Marie Reed

Arthur L. Williams • Harland D. Embree • Harold J. DeBey

SAN JOSE STATE UNIVERSITY

INTRODUCTION TO

CHEMISTRY

THIRD EDITION

ADDISON-WESLEY PUBLISHING COMPANY

Reading, Massachusetts • Menlo Park, California

London • Amsterdam • Don Mills, Ontario • Sydney

THIS BOOK IS IN THE **ADDISON-WESLEY SERIES IN CHEMISTRY**

SPONSORING EDITOR: ROBERT ROGERS
PRODUCTION EDITOR: MARY CAFARELLA
DESIGNER: MARSHALL HENRICHS
ART COORDINATOR: KRISTIN KRAMER
ILLUSTRATOR: ANCO (Boston)
COVER DESIGN: RICHARD HANNUS

Library of Congress Cataloging in Publication Data

Williams, Arthur L.   1903–1979
  Introduction to chemistry.

  Includes index.
  1. Chemistry. I. Embree, Harland D., joint
author. II. DeBey, Harold J., joint author.
III. Title.
QD31.2.W56 1981   540   80-26581
ISBN 0–201–08726–X

CONTENTS

PART TWO ORGANIC CHEMISTRY

PART THREE BIOCHEMISTRY

PREFACE

In preparing a preface to the third edition of this book we can see little reason to make major changes in the preface that was written for the first edition in 1968.

Human beings are curious and have an intense urge to explore the secrets hidden in a grain of sand, a pulsating heart, or a fragrant flower. We often feel we must interpret creatively the discoveries we make. The creative scientist constructs visions of whirling electrons, charged ions, and double spirals of DNA. Out of the visions there emerges an awe-inspiring pattern of order and unity in the very essence of many seemingly unrelated substances in the universe. The concern with the fundamental structure and composition of all these substances is the part of human knowledge we call chemistry.

This book was written for any person who wants to experience this fascinating view of our universe, either for the pleasure of expanding his or her own understanding or for the additional goal of completing a specific course. We have emphasized those topics necessary if readers are to understand chemistry and to interpret more intelligently the environment—both physical and, to a lesser extent, social—in which they find themselves. Not all statements in the book will have immediate, practical applications to daily life. Such a collection of facts would train students only for today, rather than educate them. It would leave them with no historical perspective and little ability to understand the new discoveries and new interpretations that will surely develop from our current facts and theories.

We have attempted to describe the principles of chemistry in a logical manner and amplify them with meaningful examples. We have made a particular effort to show how scientific theories are developed from observations of natural events. A textbook or course that presents only a skeletal framework, beautifully articulated as it may be, is no more inspiring than the skeleton of a famous race horse or of a beautiful woman. Equally uninspiring is a shapeless mass of material without a supporting skeleton. The skeleton of principles is definitely present throughout this book, rounded out with facts that make chemistry live.

As often as possible, observations familiar to the reader, or well-known historical experiments, have been used as a foundation for developing concepts. We have presented logical and understandable methods for writing chemical equations and performing simple calculations. Rarely have we suggested that a mathematical formula be memorized. We have emphasized the use of the Periodic Table for predicting properties and valences, as opposed to memorizing such facts.

We have assumed that readers already have some acquaintance with both the physical and biological sciences, but we have not assumed that they have had any previous course in chemistry. We have limited mathematical calculations to those requiring only arithmetic and simple algebra.

This book differs from most introductory chemistry texts in that it not only treats inorganic chemistry, but includes extensive sections of organic chemistry and bio-chemistry. The growing knowledge and challenges in these fields make it essential to provide a balanced coverage of the topics for the many readers who use this book for their first study of chemistry. We have included the customary fundamentals of general chemistry. The important classes of organic compounds are surveyed, with the concept of structure providing the central theme. Systematic names for organic compounds are emphasized. Many common names are included for reference, but are used only when there seems to be a compelling reason for doing so. The chapter on polymers should be of special interest to many readers, since it describes the manner in which numerous well-known synthetic fibers and plastic materials are formed. The biochemistry section includes—in addition to the traditional discussion of carbohydrates, proteins, and lipids—chapters on the chemistry of heredity, on disease and therapy, and on nutrition.

More and more colleges are offering a well-balanced, terminal course which provides a significant understanding of chemistry to students majoring in the liberal arts or in such fields as nursing, business, or home economics. We have had such a course at San Jose State College for many years. The need for a satisfactory text prompted us to write this book.

This book is suitable for a one-year course which includes one semester of general chemistry and one of organic and biochemistry. In colleges operating on the quarter system, one quarter could be devoted to each of the three fields. For courses taught in two quarters or one semester, several chapters and parts of chapters may be omitted without loss of continuity; we have been careful not to include in any of the optional sections those fundamental concepts that will be required later in the book. Courses for nurses or dieticians may concentrate on the organic and biochemistry sections, especially if high school chemistry is prerequisite to such programs. Although this book is not intended as a text for chemistry majors, it can provide a useful introduction for any college science majors who have not had chemistry in high school.

In this third edition we have extensively rewritten and reorganized the section on general chemistry. We have increased the level of mathematical and theoretical dis-cussions of equilibrium, pH and buffers, and the heat of reaction. The chemistry of most of the groups of the Periodic Table of the Elements has been expanded while some of the more industrially oriented descriptive materials have been omitted. Discussions of metals, nonmetals, the chemistry of colloids, chemical compounds, and chemical reactions have been extensively reorganized. Terminology and discussion of concepts have been updated throughout the book.

New information and interpretations have been added to the organic and bio-chemistry sections but these have been changed less than the first sections of the book. A chapter on the biochemistry of body fluids which includes a discussion of acid–base, electrolyte, and water balances has been added. The chapter on plant biochemistry has been eliminated and material from the chapter on the chemistry of the environment has been distributed in other chapters; for example, the chapter on water is now titled "Water, Its Properties, Pollution, and Purification."

Since we realize that the book includes more material than can be covered in any single course, we have provided the opportunity for selection by keeping discussions that are optional in discrete sections, which may be omitted or assigned as reading only, with little lecture discussion. As before, we have written in a manner intended to make the book easy for students to read and understand. Throughout the book, we have added more examples and illustrations, particularly in the areas of personal and health-related subjects.

The requirement of keeping the book at a reasonable size has meant that we have had to decrease the number of references for suggested reading at the end of chapters. A *Study Guide* for students contains a newly written section on problem-solving and mathematical operations and additional reading lists.

We are unhappy to report that our co-author and colleague, Dr. Arthur Williams, died during the preparation of this edition. Because of his importance in the conception and preparation of the first two editions of the book, we want to keep his name as senior author. His spirit and philosophy of teaching permeates this edition and will continue to do so.

Now, a personal note to our readers: We believe you will discover that chemistry is a vital, fascinating subject which overflows the printed page. It will stimulate a new way of looking at the universe and should make life more meaningful by helping you understand and interpret the natural and technical world around you and the complex world of your own body chemistry. We hope you find chemistry as dynamic and intriguing as we do.

San Jose, California   H. D. E.
October, 1980   H. J. D.

PART ONE

INORGANIC CHEMISTRY

1

The Science of Chemistry

1.1
**WHAT IS
CHEMISTRY?**

On any particular day we are likely to hear comments about chemistry and chemicals. They are mentioned in newspapers and magazines, in books and television programs, in conversations overheard in stores and restaurants. Chemistry is utilized in medical tests, in making perfumes, in the conversion of petroleum to plastics, and in the manufacture of steel, glass, and agricultural fertilizers. Chemists have discovered new medicines to treat high blood pressure and glaucoma. Chemistry is involved in the formation of "smog." Chemicals have been developed to prevent insects from destroying valuable crops. But these chemicals also may have harmful effects on the environment and on the people who handle the chemicals. Some chemicals successfully control certain types of cancer and other chemicals cause cancer. But just what *is* chemistry? How can it be related to all of these very different kinds of activity?

Briefly stated, chemistry is the science that deals with the composition, or make-up, of substances and the changes that they undergo. Although we will soon elaborate this definition, it will serve well enough for now.

Probably you would say that lime, ammonia, sulfur, and turpentine are all chemicals. Turpentine is produced by trees, but so are maple sugar, rubber, and olive oil. Are these chemicals too? Is wool a chemical? Aluminum? Aspirin? Caffeine? Vitamin C? Yes, they are, because whenever you ask a question such as, "What is caffeine made of?" you are really inquiring about the chemical nature of that substance. So in this sense *any* material is a chemical; it could be rare or common, edible or poisonous, flammable or inert. There is nothing necessarily unusual about a chemical. Bottles of salt, sugar, vinegar, and baking soda are found in chemistry laboratories as well as in homes.

We have said that chemistry is concerned not only with the composition of substances, but also with the ways that they may change. Chemical change is readily observed whenever an iron tool is left outside during a rainstorm. The tool soon has a coating of red rust. What does water do to the iron? Is the rust still iron? Why does the color change? If sugar crystals are spilled onto a hot stove, they become liquid for a moment and then turn to a black shapeless mass. The black material is no longer sweet. Is it still sugar? Was the clear liquid sugar? Many of us have been told that if we eat sugar, our bodies burn it for energy. How do we "burn" sugar? How does this process change the sugar? How does it produce energy? These questions illustrate chemistry's concern with the way substances change and with the energy released or absorbed during the changes.

From everyday experiences you already know more about chemistry than you may realize. If your knowledge is coupled with a desire to know even more about what things are made of and how they change, you will find chemistry a fascinating study. We all have quite a lot of curiosity, and that is fortunate, because the urge to know is the motivation responsible for most of the achievements of science.

**1.2
THE
SCIENTIFIC
APPROACH**

A scientist, like a nonscientist, attempts to explain the world around us on the basis of observations and of generalizations based on those observations.

Humans began making predictions based on observations long before recorded history. The periodic flooding of the Nile had important consequences for farmers in ancient Egypt. When they observed the relation of the flooding to a certain pattern of stars in the sky, at a certain time of the night, they could predict the coming of the floods. It also seemed logical to conclude that the stars controlled the flooding. Centuries passed before this generalization was shown, by another observation, to be false. Someone discovered that heavy rains in distant regions were the actual cause of the floods.

Very often it is necessary to wait for additional observations to be sure that a generalization is valid. Scientists are not usually content, however, with merely waiting for something to happen and hoping to be present to observe the incident. Instead, whenever possible they will perform experiments that can yield information related to the generalization (which may also be called a hypothesis or a theory). Indeed, one of the most difficult aspects of research work in chemistry and in other sciences is devising experiments or planning observations that will either support or refute a theory. Thus, for example, Einstein's theories of relativity had to wait for years before a way was found to test them.

The value of experimentation began to be recognized relatively late in history (A.D. 1500–1700). The ancient Greek scientist-philosophers (600–400 B.C.) were reasonably accurate observers, excellent compilers of observations, and outstanding in proposing explanations. But they lacked one attitude which we now consider necessary for a scientist; they did not experiment. The Greeks were satisfied if an explanation or theory was logical on the basis of the facts from which they argued. Aristotle, for instance, stated that a barrel full of ashes would hold as much water as would an empty barrel. Apparently no one was interested in testing this by a simple experiment.

The process of observation, generalization, and verification is the way that most of the important subject matter of chemistry, or of any science, has been established. Therefore, in this book, whenever possible we will first present observations, either those the reader can make or those that have been made historically. These will be followed by explanations and further applications. We hope that this arrangement will give the reader a more meaningful picture of chemistry and will provide some of the sense of excitement a scientist feels when seeing a new relation. In many instances, however, this method of presentation is neither possible nor practical because of the limitations of space and the difficulty of presenting the significant details in a manner understandable to readers inexperienced in the laboratory techniques of chemistry. Thus, if we occasionally present facts as though they were dictated by authority or were divinely revealed, we do so for expedience only. This is not, of course, the way scientific knowledge develops.

Today we have available a certain group of facts, and we accept theories that explain these facts consistently. But in the future there will surely be new facts, and new or modified theories. Science is a living, growing thing, not just a series of facts to be memorized. However, you obviously cannot understand chemistry without learning some facts. Let us hope that by learning both—some of the current facts *and* theories— you will begin to see chemistry as a fascinating way to understand the behavior and composition of the materials in the world around us and within us.

SUGGESTED READING

For those who wish to find further information about some of the topics in this chapter, the following suggestions are offered. The symbol (pb) indicates that the publication is available in paperback form.

Goldstein, M., and I. E. Goldstein, *How We Know*, Plenum Press, New York, 1978. Subtitled "An Exploration of the Scientific Process," this interesting book contains case histories of research in basic science and in medicine and psychiatry, and presents details that illustrate and support some of the generalizations given in this chapter.

Jaffe, B., Crucibles: *The Story of Chemistry*, 4th Revised Ed., (pb). Dover Publications, New York, 1976. This easy-to-read book is available again in paperback form. It has been updated to include developments in nuclear chemistry.

Additional references are listed in the *Study Guide* which is designed to accompany this text.

2

Methods of Measurement

2.1
INTRODUCTION
Having the skills and the instruments to make accurate, precise measurements is essential for science—both practical and theoretical. More accurate or precise measurements often lead to changes in basic theories. For example, Newton's theory of the nature of gravity and of basic physics served to explain our observations of the universe for many years. However, when more precise measurements were made, Einstein's theories of relativity extended Newton's theories.

The number of extremely fine measurements involved in the construction, launching, control, and accurate landing of a space vehicle is almost fantastic. These quantities involve calculations that are so complex and that must be done so rapidly that computers must be used. More mundane measurements are often important to us in our daily lives. These may include setting the spark gap in the spark plugs of the cars we drive, measuring the amount of sodium in the blood of a victim of heart disease, or using a strobe light to set the speed of a phonograph turntable to get exactly the right pitch.

To conduct trade we need both a *number system* and a kind of *measure*. If things to be traded come in simple units such as one sheep, we only need to count. It is probable that ancient societies (as primitive societies still found on the earth still do) originally "counted" using only words for "one," "two," "few," and "many." More complex numbers and number systems then developed, not all of them based on ten as is the one we use. When we want to measure more precisely, as when we want to indicate the difference between one large sheep and one small one, we must go to some unit of weight such as the pound for indicating differences. A major problem arises when different groups of people using different systems of measurement, perhaps different number systems, and probably different monetary systems want to buy and sell various things.

Accurate and precise measurements are the basis for developing scientific theories and also for determining whether theories are correct. This chapter tells how some of our methods of measurement developed and were changed and refined to make them useful, not only for commercial purposes but also for the purposes of science.

2.2 EARLY MEASUREMENTS AND THE DEVELOPMENT OF THE BRITISH SYSTEM

Early measurements of distance were made by counting the number of paces or steps between two points or, for longer distances, by counting the number of days or moons (months) it took to walk the distance. Measures smaller than either of these were based on some part of the human body. Although such standards were better than none at all, they tended to vary with the size, judgment, and honesty of the person making the measurement. The size of servings to people in a cafeteria line as well as the size of the fish that got away are subject to variance.

In the system of measurements that developed in England and was used throughout the British empire, the *digit* was the breadth of a middle finger; the *inch*, the breadth of the thumb; the *foot*, the length of a man's foot; the *yard*, the distance (traditionally established by the measurements of King Henry II) from the tip of the nose to the thumb at the end of an outstretched arm; and the *fathom*, the distance from fingertip to fingertip of the outstretched arms. Less standard were the units for measuring weight—the *grain*, based on a kernel or grain of wheat, and the *stone*, which is equal to 14 pounds. Volume was measured in *mouthfuls, jiggers* (two mouthfuls), the *jackpot* (a double jigger), and finally the *cup, pint, quart*, and *gallon*. Although some of these measures have persisted to the present day, they are now defined in terms more precise than body equivalents.

The beginning of this process of definition in England was probably the Magna Carta of 1215, which promised, among other reforms, that there should be but "one weight, one measure" for the kingdom. This was not realized for some time, however. For instance, from the time of the Magna Carta to the present there have been not one, but two gallons in the British system. The smaller, which is the gallon used in the United States, contains 231 cubic inches and was formerly known as the beer or wine gallon. The larger gallon is the British or Imperial gallon, which contains 277.42 cubic inches. This is slightly larger than the older unit known as the ale or corn gallon.

The *cubit* is one of the oldest units of length and has persisted through the ages. In the book of Genesis it is recorded that Noah's Ark was 300 cubits long, 50 cubits wide and 30 cubits high. The cubit was established as the distance from the tip of the elbow to the end of the middle finger. If we check this with the average adult male of today, we will find that it equals 18 to 19 inches. The cubit is presently a part of the British system and is defined as 18 inches. With this information it can be estimated that the Ark was about 450 feet long, 75 feet wide, and 45 feet high, a sizable craft for the time. The cubit was also the unit of measure used in building the Egyptian pyramids.

By the eighteenth century, English-speaking countries had developed a standard system of measurement that was suitable for trade and was consistent in its units. A foot was exactly twelve inches; a yard, exactly three feet; and a fathom, six feet. Smaller units were expressed in fractions such as 1/2, 1/4, and 1/16 of the basic units. While the system depended on defining basic units instead of measuring the width of a man's thumb, it had two shortcomings—the awkwardness of conversion from one unit to another, and the need to use fractions for precise measurements of small quantities.

The discussion above applies largely to the English-speaking countries. Other countries and sometimes even smaller groups within countries developed different methods of measurement, often based on similar quantities, such as a grain of wheat or some human dimension. The development of trade and more especially of science—which requires accurate and precise measurements that do not vary from one country to another—demanded a better system of measurement.

2.3
THE METRIC
AND INTER-
NATIONAL
SYSTEM (SI)
METHODS OF
MEASURE-
MENT
In 1790 (during the French Revolution) the French National Assembly authorized the establishment of a committee to develop a new system of weights and measures. As a result, in 1799 a logical new system called the **metric system** was completed and adopted. This system had the major advantage of being a decimal system like our monetary system, in which units are cents and dollars. Each unit was 10 times larger than or 1/10 as large as the next smaller or larger unit. It also based measurements on a single unit for each dimension to be measured—the **meter** for length, the **gram** for weight, and the **second** for measuring time. Other units such as the **liter**, used for measuring volume, could be derived from these basic units. The basic unit of length, the meter, was derived from a universal measurement, presumably one that could be made by any scientist at any time. It was supposed to be 1/10 000 000 of the distance from the north pole to the equator of the earth. Actually, the value that was established was in error and the real meter became the record the committee made of the measurement—the distance between two fine scratches on a platinum–iridium bar maintained at a given temperature in the Bureau of Standards at Sèvres (a suburb of Paris) in France. The important aspect is that one basic unit, the meter, was established.

In 1960 the General Conference of Weights and Measures modified the metric system and gave it the name of **International System of Units**. The name in French is Le Système International d'Unités, abbreviated **SI**, and the general term SI is used throughout the world to describe the system. Since 1960 additional conferences have been held and minor changes in the SI have been made. It is proposed that conferences continue to be held every few years, at which time the system may be extended, improved, and refined.

The basic units of the SI are given in Fig. 2.1. Not all of these units are used in elementary chemistry, and others that are used a great deal, such as the liter, are derived from the basic units. The prefixes used to describe units larger and smaller than the basic units, as well as the standard abbreviations for these units, are given in Fig. 2.2. In the material that follows we will discuss those units of the SI that are important in basic chemistry. All the metric–SI units will assume greater importance to people who live in the United States and other countries as they convert from the British system to a system based on the metric–SI measurements. In the United States we generally spell the unit of length as "meter" rather than the recommended "metre." It is likely that in the future we will change to the international spelling "metre" (and likewise adopt "litre" instead of "liter"); however, we have retained the common spellings in this book.

unit of measurement	abbreviation	quantity being measured (dimension)
meter	m	length
kilogram	kg	mass
second	s	time
ampere	A	electric current
kelvin	K	temperature
candela	cd	light intensity
mole	mol	amount of substance

Fig. 2.1 Basic SI measurements. Mass is the quantity that is commonly called weight. For a discussion of the difference between mass and weight, see Section 2.5.

	prefixes	symbols
$1\ 000\ 000\ 000\ 000\ 000\ 000 = 10^{18}$	exa	E
$1\ 000\ 000\ 000\ 000\ 000 = 10^{15}$	peta	P
$1\ 000\ 000\ 000\ 000 = 10^{12}$	tera	T
$1\ 000\ 000\ 000 = 10^{9}$	giga	G
$1\ 000\ 000 = 10^{6}$	**mega**	**M**
$1\ 000 = 10^{3}$	**kilo**	**k**
$100 = 10^{2}$	hecto	h
$10 = 10^{1}$	deka	da
$1 = 10^{0}$		
$0.1 = 10^{-1}$	**deci**	**d**
$0.01 = 10^{-2}$	**centi**	**c**
$0.001 = 10^{-3}$	**milli**	**m**
$0.000\ 001 = 10^{-6}$	**micro**	**μ**
$0.000\ 000\ 001 = 10^{-9}$	nano	n
$0.000\ 000\ 000\ 001 = 10^{-12}$	pico	p
$0.000\ 000\ 000\ 000\ 001 = 10^{-15}$	femto	f
$0.000\ 000\ 000\ 000\ 000\ 001 = 10^{-18}$	atto	a

Fig. 2.2 Prefixes and symbols used in the SI. The most common prefixes and symbols are shown in boldface type.

2.4
THE MEASURE-MENT OF LENGTH (DISTANCE)

The basic unit for measuring length of an object or distance between two objects in the SI is the **meter** (metre). It is officially defined as 1 650 763.73* wavelengths of the orange-red light given off by krypton as measured in a vacuum. While this distance is the same as that of the meter of the metric system, it is based on a quantity that does not vary and can be measured by individual scientists having the necessary equipment much more simply than by measuring 1/10 000 000 of the distance from the north pole to the equator. It also means that the measurement can be made by individuals rather than having to compare to a standard (such as the standard meter at Sèvres, France) that could change or be destroyed.

The meter, which is slightly more than a yard, can be used for measuring the size of rooms and short distances. Longer distances are generally measured in **kilometers**. A kilometer, which is 1 000 meters, is approximately 0.6 mile. Most scientific measurements involve quantities smaller than the meter, and centimeters and millimeters are used. Other scientific measurements use the micrometer, sometimes called the *micron*. The angstrom (Å), which is usually defined as 10^{-8} centimeter (10^{-10} meters), is still commonly used to measure the wavelengths of different kinds of light; however, the nanometer, 10^{-9} meter, is the preferred unit for such measurements.

In many instances (in the United States) we still measure distances in miles, yards, feet, and inches, and will probably continue to do so for some time; however, for smaller measurements, especially where precision is necessary, we do not use fractions of inches but use a decimal scale and record measurements in tenths, hundredths, and thousandths

---

* The SI also recommends that (because in Europe a comma is used as a decimal point) large numbers be written with spaces, not commas, separating thousands, millions, and other large units. The number above would be written conventionally as 1,650,763.73.

of inches. Although the medical profession still uses feet and inches for measuring height or other anatomical dimensions, these, like all other scientific measurements, should soon be expressed only in SI units of centimeters or similar units. Table 2.1 shows the relationship of units of length in the SI and British systems.

TABLE 2.1   Units of length in the SI and British systems

SI

1 kilometer (km) = 1 000 meters (m)
1 meter = 100 centimeters (cm)
1 meter = 1 000 millimeters (mm)
1 millimeter = 1 000 micrometers (μm)

British

1 mile (mi) = 5280 feet (ft)
1 yard (yd) = 3 feet
1 foot = 12 inches (in)

Conversion Factors*

1 kilometer = 0.60 miles
1 meter = 39.37 inches
1 centimeter = 0.40 inches

1 mile = 1.609 kilometers
1 yard = 0.915 meters
1 inch = 2.54 centimeters

* All conversion factors are approximate.

2.5 THE MEASUREMENT OF MASS

The measurement of mass in the SI uses the unit of the **kilogram**. The kilogram is the mass of a man-made object (an artifact) kept at the French Bureau of Standards. This platinum–iridium object was originally intended to be the mass of 1 000 cubic centimeters of water and it is nearly that; however, the standard is the artifact. The kilogram is the only basic SI unit that must be compared to a standard and is not defined as a quantity that can be measured by individual scientists.

Although we measure mass by weighing an object, **mass** and **weight** are not exactly the same. Both refer to the amount of matter in a given object or sample but, more precisely, weight is what we measure when we use an instrument such as a balance or a spring scale. Mass is a measure of the amount of resistance to a change of motion. Because radiation and other forms of nonmaterial energy also resist a change of motion they can be said to possess mass, but since they do not affect a balance or spring scale, they have no weight.

You may have heard it said that although astronauts are "weightless" when in outer space and weigh only 1/6 as much on the moon as they do on earth, they have the same mass under all conditions. The accuracy of this statement depends on the device we use to measure their weight or mass. The scales used for weighing produce at the market or loads of gravel, as well as the bathroom scale you use to check your weight, use springs or similar devices. Such scales would give differing weights depending on the gravity.

The most common way of making measurements of weight in the laboratory is comparing the weight of an object with the weight of standard objects in the same gravity field. Whether we use a two-pan balance, a beam balance, or some other top-loading balance, we are comparing the resistance to a change in motion of the substance we are "weighing" to a set of standard objects (weights). Thus we are really obtaining the mass of the object being "weighed." Using such devices we would find that an astronaut had the same "weight" as well as mass no matter where the determination was made.

It is common practice in chemistry to describe the process of determination of mass as weighing and to call the result a weight. It is just as correct to call this process determining the mass and the result a mass. Since both "mass" and "weight" are in common usage, we will use both throughout this book.

Although the kilogram is the unit of mass in the SI, the name is obviously based on the gram, which was formerly the basic unit. We use the **gram** to measure small quantities of substances ordinarily used in the chemical laboratory. A nickel coin weighs about 5 grams, a dollar bill weighs one gram, and a pound is approximately 454 grams. We use the **milligram** and **microgram** for measuring very small quantities. For large quantities we use the kilogram (2.2 pounds) and the **megagram** (10^6 grams), which is commonly known as the tonne or the **metric ton**. The metric ton is equal to 2 205 pounds. Thus it is intermediate between the **short ton** of 2 000 pounds, commonly used in the

TABLE 2.2   Units of mass in the SI and British systems

SI

1 ton (metric) = 1 000 kilograms (kg)
1 kilogram = 1 000 grams (g)
1 gram = 1 000 milligrams (mg)
1 milligram = 1 000 micrograms (μg)
1 microgram = 1 000 nanograms (ng)

British (Avoirdupois)

1 pound (lb) = 16 ounces (oz)
1 ounce = 480 grains (gr)
1 long ton = 2 240 pounds (used in Great Britain)
1 short ton = 2 000 pounds (used in United States)

Conversion Factors*

1 kilogram = 2.205 pounds
1 pound = 454 grams
1 gram = 15.4 grains
1 ounce = 28.4 grams
1 carat† = 200 milligrams
1 metric ton = 2 205 pounds
1 long ton = 1 016 kilograms
1 short ton = 907 kilograms

* All conversion factors are approximate.
† Used to weigh precious stones.

United States, and the **long ton** of 2 240 pounds, used by the British. Such large units are necessary when talking of world food supplies or production of industrial products.

We still retain the pound and ounce for most everday measurements; however, many physicians and researchers record the weight of people in kilograms, and some cooks (particularly those with dietetic training) measure the ingredients in gram weights when making a cake. Daily vitamin requirements are generally stated in milligrams or micrograms. Table 2.2 shows the relationship of units of mass in the SI and British systems.

2.6 THE MEASUREMENT OF VOLUME

The metric unit of volume that is commonly used in science is the **liter** (litre). The SI uses units based on measurements of length such as cubic centimeters (cc or cm^3) to measure volume. A liter is defined in the SI as 1 000 cm^3. It is also referred to as a cubic decimeter (dm^3)—a cube that is one decimeter (1/10 of a meter or 10 cm) in each dimension.

A liter is approximately a quart. This is a relatively large quantity, and scientists more often use the **milliliter** (or cubic centimeter), which is approximately 20 drops, for common measurements. The **microliter** is used to measure very small quantities. For many commercial products in the United States, contents of cans and bottles are now specified in terms of liters or milliliters. Confusion can arise when a measurement is given in ounces, since the ounce is used in the British system as a measure both of weight and of volume. One ounce of weight (sometimes called an avoirdupois ounce) is equal to approximately 28 grams, while an ounce of volume (fluid ounce) is approximately 30 milliliters. The measures are not too different if a product is primarily water, since one gram of water has a volume of one milliliter, but it is grossly misleading or confusing with other substances. Table 2.3 shows the relationship of units of volume in the SI and British systems.

TABLE 2.3　Units of liquid volume in the SI and British systems

SI

1 liter* = 1 000 cubic centimeters (cm^3)
1 liter = 1 cubic decimeter (dm^3)
1 liter = 1 000 milliliters (ml)

British

1 gallon (gal) = 4 quarts (qt)
1 quart = 2 pints (pt)
1 pint = 16 ounces
1 cup = 8 ounces

Conversion Factors†

1 liter = 1.06 quarts
1 gallon = 3 785 cubic centimeters (cm^3)
1 quart = 946 cubic centimeters (cm^3)
1 pint = 473 cubic centimeters (cm^3)
1 cup (8 oz) = 237 milliliters (ml)
1 ounce = 29.6 cubic centimeters (cm^3)

* The term liter is a metric unit, not a basic SI unit. The SI defines the liter as indicated.
† All conversion factors are approximate.

2.7
THE MEASURE-
MENT OF TIME

The basic unit of time in the SI is the second. Although it was originally based on the earth's rotation, the second is now defined as the duration of 9 192 631 770 cycles of a specific radiation given off by cesium-133. The value of the second has not changed appreciably, but it has been redefined (as was the meter) in terms that are more readily measured. The new definition is also more accurate, since the earth "wobbles" a bit and is slowing down very slightly with time.

The measurement of time in seconds, minutes, and hours developed relatively late in history. Although water clocks and hour glasses (similar to the three-minute egg timers we sometimes use) were used by the ancients, Galileo is reported to have used his own pulse to time the oscillations of a swinging chandelier in the Cathedral of Pisa, from which he deduced the law of the pendulum in 1541. Reliable mechanical clocks and watches were not developed until the seventeenth century.

We follow the metric system in designating fractions of the second and we speak of milli-, micro-, and nanoseconds when studying rapid chemical reactions. Units larger than the second, which we use for example, in specifying the half-life of radioactive isotopes, are based on the hours, days, and years of common systems. The use of a value of 60 for the number of seconds in a minute and the number of minutes in an hour probably reflects back to an earlier number system based on 60 instead of 10. The division of the year into 365 (actually $365\frac{1}{4}$) days is based on the revolution of the earth around the sun and is thus a natural measurement. However, the division of the day, which is itself a natural unit based on the earth's rotation, into 24 hours is arbitrary. The decimal system is used for larger units of time—the decade (10 years), century (100 years), and millennium (1 000 years). Thus we measure time using natural, defined, and derived units. It is interesting to speculate that at some time in the future we may measure time in kiloseconds and megaseconds rather than minutes and hours.

2.8
THE MEASURE-
MENT OF
TEMPERATURE

The ancient Greek philospher-scientist Aristotle was much concerned with heat and cold. Were they two different things, or merely extremes of one thing? At one point, his writings indicate, he came close to the idea of an instrument that would measure the intensity of heat. However, like the other Greek philosophers of his day, he thought in terms of qualities more than of quantities. The thermometer was not invented until much later.

Once the idea of temperature was developed, the problem became one of constructing a device to measure it. Although others had previously worked on this problem, we owe our present thermometers to Fahrenheit and Celsius, who developed special types of thermometers in 1724 and 1742, respectively. We can reconstruct their measurements by the following procedure.

First, we partially fill two glass tubes with a liquid that will expand when heated and contract when cooled, and then seal the tubes. We place these tubes in boiling water, and indicate the height of the liquid (mercury is most commonly used) when the tube is at the temperature of the boiling water. Then, after allowing the tubes to cool, we immerse them in a mixture of ice and water and make another mark on each tube to indicate the position of the liquid at this temperature—the freezing point of water or the melting point of ice, depending upon which way we choose to regard it. Now the problem becomes one of deciding what values to give to these points. Celsius set the freezing point of water equal to 0° and the boiling point equal to 100° on the scale that officially bears his name (formerly known as the centigrade scale). Fahrenheit, apparently in an attempt to avoid negative numbers for reporting temperatures on cold days, selected the lowest temperature he could produce with an ice-salt mixture as his zero temperature. He then set body temperature 96 units above the minimum temperature, at 96°. Later he adjusted the scale slightly so that the freezing point of water was 32° and the boiling point of water 212°, exactly 180° above the freezing point. On this scale, body temperature was 98.6°.

It is interesting to note that both scales are based on values that are relatively easy to produce anywhere on earth. We now know, however, that the boiling point of water is affected by elevation and that the values of 100°C or 212°F are correct only at sea level. We also know that dissolving substances in water changes both the boiling and freezing points, so certain precautions must be taken if we are to make accurate thermometers. While the two systems use similar reference points, the size of the degree is different. The difference between boiling and freezing is 100 degrees in the Celsius system and 180 degrees in the Fahrenheit. Thus 100°C = 180°F and 1°C = 1.8°F. The Celsius degree is nearly twice as large as the Fahrenheit degree.

The SI unit of temperature is the **kelvin**, which is defined as 1/273.16 of the freezing point of water. The temperature of zero on the Kelvin scale is "absolute zero," which is apparently the lowest temperature that can be achieved. On this scale water freezes at 273 degrees and boils at 373. Although we use the Kelvin scale for calculations involving volumes and pressures of gases, scientists most often use the Celsius scale. The size of the units (degrees) is the same in both scales. The Fahrenheit scale is still commonly used in the United States to report the daily temperature or to determine whether a patient in a hospital has a fever, but both of these values are slowly being replaced by the Celsius scale. (See Fig. 2.3.) We use the degree symbol (°) when reporting temperatures on the Fahrenheit or Celsius scale but not with Kelvin measurements.

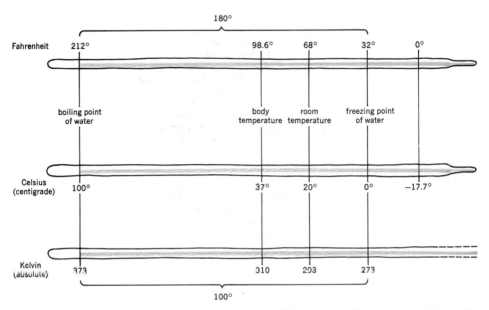

Fig. 2.3 Comparison of temperature scales. Zero Kelvin (not indicated on scale here) is equal to −273°C and −460°F. This temperature is known as absolute zero.

2.9	
THE MEASURE-	
MENT OF	
AMOUNT OF	
HEAT	

2.9 THE MEASUREMENT OF AMOUNT OF HEAT

A thermometer measures the intensity of heat or the temperature, but this is only one aspect of heat. The other measurement we use in dealing with heat is the **calorie**. *A calorie is the amount of heat required to raise the temperature of one gram of water one degree Celsius.* The SI specifies that the unit used to measure heat should be the joule and defines the calorie as 4.184 joules. In this book, we will use the calorie and kilocalorie when discussing quantities such as heats of reaction and the energy contained in foods. Most reference tables still give values in calories, although some conversions to joules (and kilojoules) are being made in tables and in reporting current research.

Since the calorie is a small unit, the **kilocalorie** (1 000 calories) is more commonly used to denote the energy lost or gained in chemical reactions. The larger unit is also the unit used to specify the calories in food. We use cal as the abbreviation for calorie and either kcal or Cal for the kilocalorie.

2.10 UNITS OF MEASUREMENT IN MEDICINE

Although the *U.S. Pharmacopoeia*, which is a standard reference in medicine, uses SI and metric units, many medications are still prescribed by doctors and dispensed by pharmacists in British units, also known as apothecaries' units. During the period of transition it will be necessary for people in the health professions to be aware of some of these units. (See Table 2.4.)

Most solid medications (pills or capsules) are measured in grams or milligrams; however, some are still compounded in terms of grains. We normally buy aspirin in 5-grain tablets. A grain is about 65 milligrams, and a gram is commonly thought of as containing 15 grains.

TABLE 2.4   Measurements used in medicine

Mass

1 pound (lb) avoirdupois = 16 ounces (oz) (avoirdupois)
1 ounce (avoirdupois) = 16 drams (avoirdupois)
1 dram = 438 grains
1 pound (apothecaries') = 12 ounces (apothecaries')
1 ounce (apothecaries') = 8 drams (apothecaries')
1 dram (apothecaries') = 60 grams
1 ounce (avoirdupois) = 28.4 grams
1 ounce (apothecaries') = 31.1 grams
1 grain (avoirdupois) = 1 grain (apothecaries')
1 grain (either) = 64.8 milligrams

Volume

1 pint = 16 (fluid) ounces
1 fluid ounce = 8 fluid drams
1 fluid dram = 60 minims
1 fluid ounce = 29.6 milliliters
1 fluid dram = 3.70 milliliters

Approximate Equivalents

1 teaspoon = 5 milliliters
1 tablespoon = 3 teaspoons = 15 milliliters
1 minim = 1 drop
1 gram = 15 grains

The grain is the same amount of material in both the avoirdupois and apothecaries' system even though the number of grains per dram and drams per ounce varies. To avoid confusing the abbreviation for gram (g) with that for grain (gr), people using both units often use gm as the abbreviation for gram.

Liquid medications administered in a hospital or other similar setting may be dispensed in ounces or fluid drams. Liquids taken by a person at home are generally measured in drops, teaspoons, or tablespoons. These are not very accurate measures. A teaspoon contains about 5 milliliters, and a tablespoon, 15 milliliters. In general, one milliliter is equivalent to about 16–20 drops, but this varies greatly with the size of the medicine dropper.

As in other disciplines the conversion to metric or SI is proceeding rapidly in the medical and health sciences. However, many people, particularly those who received their training many years ago, continue to use the older common units. Home administration of medication will probably continue to use common units for some time.

**2.11
CONVERSIONS
FROM ONE
UNIT AND
SYSTEM TO
ANOTHER
Conversions
within The SI**

Conversions from one unit to another (e.g., from centimeters to millimeters) within the SI involve only multiplication or division by a power of 10 and the change of a decimal point (sometimes with the addition of zeros to mark the new decimal point). Often we can do these conversions simply, as when we are asked how many cents there are in $2.56. We merely move the decimal point and say there are 256 cents (each equivalent to 1/100 or 10^{-2} dollars) in $2.56. We use the identical method when asked how many centimeters there are in 2.56 meters, and give the answer, 256. However, many people have problems knowing whether to divide or multiply—whether to move the decimal point to the left or right—especially when dealing with unfamiliar units. Such problems are easier to solve and mistakes avoided by using the units of each numerical value and setting up a calculation so that the units cancel. (See the Study Guide for a further discussion of this and other simple calculations.)

For instance, if we want to solve the problem asking the number of centimeters in 2.56 meters, we first must realize that, since we are given the quantity in meters and want an answer in centimeters, we will need a conversion factor relating meters and centimeters. Our calculation will be

2.56 meters × conversion factor = ? centimeters

To find our conversion factor, we consult Table 2.1 and find that 1 meter = 10^2 or 100 centimeters. We can express this relationship by means of two possible conversion factors that are stated as fractions:

$$\frac{1 \text{ meter}}{10^2 \text{ centimeters}} \quad \text{and} \quad \frac{10^2 \text{ centimeters}}{1 \text{ meter}}$$

The first factor indicates that we divide by 10^2 and the second indicates multiplication by 10^2. We choose the factor that will leave the correct units in our answer after other units have been cancelled. Our calculation becomes

$$? = 2.56 \text{ m} \times \frac{10^2 \text{ cm}}{1 \text{ m}} = 256 \text{ cm} \quad \text{or } 2.56 \times 10^2 \text{ cm}$$

The units of "meter" cancel since one is in the numerator and the other in the denominator. If we had used the other factor, the calculation would be

$$2.56 \text{ m} \times \frac{1 \text{ m}}{10^2 \text{ cm}} = 2.56 \times 10^{-2} \text{ m}^2/\text{cm}$$

This gives the irrational unit "meters² per centimeter" instead of the unit we wanted, centimeters, and consequently also the wrong numerical value.

It may be necessary to use more than one conversion factor if the conversion involves great differences in units. For example if we wanted to know how many 1-microgram tablets of a medication could be made from 36 kilograms of vitamin B_{12}, we would set up the calculation

$$36 \text{ kg} \times \frac{10^3 \text{ g}}{1 \text{ kg}} \times \frac{10^6 \text{ } \mu g}{1 \text{ g}} \times \frac{1 \text{ tablet}}{\mu g} = 36 \times 10^9 \text{ tablets} \qquad \text{or } 3.6 \times 10^{10} \text{ tablets}$$

If we had wanted tablets of 15 micrograms each, we would merely change the last factor in our calculation to get

$$36 \text{ kg} \times \frac{10^3 \text{ g}}{1 \text{ kg}} \times \frac{10^6 \text{ } \mu g}{1 \text{ g}} \times \frac{1 \text{ tablet}}{15 \text{ } \mu g} = 2.4 \times 10^9 \text{ tablets}$$

Conversions Involving the SI and the British System

Conversions from units in the British system to the SI are common, particularly now, when both systems are in use. In such conversions we proceed exactly as we did in conversions within the SI. The necessary conversion factors are listed in (or can be derived from) Tables 2.1 to 2.5. Again we will need to put relationships in the form of fractions and make certain that units cancel. Sometimes a problem may require both conversion from one system to another and conversion from one unit to another within a system.

Example

What is the distance in kilometers between two cities that are 25 miles apart? We are given miles and want the answer in kilometers; thus our calculation becomes

$$25 \text{ mi} \times \frac{1.6 \text{ km}}{1 \text{ mi}} = 40 \text{ km}$$

When we have solved a problem we should not only make certain that our answer is in the required units and that other units cancel, but also check our answer for logical sense. The kilometer is a shorter distance than a mile, so the answer should be greater than 25, and it is.

Example

How many servings of fruit punch can we make from two gallons of concentrate if it takes 20 ml of concentrate to make one serving of the fruit punch? We are given gallons and want our answer in glasses (or individual servings). Our calculation becomes

$$2 \text{ gal} \times \frac{4 \text{ qt}}{1 \text{ gal}} \times \frac{1 \text{ liter}}{1.06 \text{ qt}} \times \frac{10^3 \text{ ml}}{\text{liter}} \times \frac{1 \text{ serving}}{20 \text{ ml}} = 3.8 \times 10^2 \text{ servings or 380 servings}$$

The unit cancellation method has again been used above. The use of ratios in solving problems of this type is discussed further in the Study Guide.

Temperature Conversions

The conversion factors used in converting temperatures from one scale to another are given in Table 2.5. Since $1°C = 1 K$, the only adjustment we need to make when converting between Celsius and Kelvin involves the difference in the zero value. To change reading from Celsius to Kelvin, we simply add 273. So

$$20°C = 20 + 273 = 293 \text{ K}$$

and

$$300 \text{ K} = 300 - 273 = 27°C$$

TABLE 2.5   Temperature conversions

Conversion of zero points

To convert from degrees Celsius to Kelvin, add 273.

To convert from Kelvin to degrees Celsius, subtract 273.

In conversions between Celsius and Fahrenheit, 32° must be added to or subtracted from the Fahrenheit reading.

Equivalence of degrees

One degree (1°) Celsius = 1 kelvin
$\qquad$ 1° Celsius = 1.8° Fahrenheit

Conversions involving Celsius and Fahrenheit scales require both the adjustment of zero and a conversion to compensate for the difference in units. Thus, to convert a normal body temperature of 98.6° Fahrenheit to a Celsius reading we must first subtract 32 from the Fahrenheit reading and then multiply by the conversion factor:

$$(98.6 - 32)°\cancel{F} \times \frac{1°C}{1.8°\cancel{F}} = 37°C$$

Conversely, to convert a temperature of 25°C to Fahrenheit, we must add after we multiply:

$$25°\cancel{C} \times \frac{1.8°F}{1°\cancel{C}} + 32°F = 77°F$$

It may be helpful to remember that 32 represents a number of degrees on the Fahrenheit scale, not the Celsius, and we must always add or subtract the 32 from a temperature that is in the Fahrenheit scale. Adding or subtracting 32 from a reading on the Celsius scale is like addition or subtraction of inches from centimeters.

To convert Fahrenheit temperatures to Kelvin units, we simply convert to Celsius units and then add 273. For example, if we want to find the boiling point of water (212°F) on the Kelvin scale our calculation is

$$(212 - 32)°\cancel{F} \times \frac{1°C}{1.8°\cancel{F}} + 273 = 373 \text{ K}$$

SUMMARY

In this chapter we have discussed how systems of measurement evolved, from the arbitrary and variable units of antiquity to the precise definitions of the metric system. In addition to its simple set of basic standards, the metric system is consistent with our numerical system, in being based on 10. This usage greatly simplifies conversions from one measurement to another.

Science without accurate measurements is limited to a discussion of theories, not unlike the proverbial bull sessions. With precise measurements we can now develop replaceable parts for motors of automobiles and perform other similar feats of technology. With more accurate measurements we can develop better theories, describe phenomena more accurately, and predict with greater confidence. Prediction of the time and area to be darkened in eclipses of the moon and sun are now routine. We are beginning to hear of computers that can be programmed with the results of a series of clinical tests (each involving several accurate measurements) to report the many possible diagnoses of a patient's illness. Without measurements science could not have developed. Moreover, we can safely predict that more accurate measurements and the measurement of things yet unmeasured will lead to further progress. Yet measurements are not an end in themselves: they require interpretation.

One real advantage of the SI is that it is an international system and virtually all countries have adopted it. At the present time, conversion from British to SI units is occurring at a very slow pace in the United States. The conversion involves not only education of the public, but the manufacture of millions of new tools, scales, and other measuring devices. For those people who have training in the sciences, however, the conversion will be relatively easy.

IMPORTANT TERMS

calorie	kilo-	meter
Calorie (kilocalorie)	kilogram	micro-
Celsius	kelvin	milli-
centi-	Kelvin	ton (metric)
deci-	liter	ton (short)
Fahrenheit	mass	ton (long)
gram	mega-	weight

WORK EXERCISES

The mathematical operations needed for solving many of the exercises below are discussed in the *Study Guide*, and you should familiarize yourself with that material before proceeding. Conversion Factors can be obtained from Tables 2.1 through 2.5. All answers should be reported to a proper number of significant figures and should contain appropriate units.

1. Write the following numbers in exponential notation.
   a) 2 345
   b) 16
   c) 1
   d) 0.009 47
   e) 0.835
   f) 1/1 000
   g) 1 000 000
   h) 92 876 000 000

2. Write the following in ordinary numbers.
   a) 2.345×10^3
   b) 2.345×10^{-3}
   c) 6.89×10^5
   d) 10^5
   e) 10^{-3}
   f) 1.65×10^0
   g) 7.00×10^1
   h) 3.60×10^{-4}

3. How many significant figures are there in each of the following numbers?
   a) 2 345
   b) 0.002 34
   c) 20
   d) 20.0
   e) 1 000
   f) 0.100 0
   g) 6.89×10^5
   h) 6.89×10^{-5}

4. Rewrite the following numbers so that they contain the number of significant figures indicated in parentheses.
   a) 2 345 (3)
   b) 2 345 (1)
   c) 6.89×10^5 (2)
   d) 4 000 (3)
   e) 7 394.25 (3)
   f) 5.555 (3)
   g) 29.998 (4)
   h) 29.998 (2)

5. Make the following conversions.
   a) 25.5 m to cm
   b) 150 cm to m
   c) 50 cm to mm
   d) 6 m to mm
   e) 2 500 m to km
   f) 16 km to m
   g) 750 mm to cm
   h) 32 cm to in.
   i) 12 in. to cm
   j) 55 mi to km
   k) 150 km to mi
   l) 1 mi to m

6. Make the following conversions.
   a) 500 g to kg
   b) 500 g to mg
   c) 6 893 μg to mg
   d) 2.34 metric tons to g
   e) 3.0 lb to g
   f) 5.0 lb to kg
   g) 2.5 kg to lb
   h) 2.0 oz to g

7. Make the following conversions.
   a) 150 cm³ to ml
   b) 2.5 cm³ to ml
   c) 250 ml to dm³
   d) 874.6 ml to liters
   e) 3.25 pt to liters
   f) 750 ml to gallons
   g) 2 cups to liters
   h) 6 oz to ml

8. Make the following temperature conversions.
   a) 100°F to Celsius
   b) 50°C to Fahrenheit
   c) −10°C to Fahrenheit
   d) 100°C to Kelvin
   e) −40°C to Fahrenheit
   f) 0° Fahrenheit to Celsius
   g) 98.6° Fahrenheit to Kelvin
   h) 373 Kelvin to Fahrenheit

9. What is your weight in kilograms?

10. A horse weighs 400 kg. What is its weight in pounds?

11. A stone weighs 350 lb. What is its weight in kilograms?

12. How many kilograms are there in 1 short ton?

13. As you approach a city in Mexico, you observe a sign reading 40 km/hr. What is the corresponding speed in miles/hour?

14. Suppose you are driving a European car with a speedometer registering kilometers/hour. What speed will be indicated if you are traveling 50 mph?

15. How many 5-grain aspirin tablets can be made from a pound of aspirin?

16. Measured in yards, how much farther does a sprinter run when competing in the 100-meter dash rather than the 100-yard dash?

17. How many meters does a sprinter cover when running the 440-yard dash?

18. The muzzle velocity of a bullet is 2 700 ft/sec. What does this equal in meters/second?

19. In many countries gasoline is sold by the liter. If gasoline in Mexico sells for 40 cents per liter, what is the cost per gallon?

20. One lap on the Indianapolis speedway is 2.5 mi long.
    a) How many kilometers is this?
    b) The Memorial Day races are for 200 laps or 500 mi. How many kilometers does this equal?
    c) The record average speed for the distance made in 1978 was 161.4 mi/hr. What does this equal in kilometers/hour?

21. An imported beverage costs $6.00 per liter. What will this cost per pint?

22. What is the area of a room that is 10.2 m by 15.3 m?

23. Materials for covering roofs are available in rolls that are 1 m wide and 25 m long. Assuming that there is an overlap of 10 cm on each strip of roofing, how many rolls of roofing materials would be required to cover a flat garage roof that is 10.5 m long and 7.8 m wide?

24. What is the volume of a bar of metal that is 22 cm long, 6.3 cm wide, and 2.3 cm thick?

25. How many liters of water could be contained in a tank that is 2 m in width and length and is 5 m high?

26. What is the volume in liters and in gallons of a water supply tank that has a diameter of 2.4 m and a height of 2.0 m? (volume of a cylinder = radius$^2 \times$ pi $\times$ height)

27. How high above the base is the 100-cm^3 mark on a graduated cylinder that has a diameter of 4.0 cm? (Use formula of Problem 26.)

28. Calculate the weight in kilograms, pounds, and short tons of a water bed filled with water if it is 7.0 ft long, 6.0 ft wide, and 1.0 ft. thick. (1 000 cm^3 of water = 1 kilogram)

29. Assuming that 1 kilocalorie of heat is required to heat one kilogram of water 1°C, how much heat is required to warm the water in the water bed

(above) from room temperature (20°C) to body temperature (37°C)?

30. Find the current quotation for the value of gold per ounce and convert this to price per gram; per kilogram.

31. What is the weight in milligrams of a $\frac{1}{2}$-carat diamond? If a point is defined as being 2 mg, what is the weight (in points) of the diamond?

32. Your aunt who lives in Germany has sent you a recipe for Bienenstich (Bee sting bars). The recipe calls for 100 grams of butter and 200 grams of sugar and specifies a temperature of 180°C for baking. Convert these values to United States values.

33. If you were to reply by sending a favorite recipe of yours to your aunt, how would you express the following?
    a) 2 cups of sugar
    b) 1 teaspoon of baking soda
    c) 350° Fahrenheit

Exercises 34–36 involve medical measurements. Use Table 2.4 for conversion factors.

34. Make the following conversions.
    a) 25 ml to fluid drams
    b) 50 gr to drams (Apothecaries')
    c) 10 oz apothecaries' to ounces avoirdupois
    d) 250 mg to grains
    e) 25 ml to teaspoons (approximate)
    f) 0.5 fluid drams to drops (approximate)

35. Using apothecaries' mass measurements, how many 5-dram medications could be derived from one pound of substance?

36. If 2 fluid ounces of a medication is prescribed and you have only a medication dispenser calibrated in milliliters, how many milliliters of the medication should you administer?

37. If a person is breathing 12 times per minute and exchanging 500 ml of air with each breath, how many liters of air are being taken in (and breathed out) in 8 hours?

38. Vitamin B_{12} is required at a level of about 1.5 μg per person per day. If you wanted to make 1 000 vitamin tablets containing twice this level, how many pounds of vitamin B_{12} would be required?

39. List some advantages of using the metric system instead of the British (United States) system.

40. List some differences between SI units and the metric system as used for scientific measurements.

41. Why do some scientists still use the unit calories instead of joules?

42. Distinguish between weight and mass.

43. Give some instances in which common or British units use a decimal system for expressing smaller or larger divisions of units.

44. What is the difference in meaning of the words "heat" and "temperature"?

SUGGESTED READING

Adamson, A. W., "SI units? A camel is a camel," *Journal of Chemical Education* **55**, 634 (1978). The author, who is a professor at the University of Southern California, expresses his objections to the SI. Although a bit complex for readers of this text, the article presents an interesting dissent.

Astin, A. V., "Standards of measurement," *Scientific American*, June 1968, p. 50. The goal is to define standards of length, mass, time, and temperature that are both precise and reproducible. Of the four, only mass still lacks a reference in nature.

"Brief History of Measurement Systems with a Chart of the Modernized Metric System," National Bureau of Standards, Special Publication 304, October 1972. This brief discussion and chart are handy references. They may be ordered from the U.S. Government Printing Office. A more complex reference is another National Bureau of Standards Publication, "Units of Weight and Measure. International (Metric) and U.S. Customary," Miscellaneous Publication 286, 1967.

Klein, H. A., *The World of Measurements*, Simon and Schuster, New York, 1974. This large book (over 700 pages) tells the history and controversies associated with the development of units of measurements. It is well written and interesting.

Lord Ritchie-Calder, "Conversion to the metric system," *Scientific American*, July 1970, p. 17. A report on Britain's program of "metrification."

Paul, M. A., "International System of Units (SI)," *Chemistry* **45**, No. 9, 14, October 1972. This article describes the SI and is illustrated with photographs of stamps issued by various countries to honor the metric system. The article lists the older units that are to be used for a limited time and those units that are generally deprecated.

Royal Society Conference of Editors, "Metrication in scientific journals," *American Scientist* **56**, 159 (1968). A discussion and listing of units used in the SI. A good reference.

Van Allen, J. A., *et al.*, "Heat," *Scientific American*, Sept. 1954. This issue contains nine articles covering as many different aspects of heat.

3

Fundamental Concepts of Chemistry

3.1 INTRODUCTION   Chemists, along with many other scientists, are concerned with the material of the universe and how it changes. The first questions they ask are concerned with what things are made of and how they are put together. There are many ways of answering the questions. The ancient Greek philosophers tried to simplify the problem. Thales (600 B.C.) advanced the idea that everything was really water: gold, iron, rust, bread, and man himself were merely different arrangements of water. It is interesting to note in passing that Thales lived on a small island and was no doubt aware of the extreme importance of water. Shortly after Thales, two other philosophers, Anaximenes and Heraclitus, proposed that air and fire, respectively, were the elements of which all things were made. Since each of the theories that proposed only one basic substance were probably too simple to account for what could be observed, they were developed into a theory generally attributed to Empedocles, which said that there were four elements—water, air, fire, and earth. According to the four-element theory, the different metals just contained different amounts of the various elements, and people reasoned that by altering the proportions it should be possible to change one metal (preferably a cheap one) into another (gold).

The four-element explanation was generally accepted for almost two thousand years. If we substitute the term liquid for water, gas for air, and solid for earth, and realize that fire was somehow related to the tendency of things to burn or react, the theory makes relatively good sense even to a twentieth-century person. We must realize, of course, that the terms apply to *states* and not *kinds* of matter.

In addition to discussions about the kinds of things that made up the universe, there was also the question of the units into which this "stuff" of the universe was organized. About 400 B.C. Democritus developed a theory that the elements were made of atoms. Thus you could have an atom of earth, a different kind of atom in water, and so forth. Plato proposed that these atoms were actually different simple solid shapes, those of the earth being cubes; fire, tetrahedrons; air, octahedrons; and water, twenty-sided solids, eicosahedrons. The primary idea about the atom, however, was that it could not be subdivided. The word atom comes from the Greek *tomos*, meaning to cut, and the prefix *a*, indicating a negative property, so that "atom" meant "uncuttable."

Before we begin to discuss modern ideas about the composition and structure of matter, we must give some basic definitions and discuss them. All organized knowledge begins this way. Let us first establish certain terms without discussing them in too much detail, and then discuss the precise details and meanings of the terms.

3.2
DEFINITION OF
CHEMISTRY

Chemistry is the science concerned with the composition and structure of all the substances of the universe, the properties of these substances, the changes they undergo, and the energy relationships involved in these changes.

Any such definition, of course, can indicate only the main subjects of interest to chemists. It is useful, however, because it suggests the limits on the areas which a chemist ordinarily investigates. In our modern, complex world some limitation of interests is necessary if a subject is to be approached in depth.

This chapter and many of those that follow use the organization implied in the definition given above. We first discuss the composition and structure of matter, or a particular type of matter, then examine the properties of the substances. After this we focus on chemical changes and the energy involved in these changes.

3.3
MATTER,
ATOMS,
ELEMENTS,
COMPOUNDS,
MOLECULES,
AND MIXTURES

When asked what things are made of and how they are put together, the modern scientist would begin by saying that the universe is composed of **matter**, which is defined as anything that has *mass* and occupies *space*. The simplest *particle* of matter that exists for any appreciable length of time is the **atom**. The simplest *kind* of matter is an **element**. Gold, oxygen, sulfur, and helium are all elements. An **element** is a kind of matter that contains only one kind of atom.

Two or more different elements can combine chemically to make a substance called a **compound**. Table salt, sugar, and water are compounds. Table salt is composed of the elements sodium and chlorine; sugar, of carbon, oxygen, and hydrogen; and water, of hydrogen and oxygen. When atoms combine with each other they form particles called **molecules**. Thus the smallest particle of a compound is a **molecule**. A molecule of water contains two atoms of hydrogen and one of oxygen and has the formula H_2O. Simple sugars contain 6 atoms of carbon, 12 of hydrogen, and 6 of oxygen in each molecule. Their formula is $C_6H_{12}O_6$.

While most molecules contain two or more different kinds of atoms, molecules are also formed when two atoms of the same element are bound together. The oxygen present in the air is ordinarily present as **diatomic molecules**, which we represent as O_2. Some oxygen is also present in air as ozone molecules, which have the formula O_3. For illustrations of atoms and molecules see Fig. 3.1.

Both elements and compounds are called pure substances, or often just **substances**. Most of the things we know about and work with in our daily life are **mixtures** of pure substances. While we are quite aware that cakes, concrete, cabbages, and kings are mixtures, it is not so obvious that air is also a mixture. Historically, air was thought of as being one of the basic elements, and it was not until the eighteenth century that scientists began to realize that it was a mixture. We now know that it contains the elements oxygen and nitrogen, the compounds water and carbon dioxide, and many other substances.

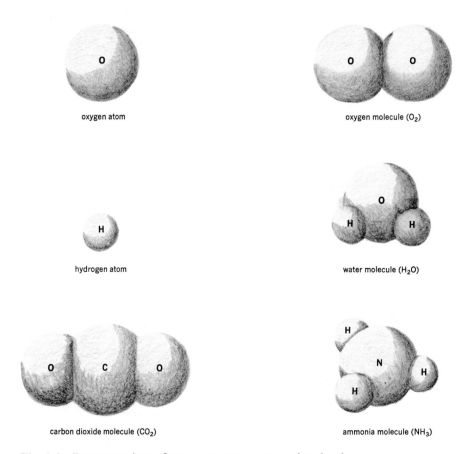

Fig. 3.1 Representations of some common atoms and molecules.

Like all mixtures, the composition of air varies, particularly with respect to the amount of water, carbon dioxide, and pollutants it contains.

Once the basic definitions have been given, the curious person will ask questions such as: "How do we distinguish pure substances from mixtures, elements from compounds, and one element or compound from another?" It is relatively easy to distinguish mixtures from pure substances. Compounds (and elements) always have the same composition. Thus, pure table salt is always 39% sodium and 61% chlorine (by weight), pure water is always 11% hydrogen and 89% oxygen (by weight). By contrast, as we analyze the mixture we call air, we find it is usually about 21% oxygen and 79% nitrogen (by volume), but that its composition may vary, especially if it contains pollutants. Thus we say that air is a mixture, not a compound. Each component of a mixture, even of polluted air, is itself an element or a compound with a definite composition.

Compounds differ from elements in that they are made of two or more kinds of atoms, while elements contain only one kind. Sometimes we know a substance is a compound because we can separate it into the elements of which it is composed. In other cases we know a substance contains two or more elements because we have observed the combination of the elements to form the compound.

Each element is different in some way from every other element, and each compound from all other compounds, just as each person is different from every other. Consider, for example, silver, gold, platinum, and mercury. Although they are all metals, only mercury is a liquid at normal temperatures. Gold is a different color from silver and platinum. We have greater difficulty in distinguishing silver from platinum and would have to measure some characteristic such as melting point or density to tell one from the other. If we can find no test that will distinguish one sample of material from another, we conclude that they are merely different samples of the same element or compound. Such tests are the basis of determining the identity or purity of an element or compound. They are very important in chemistry.

3.4
PHYSICAL AND
CHEMICAL
PROPERTIES

The characteristics of an element or compound that distinguish it from others are called its *properties*. Some properties of a substance can be determined by simple physical observations. The physical state (solid, liquid, gas), the color, and the odor are obvious **physical properties**. Chemists also include in their list of physical properties other properties that can be measured without changing the composition of the substance. Thus the melting and boiling points, the solubility, and the density are also physical properties that help us to recognize or identify a substance.

We know that gasoline is flammable and that iron rusts. These are important properties of these substances. However, when gasoline burns or iron rusts, the original substances are converted to new ones. There has been a change in composition, and we say *a chemical reaction has occurred*. A property that is seen when a substance reacts is a **chemical property**. The general degree of reactivity is also an important feature of chemical properties. We know that gold is inactive—it does not decompose and reacts with very few other substances. This property is the basis of the high value placed on this metal. White phosphorus, on the other hand, is a very active substance. When exposed to the air it bursts into flame. The degree of reactivity is also influenced by temperature. Copper does not react with oxygen at room temperature, but does when it is heated.

TABLE 3.1   Some properties of water, alcohol, and benzene

Observation or Test	Water	Alcohol	Benzene
color	colorless	colorless	colorless
taste	tasteless	characteristic burning taste	acrid
odor	odorless	sweet	definite odor
state at room temperature	liquid	liquid	liquid
freezing point	0°C	−117°C	5.5°C
boiling point	100°C	78°C	80.1°C
solubility	soluble in alcohol	soluble in water	insoluble in water; soluble in alcohol
density	1.0 g/ml	0.78 g/ml	0.80 g/ml
flammability	nonflammable	flammable	flammable

If you were given three liquids—water, ethyl alcohol, and benzene—in three separate bottles you would probably be able to identify which of the liquids was in which bottle. You would do this by observing certain properties, primarily the odor. There are many other properties that can be used to distinguish between these three substances. Some of them are given in Table 3.1.

The properties listed in Table 3.1 are sufficient to indicate which of three liquids is in which bottle if you know that only three possible compounds are present. It would be much more difficult, however, to prove beyond any doubt that, of all the substances on earth, a certain bottle contained ethyl alcohol. There are many liquids that have the same freezing points and boiling points. Therefore properties other than those listed here would be necessary for the identification of an unknown compound, especially an organic compound. Analysis of the components of a natural product such as strawberries, wood, or the human liver requires an extensive examination of the properties of each component.

| 3.5 STATES OF MATTER— SOLIDS, LIQUIDS, AND GASES | Its physical state at a given temperature is an important physical property of any pure substance, and we need to define and describe the possible states carefully in discussing basic principles of chemistry. |

3.5
STATES OF
MATTER—
SOLIDS,
LIQUIDS,
AND GASES

Its physical state at a given temperature is an important physical property of any pure substance, and we need to define and describe the possible states carefully in discussing basic principles of chemistry.

A **solid** has definite shape and volume. A **liquid** has definite volume but no definite shape; it will take the shape of the container which holds it. A **gas** has neither definite shape nor definite volume. Gases can be compressed from larger to smaller volumes and will spontaneously fill any container in which they are placed. Because both tend to flow, liquids and gases are sometimes called **fluids**. Gases are also known as **vapors** and the conversion of a liquid or a solid to a gas is **vaporization**. A substance that is easily vaporized is said to be **volatile**.

In a solid the atoms or molecules are close together; in a liquid, a bit farther apart; and in gases, there are few particles and a great deal of empty space. This is shown by the observation that a given volume of gas weighs less than the same volume of a liquid, and the liquid, in general, weighs less than the same volume of a solid. There are, of course, exceptions to the latter statement: ice, for example, is less dense than liquid water. This exception to the rule is most important since it means that ice floats when it forms on a lake or stream and eventually forms an insulating layer which protects the water at the bottom of the lake or stream from freezing. This is of great importance in allowing fish to live through the winter in colder climates.

With certain exceptions, matter may exist in any of the three states, and may be changed from one state to another under the influence of changes in temperature and pressure. The elements can exist in any of the three states, although the light gases hydrogen and helium must be cooled to very near absolute zero before they solidify, and very high temperatures are required to vaporize some of the heavier elements. While some compounds can exist in all three states, others decompose when heated. We know, for instance, that water exists in the three forms, as solid ice, as liquid water, and as a gas or water vapor. Below 0°C water exists as ice and above 100°C it is steam. Steam may be condensed to water either by cooling or by increasing the pressure.

How does a change in temperature and pressure influence a change in state of molecules or other particles? We think of all atoms and molecules (except those at absolute zero) as being in motion. If we heat such particles, they acquire more energy, which is expressed as an increase in speed of motion. The particles of a gas are moving faster than those in a solid. Although particles always tend to attract one another, the speed of the particles in a gas keeps them far apart and the attractive force is relatively unimportant. If we confine the particles in a smaller volume, as we do when we increase the outside pressure, they must come closer together and the importance of the attractive force increases. In some cases the force is sufficient to make the particles condense to form a liquid. If we remove energy by cooling the particles, this also slows them down and condensation becomes more likely. Either an increase in external pressure or cooling can cause condensation of a gas to a liquid. A combination of cooling and an increase in external pressure is very likely to cause condensation.

We can think of an atom or molecule in the gaseous state as moving rapidly and having a relatively large space of its own. It is a long way from other particles and little influenced by them—although it occasionally collides with another particle. Particles in a liquid are closer together and are moving more slowly than those in a gas. The attraction between particles in a liquid is relatively great. Particles in a solid are only slightly closer together than in a liquid (in some instances they may be slightly farther apart) but they are moving even more slowly and may form solid structures known as crystals. Cooling a liquid converts it to a solid, but pressure has little effect on either liquids or solids, since their particles are already very close together.

3.6
BOILING AND
MELTING
POINTS

The **boiling point** of a liquid and the **melting point** of a solid are important physical properties. As discussed in Chapter 2, our temperature scales are based on the boiling point and freezing point of water (the latter is the same as the melting point of ice, of course). Often we can distinguish between two compounds that look very similar by determining their melting points. Although melting and boiling points are sometimes referred to as constants, meaning that they do not change, we know that the boiling temperature of water does change with altitude. Why? To explain we need to consider, on a molecular basis, what happens when a substance boils or freezes.

To understand what happens when water or any other liquid boils, we need to recall that the molecules of the liquid are relatively close together and are attracted to one another. The molecules are also moving, but not all at the same speed. They are similar to automobiles in a busy parking lot, except that there are no traffic rules for molecules—only physical laws. If we say that molecules have an average speed it does not mean that they are all moving at the same speed. There are some slow ones that may be bumped by those that are going faster than the average. Those molecules having a speed sufficiently greater may actually escape from the surface of the liquid and **evaporate**. Observation tells us that all liquids in open containers evaporate, even at temperatures much below the boiling point of the liquid. We explain this by saying that some molecules have sufficient energy to overcome the attraction for other liquid molecules and that they enter the gaseous state or evaporate.

When we heat a liquid, we increase the rate at which molecules of the liquid are converted to gas. Not only does this happen at the air–liquid surface (as in normal evaporation), but we see that bubbles of gas are formed throughout the liquid, particularly near the source of heat. We find that when these bubbles are forming very rapidly, the temperature—which has been increasing as we heated the liquid—suddenly stops increasing and remains constant. This temperature is the boiling point of the liquid.

The tendency of molecules to evaporate is influenced not only by the temperature (which is merely a measure of the motion of the molecules) but also by the pressure of molecules in the gas above the liquid. What happens if we decrease that pressure? If we place a beaker of water in a vacuum (which means we remove some of the air molecules above the water) it will boil even at room temperature. Conversely, if we increase the pressure above a liquid, as we do when we use a pressure cooker to prepare foods, the boiling point is increased. We can define the **boiling point** as *the temperature at which the vapor pressure of a liquid equals the atmospheric pressure above the liquid*. At increased pressures the boiling point is increased, and consequently foods cook in less time. At reduced atmospheric pressures such as those found at high elevations, the boiling point of water, or any liquid, is decreased and it takes longer for "cooking" to occur. From this we realize that when we give a boiling point for a substance we need to give the atmospheric pressure at which we measured the boiling point. Many boiling points are so listed; for those that aren't, we assume that the pressure was the normal pressure at sea level. Any nonvolatile material dissolved in a liquid also increases the boiling point, and we take advantage of this when we use coolants in the radiators of automobiles (see Section 10.8 for a further discussion of this).

When a molecule evaporates, it joins other molecules—both air molecules and molecules of the liquid that have previously evaporated—in the gaseous state. Here the molecules are moving much faster than in liquids and are so far apart that they have almost no attraction for each other. If they are cooled and move slower, the attractive force between molecules becomes more important and the molecules **condense** into the liquid state. **Condensation** and **evaporation**, like freezing and melting, are opposite processes.

When a substance melts, it changes from a solid to a liquid. In solids the molecules or other particles are arranged in some sort of structure, which is lost as the molecules gain more energy and become liquids. The solids thus lose their shape, and the molecules of the liquids assume the "shape" of the containers in which they are found. Gases, of course, are shapeless although they may be confined within a balloon or other container.

The **melting point** of a solid, or the **freezing point** of the same substance in the liquid state, is the temperature at which we have both solid and liquid in the same container. When we heat a solid, we find that its temperature increases until it starts to melt. Then, as we noted with boiling liquids, the temperature does not further increase until all of the solid has melted. Similarly, when a liquid is cooled and starts to solidify, the temperature remains constant until all of the liquid has frozen. In contrast to boiling of liquids, however, the melting point of a solid (or the freezing point of a liquid) is not influenced by atmospheric pressure. It can be altered by the presence of impurities or by adding other substances. (See Section 10.8.)

If we observe dry ice (solid carbon dioxide), we see that at room temperature it does not melt, and yet its volume decreases rapidly. The carbon dioxide is passing directly from the solid state to a gas in a process known as **sublimation**. Many other substances such as paradichlorobenzene, which is used to control moths, and the element iodine also sublime. Ice can sublime and go directly from a solid to a gas when the temperature is below freezing. This process occurs naturally when ice is exposed to a cold, dry wind. Solids with a small number of atoms in each molecule have a tendency to sublime. It is this property that is responsible for the odors of such solids.

**3.7
SPECIFIC HEAT**   We have stated that the amount of heat required to raise the temperature of one gram of water one degree Celsius is one calorie. While this is true for water, it is not true for most other substances. One calorie will raise the temperature of alcohol nearly two degrees Celsius; to raise its temperature one degree requires only 0.535 cal. Each different substance has a **specific heat**, which is defined as *the amount of heat, in calories, required to change the temperature of one gram of a substance 1°C*. A list of the specific heats of some

common substances is given in Table 3.2. These specific heats are important physical properties.

TABLE 3.2 The specific heats of common materials*

	cal/g		cal/g
lead	0.0297	aluminum	0.208
mercury	0.033	ammonia	0.502
iron	0.107	ethyl alcohol	0.535
glass (flint)	0.117	water	1.000
granite	0.192	steam	0.48
		ice	0.44

* The specific heat is the number of calories required to change the temperature of one gram of a substance 1°C. These are average values. More precise values are known for narrow ranges of temperature.

From the table we can see that the specific heats of metals are low and that water has a high specific heat. One calorie of heat will raise the temperature of a gram of aluminum about five degrees, of a gram of alcohol about two degrees, and of a gram of water only one degree. On the other hand, water can *absorb* large amounts of energy from the sun or can *release* energy to warm large masses of cold air with only minor changes in the temperature of the water. The high specific heat of water accounts for the moderate climates of many areas of the world. Perhaps the best known example of this is seen in islands such as the Hawaiian Islands, where temperatures generally approximate the little-changing temperature of the ocean about them. In general, the temperatures of the coastal areas of the world are more moderate than those farther inland, due to the high specific heat of water in the nearby ocean.

**3.8
HEAT OF
VAPORIZATION
AND HEAT OF
FUSION
Heat of
Vaporization**

When we heat water or any pure liquid, we note that the temperature increases until the liquid begins to boil. Once the boiling point has been reached the temperature stays the same until all of the liquid has been converted to a gas. Since we continue to heat but do not increase the temperature, it is logical to conclude that the energy is being used to vaporize the liquid. We find that it takes 540 calories of heat to change one gram of water at 100°C to steam at 100°. *The amount of energy required to convert one gram of a liquid to a gas at its boiling point is called the* **heat of vaporization**. The same amount of energy is released to the surroundings when a gas condenses into a liquid. The heats of vaporization of some common substances are given in Table 3.3. Again we note the high value for water, which has the highest heat of vaporization of any substance of a similar molecular size.

TABLE 3.3 Heats of vaporization of some common substances at their boiling points

	cal/g		cal/g
chloroform	59	ethyl alcohol	204
mercury	71	ethyl chloride	93
benzene	94	ammonia	327
		water	540

In order for a substance to evaporate, it must acquire the necessary amount of heat. If the heat is not supplied from a flame or other heating device, it must be drawn from materials in the surroundings. This is the principle involved in refrigerators and air coolers. A substance evaporates and draws the heat necessary for the process from the surroundings, thereby cooling them. The human body uses the evaporation of perspiration to help keep its temperature normal on a hot day. Conversely, the temperature of the air often rises slightly when water vapor (a gas) is condensed into rain or snow.

Heat of Fusion   If we heat a beaker containing ice and water there is no change of temperature until the ice has all melted. This is due to the fact that energy is required merely for the change of state from a solid to a liquid. Since the old term for melting was fusion, we call this energy, which we define as *the energy required to convert a solid to a liquid*, **heat of fusion**. Again, water has a value that is high, but other substances have even higher values. The heats of fusion of some common substances are given in Table 3.4. The heat of fusion is also the amount of heat that must be removed in order for a substance to change from the liquid to the solid state, a process we commonly call freezing. As with the high specific heat of water, the heat of fusion of water is responsible for moderating some climates, since a great deal of cooling is required before water loses 80 calories per gram and freezes. Similarly, a relatively large amount of heat is required to melt ice once it has formed.

TABLE 3.4   Heats of fusion for various solids

	cal/g		cal/g
lead	5.5	glycerol	47.5
ethyl alcohol	24.9	copper	49
benzene	30	water	80
paraffin	35	aluminum	94

The curious person might ask at this point, "Why is there such a difference in the heats of vaporization and fusion and of specific heats of various substances?" The answer lies in the attraction of similar molecules for each other. Obviously, water molecules are very much attracted to one another, and great amounts of energy are required to melt ice or boil water. The molecules of some other substances have lesser attraction. We will discuss these matters further in Sections 6.2 and 6.3.

Calculations Involving Heating, Vaporization, and Melting   Calculations involving the amounts of heat involved in melting, freezing, boiling, condensing, and heating liquids, solids, and gases have many applications. We will confine our examples to relatively simple calculations.

Example 1   How many calories are required to change 50 g of water at 20°C to steam at 100°C? It requires 1 cal to raise the temperature of water 1°C, and we want to raise the temperature 100° − 20° or 80 degrees.

$$50 \text{ g} \times \frac{1 \text{ cal}}{1 \text{ g} \times 1 \text{ degree C}} \times 80 \text{ degrees C} = 4\,000 \text{ calories}$$

To change one gram of water at 100°C to steam at 100°C requires 540 cal, so

$$50 \text{ g} \times \frac{540 \text{ cal}}{1 \text{ g}} = 27\,000 \text{ cal}$$

Adding, we get

4 000 cal to heat from 20°–100°C
27 000 cal to convert to steam

———

31 000 cal required

Example 2   What will be the change in temperature (in degrees C) when 250 cal of heat are applied to 100 g of ethyl alcohol?

    In this case, we find the specific heat from Table 3.2. The value for ethyl alcohol is 0.535 cal/g per degree C. Since we know calories and want to find degrees change in temperature, we set up the following calculation.

$$250 \text{ cal} \times \frac{1°C \times 1\text{ g}}{0.535 \text{ cal}} \times \frac{1}{100 \text{ g}}$$

and our answer is 4.67°C.

Example 3   How many calories of heat must be absorbed if a tray of water at 20°C that is 9.0 cm by 28 cm by 3 cm is converted to ice cubes at −10°C. (Neglect the volume occupied by the divider in the ice-cube tray.)

    The tray contains 3.0 cm × 9.0 cm × 28 cm = 760 cm³ of water, and this water will weigh 760 g. To cool the water from 20°C to 0°C absorbs

$$760 \text{ g} \times 20°C \times \frac{1.0 \text{ cal}}{1 \text{ g} \times 1°C} = 15\ 000 \text{ cal}$$

Freezing 760 g of water absorbs 80 cal/g (from Table 3.4), and

$$760 \text{ g} \times \frac{80 \text{ cal}}{1 \text{ g}} = 62\ 000 \text{ cal}$$

Cooling 760 g of ice that has a specific heat of 0.44 (Table 3.2) from 0°C to −10°C (10 degrees) absorbs

$$760 \text{ g} \times \frac{0.44 \text{ cal}}{1 \text{ g} \times 1°C} \times 10°C = 330 \text{ cal}$$

The total heat absorbed is 15 000 + 62 000 + 330 calories, or 77 330 cal, which rounds to 77 000 cal.

**3.9
DENSITY**   If you were asked to describe the difference between the elements aluminum and lead, you might say that lead is heavier, but this would not really be accurate. Of course, the contrast between lead, water, and cork is well known to fishermen, who use lead sinkers and cork floats when they try to catch fish from the water. Nevertheless, it is inaccurate to say that one is heavier than the other. It is better to say that equal volumes of these three have different masses, as illustrated in Fig. 3.2. The property we are dealing with is **density**, which is defined as mass per unit volume. This can be expressed as

$$\text{density} = \frac{\text{mass}}{\text{volume}}$$

Since mass is commonly expressed in grams and volume in milliliters or cubic centimeters, the units of density are grams/milliliter or grams/cubic centimeter. (We tend to use milliliters if we have measured a value in milliliters, and cubic centimeters when we have measured the dimensions of an object in centimeters and calculated its volume.) We can also use the units kilograms/liter (which will give the same numerical value, since both

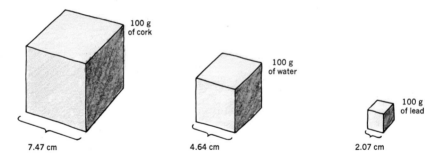

Fig. 3.2 Relative volumes of equal weights of cork, water, and lead.

of these units are 1 000 times the smaller ones). British units have also been used for density; thus we can use pounds/gallon or similar units.

To calculate the density of a substance, such as lead, we merely have to measure the mass and volume of a sample. From Fig. 3.2 we see that 100 g of lead has a volume of $(2.07 \text{ cm})^3$ or 8.87 cm^3, so its density is

$$\text{density} = \frac{100 \text{ g}}{8.87 \text{ cm}^3} = 11.3 \text{ g/cm}^3$$

Similarly, the density of cork is

$$\text{density} = \frac{100 \text{ g}}{417 \text{ cm}^3} = 0.24 \text{ g/cm}^3$$

We can find the density of water from Fig. 3.2 or can realize that a gram of water has a volume of 1 ml and since 1 ml = 1 cm^3, the density of water is 1.0 g/ml. Whichever approach you use, it is very important to know these relationships and to know that the density of water is 1 g/ml.

One very practical application of the concept of density is that things less dense than a given liquid will float on it and that things more dense will sink. Although we know that wood floats on water and metals sink, most metals will float on the liquid mercury. It's a matter of differences in density. Iron ships float because they contain great volumes of air; if they are filled with water or too much cargo, the air is replaced with more dense substances and they sink. Like the boiling point and other "constants," the density of water is not exactly constant but varies with the temperature. Water is most dense at 4°C. At this temperature its density is exactly 1.000 g/ml.

We can calculate the density of liquids as shown in the following example.

Example   Determine the density of alcohol, given that 90 milliliters weigh 71 g.

$$\text{density} = \frac{71 \text{ g}}{90 \text{ ml}} = 0.79 \text{ g/ml}$$

The densities of gases may also be calculated. The units generally used for expressing the density of a gas are either milligrams/cubic centimeter or grams/liter.

3.10
SPECIFIC
GRAVITY

It is sometimes easier to compare the mass of equal volumes of two substances rather than to determine the density of each. When we do this we determine a quantity known as the *relative density*. In particular, if water is the standard against which some other liquid is compared, we have the **specific gravity**. We must, of course, be certain that both

Fig. 3.3 Relative weights of equal volumes of liquids.

masses are determined at the same temperature; otherwise our value will not be accurate. The specific gravity is commonly given as a physical property of liquids or solutions, and density is usually given for solids and gases as well as for liquids.

A comparison of the weights of a gallon of water, of glycerol, and of gasoline are given in Fig. 3.3. To calculate the specific gravity of glycerol we set up the following:

$$\text{specific gravity} = \frac{10 \text{ lb}}{8 \text{ lb}} = 1.25$$

The general formula for calculating the specific gravity of any substance is

$$\text{specific gravity} = \frac{\text{mass of given volume}}{\text{mass of an equal volume of water}}$$

You will note that since the units in numerator and denominator must be the same, they cancel in the calculation. Therefore there are no units for a specific gravity value. This contrasts with density, where units are always involved and must be stated. However, in practical terms, if the specific gravity is calculated by comparison with water at 4°C,

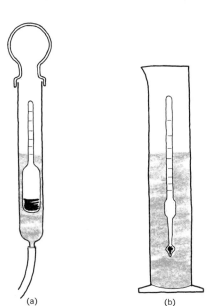

Fig. 3.4 Hydrometers used for determining specific gravity. The greater the density of the liquid being tested, the higher the hydrometer tube will float. Calibration marks on the float may indicate specific gravity, percentage of a given component, freezing point of the liquid, or charge of a battery. (a) A type of hydrometer that may be used to withdraw liquids such as those in an automobile battery or radiator. (b) Hydrometer used to test urine or other fluids.

which is often done, the specific gravity is numerically identical to density in grams/milliliter. For this reason handbooks of chemistry and physics often show specific gravity or density in the same list.

One advantage of citing the specific gravity as one of the physical properties of a substance in the liquid state is that the specific gravity is easily measured by a hydrometer (see Fig. 3.4). Since the coolant (antifreeze) in the radiator of an automobile is a mixture of two substances (water and glycol) which have different densities, a determination of the specific gravity will tell how much of each component is present, and a table of conversion can relate this specific gravity to the freezing point of the water–glycol mixture. A similar procedure is used to check the specific gravity and the charge of a storage battery such as those in automobiles. The liquid in a fully charged storage battery has a specific gravity of 1.30. When the specific gravity is less than 1.20 the battery needs to be recharged.

Hydrometers are also used to measure the specific gravity of urine (Fig. 3.4b). Here the specific gravity is influenced by the amount of solids dissolved in the urine. A low specific gravity (one that is only slightly greater than 1.0) is found after a person has been drinking a great deal of liquids. A higher specific gravity indicates greater excretion of dissolved solids or low water consumption. The specific gravity is often correlated with a disease or can be used to measure how well the kidneys are functioning.

3.11
PHYSICAL AND CHEMICAL CHANGES

We have been discussing changes that occur when solids melt and liquids boil. The question might arise, "Are these really changes, since we can usually reverse the change by cooling or heating?" The answer is "Yes, these are changes, but only of state, not of composition." We call these changes **physical changes**. They do not involve any alteration in the kinds or numbers of atoms in the molecules we are heating or cooling—only some alteration in the forces between particles or the size of the collection of molecules we are dealing with. Other examples of physical changes are processes such as crushing salt crystals into a fine powder or dissolving sugar in water.

Although chemists are interested in physical changes, they are even more interested in *changes in composition*, which are called **chemical changes**. When iron rusts, we no longer have iron but iron oxide. Burning gasoline (a compound of carbon and hydrogen) forms water (H_2O) and carbon dioxide (CO_2). These chemical reactions involve changes in the kinds of atoms in molecules and are examples of chemical changes. Chemical changes are accompanied by physical changes, as when liquid gasoline (and oxygen gas) is converted into the gases H_2O and CO_2. The reverse is not true—physical changes do not involve chemical changes. Both kinds of changes, however, involve input and output of energy.

3.12
ENERGETICS

When wood burns, heat and light are given off. We say that chemical compounds of the wood have united with oxygen from the air to form new compounds and, in the process, energy has been released.

Where did the wood get the energy? The simplest answer is that it came from the sun, was trapped by the green leaves of a tree, and was converted to the chemical energy in the wood. In the process energy was converted from that found in sunlight (radiant energy) into a form of stored or **potential energy** (chemical energy) and when the wood burns the stored energy is converted into heat and radiant energy.

Energy is defined as the ability to do work. We have listed some forms of energy—radiant, chemical, and heat. Other reactions may involve other types of energy. Explosions release sonic energy as well as heat and light. Sonic energy can initiate both physical changes, such as breaking a window, or chemical changes, as when we use sonic or ultrasonic energy to break chemical bonds in the walls of a living cell. Electrical energy can be used to decompose water into hydrogen and oxygen. It is produced in the reactions of batteries and by some animals such as the electric eel. The conduction of an impulse by a nerve involves electrical energy. Nuclear energy is released in the kinds of reactions we classify as nuclear reactions. These are discussed in Chapter 17.

**3.13
ENERGY IN
CHEMICAL
REACTIONS**

In addition to the well-known chemical reactions that occur when things burn or explode, there are many other, less obvious reactions such as those involved in moving a muscle or digesting food. All of these reactions involve changes in energy. Reactions that release energy are said to be **exergonic**. If the energy released is in the form of heat, they are said to be **exothermic**. Reactions that absorb energy are **endergonic**. Those that absorb heat are both endergonic and endothermic reactions. The decomposition of water into hydrogen and oxygen gases by the passage of a current of electricity through the water is called **electrolysis**. This is an endergonic reaction, and whenever the source of electricity is shut off, the reaction stops. When hydrogen and oxygen are mixed and a small amount of energy is provided in the form of an electrical spark or a flame, a violent explosion results. Although it takes special instruments to measure it, the amount of energy given off in the explosion of hydrogen and oxygen to form 1 g of water is exactly the amount required to electrolyze 1 g of water into hydrogen and oxygen gas.

Almost all chemical processes require an input of energy to start the reaction. Matches do not "light" until we supply some energy by striking them. Gasoline in the cylinders of automobiles requires an electrical spark to ignite it. This energy input is called the **energy of activation**. Only a few reactions are really spontaneous, and for these reactions we say that the energy of activation is sufficiently low that it can be supplied by the surrounding temperature. Endergonic reactions require a constant input of energy as well as an energy of activation. The importance of energy of activation for biological reactions is discussed in Section 34.4.

The primary source of some of the most important energy transformations is the energy of the sun. Light energy radiating from the sun travels approximately 93 million miles to the earth, where a substantial portion of it is converted to heat energy. Another portion of the sun's radiant energy is absorbed by green plants. Through the process of photosynthesis the light energy is transformed to stored chemical (potential) energy. We are indebted to this process, directly or indirectly, for all the natural fuels we have and all the energy in the foods we eat. For example, coal, petroleum, and natural gas, which meet most of the energy needs of industry, were at one time green plants. Furthermore, the warming of the earth by the sun causes water to evaporate into the atmosphere; variations in temperature cause this water to return to earth, forming streams which may be impounded behind dams. This potential energy due to the position of the water can be transformed into electrical energy, which in turn produces light, heat, chemical energy, mechanical power, and all the other things for which electricity is used.

3.14
CONSERVATION
OF ENERGY
AND MATTER

When discussing conversion of energy and matter from one state to another we might well ask about the efficiency of this conversion. Do we lose some energy in the process? What happens to the matter in a candle when it burns?

Careful experiments in which water is changed from ice to water to steam, and then back to water and ice, show that not only does the mass of water *not* change, but also that the amount of heat required to melt ice is released—in fact, it must be removed—when water freezes.

When we burn a candle it seems as if the matter of the candle has actually been destroyed, but this is not true. In fact, if we trap the gases given off by the candle we will find that they weigh more than the original candle did. The increase in weight is due to the oxygen from the air which reacted with the carbon and hydrogen atoms in the wax of the candle. When a candle burns, stored chemical energy is released as light and heat, and the carbon and hydrogen atoms are converted to the gases carbon dioxide and water. While compounds are created and destroyed in the reaction, the same number of atoms are present before and after the reaction. The amount of energy also has not been changed; it has merely been converted into different forms.

On the basis of many such experiments, we can formulate the **laws of conservation of matter and energy.** These laws state that *matter is neither created nor destroyed,* and that *energy is neither created nor destroyed in reactions.* Until recently there were no known exceptions to these laws. The advent of the atomic bomb, however, proved dramatically that matter can be converted into energy. Therefore, a more general law of conservation states that the **sum of the mass and energy is not changed during reaction.**

We now believe that even when a match burns, an extremely small amount of matter is converted into energy. Although the exact amount of matter so converted cannot be measured with even our most precise balances, it can be calculated. In a nuclear reaction such as that in an atomic bomb or in a nuclear-reactor power generator, an appreciable amount of matter is converted into energy. In this case, atoms are transformed into new kinds of atoms, but the new atoms always weigh slightly less than the old ones. Speaking strictly then, we must change the laws of conservation of matter and of energy into the new statement that indicates that matter may be converted into energy. Theoretically, the opposite process, the conversion of energy into matter, should also occur, but evidence for this has not yet been presented. It has been postulated that processes of this kind are occurring in some stars.

Since the amount of matter destroyed in ordinary *chemical* reactions is negligible, we shall treat all chemical calculations as if the old laws were true. However, the very nature of *nuclear* reactions is such that matter *is* converted to energy, and for some calculations we must recognize this fact.

SUMMARY

In this chapter we have defined and discussed some of the terms basic to chemistry. We have specified that matter exists as either elements, compounds, or mixtures, and that these are made of particles called atoms and molecules. We have discussed the meaning of physical and chemical changes and properties. We have also discussed the energy that is important in chemical reactions. Perceptive readers will realize that we have not really told them what an atom or molecule is, and while defining the terms element and compound, we have not talked about the reasons why one element is different from another, what causes two atoms of different elements to combine to form a molecule of a compound, or what forces hold them together once they have combined.   The next several chapters will discuss these questions.

IMPORTANT TERMS   *Learn these terms*

atom	energy of activation	mixture
boiling point	evaporation	physical change
chemical change	exergonic	physical properties
chemical properties	exothermic reaction	potential energy
condensation	fluid	reactivity
chemistry	freezing point	solid
compound	gas	specific gravity
density	heat of fusion	specific heat
electrolysis	heat of vaporization	sublimation
element	liquid	substance
endergonic	matter	vapor
endothermic reaction	melting point	vaporization
energy	molecule	volatile

WORK EXERCISES

1. Calculate the amount of heat (in calories or kilocalories) required or absorbed in the following processes.
   a) 48 g of water is heated from 12°C to 22°C.
   b) One kilogram of aluminum is heated from 20°C to its melting point of 660°C. (Assume the specific heat stays constant over this range of temperatures.)
   c) 50 g of ice at −15°C is changed to steam at 120°C.
   d) 1.00 liters of a liquid having a specific gravity of 0.80 and a specific heat of 0.45 is heated 25°C.
   e) The ice on a lake that is 550 m long by 320 m wide, with a layer of ice 10 cm thick, is melted. Assume that the density of the ice is 0.90.

2. Many volatile substances are used as local anesthetics because their heat of vaporization cools the skin and produces insensitivity to pain.
   a) Calculate the amount of heat absorbed when 4 g of ethyl chloride evaporates.
   b) If at 36°C this amount (4 g) of ethyl chloride affects about 8 g of skin tissue, which has a specific heat of 0.90 cal/gram, what will be the final temperature of the skin?
   c) If insensitivity to pain occurs when skin temperature drops to 50°F, is this application of ethyl chloride sufficient to produce anesthesia (insensitivity to pain)?

3. If a sample containing 1.05 g of food is completely oxidized and the energy released raises the temperature of 855 g of water 3.26°C, what was the amount of energy contained in 1 g of this food?

4. It is estimated that a human loses 300 ml of water a day by evaporation from the skin.   How much heat does this remove from the body per day?

5. Given that 50.0 ml of mercury weighs 677.3 g, what is the density of mercury in grams/milliliter?

6. An aluminum rod with a volume of 40 ml weighs 108 g.   What is the density of aluminum in grams/milliliter?

7. A gold nugget weighing 273 g was found to have a volume of 14.14 ml.   What is the density of the nugget in grams/milliliter?

8. A cube of magnesium metal 5 cm on a side weighs 217.5 g.   What is the density of magnesium in grams/cubic centimeter?

9. A bottle which holds 200 g of water can hold only 142.7 g of ether.   What is the specific gravity of ether?   What is the density of ether in grams/milliliter?

10. Wood alcohol has a density of 0.796 g.   What will 25 ml weigh?

11. The weight of water and ice are 62.5 and 57.2 lb/cu ft respectively.   What will 25 ml of each of these weigh?

12. A piece of jade, which has a specific gravity of 3.5, weighs 640 g.   What is its volume in milliliters?

13. The acid in an automobile battery has a specific gravity of 1.2.   What will 200 ml of the acid weigh?

14. A cube 4 in. on a side weighs 4 000 g.   What is its density in grams/cubic centimeter?

15. A steel shaft 6 ft long and 3 in. in diameter has a density of 7.7 g cc.
    a) What is its weight in kilograms?   (Use 1 in. = 2.5 cm and 1 kg = 2.2 lb.)

b) What is its weight in pounds?

16. Calculate the density of a gas (grams per liter) if a 250-ml balloon filled with the gas weighs 320 mg.

17. Air has a density of about 1.286 mg/cm^3 at 0°C. A 5-liter container of carbon dioxide contains 9.80 g of CO_2. Is carbon dioxide more or less dense than air? Would you expect the carbon dioxide in a poorly ventilated, crowded room to be more concentrated near the ceiling or near the floor?

18. Contrast or distinguish between the physical characteristics of the three states of matter.

19. What could you do to bring about the following changes?
    a) Conversion of a gas to a liquid
    b) Conversion of a liquid to a gas
    c) Conversion of a liquid to a solid
    d) Conversion of a solid to a liquid

20. State four physical properties of aluminum metal.

21. State some of the properties of gasoline and indicate whether they are physical or chemical.

22. Distinguish between physical changes and chemical changes.

23. Why is the statement "below the freezing point, water is a solid" not completely correct?

24. Why must chemical change always be accompanied by a physical change, whereas a physical change is not necessarily accompanied by a chemical change?

25. Why are the terms exothermic and endothermic not always appropriate in describing the energy relationships of chemical reactions?

26. Why has it been necessary to alter the original laws which pertain to the conservation of matter and energy?

27. Pure gold is labeled 24 karat. An alloy that is 75% gold is 18 karat. What is the density of 18K gold if the other 25% of the alloy is copper? (Density of gold is 19.3; of copper, 8.9.)

28. A typical dental alloy used for filling teeth has the following composition: silver 71%, tin 26%, and copper 3%. Calculate the density of this alloy. (Density of silver is 10.5; of tin, 7.3; and of copper, 8.9.)

29. A gold chain is marked 18K. The remainder is said to be silver. If the chain has a volume of 3 cm^3 and weighs 48 g, is this chain correctly labeled?

30. Using a Chemical Handbook, consult a table of the Physical Properties of the Elements and list all elements that have a density equal to or greater than that of gold.

SUGGESTED READING

Apfel, R., "The Tensile Strength of Liquids," *Scientific American*, December 1972, p. 58. Although liquids generally are not thought of as having tensile strength, this property and its measurement are discussed in this article.

4

The Composition and Structure
of Matter

**4.1
INTRODUCTION**

In previous chapters we not only defined terms, but we discussed some of them so that the reader might gain understanding instead of merely memorizing definitions. Although we said that particles known as atoms are the basic building blocks of all kinds of matter, we have not yet really discussed what an atom is. This chapter provides an answer, neither final nor detailed, but nevertheless an answer to the question, "What is an atom?" Along the way, we will also discuss the nature of molecules.

If we ask the question, "How small can a piece of gold be made?" there are two possible answers. One is that there is no limit: theoretically, you could go on cutting a piece of gold, or any other element, indefinitely if you had the proper tools or instruments. (Since this is purely a theoretical argument, we will assume that the finer and finer tools exist.) You would never arrive at a particle that could not be further divided. The other possible answer is that the process of division is not infinite, but that eventually you would come to a particle that could not be cut. This fundamental, indivisible particle is what Democritus called an atom (Section 3.1). Once the two answers are proposed we could argue indefinitely about whether we wish to think that matter is infinitely divisible or not. This is the situation that existed for nearly two thousand years after Democritus proposed his theory. Some evidence was brought up to support each side, but common sense seemed to favor infinite divisibility. What was needed to settle the argument was experimentation, a deliberate effort to uncover enough new facts to settle the argument one way or the other.

In 1803, John Dalton, an English school teacher, published a series of statements regarding the nature of atoms and molecules. These statements, which we call *Dalton's atomic theory*, were based on some of his own experiments and on those of others who had gone before him. **Dalton's theory** can be stated as follows:

1. Matter is composed of tiny, indivisible particles called atoms.

2. Atoms of the same element are all alike.

3. Atoms of different elements are different.

4. In chemical reactions atoms combine, separate, or regroup. They are the units that enter chemical reactions, forming "compound atoms" (molecules and similar particles).

5. In chemical reactions whole atoms are always involved. They may combine in a one-to-one ratio, a one-to-two ratio, or a one-to-three ratio, etc.

6. The weight of an atom never changes as a result of a chemical change.

These statements are simple and definite. Furthermore, there were many experiments that could be done to test their truth. The results removed Dalton's statements regarding the divisibility of matter and, in fact, the basic structure of matter from the level of untested assertion to that of a theory based on and supported by experiments. In general, we still believe the principles of Dalton's theory. In common with almost all theories, it has been modified slightly to conform with the results of subsequent chemical observations and experiments. In this book we cannot give all the background that led to Dalton's proposals, but the following discussion of the laws of Definite and Multiple Proportions serves to illustrate two laws derived from experiments which support Dalton's atomic theory. Without evidence of the operation of these two laws, Dalton's theory would have remained only Dalton's speculations.

**4.2
LAW OF
DEFINITE
PROPORTIONS**

Careful studies of the manner in which elements combine and of the composition of the resulting compounds produced evidence that in a pure substance or compound the percentage of each element is always the same. For instance, the percentages by weight of the elements in methane are always 75% carbon and 25% hydrogen. In water they are 88.8% oxygen and 11.2% hydrogen. This information led to the formulation of the **Law of Definite Proportions**, which states that *in a specific chemical compound the elements are always present in a definite proportion by weight.*

From the percentages of the elements in methane one might be led to the (erroneous) conclusion that methane contains three carbon atoms and one hydrogen atom, since 75% is three times 25%. This conclusion is based on the false assumption that all atoms are of equal weight. Atoms do, in fact, vary greatly in weight. When the number of grams of carbon, 75 g, contained in 100 g of methane (75% carbon) is divided by 12, the atomic weight* of carbon, the result is 6.25:

$$\frac{\text{grams of carbon (per 100 g of methane)}}{\text{atomic weight of carbon}} = \frac{75}{12} = 6.25$$

When the number of grams of hydrogen, 25 g, contained in 100 g of methane is divided by 1, the atomic weight of hydrogen, the result is 25:

$$\frac{\text{grams of hydrogen (per 100 g of methane)}}{\text{atomic weight of hydrogen}} = \frac{25}{1} = 25$$

The above calculations indicate that 100 g of methane contains 6.25 atomic weights of carbon and 25 atomic weights of hydrogen, a ratio of 4 hydrogen atoms for each carbon atom, since

$$\frac{25}{6.25} = 4$$

---

* The establishment of atomic weights was a long and involved process. It is discussed in Section 4.7.

The **empirical formula** (which is the simplest ratio of atoms in a molecule) for methane is therefore CH_4. Similarly for water, 11.2 divided by 1 is 11.2, and 88.8 divided by 16 is 5.6, so that the actual ratio (11.2/5.6) is expressed as the familiar formula H_2O.

4.3
LAW OF
MULTIPLE
PROPORTIONS

While the Law of Definite Proportions was found to be true for most compounds, there were exceptions. For example, there are two compounds of carbon and oxygen that have different percentage compositions. One is 27% carbon and 73% oxygen (we know it as carbon dioxide or CO_2), and the other (which we know as carbon monoxide) is 43% carbon and 57% oxygen. Dalton worked with two compounds of carbon and hydrogen that we now call methane (CH_4) and ethene (C_2H_4).

As a result of his studies and thinking, Dalton formulated a second law, the **Law of Multiple Proportions**. That law says that *two kinds of atoms may unite in a number of combinations forming different kinds of molecules with different percentages of composition, but their combining ratios will always be in the ratio of small whole numbers.* As has happened many times in the history of science, the original statement of a law—here, the Law of Definite Proportions—had proved to be too simple to describe what was found by experiment. A second law (or a modification of the first) was necessary to explain nature. It is interesting to speculate whether some of the things we now accept as laws may have to be modified or perhaps replaced in the future. The famous chemist Joel Hildebrand, who was born in 1881 and received his bachelor's degree in chemistry in 1903, has said that 90% of the chemistry he learned in his first college course is not now believed to be true.

Given experimental confirmation of the Laws of Definite and Multiple Proportions, most scientists became convinced of the reality of tiny, indivisible particles having a definite weight, and the atomic theory flourished. However, as recently as 1900 there were important scientists who did not accept the existence of atoms. And, while we still believe in much of what Dalton said, his theory has had to be modified in the light of more recent findings.

4.4
SUBATOMIC
PARTICLES

In the latter years of the nineteenth century, evidence began to accumulate that the atoms which Dalton had thought were indivisible, stable particles could not only be broken down under extreme conditions, but, in some cases, would spontaneously fall apart. The evidence came from a variety of sources.

It had been known for a long time that electrical discharges would travel between electrodes more readily in an evacuated (vacuum) tube. An English physicist, William Crookes, found that the rays in such a tube (Fig. 4.1) could be deflected by a magnet. The direction of deflection indicated that these rays were negative in character. In 1897 another English physicist, J. J. Thompson, found that these rays, which were named **electrons,** were tiny particles bearing the elementary negative unit of electricity. Later it was found that their mass equaled 1/1837 that of a hydrogen atom. These discoveries proved definitely that the atom is not the smallest unit of matter, as had been postulated by Dalton.

Since matter is normally neutral in character, it was reasonable to speculate that if matter consists in part of negatively charged particles, there must also be positive particles. The search for such a particle was culminated in 1919 by Rutherford. He identified a particle, the **proton,** which carried a unit of positive charge equal in magnitude to that of the electron but opposite in character. Its mass was found to be 1836 times that of the

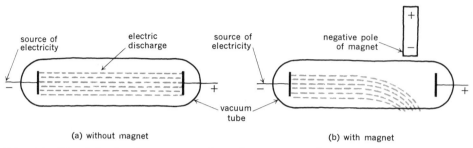

(a) without magnet   (b) with magnet

Fig. 4.1   The deflection by a magnet of rays in a vacuum tube.

electron and almost identical to that of the hydrogen atom. By 1920 it had been suggested that an elementary uncharged particle was also present in atoms. The existence of this particle, named the **neutron**, was confirmed by Chadwick in 1932. The neutron has a mass essentially equal to that of the proton or the hydrogen atom.

Discoveries during this period of active research aided and stimulated the many workers in this field. Henry Becquerel discovered radioactivity in 1895, and Marie Curie discovered the radioactive element radium in 1898. It was found that radium atoms slowly decompose, giving off three types of radiation: negative *beta* (β), which are electrons; positive *alpha* (α), which are positively charged particles similar to helium atoms; and neutral *gamma* (γ), which are considered to have zero rest mass. Gamma rays are comparable to x-rays. (A more detailed discussion of radium and radioactivity is given in Chapter 17 on nuclear chemistry.)

**4.5
COMPOSITION
AND
STRUCTURE OF
ATOMS**

At the time that some scientists were discovering various subatomic particles—the electron, the proton, and the neutron—other scientists were already performing experiments that indicated the ways these particles were arranged in atoms. In 1911 Rutherford performed a unique experiment. It had been found that when alpha particles (charged helium atoms) struck a certain type of sensitized surface, flashes of light were visible. A piece of radium (an alpha-particle emitter) was placed in a lead tube closed at one end so that the only alpha particles escaping could be aimed toward a target. Under these conditions all the strikes were concentrated in one area. When a very thin section of gold foil, 1 000 to 2 000 atoms thick, was placed in the path of the alpha particles, about 1 particle in 100 000 was deflected while the rest passed through and hit the target as if nothing were in the way. (See Figs. 4.2 and 4.3.) The results of this experiment led to the belief that very little of the space in an atom was occupied by matter. In fact, it is commonly said that an atom is as porous as a solar system, and that if we considered the likelihood of a hit by a missile shot at random through our solar system, we would have an idea of the structure of matter.

Fig. 4.2   Alpha particles striking a solid target.

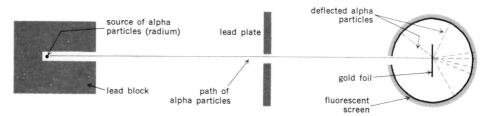

Fig. 4.3 Rutherford's experiment, showing paths of alpha particles after striking gold foil.

As the result of Rutherford's experiment, we now believe that most of the matter within the atom is located in a very small space called the **nucleus**. It has been calculated that the diameter of the nucleus is somewhere between 1/10 000 and 1/100 000 that of the atom. By way of illustration: if an atom could be enlarged until the nucleus were the size of BB shot, the atom would approximate the dimensions of a large football stadium.

In Rutherford's experiment described above, most alpha particles went straight through the foil; however, some were deflected at right angles, indicating a hit, while others were deflected only slightly, indicating that they were repelled by an electromagnetic force. Since the alpha particles are positively charged, one would expect them to be repelled by other positively charged bodies. It was reasoned, therefore, that the nuclei of the atoms must also be positively charged.

These experiments suggested that the positive and heavy particles are in the nucleus of the atom. The next problem was to pinpoint the location of the negative particles, the electrons. The most widely accepted theory became that of Rutherford and the brilliant Danish physicist Niels Bohr. They envisioned a very dense, compact nucleus surrounded by electrons traveling at exceedingly high speeds far from the nucleus, somewhat like planets traveling about the sun. This picture of an atom is a model, often known as the "Bohr atom." Bohr's model does not explain some properties of atoms, and other models and more elaborate theories have been developed. Some of these will be discussed in Chapter 5.

Research into atomic structure has indicated that there are many subatomic particles, most of which do not exist for any appreciable time when alone. The list includes the positron, the neutrino, the pions, the muons, and the newest family of particles, the quarks. These particles are observed only when atoms are disrupted by powerful forces, and they tend to combine in microseconds to reform atoms or other more stable particles. While a knowledge of such particles is important for a detailed understanding of the structure of atoms, most fundamental chemistry is understandable on the basis of only three subatomic particles: the electron, the proton, and the neutron. The other particles, particularly the quarks, are discussed in articles listed in the Suggested Reading for this chapter and in the *Study Guide*.

As the foregoing account suggests, our present conception of the atom differs significantly from Dalton's. Instead of a small, hard, indivisible particle, we think of the atom as being largely empty space containing a tiny nucleus surrounded by orbiting electrons. The nucleus is composed of protons and neutrons. The simplest atom, hydrogen, contains only one proton and one electron. Other atoms contain both protons and neutrons in their nuclei. In the smaller atoms the numbers of each are approximately equal. In the larger atoms there are more neutrons than protons. In a complete atom of any element there are always as many electrons surrounding the nucleus as there are protons in the nucleus, and the atom is electrically neutral.

After being told about atoms you might ask "How many different kinds of atoms are there?" The question is similar to "How many different elements are there?", since by definition each element contains only one kind of atom and each element is different from every other one. The difference between atoms is based on the number of protons in the nucleus of an atom—or the number of electrons orbiting around it, since the numbers are the same. The simplest atom (element) is hydrogen, which has one proton in its nucleus and one electron circling about it. The most complex atom that occurs naturally to any appreciable extent is uranium, which has 92 protons and electrons. *The number of protons in the nucleus of an atom is its* **atomic number.** The atomic number of hydrogen is 1 and of uranium is 92. Except for three elements which are so unstable that they are not found naturally, all atomic numbers from 1 to 92 are represented in the elements we find on the earth. Many elements found on earth have also been detected on the sun or in other objects in the universe. The composition of some common atoms is shown in Table 4.1.

TABLE 4.1   The composition of a few common atoms

	Number of Protons (Atomic Number)	Number of Neutrons	Atomic Weight	Number of Electrons
hydrogen	1	0	1.008	1
helium	2	2	4.003	2
carbon	6	6	12.011	6
oxygen	8	8	15.999	8
fluorine	9	10	18.998	9
sodium	11	12	22.990	11
chlorine	17	18	35.453	17

The number of protons is the most important characteristic of a given atom. If the number of protons in two atoms is different, they represent two different elements. The same is not true of electrons or neutrons. If an atom loses or gains electrons, it acquires an electrical charge and some different properties; however, it is still the same element. The number of neutrons can also vary without a major change in properties of an atom, as we shall see when we discuss isotopes in Section 4.9.

Dalton's theory of atoms was based on the weight or mass of individual atoms and the relative combining weights of elements as they formed compounds. It was not too difficult to determine the *relative* weights of elements that entered into combinations to form compounds or that could be produced by decomposing compounds. Water always contained 88.8% oxygen and 11.2% hydrogen. However, as we saw in Section 4.2, to find the number of *atoms* of each substance in a molecule it is necessary to know atomic weights; but at the same time, without knowing the formulas of at least a few compounds, early chemists could not be sure about atomic weights.

As is often done in such cases, one element was assigned a given weight and then the others were defined in terms of that element. As chemistry developed, the relative weight of the lightest element, hydrogen, came to be accepted as being 1.0. When the proton and neutron were discovered it followed that they also had a mass very near to 1.0.

Today, we use carbon as the basic reference, and the atomic mass unit (amu) is now defined as 1/12 the mass of a carbon atom having a mass of 12. One amu is equal to 1.66×10^{-27} kg. Using this standard definition of the amu, protons and neutrons have values slightly different from 1.0. (See Table 4.2.) However, for the purposes of this text, the value of 1.0 will be used for the mass of the proton, the neutron, and ordinary hydrogen.

TABLE 4.2   The mass of particles

| | Mass | |
Particle	grams	amu★
electron	9.110×10^{-28}	5.4858×10^{-4}
proton	1.6726×10^{-24}	1.0073
neutron	1.6750×10^{-24}	1.0086
hydrogen atom	1.674×10^{-24}	1.008
carbon-12 atom	1.993×10^{-23}	12.000
uranium-235 atom	3.903×10^{-22}	235.04

★ 1 amu $= 1.6605 \times 10^{-24}$ g.

The **mass number** *of an atom is equal to the sum of protons and neutrons in its nucleus.* Although there are as many electrons as there are protons in any atom, each electron has only 1/1836 the mass of a proton and in most calculations has an insignificant effect upon the mass of an atom. Even elements having 100 electrons would have only $100 \times 1/1836$ or 100/1836 amu due to the electrons.

If we want to indicate the mass of an atom, we can do so by writing a number to the upper left of the atomic symbol; thus ^{12}C indicates a carbon atom with a mass of 12. We also use the expression "carbon-12" to indicate the same atom. We can include the atomic number as well and write $^{12}_{6}$C, indicating that the atomic number, the number of protons, in carbon is 6. Of course, we do not really need the latter number (6), since the symbol C is sufficient to indicate the atom with 6 protons. However, in writing nuclear reactions, and for the purposes of being complete, we often use both the atomic number and the symbol.

Historically, when hydrogen was established as having a weight of one unit, atomic weights for other elements were based on the weights of various elements that were found to combine with each other to form compounds. For instance, the ratio of oxygen to hydrogen that combined to form water was always 8 to 1. Thus oxygen would be assigned a relative weight of 8. When it was found that there were two hydrogen atoms in a molecule of water, the ratio was better expressed as 16:2, so that each hydrogen atom would still have a relative weight of 1, thus giving oxygen a relative weight of 16. At the time these relative atomic weights were established, there was no idea of the size or weight of a single atom. Instead a weight in grams which was equal to the relative atomic weight was used, and this quantity was called the **gram atomic weight**. Similarly, a number of grams equal to the relative molecular weight of a compound was called a **gram molecular weight**. The term **gram formula weight** is used to indicate either a gram atomic weight or a gram molecular weight.

Theories that led to the establishment of the mass of just one atom and the number of atoms in a gram atomic weight start with Amedeo Avogadro, who lived from 1776 to

1856. He theorized that *there are equal numbers of particles* (which we would now call atoms or molecules) *in equal volumes of all gases.* One liter of oxygen gas is 16 times as heavy as one liter of hydrogen. From his theory it follows that a certain volume of a gas will have a weight equal to the atomic (or molecular) weight, and that this volume is the same for all gases at standard temperature and pressure (273 K temperature and 1 atmosphere of pressure). By various methods investigators determined that this standard volume of a gas—22.414 liters—contains 6.022×10^{23} particles (either atoms or molecules). This number is known as **Avogadro's number**, although he did not determine it. The official value of Avogadro's number changes from time to time as more precise measurements are made.

While Avogadro's theory originally applied only to gases, we now know that one gram formula weight of any substance, whether in the gaseous, liquid, or solid state, contains Avogadro's number of particles. Thus, to find the mass of one atom of any element, we can divide the atomic weight by Avogadro's number. For instance, an atom of uranium-238 has a mass of $238/6.022 \times 10^{23}$ or 39.5×10^{-23} grams, which should be written 3.95×10^{-22} grams.

4.8 **LISTS AND** **TABLES OF** **ELEMENTS**	The List of Elements that is found inside the front cover of this book gives the name, symbol, atomic number, and atomic weight for all known elements. The elements are listed alphabetically so you can find them easily.

Similar information is shown in the Periodic Table that is inside the back cover of the book. This Table is more difficult for the beginning chemist to use, but after the discussion of its development and use in Chapter 5, you will find that it gives much more information than the List of Elements. In addition to giving the symbol, atomic number, and atomic mass, it tells a great deal about the structure of atoms and the properties of the elements.

While the terms atomic mass and atomic weight are often used interchangeably, it is most logical to speak of the mass when referring to a single atom and to use the term atomic weight when referring to an amount of a substance that can actually be weighed. (For the difference between mass and weight, see Section 2.5.)

4.9 **ISOTOPES**	When we examine a List of Elements or a Periodic Table, we are immediately struck by the fact that the atomic weight of most elements is not a whole number. And, although many have weights that are nearly a whole number, chlorine has a weight of 35.453 and copper a weight of 63.54. Since the weight of atoms is due primarily to protons and neutrons, each of which has approximately one unit of mass, we are faced with the problem of explaining the uneven weights (masses), just as early chemists were.

The explanation came from experiments using a mass spectrometer, an instrument developed by the British physicists J. J. Thompson and F. W. Aston (Fig. 4.4). When particles (either atoms or molecules) are placed in this instrument, they tend to lose or gain electrons and become charged particles known as ions. Charged particles having different masses will behave differently when placed in a magnetic field. The particles with greater masses are deflected less than ligher ones, much as walnuts and leaves are separated by the wind and gravity (Fig. 4.5).

When neon gas was examined by the mass spectrograph it was found to be separated into two groups of ions with distinctly different masses, indicating that there were two varieties of neon ions. To the surprise of the investigators, they found many elements in

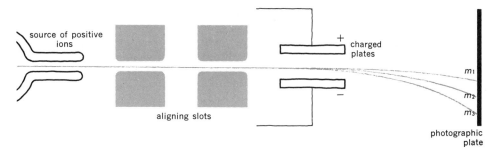

Fig. 4.4 Schematic drawing of a mass spectrograph. The charged particles produced at the left are deflected by the charged plates, and isotopes of different masses (m_1, m_2, and m_3) hit different parts of the photographic plate.

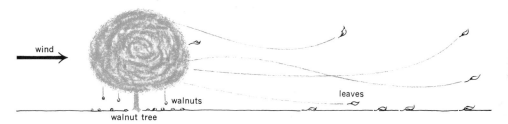

Fig. 4.5 The effect of an equal force on objects of different masses. Although the wind will blow leaves a distance from a tree, it will have little effect on walnuts.

addition to neon that gave indications of having atoms of different masses. It has since been determined that hydrogen also exhibits this property. Three kinds of hydrogen atoms exist. The simplest and most common contains 1 proton and is sometimes called *protium* to distinguish it from the other forms. The second, *deuterium*, has a nucleus containing one proton and one neutron, while the third, *tritium*, has one proton and two neutrons. It must be remembered that each of the three forms of hydrogen has only one proton and each has one electron. They all have the fundamental properties of hydrogen.

Different forms of the same atom that differ only in atomic mass are called **isotopes**. The difference in mass is explained by the presence of different numbers of neutrons in the nucleus.

From experiments, we know that there are two common isotopes of chlorine, having masses of 35 and 37, which we can write as ^{35}Cl and ^{37}Cl or as chlorine-35 and chlorine-37. All chlorine atoms must have 17 protons; therefore, one isotope has 18 neutrons (which we get by subtracting the number of protons, 17, from the atomic mass of 35) and the other isotope has 20 (37 − 17) neutrons. Further experiments tell us that approximately 75% of the chlorine atoms found on the earth are chlorine-35 and 25% are chlorine-37. Table 4.3 lists some common isotopes and gives the percentage of each found naturally.

If we know the percentage of each isotope present on earth, we can find the average atomic weight by multiplying the mass of each isotope by the percentage and then adding all the values obtained.

Example Given the values for the occurrence of the different isotopes of chlorine in Table 4.3, calculate the average atomic weight of chlorine.

TABLE 4.3   Some common isotopes

Isotope	Atomic Number	Protons	Neutrons	Mass Number	Atomic Weight	Percentage of Isotopes Found Naturally
hydrogen-1	1	1	0	1 ⎫		99.984
hydrogen-2	1	1	1	2 ⎬	1.008	0.016
hydrogen-3	1	1	2	3 ⎭		trace
carbon-12	6	6	6	12 ⎫		98.892
carbon-13	6	6	7	13 ⎬	12.011	1.108
carbon-14	6	6	8	14 ⎭		trace
fluorine-19	9	9	10	19	18.998	100.00
chlorine-35	17	17	18	35 ⎫		75.4
chlorine-37	17	17	20	37 ⎭	35.453	24.6

From the table we find that 75.4% (0.754) of chlorine is chlorine-35 and 24.6% (0.246) is chlorine-37, so

$$35.0 \times 0.754 = 26.39$$
$$37.0 \times 0.246 = \underline{9.10}$$
$$\text{Total} \qquad = 35.49$$

This value is somewhat different from that given in Table 4.3 and in lists of atomic weights because the starting figures used here are not as precise as those used in compiling such lists; however, the example does illustrate the method by which average atomic weights are obtained.

The occurrence of isotopes explains the reason for atomic weights that are not whole numbers. *The List of Elements gives the average weights of the elements as found on the earth, and these weights include isotopes of several different masses.* We refer to an individual isotope as a **nuclide.** Thus ^{14}C, ^{12}C, ^{4}He, and ^{35}Cl represent different nuclides, but only ^{14}C and ^{12}C are isotopes of carbon. Note that for calculations involving atomic weights, we usually round off to three significant figures. Thus aluminum, with a weight of 26.9815, becomes 27.0, while chlorine becomes 35.5 and copper, 63.5.

4.10
UNSTABLE
ATOMS

As indicated in Section 4.4, the atoms of many elements such as radium are unstable, and they break down in a process called *radioactive decay* to give different kinds of atoms and different elements. You will note, in the List of Elements, that some atomic weights are given in parentheses and that these weights are whole numbers. The parentheses are used to indicate that all nuclides of the element are unstable and the weight given in parentheses is that of the most stable isotope. Since the weight is due to only one nuclide, it is a whole number.

A closer look at the List of Elements shows that such parenthetical values occur for all elements having an atomic number greater than 93 (neptunium). Even neptunium is found only in very small quantities on the surface of the earth, and, although scientists have synthesized atoms with atomic numbers greater than 93, those atoms are all unstable. For the elements with numbers 94–103, we are relatively certain that at least one nuclide of each has been made; therefore, these elements are given names and included in the List of Elements and the Periodic Table. There is also some evidence that elements with atomic numbers 104, 105, and 106 have been synthesized, but there is not yet international agreement on them. American workers believe their evidence to be sufficiently good to name element 104 rutherfordium (in honor of Ernest Rutherford) and 105 hahnium (for Otto Hahn, who was one of the first to observe nuclear fission). Soviet scientists have also reported evidence that they synthesized the elements 104 and 105, and they have named element 104 kurchatovium (to honor the famous Soviet physicist Igor Kurchatov). Since the data of the various investigators are different, further research will be required before we can include names for elements 104–106 in our lists and tables.

Three of the elements having atomic numbers less than 93 (technetium, 43; astatine, 85; and francium, 87) are so unstable that they are not found on the earth. Many other elements especially those having numbers between 82 and 92, are also unstable and are continually decaying by emitting α, β, and γ radiation; however, this decay takes a long time and we still find them on the earth. We find only one stable isotope of gold. It is interesting to speculate how the history of our civilization would have differed if there had been no stable nuclide of this metal that has assumed so much commercial and artistic importance.

SUMMARY

This chapter has been concerned with atoms. On the basis of his own experiments and those of many other scientists, Dalton was able to set down a series of statements about atoms and molecules. These were based on the atomic theory originally attributed to Democritus. Less than a hundred years after Dalton proposed his theory, it was found that atoms were not the ultimate particles but were made up of simpler particles called electrons, protons, and neutrons. Further research has indicated that there are many other subatomic particles. Subatomic particles, however, do not exist as such for any appreciable time as atoms do. In fact, we might say that the only way electrons, protons, neutrons, and other subatomic particles gain real stability is by joining together into a group called an atom. We can also say that each substance is composed of the same discrete particles or building units. An atom is the smallest possible particle of an element, and a molecule is the smallest particle of a compound which still retains the typical properties of that substance. Thus we can answer the question "Can an atom be divided?" only by saying, "Yes it can , but if it is divided it isn't really an atom." Can protons, neutrons, and electrons be split? The answer is "Yes", and again we are left with the basic question about the divisibility of matter, except that we are talking about particles smaller that the atom. These smaller subatomic particles no longer have the properties of the substance, such as gold, which we originally started dividing.

In this chapter we have been concerned primarily with the units that make up an atom, but we have not specified exactly how these units are arranged and just what it is that makes the properties of one atom different from those of another. The differences and the similarities of atoms, both in general terms and with respect to specific atomic structure, will be the subject of Chapter 5.

IMPORTANT TERMS

alpha particle	atomic weight	gram atomic weight	neutron
atom	beta rays	ions	nuclide
atomic mass unit (amu)	electron	isotopes	proton
atomic number	gamma rays	mass number	

WORK EXERCISES

1. In what way did Dalton's atomic theory differ from that proposed by Democritus?
2. Distinguish between the Law of Definite Proportions and the Law of Multiple Proportions.
3. Applying the method shown in Section 4.2, determine the correct formula for a compound containing 36% calcium and 64% chlorine.
4. Using the same method as above, determine the formula of a compound containing 38.7% potassium, 13.8% nitrogen, and 47.5% oxygen.
5. A gas was found to be composed of 27.3% carbon and 72.7% oxygen. What it its simplest formula?
6. A compound contains 40% carbon, 6.7% hydrogen, and 53.3% oxygen. What is its empirical formula?
7. In what ways has it been necessary to modify Dalton's atomic theory?
8. What discoveries established the fact that atoms are not the smallest particles of matter?
9. Explain how Rutherford's experiment with the gold foil proved the porosity of matter.
10. What is the relationship between the size of an atom and the size of its nucleus?
11. Distinguish between the atomic number and the atomic weight of an element.
12. Why does the number of mass units in an atom not necessarily equal the atomic weight of the element?
13. Show diagrams representing the two isotopes of chlorine, atomic number 17, having mass numbers of 35 and 37.
14. Which of the chlorine isotopes (Table 4.2) is more abundant?
15. How many carbon atoms are there in each of the following?
    a) 12 g of carbon   b) 6 g of carbon
    c) 24 g of carbon

16. What is the weight in grams of each of the following?
    a) one atom of hydrogen
    b) one atom of carbon
    c) one diatomic molecule of oxygen
17. How many hydrogen atoms are there in each of the following?
    a) $\frac{1}{2}$ g of hydrogen (assume 1 for the atomic weight)
    b) 5 g of hydrogen
    c) 16 g of methane, CH_4
18. Using symbols of the form $^{12}_{6}C$, represent the following.
    a) the most common isotope of fluorine
    b) carbon-14   c) uranium-235
    d) two common isotopes of chlorine
19. Using the List of Elements or the Periodic Table of the Elements, complete the following table.

Nuclide	Number of		
	Protons	Electrons	Neutrons
carbon-14	(a)	(b)	(c)
sodium-23	(d)	(e)	(f)
(g)	19	19	20
(h)	20	20	20
(i)	88	88	138

20. There are two isotopes of copper: 69.1% of copper is copper-63 and 30.9% is copper-65. Calculate the atomic weight of copper.
21. There are two isotopes of gallium: 60.2% is gallium-69 and 39.8% is gallium-71. Calculate the atomic weight of gallium.
22. There are three isotopes of potassium: 93.22% is potassium-39, 0.0118% is potassium-40, and 6.77% is potassium-41. Calculate the atomic weight of potassium.

SUGGESTED READING

Andrade, E. N. da C., "The birth of the nuclear atom," *Scientific American*, November 1956, p. 93. A review of Rutherford's contributions to the knowledge of the structure of the atom.

Cline, D. B., A. K. Mann, and C. Rubbia, "The search for new families of elementary particles," *Scientific American*, January 1976, p. 44. The topic is complex but this article helps make it understandable. There is a good summary of the various types of subatomic particles.

Feinberg, G., *What is the World Made Of?* (pb), Anchor Press/Doubleday, 245 Park Ave., New York, 10017; 1978. This is a good, up-to-date book discussing atoms, leptons, quarks, and other tantalizing particles. It's a bit complex in spots but generally understandable.

See the *Study Guide* for other readings about atomic structure.

5

The Elements and the Periodic Table

5.1
INTRODUCTION

In order to understand the chemical nature of our universe, we must know something about the elements and compounds that compose it. Just as the study of words begins with a knowledge of the alphabet, it is best to begin the study of chemistry with a discussion of the elements. This is true even though, historically, the science did not always develop this way. For example, people learned a great deal about the compound known as table salt without knowing that it is made of sodium and chlorine. However, an understanding of *why* sodium chloride possesses various properties and *why* it reacts as it does depends on understanding its components.

Although scientists have made some modifications in the details of Dalton's atomic theory, it continues to provide a firm basis for explaining chemical properties. As we noted in Chapter 4, there appear to be at least 105 kinds of atoms, which are different because they have different atomic numbers. They constitute 105 kinds of matter that should be classified as elements. At first glance, the task of learning about that many different elements, to say nothing of the more than four million known compounds, may seem almost impossible. But before you throw up your hands at the prospect of ever understanding chemistry, let us consider the matter further.

A good approach to understanding a large mass of information is to try to classify or organize it. This is done on the basis of properties or characteristics. If you tell a friend you have a cat, your friend gets a mental picture of your pet which is quite different from what would emerge if you had said dog or hamster. Of course, the picture is still not distinct. If your friend is interested, you can supply more detailed characteristics: the cat's age and color; whether it is male or female; Persian, Manx, or Siamese, etc. Eventually, your friend can generate a fairly precise image of your pet on the basis of its characteristics.

Elements also have characteristics. Even before anyone realized that silver, gold, and copper are elements, it was recognized that they are similar in some ways, that they are slightly different in others, and that all three are very different from sulfur and charcoal. We shall begin this chapter by describing and classifying several of the elements. Then we shall examine the structure of the atoms that make up the elements. By *structure* we mean the definite arrangement in which the protons, neutrons, and electrons are assembled in an atom. It is the structure of the atoms that determines their properties.

**5.2
DISTRIBUTION
OF COMMON
ELEMENTS**

Given that there are 105 elements, which ones are most important to study? One approach is to consider the most common elements. Among the most familiar materials on the surface of the earth are rocks, sand, and soil. Also common are wood and metal. And there are immense quantities of water. But of what are these materials made? In other words, which *elements* are present in each one? What is the composition of water? What is in sand? Which elements are found in wood?

There have been many studies to determine which of the elements are most abundant in our immediate surroundings. For this purpose, the "surface" of the earth is often considered to be the atmosphere, the oceans, and the topmost portion (to a depth of about ten miles) of the dry land areas. When the substances found in these regions are analyzed, just twelve elements are found to make up more than 99% of the material present. These elements are listed in Table 5.1. The list includes the elements found in compounds as well as those few which may exist as free elements.

TABLE 5.1   Estimated composition of the surface of the earth*

Element	Percent by Weight	Element	Percent by Weight
oxygen	49	magnesium	1.9
silicon	26	hydrogen	0.88
aluminum	7.5	titanium	0.58
iron	4.7	chlorine	0.19
calcium	3.4	carbon	0.09
sodium	2.6	all other elements	0.56
potassium	2.4		

* Including oceans and atmosphere.

Oxygen, silicon, and aluminum constitute the largest portion of the earth's surface. One reason is that silicon and oxygen are the major components of sand and of many rocks. Also, clay soils are mostly compounds made up of silicon, aluminum, and oxygen. Oxygen is important in the oceans, too, since most of the weight of water is due to oxygen; the remainder is hydrogen.

Another way of deciding which elements are of greatest importance is to think of the composition of the human body. The most abundant elements in the human body are listed in Table 5.2. At least 25 others are also found. In the body, too, most of the elements exist in combined form, in compounds. It is interesting to note that ten of the twelve elements listed in Table 5.1 also appear in Table 5.2.

A large percentage of the human body is oxygen, because much of our body weight is due to water. The rest of our body materials are chiefly organic substances, which are all compounds of carbon. Significant amounts of hydrogen are present because it is the second most abundant element in organic compounds and is also a component of water.

TABLE 5.2   Composition of the human body

Element	Percent by Weight	Element	Percent by Weight
oxygen	65	chlorine	0.20
carbon	19	sodium	0.15
hydrogen	10	magnesium	0.05
nitrogen	3.3	iron	0.004
calcium	1.5	iodine	trace
phosphorus	1.0	fluorine	trace
potassium	0.35	silicon	trace
sulfur	0.25		

The fourth most abundant element in the organic compounds of the body is nitrogen, found largely in proteins.

In the paragraphs that follow we shall discuss briefly the elements that seem most important to human existence: those present in our bodies, those found in plants and animals, and those most abundant at the surface of the earth.

Oxygen   Oxygen, which is the most abundant element on the surface of the earth, is present in its elemental form as a gas in the atmosphere, where it makes up 23% of the total weight of dry air. (Most of the remainder of the atmosphere is nitrogen—which, however, is not among the top twelve elements in overall abundance.) Oxygen is also present in water, the most abundant *compound* found at the earth's surface. As we saw in Chapter 4, oxygen makes up 89% of the weight of water, with hydrogen accounting for the other 11%. A very large variety of other compounds, both mineral and organic, also contain oxygen. It is so important that much of Chapter 8 is devoted to oxygen and its compounds.

Silicon   The second most common element at the surface of the earth is silicon. Its most prevalent compound is silicon dioxide, the main constituent of sand. Large quantities of sand are used to manufacture glass. Some of the most prolific microscopic algae in the oceans are the diatoms. These organisms have exoskeletons ("shells") made up of a form of silicon dioxide. Pure quartz is also silicon dioxide. In nature quartz sometimes contains traces of metallic compounds, which create the lovely colored variations known as amethyst, agate, flint, and jasper. In another combination with certain metals and oxygen, silicon forms silicates, the primary constituents of granite, asbestos, mica, feldspar, and clay. The prevalence of these materials makes it easy to see why silicon is so abundant. Although the element silicon does not occur free in nature, it can be prepared in a laboratory or factory. In recent years elemental silicon has been used extensively in the production of transistors, solar cells, and many other electronic devices.

Iron   On the earth's surface, iron as a free element is found only in meteorites. Compounds of iron occur widely in nature, however. The reddish color of many rocks and the red soils of the world are due to traces of iron compounds present. The color of blood is also caused by iron in hemoglobin.

The principal ore of iron is the oxide hematite, found in concentrated deposits in various parts of the world. It is believed that methods of separating iron from its ores were discovered shortly before the rise of Greek civilization. Before that time weapons had been made of bronze; the change to the harder iron, which could be sharpened, provided a new aspect to warfare. Homer's *Iliad* speaks of iron as a precious metal.

The amount of iron metal now produced from ore each year surpasses by about fifteen times the combined amounts of all other metals. However, the number of tons of ore consumed each year is not an adequate index of the amount of iron used; many millions of tons of scrap iron are utilized to manufacture new products. The car you drive may contain iron atoms from old tractors, old battleships, or several generations of automobiles. Although iron ranks only fourth in amount among the elements on the earth's surface, the core of the earth is estimated to be largely iron, with only small amounts of a few other elements.

Calcium   This is an important component of the bones of animals and fishes and the shells of mollusks (clams, oysters, etc.). Calcium also fills an important role in body chemistry; for example, it plays a part in promoting muscle contractions and participates in the complex reactions causing blood coagulation.

Marble, limestone, and gypsum are all compounds of calcium. Across vast areas of land the topsoil is supported by a layer of limestone (which is 40% calcium) extending to depths of several thousand feet. It is obvious, then, why calcium is one of the abundant elements of the surface of the earth. It is never found as the free element, and from this we can conclude that calcium is a very active substance which readily enters into combination with other elements.

Magnesium   This is another element never found uncombined in nature. Magnesium compounds occur in considerable quantities in the oceans and as a component of some limestone deposits. The chlorophyll of green plants is essentially like the heme portion of animal hemoglobin, except that chlorophyll contains magnesium instead of iron. When isolated as the free element, magnesium is a metal. The metal burns with extreme brilliance and is used in some types of flares and fireworks. It has a very low density, and when alloyed with aluminum causes the aluminum to become less dense but stronger. Large amounts of this alloy are used in the airplane industry.

Sodium and potassium   Sodium and potassium are not found as free elements either, and again we can assume that they are quite active. In fact, if we place a sample of either of these elements in water, a violent reaction occurs. Sodium and potassium are similar in almost all their properties. They occur throughout the surface of the earth in mineral deposits, in the soil, and dissolved in the ocean waters. Ordinary salt, a compound of 39% sodium and 61% chlorine by weight, constitutes almost 3% of the ocean. If we consider that the oceans hold enough water to cover the whole earth to a depth of nearly 10 000 feet, we can only attempt to imagine the vast amount of salt present. Great deposits of salt, in some cases hundreds of feet in depth, are found in many parts of the world. They are the result of the drying up of ancient seas. Although sodium compounds have been dissolved from the rocks and carried to the seas, potassium has been retained to a greater degree in the rocks and soils of the earth. Consequently only small amounts of potassium are found in the oceans. Both sodium and potassium are essential to animals. Although potassium is also essential to green plants, only marine algae and a few other plants make use of sodium. For this reason the grass or hay fed to cattle contains little sodium and often must be supplemented with salt.

Carbon   Although carbon makes up a very small fraction of the earth's crust, it is extremely important to all living organisms. Plants use carbon dioxide from the air to make sugars, and from these build all the other organic compounds they need. Plants are the ultimate source of organic compounds for animals, which eventually return some of the carbon dioxide to the atmosphere, when they oxidize part of the organic material to obtain energy. Much of the organic carbon of the world is present in cellulose, the woody material of plants. Petroleum is a mixture of carbon compounds resulting from the decay of formerly living marine organisms. Coal consists mainly of elemental carbon, together with small amounts of carbon compounds. Coal was formed when thick layers of dead vegetation were buried under soil and rocks millions of years ago. Carbon is a component of the calcium carbonate found in the shells of shellfish, in many ocean sediments, and in limestone and marble.

Chlorine   Chlorine is never found in nature as a free element. Its most common compounds are called chlorides (sodium chloride, calcium chloride, etc.). Nearly all of these compounds are quite soluble in water. Although chlorine makes up a very small percentage of the earth's surface, in the chloride form it is the most abundant element in ocean water. Chlorine is also the most abundant nonmetallic element in body fluids.

Nitrogen   The very small quantities of nitrogen found on earth are largely in the atmosphere, in the form of elemental nitrogen gas. However, as suggested by Table 5.2, nitrogen is essential for humans. It is equally necessary for other animals and for plants. In living organisms nitrogen is a key element in all amino acids and proteins, in the nucleic acids which control hereditary traits and the formation of proteins, and in a variety of other less abundant, but important, substances.

Phosphorus and sulfur   These two elements are also necessary for the existence of all plants and animals. Phosphorus is found chiefly in various phosphate compounds, including the nucleic acids and certain compounds which release energy. Sulfur is vital in the structure and functioning of certain proteins.

5.3
SYMBOLS OF
THE ELEMENTS

Scientists often use a symbol to represent a chemical element, rather than writing its full name. Historically, some of the metals had symbols almost as soon as they had names. Gold became associated with the sun and was given the astrological symbol for the sun, which was a circle, sometimes with a dot inside. Silver was given the symbol associated with the moon, a crescent. Ancient symbols for other metals were often related to some characteristic or myth. Iron, Mars, and war were associated, as were copper, Venus, and love. Several symbols used by the medieval alchemists are shown in Fig. 5.1.

The modern symbols for the elements are much more prosaic but also somewhat more logical. The first letter of the name of the element is used as the symbol for many elements; we use C, H, O, N, S, and P for the elements carbon, hydrogen, oxygen, nitrogen, sulfur, and phosphorus. However, since there are 105 elements and only 26 letters in our alphabet, most elements have symbols consisting of two letters, only the first of which is capitalized. In general, the letters are the first two letters of the name of the element, but there are a number of exceptions. For example, because Ca was taken as the symbol for calcium, cadmium became Cd and californium became Cf. In a few cases the symbols are derived from the Latin names of the elements. The symbol Cu for copper comes from the Latin word *cuprum*, and Na for sodium comes from the Latin *natrium*.

Fig. 5.1 Ancient alchemical symbols for some common elements.

For the symbols having two letters, it is especially important to use a capital letter *only* for the first letter. Co is the symbol for cobalt, an element; but CO is the formula for carbon monoxide, a compound. Sn represents tin (Latin, *stannum*). The symbol SN would mean a combination of sulfur and nitrogen (but no such compound exists).

TABLE 5.3   Common elements and their symbols

Metals			Nonmetals	
aluminum	Al		carbon	C
calcium	Ca		chlorine	Cl
copper	Cu	(*cuprum*)	fluorine	F
iron	Fe	(*ferrum*)	helium	He
lead	Pb	(*plumbum*)	hydrogen	H
magnesium	Mg		iodine	I
manganese	Mn		nitrogen	N
potassium	K	(*kalium*)	oxygen	O
silver	Ag	(*argentum*)	phosphorus	P
sodium	Na	(*natrium*)	sulfur	S
zinc	Zn			

The chemical symbols now in use are internationally accepted. Scientists throughout the world know what is meant by NaCl even though in daily life they may use quite different names for the substance we call salt or table salt. The most common symbols appear in Table 5.3, and should be learned. Other names and symbols will be introduced from time to time; through usage they will become familiar to you.

5.4 CLASSIFICATION OF THE ELEMENTS

Most of you knew, before starting to read this book, that aluminum, iron, and magnesium are metals. You also may have known that calcium, sodium, potassium, and titanium are metals. There is even a clue in the names of these elements since, except for iron, they all end in the suffix -*ium* or -*um*, and if we substitute the Latin name of iron, *ferrum*, it falls into the same pattern. Apparently those who named these substances were aware of similarities among them. Hydrogen, nitrogen, oxygen, and sulfur also have some similarities; these elements are termed nonmetals.

What is a metal? A general description is not simple. We tend to recognize metals by their physical properties. **Metals** have a characteristic luster. They are usually good conductors of heat and electricity; they can be melted and cast into desired shapes; and they can be hammered or rolled into sheets or drawn into wire. All metals are opaque. Although we think of metals as being solids, one exception is mercury, which is a liquid at normal temperatures.

The **nonmetals** are not as uniform in their properties. Under ordinary conditions half of them are gases; one, bromine, is a liquid; the rest are solids. Few of the nonmetals have any of the typical properties of metals. In general, they are poor conductors of heat and electricity, cannot be drawn into wire, and do not have a characteristic luster.

These distinctions between metals and nonmetals are based on physical properties, most of which are easily recognized. Unfortunately, although this approach is quite satisfactory for many examples, it does not work for all. A more useful approach would be to find some fundamental trait that explains *why* certain elements behave as metals and others do not. In Section 5.9 we shall define metals and nonmetals chemically in a manner that is brief and simple, but much more precise than the physical descriptions just given.

5.5 PERIODIC PROPERTIES OF THE ELEMENTS

While a classification of elements as metals or nonmetals was helpful, many scientists looked for, and found, additional properties common to certain groups of elements. These efforts culminated in the work of Dmitri Mendeleev, a Russian chemist, who in 1869 published the first workable periodic table, a chart organized so as to group together elements having similar properties. We can recreate his work. Let us start by writing the names (or symbols) of a few of the elements in the order of increasing atomic weight, and then writing the properties—the physical and, more important, the chemical properties—for each of the elements. We begin to see a certain pattern in the list; periodically there is a repetition of properties.

If we start with lithium we obtain the list shown in Table 5.4. We see that lithium is an active metal, beryllium is somewhat less active but still a metal, boron and carbon are nonmetallic solids, and the next three (nitrogen, oxygen, and fluorine) are gases whose activity increases with increasing atomic weight. Sodium is a very active metal similar to lithium; magnesium is similar to beryllium; aluminum is somewhat different from boron; but silicon is similar to carbon. While phosphorus and sulfur are not gases, they are distinctly nonmetallic and their chemical properties are similar to those of

TABLE 5.4 Characteristic properties of low atomic weight elements

Atomic Weight	Element	Properties
7.0	lithium	active metal
9.0	beryllium	metal
10.8	boron	nonmetallic solid
12.0	carbon	nonmetallic solid
14.0	nitrogen	inactive gas
16.0	oxygen	moderately active gas
19.0	fluorine	very active gas
23.0	sodium	very active metal
24.3	magnesium	active metal
27.0	aluminum	metal
28.0	silicon	nonmetallic solid
31.0	phosphorus	nonmetallic solid; chemical properties similar to nitrogen
32.0	sulfur	nonmetallic solid; chemical properties similar to oxygen
35.5	chlorine	gas; chemical and physical properties similar to fluorine
39.0	potassium	very active metal similar to sodium and lithium
40.0	calcium	active metal similar to magnesium

nitrogen and oxygen; finally, chlorine is very similar to fluorine. If we continue with potassium and calcium, their similarity to sodium and magnesium and thus to lithium and beryllium is enough to make the repetition of properties apparent.

TABLE 5.5 Mendeleev's table of elements (in part)

lithium	beryllium	boron	carbon	nitrogen	oxygen	fluorine
sodium	magnesium	aluminum	silicon	phosphorus	sulfur	chlorine
potassium	calcium	. . .	. . .	. . .	. . .	. . .

On the basis of these patterns Mendeleev developed the arrangement of elements shown in Table 5.5. These are the same elements that are listed in Table 5.4, but they are arranged in rows so that elements having similar properties fall in the same vertical column. As we read across the first row of Table 5.5, from lithium to fluorine, the elements show a gradual change in properties from very metallic to very nonmetallic (see Table 5.4). This series, or **period**, ends with fluorine. Then there is an abrupt change of properties, for the next element, sodium, is a very active metal. Therefore, Mendeleev placed it beneath lithium; consequently magnesium fell beneath beryllium, and so on. As we proceed to the right in this second row there is again a gradual trend in properties. This row, or period, ends with chlorine, which is much like fluorine. Once again this signals the beginning of a new period; the element following chlorine is potassium, a metal similar to sodium and lithium. Although this table is only a portion of Mendeleev's periodic table of the elements, it is sufficient to show the significance of his development. On the basis of his periodic table, Mendeleev was able to formulate his Periodic Law, which states that when elements are arranged in the order of their increasing atomic weights, they exhibit a periodic recurrence of properties.

Mendeleev's original table contained 63 elements. He was perceptive enough to realize that the known elements could be grouped according to similar properties only if he left some blank spaces in his table. The spaces represented elements not yet discovered. By carefully studying the properties of the elements around the blank spaces, Mendeleev and others who came after him were able to predict the properties of the missing elements. In several instances, these predictions helped researchers find in nature the previously unknown elements. There are no blanks in modern periodic tables, the last ones having been filled in the 1940s. New elements have been added at the end of the table, and it is possible that elements beyond 105 may be added in the future.

One entire group of elements was unknown at the time Mendeleev did his work, and was omitted from periodic tables for several decades. This group, the noble gases (helium, neon, argon, krypton, xenon, and radon), is now placed in the extreme righthand column of modern periodic tables. The concept of atomic numbers was not developed until almost fifty years after Mendeleev's work, and thus the omission of the noble gases from the table was not realized.

Mendeleev worked entirely with atomic weights, and in at least one case he had to depart from his order of increasing weights in order to make the table consistent. Tellurium is heavier than iodine, but because of its properties should come *before* iodine in the chart. Mendeleev believed that this inconsistency was due to inaccurate atomic weight determinations. We now know that natural samples of tellurium actually do have a greater atomic weight than those of iodine, due to the fact that a higher percentage of heavy isotopes is present in tellurium (Section 4.9). A similar discrepancy exists with argon and potassium, and a few others.

As often happens with scientific theories, these apparent discrepancies led to a better understanding. Eventually chemists realized that the periodic repetition of properties of the elements is related to atomic numbers, not atomic weights. The most fundamental properties, especially chemical properties, of an element depend on the number of protons in its nucleus and the number of electrons surrounding the nucleus. This number, as we found in Section 4.6, is termed the atomic number, and is distinct for each element. Therefore, a modern version of the **Periodic Law** states that *when elements are arranged in the order of their atomic numbers, they exhibit a periodic recurrence of properties*. In Sections 5.7 and 5.8 we shall explore in some detail this relation between atomic structure and chemical properties. But first, let us summarize some of the important features of a periodic table.

**5.6
THE PERIODIC
TABLE**

A periodic table (also called a periodic chart) is probably the most important document available to students of chemistry. It lists all the known elements and displays some of their important relationships. In addition, nearly all periodic tables show the atomic number and the atomic weight of each element, and many tables give still other data. Periodic charts hang in almost all laboratories and chemistry classrooms and are included in most chemistry texts. A modern version of the periodic table is found inside the back cover of this book. In the paragraphs that follow we shall discuss the arrangement of the periodic table and some of the information it reveals. As each statement is made, you should compare it with the table in the back of this text, so that you develop a good understanding of the important features of a periodic chart. The more familiar you become with the table the more quickly you will be able to locate facts there, without having to review these discussions.

Metals and nonmetals   The most distinctly metallic elements are located at the extreme lefthand side of the table and the most nonmetallic elements are at the far right (excluding the noble gases). Between these extremes there is a gradual shift in properties from left to right. Several properties, not just one, serve to distinguish metals from nonmetals. Some elements are good examples of metals in many respects, while others are typical nonmetals; but some, the metalloids, have a few characteristics of both groups. A zigzag line can be drawn on the chart to separate metals from nonmetals. Several elements bordering this line are called *metalloids*.

Periods of the table   The horizontal rows of elements, in order of their increasing atomic numbers, are called series or periods. The first period contains only two elements, the second and third periods contain eight, the fourth and fifth periods contain 18 elements, and the sixth and seventh periods contain 32. The lanthanides and actinides, which are located at the bottom of the table, really belong in Periods 6 and 7, but are included at the bottom only for convenience, so the table will be more compact.

Groups of the table   There are eighteen vertical *columns* or *groups*, each consisting of a family of elements having similar properties. For example, the elements of group IIA (beryllium, magnesium, etc.) have remarkably similar chemical properties. They differ mainly in degree of chemical reactivity, and their physical properties show a regular trend from the top of the group to the bottom. All together, the elements in this family are much more like each other than they are like the elements of group IA (lithium, sodium, etc.). Thus, comparing adjacent elements, we see that calcium is more like magnesium than it is like potassium.

Representative elements   The elements in the eight main groups of the periodic table (those with the suffix A) are called *representative elements*. The elements within any particular A group display very characteristic properties which together reveal a pronounced family relationship. The differences between the groups are quite distinct. If we consider the representative elements in Groups IA through VIIIA, we see the most regular, step-wise, progression from metallic to nonmetallic properties as we proceed from left to right in each period.

Transition elements   Those elements in groups having the suffix B are known as the transition elements. They are all metals. As one views these elements from left to right, the progression of properties is more gradual and less distinct than it is for the A groups. This is partly because the series is a long one, but also because of the way it is built up, which will be discussed in Section 5.8. However, certain of the B group elements display a relationship to a similar A group. For instance, some properties of copper, silver, and gold, in Group IB, are similar to those of the metals in IA. Zinc, cadmium, and mercury, in IIB, are even more like the IIA elements (calcium, etc.).

Noble gases   The noble gases are placed in the column at the righthand side of the table. Until recently they were called the "inert gases," since it was believed that they would not react to form chemical compounds. Although it is true that they do not react readily, since 1962 several compounds containing these elements have been prepared. However, noble gas compounds are formed under extreme conditions and with only a few other elements.

**Use of the
Periodic Table**

The material in this section and the following three sections is very helpful for understanding chemistry. Because the periodic table shows general patterns, it will simplify much of the important material in the chapters that follow, so that the information will not remain just a mass of apparently unrelated facts. The periodic table will often be the place you can most quickly find a specific fact you need, to do a calculation or to answer a question about chemical behavior.

The periodic table is not perfect. In common with most attempts to classify our observations of nature, it does not neatly accommodate all the data. We have already noted that in some cases a heavier atom (such as argon, 40 amu) comes before a lighter one (potassium, 39 amu). Hydrogen and helium really do not fit well in the table. The location of the transition elements is somewhat awkward, particularly for the elements of group VIIIB. Nevertheless, despite its imperfections the periodic table is an extremely valuable aid in chemical work. We shall see in the following sections that there are sound reasons even for these "discrepancies."

**5.7
THE PERIODIC
TABLE AND
ATOMIC
STRUCTURE**

So far we have presented many statements about the periodic table without much explanation. You may be asking, "Why do the elements fall into periods of 8 or 18?" or "Why are the most active metals at the left?" Such questions puzzled scientists for many years. Rutherford and others had discovered protons, electrons, and neutrons. When these ideas of atomic composition were combined with a study of Mendeleev's table, a concept of atomic *structure* began to develop. If the difference between elements is due to the difference in the number of protons and electrons present in their atoms, those elements that have similar properties must also have some similarity in the arrangement of those electrons or protons. Speculations such as these culminated in the idea of atomic structure developed in 1913 by the Danish scientist Niels Bohr.

Bohr assumed that there is a small, heavy nucleus containing the protons and neutrons of the atom (Section 4.5). He said that the electrons, which have negligible mass, are found some distance away, moving about the nucleus much as the planets revolve around the sun. Even more important, he stated that a particular set of electrons, traveling a certain distance from the nucleus, are in one orbit. Another group of electrons, at a greater distance from the nucleus, are in a second orbit, and so on. This idea that atoms have a particular organized structure is shown by the diagrams in Fig. 5.2.

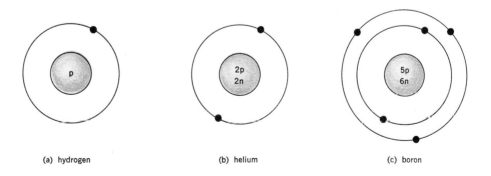

(a) hydrogen   (b) helium   (c) boron

Fig. 5.2 Sketches sometimes used for the structure of atoms. A proton is represented by p, a neutron by n, and an electron by a dot.

Bohr's picture, or model, of atomic structure has been modified in several ways for the sake of greater accuracy. For one thing, it is no longer believed that electrons travel in simple circular or elliptical paths as the planets do. Instead, one particular group of electrons is thought to occupy a certain *region* of space around the nucleus. A different group of electrons occupies another region which, on the average, is farther from the nucleus. It is as though the groups of electrons are arranged in concentric layers, or shells, around the nucleus. Therefore the term "shell" is somewhat more satisfactory than "orbit." (Even this picture is not entirely correct.)

Diagrams of the type shown in Fig. 5.2 are still useful to represent the structure of atoms in a general way. For instance, the sketch of the boron atom (c) shows five protons and six neutrons in the nucleus. Also, boron has two electrons in one region and three other electrons farther away from the nucleus. The electrons are marked on circles, which Bohr would have called orbits. However, now it is better to think of the circles as merely indicating different groups, or shells of electrons.

The later refinements in Bohr's theory in no way detract from the importance of his contributions to science. His award of the Nobel Prize for physics in 1922 was certainly deserved. Bohr himself emphasized that his imaginary model of the atom was tentative and symbolic. He was convinced that more elaborate theories would soon modify the picture. Indeed, he and some of his associates contributed to the revisions of the theories.

During the early 1920s Bohr saw a relationship between the structures of the different atoms and the periodic table. He proposed that the clustering of electrons into various groups (which we will call shells) explained the arrangement of the elements in the periodic table. He stated that the reason certain atoms have similar properties is that they have the same number of electrons in an outer shell. Sodium is similar to lithium because, despite the fact each has a different total number of electrons, both have one electron in the outer shell. Bohr also recognized that if there are only two elements in the first period of the table it must mean that only two electrons can fit into the first shell; there are eight elements in the second period because the second shell can hold no more than eight electrons. The third shell can have up to 18 electrons and larger shells as many as 32. These same numbers resulted from Bohr's theoretical calculations and coincided very nicely with the numbers of elements in each period of the table.

Using Bohr's theory, let us practice drawing the structures of some of the smaller atoms, using lithium for the first example. From the List of the Elements (inside the front cover of this book) we find that lithium has atomic number 3 and an atomic weight of 7 (rounded off). Therefore it has 3 protons and 4 neutrons. So we sketch its structure by showing these 3 protons and 4 neutrons in the nucleus. Because lithium has 3 protons, it must also have a total of 3 electrons outside the nucleus. Since no more than 2 electrons can be in the first shell, the third electron must be in the second shell. The completed sketch becomes this:

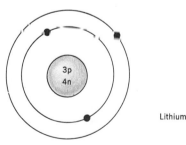

Lithium

If numbers, rather than dots, are used for the electrons, the structure of the fluorine atom, atomic number 9 and atomic weight 19 (both from the List of the Elements), can be represented thus:

2   7   Fluorine

The structure of neon and of sodium, the next two in the List of the Elements, can be summarized as follows:

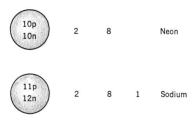

2   8   Neon

2   8   1   Sodium

Neon, like helium, has a completed outer shell of electrons; the same is true for argon. (You may wish to draw the structure of argon to verify this.) It is this property of having the outer electron shell filled that places helium, neon, and argon in the family of noble gases. This is a particularly stable condition and is the reason the noble gases are unreactive; they have no need to seek additional electrons to achieve a completed, stable, outer shell.

Also, note carefully the structure of sodium compared with that of neon. Since 10 electrons fill to capacity both the first and second shells, the last electron of sodium must be located in a third shell.

Now let us compare the position of the elements in the periodic table with the structures we have just sketched for their atoms. Lithium and sodium, in Group IA, have one electron in the outer shell. Fluorine, with seven electrons in its outer shell, is found in Group VIIA. Neon and argon, each having eight electrons in the outer shell, are in group VIIIA. This is a general principle for any A group element: *the number of electrons in its outer shell corresponds to its group number in the periodic table.* In this book, the diagrams across the top of the periodic table also show pictorially the number of electrons in the outer shell of the elements in each group. The only exception to this is helium, which is located in Group VIIIA of the table although its single electron shell can hold only two electrons.

The noble gas neon, having an outer shell filled with eight electrons, is placed in Group VIIIA and marks the end of the second period on the periodic chart. But neon can also be regarded as the *beginning* of the third period on the chart. Neon has *no* electrons in the third shell; for this reason, on many periodic charts the noble gases are labelled Group 0, rather than VIIIA.

We see now that Bohr's idea that electrons are arranged in groups, or shells, surrounding the nucleus provides a fundamental explanation for the arrangement of elements in the periodic table. This concept accounts for both the position of an element in the table and, more important, its properties. *The most distinctive chemical properties of an atom are determined by the number of electrons in its outer shell.* It is because lithium, sodium, and potassium each have just one electron in the outer shell that they are very reactive metals. Given this fundamental principle, it is often possible to predict the general behavior of an element merely by noting the group of the periodic table to which it belongs. For instance, nitrogen is located in Group VA. That number, and the diagram just above it, immediately tell you that there are five electrons in the outer shell of a nitrogen atom. In the next chapter we shall see that this simple fact reveals a great deal about the kind of chemical compounds that nitrogen will form. For such predictions it is not even necessary to sketch the complete structure of the atom; the main concern is with the electrons in the *outer* shell. This is the reason the outer-shell diagrams at the head of each group in the periodic table are useful.

You will soon become familiar with several of the common elements and able to locate them quickly in the periodic table. The information found there is sufficient to draw the structures of their nuclei and electron shells. For example, aluminum is the second one down in Group IIIA. Its atomic number is listed as 13 and its weight as 27, so we arrive at this sketch:

By filling the first shell with 2 electrons and the next shell with its maximum of 8, we are left with 3 for the outer shell. A quick check back to the periodic table assures us that this is correct: aluminum is in Group IIIA and *should* have three electrons in its outer shell.

With a little practice you should be able to diagram, in the manner just shown, the structure of any of the first 20 elements, all of which are representative elements. The elements most important for discussion in this book are the representative elements, plus a few of the transition elements of Period 4 (Cr, Fe, Ni, etc.) and Groups IB and IIB.

5.8
ELECTRON
DISTRIBUTION:
A CLOSER
LOOK

The principles described in the previous section show that there is a relationship between the structure of atoms and the position of the *representative* elements in the periodic table. However, a few additional principles are necessary to describe the rest of the periodic table and the atomic structure of the transition elements.

Subshells and
Orbitals

We have indicated that each different shell, or group, of electrons is found at a different distance from the nucleus. Another fact that is perhaps even more fundamental is that the electrons of one shell have different energies than do those of another shell. Electrons in the second shell, which are at a greater distance from the nucleus than those in the first shell, have greater energies than those of the first shell. Correspondingly, an electron in the third shell has an even greater energy than one in the second shell.

This idea of energy levels can be refined even further. The electrons populating a certain energy level (that is, those belonging to a given shell) do *not* all have exactly the same energy. For instance, at the second main energy level, which means the second shell, we find two different groups of electrons having slightly different energies. These small subdivisions within a main energy level are called **subshells**. Within the second shell, the electrons of slightly lower energy are called **s** electrons and those of slightly higher energy are called **p** electrons. There can be a maximum of two electrons in the **s** subshell, but the **p** subshell can hold as many as six.

Another important property of electrons in an atom is that they are able to pair off so that two can exist in the same territory. An **orbital** *is a region of space which can be occupied by no more than two electrons.* Most electrons are paired, but in many atoms a few remain unpaired. Each pair and each single electron occupies a separate orbital (territory) of its own.

Although electron pairs exclude other electrons, either paired or single, and exist in separate orbitals, several electrons can be at the same energy level. If so, they are members of the same subshell. Figure 5.3 shows the number of orbitals possible for each subshell within the *fourth* main shell. The figure indicates that the number of orbitals possible for each subshell is represented by a series of odd numbers: 1, 3, 5, 7. Since each orbital is filled when it has two electrons, the total number of *electrons* possible for each of the subshells is 2, 6, 10, 14.

Shells smaller than the fourth one have fewer orbitals; but their arrangement, as far as they go, is the same as in Fig. 5.3. The first shell, nearest the nucleus, can have only one orbital, called a **1s** orbital. The second shell is farther out and has more space, so it can accommodate as many as four orbitals. One is a **2s** orbital and three are **2p** orbitals. All together these provide spaces for as many as eight more electrons, and this corresponds to the eight elements in the second period of the periodic table. In the third shell, the filling of the **3s** and **3p** orbitals again gives the familiar total of eight and a particularly stable arrangement. This accounts for the eight elements in the third period of the table.

Some of the larger shells can also accommodate **d** orbitals (10 electrons), making possible a total of $2 + 6 + 10$, or 18 electrons. This accounts for one of the "long rows" of the periodic table, consisting of 8 representative elements and 10 transition elements.

The Order of Filling Subshells (Orbitals)

Now it is necessary to mention one other important idea. At the higher main energy levels, some of the sublevels overlap. For example, electrons in **3d** orbitals, which are at the *highest* energy sublevel of the third shell, have slightly more energy than do electrons in the **4s** orbital, which is the *lowest* sublevel of the fourth shell. The relative energy levels of all the various subshells are depicted in Fig. 5.4.

Now, these energy levels are important because an atom is most stable when its electrons have the least energy possible. An atom of boron, for example, will have its five electrons arranged in orbitals (regions of space) according to the pattern shown in

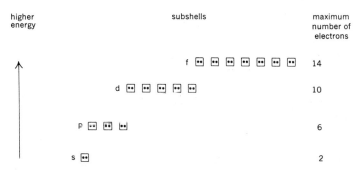

Fig. 5.3 Schematic representation of the placement of electrons in subshells of the fourth main shell. Each box accounts for one orbital, which can hold a maximum of two electrons.

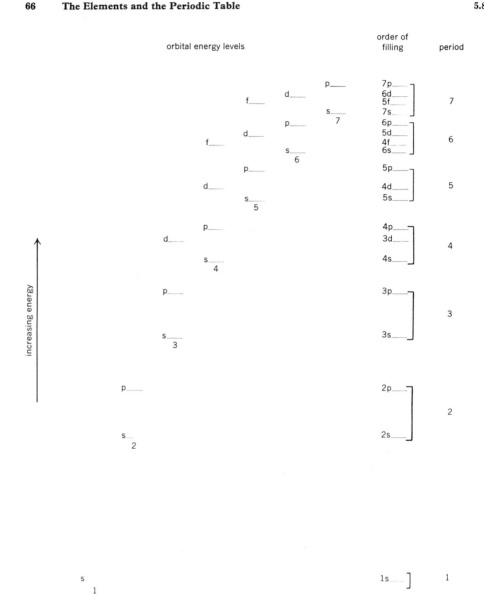

Fig. 5.4   Relative energy levels of electron orbitals, showing the order of filling by electrons.

Fig. 5.4. Two of boron's electrons will be in the **1s** orbital, two more will be in the **2s** orbital, and the "last" one must occupy a **2p** orbital.

The total structure of each kind of atom (such as boron) is arranged a certain way, according to the laws of nature. But it is convenient for humans to imagine that structure being "built up" by "placing" electrons, one by one, in the regions of space around the nucleus. As well as we can understand from experimental observations, the "build-up" should follow the rules summarized by Fig. 5.4. Each electron enters an orbital which is at the lowest energy level still available. Thus electrons first occupy the **1s** orbital, then the **2s**, then the three **2p** orbitals, and so on. The two columns at the far right in

Fig. 5.4 show the orbitals occupied by electrons for the elements in each period of the periodic table.

We presented Fig. 5.3 in order to discuss the concept of orbitals and the number available in each subshell. Figure 5.4 was provided to illustrate the idea that electrons will occupy first the orbitals of lowest energy. However, it is not necessary to memorize the patterns displayed in Fig. 5.4. The same information is provided by Fig. 5.5, in a form which shows directly the relationship between the location of elements in the periodic table and the distribution of electrons in their atoms. To make use of Fig. 5.5, compare it with the complete periodic table shown inside the back cover of this book.

Figure 5.5 shows the orbitals at the *highest* energy sublevel occupied by the electrons of a particular element. When using the diagram, remember that electrons first occupy all available orbitals of lower energy. To list the orbitals occupied by the electrons of a certain element, start with the first period in Fig. 5.5 and proceed in order from left to right through each period, stopping at the position of that element on the chart.

For example, one can see that silicon, atomic number 14, has two electrons in the **1s** orbital; the second shell is also filled with two **s** electrons and six **p** electrons. In the third shell, silicon has two **s** electrons and only two **p** electrons.

This description of the *electron distribution* of a silicon atom can be summarized briefly as follows:

Si $1s^2$ $2s^2 2p^6$ $3s^2 3p^2$

In this notation, the superscripts indicate the number of electrons in each of the orbitals listed. One can check to be sure no error has been made; in this example the total number of electrons listed is 14, which does correspond to the atomic number of silicon.

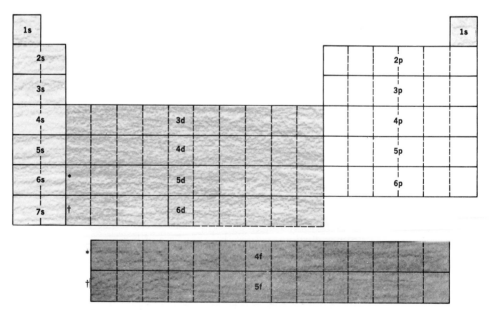

Fig. 5.5. Diagram of the periodic table, showing the last orbitals to be occupied by the electrons of each element.

Here are two more examples of writing the electron distribution of an element, with the aid of a periodic table and Fig. 5.5:

V $1s^2$ $2s^22p^6$ $3s^23p^63d^3$ $4s^2$ (a transition element)
As $1s^2$ $2s^22p^6$ $3s^23p^63d^{10}$ $4s^24p^3$

After a little practice you will find it rather easy to remember the location of the "blocks" in Fig. 5.5 and, even without having it to look at, you will be able to visualize it super-imposed upon a periodic chart. One thing to remember in particular is that the *transition elements* arise from the filling of **d** orbitals and that the order of filling the **d** orbitals is "dropped down" one row. That is, the **3d** orbitals are occupied during the build-up of the elements in the *fourth* period; and the **4d** orbitals, during build-up in the fifth period. The order of filling orbitals is also shown, for each period, in the column at the far right in Fig. 5.4.

To summarize this build-up of the transition elements: Although the filling of electrons into an outer shell may be delayed while electrons are added to a shell just beneath it, *the outer shell never has more than eight electrons.* This is the reason for the eight major groups (the A groups) of the periodic table, with rows of ten transition elements (the B groups) inserted within some periods.

Lanthanide and Actinide Elements

Although there will be no further need to discuss the lanthanide and actinide elements in this book, their structure should be mentioned briefly in order to complete the picture of the periodic table. The sixth period has not only 18 elements (10 plus 8), but within the period an *additional series* of 14 elements is inserted. This is the *lanthanide series.* The seventh period includes a similar row of 14 elements, the *actinide series.* So that the periodic table will not be too cumbersome, the lanthanide series and actinide series are printed just below the main part of the table. The proper locations for these two series are indicated in column IIIB.

The reason for the series of 14 lanthanide elements is suggested by Fig. 5.5. In the sixth period, after the **6s** orbital has been occupied (Cs and Ba), the next electrons enter **4f** orbitals. While this expansion of the fourth shell continues, *two* outer shells remain constant, the fifth with eight electrons and the sixth with two electrons. This is illustrated by the electron distribution in promethium, atomic number 61:

Pm $1s^2$ $2s^22p^6$ $3s^23p^63d^{10}$ $4s^24p^64d^{10}4f^5$ $5s^25p^6$ $6s^2$

After the fourth shell has acquired its last 14 electrons (in **4f** orbitals), filling of the fifth shell (next to the outside) resumes, due to electrons entering the **5d** orbitals. This corresponds to another series of ten transition elements (numbers 71 through 80). Tungsten, number 74, has this arrangement:

W $1s^2$ $2s^22p^6$ $3s^23p^63d^{10}$ $4s^24p^64d^{10}4f^{14}$ $5s^25p^65d^4$ $6s^2$

After the fifth shell has gained the ten **d** electrons, build-up of electrons in the **6p** orbitals finally proceeds, as in the case of lead:

Pb $1s^2$ $2s^22p^6$ $3s^23p^63d^{10}$ $4s^24p^64d^{10}4f^{14}$ $5s^25p^65d^{10}$ $6s^26p^2$

All together, the total number of elements in the sixth period (atomic numbers 55 through 86) is 32; that is, $2 + 14 + 10 + 6$. The seventh period likewise consists of 8 representative elements plus 10 transition elements plus the 14 actinide elements, for a total of 32. The elements in the lanthanide series and actinide series are often called *inner transition* elements, because they are formed by a filling of electrons into an inner shell, even deeper than a shell used for transition elements.

**5.9
THE PERIODIC
TABLE AND
CHEMICAL
PROPERTIES**

In Section 5.7 we developed the concept that each element was placed in a certain group of the periodic table because of its properties, which in turn are due to the number of electrons in the atom's outer shell. Yet to be explained is the difference in activity of substances having a similar number of outer-shell electrons. Why is potassium more reactive than sodium? Why are nickel and palladium, which possess two electrons in an outer shell, very inactive, whereas calcium, also with two in the outer shell, is very active? Why do we find the most active nonmetals near the top of the table and the most active metals near the bottom? We also need an explanation for the distinctive properties of each group, as compared with other groups. Why is calcium less active than potassium? Why is calcium a metal at all? Why are the noble gases almost entirely inactive?

If we take our clue from the noble gases we can postulate that there is something about having a completely filled outer shell of electrons that confers great stability. Therefore if a lithium atom (Fig. 5.6) could somehow get rid of the electron in its outer shell, it would be left with just one shell containing two electrons. Thus it would have an electron shell like that of helium and should be more stable. (See the diagram for helium in Fig. 5.2.) However, the lithium atom, which now has three protons (positive) and only two electrons (negative), has developed a positive charge. An electrically charged atom is called an **ion**.

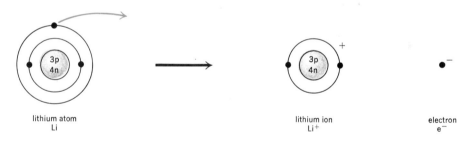

lithium atom
Li

lithium ion
Li$^+$

electron
e$^-$

Fig. 5.6   Formation of a lithium ion.

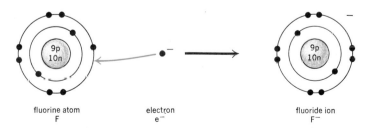

fluorine atom
F

electron
e$^-$

fluoride ion
F$^-$

Fig. 5.7   Formation of a fluoride ion.

If we consider a fluorine atom (Fig. 5.7) it seems unlikely that the seven electrons in its outer shell could all be pulled away from the attraction of the nine protons in its nucleus, even if this did leave it with a shell like helium. Instead, it would be much easier for fluorine to find *one* electron somewhere which it could add to its second shell. This would give it the same stable arrangement of eight in the outer shell that neon has. Because of the one extra electron, fluorine would become negatively charged; it too would be an ion.

We have just postulated that lithium would become more stable if it could lose an electron and fluorine would be more stable if it could gain one. This immediately leads us also to imagine how and why atoms might react with each other. If a lithium atom encountered a fluorine atom there would be a place for the lithium to get rid of one electron; the fluorine atom would readily accept it and both atoms would promptly become more stable. Figure 5.8 illustrates this behavior.

This is exactly what happens if we mix fluorine (a nonmetal) with lithium (a metal). An immediate, violent reaction occurs, releasing a great deal of energy. The two elements combine to become a compound, lithium fluoride, which is made up of lithium *ions* and fluoride *ions*. The formation of these electrically charged ions indicates that the reaction is due to a transfer of electrons from lithium atoms to fluorine atoms. The release of energy is a sign that the ions are much more stable than the original atoms. This example illustrates several fundamental principles:

1. The transfer of electrons from one atom to another is the basis of all chemical reactions. In the future we shall encounter many different kinds of chemical reactions but, in one way or another, all of them will involve movement of electrons.

2. Elements most often react in a way that converts their atoms to a noble gas structure, that is, to the stable arrangement of an outer electron shell having a full set of eight electrons.

3. Metals and nonmetals can now be defined in a more precise manner than they were in Section 5.4. A **metal** is an element that loses electrons and a **nonmetal** is an element that gains electrons, during a chemical reaction.

4. Whenever an atom gains or loses electrons to achieve a complete outer shell, it becomes electrically charged because it no longer has an equal number of protons and electrons. Metals form positive ions, since they lose negatively charged electrons, and nonmetals form negative ions.

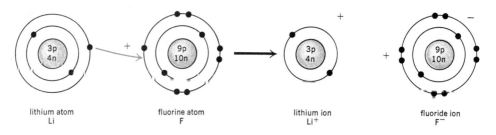

lithium atom   fluorine atom   lithium ion   fluoride ion
      Li   F   Li+   F−

lithium fluoride, LiF

Fig. 5.8   Reaction of lithium with fluorine.

Now it is possible to summarize the important generalities related to the periodic table. Elements in Group IA most readily achieve a stable outer shell by losing one electron; elements in Group IIA can lose two electrons; elements in Group IIIA can lose three electrons. Elements in Group VIIA can most easily attain a stable outer shell of eight electrons by gaining one; those in Group VIA can gain two electrons; elements in Group VA can gain three electrons. Thus, the elements toward the left of the table, which are metals, tend to lose electrons; elements toward the right of the table, the non-metals, tend to gain electrons. Furthermore, it is easier to lose one electron than two. So potassium is more active than calcium; in terms of chemical behavior we say potassium is more extremely metallic. Likewise, the most nonmetallic elements are the farthest to the right in the table (excluding the noble gases). Chlorine atoms gain electrons more readily than do sulfur atoms; chlorine reacts with magnesium rapidly, but sulfur reacts with magnesium slowly.

Atoms in the center of the periodic table, such as carbon and silicon (Group IVA), find it difficult to completely lose four electrons and equally difficult to gain four. Instead, they tend to *share* electrons with other atoms. In the next chapter we shall see how this makes it possible for carbon or silicon to have eight electrons in its outermost electron shell. Other elements, such as boron or nitrogen, also form compounds by sharing electrons.

In the transition series of Period 4, the elements have the **4s** orbital occupied but differ in the number of electrons in the **3d** orbitals. Therefore the two outer electrons of nickel or palladium are affected by the inner shell of **3d** electrons and are less active than the two electrons of calcium. Furthermore, adjacent elements in the transition series have quite a few properties in common. For example, manganese, iron, cobalt, and nickel can each lose the two electrons from the outer shell to become the ions Mn^{2+}, Fe^{2+}, Co^{2+}, and Ni^{2+}; they differ instead in the number of electrons in the next-to-outer shell. There are also more gradual changes, from left to right, among the transition elements than among the representative elements. Indeed, the transition elements are so named because they form a transition from the extremely metallic elements toward the more nonmetallic.

The properties of the elements of an inner transition series (lanthanides or actinides), and of the compounds they form, are not at all distinct from each other. These elements all have *two* identical outer electron shells and differ in the number of electrons of a deeper shell.

If we consider the outer electron attached to a lithium atom and the one attached to a potassium atom, we can see that the outer electron in potassium is much farther from the nucleus. Now, the forces holding the electron to its atom are due to the positive charge of the nucleus, and the greater the distance between the charges, the less is the attraction. As a result, is is easier to lose an electron from the larger potassium atom than from the smaller lithium atom, and thus potassium is more active than lithium. The same reasoning applies to all metals within a certain group of the periodic table: the greater the atomic size, the more easily an outer electron escapes. Consequently the most active metals of all are cesium and francium, to the left and at the bottom of the table.

The situation is just the opposite for nonmetals; the most reactive are in the upper right corner. Within Group VIIA, fluorine is the most reactive and iodine the least. Since these atoms try to gain an electron to complete the outer shell, the added electron will be able to come closer to the positive nucleus of the small fluorine atom and will be held more tightly.

Because hydrogen is in the first, very short period of the table, it is unusual. In many reactions with other elements it *loses* its only electron to become an H^+ ion. For this reason hydrogen is usually placed in Group IA, but in nearly all other properties it is *not* like the metals of that group. Although it happens rarely, hydrogen can also *gain* one electron, which would give it a complete shell of two, like helium. In this case the ion :H⁻ would result. (Hydrogen can gain an electron only from an active metal such as lithium, potassium, or strontium.) Keep in mind that hydrogen has a half-filled shell; in this respect it is like carbon. Therefore hydrogen forms very many important compounds with other elements by *sharing* its electron, rather than completely losing it or gaining another one.

In the next chapter we shall explore more thoroughly the ways in which elements react with each other to form compounds. Then it will be possible to make still more predictions by relying on the location of the elements in the periodic table.

SUMMARY

In this chapter, we started with a discussion of the occurrence and properties of some of the elements. We then indicated how similarities and differences in the properties of the elements led to the development of the law of periodicity and the periodic table. The periodic table is important, since it helps to summarize much information and also aids in predicting the properties of elements which have not been studied thoroughly. If you know the properties of one element in a group (such as sodium in Group IA) you can predict the properties of another element in the same group (for example francium) even before experimenting with it or reading about the experiments of somebody else. The generalizations and predictions must, of course, be verified by experimentation.

From the properties of protons, neutrons, and electrons observed by many experimenters, Bohr proposed a model for the structure of atoms which he could relate to the periodic table. This summarized many previously known facts and gave scientists an even better basis on which to understand and predict the properties of elements. In this connection, a very fundamental principle was discussed: the structure of a substance determines its properties.

The distribution of electrons outside the nucleus of an atom has been described in some detail, because this arrangement of electrons is the portion of structure which largely determines chemical properties. To understand the electron distribution of an atom you must know what an orbital is, how many orbitals and electrons are possible within each shell, and the order in which electrons occupy the available orbitals. Once you understand the periodic table it is easy to write out the electron distribution of an element.

Finally, in discussing the tendency of elements to lose or gain electrons from the outer shell we gave you the first glimpse of how two substances can react with each other to produce a different substance. This is the second major concern of chemistry: reactions which change one substance to another.

IMPORTANT TERMS

atomic structure	noble gases	periods
electron distribution	nonmetals	representative elements
groups (of the periodic table)	orbital	shell
ion	Periodic Law	subshell
metals	periodic table	transition elements

WORK EXERCISES

1. a) Make a list of the metals with which you were familiar previous to reading this book.
   b) Now list the nonmetals you knew previously and compare the size of the two lists.

2. State some of the properties of metals which are not common to nonmetals.

3. In some calculations the oceans and atmosphere are not included as part of the surface of the earth. Will the percentage of oxygen in these calculations be higher or lower than in Table 5.1? Justify your answer.

4. List at least four ways in which the periodic table will be useful in your study of chemistry.

5. In what other way has the periodic table been of help to chemists?

6. List four irregularities in the periodic table.

7. Draw a diagram shaped approximately like the periodic table. Then, without looking at the one in the book, mark the following on your diagram.
   a) the columns or rows called groups
   b) the columns or rows called periods
   c) the locations of the representative elements
   d) the noble gases
   e) the transition elements
   f) the side of the table where the most metallic elements are found
   g) the side of the table where the most non-metallic elements are found
   h) the zigzag line dividing metals from nonmetals

8. Make the following predictions by considering the positions of the elements in the periodic table.
   a) Would you expect potassium or cesium to react more vigorously with oxygen? Explain.
   b) Would sodium react faster with sulfur or with bromine? Explain.
   c) Would magnesium react more easily with bromine or chlorine? Why?
   d) Would sulfur react more vigorously with aluminum or calcium? Explain.

9. Make a sketch, like the one for sodium in Section 5.7, to represent the structure of each atom listed below. To begin, find information in the List of Elements (inside the front cover).
   a) Si   b) N   c) K
   d) Cl   e) Ti   f) B
   g) Ne   h) Mg   i) Mn

10. In the periodic table find each element for which you sketched a structure in Exercise 9.
    a) For which ones do you see a relation between the group number and the number of outer-shell electrons? What is the relationship?
    b) Which element(s) in Exercise 9 belong(s) to a **B** group?

11. Sketch the structure of each atom below, using only information available from the periodic table.
    a) O   b) Na   c) Ca
    d) C   e) F   f) Be
    g) P   h) V   i) S

12. From the periodic table, predict for each element below (1) whether it would gain electrons or lose them; (2) how many electrons it would gain or lose; (3) whether it is a metal or nonmetal.
    a) Rb   b) I
    c) Ba   d) Al
    e) S   f) Fr
    g) Rn   h) N
    i) Ga

13. For each element in Exercise 12 write a formula for the ion that results when electrons are gained or lost.

14. When we read across from left to right in the periodic table we expect differences in properties from one element to the next.
    a) Are these differences more pronounced among transition elements or among representative elements?
    b) In what way does the structure of transition elements relate to your answer for (a)?

15. Distinguish between an atom and an ion.

16. In Section 5.2 we stated that calcium, chlorine, potassium, and sodium (among others) do not exist in nature as free elements; they are found only in chemical combination with other elements. From the position of these elements in the periodic table, offer an explanation for this behavior.

17. List the complete electron distribution for each element shown below, as was done for Si in Section 5.8.
    a) N   b) Na   c) Cl
    d) Al   e) Cr   f) Zn
    g) Br   h) Sr   i) Co

18. Write the correct chemical symbol for each element.
    a) carbon   b) oxygen
    c) silver   d) fluorine
    e) copper   f) magnesium
    g) hydrogen   h) potassium
    i) chlorine   j) nitrogen
    k) lead   l) zinc
    m) phosphorus   n) sodium
    o) calcium   p) sulfur
    q) aluminum   r) iron

SUGGESTED READING

Asimov, Isaac, *The Search for the Elements*, Basic Books, New York, 1962.

Day, F. H., *The Chemical Elements in Nature*, Reinhold, New York, 1963.

Frieden, E., "The Chemical Elements of Life," *Scientific American*, July 1972, p. 52. Until recently it was believed that living matter incorporated 20 of the natural elements. Now it has been shown that a role is played by four others: fluorine, silicon, tin, and vanadium.

Weeks, M. E., *Discovery of the Elements*, 6th ed., Chapter 24, "The Periodic System of the Elements," Journal of Chemical Education Publishing Co., Easton, Pa., 1956.

Wolfenden, J. H., "The Noble Gases and the Periodic Table," *J. Chem. Ed.*, **46**, 569 (1969).

6

Compounds and Chemical Bonds

6.1
INTRODUCTION
In the previous chapter we described the structure of the atoms which constitute the individual particles of an element. We showed that each different element is unique because its atoms have a particular structure that makes them unlike the atoms of any other element. It is important to understand the nature of atoms, for they are the building units that make up compounds. Indeed, the various elements on our earth almost never exist as individual atoms. The atoms of different elements have a strong tendency to combine with each other, in specific ways, to form compounds. Water is formed by the union of atoms of hydrogen and oxygen. Table salt is sodium combined with chlorine. Compounds vary greatly in their complexity. Baking soda contains atoms of sodium, hydrogen, carbon, and oxygen. A typical protein molecule consists of thousands of the atoms of several different elements. Regardless of whether a compound is simple or complex, its existence depends on chemical bonding.

The combining of elements to form bonds and make compounds is due to the tendency of atoms to establish stable outer shells of electrons. In most cases an outer shell is stable when it has eight electrons. In Section 5.9 we stated that atoms could achieve a stable outer shell of eight electrons by either gaining or losing electrons. This shift of electrons from one atom to another results in a chemical bond—that is, an attractive force which holds together the atoms in the compound. In some instances the electrons are completely transferred from one atom to another, causing the formation of an *ionic bond*. In other instances, the atoms obtain stable outer shells by sharing electrons with each other; this combination produces a *covalent bond*. These two fundamental bond types, the formulas of the compounds which result, and the effect of the bond types on physical properties will be discussed in this chapter.

6.2
IONIC BONDS
The formation of the compound lithium fluoride, from the reaction of a lithium atom with a fluorine atom, was illustrated in Fig. 5.8. The atoms reacted in this manner in order to obtain stable outer shells of electrons. Lithium achieved a stable "noble gas"

structure by ejecting the one electron from its outer shell. Fluorine, which has seven electrons in its outer shell, accepted the electron from lithium. Consequently, the outer electron shell of fluorine became filled with eight electrons, providing stability. The reaction of a sodium atom with a chlorine atom, shown in Fig. 6.1, is another example of compound formation. Sodium loses the electron from its outermost shell; its next-lower shell, which is then exposed, has the stable configuration of eight electrons. Chlorine adds the electron to its outer shell, providing it, too, with the stable number eight.

In this process of gaining a negatively charged electron the chlorine atom has become a chloride *ion* having a charge of $1-$. This can be verified by noting that in the chloride ion the nucleus, containing 17 protons, is surrounded by 18 electrons. The sodium *ion*, with only 10 electrons surrounding its 11 protons, has a charge of $1+$.

Because of the opposite electrical charges on the Na^+ ion and the Cl^- ion, they are strongly attracted to each other; it is this attractive force that binds them together. This type of chemical bond, *due to the attraction of oppositely charged ions*, is called an **ionic bond**. Although the individual ions are charged, the compound NaCl as a whole is electrically neutral.

A diagram such as Fig. 6.1 helps us visualize how sodium atoms can react, one by one, with chlorine atoms; each sodium atom can get rid of an electron only if there is a chlorine atom present to accept the electron. Therefore the formula NaCl is an appropriate symbol for the compound sodium chloride. The formula indicates that there is one sodium ion for each chloride ion, no matter how large or small a particular granule of solid salt happens to be. In a crystal of sodium chloride, Fig. 6.2, the positive sodium ions and negative chloride ions are stacked together one after the other, in all three dimensions. But there are always equal numbers of Na^+ ions and Cl^- ions.

We have seen that the reaction of one atom with another depends mainly on the number of electrons in the outer shell of each atom. These are often called valence electrons, and *the outer shell is termed the* **valence shell**. *Valence refers to combining ability.* We can usually describe the chemical behavior of an atom by focusing our attention on the valence shell. For this reason it is not always necessary to draw diagrams showing *all* electron shells, as in Fig. 6.1. Instead symbols can be written showing only the valence electrons, as in Fig. 6.3, which represents the same reaction as Fig. 6.1, but in briefer form. In Fig. 6.3 the seven valence shell electrons of the chlorine *atom* are shown as dots; the symbol Cl represents not only the nucleus with its protons and neutrons, but also the first two filled electron shells. In the chloride *ion* the outer, valence shell has eight electrons. The representation of the sodium *atom* shows the one electron

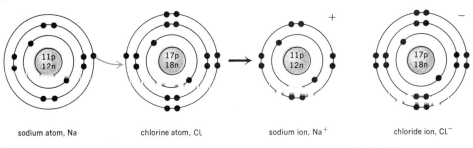

sodium atom, Na   chlorine atom, Cl   sodium ion, Na^+ chloride ion, Cl^-

sodium chloride, NaCl

Fig. 6.1 Formation of the compound sodium chloride.

Fig. 6.2 The arrangement of sodium ions and chloride ions in a crystal of sodium chloride. The larger ions are the chloride ions.

sodium atom   chlorine atom   sodium ion   chloride ion

Fig. 6.3 Formation of sodium chloride as represented by electron-dot structures. (Compare with Fig. 6.1.)

(a dot) in its valence shell and the symbol Na^+ is used for the sodium *ion*. It is important to remember that when the symbol Na is used in this way it represents the nucleus and the inner electron shells (all those beneath the valence shell). Thus, the sodium ion shown as Na^+ in Fig. 6.3 means exactly the same thing as the more complete diagram given in Fig. 6.1. Similarly, when we see an electron-dot diagram for a chloride ion, as in Fig. 6.3, we should realize that it includes all the electron shells and the nucleus, as shown in Fig. 6.1.

Figure 6.4 illustrates, with electron-dot structures, the formation of the compound calcium chloride. Calcium has two valence electrons. It can most readily achieve the stability of an exposed shell of eight electrons by losing the two from its outer shell. Since each chlorine atom can accept just one electron, two chlorine atoms must be present to accommodate the two electrons released by calcium. For this reason a ratio of two chloride ions for one calcium ion is found in any sample of calcium chloride, and $CaCl_2$ is used as its formula. The calcium ions and chloride ions are held in the compound by ionic bonds.

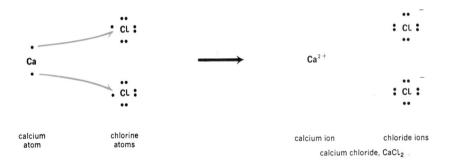

Fig. 6.4   Formation of calcium chloride.

Magnesium (Fig. 6.5) is another metal that has two electrons in its valence shell. It could find stability by transferring these electrons to some atom which needs electrons to complete an almost-filled shell; oxygen would be an example. Because oxygen has six valence electrons, it can accept both of the electrons lost from magnesium. This results in the formation of one magnesium ion and one oxide ion, and the correct formula becomes MgO.

In general, we can predict that elements from the far sides of the periodic table will tend to react by completely transferring electrons from one atom to another. Thus, any of the metals found in groups IA or IIA of the periodic table will react with nonmetals of VIA or VIIA, by a transfer of electrons, to form compounds held together by ionic bonds. Because the IA and IIA elements form positive ions as a result of chemical reactions, they are said to be *electropositive*; the VIA and VIIA elements are said to be *electronegative* because they can form negative ions.

Fig. 6.5   Formation of magnesium oxide.

One more example, Fig. 6.6, will illustrate this generalization. Sodium, with one valence electron, will have a great tendency to transfer it to some kind of atom such as oxygen, which can readily accept an electron. Such a reaction actually takes place whenever metallic sodium is exposed to oxygen. Since oxygen requires two additional electrons to build up its valence shell from six up to the stable eight, *two* sodium atoms must react with each oxygen. There results a white, crystalline compound, sodium oxide, formula Na_2O. In each crystal of sodium oxide the sodium ions and oxide ions are held together by ionic bonds. In either a large crystal or a microscopic one there are always two sodium ions for each one of oxygen.

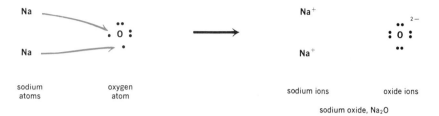

Fig. 6.6 Formation of sodium oxide.

<table>
<tr><td>6.3
COVALENT
BONDS</td><td>The natural gas widely used as a fuel consists almost entirely of one compound, methane. The mere fact that it is a gas suggests that its atoms are held together in a manner that differs from the ionic bonding found in sodium chloride, which is a hard, crystalline solid. Analysis of methane reveals the presence of four hydrogen atoms for each carbon atom. How do these facts relate to a possible formula for methane and to the structure of the carbon and hydrogen atoms?</td></tr>
</table>

Carbon, in Group IVA, is in the middle of the periodic table. There is space in carbon's valence shell for four more electrons. However, the positive nucleus of carbon does not exert enough attraction to be able to hold onto those four additional electrons by completely taking them away from some other atoms. Instead, a carbon atom can *share* electrons with hydrogen atoms to produce methane, Fig. 6.7. Hydrogen has one electron which it could share. When carbon contributes one electron as its share, the result is a pair of electrons held between the carbon atom and the hydrogen atom. All together, the carbon atom shares four pairs of electrons with four hydrogen atoms. For each pair, the carbon atom donates one electron and the hydrogen atom donates one.

Sharing of electrons is thus an alternative way in which an atom can reach the stable condition of having a completed outer shell of electrons. Figure 6.7 shows that in a molecule of methane there are eight electrons in the outer shell around the carbon atom. Even though these electrons are shared with hydrogen atoms, the shell of eight provides stability for the carbon. At the same time, there are two shared electrons just outside each hydrogen nucleus, which is a stable condition for it.

A bond created by the sharing of a pair of electrons between two atoms is a **covalent bond.** The sharing holds the atoms together. In methane, for example, the carbon atom and the hydrogen atoms are bonded together into a particle called a molecule. The

Fig. 6.7 Formation of methane, CH_4, a covalent compound.

structure of methane shown in Fig. 6.7 requires that a correct formula be written CH_4; this also agrees with the analytical results, which show four hydrogen atoms for each carbon atom.

The formation of compounds by covalent bonding occurs among all of the non-metallic elements of the periodic table, and between many of the metalloids and non-metals. Examples are the three common substances illustrated in Fig. 6.8 with electron-dot structures. To write such formulas, first write the symbol for each element involved and then, from the periodic table, determine the number of valence electrons to place around it. In each case think of "building" the molecule by using one electron from each atom to form each shared pair and do so in a way that will finally create a stable electron shell around each atom (i.e., eight electrons around every atom except hydrogen, which can have only two). Thus the oxygen atom, having six valence electrons, will share electrons with two hydrogen atoms, forming a molecule of water. Similarly, three covalent bonds are formed in ammonia because nitrogen starts with five valence electrons. In the case of carbon tetrachloride, four chlorine atoms are necessary to provide a shell of eight around carbon. In the structure of CCl_4 we see that not only the carbon atom, but each chlorine atom is surrounded by eight electrons.

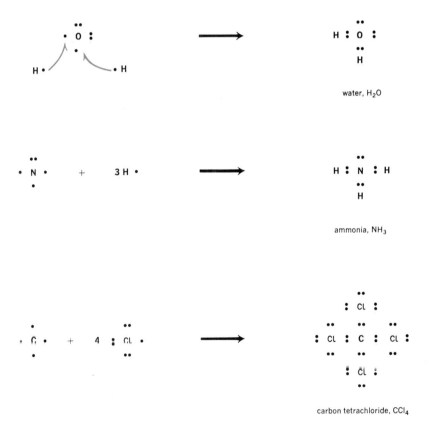

water, H_2O

ammonia, NH_3

carbon tetrachloride, CCl_4

Fig. 6.8 Formation of molecules of water, ammonia, and carbon tetrachloride. In each case, electron pairing creates covalent bonds.

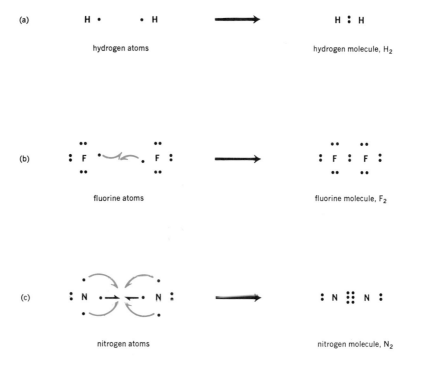

Fig. 6.9 Formation of carbon dioxide, CO_2.

Two atoms can share more than one pair of electrons between them, if this will lead to the important number of eight for the valence shell of each. An example is the covalent bonding of carbon and oxygen to form carbon dioxide, Fig. 6.9. Both carbon and oxygen donate an electron to each shared pair, but finally an oxygen atom shares two pairs with the carbon atom. The oxygen is held by two covalent bonds, the same number of bonds it has in H_2O. The carbon in CO_2 has four covalent bonds, as it does in CH_4 or CCl_4.

The tendency of atoms to achieve stable outer electron shells is so great that covalent bonds will form even between atoms of the same element, if no different element is present for reaction. For instance, the element hydrogen does not exist as separate, individual atoms (except when disrupted by a large amount of energy such as from high termperatures). Instead, two hydrogen atoms share their electrons, thereby forming a covalent bond that ties them together in a hydrogen *molecule*, H_2. Figure 6.10(a) shows the electron pairing by means of electron-dot structures.

Fig. 6.10 Covalent bonding in molecules of (a) hydrogen, (b) fluorine, and (c) nitrogen.

Fluorine is another element that exists as a diatomic (two-atom) molecule, F_2, rather than as individual atoms; see Fig. 6.10(b). The other elements of Group VIIA also form diatomic molecules.

Nitrogen often exists in compounds, bonded to other elements. But on earth there is also a significant amount of the separate element nitrogen. Even this "free" nitrogen does not remain as individual atoms. The atoms are held together in pairs by covalent bonds, so we write the formula N_2 (rather than N) when nitrogen exists as an element. The electron-dot structures in Fig. 6.10(c) show that the N_2 molecule is held together by *three* covalent bonds.

Of the nonmetallic elements that normally exist as diatomic molecules, the most important are H_2, N_2, O_2, and the Group VIIA elements, commonly called the halogens (F_2, Cl_2, Br_2, I_2). In general, the molecular formulas (rather than H, N, etc.) should be used to represent these elements in most situations. We shall see that the molecular formulas (O_2, etc.) are especially important in writing chemical equations (Section 7.5) or in performing calculations (Section 7.7).

The electron-dot structures we have been writing are often called Lewis structures, in honor of the United States chemist G. N. Lewis, who first used them. In 1916 he proposed the idea that a covalent bond is due to a pair of shared electrons. Before that time chemists realized that carbon had a "valence" (combining ability) of four, hydrogen of one, oxygen of two, and so on, which "explained" the formulas CH_4 and CO_2. But no one understood what "valence" really meant—that is, why the atoms should be held together and why the combining number should be four for carbon and two for oxygen, etc. Lewis also emphasized the "rule of eight," the idea that atoms strive to complete an outer shell of eight electrons. These two proposals by Lewis revolutionized the thinking about chemical compounds and greatly enhanced our ability to understand the structure and properties of more complex molecules.

In writing formulas for covalent compounds it is often convenient to use a dash, rather than a pair of dots, to indicate the bond. Accordingly, we can also write formulas of the type shown below. But whenever we do, we need to remind ourselves that each bond line really represents a pair of electrons.

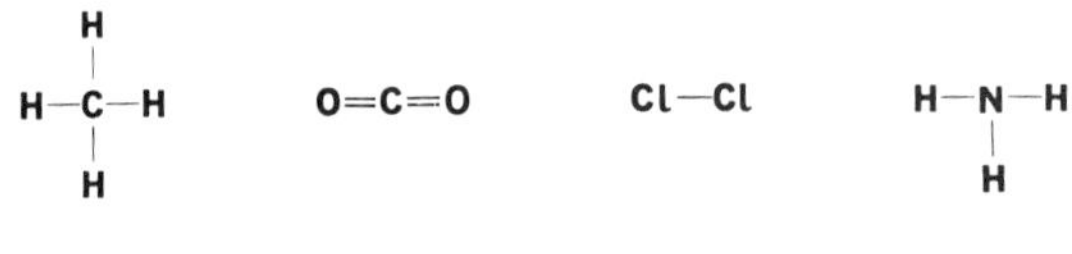

$$
\begin{array}{c}
\text{H} \\
| \\
\text{H}-\text{C}-\text{H} \\
| \\
\text{H}
\end{array}
\qquad
\text{O}=\text{C}=\text{O}
\qquad
\text{Cl}-\text{Cl}
\qquad
\begin{array}{c}
\text{H}-\text{N}-\text{H} \\
| \\
\text{H}
\end{array}
$$

Formulas of this type are often called valence-bond formulas.

The formation of an H_2 molecule was shown in Fig. 6.10. Because both hydrogen atoms are identical, they would have identical tendencies to either hold or release their electrons. Therefore the electrons in the covalent bond of H_2 are shared exactly equally by both hydrogen atoms. Now, if two atoms are different, when they form a covalent bond by sharing their electrons, the sharing may or may not be equal; one atom may hold the electron pair closer to itself. Such is the case in a molecule of hydrogen chloride, Fig. 6.11(a). The pair of electrons shared by hydrogen and chlorine is shifted nearer to chlorine than to hydrogen. This is *not* a complete transfer of an electron from hydrogen to chlorine, as in the formation of an ionic bond. In this covalent bond, however, the chlorine atom has acquired a little more than its share of electrons and has a *partially*

H : Cl :   H————Cl
                                                         δ+   δ−

(a)   (b)

Fig. 6.11 The polar covalent bond of hydrogen chloride. (a) Lewis electron-dot structure, showing the bonding pair of electrons closer to Cl than to H. (b) Positive and negative poles of the covalent bond shown on a valence-bond structure.

developed negative charge. Consequently, the hydrogen atom has a partial positive charge. The partial charges at the ends of the molecule are often shown on a valence bond structure as in Fig. 6.11(b); the Greek letter delta (δ) signifies "partial." Since the bond in the HCl molecule has a positive pole and a negative pole, it is termed a *polar covalent bond*.

In contrast, a bond between two identical atoms, as in H_2 or Cl_2 molecules, is always a nonpolar bond. Note, too, that in many covalent bonds between *different* atoms the electrons are equally shared because the two atoms have an approximately equal attraction for electrons. Such a bond is also nonpolar. A very common example of a nonpolar covalent bond between unlike atoms is the C—H bond, as in CH_4.

Finally, we should emphasize that *only two fundamental bond types are found in compounds, ionic and covalent*. Even though some covalent bonds are polar and others are nonpolar, every covalent bond consists of a pair of electrons. Whether the covalent bond is more or less polar depends on how unequally the electron pair is shared.

**6.4
ELECTRO-
NEGATIVITY**

The ability of an atom to attract electrons into its valence shell is termed its **electronegativity.** Within each period of the periodic table, electronegativity increases from left to right; oxygen is more electronegative than nitrogen. And within each group, electronegativity increases from bottom to top; chlorine is more electronegative than iodine. As we have seen, this is because the valence shell of iodine is farther from the positive nucleus, so that iodine attracts and holds another negative electron less strongly than does chlorine. Fluorine—at the upper righthand corner of the table—is the most electronegative of all elements. These trends in electronegativity are shown graphically in Table 6.1.

TABLE 6.1   Electronegativity trends within the periodic table

increasing negativity ⟶

			H				
Li	Be	B	C	N	O	F	
Na	Mg	Al	Si	P	S	Cl	increasing
K	Ca				Se	Br	negativity
Rb	Sr					I	
Cs	Ba						

The comparative electronegativity of two atoms determines whether a covalent bond between them will be polar or nonpolar. In very general terms, we can expect to find polar covalent bonds between two atoms from different groups of the periodic table, especially if they are rather far apart. A covalent bond between carbon and oxygen or between phosphorus and chlorine should be quite polar, but a bond between carbon and nitrogen or nitrogen and oxygen should be only slightly polar. Of course, for atoms that are extremely different in their ability to hold valence-shell electrons, such as sodium and chlorine, we can expect a complete transfer of electrons, resulting in an ionic bond. For our purposes these general predictions are sufficient to understand a number of cases in which polar covalent bonds may have an important effect on chemical properties or solubilities. To predict more accurately how polar a certain covalent bond would be, it is necessary to consult a more advanced text showing specific numerical values for electronegativity.

In terms of electronegativity, the usual position of hydrogen in the periodic table is misleading. As we have seen, hydrogen forms a polar covalent bond with chlorine, not an ionic bond as does lithium or sodium. We need to remember that one electron of a hydrogen atom gives it a half-filled valence shell. In this regard we might think of hydrogen as being halfway across the periodic table, like carbon; see Table 6.1. Indeed, hydrogen forms nonpolar covalent bonds with carbon, slightly polar bonds with nitrogen, and quite polar covalent bonds with oxygen or chlorine. Thus, the covalent bonds of water, H—O—H, are highly polar.

6.5
OXIDATION
NUMBERS

We have stated that elements may combine to produce compounds having either ionic bonds or covalent bonds. In either case, a distinct number of electrons will be gained, lost, or shared, and the number determines the formula of the compound. It is not always desirable to draw elaborate diagrams such as Fig. 6.1, or even Fig. 6.3, just to decide that the correct formula of salt is NaCl and not Na_2Cl or $NaCl_3$. Instead, oxidation numbers can be used to write formulas. In this section we will show how to select the correct oxidation numbers, and in Section 6.6 we will apply the numbers to the writing of formulas.

Figure 6.5 shows that magnesium reacts by giving away two electrons (negative charges) to oxygen and becoming an Mg^{2+} ion. Because of the reaction with oxygen it is said that magnesium was oxidized, and that in the compound MgO magnesium has an oxidation number of 2+. When oxygen reacts with sodium, Fig. 6.6, the sodium becomes an Na^+ ion; the oxidation number of sodium in sodium oxide is 1+. Because the original metals, magnesium and sodium, had all their electrons and had no electrical charge, in their elemental form they are assigned an oxidation number of 0.

The **oxidation number** (also called *oxidation state*) *corresponds to the number of electrons gained or lost by the original atom.* Gain of electrons, which have negative charges, produces a negative oxidation number. Loss of electrons produces a positive oxidation number. For an ion, *the oxidation number is the charge on the ion.*

In the process of being oxidized, magnesium and sodium lost electrons. The term oxidation has been generalized and applied to any case in which electrons are lost, even though the reactions may not involve oxygen. **Oxidation** *is a loss of electrons.* It corresponds to a numerical increase in oxidation number; thus, for example, magnesium goes from oxidation number 0 to 2+ when it is oxidized. Conversely, during reaction with either of these metals, oxygen gains electrons. The oxidation number of oxygen changes from 0 to 2−, which is a numerical decrease. Since its oxidation number has been reduced in numerical value, it is said that oxygen has been reduced. Chemically, then, **reduction** *is a gain of electrons*. It is accompanied by a decrease in oxidation number. If, by some different process, electrons were added to Mg^{2+} ion so that it was converted back to the metal, Mg^0, we would say it had been reduced; there was a decrease in its oxidation number from $2+$ to 0. It is also important to note that *any chemical reaction involving oxidation must be accompanied by reduction*. This, of course, is because electrons cannot be lost from one atom unless another atom is present to accept the electrons.

Now let us apply these general terms to an example which does not actually include oxygen. Figure 6.3 shows that during the formation of $NaCl$ the sodium loses an electron and goes from 0 to $1+$ in oxidation number; sodium has been oxidized. During the same reaction chlorine gains an electron and changes from Cl^0 to Cl^-. Chlorine has been reduced. In general it is apparent that for simple ions the oxidation number corresponds to the charge on the ion. Fundamentally it arises from the number of electrons in the atom's valence shell and consequently the number of electrons it will tend to gain or lose.

Because oxidation numbers depend on valence electrons, the periodic table (and the valence shell diagrams across the top of the table) can be used to predict oxidation numbers. Below are listed some rules for establishing oxidation numbers. The rules apply to the representative elements and to a large extent to other elements as well.

1. The maximum oxidation number of an element corresponds to its periodic group number. (Magnesium $2+$, carbon $4+$, nitrogen $5+$, etc.)
2. The negative oxidation number (the minimum) of a nonmetal equals the number of electrons required to fill its valence shell to eight. (Nitrogen $3-$, oxygen $2-$, chlorine $1-$.)
3. The oxidation number of an uncombined element is always 0.
4. The oxidation number of hydrogen is $1+$ in nearly all compounds.
5. The oxidation number of oxygen is $2-$ in nearly all compounds.

Among the representative elements there are two exceptions to these rules, which, fortunately, are not frequently encountered. First, when hydrogen forms a compound with an active metal (Na, Ca, etc.) the metal, being more electropositive, loses electrons which hydrogen must accept. Therefore hydrogen has an oxidation number of $1-$ in compounds such as calcium hydride (CaH_2) or sodium hydride (NaH). Second, oxygen is assigned an oxidation number of $1-$ in peroxides, such as H_2O_2.

In a great many cases one can select the correct oxidation number by using the rules just given. However, as a time-saver the oxidation numbers of some of the most frequently encountered elements are listed in Table 6.2. You will find it helpful to become familiar with the list.

All of the elements in Table 6.2 fit the rules except for copper, iron, and lead. In general, the transition elements, because of their atomic structure, require more elaborate rules; but, since those rules will not be needed in later chapters, we will not list them here. In the case of copper, iron, and lead, however, we should take the time to see how they depart from the simpler oxidation-number rules.

The $1+$ oxidation number of copper results from the loss of its one outer-shell electron and corresponds to its group, IB. However, copper forms many other stable compounds in which it has oxidation number $2+$ due to loss of a second electron, from an inner shell. The $4+$ oxidation number of lead also relates to its group, IVA,

TABLE 6.2   Oxidation numbers of some common elements

Element	Oxidation Number	Element	Oxidation Number
aluminum, Al	3+	lead, Pb	2+ (4+)
barium, Ba	2+	magnesium, Mg	2+
bromine, Br	1−	nitrogen, N	5+, 3−
calcium, Ca	2+	oxygen, O	2−
copper, Cu	(1+) 2+	potassium, K	1+
carbon, C	4+, 4−	silver, Ag	1+
chlorine, Cl	1−	sodium, Na	1+
hydrogen, H	1+	sulfur, S	6+, 2−
iron, Fe	(2+) 3+	zinc, Zn	2+

but is exhibited in only a very few lead compounds. The oxidation number of lead is 2+ in most cases. Iron, in group VIIIB of the periodic table, has two electrons in its outermost shell. Loss of these two would produce Fe^{2+}, and in several compounds iron does exist in this condition. However, in most compounds, actually the more stable ones, iron has lost a total of three electrons and thus has oxidation number 3+. To summarize, unless other information is provided, *use 2+ for the oxidation number of copper, 2+ for that of lead, and 3+ for that of iron.*

Table 6.2 also indicates that some of the representative elements can produce different compounds in which they have different oxidation numbers; for example, nitrogen has 5+ or 3−. This results because either rule 1 or rule 2 can apply, depending on whether nitrogen reacts with an element that is more electropositive or more electronegative. This behavior is observed especially among the nonmetals of Groups IVA, VA, and VIA. A few elements may also display additional, intermediate oxidation numbers in certain compounds; nitrogen can also be 3+ and sulfur 4+.

6.6
FORMULAS OF
COMPOUNDS

In this chapter we have begun to write formulas to represent compounds, for instance, $CaCl_2$ for calcium chloride. A **formula** *tells the kind and number of atoms present in a compound.* Any complete compound is electrically neutral—one atom gains only as many electrons as another releases. Therefore, in the correct formula of a compound, the algebraic sum of the oxidation numbers within the compound must be zero. Thus sodium chloride is Na^+Cl^- and magnesium oxide is $Mg^{2+}O^{2-}$. We shall now see how oxidation numbers can be used to write formulas correctly.

How to Write a
Formula

The formula of a compound can be written by following these steps:

1. Write the symbol of the atoms present in the compound, with the more electropositive atom written first.
2. Write (temporarily) the oxidation number of each atom above its symbol.
3. At the lower right of the atom symbols write the numbers (called subscripts) required to balance the plus and minus oxidation numbers.
   a) This step is necessary only if the oxidation states of the atoms are numerically different (for example, 3+ and 2−, or 1+ and 3−).
   b) An easy method to select the correct subscripts is to "cross over" the oxidation numbers from above the atom symbols (but without + and − signs).

     c) If the two oxidation numbers are 4 and 2, the subscripts 1 and 2 are used. (The *smallest* numbers required to balance the charges should be used.)

4. For the final formula of a compound, omit the oxidation numbers temporarily used in step 2.

Several examples will illustrate the use of these steps and of the oxidation numbers previously discussed.

Example 1   Suppose lithium metal reacts with oxygen. How should the formula of the resultant compound be written?

1. LiO

2. $\overset{(1+)}{Li} \overset{(2-)}{O}$

3. $\overset{(1+)}{Li_2} \overset{(2-)}{O_1}$

    The symbols Li_2 indicate two lithium ions. Therefore the total positive charge is $2+$, which balances the $2-$ of oxygen.

4. Li_2O is the final correct formula; it means the compound consists of two lithiums and one oxygen. The subscript 1 is not written in a final formula.

Example 2   What is the formula of the compound produced by bromine and aluminum?

1, 2. $\overset{(3+)}{Al} \overset{(1-)}{Br}$

3. $\overset{(3+)}{Al_1} \overset{(1-)}{Br_3}$

4. $AlBr_3$

    This means the compound consists of one aluminum and three bromines.

Example 3   What compound is produced by reaction of oxygen and calcium?

1, 2. $\overset{(2+)}{Ca} \overset{(2-)}{O}$

3. In this example both oxidation numbers are numerically 2. Step (3) is not necessary because the charges are already balanced.

4. CaO is the correct formula.

Example 4   Magnesium + nitrogen $\rightarrow$?

1. Mg   N

2. $\overset{(2+)}{Mg} \overset{(3-)}{N}$

    Nitrogen can have oxidation number $5+$ or $3-$. In this case it is combining with magnesium, which is a metal, is more electropositive than nitrogen (a nonmetal), and always has oxidation number $2+$. Therefore we must use a negative oxidation number for nitrogen. To say it another way, magnesium and nitrogen cannot both use a positive oxidation number.

$$(2+) \quad (3-)$$

3. $Mg_3 \quad N_2$

Notice, in this example, that when we use 3 magnesiums and 2 nitrogens, the net charge of the compound will be:

$$3(2+) + 2(3-) = (6+) + (6-) = 0$$

4. Mg_3N_2 is the correct formula.

The four examples just shown have concerned ionic compounds, but the rules can also be applied to combinations of elements that we might expect would produce covalent compounds. The indication of a plus or minus value in the oxidation number does not mean that the atom *must* exist as an ion, with a completely developed charge. Formulas of covalent compounds are illustrated by the next few examples.

Example 5   Silicon and oxygen $\rightarrow$?

1. Si   O

$$(4+) \quad (2-)$$

2. Si   O

Oxygen should be $2-$; also it is more electronegative than silicon (oxygen is farther right in the periodic table than silicon), so silicon should be positive in this compound. Silicon is not listed in Table 6.2, but from rule 1 and the location of silicon in the periodic table (Group IVA) we decide upon $4+$.

$$(4+) \quad (2-)$$

3. $Si_2 \quad O_4$

4. SiO_2

Because the subscripts shown in step (3) are 2 and 4, in the final formula they have been reduced to 1 and 2. Short cuts can be taken in a case like this. Since the oxidation numbers in step (2) are $4+$ and $2-$, it is easy to see that doubling the amount of oxygen would give $4-$ and balance the charges. So we can immediately write the final formula, SiO_2.

Example 6   Hydrogen + oxygen $\rightarrow$ water; what is the formula of water?

$$(1+) \quad (2-)$$

1, 2. H   O

$$(1+) \quad (2-)$$

3. $H_2 \quad O$

4. H_2O

The rules have provided us with the correct formula for water; it really doesn't matter that we know from the start whether water is ionic or covalent. Actually, the bonds are covalent (Fig. 6.8) but very polar. Because hydrogen is fairly electropositive and oxygen is highly electronegative, the electrons in the covalent bonds are shifted distinctly toward oxygen. Since the electron that hydrogen contributed to the covalent bond has been shifted *away from* hydrogen, in this sense it has been oxidized. *Oxidation amounts to a loss of electrons, whether due to complete removal or merely a shift away from the atom being oxidized.*

Example 7   Nitrogen + hydrogen $\rightarrow$ ammonia; what is the formula of ammonia?

$$
\begin{array}{lll}
 & (3-) & (1+) \\
1, 2. & N & H \\
 & (3-) & (1+) \\
3. & N & H_3 \\
4. & NH_3 &
\end{array}
$$

The main problem in this example is to decide which atom has a positive oxidation number and which has a negative number. Nitrogen is farther left in the periodic table and should be more electronegative than hydrogen (even if we think of hydrogen as being halfway across the periodic table); also, rule 4 states that hydrogen should be $1+$. Therefore we select $3-$ for the oxidation number of nitrogen in this compound and arrive at the correct formula, NH_3. Incidentally, this is one case in which the practice of writing the more electropositive atom first in the formula is usually not followed. Because of long-standing custom, ammonia is most often written NH_3; however, H_3N is certainly not wrong.

Finding an Oxidation Number From a Formula   So far, oxidation numbers have been used to determine the correct formula of a compound. The reverse can be done also; if a formula is known, the oxidation number of some element in the compound can be calculated. For example, under different conditions sulfur can combine with oxygen to produce two different gaseous compounds. Careful analysis of the compounds shows that in one there are two oxygen atoms and one sulfur atom, while in the other there are three oxygens for one sulfur. The respective formulas must be SO_2 and SO_3 and we can ask, "What is the oxidation number of sulfur in each?" Since the sum of the oxidation numbers in a compound must total zero, and since we can reliably assume that oxygen will be $2-$, we can write:

$$
\begin{array}{ll}
? & 3(2-) \\
S & O_3
\end{array}
$$

The total for oxygen is $3(2-) = 6-$. Therefore sulfur must be $6+$ in this compound; this is an oxidation number we might have expected from sulfur in periodic group VIA.

Following a similar procedure for SO_2 we write:

$$
\begin{array}{ll}
? & 2(2-) \\
S & O_2
\end{array}
$$

For oxygen, $2(2-) = 4-$; so sulfur is $4+$, a less common oxidation number for sulfur which we could not predict from the periodic table. Additional illustrations of calculating an oxidation number from a formula follow.

Example 1   In a chemical reference book we might encounter a discussion of the compound PCl_3; we could calculate the oxidation number of phosphorus by assuming that chlorine has its familiar oxidation number of $1-$. Since the three chlorines will total $3-$, phosphorus must be $3+$ in PCl_3.

Example 2   We can calculate the oxidation number of bromine in HBrO by making the usual assumption that hydrogen is $1+$ and oxygen is $2-$. In order for the compound as a whole to have charge zero, obviously bromine must be $1+$.

Example 3   A slightly more complicated example is to calculate the oxidation number of sulfur in magnesium sulfate, $MgSO_4$, shown as follows:

$$2+ \quad ? \quad 4(2-)$$
$$Mg \quad S \quad O_4$$

We can safely assume that magnesium is $2+$ and, as usual, that each oxygen is $2-$. The total negative charge from oxygen is $8-$, which must be balanced by a total of $8+$ from other atoms. Since magnesium accounts for only $2+$, the remaining $6+$ must be from sulfur.

6.7
THE MEANING OF A FORMULA

Formulas of compounds have been discussed at some length. But it might be a proper question to ask, "What do the formulas mean in terms of the real substances, such as a teaspoon of salt, a bottle of liquid carbon tetrachloride, or a tank of methane gas?"

Covalent Compounds

For a covalent compound such as carbon tetrachloride the formula CCl_4 means not only that there are four chlorine atoms for each carbon, but quite specifically that all these atoms are held together (by covalent bonds) in a definite particle. This particle is called a molecule and its structure is represented in Fig. 6.8. The molecules are so tiny that in a small bottle of the liquid there would be billions of them. But each molecule is a separate particle of CCl_4; although its atoms are bonded to each other they are not bonded to the atoms of any other molecule. Since even the tiniest particle of CCl_4, one molecule, consists exactly of four chlorines and one carbon, this formula also means that in any sample of carbon tetrachloride large enough for us to handle there will always be four times as many chlorines as carbons.

These same principles apply to all covalent compounds. The formula tells the number and kind of atoms in one molecule; it also says that the same proportions of these atoms will exist in any quantity of the compound. In a tank of methane gas, all of the billions of particles bouncing around, colliding with each other, consist of separate CH_4 molecules (Fig. 6.7). Within the whole tank, for each billion carbon atoms there are four billion hydrogen atoms. Similarly, the formula H_2O means that each molecule (particle) of water is made up of two hydrogens and one oxygen bonded together (Fig. 6.8). In a teaspoon of water, a glassful, or a lakeful, there will always be exactly twice as many hydrogen atoms as oxygen atoms.

Ionic Compounds

For an ionic compound such as salt, the formula $NaCl$ has the same meaning, in terms of relative composition, as does the formula for a covalent compound. In one small crystal of salt or in a whole teaspoonful there will always be equal numbers of sodium ions and chloride ions.

However, in terms of the individual particles found in the compound, the situation in an ionic compound is not at all like that in a covalent compound. In an ionic compound there are no molecules; instead, the individual particles present are ions. The ions are "building blocks" that are stacked together to make a crystal large enough to see. In the case of sodium chloride the crystal arrangement is shown (greatly magnified) in Fig. 6.2. In the crystal we cannot pick out one certain pair of Na^+ and Cl^- ions and call it a molecule; each Na^+ ion is surrounded by six Cl^- ions and is equally attracted to all of them.

The individual nature of the ions is emphasized even more dramatically by a solution of salt in water. As you know from common observation, when salt is placed in water it seems to disappear into the water. The water breaks down (dissolves) the salt crystals and this allows all of the separate Na^+ and Cl^- ions to move about in the water. One certainly cannot say that one particular Na^+ ion "belongs" to a certain Cl^- ion. All that is certain is that there are equal numbers of Na^+ and Cl^- ions in the water.

From these discussions it is evident that the formula of a compound (a) tells exactly the number and kind of atoms present, and (b) has a specific meaning in terms of the particles (ions or molecules) of the compound. In Section 7.2 we shall discuss a third, equally important, meaning related to the formula of a compound; it can also represent the mass or weight of the compound.

6.8 MULTIPLE OXIDATION NUMBERS

In Section 6.6 we found that sulfur has oxidation number $6+$ in $MgSO_4$ and in SO_3, but that in SO_2 it has $4+$. In another compound, Na_2S, sulfur must have oxidation number $2-$, because sodium can have only $1+$. It is apparent from Table 6.2 that some of the common elements can exist in more than one oxidation state, and that many, especially the transition elements, have multiple oxidation numbers.

Nitrogen is a familiar element that has multiple oxidation states. In Section 6.6 nitrogen was shown having oxidation state $3-$ in NH_3. Two other well-known compounds of nitrogen are HNO_3 (nitric acid) and HNO_2 (nitrous acid); in these compounds the oxidation numbers of nitrogen are $5+$ and $3+$, respectively. These numbers can be calculated from the formulas by assuming the usual $1+$ for hydrogen and $2-$ for oxygen. But the oxidation states can also be observed, in a more fundamental way, in the electronic structures of the compounds. This is displayed for nitrous acid in Fig. 6.12. The formula is written HNO_2, due to the custom of placing first the elements of positive oxidation number and last the negative ones (oxygen in this case). However, the structure is H—O—N=O, with hydrogen bonded to oxygen. Figure 6.12 shows that nitrogen shares three of its electrons in covalent bonds with oxygen. Since oxygen is more electronegative than nitrogen, the electrons in the covalent bonds will be pulled more toward oxygen. Nitrogen has donated three of its electrons *toward* oxygen and has been oxidized to a $3+$ state.

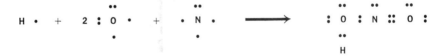

Fig. 6.12   The structure of nitrous acid, HNO_2.

Although for our purposes it is not necessary to do so, an examination of the covalent structure of HNO_3 would show that in it the nitrogen atom had donated all five of its electrons toward oxygen. In this case the nitrogen has essentially been oxidized to a $5+$ state. Similar analyses of structure can account for multiple oxidation numbers of still other elements.

6.9 IONIC GROUPS

The structure of nitrous acid has been shown as covalent (Fig. 6.12). In it, the hydrogen–oxygen bond is very polar. Under certain conditions, such as in a water solution, the highly polar H—O bond may actually become ionic; that is, hydrogen departs to become a separate H^+ ion. This means that a negative ion, a nitrite ion ($^-$O—N=O), is also

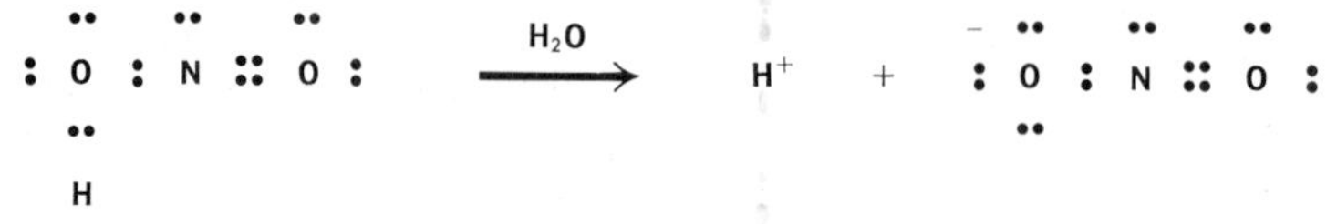

$$\ddot{\ddot{O}} : N :: \ddot{O} : \quad \xrightarrow{H_2O} \quad H^+ \; + \; {}^{-}\ddot{\ddot{O}} : N :: \ddot{O} :$$

H

$$HNO_2 \quad \xrightarrow{H_2O} \quad H^+ \; + \; NO_2^{-}$$

Fig. 6.13 In water solution, a covalent molecule of nitrous acid can break apart (ionize) to form H^+ and an ionic group, NO_2^-.

formed. Such a process is called *ionization* of a covalent compound. (Ionization is discussed in Section 11.2.) The resultant hydrogen ion and nitrite ion are represented in Fig. 6.13 by structural formulas.

The structure of the nitrite ion illustrates an important general principle. The atoms of the nitrite ion are held together as a *group* by covalent bonds. But at the same time the group has a portion of ionic structure; one of the oxygen atoms has a negative charge.

In a solution of nitrous acid in water, the ionic charge of the nitrite group would be associated with an H^+ ion. However, in other situations the nitrite ion could be associated with a different positive ion such as Na^+ or K^+. For example, when nitrous acid, HNO_2, reacts with sodium hydroxide, an Na^+ ion replaces an H^+ ion to produce a new compound, sodium nitrite, $NaNO_2$. Not only in this reaction but during numerous other chemical changes the nitrite group persists as a definite structure. Thus, an NO_2^- group is a particular ion with its own properties and existence, just as a Cl^- ion has its own existence.

The behavior just described is found also among many other ionic groups. The atoms of a group remain bonded together as a distinctive cluster. The group maintains that structure even while being involved in various chemical reactions. Always, the group is held together by covalent bonds, while one (or more) of the atoms also bears an ionic charge.

TABLE 6.3   Common ionic groups with illustrative compounds

Formula of Group	Name of Group	Illustrative Compound	Name of Compound
OH^-	hydroxide	$NaOH$	sodium hydroxide
NO_3^-	nitrate	KNO_3	potassium nitrate
NO_2^-	nitrite	KNO_2	potassium nitrite
SO_4^{2-}	sulfate	$CaSO_4$	calcium sulfate
SO_3^{2-}	sulfite	$CaSO_3$	calcium sulfite
CO_3^{2-}	carbonate	Na_2CO_3	sodium carbonate
HCO_3^-	hydrogen carbonate (bicarbonate)	$NaHCO_3$	sodium hydrogen carbonate (sodium bicarbonate)
PO_4^{3-}	phosphate	Na_3PO_4	sodium phosphate
CN^-	cyanide	KCN	potassium cyanide
CrO_4^{2-}	chromate	$PbCrO_4$	lead chromate
MnO_4^-	permanganate	$KMnO_4$	potassium permanganate
NH_4^+	ammonium	NH_4Cl	ammonium chloride

Many ionic groups are known. Those most frequently encountered are listed in Table 6.3. Although electron-dot formulas can be used to determine the structures and charge of each group, it is more efficient to simply memorize the formula, name, and charge of all the groups in Table 6.3, in order to use them to write formulas of compounds.

When the correct formula of a compound requires more than one group, the entire group symbol is enclosed within parentheses. Thus $(NH_4)_2CO_3$ is written for ammonium carbonate and $Al(OH)_3$ for aluminum hydroxide. Further examples of formulas containing ionic groups are provided in Table 6.4.

TABLE 6.4 Examples of names for compounds

Formula	Name
MgO	magnesium oxide
$(NH_4)_2S$	ammonium sulfide
$FeBr_2$	iron(II) bromide (ferrous bromide)
$Fe(OH)_3$	iron(III) hydroxide (ferric hydroxide)
Na_2SO_4	sodium sulfate
$Cu(NO_3)_2$	copper(II) nitrate (cupric nitrate)
$CuCl$	copper(I) chloride (cuprous chloride)
BF_3	boron trifluoride
CCl_4	carbon tetrachloride
CO	carbon(II) oxide (carbon monoxide)
CO_2	carbon(IV) oxide (carbon dioxide)
SO_3	sulfur(VI) oxide (sulfur trioxide)

6.10 NAMES OF COMPOUNDS

Over the years numerous methods for naming chemical compounds have been devised. At present most chemists throughout the world use the rules agreed upon by the International Union of Pure and Applied Chemistry (IUPAC). A brief version of these rules, adequate for naming a great many compounds, follows.

1. Write the name of the element that appears first in the formula of the compound (the more positive element). Use a roman numeral in parentheses after the name of the element if it has more than one oxidation state.

2. Then write the name of the more negative portion of the compound. If the negative portion is an ionic group, use the name given in Table 6.3. If the negative portion is a single element, change its name ending to *-ide*.

3. When a metalloid or nonmetal forms a compound with another nonmetal, the compound can be named by including a Greek prefix with the more negative element's name.

mono, one   *tri*, three   *penta*, five
di, two   *tetra*, four

The rules are illustrated by examples in Table 6.4. The names of several hydrogen compounds also follow the rules; for example, HCl is named hydrogen chloride. For a few hydrogen compounds, however, common names have been adopted as proper, systematic names: NH_3 is called ammonia, CH_4 is methane, and H_2O is water.

TABLE 6.5   Alternative names for some familiar metallic ions

Ion	Systematic Name	Common Name	Latin Name Origin
Hg^{2+}	mercury(II)	mercuric	Mercurius
Hg^+	mercury(I)	mercurous	
Cu^{2+}	copper(II)	cupric	cuprum
Cu^+	copper(I)	cuprous	
Fe^{3+}	iron(III)	ferric	ferrum
Fe^{2+}	iron(II)	ferrous	
Sn^{4+}	tin(IV)	stannic	stannum
Sn^{2+}	tin(II)	stannous	

Old common names are also still in use for some familiar metals having more than one oxidation state. The metal names are listed in Table 6.5, and illustrations of their use are included in Table 6.4. In each case, an *-ic* ending is used for the higher oxidation state and an *-ous* ending for the lower oxidation state. The main portion of each common name, which dates back to the Dark Ages, was derived from a Latin word.

Each of these ions has its own unique chemical properties; for instance, ferric ions, Fe^{3+}, behave differently from ferrous ions, Fe^{2+}. Such differences can also be important biologically. In hemoglobin, iron is in the Fe^{2+} condition, not Fe^{3+}. Dietary supplements to help alleviate anemia should be compounds such as ferrous sulfate or ferrous gluconate; these are more soluble and are more readily absorbed than are ferric compounds. Similarly, stannous fluoride is used in toothpastes, not stannic. Mercurous chloride has had certain medical uses, but mercuric chloride is too poisonous.

6.11 PHYSICAL PROPERTIES OF SUBSTANCES

One of the most important principles of chemistry is that the structure of a substance dictates its properties. We have already given several illustrations of how structure determines chemical properties. For example, a sodium atom reacts with chlorine to produce an ionic bond, but due to a different atomic structure carbon forms a covalent bond with chlorine.

Most of the rest of this book will be devoted to discussions of *chemical* properties, all of which ultimately depend on structure. Not surprisingly, however, the *physical* properties of substances are also determined by their structure. Each different type of substance has its own distinctive physical properties, which are summarized in Table 6.6 and discussed in the following paragraphs. (See also Sections 3.5 and 3.6.)

Ionic Compounds

Common salt, sodium chloride, is a very typical example of an ionic compound. The particles found in a sample of salt are Na^+ ions and Cl^- ions. These particles are very strongly attracted to each other in a crystal (Fig. 6.2) because of their opposite electrical charges. The crystal is hard and brittle; it breaks only if struck with considerable force. A high temperature (800°C) is required to melt the crystal; that is, a great amount of heat energy is necessary to overcome the attractions between the ions so that they can move about in a liquid condition rather than being held in the rigid crystal pattern. A

TABLE 6.6   Physical properties of substances

Property	Ionic Compounds	Covalent Compounds	Metals
Type of particles	ions	molecules	atoms
Attraction between particles	very strong	weak	moderate to strong
Melting temperature	high (500–3 000°)	low (within 250° above or below room temperature)	moderate to high (50–2 500°)
State at room temperature (20°)	solid	gas or liquid or solid (easily melts)	solid
Conditions of the solid state	hard, brittle, translucent	soft, translucent	soft or medium hard, opaque, malleable, metallic luster
Conditions of liquid and gas states	long liquid range (500–1 000°) nonvolatile	short liquid range (50–200°), easily vaporized, often has odor	long liquid range (1 000–2 000°) nonvolatile
Most soluble in:	polar compounds (water)	nonpolar compounds (oil, gasoline, ether, etc.)	insoluble
Electrical conductivity	conducts only when melted	nonconducting	excellent
Heat conductivity	poor	poor	very good
Example	sodium chloride, NaCl	octane, C_8H_{18} (gasoline)	copper, Cu

very high temperature (1 410°C) is required to make the ions move so fast that they can escape from the liquid and pass into the gaseous state.

Because of its rigid crystal structure, salt is a poor conductor of heat and does not conduct electricity. However, once it melts (becomes liquid), so that the ions can move about, the charged particles do conduct an electric current.

Water easily dissolves salt. Water is one of the most polar covalent compounds known. Its positive and negative poles exert enough attraction for charged ions that the ions break loose from the crystal and dissolve. The water molecules literally insulate Na^+ and Cl^- ions from each other, thus keeping the salt dissolved (Fig. 9.4). In general, ionic compounds (if soluble at all) will dissolve in water. In comparison, ionic compounds are unattracted to and are insoluble in oils, which have nonpolar covalent structures.

Covalent Substances   The particles of covalent substances are molecules. The protons and electrons of one atom in a molecule are occupied almost entirely in the bonding which holds the atom to other atoms in the same molecule. However, the atoms of one molecule do have a very weak attraction to atoms in a different molecule, whenever the two molecules come quite close. This weak, close-range attraction means that not much energy is required to separate one molecule from another.

Octane, a constituent of gasoline, is a liquid. Even the modest energy provided by "ordinary" temperatures is sufficient to keep octane molecules moving in a fluid state.

A slight increase in temperature to 125°C, the boiling point of octane, supplies enough additional heat energy to overcome all attractions between the molecules; above this temperature they exist as gas. The melting point of octane is −57°C. At this temperature there is enough energy to break down solid octane, allowing it to flow and become liquid.

In general, among covalent compounds of similar type, one having larger molecules will have a higher melting point and boiling point because its larger surface allows greater attraction between molecules. Paraffin wax is a familiar material which is solid at room temperature. Structurally it is like octane, but most of the molecules contain 25 or 30 carbons (plus hydrogens), rather than eight as in octane, so paraffin has a higher melting point, about 55°C. Even this is very low compared with 800°C for sodium chloride. Similarly, very little physical force is required to disrupt solid paraffin wax. The shape can be easily deformed or broken, but it is not brittle.

Another generality is that "like dissolves like." Oil, grease, and paraffin wax are soluble in gasoline (all are nonpolar), but not in water, which is highly polar. Solubility will be further discussed in Chapter 10.

Only covalent substances—for example, octane—have odors. The odor sensation occurs only if some of the molecules can be in the gas state at room temperature and can reach our noses. As we have seen, many covalent compounds are gases. Moreover, an appreciable number of molecules can escape even from liquid compounds. For example, we readily observe much vapor escaping from liquid octane when it is poured, even though the temperature is less than its boiling point. Of course, not all molecules that reach our olefactory nerve-ends have an odor; carbon dioxide, nitrogen, and oxygen do not.

A few covalent substances (Cl_2, O_2, N_2, etc.) are gases which are elements rather than compounds. Nevertheless, because they are covalent they have the same general physical properties as those described for covalent compounds.

It is important to emphasize that these properties typical of covalent substances are due to weak attractions *between* molecules. This does not mean the covalent bonds *within* the molecules are weak. Indeed, some covalent bonds are exceedingly strong. Diamond provides one striking example. Diamond consists entirely of carbon atoms, each one covalently bonded to four others in a vast three-dimensional network of extreme rigidity. Diamond is the hardest solid known and its melting temperature is above 3 600°C. Only a few other substances have such a structure. Silison carbide, SiC, has a diamondlike structure because carbon and silicon are both in Group IVA. However, largely due to the difference in size between carbon and silicon, the atoms do not fit into the network as rigidly as carbon atoms alone do in diamond. Thus silicon carbide is not quite as hard as diamond and has a lower melting point (2 700°C).

Metals   The nature of possible attractions or bonding in metals has not yet been described. A metal atom has so few valence electrons that it can rarely fill the valence shell by forming covalent bonds—that is, by sharing specific pairs of electrons with other atoms. But metals do achieve a kind of stability when a few electrons (often as few as one or two) from each atom participate in a large "pool" of electrons shared by all the atoms. This creates a moderate attraction between the atoms in any sample of metal. Any given electron in the pool is not associated exclusively with a specific metal atom, so the electron can move about rather freely. The metal atoms can also move past each other without appreciably disturbing the attractive forces among all the atoms.

If we apply force to a metal, for example by pounding it, the metal can change shape but remain solid. We say it is *ductile* and *malleable*. It does not break, as would the brittle crystal of an ionic compound when enough force is applied to disturb the ionic bonds. If a moderate amount of heat energy is applied, metal atoms can move past each other more easily; the metal melts and becomes liquid. On the average, metals have lower melting points than do ionic compounds. Some metals are quite soft and have quite low melting points (potassium, 63°C). But some transition metals, such as chromium, tungsten, and iron, have high melting temperatures and are very hard; however, they are tough rather than brittle.

The fluid pool of electrons in a metal makes it an excellent conductor of electric current. If an electron from a current enters one end of a copper wire, many electrons down the length of the wire can each shift slightly, and almost instantly an electron can be "pushed out" the other end of the wire.

Properties of Atoms Compared with Those of Compounds

Table 6.6 indicates that different types of compounds have different properties. It is also very important to recognize that an atom in the elemental form has strikingly different properties than does the same kind of atom when it is a member of a compound. For instance, *atoms* of the element calcium form a soft, malleable solid and react very easily with oxygen or with water. But in the compound calcium phosphate, the calcium atoms exist in the form of *ions*. Calcium phosphate, found in rocks, bones, and teeth, is a very hard, somewhat brittle solid and obviously does not react chemically with oxygen or with water.

Likewise, many other metallic elements when found in compounds have physical, chemical, and biochemical properties different from those of the same element when it exists as uncombined atoms. It is ferrous ions that function in hemoglobin, not atoms of iron metal. Similarly, magnesium ions function in chlorophyll, not magnesium atoms. Sodium ions are very soluble and stable in water, and are found in all body fluids. In contrast, atoms of sodium react violently with water.

Comparable profound differences between elements and compounds are found among the nonmetals and metalloids. The element chlorine is a pale green, poisonous gas, almost insoluble in water. But chloride ions are colorless, very soluble in water, and essential to living organisms. Also essential are the protein compounds, which contain covalently bonded carbon atoms that are totally different in behavior from the elemental carbon atoms of soot or charcoal.

SUMMARY

The central theme of this chapter has been the idea that atoms, in order to achieve stable valence shells, combine to form compounds. How the atoms accomplish this determines whether the resultant compound is held together by ionic bonds or covalent bonds. We have also discussed which combinations of atoms are most apt to lead to ionic structures and which to covalent structures, and we have defined oxidation, reduction, and oxidation numbers.

Just as symbols are used to represent elements, collections of symbols, called formulas, are written for compounds. This chapter presented a method for correctly writing formulas, discussed the meanings of the formulas, and illustrated the procedures for naming compounds.

The last concept presented in this chapter is that the structure of a substance—ionic, covalent, or metallic—has a profound effect on its physical properties. In the next chapter we shall see that the structure of a substance likewise determines its chemical properties; just as elements combine to produce compounds, the compounds may undergo chemical changes to yield new substances. The new substances have different groupings of atoms from those found in the old substances, but the total number and kind of atoms remain the same during these chemical changes. This must be so because, as we have seen, the fundamental particles of elements are atoms, and atoms are neither created nor destroyed during chemical changes. An atom merely changes its condition when it participates in a compound. The atom may become ionic, or in a covalent compound it will be bonded.

IMPORTANT TERMS

covalent bond	ionic bond	oxidation number	valence shell
electronegativity	ionic group	polar bond	
formula	oxidation	reduction	

WORK EXERCISES

1. Which of the following would have ionic bonds and which would have covalent bonds?
   a) $MgBr_2$ b) O_2 c) K_2S
   d) CaO e) CH_4 f) Br_2
   g) Na_3N h) $RbCl$ i) NH_3
   j) H_2O k) PCl_3

2. Among the compounds in Exercise 1 that you identified as covalent, which ones have quite polar bonds and which have essentially nonpolar bonds?

3. On the basis of the periodic table, what oxidation number would you expect for each element below? Check your answers in Table 6.2.
   a) Cl   b) H   c) O   d) K
   e) Mg   f) Al   g) Na   h) Ba
   i) Br

4. Each of the following nonmetals can have more than one oxidation number. Using the periodic table, what two numbers could you predict for each?
   a) S   b) C   c) N   d) Cl
   e) Se   f) P

5. Write the symbol and the common oxidation number for each of these elements.
   a) lead   b) silver   c) iron   d) zinc
   e) copper

6. Write the correct formula for the compound formed by each of these pairs of elements.
   a) sodium, bromine   b) chlorine, magnesium
   c) oxygen, barium   d) calcium, nitrogen
   e) carbon, oxygen   f) calcium, sulfur
   g) oxygen, aluminum   h) boron, fluorine

7. Using electron-dot structures, show the formation of compounds (as in Fig. 6.3) for these combinations of elements:
   a) Na + I   b) K + O   c) Ca + S
   d) C + F   e) C + S   f) Al + Cl

8. Each of these compounds is covalent. Write a Lewis (electron-dot) structure for each.
   a) H_2 b) Cl_2 c) CH_4 d) H_2S
   e) I_2 f) H_3CBr g) NH_3 h) PBr_3

9. In which compound are the bonds more polar? State briefly why.
   a) H_2O or H_2S b) NH_3 or H_2O
   c) BF_3 or BBr_3 d) PCl_3 or $SiCl_4$

10. The formulas shown below are correct. What is the oxidation number of the indicated element in each compound?
    a) $\overset{?}{Sr}Cl_2$ b) $Na_2\overset{?}{S}$ c) $\overset{?}{C}O_2$
    d) $Cu\overset{?}{O}$ e) $\overset{?}{N}H_3$ f) $Ca\overset{?}{S}O_3$
    g) $Na_3\overset{?}{P}O_4$ h) $\overset{?}{N}O_2$ i) $\overset{?}{Fe}S$
    j) $K_2\overset{?}{Cr}O_4$ k) $Li\overset{?}{Mn}O_4$ l) $\overset{?}{N}I_3$

11. Name each of the compounds shown in Exercise 10.

12. Name each compound.
    a) Al_2O_3 b) LiF
    c) $AgNO_3$ d) CBr_4
    e) $NaOH$ f) NH_4NO_3
    g) $Ba(NO_2)_2$ h) $(NH_4)_2CO_3$
    i) KCN j) $KMnO_4$
    k) PbS l) $Mg(HCO_3)_2$
    m) $CaCl_2$ n) $BaSO_4$
    o) $Fe_2(CO_3)_3$ p) K_3PO_4
    q) Ag_2CrO_4 r) CrO_3
    s) $FeSO_4$ t) $Sn(NO_3)_4$
    u) P_2O_5 v) Hg_2O
    w) NO

13. Write the formula for each of the following.
    a) potassium bromide
    b) copper(II) sulfide
    c) silver oxide
    d) sodium nitrite
    e) aluminum chloride
    f) magnesium hydroxide
    g) zinc carbonate
    h) barium permanganate
    i) lead sulfate
    j) ferric nitrate

k) calcium hydrogen carbonate
l) aluminum sulfate
m) potassium nitrite
n) lithium bicarbonate
o) sodium sulfite
p) silver phosphate
q) ammonium iodide
r) mercuric sulfide
s) cuprous carbonate
t) calcium cyanide
u) barium phosphate
v) ammonium sulfite
w) aluminum chromate
x) stannous chloride

14. Compare covalent compounds with ionic compounds in terms of the following properties.
a) solubility in water
b) melting temperature
c) ability to conduct electricity
d) solubility in mineral oil (a nonpolar liquid)
e) ability to conduct heat
f) the hardness of the compound when solid (the crystals)

15. Give a brief explanation of your answer for each pair below.
a) Which has a lower melting temperature, PCl_3 or KCl?
b) Which is a better conductor of electricity, $CaCl_2$ or Cl_2?
c) Which is a gas at room temperature, Na_2O or SO_2?
d) Which is a liquid at room temperature, CCl_4 or $FeCl_3$?
e) Which is a better conductor of electricity, K or P?
f) Which has a higher melting temperature, Fe or I_2?

SUGGESTED READING

Fernelius, W. C., K. Loening, R. M. Adams, "Notes on Nomenclature," *J. Chem. Ed.* **48**, 730 (1971).

IUPAC, "Reports on Symbolism and Nomenclature," *J. Amer. Chem. Soc.* **82**, 5517 (1960).

7

Chemical Reactions

The universe around us is continually changing. While most of us are unaware of changes in the stars, we are aware of changes in our immediate environment. The melting of ice, the burning of wood, the growing of plants, and the decay of dead materials are all common changes. Some of these changes are physical, and others are chemical. The chemical changes, as the reader is now aware, are referred to as *chemical reactions*. In the previous chapter we discussed reactions of elements to form compounds. After compounds have been formed, many other kinds of chemical changes are possible; the compounds can react with each other and with other elements to produce still different substances. During all of these reactions the substances participating undergo not only a change in composition but also a change in energy content. Several of the most common chemical reactions will be described in this chapter, and the accompanying energy changes will be briefly discussed.

For ease of discussion, we nearly always represent a chemical reaction by a chemical equation, which is a type of shorthand utilizing chemical symbols. In the previous chapter we showed how elemental symbols could be used to write formulas for compounds. Now we shall see how the formulas can be used to write chemical equations. A chemical equation indicates what substances react with each other and what new substances are produced. Even more important, the chemical equation has a numerical or quantitative meaning and can therefore be used to calculate the amount of each particular substance involved. As a prelude to such calculations, we shall begin the chapter by discussing some important numerical meanings related to the formula of a single compound.

When we use a chemical formula to represent a compound, the formula tells precisely the number of each kind of atom present and also signifies the type of particles existing in the compound (Section 6.7). In addition, we shall see now that the formula can represent the mass, or weight, of the compound. Since atoms are the units that combine to make compounds, the mass of a compound is the sum of the masses of all its constituent atoms. Because atoms are so small, for practical purposes the gram atomic weights (Section 4.8) are most often used.

For example, given the rounded atomic weights from the periodic table, the formula H_2O can be used to calculate the mass of one molecule of water. The same number expressed in grams will provide a unit quantity of water large enough for laboratory measurement.

mass of one molecule	*bulk mass* (*weight of 6.02×10^{23} molecules*)
2H = 2 × 1.0 amu = 2.0 amu	2 × 1.0 g = 2.0 g
1O = 1 × 16.0 amu = 16.0 amu	1 × 16.0 g = 16.0 g
molecular mass = $\overline{18.0\ amu}$	mole weight = $\overline{18.0\ g}$

For water, the sum of the gram atomic weights, 18.0 g, is called its *mole weight*. Until recently this quantity has generally been called the "gram molecular weight." In this book we shall nearly always use the modern term, mole weight. The word "mole" comes from the Latin *moles* meaning heap or pile.

A similar calculation for sodium chloride, NaCl, is this:

$$Na = 23.0 \text{ g}$$
$$Cl = 35.5 \text{ g}$$
$$\text{mole weight} = \overline{58.5 \text{ g}}$$

Again, the sum of the atomic weights in grams, calculated from the formula, is properly called the mole weight. However, it would not be accurate to use the term "gram molecular weight" for this ionic compound because, as explained in Section 6.7, there are no discrete *molecules* of sodium chloride. Difficulties of this sort are avoided by the use of the term *mole*, which has been defined in the *Système International* to include all types of substances.

The fundamental SI unit used to measure the quantity of a substance is the mole (abbreviated *mol*). The **mole** *is the amount of substance which contains Avogadro's number of elementary entities.* The amount of substance is almost always expressed in *grams*. The formula of the substance designates the elementary entity, which may be an atom, a molecule, an ion, or an assembly of ions. Thus for H_2O the elementary entities are molecules, but for sodium chloride the entities indicated by the formula are assemblies of ions, Na^+ and Cl^- ions. This means that in one mole of NaCl, which is 58.5 g, there is one mole of Cl^- ions (that is, 6.022×10^{23} ions) and one mole of Na^+ ions (also 6.022×10^{23} ions). Magnesium metal is an element, so the elementary entities present are atoms and its mole weight is 24.3 g/mol. This of course is *numerically* the same as the relative mass of the magnesium atom, 24.3 amu, but the units differ. The advantage of the term mole is that it can be used for both elements and compounds (and other entities) and we need not be concerned whether the compounds are ionic or covalent. Figure 7.1 demonstrates one mole of each of the three substances just discussed.

A few more examples will help to illustrate the concept of the mole and some of the ways it can be used in calculations.

Example 1   For the compound sodium sulfate: (a) What is the mole weight? (b) How many moles of sodium ions are present in one mole of sodium sulfate?

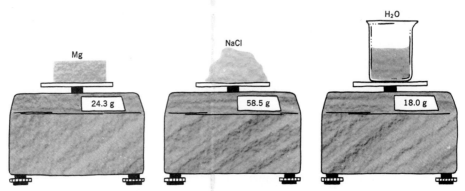

Fig. 7.1 Weights (in grams) of one mole of several substances. (a) Magnesium, 6.02 × 10^{23} atoms; (b) sodium chloride, 6.02 × 10^{23} chloride ions (35.5 g) plus 6.02 × 10^{23} sodium ions (23 g); (c) water, 6.02 × 10^{23} molecules. [Adapted with permission from W. T. Lippincott, A. B. Garrett, and F. H. Verhoek: *Chemistry, A Study of Matter*, 3rd Edition, John Wiley & Sons, Inc., New York, 1977.]

Solutions   a) The formula of sodium sulfate is Na_2SO_4 (see Table 6.3), so we write:

$$
\begin{aligned}
2Na &= 2 \times 23.0\ g = \quad 46.0\ g \\
1S &= 1 \times 32.0\ g = \quad 32.0\ g \\
4O &= 4 \times 16.0\ g = \quad 64.0\ g \\
\hline
\text{mole weight} &= \overline{142.0\ g/mol}
\end{aligned}
$$

b) By inspecting the formula we can see that two moles of Na^+ ions are present in one mole of Na_2SO_4. Or, to say it more explicitly, one mole of sodium sulfate contains $6.02(10^{23})$ of the entities Na_2SO_4. Consequently this amount contains 2 × $6.02(10^{23})$ Na^+ ions, or 2 moles of Na^+ ions.

Example 2   In the laboratory is a bottle containing chips of aluminum metal. (a) How many grams of these chips must be weighed out to obtain two moles of aluminum? (b) How many atoms are present in the two moles?

Solutions   a) The mole weight of aluminum is 27.0 g/mol, therefore:

$$
\frac{27.0\ g}{mol} \times 2\ mol = 54.0\ g
$$

b) $2\ mol \times 6.02(10^{23})\ \dfrac{atoms}{mol} = 12.04(10^{23})$ or $1.20(10^{24})$ atoms

Example 3   If a gold nugget weighs 42.8 g, how many moles is this?

Solution $42.8\ g \times \dfrac{1\ mol}{197.0\ g} = \dfrac{42.8}{197.0}\ mol = 0.217\ mol$

Here the units help us decide whether to multiply or divide. If the factor 1 mol/197.0 g is used in the calculation, the g units cancel.

Example 4   For copper(II) nitrate: (a) What is the mole weight? (b) How many grams would make three moles? (c) How many grams of copper are present in three moles of the compound? (d) How many copper ions are present in 0.50 mol of copper(II) nitrate?

Solutions   a) The correct formula is $Cu(NO_3)_2$; see Table 6.3.

$$1Cu = 1 \times 63.5 \text{ g} = \quad 63.5 \text{ g}$$
$$2N = 2 \times 14.0 \text{ g} = \quad 28.0 \text{ g}$$
$$6O = 6 \times 16.0 \text{ g} = \quad 96.0 \text{ g}$$
$$\text{mole weight} = \overline{187.5 \text{ g/mol}}$$

b) $3 \text{ mol} \times \dfrac{187.5 \text{ g}}{\text{mol}} = 562.5 \text{ g}$

c) The formula shows that one mole of copper(II) nitrate contains one mole of copper. Since the mole weight of copper is 63.5 g/mol, we write:

$$3 \text{ mol } Cu(NO_3)_2 \times \frac{1 \text{ mol } Cu}{1 \text{ mol } Cu(NO_3)_2} \times \frac{63.5 \text{ g } Cu}{\text{mol } Cu} = 190.5 \text{ g } Cu$$

d) $0.5 \text{ mol } Cu(NO_3)_2$

$$\times \frac{1 \text{ mol } Cu^{2+}}{1 \text{ mol } Cu(NO_3)_2} \times \frac{6.02(10^{23}) \ Cu^{2+} \text{ ions}}{1 \text{ mol } Cu^{2+}} = 3.01(10^{23}) \ Cu^{2+} \text{ ions}$$

7.3
PERCENTAGE
COMPOSITION
OF A
COMPOUND

When a new compound is discovered it is analyzed to determine which elements it contains and the percentage of each. From the percentage composition a formula can then be calculated (Section 4.2). However, since many thousands of compounds are now well known, it is not necessary to again perform a time-consuming analysis in the laboratory. If we know we have a specific compound and that it is pure, we can calculate from its formula the exact percentage of each element present. The following examples demonstrate the method of calculation.

Example 1   The dark red ore known as cinnabar is the chief source of mercury. It has been established that cinnabar is mercury(II) sulfide. What percentage of cinnabar is mercury?

Solutions   First, we need the correct formula of cinnabar. We could find it in a chemical handbook or, more quickly, figure it out by means of the naming rules (Section 6.9). It is HgS. Now from the formula we determine the weight of each element present and the total weight (the mole weight). Then the percentage of any particular element can be calculated in the usual manner.

$$Hg = 201 \text{ g} \qquad \% \ Hg = \frac{201 \text{ g}}{233 \text{ g}} \times 100\% = 86.3\%$$

$$S = \frac{32 \text{ g}}{233 \text{ g}}$$

Example 2   The pigment called chrome yellow is $PbCrO_4$. Calculate the percentage of each element present.

Solutions $Pb = 207 \text{ g} \qquad \% \ Pb = \dfrac{207 \text{ g}}{323 \text{ g}} \times 100\% = 64.1\%$

$Cr = 52 \text{ g} \qquad \% \ Cr = \dfrac{52 \text{ g}}{323 \text{ g}} \times 100\% = 16.1\%$

$4O = \dfrac{64 \text{ g}}{323 \text{ g}} \qquad \% \ O = \dfrac{64 \text{ g}}{323 \text{ g}} \times 100\% = 19.8\%$

$$\text{Total} = \overline{100.0\%}$$

Example 3   Niacinamide (sometimes called vitamin B_3) has the formula $C_6H_6N_2O$. What percentage of its weight is carbon and what percentage is nitrogen?

Solutions $6C = 6 \times 12.0 \text{ g} = 72.0 \text{ g}$ $\% \text{ C} = \dfrac{72.0 \text{ g}}{122 \text{ g}} \times 100\% = 59.0\%$

$6H = 6 \times 1.0 \text{ g} = 6.0 \text{ g}$

$2N = 2 \times 14.0 \text{ g} = 28.0 \text{ g}$ $\% \text{ N} = \dfrac{28.0 \text{ g}}{122 \text{ g}} \times 100\% = 23.0\%$

$1O = 1 \times 16.0 \text{ g} = \dfrac{16.0 \text{ g}}{122 \text{ g}}$

Organic chemists often use the percentage composition to help prove that they have synthesized a particular compound. First they must be certain that the expected chemical elements are present and then that the amounts are correct.

Example 4   A chemist believes he has synthesized niacinamide (see above). When 9.8 mg of the sample is analyzed it is found to contain 3.1 mg of nitrogen. Is it likely that the sample submitted for analysis is pure niacinamide?

Solution $\% \text{ N} = \dfrac{3.1 \text{ g}}{9.8 \text{ g}} \times 100\% = 32\%$

Since this percentage differs significantly from 23% nitrogen, either the synthetic sample is impure or it is not niacinamide.

7.4
CHEMICAL
CHANGES AND
EQUATIONS

The element chlorine is a pale green, poisonous gas. Sodium is a soft metal (about as hard as paraffin wax) but, like many metals, has a shiny, silver-gray appearance. If some chlorine gas is passed over a piece of sodium, a vigorous reaction occurs which evolves considerable heat. A completely new substance, sodium chloride, has been formed. It has all the familiar properties of table salt: it is a hard, white solid, has a salty taste, is edible rather than poisonous, etc. This example demonstrates one of the most important aspects of chemical reactions. Because reactions involve changes in composition and structure, the new substances always have properties different from the original ones. Very often, as in this case, the differences are dramatic.

Much of the rest of our study of chemistry will be concerned with chemical changes and the accompanying changes in the properties of the substances involved. In order to describe what happens when substances react it is often convenient to write a very brief summary of the chemical reaction. For instance, one sentence can be used for the spectacular reaction just described:

"Sodium combined with chlorine produces sodium chloride."   (7-1)

This statement describes the changes in composition resulting from this particular chemical event, by telling which substances reacted and which new substance was created. Here is another example.

"Magnesium and sulfur react to form magnesium sulfide."   (7-2)

This word statement can be made even more brief by translating it into a **chemical equation** in which we use the standard symbols for the atoms:

$$Mg + S \rightarrow MgS \qquad (7\text{-}3)$$

In a chemical equation, instead of an equals sign an arrow is used; the arrow means "produces," "forms," "yields," or "is converted to." The + sign does not mean "add" in the strict mathematical sense, but "and," "combined with," or "reacts with." The newly formed substance, MgS in Eq. (7–3), is called a **product**. The original substances, Mg and S in this example, are called **reactants**.

7.5
BALANCING
CHEMICAL
EQUATIONS

A properly written chemical equation is also a quantitative statement; it tells not only *which* substances are involved, but also accounts for *the number* of particles (atoms, ions, or molecules) of each substance. The accounting is done by "balancing" the equation. There is no significant change of mass during a chemical reaction and no elements are created or destroyed. Elements may be associated in different combinations after a chemical reaction, but the total number of each type of atom will be the same after the reaction as before. For instance, Eq. (7–3) indicates that every time one magnesium atom reacts, one magnesium sulfide is produced. After the reaction there is still one magnesium atom, even though it is now in ionic form. The same is true for sulfur atoms in this reaction.

To write a chemical equation, three distinct steps are required:

1. Determine what will be the product(s) of the reaction.
2. Write the correct formula for each substance.
3. Balance the equation by inserting the necessary numbers before the formulas.

Each of these steps must be completed in the order shown.

To begin, how do we know what product(s) to expect from a certain reaction? Experimental observation is the original source of this information. Only from a laboratory analysis can we be sure that sodium chloride really is the white solid produced by the interaction of sodium metal and chlorine gas, so that we can make the statement represented by Eq. (7–1). But it would be tedious, indeed, if we had to perform an analysis each time we needed to predict the products from a chemical reaction mixture. Instead, we nearly always find the information in a reference book. We should recognize, however, that the facts are there only because someone has already studied the reaction, analyzed the product, and reported the results in a chemical journal. For your immediate use, many of the facts that help us predict the products of various chemical reactions are summarized in Section 7.6. The results of still other chemical reactions will be described in the remainder of the book.

For the second step in completing an equation, the correct formula of each compound can be written by following the procedure described in Section 6.6. Balancing, the third step in writing an equation, will be illustrated by the examples that follow.

Example 1   If we wish to translate the word statement about the reaction of sodium and chlorine, Eq. (7–1), into a chemical equation, we already have the information for step 1: the product is sodium chloride. We can proceed to step 2 and replace the words with correct formulas:

$$Na + Cl_2 \rightarrow NaCl \tag{7-4}$$

Here we must use the symbol Cl_2 for the element chlorine because it actually exists in that form, not as individual atoms (Section 6.3).

Now we proceed to step 3, the numerical balancing. We can see at a glance that in Eq. (7–4) there are two chlorines on the left side of the arrow but only one on the right. This violates the law of conservation of matter; each chlorine must be accounted for. The situation can be remedied by placing a 2 before the formula NaCl:

$$\text{Na} + \text{Cl}_2 \rightarrow \text{2NaCl} \tag{7-5}$$

The symbols 2NaCl mean two times the whole formula; they signify "two sodium chlorides" (not two Na and one Cl).

Equation (7-5) accounts for the chlorines but now the sodiums are not balanced. This can be corrected if we write a 2 before the Na symbol on the left side:

$$\text{2Na} + \text{Cl}_2 \rightarrow \text{2NaCl} \tag{7-6}$$

Equation (7-6) is completely balanced. We can interpret this chemical equation to mean, "Two sodium *atoms* react with one chlorine *molecule* to produce two sodium chlorides."

It should be stressed that the three steps for completing an equation must be done in order. The correct formulas must be written *first* and then must not be changed. For instance, to balance Eq. (7-4) it might be tempting to account for the two chlorines by simply changing the NaCl to NaCl_2. However, this would violate the experimental fact that sodium chloride never exists as NaCl_2; its correct formula is NaCl, so this is what we must use. Finally, we must never lose sight of the fact that an equation is merely a convention for expressing the changes in composition during a chemical reaction. The reaction is the real event; the equation is only one way of representing that reality.

Example 2 The reaction of lithium metal with oxygen gas provides another illustration for writing a chemical equation. First we might reasonably expect that this reaction mixture would produce lithium oxide. (See the discussions in Section 6.2.) Second, to write the correct formulas we must remember that oxygen exists as molecules of O_2 rather than as atoms; and that the formula of lithium oxide can be written using the procedure described in Section 6.6. Thus we write:

$$\text{Li} + \text{O}_2 \rightarrow \text{Li}_2\text{O} \tag{7-7}$$

We can begin the third step by showing two lithium oxides which will balance the two oxygens on the left:

$$\text{Li} + \text{O}_2 \rightarrow \text{2Li}_2\text{O} \tag{7-8}$$

Equation (7-8) indicates that the product ($\text{2Li}_2\text{O}$) from the reaction of one O_2 molecule contains a total of four lithiums; therefore we must write 4 in front of the symbol Li on the left of Eq. (7-9). In terms of the real chemical reaction, this means that each time an O_2 molecule reacts there must be four lithium atoms available for it to combine with. The balanced equation is:

$$\text{4Li} + \text{O}_2 \rightarrow \text{2Li}_2\text{O} \tag{7-9}$$

Example 3 A slightly more complicated example will help clarify the procedure for balancing an equation. Although aluminum is supposedly a reactive metal, our daily observations tell us that it is quite stable in the presence of oxygen; stepladders and frying pans made of aluminum last for years. However, if aluminum is ground into small chips and heated to a very high temperature it actually burns and becomes completely oxidized to aluminum oxide. Let us develop a balanced equation to represent this chemical reaction.

Once again, we write O_2 for the reactant oxygen and use the method of Section 6.6

to decide upon the correct formula of aluminum oxide, the product. The unbalanced equation becomes:

$$Al + O_2 \rightarrow Al_2O_3 \tag{7-10}$$

In this case the last step, balancing, will be a bit more involved. In Eq. (7–10) there are two oxygen atoms on the left and three on the right. The least common multiple of 2 and 3 is 6; so to have 6 oxygens on each side we must multiply O_2 by 3 and Al_2O_3 by 2, to obtain:

$$Al + 3O_2 \rightarrow 2Al_2O_3 \tag{7-11}$$

Finally, to complete the balancing we must show 4 aluminum atoms on the left:

$$4Al + 3O_2 \rightarrow 2Al_2O_3 \tag{7-12}$$

A chemical equation is extremely important in many discussions. Therefore the rules described in this section for completing the equation are equally important. The rules apply to any chemical reaction, although, as we shall see later, for reactions involving reduction and oxidation ("redox" reactions) a few additional procedures are helpful.

7.6
TYPES OF
CHEMICAL
REACTIONS

It has become apparent from previous sections that a chemical equation is a particularly useful way to represent a chemical reaction. The equation enables us to discuss the chemical change taking place and (as we shall see in the next section) to calculate the relative *amounts* of the substances involved in the change. The first essential step in understanding a chemical change, and writing an equation for it, is to be able to predict the products expected from the reaction of a certain mixture of substances. In this section the most common inorganic chemical reactions will be summarized. The generalities described will make it possible to predict the products of a great many reaction mixtures. Similar discussions of chemical reactions occurring among organic compounds and in biochemical situations will be presented later in the text.

Most reactions of elements and of inorganic compounds correspond to one of the types in the following list. It is a short list, comprising only four main categories, and you will soon discover that by using it you can predict the outcome of surprisingly many chemical reactions.

A.
Combination or
Addition
Reactions

This is a simple kind of reaction in which two substances, either elements or compounds, join together by forming chemical bonds. The spontaneous formation of new chemical bonds (other than those that may be present already) invariably leads to greater stability and a release of energy. Therefore combination reactions will be *exothermic* (Section 3.13).

1. Combination of two elements Two familiar examples involving combinations of elements are shown below. Equations (7–3) and (7–12) represent still other reactions of this type.

$$2Na + Cl_2 \rightarrow 2NaCl \tag{7-13}$$

$$2H_2 + O_2 \rightarrow 2H_2O \tag{7-14}$$

Equation (7–13) shows a metal and a nonmetal combining, while Eq. (7–14) shows two nonmetals. The generalities concerning which metals and nonmetals usually react with each other were discussed in Sections 6.2 and 6.3.

2. Combination of two compounds Two compounds may combine to produce a larger molecule. The following reactions, which will be discussed in greater detail later, are of this type. For the first example, we find that oxides of active metals combine with carbon dioxide to form carbonates:

$$CaO + CO_2 \rightarrow CaCO_3 \tag{7-15}$$

Oxides of active metals also combine with water to produce hydroxides; a hydroxide is one kind of base (Section 11.5.)

$$CaO + H_2O \rightarrow Ca(OH)_2 \tag{7-16}$$
$$\text{(calcium hydroxide, a base)}$$

$$Na_2O + H_2O \rightarrow 2NaOH \tag{7-17}$$
$$\text{(sodium hydroxide, a base)}$$

Yet another general type of reaction is the combination of oxides of nonmetals with water to produce acids (Section 11.4).

$$SO_2 + H_2O \rightarrow H_2SO_3 \tag{7-18}$$
$$\text{(sulfurous acid)}$$

The three kinds of reactions just mentioned will allow the reader to recognize many combination reactions involving two compounds. It is also helpful to note that at least one of the two compounds participating in a combination reaction is a *covalent* molecule. Some substitution reactions (type C-2, below) also involve two compounds; in that case, however, both are *ionic* compounds.

B. Decomposition Reactions   Many compounds can be made to break down, or *decompose*, to give smaller compounds or elements, if sufficient energy is provided. Thus decomposition reactions are nearly always *endothermic* (Section 3.13) and, in a real sense, decomposition reactions are the reverse of combination reactions. (These statements concerning energy changes are generalities; there may be occasional exceptions. It is true that the breaking of chemical bonds requires energy, whereas bond formation releases energy. But to calculate the net energy change for a complete reaction, both the bonds made and those broken must be considered.)

1. Decomposition to yield elements Moderate heating (a gas-burner flame is sufficient) decomposes mercury(II) oxide into metallic mercury and oxygen gas. In this case, heat is the energy that breaks the chemical bonds.

$$2HgO \xrightarrow{\Delta} 2Hg + O_2 \uparrow \tag{7-19}$$

The upward-pointing arrow at the end of this equation symbolizes a gas (O_2) escaping. The triangle beneath the arrow indicates heat; the symbol originated in the days of alchemy and represents a flame.

Water can be decomposed by supplying energy in the form of an electric current.

It is evident that Eq. (7–20) is the reverse of Eq. (7–14).

$$2H_2O \xrightarrow[\text{electricity}]{} 2H_2 \uparrow + O_2 \uparrow \tag{7-20}$$

2. Decomposition to yield compounds Lime, CaO, can be obtained by heating limestone, $CaCO_3$; Eq. (7–21) is the reverse of Eq. (7–15).

$$CaCO_3 \rightarrow CaO + CO_2 \uparrow \tag{7-21}$$

Some compounds are said to be "unstable." Usually this just means that the compound decomposes so easily that even ordinary temperatures provide sufficient heat energy to cause the breakdown. One such compound is carbonic acid:

$$H_2CO_3 \rightarrow H_2O + CO_2 \uparrow \tag{7-22}$$

Carbonic acid is present in "carbonated" beverages, and reaction (7–22) helps explain the rapid loss of CO_2 gas bubbles from a bottle of beverage that has no cap and has become warm. The increase of heat energy hastens the reaction, and the escape of CO_2 gas from the uncapped bottle also promotes the reaction. If instead, the bottle is capped so that the CO_2 gas is kept in, the CO_2 can recombine with water to cause the reverse of Eq. (7–22). Potentially reversible reactions such as this, which can often be made to proceed in either direction, will be discussed in Chapter 12.

In general, decomposition reactions can be identified by the fact that the reactant is only one compound and that a source of energy is required.

**C.
Substitution or
Displacement
Reactions**

In a great many chemical reactions, one atom or a group of atoms takes the place of another atom or group in a compound. The substitution may be due to one element displacing another from a compound or to an exchange between two compounds.

1. Single substitution (one element and one compound) If we study the effects of placing copper wire in a colorless solution of silver nitrate, we find that the solution becomes blue. We recognize that the blueness is due to the presence of copper ions, and we see a beautiful deposit of crystals forming on the surface of the copper wire. When analyzed these crystals prove to be silver metal. We can write the reaction as follows:

$$2AgNO_3 + Cu \rightarrow Cu(NO_3)_2 + 2Ag \downarrow \tag{7-23}$$

We say that copper has displaced the silver in the compound.

Similarly, if we place iron in a solution of copper sulfate we find copper metal being deposited and the iron going into solution. We can write the equation:

$$Fe + CuSO_4 \rightarrow Cu \downarrow + FeSO_4 \tag{7-24}$$

If we reverse the situation and place some copper metal in a solution of iron sulfate, we find no evidence of reaction so we write the following:

$$Cu + FeSO_4 \rightarrow \text{no reaction} \tag{7-25}$$

In single substitution there must be some way of predicting whether a reaction will proceed or not. Of course, we could try each reaction and check the evidence. Actually, such experiments have been done by many scientists, and as a result we can make some generalizations. We can say that in order for an element to replace another in a compound it must be more active than the element in the compound. Thus, on the basis of the equations given above, we can say that copper is more active than silver but less active than iron. On the basis of many experiments we can develop an activity series or displacement table (see Table 13.3). Such reactions form part of the evidence for statements made in Chapter 5 about the activity of metals.

We also know from experiments that very active metals react with water, displacing hydrogen gas and forming hydroxides:

$$2Na + 2HOH \rightarrow 2NaOH + H_2 \uparrow \tag{7-26}$$

$$Ca + 2HOH \rightarrow Ca(OH)_2 + H_2 \uparrow \tag{7-27}$$

Single substitution reactions occur among nonmetals, too; for instance, when we pass chlorine gas through a solution of potassium iodide we find that iodine is displaced.

$$2KI + Cl_2 \rightarrow 2KCl + I_2 \tag{7-28}$$

A single substitution reaction may be identified by having one element and one compound as reactants. Either the periodic table or an activity series (Table 13.3) is required to predict whether or not a single substitution will take place.

2. Double substitution (two compounds)   Two ionic compounds may react by substitution to form two new compounds. The positive ion of one compound combines with the negative ion of the other compound, while the two remaining ions also pair off. There must, of course, be some reason for the compounds to exchange ions. For instance, without experimental observation we could not predict whether anything would happen if we mixed a water solution of silver nitrate with a solution of sodium chloride:

$$Ag^+NO_3^- + Na^+Cl^- \rightarrow ?$$

Because we know that these are ionic compounds and are very soluble in water, we have shown them as individual ions. In the laboratory we find that as soon as the two solutions are mixed a thick, white precipitate settles out. Upon analysis this solid proves to be silver chloride. We conclude that the formation of the insoluble solid resulted from a chemical change. Although in the solid silver chloride, ions are still present, they are held together so strongly that they are not free to move out into the water and exist as individual, dissolved ions. Therefore in the completed equation we represent silver chloride as unseparated ions by omitting the charges, and use a downward arrow to indicate that it is insoluble (a precipitate):

$$Ag^+NO_3^- + Na^+Cl^- \rightarrow AgCl \downarrow + Na^+NO_3^- \tag{7-29}$$

In contrast to reaction (7–29), if a potassium nitrate solution is mixed with a sodium chloride solution, no change is visible. Careful analysis shows that all four soluble ions are still present in the water solution. The situation could be represented by this equation:

$$K^+NO_3^- + Na^+Cl^- \rightleftharpoons K^+Cl^- + Na^+NO_3^- \tag{7-30}$$

In this case we might think there has been a double substitution yielding the products on the right (K^+Cl^- and $Na^+NO_3^-$). But since those products are *soluble* ionic compounds there is no reason for them to remain as K^+Cl^- and $Na^+NO_3^-$; they can just as readily exchange ions to give back $K^+NO_3^-$ and Na^+Cl^-. The situation is symbolized by arrows pointing in both directions in Eq. (7–30); the arrows show that the reaction is completely reversible. All we know about the reaction mixture is that all four ions are present, and we must imagine any of the possible combinations. (Combinations of the two positive ions or the two negative ions are, of course, impossible; the like charges would repel each other.) It is interesting to discover that if we evaporate the water away from the solution represented by Eq. (7–30), we are left with crystals of all four compounds: KNO_3, $NaCl$, KCl, and $NaNO_3$.

For a mixture such as (7–30) we often say "no reaction" because the condition of the individual ions has not changed, whereas in Eq. (7–29) one pair of ions, AgCl, is in a distinctly different condition: the silver and chloride have separated themselves from the other ions in the solution. In general, whenever one of the products is an insoluble precipitate, a double substitution reaction can occur. The rules needed to predict which compounds will be insoluble are discussed in Section 11.7.

A double substitution can also take place if one combination of the ions produces a *covalent* compound. For instance, the most common acid–base reactions occur because hydrogen ions combine with hydroxide ions to produce HOH, a covalent molecule of water:

$$H^+Cl^- + Na^+OH^- \rightarrow HOH + Na^+Cl^- \tag{7-31}$$

The compounds H_2CO_3, H_2SO_3, and HCN are three others that exist chiefly as covalent molecules rather than as ions; they would therefore promote completion of a substitution reaction. These three compounds are types of acids and will be further discussed, along with reactions such as (7–31), in Chapter 11.

A third condition permitting a double substitution reaction to proceed is the formation of an escaping gas. An example is the reaction of sulfuric acid (H_2SO_4) with zinc sulfide.

$$2H^+, SO_4^{2-} + Zn^{2+}S^{2-} \rightarrow H_2S\uparrow + Zn^{2+}SO_4^{2-} \tag{7-32}$$

In fact, in this case the H_2S is not only a gas that escapes but is also a covalent molecule. There are not many reactions of this type; a few will be discussed later.

To summarize, two *ionic* compounds undergo a double substitution reaction if a new compound can result that is: (a) an insoluble solid or (b) a covalent molecule or (c) a gas that escapes. If one of the two *reactant* compounds is covalent, rather than ionic, a substitution reaction will *not* occur, but a combination reaction may; compare A(2). (This rule does *not* apply to reactions of *organic* compounds to be discussed later.)

**(D.)
Reduction–
Oxidation
(Redox)
Reactions**

Don't worry

Three types of reactions have just been discussed: combination, decomposition, and substitution. It is convenient to think of chemical changes according to these groups simply because it helps us to know what to expect of certain reaction mixtures. In thinking about chemical reactions, sometimes it is also important to note whether oxidation and reduction have taken place. Considered in this manner, all chemical reactions can be placed in one of two very broad groups: (1) *reduction–oxidation reactions*—reactions in which the oxidation numbers of two or more elements change (possibly accompanied by other chemical changes), and (2) *transposition reactions*—reactions in which the oxidation numbers of all elements remain the same, but the atoms are transposed, or rearranged, into different compounds.

The classification of redox versus transposition reaction can be applied instead of, or in addition to, the previous classification of combination, decomposition, and substitution. For instance, we classified Eq. (7–13) as a combination reaction, but we can also regard it as a redox reaction because the sodium was oxidized and the chlorine was reduced. In fact, the first category within each of the three types of reactions previously mentioned (A(1), B(1), C(1)) involves redox changes. Such changes can be quickly identified by the fact that at least one individual *element* is involved as either reactant or product. The second categories (A(2), B(2), C(2)) concern only compounds, and no reduction–oxidation occurs. Unfortunately, in some other very important chemical (and biochemical) redox reactions, the elements being reduced and oxidized are all present in the form of compounds, with none as an individual element. How, then, can we reliably identify a redox reaction? The key is to look for a change in oxidation number of some element (including an element in a compound).

Let us use Eq. (7–24) as an example and, for convenience, omit the SO_4^{2-} ion, since it is unchanged by the reaction. By so doing we can write a briefer equation:

$$Fe + Cu^{2+} \rightarrow Cu \downarrow + Fe^{2+} \tag{7-33}$$

In Eq. (7–33) we readily note that the element Fe on the left side must have oxidation number 0, whereas the Fe^{2+} on the right has oxidation number $2+$. This increase in oxidation number means that the iron atom has been oxidized, since it has lost two electrons. In Eq. (7–33) the oxidation number of copper decreases from $2+$ to 0, which amounts to reduction and is due to a gain of two electrons. When we visualize reaction (7–33) we should remember that all the electrons lost by Fe must be acquired by the Cu^{2+} ion; the equation can be rewritten to show the number of electrons changing possession:

$$\overset{\text{two electrons gained}}{Fe^0 + Cu^{2+} \rightarrow Fe^{2+} + Cu^0} \tag{7-34}$$
$$\underset{\text{two electrons lost}}{}$$

At this point we should mention two other terms. An **oxidizing agent** is a substance that *causes* oxidation of some other element (whether an atom, an ion, or a part of a compound). A **reducing agent** is a substance that causes some other element to be reduced. In Eq. (7–34), Cu^{2+} ion is the oxidizing agent because it takes two electrons from Fe^0, thereby oxidizing the iron. Note also that the oxidizing agent, Cu^{2+}, is itself reduced. Similarly, Fe^0 acts as the reducing agent, but is itself oxidized to Fe^{2+}.

A similar examination of Eq. (7–23) reveals that it, too, is a reduction–oxidation reaction. If the equation is rewritten to exclude the NO_3^- ions and the electron changes are shown, we have:

$$\overset{\text{two electrons lost}}{Cu^0 + 2Ag^+ \rightarrow Cu^{2+} + 2Ag^0} \tag{7-35}$$
$$\underset{\text{one electron gained by each } Ag^+}{}$$

The copper was oxidized and the silver was reduced. It should be observed that two silver ions (each gaining one electron) were required for each copper atom, which lost two electrons.

One more example will be sufficient to emphasize that in any redox reaction the total number of electrons lost from one element must be gained by another:

two electrons lost

$$2Fe^{3+} + Sn^{2+} \rightarrow 2Fe^{2+} + Sn^{4+} \qquad (7\text{-}36)$$

one electron gained by each Fe^{3+}

Equation (7–36) does not fit into category C(1); it illustrates a redox reaction in which both elements being oxidized or reduced exist in compounds. The situation is more apparent if we replace the "missing" ions and write a complete equation:

$$2FeCl_3 + SnCl_2 \rightarrow 2FeCl_2 + SnCl_4 \qquad (7\text{-}37)$$

The procedures for balancing a redox equation are much more involved than those for other chemical equations and we shall not attempt to describe the necessary rules. Although some very important reactions to be discussed later in this text depend on oxidation and reduction, the balanced equations will be given without demonstrating how they were balanced.

7.7 WEIGHT RELATION-SHIPS IN CHEMICAL REACTIONS

So far, in this chapter, two fundamental ideas have been discussed. First, the formula of an element or compound can represent its weight (Section 7.2). Second, a balanced chemical equation accounts for the exact number of each kind of atom involved (Section 7.5). By combining these two ideas we can use a chemical equation to calculate the weights of any of the substances produced or consumed by the reaction.

The quantitative significance of a chemical equation will be illustrated first by the formation of methane. Under the proper conditions hydrogen gas combines with carbon to produce methane. This reaction is represented by Eq. (7–38), beneath which are shown some of the important numerical meanings related to it.

	C	+	2H$_2$	→	CH$_4$	(7-38)
Particles:	1 atom of carbon	+	2 molecules of hydrogen	→	1 molecule of methane	
	12.0 amu	+	2 × 2.0 amu	→	16.0 amu	
Bulk mass:	1 mole of carbon	+	2 moles of hydrogen	→	1 mole of methane	
	12.0 g	+	4.0 g	→	16.0 g	

When we see an equation such as (7–38) we can interpret it to mean "One atom of carbon (which is 12.0 amu) reacts with two molecules of hydrogen (or 4.0 amu) to produce one molecule (16.0 amu) of methane."

For handling bulk mass in the laboratory we can use the same mass numbers but weigh them in grams; thus we will be using Avogadro's number, 6.02×10^{23}, of carbon atoms, not just one carbon atom, and so on. On this scale we can read Eq. (7–38) as "One mole of carbon plus two moles of hydrogen give one mole of methane" or "12.0 g of carbon plus 4.0 g of hydrogen give 16.0 g of methane."

In practical manipulations we do not even need to confine ourselves to the use of moles or grams; the chemical equation, such as (7–38), indicates that the amounts of the substances will always be in the same proportions regardless of the units we use. For calculating amounts, the most general number of all is the formula weight of a substance. The **formula weight** *is numerically the same as the mass in amu corresponding to the formula, but without units.* Any units convenient for a particular purpose may be used with a formula weight. Thus the formula weight of carbon is 12.0; depending on the situation, we could use 12.0 oz, 12.0 kg, or 12.0 tons. The formula weight of methane is 16.0, which could be 16.0 oz, kg, or tons, or any other unit of weight.

Let us apply these principles to answer the question, "How many lb of sodium metal would be consumed during the formation of 50.0 lb of sodium chloride?" To begin, we should always write a balanced equation and show beneath it the weight relationship of each substance:

$$2Na \quad + \quad Cl_2 \quad \rightarrow \quad 2NaCl \tag{7-39}$$

Formula weight: 23.0   71.0   58.5
Reaction weight: $2 \times 23 = 46.0$ 71.0 $2 \times 58.5 = 117$

We shall show two methods for performing the mathematical calculation. (The instructor may recommend one method or the other; if not, the student can make the choice that seems most suitable.) By one approach, the calculation can be done using simple proportions. Comparing the question with the weight relations in Eq. (7–39) we can think of the proportions as follows: "The unknown amount of sodium compared to a weight of 46.0 is the same proportion as 50.0 lb of sodium chloride compared to 117." The calculation is even more easily visualized if we set it up above the chemical equation and include (for this example) the unit "lb" with the "reaction weight":

Problem: ? wt. Na   50.0 lb

$$2Na \quad + \quad Cl_2 \quad \rightarrow \quad 2NaCl \tag{7-40}$$

Reaction weight: 46.0 lb   71.0 lb   117 lb

The mathematical equation can be correctly written if the factors are kept in the same relative positions as in the chemical equation:

$$\frac{\text{wt. Na}}{46.0 \text{ lb Na}} = \frac{50.0 \text{ lb NaCl}}{117 \text{ lb NaCl}} \tag{7-41}$$

Now we can rearrange Eq. (7–41) so that "46.0 lb Na" appears in the numerator on the opposite side. Since "lb NaCl" has been cancelled, the final answer, Eq. (7–42), contains the correct units, "lb Na."

$$\text{wt. Na} = \frac{50}{117}(46.0 \text{ lb Na}) = 19.7 \text{ lb Na} \tag{7-42}$$

This problem can just as well be solved using only dimensional analysis as a guide. We can start by restating the problem as "The amount of sodium consumed will equal 50.0 lb sodium chloride times some factor," and translate this into a mathematical equation:

$$\text{wt. Na} = 50.0 \text{ lb NaCl} \times \; ? \tag{7-43}$$

Now, chemical equation (7–39) states that 46.0 lb of Na are used per 117 lb NaCl formed; this translates into the factor 46.0 lb Na/117 lb NaCl. Alternatively, the chemical equation can be read to mean 117 lb NaCl are produced per 46.0 lb Na consumed; the corresponding factor is 117 lb NaCl/46.0 lb Na. By analyzing the dimensions, or units, of weight we select the first factor because it will give the correct dimensions in the final answer:

$$\text{wt. Na} = 50.0 \text{ lb NaCl} \times \frac{46.0 \text{ lb Na}}{117 \text{ lb NaCl}} = 50.0 \left(\frac{46.0}{117}\right) \text{ lb Na} = 19.7 \text{ lb Na} \tag{7-44}$$

Of course, Eq. (7–44) is mathematically equivalent to Eq. (7–42) and has given exactly the same final answer.

One more chemical reaction will serve to illustrate the use of a balanced equation to make calculations.

Example   The natural gas burned to provide heat in many houses is predominantly methane, CH_4. In one year a typical two- or three-bedroom house in a moderate climate consumes about $7(10^4)$ liters, or 50 kg of methane. For this amount let us calculate: (a) the weight, in kilograms, of oxygen removed from the air, (b) the weight, in kilograms, of carbon dioxide added to the atmosphere, and (c) the volume, in liters, of this amount of carbon dioxide, given that there are 509 liters of CO_2 per kilogram.

Solutions   a) First it is absolutely essential to write a *balanced* equation for this oxidation reaction. Then, using formula weights, we can write the reacting weights beneath each substance. If this problem is to be solved by the proportion method we will indicate the question above the chemical equation, as was done with Eq. (7–40) previously:

$$\begin{array}{lll}
50.0 \text{ kg} & \text{? wt. } O_2 & \\
CH_4 \ + \ 2O_2 & \rightarrow \ CO_2 \ + \ 2H_2O & \text{(7–45)}\\
16.0 \text{ kg} \quad 2 \times 32.0 = 64.0 \text{ kg} & 44.0 \text{ kg} \quad 2 \times 18.0 = 36.0 \text{ kg} &
\end{array}$$

From this, the calculation can be set up and completed:

$$\frac{50.0 \text{ kg } CH_4}{16.0 \text{ kg } CH_4} = \frac{\text{wt. } O_2}{64.0 \text{ kg } O_2}$$

$$\text{wt. } O_2 = \frac{50.0}{16.0}(64.0 \text{ kg } O_2) = 200 \text{ kg } O_2 \tag{7–46}$$

By dimensional analysis this problem is solved as follows. The question means, "The weight of O_2 consumed equals 50.0 kg of CH_4 times a factor":

$$\text{wt. } O_2 = 50.0 \text{ kg } CH_4 \times \text{ ?} \tag{7–47}$$

The weights beneath Eq. (7.45) indicate that 64.0 kg of O_2 reacts for each 16.0 kg of CH_4 reacting, so by inserting this factor into Eq. (7–47) we obtain:

$$\text{wt. } O_2 = 50.0 \text{ kg } CH_4 \times \frac{64.0 \text{ kg } O_2}{16.0 \text{ kg } CH_4} = 50.0 \left(\frac{64.0}{16.0}\right) \text{kg } O_2 = 200 \text{ kg } O_2 \tag{7–48}$$

Again, this method has given the same answer as the proportion method, Eq. (7–46).
b) The weight of CO_2 produced by 50 kg of CH_4 is calculated as follows, using the proportion method:

$$\begin{array}{lll}
50.0 \text{ kg} & \text{? wt. } CO_2 & \\
CH_4 + \ 2O_2 & \rightarrow \quad CO_2 \ + H_2O & \text{(7–49)}\\
16.0 \text{ kg} & 44.0 \text{ kg} &
\end{array}$$

$$\frac{50.0 \text{ kg } CH_4}{16.0 \text{ kg } CH_4} = \frac{\text{wt. } CO_2}{44.0 \text{ kg } CO_2}$$

$$\text{wt. } CO_2 = \frac{50.0}{16.0}(44.0 \text{ kg } CO_2) = 137.5 \text{ kg } CO_2 \tag{7–50}$$

By the dimensional-analysis method the solution becomes:

$$\text{wt. } CO_2 = 50.0 \text{ kg } CH_4 \times \text{ ?} \tag{7–51}$$

For this problem we can select from beneath Eq. (7–45) the factor 44.0 kg CO_2/16.0 kg CH_4 and arrive at a completed equation:

$$\text{wt. } CO_2 = 50.0 \text{ kg } CH_4 \times \frac{44.0 \text{ kg } CO_2}{16.0 \text{ kg } CH_4} = 50.0 \left(\frac{44.0}{16.0}\right) \text{kg } CO_2 = 137.5 \text{ kg } CO_2 \tag{7–52}$$

c) The addition of 138 kg of CO_2 to the atmosphere may not seem to be a large amount at first glance. However, gases are so diffuse that a small amount of weight occupies a large volume at normal pressure and temperature. The problem states that for CO_2 this amounts to 509 liters per kilogram so in this case we convert weight to volume as follows:

$$1.38(10^2)\,\text{kg}\ CO_2 \times \frac{5.09(10^2)\ \text{liters}}{1\ \text{kg}} = 7.02(10^4)\ \text{liters}\ CO_2 \tag{7-53}$$

$$= 70\ 200\ \text{liters}\ CO_2$$

This 70 200 liters would fill about one-quarter of the "typical" house with *pure* carbon dioxide.

There has been some debate as to whether the continued burning of fossil fuels will increase the amount of carbon dioxide in the atmosphere enough to cause a climatic change. Whatever the final verdict, it does appear that such fuel-burning places a considerable burden on the atmosphere to dilute the amount of carbon dioxide produced. Since air contains approximately 0.03% carbon dioxide, the one-quarter houseful of pure carbon dioxide would need to be diluted by about 800 housefuls of air to maintain that 0.03% concentration of carbon dioxide unchanged. When we consider that this effect was from *one* house for *one* year and then think of the carbon fuels burned in all the houses, factories, planes, trains, trucks, and autos, the amounts of carbon dioxide added to the atmosphere are impressive.

Hereafter we shall encounter numerous different chemical reactions. For any reaction, as long as we write a complete, balanced equation, it is possible to calculate the quantities of the various substances involved. The methods described in this section, or slight variations thereof, are sufficient to complete the calculations.

7.8

NET IONIC EQUATIONS

Sometimes when a chemical reaction involves ionic compounds it is desirable to represent it by an even shorter equation that shows only the substances actually participating in the reaction. Such an equation is called a **net ionic equation**. Equation (7–54) is an example of one very common type of chemical change, the reaction of an acid (HCl) with a base (NaOH). Since both reactants are ionic compounds we shall write them as ions:

$$H^+ + Cl^- + Na^+ + OH^- \rightarrow HOH + Na^+ + Cl^- \tag{7-54}$$

The products are the covalent compound water, HOH, and the ionic compound sodium chloride, Na^+Cl^-. Here the important chemical event, the only thing causing any change of composition, is the combining of a hydrogen ion (H^+) with a hydroxide ion (OH^-) to produce a covalent water molecule. The Na^+ ion and the Cl^- ion really have nothing to do with the main event taking place, and they appear in the same form on both sides of the equation. They are often referred to as **spectator ions**. If we rewrite Eq. (7–54), leaving out the spectator ions, we obtain

$$H^+ + OH^- \rightarrow HOH \tag{7-55}$$

Equation (7–55) is a *net ionic equation*. It has the advantage that it focuses our attention on the key event occurring.

A net ionic equation also represents a general reaction which can have numerous variations. If we mixed any compound supplying hydrogen ions with any compound supplying hydroxide ions, reaction (7–55) would occur. For instance, if we used hydro-bromic acid and potassium hydroxide, the complete equation would be:

$$H^+ + Br^- + K^+ + OH^- \rightarrow HOH + K^+ + Br^- \tag{7-56}$$

In this example the spectator ions are K^+ and Br^-, and the net ionic equation is again represented by Eq. (7–55).

In Eqs. (7–54) and (7–56) all the ions are shown, including spectator ions. Such equations are called *total ionic equations*. This does not mean all the substances present are ionic; there may be at least one covalent compound, such as the HOH in these examples.

Net ionic equations are written for *reactions of ions in a water solution*. A few simple rules suffice for converting a complete equation to a net ionic equation:

1. First, write a total ionic equation, but in doing so, do *not* show as ions:
   a) Covalent compounds. The most common examples are H_2O, NH_3, certain gases (such as CO_2, SO_2) and "weak" acids (such as H_2SO_3, H_2CO_3, HCN; to be discussed in Section 11.8).
   b) Insoluble substances. Some insoluble compounds such as $CaCO_3$ are actually ionic in the solid state, but the ionic equation should show only those ions that exist *in solution*.

2. Eliminate from the complete equation only the ions that are exactly the same on both sides; these are the spectator ions.

3. After writing the net ionic equation, check it to see that the algebraic sum of the charges is the same on both sides of the equation. (Like any chemical equation, a net ionic equation must be balanced, including electrical charges.)

Here are two illustrations of writing a net ionic equation from a complete equation:

Example 1   Treatment of magnesium metal with hydrochloric acid.

1. $Mg + 2HCl \rightarrow MgCl_2 + H_2\uparrow$
$Mg + 2H^+ + 2Cl^- \rightarrow Mg^{2+} + 2Cl^- + H_2\uparrow$

2. $Mg + 2H^+ + \cancel{2Cl^-} \rightarrow Mg^{2+} + \cancel{2Cl^-} + H_2\uparrow$

3. $Mg + 2H^+ \rightarrow Mg^{2+} + H_2\uparrow$

The final net ionic equation is balanced. There is one magnesium on each side, even though it has changed from a magnesium atom to an ion. There are two hydrogens on each side. There are two + charges on each side.

Example 2   Mixing a solution of barium nitrate with a solution of sodium sulfate.

1. $Ba(NO_3)_2 + Na_2SO_4 \rightarrow BaSO_4\downarrow + 2NaNO_3$
$Ba^{2+} + 2NO_3^- + 2Na^+ + SO_4^{2-} \rightarrow BaSO_4\downarrow + 2Na^+ + 2NO_3^-$

2. $Ba^{2+} + \cancel{2NO_3^-} + \cancel{2Na^+} + SO_4^{2-} \rightarrow BaSO_4\downarrow + \cancel{2Na^+} + \cancel{2NO_3^-}$

3. $Ba^{2+} + SO_4^{2-} \rightarrow BaSO_4\downarrow$

A check of all atoms and charges shows that the final equation is balanced.

Although for convenience we may write a net ionic equation, it is important to bear in mind that the spectator ions are always present in solution, as shown by Eq. (7–56). It is not possible to prepare a solution containing only OH^- ions; their negative charges must be balanced by positive ions, such as K^+ ions. Similarly, after reaction (7–56) is complete, the K^+ ions must be present to balance the Br^- ions.

**7.9
ENERGY
CHANGES
DURING
CHEMICAL
REACTIONS**

All chemical reactions are accompanied by energy changes, very often in the form of heat energy. The net change in heat content during a chemical reaction is termed the *heat of reaction*. Frequently the energy change associated with a reaction is as important to our daily lives as is the change in composition. For instance, methane (natural gas) is burned [Eq. (7–45)], for the sake of the heat released by the reaction. Specifically, 213 kcal are produced by the oxidation of one mole of methane gas, so a more complete equation would be:

$$CH_4 + 2O_2 \rightarrow CO_2 + H_2O + 213 \text{ kcal} \qquad (7–57)$$

The amount of methane used as a raw material for manufacturing is very small compared with the quantities burned for heat energy.

Experimentally, a heat-of-reaction value can be determined in an apparatus called a calorimeter (Section 8.7). The heat of reaction for a particular chemical change is a definite, constant value, just as the mass of a molecule is specific. For example, the heat of reaction for the formation of liquid water has been found to be 68 kcal per mole. Since a balanced equation shows *two* moles of water, the equation should be written:

$$2H_2 + O_2 \rightarrow 2H_2O + 136 \text{ kcal} \qquad (7–58)$$

The energy changes accompanying various chemical reactions will be discussed in more detail in the next chapter. Here, we shall merely point out that some of the most fascinating energy changes are those associated with biochemical reactions. In the presence of sunlight, green plants combine carbon dioxide and water to produce oxygen and glucose ("grape sugar," "blood sugar"). Simultaneously, light energy is stored as potential energy in the bonds of the glucose molecules. In our bodies, when glucose is oxidized back to carbon dioxide and water, the stored energy is released to warm the body, to cause muscle contractions, to provide energy for the synthesis of other compounds, and so on.

SUMMARY

The formulas and equations of chemistry are comparable to the words and sentences of a language. The more familiar they are, the easier it is to think about and express ideas and to understand the statements of others. In this chapter we have described what chemical equations mean and how to write them. Throughout the rest of the book they will be used routinely, as statements of what occurs during chemical changes.

The quantitative meaning of a formula or an equation is an inherent part of what it symbolizes. We have explored the numerical meaning of a formula and the ideas of the quantities related to a chemical equation. A complete, balanced equation can tell not only the *kinds* of particles and substances involved, but also the *amounts* of mass, of electrical charge, of heat, and of electrons being transferred. Because of these precise quantitative meanings, an equation can be used to calculate the amount of compound A that will react with B and the quantity of product C one can expect as a result. When desired, one can also calculate the amount of heat energy absorbed or released by a reaction.

All in all, a chemical equation is an extremely useful device for representing and discussing a chemical reaction. With these fundamental concepts established we can proceed, in the remaining chapters, to investigate the chemical nature of oxygen, water, salts, petroleum, vitamins, proteins, and a host of other substances vital to our existence.

IMPORTANT TERMS

formula weight	mole weight	product	spectator ion
mole	net ionic equation	reactant	

WORK EXERCISES

1. Calculate the mole weight of each of the following.
   a) NaBr   b) I_2 c) ZnS
   d) $MgBr_2$ e) $BaCO_3$ f) SO_2
   g) Fe_2O_3 h) $HBrO_4$ i) K_2HPO_4

2. How many moles is each of the following?
   a) 110 g of Li_2SO_4 b) 7.00 g of N_2
   c) 86.5 g of $CHCl_3$ d) 32.7 g of Cu
   e) 46.8 g of $NaNO_3$ f) 53.6 g of NiS
   g) 73.2 g of MnO_2 h) 218 g of $(NH_4)_2SO_4$
   i) 15.4 g of C_3H_8

3. How many atoms are present in each of the following?
   a) 153 g of magnesium metal
   b) 22.5 g of tin
   c) 15.0 g of sulfur

4. A block of "dry ice" (solid CO_2) weighing 4.00 ounces contains how many moles?

5. If a volleyball or basketball were inflated with natural gas (CH_4) instead of air, it would be very firm if only 18.0 g of the gas were pumped in. How many molecules of CH_4 would that be?

6. A 50.0 g sample of calcium chloride contains:
   a) how many moles of compound?
   b) how many moles of calcium ions?
   c) how many moles of chloride ions?
   d) how many grams of chloride ions?
   e) how many calcium ions?
   f) what percentage of calcium?

7. A 28.3 g sample of sodium sulfate contains:
   a) how many moles of sodium ions?
   b) how many sulfate ions?
   c) what percentage of oxygen?
   d) what percentage of sodium?

8. Calculate the percentage of each element in the following compounds.
   a) sodium chloride   b) carbon dioxide
   c) water   d) calcium phosphate
   e) iron(II) hydroxide   f) stannous chloride
   g) butane, C_4H_{10} h) glucose, $C_6H_{12}O_6$
   i) monosodium glutamate (MSG), $NaC_5H_8NO_4$
   j) the dental anesthetic lidocaine chloride, $C_{14}H_{23}N_2OCl$

9. Calculate the percentage of each metal present in these ores.
   a) horn silver, AgCl
   b) zinc blende, ZnS
   c) aluminum oxide
   d) chalcopyrite, $CuFeS_2$
   e) cassiterite, tin(IV) oxide
   f) cerrusite, lead(II) carbonate

10. For each equation below (1) tell which are products and which are reactants; (2) translate the equation into a word sentence that is *qualitative* (that is, tells only *what* is involved); (3) translate it into a *quantitative* sentence (which tells *how much*), in terms of atoms or molecules and in terms of moles.
    a) $C + 2S \rightarrow CS_2$
    b) $N_2 + 3H_2 \rightarrow 2NH_3$
    c) $2H_2S + 3O_2 \rightarrow 2SO_2 + 2H_2O$

11. In the following equations, change the names to formulas, complete the equations and balance them. Name the products. (Assume that a reaction does occur in each case.)
    a) potassium chloride + silver nitrate $\rightarrow$
    b) solium sulfate + barium chloride $\rightarrow$
    c) sodium hydroxide + iron(III) chloride $\rightarrow$
    d) ammonium sulfate + barium nitrate $\rightarrow$
    e) sodium hydroxide + nitric acid $\rightarrow$
    f) lithium + oxygen $\rightarrow$
    g) sulfur trioxide + water $\rightarrow$
    h) aluminum + bromine $\rightarrow$
    i) potassium oxide + water $\rightarrow$

12. Complete and balance these combination reactions.
    a) $C + O_2 \rightarrow$ b) $Mg + Cl_2 \rightarrow$
    c) $K + S \rightarrow$ d) $Al + S \rightarrow$
    e) $CaO + CO_2 \rightarrow$ f) $Na_2O + H_2O \rightarrow$
    g) $H_2O + SO_2 \rightarrow$ h) $H_2O + SO_3 \rightarrow$

13. Complete and balance these decomposition reactions.
    a) $HgO \xrightarrow{\Delta}$ b) $CaCO_3 \xrightarrow{\Delta}$
    c) $Ca(OH)_2 \xrightarrow{\Delta}$ d) $H_2CO_3 \rightarrow$

e) $Na_2CO_3 \xrightarrow{\Delta}$

f) $H_2O \xrightarrow{\text{electricity}}$

g) $H_2SO_3 \rightarrow$

h) $Al(OH)_3 \xrightarrow{\Delta}$

h) "milk of magnesia," $Mg(OH)_2$
i) the amino acid alanine, $C_3H_7NO_2$
j) acetaminophen (Tylenol), $C_8H_9NO_2$
k) thiamine (Vitamin B_1), $C_{12}H_{17}ClN_4OS$

14. Balance the following equations.
a) $NaOH + H_2SO_4 \rightarrow Na_2SO_4 + HOH$
b) $Fe_2(SO_4)_3 + Ba(OH)_2 \rightarrow Fe(OH)_3\downarrow + BaSO_4\downarrow$
c) $Al + H_2SO_4 \rightarrow Al_2(SO_4)_3 + H_2\uparrow$
d) $BaCl_2 + (NH_4)_2CO_3 \rightarrow BaCO_3\downarrow + NH_4Cl$
e) $KOH + H_3PO_4 \rightarrow K_3PO_4 + HOH$
f) $Fe + H_2O \xrightarrow{\Delta} Fe_3O_4 + H_2\uparrow$
g) $CaCl_2 + AgNO_3 \rightarrow Ca(NO_3)_2 + AgCl\downarrow$
h) $NaOH + H_2CO_3 \rightarrow Na_2CO_3 + HOH$
i) $Mg(OH)_2 + HNO_3 \rightarrow Mg(NO_3)_2 + HOH$
j) $Ca(OH)_2 + H_3PO_4 \rightarrow Ca_3(PO_4)_2\downarrow + HOH$

15. Rewrite the equations in Exercise 14 as net ionic equations.

16. Complete and balance these substitution reactions.
a) $KCl + AgNO_3 \rightarrow$ KNO3 + AgCl
b) $Na_2SO_4 + BaCl_2 \rightarrow$ NaCl + BaSO4
c) $NaOH + FeCl_3 \rightarrow$ 3 NaCl + Fe(OH)3
d) $NaOH + HNO_3 \rightarrow$ NaNO3 + HOH
e) $Mg + NiCl_2 \rightarrow$ MgCl2 + Ni
f) $CaO + HNO_3 \rightarrow$ Ca(NO3)2 + H2O
g) $ZnO + HCl \rightarrow$ ZnCl2 + H2O
h) $BaCl_2 + H_2SO_4 \rightarrow$ BaSO4 + 2HCl
i) $K_2SO_3 + HBr \rightarrow$ 2KBr + H2SO3
j) $Ag_2SO_4 + Zn \rightarrow$ ZnSO4 + 2Ag
k) $Na_2CO_3 + H_2SO_4 \rightarrow$ Na2SO4 + H2CO3
l) $KOH + H_3PO_4 \rightarrow$ K3PO4 + 3HOH
m) $Cu(NO_3)_2 + Al \rightarrow$ 2Al(NO3)3 + 3Cu
n) $Al_2(SO_4)_3 + Ba(NO_3)_2 \rightarrow$ 2Al(NO3)3 + 3Ba(SO4)2

17. Rewrite the equations in Exercise 16 as net ionic equations.

18. Complete, balance, and identify the type of equation.
a) $Zn + S \rightarrow$
b) $MgO + HCl \rightarrow$
c) $Zn + CuSO_4 \rightarrow$
d) $CuCO_3 \xrightarrow{\Delta}$
e) $Ca(OH)_2 + H_3PO_4 \rightarrow$
f) $Al_2O_3 + H_2SO_4 \rightarrow$
g) $K + Cl_2 \rightarrow$
h) $Na_2O + CO_2 \rightarrow$
i) $Mg + Co(NO_3)_2 \rightarrow$
j) $C + Cl_2 \rightarrow$

19. Calculate the formula weight of each compound.
a) Br_2 b) H_3PO_4 c) $BaCO_3$
d) K_2SO_4 e) $AgNO_3$ f) $CaCl_2$
g) baking soda, $NaHCO_3$

20. How many grams of oxygen are required to completely burn 30.5 g of carbon?

21. How many grams of carbon dioxide will be produced by the complete oxidation of 15.2 g of carbon?

22. How many grams of carbon, completely oxidized, will be required to produce 33 g of CO_2?

23. This equation shows the reaction used to convert limestone to lime, CaO:
$CaCO_3 \xrightarrow{\Delta} CaO + CO_2$

a) How many grams of CO_2 are released by heating 50.0 g of $CaCO_3$?
b) How many tons of CaO will be produced when 10.5 tons of $CaCO_3$ are heated?
c) How many moles of $CaCO_3$ are required to make 88 g of CaO?

24. For the complete combustion of natural gas (see Eq. 7–45) calculate the following.
a) the weight of CO_2 in grams released when 22 g of CH_4 burns
b) The weight of O_2, in kilograms, required to burn 745 g of CH_4.
c) The weight of H_2O, in pounds, produced when 2.5 lb of CH_4 burns

25. Consider the following equation:
$2KClO_3 \xrightarrow{\Delta} 2KCl + 3O_2$

a) How many moles of O_2 will be produced from the decomposition of 2 moles of $KClO_3$? Of 5 mol $KClO_3$? Of 0.60 mol $KClO_3$?
b) How many grams of $KClO_3$ are required to produce 10 g of O_2?
c) How much O_2 will be produced by the decomposition of 5 g of $KClO_3$?

26. Complete and balance this substitution reaction and then do the calculations requested:
$NaNO_3 + H_2SO_4 \rightarrow$
a) How many moles of HNO_3 are produced by 3 moles of H_2SO_4?
b) How many moles of HNO_3 are produced from each mole of $NaNO_3$ used?
c) How many grams of $NaNO_3$ are required to react completely with 26.7 g of H_2SO_4?
d) How many pounds of H_2SO_4 would be needed to react with 1.35 lb of $NaNO_3$ and how many pounds of HNO_3 would this give?

SUGGESTED READING

Kieffer, W. F., *The Mole Concept in Chemistry*, Reinhold, New York, 1962. A discussion of the mole and how it is employed in chemical calculations.

8

Oxygen, Nitrogen, the Noble Gases

8.1
INTRODUCTION

Now that we have presented the basic theories about the nature of matter and its reactions, let us focus our attention on some of the individual elements and groups of elements. This subject matter is often called descriptive chemistry. Such discussions generally begin with a consideration of the forms in which an element is found and the ways it can be prepared (isolated) in a pure state. This is followed by a description of the physical and chemical properties of the element. The important compounds of the element and their uses are usually discussed last. That traditional format is the one we shall follow here.

In this chapter we shall discuss those nonmetals that are found in the atmosphere—oxygen, nitrogen, and the noble gases. The compound water is discussed in Chapter 9. In Chapters 10–13 we shall take time out to consider solutions, acids, bases, salts, and the topic of equilibrium, after which we can return to the descriptive chemistry of the metals and other nonmetals. Descriptive chapters may also introduce new theories and important new terms. In this chapter we introduce the concept of catalysis, heat of reaction, and allotropic forms of elements.

8.2
OCCURRENCE
OF OXYGEN

Oxygen in one form or another is the most prevalent of the elements that make up the portion of the earth with which we are familiar. It is, in fact, estimated that approximately 50% by weight of the so-called crust (the atmosphere, oceans, and outer layer of dry land) of the earth is composed of oxygen. In terms of weight about 23% of the atmosphere, nearly 90% of water, and more than 50% of sand is oxygen. In the atmosphere oxygen exists in the free, uncombined state, while in sand and water it is in a combined form. Many elements are essential for life, but oxygen is the one that is the most continuously required. Under most conditions we can go for weeks without food, for days without water, but only minutes without oxygen. It may seem strange that so common an element was unidentified until less than 200 years ago.

8.3
THE
DISCOVERY OF
OXYGEN

Joseph Priestley, an English clergyman, teacher, and experimenter, is generally credited with being the first to prepare oxygen in a pure state, and certainly he was the one who announced to the world many of its interesting properties. He demonstrated experimentally that a candle would soon stop burning when confined in a closed container of air. Afterward, if plants were grown in this atmosphere, a regeneration took place and the candle would burn again. In his experiments he also observed the effects of oxygen on animals, usually mice. Carl Scheele, a Swedish druggist, prepared oxygen a few years earlier than Priestley but apparently did not realize the importance of the discovery and made no mention of it until after Priestley's work in 1774. Priestley prepared the "new air," as he referred to it, by heating mercuric oxide.

$$2HgO \rightarrow 2Hg + O_2 \tag{8-1}$$

He used a magnifying glass to intensify the heat of the sun and collected the gas over mercury. A comparison of Priestley's apparatus and that used today is shown in Fig. 8.1. The oxygen that Priestley prepared over mercury was transferred to an apparatus in which water was substituted for the mercury, making experiments with animals more feasible.

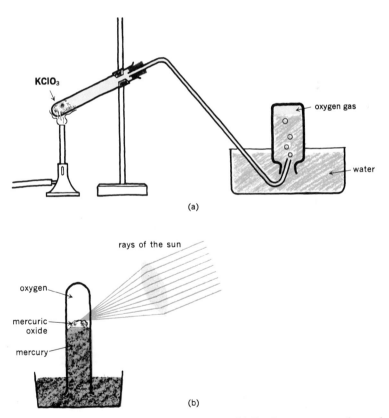

Fig. 8.1 Apparatus for collecting oxygen. (a) Equipment presently used in the laboratory; (b) equipment used by Priestley.

Priestley compared pure oxygen (which he called dephlogisticated air) with other gases as well as experimenting with oxygen. The following paragraphs from his writings indicate his methods and interests.

"For the purpose of these experiments it is most convenient to catch the mice in small wire traps, out of which it is easy to take them, and holding them by the back of the neck, to pass them through the water into the vessel which contains the air. If I expect that the mouse will live a considerable time, I take care to put into the vessel something on which it might conveniently sit, out of reach of the water. If the air be good, the mouse will soon be perfectly at its ease, having suffered nothing by its passing through the water. If the air be supposed to be noxious, it will be proper (if the operator be desirous of preserving the mice for further use) to keep hold of their tails, that they may be withdrawn as soon as they begin to show signs of uneasiness; but if the air be thoroughly noxious, and the mouse happens to get a full inspiration, it will be impossible to do this before it will be absolutely irrecoverable . . .

"From the greater strength and vivacity of the flame of a candle, in this pure air, it may be conjectured that it might be peculiarly salutary to the lungs in certain morbid cases, when the common air would not be sufficient to carry off the phlogistic putrid effluvium fast enough. But, perhaps, we may also infer from these experiments, that though pure dephlogisticated air might be very useful as a *medicine*, it might not be so proper for us in the usual healthy state of the body: for as a candle burns out much faster in dephlogisticated than in common air, so we might, as may be said, *live out too fast*, and the animal powers be too soon exhausted in this pure kind of air. A moralist, at least, may say that the air which nature has provided for us is as good as we deserve.

"My reader will not wonder that, after having ascertained the superior goodness of dephlogisticated air by mice living in it, and the other tests above mentioned, I should have the curiosity to taste it myself. I have gratified that curiosity by breathing it, drawing it through a glass siphon, and, by this means, I reduced a large jar full of it to the standard of common air. The feeling of it to my lungs was not sensibly different from that of common air; but I fancied that my breath felt peculiarly light and easy for some time afterward. Who can tell but that, in time, this pure air may become a fashionable article of luxury? Hitherto only two mice and myself have had the privilege of breathing it."★

The terms phlogistic and dephlogisticated used in the foregoing quotation refer to the Phlogiston Theory of combustion, which was commonly accepted during the eighteenth century. According to this theory, any material capable of burning contained an element or substance called phlogiston. Phlogiston was supposed to be driven off during a combustion; the remaining ash was therefore "dephlogisticated." Air in which nothing had been burned did not contain phlogiston. This theory gave a reasonably good explanation of what happened when a thing burned or a mouse breathed. Although the theory was replaced by our modern theory of oxidation during Priestley's lifetime, he continued to believe in phlogiston. In spite of this, it is interesting to note Priestley's anticipation of the medical use of oxygen and his concern about the effects of breathing pure oxygen.

Antoine Lavoisier, a French scientist and statesman, duplicated Priestley's experiments, then continued with others. He established the fact that Priestley's "new air" constituted approximately one fifth of the atmosphere. He conceived the idea that combustion is usually a combination of a fuel with this portion of the air, and he rather than Priestley is credited with naming the gas oxygen, from the Greek meaning "acid former." Lavoisier erroneously believed that all acids contained oxygen. Of course, many acids do, but we now know others, such as HCl, that do not.

★ F. J. Moore, *A History of Chemistry*, McGraw-Hill Book Company, Inc., 1918.

In the laboratory, oxygen may be made in a variety of ways. A common method previously used for producing small amounts of oxygen was the one used by Priestley, the heating of mercury oxide (mercury(II) oxide). Due to the great toxicity of mercury, this method is no longer recommended. The method presently used in the laboratory is that of heating potassium chlorate, $KClO_3$ (Fig. 8.1a), which readily gives up all its oxygen, according to Eq. (8–2).

$$2KClO_3 \xrightarrow[\Delta]{} 2KCl + 3O_2 \qquad\qquad (8\text{–}2)$$

This reaction requires a relatively high temperature, but in the presence of manganese dioxide, MnO_2, the rate of reaction is greatly increased, even at much lower temperatures. The MnO_2 does not lose its oxygen and may be recovered unchanged when the reaction is complete. Substances that act in this manner are called **catalysts**. A simple definition of a catalyst would be *a substance that alters the rate of a chemical reaction without being consumed in that reaction*. The presence of a catalyst during a chemical reaction is indicated by placing it over the arrow of the equation rather than on either side.

$$2KClO_3 \xrightarrow[\Delta]{MnO_2} 2KCl + 3O_2 \qquad\qquad (8\text{–}3)$$

Not only are catalysts commonly used in industry, they are responsible for many of the chemical reactions taking place in living plants and animals. The laboratory instructor might be called a catalyst since his presence usually results in more work, and it is hoped that he will not be much changed at the end of the period. Substances that decrease the rate of chemical reactions are called *inhibitors*.

For demonstrations, oxygen is often prepared by the *electrolysis of water*, using an apparatus (Fig. 8.2) filled with water containing a little sulfuric acid. A direct current of electricity, such as that produced by a battery, is passed through the solution. Oxygen is liberated at the positive electrode, the **anode** (where electrons leave the solution), and hydrogen at the negative electrode, the **cathode** (where electrons enter the solution). The sulfuric acid is added to increase the electrical conductivity of the water.

$$2H_2O \xrightarrow[\text{current}]{\text{direct}} 2H_2 + O_2 \qquad\qquad (8\text{–}4)$$

This method produces very pure oxygen, but is too expensive to be practical on a commercial scale.

Industrially, nearly all oxygen is obtained from the atmosphere. Air is liquefied and cooled to about $-200°C$, using principles essentially the same, but in a higher degree of sophistication, as those used in a household refrigerator. When the liquid is warmed to $-196°C$, the point at which nitrogen boils at atmospheric pressure, the temperature will remain almost constant until nearly all the nitrogen has been removed. Further warming to $-183°C$, will ensure that the remaining liquid is almost pure oxygen (see Fig. 8.3). For permanent storage, the oxygen may be vaporized and pumped under pressure into steel cylinders. Oxygen used for making steel is often piped directly from the plant where it is made to a steel mill.

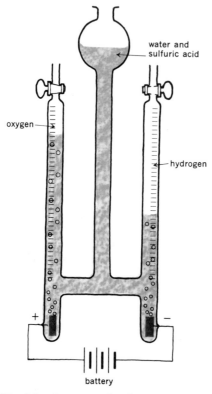

Fig. 8.2 Apparatus for the electrolysis of water.

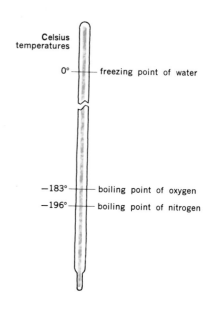

Fig. 8.3 Difference in boiling points, which allows separation of oxygen from nitrogen.

8.5
PROPERTIES
OF OXYGEN

Physical properties   Oxygen is a colorless, odorless, and tasteless gas that boils at $-183°C$ and freezes at $-216°C$. It occurs in the atmosphere in the diatomic form O_2. At room temperature and atmospheric pressure, oxygen dissolves in water to the extent of 3 ml/100 ml of water. As the temperature increases, the solubility of the oxygen decreases until, at temperatures near the boiling point of water, the solubility approaches zero. Conversely, if the water becomes colder, oxygen becomes more soluble. This greater solubility accounts in part for the greater activity of fish in cold water, and the decreasing solubility in warmer water causes sluggishness or even death of fish.

Chemical properties   Only fluorine has greater activity as a nonmetal than oxygen. In fact, oxygen forms compounds with nearly all the elements. From its position on the periodic table we expect oxygen to be electronegative and to gain two electrons in most reactions. Compounds formed by the union of oxygen with other elements are called oxides. Typical reactions of nonmetals with oxygen are:

$$C + O_2 \rightarrow CO_2 \tag{8-5}$$

$$S + O_2 \rightarrow SO_2 \tag{8-6}$$

$$2H_2 + O_2 \rightarrow 2H_2O \tag{8-7}$$

Oxygen also reacts with many metals to form oxides:

$$2Ca + O_2 \rightarrow 2CaO \tag{8-8}$$

$$2Mg + O_2 \rightarrow 2MgO \tag{8-9}$$

$$4Li + O_2 \rightarrow 2Li_2O \tag{8-10}$$

Oxygen forms oxides with almost all of the chemical elements. Many of these oxides are very insoluble in water, and we find silicon dioxide, SiO_2, as the major component of sand. Other oxides such as iron and aluminum oxides are important ores and serve as the raw material for our production of these metals. Soluble oxides have a tendency not only to dissolve but also to react with water to give acids and bases. These reactions are discussed in Sections 11.4 and 11.5.

**8.6
COMBUSTION,
EXPLOSIONS,
AND FIRE
EXTIN-
GUISHERS**

A reaction in which energy is released so rapidly that flames and light are produced is called **combustion,** and a material that will burn is said to be *combustible.* Many reactions of substances with oxygen are combustions. In the presence of a spark or a flame oxygen of the air reacts with methane, the main constituent of natural gas, to give carbon dioxide and water.

$$CH_4 + 2O_2 \rightarrow CO_2 + 2H_2O \tag{8-11}$$

The combustion of other compounds of carbon and hydrogen that occur in automobiles and other engines that burn gasoline or similar fuels leads to their being called *internal combustion engines.*

Even though many oxidations do not proceed so rapidly that they produce flames and light, *oxidations release energy.* They are exergonic reactions. Heat generally increases the rate of a chemical reaction, and sometimes, when a small amount of heat produced by a slow oxidation is not dissipated, the temperature increases until combustion occurs. Thus fires caused by storing oily rags in a closed container are not magical, but the result of normal chemical processes. The phenomenon is called **spontaneous combustion,** and the temperature at which a substance bursts into flames is called its **kindling temperature.** Normally we supply the energy necessary for reaching the kindling temperature by a match or flame, as when we strike a match to a campfire or use a pilot light to start a furnace. In the laboratory white phosphorus is the material that most readily exhibits spontaneous combustion, for it combines with the oxygen of the air and has a very low kindling temperature. Consequently, white phosphorus is stored under water; if exposed to the air at room temperature it bursts into flame.

A reaction that produces a great deal of pressure, usually accompanied by light, sound, and heat, is an **explosion.** Explosions and fires caused by spontaneous combustion do much damage and cause considerable loss of life annually. Oxidation takes place on the surface of solids and the rate of oxidation increases with a decrease in particle size. When combustible substances such as coal, flour, starch, and wood dust are finely divided into dustlike particles, their surface areas are enormously increased. Many coal mine explosions and fires in grain elevators, flour mills, and woodworking factories are caused by the explosion of small particles of coal, flour, or wood dust. Any substance that can be burned will form an explosive mixture when finely divided and suspended in oxygen or air. Many of these explosions are set off by electric sparks; others may be due to spontaneous combustion.

In a laboratory, explosions may result from heating substances that are unstable at higher temperatures, or by heating or even by grinding combustible substances which are mixed with oxidizing agents such as $KClO_3$. This type of explosion is particularly violent since solids are converted into gases at high temperatures, causing a sudden and tremendous increase in volume.

Fire and explosions cause extensive damage and loss of life each year. A knowledge of the principles involved helps us understand the use of fire extinguishers. Spraying water on a fire lowers the temperature of most burning substances below the kindling temperature; however, in some cases water will react with the burning substance, and the reaction may cause an *increase* in temperature. Therefore, a more useful principle to remember is that oxygen is necessary for the combustion involved in all ordinary fires; thus they can be controlled by taking away their supply of oxygen. For small fires a coat or blanket thrown over the burning area will "smother" the fire by cutting off its supply of oxygen.

The best type of commonly used fire extinguisher contains carbon dioxide (CO_2) under pressure. When a gas is released from pressure it becomes much cooler, and a stream of cold CO_2 directed at a flame not only reduces the temperature but also removes oxygen and surrounds the burning object with CO_2, which neither burns nor supports combustion. Another advantage of CO_2 fire extinguishers is that they do not damage other materials as water or other chemical agents might do.

Even with the CO_2 extinguisher, there are disadvantages. Carbon dioxide will react with magnesium to give carbon and magnesium oxide:

$$2Mg + CO_2 \rightarrow 2MgO + C$$

This reaction releases a great deal of energy, so CO_2 is not good for extinguishing fires involving magnesium alloys, such as aircraft fires. Furthermore, since a large amount of CO_2 in the atmosphere interferes with oxidations, it restricts the biochemical oxidations that are necessary to maintain life. Thus after extensive use of a CO_2 extinguisher in a closed area, people and other animals should be evacuated until adequate ventilation has restored a supply of oxygen.

While water and CO_2 extinguishers are sufficient for controlling most fires, fires involving many substances found in the chemistry laboratory as well as extensive use of CO_2 extinguishers require special knowledge and are best left to professionals.

8.7
HEAT OF
REACTION

When a substance reacts with oxygen, the amount of heat given off varies with the substance being oxidized. While both gasoline and sugar will burn, the amount of energy released per gram of these two substances is different. Compounds formed in reactions in which large amounts of heat are liberated do not decompose easily when they are heated. Thus water, carbon dioxide, and magnesium oxide are very stable compounds.

We can measure the amount of heat given off in an oxidation using a device known as a **calorimeter**. We place our reactive substance and an excess of oxygen under pressure in a thick-walled vessel called a *combustion bomb*. We then place the bomb in a large container of water and ignite the combustible substance by means of electricity. The increase in temperature of the water is equal to that given off by the oxidation.

When one mol of hydrogen gas (H_2) reacts with one-half mol of oxygen to form one mol of liquid water, 68.3 kcal of heat are released.

$$H_2 + \tfrac{1}{2}O_2 \rightarrow H_2O + 68.3 \text{ kcal} \qquad (8\text{-}12)$$

This equation has been written with $\frac{1}{2}O_2$ so we could state the energy *per mol* of hydrogen burned. If we write the equation in the usual fashion,

$$2H_2 + O_2 \rightarrow 2H_2O \qquad (8\text{-}13)$$

the energy released is 136.6 kcal.

The symbol H is used to represent heat content, and ΔH (**delta** H) represents a change in heat content when a reaction occurs. When heat or another form of energy is given off by a reaction, the heat content of the substances formed is less than that of the reactants and the change in heat content or ΔH is negative. We can rewrite Eq. (8-12) as

$$H_2 + \tfrac{1}{2}O_2 \rightarrow H_2O; \qquad \Delta H = -68.3 \text{ kcal} \qquad (8\text{-}12a)$$

The **heat of combustion** is defined as the heat evolved by the combustion of one mole of a substance. Thus, the quantity of heat shown in Eqs. (8-12) and (8-12a) is the heat of combustion of hydrogen. Since energy has been released, there is less energy in the water than in the oxygen and hydrogen, and we say that the heat of combustion of hydrogen is 68.3 kcal per mol. The ΔH for the reaction is -68.3 kcal per mol. The heats of combustion of several substances are given in Table 8.1.

Table 8.1   Heats of combustion of selected substances*

Substance	Formula	State	Heat of combustion (ΔH) (kcal/mol)
carbon	C	solid	-94.1
glucose	$C_6H_{12}O_6$	solid	-670
hydrogen	H_2	gas	-57.8
methane	CH_4	gas	-210
octane	C_8H_{18}	liquid	-1302
stearic acid	$C_{17}H_{35}COOH$	solid	-2711

* Reactions considered are complete combustion. Products are all in the gaseous state.

Another way to look at the reaction of oxygen and hydrogen to give water is to consider it not as a combustion of hydrogen, but as the formation of water. When we do this we call the value the **heat of formation**, which is defined as the heat either released or absorbed when a mole of a compound is formed from its elements. The heat of formation (ΔH) of water is -68.3 kcal/mol, the same as the heat of combustion for hydrogen. Although we can measure the oxidation of glucose to give CO_2 and H_2O

$$C_6H_{12}O_6 + 6O_2 \rightarrow 6CO_2 + 6H_2O \qquad (8\text{-}14)$$

and thus we can determine the heat of combustion easily, the reverse of this reaction does not give heat of formation since in this equation, as in nature, glucose is not made by combining its elements. We can, however, calculate heats of formation for those compounds that are not readily synthesized from elements. The heats of formation of some common substances are given in Table 8.2. It should be noted that we need to know the physical state of both the reactants and products when calculating heat involved in reactions, since energy is involved in changing a solid to a liquid and a liquid to a gas or is absorbed in the reverse processes.

TABLE 8.2   Heats of formation*

Substance formed	State	Heat of formation (kcal/mol)
CO	gas	-26.4
CO_2	gas	-94.1
H_2O	gas	-57.8
H_2O	liquid	-68.3
Al_2O_3	solid	-401
HCl	gas	-22
Fe_2O_3	solid	-197
O_3	gas	34.1

* Negative values indicate that heat is given off by the reaction.   These values are also known as *standard enthalpies of formation* and given the symbol ΔH_f.

When we decompose water to give hydrogen and oxygen, the amount of energy required is exactly the same as the amount released when water was formed. This is true of all chemical reactions; the reverse of a reaction involves a numerically identical amount of energy with the sign reversed. If the heat of combustion of glucose to yield carbon dioxide and water releases 670 kcal/mol glucose ($\Delta H = -670$ kcal/mol), then the synthesis of glucose from carbon dioxide and water will require 670 calories; that is the ΔH is $+670$ kcal/mol.

We can use the information given above to make calculations involving the energy consumed or released in chemical reactions.

Example 1   How much energy will be liberated during the formation of 100 g of carbon dioxide from carbon and water ?

We begin by writing the equation for the reaction and then indicating the weights and amounts of energy involved. (See Table 8.2 for energy relationships.)

$$\text{C} \;+\; \text{O}_2 \;\rightarrow\; \text{CO}_2; \quad \Delta H = -94.1 \text{ kcal/mol}$$
$$12 \text{ g} \quad 32 \text{ g} \quad\quad 44 \text{ g}$$

If we choose to solve by ratios, the equation becomes

$$\frac{\Delta\text{heat}}{100 \text{ g CO}_2} = \frac{-94.1 \text{ kcal}}{44 \text{ g CO}_2}$$

$$\Delta\text{heat} = -94.1 \text{ kcal} \times \frac{100 \text{ g CO}_2}{44 \text{ g CO}_2} = -214 \text{ kcal}$$

Therefore 214 kcal of heat would be released.

If we choose to solve by the use of dimensions and canceling units, we obtain

$$\Delta\text{heat} = 100 \text{ g CO}_2 \times \frac{1 \text{ mol}}{44 \text{ g CO}_2} \times \frac{-94.1 \text{ kcal}}{1 \text{ mol}} = -214 \text{ kcal}$$

Example 2   How much energy is required to produce 200 g of oxygen by the electrolysis of water ?

$$2\text{H}_2\text{O} \;\rightarrow\; 2\text{H}_2 \;+\; \text{O}_2; \quad \Delta H = +136.6 \text{ kcal}$$
$$36 \text{ g} \quad\quad 4 \text{ g} \quad\; 32 \text{ g}$$

Solving by use of ratios, we obtain

$$\frac{\Delta H}{200 \text{ g O}_2} = \frac{136.6 \text{ kcal}}{32 \text{ g O}_2}; \quad \Delta H = 136.6 \text{ kcal} \times \frac{200 \text{ g O}_2}{32 \text{ g O}_2} = 854 \text{ kcal}$$

Using dimensions and canceling, we obtain

$$\text{energy} = 200 \,\cancel{\text{g } O_2} \times \frac{1 \,\cancel{\text{mol } O_2}}{32 \,\cancel{\text{g } O_2}} \times \frac{136.6 \text{ kcal}}{1 \,\cancel{\text{mol } O_2}} = 854 \text{ kcal}$$

Example 3   How much energy can be released by the oxidation of 1.00 g of glucose to give CO_2 and H_2O in the human body?

The energy relationships are the same whether a substance is burned in a calorimeter or in a living organism. Thus

$$C_6H_{12}O_6 + 6O_2 \rightarrow 6CO_2 + 6H_2O; \quad \Delta H = -670 \text{ kcal}$$
$$\quad\; 180 \text{ g} \quad\;\; 192 \text{ g} \quad\;\; 264 \text{ g} \quad 108 \text{ g}$$

$$\frac{\Delta H}{1.00 \text{ g glucose}} = \frac{-670 \text{ kcal}}{180 \text{ g glucose}}$$

$$\Delta H = -3.72 \text{ kcal}/100 \text{ g glucose}$$

Solving by the use of dimensions yields

$$\Delta H = 1.00 \,\cancel{\text{g glucose}} \times \frac{1 \,\cancel{\text{mol glucose}}}{180 \,\cancel{\text{g glucose}}} \times \frac{-670 \text{ kcal}}{1 \,\cancel{\text{mol glucose}}}$$

$$= -3.72 \text{ kcal}/100 \text{ g glucose}$$

8.8
OXYGEN
AND LIFE

At the beginning of this chapter the continuous need of the body for oxygen was mentioned. The average active adult inhales at least 10 000 liters of air/day (more than half the volume of a large gasoline transport truck). From this amount of air some 500 liters of oxygen will be removed and more than 400 liters of carbon dioxide, produced by various body reactions, will be exhaled. While this may appear to be quite a task, we unconsciously accomplish it day after day throughout our lives. In serious illnesses such as pneumonia, when the lungs are incapacitated, the process of breathing may become laborious. Under these circumstances a patient may be placed in an oxygen tent where the percentage of oxygen will be increased from the normal 21% to 40–60%, considerably reducing the energy required in his breathing. Higher concentrations of oxygen, when inhaled over a period of time, produce undesirable effects. Smoking or any type of open flame is not permitted in a room where an oxygen tent is in operation because of the greatly increased rate at which fires burn in oxygen-enriched atmospheres.

When breathing becomes extremely difficult, helium rather than nitrogen can be used to dilute the oxygen. Helium, which is much lighter than nitrogen, diffuses through small openings of the lungs at almost twice the rate that nitrogen does (Section 16.6). This easy passage in and out of the lungs further decreases the effort of breathing.

The atmosphere diminishes in density with increases in altitude at such a rate that at approximately 17 500 ft its density is only one-half that at sea level. Above this altitude the oxygen concentration is so low that breathing becomes difficult and supplementary oxygen must be used unless a person has become acclimated to the higher elevation. Mountain climbers use base camps at high elevations for this purpose. A lack of oxygen is called **anoxia** and is present in many forms of asphyxiation. Unfortunately, the body has no satisfactory warning system to alert us to an impending state of anoxia. The gasping for breath after strenuous exercise is brought about by an oversupply of carbon dioxide produced by the increased demand for energy and not by any deficiency of oxygen in the bloodstream. In fact, pressure chamber experiments show that a person loses consciousness from lack of oxygen without experiencing any discomfort. In the early days, passengers and crewmen of aircraft wore oxygen masks on high-altitude flights. While the pressurized cabins of modern aricraft have eliminated this necessity, the law requires that oxygen equipment, including masks, be available in the event of an emergency.

Deep-Sea Diving   The combination of oxygen and helium as a breathing mixture is used to advantage in deep-sea diving. As the diver descends, the water pressure increases approximately 15 psi for every 34 ft. At slightly more than 200 ft the pressure of the water is 100 psi. Breathing at this pressure causes no inconvenience, and divers are comfortable as long as they remain under pressure. The solubility of gases in liquids is increased by pressure; therefore, more oxygen and nitrogen will be dissolved in the bloodstream when the pressure is great. If the diver surfaces quickly, the nitrogen becomes less soluble, forming bubbles in the bloodstream in much the same manner that carbon dioxide bubbles form when the cap is removed from a bottle of carbonated beverage. Within the bloodstream small bubbles of nitrogen collect to form larger bubbles that may block the flow of blood to a portion of the body. This malady, known as the "bends," is very painful and sometimes fatal. If divers are brought up very slowly, they are decompressed slowly and suffer no ill effects. This is a monotonous procedure since an ascent from 200 or more feet would require many hours. In an emergency situation requiring rapid ascent, there are two alternatives. One is to bring the divers up quickly and put them into a decompression chamber where the pressure is rapidly increased and then slowly allowed to decrease to normal values. The second choice is to start forcing a mixture of oxygen and helium through the air lines a few minutes before the ascent. Helium is practically insoluble in the blood plasma, and in the absence of atmospheric nitrogen, the nitrogen in the blood is rapidly eliminated. The diver may then be brought quickly to the surface with no ill effects.

In the fall of 1965 the astronaut Scott Carpenter and his associates spent thirty days in the Sealab more than 200 ft below the surface of the ocean off the coast of La Jolla, California. The atmospheric pressure in the Sealab was roughly seven times that at the surface. The artificial atmosphere in the Sealab consisted primarily of oxygen and helium. This was a precaution taken in the event that an emergency surfacing became necessary. An unusual effect of the presence of helium was the odd sound of the voices of persons in the Sealab, caused by the decreased density of the atmosphere inside the Sealab.

8.9 INDUSTRIAL USES OF OXYGEN   In addition to the ever-increasing demand for oxygen in hospitals and in the aircraft industry, there are many other important uses. Large quantities are used in the oxyacetylene torches for cutting and welding steel. Almost every garage and machine shop in the country uses oxygen for such purposes. In machine shops where large quantities of oxygen are used it has been found that liquid oxygen is the more economical. For such purposes the liquid oxygen is stored in specially designed Thermos-bottle-type containers.

Relatively pure oxygen (95–99.5%) is being used in ever-increasing amounts in the manufacturing of iron and steel. With the use of nearly pure oxygen, comparable equipment can produce iron and steel several times faster than conventional processes that use air.

During the past decades another field has opened for the use of oxygen. The rocket program uses such large amounts of liquid oxygen that many great industrial plants for producing it are located on or near the rocket bases. Thus there is an ever-ready source with minimum delivery costs. The only material other than electricity needed for such a plant is air, which is, of course, available at all locations.

8.10
OZONE

Oxygen, like most other gaseous nonmetals, is found in the elementary state as diatomic molecules. It may also exist in the triatomic molecule O_3, called ozone. Oxygen and ozone are called **allotropic forms** of oxygen. Allotropic forms or allotropes of other elements are also known.

Energy is required for the conversion of oxygen to ozone.

$$3O_2 + energy \rightarrow 2O_3 \tag{8-15}$$

At high altitudes, ultraviolet radiation provides the energy for the formation of an ozone layer. This process also absorbs much of the ultraviolet radiation from the sun and prevents it from reaching the surface of the earth. Small amounts of ozone are formed nearer the earth's surface, where it is regarded as a pollutant because of its destructive effects on living tissues.

If ozone is used in place of oxygen in chemical reactions, more energy is liberated or is made available for driving the reaction. For instance, the reaction

$$3C + 2O_3 \rightarrow 3CO_2 \tag{8-16}$$

liberates almost 25% more energy than

$$3C + 3O_2 \rightarrow 3CO_2 \tag{8-17}$$

even though in each reaction the same amount of carbon is oxidized to carbon dioxide. Industrially, ozone is sometimes used to advantage in place of oxygen in the preparation of chemicals.

8.11
NITROGEN

Nitrogen is the major component (79% by volume) of air, where it occurs in the free, uncombined state. It is also found as nitrates, nitrites, and ammonium compounds in the soil. Many organic compounds both in plants and animals, including the amino acids, proteins, and nucleic acids, contain appreciable amounts of nitrogen in their structures.

Nitrogen is produced commercially by cooling air to a liquid state and then separating the nitrogen from the oxygen and other liquefied gases by distillation. We can attribute much of the nitrogen we find in the air to the action of denitrifying bacteria that produce free nitrogen gas from nitrates and nitrites.

The reactions in which free nitrogen from the air is combined with other elements to form compounds are called **nitrogen fixation**. Much of the nitrogen in nitrites, nitrates, and other nitrogen-containing compounds found in the soil has been fixed by bacteria. Since nitrogen-containing compounds are essential for plant growth, nitrogen-fixing bacteria are very important. The reaction of nitrogen and oxygen either due to lightning or in an internal combustion engine is also a form of nitrogen fixation.

Some of the different forms of nitrogen and their relation to each other are shown in Fig. 8.4. Elemental nitrogen is very stable. In contrast to many other elements, a great deal of energy must be provided to form compounds of nitrogen. At high temperatures and under pressure, nitrogen and oxygen combine to form a variety of oxides. Dinitrogen oxide (N_2O), commonly called *laughing gas*, is used as an anesthetic, and NO and NO_2, which are formed in internal combustion engines and other places, are air pollutants.

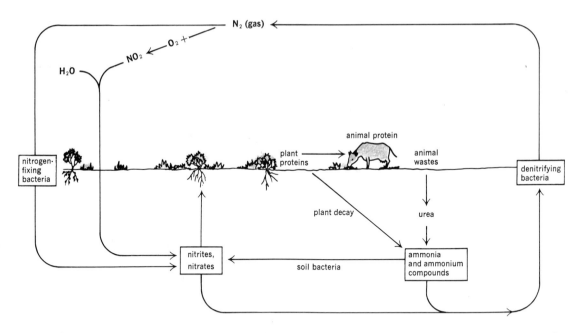

Fig. 8.4 The nitrogen cycle. Nitrogen gas in the atmosphere is fixed by nitrogen-fixing bacteria on the roots of some plants, to form nitrites and nitrates. Nitrogen may also be fixed by reacting with oxygen to produce oxides of nitrogen, which then react with water to give nitrites and nitrates. Nitrates are absorbed through the roots of green plants and made into plant protein. When the plant dies and decomposes, or when it is eaten by an animal that excretes nitrogen-containing wastes or dies and decomposes, the nitrogen is returned to the soil and is converted to ammonia or ammonium compounds. Denitrifying bacteria act on nitrates or ammonium compounds and release nitrogen gas into the atmosphere.

When nitrogen dioxide reacts with water, nitric acid is produced.

$$3NO_2 + H_2O \rightarrow 2HNO_3 + NO \tag{8-18}$$

Nitrogen combines with hydrogen to form ammonia. The reaction, which is called the **Haber process**, requires a pressure of 200 atmospheres or more and a temperature of 550°C, as well as a catalyst.

$$N_2 + 3H_2 \xrightarrow{\text{catalyst}} 2NH_3 \tag{8-19}$$

The large amounts of energy required to form compounds of nitrogen are released when these compounds decompose, and many compounds of nitrogen such as TNT (trinitro-toluene) and nitroglycerine are explosives. The compound ammonium nitrate will also explode.

Liquid elemental nitrogen is used industrially as a coolant, especially in the frozen-foods industry. Even greater amounts of nitrogen gas are used to provide "blanketing atmospheres" that exclude oxygen in the processing of metals and the production of transistors and other electronic components. These uses take advantage of the chemical inactivity of elemental nitrogen. The use of compounds of nitrogen for explosives again uses the same properties, but in reverse—the tendency of compounds of nitrogen to decompose to the elemental form.

Due to the large amounts of nitrogen in many important compounds in green plants, it must be present in soils. Nitrogen compounds, particularly ammonia, are necessary as fertilizers for many crops. Sometimes the ammonia is drilled into the soil and other times it is bubbled into water used for irrigation. The increasing cost of ammonia fertilizers, due both to the energy required for its synthesis and to the fact that expensive petroleum is the source of the hydrogen used in the Haber process, is one factor that contributes to the world's food crisis.

Although many organic compounds of nitrogen are essential for life, others are toxic. Exposure of humans to nitrobenzene and similar compounds is controlled by government agencies. When nitrites react with some organic compounds they can produce nitroso-amines, which have been shown to cause cancer. Many people are now concerned about the amounts of nitrites and nitrates in foods such as ham and bacon. These compounds are added to prevent the growth of bacteria that cause the deadly botulism poisoning; however, the nitrites may react with food components in the stomach to produce nitros-amines. In the light of current research, it would seem that we should use other methods for preventing botulism. Some countries have already done so and have banned the addition of nitrites and nitrates to foods.

8.12 THE NOBLE GASES

There is one group of elements found in the atmosphere that are so unreactive that they were not discovered on the earth until 1894, much later than most other nonradioactive elements were found. These gases are helium, neon, argon, krypton, xenon, and radon. The evidence for the existence of helium in the atmosphere of the sun was first observed, by spectroscopic methods, in 1868. The element was named *helium*, from the Greek name for the sun, *helios*. It was not discovered on the earth until 1895, when it was found in uranium-containing ores. Today it is obtained commercially from the natural petroleum gases occurring in some of the oil–gas fields of Kansas, Oklahoma, and Texas.

A hundred years before the discovery of helium, it was suspected that the atmosphere contained gases other than oxygen and nitrogen, but it was not until 1894 that such a gas, which was named *argon*, was identified. The name argon was suggested since it is the Greek word for "inert." This was indeed a logical name since all the members of this family are so inactive they do not naturally react with other elements. For more than sixty years they were called the *inert gases*. Subsequently, other members were isolated and identified. Some of their physical properties are shown in Table 8.3.

TABLE 8.3   Some of the properties of the noble gases

	Atomic weights	Boiling point, °C	Melting point, °C
helium	4.0	268.9	−272.0
neon	20.0	−245.9	−248.8
argon	40.0	−185.8	−189.2
krypton	83.8	−152.9	−156.6
xenon	131.3	−108.1	−111.9
radon	(222.)	− 61.8	− 71.0

As their former name implies, the noble gases do not react readily. In fact, they do not even form diatomic molecules. Thus, unlike most other gases they are found as atoms, not molecules.

In 1933, Professor Linus Pauling stated that it should be possible for a so-called inert gas to form compounds. Despite the fact that the outer electron shell of such an atom has a completed set of eight electrons, he predicted that, under the proper circumstances, these electrons could be used for bonding to other elements. However, it was not until 1962 that chemists succeeded in making any of the compounds. Since that time, several compounds have been prepared, by the reaction of the noble gases with fluorine or oxygen. Xenon tetrafluoride, XeF_4, is a typical example (Fig. 8.5). In general, it is still proper to regard these gases as inert, for they do not participate in most of the common chemical reactions.

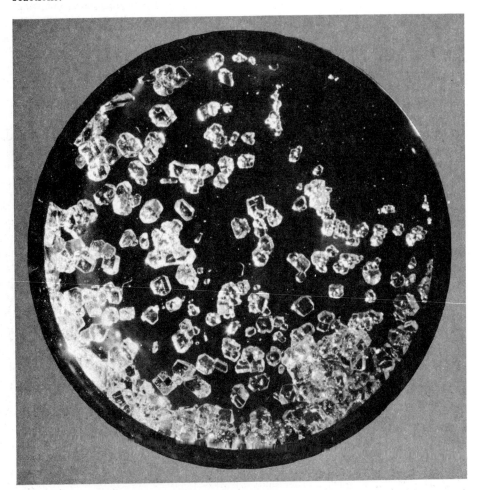

Fig. 8.5 Xenon tetrafluoride crystals, prepared at the Argonne National Laboratory in 1962 and first appearing as a cover photograph on *Science*, Vol. 138, 12 October 1962. Copyright 1962 by The American Association for the Advancement of Science. [Courtesy of the Argonne National Laboratory and *Science*.]

The discovery of the noble gases posed a problem in that there appeared to be no ideal place for them in the periodic table. They have been justifiably placed in either of two locations in the table. Since it appeared for many years that they were inert, they were considered to have a valence of zero and were therefore placed before Group I under a heading of "0" valence. However, it was established that the inertness of the noble gases was due to the fact that the outside shell of electrons for each was complete; therefore they were also placed at the extreme righthand side of the chart and sometimes indicated as Group 0. In modern periodic tables, they are usually placed at the right and are designated Group VIII, as in the chart inside the back cover of this book.

Uses   The noble gases have many uses, some of which derive from their inactive characteristics. Argon is used in electric lightbulbs to replace air, which would destroy the hot filament by oxidation. Electricity, when passed through an evacuated glass tube containing neon gas at low pressure, produces the characteristic "neon light."

Helium has many uses. The most important one for many years, after it was obtained in quantity, was in balloons and other lighter-than-air craft. Helium eliminated the fire hazard always present when hydrogen was used. As we noted in Section 8.8, helium has been used by deep sea divers to prevent the "bends" and in hospitals to dilute the oxygen needed to treat respiratory diseases. Helium is also used by the welding industry when oxygen-free atmospheres are necessary. The National Aeronautics and Space Administration is presently one of the heavy users. Since helium has the lowest boiling point of any known substance, $-268.9°C$, it is used as the cooling agent when extremely low temperatures are necessary. Because of the rarity of the other noble gases, there are few practical applications of them.

SUMMARY

In this chapter we have described and discussed the occurrence, properties, uses, and significance of some of the elements that are commonly found in the atmosphere. Although all occur as gases, in chemical properties they range from reactive oxygen, through relatively inactive nitrogen, to the extremely unreactive noble gases. Oxygen and nitrogen are found both in the free state and in various compounds. Oxygen gas and many compounds of both oxygen and nitrogen are essential to life. Compounds of nitrogen are especially important as fertilizers for green plants.

Formation of chemical compounds of the noble gases provides an example of experimental evidence challenging an important chemical idea and requiring that it be modified. The idea that the noble gases and, indeed, any element with a completed outer shell of electrons would not form compounds is still correct in most cases. However, we now know that, under unusual circumstances, the outer-shell electrons of noble gases can be used for bonding. The heat of reaction, especially the heat given off by oxidations, was discussed and sample calculations dealing with this important physical constant were presented. The allotropic form of oxygen, ozone, was discussed briefly.

The importance and interactions of oxygen, ozone, nitrogen, the rare gases, and other substances in the atmosphere are discussed in Chapter 15. Hydrogen and some other nonmetals and metalloids are discussed in Chapter 14. The following chapter discusses that most important compound—water.

IMPORTANT TERMS

allotropic form	combustion	heat of reaction
anode	delta H (ΔH)	kindling temperature
anoxia	explosion	nitrogen fixation
calorimeter	Haber process	noble gas
catalyst	heat of combustion	ozone
cathode	heat of formation	spontaneous combustion

WORK EXERCISES

1. Write the equation for the electrolysis of water.
2. Write chemical equations for the reaction between (a) oxygen and three different nonmetals and (b) oxygen and four metals from Groups IA or IIA.
3. Why do finely divided combustible materials burn more rapidly than larger particles of the same material?
4. Name two methods of controlling fires.
5. Distinguish between oxygen and ozone.
6. Is ozone an isotope of oxygen?
7. What term is employed to describe the relationship of ozone and oxygen?
8. Why are carbon-dioxide-filled fire extinguishers preferred to water and water solutions for some types of fire?
9. List the advantages and disadvantages of living in an atmosphere containing 50% or more oxygen.
10. Why is there less danger that deep-sea divers will have the "bends" if they breathe a helium–oxygen mixture instead of ordinary air, which contains primarily oxygen and nitrogen?
11. Would you predict that Indians living at high elevations (17 000 ft) in the Andes would have relatively smaller or larger lungs than people who live near sea level? Why?
12. List several uses for oxygen.
13. According to Eq. (8–2):
    a) How many moles of oxygen are produced by the thermal decomposition of one mole of $KClO_3$?
    b) How many grams of oxygen are produced by decomposition of 2 moles of $KClO_3$?
    c) How many grams of oxygen are produced by the decomposition of 10 g of $KClO_3$?
    d) How many grams of $KClO_3$ are required to produce 24 g of oxygen?
    e) How many grams of KCl will be produced by the decomposition of 100 g of $KClO_3$?
14. According to Eq. (8–11):
    a) How many grams of carbon dioxide will be formed by the oxidation of 32 g of methane (CH_4)?

b) How many grams of water will be formed by the complete oxidation of 100 g of methane?
c) How many grams of oxygen are required to completely burn 100 g of methane?

15. According to the equation $C_8H_{18} + 12\frac{1}{2}O_2 \rightarrow 8CO_2 + 9H_2O$:
    a) How many pounds of oxygen will be required for the complete oxidation of 100 pounds of octane (C_8H_{18})? This is approximately the equivalent of 18 gallons of gasoline. (Any unit weight may be substituted for grams in chemical equations.)
    b) How many pounds of carbon dioxide will be produced by the complete oxidation of 100 pounds of octane?
    c) How many pounds of water will be produced by the complete oxidation of 100 pounds of octane?
16. The heat of formation of water is -68.3 kcal/mol. How many kilocalories of energy will be liberated by the formation of one liter of water produced by the combining of hydrogen and oxygen?
17. The heat of formation of magnesium oxide is -143.8 kcal/mol. According to Eq. (8–9), how many kcal of energy will be liberated by the oxidation (burning) of 10 g of magnesium?
18. Using the information given in the preceding question, find out how many grams of magnesium, when oxidized, are needed to produce 100 kcal of energy.
19. In an experiment to determine the heat of combustion of glucose ($C_6H_{12}O_6$), a 1.25-g sample was completely oxidized. The heat liberated raised the temperature of 900 ml of water 5.17°C. As measured in this experiment, what is the heat of combustion of one gram of glucose? of one mole of glucose?
20. Describe three ways in which nitrogen of the air is fixed.
21. Can you correlate the fact that it is difficult to form compounds of nitrogen and the fact that nitrogen compounds are used in many explosives?

22. What is the economic and nutritional significance of the Haber process?

23. List some of the important uses of nitrogen-containing compounds.

24. Why are the noble gases unreactive chemically?

25. What substances might be used to produce an inert atmosphere that would prevent corrosion of materials?

SUGGESTED READING

Brill, W. J., "Biological nitrogen fixation," *Scientific American*, March 1977, p. 68 (Offprint #922). Only a few bacteria and simple algae have the cellular equipment needed to "fix" the nitrogen of the atmosphere into ammonia. They are the major suppliers of this limited agricultural resource.

Cload, P., and A. Gibor, "The oxygen cycle," *Scientific American*, September 1970, p. 110 (Offprint #1192). Oxygen occurs in many compounds and its recycling is complex; this article describes the compounds and processes involved.

Delwiche, C. C., "The nitrogen cycle," *Scientific American*, September 1970, p. 136 (Offprint #1194). From the fixing of atmospheric nitrogen to its release by denitrification a complex series of reactions is involved. These reactions and the intermediate compounds are discussed in this article.

9

Water: Its Properties, Pollution, and Purification

Water is part of everyone's experiences. In childhood, water in puddles presented to many of us an irresistible urge to wade in them, while water in a bathtub was to be avoided. On hot days a drink of cool water or a plunge into a pool or lake, water skiing or other sports, waiting for rain after a dry, dusty drought, the peaceful feeling brought about by a calm body of water, and the exhilaration or fear of a stormy ocean—these and many other relationships to water are part of people's lives.

While these emotional experiences are important, many scientists find that a study of the chemistry of water produces even more wonder and satisfaction. Not only is water everywhere on our planet, but the properties of this common compound are so unusual that it is difficult to imagine this blue sphere, the earth—or life itself—without water.

Water is unusual in that it is commonly found in all three states: as a solid (ice), a liquid (water), and a gas (water vapor). There is a constant recycling of water as it evaporates from oceans, lakes, and streams to form molecules of water vapor, which then condense as dew or in clouds and eventually fall as precipitation. The precipitation absorbed by the soil may be picked up by a green plant, where it performs a variety of functions before it is incorporated into the leaves, stems, or roots of the plant or is returned to the air by evaporation. Water may also trickle down through the soil until it reaches an impervious layer of rock and forms a pool from which it may be pumped up through a well, where it again enters the general circulation. Animals as well as plants need water, which they use for a variety of purposes and then excrete to help cool the animal or to carry waste products to the outside. In temperate areas, the freezing of water to make ice and the melting of ice to give water happens almost daily during winter, and in areas near the north and south poles it happens throughout the year. Except for glaciers, icebergs, and floating blocks of ice, water is relatively stationary when in its solid form.

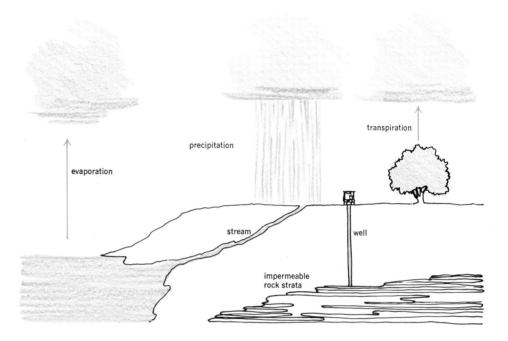

Fig. 9.1   The water cycle, showing three of the pathways by which water is recycled. When water falls from clouds, some of it runs into streams and rivers that eventually empty into an ocean or sea, where the water evaporates and again forms clouds. Water absorbed by the soil may migrate downward until it is stopped by layers of impervious rocks; it may then be brought back to the surface from a well. It may also be taken up from the soil by trees and other plants, which release large amounts of water into the atmosphere by a process called transpiration.

Throughout all these changes (see Fig. 9.1) water remains in the molecular form, and cycling from one situation to another is primarily a series of physical changes brought about by the addition or removal of energy. An exception to this is that some reactions in plants and animals actually result in the breaking of the bonds between the hydrogen and oxygen atoms. The hydrogen is incorporated into organic compounds, which are later oxidized to again yield water; the oxygen that is released by the reactions goes into the atmosphere. It recombines with hydrogen to form water when organic compounds are burned or metabolized by animals.

$$H_2O \longrightarrow \substack{O_2 \\ \text{''H''} \rightarrow \text{Organic compounds} \\ O_2}$$

When we consider the widespread occurrence of water and the stability of the water molecule it is easy to see why ancient people considered it to be one of the elements and not a compound. It is estimated that 97% of the water molecules of the earth are present as liquids in the oceans and seas, 2% is found in the solid ice caps at the poles, and the remaining 1% makes up the clouds, rivers, water in living things, and water in ground water.

Many of the foods we eat are primarily water (Table 9.1). Even "dried" foods contain considerable amounts of water. Two-thirds of our body weight is water, and all animals contain large amounts of this compound. Plants are somewhat more variable than animals, but while alive, they contain appreciable percentages of water. Even rocks and minerals often contain water, which is sometimes held in a particular form known as *water of crystallization.*

TABLE 9.1 Water content of foods

Food	Water content (%)
apples	84
bread (fresh)	35
broccoli	90
cheese, cheddar	37
cucumbers	96
dates (dried)	20
eggs	74
fish (cod) uncooked	83
fish (tuna) canned	60
grapes	82
honey	20
macaroni (dry)	12
milk	87
peanuts, roasted	3
potatoes	78
raisins, dry	24
steak, round	74
watermelon	92

9.3
PHYSICAL
PROPERTIES
OF WATER

Because of its physical and chemical properties, water is probably the most important compound on our earth. Indeed, life could not exist without it. In discussing its properties, we generally start by saying that water is colorless, odorless, and tasteless—but we know that deep water seems to be blue because it reflects this color to our eye, and we realize that even if it had an odor or taste, we would not be aware of it, since we have it constantly in our mouth and nostrils.

Perhaps the most interesting property of water is that it exists as a liquid at normal temperatures. The physical state of a covalent compound is related to its molecular weight, with small molecules being gases, intermediate ones liquids, and large ones solids. This is logical when we consider that the attraction between molecules depends upon their relative sizes. Thus, when we compare water with similar substances (Table 9-2), we would expect it to be a gas and not a liquid at room temperature. We note not only that the melting point of ice and the boiling point of water are high, but also that the heat of fusion and of vaporization (see Section 3.8 for discussion of these terms) are large values. These physical constants led scientists to look for some force that attracts one water molecule to another. Studies show that there is such a force and that it is due to the shape and distribution of charges of the atoms of the water molecule. We find from x-ray studies

TABLE 9.2 Physical constants of water compared with other compounds of low molecular weight

Substance	Formula weight	20°C	Boiling point °C	Freezing point °C	Heat of vaporization (cal/gram)
H_2O	18	liquid	100	0	540
H_2S	34	gas	− 61	− 86	132
NH_3	17	gas	− 33	− 78	327
CH_4	16	gas	−162	−183	122
HF	20	gas	20	− 83	360
PH_3	34	gas	− 88	−134	103
CO_2	44	gas	− 78	− 57	87
C_3H_8	44	gas	− 42	−188	102

that the water molecule is not linear or in a right-angle form as might be expected, but that the bond angles between the oxygen and hydrogen nuclei are about 104.5° (see Section 6.12).

H—O—H   O—H   O
   180°   H 90°   H   H
                           104.5°

(a) linear water   (b) right angled   (c) actual water
    molecule   molecule   molecule

In terms of electron-dot structures, the actual structure for water is intermediate between (a) and (b) below, somewhat closer to (b) than to (a).

H : O : H   : O : H
                    H

   (a)   (b)

Because the covalent bonds of water molecules are polar (Sections 6.3, 6.4) there is an uneven distribution of electrical charge; the oxygen atom has a slight negative charge and the hydrogen atoms have a slight positive charge. We symbolize this partial charge as δ^+ or δ^- to indicate that it is much less than the ordinary unit charge such as that on sodium or chloride ions. Two representations of polar water molecules are shown below.

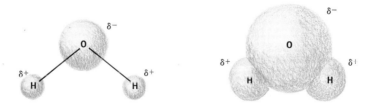

The fact that water molecules are polar means that the positive part of one molecule has a strong attraction for the negative part of another molecule, as shown at the top of page 144.

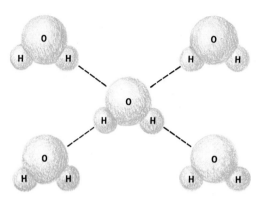

This attraction between water molecules explains why so much heat is required to separate molecules of liquid water and convert them to the gaseous state. It also explains the high specific heat and heat of fusion we find for water. Measurements indicate that there are five or six water molecules bound together at a temperature of 20°C, and perhaps the correct formula for water should be $(H_2O)_{5-6}$. But the number of molecules in a cluster changes with temperature, the number decreasing as the temperature increases. This is because an increase in temperature increases the motion of the individual molecules so that they are less likely to be bound together. In the gaseous state (steam or water vapor), water molecules tend to be single molecules, not clumps.

Most substances contract as they freeze or solidify, but again, water is an exception. As ice forms, there is an expansion; thus ice occupies more space, or is less dense, than cold water. This may cause an annoyance, as when water in pipes or automobiles freezes and cracks the pipe or radiator. The process is important, however, in breaking rocks into soil. The lesser density of ice is even more important to organisms that live in water.

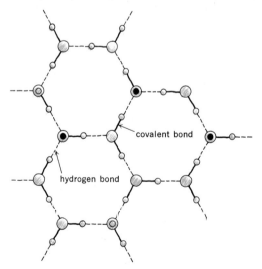

Fig. 9.2 Vertical view of arrangement of water molecules in an ice crystal: ◉ represents a hydrogen atom above an oxygen atom; ◎ represents a hydrogen atom below an oxygen atom.

When ice forms, it remains at the top of the body of water, and often forms a solid layer of ice that insulates the water below from the air. Many plants and animals can survive even in very cold water, provided they are not frozen. Although small ponds may freeze solid, large bodies of water rarely do. If ice were more dense than water, it would sink and even large bodies of water would often freeze solid.

Water expands as it freezes because it forms a crystalline structure influenced by the distribution of polar molecules (Fig. 9.2). This structure contains vacant spaces, which account for the expansion. Water molecules in the liquid state are much more tightly packed than those in ice. Of course, most ice also contains trapped air, which further decreases its density.

The polar nature and the mutual attraction of water molecules also explains why water has a tendency to form drops rather than spreading out over a flat surface, particularly if the surface is nonpolar, as when it contains a coating of wax (Fig. 9.3). We describe this tendency by saying that water has a high *surface tension*. The ability of small dense objects such as a greased needle to float on the surface of water also illustrates this property. When we add a surface-active agent, a detergent, we decrease the surface tension of water and it tends to spread out instead of forming drops. Some water birds sink rather than remain afloat when the oils around their feathers are removed by detergents. The human lungs are always coated with water, and because of the attraction of water molecules for one another, the lungs will tend to collapse unless a substance called *surfactin* is present. It is postulated that a decrease in surfactin is the cause of diseases such as emphysema. The action of soaps and detergents is discussed in Sections 24.6 and 24.7. The importance of bile in digestion is included in Section 37.2.

Another phenomenon that is due to the attraction of water molecules, this time both to each other and to the surface of a clean glass cylinder, is the *meniscus* or curved surface we find when we measure water solutions in a graduated cylinder (Fig. 9.3). Water molecules are attracted to the surface of glass because it is also highly polar.

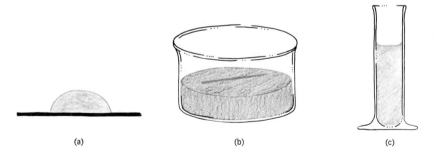

(a)   (b)   (c)

Fig. 9.3 Some effects of surface tension of water. (a) A drop of water on a nonwettable surface such as a waxed automobile. (b) A waxed or greased needle floating on water. (c) The meniscus formed when water adheres both to other water molecules and to the glass surface of a graduated cylinder.

9.4
HYDROGEN
BONDS

The attraction between two molecules of water is called a **hydrogen bond.** Although each hydrogen nucleus is held to an oxygen atom by a covalent bond, it is also attracted, by polarity, to the oxygen atom of another water molecule (one that is already covalently bonded to two hydrogen atoms of its own). Hydrogen bonds are commonly represented as dashed lines to distinguish them from covalent bonds (solid lines).

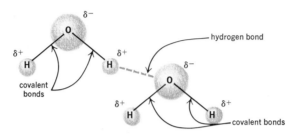

Hydrogen bonds are much weaker than covalent bonds. It generally requires only 5 kcal to rupture a mole (6.02×10^{23}) of hydrogen bonds, while it takes 90–100 kcal to rupture a mole of covalent bonds.

It should be emphasized that a hydrogen "bond" is a force acting *between* molecules, not within them—it is due to the attraction of the positive and negative poles of two water molecules, and thus is quite distinct from covalent bonding within a molecule. If hydrogen were covalently bonded to two oxygens, there would be four electrons around hydrogen, and this is an impossibility.

Hydrogen bonding occurs in many compounds containing a hydrogen atom that is covalently bonded to oxygen, nitrogen, or sulfur nuclei. Hydrogen bonding is found to increase the boiling points of some organic compounds, such as the alcohols, and hydrogen bonds are even more important in the structures of proteins, nucleic acids, and other biochemical molecules. Hydrogen bonding in compounds with nitrogen or sulfur is weaker than in those with oxygen because of the lesser electronegativity of nitrogen and sulfur. See Table 9.2 for the relative effects of the hydrogen bonds in H_2O, H_2S, and NH_3.

**9.5
CHEMICAL
PROPERTIES
OF WATER**

Water reacts with other substances in a variety of ways. It is often characterized as the universal solvent. Although this term is misleading (there are many substances that do not dissolve in water), it does dissolve a large number of substances. When a substance dissolves we can think of the particles (molecules, ions) of the substance as being attracted to the water molecules and vice versa (Fig. 9.4). This is similar to the attraction of water molecules for each other and is found whenever water reacts with ions or with polar or semipolar molecules. Solutions are discussed in detail in Chapter 10.

In some cases water remains trapped in a solid compound that exists in the crystalline state. For instance, solid, crystalline copper sulfate has five molecules of water for every one of copper sulfate. We use a dot to indicate the formulas for compounds such as this, which we call **hydrates**.

$CuSO_4 \cdot 5H_2O$

Many crystalline substances are found as hydrates, as shown in Table 9.3.

While the number of molecules of water associated with different compounds varies, the amount associated with a given compound generally does not. There are five molecules of H_2O per molecule of $CuSO_4$ and seven molecules of H_2O per molecule of $MgSO_4$ in its hydrated form. A few compounds have two or more hydrated forms, in which different numbers of molecules of water are associated with each molecule of the compound.

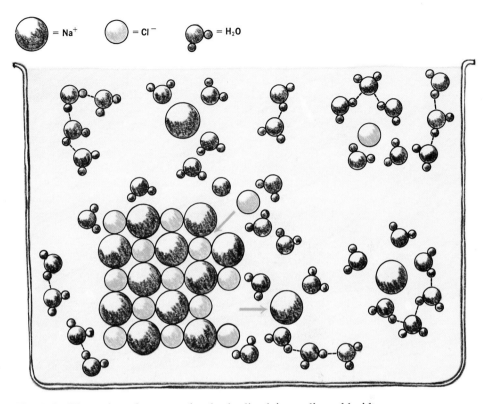

Fig. 9.4 The action of water molecules in dissolving sodium chloride.

Water of hydration may be removed by heating:

$$\text{CuSO}_4 \cdot 5\text{H}_2\text{O} \xrightarrow{\Delta} \text{CuSO}_4 + 5\text{H}_2\text{O} \tag{9-1}$$

The CuSO_4 produced in this way is an anhydrous, noncrystalline (or amorphous) powder. In the case of copper sulfate the anhydrous form is a white powder, whereas the crystals are a brilliant blue.

TABLE 9.3 Some common hydrates

Name	Formula of Hydrate
asbestos	$3\text{MgO} \cdot 2\text{SiO}_2 \cdot 2\text{H}_2\text{O}$
barium bromide	$\text{BaBr}_2 \cdot 2\text{H}_2\text{O}$
borax	$\text{Na}_2\text{B}_4\text{O}_7 \cdot 10\text{H}_2\text{O}$
clay	$\text{Al}_2\text{O}_3 \cdot 2\text{SiO}_2 \cdot 2\text{H}_2\text{O}$
copper sulfate, crystalline	$\text{CuSO}_4 \cdot 5\text{H}_2\text{O}$
epsom salts	$\text{MgSO}_4 \cdot 7\text{H}_2\text{O}$
glauber's salt	$\text{Na}_2\text{SO}_4 \cdot 10\text{H}_2\text{O}$
gypsum	$\text{CaSO}_4 \cdot 2\text{H}_2\text{O}$
talc (soapstone)	$3\text{MgO} \cdot 4\text{SiO}_2 \cdot \text{H}_2\text{O}$

When exposed to air, particularly dry air, many crystals tend to lose their water of crystallization spontaneously in a process known as **efflorescence**. Other substances tend to pick up water from the atmosphere; these are called **hygroscopic** substances. If they absorb sufficient quantities of water from the air to actually dissolve, they are said to be **deliquescent**. A hygroscopic substance such as $CaCl_2$ is sometimes used to remove moisture from the air in an enclosed space (for example, a damp basement room); it is then called a **desiccant**. Other hygroscopic substances, such as glycerol and some sugar-like compounds, are used to help retain the moisture in a food or a similar product (for example, pipe tobacco). Such substances, called **humectants**, are among the common food additives. Humectants are also included in lotions used to keep the skin moist.

We make use of two different hydrated forms of a compound when we use plaster of Paris in making an art or dental casting or in constructing a cast for a broken or injured limb. Calcium sulfate is found as the transparent, crystalline dihydrate called gypsum. Upon relatively gentle heating (125°C), gypsum loses water to form plaster of Paris, which is a hemihydrate.

$$2CaSO_4 \cdot 2H_2O \xrightarrow{\Delta} (CaSO_4)_2 \cdot H_2O + 3H_2O \qquad\qquad (9\text{-}2)$$

Gypsum   Plaster of Paris

When water is added to plaster of Paris the reaction is reversed to give gypsum, which quickly "sets" or crystallizes to form a strong solid (see Fig. 9.5). Setting also involves a slight expansion, which must be taken into consideration in artistic or medical and dental procedures.

The reactions already considered involve the action of whole water molecules, and although they can be regarded as physicochemical reactions, since they involve both

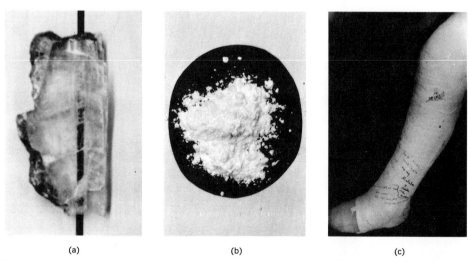

(a)   (b)   (c)

Fig. 9.5 Hydrated forms of calcium sulfate: (a) is a naturally occurring transparent hydrated mineral, $CaSO_4 \cdot 2H_2O$ (gypsum). The gypsum when heated is changed to a white powder (b) with the formula $(CaSO_4)_2 \cdot H_2O$. The powdered form is commonly called plaster of Paris. Plaster casts (c) are formed by adding the proper amount of water and molding into the desired form or shape.

physical and chemical changes, the water is not transformed into other substances. Other changes involve definite chemical reactions because the water molecules are converted into new substances. Some reactions are due to the fact that water has a very slight tendency to dissociate or ionize to produce hydrogen and hydroxide ions.

$$\text{HOH} \rightleftharpoons \text{H}^+ + \text{OH}^- \tag{9-3}$$

The degree of dissociation is very small and most water is always present in the molecular form; however, if one of the ions is consumed in a reaction, another molecule of water ionizes. The significance of the slight ionization of water will be discussed further in Chapters 11 and 12.

Water reacts with oxides to yield new compounds in a combination reaction:

$$\text{Na}_2\text{O} + \text{HOH} \rightarrow 2\text{NaOH} \tag{9-4}$$

$$\text{SO}_3 + \text{HOH} \rightarrow \text{H}_2\text{SO}_4 \tag{9-5}$$

In other reactions, a water molecule and some other molecules are split and recombined into two new compounds. This type of double substitution reaction (Section 7.6C) is called hydrolysis, from the prefix *hydro-* for water and *-lysis* indicating a breaking apart.

$$\text{Na}_2\text{CO}_3 + \text{HOH} \rightleftharpoons \text{NaOH} + \text{H}_2\text{CO}_3 \tag{9-6}$$

Many hydrolyses are reversible and the reaction often favors the production of the salt and water. Many covalent organic compounds also undergo hydrolysis. Hydrolyses are very important in biochemistry and, under the influence of enzymes as catalysts, proceed rapidly and primarily in the direction of hydrolysis. The digestion of food is a hydrolytic reaction:

$$\text{proteins} + \text{HOH} \rightarrow \text{amino acids} \tag{9-7}$$

$$\text{starch} + \text{HOH} \rightarrow \text{simple sugars} \tag{9-8}$$

These reactions are used in reverse by plants and animals to synthesize the complex molecules of proteins and starch. Water is a byproduct of the synthesis. Hydrolyses are generally exergonic, and the reverse, or synthetic, reactions require energy.

**9.6
IMPORTANCE
OF WATER
TO LIFE**

Life as we know it cannot exist without water. While it is interesting to speculate that perhaps life on other planets could use H_2S or NH_3 for some of the purposes for which life on earth uses water, neither of these related compounds has all of the properties that make water so vital to the life of our plants and animals.

We know that animals or plants die when deprived of water. While some plants and animals can survive on little water, no living organism can exist in the total absence of water. The question naturally arises, "Why is water vital?" The answer, as we have already suggested, lies in the amazing properties of water.

Plants and animals depend upon chemical reactions for their life and growth. In chemical reactions, the reacting particles must be able to collide. While collisions are routine for gas molecules and present no real difficulty in molecules of a liquid, molecules of solids must generally be in solution in order to react. In some cases, the molecules must also dissociate or ionize before they can react. Water is the best solvent known, and because of its polar nature, it promotes ionization of substances that dissolve in it. The extremely large protein molecules, which are so necessary for life, are kept in solution by being surrounded by hundreds of water molecules. The elimination of waste products as water solutions (especially in the urine but also in perspiration) represents another important use of water as a solvent.

In addition to the function of water as a neutral, nontoxic solvent, it is estimated that water participates as a reactant or product in 50% of the reactions of living organisms. Digestion of food is primarily a series of hydrolysis reactions. The process of photosynthesis involves water as a reactant:

$$6H_2O + 6CO_2 \rightleftharpoons C_6H_{12}O_6 + 6O_2 \qquad\qquad (9\text{-}9)$$
$$\text{(a sugar)}$$

The reverse of this reaction is the oxidation of sugar, which requires oxygen and yields water and carbon dioxide.

One of the major problems of the higher animals is temperature regulation. If their body temperature varies only a few degrees, serious consequences result. If body temperature is sufficiently elevated, brain damage and eventually death may occur. Since many of the reactions taking place in animals release heat, it is necessary that cooling processes be employed. Water is well suited to act as a coolant. It has a high specific heat. Consequently small volumes of it can absorb relatively large quantities of heat without a large increase in temperature. Water also has a high heat of vaporization. This property means that evaporation from the skin or lungs consumes large amounts of heat. Furthermore, water is a reasonably good heat-transfer agent, and, as such, serves to help transfer the heat from the places where it is produced to places where it can be dissipated.

When we consider the importance of water, we wonder how some animals manage to survive with almost no water intake. The desert rat (kangaroo rat) apparently never drinks water. While other animals get appreciable amounts of water from their foods (as is obvious from the water content of foods listed in Table 9.1), the desert rat's main food is dry seeds, which contain only small quantities of water. The desert rat derives a great deal of its water from the oxidation of hydrogen-containing substances in its food. Water that is produced as the result of reactions within an animal is called **metabolic water** (metabolism is a general term used to describe all the reactions of an animal or plant). Seeds contain relatively large amounts of fats, which are rich in hydrogen and thus are a good source of metabolic water.

The desert rat conserves its meagre supply of water in many ways. Although it lives on the hot, dry desert, where it would be expected that it would have to use large amounts of water for cooling purposes, it remains in burrows where it is relatively cool (60–75°F) and relatively humid (15–40% relative humidity) during the day, and it confines its activities to the night hours when the desert is cool. Whereas humans cannot excrete urine that contains more than 1.75% NaCl,* the desert rat can excrete urine that is 7% NaCl. It excretes other waste products as concentrated solutions, also. By these various adaptations, it is able to survive under conditions that provide little water. It maintains a nearly constant body temperature, and water always makes up 66% of its weight.

Many reptiles use some of the processes available to the desert rat, but they have an advantage (at least as far as desert life is concerned) in that they do not need to maintain a constant body temperature; it fluctuates with the surroundings. While it is tempting to assume that these animals learned how to get along with little water, it is much more likely that their

---

* This explains why people excrete additional water and dehydrate themselves when they drink sea water containing 3.5% NaCl.

inherited behavior patterns and inherited biochemical patterns have enabled them to live in areas where the competition for survival is less rigorous.

The camel survives desert conditions by various adaptations, some of which are different from those of the desert rat. The camel certainly produces metabolic water to supplement what it drinks. Its hump is largely fat, a good source of metabolic water. Camels also excrete a concentrated urine. The unusual adaptations of camels are that their body temperature can vary from 93° to 105°F without serious consequences, and they can lose up to 40% of their body water without dying. (Humans, on the contrary, become delirious when 10% of their body water is lost, and cannot survive a 20% loss of body water.) In addition, the camel can replenish its water supply in a short time. It can drink an amount of water equivalent to one third of its body weight in 10 minutes!

Sea birds face a problem not of the lack of water, but of the lack of fresh, unsalty water. They solve their problem by drinking salt water and then excreting a solution containing 5% NaCl through special salt glands in their beaks.

The animals discussed above represent exceptions to general rules about requirements for fresh water. The majority of animals need large amounts of the amazing substance we call water for the maintenance of life.

**9.7
WATER
POLLUTION**

Absolutely pure water is extremely rare. As soon as water is purified it starts dissolving small amounts of gases from the air. Even the purest spring water contains at least traces of dissolved ions of many metals, nonmetals, and ionic groups. Although such water might be considered "polluted", we generally use this word only when the water contains substances that make it unsuitable for drinking or swimming and other recreational uses, and that have harmful effects on fish, water plants, and other forms of life. The major types of pollutants are discussed in the following sections.

**Inorganic
Pollutants**

As water flows over and through rocks and soil it dissolves many substances. Streams and rivers continually flow into the oceans and seas of the world. As the water evaporates from these large bodies of water, the dissolved substances are left behind. Thus the oceans of the earth are getting saltier as they accumulate ions of sodium, potassium, calcium, magnesium and other cations, and the chlorides, sulfates, nitrates, phosphates, and other anions. Dried lake beds as well as the oceans are the sources of various inorganic compounds that are used commercially. Even in some rivers the total amount of dissolved ions is so great that the water cannot be used for irrigation. Nitrate ions, which accumulate in wells located near heavily fertilized fields, are toxic to animals. The phosphates, both from fertilizers and from some detergents, may combine with other ions to cause a great increase in the tiny green plants, the algae, in lakes and other bodies of water. This may seem to be a good thing, since algae produce oxygen and provide food for fish and other animals; but when there are more algae than can be consumed for food, they are decomposed by bacteria and other microorganisms, and this requires large amounts of oxygen. The overproduction of algae, which is called *eutrophication*, results in the uptake of so much oxygen that the fish die (see Fig. 9.6). Eutrophication has been a serious problem in many agricultural regions, particularly in the midwestern part of the United States. The problem is being solved by virtually eliminating phosphates from detergents and by being more careful in application of fertilizers.

Other inorganic substances, such as the metal ions lead, mercury, cadmium, and the anion cyanide, which is used in metal-plating processes, are highly toxic. The toxicity of

Fig. 9.6 Fish killed as a result of eutrophication. This process occurs when large amounts of nutrients, added to bodies of water stimulate the growth of algae. If the algae are not eaten by fish, but instead fall to the bottom of the body of water, their decay consumes large amounts of oxygen. When the water is deprived of oxygen, the fish die. [Photo courtesy of the Environmental Protection Agency.]

these metals, which are found not only in water but in the air and soil, are discussed in Section 13.13. If water contains too many small suspended particles of sand, clay, and other essentially nontoxic substances, it can also be considered to be polluted since some fish cannot tolerate muddy water and since growth of underwater plants is decreased because less sunlight penetrates this water. Some water supplies also contain radioactive substances, which must be considered dangerous pollutants.

Organic Pollutants Leakages from ships carrying petroleum have made us aware of the deadly effects of oil spills on birds and other forms of sea life. Scientists have insufficient evidence to say certainly whether the damage is due to the organic compounds present in petroleum, to oxidation products of these compounds, or perhaps to the detergents used to help disperse the oil. A few kinds of microorganisms can actually use petroleum as a food and they help clean up oil spills, but many more seem to be killed by oil.

Even more toxic than petroleum are the many organic compounds that are manufactured and eventually find their way into water. In most instances organic compounds are not very soluble; however, only a few parts per million may be sufficient to poison fish or shellfish. Particularly toxic are the polychlorinated biphenyls (PCB's), which are used industrially but are finding their way into our fresh-water supplies or even into the ocean and the Great Lakes. Insecticides are often found in water supplies following spraying, and these are toxic to fish and other forms of life as well as to insects. Many insecticides are not broken down readily and persist for long times in water supplies or in sediments at the bottom of lakes. A lesser problem has been that of detergents that are not broken down by microorganisms (we say they are not *biodegradable*). After the harm done by such detergents became evident the formulas were changed, and now almost all of the detergents that are used and eventually are found in sewage provide food for bacteria.

Many organic compounds that are present in wastes from food-processing plants and even from sewage serve as food and can promote growth of microorganisms and other forms of life. In the process, most of these nontoxic organic compounds are eventually oxidized. If plenty of oxygen is available, as in a shallow stream running and splashing over rocks, the organic compounds create little problem. On the other hand, the combination of a great amount of organic compounds and little oxygen results in eutrophication or sometimes merely in the presence of bad odors. For example, when sufficient oxygen is present, sulfur-containing compounds are converted to the odorless sulfates, but a lack of oxygen results in the production of the vile-smelling H_2S and other sulfides. Some of the products of both oxidizing and nonoxidizing (reducing) environments are given in Table 9.4. Many of the reduced compounds are produced by the bacteria we call anaerobic, those that do not use oxygen.

TABLE 9.4 Oxidized and reduced compounds or ions of common elements

Element	Oxidized forms	Reduced forms
carbon	CO_2	CH_4
sulfur	SO_2, SO_4^{2-}	H_2S
nitrogen	NO_2, NO_3^-	NH_3
phosphorus	P_4H_{10}, PO_4^{3-}	PH_3

If we analyze the amount of oxygen required to oxidize the organic compounds in any sample of water, we can express our results in terms of the **Biochemical Oxygen Demand** or the BOD. Water having a BOD of one part per million (ppm) is considered pure, that with a value of 5 is doubtful, and values over 20 ppm mean the water must be purified for most uses. Since water at 20°C can hold only 9 ppm of oxygen, and 5 ppm of oxygen is required for most animals living in warm water, a sewage effluent with a BOD of 1 000 (which is not rare) can cause serious problems unless the sewage is immediately diluted with large amounts of water containing oxygen, or is otherwise exposed to oxygen.

Other water pollutants that might be considered organic are the bacteria and viruses that can cause typhoid fever, dysentery, cholera, infectious hepatitis, and other diseases. Cases of such diseases are very often traced back to a polluted water supply. The inspection of wells and treatment of sewage or of drinking water with chlorine gas have greatly decreased the incidence of most water-borne diseases in the developed countries of the world. These diseases still affect many people and claim many lives in the less-developed countries. Purer water supplies would help control these serious diseases.

Heat Pollution Modern technology adds an appreciable amount of heat to natural water supplies. Although large amounts of heat obviously kill plants and animals, even moderate heating of rivers and streams may make the water less able to support life and thus pollute the water. The major problem arises from the fact that oxygen is less soluble in warmer water. The use of water from a river for cooling purposes in a factory or power plant can result in the elevation of the temperature by several degrees. This is especially true if the water source is a small stream. Regulations now require that some factories pass water through cooling towers before returning it to rivers or streams. Many nuclear reactors are built near the oceans and use sea water for cooling. There is little danger of warming the ocean significantly, but local changes in the types of ocean life are possible. It has been speculated that sharks, which prefer warmer water, may extend their territory if we warm coastal waters.

| 9.8 WATER PURIFICATION | Various methods are used to purify water; the method depends upon the extent of pollution, the nature of the pollutants, and the purposes for which the purified water is to be used. We require very pure water for making solutions of medications or nutrients that are to be injected into sick patients, and less pure water for studying chemical reactions or for filling a storage battery or a steam iron. Still lesser degrees of purity are required for water to be used for drinking, for filling a swimming pool, or for irrigation of crops. Many cities depend for their water supplies on streams that contain the sewage of cities upstream; in such cases, purification of sewage by the upstream cities and of drinking water by downstream cities is essential. Due to a variety of causes, even water from mountain streams is often polluted, and back-packers far from cities must be concerned about some method of water purification in many areas. Water is often boiled to remove pollutants, and although boiling is effective in destroying bacteria, most other pollutants are not changed by this process. It is not practical to purify water commercially by boiling. |

Settling, Filtering and Flocculation

The simplest method of purification of water to be used for drinking, and the first step in treatment of waste water, is merely letting the water stand or filtering it so that large particles or sediments settle out. If insoluble substances rise, they may be skimmed off. Such procedures are known as primary treatment when applied to sewage. Settling and filtration, particularly of water to be used for drinking, is sometimes aided by a process known as **flocculation**. In this process slaked lime, $Ca(OH)_2$, and aluminum sulfate, $Al_2(SO_4)_3$, are thoroughly mixed in the water to be purified. The reaction of these two substances produces the flocculating agent aluminum hydroxide, $Al(OH)_3$, which is a gelatinous solid that tends to entrap suspended particles as it settles to the bottom of a sedimentation tank.

$$3Ca(OH)_2 + Al_2(SO_4)_3 \rightarrow 2Al(OH)_3 + 3CaSO_4 \tag{9-10}$$

The calcium sulfate also tends to settle out, but is not as good a flocculating agent as aluminum hydroxide. Water that has been treated with flocculating agents, as well as that which has not, may also be filtered through sand or gravel. Further purification is effected if water is filtered through a bed of charcoal, which adsorbs many impurities.

Oxygenation

The oxidation of biological substances in water is accomplished in many ways. Water for drinking is often sprayed into the air, which not only adds oxygen but often results in the evaporation of substances that have undesirable odors or tastes. Some oxygenations involve the addition of ozone (O_3), which also serves to destroy bacteria. Sewage or other wastes with a high BOD generally have air bubbled through them and are passed through tanks known as *digesters*, where bacteria are added that aid in the process of oxidation. Addition of bacteria and air are generally known as secondary treatment of waste water. Most cities in the United States now use primary treatment and many use secondary treatment for their waste water. (See Fig. 9.7.)

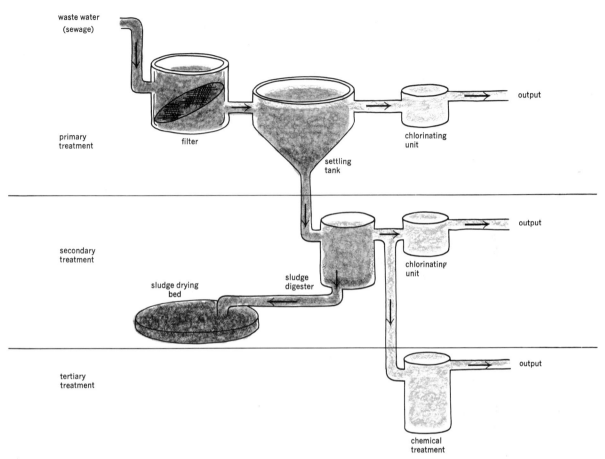

Fig. 9.7   Treatment of waste water. In primary treatment, waste water is filtered and allowed to settle to remove impurities. Sometimes there is more than one filtration or settling tank. Some primary treatment also includes chlorination. Secondary treatment consists in digesting the product of primary treatment to oxidize organic compounds. Water is usually chlorinated before being released from a secondary treatment plant. In tertiary treatment, the effluent from secondary treatment is subjected to various other processes, such as the addition of calcium salts to precipitate phosphates.

<table>
<tr><td>**Chlorination**</td><td>Both drinking water and sewage are often treated with chlorine gas to destroy bacteria. Chlorination must follow secondary treatment of sewage, and chlorinated water must be allowed to stand to allow excess chlorine to escape before it is returned to a river or other body of water; otherwise it will kill fish and other forms of life. Scientists are currently concerned about the chlorine-containing organic compounds found in the water of New Orleans and other communities that reuse river water. They are concerned that these compounds, which have been shown to cause cancer in some animals, may be produced during chlorination.</td></tr>
<tr><td>**Removal of Soluble Substances**</td><td>Neither settling, filtration, nor oxidation removes soluble substances such as inorganic ions from water. For many purposes these dissolved inorganic ions present no problem and are allowed to remain in water. However, they can cause eutrophication, they may react</td></tr>
</table>

with soaps to produce a scum, or they may form mineral deposits in pipes and boilers known as *scale*, particularly when heated. Some ions interfere with chemical reactions, including those in lead storage batteries or the reactions of living systems. We can remove ions in a variety of ways.

Water that has passed through soil dissolves soluble substances. For this reason, sodium, potassium, calcium, and magnesium ions, as well as a variety of negative ions, are found not only in the oceans but in many water sources. Water containing less than 50 parts per million (ppm) of dissolved solids is considered to be soft water, and samples containing more than 50 ppm are considered hard. It is not unusual to find water with 500–1 000 ppm of dissolved solids, particularly in some areas of the southwestern United States. On the other hand, mountain streams that have run only over hard, insoluble rocks such as granite may have only 10–20 ppm dissolved solids. If the hydrogen carbonate (or bicarbonate) ion, HCO_3^-, is present in waters containing magnesium and calcium ions, boiling the water results in the formation of insoluble carbonates of magnesium and calcium.

$$Mg^{2+} + 2HCO_3^- \rightarrow MgCO_3\downarrow + H_2O + CO_2 \tag{9-11}$$

$$Ca^{2+} + 2HCO_3^- \rightarrow CaCO_3\downarrow + H_2O + CO_2 \tag{9-12}$$

Such bicarbonate-containing waters are often said to be only temporarily hard. However, even this type of hardness may cause much inconvenience and damage, as when precipitates of $MgCO_3$ and $CaCO_3$ form a hard crust inside teakettles or plug up hot-water pipes and water heaters. Removal of dissolved solids by precipitation or other methods is referred to as softening the water.

Another method of removing the calcium and magnesium ions is by adding a substance that will form a precipitate with these ions. Sodium carbonate and phosphate are used as softeners because they form precipitates with ions that cause hardness. If the water contains sufficient amounts of the dissolved ions, the amount of precipitate may be so large that it must be removed by filtration. The softening of water by addition of carbonate and phosphate ions is shown below.

$$Ca^{2+} + CO_3^{2-} \rightarrow CaCO_3 \tag{9-13}$$

$$3Mg^{2+} + 2PO_4^{3-} \rightarrow Mg_3(PO_4)_2 \tag{9-14}$$

Since the magnesium and calcium ions, and to some extent the iron ions, found in water are responsible for most of the undesirable reactions of hard water, we usually consider water to be soft when they are removed even if sodium and potassium ions remain. However, large concentrations of even these ions will make a water hard, and truly purifying water requires the removal of all ions.

Water
Purification
by Ion Exchange

There are many substances, both natural and synthetic, that will absorb certain ions and release others. We call these **ion-exchanging substances**. One group of such substances that is widely used for softening water is the hydrated sodium–aluminum silicates, a group commonly known as *zeolites*. The zeolite crystal has a large, regular pore structure that can absorb certain types of ions and holds some more firmly than others. When zeolite that contains sodium ions is exposed to water that contains calcium and magnesium ions, the sodium is released and the calcium and magnesium ions are absorbed. Using the word zeolite for the complex aluminum silicates, we can write the following equation:

$$2Na \cdot zeolite + Ca^{2+} \rightarrow Ca(zeolite)_2 + 2Na^+ \tag{9-15}$$

Thus if we pass hard water through a bed of the sodium form of zeolite, the water is softened. However, the zeolite eventually becomes saturated with calcium and/or magnesium ions and no longer removes these ions from hard water. When this happens, we can reverse the process by adding a large excess of NaCl to reconvert the zeolite to its sodium form in a process we call regeneration.

$$2Na^+ + 2Cl^- + Ca(zeolite)_2 \rightarrow 2Na \cdot zeolite + Ca^{2+} + 2Cl^- \tag{9-16}$$

The calcium and magnesium ions as well as excess NaCl are flushed away and the water softener is again effective.

Looking at the above equations, it is obvious that removing calcium and magnesium ions increases the number of sodium ions in the softened water. Many people with heart conditions must severely restrict their intake of sodium ions and these people should not drink water softened by this type of ion exchange.

Other ion-exchanging substances (often called *resins*) will exchange either cations or anions, and we can use a combination of two resins to completely "deionize" water. One resin is a cation exchanger, which removes any positive ions and replaces them with hydrogen ions.

$$Mg^{2+} + 2H \cdot cation\ exchanger \rightarrow 2H^+ + Mg \cdot cation\ exchanger \tag{9-17}$$

A second resin releases OH^- ions in exchange for negative ions (anions). When water is passed through both exchangers, it contains H^+ and OH^- ions, which react to form water molecules, and thus we have effectively removed all ions from the water, adding back only water molecules. If the two resins are kept separated, they can be regenerated with substances containing hydrogen and hydroxide ions, respectively. For many purposes, the resins are mixed together in a resin-filled cartridge that will purify water; however, these resin mixtures cannot be regenerated. Such cartridges are sold for purifying water for steam irons, for purifying drinking water, and for use in laboratories when a relatively small amount of very pure water is required. Deionized water is relatively inexpensive.

Medications that are to be given orally (by mouth) are often dissolved in deionized water. However, if a medication is to be injected, the water used to dissolve or dilute it must be purified by distillation. This is because deionization does not remove bacteria and if deionized water is heated to sterilize it, the bacteria that are present decompose to produce substances that cause fever and other harmful effects.

Distillation Water as well as many other liquids can be purified by distillation. In this process the liquid is heated to the boiling point and the vapor transferred to another part of the distillation apparatus, where it condenses (see Fig. 9.8). Not only does the distillation of water separate it from most impurities, but the boiling also destroys and separates any bacteria present. Distillation will not separate substances having similar boiling points.

Ordinary distilled water may contain traces of metal ions or other impurities if it has been distilled from metal stills. Such ions must be removed (or the water must be

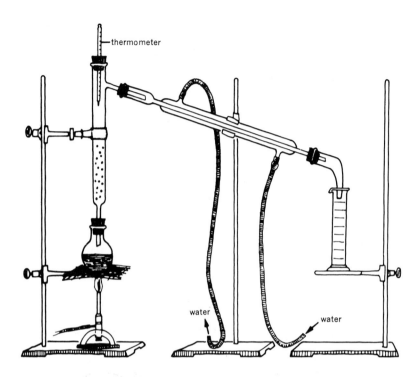

Fig. 9.8   A laboratory distillation apparatus.   The substance to be distilled is heated in a flask. Vapors from the heated flask are condensed by the cooling action of water circulating through the outer jacket of a condenser and are collected in a graduated cylinder or other container.   The thermometer shows the boiling point of the substance being distilled.

distilled from glass vessels) for some procedures, such as the growth of cultures of single mammalian cells. It is reported that much of the illegally distilled alcohol produced in the United States contains appreciable quantities of the poisonous metal lead, which comes from the automobile radiators used as condensers in these operations.

9.9
DESALINATION
OF SEA WATER

Many areas of the world suffer from a shortage of water pure enough to be used for drinking or irrigation. Many of these desert regions are located near oceans, and the possibility of desalination of sea water (removing its salt) presents great hope for increased water supply and food production. We know several methods for purifying sea water. The major problem is that they are all expensive, either in terms of materials or in terms of the amount of energy required.

Distillation is the simplest method for purifying sea water. Ships have used this process to supplement the supplies of fresh water they normally carry. The difficulty with distillation is that, due to the high heat of vaporization of water, large amounts of energy are required to boil it. There is no way to decrease the energy input required for distillation. For some desert countries, those with a plentiful supply of petroleum, the energy required to desalinate water by distillation is not a drawback. For other countries, it is possible that solar-powered distillation plants can be developed. These would use the sun, a cheap and readily available energy source.

Ion exchange can be used to desalinate sea water, but here there is a high cost both for the resins and for the chemical substances required to recharge the resins. Energy requirements, however, are not great. Other processes for desalination include *electrodialysis* and *reverse osmosis*. The basis for these processes is discussed in Sections 10.13 and 10.14.

At the present time all methods of desalination of sea water for irrigation purposes are too expensive to be commercially feasible. Research and development of new technology may change the situation, particularly if solar energy can be used or if atomic fusion is developed and produces relatively inexpensive energy supplies. One additional problem with desalination is the disposal of the large amounts of the various salts that will be produced.

9.10 **SOLUTIONS TO** **THE PROBLEM** **OF WATER** **POLLUTION**	As with other forms of pollution, probably the best method of control of water pollution is to keep pollutants from entering water supplies. Industries that produce toxic substances should remove them before water used in a factory or power plant is returned to a lake or stream. Farmers must prevent excessive runoff of fertilizers and ordinary citizens must become more conscious of their actions that contribute to pollution. The removal of phosphates and nonbiodegradable substances from detergents, largely because of citizen concern and government regulation, represents one victory in the war against pollution.

In many instances, particularly the removal of biological wastes, the solution lies in treatment of the wastes before releasing the used water. (See Fig. 9.7.) It is estimated that only 33% of municipal waste-treatment plants in the United States were meeting standards for secondary treatment of waste water as of July 1977. Increasing the number and efficiency of treatment plants should have a high national priority. As with all processes, water treatment is expensive and becomes more so as we remove more and more substances. Some communities have spent the funds necessary to initiate tertiary treatment of waste water to remove particularly the soluble impurities such as phosphate ions. These ions are generally removed by precipitation. Tertiary treatment is used at Lake Tahoe on the California–Nevada border, where the primary purpose of tertiary treatment is to prevent eutrophication of clear, blue Lake Tahoe.

The primary cause of water pollution is that we are discharging too many substances into limited amounts of water. Even if we eliminate the truly toxic substances such as the cyanides, pesticides, and some metals, the load of relatively nontoxic substances is too great for natural processes to purify the water. As with air pollution, our society must consider whether we need all of the pollution-producing things we now have and it must be willing to allot the money for removing those wastes, such as wastes from food processing and human body processes, which are inevitable. Progress is being made in cleaning up water pollution in the United States, and fish and other forms of life are returning to many streams and lakes that were formerly highly polluted. Greater awareness and willingness to take the necessary steps will make still more bodies of water pleasant places for people to swim in and otherwise enjoy instead of being streams of pollution.

SUMMARY

This chapter has been devoted to that most remarkable compound, water. We have described and discussed its occurrence, properties, uses, and significance. We have shown how the polar nature of water molecules and the hydrogen bonds between molecules are responsible for many of the properties that make water one of the most important substances for life on the planet earth.

The types of pollutants found in the waters of rivers, streams, lakes, and oceans were discussed and, on the basis of the types of pollution present and the uses intended for water, appropriate methods of water purification have been presented. As with other forms of pollution, an understanding of the nature of the pollution often suggests methods for the chemical and physical processes necessary for purification. As usual, both the processes and the energy required for them prove to be important.

In the chapters that follow, the properties of water solutions and some of the aspects of the ionization of water will be discussed. Reactions involving water will be discussed throughout the book, since it is a reactant or product in many of the reactions of organic chemistry and biochemistry.

IMPORTANT TERMS

anhydrous
Biochemical Oxygen Demand (BOD)
deliquescence
desalination
desiccant
distillation

efflorescence
eutrophication
flocculation
hard water
humectant
hydrate

hydrogen bond
hygroscopic
ion exchange
metabolic water
zeolite

WORK EXERCISES

1. Why do inland bodies of water, typically the Great Salt Lake and the Great Lakes, differ so much in mineral content?

2. Illustrate the structure of a typical unit of water as it occurs in the liquid state.

3. Most liquids decrease in volume when changing to a solid. How do you account for the fact that ice does not follow this pattern?

4. What is the explanation for the comparatively high heats of fusion and vaporization of water? Might the same explanation apply to the specific heat of water?

5. While the density of ice is less than one, because of its crystal structure, it also contains bubbles of gases from the air that further decrease its density. Why are these gases found in ice? Do you think the top millimeter of ice will contain more or fewer bubbles than ice that is 30 m below the surface?

6. Complete and balance the following equations.

   a) $Ca(HCO_3)_2 \xrightarrow{\Delta}$

   b) $CaCl_2 + Na_2CO_3 \rightarrow$

   c) $MgSO_4 + Na_3PO_4 \rightarrow$

   d) $Ca(HCO_3)_2 + Na_3PO_4 \rightarrow$

   e) $2Na \cdot zeolite + Ca^{2+} \rightarrow$

   f) $Ca(OH)_2 + Al_2(SO_4)_3 \rightarrow$

7. How do the small animals that live in desert areas survive on so little water?

8. How is metabolic water produced in the animal body?

9. One of the most important ores from which borax, $Na_2B_4O_7 \cdot 10H_2O$, is obtained is *kernite*, $Na_2B_4O_7 \cdot 4H_2O$.

   a) How much borax can be produced from one ton of the pure kernite ore?

   b) What is the percent of water in kernite?

   c) What is the percent of water in borax?

10. How many grams of metabolic water are produced when 50 g of a fat with a formula of $C_{57}H_{110}O_6$ are oxidized in the body? How many grams of carbon dioxide are produced in the above?

11. What is the percent of water in each of the following hydrates?

    a) $CaSO_4 \cdot 2H_2O$
    b) $Na_2CO_3 \cdot 10H_2O$
    c) $MgSO_4 \cdot 7H_2O$
    d) $BaCl_2 \cdot 2H_2O$
    e) $Al_2O_3 \cdot 2SiO_2 \cdot 2H_2O$
    f) $CuCl_2 \cdot 2H_2O$
    g) $MnSO_4 \cdot 3H_2O$
    h) $FeSO_4 \cdot 5H_2O$
    i) $LiNO_3 \cdot 3H_2O$
    j) $KAl(SO_4)_2 \cdot 12H_2O$

12. List some metallic ions that would be considered water pollutants; list other metal ions that would not normally be considered pollutants.

13. List the compounds in which we find the elements carbon, sulfur, and nitrogen in oxidized and reduced forms.

14. Why does the addition of warm water to a river exert harmful effects on fish?

15. Tell how you would remove (or destroy) the following substances found in water.
   a) calcium ions (2 methods)
   b) phosphate ions
   c) bacteria (2 methods)
   d) petroleum
   e) sodium ions
   f) sodium chloride
   g) dissolved $Ca(HCO_3)_2$

SUGGESTED READING

"Environmental Quality—1977," Stock No. 041-011-00035-1. A government publication discussing air pollution, water pollution, energy and natural resources of the United States. New reports are issued each year. An authoritative report, filled with facts and figures.

Penman, H. L., "The water cycle," *Scientific American*, September 1970, p. 98 (Offprint #1191). The way water molecules are recycled is discussed in detail.

Stoker, H. S., and S. L. Seager, *Environmental Chemistry: Air and Water Pollution*, 2nd Ed., Glenview, Ill., and London: Scott, Foreman and Co., 1976. This is an excellent paperback on the topics listed in the title. It gives the facts necessary for conclusions and generalizations in this area.

10

Solutions and Suspensions

Life exists only in the presence of water. It is believed that life originated in an ancient ocean, and although some forms of life were able to escape the necessity of being continuously bathed in water, individual cells of these organisms still require a watery environment. Why is water so important to living things? We could perform an experiment that would show one important reason. If we mix some sodium hydrogen carbonate with some crystals of salicylic acid in a dry test tube, nothing happens. If we then add water to the dry mixture, we immediately see evidence of a reaction in the form of bubbles, which fill and perhaps overflow from the tube. What has the water done so that the two substances react? It has dissolved the two solids, allowing the molecules and ions of the two substances to come close enough to react with each other. Most of you have probably observed this type of reaction when you dropped a seltzer tablet into a glass of water. The mixture of substances (usually a carbonate and a dry acid plus a medication or sugar and flavoring) dissolved in the water to produce carbon dioxide gas in a reaction we describe as "fizzing." The many chemical reactions that occur in living plants and animals require that the reactants be dissolved or at least suspended in water so that particles can come close to each other and react. In addition, solutions are used to transport important substances from one place to another within the organism.

Any time we dissolve one substance in another we make a *solution*. While we commonly think of solutions as solids dissolved in liquids, if we extend our thinking, we can understand that the air we breathe is a mixture of several gases, primarily molecules of nitrogen, oxygen, argon, and carbon dioxide, and often contains other substances we classify as pollutants. Air is generally classified as a solution. Mixtures of metals such as bronze, stainless steel, and solder, which are called alloys, are also solutions although the components are solids. Salt dissolved in water, gases in air, and metals in alloys are very different, yet all are called solutions. How can we define the word *solution* so that it includes all of these examples? The definition generally given is that *a* **solution** *is a homogeneous mixture of two or more kinds of atoms, molecules, or ions.* Some examples of kinds of solutions are given in Table 10.1.

TABLE 10.1 Example of solutions

Type	Example
solid in liquid	syrup; sugar in water*
liquid in liquid	alcoholic beverages; alcohol in water
gas in liquid	carbonated beverages; carbon dioxide in water
solid in gas	air near moth preventives; naphthalene or paradichlorobenzene in air
liquid in gas	water vapor; water in air
gas in gas	the atmosphere; oxygen and other gases in nitrogen and each other
solid in solid	metallic alloys; one metal in another
liquid in solid	skin creams; perfumes and other liquids in a fat

* The solute is listed first and the solvent second.

If sugar crystals are stirred into water, the sugar is dissolved by the water and a **solution** is formed. Water is the **solvent** and sugar is the **solute**. In any solution involving a solid and a liquid, we call the solid the solute and the liquid the solvent. If both components of a solution are liquids, as in an alcohol–water solution, we say that the component present in the greater amount is the solvent. This, of course, creates a problem when both components are present in equal amounts. In general, the rule for distinguishing between solvent and solute is somewhat arbitrary and should be considered a guideline rather than an absolute criterion. When we talk of the components of complex solutions such as the air and alloys, we are not concerned with distinguishing between solutes and solvents.

Since the solute particles may be considered to be scattered or dispersed in a solution, sometimes we say that *a solvent is a dispersing medium or phase* and that *the solute is the dispersed phase*. Using these terms *a solution is a permanent dispersion of one substance in another*. The term dispersion is also used in describing mixtures such as oil and water, which are not solutions but only temporary dispersions. In such a case, the dispersed phase consists of particles that are larger than molecular size or may contain really giant molecules (macromolecules), such as the protein molecules of casein that are dispersed in water in the relatively permanent suspension we call milk.

In a solution the solute and solvent do not separate. If we dissolve sugar in tea, the sugar stays dissolved. If we were to pour sweetened tea through a filter paper, the filtrate would still taste sweet and we could conclude that the sugar and tea had not been separated by filtration. *This is true of all solutions—they are not separated by filtration*. This is logical since the solute particles are the size of atoms, ions, or small molecules. Such small particles cannot be trapped by filter paper. However, the larger particles that are usually found in tea and coffee can be trapped, since they are not true solutes but only suspended solids. Actually these large particles would eventually settle out and form a sediment. The filter paper has only hastened the separation.

If we dissolve some crystals of the blue compound copper sulfate, we obtain a beautiful clear blue liquid. If we filter this, the blue color goes through the filter paper. Since the blue color will not separate from the water on standing, we can say we have a solution of copper sulfate. Thus, *solutions do not need to be colorless, but they must be clear*. This, again, is logical, since any suspended particles that are sufficiently large to make the solution look cloudy are certainly too large to be dissolved. In some cases, a dispersion that appears clear will show the path of a beam of light. If so, it is not a true solution but is a colloidal dispersion (see Section 10.9).

If two liquids form a homogeneous mixture or solution, they are said to be **miscible**. For example, glycerine and water are miscible; any proportion of the two components will give a clear solution that does not separate upon standing. On the other hand, if we try to mix water and salad oil, we find that they form only a temporary suspension, which soon separates into two liquid layers; we say the two liquids are **immiscible**. In some instances the substances are partially miscible (that is, partly soluble). We can dissolve a very small amount of ether in water and a small amount of water in ether, but generally we say the two are immiscible.

10.2 FACTORS AFFECTING SOLUBILITY   *Solubility is generally given in terms of the amount (in grams) of a solute that can be dissolved by 100 grams of the solvent at a given temperature.* Many factors influence the solubility. As will be illustrated in the following, common experience is a good guide in making generalizations about solubilities.

Solids in liquids   From the observation that more sugar can be dissolved in hot water than in an equal volume of cold water, we could conclude that *the solubility of solids in liquids increases as the temperature of the solvent is increased.* This is generally true, but there are exceptions. Sodium chloride is only slightly more soluble in hot water, and a few substances are *less* soluble in hot solvents. Increasing or decreasing the pressure above a solution does not materially affect the solubility of solids in liquids.

Gases in liquids   Most of us have opened warm cans or bottles of carbonated beverages and know that the contents are likely to spurt out violently. When we compare this to what happens when we open a cold container of the same beverage, we can conclude that the gas (CO_2) is more soluble in a cold solvent. From this we can generalize further that *the solubility of gases in liquids decreases with an increase in temperature.* Observation of a variety of gases in a variety of liquids would show that this generalization is almost always true. If we went on to test the effect of pressure on solubility of gases in liquids, we would discover that pressure is just as important as temperature—but that it operates in the opposite manner. *An increase in pressure tends to increase the solubility of gases in liquids.* In preparing solutions of gases we generally force the gas into solution under pressure and then seal the container with a lid of some sort to maintain that pressure and thus keep the gas in solution.

Another example of the effect of temperature on the solubility of gases in liquids is provided by the observation that when water is heated, bubbles of dissolved gases form on the side of a container long before boiling occurs. Divers get the "bends" when gases (primarily nitrogen) that have dissolved in the blood under increased pressure suddenly come out of solution. This causes pain and may cause death if the bubbles of gas block normal circulation of the blood.

Nature of Solvent and Solute   So far we have been discussing water as a solvent, and indeed it is the most common solvent. However, if you have attempted to remove a grease spot from your clothing, you are aware that water is ineffective. To dissolve grease we must use solvents such as carbon tetrachloride. On the other hand, for stains that are water-soluble, carbon tetrachloride is not a good solvent. We find that water, which is a polar molecule (one that has a slight separation of electrical charges—see Section 6.5), is a good solvent for those molecules that consist of charged particles (ions) or that consist of partly charged particles, that is, polar covalent molecules. Oils and greases, which are nonpolar molecules, are soluble in the nonpolar solvents such as carbon tetrachloride. In order for solids to dissolve, the solvent molecules must exert an attraction on a molecule or other small particle of the solid that is greater than its attraction for other solid particles.

We sometimes say "*like dissolves like*" in talking about solutes and solvents. A slightly more precise statement is that *polar solvents dissolve polar solutes and nonpolar solvents dissolve nonpolar solutes.*

It is interesting to observe that people we know as "dry cleaners" actually use liquid solvents, usually of the nonpolar variety, for cleaning clothes. They also need to have a great deal of knowledge about which solvents will dissolve or remove various kinds of stains and spots.

10.3 FACTORS AFFECTING RATES OF SOLUTION

When you add sugar to coffee, tea, or lemonade, you generally stir the mixture. Why do you do this? Obviously to hasten the process of solution. If you do not stir, the sugar settles to the bottom of the cup and is soon surrounded with a layer of solution that contains dissolved sugar. *Stirring (agitation) brings the undissolved solute into contact with a different portion of solvent which contains much less dissolved material. This helps the process of solution.* In the laboratory you will often place a stopper in a test tube and shake the contents to speed up the process of dissolving a solid in a liquid, or to help in determining whether or not one liquid is soluble in another.

If you compare the rate of solution of one gram of coarse salt crystals with that of one gram of very fine salt crystals, you will observe that the fine crystals dissolve first. This leads to the generalization that particle size is important in determining the rate of solution. *Small particles, which have a greater surface area per unit of weight, dissolve faster than large particles.*

Another factor that influences the rate of solution as well as the total solubility is temperature. *The rate of solution is generally increased by heating.* In chemistry service centers where technicians must prepare large volumes of many different kinds of solutions, they usually start with a finely powdered solute (grinding it first if necessary) and prepare the solution by heating the solvent and stirring the mixture with a power-driven stirrer. This is much more efficient than taking time to shake large crystals in a small container of cold solute.

10.4 CONCENTRATION OF SOLUTIONS

If we dissolve a small amount of solute in a solvent we have a **dilute solution**. If we add more solute, we produce a **concentrated solution**. The terms *dilute* and *concentrated* are relative terms, and we cannot define precisely when a solution changes from dilute to a concentrated solution.

If we add solute to a solution until no more can dissolve, we produce a **saturated** solution. This means that, although we can add more solute, no greater amount of it will go into solution. If we were to add some radioactive sodium chloride to a saturated solution of sodium chloride in water, we would find that some of the radioactive particles would go into solution, but for every particle that goes into solution, another particle would settle out of solution, so that the total amount dissolved at any moment remains the same. We cannot force a greater amount of salt into a saturated solution. Since some solutes are not very soluble in some solvents, a saturated solution may be a relatively dilute solution.

There is a way, of course, to get more solute to dissolve in a saturated solution, and that is to heat it. This is particularly true of certain compounds such as the salt sodium acetate. Actually, when we heat solvents or solutions we are merely changing conditions (temperature) and thus increasing the amount of compound necessary to saturate the solution. If the temperature of a hot saturated solution of sodium acetate is decreased, normally we see the sodium acetate settling out of solution, since the amount of solute required to saturate the solution at a lower temperature is less than that required at the higher temperature.

If we carefully cool a saturated solution, sometimes we find that no solute is settling out or precipitating, even though this should happen because of a decrease in solubility. Such a solution, one that has more solute than it can normally hold at a given temperature, is called a **supersaturated solution**. This is an unstable condition, and if a single crystal of the solute is added to a supersaturated solution, crystallization begins immediately. With highly supersaturated solutions of sodium acetate, addition of a single crystal often causes a striking transformation, as almost all of the solution in the test tube instantly turns to a solid. The transformation is accompanied by a release of heat—which is logical, since heat was required to produce a supersaturated solution, and the heat is released when the solute precipitates.

Clouds are often supersaturated with water. Adding ice crystals or even crystals of silver iodide to the cloud or suddenly chilling it by adding small particles of dry ice can cause the moisture to form drops of rain or snow, which then fall as precipitation. As with the supersaturated solution of sodium acetate, the temperature often rises when precipitation starts to fall from supersaturated air.

**10.5
USES OF
SOLUTIONS**

We have already indicated that it is necessary to have substances in solution for most reactions to take place. Solutions have many other uses in chemistry. Many reactions or processes require a definite amount of a substance, and it is generally easier to measure the volume of a solution than it is to weigh a given amount each time we want to make an addition. Canneries add sugar to fruits to sweeten them and in some cases to act as a preservative. It is usually easier to add a measured volume of a sugar solution (sugar syrup) to each batch of fruit than to weigh out the desired amount of solid sugar. Addition of a solution also means that the solute is already dissolved and reactions can begin immediately. We use solutions of glucose (dextrose) for intravenous injection when patients are so ill they cannot or should not eat solid food. Having the glucose in solution means it can circulate and be metabolized rapidly.

When we use solutions for specific purposes and especially when we use them in processes involving chemical reactions, we need to indicate the concentration of the solutions. The terms dilute, concentrated, and saturated are generally not precise enough to specify concentration: we need to use other terms such as percentage composition and molarity. The following sections describe these types of solutions. *Unless another solvent is specified, it is assumed that all solutions are made using water as the solvent.* Such solutions are sometimes referred to as *aqueous solutions.*

**10.6
CONCENTRA-
TION OF
SOLUTIONS
EXPRESSED BY
PERCENTAGE**

A common method for indicating the concentration of a solute in a solution is by using percentage to indicate the parts of solute per hundred parts of a solution. Both the *amount of solute* and that of the *total solution* must be measured. This can be done either by *weighing* a solid or liquid solute on a balance or by *measuring the volume* of a liquid (solute or solvent) in a graduated cylinder or a volumetric flask. In discussing such solutions we should always specify which type of measurement has been used, indicating first the method of measuring the amount of solute and second the method used for measuring the 100 parts of the solution. Thus we have weight:volume (w/v), weight: weight (w/w), and volume:volume (v/v) solutions. The following examples illustrate differences in preparing these three types of solutions.

Solutions in which the solute is weighed and the amount of solution measured in terms of volume are probably the most commonly used in chemistry. To make a 5% solution (w/v), we weigh 5 g of solute, place it in a flask or cylinder with a mark indicating 100 ml and add solvent. We first add sufficient solvent to dissolve the solute and then add solvent up to the 100-ml mark. This gives us 100 ml of a solution containing 5 g of solute; thus each milliliter of the solution contains 0.05 g of solute. Solutions used for intravenous injections and most solutions commonly used for medical purposes are weight/volume solutions.

Example   How much glucose is needed to make a liter of 10% glucose (w/v) solution?

$$\frac{10 \text{ g glucose}}{100 \text{ ml solution}} \times 1\ 000 \text{ ml solution} = 100 \text{ g of glucose}$$

In solving the problem we have taken the ratio of solute to solvent and multiplied by the total volume of the solution.

Volume:volume solutions are commonly used when the solute is a liquid. To make a 70% solution (v/v) of ethyl alcohol we measure 70 ml of ethyl alcohol and add it to a 100-ml volumetric flask or graduated cylinder. We then add water to the 100-ml mark. If we start with only 95% ethyl alcohol (which is the most common form) we must multiply by the factor 100/95 to find out how much 95% alcohol to start with.

Example   Tell how you would prepare 500 ml of a 50% solution (v/v) of ethyl alcohol starting with a solution that is 95% ethyl alcohol.

$$\frac{50 \text{ ml alcohol}}{100 \text{ ml alcohol}} \times \frac{100\% \text{ alcohol}}{95\% \text{ alcohol}} \times 500 \text{ ml solution} = 263 \text{ ml}$$

Our answer is 263 ml of 95% alcohol. This is dissolved to make 500 ml of solution.

While it is easier to measure a volume than to weigh a substance, determinations of volumes are, as a rule, less accurate. Further error is introduced when liquids expand with increases in temperature. Thus a change in temperature will cause a change in concentration. While these changes are not significant for most purposes, for many very accurate measurements the solvent is weighed and we have weight:weight (w/w) solutions. Such solutions are used for some chemical measurements.

Thus to make a kilogram of 10% sodium chloride solution (w/w) we would weigh 100 g of sodium chloride and 900 g of water and then mix the components.

Example   Describe the preparation of 50.0 g of a 2.50% solution (w/w) of glucose in water.

$$\frac{2.50 \text{ g glucose}}{100 \text{ g solution}} \times 50.0 \text{ g solution} = 1.25 \text{ g glucose}$$

We can find the weight of water by subtracting 1.25 from 50.0 to get 48.75 g of water. To prepare the solution we would mix 1.25 g of glucose and 48.75 g of water.

In addition to preparation of solutions, calculations are sometimes needed to determine other values. Some examples follow.

Example 1   A chemical reaction calls for 1.2 g of sodium chloride. How many milliliters of a solution that is 5% NaCl will be required? We know the grams required and need a factor involving the relationship of grams of salt to volume of solution. Thus our calculation becomes:

$$1.2 \text{ g NaCl} \times \frac{100 \text{ ml solution}}{5 \text{ g NaCl}} = 24 \text{ ml of solution}$$

Example 2   Glucose provides about 4.0 kcal of energy per gram. How many milliliters of a 10% solution of glucose should be administered to provide 1 250 kcal of energy?

$$1 250 \text{ kcal} \times \frac{1 \text{ g glucose}}{4 \text{ kcal}} \times \frac{100 \text{ ml solution}}{10 \text{ g glucose}} = 3 125 \text{ ml}$$

Example 3   When 250 g of a solution of KCl was evaporated to dryness, the residue (KCl) was found to weigh 15.0 g. What was the percentage of KCl in the solution?

$$\frac{15 \text{ g KCl}}{250 \text{ g solution}} \times 100 \text{ g solution} = 6 \text{ g KCl}$$

and 6 g KCl per 100 grams of solution is a 6% solution.

Example 4   How many grams of KNO_3 are present in 40 g of a 15% solution?

$$40 \text{ g solution} \times \frac{15 \text{ g KNO}_3}{100 \text{ g solution}} = 6.0 \text{ g KCl}$$

When we make a solution, the density is usually greater than that of the solvent alone. We can often determine the concentration of a substance in solution by measuring the density and looking in a chemical handbook for a table relating density to concentration. It is relatively easy to determine densities by weighing and measuring volumes or by the use of a hydrometer (see Fig. 3.4). These methods of finding the concentration are simpler than evaporating the solvent. Measuring densities may also be used to determine concentrations of volatile solutes such as ethanol or ammonia. In the latter two cases the densities of the solutes are less than that of water and the density of the solutions is less than 1.0 g/cm³. The density or specific gravity of a wine is often used to determine its alcohol content, especially during the fermentation process.

Example   A 60.0-ml sample of a solution of NaOH weighs 67.8 g. (a) What is its density, (b) what is the percentage composition of NaOH in the sample, and (c) how many grams of NaOH are there in the sample?

(a) Density $= \dfrac{67.8 \text{ g}}{60.0 \text{ ml}} = 1.13$ g/ml

(b) By consulting a handbook of chemistry we find that a solution of NaOH having a density of 1 13 g/ml contains 12% NaOH.

(c) $67.8 \text{ g solution} \times \dfrac{12 \text{ g NaOH}}{100 \text{ g solution}} = 8.14 \text{ g NaOH}$

As instruments become capable of measuring very small amounts of substances, it it sometimes convenient to express concentrations in parts per million (ppm). These are weight:weight solutions. The procedures in making such solutions are similar to percentage except that the number 1 000 000 (10⁶) is used instead of 100.

Example 1   Describe how you would make 3.0 liters of a solution containing 3.5 ppm of substance Z.

3.5 parts per million means 3.5 g per million (10^6) grams of solution. Since such a dilute solution would have a density very near that of water we could assume that 1 ml of solution is equal to 1 g of solution.

$$\frac{3.5 \text{ g Z}}{10^6 \text{ g solution}} \times 3 \times 10^3 \text{ ml solution} \times \frac{1 \text{ g solution}}{1 \text{ ml solution}} = 10.5 \times 10^{-3} \text{ g Z}$$

or 10.5 mg of the substance Z would be mixed and dissolved in a total of 3 liters of solution.

Example 2   If the tolerance level of an insecticide on lettuce is established as being 5 ppm, does a 100-g sample of lettuce containing 0.75 mg of insecticide exceed the tolerance level?

$$\frac{0.75 \times 10^{-3} \text{ g}}{10^2 \text{ g}} \times 10^6 = 0.75 \times 10^1 = 7.5 \text{ ppm}$$

The sample exceeds the tolerance.

10.7
MOLAR
SOLUTIONS

While specifying concentrations in terms of percentage is sufficient for many purposes, chemists use other units when dealing with chemical reactions. For example, if we add a 5% solution of HCl to an equal volume of 5% KOH, the acid and base will react but at the end of the reaction there will still be some acid that has not reacted. HCl and KOH do not react with each other in terms of an equal number of grams but instead in terms of formula weights. From the equation

HCl + KOH → KCl + HOH

we know that one formula weight of HCl (36.5 units) reacts with one formula weight of KOH (56.1 units). The unit, the mole, has previously been defined as being equal to the formula (molecular, atomic) weight of a substance in grams. It is in terms of this unit—the mole—that we express the concentrations of solutions that are used for chemical reactions. *A solution that contains one mole of a solute per liter of solution is called a 1 molar (1 M) solution.* In the SI, the unit of volume equal to the liter is the cubic decimeter (dm^3) and this unit of volume is now used in some cases.

The molecular weight of common table sugar, sucrose, is 342 amu. If we place 342 g of sucrose in a flask and add water to the 1 liter (1 dm^3) mark, we have a 1 M solution of sucrose. If we dissolve 684 (2 × 342) grams of sucrose in a liter of water, we have a 2 M solution. A solution that contains 34.2 g of sucrose per liter is a 0.1 M solution. In terms of the categories we used for percentage solutions, molar solutions are weight: volume solutions.

The following examples illustrate the use of molarity in expressing concentrations of solutions.

Preparing solutions of given molarity

Example 1   Describe the preparation of a liter of a 0.20 M solution of calcium hydroxide.

First we determine the molecular weight of $Ca(OH)_2$.

$$
\begin{aligned}
1 \text{ Ca} &= 40 \\
2 \text{ O} &= 32 \\
2 \text{ H} &= 2 \\
\hline
Ca(OH)_2 &= 74
\end{aligned}
$$

We want a solution that contains 0.20 mole Ca(OH)$_2$ in 1 liter of solution. The solution is to be made from calcium hydroxide, and we have just found that there are 74 g of Ca(OH)$_2$ in 1 mole of Ca(OH)$_2$. Thus we multiply the number of moles by the factor that will convert grams to moles and get

$$\frac{0.20 \text{ mole Ca(OH)}_2}{1 \text{ liter of solution}} \times \frac{74 \text{ g Ca(OH)}_2}{1 \text{ mole Ca(OH)}_2} = \frac{14.8 \text{ g Ca(OH)}_2}{1 \text{ liter of solution}}$$

So we would weigh 14.8 g of calcium hydroxide and add it to a volumetric flask, dissolve, and dilute to a total volume of 1 liter.

Example 2   How would you prepare 250 ml of a 0.150 M solution of KCl?
   The molecular weight of KCl is 74.5. The solution is similar to Example 1 except that we must add a factor to give us an answer for 250 ml instead of a liter (1 000 ml).

$$\frac{0.150 \text{ mole KCl}}{1\,000 \text{ ml solution}} \times \frac{74.5 \text{ g KCl}}{1 \text{ mole KCl}} \times 250 \text{ ml solution} = 2.79 \text{ g KCl}$$

The result of the calculation is 2.79 g of KCl. We would weigh this amount of the salt and place it in a 250 ml volumetric flask, dissolve, and dilute to the 250-ml mark.

Determining molarity when amounts of solute and solvent are known

Example 1   Calculate the molarity of a solution made by dissolving 50.0 g of sodium carbonate (Na$_2$CO$_3$) and diluting the solution to a total volume of 800 ml.
   Here we know grams and want to find molarity or moles/liter. To find moles we multiply grams by a factor that will give us moles, in this case

$$\frac{1 \text{ mole Na}_2\text{CO}_3}{106 \text{ g Na}_2\text{CO}_3}$$

Thus we obtain

$$50 \text{ g Na}_2\text{CO}_3 \times \frac{1 \text{ mole Na}_2\text{CO}_3}{106 \text{ g Na}_2\text{CO}_3} = 0.472 \text{ mole Na}_2\text{CO}_3$$

This 0.472 mole is contained in 800 ml:

$$\frac{0.472 \text{ mole}}{800 \text{ ml solution}}$$

Since we want to know the moles in a liter (1 000 ml), we must then multiply by 1 000; our total calculation becomes

$$50 \text{ g Na}_2\text{CO}_3 \times \frac{1 \text{ mole Na}_2\text{CO}_3}{106 \text{ g Na}_2\text{CO}_3} \times \frac{1\,000 \text{ ml}}{800 \text{ ml}} = 0.59 \text{ mole Na}_2\text{CO}_3$$

The solution is 0.59 M.

Example 2   What is the molarity of a 10% solution of glucose (C$_6$H$_{12}$O$_6$)?
   The calculation is:

$$\frac{10 \text{ g glucose}}{100 \text{ g solution}} \times \frac{1 \text{ mole glucose}}{180 \text{ g glucose}} \times 1\,000 \text{ ml} = 0.56 \ M$$

The Use of Molarity in Chemical Reactions   Solutions having concentrations expressed in molarity are used primarily for processes that involve chemical reactions. The illustrations that follow show some examples of such uses.

Example 1   How many milliliters of 6.0 M HCl would be required to react with 10 g of zinc?

We begin by writing the equation for the reaction.

$$Zn\ + 2HCl \rightarrow ZnCl_2 + H_2$$
(65.4 g)

The equation tells us that one mole (65.4 g) of zinc reacts with two moles of HCl.

To find the number of moles of zinc in 10 g of zinc,

$$10\ g\ Zn \times \frac{1\ mole\ Zn}{65.4\ g\ Zn} = 0.153\ mole\ Zn$$

Since each mole of zinc reacts with 2 moles of HCl, 0.306 mole of HCl is required. If our solution contains 6.00 moles of HCl per liter, we find the volume that contains 0.306 mole of HCl by

$$0.306\ mole\ HCl \times \frac{1\ 000\ ml\ solution}{6.00\ moles\ HCl} = 51\ ml\ HCl\ solution$$

The complete calculation is

$$10\ g\ Zn \times \frac{1\ mole\ Zn}{65.4\ g\ Zn} \times \frac{2\ moles\ HCl}{1\ mole\ Zn} \times \frac{1\ 000\ ml\ HCl\ solution}{6.00\ moles\ HCl} = 51.0\ ml\ HCl\ solution$$

Example 2   It took 40 ml of a 2.0 M solution of NaOH to neutralize 25 ml of a solution of HCl. What was the molarity of the HCl?

$$HCl + NaOH \rightarrow NaCl + HOH$$

Since the equation tells us that one mole of NaOH reacts with one mole of HCl, our calculation becomes

$$40\ ml\ NaOH\ solution \times \frac{2.0\ moles\ NaOH}{1\ 000\ ml\ NaOH\ solution} \times \frac{1\ mole\ HCl}{1\ mole\ NaOH}$$

$$\times \frac{1\ 000\ ml\ HCl\ solution}{25\ ml\ HCl\ solution} = 3.2\ moles\ HCl$$

The solution is 3.2 M.

We can also solve problems of this sort by setting up an equation and then substituting in the equation. If we write the equation in a general form

$$A + B \rightarrow C + D$$

we are saying that one mole of A reacts with, or is equivalent to, one of B, and when the reaction is complete

moles of A = moles of B

We find moles of a substance by multiplying the molarity of a solution by the number of liters of the solution. For example, 1 liter of a 1 M solution contains 1 mole; 0.4 liter of a 2.0 M solution contains 0.8 mole. Thus we can write our equation

liters of A $\times$ molarity of A = liters of B $\times$ molarity of B

In fact, when we multiply milliliters by molarity we call the quantity a millimole, which we abbreviate mmol. Solving the problem in Example 2 by this method, we let A be HCl and B represents NaOH:

ml of HCl $\times$ concentration of HCl = ml of NaOH $\times$ molarity of NaOH

concentration of HCl $\times$ 25 ml = 2.0 M $\times$ 40 ml

$$concentration\ of\ HCl = 2.0\ M \times \frac{40\ ml}{25\ ml} = 3.2\ M$$

This gives the same results as the previous solution but represents a method of solution that many people find simpler to use. This sort of calculation will be extended to deal with the use of normality and milliequivalents in Chapter 11.

Calculations Involving Dilutions of Solutions

Chemists often have a solution of one concentration, usually a concentrated one, and need to make a solution of different concentration. Problems of this sort can be solved by using an approach similar to that given above. We realize that when we dilute a solution, the amount of solute will be the same in both the concentrated and the dilute solution; water (or other solvent) is all that is added. We can state this in the form of an equation:

amount of solute in concentrated solution = amount of solute in dilute solution

We can determine the amount of solute by multiplying the concentration of the solution by the volume of the solution. For a molar solution, the equation becomes

ml of conc. solution × molarity of conc. solution =
                    ml of dilute solution × molarity of dilute solution

In solutions whose concentration is expressed in terms of percentage the equation is

ml of conc. solution × % composition = ml of dilute solution × % composition

Example 1

Concentrated sulfuric acid is about 18 M. How much concentrated solution would be required to make a liter of 2.5 M H_2SO_4?

M of conc. H_2SO_4 × ml of conc. H_2SO_4 = M of dilute H_2SO_4 × ml of dilute H_2SO_4

amount of conc. $H_2SO_4 = \dfrac{2.5\ M}{18\ M} \times 1\ 000\ \text{ml} = 140\ \text{ml}$

If this dilution were to be made, great care would have to be taken, since the addition of concentrated sulfuric acid to water produces a great deal of heat, and the acid may spatter if added too rapidly. Note also that as a rule we add the acid to water rather than water to the acid.

Example 2

How much water would have to be added to 50 ml of 95% ethyl alcohol to give a solution that is 50% ethyl alcohol?

ml of 95% ethyl alcohol × 50 ml = ml of 70% ethyl alcohol × amount of water

amount of water = $\dfrac{95\%\ \text{ethyl alcohol}}{70\%\ \text{ethyl alcohol}} \times 50\ \text{ml} = 68\ \text{ml of 70\% alcohol}$

Since the final volume is 68 ml and we started with 50 ml, we would have to add (68 − 50) or 18 ml of water.

10.8 FREEZING AND BOILING POINTS OF SOLUTIONS

We know that sea water is less likely to freeze than water in lakes and streams. While this is due partly to the fact that sea water is constantly in motion, it is also due to the fact that the *salts dissolved in the water of the oceans decrease the freezing point of the solution* (*sea water*). We take advantage of the lower freezing point of salt solutions when we make home-made ice cream; we add salt to the ice in the freezer which eventually forms a salt water solution.

From experimentation it has been found that adding one mole of a solute to 1 000 g of water lowers the freezing point to $-1.86°C$. When we add the liquid ethylene glycol (an antifreeze) to water in the cooling system of automobiles, we find that adding one mole of ethylene glycol (62 g) per 1 000 g (1 liter) of water depresses the freezing point by 1.86°C. When we add two moles per kilogram of water, the freezing point is depressed twice as much or 3.72°C. If we add one mole (58.5 g) of sodium chloride per kilogram, we find that the freezing point is depressed 3.72°C, or twice as much per mole as it was with ethylene glycol. We note that a mole of sodium chloride produces two moles of particles and conclude that *it is the number and not the size of the particles that is reponsible for lowering the freezing point.*

It might seem that sodium chloride or even calcium chloride, which can produce three particles per molecule, would be the best antifreezes. Unfortunately, these salts are too corrosive for automobiles, but they are used in commercial refrigeration plants.

There are fewer commonly known examples, but we find that *solutions boil at temperatures higher than that of the pure solvent. One mole of solute per kilogram of solvent increases the boiling point 0.51°C.* Thus a solution containing 342 g of sucrose added to 1 kg (1 liter) of water will boil at a temperature of 100.51°C. As with depression of the freezing point, a mole of sodium chloride is twice as effective as one of sucrose. *The boiling-point elevation and freezing-point depression are greater for electrovalent compounds than for covalent compounds. In fact, we can determine the amounts of dissociation or ionization of compounds by measuring their effects on boiling and freezing points.*

If you have ever made candy at home and used a candy thermometer, you have taken advantage of the increase in boiling point of a solution. Here you measure the boiling point of the solution and note changes as water is evaporated (boiled off) from the solution. At a given temperature, you have a given concentration of sugar, and this is related to the ability of the sugar solution to harden. In an automobile, we use ethylene glycol not only as an antifreeze, but also as an anti-boiling agent, since such solutions have a boiling point higher than pure water.

Why does addition of solutes increase the boiling point of liquids? The boiling point is the temperature necessary to increase the vapor pressure of the solvent to that of the atmosphere. If there are solute molecules at the surface of a liquid, there are fewer solvent molecules at that surface, and the solvent has a lower vapor pressure. The temperature must therefore be raised in order to have the vapor pressure of the solvent equal to the vapor pressure of the atmosphere, which is necessary for a substance to boil. Of course the solvent must be nonvolatile, or appreciably less volatile than the solvent, for the boiling point to be raised. Solutions of ethanol (boiling point 79°) in water have a boiling point less than that of water, but greater than that of ethanol. We can explain freezing-point depression by saying that the solute particles prevent solvent molecules from condensing with other solvent molecules and solidifying until a lower temperature is reached.

In solving problems involving boiling-point elevation and freezing-point depression, we always talk about the amount of solute dissolved in a given weight of solvent (and we have weight:weight solutions). A solution containing one mole of solute per 1 000 g of solvent is called a **molal** solution. A solution containing 1 mole per 250 g of solvent is a 4-molal (4 m) solution. Note that we use the small letter m for molality to distinguish it from molarity, where we use a capital M.

Example 1   How much solute should be added to 250 g of a solvent to make a 3.0 m solution of a covalent substance?

A 3.0 m solution contains 3.0 moles of solute and 1 000 g of solvent, so

$$\frac{3.0 \text{ moles solute}}{1\,000 \text{ g solvent}} \times 250 \text{ g solvent} = 0.75 \text{ mole solute}$$

Dissolve 0.75 mole of solute in 250 g of solvent.

Example 2   What would be the freezing point of a solution containing 124 g of ethylene glycol ($C_2H_6O_2$) added to 1 kg of water?

The molecular weight of glycol is

$$
\begin{aligned}
2C &= 24 \\
6H &= 6 \\
2O &= 32 \\
\hline
&62
\end{aligned}
$$

The freezing point is depressed by

$$\frac{1.86°}{1 \text{ mole}} \times \frac{1 \text{ mole}}{62 \text{ g}} \times 124 \text{ g} = 3.72°$$

Thus the freezing point is $-3.72°C$.

Example 3   If a solution that is 2.0 m increases the boiling point of water to 102.04°, how many particles are formed from each molecule of the solute? Is the solute a covalent or electrovalent compound?

A solution that is 2.0 m would be expected to increase the boiling point by $0.51 \times 2 = 1.02°$. The increase ($102.04° - 100°$) is 2.04°, which is twice that expected, so the solute forms two particles per molecule. It is an electrovalent compound.

10.9 SUSPENSIONS

So far, our discussion has been about soluble substances and solutions. However, we are all aware that some substances are not very soluble in water or other solvents. Actually, almost everything is at least very, very slightly soluble, but for all practical purposes we regard some substances as insoluble. If insoluble substances are finely powdered, they may be suspended in water or another liquid to produce a **suspension**. Small particles remain suspended as long as we shake the mixture, but if left alone, they will soon settle to the bottom of a container. We can then remove the liquid by pouring it off in a process called **decantation**, or we can separate the solid from the liquid by **filtration**. If a chemical reaction between two soluble substances such as sodium chloride and silver nitrate produces an insoluble substance (silver chloride), it will settle out of solution in a process called **precipitation**. The insoluble substance formed this way is a **precipitate**.

The chemical and physical processes that might be expected to produce precipitates or temporary suspensions sometimes do not. Although large particles are involved, they are permanently suspended, or at least suspended for a long period of time. *We call these permanent suspensions* **colloidal dispersions** *or* **colloidal suspensions**. Sometimes we use the term *colloids* to describe the suspension although, strictly speaking, we should refer only to the substance that is being dispersed as the colloid. Soluble substances are sometimes called crystalloids. See Table 10.2 for a list of typical colloids.

We can usually see that a colloidal dispersion such as milk or muddy water is cloudy or even opaque. Sometimes the colloidal dispersion is dilute and may appear to be clear like a solution; we may then need to perform a test to distinguish whether it is a colloidal dispersion of a solution. A beam of light passed through the solution at right angles to

TABLE 10.2 Examples of colloidal dispersions

Type	Example
solid in liquid	starch suspension; starch in water*
liquid in liquid	salad dressings; oil in vinegar and water
gas in liquid	soapsuds; air in soap solution
solid in gas	smoke or polluted air; various particles in air
liquid in gas	clouds; water in air
solid in solid	gemstones; various particles in complex silicates
liquid in solid	cheese; water in milk proteins
gas in solid	styrofoam; air in styrene polymer

* The colloidal substance is listed first and the dispersing medium second.

the observer leaves a visible path in a colloidal dispersion and none in a solution (Fig. 10.1). This procedure can be used to distinguish true solutions from clear colloidal suspensions. Light scattering by dispersion is known as the *Tyndall effect*, after John Tyndall, who described it about a hundred years ago. We can sometimes determine the size of colloidal particles by measuring the light scattering of the dispersion.

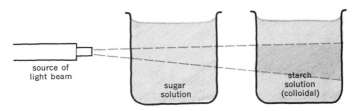

source of light beam

sugar solution

starch solution (colloidal)

Fig. 10.1   The Tyndall effect. The beaker on the left contains a sugar solution; the one on the right contains a colloidal dispersion of starch. When the light beam passes through the true solution of sugar it leaves no perceptible path; however, when it passes through the starch solution, where the starch particles are large enough to reflect light, the beam produces a visible path.

Many substances that are important for life are very large molecules. These substances must react and in some cases must circulate throughout the body. They are generally present as colloidal suspensions.

**10.10
KINDS OF
COLLOIDAL
DISPERSIONS**

Dispersions in gases   You have probably seen the beam of light formed by a searchlight piercing the air. What we actually see is the light reflected by the countless tiny particles of solids and liquids suspended in the air. We usually call the suspension of solids *dust* and that of liquids *fog*. Although we often use the word **aerosol** to describe the spray from a pressurized can, *the word is used scientifically to describe any suspension of solids or liquids in a gas.* Since the suspended particles are larger than those that form solution, the permanent suspensions of this type are examples of colloidal dispersions. Although most aerosols tend to settle out of air, dust from great volcanic disturbances and radioactive particles from atomic explosions tend to stay dispersed for long times. Often these particles are precipitated along with rain or snow.

While we tend to regard any air that is not perfectly clear as being polluted, fog consisting only of water droplets is very important for growth of redwood trees and this foggy air is not necessarily polluted. Many types of trees (pines especially) create a blue haze by giving off substances known as *terpenes* which are not harmful. It is important to know the composition of the particles that are often reported as "particulates" in air-pollution reports. Some, such as small particles of asbestos or coal dust, can be extremely damaging to our lungs; others (such as water) are harmless or perhaps even beneficial.

Dispersions in liquids   Starch is not soluble in water, but if we heat a starch suspension, it will form a colloidal dispersion. Other common suspensions of solids in liquids are those of clay to make muddy water and the water from glaciers, which contains sediments that cause it to look like milk even when the particles have had a great deal of time to settle out. We suspend a gas (air) in a liquid when we make whipped cream or meringue. Foams of various sorts are also examples of this type of colloidal dispersion.

Some of the commonest colloidal dispersions are those of one liquid in another. Such dispersions are called **emulsions** or emulsoids. Salad dressings, such as French dressings and mayonnaise, are emulsions. Cream is a suspension of fat droplets in water. We generally need to use an emulsifying agent to permanently suspend one liquid in another. Although we call the dispersion of silver particles in gelatin and water a photographic emulsion, it is technically not an emulsion.

Dispersions in solids   Both gases and liquids can be suspended in solids. Although the product known as Styrofoam has gas particles (bubbles) that are too large to qualify as a colloidal suspension, it is an example of a gas suspended in a solid. Much more important to living things are suspensions of liquids (usually water, sometimes an aqueous solution) in a solid. We call such a suspension a **gel**. An apple is about 90% water, yet we think of apples as being solid. In an apple and many other fruits, water is suspended in the molecular network of substances called **pectins**. We use pectins to provide a solid or semisolid form for a water solution of sugar and flavors when we make jams and jellies. Gelatin is a solid that can hold a great deal of water in a semisolid state. More important to the survival of animals is the substance *fibrin*, which forms a network that holds red blood cells and other materials in a blood clot. Studies of the change from the colloidally dispersed fibrinogen to the precipitated network called fibrin are very important in medicine. If we understand blood clotting, we can help it occur in some cases and prevent it in others. An unwanted blood clot (for example, in the vessels of the heart or brain) can cause death.

**10.11
WHY DO
COLLOIDS
REMAIN IN
SUSPENSION?**

Since the particles in a colloidal suspension are larger than those in a solution, and occasionally we can observe them precipitating out of suspension, the question arises: Why do they stay in suspension at all? As usual in chemistry, there are several answers to the question.

We believe that molecules are constantly in motion. The observation that led to this conclusion was made in 1827 by Robert Brown, who observed flashes of light being reflected from a colloidal dispersion and concluded that the particles were being bumped by molecules in constant motion. We sometimes call this phenomenon *brownian motion*. Just as a group of people can keep a volleyball suspended in the air by constantly hitting it, the constantly moving air molecules or even the slower moving molecules of a liquid keep a colloid from settling out. We can make a colloidal dispersion or help to maintain one by decreasing the size of the particles so they will stay in suspension. This is done when cream globules are forced through small openings to break them up and produce homogenized milk.

The repulsion between particles of like electrical charge is another factor that tends to keep particles in suspension. For example, if all soot particles in the smoke from a furnace are positively charged, they repel each other so that particles large enough to precipitate are not formed. However, if we want to precipitate smoke or other particles, we can use an electrostatic precipitator. By placing a charge on the precipitator, or by channeling the smoke alternately through positively and negatively charged areas, charged particles can be precipitated (see Fig. 15.7). If we incorporate some kind of collector to remove the precipitated particles, not only are they removed from the air, but many times they can be put to some use. The smoke from metal smelters contains valuable metals that can be reclaimed by using electrostatic precipitators in the smoke-stacks. A different kind of example is the colloidal silt (clay) particles in river water. The charges on these particles are neutralized when rivers enter the sea, so that particles settle out at that point. It is this natural process that has, over many centuries created the great fertile deltas at the mouths of the Nile and Mississippi rivers.

Another way in which electrical charges help to keep large particles in suspension is important for the action of protein and nucleic acid molecules in living systems. The molecules of protein and of the nucleic acids (DNA and RNA) are extremely large, and we would not expect them to be permanently suspended. However, they also contain many areas that carry positive and negative charges. These charged areas attract (or are attracted to) water molecules (Fig. 10.2). Since the water molecules are attracted both to the large molecule and to other water molecules, a permanent suspension is formed. We can precipitate nucleic acids or proteins by neutralizing the charges on the molecules. When milk becomes sour (due to the production of lactic acid by bacteria), the milk protein, casein, settles out of solution. Although we generally do not like to drink sour or precipitated milk, it is used for making yogurt and cheese. Similarly, gamma globulin can be separated from blood plasma by a series of precipitations. Gamma globulin is a valuable source of antibodies and is often given to people exposed to infectious hepatitis and other diseases.

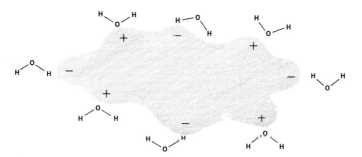

Fig. 10.2 Representation of part of a protein molecule surrounded by water molecules. Protein molecules have both positive and negative charges. The oxygen atom of a water molecule has a slight negative charge and thus is attracted to the positively charged areas on the protein. The hydrogen atoms, having a slight positive charge, are attracted to the negatively charged areas.

Emulsions can be formed and maintained by substances we call emulsifying agents. Soap is a good emulsifying agent and lecithin, found in egg yolks, is another. Emulsifiers such as alginates and carrageenan are used commercially to help produce and stabilize the colloids in ice cream, mayonnaise, and other foods. Emulsifying agents generally contain relatively large molecules in which one end is polar (soluble in water) and the other is nonpolar (soluble in fats and oils). The way these agents work is discussed further in Section 24.6. Gelatin and the alginates function by surrounding small particles with thin films. This prevents the particles from aggregating (coming together) and forming precipitates or larger crystals. Emulsifiers in ice cream not only keep fats suspended but prevent formation of large ice crystals that would give the ice cream a grainy texture.

10.12
COLLOIDS AS
ADSORBENTS
AND
CATALYSTS

Although the particles in a colloid are larger than those in a solution, they are still very small. These small particles have a much larger surface area than normal solids. For instance, a cube that is 10 cm on each side has six sides, each having a surface area of 10×10 or 100 cm^2, and the total surface area is 600 cm^2. If we divide this cube into cubes that are 1 cm on a side we will have 1 000 cubes each having a surface area of 6 cm^2 and we will have a total surface area of $6 000 \text{ cm}^2$. Although colloidal particles are not cubes, they are very small and have a large surface area per unit of weight. This makes them useful in reactions which occur on surfaces. Such reactions involve adsorption— attraction to surfaces as contrasted to absorption, in which substances penetrate into solids.

We take advantage of this large surface area to make catalysts such as the platinum or nickel catalysts that can be used in the hydrogenation of vegetable oils or the catalysts in automobile exhaust systems that promote the oxidation of potential air pollutants.

If we carefully prepare carbon from coconuts, peach pits, or some woods so that it has a large surface area, we produce a substance known as *activated charcoal*. This substance is a good adsorbent and can be used for a variety of purposes, from adsorbing colored impurities in a manufacturing process to absorbing poisonous gases in the cannister of a gas mask.

10.13
OSMOSIS

The process of osmosis, which is essential to all living things, involves substances both in solution and in colloidal suspension. **Osmosis** consists of the passage of substances through a semipermeable membrane. Most membranes in plants and animals are semipermeable; that is, these thin structures (membranes) have openings of definite size, large enough to permit the passage of some molecules but not others. Most biological membranes permit the passage of water and some of the small ions. Many membranes will also permit the passage of the sugar glucose (which has a molecular weight of 180). However, the passage of large molecules such as proteins is not possible.

Red blood cells contain water, hemoglobin and other proteins, and salts enclosed in a delicate membrane. If we place some red blood cells (commonly called *red corpuscles*) in a solution of distilled water, they swell and burst (Fig. 10.3a). We can explain what has happened by saying that more water molecules went into the cell than came out. This happened because the water moved freely into the cell, but its movement out of the cell was impeded, due largely to the presence of proteins and other large molecules. Not only do these large molecules attract and hold some water molecules, but more important, their presence at the inner cell membrane means that the concentration of water at the inner membrane is less than at the outer membrane. Since the concentration

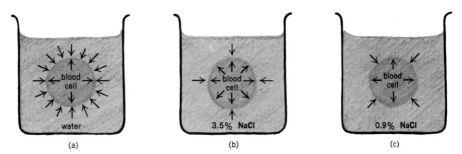

Fig. 10.3   Factors controlling the direction of osmosis in blood cells.

of water molecules outside the cell is greater than that inside, there is a net movement of water into the cell. While there is a slight movement of salt out of the cell, this is not enough to prevent the membrane from bursting.

If a red blood cell is placed in sea water, which is about 3.5% salt, the cell will shrink. In this case more molecules of water have left the cell than have entered it. The relatively high concentration of salt outside the cell means that water molecules are in lower concentration at the outer surface than at the inner surface, so there is a net movement of water out of the cell.

If a red blood cell is placed in a solution that is 0.9% sodium chloride, there is neither swelling nor shrinking. The normal salt content of the blood (mostly sodium chloride) is 0.9%, and this is the environment that is best for the normal functions of the red blood cells as well as other cells of the human body.

We use the term **isotonic** to describe a solution that contains the same amount of salt as normal blood. In such a solution, red blood cells maintain their normal size and can function normally. A solution that contains less than 0.9% salt is **hypotonic** (*hypo* = below) and one that contains more than 0.9% salt is **hypertonic** (*hyper* = above). Medications and solutions containing glucose, vitamins, and other nutritious substances are often given to sick persons by injecting large volumes (often several liters) of the solution into a blood vein. From the above discussion it is obvious that, for intravenous feeding, it is essential that salt concentrations be carefully adjusted to prevent rupture or shrinking of red blood cells as well as other cells of the body.

Differences in pressure due to osmosis can be demonstrated by placing a concentrated solution of sugar in a tube that is covered by a synthetic membrane (Fig. 10.4). When the tube is placed in water, the water molecules readily pass through the membrane and the sugar molecules do not, so the level of solution in the tube rises. The difference between the level of water in the tube and in the container is due to the osmotic pressure. If we wish to measure the osmotic pressure we must determine the pressure necessary to have the same water level in both the tube and the larger container of water (Fig. 10.5). Another demonstration of osmotic pressure uses an egg with part of the shell removed but having the membrane around the egg carefully preserved (Fig. 10.4). In fact, *we see the difference in volume and/or pressure any time two solutions having different concentrations of different-sized molecules are separated by a membrane that is more easily permeated by one of the components than others. We generally say that water travels in a direction that will dilute the more concentrated solution.* There is nothing mysterious about this; it is merely diffusion of the various substances to establish a situation (equilibrium) in which concentrations of all substances on both sides of a membrane are equal.

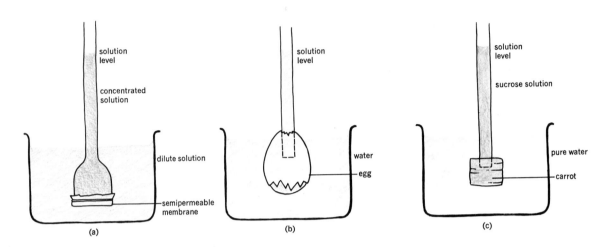

Fig. 10.4 Osmosis. (a) Water passing through a semipermeable membrane from a dilute solution to a concentrated solution. (b) An egg with part of the shell removed but with membrane intact, showing the rise of the egg white solution in a tube inserted through both shell and membrane. (c) A carrot serving as a semipermeable membrane. In this case the pure water will eventually become a dilute sugar solution and the levels of the solutions inside the tube and in the outer vessel will be equal.

The osmotic pressure generated by a solution depends on the number, and not the size, of the particles that do not freely cross the membrane. Thus, as with freezing- and boiling-point changes, a mole of sodium chloride is twice as important in determining osmotic pressure as is a mole of glucose. However, this is true only if sodium and chloride particles do not readily cross the membrane. We sometimes speak of the osmolarity of a solution or colloidal suspension. The osmolarity is related to the total number of non-diffusible particles in a specific amount of solvent.

Hypertension or high blood pressure is a common condition of many people. It is related to heart attacks, strokes, and kidney diseases, which are the leading causes of death in the United States and other developed countries. One of the possible causes of hypertension is that there are too many sodium ions in the blood and this increases the

Fig. 10.5 Osmotic pressure. In (a) the two solutions separated by a permeable membrane have different levels. In (b) a weight has been applied so that the two solutions have equal levels. The pressure applied by the weight in (b) is analogous to osmotic pressure.

osmotic pressure. (Although sodium ions are small they do not cross some membranes.) Most people with hypertension are told to decrease the amount of salt and other sodium-containing compounds in their diet. Sometimes they are given medications known as diuretics, which help the kidney to excrete more sodium ions from the body.

10.14 DIALYSIS

The process called **dialysis** also makes use of a semipermeable membrane. If we place a solution containing proteins, sugar, and salts in a tube or sack made from a semipermeable membrane and immerse this bag in water, small molecules tend to leave the bag, while the larger ones remain. If we keep changing the water in the container (Fig. 10.6) we will soon remove all the small molecules and will have only the larger molecules in the bag. This process, known as **dialysis**, can be used to separate small molecules from larger ones and thus purify the large molecules. It is used to purify proteins and nucleic acids.

The kidney uses a process similar to dialysis to eliminate wastes from the body. Blood is filtered in the kidney by a semipermeable membrane and small molecules of impurities are forced out into the urine, while the proteins and other large molecules are retained. Since some of the small molecules filtered out are beneficial, the kidney also has a system of "reverse dialysis" that allows small molecules such as glucose, which are originally filtered out, to be reabsorbed. A great deal of water is also absorbed, drawn back in by osmosis. (See Section 40.9 for a further discussion of kidney function.)

Because we understand some of the processes of the kidney, we can make artificial kidneys available for those people who have badly injured or diseased kidneys. These machines are called *dialysis units* and the process is referred to as **hemodialysis** (*hemo* = blood), since it is the blood that is being dialyzed. In hemodialysis, blood is taken out from

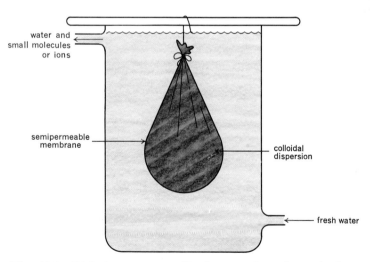

water and small molecules or ions

semipermeable membrane

colloidal dispersion

fresh water

Fig. 10.6 Dialysis apparatus. Fresh water flows in at the lower right, and when it flows out at the upper left it carries away the small molecules that have diffused through the semipermeable membrane. Large molecules in colloidal dispersion remain inside the membrane and are purified.

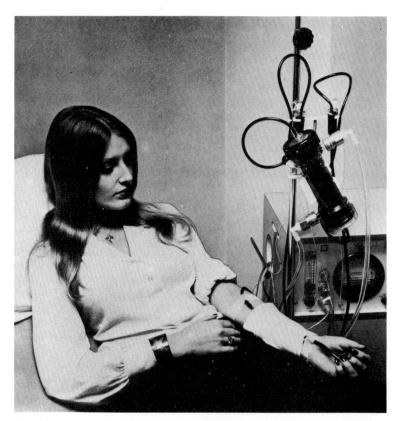

Fig. 10.7   An artificial kidney or hemodialysis unit. Blood is withdrawn from an artery (upper tube) and enters the hemodialysis unit at the top. Fluid supplied by the machine enters the unit at the lower left and leaves at the upper right. The fluid and blood exchange small molecules through a semipermeable membrane in the hemodialysis unit. The blood leaves the unit through a tube at the bottom and is returned to a vein. [Photo courtesy of the Cordis Dow Corporation.]

a convenient artery and passed through a series of semipermeable membranes before being returned to the body. (See Fig. 10.7.) This removes the small molecules—both those we classify as wastes, such as urea, and glucose, which can still be used for energy. We compensate for this loss by having the solution outside the membranes contain glucose and other substances that we want to retain in the blood.

The conduction of nerve impulses involves the passage of substances (mainly sodium and potassium ions) through the membrane of a nerve cell. In this instance, passage is reversed when the nerve is getting ready to carry another impulse. This reversal probably involves changes in the permeability of the membrane. It happens in a small fraction of a second.

In one form of serious condition known as shock, many of the symptoms appear to be due to an increased permeability of blood capillaries. This allows larger particles such as proteins to leave the capillary. When these large molecules are lacking, the osmotic pressure of the blood is decreased, less water flows back into the blood, and more water remains in the tissues. This means there is less volume of blood and the blood pressure drops. If the drop is sufficient, death may result. Inflammation of tissues which follows an injury also involves a change in permeability of capillaries.

SUMMARY

This chapter has been concerned with the characteristics of solutions and suspensions. Several methods of expressing the concentration of solutions were given and the relation of concentration to boiling and freezing points was discussed. Suspensions, particularly colloidal suspensions, which are not true solutions, were also described and their properties were discussed.

Most chemical reactions, including those that occur in living organisms, occur in solutions or suspensions. Such mixtures are necessary to ensure that particles can come close enough to other particles for a reaction to occur. Precipitation of substances generally stops a chemical reaction. The size of suspended particles and their relations to semipermeable membranes are important in the processes of osmosis and dialysis. These two processes are extremely important in living systems.

Osmosis and dialysis are terms used to describe similar processes that involve membranes that are permeable to some molecules and impermeable to others. The term osmosis is generally used to describe naturally occurring processes in living systems—*in vivo* processes. The term *dialysis* is used to describe procedures in which semipermeable membranes are used *in vitro*, or outside the living system, to purify various substances. While some of the substances purified by dialysis, such as proteins or the blood itself, are derived from living organisms, the process also occurs outside the organism. With the use of suitable membranes, dialysis can also be used to purify (desalinate) sea water.

The following chapter discusses the properties of solutions of acids, bases, and salts. This leads to a further study of chemical reactions and a consideration of their reversibility and equilibrium situations.

IMPORTANT TERMS

aerosol	hemodialysis	osmosis
colloid	hypertonic solution	precipitate
colloidal suspension	hypotonic solution	precipitation
concentrated solution	immiscible	saturated solution
decant	isotonic solution	solute
dialysis	miscible	solution
dilute solution	molal solution	solvent
emulsion	molar solution	supersaturated solution
filtration	molarity	suspension

WORK EXERCISES

1. Define the term "solution."
2. Name and give illustrations of three common kinds of solutions.
3. Make at least three distinctions between "solvent" and "solute."
4. State the effects of temperature and pressure on the solubility of gases in liquids.
5. Does pressure materially affect the solubility of solids in liquids? How?
6. Does temperature materially affect the solubility of solids in liquids? How?
7. List three factors that increase the rate of solution.
8. How does the polar nature of the solute and solvent molecules (or ions) influence solubility?
9. List eight ways of specifying the concentration of solutions. Which terms are only general and which ones can be used for precise descriptions of concentration?

10. How could you tell whether a clear liquid is a suspension or a solution?
11. How could you tell whether a clear liquid is a solution or merely a pure liquid?
12. List some reasons why colloids remain in suspension.
13. Tables 10.1 and 10.2 list solutions and suspensions according to type. Give one more example of each of the categories listed in these tables.
14. Give the amount (in appropriate units) of both solute and solvent required to make the following solutions. Unless otherwise specified, the solvent is water.
    a) 500 ml of a 0.9% NaCl solution (w/v)
    b) 1 dm^3 of a 70% solution of ethyl alcohol (v/v)
    c) 25 g of a solution of $CuSO_4 \cdot 5H_2O$ that is 3.4% (w/w)
    d) 1 liter of a solution of EDTA that contains 3.5 ppm (w/w)
    e) a 10% solution of iodine in ethyl alcohol (w/v)
    f) 1 liter of a solution that contains 10% glucose in 0.9% saline (NaCl) (w/v)
    g) 1.50 liters of a solution of NaOH that is 0.250 M
    h) 1.50 liters of a solution of ethylene glycol (mol wt. 62) that is 0.250 m
    i) 100 ml of a 1% $AgNO_3$ solution (w/v)
15. Calculate the number of grams of solute required to prepare the following molar solutions.
    a) 2 liters of 2 M NaOH
    b) 3 dm^3 of 0.1 M $AgNO_3$
    c) 10 liters of 0.5 M NaCl
    d) 500 ml of 1.0 M H_2SO_4
    e) 100 ml of 0.5 M HCl
    f) 125 ml of 1.0 M $CuSO_4 \cdot 5H_2O$
16. Calculate the molar strength of solutions containing the following.
    a) 60.0 g of NaOH in 1 liter
    b) 200.0 g of NaOH in 2 liters
    c) 4.0 g of NaOH in 1 dm^3
    d) 50.0 g of HNO_3 in 4 dm^3
17. How many grams of HCl are contained in 130 ml of hydrochloric acid which has a specific gravity of 1.19 and contains 37.2% of HCl by weight?
18. A solution of KCl weighing 150 g, on being evaporated to dryness, left a residue of 18 g of KCl. What was the percent of KCl in the solution?
19. A sample of sulfuric acid has a specific gravity of 1.80 and is 87% by weight sulfuric acid.
    a) What will 1 liter of the acid weigh?
    b) How many grams of sulfuric acid are there in 1 liter of the acid?
20. The sulfuric acid in the conventional automobile storage battery weighs 1.2 g/ml.
    a) What will 1 liter of the acid weigh?

b) Consult a chemistry handbook for the percent H_2SO_4 in the acid.
21. The density of an aqueous solution of ethanol, which is used as an antifreeze in automobile radiators, is less than that of water. A 26.7% (by weight) alcohol–water solution has a density of 0.959 g/ml and freezes at $-16°C$. How many grams of ethanol are present in 1 liter of the solution?
22. a) Glycol, the most common permanent antifreeze, has a density greater than 1; therefore aqueous solutions of it have densities greater than that of water. A solution containing 40% glycol (by weight) has a density of 1.053 g/ml and freezes at $-22°C$. How many grams of glycol are present in 1 liter of the solution?
    b) How many grams of glycol are there in one quart of the solution?
23. How many grams of glycol, $C_2H_4(OH)_2$, must be added to 1 liter of water to lower the freezing point to $-15°C$?
24. What will be (a) the molality and (b) the freezing point of the solution in a radiator that contains 4 liters of ethanol (molecular wt. 46 and density 0.79 g/ml) mixed with 4 liters of water?
25. How many ions are produced per molecule of a substance if 2.0 moles dissolved in 500 g of water elevates the boiling point of the solution 4.2°C?
26. How many milliliters of a 0.10 M solution of KCl will be required to provide the following?
    a) 150 mmol of KCl
    b) 6 g of KCl
    c) complete reaction with 5.0 ml of 0.25 M $AgNO_3$
27. How many ml of a 0.30 M solution of H_2SO_4 will be required to convert 10 g of $BaCl_2$ to $BaSO_4$? The reaction is $H_2SO_4 + BaCl_2 \rightarrow BaSO_4 + 2HCl$
28. Given the reaction $H_2SO_4 + 2NaOH \rightarrow Na_2SO_4 + 2HOH$.
    a) How many moles of NaOH would be required to react completely with 1.5 moles of H_2SO_4?
    b) If 38 ml of 2.5 M NaOH reacts completely with 13 ml of H_2SO_4, what is the molarity of the H_2SO_4?
29. In the reaction $H_2SO_4 + Ca(OH)_2 \rightarrow CaSO_4 + 2HOH$, how many ml of a 4.6 M solution of calcium hydroxide will be required to react completely with 50 ml of 6.0 M H_2SO_4?
30. Describe how you would make the following solutions from those given as starting materials.
    a) 1.0 dm^3 of 0.10 M H_2SO_4, starting with 6.0 M H_2SO_4

b) A 1% AgNO₃ solution, starting with 10% AgNO₃ solution

c) 500 ml of a 70% ethyl alcohol solution, starting with 95% ethyl alcohol

d) 1 liter of a solution containing 10 ppm of a substance, starting with a 1% solution of the substance

31. Since water solutions of vitamins will deteriorate on standing, people who make up vitamin mixtures for animal feeding experiments often make a "solid solution" by mixing vitamins with table sugar (sucrose).

a) How would you make a vitamin mixture to contain 120 mg of thiamin (Vitamin B₁) and 150 mg of riboflavin (Vitamin B₂) in 100 g of mixture?

b) How much of the mixture would be needed to supply 3 mg of thiamin?

c) What would be the riboflavin content of the sample in (b)?

SUGGESTED READING

Debye, P., "How giant molecules are measured," *Scientific American*, September 1957, p. 90. The light-scattering ability of colloidal particles as an indication of their size.

11

Acids, Bases, Salts, and Ionization

Since the days of the first experiments with electricity, it has been known that solutions of some substances in water would conduct an electric current, whereas solutions of other substances in water would not. The former were called **electrolytes** and the latter **nonelectrolytes**. It was later observed that electrolytes invariably belonged to one of three classes of compounds called *acids*, *bases*, and *salts*.

There were other distinguishing differences between electrolytes and nonelectrolytes. When substances are dissolved in water, the freezing point of the solution is lower than that of pure water, and the depression of the freezing point is in proportion to the number of particles present in a given volume of water (Section 10.8). It was known that solutions of nonelectrolytes such as sugar, alcohol, and glycerol all have the same freezing point, provided the number of moles of solute is the same in each solution. However, the same number of moles of an acid, base, or salt always gives a water solution having a lower freezing point than solutions of nonelectrolytes. Furthermore, the abnormal depression of the freezing point occurs in whole-number multiples; that is, the lowering is two, three, or four times the expected value. The boiling point of a water solution is elevated in a similar abnormal manner when the solute is an electrolyte.

For many decades these phenomena baffled chemists and physicists. Although various theories to explain the observations were advanced, none stood up under close scrutiny.

In 1884 the Swedish chemist Svante Arrhenius proposed the *theory of ionization*, which, with certain modifications, is still considered to be essentially correct. Arrhenius stated that when an electrolyte is placed in water the molecules separate into particles with positive charges and particles with negative charges. Such particles are called ions (from the Greek word for wanderer). These charged ions account for the ability of solutions of electrolytes to conduct electricity. Furthermore, since dissociation in solution yields two or more particles from each "molecule," the freezing point would be "abnormally" lowered by exactly twice or three times the anticipated change.

At first Arrhenius received little support for his theory; even his professors were skeptical of its merits. He used it as a portion of his Doctor of Philosophy dissertation and was given the lowest possible passing grade for it. Twenty years later, for the same work, he was awarded the Nobel Prize in chemistry. Outstanding scientists, like great artists, musicians, and poets, often have to wait many years before their works are fully understood and appreciated.

Arrhenius proposed that sodium chloride in water would form two ions:

$$NaCl \xrightarrow{H_2O} Na^+ + Cl^- \qquad\qquad (11\text{-}1)$$

Copper(II) chloride dissolved in water would produce three ions:

$$CuCl_2 \xrightarrow{H_2O} Cu^{2+} + 2Cl^- \qquad\qquad (11\text{-}2)$$

An electric current can flow through a water solution of an electrolyte such as $CuCl_2$, but not through pure water (unless the voltage is very high). The dissolved ions of the electrolyte can move through the solution and, being electrically charged, form a conducting "bridge" between the electrodes. The situation is illustrated by Fig. 11.1. It is interesting to see that the electrical current flows through the wire because the dissolved ions are undergoing a chemical change at the electrodes. Electrons can enter the solution from the cathode because Cu^{2+} ions are there to accept the electrons; by gaining electrons the ions are reduced to Cu atoms, which plate out as a coating of copper metal on the electrode. Using e^- to represent an electron, we can write an equation for the reaction at the cathode:

$$Cu^{2+} + 2e^- \rightarrow Cu \qquad\qquad (11\text{-}3)$$

Since a positive copper ion is attracted to the cathode, it is called a *cation*. This is a general term. A **cation** *is any positive ion.*

Meanwhile, a chemical reaction is also taking place at the positive electrode, the anode. The negative chloride ions are attracted to it. As soon as the Cl^- ions reach the

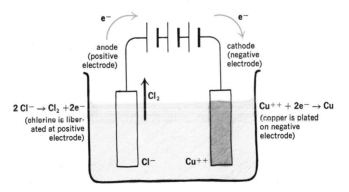

Fig. 11.1 Conductivity of copper chloride in water, due to the presence of ions.

anode they release electrons and are thereby oxidized to Cl atoms, which pair off and escape as Cl_2 gas. The equation for the reaction at the anode is:

$$2Cl^- \rightarrow Cl_2 + 2e^- \qquad (11\text{--}4)$$

The Cl^- ion or *any negative ion is called an* **anion**.

Arrhenius' theory implies that an electrolyte such as $CuCl_2$ or $NaCl$ does not form ions until it dissolves in water. This is one part of his theory that has been modified. As shown in Fig. 6.2, it is now known that sodium chloride already exists as Na^+ ions and Cl^- ions even in a crystal. The change occurring when the salt crystal dissolves in water is that the ions are separated from each other. This state of affairs can be represented by altering Eq. (11–1) as follows:

$$Na^+Cl^- \xrightarrow{H_2O} Na^+ + Cl^- \qquad (11\text{--}5)$$
(solid)

When you see an equation such as (11–5), keep in mind that each dissolved Na^+ or Cl^- ion has several water molecules clustered around it, as illustrated by Fig. 9.4.

Some electrolytes do conform with Arrhenius' original theory, in that they do not ionize until forming a solution. For instance, a sample of pure hydrogen chloride, a gas at ordinary temperatures, is not ionized. A molecule of hydrogen chloride is covalent, although the bond is highly polar.

$$\overset{\delta^+ \quad \delta^-}{H-Cl}$$

[handwritten: electrons shared between atoms]

Hydrogen chloride gas, when dissolved in a nonpolar solvent such as benzene, remains nonionized; the solution does not conduct an electric current. In water, hydrogen chloride is very soluble, and the resulting solution is an excellent conductor of electricity. The very polar water molecules induce the hydrogen chloride molecules to split into ions. In this case there is a chemical reaction between one HCl molecule and one H_2O molecule:

$$H:\overset{..}{\underset{..}{O}}: \quad H:\overset{..}{\underset{..}{Cl}}: \longrightarrow H:\overset{..}{\underset{..}{O}}:H^+ + :\overset{..}{\underset{..}{Cl}}:^- \qquad (11\text{--}6a)$$
$$\overset{|}{H} \qquad\qquad\qquad \overset{|}{H}$$

$$H_2O \quad + \quad HCl \xrightarrow{H_2O} H_3O^+ \quad + \quad Cl^- \qquad (11\text{--}6b)$$
(gas)

The H_3O^+ ion is called a **hydronium ion**; in its structure the plus charge is actually on oxygen rather than on hydrogen. Additional water molecules, acting as solvent, surround each H_3O^+ ion and each Cl^- ion, in the same way that water molecules keep Na^+Cl^- dissolved.

Although it is this action of water, Eq. (11–6), that causes ionization of hydrogen chloride, for many purposes it is more convenient to represent the reaction by using an equation that shows the water molecules only as solvent:

$$HCl \xrightarrow{H_2O} H^+ + Cl^- \qquad (11\text{--}7)$$

When we use an equation such as (11–7) or speak of "hydrogen ions" in solution, it must be understood that the ions are really hydronium ions. Although the correct name for HCl alone is hydrogen chloride, the water solution containing H^+ (or H_3O^+) ions and Cl^- ions is called *hydrochloric acid*.

In a similar fashion the ionization of nitric acid, when dissolved in water, may be represented in either of two ways:

$$H_2O + HNO_3 \longrightarrow H_3O^+ + NO_3^-$$ (11–8)

$$HNO_3 \xrightarrow{\text{H}_2\text{O}} H^+ + NO_3^-$$ (11–9)

In modern terms, **ionization** *is the action of water (or some other solvent) upon a solute which produces ions in solution.* In general, there are two ways this can happen. The water may merely "set free" from a crystal the ions that already exist, as in the case of sodium chloride, Eq. (11–5); or, in the case of a polar covalent molecule like HCl, Eq. (11–6b), no ions are produced *until* the molecule dissolves in (and reacts with) water. Now that electrolytes have been described as fundamentally ionic in solution, we can turn our attention to the types of compounds that have such properties.

11.3 RELATIONSHIPS OF ACIDS, BASES, AND SALTS

Just as chemical reactions were classified into certain types (Section 7.6), it is convenient to think of compounds as being of different types. The terms *acid*, *base*, and *salt* have been used since ancient times to describe particular kinds of materials. The classifications originally were based primarily on physical properties, and the name "salt" still suggests the taste of many (but not all) compounds of that type. In similar fashion the word "acid" is associated with a sour taste, vinegar and tart fruit juice being the oldest examples. More satisfactory and reliable concepts for distinguishing between acids, bases, and salts were not developed until after Arrhenius proposed his theory of ionization.

Descriptions or definitions of acids and bases have evolved ever since about 1890, with valuable ideas being contributed by several different chemists. It is not essential that we retrace the development of acid–base concepts. Rather, we shall state definitions that have the most general application to a variety of chemical situations.

The typical *acid and base react with each other to produce a salt.* If the base is a hydroxide, water is formed also. Such behavior is illustrated by Eq. (11–10).

an acid		a base		a salt		
HCl	+	KOH	→	KCl	+	H_2O
hydrochloric acid		potassium hydroxide		potassium chloride		water

(11–10)

This type of chemical reaction is often termed **neutralization** because the strongly corrosive properties of both the acid and the base have been cancelled out and in their place we have a salt, which in general is regarded as "neutral." The salt certainly is much less corrosive and irritating, although it may not be entirely neutral in the same sense that water is.

The definitions of acids and bases need not be complicated for most purposes. An **acid** *is a proton donor.* A **base** *is a proton acceptor.* Or, to say it in different words, an acid is any molecule or ion that gives up hydrogen ions (protons), while a base is any molecule or ion that combines with protons. Note that the terms "hydrogen ion" and "proton" are equivalent. In Section 4.6 we saw that a hydrogen atom consists of one proton and one electron. During certain reactions a hydrogen atom may give up its electron; the H^+ ion that results is only a proton. (The chemical meaning of the term *salt* will be described in Section 11–6.)

In Eq. (11–10) hydrochloric acid, HCl, donates a proton. The base that accepts the proton is KOH or, more precisely, the OH^- ion. This behavior is even more apparent if we rewrite the equation to show the electrolytes in their ionic form:

$$H^+ + Cl^- + K^+ + OH^- \xrightarrow[\text{solution}]{\text{water}} K^+ + Cl^- + HOH \qquad (11\text{--}11)$$

Here it is seen that neutralization occurs when the H^+ ion from the acid combines with the OH^- ion, the base. This particular type of acid–base reaction produces a covalent water molecule, and the remaining ions, K^+ and Cl^-, constitute a salt. In the next three sections the most typical acids, bases, and salts will be discussed in more detail.

11.4
ACIDS

The term **acid** is from the Latin word *acidus*, meaning sour. All water-soluble acids taste sour unless they are very weak. Furthermore, only acids taste sour. (*Caution:* Although the acids in foods such as lemonade or pickles may be eaten, in general it is *not* safe to taste acids, or any other chemicals.) If we examine the formulas of a few common acids,

| hydrochloric | HCl | carbonic | H_2CO_3 | sulfuric | H_2SO_4 |
| nitric | HNO_3 | boric | H_3BO_3 | phosphoric | H_3PO_4 |

we note that the only element common to all is hydrogen. However, not all compounds containing hydrogen are acids; for instance, water (H_2O), ammonia (NH_3), methane (CH_4), and sucrose (common sugar, $C_{12}H_{22}O_{11}$) are not acids in water solution. In order to be an acid, a compound must have one or more hydrogen atoms bonded to other atoms in such a manner that hydrogen *ions* can be released at the moment of reaction. This means that the "acidic" hydrogen must be held either by an ionic bond or by a very polar covalent bond that can easily *become* ionized.

It is the hydrogen ions released by an acid that cause the sour taste. The hydrogen ions also change the color of an indicator; this provides a much safer and more accurate method for detecting the presence of an acid. An indicator is an organic compound that has a certain color when placed in a basic solution but has a different color in an acidic solution. There are many such compounds, but one of the most common is litmus, a vegetable substance obtained from lichens. In basic solution litmus is blue; in acidic solution it is pink. A piece of paper impregnated with litmus is often used for testing. If a drop of solution placed on a blue litmus paper changes the color to pink, we conclude that the solution contains an acid.

Nomenclature of Acids and their Salts

Some common acids are called **binary acids** because they consist of just two elements, hydrogen plus either sulfur or a halogen (Group VIIA element). The names of binary acids have the prefix *hydro-* and the ending *-ic*. A salt derived from such an acid has the name-ending *-ide*:

| HBr | hydrobromic acid | H_2S | hydrosulfuric acid |
| KBr | potassium bromide | Na_2S | sodium sulfide |

Note that these salt names correspond to the rules given in Section 6.10.

Other acids are made up of three elements. Each of the nonmetals and several metalloids will form compounds with hydrogen and oxygen known as **oxyacids** or **ternary acids**. The most common examples are listed in Table 11.1, together with the

TABLE 11.1   Names of some common oxyacids and their salts

Acids		Salts	
Formula	Name	Formula	Name
H_2SO_4	sulfuric acid	$CaSO_4$	calcium sulfate
H_2SO_3	sulfurous acid	$MgSO_3$	magnesium sulfite
HNO_3	nitric acid	$Ba(NO_3)_2$	barium nitrate
HNO_2	nitrous acid	$LiNO_2$	lithium nitrite
H_3PO_4	phosphoric acid	Na_3PO_4	sodium phosphate
H_3PO_3	phosphorous acid	K_3PO_3	potassium phosphite

names of the corresponding salts. (Oxyacids of the halogens will be discussed in Chapter 14.) In the formula of an oxyacid, hydrogen is customarily written first and the oxygens last, as in HNO_2. But structurally the oxygens are bonded to the "central" atom (N in this example) and the hydrogen is in turn bonded to an oxygen. An example is the structure of nitrous acid shown in Fig. 6.12.

Most of the nonmetals that form oxyacids can have more than one oxidation state, with the highest oxidation number corresponding to the periodic group number. The oxyacid containing the central element in its highest oxidation state has a name ending in *-ic*, as a rule, while an oxyacid involving a lower oxidation number has a name ending in *-ous*:

sulfur*ic* acid, H_2SO_4 sulfur (Group VIA) has oxidation number 6
sulfur*ous* acid, H_2SO_3 sulfur has oxidation number 4

It is helpful to remember that an *-ous* acid always has one less oxygen than an *-ic* acid of the same central element. As shown in Table 11.1, an oxyacid having an *-ic* name-ending forms a salt having the name-ending *-ate*, whereas an *-ous* acid forms a salt with an *-ite* name.

An oxyacid that has more than one hydrogen is called a **polyprotic acid**. The terms *monoprotic*, *diprotic*, or *triprotic* are used to designate the number of protons (hydrogen ions) a certain acid can donate to a base:

HNO_3 monoprotic H_2SO_4 diprotic H_3PO_4 triprotic

Diprotic and triprotic acids ionize in two or three steps, respectively:

$H_2SO_4 \leftrightharpoons H^+ + HSO_4^-$ $H_3PO_4 \leftrightharpoons H^+ + H_2PO_4^-$

$HSO_4^- \leftrightharpoons H^+ + SO_4^{2-}$ $H_2PO_4^- \leftrightharpoons H^+ + HPO_4^{2-}$

$HPO_4^{2-} \leftrightharpoons H^+ + PO_4^{3-}$

Occurrence and Preparation of Acids

Acids occur throughout nature in both living and nonliving substances. Some common acids and their sources are listed below:

Acid	Source	Acid	Source
citric	citrus fruits	carbonic	all natural waters
tartaric	grapes	hydrochloric	
oxalic	rhubarb	hydrofluoric	volcanic gases
acetic	vinegar	sulfuric	
hydrochloric	stomachs of animals		

Aside from the natural sources, many acids can be prepared in a laboratory or factory. One simple method to obtain a volatile acid is to treat one of its salts with a nonvolatile acid, such as sulfuric:

$$H_2SO_4 + 2NaCl \rightarrow Na_2SO_4 + 2HCl \uparrow \tag{11-12}$$

$$H_2SO_4 + FeS \rightarrow Fe_2SO_4 + H_2S \uparrow \tag{11-13}$$

$$H_2SO_4 + KNO_3 \xrightarrow{\Delta} KHSO_4 + HNO_3 \uparrow \tag{11-14}$$

HCl and H_2S are gases that escape from the reaction mixture and can be captured in another vessel. HNO_3 is a liquid, so additional heat must be applied to make it all boil out of the mixture. (The reaction itself supplies some heat.)

Water-soluble oxides of nonmetals react with water to form acids:

$$SO_2 + H_2O \rightarrow H_2SO_3 \tag{11-15}$$

$$SO_3 + H_2O \rightarrow H_2SO_4 \tag{11-16}$$

$$P_2O_5 + 3H_2O \rightarrow 2H_3PO_4 \tag{11-17}$$

Since the addition of water to the *soluble* oxide of a nonmetal produces an acid, the nonmetallic oxide is called an **acid anhydride** (*anhydro* means "without water"). However, the oxides of some nonmetals are *not* soluble in water. Silicon dioxide, SiO_2, is the major constituent of sand and an important part of all granite-type rocks. This nonmetallic oxide is very insoluble, and under ordinary conditions will *not* react with water to form detectable amounts of acid. A sand pile does not become acidic after a rainstorm!

Industrially, sulfuric acid is produced by reaction (11–16). Most of the SO_3 required is obtained by oxidation of elemental sulfur, but considerable amounts can be recovered at those metal refineries that process sulfide ores. The refining first produces SO_2 gas, which if allowed to escape, not only would be wasted but would severely pollute the air. The SO_2 can be captured, further oxidized to SO_3, and then converted to sulfuric acid, a useful product. The total United States production of sulfuric acid in 1979 was about 42 million tons.

Chemical Properties of Acids

Acids react with hydroxides of metals to form salts and water. As we have seen, metallic hydroxides are basic compounds.

$$2NaOH + H_2SO_4 \rightarrow Na_2SO_4 + 2HOH \tag{11-18}$$

$$Ca(OH)_2 + 2HNO_3 \rightarrow Ca(NO_3)_2 + 2HOH \tag{11-19}$$

$$Al(OH)_3 + 3HCl \rightarrow AlCl_3 + 3HOH \tag{11-20}$$

Acids react with oxides of metals to form salts and water:

$$Na_2O + H_2SO_4 \rightarrow Na_2SO_4 + HOH \tag{11-21}$$

$$CaO + 2HNO_3 \rightarrow Ca(NO_3)_2 + HOH \tag{11-22}$$

It is interesting to compare Eq. (11–21) with (11–18) and Eq. (11–22) with (11–19). Metallic oxides are basic anhydrides (Section 11.5) and have the same relation to hydroxides that acid anhydrides have to acids.

Acids react with a number of metals to produce salts and hydrogen gas:

$$Zn + H_2SO_4 \rightarrow ZnSO_4 + H_2 \uparrow \qquad (11-23)$$

$$Fe + 2HCl \rightarrow FeCl_2 + H_2 \uparrow \qquad (11-24)$$

Metals that react with an acid are often called "active" metals; the principle will be discussed further in Section 13.6. Other metals, such as copper, silver, or gold, do not react with an acid to release hydrogen.

Acids react with carbonates to form salts, water, and carbon dioxide. Actually, two separate reactions occur, represented by the next two equations:

$$2HCl + Na_2CO_3 \rightarrow 2NaCl + H_2CO_3 \qquad (11-25)$$

The carbonic acid (H_2CO_3) that forms is unstable and promptly decomposes to yield water and carbon dioxide:

$$H_2CO_3 \rightarrow H_2O + CO_2 \uparrow \qquad (11-26)$$

The two preceding equations can be combined into one equation that represents the overall chemical change:

$$2HCl + Na_2CO_3 \rightarrow 2NaCl + H_2O + CO_2 \uparrow \qquad (11-27)$$

Another example of the reaction between an acid and a carbonate is the action of nitric acid on limestone (calcium carbonate):

$$2HNO_3 + CaCO_3 \rightarrow Ca(NO_3)_2 + H_2O + CO_2 \uparrow \qquad (11-28)$$

Geologists use this reaction to identify carbonates in mineral specimens. If gas bubbles are observed when an acid is dropped on a rock, it is good evidence that the specimen contains a carbonate.

**11.5
BASES**

Compounds that are bases in water solution have several distinctive properties in common. They have a bitter taste; turn pink litmus paper to blue; have a slippery, soapy feeling; and neutralize acids. Bases range from extremely caustic substances (NaOH, KOH) capable of destroying flesh and fabrics, to mild compounds (NH_3) which may be used in household cleaning. Magnesium hydroxide ($Mg(OH)_2$, milk of magnesia) may be taken internally. The basic strength of most metallic hydroxides depends on their solubility, since this determines the concentration of hydroxide ions released into solution. Sodium and potassium hydroxides are very soluble in water, are completely ionized, and are the strongest of the common bases. Such bases (hydroxides of Group IA metals) are also called *alhalis*.

We have defined bases as being *proton acceptors*. In a metallic hydroxide, such as NaOH, the proton acceptor is really the hydroxide ion:

$$Na^+ + OH^- + H^+ + Cl^- \rightarrow HOH + Na^+ + Cl^- \qquad (11-29)$$

Fundamentally, a hydroxide ion is a base because it has a nonbonded pair of electrons available which it readily shares with a proton:

$$H:\overset{..}{\underset{..}{O}}:^- + H^+ \rightarrow :\overset{\overset{\textstyle H}{..}}{\underset{..}{O}}:H$$

It is not necessary for the base to have a negative charge; having an electron pair is the important feature. For example, NH_3 is a neutral molecule, which is a base because it has an electron pair to be shared with a proton. Thus, when ammonia gas is mixed with hydrogen chloride gas they form a white salt, ammonium chloride; tiny crystals of the solid salt settle out, much like a snowfall.

$$H:\overset{\overset{\textstyle H}{..}}{\underset{\underset{\textstyle H}{..}}{N}}: \quad H:\overset{..}{\underset{..}{Cl}}: \rightarrow H:\overset{\overset{\textstyle H}{..}}{\underset{\underset{\textstyle H}{}}{N}}\overset{+}{\!}:H \qquad :\overset{..}{\underset{..}{Cl}}:^- \tag{11–30}$$

ammonia   hydrogen   ammonium chloride
           chloride NH_4Cl

In reaction (11–30) hydrogen chloride is the proton donor; ammonia is the proton acceptor. If a solution of NH_3 in water were mixed with a water solution of HCl, the same acid–base reaction would occur, with just two minor differences: (1) The HCl would already be ionized into separate H^+ and Cl^- particles because of the influence of the water, and (2) the white cloud of small NH_4Cl crystals would not be seen because the salt would already be dissolved in water.

Acid–base reactions of the type shown in Eq. (11–30) are essential in all living organisms. Organic bases called *amines* are derivatives of an ammonia molecule in which one or more hydrogen atoms have been replaced by hydrocarbon groups. Nonetheless, the nitrogen with its pair of electrons has the same basic character that NH_3 does.

$$C_2H_5NH_2 + H^+Cl^- \rightarrow C_2H_5NH_3{}^+Cl^- \tag{11–31}$$

ethylamine   ethylammonium
                        chloride

Amino groups are present in amino acids, which in turn form protein material present in all living matter. In addition to many other important functions, they help control the hydrogen-ion concentration within the plant or animal body.

Occurrence
and Preparation
of Bases

A tremendous variety of amines, the ammonia-type bases, are produced by plants and animals; however, many are toxic and large concentrations of an amine are rarely found in any one organism. Certain amine compounds found in many plants are called *alkaloids*, meaning alkali-like, because of their distinctive basic properties. Nicotine, quinine, and morphine are alkaloids.

Among the inorganic or mineral examples we seldom find, in nature, hydroxides of typical metals, especially Group IA or IIA metals; such hydroxides are so basic that they would have been neutralized by CO_2 from the air or by contact with some acid. But industrially some hydroxides are manufactured for a variety of uses; the current annual production of sodium hydroxide in the United States is approximately 12 million tons. Calcium hydroxide, "slaked lime," is manufactured by the reaction shown in Eq. (11–33); it is used agriculturally to help neutralize acidic soils and for numerous other purposes. About 19 million tons of lime are produced in the United States each year.

Hydroxides are most often prepared by treating the oxide of a metal from Group IA or IIA with water:

$$\textbf{Na}_2\textbf{O} \quad + \textbf{H}_2\textbf{O} \rightarrow \quad \textbf{2NaOH} \tag{11-32}$$
sodium oxide   sodium hydroxide

$$\textbf{CaO} \quad + \textbf{H}_2\textbf{O} \rightarrow \quad \textbf{Ca(OH)}_2 \tag{11-33}$$
calcium oxide   calcium hydroxide
"lime"   ("slaked lime")

$$\textbf{MgO} \quad + \textbf{H}_2\textbf{O} \rightarrow \quad \textbf{Mg(OH)}_2 \tag{11-34}$$
magnesium oxide   magnesium hydroxide
("milk of magnesia")

Oxides of metals are called **basic anhydrides,** or sometimes basic oxides, because they combine with water to yield bases.

The other economically important type of base is ammonia, which is synthesized directly from the gaseous elements:

$$\textbf{N}_2 + \textbf{3H}_2 \xrightarrow{\text{catalyst}} \textbf{2NH}_3 \tag{11-35}$$

At present, about 18 million tons of ammonia are produced in the United States each year. Much of this is used in agriculture as a source of nitrogen fertilizer. Because ammonia is a mild base, it can be added directly to irrigation water. Or, depending on the needs of the particular crop, the ammonia can be treated with CO_2 and converted to $(NH_2)_2CO$, urea, which is also water-soluble and usable by plants. For still other purposes the ammonia can be neutralized with nitric acid; both ions $(NH_4{}^+, NO_3{}^-)$ of the resultant salt can be assimilated and used by plants. Additional quantities of ammonia are used in the production of certain plastics and synthetic fibers (nylon and acrylic).

Chemical Properties of Bases As previously stated, a common reaction of a base is that of neutralizing an acid:

$$\textbf{NaOH} + \textbf{HNO}_3 \rightarrow \textbf{NaNO}_3 + \textbf{HOH} \tag{11-36}$$
$$\textbf{2KOH} + \textbf{H}_2\textbf{SO}_4 \rightarrow \textbf{K}_2\textbf{SO}_4 + \textbf{2HOH} \tag{11-37}$$
$$\textbf{NH}_3 + \textbf{HCl} \rightarrow \textbf{NH}_4\textbf{Cl} \tag{11-38}$$
$$\textbf{2NH}_3 + \textbf{H}_2\textbf{SO}_4 \rightarrow \textbf{(NH}_4\textbf{)}_2\textbf{SO}_4 \tag{11-39}$$

Hydroxide bases react with acidic oxides (nonmetallic oxides) to form salts and water:

$$\textbf{Ca(OH)}_2 + \textbf{CO}_2 \rightarrow \textbf{CaCO}_3 + \textbf{HOH} \tag{11-40}$$
$$\textbf{2KOH} + \textbf{SO}_2 \rightarrow \textbf{K}_2\textbf{SO}_3 + \textbf{HOH} \tag{11-41}$$
$$\textbf{Ca(OH)}_2 + \textbf{SO}_3 \rightarrow \textbf{CaSO}_4 + \textbf{HOH} \tag{11-42}$$

Amphoteric Compounds We have seen that *nonmetals* can combine with hydrogen and oxygen to form oxyacids, whereas active *metals* combine with hydrogen and oxygen to yield hydroxides, which are bases. It is reasonable, then, that the hydroxides of some metals and metalloids near the center of the periodic table should exhibit both acidic and basic traits. Zinc, tin, and aluminum are typical examples. The hydroxides of these metals, which are mainly covalent, react with *either* strong acids or strong bases to form salts. Such a hydroxide must, therefore, be regarded as both an acid and a base and is said to be **amphoteric**. (The term comes from the Greek *amphoteros*, meaning both; it is closely related to the word amphibian.)

Tin hydroxide, for example, reacts as follows:

$$\text{Sn(OH)}_2 + 2\text{HCl} \;\rightarrow\; \text{SnCl}_2 + 2\text{HOH} \tag{11-43}$$

$$\text{Sn(OH)}_2 + 2\text{NaOH} \rightarrow \text{Na}_2\text{SnO}_2 + 2\text{HOH} \tag{11-44}$$

When tin hydroxide reacts as an acid, Eq. (11–44), its formula sometimes is written H_2SnO_2 to make it "look" like an acid. Its covalent structure is H—O—Sn—O—H. Whether it reacts as acid or base depends on whether H^+ or HO^- is removed by some other reactant. We found a similar structure in the oxyacids (for example H—O—N=O, Fig. 6.13), the main difference being a greater tendency for the common oxyacids to release H^+ ions, but not OH^- ions.

11.6 SALTS   *Salts are ionic compounds having positive ions other than hydrogen ions and negative ions other than hydroxide ions.* *

The **cation** may be an ammonium ion or a metal ion from:

Group IA,   Group IIA,   or a transition element

The most common are: NH_4^+, Na^+, K^+, Ca^{2+}, Ba^{2+}.

The **anion** may be from:

A nonmetal of Group VIA or VIIA,   or an ionic group (Tables 6.3 and 11.1)

The most common are: Cl^-, Br^-, S^{2-}, NO_3^-, SO_4^{2-}, PO_4^{3-}, CO_3^{2-}.

It should be emphasized that the existence of salts does not depend only on neutralization reactions. Salts are produced by many different chemical reactions, as we shall see in the next few paragraphs.

Occurrence and Preparation of Salts   Salts are widely dispersed over the surface of the earth and are found, in low concentrations, in the tissues of plants and animals. Great quantities of the water-soluble salts have accumulated in the oceans, or in rock strata that were formerly on the bed of an ocean or lake. The less soluble salts and small amounts of some soluble ones are found in other rocks and in soil.

The processing or mining of ordinary salt, sodium chloride, is one of the most ancient and most widely distributed industries on earth. Salt may be isolated by evaporation of seawater or of natural brines (concentrated solutions) trapped within the earth's crust. Salt can also be mined from underground deposits of "rock salt" crystals and from salt domes, all of which were covered by a layer of impervious rock and were therefore protected from water. In recent times, among the developed nations of the world, the processing of sodium chloride is often accompanied by the isolation of other valuable salts, especially of potassium salts needed for agriculture. The three major nutrients essential for plant growth are nitrogen, potassium, and phosphorus; commercial fertilizers contain various amounts of these, depending on the intended application.

---

* According to the most general definitions, basicity is entirely relative. The anion of *any* salt (not just an OH^- ion) can be regarded as a base because it has some ability to capture an H^+ ion. If the tendency to acquire a proton is great, the anion is considered a strong base; if the tendency is small, it is a weak base. In this sense, NaOH may be called either a base or a salt.

For small-scale laboratory purposes, salts may be obtained from a variety of chemical reactions; the types in the following list have all been mentioned previously in this text. Some of these reactions occur in nature as well as in the laboratory.

Reaction between an acid and a base:

$$H_2SO_4 + Mg(OH)_2 \rightarrow MgSO_4 + 2H_2O \qquad (11\text{--}45)$$

sulfuric   magnesium   magnesium
acid   hydroxide   sulfate

$$HBr + NH_3 \rightarrow NH_4Br \qquad (11\text{--}46)$$

hydrobromic   ammonia   ammonium
acid   bromide

Reaction between an acid and an active metal:

$$2HCl + Fe \rightarrow FeCl_2 + H_2 \uparrow \qquad (11\text{--}47)$$

hydrochloric   iron   iron (II)
acid   chloride

Reaction between an acid and a metallic oxide:

$$2HNO_3 + CaO \rightarrow Ca(NO_3)_2 + H_2O \qquad (11\text{--}48)$$

nitric   calcium   calcium
acid   oxide   nitrate

Reaction between a base and an acidic oxide:

$$2NaOH + CO_2 \rightarrow Na_2CO_3 + H_2O \qquad (11\text{--}49)$$

sodium   carbon   sodium
hydroxide   dioxide   carbonate

Direct union of elements:

$$2Na + Cl_2 \rightarrow 2NaCl \qquad (11\text{--}50)$$

sodium   chlorine   sodium chloride

Formation of an insoluble salt from two soluble salts:

$$AgNO_3 + NaCl \rightarrow AgCl \downarrow + NaNO_3 \qquad (11\text{--}51)$$

silver   sodium   silver   sodium
nitrate   chloride   chloride   nitrate

Chemical Properties of Salts

The major reactions of salts fall into some of the categories described in Section 7.6.

Hydrogen Salts, Hydroxysalts, and Normal Salts

The diprotic and triprotic acids we encountered in Section 11.4 can be neutralized step by step. For instance, if we add one mole of sodium hydroxide to one mole of sulfuric acid, only half of the hydrogen ions will be neutralized:

$$NaOH + H_2SO_4 \rightarrow NaHSO_4 + HOH \qquad (11\text{--}52)$$

sodium hydrogen
sulfate

The product, sodium hydrogen sulfate, is a salt because it has a metallic ion (Na^+), but it also still has an acidic hydrogen ion. Such a salt is properly called a **hydrogen salt**; the term *acid salt* has been used also. One mole of sodium hydrogen sulfate can, of course, still neutralize one mole of sodium hydroxide:

$$NaOH + NaHSO_4 \rightarrow \quad Na_2SO_4 \quad + HOH \tag{11–53}$$
$$\text{sodium sulfate}$$

In a similar fashion, dihydroxy or trihydroxy bases can be neutralized by acids stepwise:

$$HCl + Ca(OH)_2 \rightarrow \quad Ca(OH)Cl \quad + HOH \tag{11–54}$$
$$\text{calcium}$$
$$\text{hydroxychloride}$$

A salt of the type shown in Eq. (11–54) is called a **hydroxysalt**, or sometimes a *basic salt*. By adding to it one more mole of hydrochloric acid we could complete the neutralization:

$$HCl + Ca(OH)Cl \rightarrow \quad CaCl_2 \quad + HOH \tag{11–55}$$
$$\text{calcium chloride}$$

A **normal salt** is one that has neither hydrogen ions nor hydroxide ions. Examples are calcium chloride, Eq. (11–55); sodium sulfate, Eq. (11–53), and all the other salts shown in the previous sections of this chapter.

11.7
SOLUBILITY OF ACIDS, BASES, AND SALTS

In order to prepare solutions, to predict the outcome of a reaction (Section 7.6), or for other purposes, it is often important to know how soluble a compound is in water. Although the information can be found in a chemical handbook, it is very helpful to know the solubilities of the most common compounds. Fortunately, a few simple generalities summarize many solubility facts. The "rules" are the following.

Acids

All of the acids mentioned in this chapter are soluble in water; most of them are very soluble. We should note, however, that HCN or H_2S will escape rapidly from a *heated* solution (or slowly from a cold one) as a gas; both are also very poisonous. HNO_2 and H_2CO_3 decompose in warm solutions.

Bases

Ammonia is soluble. The hydroxides of the metals of Groups IA and IIA are soluble, except for beryllium and magnesium. Calcium hydroxide is moderately soluble. Other metallic hydroxides are insoluble.

Salts

The following salts are soluble:

1. All nitrates.
2. All salts of sodium, potassium, and ammonium.
3. All chlorides except those of silver, lead, mercury(I), and copper(I).
4. All sulfates except those of strontium, barium, and lead.

We can assume that all other salts are insoluble. Note especially that most carbonates and phosphates are *insoluble*; only those of the Group IA metals and ammonium are soluble (compare rule 2).

**11.8
STRONG AND
WEAK
ELECTROLYTES**

We have stated that acids, bases, and salts are electrolytes (Section 11.1) because the ions they provide in a water solution will conduct an electric current (Section 11.2). However, some of these compounds conduct a current poorly because they generate only a few ions in solution. With the apparatus shown in Fig. 11.2 one can easily determine, qualitatively, the extent of ionization of a compound dissolved in water. If the lamp does not glow at all, the compound is a nonelectrolyte. If the lamp glows very brightly, the compound is completely ionized and we call it a *strong electrolyte*. If the lamp glows weakly, it means only a few ions have been released into solution, and we say the compound is a *weak electrolyte*. A medium glow indicates a moderate electrolyte. For a quantitative measurement of the conductivity the lightbulb can be replaced by some sort of meter.

Salts

Since a salt is by nature ionic in the crystalline state, its degree of ionization depends upon its solubility at a given temperature. Any soluble salt provides many ions in solution and is a strong electrolyte. An insoluble salt, such as calcium carbonate, is a nonelectrolyte. Calcium sulfate is an example of a salt that is only slightly soluble and is therefore a weak electrolyte.

Bases

The same principles apply to the hydroxide bases, all of which are ionic in structure. The very soluble ones are strong electrolytes; the insoluble ones are nonelectrolytes. Some, of borderline solubility, would be weak or moderate electrolytes.

Ammonia, NH_3, is the other familiar base. Although it is soluble in water, when the solution is tested with a conductivity apparatus (Fig. 11–2) we find it is a *weak* electrolyte. If we replace the bulb in the apparatus with an appropriate meter, we can calculate from the dial reading that ammonia is only one-percent ionized in water solution. The molecules of NH_3 are electrically neutral and nonconducting. Since ammonia is a weak base,

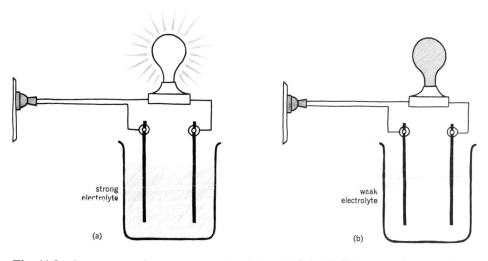

Fig. 11.2 Apparatus to demonstrate conductivity. (a) Salt (NaCl) in water is an excellent conductor. (b) Ammonia (NH_3) in water is a poor conductor.

only about one NH_3 out of every hundred is able to take a hydrogen ion away from a water molecule, thus creating some ions in the solution:

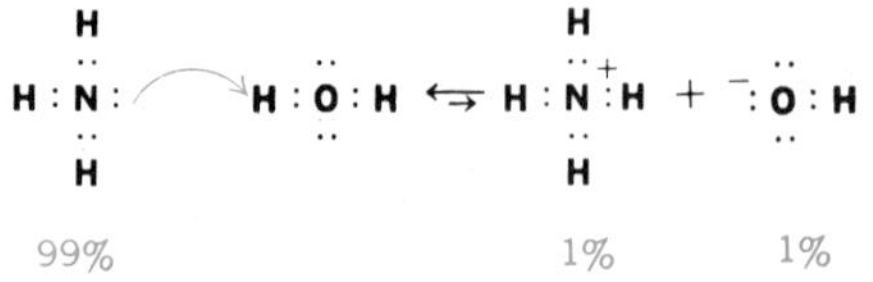

$$
\begin{array}{c}
H \\
\ddots \\
H : N : \qquad H : O : H \rightleftharpoons H : N : H \;+\; \overset{-}{:} O : H \\
\ddots \qquad\qquad\qquad\qquad\qquad \ddots \\
H \qquad\qquad\qquad\qquad\qquad\quad H
\end{array}
\qquad (11\text{--}56)
$$

99%   1%   1%

It is interesting to note that in this reaction a water molecule must serve as the acid, the proton donor. Because ammonia is a weak base, it is difficult for it to take a proton from water. The important principle here is that, if a base is a weak electrolyte (a poor conductor), automatically it is poorly ionized and is a **weak base**. Even so, if the proton donor is a strong acid, one mole of ammonia can react completely with one mole of the acid, Eq. (11–30).

Acids   Hydrogen chloride is a strong acid. When it dissolves in water, essentially all the HCl becomes ionized, Eq. (11–6), so it is a strong electrolyte. However, some other acids are weak electrolytes, despite the fact that they are water-soluble. As with ammonia, this depends not on their solubility but on the fact that they are weakly ionized in solution. This happens because they are weak acids; they are less able to force their hydrogen ions onto water molecules or other substances. An example is hydrocyanic acid (hydrogen cyanide):

$$
HCN \;+\; H_2O \overset{\star}{\rightleftharpoons} \; H_3O^+ \;+\; CN^- \qquad (11\text{--}57)
$$

hydrocyanic   hydronium   cyanide
   acid   ion   ion
99.999%   0.001%   0.001%

If we use a simpler equation, showing dissolved hydrogen ions as H^+ rather than H_3O^+, we can write:

$$
HCN \overset{H_2O}{\rightleftharpoons} H^+ + CN^- \qquad (11\text{--}58)
$$

The characteristic of an acid is its ability to generate hydrogen ions (or hydronium ions). Because hydrocyanic acid does this very poorly when alone in solution, it is designated as a *weak* acid. Equation (11–57) shows that at one particular moment nearly all the dissolved HCN molecules are covalent and only a few are ionized into H^+ and CN^- particles. It is important to distinguish this trait of a weak acid from its total *capacity* to provide hydrogen ions. If we add NaOH or some other strong base to the solution of HCN, the OH^- ions remove the H^+ ions from *all* of the HCN molecules. Thus one mole of HCN has the capacity to react with one mole of NaOH.

Among the common acids, the **strong acids** are:

HNO_3 H_2SO_4 HCl   HBr   HI   (H_3PO_4 is moderately strong)

$\star$ The longer arrow pointing toward the left is used to indicate that most of the substances in the solution are those shown on the left.

Some of the common **weak acids** are:

hydrocyanic acid	HCN	boric acid	H_3BO_3
hydrosulfuric acid	H_2S	carbonic acid	H_2CO_3
sulfurous acid	H_2SO_3	acetic acid*	CH_3COOH

Acetic acid, the sour compound in vinegar, is an organic substance. There are several hundred organic acids, nearly all of which are weak acids. The covalent structure of an organic acid can be represented by a general formula:

$$R-\underset{\underset{O}{\|}}{C}-O-H$$

With this structure it is important to note that ionization in water breaks the O—H covalent bond, *not* the C—O bond. So despite its appearance, an organic acid does not release OH^- ions.

There can be many variations in the R portion of the molecule. It may contain only one carbon atom, as in acetic acid, or as many as 50 or more; the carbon skeleton also holds hydrogen atoms and sometimes oxygen or nitrogen atoms. Often the formula for organic acids is written more briefly as RCOOH. This condensed structural formula is intended to show that two different oxygen atoms are bonded to the carbon and that one of the oxygens holds the "acidic" hydrogen atom. In an organic acid, only the hydrogen attached to an oxygen is potentially available to become an H^+ ion. Acetic acid, CH_3COOH, has only one acidic hydrogen; the three hydrogens bonded to carbon do not react with a base even as strong as NaOH. It is significant that organic acids are weak acids. Living organisms can hardly ever tolerate strong acids.

One curious exception is the digestive fluid in the human stomach, which contains 0.1 M hydrochloric acid. This can become irritating if, occasionally, the stomach produces too much acid. The antacid pills available from drug stores counteract much of the acid without making the stomach fluid completely neutral. And, of course, the compound used must not be a strong base, such as NaOH, which itself would damage the throat and stomach. A typical ingredient of the pills is milk of magnesia, $Mg(OH)_2$.

The same principle applies to first-aid treatments; use a *weak* base for neutralization. If a strong acid (battery acid, H_2SO_4, or swimming pool acid, HCl) is spilled on the skin, treat with ammonia solution or with baking soda ($NaHCO_3$). If a strong base (lye, Drano, etc.) is spilled, treat with vinegar (dilute acetic acid). For internal use, drink sour fruit juice or diluted vinegar. Milk is an excellent antidote. It soothes irritated membranes and the proteins it contains are buffers (Section 13.12) capable of neutralizing either acids or bases.

Acid–base balance is extremely important in all living organisms and in the practice of medicine. Most body fluids must be maintained near neutrality, with only very low concentrations of even a weak acid (acetic acid, etc.) or of a weak base (ammonia or an amine). In cases where a higher concentration of a weak acid, for example, is found in nature, usually it is confined to certain tissues. For instance, citric acid is largely in the lemon fruit, with only traces throughout the rest of the tree. Failure to maintain a proper acid base balance can be serious for an organism. If diabetes is not controlled the resultant acidosis (accumulation of acids in the body) is one of the factors leading to collapse or death.

---

* Because acetic acid is the most frequently discussed weak acid, it is often represented by the abbreviated formula AcOH, or HOAc. In this formula, the O holding the acidic hydrogen is shown and the Ac stands for the CH_3CO portion of the structure. Thus we have:

     HOAc acetic acid OAc^- acetate ion   NaOAc sodium acetate

The next chapter will include discussions of pH values, used to indicate acid or base concentrations; factors affecting the ionization of weak acids or bases; and exact mathematical terms for expressing the strength of a weak acid or base.

**11.9
TITRATION OF
ACIDS AND
BASES**

There are many occasions when it is necessary to know the amount of acid or of base present in a certain material. The food processor wants to determine the amount of acid in vinegar, tomato juice, or fresh fruits. The wine master needs to know how tart each batch of grape juice is. The medical technologist is asked to measure the amount of acid in the gastric fluid of a suspected ulcer patient. And the chemist needs to determine the amount of base or acid present in many different kinds of materials.

These determinations can be made by the **titration** method. The apparatus used is shown in Fig. 11.3. A burette is a long glass tube, accurately marked to show how many milliliters of solution it contains. If we read the scale to see how much solution is present, drain some into a flask, and then again read the scale, we can determine how much solution was drained into the flask. If we wish to know the amount of acetic acid in a vinegar solution, for example, we use one burette to measure out a definite amount of the vinegar into a flask. In the other burette we have a solution of base of known strength. We allow this to drip slowly into the vinegar in the flask until the acid is just neutralized, Eqs. (11–10) and (11–11). By observing the amount of base solution we needed to add, we can calculate the amount of acid in the vinegar sample. The condition at the end of the titration, when an equal number of H^+ ions and OH^- ions have been added to the flask, is called the *equivalence point* or the *endpoint*.

Of course, we need to have some way of knowing when the endpoint of a titration is reached. One convenient way is to use an indicator. For this purpose litmus is not

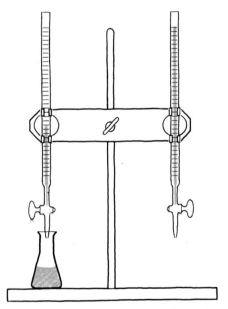

Fig. 11.3 Burettes are used to determine the strengths of acid or base solutions.

satisfactory because its change from blue to red gives shades of purple, making it difficult to distinguish the exact point of equivalence. Instead we can use the compound phenolphthalein. It is colorless in an acid solution; but as soon as the endpoint is reached, one small drop of excess base turns the phenolphthalein red.

Now let us see how we can calculate the results from a titration. In Section 11–3 a base was defined as an acceptor of protons (H^+ ions). Since one OH^- ion can accept exactly one H^+ ion, it is evident that one *mole* of NaOH would react with one *mole* of HCl, or that 0.25 mole of each would exactly neutralize each other, and so on. To put this statement into equation form we can say that, at the equivalence point:

Moles of H^+ donated by acid = Moles of acceptors in base   (11–59)

The proton acceptors could be, for example, OH^- ions in a sodium-hydroxide solution or NH_3 molecules in an ammonia solution.

We should pause to note that, because small quantities of solutions are required for titrations, the volumes are measured in *milli*liters; consequently the amount of acid or base present can be expressed conveniently in *milli*moles. Although molarity was originally defined in terms of moles per liter (Section 10.7), the concentration is numerically the same if the units used are millimoles (mmol) per milliliter (ml):

$0.1\ M = 0.1\ \text{mol/liter} = 0.1\ \text{mmol/ml}$

It is essential to consider the *concentration* of the solutions used for titration. Let us compare the two following examples of titrations. At the neutralization point:

	This much acid:	Requires this much base:
I	30 ml of 0.10 M HCl	30 ml of 0.10 M NaOH
II	30 ml of 0.20 M HCl	60 ml of 0.10 M NaOH

This is logical. All it amounts to is that in titration II the *concentration* of the HCl solution is twice as much as in I. Therefore, if we use the same *concentration* of NaOH solution, in example II we will need to use twice the *volume* of NaOH solution in order to obtain as many of the OH^- ions as we have of H^+ ions.

Although we may not consciously realize it at first, the reason this seems logical is that we are really applying Eq. (11–59); that is, we are making sure the moles of H^+ equal the moles of OH^-. Let us see how this works out mathematically for titration II. The number of moles of acid present is found by multiplying the concentration of the acid, M_A, by the volume of acid, V_A:

moles $H^+ = M_A \times V_A$ (11–60)

In titration II,

$$\text{moles } H^+ = \frac{0.2\ \text{mmol}}{\text{ml}} \times 30\ \text{ml} = 6.0\ \text{mmol}$$

Similarly, the number of moles of base would be:

moles $OH^- = M_B V_B$ (11–61)

In titration II,

$$\text{moles } OH^- = \frac{0.1\ \text{mmol}}{\text{ml}} \times 60\ \text{ml} = 6.0\ \text{mmol}$$

We can see that since we had 6.0 mmol of H^+ ions in titration II, we needed to supply an equivalent amount of OH^- ions, 6.0 mmol, to reach the neutral point.

Since Eq. (11–59) shows that the moles of H^+ ions must equal the moles of OH^- at the equivalence point, we can combine Eqs. (11–60) and (11–61) to obtain a general equation:

$$M_A \times V_A = M_B \times V_B \tag{11–62}$$

As shown by the following examples, this equation can be used for titrations involving any solutions of acids and bases, provided that the acid is one such as HCl, which releases just one mole of H^+ ions from each mole weight of acid, and that one mole weight of base (such as NaOH) supplies one mole of OH^- ions.

Example 1   During a titration 23.8 ml of 0.114 M NaOH solution was withdrawn from one burette. To neutralize this, what volume of HCl solution must be added from the other burette, if the HCl is 0.0972 M?

Solution   First we should write down what information we have and what we need to find:

$M_B = 0.114$ mmol/ml $V_B = 23.8$ ml
$M_A = 0.0972$ mmol/ml $V_A = ?$

Now we can substitute these values into Eq. (11–62):

$$\frac{0.0972 \text{ mmol}}{\text{ml}} \times V_A = \frac{0.114 \text{ mmol}}{\text{ml}} \times 23.8 \text{ ml}$$

Transposing gives:

$$V_A = \frac{0.114 \text{ mmol}}{\text{ml}} \times 23.8 \text{ ml} \times \frac{\text{ml}}{0.0972 \text{ mmol}} = 28.0 \text{ ml}$$

Example 2   A vinegar solution of unknown strength was placed in a burette; 24.5 ml of the vinegar was measured out and titrated with 1.03 M NaOH. The endpoint was reached when 20.1 ml of the NaOH solution had been added. What was the concentration of acid in the vinegar?

Solution   We can list these facts:

$M_B = 1.03$ mmol/ml $V_B = 20.1$ ml
$M_A = ?$ $V_A = 24.5$ ml

Using Eq. (11–62) we obtain:

$$M_A \times 24.5 \text{ ml} = \frac{1.03 \text{ mmol}}{\text{ml}} \times 20.1 \text{ ml}$$

$$M_A = \frac{1.03 \text{ mmol}}{\text{ml}} \times 20.1 \text{ ml} \times \frac{1}{24.5 \text{ ml}} = \frac{0.845 \text{ mmol}}{\text{ml}} = 0.845 \ M$$

Example 3   If 22.6 ml of an unknown potassium hydroxide solution required 25.3 ml of 0.0972 M HCl solution at the equivalence point, what was the concentration of the potassium hydroxide solution?

Solution   In this case we have:

$M_A = 0.0972$ mmol/ml $V_A = 25.3$ ml
$M_B = ?$ $V_B = 22.6$ ml

Our calculation becomes:

$$\frac{0.0972 \text{ mmol}}{\text{ml}} \times 25.3 \text{ ml} = M_{\text{B}} \times 22.6 \text{ ml}$$

$$\frac{0.0972 \text{ mmol}}{\text{ml}} \times 25.3 \text{ ml} \times \frac{1}{22.6 \text{ ml}} = M_{\text{B}}$$

$$0.109 \ M = M_{\text{B}}$$

In this section we have described how titrations are carried out, how to calculate the results, and how to apply the method to monoprotic acids (HCl, HNO_3) and monohydroxy bases (NaOH, KOH). In the next two sections we shall see that, by modifying the calculations, we can use the titration method with an even greater variety of acids and bases.

**11.10
EQUIVALENT
WEIGHTS**
When an acid is titrated with a base, at the endpoint one mole of H^+ ions has reacted with, or is chemically equivalent to, one mole of OH^- ions. However, for an acid such as H_2SO_4 the number of moles of sulfuric acid used in the reaction will *not* be the same as the number of moles of H^+ ions it supplies. From the formula we can see immediately that each molecule of H_2SO_4 supplies *two* H^+ ions, or that *one* mole of H_2SO_4 (6.02 $\times$ 10^{23} molecules) supplies *two* moles of H^+ ions (2 $\times$ 6.02 $\times$ 10^{23} ions). In other words, in a titration one mole of OH^- ions will be chemically equivalent to one mole of H^+ ions, but not equivalent to one mole of the whole compound, H_2SO_4.

For this reason it is convenient to define another term: an **equivalent** *is one mole of reacting units.* For an acid one equivalent is one mole of H^+ ions; for a base one equivalent is one mole of OH^- ions or one mole of NH_3 molecules. The number of equivalents supplied to a reaction by a compound can be determined from its formula. Thus, for H_2SO_4 there are two equivalents (eq) per mole of compound; or in some situations we may wish to think of it as 1 mol/2 eq. Likewise, $Ca(OH)_2$ provides 2 eq/mol, or 1 mol/2 eq. In either HCl or NaOH there is 1 eq/mol. The convenience of the term equivalent is that we can always be sure that 1 eq of NaOH will react with 1 eq of HCl, or of H_2SO_4, or of *any* acid (even including an acid whose formula we do not know). Again, for convenience in working with small quantities, we may choose to express them in milliequivalents (meq).

When we are concerned with the number of equivalents of a compound taking part in a reaction, we will want to measure it in terms of its equivalent weight, not its mole weight. The **equivalent weight** *is the number of grams per equivalent,* that is, the weight of compound (in grams) that provides one mole of reacting units. The definition of equivalent weight is also expressed by Eq. (11–63), and Eq. (11–64) shows how it can be calculated from the mole weight of the compound:

$$\text{eq wt.} = \frac{\text{g}}{\text{eq}} = \frac{\text{mg}}{\text{meq}} \tag{11–63}$$

$$\text{eq wt.} = \frac{\text{mole wt.}}{\text{eq}} = \frac{\text{g}}{\text{mol}} \times \frac{\text{mol}}{\text{eq}} = \frac{\text{g}}{\text{eq}} \tag{11–64}$$

Thus to determine the equivalent weight of H_2SO_4 we first calculate its mole weight to be 98 (see Section 7.2) and note from its formula that there is 1 mol/2 eq. Then using Eq. (11–64) we have:

$$\text{eq wt.} = \frac{98.0 \text{ g}}{\text{mol}} \times \frac{1 \text{ mol}}{2 \text{ eq}} = \frac{49.0 \text{ g}}{\text{eq}} \quad \text{(or 49.0 mg/meq)}$$

Additional illustrations of calculating an equivalent weight are the following.

Example 1   For calcium hydroxide determine (a) the number of equivalents per mole and (b) the equivalent weight.

Solution   First we must write the correct formula, $Ca(OH)_2$. Now we can work on the questions. (a) The formula shows two OH^- ions, so there are 2 eq/mol. (b) The mole wt. = 40.0 + 2(16.0) + 2(1.00) = 74.0 g/mol, so

$$\text{eq wt.} = \frac{74.0 \text{ g}}{\text{mol}} \times \frac{1 \text{ mol}}{2 \text{ eq}} = \frac{37.0 \text{ g}}{\text{eq}} \quad \text{(or 37.0 mg/meq)}$$

Example 2   For nitric acid, what is (a) the number of equivalents per mole and (b) the equivalent weight?

Solution   The formula is HNO_3, so for (a) we see 1 eq/mol. (b) Mole wt. = 1.00 + 14.0 + 3(16.0) = 63.0 g/mol, so

$$\text{eq wt.} = \frac{63.0 \text{ g}}{\text{mol}} \times \frac{1 \text{ mol}}{1 \text{ eq}} = \frac{63.0 \text{ g}}{\text{eq}} \quad \left(\text{or } \frac{63.0 \text{ mg}}{\text{meq}}\right)$$

In other words, for a compound such as this, which has 1 eq/mol, the equivalent weight is the same as the mole weight.

**11.11
NORMAL
SOLUTIONS**
For many purposes we have expressed the concentration of solutions in terms of mol/liter, or molarity. But for titrations of acids and bases, as well as other chemical reactions, it is particularly convenient to use **normality**, which *is equivalents per liter of solution.* An acid or base solution that contains one equivalent per liter is said to be one normal (1 N) in concentration. Or, if the solution contains 0.5 eq/liter we have:

$$0.5 \, N = \frac{0.5 \text{ eq}}{\text{liter}} = \frac{0.5 \text{ meq}}{\text{ml}}$$

When we use normality to express the concentration of the solutions we employ, we can be sure that equal volumes of solutions having the same normality will contain the same number of reacting units (the same number of equivalents).

In a titration, if we represent the normality of the acid solution by N_A and of the base solution by N_B, we can say that:

equivalents of acid = $N_A \times V_A$ (11–65)
equivalents of base = $N_B \times V_B$ (11–66)

Therefore:

$$N_A \times V_A = N_B \times V_B \tag{11-67}$$

The following examples apply the equations above.

Example 1   How many milliequivalents are contained in 24.0 ml of 1.27 N sodium hydroxide solution?

Solution   Eq. of base = $\dfrac{1.27 \text{ meq}}{\text{ml}} \times 24.0 \text{ ml} = 30.5$ meq

Example 2   What would be the answer if the base in Example 1 were calcium hydroxide?

Solution   The result would be exactly the same, even though one is NaOH and the other is $Ca(OH)_2$, because 1.27 N means 1.27 milli*equivalents* per ml, regardless of what the base is or what its mole weight is.

Example 3   What weight of base is present in the sample of solution (a) from Example 1 and (b) from Example 2?

Solution   The *weight* of compound present *will* depend upon its mole weight and number of equivalents.

(a) $\text{Wt.} = 30.5 \, \cancel{\text{meq}} \times \dfrac{1 \, \cancel{\text{mmol}}}{1 \, \cancel{\text{meq}}} \times \dfrac{40.0 \, \text{mg}}{\cancel{\text{mmol}}} = 1\,220 \, \text{mg} = 1.22 \, \text{g}$

(b) $\text{Wt.} = 30.5 \, \cancel{\text{meq}} \times \dfrac{1 \, \cancel{\text{mmol}}}{2 \, \cancel{\text{meq}}} \times \dfrac{74.0 \, \text{mg}}{\cancel{\text{mmol}}} = 1\,128 \, \text{mg} = 1.13 \, \text{g}$

Example 4   If 28.3 ml of 0.105 N HCl solution is required to titrate 25.9 ml of KOH solution of unknown strength, what is the concentration of the KOH solution?

Solution $N_A = 1.05$ meq/ml $V_A = 28.3$ ml
$N_B = ?$ $V_B = 25.9$ ml
$N_A \times V_A = N_B \times V_B$

$\dfrac{1.05 \, \text{meq}}{\text{ml}} \times 28.3 \, \text{ml} = N_B \times 25.9 \, \text{ml}$

$\dfrac{1.05 \, \text{meq}}{\text{ml}} \times \dfrac{28.3 \, \cancel{\text{ml}}}{25.9 \, \cancel{\text{ml}}} = N_B$

$N_B = \dfrac{1.15 \, \text{meq}}{\text{ml}} = 1.15 \, N$

It is also possible to do a titration when one of the reactants, either the acid or the base, is a solid compound. For instance, suppose we have a solid acid; a portion of it can be weighed, dissolved in water or alcohol, and then titrated with a base solution measured out from a burette. The number of equivalents of base used can be calculated from Eq. (11–66) and we know that there must be the same number of equivalents present in the sample of acid used. The next example illustrates this method.

Example 5   A sample of benzoic acid (an organic acid) weighing 269 mg was titrated with 0.106 N NaOH solution. The endpoint was reached when 20.8 ml of base had been added. (a) This sample of benzoic acid contained how many equivalents? (b) What is the equivalent weight of benzoic acid? (c) It is known that benzoic acid is a monoprotic acid. With this information, can you determine its mole weight?

Solution   (a)   The equivalents of base used for the titration will be given by Eq. (11–66) and we can write:

amount of benzoic acid = equivalents of base used

$$= \dfrac{0.106 \, \text{meq}}{\text{ml}} \times 20.8 \, \text{ml} = 2.205 \, \text{meq}$$

(b) In this experiment the amount of benzoic acid used, 269 mg, was found to contain 2.205 meq, therefore:

$$\text{eq wt.} = \frac{269 \text{ mg}}{2.205 \text{ meq}} = 122 \text{ mg/meq}$$

(c) If benzoic acid is monoprotic, one mole of it supplies one mole of H^+, or 1 eq/1 mol. Therefore the mole weight will be the same as the equivalent weight:

$$\text{mole wt.} = \frac{122 \text{ mg}}{\text{meq}} \times \frac{1 \text{ meq}}{1 \text{ mmol}} = 122 \text{ mg/mmol}$$

In these last three sections we have discussed one of the most common applications of the titration method, for reactions of acids with bases. Titration can also be used for quantitative measurements of many other kinds of chemical reactions, including reduction-oxidation reactions and precipitations in double substitution reactions. In these applications, too, it is most satisfactory to calculate the concentration of the solutions in normality units, and to consider the number of equivalents of reactant supplied by each mole of compound used. A solution of $AgNO_3$ could react, for instance, with either NaCl or $CaCl_2$.

$$Ag^+, NO_3^- + Na^+, Cl^- \rightarrow AgCl\downarrow + Na^+, NO_3^- \tag{11–68}$$

$$2Ag^+, NO_3^- + Ca^{2+}, 2Cl^- \rightarrow 2AgCl\downarrow + Ca^{2+}, 2NO_3^- \tag{11–69}$$

In Eq. (11–68) we find 1 eq Cl^-/1 mol NaCl and in Eq. (11–69) we see 2 eq Cl^-/1 mol $CaCl_2$.

SUMMARY

In this chapter we have seen that Arrhenius' ideas of ionization helped to establish a firm basis for understanding the behavior of electrolytes and for classifying these compounds as acids, bases, or salts. The solubility of salts and the observation that some compounds are good electrolytes and others are poor electrolytes helps explain why some reactions occur while others do not, and why some acids or bases are strong while others are weak. Later in this book we shall see that many of the compounds important in organic chemistry and biochemistry are also acids, bases, or salts, but that in addition there are several other types.

The name of a compound usually reveals whether it is an acid, a base, or a salt. From the name it is also possible to write the formula and to predict some of the important properties of the compound. Rules for naming the common acids, bases, and salts have been mentioned in this chapter.

Previously, in Section 7.7, we showed that definite amounts of reactants participate in chemical reactions and that when you represent a reaction by a balanced equation you can use the equation to calculate the amount of some reactant or product in terms of either moles or weight (g, lb, etc.). In the present chapter the method of titration has been described. It is one of the most versatile and accurate methods for measuring, in the laboratory, the amount of some substance participating in a chemical reaction. In this connection the useful concept of equivalent weight has been defined; in many of these measurements the equivalent weight is a more satisfactory unit of quantity than is the mole weight.

IMPORTANT TERMS

acid
acid anhydride
amphoteric
anion
base
basic anhydride
binary acid
cation
electrolyte

equivalent
equivalent weight
hydronium ion
hydrogen salt
hydroxysalt
ionization
neutralization
nonelectrolyte
normality

oxyacid
polyprotic acid
proton donor
salt
strong electrolyte
ternary acid
titration
weak electrolyte

WORK EXERCISES

1. What observations were made that set the stage for Arrhenius' development of the theory of ionization?

2. Illustrate the role of water molecules in the ionization of hydrogen chloride, HCl.

3. Why is an H^+ ion often called a proton?

4. Distinguish between a hydrogen ion and a hydronium ion.

5. Write equations for the action of sodium hydroxide with each of the acids listed in Table 11.1 and name the products.

6. Ammonia, NH_3, has been called a base. Explain, with an appropriate structural formula, *how* it can react with an acid.

7. What is the principal distinction between strong and weak acids?

8. Name and write the formulas of three polyprotic acids.

9. Name each of the salts formed in Eqs. (11–36) through (11–42).

10. Contrast the properties of acids and bases.

11. What should you look for in the formula of a compound to decide whether it is a salt?

12. Which of the following are salts: KI, CH_4, LiOH, $BaSO_4$, $CuBr_2$, SO_3, MgS, HNO_3?

13. Write the chemical equations for four reactions that produce salts.

14. Review the rules for the solubility of salts and write the names and formulas for two salts in each category.

15. List all the possible ions that might be found in the water solutions of each acid listed in Table 11.1.

16. What chemical reaction takes place when water containing carbonic acid, H_2CO_3, is warmed?

17. List oxides of two different elements that would react with water to form acids. Write the equations.

18. List oxides of four different elements that would react with water to form bases. Write the equations.

19. List three different oxides that would react with an acid. Write the equations.

20. List two different oxides that would react with a base. Write the equations.

21. Complete the following ionic equations, and name the products other than water. Indicate the insoluble substances formed.
    a) $NH_4^+ + Cl^- + Ca^{2+} + 2OH^- \rightarrow$
    b) $Cu^{2+} + SO_4^{2-} + Ba^{2+} + 2Cl^- \rightarrow$
    c) $2H^+ + SO_4^{2-} + K^+ + CN^- \rightarrow$
    d) $H^+ + Cl^- + Ba^{2+} + 2OH^- \rightarrow$
    e) $Al^{3+} + 3Cl^- + Na^+ + OH^- \rightarrow$
    f) $Ca^{2+} + S^{2-} + H^+ + Cl^- \rightarrow$
    g) $Ag^+ + NO_3^- + Al^{3+} + 3Cl^- \rightarrow$
    h) $3H^+ + PO_4^{3-} + Fe(OH)_3 \rightarrow$
    i) $2K^+ + CO_3^{2-} + H^+ + Br^- \rightarrow$

22. Complete and balance the following as molecular equations, name the salts formed, and then rewrite as total ionic equations.
    a) $NaOH + HCl \rightarrow$
    b) $KOH + H_2SO_4 \rightarrow$
    c) $Ca(OH)_2 + HNO_3 \rightarrow$
    d) $Ba(OH)_2 + H_3PO_4 \rightarrow$
    e) $NH_3 + HNO_3 \rightarrow$
    f) $NH_3 + H_2SO_4 \rightarrow$
    g) $Al(OH)_3 + H_2SO_4 \rightarrow$
    h) $Zn(OH)_2 + HNO_3 \rightarrow$
    i) $H_2SO_3 + LiOH \rightarrow$
    j) $Zn(OH)_2 + H_2S \rightarrow$

23. Complete the following, and name the products.
    a) $HCl + Na_2CO_3 \rightarrow$
    b) $CaCO_3 + HNO_3 \rightarrow$
    c) $H_2O + SO_3 \rightarrow$
    d) $H_2SO_4 + NaCl \rightarrow$
    e) $CaO + H_2O \rightarrow$
    f) $MgO + HNO_3 \rightarrow$
    g) $Al_2O_3 + H_2SO_4 \rightarrow$
    h) $K_2O + HNO_3 \rightarrow$
    i) $H_2O + CO_2 \rightarrow$
    j) $H_3PO_4 + Na_2O \rightarrow$

24. What is a diprotic acid?
25. One mole of HCN can neutralize one mole of NaOH, yet HCN is called a weak acid. Explain.
26. Calculate the number of grams of solute required to prepare the following solutions:
    a) 2 liters of 0.5 N NaOH
    b) 5 liters of 2 N NaOH
    c) 2 liters of 0.5 N H_2SO_4
    d) 6 liters of 1 N H_3PO_4
    e) 4 liters of 0.1 N $Ca(OH)_2$
    f) 200 ml of 0.1 N NaOH
27. What is the normality of solutions containing the following amounts of solute?
    a) 10 g of NaOH/liter
    b) 49 g of H_2SO_4/liter
    c) 0.04 g of NaOH/ml
    d) 0.5 g of NaOH/10 ml
    e) 0.0365 g of HCl/2 ml

28. Twenty-five ml of 0.6 N NaOH is exactly neutralized by 30 ml of a solution of HCl. What is the normality of the HCl.
29. Fifteen milliliters of 0.6 N HCl is neutralized by 10 ml of NaOH. What is the normality of the base?
30. Solve the normality of the following solutions.
    a) 24 ml of 0.8 N H_2SO_4 = 48 ml of NaOH
    b) 16 ml of 3 N NaOH = 48 ml of HCl
    c) 18 ml of 0.1 N $Ca(OH)_2$ = 30 ml of H_3PO_4
    d) 18 ml of 0.1 N $Ca(OH)_2$ = 12 ml of H_2SO_4
31. How many grams of solute are present in each of the following?
    a) 1 liter of 0.30 N NaOH
    b) 1 liter of 2.0 N H_2SO_4
    c) 500 ml of 0.60 N NaOH
    d) 250 ml of 2.0 N HCl
    e) 10 ml of 0.50 N H_3PO_4

12

Equilibrium

When we observe a physical or chemical process we think of it as going in a certain direction. Sugar dissolves in water, snow precipitates from clouds, and acids are neutralized by bases. However, the previous chapters have given examples of situations in which two opposite processes were occurring at the same time. When we add salt crystals to a solution already saturated with the salt, some salt is always dissolving and some precipitating. Weak acids such as acetic acid ionize in water to give only a few hydrogen ions and acetate ions, and these two ions react with each other to give acetic acid. Upon further study you will find that most chemical and physical processes do not go completely in one direction, but that a balance point is reached where reaction in one direction and in the reverse direction—solution and precipitation, dissociation and association—are both occurring. The point at which the two processes are occurring at equal rates is the point of **equilibrium**.

One of the common examples of an equilibrium situation is a seesaw. If two people of equal weight are on the seesaw and they sit perfectly still, a point of equilibrium can be reached. A chemical equilibrium is different, it is never static. Although you may weigh the same amount from day to day, you are aware that some molecules are entering your body and some leaving. Although the amount of a solute in a saturated solution and the amount that remains as undissolved solid do not change after an equilibrium has been reached, individual solute particles are continually entering and leaving solution. If we heat the solution and this increases the solubility, we will disturb the equilibrium, but if there is still some undissolved solute present, a new point of equilibrium will be established. We can illustrate these principles further in the following.

Suppose we take a large bottle about half full of water, add some sulfur-dioxide gas, close the bottle with a stopper, and then allow it to stand for a period of time (Fig. 12.1). We may even shake the stoppered bottle to hasten any processes that might be occurring. If

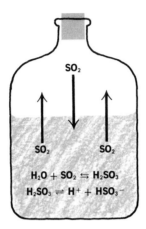

Fig. 12.1 Equilibria in a closed bottle containing water and sulfur dioxide.

we then remove the stopper, we can detect the odor of sulfur dioxide. On first thought, we might conclude that sulfur dioxide does not dissolve in water. However, if we remove some of the water and heat it, we find that sulfur dioxide is released. We can conclude, therefore, that sulfur dioxide is soluble in water, but that sulfur-dioxide molecules also tend to escape from a water solution and that the latter process is increased by heating. We would assume that, when a bottle containing sulfur dioxide and water is allowed to stand for a period of time, an equilibrium is established and sulfur dioxide molecules are entering the solution at the same rate as that at which others are escaping.

If we test the water in the container with litmus paper we find it is acid. The hydrogen ions in our solution of sulfur dioxide shows that it not only dissolves in the water, but also reacts with it to form sulfurous acid:

$$HOH + SO_2 \rightarrow H_2SO_3 \tag{12-1}$$

The sulfurous acid then dissociates to produce hydrogen and hydrogen sulfite ions:

$$H_2SO_3 \rightarrow H^+ + HSO_3^- \tag{12-2}$$

If we remove the stopper from our bottle of sulfur dioxide and water that has reached equilibrium and allow molecules of sulfur dioxide gas to escape from the bottle, and especially if we apply heat or reduce the pressure above the liquid by use of a vacuum pump, we find that the water no longer turns blue litmus red. From this we can conclude that there was more than one equilibrium involved in our bottle of sulfur dioxide and water. The sulfur-dioxide molecules in the space above the solution are in equilibrium with the dissolved sulfur-dioxide molecules. There is another equilibrium in which dissolved sulfur-dioxide molecules and water molecules combine to form molecules of sulfurous acid, which can in turn decompose to give back sulfur dioxide and water, Eq. (12–1). There is a further equilibrium between molecules of sulfurous acid and hydrogen and hydrogen-sulfite ions, Eq. (12–2).

When we remove the stopper from the bottle, the equilibrium between dissolved and gaseous molecules of SO_2 shifts (Fig. 12.2). The sulfur-dioxide molecules tend to leave

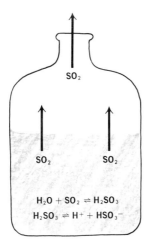

Fig. 12.2 Effect of removing stopper from container of water and sulfur dioxide.

the solution, which in turn causes the decomposition of some of the sulfurous-acid molecules. Consequently, fewer hydrogen and hydrogen-sulfite ions remain in the solution. Continuation of these processes results in the formation of pure water and the escape of all the SO_2 as gas.

Thus in one experiment three types of equilibria have been illustrated. These are: (1) an equilibrium involving the solubility of a substance; (2) an equilibrium involving a reversible chemical reaction; and (3) an equilibrium that involves the dissociation of a substance into ions. These are three major types of equilibria encountered in chemistry. Some equilibria that are of importance to living systems are the equilibrium between oxygen (and CO_2) in the air and that dissolved or bound to hemoglobin in the blood. Another set of equilibria involves the reversible reactions and dissociations that are observed when CO_2 reacts with water to form H_2CO_3, which then dissociates to give hydrogen and hydrogen-carbonate ions in the human body. Such reactions are important in maintaining a correct acid–base balance.

The illustration of the behavior of sulfur dioxide and water has also shown how altering conditions such as pressure and temperature may effect the point of equilibrium. Similar equilibria are involved when people live under lower atmospheric pressure at high altitudes and have to compensate for a decreased oxygen concentration in the air. Adjustment to different temperatures is very important, especially in the cold-blooded animals.

In discussing chemical reactions, we often use the letters A and B to indicate the substances that react (the reactants) and C and D to represent the products. Thus a generalized chemical reaction becomes

$$A + B \rightarrow C + D$$

If we allow A to react with B and find after a time that some C and D have been produced but that there are still considerable amounts of A and B present, we say that the reaction is *reversible* or that an equilibrium has been established. We use two sets of arrows to indicate this equilibrium

$$A + B \rightleftarrows C + D$$

Thus we can rewrite Eqs. (12–1) and (12–2) to show that they are reversible and really represent equilibrium situations.

$$HOH + SO_2 \rightleftharpoons H_2SO_3 \tag{12-3}$$

$$H_2SO_3 \rightleftharpoons H^+ + HSO_3^- \tag{12-4}$$

If an equilibrium is established in which there is more of A and B than of C and D, we sometimes use arrows of different lengths to indicate this fact.

$$A + B \underset{\longleftarrow}{\overset{\longrightarrow}{\rightleftharpoons}} C + D$$

It is believed that after any reaction there are at least a few particles of A, B, C, and D, so that theoretically all reactions are reversible; however, the amounts of A and B are often so small that they cannot be detected. In such cases they can be disregarded and the equation can be thought of as going almost completely to the right. If the opposite is true—if the amounts of A and B are almost unchanged and the amounts of C and D are extremely small—we say that no reaction has occurred.

Some reactions involve only one reactant

$$A \rightarrow C + D$$

or one product

$$A + B \rightarrow C$$

Obviously these reactions are just simplifications of the general reaction involving two reactants and two products.

**12.3
EFFECTS OF
TEMPERATURE
AND
PRESSURE ON
EQUILIBRIUM**

For a chemical reaction to occur, we must have a collision of two (or perhaps more) particles (ions, atoms, molecules) that are in a reactive state. The probability that any reaction will occur (that A will react with B, or C with D) depends on the concentration of reactive particles in the space in which the reaction may take place. When we combine these ideas with those of equilibrium, we arrive at a better explanation of the effects of temperature and pressure on shifting an equilibrium in one direction or another, particularly in reactions of gases.

If the temperature of a sulfurous-acid solution is increased, the energy, and therefore the speed of the particles, also increases. Since high-speed molecules are more likely to be in the gaseous state, the increased velocity of the molecules increases the rate at which the sulfur-dioxide molecules escape from the solution and decreases the concentration of SO_2 molecules that are available to react with water; the equilibrium in Eq. (12–3) therefore shifts to the left. Conversely, a decrease in temperature causes sulfur-dioxide molecules to slow down, and when they reenter the water very few of them have sufficient velocity to escape from the solution. Since there are now more molecules of SO_2 in the water, the possibility that they will react with a water molecule is increased, and the equilibrium in Eq. (12–3) is shifted to the right (Fig. 12.3). Temperature has a similar effect on the

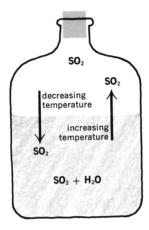

Fig. 12.3 Effect of temperature on the solubility of sulfur dioxide in water.

solubilities of most gases in liquids, although not all gases react with their solvents as SO_2 does with water.

Since gases are compressible, an increase in pressure causes an increase in the concentration of the molecules of a gas above a solution. Because there are more molecules in a given volume, the chance of their being captured by the liquid solvent is increased. Consequently, an increase in the pressure of a gas above a liquid increases the solubility of the gas in a liquid, and a decrease in pressure decreases the solubility.

The shift in equilibrium due to a change in conditions was described by Henri Le Châtelier in 1888 and is known as Le Châtelier's principle. We can state the principle as: *When a system in equilibrium is subjected to a stress, there is an adjustment to relieve the stress.* Knowledge and application of this principle enabled Haber to specify conditions for the important process for making ammonia from hydrogen and nitrogen.

$$N_2 + 3H_2 \rightleftharpoons 2NH_3 \tag{12-5}$$

Since the four molecules (one of nitrogen and three of hydrogen) on the left occupy more space than two molecules of ammonia, an increase in pressure (a stress) will drive the reaction to the right and increase the amount of ammonia synthesized. This synthesis of ammonia is very important in making fertilizers to increase the world's food supply.

A further application of the principle is given in the law of mass action, which says that *the rate of a chemical reaction is proportional to the active masses of the reacting substances.* The active mass refers to the number of particles available for chemical reaction. Thus a reaction that produces an insoluble precipitate as a product (C) will not proceed to produce more A and B because, with most of C not available for reaction, the rate of reaction of C and D is decreased.

$$A + B \rightarrow C\downarrow + D$$

**12.4
REACTIONS
THAT GO
PRIMARILY IN
ONE DIRECTION**

While reversible reactions that reach an equilibrium at which there are appreciable quantities of both reactants and products are important in many instances, most chemical reactions are performed to obtain large amounts of the products. To do this chemists must choose reactions that go primarily in one direction. Such reactions are virtually irreversible if the products do not react because they are either insoluble or not ionized. For instance, in the reaction of silver nitrate and sodium chloride. The silver chloride is insoluble and precipitates.

$$AgNO_3 + NaCl \rightarrow AgCl \downarrow + NaNO_3 \qquad (12\text{-}6)$$

The reaction of zinc with hydrochloric acid produces another type of insoluble substance, a gas.

$$Zn + 2HCl \rightarrow ZnCl_2 + H_2 \uparrow \qquad (12\text{-}7)$$

Neutralization of acids with bases produces water that is only slightly ionized, and we can write the reaction as going in one direction only.

$$2KOH + H_2SO_4 \rightarrow K_2SO_4 + 2HOH \qquad (12\text{-}8)$$

Actually, we will see later that the very small amount of ionization of water is very important for many purposes; however, it is correct to write the equation as shown above.

Another way of describing reactions that go primarily in one direction is to say that the active mass of one (or both) of the products of the reaction is very small and the law of mass action predicts that the reverse reaction will not proceed or will proceed very slowly.

**12.5
THE
EQUILIBRIUM
CONSTANT**

If we want to describe more precisely the situation at equilibrium we can use an **equilibrium constant**, abbreviated K_{eq}. We get this constant by multiplying the concentrations of the two products of the reaction and dividing by the number we get when we multiply the concentrations of the reactants. We express this mathematically as

$$K_{eq} = \frac{\text{concentration of C} \times \text{concentration of D}}{\text{concentration of A} \times \text{concentration of B}} \qquad (12\text{-}9)$$

We generally use brackets [ ] to indicate "concentration of," and the equation becomes

$$K_{eq} = \frac{[C][D]}{[A][B]} \qquad (12\text{-}10)$$

Example 1   What is the equilibrium constant for a reaction which at equilibrium contains 2 moles/liter of the products C and D and 0.5 moles/liter of reactants A and B?

$$K_{eq} = \frac{\dfrac{2 \text{ moles}}{\text{liter}} \times \dfrac{2 \text{ moles}}{\text{liter}}}{\dfrac{0.5 \text{ moles}}{\text{liter}} \times \dfrac{0.5 \text{ moles}}{\text{liter}}} = \frac{2 \times 2}{0.5 \times 0.5} = \frac{4}{0.25} = 16$$

Example 2 (a)   One molecule of ethyl acetate reacts with one of water in the presence of acid as a catalyst to give one molecule of ethyl alcohol and one of acetic acid.

ethyl acetate + water $\rightleftharpoons$ ethyl alcohol + acetic acid

If we start with 20.0 mmol of water and 5.00 mmol of ethyl acetate and find by titration at equilibrium that there are 2.75 mmol of acetic acid, what is the equilibrium constant for the reaction of ethyl acetate with water?

$$K_{eq} = \frac{[\text{ethyl alcohol}][\text{acetic acid}]}{[\text{ethyl acetate}][\text{water}]}$$

Concentration of acetic acid is 2.75 mmol. Concentration of ethyl alcohol is also 2.75 mmol, since one mole of ethyl alcohol is formed for each of acetic acid. Concentration of water = $20.0 - 2.75 = 17.25$ mmol. Concentration of ethyl acetate = $5.00 - 2.75 = 2.25$ mmol.

$$K_{eq} = \frac{2.75 \text{ mmol} \times 2.75 \text{ mmol}}{2.25 \text{ mmol} \times 17.25 \text{ mmol}} = 0.195$$

(b) Using the data above, what is the K_{eq} for the reaction of acetic acid and ethyl alcohol to form ethyl acetate?

Since the reaction is ethyl alcochol + acetic acid $\rightarrow$ ethyl acetate + H_2O, the K would be determined by transferring

$$K_{eq} = \frac{[\text{HOH}][\text{ethyl acetate}]}{[\text{acetic acid}][\text{ethyl alcohol}]} = \frac{17.25 \times 2.25}{2.75 \times 2.75} = 5.13$$

This reaction is the reverse of that in 2(a) and we find that the K values for a reaction and the reverse of the reaction are reciprocals of each other, i.e.,

$$1/5.13 = 0.195$$

(c) Does the equilibrium favor formation of ethyl acetate or ethanol and acetic acid?

Since the K_{eq} for the formation of ethyl acetate is 5.13, which is larger than that for the reverse reaction of 0.195, the equilibrium favors the formation of ethyl acetate.

The values we obtain for K_{eq} depend on temperature, pressure, and other variables and might be different if there were changes in these conditions. More complex calculations are required when one molecule of A reacts with more than one of B ($A + 2B \rightarrow 2C + D$) or when more than one particle of the product is formed.

Many of the chemical reactions used in chemical industries involve tons of particles. They usually reach an equilibrium in which there are a great many more particles of products than of reactants. We sometimes speak of such reactions as those that go to the right as irreversible reactions. Alternatively we could say the K_{eq} for these reactions is a very large number.

12.6
IONIZATION OF WEAK ACIDS

Some of the most important equilibria studied by chemists are the ionizations of weak acids. Acetic, carbonic, and similar acids are characterized as weak acids because they do not ionize to any appreciable extent. Although this means they are not useful in many chemical processes, it is their low degree of ionization that makes them especially important in other reactions, including those that take place in living organisms. We can write the ionization of acetic acid as follows.

$$H(C_2H_3O_2) + H_2O \rightleftharpoons H_3O^+ + C_2H_3O_2 \tag{12-11}$$

If we wish to express the ionization with a constant as we did for other equilibria, we use K_i as the symbol for the ionization constant and write the following:

$$K_i = \frac{[H_3O^+][C_2H_3O_2{}^-]}{[H_2O][HC_2H_3O_2]} \tag{12-12}$$

To simplify calculations we generally combine the concentration of water with K_i, since it is nearly constant for dilute solutions of acids. This gives us a value known as K_a. We also use H^+ in place of the hydronium ion H_3O^+ to get

$$K_i[H_2O] = K_a \quad \text{and} \quad K_a = \frac{[H^+][C_2H_3O_2^-]}{[HC_2H_3O_2]} \tag{12-13}$$

The K_a for acetic acid at 25°C is 1.8×10^{-5}.

We often use HA to represent any acid and write the general dissociation as

$$HA \rightarrow H^+ + A^-, \quad \text{with} \quad K_a = \frac{[H^+][A^-]}{[HA]} \tag{12-14}$$

Problems involving ionization of acids

Example 1   What is the hydrogen-ion concentration of a solution that contains 0.100 M acetic acid ?

$$K_a = \frac{[H^+][C_2H_3O_2^-]}{[HC_2H_3O_2]}$$

Since the number of hydrogen ions is equal to the number of acetate ions, we may say that each is equal to X and

$$K_a = \frac{(X)(X)}{[HC_2H_3O_2]} \quad \text{or} \quad X^2 = K_a[HC_2H_3O_2]$$

The number of acetic-acid molecules remaining is $0.100 - X$. However, X is so very small when compared to the number of undissociated acetic-acid molecules that subtracting it from 0.100 would not significantly change 0.100. In fact, the number subtracted would be smaller than the number of significant figures involved in the calculation. Therefore, we use 0.0100 as the concentration of $HC_2H_3O_2$ and the equation becomes

$$X^2 = K_a(0.0100)$$

Since we know that $K_a = 1.8 \times 10^{-5}$,

$$X^2 = (1.8 \times 10^{-5})(10^{-1}) = 1.8 \times 10^{-6}$$

The square root of 1.8 is 1.34 and that of 10^{-6} is 10^{-3}, so $X = 1.34 \times 10^{-3} =$ hydrogen-ion concentration of a solution of 0.100 M acetic acid.

The K_a gives an indication of the relative strength of an acid. If they are not already known or given in a problem, K_a values can be obtained from tables in handbooks. Such constants are found by calculations like the following.

If an acid were 99% ionized, the concentration of A^- and H^+ would be 0.99 and that of HA would be 0.01. The K_a would be

$$K_a = \frac{0.99 \times 0.99}{0.01} = 98$$

Thus strong acids have K_a values near 100. On the other hand the K_a for acetic acid is 1.8×10^{-5}, a very small number. Acids with small K_a values are weak, and the lower the K_a the weaker the acid. An acid with a K_a of 2×10^{-6} is weaker than one with a K_a of 2×10^{-5}.

Example 2   What is the K_a for an acid if a 0.200 M solution of the acid is found to contain 10^{-4} moles of hydrogen ions per liter?

$$\text{HA} \rightarrow \text{H}^+ + \text{A}^- ; \qquad K_a = \frac{[\text{H}^+][\text{A}^-]}{[\text{HA}]}$$

$$K_a = \frac{[10^{-4}][10^{-4}]}{2.00 \times 10^{-1}} = \frac{10^{-8}}{2 \times 10^{-1}} = \frac{10 \times 10^{-7}}{2 \times 10^{-1}} = 5 \times 10^{-6}$$

**12.7
IONIZATION
OF WATER**

We ordinarily think of water as not being ionized, and this is shown by the fact that pure water does not conduct an electric current. However, there is a slight ionization of water molecules in any aqueous solution.

$$\text{HOH} \rightleftharpoons \text{H}^+ + \text{OH}^- \tag{12-15}$$

We can write the constant for this ionization as

$$K_i + \frac{[\text{H}^+][\text{OH}^-]}{[\text{HOH}]} \tag{12-16}$$

As we did with the dissociation of acids, we can combine K_i with the concentration of water (since it is 55.55 moles/liter in solutions that are dilute, i.e., principally water). When we multiply K_i by 55.55 we obtain a value known as **K_w, the ionization constant** (sometimes called the ion product) **for water.**

$$K_i[\text{HOH}] = K_w = [\text{H}^+][\text{OH}^-] \tag{12-17}$$

We can determine experimentally that there are 10^{-7} or $1/10\,000\,000$ moles of hydrogen ions in a liter of pure water at 25°C and, of course, one hydroxide ion is produced for each hydrogen ion, so the concentration of hydroxide ions is also 10^{-7}. Knowing this, we can calculate K_w.

$$K_w = [\text{H}^+][\text{OH}^-] = 10^{-7} \text{ moles/liter} \times 10^{-7} \text{ moles/liter}$$
$$= 10^{-14} \text{ moles}^2/\text{liter}^2 \tag{12-18}$$

Since K_w is a constant it applies to the concentrations of hydrogen and hydroxide ions in any solution and not only to pure water.

Example 1   What is the concentration of hydroxide ions in a solution that contains 10^{-3} moles of hydrogen ions per liter? (It may be necessary to review the discussion of the use of powers of 10 in the *Study Guide*.

$$K_w = 10^{-14} = [\text{H}^+][\text{OH}^-]$$

$$[\text{OH}^-] = \frac{10^{-14} \text{ moles}^2/\text{liter}^2}{[\text{H}^+]} = \frac{10^{-14} \text{ moles}^2/\text{liter}^2}{10^{-3} \text{ moles/liter}} = 10^{-11} \text{ moles/liter}$$

The hydroxide-ion concentration is 10^{-11} moles/liter.

Example 2   What is the hydrogen-ion concentration of a solution of 0.0100 M KOH (assuming that the base is completely dissociated)?

$$[\text{OH}^-] = 0.010 \; M = 10^{-2} \text{ moles/liter}$$
$$[\text{H}^+][\text{OH}^-] = 10^{-14} \text{ moles}^2/\text{liter}^2$$

$$[\text{H}^+] = \frac{10^{-14} \text{ moles}^2/\text{liter}^2}{10^{-2} \text{ moles/liter}} = 10^{-12} \text{ moles/liter}$$

The hydrogen ion concentration is 10^{-12} moles/liter.

12.8
pH

Since it is more convenient to work with positive numbers, scientists usually speak of small concentrations of hydrogen and hydroxide ions in terms of their negative logarithms, or more precisely, as the negative of their logarithms.

If the hydrogen-ion concentration is 10^{-3} moles/liter, the -3 represents the logarithm of the concentration of the H^+ ion. The negative of this value is $+3$, or merely 3. The common symbol for the negative logarithm is the letter "p," so the negative log of the H^+ content becomes pH and we say that a solution containing 10^{-3} moles of H^+ per liter has a pH of 3. Similarly, pOH is the negative log of the hydroxide-ion content. The use of pH is very important in chemistry, particularly in biochemistry.

neutral

increasing acidity

increasing basicity

pH	0	1	2	3	4	5	6	7	8	9	10	11	12	13	14
hydrogen ion concentration (moles/liter)	10^0	10^{-1}	10^{-2}	10^{-3}	10^{-4}	10^{-5}	10^{-6}	10^{-7}	10^{-8}	10^{-9}	10^{-10}	10^{-11}	10^{-12}	10^{-13}	10^{-14}
hydroxide ion concentration (moles/liter)	10^{-14}	10^{-13}	10^{-12}	10^{-11}	10^{-10}	10^{-9}	10^{-8}	10^{-7}	10^{-6}	10^{-5}	10^{-4}	10^{-3}	10^{-2}	10^{-1}	10^0
pOH	14	13	12	11	10	9	8	7	6	5	4	3	2	1	0

Fig. 12.4   A comparison of ways of expressing acidity or basicity of solutions.

Figure 12.4 shows the relationship of pH, H^+ ion concentration, pOH, and the concentration of OH^- ions. Solutions having a pH of 7 contain equal numbers of OH^- and H^+ ions and are neutral. (Of course, we should not forget that such solutions always contain much greater quantities of undissociated water molecules.) Solutions with pH values less than 7 are acid and the pH decreases as the hydrogen ion concentration increases. An example will illustrate this principle.

Example 1 Compare the hydrogen-ion concentration of a solution with a pH of 4 with that of one having a pH of 2.

pH = 4; $[H^+] = 10^{-4} = 1/10\,000$ moles of H^+ per liter
pH = 2; $[H^+] = 10^{-2} = 1/100$ moles of H^+ per liter

A solution of pH 2 contains 100 (10^2) times as many hydrogen ions per liter as one of pH 4. It is 100 times as acid.

Thus solutions of pH 1 are very acid and those with a pH between 1 and 6 have decreasing acidity. Solutions with pH values of 8–9 are basic and a solution of pH 14 is very basic.

Example 2 Find both the hydrogen and hydroxide-ion concentrations of a solution with a pH of 10. What is the pOH of this solution?

$$K_w = [H^+][OH^-] = 10^{-14} \text{ moles}^2/\text{liter}^2$$

At pH 10,

$$[H^+] = 10^{10} \text{ moles/liter}$$

Thus

$$[OH^-] = \frac{10^{-14} \text{ moles}^2/\text{liter}^2}{10^{-10} \text{ moles/liter}} = 10^{-4} \text{ moles/liter}$$

$$pOH = 4$$

Since the pH increases as the pOH decreases and their product is always equal to 14, we can say that

$$pOH + pH = 14 \tag{12–19}$$

So far our discussion has involved only pH values in whole numbers, whereas many substances have pH values that are 3.2, 7.8, or other intermediate values. We can deal with such solutions using logarithms. You may need to review the use of logarithms in the *Study Guide* before proceeding.

Example 1 Find the pH of a solution having a hydrogen ion concentration of 5×10^{-4}.

log of 5 = 0.69897 = 0.70
log of 10^{-4} = -4.00

The log of 5×10^{-4} is

-4.00
$+0.70$
—————
-3.30 = log of $[H^+]$

Therefore, pH = 3.30.

Example 2 What is the hydrogen ion concentration of a solution having a pH of 5.40?
We know that the $[H^+]$ is $10^{-5.40}$. This represents a value between 10^{-5} and 10^{-6}, and -5.40 is obtained by subtracting a positive number from -6.00;

$-5.40 = -6.00 + 0.60$ or -6.00
                                    $+0.60$
                                    ————
                                      5.40

The hydrogen-ion concentration is a number that will be written in exponential form with the exponent of 10 being -6. The antilog of 0.60 is about 4 and we write the hydrogen ion concentration as 4×10^{-6}.

Although we have been discussing pH in terms of solutions of acids and bases, all solutions in which water is the solvent have a pH and pOH value. Human blood has a pH of 7.35, orange juice is a great deal more acid, having a pH between 3 and 4, fresh egg white (which is a solution of egg albumin and other substances in water) has a pH that is slightly alkaline. The values of the pH of some solutions as well as the juice from common foods are given in Table 12.1.

TABLE 12.1   Approximate pH values of some common substances

Substance	pH	
hydrochloric acid	0	very acid
gastric juice	1.0–1.5	
lemons	2.2–2.4	
vinegar	2.4–3.4	
apples	2.9–3.3	
raspberries	3.0–3.5	
oranges	3.0–4.0	
grapes	3.5–4.5	
tomatoes	4.0–4.4	
boric acid solutions	5	
cabbage	5.2–5.4	
asparagus	5.4–5.8	
potatoes	5.6–6.0	
peas	5.8–6.4	
corn	6.0–6.5	
milk	6.3–6.6	
pure water	7.0	neutral
blood	7.35	
egg white	7.6–8.0	
ammonia solutions	11	
solutions of Na_2CO_3	12	
solutions of NaOH	13	very basic

12.9
MEASUREMENT
OF pH

There are various ways we can measure pH. Litmus paper will turn pink in an acid and blue in a base; thus it measures pH and belongs to the class of compounds known as indicators. Phenolphthalein is an indicator that is colorless in acid solutions and is a bright red-violet at pH values higher than 8.3. We often use this indicator to follow the course of titration. There are more than 75 compounds that change color within a given pH range (see Table 12.2). These compounds can be used in titrations or may be used to measure the pH of solutions of biological interest. To obtain values more accurate than those we get by observing the color with our eyes, we can measure the intensity of the color at a specific wavelength in a colorimeter or a spectrophotometer.

TABLE 12.2   Common acid–base indicators

Indicator	pH Range	Color change on proceeding from low to higher pH
thymol blue	1.2–2.8	red–yellow
methyl orange	3.1–4.4	red–yellow
methyl red	4.2–6.3	pink–yellow
bromocresol purple	5.2–6.8	yellow–red
litmus	4.7–8.2	red–blue
phenolphthalein	8.3–10.0	colorless–pink
thymolphthalein	9.3–10.5	colorless–blue
thymol blue	8.0–9.6	yellow–blue
alizarin yellow	10.1–12.1	yellow–red

Fig. 12.5 A pH meter. The digits on the instrument's display screen indicate the pH of the solution into which the electrodes are inserted. (Reproduced by permission of Beckman Instruments, Inc., Fullerton, California.)

Chemists generally use an instrument known as the pH meter (see Fig. 12.5) to measure pH. Such instruments give pH values that are both more accurate and more precise than we can obtain with indicators. We do not have to use a clear, colorless solution when measuring pH with a pH meter; thus a pH meter can be used to measure the pH of blood without removing the red-colored blood cells. Indicators are still used, however, where great precision is not required or to analyze samples obtained in field studies or in other situations in which electricity or relatively expensive instruments are not practical.

12.10 **WHO USES pH?**	Knowing the pH of a solution can tell the chemist when a reaction is complete or, in those reactions that proceed better at one pH than at another, whether the pH is at the level that is necessary to obtain the maximum amount of products in the shortest time. The pH of human blood normally remains constant at about 7.35, and even small changes have serious consequences, so clinical laboratories are concerned with measuring the pH of blood. Urine, gastric juice, and other body fluids also have characteristic pH values, and knowing their acidity can often help in diagnosing an illness. (See Chapter 40.)

Plants require soils having a certain pH value for optimum growth. Some grow or produce better under slightly acid conditions, while others do better when planted in soil that is slightly alkaline. Some plants are tolerant of a fairly wide range of pH values;

others are not. Many plants produce flowers when acids (weak ones, of course) are added to the solutions used to water the plants. Soils containing a great deal of humus, such as those found in Iowa and neighboring states, are too acid for many plants, and an alkaline substance or one that produces an alkali must be added for maximum crop production. Soils in the western United States, on the other hand, are often too alkaline and need acid fertilizers for growth of crops. Soil chemists analyze for soil pH as well as for content of mineral elements. Well-informed nurserymen can tell customers the optimum pH of soil for each of the plants they sell, and can problably recommend and supply an appropriate fertilizer.

12.11 HYDROLYSIS OF SALTS

Since ordinary salts do not contain either H or OH groups, when we dissolve a salt in water, we expect the solution to be neither acidic nor basic, but neutral. While the solutions of many salts are, in fact, neutral, others are either slightly acidic or basic. We might be led to postulate that such salts were really acid salts or basic salts, i.e., formed by incomplete neutralization of an acid or base. Such salts are known, salts such as $NaHCO_3$ or $BiOHCl_2$. Others, such as ammonium chloride and sodium acetate, which are formed by addition of exact equivalent amounts of acids and bases, also give solutions that are not neutral. Why is this so? We can get a clue that helps in explaining these observations by comparing the formulas for salts that produce acidic, neutral, and basic solutions.

acid solutions	NH_4Cl, $AlCl_3$, $Pb(NO_3)_2$, $(NH_4)_2SO_4$
neutral solutions	KCl, $NaCl$, $NH_4C_2H_3O_2$, KNO_3, Na_2SO_4
basic solutions	$NaC_2H_3O_2$, K_3PO_4, Na_3PO_4, $NaCN$

We see that the salts producing acid solutions are formed from the combination of a strong acid and a weak base, those producing alkaline solutions are formed from the combination of a weak acid and a strong base, and the neutral salts are formed from the combination of either a strong base and a strong acid or a weak base and a weak acid. Experiments in dissolving other salts would support our conclusions.

Once we have made this generalization, we must look for an explanation for the presence of excesses of H^+ or OH^- ions in solutions of salts. The concept of equilibrium and our definitions of weak acids and bases help us reach an explanation. We assume that most salts dissociate to give ions in water solutions. So a solution of NaCN would give Na^+ and CN^- ions.

$$NaCN \rightleftharpoons Na^+ + CN^- \tag{12-20}$$

Now, CN^- ions have a tendency to form covalent, nondissociated bonds with hydrogen ions. (This is why we classify HCN as a weak acid.) Since water always contains small amounts of OH^- and H^+ ions, some of the H^+ ions are bound to the CN^- group, giving undissociated HCN:

$$H^+ + CN^- \rightarrow HCN \tag{12-21}$$

While OH^- ions may be attracted to the Na^+ ions formed by the dissociation of NaCN, there is almost no tendency for Na^+ and OH^- ions in water solutions to form a compound, so we are left with both Na^+ and OH^- in the ionic form. We can write the reaction:

$$Na^+ + CN^- + H^+ + OH^- \rightleftharpoons HCN + Na^+ + OH^- \tag{12-22}*$$

On the righthand side of the equation we have an excess of OH^- ions, and since the solution is basic when tested with an indicator, we assume that the equilibrium is displaced somewhat to the right. This reaction of salts with water to produce solutions that are not neutral is called **hydrolysis**. The term hydrolysis is also applied to other reactions, but all involve the lysis (splitting) of the compounds by water. Another way of looking at the same phenomenon is as the splitting of the H^+ and OH^- of water, whereby the water is "lysed."

On examination, we see that Eq. (12-22) is just the reverse of the expression for the neutralization of HCN with NaOH. In general, we would think of Eq. (12-22) as going primarily to the left (as written here). In fact, if water is removed, the reaction will go to completion. In the presence of water, however, an equilibrium situation exists in which a small amount of the salt is hydrolyzed to give a slight reversal of the neutralization reaction.

A solution of aluminum chloride in water is found to be slightly acidic. The equations for the hydrolysis of aluminum chloride can be written:

$$AlCl_3 \rightleftharpoons Al^{3+} + 3Cl^- \tag{12-23}$$

$$Al^{3+} + 3Cl^- + 3H^+ + 3OH^- \rightleftharpoons Al(OH)_3 + 3H^+ + 3Cl^- \tag{12-24}$$

Since HCl is a strong acid, we know it is almost totally dissociated. But aluminum hydroxide is a weak base, so the Al^{3+} and OH^- ions tend to associate, leaving a slight excess of H^+ ions and therefore an acidic solution.

The salts of strong acids and bases tend to be completely dissociated and therefore do not react with H^+ or OH^- of water to give nonionized compounds:

$$K^+ + Cl^- + H^+ + OH^- \rightleftharpoons H^+ + Cl^- + K^+ + OH^- \quad \text{(no reaction)} \tag{12-25}$$

If a salt such as aluminum acetate has been formed from a weak acid and a weak base, both of which have about equal tendencies to dissociate, the salt will produce a neutral solution of water. Its ions will associate with approximately equal quantities of H^+ and OH^- ions. (Here we use OAc^- to represent the acetate ion, $C_2H_3O_2{}^-$, and HOAc for acetic acid:

$$Al^{3+} + 3OAc^- + 3H^+ + 3OH^- \rightleftharpoons Al(OH)_3 + 3HOAc \tag{12-26}$$

---

* Actually, most of the water is present as undissociated molecules (HOH). It is also true that HCN has some tendency to dissociate. It is more nearly correct, although much more complex, to write Eq. (12-22) as follows:

$$H^+ + CN^-$$

$$Na^+ + CN^- + H^+ + OH^- \xrightleftharpoons{} HCN + Na^+ + OH^-$$

$$HOH$$

We should not forget, however, that solutions of sodium cyanide are quite basic, and do contain an excess of the hydroxide ion. Also, we should remember that the cyanide ion is very poisonous. Cyanides are good for illustrations, but actual experiments with hydrolysis use less toxic subtances. HCN is among the most toxic gases known.

We can formulate the rules governing the hydrolysis of salts as follows.

Salts formed by the reaction of	Produce
a) a strong acid and a weak base	acid solutions
b) a strong base and a weak acid	base solutions
c) a strong acid and a strong base	neutral solutions
d) a weak acid and a weak base	depends on relative strength of acid and base

We use our knowledge of the product of hydrolysis of salts in various ways. If we want to compensate for a soil that is too acid (see previous section) we can add a weak base, lime [CaO, which becomes $Ca(OH)_2$], or we can use the less expensive substance limestone, $CaCO_3$, which is the salt of a strong base and a weak acid. Similarly, ammonium-sulfate fertilizers not only supply nitrogen from the ammonium cation and sulfur from the anion, but will increase the acidity of soil.

Another important use of salts that hydrolyze is the preparation of buffers, which are discussed in the following section.

12.12 BUFFERS

Most chemical reactions proceed best at a certain pH. This is especially true of the reactions that occur in the human body and in other living things. We often need to maintain a relatively constant pH in certain systems in spite of the fact that acids or bases may be formed in the system. In animals, the problem is greater because much of the food they eat is also acid or basic, sometimes far removed from the pH of the body or organ into which the food is placed.

While it is impossible to prevent changes in pH, it is possible to use substances or mixtures that resist changes in pH. *Substances that act to minimize the pH changes caused by addition of acids or bases are known as* **buffers**. The action of a simple buffer is illustrated in Fig. 12.6.

A buffer needs to react with either hydrogen or hydroxide ions to remove them from solutions so they do not change the pH. One of the best types of buffers is a mixture of a weak acid and a salt containing the anion produced when the acid dissociates. A typical buffer solution contains acetic acid and acetate ion (from either sodium or potassium acetate). We will use the symbol OAc to represent the acetate group $(C_2H_3O_2^-)$, so acetic acid becomes HOAc, and acetate ion is OAc^-. Thus the buffer contains HOAc, OAc^-, and a cation such as sodium or potassium, which is a spectator ion. If we add H^+ to the buffer, it reacts with OAc^- and we get

$$H^+ + OAc^- \rightarrow HOAc \tag{12-27}$$

The acetic acid has only a small tendency to dissociate (remember, the K_a is only 1.8×10^{-5}), so we have effectively removed most of the added H^+ from the solution. On the other hand, if we add hydroxide ion to the buffer, the OH^- will react with a molecule of acetic acid

$$OH^- + HOAc \rightarrow HOH + OAc^- \tag{12-28}$$

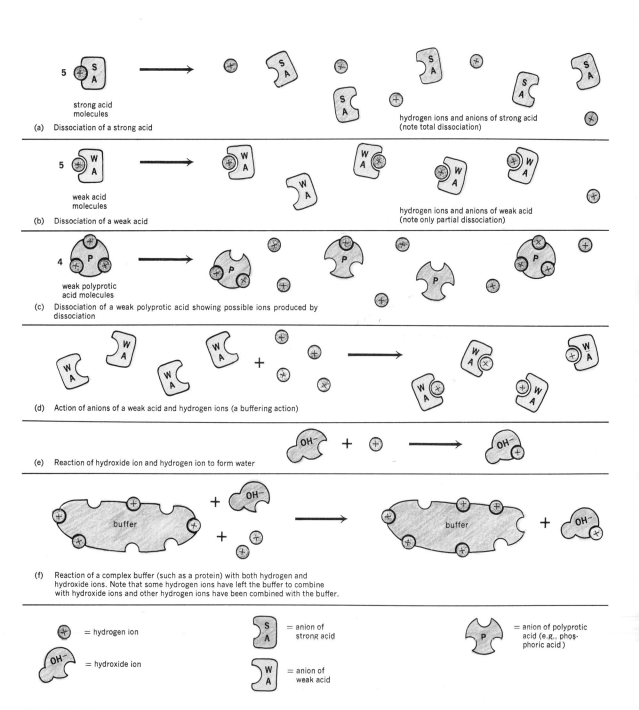

Fig. 12.6 Diagrams illustrating the dissociation of strong and weak acids, buffer action, and the association of hydrogen and hydroxide ions to produce water.

and the ion, OH^-, that could change the pH will be removed. (Alternatively, we could say that the hydroxide ion reacts with one of the few hydrogen ions produced by the dissociation of acetic acid, and that another molecule of the acetic acid dissociates to maintain the equilibrium.)

Phosphate buffers, containing mixtures of $H_2PO_4^-$ and HPO_4^{2-} ions, are also good buffers. The reactions involved are shown below.

$$H_2PO_4^- + OH^- \rightarrow HPO_4^{2-} + HOH \tag{12-29}$$

$$HPO_4^{2-} + H^+ \rightarrow H_2PO_4^- \tag{12-30}$$

With the phosphates, buffer ions react with hydrogen or hydroxide ions to generate more of the ions that serve as buffers.

Although buffers prevent major changes in pH by reacting with H^+ or OH^-, this cannot go on indefinitely. Continual addition of hydroxide ions to a buffer such as an acetate or phosphate buffer will eventually exhaust the supply of acetic acid or acid phosphate molecules (or the H^+ they produce), and eventually some hydroxide ions will be left in the solution. Similarly, addition of excessive amounts of H^+ will exhaust the supply of acetate or phosphate ions and some H^+ will remain in solution. In such cases we say that the buffering capacity of the buffer has been exceeded. This can happen in the human body and is the reason why slight changes in the pH of the blood are so serious—they indicate that the buffers are no longer capable of dealing with acids or bases.

Calculations Involving Buffers

If we want to describe a buffer solution in greater detail we can use an equation known as the Henderson–Hasselbach (H–H) equation:

$$pH = pK_a + \log \frac{[A^-]}{[HA]} \tag{12-31}$$

The H–H equation uses the term HA to represent any weak acid and A^- to represent the anion of that acid.

To use this equation we need to convert the dissociation of the acid involved in the buffer to its pK_a by a process similar to that we used to get pH from the hydrogen-ion concentration. See Table 12.3 for a listing of K_a and pK_a values for some common acids.

The pK_a can be calculated from a K_a value much as the pH can be obtained from a hydrogen-ion concentration.

TABLE 12.3 The dissociation constants (K_a) and pK_a values for some common acids

Acid		K_a	pK_a
acetic acid		1.8×10^{-5}	4.75
boric acid		5.8×10^{-10}	9.24
carbonic acid*	K_1	4.31×10^{-7}	6.37
	K_2	5.6×10^{-11}	10.25
formic acid		1.77×10^{-4}	3.75
lactic acid		1.38×10^{-4}	3.86
phosphoric acid	K_1	7.5×10^{-3}	2.12
	K_2	6.2×10^{-8}	7.21
	K_3	4.8×10^{-13}	12.32

* Polyprotic acids have more than one dissociation constant.

Example   Express the K_a of acetic acid (1.8×10^{-5}) as a pK_a.
               The log of 1.8 is 0.255 and the log of 10^{-5} is -5.000; thus the log of 1.8×10^{-5} is

$$
\begin{array}{r}
-5.000 \\
+0.255 \\
\hline
-4.745
\end{array}
$$

The negative log is then 4.75, which rounds to 4.8 and is the pK_a of acetic acid.

While the H–H equation can be used for calculations, it can also be used to understand the nature of buffer action. The H–H equation indicates that the pH of a buffered solution containing a weak acid and anions of its salt depends on the pK_a of the acid and on the ratio of the concentrations of the acid and its salt.

The importance of the ratio of the amount of anion to acid is shown below. For instance, if [A$^-$] is 10 and [HA] is 1;

$$\frac{[\text{A}^-]}{[\text{HA}]} = 10$$

Since the log of 10 is 1, the equation becomes

$$\text{pH} = \text{p}K_a + 1$$

Thus the pH is greater than the pK_a, and the solution is more alkaline when there is an excess of the anion. Conversely, if [A$^-$] is 1 and [HA] is 100,

$$\frac{[\text{A}^-]}{[\text{HA}]} = \frac{1}{100} = 10^{-2}$$

The log of 10^{-2} is -2, and the pH is thus 2 units less (that is, 2 units more acid) than the pK_a. If the amounts of A$^-$ and HA are equal, the ratio is 1; since the log of 1 is zero, pH = pK_a. This is the best buffer since it has equal amounts of the acid and its anion. Addition of either acid or base will not alter the ratio much. When the ratios are as low as 1/10 or as high as 10/1 (pH = ± 1) the buffer is less effective, and small additions will make major changes in pH. At ratios outside these ranges (± 1) a buffer is relatively ineffective. In other words, for a buffer to be effective, the pK_a must be near the pH the buffer is to maintain. [See Work Exercises 22 and 23 for examples.] This means that acetate buffers, in which the pK_a is 4.8, are not effective in maintaining the pH of blood (7.35) in the human body.

Example   What is the pH of a buffer of an acid having a pK_a of 7.0 if it contains 10 mmol per liter of A and 100 mmol per liter of HA?

$$\text{pH} = \text{p}K_a + \log \frac{[\text{A}^-]}{[\text{HA}]}$$

$$\text{pH} = 7.0 + \log \frac{10}{100}; \qquad \log \frac{10}{100} = \log \frac{1}{10} = -1.0$$

$$\text{pH} = 7.0 - 1.0 = 6.0$$

As with pH, in most calculations involving buffers the values are not even powers of 10. The calculations can be done using a table of logs, but are probably beyond the scope of this text. What is important is to grasp the principles as shown by the simple calculations in our examples.

Although acetate buffers provide simple illustrations of the principles, as stated previously they are not useful for maintaining the pH of human blood. The proteins, phosphates, and carbonates are better for this purpose. Although the carbonate/bicarbonate system has a pK_a of 6.1, which differs by more than one pH unit from that of blood ($pK_a = 7.4$), it is still useful (see Section 40.12). This buffer system uses bicarbonate ion and carbonic acid as the A^- and HA of the H–H equation:

$$H_2CO_3 \leftrightarrow HCO_3^- + H^+ \tag{12-32}$$

However, H_2CO_3 is better described as a solution of CO_2 in water or

$$H_2CO_3 \rightleftharpoons H_2O + CO_2 \tag{12-33}$$

Because of these two reactions, we can write the H–H equation for the system as

$$pH = 6.1 + \log \frac{[HCO_3^-]}{[CO_2]} \tag{12-34}$$

From this we can calculate that the ratio of HCO_3^-/CO_2 at a pH of 7.4 is 20/1 (log 20 is 1.300). More important, it tells us that if the body is in a state of **acidosis** (having too much acid), as is commonly found in people who are starving or who have untreated *diabetes mellitus*, the body should compensate by getting rid of excess acid. While this can be done by excreting urine that is more acid than normal, it can also be accomplished by excreting more of the acid component of the buffer system, CO_2, by breathing faster. It was observed long ago that untreated diabetics breathe faster than normal people. We say they *hyperventilate*, and we see this tendency any time a person is in the state we call acidosis. People in a state of **alkalosis** breathe slowly, conserving CO_2, and excrete more bicarbonate or other alkaline substances in their urine.

The Henderson–Hasselbach equation can be used to calculate the amounts of substances to be be used to make a buffer and to explain how the acid and anions in a buffer resist changes in pH. It can also be seen as a model for control of pH in a living system. Our bodies cannot control pH by changing the pK_a—that is a constant. Instead, we respond to changes in internal pH by changing the ratio of A^- to HA. If too much acid is being ingested or being formed, we tend to excrete more of the HA or retain the A^- if this is possible. The response to an excess of base is excretion of more of the anion (A^-) or retention of more of the acid (HA).

We can speculate that the reason animals have retained the bicarbonate/carbonate acid system, in spite of the fact that its pK_a is really too far from the pH to be maintained, is the rapidity of excretion of the acid component of the buffer pair, carbon dioxide, through the lungs.

There are many organic molecules, from the simple amino acids to the macromolecules of proteins and nucleic acids, that serve as buffers. These substances contain groups we can describe as A^- and HA units on a single molecule. Many proteins, especially the red blood protein hemoglobin, act as buffers to control pH in the human body. The way in which these substances act as buffers is described in Chapter 31, Chemistry of the Proteins.

SUMMARY

Chemical reactions that go predominantly in one direction, or "irreversible" reactions, are important for many chemical processes. We can choose such reactions by selecting those that give insoluble or nonionized products. Alternatively, we can sometimes apply a stress such as increased pressure or an increase in temperature to force a reaction in the direction we want it to go.

With many other chemical reactions, a state of equilibrium is reached. At this point of equilibrium, the forward and reverse reaction are proceeding at equal rates. Although the number of each kind of particle remains constant, individual particles are constantly reacting.

Very weak electrolytes such as water and weak acids ionize only slightly, and at equilibrium there are always many more undissociated molecules than ions. However, these few ions are extremely important for some chemical processes, including those that occur in living organisms.

We describe the acidity produced by acids in aqueous solutions in terms of hydrogen-ion concentration or pH. These values are useful in specifying conditions for chemical reactions. The pH can be controlled by substances or mixtures of substances called buffers. Buffers are very important in providing proper conditions for chemical rections that are necessary for the proper functioning of living organisms.

IMPORTANT TERMS

acidosis	equilibrium	ionization constant (K_i)	pH
alkalosis	equilibrium constant (K_{eq})	K_a	pKa
buffer	hydrolysis of salts	K_w	

WORK EXERCISES

Note: For practice in the use of logarithms, see Exercises 28 and 29.

1. Describe three different situations or conditions in which some type of equilibrium is involved.
2. Write three equations for the above, using arrows to indicate that in each equation an equilibrium is involved.
3. Give three factors that bring about changes in equilibrium.
4. Write an equation for a reaction that goes almost entirely to completion (use appropriate arrows).
5. Give four factors that can cause reactions to go toward completion.
6. What evidence do we have that water is only slightly ionized?
7. According to the information shown in Table 12.1, approximately how much greater is the hydrogen-ion concentration in lemons than in cabbage?
8. Explain why solutions having a relatively high concentration of hydrogen ions always have low OH$^-$ ion concentrations.
9. List three indicators of pH commonly used in the laboratory, and state the pH ranges of each.
10. List five different professions in which some knowledge of pH is essential.
11. Which is the stronger acid, one having a pK_a of 6.5 or one with a pK_a of 4.5? What is the pK_a of an acid that is 100% dissociated?

12. Why do some salts, when dissolved in water, form solutions that are not neutral?
13. Referring to the discussion of strong and weak acids in Chapter 11, indicate whether solutions of the following salts would be acidic, basic, or neutral.
    a) NaCl   b) Na_2CO_3
    c) $K(C_2H_3O_2)$ d) $(NH_4)_2SO_4$
    e) KNO_3 f) Na_2SO_3
    g) Na_2S
14. Calculate the equilibrium constants (K_{eq}) for the following concentrations of substances at the equilibrium point of the general reaction A + B $\rightleftarrows$ C + D.
    a) A = 1.00; B = 1.00; C = 0.50; and D = 0.50
    b) A = 0.50; B = 0.25; C = 1.00; and D = 2.00
    c) A reaction starts with 1.00 mole of both A and B and at equilibrium 0.30 mole of C is found to be present.
15. Which of the reactions given above (Exercise 14) could be considered as proceeding primarily to the right?
16. What is the hydrogen-ion concentration in the following instances?
    a) We have a 0.100 M solution of an acid having an ionization constant of 9.0×10^{-7}.
    b) We have a 0.050 M solution of an acid having an ionization constant of 3.8×10^{-6}.

17. What is the acid dissociation constant K_a of the following? (Assume that the final concentration of HA is not significantly different from the molarity.)
   a) An acid solution that is $1.00\ M$ has a hydrogen-ion concentration of 2.50×10^{-4}.
   b) An acid solution that is $0.100\ M$ has a pH of 4.30.

18. What is the pH of the following solutions?
   a) a solution having a hydrogen-ion concentration of 1×10^{-5}
   b) a solution having a hydrogen-ion concentration of 3.0×10^{-4}
   c) a solution having a hydroxide concentration of 1×10^{-5}
   d) a solution having a hydrogen-ion concentration of 2×10^{-9}

19. What is the hydrogen-ion concentration of the following solutions?
   a) a solution with a pH of 4.00
   b) a solution with a pH of 5.29
   c) a blood specimen that has a pH of 7.35
   d) a urine specimen that has a pH of 6.82
   e) a solution with a pOH of 5.00
   f) a solution with a pOH of 3.50
   g) a 0.10 M solution of a strong acid
   h) a 0.100 M solution of a weak acid having a K_a of 4.0×10^{-5}

20. Calculate the pK_a for acids having the following K_a values.
   a) 1×10^{-6} b) 3.8×10^{-8}
   c) 5.6×10^{-5}

21. What is the pH of the following buffers? (pK_a for acetic acid is 4.8.)
   a) a buffer that contains equal amounts of acetate ions and acetic acid molecules
   b) a buffer that contains 10 times as many acetate ions as acetic acid molecules
   c) a buffer that contains 10 times as many acetic acid molecules as acetate ions
   d) a buffer that contains twice as many acetate ions as acetic acid molecules

22. What would be the effect of adding 5 millimoles of a strong acid (5 mmol of H^+) to the following buffers?
   a) a buffer having a pK_a of 6.0 containing 11 mmol of A and 11 mmol of HA in a liter of solution
   b) a buffer having a pK_a of 6.0 containing 2 mmol of A and 20 mmol of HA in a liter of solution

23. What would be the resulting hydrogen-ion concentration of the solution in Exercise 22(b) after addition of the acid?

24. What ratio of A^- to HA is required to make a buffer having a pH of 7.40 from an acid (HA) having a pK_a of 7.00?

25. Why are acetic-acid buffers ineffective in maintaining the pH of the human body?

26. One of the pK_a values for arsenic acid (H_3AsO_4) is 7.08. Why is this acid and its salt not used to make buffers having a pH of 7.4 for use in medications?

27. In a chemistry handbook check the K_a and pK_a values for acids that could be used for maintaining a pH near the body pH of 7.35.

Use an electronic calculator or the table of logarithms in the *Study Guide* to find answers to the following. (Remember that a logarithm has only one more significant figure than the number that it represents.)

28. Give the logarithms (base 10) of the following numbers.
   a) 100   b) 386
   c) 38.6   d) 3.86
   e) 0.010   f) 7.47×10^2
   g) 7.47×10^{-2} h) 0.038 6
   i) 186 000   j) Avogadro's number

29. Give the antilog of the following.
   a) 2.00   b) -2.00
   c) 0.313 9   d) 2.313 9
   e) 0.916 5   f) 9.165 0
   g) 2.721 0   h) $-1.279\ 0$
   i) -3.600 j) $-22.220\ 0$

SUGGESTED READING

Morris, D. L., "The carbonates: An entertainment in equilibrium," *Chemistry* 47, No. 7 (July–August 1974) p. 6. An interesting article that discusses the equilibria of carbonate ions, including the use charlatans made of at least one of them, as well as modern industrial applications.

13

The Metals

The metals gold and silver were known and valued in prehistoric times. Ancient tombs contain articles made of these metals, and the earliest writings mention silver and gold. Both metals were used for coins and ornaments in almost all civilizations. To a great extent the value placed on gold and silver is due to their lack of reactivity. In this world where all things perish and pass away, gold—and to a lesser extent, silver—remains. It is easy to see how gold was associated with gods and immortality. The classical Greek civilization knew of seven metals, gold, silver, copper, iron, tin, lead, and mercury. Antimony was also known even before the Greek era but was not among the seven commonly mentioned by them. Gold and silver have been used for ornamental purposes, but never for tools.

Gold, silver, and copper occur as free metals, and free iron occurs in some meteorites, although this is not a significant source of the metal. The other metals known to ancient civilizations are easily extracted from their ores by heating. Metallurgy was probably established when observant people noted that certain stones around a campfire had been "changed" into a metal. If the stone was blue (as are most copper ores) and it gave copper, the relationship between the color and the product should have been obvious. We know that copper was plentiful in the Sinai area and that mines had been developed there as early as 2500 B.C. Impetus was added to the mining and use of copper when someone learned how to combine tin with copper to make bronze. When this happened civilization changed. A similar major change came when methods for extracting iron from its ores were developed.

The archeological classification of civilization as the Stone Age, Bronze Age, and Iron Age depends primarily on the kind of tools used. Civilizations in the Americas, although they certainly used silver and gold, remained in the stone age until the time of Columbus. The bronze age occurred at different times in different cultures of the Mediterranean area. Bronze is an alloy of copper and tin that is much harder than pure copper and much easier to cast into molds. Once bronze became available, stone implements and tools were obsolete. The bronze age did not last long in most civilizations, since the people soon learned how to produce iron which was much harder than the bronze (at least in the degree of purification they could achieve) and could be sharpened to make better tools and swords.

Iron was probably first produced by the Hittites about 1500 B.C. This metal cannot be separated from its ores by heating; extraction requires the presence of carbon. Perhaps the combination of coal, or more likely of charcoal, with iron ore, probably again in a campfire, led to the discovery of the process for purifying iron. This accidental purification of iron could have happened hundreds of times before some prehistoric "scientist" observed and interpreted the facts carefully and the Iron Age began. The knowledge of iron was carried from one civilization to another largely by men with iron weapons who were conquering other peoples. This fusion of different peoples and knowledge resulted in new, more powerful societies, much as combining metals in alloys gives products with properties surpassing those of the individual components.

From the time of the Greeks until about 1700, no new metals were discovered and used. Since that time more than fifty metals have been recovered from their compounds, and in the past few decades several synthetic metals have been produced in atomic accelerators.

The transition from the stone age through the bronze age to the iron age took centuries. It is difficult to classify our present age; perhaps we are still living in an age of steel and aluminum, or perhaps we are in an age of plastics. We are much too close to the most recent developments to have historical perspective. Another way of classifying ages is based on sources of power rather than on materials used, and we are said to be entering the atomic age in this respect. How historians of the future eventually classify our civilization is not very important to us, but the metals themselves are of great significance.

Chapters 3 and 5 provided an introduction to some of the metals. In this chapter we will discuss in greater detail some important groups of metals as well as some individual metals that are of special significance.

**13.2
METALS AND
NONMETALS**

The word "metal" is widely used and to most people it means a hard, durable solid that can be polished. As indicated in Sections 5.6 and 5.8, the word has many other meanings

TABLE 13.1   **Comparison of physical and chemical properties of metals and nonmetals**

Metals	Nonmetals
Physical properties	
1. Good conductors of heat and electricity	1. Poor conductors or nonconductors
2. Malleable and ductile	2. Brittle
3. Most are solids	3. Occur as gases, liquids, and solids
4. Relatively high density	4. Relatively low density
5. Metallic bonds between atoms	5. Covalent bonds between atoms
Chemical properties	
1. Oxides + water give bases	1. Oxides + water give acids
2. Have one to four electrons in valence shell	2. Have four to eight electrons in valence shell
3. Tend to form cations by losing electrons	3. Tend to form anions by gaining electrons
4. Oxidation numbers are positive	4. Oxidation numbers either negative or positive
5. Reducing agents	5. Oxidizing agents

to a chemist. Some of the properties of those elements we classify as metals, as well as those of the nonmetals, are given in Table 13.1

As with most other classifications, the distinction between **metals** and **nonmetals** is not always precise. Some elements have properties of both metals and nonmetals; we call these elements **metalloids**. As indicated in Chapter 5 and as shown in the Periodic Table of the Elements inside the back cover of this book, the metals are found to the left side and at the bottom of the Periodic Table, and the nonmetals are shown at the right and near the top of the Table. A line can be drawn to separate metals and nonmetals, with elements near the line being classified as metalloids. Approximately three-fourths of the elements are classified as metals.

13.3
OCCURRENCE
OF METALS

Some metals—for example, silver, gold and platinum—are found as elements in the free uncombined state as well as in compounds. Others such as sodium and potassium are always found naturally as part of a compound. The oxides, sulfides, and silicates of most metals are insoluble, and the best natural sources of metals are these compounds. Soluble compounds of metals tend to accumulate in the oceans, so the oceans of the world are good sources of many metals—generally in the ionic form, of course.

A material that can be used commercially as a source of a metal is called an **ore**. The common ores and sources of representative metals are given in Table 13.2. The most common ores are oxides, sulfides, halides, carbonates, and sulfates. Although metals are widely found as silicates, it is difficult to separate the metals from these compounds and they are not good ores.

13.4
PREPARATION
OF METALS

In preparing or isolating metals, we usually begin by using a physical separation to remove substances of different solubility or density from the ore. If an ore (or an impurity) is magnetic, a magnet can be used to separate the ores from impurities. Grinding or pulverizing the ore may also be necessary before chemical processes are used.

In many instances, an ore is first converted to the oxide of the metal by heating in air, a process commonly known as roasting.

$$2ZnS + 3O_2 \xrightarrow[\text{heat}]{} 2ZnO + 2SO_2 \tag{13-1}$$

$$PbCO_3 \xrightarrow[\text{heat}]{} PbO + CO_2 \tag{13-2}$$

The oxides of a few metals are so unstable that heating will give the pure metal.

$$2HgO \xrightarrow[\text{heat}]{} 2Hg + O_2 \tag{13-3}$$

The oxygen of most ores, however, must be removed by a simple displacement reaction with another element. Because it is inexpensive, carbon (usually in the form of coke or charcoal) is often used for preparation of metals.

$$ZnO + C \rightarrow Zn + CO \tag{13-4}$$

Since the carbon removes oxygen from the oxide, we call it a reducing agent. It is not necessary to use carbon as the reducing agent; any substance that will combine with

TABLE 13.2   Common metals: their ores, sources, and important uses

Metal	Common Ore	Source	Uses
aluminum	Al_2O_3	Australia, Jamaica, Guinea, Surinam	lightweight alloys
antimony	Sb_2S_3	China, Bolivia, Chile	primarily in alloys (see Table 13–4)
bismuth	Bi_2S_3, Bi_2O_3	(by-products of Pb, Cu, Sn, Ag ores)	type metal, low melting alloys
cadmium	CdS	(by-product of Zn, Cu, Pb ores)	batteries, electroplating
chromium	$FeO \cdot Cr_2O_3$	Southern Rhodesia, Cuba, Philippines	alloys, electroplating, pigments
cobalt	$CoAsS$, $CoAs_2$	Ontario (Canada)	magnets, alloys
copper	CuS, CuO, $CuCO_3$	United States, Chile, Canada, Zambia, Zaire	conductors, coins
iron	Fe_2O_3	USSR, Australia, United States, Brazil	steel for structures
lead	PbS	Oklahoma, Kansas, Missouri	paints, storage batteries
magnesium	$MgCO_3$, $MgCl_2$	solid ores—Austria, USSR, Greece brines, sea water	lightweight alloys
manganese	MnO_2, $MnCO_3$	ocean floor, USSR, United States	alloys
mercury	HgS	Spain, Italy, California	scientific instruments, vapor lamps
molybdenum	MoS_2	Colorado	electrical resistance wire, alloys for withstanding high temperatures
nickel	sulfides	Canada, Cuba, Norway	alloys, structural steel, coin metal
silver	Ag, Ag_2S	Mexico, United States	coin metals, photographs, jewelry
tin	SnO_2	Bolivia, East Indies	alloys, coating for tin cans, solder
tungsten	$(Fe,Mn)WO_4$	Nevada, California, North Carolina	electric filaments, alloys for high speed cutting tools (highest melting point of all metals, 3410°C)
zinc	ZnS, $ZnCO_3$	Oklahoma, Kansas, Missouri	ZnO (adhesive tape), alloys, galvanized coating

oxygen can be used. Both carbon and carbon monoxide, which is produced by the partial oxidation of coke, are used as reducing agents in the purification of iron.

$$2C + O_2 \rightarrow 2CO \tag{13-5}$$

$$Fe_2O_3 + 3CO \rightarrow 2Fe + 3CO_2 \tag{13-6}$$

Any element that is more active than the metal that is to be produced may be used as a reducing agent; however, since these active elements are more expensive than carbon, metals produced in this way will be more expensive.

$$Cr_2O_3 + 2Al \rightarrow 2Cr + Al_2O_3 \tag{13-7}$$

$$WO_3 + 3H_2 \rightarrow W + 3H_2O \tag{13-8}$$

In some cases, as in the production of tungsten above, the additional expense may be warranted.

Metals may also be separated from their ores by electrolytic processes. Sodium is prepared by the electrolysis of molten (fused) sodium chloride:

$$2NaCl \xrightarrow[\text{electricity}]{} 2Na + Cl_2 \tag{13-9}$$

We can also write this equation as

$$Na^+ + e^- \rightarrow Na^\circ \tag{13-10}$$

Aluminum is produced by the electrolysis of aluminum oxide (commonly known as bauxite) dissolved in fused cryolite (Na_3AlF_6). Electrolytic reactions are often used in the last step of purification of metals. Copper that is 99.955% pure can be made this way.

Distillation (see Section 9.8) can also be used to purify metals that have relatively low boiling points, such as mercury, zinc, and cadmium. This process is effective because compounds of metals generally have boiling points much higher than the pure metal.

13.5 PHYSICAL AND CHEMICAL PROPERTIES OF METALS

The common physical properties of the metals were given in Table 13.1. Although most are solids at room temperature (20°C), three—cesium, gallium, and mercury—are liquids, and others, such as potassium and sodium, melt at relatively low temperatures. While most metals have relatively high densities, they vary from lithium with a density of 0.53, which is less than that of water, to lead with a density of 11.3, and gold and tungsten with a density of 19.3. Platinum and osmium are even more dense, having densities of 21.5 and 22.6, respectively. Metals also vary in their conductivity of heat and electricity, although all serve as conductors due to the fact that valence electrons pass easily from one atom to another in a sample of a metal. In fact we can say that the electrons are shared thoughout the sample and that this sharing helps hold the atoms together in a **metallic bond**. Small amounts of impurities may change the conductivity, especially in the metalloids we use in semiconductor devices such as transistors. The variance of other physical properties with groups of metals and individual metals will be discussed in the materials which follow.

While metals share many chemical properties, these vary even more than their physical properties. Almost all metals combine with sulfur.

$$Zn + S \rightarrow ZnS \tag{13-11}$$

Even the metals silver and gold, which do not combine to any extent with oxygen, form sulfides in a reaction we call *tarnishing*.

$$2Au + 3S \rightarrow Au_2S_3 \tag{13-12}$$

The more active metals form compounds with a variety of nonmetals.

$$Cu + Cl_2 \rightarrow CuCl_2 \tag{13-13}$$

$$2Ca + O_2 \rightarrow 2CaO \tag{13-14}$$

The very active metals react with water to give a hydroxide and hydrogen gas.

$$2K + 2HOH \rightarrow 2KOH + H_2 \tag{13-15}$$

Less active metals do not react at an appreciable rate with water but will react with steam.

$$Mg + HOH_{(gas)} \rightarrow MgO + H_2 \tag{13-16}$$

If sufficient water is present the oxide can then react with water to form the hydroxide.

$$MgO + HOH \rightarrow Mg(OH)_2 \tag{13-17}$$

Most metals do not react with water.

Another common reaction of metals is with strong acids to yield a salt and hydrogen.

$$Ni + H_2SO_4 \rightarrow NiSO_4 + H_2 \tag{13-18}$$

The reaction of active metals with strong acids is explosive, while the less active metals will not react with acids.

Free metals may react with compounds of other metals in a simple displacement reaction:

$$2AgNO_3 + Cu \rightarrow Cu(NO_3)_2 + 2Ag \tag{13-19}$$

This reaction may also be written

$$2Ag^+ + Cu^0 \rightarrow Cu^{2+} + 2Ag^0 \tag{13-20}$$

If this reaction is performed by immersing a coil of copper wire in a solution of silver nitrate, beautiful silver crystals will form and hang from the coil of copper while the solution turns blue due to the formation of copper ions.

As indicated previously, this sort of reaction may be used to produce a metal when we have another (usually less expensive) active free metal and a compound of the metal we want to prepare in a pure form. Although such reactions go primarily in one direction when the free metal is more active than the combined metal, some reactions reach an equilibrium in which both free metals as well as ions of both are present. If the free metal is less active than the one in the compound, no reaction occurs. For instance, we get no reaction if we add silver to a solution of copper ions.

$$Ag + Cu(NO_3)_2 \rightarrow \text{no reaction} \tag{13-21}$$

13.6
THE ACTIVITY
SERIES OF
METALS

In order to classify the relative reactivity of metals as well as to be able to predict whether a given metal will displace another, we can construct an activity series of metals such as that in Table 13.3. The most active metals are at the top of this series, and each element will displace any element below it. The least active elements are at the bottom of the table. These inactive metals are often found in the free state and can be prepared by heating their oxides. For these reasons they are the metals that were known to ancient people.

TABLE 13.3   Activity series of metals. The most active metals are listed at the top, with any metal being capable of displacing any metal below it. Typical reactions are given using the symbol M to represent a metal with an oxidation number of 2.

Li
Rb
K
Ba   react readily with cold water to
Sr   give hydrogen
Ca $M + 2HOH \rightarrow M(OH)_2 + H_2$
Na

Mg
Be
Al
Mn   heated metals react with steam
Zn   to give hydrogen
Cr $M + H_2O \rightarrow MO + H_2$
Fe
Cd
Co
Ni

Sn
Pb   react slowly with acids

react with acids to give hydrogen
$M + H_2SO_4 \rightarrow MSO_4 + H_2$

H
Sb
As
Bi
Cu
Hg(I)
Ag
Pd
Hg(II)
Pt
Au

will not react with acids to give hydrogen

The activity series of metals was developed by studying reactions of various metals. Some of the reactions that can be used to group metals are their tendency to react with water, with steam, and with acids (see Table 13.3). The exact placement of metals in the series can be determined by studying their tendency to displace one another in compounds. For instance, if we place copper in a solution of mercury(II) nitrate,

$$Cu + Hg(NO_3)_2 \rightarrow Cu(NO_3)_2 + Hg \qquad\qquad (13\text{-}22)$$

we soon detect a blue color in the solution and may see drops of mercury forming, indicating that copper is above mercury in the activity series. There would be no reaction if mercury were added to a copper compound.

Since the ability to displace another metal is related to the tendency to lose electrons (the most active metals having the greatest tendency to lose electrons), we can also determine the activity of metals by measuring their tendency to lose electrons. When we measure this tendency, known as the electromotive force, we get a table of activities which we call the electromotive force table. Such tables are also known as tables of oxidation potentials, which is a logical classification when we consider that loss of electrons is oxidation. In most, but not all, cases the tables so constructed agree with tables constructed from studying displacement reactions in solution. The situation is complicated when different oxidation states of a metal are present. Note that the two states of mercury occur at different points in the series given in Table 13.3.

From activity series we can also predict the tendency of metals to release hydrogen from water or steam, and can use this information when extinguishing fires. If we add water to a fire involving burning sodium, we only increase the production of hydrogen which is being burned. Many alloys of magnesium are used in aircraft, and if the magnesium is burning, water only aggravates the fire (Eq. (13–16)).

**13.7
USES OF
METALS**

The development of a civilization is greatly influenced by the metals it uses. While our civilization uses a few pure metals, a much greater use is made of **alloys**, the mixtures of metals we sometimes call solid solutions. Adding a small amount of another substance to a metal, as when we add carbon to iron to produce steel, almost always makes the alloy harder than the pure metal. Too much of an impurity may increase brittleness—an undesirable property for most uses made of metallic alloys. Additions generally decrease the conductivity of a metal. It is more difficult to predict the other effects of composition on alloys, and many alloys are made and their properties determined later. Some of the alloys that are used today are given in Table 13.4.

To illustrate some uses of alloys, let us consider some of the uses of various kinds of iron (steel) in an automobile. The engine is made of cast iron, which is soft enough to be machined and polished yet hard enough to resist wear. The springs must be flexible and resilient to the extent that they maintain their original shape after millions of bumps. The axles must be rigid and tough enough to withstand the pressures exerted by today's high-powered engines. The gears also require special alloys that are strong and resistant to wear. The body is composed of sheet metal, which can be rolled back to the original shape if it has been dented.

Although iron-containing or **ferrous** alloys such as the various types of steel are widely used, the lighter-weight alloys of aluminum have made possible the construction of high-rise buildings. Without these lightweight alloys, airplanes, lightweight trains, and rapid transportation, which is so much a part of our lifestyle, would be impossible or would require much more energy. While we do not generally call them alloys, many elements, particularly the metalloids, have small amounts of impurities added in a process called "doping." These "mixtures" are then used in transistors and other solid-state devices which are significant factors in the lifestyle of individuals and in the civilization of today's world.

**13.8
THE METALS
OF GROUP IA**

The elements hydrogen, lithium, sodium, potassium, rubidium, cesium, and francium, which comprise the elements of Group IA, all have one electron in their outer shell. Hydrogen, which is usually also included in this group, has the properties of a nonmetal and is discussed in Chapter 14. The remainder of the elements of the group have metallic properties and are known as the **alkali metals**. Some of the significant properties of these metals are given in Table 13.5. Examination of the table shows that most of the physical properties of these elements change as atomic weight and size (radius of the atom) increase. We will see in the sections that follow that changes in boiling and melting

TABLE 13.4 Common alloys

Name	Composition Percentage	Properties	Uses
steel	Fe 99, C 1.0	hard, durable	structures of steel, rails, etc.
steel, stainless*	Fe 74, Cr 18, Ni 8	corrosion resistant	chemical equipment, tanks for trucks, railroad cars
steel, high speed	Fe 77.7, W 18, Cr 4, Mn 0.3	very hard retains temper at very high temperatures	drill bits, cutting tools
steel, molybdenum	Fe 97–99, Mo 1–3	hard, heat resistant	axles for autos
duralumin*	Al 94.6, Cu 4.0, Mg 0.8, Mn 0.8	lightweight, durable	structures and vehicles
alnico*	Fe 50, Al 20, Ni 20, Co 10	highly magnetic	magnets
brass	Cu 85, Zn 15	easily machined, corrosion resistant	fittings, hardware, radiator cores
brass, cartridge	Cu 70, Zn 30	easily shaped	shell cases
brass, naval	Cu 60, Zn 39, Sn 1	resistant to seawater and corrosion, polishes well	ship fittings
bronze	Cu 70–95, Zn 1–25, Sn 1–18	harder than brass	fittings, statuary
nickel coinage	Cu 75, Ni 25		
silver, sterling	Ag 92.5, Cu 7.5	durable, beautiful	silverware and service, jewelry
silver, coinage	Ag 90, Cu 10		
gold, coins	Au 90, Cu 10		
gold, 12-carat	Au 50, Cu 35, Ag 15		jewelry
gold, 18-carat	Au 75, Ag 10–20, Cu 5–15		jewelry
Monel metal	Ni 60, Cu 33, Fe 7	strong, durable, bright	marine shafts, vats for foods and chemicals
pewter	Sn 85, Cu 6.8, Bi 6, Sb 1.7	soft, corrosion-free	decorative vases, pitchers, etc.
solder, plumbers'	Pb 67, Sn 33	low melting point	soldering
type metal	Pb 70, Sb 18, Sn 10, Cu 2	low melting point	printers' type
Wood's metal	Bi 50, Pb 25, Sn 12.5, Cd 12.5	very low melting point (70°C)	soft plugs in automatic sprinklers
Nichrome	Ni 80–85, Cr 15–20	high melting point, low electrical conductivity	heating elements

* Example of one of many alloys given this name.

TABLE 13.5   Properties of metals in Group IA

Element	Atomic Number	Atomic Weight	Atomic Radius ($\times 10^{-11}$ m)	Melting Point (°C)	Boiling Point (°C)	Density (at 20°C) g/cm^3	Common Oxidation States	Hardness
lithium	3	7.0	15.2	179	1 336	0.54	1	0.6
sodium	11	23.0	15.4	97.8	883	0.97	1	0.4
potassium	19	39.0	22.7	63.5	758	0.87	1	0.5
rubidium	37	85.5	24.8	39	700	1.53	1	0.3
cesium	55	133	26.5	28.6	670	1.9*	1	0.2
francium	87	(223)	27.	(27)	—	—	1	—

* Value determined at 15°C.

points, density, and other properties are also found in other families of elements, although nowhere are they so regular and definite as in the elements of Group IA. As with many other groups, the heaviest atom, in this case francium, is radioactive. Since there are no stable isotopes of francium, we know less about its physical and chemical properties than about those of the other elements of the group.

The tendency of a metal to react is in direct relationship to the ease with which it can lose electrons. The outer or valence electron is relatively close to the positive nucleus in sodium atoms and, although it is active, sodium is much less active than the other metals of Group IA. Rubidium and cesium (and presumably francium) are the most active metals known, and sodium and potassium are the most active metals normally found in a chemistry laboratory. They react violently with water to produce hydrogen gas and a base.

$$2Na + 2HOH \rightarrow 2NaOH + H_2 \qquad\qquad (13\text{-}23)$$

The heat of the reaction is sufficient to melt the metals and, as the reaction proceeds, we note a small sphere of sodium or potassium darting across the surface of the water with which it is reacting. Sometimes the hydrogen that is being formed catches fire and burns. Lithium reacts slowly with water, and cesium and rubidium react explosively.

Due to their extreme reactivity, the alkali metals are never found free in nature. Since most compounds of the alkali metals are soluble, we find large amounts of them in oceans and in dried up lake beds. A source (ore) of any one of the metals in this group is likely to contain appreciable amounts of the others. Most of the alkali metals are prepared in pure form by electrolysis of their salts. Sodium and potassium, the most abundant and the most important of the alkali metals, are discussed in detail in the following sections.

Sodium   Sodium is the sixth most abundant element on earth, and the yellow light it gives off reminds us that it is also a prominent element in our sun. Sodium is ninth in abundance in the human body. Compounds of sodium such as $NaCl$ and Na_2CO_3 have been known and used for centuries, but it was not until 1807 that the pure metal was produced.

Sodium metal has a low density and will float on water, especially when buoyed up by bubbles of hydrogen gas produced by its reactions with water. Because of its extreme reactivity, sodium metal must be handled carefully. It is generally stored under an unreactive liquid such as light mineral oil or kerosene.

Sodium metal is used in some alloys, and a mixture of sodium and potassium metals is used in some nuclear reactors to transfer heat from one part of the reactor to another. The yellow light given off when sodium or its compounds are heated to incandescence helps us identify the element in the laboratory. The intensity of this light makes sodium vapor lamps especially useful in foggy weather. The many compounds of sodium have a variety of uses. NaCl is used for seasoning food and in high concentrations it serves as a preserving agent. Sodium hydroxide is used in many manufacturing processes. $NaHCO_3$, which we know as baking soda or sodium bicarbonate, has many uses. Its use in baking depends on its reaction with acids, either those solid acids contained in baking powder or those from sour milk or cream, to produce carbon dioxide, which serves as a leavening agent.

$$NaHCO_3 + H^+ \rightarrow Na^+ + H_2O + CO_2 \uparrow \tag{13-24}$$

Sodium is an important nutrient for all animals, some of which will travel miles to natural salt deposits we call *salt licks*. The sodium ion is the major cation of the fluids of the human body and it, along with potassium ions, is involved in the conduction of impulses by nerves. People with heart disease may have difficulty eliminating sodium ions and are placed on diets low in sodium content or are given diuretics to aid in the elimination of both sodium ions and water.

Potassium The metal we know as potassium is present in ashes from burned wood and other plants; in fact, its name is derived from the words "pot ashes," and mixtures of salts of the metal are known commonly as *potash*. Potassium is just slightly less abundant than sodium, ranking seventh in abundance on the earth. It is also seventh in abundance in the human body. The oceans contain much less potassium than sodium, probably because many minerals that contain potassium are complex compounds that are insoluble in water, such as feldspar ($K_2O \cdot Al_2O_3 \cdot 6H_2O$).

Due to its great reactivity potassium was not isolated until 1807, when Sir Humphrey Davy (see Section 13.10) produced both sodium and potassium by electrolysis of their hydroxides.

Because sodium and potassium compounds are very similar, for most chemical and industrial purposes we can use a compound of either sodium or potassium. Plants and animals need both sodium and potassium; they use them interchangeably for some functions but in very different ways in other processes, such as the conduction of impulses by nerves. People who take diuretics to help them eliminate sodium ions from their bodies also usually excrete larger than normal amounts of potassium ions. These people are generally told to consume foods such as bananas or orange juice, which are high in potassium, to keep from developing a potassium deficiency. Plants require large amounts of potassium, which is therefore included in many fertilizers. While some plants can live in sea water or in soils containing high levels of sodium chloride, many will not.

Lithium, Lithium has the distinction of being the least dense of all the metals and having the
Rubidium, highest specific heat. It is used in some alloys and in making ceramics and special kinds
Cesium, and of glass.
Francium Rubidium and cesium are extremely reactive, burning spontaneously when exposed to air and reacting explosively with water. Due to the ease with which they ionize, they have been used experimentally in engines functioning by ion propulsion and may find use as propellants for space travel. Since they are less abundant than sodium and

potassium, their compounds are generally expensive. The less expensive sodium and potassium compounds are often used in place of compounds of rubidium and cesium.

Twenty isotopes of francium are known; however, all are radioactive, with the most stable having a half-life of only 22 minutes. Although very small amounts of francium are produced by decay of other isotopes, there is probably less than 30 grams of this element on the earth at any one time. The amounts of francium that have been prepared are extremely small and its chemical properties have not been studied extensively; however, it is probable that it is the most active metal. Any possible chemical uses for it are negated by its radioactivity and short half-life.

**13.9
THE METALS
OF GROUP IIA**

Beryllium, magnesium, calcium, strontium, barium, and radium, which are classified in Group IIA of the periodic table, are known as the **alkaline earths**. Like the elements of Group IA, they show a variation of properties that correlates with their atomic weight, atomic radius, and position on the periodic table (see Table 13.6). All have a common oxidation state of 2^+. Metallic properties as well as activity increase as atomic weight and size increase. However, beryllium is best classified as a metalloid. Beryllium, magnesium, and calcium do not react appreciably with oxygen at room temperature. Magnesium must be strongly heated before it starts to burn, while calcium burns when heated slightly. Strontium oxidizes rapidly at 20°C and barium undergoes spontaneous combustion in moist air at this temperature. All isotopes of the largest and heaviest element, radium are radioactive (as we found with francium).

None of the alkaline earth elements are found free in nature. The free elements are generally produced by electrolysis of fused salts. Four of the elements, barium, strontium, calcium, and magnesium, were first prepared by electrolysis in 1808 by Davy (see Section 13.10).

The similarity in chemical properties of the alkaline earths means that many chemical or industrial processes can use compounds of the different elements for similar purposes. Since the free elements magnesium and calcium are active and relatively inexpensive, they are used to prepare elements such as uranium by simple displacement reactions. Although calcium and magnesium are abundant and have many functions in plants and animals, beryllium and barium compounds are poisonous.

Magnesium

Magnesium, which is the eighth most abundant element on earth and eleventh in the human body, is found in a variety of minerals. The carbonate of magnesium is known as

TABLE 13.6   Properties of metals in Group IIA

Element	Atomic Number	Atomic Weight	Atomic Radius ($\times 10^{-11}$ m)	Melting Point (°C)	Boiling Point (°C)	Density (at 20°C) g/cm³	Common Oxidation States	Hardness
beryllium	4	9.0	11.1	1 285	2 970	1.9	2	—
magnesium	12	24.3	16.0	650	1 117	1.74	2	2.0
calcium	20	40.0	19.7	851	1 487	1.55	2	1.5
strontium	38	87.6	21.5	774	1 366	2.6	2	1.8
barium	56	137	21.7	850	1 537	3.6	2	—
radium	88	(226)	22.0	(700)	(1 525)	(6)	2	—

magnesite. Dolomite is a mixture of the carbonates of magnesium and calcium. Talc (which we use in talcum powder) is $Mg_3Si_4O_{10}(OH)_2$, and asbestos contains compounds that have been assigned the formulas $H_4Mg_3Si_2O_9$ and $Ca_2Mg_5(Si_4O_{11})_2(OH)_2$. Chlorophyll, the green, light-absorbing matter of plants, contains magnesium in a complex organic structure.

Magnesium metal is usually available in chemistry laboratories in the form of a ribbon. When magnesium ribbon is heated it burns with a dazzling white flame. In the presence of air, both the oxide and the nitride are formed.

$$2Mg + O_2 \rightarrow 2MgO \tag{13-25}$$

$$6Mg + 2N_2 \rightarrow 2Mg_3N_2 \tag{13-26}$$

Magnesium metal reacts very slowly with water to produce magnesium hydroxide.

$$Mg + 2HOH \rightarrow Mg(OH)_2 + H_2 \tag{13-27}$$

The reaction proceeds better with steam (Eq. 13–16), giving the oxide instead of the hydroxide.

In spite of its activity, magnesium can be stored in air because a coating of inactive $MgCO_3$ forms by the reaction of MgO with CO_2 from the air. This coating prevents further oxidation. At higher temperatures magnesium metal reacts with carbon dioxide.

$$2Mg + CO_2 \rightarrow 2MgO + C \tag{13-28}$$

Because of the brilliance of the flame produced when it is oxidized, magnesium metal is used in photographic flash devices, for flares, and in fireworks. Its relative inactivity and low density make magnesium useful as a component of lightweight alloys. A suspension of magnesium hydroxide in water, known as milk of magnesia, is used as an antacid and laxative, and Epsom salts ($MgSO_4 \cdot 7H_2O$) is used as a strong laxative or purgative, particularly in veterinary medicine. Asbestos is widely used for brake linings and in fireproofing and insulating materials. However, it now appears that exposure to asbestos is linked to a high incidence of lung cancer, which may not develop until many years after exposure. Apparently, the lung tumors develop as a reaction to the crystalline shape of asbestos, not to the elements present. Magnesium (in its ionic form) is required for both plant and animal life.

Calcium   Calcium is fifth in abundance not only in the earth but also in the human body. It is a prominent substance in sea water. Calcite, chalk, and marble are all forms of calcium carbonate, and gypsum, which we use in plaster of Paris, is a hydrated calcium sulfate. Coral, as well as the shells of many organisms that live in both fresh and salt water, contains large amounts of calcium carbonate. In fact, deposits of chalk and other calcium carbonates probably formed from shells of marine organisms. Calcium ions circulate in human blood and are involved in blood clotting as well as muscle function. There are insoluble calcium compounds in the bones and teeth. Both milk, which is produced by all mammals, and the egg shells of birds contain calcium compounds.

When calcium carbonate is heated, carbon dioxide is driven off and calcium oxide, which is commonly called lime or quicklime, is produced.

$$CaCO_3 \xrightarrow{\text{heat}} CaO + CO_2 \tag{13-29}$$

limestone   quicklime

Addition of water to quicklime produces calcium hydroxide, which is called slaked lime.

$$CaO + HOH \rightarrow Ca(OH)_2 \qquad\qquad\qquad (13\text{-}30)$$
$$\text{slaked lime}$$

Mortar, which is used between bricks and other building materials, is a mixture of slaked lime, sand, and water. As the mortar sets, it absorbs CO_2 from the air and loses H_2O to form insoluble $CaCO_3$.

$$Ca(OH)_2 + CO_2 \rightarrow CaCO_3 + H_2O \qquad\qquad\qquad (13\text{-}31)$$

Some mortar also contains cement. Calcium hydroxide is one of the least expensive bases and is widely used as such in industrial processes. Lime is often used to neutralize acid soils, as in the midwestern United States, to make them suitable for growing crops such as corn.

Plaster and Portland cement contain calcium aluminum silicates. When plaster sets or cement and sand solidify to form concrete, complex crystalline structures form. This process continues for a long time, and the exact chemical and physical changes involved are not fully understood. The setting of plaster of Paris was discussed in Section 9.5; the formation of caves in limestone is explained in Section 14.18.

Because of the many functions it serves in all living things, calcium is a necessary component of the soil and of food. The requirement for calcium in the diet of humans (800 mg/day) is greater than that of any other mineral element. The need is even greater during a period of growth when new bones are being formed and during pregnancy and lactation.

Barium, Strontium, Radium, and Beryllium

Barium is a toxic element, but barium sulfate is given to patients by mouth or enema before x-rays are taken. Since it is opaque to radiation it helps make the digestive tract visible on the x-ray picture. The reason $BaSO_4$ can be used is that it is extremely insoluble in water and even in the acid (HCl) of the stomach; thus it is not absorbed by the body. It is the barium ion that is toxic. Water- or acid-soluble compounds of barium would, of course, ionize to produce barium ions, but barium sulfate does not.

Strontium is a relatively rare element. Many of the practical uses of it are based on the fact that, when compounds containing strontium are heated, a brilliant red flame is produced. Compounds of strontium are used in fireworks and in the emergency flares we carry in our automobiles to warn others of a wreck or other highway hazard. The radioactive isotope ^{90}Sr is produced in nuclear explosions. When it falls to earth, plants and animals absorb it and treat it much as they do calcium. Thus cattle grazing on grass that contains strontium-90 incorporate the radioactive isotope in their milk. People who drink this milk tend in turn to deposit ^{90}Sr along with calcium in their bones. Since the red blood cells are produced in bone marrow, the presence of the radioisotope can disrupt normal blood cell synthesis as well as possibly causing bone cancer.

The use we make of radium depends much more on its radioactivity than on chemical or physical properties. When radium is enclosed in metallic "needles" and placed near a cancerous tumor, it can help to destroy the tumor. Although radiation can cause as well as treat cancer, the possibility of destroying the tumor is greater (and of more immediate importance) than the lesser probability that radioactivity will itself produce a cancer. The current trend in developed countries is to replace radium with other sources of radiation for tumor treatment. Care must be taken not to ingest radium, since it, like strontium, is a "bone-seeking" element.

Although beryllium belongs to Group IIA, it has many properties in common with the first element in Group IIIA, aluminum, as well as with the metalloids. Emeralds and aquamarines are slightly impure silicates of beryllium and aluminum. Like aluminum, beryllium is used in many lightweight alloys. Modern industrial users of beryllium must take into consideration the toxic nature of this element.

**13.10
THE DISCOVERY
OF THE
ELEMENTS OF
GROUPS IA
AND IIA**

We generally credit a person with discovering a chemical element when he or she first prepares or isolates the element in its pure form or otherwise shows that a new substance is present. Sometimes, as we shall see, an individual or a small group of people has been responsible for the discovery of more than one element. In most cases the discovery or preparation has depended on having pure compounds of the element to work with and a new method either of producing or of identifying the element.

The English scientist Sir Humphrey Davy is credited with the discovery of potassium, sodium, barium, strontium, calcium, and magnesium. He had relatively pure compounds of these elements to work with and a new device that made their isolation in the pure state possible—a strong battery containing more than 250 plates. When current from this battery was passed through the melted hydroxides or salts of the alkali or alkaline earth metals, the pure elements were produced. Potassium was isolated on October 6, 1807, and a week later, sodium metal was prepared for the first time. In 1808 the four alkaline earth elements were isolated. Thus, although compounds of these elements had been known for centuries, a powerful battery made is possible to isolate all of them in a free state within a period of about one year. Amazingly, this was done by one man (probably with the help of laboratory assistants).

The German chemist Robert Bunsen is best known for his development of a gas burner which admitted air at the bottom of a tube to produce an almost invisible flame. He and a fellow German, Gustav Kirchhoff, developed the spectroscope, in which light from a substance being burned passed through a prism, where it was sorted out into light of different colors, which appeared as lines. They helped develop the theory that each element when heated gave off light of definite colors—that it had a characteristic spectrum. A sodium flame was an intense yellow-orange, and when the light from the sun was analyzed by a spectroscope, lines in the identical part of the spectrum indicated the presence of sodium in the sun.

Bunsen and Kirchhoff studied the spectra given off by different pure compounds. In 1860 they found that a sample of mineral water produced a new line in the blue part of the spectrum, one that was not produced by any known element. They concluded there was a new element present, and named it cesium. The name comes from the Latin *caesius*, meaning sky blue. In 1861 a deep red line in the spectrum of the mineral lepidolite helped them identify rubidium, which derives its name from the Latin *rubidis*, meaning deepest red.

Radium, which was first produced in 1910 by the electrolysis of radium chloride by Marie Curie and André Debierne, had previously been identified by its radioactivity. Credit for its discovery goes to Madame Curie and her husband Pierre. The crucial step in the isolation of radium was the purification of radium chloride. There is about 1 gram of radium in 7 tons of the mineral pitchblende. The prodigious efforts of Marie Curie in purifying radioactive substances from pitchblende resulted not only in the discovery of the radioactive element polonium (in July 1898) and radium (in December 1898) but also in her exposure to a great deal of radiation. Both polonium (named for Madame Curie's native country, Poland) and radium (named for the radiation it produced) were detected after Pierre Curie developed a device that measured radioactivity. Thus both a purification and a new instrument were responsible for identifying the elements radium and polonium. Marie Curie, who is the only person to be awarded two Nobel Prizes in science, died from leukemia, which was probably brought about by exposure to radiation. André Debierne, who helped Madame Curie prepare radium, was a close friend as well as co-worker with the Curies and is credited with the discovery, in 1899, of the element actinium. Although Pierre Curie worked with his wife during their experiments, his death in 1906 prevented his sharing the Nobel Prize for the discovery of radium and polonium.

Francium, as well as the other elements that do not exist in any appreciable amounts on earth, was synthesized by bombardment of heavier elements in cyclotrons and similar devices. The identification of these rare elements depends on analyzing their radioactivity, since most have not been produced in visible or weighable quantities. Francium can be formed by radioactive decay of actinium or proton bombardment of thorium. It was discovered by Marguerite Perey, working at the Curie Institute in Paris. Obviously it was named for the country France.

13.11
COPPER,
SILVER, AND
GOLD

It is interesting to consider the chemistry of the elements of Group IB—copper, silver, and gold—and especially to compare their properties with those of the elements of Group IA. Although the elements of both groups have one electron in their outer shells and we would predict they would be similar in properties, we know this is not true. The elements of Group IA—lithium, sodium, potassium, cesium, rubidium, and francium—are the most active metals known, and those of Group IB are among the least active. In fact, the uses and value of the elements of Group IB are primarily based on their inactivity. Some of the important properties of the metals of Group IB are given in Table 13.7.

One property that all elements in Groups IA and IB have in common is that they are good conductors of electricity and in most instances have oxidation states of 1. Copper and gold also have common oxidation states of 2 and 3, respectively. We explain these oxidation states by saying that electrons are pulled from the inner shell when they show these oxidation states.

We also explain some of the differences in reactivity by looking to the electron shell just closer to the nucleus than the valence shell. In Group IA there are 8 electrons in this shell, while there are 18 in the elements of Group IB. While one reason for the lack of reactivity of copper, silver, and gold involves the effect of the underlying electrons on the loss of the valence electron, probably the major reason for the difference in activity is that the elements of Group IB have much smaller atomic radii. Thus their positive nuclei are nearer the valence electron and can hold it more tightly. The atomic radius

TABLE 13.7 Properties of metals in Group IB

Element	Atomic Number	Atomic Weight	Atomic Radius ($\times 10^{-11}$ m)	Melting Point (°C)	Boiling Point (°C)	Density (at 20°C) g/cm^3	Common Oxidation States	Hardness
copper	29	63.5	12.8	1 083	2 582	8.9	1, 2	2.5–3.0
silver	47	108	14.4	960	2 177	10.5	1, 2	2.5–4.0
gold	79	197	14.0	1 063	2 707	19.3	1, 3	2.5–3.0

is much smaller than in Groups IA and IIA, not only in the elements of Group IB, but in all the elements indicated as transition elements on the Periodic Table. Thus the transition elements are much less active than the alkali and alkaline earth metals. These elements also show multiple oxidation states, which are due to loss of electrons from inner shells.

Gold and Silver Gold and silver are found both in the free state and in compounds. Production of the free elements from compounds is relatively easy due to the inactivity of the elements. It might be said that the chemistry of silver and gold is of little significance because they form so few compounds and because both elements are relatively rare in nature. It is their physical properties, especially their malleability, ductility, and resistance to chemical change, that make them useful. Furthermore, we cannot discount the social effect of the relative scarcity of these elements. We tend to value that which is rare or hard to obtain.

In addition to uses for coins, ornaments, and as an item of commerce, gold and silver can be used for filling or capping teeth. Both silver and gold are good conductors of electricity, but they are too expensive to be used extensively for this purpose. Since pure gold is too soft for many purposes, it is generally alloyed with other materials, both to harden it and to make it less expensive. Silver alloys are also widely used (see Table 13.4). Silver compounds are used in photography, and diluted solutions of silver nitrate are routinely applied to the eyes of newborn babies to destroy bacteria they may have acquired in the process of being born.

Copper Although copper is found in the free state, it is more commonly found in the form of sulfides, carbonates, and oxides. A copper-containing blue pigment is used by lobsters and some other marine animals as an oxygen-carrying substance similar to hemoglobin.

Copper is produced by the common methods of purifying metals from their ores (Section 13.4); electrolysis is usually needed as the final step in order to produce the pure copper needed for most uses. Copper is more active than either silver or gold. When copper "weathers" it produces a green compound, $Cu_2(OH)_2CO_3$, which is similar in structure to the copper ore malachite. In fact, most compounds of copper are colored, with the colors ranging from yellow to green to blue to black. The copper(II) ion is a brilliant blue color. Copper is one of the few metals that has a color, and the color is so distinctive that we use the word copper to describe other objects having a similar hue.

Although silver is a slightly better conductor, copper metal is the most widely used electrical conductor in the world today. For this use copper must be very pure, since small amounts of impurities alter its conductivity. Bronze and brass (Table 13.4) are widely used alloys of copper, and all United States coins now contain copper alloys. Reagents such as Benedict's and Barfoed's reagents, which are used for testing for sugars, contain copper salts. In tests with these reagents the bright blue copper(II) ion is changed to a yellow/orange precipitate when copper is reduced to the Cu(I) form. Copper sulfate is used to control the growth of algae in bodies of water; however, it can poison other forms of life also. Very small amounts of copper are needed by the human and other animals, while larger amounts are poisonous.

13.12
IRON AND
ALUMINUM
Many of the metals we have discussed are important to our civilization, but iron and aluminum are generally regarded as the two most important metals, with copper being third. While it is not always true that substances used in the greatest amount are the

most important, it is difficult to imagine a modern technological society without iron and aluminum. Aluminum is the most abundant metallic element on the earth (and the third most abundant among all elements), while iron is just behind it, ranking fourth in overall abundance and second among the metallic elements. More iron is used in the United States than all other metals combined. Although the metals nickel and cobalt have properties similar to iron and might be used in place of it, they are relatively rare and are much more difficult to mine and refine; thus they are much more expensive. (See Table 13.8 for the properties of aluminum, iron, and similar metals.)

Both aluminum and iron are active and found mainly in the form of compounds, particularly their oxides. While almost every soil or rock sample you might find would contain iron and aluminum oxides, good ores—samples containing high percentages of these metals—are relatively rare. In most industrially developed countries these ores have been extensively mined. Since much of the best iron ore (hematite) has been mined from the Mesabi range in northern Minnesota, we now use the taconite ore, a poorer source of iron. The mining of bauxite (a hydrated aluminum oxide) in the United States has decreased appreciably because the richer ores have been exhausted.

Very little iron is found in the free state. The small amounts which reach the earth in the form of meteorites, alloyed with nickel, are thought to be fragments of the interior of a disintegrated member of the solar system. Free iron of terrestrial origin has been found in rather rare instances, apparently forced up from the interior of the earth, which we believe is primarily iron and nickel. Iron is too active chemically to remain in a pure state for long periods at the earth's surface unless it is protected from weathering. While there are many naturally occurring minerals or compounds of iron, only a few are used as ores. The most widely used are hematite, Fe_2O_3, limonite, a hydrated form of hematite, $2Fe_2O_3 \cdot 3H_2O$, and magnetite, Fe_3O_4. This formula can also be written as $Fe_2O_3 \cdot FeO$, indicating a compound containing both common oxidation states of iron.

The name hematite is of Greek origin, from *haimatites*, blood-like. The ore is red in color and is sometimes used as the pigment in red paint. Magnetite has magnetic properties which are utilized in separating it from impurities in its ores. When the molecules in a specimen of magnetite are properly oriented (polarized) the mineral will attract a compass needle. Such rocks are called lodestones or natural magnets.

Iron is purified from its ores by the use of the reducing agent carbon monoxide, which is produced from coke. Coke, which is mostly carbon, is produced by heating

TABLE 13.8 Properties of commercially important and poisonous metals

Element	Atomic Number	Atomic Weight	Group in Periodic Table	Atomic Radius ($\times 10^{-11}$ m)	Melting Point (°C)	Boiling Point (°C)	Density (at 20°C) g/cm³	Common Oxidation States	Hardness
aluminum	13	27.0	IIIA	14.3	660	2 447	2.70	+3	2–2.9
iron	26	56.0	VIII	12.4	1 530	3 000	7.9	2, 3, 4, 6	4–5
cobalt	27	59.0	VIII	12.5	1 495	3 550	8.9	2, 3	—
nickel	28	58.7	VIII	12.5	1 455	2 840	8.9	0, 1, 2, 3	—
mercury	80	201	IIB	16.0	−38.9	356.6	13.6	1, 2	—
lead	82	207	IVA	17.5	327	1 751	11.3	2, 4	1.5

coal in the absence of air. Some carbon also combines with iron to produce the compound cementite, Fe_3C.

$$3Fe + C \rightarrow Fe_3C \tag{13-32}$$

This reaction is reversible and the amount of carbon present as the element in an alloy depends on the temperature and rate at which the mixture has cooled. The initial purification of iron from iron ore produces a mixture called *pig iron*, containing 3–4% carbon and lesser amounts of silicon, phosphorus, and magnesium. Most of the silicon, which is present in iron ores as silicon dioxide (sand) and silicates, is removed by adding limestone, which releases CO_2 to form CaO, which then combines with SiO_2 and the silicates to form calcium silicate.

$$CaO + SiO_2 \rightarrow CaSiO_3 \tag{13-33}$$

Calcium silicate (along with some other substances) is called *slag*. Since molten slag will float on molten iron it can be separated. Slag is used in making Portland cement and glass.

The iron and steel industry consumes large quantities of limestone and coke. In general, 50 carloads of iron ore require 25 carloads of coke and 15 of limestone for processing. Steel mills are often located near supplies of coal, which is used to make coke, rather than near deposits of iron ore. Indeed, the Industrial Revolution, which depended so much on the production of iron and steel, depended on coal as much as on iron ore.

As with most metals, pure iron is soft, and many physical properties are enhanced in its alloys, the steels. Addition of small amounts of carbon increases the hardness of iron. but larger amounts make the alloy brittle. The purification of the hard but brittle pig iron and addition of other substances (mostly other metals) is the basis of the complex steel industry. A listing of some of the important kinds of steel is given in Table 13.4.

Billions of dollars are lost each year due to the reaction of iron to produce iron oxide, which is known as rust. Neither oxygen (in the air) nor water alone will react with iron to produce rust; it is a combination of the two that is needed. A variety of methods are used to prevent rusting. In some cases we coat iron with a paint, lacquer, or ceramic material, as in the porcelain used in sinks and bathtubs. In other cases the coating is a less active metal, for example, tin (in "tin" cans) or zinc (in galvanized iron or steel). Sometimes, also, iron is exposed to superheated steam and the iron oxide which forms adheres to protect the iron underneath from further corrosion. We avoid the problem when we use noncorroding alloys such as stainless steel, a tough, durable alloy that does not rust.

Most iron compounds are colored and many of the colors of natural rock formations, particularly the rust reds, are due to iron. Other iron compounds, such as those used in making blueprints and inks, are blue or black.

Iron is an essential compound of the large organic molecule hemoglobin, which carries oxygen in the human blood. Although the human body has a complex recycling system that promotes reuse of iron, small amounts of iron should be included in the diet. Plants also need iron, and deficiencies are generally seen as yellowing of the leaves. Since iron is so widespread, the deficiency seen in a plant is generally not of iron itself but of iron in a soluble form that the plant can absorb.

Aluminum　It is interesting to note that aluminum, now the second most common metal in terms of quantities produced, is so new that it is almost a twentieth-century metal. It was first produced in 1828, but for many years was primarily an expensive curiosity. It is reported that Emperor Napoleon III of France was so fascinated by this light and beautiful metal that he had a set of tableware made of it to be used in place of gold for his most important banquets. The unusual lightness, strength, and durability of aluminum indicated that it should have great utility, but the price remained so high, several dollars per pound, that its uses were limited.

The story is told that one day in 1885 Charles Martin Hall, a student at Oberlin College, was listening to his chemistry professor, who made the statement that fame and fortune awaited the man who could find an inexpensive method of making aluminum. Hall turned to his laboratory partner and said, "I'm going after that metal." He started his work in a back-yard laboratory, and in February, 1886, he produced small amounts of aluminum by a unique method of electrolysis, a method still being used. He dissolved aluminum oxide in molten cryolite, Na_3AlF_6, and using carbon electrodes passed a strong current of electricity through the molten liquid. Hall did become famous and wealthy, but much effort was required to convince the world of the advantages of this unusual metal.

Aluminum is a silvery metal with a density of 2.7 g/cm^3, which is approximately one-third that of iron. It is a soft malleable, and ductile metal that is an excellent conductor of heat and electricity. Industrial users take advantage of some or all of these properties. It is hammered or rolled into many useful objects. It is used, for example, in cooking utensils where a transfer of heat is essential, and it is drawn into wires for conducting electricity.

Aluminum is quite active chemically, and were it not for one peculiar characteristic, it would be useless, or used for much different purposes. It reacts with oxygen to form a very thin, invisible film of Al_2O_3 which serves as a protective coating that renders the surface inactive. The fact that aluminum forms a wide variety of alloys with greatly different properties has added materially to its importance. It seems doubtful that one of our largest industries—aviation—could have reached any degree of importance without the lightweight aluminum and its alloys for the structure and the skin of large aircraft. Aluminum is also used in structural beams, window frames, doors, canoes, travel trailers, mobile homes, siding for buildings, paint pigments, pots, pans, foil, and countless other items.

Aluminum forms some interesting compounds. Corundum, Al_2O_3, is an important abrasive. Sapphires and rubies are also naturally occuring oxides of aluminum containing small amounts of impurities. In recent years synthetic sapphires and rubies of high quality have been made. They are used for jewels in watches and as gems. Aluminum has no known function in the human body.

13.13
THE
POISONOUS
ELEMENTS

The ions of approximately 25 elements are highly toxic to humans and other animals. Among the most toxic are the metals lead, mercury, beryllium, cadmium, chromium, and nickel and the metalloid arsenic. The nonmetal fluorine is apparently needed in small concentrations, but it becomes toxic at higher levels. Although these elements are widely distributed, they are relatively rare. They are usually found in compounds that are insoluble and therefore not likely to poison humans. It is when we concentrate these elements and make them soluble that problems may arise. Many cases of acute poisoning have been caused by ingestion of relatively small quantities of metals, but more insidious are death and disabilities due to long-term ingestion of very small doses. Since the animal body does not excrete these substances very well, levels tend to increase: we say that these are cumulative poisons. Many of the effects of these poisons are similar to those of other diseases and disorders, and it is difficult to be certain whether they are the actual causes of death. We know from studies with experimental animals that deaths do occur after chronic poisoning, but we must make a test of the content of these elements in blood, bone, or other tissue before we can determine the extent to which they are responsible for death or illness of humans.

Lead Lead was one of the elements known to ancient peoples. The ease with which it was separated from ores, its low melting point, the ease with which it was shaped, and its resistance to corrosion made it useful. Although the Greeks knew about the toxicity of lead, apparently the Romans did not. They used it for water pipes and for lining wine-storing vessels (in fact, the word *plumbing* and the chemical symbol Pb are derived from the Latin word for lead, *plumbum*). Examination of bones from Roman tombs shows concentrations of lead sufficiently high that we can be certain they suffered from lead poisoning. Since lead poisoning is known to affect mental processes, many have speculated that the erratic behavior of some of the latter-day Romans may have been due to lead poisoning, and others even go so far as to suggest that part of the reason for the fall of the Roman Empire was widespread lead poisoning among the ruling classes. Poor people did not use the expensive lead containers and were not so likely to be poisoned.

We have used lead for a variety of purposes in our civilization. While we no longer use lead pipes to carry drinking water, this change is comparatively recent. The insecticide lead arsenate (which contains arsenic in addition to lead) was widely used, particularly on tobacco plants, until comparatively recently. Both of these toxic elements were absorbed by plants and probably still persist in soils. Until about 1940 lead was widely used as a pigment in paints. It is used for glazes on pottery, and as recently as 1970 there were many episodes of lead poisoning due to drinking weakly acidic liquids, such as apple juice and cola beverages, that had been stored in glazed pottery. Perhaps the greatest source of lead in our environment today is the tetraethyl lead which continues to be used as an antiknock ingredient in high octane gasoline. Examination of ice and snow in the region of the North Pole shows a steady increase in lead content from the beginning of the Industrial Revolution and an increase of about 300 percent between 1940 and the present. The presence of lead in such a remote place is a clear indication of the global nature of the problem.

Acute lead poisoning results in brain damage, mental deficiency, and serious behavior problems. Kidney damage is seen both in acute and chronic cases of lead poisoning. Acute poisoning is observed in some people who work in industries which expose them to high concentrations of lead. Even more tragic is the acute lead poisoning of children who live in urban slum areas. These children often eat chips of paint that contain lead as a pigment. Although titanium dioxide has replaced lead pigments in most white paints today, older houses may still have the older toxic paints, perhaps a layer of two below the current paint. Surveys in major United States cities show that many children who live in the cities have levels of lead in their blood that are one-half those found in children who have definite symptoms of lead poisoning. For an excellent discussion of this and other aspects of lead poisoning, see the article by Chisholm in the list of readings at the end of this chapter. Lead poisoning is a condition that can be prevented with sufficient knowledge and care.

We do not know the effects on the general population of exposure to lead. As with ionizing radiation, none of our senses warn us that we are being poisoned. However, chemical analysis shows us that people accumulate more lead in their bones as they grow older, and thus lead can be considered a cumulative poison. Since the symptoms of chronic lead poisoning are similar to those caused by many diseases and other poisons, it is difficult to assess the long-term effect of low levels of exposure. Tests for the presence of lead in body tissues will have to be widely made and the results correlated with the cause of death in a large number of people before we can give definite answers. In contrast to the situation with the ancient Romans, in our civilization it is the poor who live in old houses near a great deal of traffic that are most likely to be poisoned.

Even in the absence of proof of the influences of low levels of lead, it seems logical to decrease the amount in our environment as much as possible. The widespread use of lead in plumbing, insecticides, and in paints is probably a thing of the past. It is clear that we must become more aware of the content of pottery glazes, and perhaps widespread testing will have to be instituted. The most important current effort to reduce lead in the environment is the introduction of low-lead and lead-free gasolines. These products are a response both to concern for the amount of lead in the atmosphere and to the fact that catalytic after-burners used to decrease other types of pollution from automobile exhausts cannot function efficiently if lead is present.

The story of the use of lead illustrates how chemical knowledge of things not perceived directly by our senses can lead to improvement in health and decrease in deaths by poison. It also illustrates how facts known as long ago as the age of the ancient Greeks may be rediscovered centuries later—often only after the senseless death of many people, especially children. Unfortunately, we lack knowledge and concern about many other poisons we are exposed to every day.

Mercury   The Mad Hatter in "Alice in Wonderland" is a classic example of a person poisoned by exposure to mercury. Felt hats were treated with compounds of mercury, and the expression "mad as a hatter" indicates a widespread knowledge of the symptoms but indicates no awareness that such behavior was due to mercury.

In addition to using mercury in thermometers, where it is encased in glass, we have used a variety of mercury compounds over the years. The compound $HgCl_2$ was once used as a germicide. Various compounds of mercury are used to prevent growth of fungi on seed grains. Many chemical processes, particularly the production of chlorine and sodium hydroxide, use mercury.

In 1969, the children of a New Mexico family fell ill with a variety of symptoms. Several weeks elapsed before urinalysis was carried out which showed that they were suffering from mercury poisoning. As the medical detective work later revealed, the father of the family fed his hogs some seed grain that had been treated with mercury to retard growth of fungi. Some of the hogs died of a condition described as "blind staggers," but this fact did not prevent the butchering of hogs from this group and the use of the pork for food. The symptoms observed in the afflicted children included dizziness, loss of muscular control, loss of vision, deafness, and long periods of unconsciousness or coma. A year later one child was confined to his wheelchair unable to walk more than a step or two and unable to control the bobbing of his head; another could walk a bit; and a third had just emerged from eight months of coma, blind and unable to talk. A fourth child who was born soon after the other children were affected is blind, and hope for him was slight. The effects on the children were greater than on the parents.

From 1953 until 1960 near Minimata Bay in Japan, more than 111 persons died or were seriously handicapped as a result of long-term consumption of mercury in fish and shellfish which were taken from industrially polluted water. Five more died and 25 were injured at Nigata in Japan in 1968. In Iraq, 35 people died and 321 were injured in 1961 from mercury poisoning, and there are unconfirmed reports of another poisoning incident involving hundreds of people in Iraq.

It has been known for some time that organic compounds of mercury, such as those containing the methylmercury ion, are water-soluble, and that dimethyl mercury is fat-soluble. It is also known that both of these compounds are extremely toxic. It has been assumed, however, that mercury itself and inorganic compounds of mercury are insoluble and heavy enough to settle to the bottom of bodies of water and thus remove themselves from circulation. In addition, we reclaim much of the mercury in industrial use because it is a valuable and rare metal. However, recent findings have led to reconsideration of the safety of our disposal procedures. There is a bacterium that converts apparently insoluble inorganic mercury compounds to the toxic, soluble organic compounds. When we add to this knowledge the realization that much of the mercury used to treat seeds leaches out of the soil into streams and rivers, we have cause for concern.

These concerns led to the widespread analysis of various kinds of fish for mercury. Much of the American public was shocked in 1971 when analysis of deep-sea fish showed extensive amounts of mercury. Where are these large deep-sea animals getting the poison? Apparently the explanation involves food chains. Soluble mercury is present in water either as runoff from fields or from conversion of insoluble forms. The soluble mercury is ingested by plants or small animals, and these are eaten by larger animals. Each step can produce a greater concentration of mercury in the tissues, and it seems that the largest fish have the greatest amounts of mercury. Of course, a smaller fish or even a shellfish such as an oyster or clam could have high concentrations if it grew in water with high mercury content.

Examination of preserved fish which lived fifty to a hundred years ago show that, even then, they contained a fairly high concentration of mercury. Even fossil fish contain small amounts of this element.

With mercury as with most cumulative poisons, we do not know whether there is a level low enough to be safe or what limits should be set on foods for human consumption. We do know that the substance is poisonous and that we must be cautious in its use and consumption. Recent analyses have shown that most foods, other than fish and sea foods, have very low levels of mercury.

Although some incidents have been reported, there are at present no evidences of widespread poisoning by the elements arsenic, beryllium, cadmium, chromium, and nickel. These elements are widely used for a variety of purposes in technologically advanced societies. We know that they are toxic, but at present we have little information about levels of toxicity, amount and duration of retention in the human body, and effects of long-term exposure. Most chemical industries know of these toxicities and take care to protect their workers, but some firms are not sufficiently concerned about release of small amounts into the water or atmosphere. Less well-informed people use these elements in ways that could result in acute poisoning. Unfortunately, we usually do not become sufficiently concerned about such problems until they cause the death of several people.

Fluorine   Many ores, particularly those of aluminum, are fluorides. Chemical plants that process these ores often release fluorine in a volatile compound or small particles. Cattle, sheep, and other grazing animals have often been shown to die from fluoride toxicity. With continuous care, it is possible to remove this toxic element from both gaseous and liquid wastes of industry.

**13.14
RESOURCES
IN SHORT
SUPPLY**

While chemists must be concerned about the poisonous substances that have been introduced into the environment because they were concentrated and made soluble, there is a related problem that is the concern of all people. Our use of metals has depleted many good ores and our use of coal and petroleum in the refining and alloying of metals as well as for many other purposes presents a serious problem.

Modern industrialized civilization, with its emphasis on buildings, automobiles, automatic dishwashers, television sets, and the myriad other things we use and feel we need creates a great demand for metals. Those most critical in our society are iron, aluminum, and copper. Other metals such as mercury, zinc, lead, molybdenum, tin, tungsten, cobalt, nickel, and vanadium are needed in smaller amounts, but when the supply of these is small or located at relatively few places on the earth, they may also be critical.

At one time there were huge deposits of iron ore in the Mesabi range in Minnesota. Most of the high-grade ore there has now been used. Once copper ores containing nearly pure copper could be picked up at the surface of the earth. One hundred years ago in the United States we were mining ores that contained 5% to 6% copper. Now we have used these high-grade ores and are mining and refining ores that contain 0.4% to 0.8% of the metal. Mines once abandoned are being reexplored with the hope that they can now be worked profitably. Although there were once good deposits of high-grade aluminum ores in the United States, we are importing much of this ore today.

Almost all the industrialized nations are now importing the ores of iron, aluminum copper, and the other metals needed to maintain a high production of consumer goods. In the past, mineral supplies have led to armed invasions and conquest of mineral-rich nations. We cannot disregard the relationship of mineral resources, including petroleum, to the wars of the twentieth century.

Even if political and transportation problems could be solved, the basic problem remains. We are using mineral resources at a rapid rate. Except for deposits in some of the nonindustrialized countries, the high-percentage ores have been used. Although technology can develop ways of using minerals having lower content of the desired metal, this is done only by using an increased amount of energy to separate the metal from the source in which it is found. One aspect of this problem is that the ores we are now using come from greater depths and we must remove more dirt and rock to get to the deposits. We also remove more waste material or slag as we refine lower and lower grade ores. In order to do all this, the price for the metal must rise to compensate for the greater cost of producing it. In terms of the available reserves of energy, particularly the fossil fuels, it is questionable that we should continue to use more of these to produce metals from low-grade ores. Many critical metals might be extracted from sea water, but they exist there in extremely low levels and a great deal of energy is required to concentrate and refine them.

Combined with this shortage of many metals, we see huge lots filled with discarded automobiles, streams littered with soft drink cans, and much of our metal waste is buried in disposal areas. While there are some losses that cannot be easily prevented such as loss of iron due to rusting, we can reuse much of our metals by recycling. There is, of course, no shortage of iron atoms. It is just that most of them are widely dispersed in the earth's crust. We have concentrated some of them and used them to build skyscrapers and automobiles. If we are to build more skyscrapers and automobiles, we must increase the amount of iron that is recycled. The same is true of all the other metals we need. Except for corrosion, our use of them does not exhaust them in the same way our use of coal does; it merely requires an awareness and the development of facilities to use more effectively those atoms we have on board our spaceship earth.

Energy Resources   While this chapter has been concerned with the atoms and molecules of ores, metals, and alloys, it is also appropriate to consider briefly the sources of energy necessary for the refining and use of metals. Modern civilization has been made possible because of our exploitation of fossil fuels. So long as our sources of power were animals who ate plants or we used the actual plants themselves as wood or charcoal, the amount of energy each individual could appropriate to his use was small. While some use could be made of water power, the industrial revolution came only when coal was available. The addition of petroleum products greatly increased the amount of available power and brought us from the age of the steam locomotive to the automobile and diesel tractor. Such use of energy poses several problems. Much of the pollution in our atmosphere today is due to the burning of these fuels. They also create problems associated with the effects on global concentrations of oxygen and carbon dioxide and the subsequent effects on the temperature of the earth. These seem not to be crucial at the present, but they cannot be neglected. (See Section 15.5.)

The major problem involves our depletion of sources of energy. The amount of coal and petroleum is limited and there is no way to recycle them. Although it is possible that some coal and petroleum is being formed within the earth today, we are using these resources at a rate infinitely greater than their rate of formation. Other sources of energy, primarily that from nuclear reactions and from solar energy, must be explored. Energy from nuclear fission is now producing somewhat less than 1% of the electricity in the United States, but rapid expansion has stopped because the possibilities of pollution of the environment by radioactive substances is great. The possibility of control of nuclear fusion offers great hope for the future. On a long-term basis, we must begin to look to solar energy—more efficient ways of trapping it and ways of storing it for use on cloudy days or at a distance from the site of production.

13.15 COMMERCIALLY IMPORTANT CHEMICALS   An idea of which chemical substances are important to our industrialized civilization can be gained from Table 13.9. This table is taken from a listing of the chemicals that were produced in the greatest volume in the year 1979 in the United States. It will be noted that only one organic compound, ethylene, is among the top 12. Organic chemical substances dominate the list below this point and only the inorganic chemicals are listed for the remainder of the table. It should be realized that while the chemicals on this list are commercially the most important, this is only one way of considering importance. A few milligrams of penicillin or a crucial valve made from an organic polymer may also have great importance.

SUMMARY

The metals have found many different uses because of great variety in their properties. They range from the very light to the very dense, and from the extremely reactive to the unreactive. In general, their properties include hardness, strength, and durability. The discovery of a new metal or a new mixture of metals—an alloy—can change a civilization. The number of possible alloys is almost limitless. Thousands have already been produced, and new ones are being developed to meet the needs of today's exploration of space. Although certain synthetic organic compounds, the plastics, are replacing metals for many purposes, our civilization needs both the metals and the plastics. To understand many of the events occurring in our world and the factors bringing about change, some knowledge and understanding of both the inorganic world of metals and nonmetals, and of the organic compounds is necessary.

TABLE 13.9 Selections from the chemicals that were produced in the greatest amounts in the United States in 1979*

Rank	Name	Formula	Production (Billions of Pounds)
1	Sulfuric acid	H_2SO_4	83.98
2	Lime	CaO, $Ca(OH)_2$	38.78
3	Ammonia, anhydrous	NH_3	36.24
4	Oxygen (gas)	O_2	35.35
5	Nitrogen	N_2	29.92
6	Ethylene	C_2H_4	29.19
7	Sodium hydroxide	$NaOH$	24.78
8	Chlorine	Cl_2	24.22
9	Phosphoric acid	H_3PO_4	20.27
10	Nitric acid	HNO_3	17.13
11	Sodium carbonate	Na_2CO_3	16.51
12	Ammonium nitrate	NH_4NO_3	15.60
23	Carbon dioxide	CO_2	7.07
26	Hydrochloric acid	HCl	5.95
31	Ammonium sulfate	$(NH_4)_2SO_4$	3.94
34	Carbon black	C	3.33
37	Aluminum sulfate	$Al_2(SO_4)_3$	2.46
39	Sodium sulfate	Na_2SO_4	2.34
41	Calcium chloride	$CaCl_2$	2.03
46	Sodium silicate	Na_2SiO_3	1.55
47	Sodium tripolyphosphate	$Na_5P_3O_{10}$	1.51
49	Titanium dioxide	TiO_2	1.48

* Data selected from *Chemical and Engineering News*, Vol. 58, No. 18, May 5, 1980, page 35. Reprinted with permission of the copyright owner, the American Chemical Society.

The following chapter discusses some of the most important nonmetals. The discussion of organic compounds begins with Chapter 18, and the organic polymers which we commonly call the plastics are discussed in Chapter 27.

IMPORTANT TERMS

alkali metals	ductile	nonmetal
alkaline earths	malleable	ore
alloy	metal	pig iron
brass	metallic bond	steel
bronze	metalloid	

WORK EXERCISES

1. List the most important properties of metals.
2. List the most important properties of nonmetals.
3. What is the term assigned to the elements along the border between the true metals and the non-metals?
4. Name two important reactions typical of active metals which the less active metals do not undergo.
5. Give six steps involved in separating or purifying metals from their ores.
6. Name five types of compounds which are good ores of metals.

7. Why should a metal as prevalent in nature as iron be so rarely found in a free state?
8. Most plant-eating animals (herbivores) require an additional source of sodium ions. What does this indicate about the relative concentrations of sodium and potassium in plants and animals? Why do carnivores (animals that eat other animals) not need an additional source of sodium?
9. Given your answers to the above question, would vegetarians need to add more or less sodium chloride to their diets than others?

10. Which elements of Groups IA and IIA are poisonous to animals (in relatively low concentrations)?
11. List some of the developments that were necessary for the first isolation of the elements in Groups IA and IIA.
12. How can the differences in chemical reactivity of the metals in Groups IA and IB be explained?
13. List three uses of gold and silver other than for coins and jewelry.
14. List the most important uses of the metal copper and its compounds.
15. What is the apparent oxidation number of iron in the ore Fe_3O_4? How can a more acceptable oxidation number be assigned to the iron atoms in this ore?
16. What purpose does limestone play in the separation of iron from its ores?
17. Name some of the properties of copper which make it more desirable than iron for certain uses.
18. State some of the properties of aluminum which make it more desirable for some purposes than either copper or iron.
19. What are some of the advantages of alloys over pure metals?
20. Name some of the uses of the slag produced in blast furnaces.
21. Why were metals such as gold, silver, and copper known to ancient peoples?
22. What technical processes were necessary before the iron age developed?
23. List some methods used to prevent the formation of rust.
24. What are some of the effects on properties of metals when small amounts of impurities are added to them?
25. List those metals that are poisonous when consumed in small amounts.
26. Consult a more recent reference to find whether there have been significant changes in the figures given in Table 13.9. (*Chemical and Engineering News* generally publishes lists similar to Table 13.9 each year in May or June. The *World Almanac* is also a good source.)
27. Complete and balance the equations for which a reaction occurs, or state "no reaction."
    a) $NaCl$ + electricity $\rightarrow$
    b) $ZnO + C \xrightarrow{\Delta}$
    c) $ZnO \xrightarrow{\Delta}$
    d) $K + HOH \rightarrow$
    e) $Zn + S \xrightarrow{\Delta}$
    f) $PbCO_3 \xrightarrow{\Delta}$
    g) $ZnS + O_2 \xrightarrow{\Delta}$

h) $Fe_2O_3 + CO \xrightarrow{\Delta}$
i) $CaCO_3 \xrightarrow{\Delta}$
j) $CaO + SiO_2 \xrightarrow{\Delta}$
k) $HgO \xrightarrow{\Delta}$
l) $WO_3 + H_2 \xrightarrow{\Delta}$
m) $Cu_2S + O_2 \xrightarrow{\Delta}$
n) $Au + O_2 \xrightarrow{\Delta}$
o) $Au + S \xrightarrow{\Delta}$
p) $Cu^{2+} + Zn \rightarrow$
q) $Ag^+ + Cu \rightarrow$
r) $CuO + H_2SO_4 \rightarrow$

28. How many tons of calcium hydroxide are needed to displace one ton of magnesium from sea water?
29. An iron ore contains 60% Fe_2O_3. How many pounds of iron are there in one ton of ore?
30. Fifty grams of fine copper wire were completely oxidized, forming CuO. How much CuO was formed?
31. How many grams of copper are needed to yield on oxidation 40 g of CuO?
32. How many grams of PbO can be obtained by heating 200 g of $PbCO_3$?
33. How much ZnO can be obtained from roasting 60 g of ZnS? [See Eq. (13–1).]
34. Complete and balance the following equations. If no reaction takes place, write "no reaction."

    a) $Zn + HCl \rightarrow$
    b) $Ca + H_2SO_4 \rightarrow$
    c) $Ca + H_2O \rightarrow$
    d) $Zn + CuSO_4 \rightarrow$
    e) $Cu + FeCl_2 \rightarrow$
    f) $Cu + AgNO_3 \rightarrow$
    g) $Zn + Pb(NO_3)_2 \rightarrow$
    h) $CuSO_4 + Fe \rightarrow$
    i) $H_2O + C \xrightarrow{\Delta}$
    j) $Fe + AgNO_3 \rightarrow$
    k) $Hg + HCl \rightarrow$
    l) $Na + HOH \rightarrow$
    m) $Mg + HCl \rightarrow$
    n) $Al + H_2SO_4 \rightarrow$
    o) $Ag + HCl \rightarrow$
    p) $Cu + H_2O \rightarrow$
    q) $Mg + H_2O$ (steam) $\rightarrow$
    r) $Fe + H_2SO_4 \rightarrow$
    s) $CuO + H_2 \xrightarrow{\Delta}$
    t) $H_2 + Cl_2 \rightarrow$

35. How many grams of copper could be displaced from a solution of $CuSO_4$ by the addition of 100 g of aluminum?

36. How much copper can be precipitated by reaction with 50 g of zinc?

37. How much silver could be recovered from a silver solution by causing it to react with 10 g of copper?

38. a) How many grams of hydrochloric acid will react with 10 g of zinc?
   b) How many grams of hydrogen will be liberated?
   c) How many grams of $ZnCl_2$ will be formed?

39. How many grams of zinc are needed to liberate 10 g of hydrogen from sulfuric acid?

40. Give the number of important commercial uses indicated in the parentheses for either the metal or compounds of the metal.
   a) sodium (3)   b) potassium (2)
   c) calcium (3)   d) magnesium (4)

41. Discuss the importance to the human body of the following elements.
   a) sodium   b) calcium
   c) potassium   d) mercury
   e) lead   f) fluorine

42. Give the composition and the number of uses indicated in parentheses for the following alloys.
   a) stainless steel (2)
   b) sterling silver (1)
   c) pewter (2)
   d) Wood's metal (1)
   e) brass (naval) (1)
   f) molybdenum steel (1)
   g) alnico (1)
   h) duralumin (1)

SUGGESTED READING

Chisholm, J. J., Jr., "Lead poisoning," *Scientific American*, February 1971, p. 15 (Offprint #1211). A good, modern discussion of this killer and maimer of children—and of adults.

Cottrell, A. H., "The nature of metals," *Scientific American*, September 1967, p. 90. The gas of electrons that binds metal atoms together makes metals behave as they do. Their mechanical properties in particular come from the close-packed crystal structure favored by the metallic bond.

Dye, J. L., "Anions of the alkali metals," *Scientific American*, July 1977, p. 92 (Offprint #368). We know the alkali metals such as sodium as cations. When they gain an electron to form anions, unusual things happen.

Goldwater, L. J., "Mercury in the environment," *Scientific American*, May 1971, p. 15. An up-to-date discussion of this potentially deadly element and its significance in our lives.

Magliulo, A., "Medical chelating agents," *Chemistry* **47**, No. 1, 25, January 1974. This article discusses the chelating agents that bind to poisonous metals and help in their excretion from the body. The article is brief but has a lot of information in it.

14

Hydrogen, the Halogens, Other Nonmetals, and the Metalloids

14.1
INTRODUCTION

In this chapter we continue the presentation of descriptive chemistry which was begun in Chapter 8 and continued in Chapter 13. Here we will begin with the unique element hydrogen, after which we discuss the elements in Group VIIA of the periodic table, which are known as the halogens. We then consider the elements in Groups IVA, VA, and VIA as well as boron, the first element in Group IIIA (see Fig. 14.1). As will be seen,

Group IA		IIIA	IVA	VA	VIA	VIIA
1 1.0		5 10.8	6 12.0	7 14.0	8 16.0	9 19.0
H		B	C	N	O	F
Hydrogen		Boron	Carbon	Nitrogen	Oxygen	Fluorine
			14 28.0	15 31.0	16 32.0	17 35.5
			Si	P	S	Cl
			Silicon	Phosphorus	Sulfur	Chloride
			32 72.6	33 75.0	34 79.0	35 80.0
			Ge	As	Se	Br
			Germanium	Arsenic	Selenium	Bromine
			50 119	51 122	52 128	53 127
			Sn	Sb	Te	I
			Tin	Antimony	Tellurium	Iodine
			82 207	83 209	84 (210)	85 (210)
			Pb	Bi	Po	At
			Lead	Bismuth	Polonium	Astatine

Fig. 14.1 Location in the periodic table of the elements discussed in this chapter.

the elements in these groups having the lowest atomic weights are nonmetals, but as the atomic weight and atomic size increase, we find a transition from nonmetals to metalloids. The elements tin and lead, in Group IVA and bismuth in Group VA are classified as metals.

14.2
THE DISCOVERY AND OCCURRENCE OF HYDROGEN

The gas we know as hydrogen was given its name by the great French chemist Lavoisier. He derived the name from Greek terms meaning water-forming, because his experiments showed that water was formed when hydrogen burned in air. Hydrogen had been known as "inflammable air" for at least 200 years before Lavoisier gave it its current name. In 1670 Robert Boyle knew that the gas could be produced by adding iron to acids. Much later, the English scientist Henry Cavendish collected the gas and studied it systematically. In 1766 he read a paper before the Royal Society reporting on his experiments. Cavendish is generally credited with the discovery of hydrogen, which he called "inflammable air." As with so many scientific discoveries, the man finally credited with the discovery based his work on that of many, many other investigators.

In many ways hydrogen is unique among the elements. Throughout the universe it is by far the most abundant element. Scientists believe that approximately 90% of the atoms and 75% of the mass of the universe is hydrogen. The great mass of the sun is chiefly hydrogen that is undergoing a nuclear reaction (Section 17.7) to become helium. Tremendous amounts of energy are released in this conversion. Only a small percentage of this energy reaches the earth, but even this is sufficient to provide the heat and light necessary for life on earth.

The most prevalent material of the universe is also the simplest in atomic structure. The hydrogen molecule is composed of two hydrogen atoms, each of which contains one proton and one electron. Hydrogen combines with most of the elements to form compounds, and it is believed by some to exist in more chemical compounds than any other element. Hydrogen is the lightest known substance, being only 1/16 as dense as oxygen. It is interesting that element No. 1 should be first in so many different respects.

Hydrogen is widely distributed on earth, almost entirely in the form of compounds. The greater portion of hydrogen on the surface of the earth occurs in water, since it constitutes about 11% by weight of this substance. In petroleum compounds it is combined with carbon: e.g. methane CH_4, butane C_4H_{10}, and octane C_8H_{18}. Hydrogen is also present in compounds in all plant and animal tissues. The rocks of the earth's surface contain small amounts of combined hydrogen, present as water in the form of hydrated minerals such as gypsum, $CaSO_4 \cdot 2H_2O$. Large amounts of free hydrogen are found in volcanic gases, and small amounts are found in natural gas. Two factors account for the almost complete absence of free hydrogen in the atmosphere. One, it reacts with the oxygen in the air forming water, and two, the gravitational attraction of the earth is insufficient to hold the small, fast-traveling hydrogen molecules. Hydrogen molecules that reach "escape velocities" are continually being lost to outer space.

14.3
PREPARATION OF HYDROGEN

The most common method of preparing hydrogen in the laboratory is by the reaction of an acid on one of the moderately active metals such as zinc or iron:

$$H_2SO_4 + Fe \rightarrow FeSO_4 + H_2 \qquad (14-1)$$

$$2HCl + Zn \rightarrow ZnCl_2 + H_2 \qquad (14-2)$$

Commercial preparation depends on various processes:

a)   The "water gas" method utilizes steam which is passed over white-hot carbon,

$$H_2O + C \xrightarrow{\Delta} H_2 + CO \tag{14-3}$$

Carbon monoxide and hydrogen both burn readily in air, so this method has been used extensively for the manufacture of a gaseous fuel. If carbon monoxide and water are heated to 500°C in the presence of a catalyst, hydrogen and carbon dioxide result:

$$H_2O + CO \xrightarrow[\Delta]{catalyst} H_2 + CO_2 \tag{14-4}$$

If we combine reactions (14–3) and (14–4) the complete reaction would be:

$$2H_2O + C \rightarrow CO_2 + 2H_2 \tag{14-5}$$

The carbon dioxide can easily be removed, leaving pure hydrogen.

b)   The greatest amount of hydrogen produced in the United States is obtained by heating (cracking) hydrocarbons (Section 20.6).

$$CH_4 \xrightarrow[\Delta]{catalyst} C + 2H_2 \tag{14-6}$$

and by the reaction of methane and water.

$$CH_4 + 2H_2O \xrightarrow[\Delta]{catalyst} CO_2 + 4H_2 \tag{14-7}$$

Smaller amounts are produced by the electrolysis of water and as a by-product from electrolytic industries. Production of hydrogen in the United States is approximately 80 billion cu ft annually. Such a volume would fill a container 1 mi square and 2800 ft high.

**14.4
PHYSICAL
PROPERTIES
OF HYDROGEN**

Hydrogen is the lightest known substance. Although it occurs as diatomic molecules, these molecules have little attraction for each other, and hydrogen remains a gas until cooled to -253°C (20 K). Since liquid hydrogen freezes at -259°C (14 K), it exists only within a six-degree range. Hydrogen is colorless, odorless, and only slightly soluble in water. At 20°C and 1 atmosphere of pressure the solubilty is about 2 ml/100 ml of water.

There are three isotopes of hydrogen, each commonly known by a distinctive name. The most prevalent isotope has one proton, one electron, and no neutrons. It is sometimes called **protium**. **Deuterium** is the common name for ^2H, which contains a neutron in addition to the proton and electron. This isotope is stable and does not break down radioactively. About one atom of deuterium is found for every 5,000 atoms of protium. **Tritium**, which contains two neutrons in the nucleus, has a total atomic mass of 3 and is unstable, breaking down radioactively. Tritium is relatively rare and is constantly breaking down whenever formed. It has a half-life of 12.5 years. Compounds of deuterium and tritium with oxygen are sometimes given the formulas D_2O and T_2O, and D_2O especially is known as *heavy water*. Heavy water is used in some nuclear reactors, and both unusual isotopes are used as "tracers," especially in living systems, to study the action of compounds of hydrogen. (See Section 17.12 for a further discussion of isotopic tracers.)

14.5
CHEMICAL
PROPERTIES
OF HYDROGEN

We discovered in Chapter 5 that the structure of an atom determines its properties and that sensible patterns occur in the periodic table. Hydrogen is usually placed at the top of Group IA of the periodic table, and also commonly placed at the top of Group VIIA (see Fig. 14.2). If we consider the characteristics of hydrogen, we find that it is like sodium, potassium, and other Group IA elements, each of which may lose one electron and become an ion with a charge of 1+. Hydrogen readily combines with the electronegative elements on the righthand side of the periodic table as do other Group IA elements. However, as a gas, its physical characteristics have little resemblance to IA elements and a greater resemblance to VIIA elements. Furthermore, hydrogen commonly combines with active metals such as sodium and calcium, forming metallic **hydrides**:

NaH   **CaH$_2$**

sodium hydride   calcium hydride

In such compounds the hydrogen gains an electron to complete its outer shell and becomes negatively charged as do other Group VIIA elements. The resultant ion, H⁻, is called a hydride ion.

Thus, hydrogen can react either by losing an electron or by gaining one. Since the valence shell of hydrogen is half filled, it might be almost as logical to assign hydrogen a new position in the periodic table above carbon, which also has a half-filled valence shell. Lines connecting hydrogen to Groups IA and VIIA would indicate that it has properties relating it to both of these groups (Fig. 14.2). This position approximately halfway across the table would also be consistent with the electronegativity of hydrogen. Hydrogen forms both ionic and covalent bonds. The bonds are ionic in the hydrides of active metals (as H⁻) and in some acids (as H⁺). Other compounds of hydrogen exhibit covalence, typically:

```
       H            H            H
       ..           ..           ..
 H : C : H     : N : H      H : O :       H : Br :
       ..           ..           ..          ..
       H            H
```

Hydrogen and oxygen do not react at room temperature, but with an increase in temperature or in the presence of a proper catalyst they combine explosively:

$$H_2 + \tfrac{1}{2}O_2 \xrightarrow[\text{catalyst}]{\text{heat}} H_2O + \text{energy} \tag{14--8}$$

In the laboratory great care must be exercised in preventing the ignition of hydrogen–air mixtures. The oxyhydrogen torch takes advantage of the great amount of heat liberated in the above reaction. Hydrogen reacts with oxides of many metals, removing the oxygen

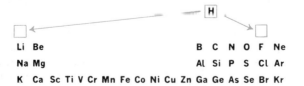

Fig. 14.2 Possible location of hydrogen in a periodic table.

and leaving the pure metal. Removal of oxygen from a substance (decreasing its oxidation number) is **reduction**. If hydrogen is passed over heated copper oxide it removes the oxygen, producing pure copper and water.

$$CuO + H_2 \xrightarrow{\Delta} Cu + H_2O \tag{14–9}$$

Since hydrogen is the agent responsible for many reduction reactions like that shown above, it is known as a **reducing agent**. Although carbon monoxide and carbon (coke) are the most commonly used reducing agents in the purification of metals from ores, in some instances [as with tungsten, shown in Eq. (13–8)] hydrogen can be used for this purpose.

The addition of hydrogen to an element or compound is also called reduction. There are many instances in biochemical reactions in which hydrogen is added to a compound. In photosynthesis, it is the hydrogen derived from water that is added to carbon dioxide to produce sugars; the process is sometimes described as the reduction of carbon dioxide. Since oxidation and reduction are regarded as opposite processes, it follows that removal of hydrogen from a compound is oxidation. Many biochemical oxidations involve the removal of hydrogen from a complex organic compound. The enzymes that catalyze these reactions are called *dehydrogenases*. Oxidations release energy and reductions require energy to proceed.

**14.6
USES OF
HYDROGEN**

Of the enormous quantity of hydrogen produced in the United States annually, by far the greater portion is used in producing ammonia synthetically. Ammonia, in turn, is used primarily as an agricultural fertilizer:

$$N_2 + 3H_2 \xrightarrow[\substack{\text{heat and} \\ \text{pressure}}]{\text{catalyst}} 2NH_3 \tag{14–10}$$

Another important use of hydrogen is in the food industries. Large quantities of vegetable oils—cottonseed, soybean, etc.,—are hydrogenated and thereby changed from liquids to solids (Section 24.4). The hydrogenated products have superior keeping qualities. Hydrogen is also used in the manufacture of hydrogen chloride.

$$H_2 + Cl_2 \rightarrow 2HCl \tag{14–11}$$

**14.7
THE
HALOGENS—
THE SALT
FORMERS**

The five elements in Group VIIA—fluorine, chlorine, bromine, iodine, and astatine—are known as the **halogens**. This name, which is derived from Greek words for salt formers, was suggested in 1825 by the Swedish chemist Berzelius. While elemental chlorine and iodine were known at that time, fluorine was known only in its compounds. Bromine was not discovered until later, probably because of its rarity and its tendency to form compounds that were so similar to chlorides that separation was difficult. Astatine was unknown until 1940. Since all the known isotopes of astatine are radioactive and have only short half-lives, astatine must be made synthetically, for example, by bombardment of bismuth with alpha particles. Small amounts of astatine may be produced naturally during the decay of other radioactive compounds (see Chapter 17).

The most important compound of the halogen elements is sodium chloride, commonly known as salt or table salt. In our present civilization because of mining, extraction of minerals from sea water, and rapid transportation, possession of sodium chloride has lost most of its urgency.   To primitive peoples, this compound was precious, and some archeologists have pointed to the development of civilizations in areas close to dried-up lakes and seas, as indicating the importance of salt to societies. Wars have been fought for the control of salt deposits.

The halogens are similar in most properties; however, they differ greatly in activity; fluorine is the most active, and astatine the least active. Although the properties of astatine have not been widely studied, it is reasonable to assume that they are similar to those of the other halogens. The evidence that is available concerning astatine confirms this assumption.

**14.8
OCCURRENCE
OF THE
HALOGENS**

Although the halogens are much too active to occur naturally in the elementary state, their compounds are widely distributed with the exception of radioactive astatine. Fluorine and chlorine are the most prevalent. It is estimated that they occur in almost equal amounts on the earth, but chlorine is more easily separated from its compounds. Most of the chlorine exists in the form of chloride ions in the oceans and in the salt deposits resulting from the evaporation of inland oceans. Fluorine, which occurs in low concentration in the ocean, is found primarily as a constituent of minerals and rocks. Table 14.1 shows the halide ion content of sea water.

TABLE 14.1   Concentration of halide ions in sea water

Ion	Milligrams/Liter	Liters of Sea Water Required to Contain 1 g of Halide Ions
chloride	19 000 (19 g)	0.052 (52 ml)
bromide	65	15.4
fluoride	1.3	769
iodide	0.05	20 000

**14.9
PHYSICAL
PROPERTIES
OF THE
HALOGENS**

The halogens are the only family of elements with members existing in the three physical states at room temperature. Fluorine and chlorine are gases. Bromine is the only non-metallic element that is a liquid, and iodine is a solid. Table 14.2 lists some of the physical

TABLE 14.2   Properties of elements of Group VII

Element	Atomic Number	Atomic Weight	Atomic Radius ($\times 10^{-11}$ m)	Melting Point (°C)	Boiling Point (°C)	Density (at 20°C) g/cm³	Common Oxidation States
fluorine	9	19.0	6.4	−220	−188	1.6×10^{-3}	−1
chlorine	17	35.5	9.9	−101	−34.05	3×10^{-3}	−1, 3, 5, 7
bromine	35	80.0	11.4	−7.08	58.8	3.1	−1, 3, 5, 7
iodine	53	127	13.3	113.6	184.4	4.7	−1, 3, 5, 7
astatine	85	(210)	14.0	302	334	—	(−1, 3, 5, 7)

properties of the halogens. With gentle heating, iodine readily sublimes, forming beautiful violet vapors. When cooled, the vapors condense to form shiny, purple-black, metallic-appearing crystals. It does not melt.

14.10
CHEMICAL
PROPERTIES
OF THE
HALOGENS

Fluorine is the most electronegative and the most active of all chemical elements. It will react with substances such as water, carbon, glass, and even the highly inactive metals. Even some of the noble gases (Section 8.12) react with fluorine. Although elemental fluorine is extremely active chemically, once it has reacted by obtaining an electron, it forms the stable fluoride ion. A great deal of energy is required to remove the electron from fluoride ions to regenerate elemental fluorine.

As the size and weight of the halogens increase, their reactivity decreases. The elemental forms become less active and the ions are more stable. Chlorine is less reactive than fluorine, and crystals of elemental iodine are relatively unreactive.

The most common reaction of the halogens with metals forms salts which, by definition, are completely ionic (Section 11.6).

$$Cu + Cl_2 \rightarrow CuCl_2 \quad \text{or} \quad Cu^0 + Cl_2^0 \rightarrow Cu^{2+} + 2Cl^- \tag{14-12}$$

With less active metals there is an increasing tendency of halogens to form covalent bonds because of the decreasing differences in electronegativity between the reacting elements. With nonmetals they form covalent bonds:

$$
\begin{array}{cccc}
 & Cl & I & F \\
 & \overset{..}{} & \overset{..}{} & \overset{..}{} \\
H:Cl & Cl:C:Cl & H:C:I & Cl:C:F \\
 & \overset{..}{} & \overset{..}{} & \overset{..}{} \\
 & Cl & I & Cl \\
\end{array}
$$

The halogens react with hydrogen to form the corresponding halogen acids:

$$H_2 + F_2 \rightarrow 2HF \qquad \text{hydrogen fluoride or hydrofluoric acid} \tag{14-13}$$

$$H_2 + Cl_2 \rightarrow 2HCl \qquad \text{hydrogen chloride or hydrochloric acid} \tag{14-14}$$

$$H_2 + Br_2 \rightarrow 2HBr \qquad \text{hydrogen bromide or hydrobromic acid} \tag{14-15}$$

$$H_2 + I_2 \rightarrow 2HI \qquad \text{hydrogen iodide or hydriodic acid} \tag{14-16}$$

As explained in Section 11.2, the hydrogen halides are all strong acids in water solutions and are often named as acids even when not in solution.

It is convenient to use X as a general symbol to represent any halogen, particularly when the reaction involved is applicable to all halogens. In the equation below, the X could be Cl or Br, or some other halogen.

$$H_2 + X_2 \rightarrow 2HX \tag{14-17}$$

The reactions of hydrogen with the halogens fluorine and chlorine are violent, liberating considerable heat. Bromine and iodine, however, do not react with hydrogen unless heated.

The difference in the reactivity of halogens means that the more active halogens can displace the less reactive ones from compounds or, speaking more precisely, can take electrons from less active ions to become ions themselves and release the less active element in its free state. For example,

$$F_2 + CaCl_2 \rightarrow CaF_2 + Cl_2 \tag{14-18}$$

which we can also write

$$2F^\circ + 2Cl^- \rightarrow 2F^- + 2Cl^\circ \qquad (14\text{–}19)$$

These displacement reactions can be used to prepare a halogen if we have one of its salts and a free halogen that is more active.

14.11
FLUORINE

The element fluorine is so reactive that it resisted the efforts of chemists to isolate it for more than half a century after the discovery of other halogens, even though its compounds were plentiful. Not until platinum vessels were utilized could the liberated element be restrained from combining with the container. Its great toxicity caused the death or serious illness of several scientists. It is very important to note, however, that the vigorously reactive *element*, fluorine, is not the same as the fluor*ide* ion which is found in rocks and in many natural water supplies used for drinking water. Fluorine occurs in a number of mineral forms, the most common being fluorspar, CaF_2, cryolite, Na_3AlF_6, and fluorapatite, $Ca_5(PO_4)_3F$. Hydrofluoric acid is made by the reaction between calcium fluoride and sulfuric acid:

$$CaF_2 + H_2SO_4 \rightarrow CaSO_4 + 2HF \qquad (14\text{–}20)$$

Because of the hydrogen bonding, the formula for hydrofluoric acid resembles that of water and is more correctly written as H_2F_2 or $(HF)_2$. It should be recalled that water exists as the monomer (H_2O) only in the vapor state, and that in the liquid state it exists as $(H_2O)_x$. For our purpose we will use the simplified notation, HF.

Hydrogen fluoride is commercially important. Because it can dissolve glass, it is used in large quantities for etching designs on glass and in the "frosting" of lightbulbs. Obviously, it cannot be kept in glass containers. For many years wax containers were used for laboratory storage; at the present time plastic bottles are being used.

Fluorine is used in the production of the well-known **Teflon**, a material made from fluorine and carbon, and for **Freon**, CCl_2F_2, which is used as a refrigerant in household refrigerators. The solvent propellents in some aerosol pushbutton dispensers are freon or similar compounds. Since it is likely that these compounds may act to destroy the ozone layer in the upper atmosphere (see Section 15.5), the use of freon and similar propellents is being regulated. Such propellents are banned in many areas. The compound UF_6, uranium hexafluoride, is used to separate the fissionable isotope uranium-235 from nonfissionable uranium-238.

Fluoride ions are highly toxic when taken in *excess*, but some form of the element is a physiological necessity. In the body it occurs primarily in the teeth and bones. Fluoride in drinking water, in the amount of one part per million (ppm) inhibits tooth decay. In areas where there are several ppm of fluoride naturally in the drinking water, a mottling or spotting of the tooth enamel occurs. Many communities in the United States now add fluoride to water supplies in amounts that prevent tooth decay but do not produce mottling. There is also considerable evidence to indicate that fluoride is important in the maintenance of good health and strong bones in elderly people.

14.12
CHLORINE

Chlorine is a greenish-yellow, poisonous gas, first prepared by Scheele in 1774. It was used in World War I, the first poison gas to be part of modern warfare.

Chlorine is made commercially by the electrolysis of a solution of sodium chloride in specially constructed electrolytic cells. In these cells chloride ions migrate to the anode, where they give up electrons and form chlorine gas:

$$2Cl^- \rightarrow Cl_2 \uparrow + 2e^- \tag{14-21}$$

At the cathode, hydrogen ions, which are always present in small amounts in water and are less active than the sodium ions, receive electrons and are discharged as hydrogen gas:

$$2H^+ + 2e^- \rightarrow H_2 \uparrow \tag{14-22}$$

The source of chlorine is usually NaCl from sea water; after removal of the H_2 and Cl_2, the remaining solution contains sodium hydroxide. The overall reaction is:

$$2NaCl + 2H_2O \xrightarrow{\text{electrolysis}} H_2 \uparrow + Cl_2 \uparrow + 2NaOH \tag{14-23}$$

In the process not only chlorine but two other highly useful substances, hydrogen and sodium hydroxide, are produced. In a separate reaction, hydrogen can be combined with chlorine to produce hydrogen chloride (Eq. (14–14)). This is one of the most important sources of hydrochloric acid.

The uses of chlorine are many. More than 10.7 million tons of chlorine were produced in the United States in 1977.

You can appreciate how much 10 700 000 tons of chlorine is if you realize that it would require a train of tank cars about 2 200 miles long (reaching from San Francisco to Chicago) to hold this quantity.

Large quantities of chlorine are used in the purification of water and in the synthesis of the vinyl chloride required to manufacture the "vinyl" plastics used as shower curtains, phonograph records, and floor coverings. Chlorine is used in the manufacture of special types of solvents that are needed for water-insoluble materials. Most paper is produced from wood pulp; without a bleach, paper would resemble wood in color. Chlorine is one of the important bleaches used for this purpose. If we inspect the labels in a chemistry laboratory or storeroom, we find that many bottles contain compounds of chlorine. Organic chemists often use chlorine-containing compounds to synthesize new compounds.

Sodium chloride is the most prevalent and most widely used halogen compound. Many millions of tons are obtained annually from underground salt deposits and by evaporation of water from the ocean or from inland seas. In some underground deposits, it is mined in much the same way as coal. In other areas, water is circulated through the salt deposits, and the salt is recovered by evaporation of water from the brine.

Chlorine is the most plentiful and therefore the cheapest of the halogens. Sodium compounds, in turn, are less expensive than those of potassium. Consequently, when metallic halogen compounds are needed, sodium chloride is most commonly used. Sodium chloride is used extensively by the chemical industries in the production of hundreds of compounds which contain chlorine. It is also used in the recharging of water softeners and the melting of ice on roads. For centuries it has been used to preserve foods.

14.13
BROMINE

Bromine is a reddish-brown, volatile liquid with a density of 3.12 g/ml. It has a pungent odor from which its name was derived; the Greek word for stench is *bromos*. Contact of bromine with the skin causes severe burns.

Bromine is manufactured from bromide salts. These are obtained from the saturated brine from which common salt has been crystallized. The bromide salts, being more soluble, remain in solution after the sodium chloride has been removed. Bromine is a relatively rare element. All known sources contain only very low concentrations. Natural brines from wells of the eastern United States were the principal sources until a satisfactory method of recovery from the ocean was perfected. Sea water contains 65 ppm of bromine. While this may seem like a small amount, calculations show that 1 cu mi of ocean water contains more than 300 000 tons of bromine (in the ionic form).

Since bromine is less active than chlorine, it can be replaced in its compounds by the action of its sister element:

$$MgBr_2 + Cl_2 \rightarrow MgCl_2 + Br_2 \tag{14-24}$$

In practice, ocean water is made slightly acid to decrease the solubility of the bromine, which is then replaced by chlorine. The free bromine is removed by aeration and collected in an alkaline solution.

The greatest need for bromine grew from the practice of adding tetraethyl lead to gasoline as an antiknock compound. Since lead fouls the ignition points in an engine, it is necessary to add something which will form a volatile compound with lead. The addition of 2 or 3 ml of $C_2H_4Br_2$/gal of ethyl gasoline results in the formation of lead bromide, $PbBr_2$, which is volatile and is removed in the exhaust gases. The use of tetraethyl lead in gasoline, which began in 1924, accelerated the demand for bromine so much that the process of extraction of bromine from the sea was developed. Since lead is not being added to gasoline for most new automobiles because it inactivates the catalysts used to control pollution, the demand for bromine is decreasing.

Bromine is used extensively in the field of photography. Silver bromide, AgBr, is the light-sensitive substance on photographic film. Bromine is also used in the manufacture of dyes, drugs, and many other organic chemicals.

14.14
IODINE
Elementary iodine is a dense solid, usually occurring as lustrous, almost black crystals. It is only slightly soluble in water. While iodine occurs only in low concentration in ocean water, it is concentrated in seaweeds. Iodine was first obtained from the ashes of seaweeds.

Early in the nineteenth century, nitrate deposits containing an unusually high concentration of iodine were discovered in Chile. This was the world's major source for many years. Some years ago it was noted that the saltwater brines occurring in some California oil wells contained considerable amounts of iodine. This source is now a major one, and broke the monopoly which had controlled the price for many years. Iodine now sells for a small fraction of what it previously cost.

Iodine is a physiological necessity for animals, including humans, and in areas where its natural occurrence is low, a dietary deficiency may exist unless a supplement is provided. Nearly all the iodine in the body is located in the thyroid glands, which are found encircling the windpipe (trachea) of most animals. A deficiency of iodine in the diet of animals lowers the ability of the thyroid glands to produce certain hormones and causes an enlargement of the thyroid glands, a condition known as goiter. Thyroid selectivity is so effective that the iodine concentration of the thyroid glands is normally 2 000 to 4 000 times that of blood. Adult humans require only 0.1 mg of iodine/day. This is such a small amount that 1 g, properly administered, would be adequate for more than 24 years. In areas near oceans, the wind-blown sprays carry iodine and deposit it on soils from which plants absorb it. Seafood also is rich in iodine. In contrast, areas within some large continents are almost without iodine. The north central part of the United

States is such an area. People living in this area are likely to develop goiters, but the inclusion of sodium iodide in table salt to give "iodized" salt prevents the condition. Farmers and stockmen who raise cattle find that their livestock are healthier if they, too, are fed iodized salt.

Iodine and iodine compounds are used extensively in the laboratory and in the preparation of organic compounds. Silver iodide is used in the making of photographic films.

**14.15
HALOGEN
ACIDS AND
THEIR SALTS**

In addition to the hydrohalide acids, such as HCl, a series of ternary, or oxy acids, can be formed with each of the halogens with the exception of fluorine. Since fluorine is more electronegative than oxygen, it does not react to form these compounds. Bromine and iodine form a series of acids and salts similar to those of chlorine listed in Table 14.3. By substituting the word stems brom- and iod- in place of chlor-, you get the names of the corresponding acids and salts. This system of naming acids and their salts is used wherever such series occur. Other metallic ions may be substituted for sodium.

TABLE 14.3 The nomenclature of the acids and salts of chlorine and the oxidation number of chlorine in each

Acid		Corresponding Salt		Oxidation Number of Chlorine
hydrochloric	HCl	sodium chloride	NaCl	$1-$
hypochlorous	HClO	sodium hypochlorite	NaClO	$1+$
chlorous	$HClO_2$	sodium chlorite	$NaClO_2$	$3+$
chloric	$HClO_3$	sodium chlorate	$NaClO_3$	$5+$
perchloric	$HClO_4$	sodium perchlorate	$NaClO_4$	$7+$

Sodium hypochlorite, NaClO, is the active ingredient in common household bleaches such as Clorox or Purex. Sodium hypochlorite can be made by passing an electric current through a dilute solution of NaCl in a specially designed cell.

The chlorates and perchlorates are powerful oxidizing agents, and under certain conditions are explosive. Great caution should be exercised to avoid mixing these substances with any easily oxidizable materials. Sodium chlorate, $NaClO_3$, is used in sprays to kill weeds.

**14.16
SULFUR AND
THE OTHER
ELEMENTS OF
GROUP VIA**

Although they share some properties, the elements of Group VIA are not as similar as are the halogens. Oxygen, which was discussed in Chapter 8, is a gas and sulfur, selenium, tellurium, and polonium are solids, although the melting points of sulfur and selenium are relatively low (see Table 14.4). While oxygen and polonium differ greatly in their chemical properties, we find similarity in the properties and the compounds formed by members of the group that are near each other in weight and size. For instance, for many organic compounds that contain oxygen, such as ethyl alcohol, C_2H_5OH, there exist analogs such as C_2H_5SH that contain sulfur. Organic selenium compounds similar to organic sulfur compounds are also known. Oxygen is considered to have only the

TABLE 14.4   Properties of elements in Group VIA

Element	Atomic Number	Atomic Weight	Atomic Radius ($\times 10^{-11}$ m)	Melting Point (°C)	Boiling Point (°C)	Density (at 20°C) g/cm³	Common Oxidation States	Classification
oxygen	8	16	6.6	−219	−183	1.3×10^{-3}	−2, 0	nonmetal
sulfur	16	32	10.4	106, 115*	445	2.0	−2, 4, 6	nonmetal
selenium	34	79	11.7	170, 217*	685	4.5, 4.8*	−2, 4, 6	nonmetal
tellurium	52	128	13.7	450	994	6.2	−2, 4, 6	metalloid
polonium	84	(210)	15.3	254	962	9.2, 9.4*	−2, 2, 4, 6	metalloid

* Different values indicate values for allotropic forms.

oxidation state of 0 (as the free element) and 2⁻. With increasing weight, the elements are more likely to be found in compounds in which their oxidation states are 2, 4, or 6.

Sulfur   Sulfur is one of the elements that is found in the free state on the earth. It and its compounds H_2S and SO_2, both of which have bad odors, are found near volcanoes and in waters of hot springs and geysers. There is evidence that there may be deposits of sulfur in one of the craters on the moon. Just as plants release O_2 to the atmosphere, some bacteria produce elemental sulfur, S, which is left in deposits, particularly near hot springs.

Because sulfur occurs as a pure solid substance, it was known to ancient people. Some people called it brimstone, and Hell is described in the Bible as a place of "fire and brimstone"—an apt name when one considers the rotten-egg smell of H_2S and the choking odor of SO_2, as well as the purple flames of burning sulfur. Sulfur is one of the components of gunpowder, which was developed by the ancient Chinese. It is reported that in medieval times sticks of sulfur were wrapped in cloth, dipped in molten sulfur, ignited, and thrown into enemy castles. These burning sulfur sticks were likely to start fires in readily combustible substances. Sulfur was also valued by the alchemists, to whom it had some of the essences of fire. Some people still remember when a mixture of powdered sulfur and molasses was given as a "spring tonic," especially to children. It is unlikely that the insoluble sulfur had many beneficial effects; however, we now know that sulfur-containing compounds are essential to life.

Sulfur has been obtained from natural deposits. Those located in Louisiana and eastern Texas are underground in formations that are not easily mined by the usual processes. The Frasch process takes advantage of the low melting point (113°C) of sulfur and forces superheated water (180°C) into shafts along with air. The underground sulfur is melted and then brought to the surface by the air pressure.

Sulfur is also found as compounds in petroleum and coal. The removal of sulfur from petroleum began as a method of purifying the petroleum, with the sulfur being discarded, but we now reclaim the sulfur. A large amount of our annual production of sulfur comes from refining petroleum and from recovery of the sulfur produced when coal and natural gas are burned. The added advantage of these processes is that they reduce air pollution.

Sulfur is a pale yellow, odorless, tasteless solid that is insoluble in water. It occurs in many **allotropic forms** from the amorphous to the crystalline. In the gaseous form sulfur has the formula S_2 and in some solid forms it is S_8.

Sulfur burns readily in the atmosphere, producing the irritating and toxic gas sulfur dioxide.

$$S + O_2 \rightarrow SO_2 \tag{14-25}$$

Sometimes the reaction goes one step further and produces sulfur trioxide, another extremely irritating gas. Industrially sulfur trioxide is produced by passing oxygen and sulfur dioxide over a catalyst (vanadium pentoxide, V_2O_5, and divided platinum are used).

$$O_2 + 2SO_2 \xrightarrow{\text{catalyst}} 2SO_3 \tag{14-26}$$

The sulfur trioxide is then treated with water to produce sulfuric acid (H_2SO_4).

$$SO_3 + H_2O \rightarrow H_2SO_4 \tag{14-27}$$

Sulfuric acid is probably the most important commercially produced chemical compound. It has been said that the degree of industrialization of a country can be determined by the amount of sulfuric acid it produces and consumes. (See Table 13.9.) Sulfuric acid is used in the manufacture of fertilizers, steel, rayon, lead storage batteries, petroleum products, explosives, dyes, textiles, and a variety of other compounds.

The sulfides are important ores of many metals. Iron sulfide, FeS, is known as iron pyrite; because of its yellow color, it is also sometimes known as "fool's gold." Lead sulfide, PbS, is called galena. The most important ore of mercury is cinnabar, HgS. Chalcocite, CuS, and chalcopyrite, $CuFeS_2$, are important sources of copper.

$$CuS + O_2 \rightarrow Cu + SO_2 \tag{14-28}$$

The "roasting" of these ores in air produces SO_2, which can be recovered and converted to sulfur or H_2SO_4. Hydrogen sulfide, H_2S, is a vile-smelling, poisonous gas (toxic level is 1 ppm in air) that is produced both in industrial processes and in the decay of plants and animals.

The sulfates of many metals are also widely found in nature. Gypsum, $CaSO_4 \cdot 2H_2O$, occurs in large deposits throughout the midwestern United States. It is used in the manufacture of plaster and plasterboard. Epsom salts, $MgSO_4 \cdot 7H_2O$, and Glauber's salt, $Na_2SO_4 \cdot 10H_2O$, are also found both in crystalline form and dissolved in water. It has been said that the effectiveness of many mineral waters is due to the laxative action of the latter two compounds.

In addition to its use in manufacturing sulfuric acid, sulfur is used in vulcanizing rubber and in making paper from wood pulp. Sulfur dioxide is used as a bleaching agent and is also used as a fumigant, since it destroys microorganisms, particularly yeasts. When grape juice is fermented to make wine, it is often treated with SO_2 or a sulfide such as $KHSO_3$ (which releases SO_2) to destroy unwanted yeasts before adding a culture of desirable yeast.

Selenium, Tellurium, and Polonium

Selenium is a rare element and is recovered chiefly as a by-product from refining of copper ores. It exists in several allotropic forms, varying from an amorphous red powder to gray and black metallic or glassy-appearing forms. Because its electrical resistance decreases with exposure to light, it is used in photocells, in exposure meters for photography, and in other applications in the solid state electronics industry. Some plants that absorb and concentrate selenium from soils have been given the name locoweed because of the strange behavior displayed by cattle or other animals that consume them. The effects can result in death of the animal. Although selenium can be toxic, very low amounts may be beneficial to animals.

Like selenium, tellurium is recovered from the "muds" produced during copper refining. Pure tellurium has a silvery white, metallic appearance. It is found as the negative component of some metallic compounds such as gold and bismuth tellurides. There is a town in Colorado whose name, Telluride, derives from these compounds. Tellurium is a p-type semiconductor and is used in electronic products. It is probably toxic.

Polonium was the first element discovered by Marie Curie and was named in honor of her native country, Poland. It is a rare element, being formed by the radioactive decay of other elements and itself decaying in a relatively short time. Although there are more isotopes of polonium (27) than any other element, they all decay rapidly and produce a high level of radioactivity. Polonium can be synthesized in nuclear reactions, but the expense of this process and the danger associated with its intense radioactivity mean that it has found few uses.

14.17 PHOSPHORUS AND THE OTHER ELEMENTS OF GROUP VA

Group VA contains the elements nitrogen, phosphorus, arsenic, antimony, and bismuth. Some of the properties of these elements are given in Table 14.5. As with Group VIA, we see that the small atoms have nonmetallic properties such as combining with metals and commonly having the oxidation state of -5, while the larger ones have metallic properties such as combining with nonmetals and having positive oxidation states in compounds. Intermediate elements are metalloids. All elements in the group have different allotropic forms. Nitrogen and phosphorus are relatively abundant and their compounds are essential to life, while the others—arsenic, antimony, and bismuth—not only are rare, but also form compounds that are poisonous even at relatively low concentrations.

The chemistry of nitrogen was discussed in Chapter 8; its role in relation to air pollution is examined in Chapter 15.

Phosphorus

Unlike nitrogen, which is found in the free state in the atmosphere, phosphorus is never found free in nature. There are major deposits of calcium phosphate, $Ca_3(PO_4)_2$, in the Soviet Union, Morocco, and the United States. It is likely that these deposits were formed from the bones of various animals, since we find calcium phosphate in bones. Other phosphates are found also, but the insolubility of calcium phosphate makes it the most likely one to persist in areas that have been exposed to rain.

Phosphorus was first prepared by Brand, one of the last of the alchemists, in 1669. He heated urine, sand, and carbon and produced a substance that glowed in the dark.

TABLE 14.5   Properties of elements in group VA

Element	Atomic Number	Atomic Weight	Atomic Radius ($\times 10^{-11}$ m)	Melting Point (°C)	Boiling Point (°C)	Density (at 20°C) g/cm³	Common Oxidation States	Classification
nitrogen	7	14	7.0	-210	-196	1.2×10^{-3}	$-3, 0$	nonmetal
phosphorus (white)	15	31	11.0	44.2	280	1.8	$-3, 5$	nonmetal
arsenic	33	75	12.1	817	613	2.0, 5.7*	$-3, 3, 5$	metalloid
antimony	51	122	14.1	631	1 640	6.7	$-3, 0, 3, 5$	metalloid
bismuth	83	209	15.5	271	1 560	9.8	3, 5	metal

* Different values indicate values for allotropic forms.

He called the substance phosphorus, meaning bearer of light. Our modern preparation substitutes calcium phosphate for the phosphate from urine, but generally follows the old preparation.

$$Ca_3(PO_4)_2 + 3SiO_2 \rightarrow 3CaSiO_3 + P_2O_5 \qquad (14-29)$$

$$P_2O_5 + 5C \rightarrow 5CO + 2P \qquad (14-30)$$

There are two common allotropic forms of phosphorus, white (yellow) and red. White phosphorus is a waxy solid that melts at 44°C. It ignites and burns when exposed to air and is commonly stored under water, in which it is insoluble. Red phosphorus is a dark red, amorphous form and is much more stable. It is the form more often used in laboratory experiments. Although high temperatures are required to ignite red phosphorus, it does burn.

The product of the oxidation of phosphorus is diphosphorous pentoxide.

$$4P + 5O_2 \rightarrow 2P_2O_5 \qquad (14-31)$$

While gaseous phosphorus is probably actually P_4 and the oxide is P_4O_{10}, we generally use the simplest formula, P_2O_5. The oxide reacts with water to form phosphoric acid.

$$P_2O_5 + 3H_2O \rightarrow 2H_3PO_4 \qquad (14-32)$$

Trying to extinguish phosphorus fires with water is not very effective since the water reacts with the oxide to produce phosphoric acid. Although this is not a strong acid, it still produces burns in flesh. Diluted phosphoric acid is sometimes added to soft drinks such as a cherry phosphate, where it provides the sour taste. Phosphorus can also be made into the compound phosphine, PH_3, which is similar to ammonia. While elementary phosphorus and phosphine are toxic, the phosphates are necessary components of both plants and animals. Fertilizer for plants and food for animals must contain adequate amounts of phosphates. Some of the most important biochemical compounds—for example, ATP (adenosine triphosphate), the nucleic acids DNA and RNA, and many proteins—contain phosphate groups.

The greatest commercial use of phosphates is in fertilizers, and these are in great demand to produce food to feed the billions of people on the earth, $Ca(H_2PO_4)_2$, called "superphosphate," is widely used because it is more soluble than normal calcium phosphate. Compounds of phosphorus are also used in making matches, in baking powder, and as the cleaning agent trisodium phosphate (Na_3PO_4), or TSP. Since phosphorus produces a dense white smoke when it burns, it is often used in bombs designed for smoke screens. The burning phosphorus also starts fires and may cause deep burns in people; thus it is one of the most unpleasant agents of chemical warfare.

Arsenic   Arsenic is found as arsenic sulfide and oxides, as well as arsenites and arsenates. It will form a compound arsine, AsH_3, similar to ammonia and phosphine. When heated, arsenic is oxidized to As_2O_3, which has an odor of garlic. Like other elements in this group, arsenic exists in many allotropic forms. A steel gray, brittle, crystalline, semi-metallic solid is one common form.

Arsenic is used in many alloys and is being used as a "doping agent" in transistors and other solid-state electronic devises. All compounds of arsenic are poisonous, and calcium and lead arsenates have been used as insecticides. This use has largely been discontinued, however, since it leaves toxic residues in the soil or on plants which may be eaten. It is possible that the poisonous nature of the compounds of arsenic, particularly the arsenates, stems from their close resemblance to the phosphates which are so necessary to life. That is, because of their similar molecular structure, plants and animals do not distinguish between phosphates and arsenates and take both of them in; but the differences in properties means that arsenates cannot carry out the functions of phosphates and thus are poisonous. This hypothesis requires further testing to determine whether it is true.

Antimony   Antimony is found as the sulfide stibnite, Sb_2S_3, as oxides, as antimonides of heavy metals such as Cu_3Sb, and as antimonates such as $Pb_3(SbO_4)_2$. It is also found to some extent as a free element. The element and many of its compounds are toxic. Antimony is used in alloys, paints, glass, pottery, and in flame-proofing compounds.

Bismuth   Bismuth is found primarily as its sulfide, Bi_2S_3, and its oxide, Bi_2O_3. Peru, Japan, Mexico, Bolivia, and Canada have deposits of relatively high-grade ores, but in the United States bismuth is generally produced as a by-product of refining other metals. Bismuth is generally classified as a metal and it has many of the properties of a metal; however, it is not a good conductor of electricity as metals are. In fact, it has a very high resistance and this resistance increases when bismuth is placed in a magnetic field. Two other unusual properties of bismuth are the basis for its important uses. Bismuth expands on solidifying, and alloys of bismuth are used to obtain castings that do not shrink when molded. The melting point of bismuth is relatively low (Table 14.5), and some bismuth-containing alloys have melting points low enough that they are used as plug materials in automatic fire sprinklers.

14.18
CARBON AND
THE ELEMENTS
OF GROUP IVA

The properties of the elements in Group IVA vary widely, as can be seen in Table 14.6. Carbon is a nonmetal and yet is very different from nitrogen and oxygen, which head Groups VA and VIA. While silicon and germanium resemble some of the other metalloids, they have extremely high boiling points. Tin and lead, which have metallic properties, were discussed in Chapter 13 with the other metals.

Carbon   Carbon occurs naturally as a free element in at least three forms—diamond, graphite, and the amorphous form we find in charcoal. There are more than three million compounds of carbon, most of which we classify as being organic compounds. These compounds are discussed in the chapters on organic chemistry and biochemistry. In this section we

TABLE 14.6   Properties of elements in group IVA

Element	Atomic Number	Atomic Weight	Atomic Radius ($\times 10^{-11}$ m)	Melting Point (°C)	Boiling Point (°C)	Density (at 20°C) g/cm³	Common Oxidation States	Classification
carbon	6	12	7.7	—	sublimes	1.8–3.5*	0, 2, 4	nonmetal
silicon	14	28	11.7	1 415	2 680	2.3	4	metalloid
germanium	32	72.6	12.2	937	2 830	5.3	2, 4	metalloid
tin	50	119	14.1	232	2 687	7.3	2, 4	metal
lead	82	207	17.5	327	1 751	11.3	2, 4	metal

* Different values indicate values for allotropic forms.

will discuss those compounds that contain carbon and oxygen but no hydrogen. The most important of these are carbon monoxide, carbon dioxide, and the carbonates.

Pure carbon in an amorphous form can be prepared by burning organic compounds in insufficient oxygen.

$$CH_4 + O_2 \rightarrow C + 2H_2O \tag{14-33}$$

We see this reaction when an automobile engine or fire "smokes," producing the carbon we often call soot.

If we subject pure carbon to various conditions we can prepare the two other allotropic forms of carbon, graphite and diamond. Production of diamonds requires high temperatures and pressures.

Diamond, which is pure crystalline carbon, is the hardest substance known. We are all aware of the beauty and durability of polished diamonds. However, most of the diamonds found are not of gem quality (see Fig. 14.3) but are used in industries for cutting and polishing metals or stones. Diamonds are found principally in Africa, with some also being found in Arkansas. It is believed that they were formed deep in the

(a)   (b)

Fig. 14.3 (a) This 530-carat diamond, approximately actual size, is the largest of the stones cut from the famous Cullinan diamond. [Courtesy of the Gemological Institute of America.] (b) Synthetic diamonds are made in a variety of forms and are superior to natural diamonds for many industrial uses. The graphite pencil point indicates the size of the diamonds. [Courtesy of General Electric Company.]

earth under great pressure and heat. We can duplicate these conditions, and 30% of the diamonds used industrially are now produced synthetically.

Graphite is an allotropic form of carbon which has soft, black, shiny, flat crystals. It occurs naturally in many parts of the world and is also produced synthetically. The minute, flat particles of graphite slide over each other like the cards in a deck, producing a slick feel when rubbed between the fingers. This property makes graphite an excellent lubricant for dry bearings such as are used in many electric motors, and in door locks. It is also used in motor oil additives. The "lead" in lead pencils is compressed graphite; when we write, a thin layer of the minute flat crystals is left on the paper. Graphite is also used as electrodes, since it is a good conductor of electricity but has little tendency to react chemically.

Coal, carbon black, charcoal, and similar products are primarily amorphous carbon. They lack the crystalline structure of diamond and graphite and usually contain impurities. These impure, amorphous forms of carbon find many uses. Coal is converted to a purer form of carbon known as coke before being used in recovering iron from its ores and making steel. Charcoal can be made by heating wood or other plant materials in a limited supply of air. Much of the charcoal we use as a fuel for outdoor barbecues is made from the pits of fruits such as peaches and apricots. Slightly more refined charcoal is used in the chemistry laboratory and in industry as an adsorbant which removes impurities, especially colored impurities, from products such as sugar. The finely divided powder we get from burning methane, as in Eq. (14–33), is called carbon black. It is compounded with rubber in automobile tires to the extent of several pounds per tire, where it adds durability to the rubber. Carbon black is also used as the pigment in printer's ink. This use consumes many tons of carbon black each year.

When carbon-containing compounds burn they may produce a variety of products. We have already discussed the formation of carbon in this manner. If slightly more oxygen is present, carbon monoxide may be produced.

$$2C + O_2 \rightarrow 2CO \tag{14-34}$$

If more air is present, carbon dioxide is formed.

$$C + O_2 \rightarrow CO_2 \tag{14-35}$$

Carbon monoxide is a very poisonous substance because it reacts with the hemoglobin in red blood cells and prevents their combining with oxygen. Prolonged breathing of carbon monoxide, even at low concentrations, can have serious effects. Some of these are listed in Table 14.7. While most of us are aware that automobile motors produce appreciable amounts of carbon monoxide and that running automobiles in a closed garage can cause death, there are other sources of carbon monoxide in our lives. Improperly installed gas heaters and even charcoal fires in enclosed places such as vans and boat cabins cause many deaths. Less obvious is the effect of smoking cigarettes. Moderate to heavy smokers (1–2 packs per day) generally have from 6–7% of their hemoglobin combined with carbon monoxide. This level is sufficient to impair the function of blood vessels, particularly those in the heart, and most people with heart diseases are advised by their physicians not to smoke.

Carbon monoxide can be converted to carbon dioxide

$$2CO + O_2 \rightarrow 2CO_2 \tag{14-36}$$

when burned in air. In fact, carbon monoxide is a good fuel, releasing a great deal of heat when it burns, but its usefulness is decreased by its toxicity.

TABLE 14.7   Effects of saturation of hemoglobin with carbon monoxide and exposure required to produce various levels of hemoglobin-CO

Percent of Hemoglobin that is Saturated with Carbon Monoxide	Physiological Effects
less than 2%	Few effects
2–5%	Impairment of vision and time discrimination
5–10%	Changes in heart and lung function
more than 10%	Headache, fatigue, drowsiness, and at high levels, respiratory failure and death

Conditions required to produce certain levels of saturation of hemoglobin with carbon monoxide

2%	Long-term exposure to air with 10 ppm CO Smoking 10 cigarettes per day
5%	Long-term exposure to air with 30 ppm CO Smoking 10–40 cigarettes per day
7%	Long-term exposure to air with 40 ppm CO Smoking more than 40 cigarettes per day
12%	Long-term exposure to air with 70 ppm CO

Data taken from publications of the United States Department of Health, Education, and Welfare.

Carbon dioxide is a colorless, odorless gas with a density 1.5 times that of air. At atmospheric pressure it liquefies at $-78°C$, or we might say, the liquid boils at that temperature. At 22°C, it is converted to a liquid if the pressure is raised to 900 psi. This is the pressure that exists inside the ordinary carbon dioxide fire extinguisher. If the pressure is suddenly released, the liquid evaporates very rapidly, and since evaporation requires heat, the temperature of the escaping gas is rapidly lowered. At atmospheric pressure, carbon dioxide sublimes, and a sudden lowering of temperature causes the gaseous carbon dioxide to turn to a solid called **dry ice** with a temperature of $-78°C$ (or $-109°F$). This solid also sublimes, changing directly from the solid to the gaseous state, which accounts for the name "dry ice." Large amounts of dry ice are used as a refrigerant. The carbonated beverage industries also provide a large market for carbon dioxide.

Carbon dioxide is nontoxic in low concentrations, and it makes up approximately 4% of our exhaled breath. It has an interesting physiological function in that the concentration of CO_2 in our blood controls the rate of our breathing. During heavy exercise the rate of carbon dioxide formation in the body is increased, and therefore we breathe more rapidly. Small amounts of carbon dioxide can be used during surgery and resuscitation to stimulate breathing. The inhalation of high concentrations of carbon dioxide over prolonged periods of time, however, causes asphyxiation.

Carbon dioxide is somewhat soluble in water, with about 1.5 g dissolving in a liter of water. In addition to dissolving, about 1% of the carbon dioxide reacts with water to form carbonic acid, H_2CO_3.

$$H_2O + CO_2 \rightleftharpoons H_2CO_3 \qquad\qquad (14\text{--}37)$$

Although carbonic acid is a weak acid, there is some ionization to produce hydrogen and hydrogen carbonate ions.

$$H_2CO_3 \rightleftharpoons H^+ + HCO_3^- \qquad\qquad (14\text{--}38)$$

A solution containing carbon dioxide under pressure is called carbonated water or soda water. While some spring waters are naturally carbonated, most carbonated beverages are made by forcing CO_2 into water under pressure. The slight sour taste of soda water is due to the hydrogen ions produced by ionization of carbonic acid. The low pH probably also prevents the growth of microorganisms in carbonated beverages. Most flavored sodas contain citric or phosphoric acid to give additional sourness.

We find many carbonates in nature. Limestone and marble are forms of calcium carbonate, usually mixed with smaller amounts of magnesium carbonate and other impurities. Calcium carbonate is relatively insoluble. It is formed when calcium ions come into contact with carbonate ions in water.

$$Ca^{2+} + CO_3^{2-} \rightarrow CaCO_3 \tag{14–39}$$

Lime (CaO) also reacts with CO_2 to give $CaCO_3$ Sodium and potassium carbonates are soluble and are found in many waters on the surface of the earth and in deposits formed in dry lake beds.

When surface water containing CO_2 (and H_2CO_3) trickles through deposits of limestone ($CaCO_3$), the soluble salt calcium hydrogen carbonate $Ca(HCO_3)_2$ is formed.

$$CaCO_3 + H_2CO_3 \rightarrow Ca(HCO_3)_2 \tag{14–40}$$

The calcium hydrogen carbonate may be carried away by underground water and in some instances this forms a cave or cavern. Later, as water containing calcium hydrogen carbonate enters the ceiling of a cave, it may lose water and carbon dioxide to reform $CaCO_3$.

$$Ca(HCO_3)_2 \rightarrow CaCO_3 + H_2O + CO_2 \tag{14—41}$$

This calcium carbonate forms deposits we call stalactites if they take shape on the roof of the cave, or stalagmites if evaporation occurs after the solution drips to the floor of the cave. As with marble, it is the impurities in cave formations that give them their interesting colors.

Silicon It has been estimated that 95% of the material on the earth's surface contains silicon. Sand is silicon dioxide, SiO_2, and many rocks and minerals contain silicates, which may be written as $CaSiO_3$ or $CaO \cdot SiO_2$ (see Table 14.8).

When minerals in a molten state cool *slowly*, they form crystals. *Granite*, which is a mixture of several different compounds, is a rock that has cooled very slowly while still underground. It contains several kinds of crystals. *Obsidian*, or volcanic glass as it is commonly called, is similar to granite in composition. However, it was forced to the

TABLE 14.8 Formulas of some compounds of silicon

Name	Uses	Formula
garnet	gems and abrasives	$3CaO \cdot Al_2O \cdot 3SiO_2$
talc	talcum powder	$3MgO \cdot 4SiO_2 \cdot H_2O$
feldspar	ceramics	$K_2O \cdot Al_2O_3 \cdot 6SiO_2$
zircon	gems	$ZrO_2 \cdot SiO_2$
clay	ceramics	$Al_2O_3 \cdot 2SiO_2 \cdot 2H_2O$
asbestos	insulation	$3MgO \cdot 2SiO_2 \cdot 2H_2O$
sand or quartz	innumerable	SiO_2

surface of the earth while still molten, and therefore cooled rapidly into a noncrystalline material we call glass. The *glass* with which we are familiar is a rapidly cooled silicate material.

While carbon is the key element of the plant and animal world, silicon is the dominant element of the mineral world. In contrast to carbon, silicon is never found in the free state but always in compounds.

Silicon may be produced by the reaction of sand with carbon.

$$SiO_2 + C \rightarrow Si + CO_2 \tag{14-42}$$

This reaction is similar to those seen with the halogens, where an element in a compound is displaced by a more active element. Amorphous silicon is a brown powder, while crystalline silicon is gray with a metallic luster.

Glass is a mixture of calcium and sodium silicates and is made by reacting sand with sodium and calcium carbonates.

$$Na_2CO_3 + SiO_2 \rightarrow Na_2SiO_3 + CO_2 \uparrow \tag{14-43}$$

and

$$CaCO_3 + SiO_2 \rightarrow CaSiO_3 + CO_2 \uparrow \tag{14-44}$$

The mixture must be rapidly cooled to form glass. Glass ordinarily has other substances mixed with it, either as impurities or as substances deliberately added to give it special properties. Boron oxide (B_2O_3) is added to make glass such as Pyrex® that is less likely to crack when heated or cooled. Iron(III) produces yellow to brown colors when added to glass; cobalt produces an intense blue; manganese(IV) produces a pink to violet color; and either colloidal selenium or gold produces a ruby red glass.

One of the newer uses for silicon is in a variety of semiconductor and solid-state electronic devices such as transistors. Silicon carbide, SiC, is used as an abrasive in grinding and polishing. High-molecular-weight polymers of silicon are known as silicones. They range from liquids to solids and have many uses such as in adhesives and bonding materials, auto polishes, coolants, and electrical insulation materials.

Diatoms, which are microorganisms found in both sea and fresh water, absorb silicates from water and use them in their skeletal structure. While small amounts of silicon are found in human bones, they are primarily protein and calcium phosphate; the silicon is probably there merely because it is everywhere around us and we cannot keep from consuming it. When too much silicon-containing dust is inhaled, as it may be by miners and stonecutters, silicosis, a serious lung disease, may develop.

Germanium   Germanium, which gets its name from the country Germany, is interesting because although it was unknown when Mendeleev developed the periodic table, he left a space for it. He gave this missing element the temporary name "eka-silicon." When germanium was discovered much later, it had the properties Mendeleev had predicted and fit into the space left for it.

Germanium compounds are often found along with ores of silver and zinc and in coal deposits. In the pure form it is a brittle, crystalline, gray-white metalloid. It, as well as silicon, is used in transistors and other solid-state electronic devices. Germanium is also used in metal alloys and is added to glass in wide-angle camera lenses and microscope objectives. Some germanium compounds have been reported to be toxic to bacteria without harming animals, and they may find use as chemotherapeutic agents.

14.19
BORON
Boron is the top element in Group IIIA and as a metalloid has properties that are related to those of the elements previously discussed. Compounds of boron have been known for thousands of years. It is reported that they were used as a preserving agent in Egyptian mummies. They are still used as preservatives in some parts of the world. For centuries metalsmiths have used boron compounds as a flux for removing oxide films from metals. Boron compounds have also long been used as antiseptics.

Boron exists both in a brown, amorphous powder (which contains some impurities) and in a pure crystalline state. It is not found in these pure states in nature but always as compounds. Boron melts st 2300°C and sublimes at 2550°C. Its density is 2.34 in the crystalline form and 2.37 in the amorphous variety. Its most common oxidation state is 3.

Boron compounds, in low concentration, are widespread throughout the world. Ores of boron are believed to be of volcanic origin; all workable deposits are in areas of volcanic action—present or past. More than 85% of the world's supply comes from California, most of it from the Death Valley area mines, Searles Lake, and from an enormous open-pit mine near Barstow, California. Apparently these deposits are associated with the past volcanic activity of the Sierra Mountains and the collecting of these minerals in lakes that later became dry. Boric acid, H_3BO_3, is found in volcanic areas; so too, are the more important mineral ores, borax, $Na_2B_4O_7 \cdot 10H_2O$, of 20-mule-team fame, and kernite, $Na_2B_4O_7 \cdot 4H_2O$.

Borax is used in the manufacturing of Pyrex® and similar types of glass which have very low coefficients of thermal expansion. It is also used in enamels and glazes. Borax is the salt of a strong base and a weak acid, and therefore it produces a basic reaction in water. The borate ion hydrolyzes as follows:

$$B_4O_7{}^{2-} + 7H_2O \rightarrow 4H_3BO_3 + 2OH^- \tag{14-45}$$

Borates react with calcium or magnesium ions, forming insoluble calcium or magnesium borates:

$$Na_2B_4O_7 + Ca^{2+} \rightarrow CaB_4O_7 + 2Na^+ \tag{14-46}$$

These properties make borax a very good water-softening agent.

Rods made of the isotope boron-10 are used in nuclear reactors, and boron is also used in instruments that detect neutrons. Some newly developed organic-boron compounds which are amino acid analogs have been tested in animals and show promise of being useful as anticancer agents and for the treatment of high blood cholesterol and arthritis. Time and human testing will be required to see whether these compounds fulfill their original promise.

SUMMARY

This chapter has described and discussed many of the elements we call the nonmetals and metalloids, along with a few metals. While the properties of these substances vary greatly, they show a pattern when we consider them in relationship to their position in the Periodic Table of the Elements. We find that the elements at the top of each group are often gases, are more likely to be found as the negative or electron-accepting components of molecules, and are generally poor conductors of electricity. All of these characteristics are related to the fact that they tend to hold tightly the electrons in their outer shells and to form compounds by acquiring electrons from other elements. These are the properties of the nonmetals.

In contrast, metals are generally solids, are positive or electron-donating parts of compounds, and are generally good conductors of electricity. We find that the elements at the bottom of Groups IIIA to VIIA are more metallic than those above them, and we also see that as we move from the right side of the table (the halogens) toward the left, metallic properties increase. This is logical if we consider that it is both the charge on the nucleus and the distance of outer electrons from the nucleus that influence the tendency to lose electrons and therefore require that the substance be classed as a metal.

Between the nonmetals and the metals are the metalloids such as silicon, germanium, arsenic, and tellurium. The metalloids have some of the properties of both metals and nonmetals. It is especially interesting that whereas metals are good electrical conductors and nonmetals are not, many of the metalloids are finding use as semiconductors, which form the basis for the solid-state electronics industry.

While various metallic ions are the principal cations in body fluids, the halides are probably the most important anions, from a quantitative standpoint, in the human body. Halogen compounds are widely used in organic reactions. Iodine is essential for animal life, and apparently fluorine is also required. No specific biological function has been found for chlorine, and it may be that it is present primarily as the anionic partner of the essential cations, sodium and potassium.

While we can often predict the properties and consequently the uses of an element from its position in the periodic table, other properties such as whether it is found free in nature and whether it is poisonous are more difficult to predict. We hope that future research, thought, and the development of new theories will help us to understand even more of the pattern of properties and uses of the elements and compounds.

IMPORTANT TERMS

allotropic forms	Freon	hydride	reduction
deuterium	graphite	metalloid	reducing agent
diamond	halogen	protium	tritium

WORK EXERCISES

1. How can you account for the fact that although hydrogen is the most abundant element in the universe, there is little of it in the atmosphere of the earth?
2. List four methods of preparing hydrogen.
3. How can you explain the very low freezing point of hydrogen?
4. Would you expect the boiling point of heavy water (water containing appreciable amounts of deuterium) to be different from that of ordinary water? How would you expect it to differ? Why?
5. Give reasons for placing hydrogen at each of the three locations on the Periodic Table, as shown in Fig. 14.2.
6. Give three uses for hydrogen gas.
7. Why is the ocean a good source of the halogens?
7. Why is fluorine more active chemically than iodine?
9. Give three uses for the element chlorine.
10. Write the formulas for the following compounds.

a) calcium hypochlorite
b) potassium iodate
c) bromous acid
d) iodic acid
e) magnesium bromite
f) sodium astatide
g) perfluoric acid
h) potassium perchlorate

11. Which elements disccussed in this chapter occur in allotropic forms? List those allotropic forms that have distinctive names.
12. List five forms in which sulfur is found naturally and give a commcrical use for each.
13. Give the number of commercial uses indicated in the parentheses for each of the following. You may use either the element itself or a compound in answering this question.

a) bromine (2)   d) phosphorus (4)
b) iodine (2)   e) arsenic (2)
c) selenium (1)   f) antimony (2)

g) bismuth (2)   i) germanium (1)
h) silicon (3)   j) boron (3)

14. Give five uses for carbon in the free, elemental form.

15. Give the number of commercial uses specified for the following compounds.
a) H_2SO_4 (5)   b) CO_2 (3)
c) NaCl (3)   d) NaClO (1)
e) HF (1)   f) $NaClO_3$ (1)

16. What is the oxidation state of the element indicated in the given compound?
a) silicon in Na_2SiO_3
b) phosphorus in H_3PO_3
c) phosphorus in PH_3
d) boron in $Na_2B_4O_7$
e) sulfur in H_2S
f) sulfur in H_2SO_3
g) sulfur in H_2SO_4
h) bromine in $NaBrO_2$
i) chlorine in $NaClO_4$
j) iodine in NaI

17. Complete and balance the equations for which a reaction will occur.
a) $H_2 + Cl_2 \rightarrow$
b) $Na^+ + F^- + Br_2 \rightarrow$
c) $K^+ + Br^- + Cl_2 \rightarrow$
d) $Cu^{2+} + 2Cl^- + Br_2 \rightarrow$
e) $Ca^{2+} + 2Cl^- + F_2 \rightarrow$
f) $H_2 + F_2 \rightarrow$
g) $Cu + Cl_2 \rightarrow$
h) $Na + Cl_2 \rightarrow$
i) $K^+ + I^- + Br_2 \rightarrow$
j) $Na^+ + F^- + Cl_2 \rightarrow$
k) $Ca + Cl_2 \rightarrow$
l) $Al + Br_2 \rightarrow$
m) $Cl_2 + Br^- \rightarrow$
n) $Br_2 + I^- \rightarrow$
o) $I_2 + Cl^- \rightarrow$
p) $Br_2 + F^- \rightarrow$

18. Complete and balance.
a) $Na^+ + Cl^- + 2H^+ + SO_4^{2-} \rightarrow$
b) $Na^+ + Cl^- + Na^+ + HSO_4^- \rightarrow$
c) $Na^+ + Cl^- + HOH \xrightarrow{\text{electrolysis}}$
d) $Ag^+ + NO_3^- + Na^+ + Cl^- \rightarrow$
e) $H_2 + X_2 \rightarrow$
f) $Ca^{2+} + 2OH^- + H^+ + ClO_3^- \rightarrow$
g) $KClO_3 \xrightarrow{\Delta}$

19. Chlorine gas was passed through a solution containing 20 g of KI until all the iodine was replaced.
a) How many grams of iodine were released?
b) How many grams of chlorine were required?
c) How many grams of KCl were produced?

20. Complete and balance.
a) $Na_2B_4O_7 + Mg^{2+} \rightarrow$
b) $CO + O_2 \xrightarrow{\Delta}$
c) $C_3H_8 + O_2 \rightarrow$
d) $CaCO_3 + SiO_2 \xrightarrow{\Delta} CaSiO_3 +$
e) $P_2O_5 + H_2O \rightarrow$
f) $SO_2 + O_2 \xrightarrow{\text{catalyst}}$
g) $FeS + H_2SO_4 \rightarrow$
h) $CO_2 + H_2O \rightarrow$
i) $C_8H_{18} + O_2 \rightarrow$
j) $Ca_3(PO_4)_2 + H_2SO_4 \rightarrow$
k) $H_2O + SO_3 \rightarrow$
l) $H_2O + SO_2 \rightarrow$

21. Determine the following percentages.
a) boron in $Na_2B_4O_7$
b) silicon in Na_2SiO_3
c) sulfur in H_2SO_4
d) phosphorus in $Ca_3(PO_4)_2$
e) iron in $FeSO_4$

22. How many tons of sulfuric acid can be made from one ton of sulfur?

23. How many pounds of phosphorus are there in 100 pounds of $Ca(H_2PO_4)_2$?

24. How much sulfuric acid is required to react with 50 pounds of $Ca_3(PO_4)_2$ to form $Ca(H_2PO_4)_2$?

25. In the above problem, how many pounds of $Ca(H_2PO_4)_2$ will be produced?

26. Why is hydrogen no longer used to fill lighter-than-air balloons?

27. Why are the chlorofluorocarbon compounds known as the Freons banned from being used in aerosol spray containers?

28. What commercial advantages have resulted from the removal of sulfur dioxide from gases given off in industrial processes?

29. Why is lead arsenate no longer used as an insecticide?

30. Discuss in a paragraph the relationship of carbon monoxide from smoking cigarettes to a person's health.

SUGGESTED READING

Goldsmith, J. R., and S. A. Landaw, "Carbon monoxide and human health," *Science* **162**, 1352 (December 20, 1968). This article is a bit technical, but contains a great deal of information.

Groth, Edward III, "Science and the fluoridation controversy," *Chemistry* **49**, No. 4, 5 (May, 1976). An excellent discussion of the controversy about adding fluoride to drinking water. The author gives 28 additional references on the subject.

Hittinger, W. C., "Metal-oxide semiconductor technology," *Scientific American*, August 1973, p. 48. The techniques described in this article make possible the placement of 10 000 electronic components on a silicon "chip" such as those used in pocket calculators.

Woodwell, G. M., "The carbon dioxide question," *Scientific American*, January 1978, p. 34 (Offprint #1376). The author discusses the problems brought about by the increase in carbon dioxide in the atmosphere. He raises the question: "Will enough carbon be stored in forests and the ocean to avert a major change in climate?" An important problem, well presented.

15

The Atmosphere and its Pollution

"The earth's atmosphere is a vast churning mixture of gases and trace quantities of liquids and solids. Held to the earth by the pull of gravity, it is the transparent envelope without which life on earth would cease to exist.

"The atmosphere is the source of oxygen essential to man and of carbon dioxide essential to plants. It is the source of rain. It is the shield that absorbs the sun's ultraviolet radiation that would otherwise destroy all life on earth. It is the barrier that dissipates most of the bombarding cosmic rays from outer space. It is the protective layer that burns up the meteors that would otherwise leave the earth as crater-pocked as the moon.

"The earth's atmosphere maintains the temperature and climate that make man's survival possible. Without the atmosphere, the temperature at the equator would reach as high as 180°F in the day and as low as −220°F at night.

"The atmosphere makes flight possible. It makes sound propagation possible. The free electrons in the upper atmosphere make conventional long-distance radio communications possible.

"Tenuous, impalpable, ceaselessly in flux, the atmosphere is as vital to man as the sun, the oceans, and the solid earth itself."*

Although we are now aware that the earth, like many other planets in our solar system, has an atmosphere surrounding it, this was not obvious to ancient people. One bit of evidence that the earth has an atmosphere is that we can measure its pressure. This was shown by Torricelli, an Italian physicist, in 1643, when he performed an experiment in which he sealed the end of a glass tube, filled it with mercury, then inverted it in an open container of mercury (Fig. 15.1). He found the top of the column of mercury fell to

---

* Taken from "Chemistry and the Atmosphere," special report, *Chemical and Engineering News*, March 28, 1966. Reprinted with permission from the copyright owner, the American Chemical Society.

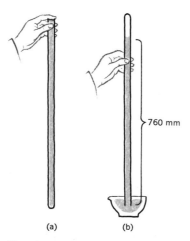

760 mm

(a)   (b)

Fig. 15.1 The first barometer. The production of a Torricellian vacuum: (a) a glass tube, more than thirty inches long, is filled with mercury and closed with the finger; (b) the tube is inverted, the open end placed in a bowl of mercury, and the finger removed. The mercury level in the tube drops until it is about 760 mm above the level of the mercury in the bowl. The top part of the tube is then empty (a vacuum).

approximately 760 mm above the level of the mercury in the container. When other investigators carried the instrument to the top of a mountain, the height of the mercury column decreased. Torricelli and other scientists decided the height of the column was proportional to the weight of the atmosphere pushing down on the mercury in the container and that the atmospheric pressure decreased as altitude above the earth increased.

We still use barometers similar to Torricelli's to measure atmospheric pressure in the laboratory. In other cases we use instruments called aneroid barometers. These contain a hollow, flat, wafer-like mechanism which compresses or expands with changes in pressure. A knowledge of small, day-to-day changes in atmospheric pressure is useful in predicting changes in weather. Barometric readings must also be made and used in calculations dealing with volumes of gases. The altimeter found in an aircraft is an aneroid barometer that is calibrated in terms of altitude rather than pressure.

A mercury barometer at sea level will, like Torricelli's instrument show a difference in height of 760 mm between the surface of the mercury in the bowl and in the tube. The pressure that this represents has been adopted as a standard unit called one *atmosphere*. In dealing with extremely high pressures, we generally express the total pressure in terms of multiples of atmospheres; a pressure of 1 520 mm of mercury is called 2 atmospheres (2 atm). When dealing with lower pressures, we generally record them in either millimeters of mercury or torrs. The **torr**, which is named for Torricelli, is a unit of pressure equal to 1 mm of mercury. Engineers routinely use pounds per square inch (lb/in² or psi) when describing pressure. Atmospheric pressure is 14.7 psi, and an automobile tire that we say contains 25 lb of pressure actually contains $14.7 + 25$ or about 40 psi of pressure. The SI recommends that pressure be measured in terms of a unit called the **pascal**, with the symbol Pa. A pressure of 1 mm of mercury is defined as 133.322 4 pascals. Normal atmospheric pressure then is 101.323 pascals or 101.3 kilo-pascals.

**15.3
COMPOSITION
OF THE
ATMOSPHERE**

Although the atmosphere is composed of ions, atoms, compounds, and colloidal particles, three elements—nitrogen, oxygen, and argon—account for more than 99.9% of its so-called fixed components. By fixed components we mean those that are the same everywhere on earth. Table 15.1 shows the percentages of the various fixed components of the atmosphere. In addition to the fixed components, water, in its gaseous or vapor form, is present in all samples of air, but its amount varies from place to place and from time to time. Since warm air can hold more water than cold air, when very moist air is cooled the water usually precipitates. There are many other substances (ions, atoms, molecules, and larger particles) found in most samples of air, particularly those lying over a city or near an industrial plant. Most of these substances are pollutants, although usually that term is used only if a substance is harmful to some form of life.

TABLE 15.1 Composition of clean, dry air near sea level

Component	Content (Percent by Volume)
nitrogen	78.084
oxygen	20.947 6
argon	0.934
carbon dioxide	0.031 4
neon	0.001 818
helium	0.000 524
krypton	0.000 114
xenon	0.000 008 7
hydrogen	0.000 05

Since the pressure of the atmosphere decreases with altitude above the earth, the amounts of each of the individual gases also decrease. The decrease in the amount of oxygen is one concern that must be met in the design of planes. It is even noticeable when persons who have lived near sea level travel to a higher elevation and exercise vigorously. They must breathe more rapidly to get the required amount of oxygen from the "thin" air. In a short time, the human body compensates for this decrease in amount of oxygen. One of the methods of adapting is the synthesis of more red blood cells that carry oxygen to all parts of the body. People living near sea level normally have about 5×10^6 red blood cells (RBC's) per mm^3 of blood, whereas it is not unusual to find RBC counts of $7 \times 10^6/mm^3$ among people living in Denver, Colorado, which is one mile above sea level. Pilots must reset altimeters and cooks must vary recipes or change cooking times if they live at altitudes appreciably above sea level.

**15.4
NORMAL COM-
PONENTS OF
THE EARTH'S
ATMOSPHERE
Recycling of
Atoms**

In the following sections we will discuss some of the most important components of the atmosphere. In discussing these, there are some general principles we must consider.

The number of atoms on the earth (which includes the atmosphere) does not change appreciably with time. An oxygen atom may be present first as free atmospheric oxygen, then be transformed into one of the oxygen atoms in a molecule of CO_2, then become part of a molecule of calcium carbonate, and after a series of transformations again be present as free atmospheric oxygen. In most instances this recycling involves not only the atmosphere but also the waters (hydrosphere) and the earth's crust. Since these three—atmosphere, hydrosphere, and crust—are places where life is found we often speak of them as constituting the **biosphere**. A given atom can go from one part of the biosphere to another as it becomes parts of different compounds and is recycled.

The earth has been compared to a spaceship. All our supplies are on board. While we should not run out of atoms, we may get into trouble if we don't have enough of them present in the right form. Too much CO_2 suffocates us, and a small amount of CO poisons us; we need to breathe free molecular oxygen, preferably at about 20% concentration.

Energy Balances

Although the specific number of atoms on the earth is fairly constant, we cannot make the same statement about the amount of energy. Every day we receive vast amounts of energy from our star, the sun. Since we know that the temperature of the earth is relatively stable, we can deduce that the earth loses energy at a rate that is approximately equal to that at which we receive energy from the sun. The relatively stable temperature is possible only because of a delicate balance, between gain and loss of energy. We know that at times this balance has been upset, and there have been drastic changes in the climate of the earth. There is evidence that there have been five or six ice ages. Careful measurements show that the average surface temperature of the earth was increasing throughout the past century but that, since 1950, a cooling trend has started, and our temperature has decreased a few tenths of a degree Celsius. Although the overall changes in temperature are slight, relatively small increases can trigger processes that have tremendous consequences. A warming of only a few degrees could melt most of the polar ice caps, raise the sea level, and flood many of the major cities of the world.

It is important to remember that although temperatures and climate have varied widely in the past, these fluctuations have been due primarily to natural causes—not the works of humans. There is little we can do about natural changes other than accept them. However, with the increase in the number of people on our planet and with growth of technological knowledge, humans now have the power to initiate drastic changes. Although there are powerful natural forces serving to keep things in a state of dynamic equilibrium, many of them are delicately balanced, and only slight disruptions of their balance can lead to greatly amplified consequences. Environmental changes have resulted in the extinction of many species of plants and animals in the past. Many people are now concerned that humans using technology may bring about the extinction of more and more species, perhaps including humans themselves.

What is causing the current decrease in temperature? Is this the beginning of another ice age? What is the effect of the large amounts of energy released when we burn fossil fuels—coals, natural gas, petroleum? What are the possibilities of oxidizing these fuels so rapidly that we will run out of free molecular oxygen and suffocate in an excess of carbon dioxide? These are important questions for everyone concerned with the continuance of life on earth. Their answers, although complex, are based on chemical and physical principles.

It has already been emphasized that chemical and physical changes release or absorb energy. The ways in which reactions and energy are related to each other and the factors that control the overall processes are important not only to the chemistry of the atmosphere but to all changes in the biosphere.

Oxygen in the Atmosphere

It is difficult to explain why we have the quantity of oxygen we find on our planet earth. We do not find this gas in the atmospheres of other planets, and apparently it was not present when the earth was formed. The oxygen we now find in the air came from various types of rocks—the silicates and perhaps carbonates, and from water. The photolysis of water, particularly in the upper atmosphere, to yield hydrogen and oxygen gases has been important and still continues. Some of the hydrogen formed in this process, is lost into space, but most of the oxygen is retained in our atmosphere. The other major source of oxygen today is the process of photosynthesis, in which plants take in water and carbon dioxide and release free oxygen into the atmosphere. We can summarize the process by the following equation:

$$6CO_2 + 6H_2O \underset{\text{oxidation}}{\overset{\text{photosynthesis}}{\rightleftharpoons}} C_6H_{12}O_6 + 6O_2$$
$$\text{(a simple sugar)}$$

The process is reversible and yields water and carbon dioxide when the products of photosynthesis are burned in an animal body or in a fire. In the distant past, huge amounts of the products of photosynthesis were converted to coal, petroleum, and natural gas, which are known as **fossil fuels**. This process involved the removal of oxygen from the molecules found in plants, leaving hydrocarbons in the case of petroleum and natural gas, and carbon in coal. Some of the oxygen in our atmosphere is there only because these fossil fuels were stored in a reduced (nonoxidized) state.

We are burning gasoline, fuel oil, natural gas, and coal at a rate infinitely more rapid than the rate at which they are being produced on the earth today. While this fact has serious consequences in terms of exhausting sources of stored energy, it could have even more serious consequences in terms of the amount of oxygen and carbon dioxide in our atmosphere. At the present time, the level of oxygen in our atmosphere seems to be staying constant at 20.947 6% in spite of the amount of fossil fuels we are burning. This balance is due in part to the fact that green plants are producing large amounts of free oxygen. In fact, as predicted by the law of mass action, the presence of increased amounts of carbon dioxide increases the rate of photosynthesis by green plants. Although we are aware of trees and grasses as green plants, we are likely to forget, unless reminded, those one-celled green plants, the algae, which are found both in fresh water and in the oceans. These unicellular plants are responsible for about a third of the oxygen produced by green plants. Thus, we decrease our supplies of atmospheric oxygen not only when we cut down forests and convert farm or grass land to housing areas and asphalt parking lots, but also when we allow the pollution of streams, lakes, and the seas to inhibit the growth of algae. Although the evidence indicates that we have not *yet* exceeded the power of natural processes to regulate the level of oxygen, it is not a matter we can ignore.

Carbon Dioxide in the Atmosphere

The carbon dioxide content of the atmosphere is extremely low, as indicated in Table 15.1. You might conclude, therefore, that it is an unimportant constituent; however, such a conclusion would be wrong. Although the concentration is low, it is in a rapid state of turnover. Millions of tons of carbon dioxide are poured into the atmosphere every day from the burning of gasoline, coal, and other fuels and from the respiration of animals. At the same time, carbon dioxide, along with water, is the major reactant that green plants use in making our food and other products. The relatively few molecules of CO_2 in any air sample represent only those temporarily released from a variety of other compounds of carbon.

In contrast to the constant level of oxygen, that of carbon dioxide in our atmosphere has been increasing at a slow but steady rate. It is estimated that the amount of carbon dioxide increased from 290 to about 330 parts per million (ppm) from 1900 to about 1950. As the result of definite measurements, we know that the level of carbon dioxide has been increasing about 0.2% or 0.7 ppm each year since 1958.

Some scientists are concerned about this increase, because it is occurring in spite of several processes that tend to keep the level constant (see Fig. 15.2). We have already mentioned the effects of the law of mass action on photosynthesis, which is one important control. Another homeostatic force is the solution of carbon dioxide in water. The oceans of the world are huge reservoirs of dissolved carbon dioxide, not because it is present in any great concentration but because the volume of the oceans is so great. The higher the concentration of carbon dioxide in the air, the more we would expect to be dissolved in the oceans. Another factor that influences the solubility of a gas in water is the temperature, and we would expect any increase in the average earth temperature to decrease the amount of carbon dioxide in the seas and a general cooling to add to the amount dissolved.

Another important property of carbon dioxide gas is that (to a far greater extent than does oxygen or nitrogen) it absorbs various kinds of energy from the sun and converts them into infrared radiation or heat. This is similar to what happens when sunlight is absorbed in a glass-enclosed area such as the interior of an automobile or a greenhouse.

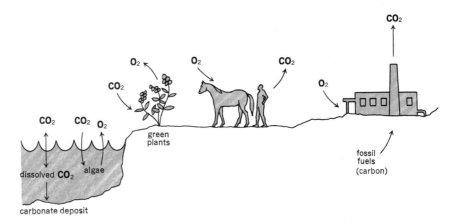

Fig. 15.2 The carbon and oxygen cycle. Many processes are active in maintaining the amounts of oxygen and carbon dioxide in our atmosphere. The most important ones are shown above. Photosynthesis in green plants on land and in the sea consumes carbon dioxide and produces free oxygen, while oxidation of plant material (wood, coal, etc.), whether in an animal's body or in a fire, consumes oxygen and releases carbon dioxide. On a global scale the amount of carbon dioxide dissolved in seawater is a significant factor; this process is highly sensitive to temperature.

This conversion to heat and the lesser tendency of the heat to radiate back outside is called the *greenhouse effect*. The fact that the earth maintains a temperature much more stable than that of the moon is due both to the fact that the earth has an atmosphere and also to the fact that there is carbon dioxide in the atmosphere to convert energy to heat which is not rapidly reradiated out into space.

Thus we would expect an increase in the amount of carbon dioxide in the air to cause an increase in temperature of the earth's atmosphere. In fact, it has been estimated that doubling the amount of carbon dioxide in the atmosphere would bring about a tropical climate throughout most of the world. This would melt the polar ice, raise sea level about 100 feet, and flood many of the major cities of the world. Significant decreases in carbon dioxide could start another Ice Age.

When we consider the implications of the solubility of carbon dioxide in water and the relationship of heat to the process we realize that a higher temperature would cause the oceans to release more carbon dioxide into the atmosphere and this, in turn, would tend to further increase the temperature, resulting in the release of more carbon dioxide. Thus we have a self-catalyzing process or amplification in which a small change could result in major consequences.

Since we know from careful measurements that the amount of carbon dioxide in the atmosphere is increasing, we would suppose that the temperature of the earth is increasing. However, careful measurements of temperature show the opposite, and this decrease in temperature is happening even though we are releasing great amounts of heat by burning fossil fuels. Therefore we must look for a compensating factor to account for the decrease in temperature. It is possible that the amount of radiation being emitted by the sun is decreasing; however, it seems more likely that the amount of this radiation being absorbed by the earth's atmosphere is decreasing. This tendency may be due to the fact that we are constantly increasing the amount of particulate matter (ash, soot) that is being discharged into the air. This not only reflects the sun's energy back into space, but also forms nuclei around which water particles can condense and form clouds. Clouds also reflect the sun's radiation back into space. One has only to recall the difference in temperature on sunny and cloudy days to become aware of the tremendous effect of cloud cover on temperature. Although we know that the amount of particulate matter in our atmosphere is increasing, we have inadequate information about the amount of cloud cover on the whole globe on an annual basis. The increasing use of weather satellites should give us more data in the near future.

There is, under natural conditions, very little circulation or deposition of particles in our upper atmosphere. It is estimated that in the stratosphere (elevations above 11 km), particles may persist for as long as five years. Since supersonic planes are now being designed to fly at these elevations, many people are concerned about the effect of the water (ice) particles as well as other exhaust substances from these high flying aircraft. Our information is inadequate to accurately estimate the results of such particles, but one possibility is that they could form long-lasting clouds that would decrease still further the amount of solar radiation that reaches the earth.

From the foregoing discussion, you can see that the levels of just two of the gases of the atmosphere, oxygen and carbon dioxide, are the result of several related processes. These relationships are represented in Fig. 15.2. Although the amount of carbon in each of its various forms is important, the rate of conversion of one compound to another is even more important. We use the concept of **feedback** in describing the rate and control of these changes (see Fig. 15.3). We say that there is positive feedback when the production of one factor increases the production of itself or of another substance. Such a relation-

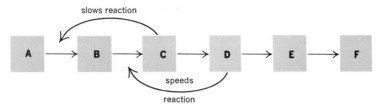

Fig. 15.3　Feedback. The basic process is the conversion of substance A to substance F. If one of the intermediates or products in the overall series of reactions inhibits a step in the process, we say there is negative feedback. If an intermediate or product helps the reaction proceed at a faster rate, we call this positive feedback, or amplification. In the illustration above, C exerts negative feedback and D positive feedback on the conversion of A to F.

ship is seen in the case of atmospheric carbon dioxide increasing the temperature of the earth's surface, which in turn increases the rate of release of carbon dioxide from sea water. An example of negative feedback is found in the fact that high levels of atmospheric oxygen depress the rate of photosynthesis. If we are to understand and perhaps control a complex series of physical and chemical changes, it is necessary to know not only the products and reactants involved but also the rate of change and the effects of many factors on these rates of change.

Many factors are controlled by negative feedback so that an increase in any substance slows the general process and leads to a state of balance. We must be more concerned about systems that involve positive feedback or amplification. These can have serious consequences. Humans have been concerned about controlling them as when they prevent spontaneous combustions or control a fever in a seriously ill patient. The increase in the amount of carbon dioxide in the atmosphere can result in amplification or positive feedback and must be an area of serious concern.

Nitrogen in the Atmosphere　　Although there is a large amount of nitrogen in the atmosphere, it is of less significance to the direct well-being of humans and other animals than oxygen and carbon dioxide. It is essentially a relatively inert component which serves to dilute the oxygen. While in some instances it can be converted by lightning or an internal combustion engine into toxic oxides of nitrogen, and thus exert harmful effects on living organisms, on a global scale these compounds are normally soon recycled to the soil in the form of nitrates and nitrites. (See Fig. 8.4.)

A supply of nitrogen, combined with oxygen as nitrates or in the form of ammonia or similar compounds in which the nitrogen is attached to hydrogen atoms, is of vital importance for the growth of plants. The processes by which nitrogen is transformed from a gaseous element into a compound that is usually a liquid or solid is called **fixation**. Much of this process is catalyzed by bacteria that grow in nodules of the roots of plants such as peas, clover, and soybeans, which we call the *legumes*. Since all plants require a source of nitrogen and most of the plants we grow for food such as wheat and corn do not support the bacteria we call nitrogen fixers, we are forced to use fertilizers if we want to produce enough food to feed a hungry world. Scientists who are sometimes called genetic engineers are now trying to change the hereditary characteristics of nitrogen-fixing organisms so they can live on roots of nonlegumes. This accomplishment would increase the amount of nitrogen present in the soil and in plants. The decrease in volume of atmospheric nitrogen would probably have little effect.

15.5
AIR POLLUTION

More than 3 000 foreign substances have been found in our air. Some of these, such as the odor of pine trees and pollen from a variety of plants, we regard as natural. Others, such as the oxides of nitrogen, carbon monoxide, and the ozone we find near the earth's surface, can be directly attributed to humans and their work—especially the internal combustion engine. Many of these substances are present in only a few parts per million even in the air over our most polluted cities on a bad day. While these levels seem low, it is known that a concentration of only 6 ppm of ozone will kill any laboratory animal so far tested, and it does so within a period of four hours.

Three air pollution disasters have received worldwide attention. In 1952, a black fog that persisted for several days over London was responsible for at least 4 000 deaths. In Donora, Pennsylvania, in 1948 and in the Meuse Valley of Belgium in 1930 scores of deaths and a great deal of intense discomfort were caused by persistent fogs that became loaded with pollutants from nearby factories. Many of the victims of these disasters were older people, and the cause of death was usually respiratory or heart failure.

Since these disasters produced immediate death and widespread discomfort, they created public concern. Probably much more dangerous are the effects of continuous breathing of polluted air over a period of months or years. These effects are difficult to attribute to pollution since people die from a variety of lung diseases and heart conditions even in the absence of air pollution. Only extended studies will show the extent to which various causes of death are influenced by long-term exposure to relatively low levels of pollutants. However, current evidence concerning the increase of the lung disease emphysema, as well as other lung disorders, is alarming. Further research will probably uncover more of the relationships between pollutants and health problems.

Ever since they first used fire, humans have polluted the air. The concentration of people in cities and the introduction of factories and automobiles have greatly complicated the problem. Another complicating factor is the presence of **inversion layers** in our atmosphere. Normally, when we produce a pollutant, it rises in the air and is scattered by wind currents because warm air is lighter and rises through the denser cold air. The pollutant is then recycled into harmless compounds. However if pollutants reach a region of **temperature inversion**, that is, a region where the air is warmer and less dense, they stop rising and concentrate below the inversion zone (see Fig. 15.4). Although there is little evidence that cities create these inversions, many cities are located in areas where they form easily. The tragic incidents in London, Donora, and the Meuse Valley were probably due to inversion layers which trapped gases being discharged into the atmosphere.

Oxides of
Sulfur

Most coal and petroleum contain sulfur. When burned, they produce a pungent gas, sulfur dioxide. Although the sulfur content of petroleum can be reduced by refining, this is not done with coal. In general, soft coal contains more sulfur than hard coal.

Sulfur dioxide is the oxide of a nonmetal, and when it is combined with water it forms sulfurous acid (H_2SO_3). If sulfur dioxide is oxidized by oxygen or other substances in air, it forms sulfur trioxide, which reacts with water to form sulfuric acid. Although the amount of this acid in the air is small, it is a strong acid capable of causing major lung damage and death, as well as the disappearance of nylon hose when women wear them on days when pollution is high. Sulfur dioxide was the major culprit in the killer fogs of

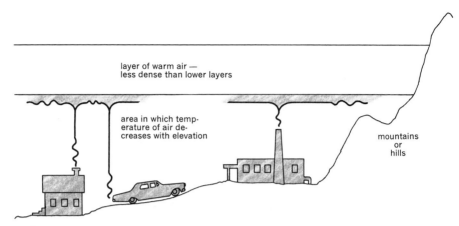

Fig. 15.4 Diagram of an inversion layer in the atmosphere. Normally, warm, polluted air rises and is scattered by the wind. When an inversion layer forms, polluted air can no longer rise and disperse. Inversion layers form only in the absence of wind and are broken up when the wind returns.

London, Donora, and the Meuse Valley. It is the major cause of "acid rain" that falls downwind from pollution sources and may destroy crops and life in lakes or streams.

Carbon Monoxide and Other Carbon-Containing Pollutants

Another by-product from the burning of coal and other carbon-containing substances is the mixture of carbon particles we call soot. Lungs of people who have lived in London or other major cities where coal has been burned are usually a dirty gray, due to the continued inhalation of soot. Along with these dark particles are other coal derivatives, many of them colorless, but nonetheless deadly. One of the first correlations of occupations with cancer was observed in chimney sweeps. Something in the soot and other debris they continually encountered appeared to be causing cancer of the skin and lungs. We are now quite certain that many of the compounds that are found not only in coal smoke but also in the smoke of cigarettes and even on charcoal-broiled steaks can cause cancer. These substances, which are commonly known as tars, do not cause cancer immediately and not everyone exposed to them will die from cancer, but the cause-effect relationship is well established.

Although coal is the source of two major pollutants, it is not the major cause of the carbon monoxide, ozone, oxides of nitrogen, and hydrocarbons we find in the air above major cities today. Here the culprit is the internal combustion engine. This engine is a device to derive power from oxidation of the hydrocarbons in gasoline and similar fuels. While it does this efficiently enough to power automobiles and trucks, it also produces carbon monoxide, oxides of nitrogen, and unburned hydrocarbons which are usually eliminated in the exhaust. When these substances react, particularly in the sunlight, with the oxygen in the air and with each other, they produce the mixture we call *smog*. The name is really inappropriate, since it should indicate a mixture of smoke and fog, neither of which is present in cities located in relatively dry, sunny areas such as Los Angeles and Denver. Whatever we choose to call this polluted air, it represents a major cause of concern for everyone living in an urban area.

Carbon monoxide is produced whenever carbon-containing substances are burned in insufficient oxygen. Carbon monoxide has an affinity for hemoglobin that is more than

200 times that of oxygen, and molecules of hemoglobin react with carbon monoxide to form **carboxyhemoglobin**, which is unable to combine with oxygen. If a sufficient percentage of molecules of hemoglobin are so tied up, death occurs. We are all aware of the deaths of people who operate automobiles in closed garages. If the automobile operates outside, the same carbon monoxide is being produced, but it is being widely scattered until it reaches a nontoxic level. If thousands of cars are producing carbon monoxide in a city under an inversion layer, we have increased the size of the "garage" but must expect bad consequences. It is probable that many automobile accidents are partially due to drivers forced to breathe too much carbon monoxide. Continued breathing of air containing 0.1% or 1 part per thousand of carbon monoxide for 90 minutes may be fatal. (In discussing toxic effects of gases, we must specify not only a level of poison but also the period of exposure, since it is the total amount of poison absorbed that is significant.)

Those who smoke cigarettes and particularly those who inhale are also exposed to relatively large concentrations of carbon monoxide and probably expose those in their immediate vicinity also. It has been shown that a person who smokes one pack of cigarettes per day and inhales deeply has about 6% of his hemoglobin tied up as carboxyhemoglobin. Experiments have shown that this level can affect eyesight and time discrimination. Other effects that have been connected with carbon monoxide toxicity are impairment of vision and time-interval discrimination, fatigue, headaches, irritability, drowsiness. If the levels of carboxyhemoglobin are high enough, coma and death result. Experiments with animals have shown definite heart damage due to relatively low levels of carbon monoxide. (See Table 14.7.)

The state of California lists 30 ppm of carbon monoxide for eight hours as an adverse level, and 120 ppm for one hour as a serious level of pollution. The latter level has been measured on the streets of Los Angeles. The present maximum allowable atmospheric concentration for occupational exposure to carbon monoxide is 50 ppm for eight hours. This limit was set at 100 ppm until 1964.

The human body gradually excretes carbon monoxide that has combined with hemoglobin and makes new hemoglobin to replace it, so that the effects of the poison can be reversed. However, the impaired function of the body operating under decreased supplies of oxygen may produce permanent damage, especially to the heart. As with many poisons, children are more susceptible to carbon monoxide than are adults.

Although carbon monoxide is oxidized to carbon dioxide in the atmosphere and certain soil organisms also remove carbon monoxide, these processes are affected by many variables. If the conversion of CO to CO_2 is decreased it is conceivable that we could begin to accumulate high levels of CO in our global air supply. This is another substance whose concentration should be monitored more carefully on both a local and global scale.

Oxides of Nitrogen The two most abundant substances in air are oxygen and nitrogen. Under ordinary conditions these two gases do not react with each other. However, when heated under pressure in the presence of catalysts—conditions present in automobile engines—they react to form nitrogen oxide (NO). This gas, commonly called nitric oxide, is readily oxidized to yield nitrogen dioxide (nitrogen(IV) oxide). These and other possible oxides of nitrogen make up another significant and deadly component of air pollution. It is nitrogen dioxide that is responsible for the brown color of much air pollution.

We can summarize the reaction of oxygen and nitrogen as shown below:

$$O_2 + N_2 \rightarrow 2NO$$
$$2NO + O_2 \rightarrow 2NO_2$$

People exposed to 500 ppm of NO_2 in the air die almost immediately. Lower concentrations produce lung damage in experimental animals; in fact, nitrogen dioxide has been characterized as the only gas that invariably produces the lung disease emphysema in rats. Scientists disagree about levels that can be tolerated by humans. The American Conference of Governmental and Industrial Hygienists gives 5 ppm as a tentative threshold (allowable) dose, whereas Russian scientists report that a level of 3 ppm has produced emphysema. Levels of nitrogen dioxide reach 0.9 ppm in Los Angeles on days of high pollution.

In addition to its primary effects, nitrogen dioxide serves as a facilitator in the production of ozone, as shown by the following reaction:

$$NO_2 + O_2 \xrightarrow{\text{ultraviolet radiation}} O_3 + NO$$

This reaction accounts for the formation of ozone and explains the fact that ozone levels are higher on sunny days than on cloudy days. The reaction is reversible and is further complicated by the fact that the nitrogen oxide formed can also react with normal oxygen to give more nitrogen dioxide which, in turn, reacts with oxygen to produce more ozone (see Fig. 15.5).

Ozone   Ozone is an interesting substance because it has some actions that benefit humans and others that are harmful. In the upper atmosphere, from 25 to 50 kilometers above the earth's surface, the reaction of oxygen with ultraviolet radiation from the sun produces ozone. This process prevents the ultraviolet radiation from reaching the earth's surface, where it would certainly increase the amount of sunburn that beach lovers get and might even destroy many forms of life.

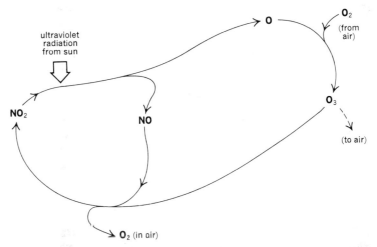

Fig. 15.5 Interrelationship of ultraviolet radiation, oxides of nitrogen, ozone, and oxygen. This diagram shows the effects of ultraviolet radiation on production of some of the components of photochemical smog. If some part of the cycle is interrupted, one of the substances (e.g., ozone) may accumulate or react with substances not included in this cycle.

Ozone is extremely toxic, with relatively large amounts producing immediate death, whereas smaller amounts cause respiratory diseases. So long as ozone is 15 miles (25 km) above the earth, only those who fly aircraft at this height need be concerned, but at the surface of the earth it becomes of great importance to all of us. It has been shown that 6 ppm of ozone kills laboratory animals in a short time. The effects of ozone are increased if the person exposed is exercising or if the temperature is high, and ozone is especially damaging to children. Ozone in the atmosphere is generally not measured as such, but it makes up about 90% of the general category of "oxidants" which are now measured in many locations on a continuing basis. In Los Angeles, school children are not allowed to play outside when the level of oxidants goes above 0.35 ppm.

Parenthetically, we should say that we make many references to Los Angeles not because conditions are necessarily worse there than in other places, but because the people there have become sufficiently concerned to find out the facts and make protective regulations. A large number of the air pollution experts in the United States work in the Los Angeles area, and it has become the source of accurate measurements which we need to speak precisely about levels and effects. Other cities in the United States and the world may have pollution as bad as or worse than Los Angeles; we just do not have the figures to cite. Tokyo, Mexico City, and Ankara in Turkey have especially bad air pollution problems. Other major industrial cities are saved only because they receive large amounts of rain, which is one of the best remedies for air pollution—probably at the expense of the streams that carry this polluted rain water.

One of the most obvious effects of air pollution is its effects on the eyes. People living in polluted areas are highly aware of eye irritation that results in pain and excessive watering. Although a variety of chemical compounds are probably responsible, peroxyacetylnitrate (PAN) is probably the main tear gas or lacrimator in smog.

$$CH_3-\overset{\overset{\displaystyle O}{\|}}{C}-O-O-NO_2$$

peroxyacetyl nitrate (PAN)

It is produced by the reaction of unburned hydrocarbons (particularly those with a double bond in them) with ozone and the nitrogen oxides. Although eye irritation is painful, it does not seem to produce long-term permanent damage. Other pollutants, some of which our senses do not detect directly, are much more damaging. PAN however, can permanently damage plants.

Particles in the Air   In addition to the many gases found in the air surrounding our cities, there are many kinds of small particles. Included in the value reported as "particulates" are particles of ash, soot, rubber, asbestos, and hundreds of other materials. Asbestos, which comes from insulating materials and from brake linings of cars, produces lung cancer. It does not do so immediately but has a latency period of several years, as is the case with many cancer-causing substances. Rubber particles from automobile tires are also suspected of producing cancer. Although ash and soot are decreasing in many cities as less coal is burned, the amounts of other particulate matter seem to be increasing.

So far we have concentrated on effects of air pollution on humans; however, widespread plant damage is also occurring near many cities. Spinach, orchids, and violets can no longer be grown in the Los Angeles area. Forests on the nearby mountains are also being affected. This damage is probably due to ozone and some of the unburned hydrocarbons, especially ethylene (C_2H_4). This sensitivity of plants is in contrast to their ability to grow in and concentrate certain other poisons. Plants growing near highways have been shown to contain high levels of lead. Although the lead does not seem to affect the plants, it can poison the animal that eats them. Other aspects of lead poisoning were discussed in Section 13.13.

15.6
CONTROL OF
AIR POLLUTION

With the exception of carbon dioxide, there is no evidence that air pollution by humans has changed the composition of the global atmosphere. Some soils have been contaminated, perhaps for long periods of time, by airborne particulates such as fluorides and radio-active particles that have settled out. However, there is no question that the air in some local areas, often in areas where many people live, is seriously polluted a great deal of the time, and there is no question that this local pollution must be controlled. Fortunately, an understanding of the chemical and physical properties of the major pollutants suggests methods of control. In many instances we control pollutants by using a physical or chemical process to prevent their release from factories or power plants.

Scrubbers

Many pollutants are soluble in water. In fact, most of the damage from the oxides of sulfur and nitrogen is due to their reacting with water in the tissues of a living plant or animal. If polluted air is bubbled through or otherwise brought into contact with water in a device commonly known as a scrubber, the soluble gases are removed (Fig. 15.6). In addition, the reaction with water or subsequent reactions with other chemical substances can produce products with commercial value. The equations below show some possible conversions.

$$SO_3 + H_2O \rightarrow H_2SO_4$$

$$3NO_2 + H_2O \rightarrow 2HNO_3 + NO$$

$$HNO_3 + NH_3 \rightarrow NH_4NO_3$$

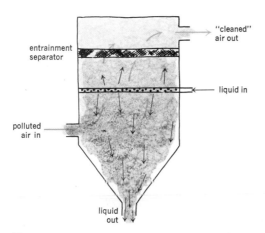

Fig. 15.6 Schematic drawing of a scrubber. Polluted air is introduced near the bottom of a spray tower, where it is treated by spraying water or other liquids through the gas. Purified air leaves the tower near the top.

Ammonium nitrate is a good fertilizer and sulfuric acid is used in many industries. In fact, more than half of the supply of sulfur we use annually in the United States now comes from recovery of sulfur from industrial wastes.

Precipitators   Many solid pollutants that occur as small particles can be removed by filters. In other instances the particles are so small that they would pass through filters, or filters are impractical. In these cases the particles are small because they are charged and will not come together to form larger particles (see Section 10.11). Knowing this, we can use devices known as *electrostatic precipitators* (see Fig. 15.7) for removing such particles. These devices place anodic and cathodic plates or other surfaces in the stream of polluted air. The anodes pick up negatively charged particles, neutralizing their charges, and the particles can then be scraped off the surface. The same process applies to the positively charged particles at the cathodes.

As with scrubbers, many valuable substances can be recovered by the use of electrostatic precipitators. They have been used for many years in the smokestacks of smelters where precious metals such as silver and gold are processed. Their use to recover less precious substances will help to preserve even more precious things—an unpolluted atmosphere and healthy humans.

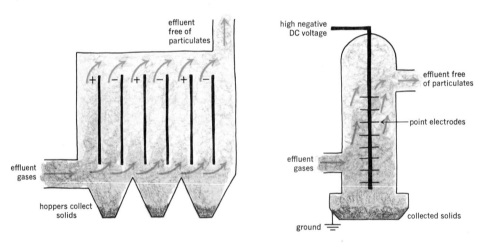

(a) plate-type electrostatic precipitator   (b) tube-type electrostatic precipitator

Fig. 15.7 Schematic drawings of two types of electrostatic precipitators. Particulates are removed from polluted air by (a) attraction to charged plates or (b) electrical discharge from point electrodes. The walls in (b) are either attached to a ground or charged with a charge opposite to that on the point electrodes.

Other Devices   While scrubbers and precipitators are useful in factories, power plants, and other stationary sources of pollution, they are so heavy that they are of almost no use for the automobile and other mobile internal combustion engines. In many instances pollutant emissions have been decreased by redesigning the engine. This has resulted in burning the fuel more completely to reduce the output of unburned hydrocarbons and carbon monoxide. Unfortunately, the increased efficiency has resulted in an increased production of the oxides of nitrogen. This has been shown by measurements of air pollution in the Los

Angeles air basin, where carbon monoxide has decreased but oxides of nitrogen have increased in recent years.

Many automobiles now have catalytic converters. These devices provide a bed of hot catalyst which removes not only unburned hydrocarbons and carbon monoxide but also the oxides of nitrogen. There is a concern that such catalysts may produce more SO_3, H_2SO_4, or sulfates, but they have not been used long enough or widely enough for us to know their effects on the level of sulfur compounds.

One of the side effects of the use of catalytic converters has been the elimination of lead from gasoline, a change made because lead "poisons" the catalyst, i.e., interferes with its action. Originally, the addition of lead (as tetraethyl lead) enabled us to use more of the components of petroleum in gasoline and prevented "knocking" in the engine. There is some research indicating that it may be possible to replace tetraethyl lead by other metallic compounds which are less toxic to catalysts and to humans. Some compounds of the elements generally known as the rare earths or lanthanides are being used experimentally at the present time.

As we have developed methods of control of air pollution, we have often seen the beneficial results lost as the number of potential polluters increased. Industrial growth means more factories and power plants, and the good life for more people often means more automobiles. Perhaps one of the best methods of controlling pollution is by changing the lifestyle of people. If we had fewer automobiles and relied more on buses and especially electric trains, not only would we have less pollution but we would use less petroleum. Since petroleum is the raw material of many synthetic substances we use in construction and for making clothing, it seems wasteful to many chemists to burn this precious material that is limited in supply. In terms of conservation of energy, riding on an electric train is much more efficient than the use of automobiles. A decrease in the number of manufactured items we seem to need (and in the number of factories and power plants) would also help conserve raw materials and energy as well as decreasing pollution.

Another important aspect of controlling pollution is that in some cases we can help natural recycling that will remove pollutants. For instance, when carbon monoxide in air is exposed to soil, the carbon monoxide is converted to carbon dioxide. We know that green plants recycle CO_2 and produce oxygen as well as other important substances. Fortunately, many cities have started planting trees, shrubs, and grass and are tearing up concrete. Fountains which probably help "scrub" the air are being built. As with many other changes, these produce not only a less polluted atmosphere but also a more beautiful setting for living.

SUMMARY

The atmosphere of the earth is one of its distinctive features and without it life would not be possible. The present composition of the atmosphere is the result of processes from the past, and its present status is controlled by a complicated network of cyclic interactions. Humans, and the level of technology they use, can now alter the atmosphere in relatively large areas and may be responsible for the increase in carbon dioxide content of the atmosphere.

There are several methods of control of atmospheric pollution. Some of them involve prevention of release of contaminants, while others require increasing the rate of recycling of pollutants. Still another approach to controlling pollution involves changes in lifestyle so that lesser amounts of pollution-producing substances are used. That approach will help conserve mineral resources and energy as well as conserving or perhaps restoring an unpolluted atmosphere.

IMPORTANT TERMS

atmosphere   carboxyhemoglobin   hydrosphere   temperature inversion
barometer   feedback   pascal   torr
biosphere   fossil fuels   smog

WORK EXERCISES

1. What evidence is there that the atmosphere becomes thinner with an increase in altitude above the earth?

2. It is estimated that the pressure of the atmosphere is halved for each 17 500 ft of altitude above the earth. Assuming this estimate is correct, find the atmospheric pressure in pounds per square inch and in millimeters of mercury at 70 000 ft.

3. What changes would likely take place if the concentration of CO_2 in the atmosphere were doubled?

4. Which of the following components of air would be considered pollutants?
   a) oxygen
   b) carbon monoxide
   c) carbon dioxide
   d) nitrogen dioxide
   e) fog
   f) sulfur dioxide

5. How can the following substances be removed from air?
   a) sulfur dioxide
   b) nitrogen dioxide
   c) asbestos particles
   d) carbon monoxide

6. What reason can you give for the fact that we find many of the noble gases in the atmosphere?

7. Why is the content of hydrogen in the atmosphere low?

8. What is the purpose of catalytic converters on automobiles?

9. List the beneficial effects and the harmful effects of ozone.

10. List three chemical reactions that result in the addition of carbon dioxide to the atmosphere; three that remove it.

11. Why is smog often worse on days when there is a great deal of sunshine?

SUGGESTED READING

"Environmental Quality—1977," Stock No. 041-011-00035-1. A government publication discussing the air pollution, water pollution, energy, and natural resources of the United States. New reports are issued each year. An authoritative report, filled with facts and figures.

Hutchinson, E. G., "The biosphere," *Scientific American*, Sept. 1970 (Offprints #1188–1198). This is only the lead article in an issue packed with ecologically important articles such as those on the cycles of carbon, oxygen, energy, water, nitrogen, and minerals. The whole September 1970 issue is highly recommended.

Lippincott, W. T., editor, *Journal of Chemical Education* **49** (1), January 1972. A major part of this issue is devoted to environmental chemistry. Methods for measuring mercury and oxides of nitrogen are discussed. Highly recommended.

Stoker, H. S., and S. L. Seager, *Environmental Chemistry: Air and Water Pollution*, 2nd Ed., Scott, Foreman and Co., Glenview, Ill., and London, 1976. This is an excellent paperback on the topics listed in the title. It gives the facts necessary for conclusions and generalizations in this area.

16

The Properties of Gases

In Section 3.5 we described briefly the three states of matter—solid, liquid, and gas. All of us have long been accustomed to these classifications and know, for example, that solids melt to give liquids and liquids boil to give gases. However, further thought may raise some questions: What is the real difference between a liquid and a gas? Why is it possible to compress gases and not liquids or solids? Why does a gas tend to fill any container into which it is placed?

Questions such as these have concerned scientists for hundreds of years. As a result of the experiments of Robert Boyle, J. A. C. Charles, Joseph Gay-Lussac, Amedeo Avogadro, Thomas Graham, and many, many others, a theory has been developed that answers the questions. It is called the *kinetic-molecular theory of gases*. Interpretations of the nature of gases made by these pioneers of science also helped to establish the atomic theory which was discussed in Chapter 4. We will now tell the story of how the present understanding of the properties of gases developed. We will include some of the kinds of calculation that were important in first establishing the theories and that are still used today in either measuring gases or predicting their behavior.

**16.2
ROBERT BOYLE
AND THE
RELATIONSHIP
OF PRESSURE
TO VOLUME**

Most of you are aware that air can be compressed into a smaller volume. After it has been compressed, it can be used for many purposes. If we encase it in an automobile tire, it supports the weight of the automobile and its springiness makes the automobile ride more smoothly. Compressed air can also be used to drive jack hammers and sand blasting machines.

Robert Boyle, who lived from 1627 to 1691, was one of the first scientists to study the effect of pressure on a gas; the gas he used was air. His experiments were made possible by the prior invention of a good vacuum pump. For his experiments he used a container from which air could be pumped or in which pressures could be varied by adding or removing mercury. A somewhat simplified representation of his experiments is shown in

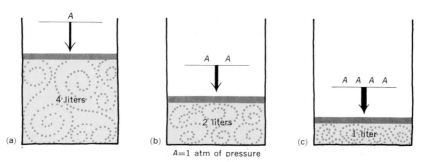

Fig. 16.1 Diagrams showing the effect of different pressures on the volume of air.

Fig. 16.1. In this diagram three cylinders, each holding 4 liters of air, are fitted with pistons which move up and down but will not allow air either to escape or to enter the cylinder. In cylinder (a) the pressure is the same inside and outside the container. It is at atmospheric pressure, or has a pressure of 1 atm. In cylinder (b) the pressure on the gas is double that in (a), and is 2 atm of pressure. In cylinder (c) the total pressure on the gas is 4 atm. The volume of air in the three cylinders is 4 liters in (a), 2 liters in (b) and only 1 liter in (c).

When we interpret the data shown in Fig. 16.1, we see that not only does increasing the pressure decrease the volume of air, but that doubling the pressure halves the volume, and that increasing the pressure by a factor of four reduces the original volume to one-fourth. The *qualitative* nature of the relation between pressure and gas volumes was probably known to many scientists who were contemporaries of Boyle, just as it is common knowledge today. The important contribution Boyle made was in stating the relationship in specific *quantities*. This quantitative relationship can be expressed as follows: the product of the volume and pressure for a given gas, at a certain temperature, is always the same number (or is constant). From Fig. 16.1, we can see that:

$$\left.\begin{array}{l} 4 \text{ liter} \times 1 \text{ atm} = 4 \text{ liter atm} \\ 2 \text{ liter} \times 2 \text{ atm} = 4 \text{ liter atm} \\ 1 \text{ liter} \times 4 \text{ atm} = 4 \text{ liter atm} \end{array}\right\} \text{a constant value}$$

In general terms, Boyle's law is expressed by the following equation, in which c stands for a constant.

$$PV = c \tag{16–1}$$

Actually, Boyle expressed his original statement in terms of the constant value obtained when pressure and volume were multiplied, and he does not seem to have been concerned with the effects of temperature. He was probably fortunate in that the gases he studied were at the same temperature throughout his experiments. The fact that temperature could affect the results was first made known by the Frenchman Edme Mariotte, and the French still designate the statement just quoted as Mariotte's law, rather than as Boyle's law.

There are many applications, both practical and theoretical, for Boyle's law. If we have a known volume of gas under a known pressure, we can calculate its volume at any other pressure, providing the temperature is the same in both cases. Thus if we increase the pressure, the volume will be decreased, and vice versa. When we change a specific sample of gas from one pressure and volume (P' and V') to a new pressure and volume (P and V), we can see from Eq. (16–1) that:

$$PV = c \quad \text{and} \quad P'V' = c$$

therefore,

$$PV = P'V' \tag{16–2}$$

By using this equation we can always calculate the new volume, V, the gas will occupy at a new pressure, P, if we know the original volume and pressure, P' and V'. For illustration, let us consider these examples.

Example 1 If we fill a perfectly elastic balloon with 500 ml of air at sea level, where the atmospheric pressure is 760 torr, and transport it to the mountains, where the atmospheric pressure is 600 torr, what will be its new volume?

$$PV = P'V'$$
$$600 \text{ torr} \times V = 760 \text{ torr} \times 500 \text{ ml}$$
$$V = \frac{760 \text{ torr}}{600 \text{ torr}} \times 500 \text{ ml} = 633 \text{ ml}$$

The answer should always be checked to see if it makes sense. We know that when the pressure decreases, as in this example, the volume will expand. So the volume we have just calculated seems to be correct, because it is larger than the original volume. At least, the answer is in the right direction and it should be correct if the arithmetic is correct. If we had mistakenly interchanged 760 torr and 600 torr when substituting for P and P', the new volume, V, would have been *smaller* than V' and reason warns us that this is wrong.

Example 2 Assuming the same atmospheric pressures, suppose we reverse the procedure—fill the balloon with 500 ml of air on the mountain and return to sea level. What will be the volume of the balloon at sea level?

Again using Eq. (16–2), we have:

$$760 \text{ torr} \times V = 600 \text{ torr} \times 500 \text{ ml}$$
$$V = \frac{600 \text{ torr}}{760 \text{ torr}} \times 500 \text{ ml} = 395 \text{ ml}$$

This answer is likewise sensible, for the balloon will be subjected to increased pressure at sea level and its volume should become smaller.

Example 3 The steel cylinders ordinarily used to transport hydrogen, oxygen, or other gases have a volume of 2 cubic feet, and they may be filled to a pressure of 2 000 psi. If such a cylinder is filled with oxygen, what volume of oxygen would we obtain by opening the valve and allowing it to expand to a pressure of 1 atm? (Assume 1 atm = 14.7 psi, Section 15.2.)

For calculations using Eq. (16 2), any units of volume and units of pressure may be used.

$$14.7 \text{ psi} \times V = 2.00(10^3)\text{psi} \times 2.00 \text{ ft}^3$$
$$V = \frac{2.00(10^3)\text{psi}}{14.7 \text{ psi}} \times 2.00 \text{ ft}^3 = 272 \text{ ft}^3$$

When we think further about the implications of the fact that doubling the pressure exactly halves the volume of a gas, we realize that the gas must really be mostly space.

The volume occupied by the molecules is so small that we don't even need to consider it. The same is true of all gases studied, not only of air with which Boyle did his experiments.

Under ordinary conditions, Boyle's law is entirely satisfactory. When a gas is compressed to a very small volume by high pressure, however, we find that we cannot properly neglect the volume of the molecules. To be precise, we must add correction factors to Boyle's law to make it truly fit the behavior of a gas under these conditions. The correction factors are necessary because when the total gas volume is extremely small the molecules themselves *do* occupy a significant part of the volume. Also, when molecules are very close to one another, there are attractions and repulsions between them.

Even though molecules are minute and occupy only an extremely small part of the volume of a gas under normal pressures, they definitely create the pressure *within* the gas which balances the external, or applied, pressure. Boyle talked about the "spring of air" when describing the pressure of a gas which counterbalances efforts to compress the gas. We see this springiness or internal pressure of a gas in our automobile tires, in the various devices that use compressed air, and in what happens when we pop a paper bag filled with air.

16.3 THE RELATIONSHIP OF VOLUME TO TEMPERATURE —CHARLES'S LAW

We are all familiar with the fact that heating a gas causes it to expand. Mariotte was one of the first to describe this fact. During the latter part of the eighteenth century (some 100 years after Boyle's work) precise measurements of the effect of temperature on the volume of gases were probably first made by the French scientist J. A. C. Charles, who did not publish his observations. They were made later, apparently independently, by another French scientist, Joseph Gay-Lussac, who did publish them, in 1802. The qualitative observation from which both men started was that hot air was lighter than cold air, and that a bag which had been inverted over a fire until it was full of heated gases would float for some time.

It is interesting to note that neither Charles nor Gay-Lussac was merely an academic scientist. When Charles heard of the bags of hot air being carried into the air, he realized that hydrogen, a newly discovered gas, would have much greater buoyancy than hot air. Charles was the first to make hydrogen balloons, and he and a coworker were the first men to go aloft. On one occasion he went more than a mile high. Gay-Lussac also went up in balloons, eventually reaching a height of four miles. He even took samples of air at that altitude and concluded that the composition of air did not vary with height. More precise measurements, however, have shown that his conclusion was not entirely correct.

One feature of Charles's observations was most significant. He established the fact that a sample of any gas at 0°C would expand by 1/273 of its volume for each degree he heated it, and would contract by 1/273 of its volume for each degree of cooling. This was true for all gases. (See Fig. 16.2 for a graphic presentation of Charles's results.) This relationship can be expressed in a statement now generally known as Charles's law (alternately as Gay-Lussac's law): The volume occupied by a gas (pressure remaining constant) is directly proportional to the absolute temperature. Mathematically, this statement becomes the equation:

$$V = cT \tag{16-3}$$

At the time of his work, Charles did not have methods for producing extremely low temperatures. Also, there was no temperature scale that could be directly applied in

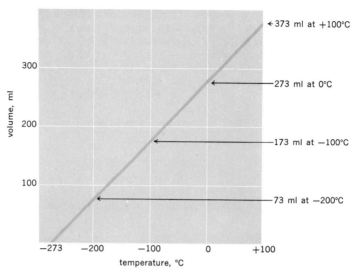

Fig. 16.2   The influence of temperature on the volume of a gas.

calculations, for some temperatures were plus and some were minus. A satisfactory temperature scale was first suggested by William Thomson, better known as Lord Kelvin, a Scottish mathematician and physicist.

Kelvin reasoned that if gases lost 1/273 of their volume at 0°C for every degree they were cooled, there must be a temperature of −273°C that was the lowest possible temperature. Theoretically, this would be the place where gases had no volume at all; he realized, of course, that gases would undoubtedly condense to the liquid state long before this temperature was reached. He took −273°C, which he called absolute zero, as the temperature at which the molecules of a gas would have lost all their energy. As a result of his reasoning, a temperature scale was developed in which zero corresponded to −273° Celsius. The new scale, now known as the *Kelvin* or *absolute scale*, is related to the centigrade scale and uses the same size of degrees. The only difference is the zero point. Celsius temperature is converted to Kelvin temperature by adding the number 273 (Table 16.1). In recent years a temperature approaching absolute zero, 0.004 K, has been reached.

TABLE 16.1   Relationship between Celsius and Kelvin temperatures

Description	Computation	°C	K
boiling point of water:	100°C + 273 = 373 K	100	373
freezing point of water:	0°C + 273 = 273 K	0	273
absolute zero:	−273°C + 273 = 0 K	−273	0

The tremendous advantage of the absolute temperature scale is that the volume of a gas is *directly* proportional to the numerical value of the temperature. That is, at 200 K a gas will have twice the volume it has at 100 K, and at 300 K it will have three times the volume. With the use of the absolute scale we can now calculate the change in volume of a sample of a gas for any change in temperature: *if the temperature is lowered, the volume will be less; if the temperature is raised, the volume will increase.*

By rearranging Eq. (16–3) we can write:

$$\frac{V}{T} = c \quad \text{and} \quad \frac{V'}{T'} = c$$

therefore,

$$\frac{V}{T} = \frac{V'}{T'} \qquad \qquad (16\text{–}4)$$

Typical calculations of the effect of temperature on the volume of a gas are the following.

Example 1   Suppose we have collected 150 ml of oxygen at 30°C and wish to know what volume it would occupy at 0°C. (Assume no change in pressure.) First, we must convert the original temperature, T', and the new temperature, T, to absolute temperatures (K):

$$T' = 30°C = (30 + 273)\,K = 303\,K$$
$$T = 0°C = (0 + 273)\,K = 273\,K$$

Now by substituting in Eq. (16–4) we have:

$$\frac{V}{273\,K} = \frac{150\,ml}{303\,K}$$

$$V = \frac{150\,ml}{303\,K} \times 273\,K = 135\,ml$$

As usual, we should check the answer to see if it is reasonable. In this example the temperature decreased, so the volume should also decrease; the answer appears to be satisfactory.

Example 2   A child is given a toy balloon inside an air-conditioned auditorium, where the temperature is 22°C (72°F). If the volume of the balloon is 918 ml, to what volume will it expand when it is taken outside, where the temperature is 36°C (97°F)?

$$T' = 22°C = (22 + 273)\,K = 295\,K$$
$$T = 36°C = (36 + 273)\,K = 309\,K$$

$$\frac{V}{309\,K} = \frac{918\,ml}{295\,K}$$

$$V = \frac{918\,ml}{295\,K} \times 309\,K = 962\,ml$$

In this case, the increased temperature should cause an increased volume. Apparently the problem was set up correctly because the new volume, V, is indeed greater than the original volume. Or, if we inspect the calculation we see that an increase of volume should result, because the original, 918 ml, is multiplied by the factor 309/295.

This method of calculation utilizing Eq. (16–4) can be used for any situation in which the volume of a gas changes because of a temperature change, while the pressure remains the same.

16.4
THE RELA-
TIONSHIP OF
TEMPERATURE
TO PRESSURE

If a gas is contained in a vessel that does not expand, such as a steel cylinder the pressure of the gas increases when heated. We can modify Charles's law and say that the *pressure of a gas is directly proportional to the absolute temperature if the volume is held constant.*

This application of Charles's law, which is also sometimes called the law of Gay-Lussac, is conveniently expressed by the equation:

$$P = cT \tag{16-5}$$

In order to calculate the change in pressure resulting from a change in temperature it is again useful to restate the fundamental equation:

$$\frac{P}{T} = c \quad \text{and} \quad \frac{P'}{T'} = c$$

from which we can write:

$$\frac{P}{T} = \frac{P'}{T'} \tag{16-6}$$

Example

Let us assume that we check our tires early in the morning before starting out on a trip that will take us across a desert in midafternoon. The pressure gauge registers a tire pressure of 30 psi when the temperature is 15°C (59°F). By midafternoon the high temperature of the desert air combined with hot pavement and tire friction will easily produce tire temperatures of 100°C (212°F). What will be the pressure in the tires at this temperature? (Assume a negligible change in the volume of the tires. Also note that the pressure inside the tire is the pressure shown by the gauge *plus* the atmospheric pressure.)

Here again, we must convert the temperature to absolute temperature, which then can be substituted into Eq. (16-6).

$$T' = 15°C = (15 + 273)K = 288 \text{ K}$$
$$T = 100°C = (100 + 273)K = 373 \text{ K}$$

The total pressure in the tire initially will be:

$$P' = 30 \text{ psi} + 14.7 \text{ psi} = 44.7 \text{ psi}$$

$$\frac{P}{373 \text{ K}} = \frac{44.7 \text{ psi}}{288 \text{ K}}$$

$$P = \frac{44.7 \text{ psi}}{288 \text{ K}} \times 373 \text{ K} = 57.9 \text{ psi}$$

Finally, the pressure *observed* on the gauge in the afternoon will be:

$$57.9 \text{ psi} - 14.7 \text{ psi} = 43.2 \text{ psi}$$

This is a reasonable answer; the increased temperature has resulted in a higher pressure. These figures have actually been observed by one of the authors of this book. Much higher tire temperatures have also been reported.

16.5
DALTON'S LAW
OF PARTIAL
PRESSURES

John Dalton (who is also known for his contributions to the atomic theory) was, throughout his life, interested in the weather. He made daily observations of weather conditions for nearly 60 years. Thus, it was natural that his early experiments involved measurements of water content (humidity) of the air. He found that when water vapor was added to dry air, the total pressure of the moist air was the sum of the individual pressures of the dry air and the water vapor. He found that the amount of water vapor present in air was related to the temperature and that gases other than air behaved similarly when they were mixed with water vapor. From these observations, he formulated the law we now know as *Dalton's law of partial pressures*, which says that the total pressure of a mixture of gases is equal to the sum of pressures of each gas when measured alone.

For example, if a liter of oxygen with a pressure of 200 torr is added to a liter of nitrogen with a pressure of 400 torr, the new pressure will be 600 torr, provided the volume is still one liter and that the temperature remains constant. We can express these relationships as

$$P_{total} = P_{oxygen} + P_{nitrogen} = 200 \text{ torr} + 400 \text{ torr} = 600 \text{ torr}$$

or, for any gas mixture:

$$P_{total} = P_1 + P_2 + P_3 \cdots \tag{16-7}$$

Dalton's law is simple; it is easy to understand and to use in calculations. But it has a profound meaning theoretically, in helping us understand that molecules are tiny, individual particles. We shall see in Section 16.9 that the concept of partial pressure is an important part of our picture (or theory) of gases and molecules.

Even now, one practical use of Dalton's law in the chemistry laboratory is in calculations involving gases collected over water. Oxygen, hydrogen, and many other gases are prepared in the laboratory and collected by displacement of water from a container. Since water has an appreciable vapor pressure (Section 3.6), any hydrogen or oxygen collected over water is not pure hydrogen or oxygen but actually a mixture of the gas being generated and water vapor. Thus, if we are to make calculations involving the pressure or volume of a gas formed in a reaction, we must correct for the water vapor present. We do this by subtracting the vapor pressure of water, P_w, at the temperature of the experiment from the total atmospheric pressure, P_{total}. The difference is the pressure of the gas, P_g, being collected. This statement put into equation form, starting from Eq. (16-7), would be:

$$P_{total} = P_g + P_w \tag{16-8}$$

or

$$P_{total} - P_w = P_g$$

TABLE 16.2   Vapor pressure of water

Temperature, °C	Vapor Pressure, torr	Temperature, °C	Vapor Pressure, torr
10	9.2	32	35.7
14	12.0	34	39.9
18	15.5	40	55.3
20	17.5	50	92.5
22	19.8	60	149.4
24	22.4	70	233.7
26	25.2	80	355.1
28	28.3	90	525.8
30	31.8	100	760.0

Example   In a laboratory experiment 400 ml of oxygen is collected over water at sea-level pressure (760 torr) and at a temperature of 22°C. What will be the pressure of the oxygen in the container?

Consulting Table 16.2, we find that at 22°C the vapor pressure of water, P_w is 19.8 torr = 20 torr. Using Eq. (16–8) we have:

$$P_{total} = P_O + P_w$$
$$760 \text{ torr} = P_O + 20 \text{ torr}$$
$$P_O = 760 \text{ torr} - 20 \text{ torr} = 740 \text{ torr}$$

16.6
DIFFUSION
OF GASES

The migration of molecules among other molecules is called *diffusion*. We are familiar with the phenomenon of diffusion of gases, since we know that if a gas with a characteristic odor is released on one side of a room, it is soon detectable on the other side. If we invert a container of hydrogen gas over a container of oxygen gas, it can soon be demonstrated that the two gases have mixed. Why should this be so, since hydrogen is much lighter than oxygen? This and similar questions concerned the Scottish scientist Thomas Graham. We can perform a demonstration similar to the experiments he did.

For our demonstration of the diffusion of gases we will use a glass tube that is 3 cm or more in diameter and about 1 m long (Fig. 16.3). Small containers, one filled with concentrated HCl and the other with a concentrated solution of NH_3 in water, are placed at the opposite ends of the tube, and stoppers are inserted at each end of the tube. Some time later a faint white cloud of smoke, which eventually forms a white band around the inside of the tube, is observed somewhat nearer the end of the tube that contains the HCl. We know that such white solid particles of NH_4Cl are formed when HCl reacts with NH_3 according to the equation:

$NH_3 + HCl \rightarrow NH_4Cl$ (16–9)

We interpret our experiment by saying that hydrogen chloride gas and ammonia gas diffuse through the tube, and where they meet, a reaction occurs. The zone of reactions is nearer the hydrogen chloride source. We know that the molecular weight of NH_3 is 17 and the molecular weight of HCl is 36.5. Our conclusion must be that small molecules move faster than large molecules and therefore diffuse faster.

Two important ideas originate from experiments such as these:

1. Gas molecules must be in constant motion, if they can mix with each other and can travel from one place to another.

2. Lighter gas molecules move faster than heavier ones.

A third important result of Thomas Graham's experimentation was that he could assign *quantitative* values to the rates of travel mentioned above. Graham's law states that the rates of diffusion of two gases are inversely proportional to the square roots of their molecular weights. Thus, in the experiment pictured in Fig. 16.3, if we place our containers of HCl and NH_3 exactly 1 m apart in the tube and measure the distance to the zone where NH_4Cl

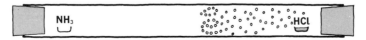

Fig. 16.3   Rates of diffusion of NH_3 and HCl gases.

forms, we will find that it is about 60 cm from the NH_3 and 40 cm from the HCl. Graham's law can predict where the white ring will form in the glass tube. The molecular weight of HCl is 36.5 and that of NH_3 is 17. Using approximations in our calculations, we find that the square roots for the two molecular weights are:

for HCl, $\sqrt{36.5} \simeq 6$; for NH_3, $\sqrt{17} \simeq 4$

Since the rates of diffusion for ammonia, R_A, and for hydrogen chloride, R_H, are inversely proportional to the square roots of their molecular weights we have:

$$\frac{R_A}{R_H} = \frac{\sqrt{\text{M.W.}_H}}{\sqrt{\text{M.W.}_A}} = \frac{\sqrt{36.5}}{\sqrt{17}} \simeq \frac{6}{4}$$

This ratio of the diffusion rates means that ammonia travels about 60 cm while hydrogen chloride travels 40 cm.

16.7 GAY-LUSSAC AND THE LAW OF COMBINING VOLUMES

Gay-Lussac, who shared with Charles the discovery of the effect of temperature on gases and an enthusiasm for ballooning, is credited with another important observation concerning volumes of gases. Gay-Lussac found that when chemical reactions involving gases take place, the reacting volumes are always in the ratios of small whole numbers. This is not so startling when one volume of hydrogen gas reacts with one of chlorine to produce two volumes of hydrogen chloride,

Hydrogen	+	Chlorine	→	Hydrogen Chloride	Hydrogen Chloride

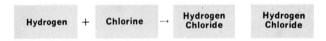

but when two volumes of hydrogen react with one of oxygen to give two of water vapor,

Hydrogen	Hydrogen	+	Oxygen	→	Water	Water

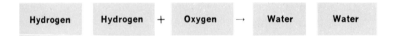

or when one volume of nitrogen gas reacts with three of hydrogen to give only two volumes of ammonia gas, some explanation is needed.

Nitrogen	+	Hydrogen	Hydrogen	Hydrogen	→	Ammonia	Ammonia

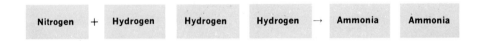

The basic observations were made by Gay-Lussac, but the explanation for them was provided by the Italian scientist Amedeo Avogadro. It is interesting that his explanation was made only a few years after Gay-Lussac reported his generalizations, but Avogadro's explanation was neglected by the scientific community for fifty years. In fact, Dalton, who lived (1766–1844) and worked at the same time as Gay-Lussac and Avogadro, never accepted Gay-Lussac's data as having any real validity, since they seemed to contradict Dalton's atomic theory. For instance, if equal volumes of gases contained equal numbers of atoms, then one atom (liter) of oxygen reacted to form two particles (liters) of water. This would imply that the oxygen atom had actually been split. Such a thing was impossible according to Dalton's atomic theory. The explanation, and the reconciliation of Dalton's atomic theory with Gay-Lussac's law of combining volumes, came from Avogadro and his fellow countryman Cannizzaro, who championed Avogadro's theory after his death.

16.8
AVOGADRO'S
LAW

It had been believed by many that equal volumes of gases (under similar temperatures and pressures) contained the same number of atoms. Gay-Lussac's experiments seemed to contradict this. If we change the statement slightly, however, and say that *equal volumes of gases at the same temperature and pressures contain an equal number of molecules*, we have a statement of **Avogadro's law**. The substitution of the word molecule (we could just as well have used particle) for atom, and the idea that there might be more than one atom in a molecule, made all the evidence consistent with one theory. According to Avogadro's theory we can write the equations for the formation of hydrogen chloride, of water, and of ammonia as follows:

$$\boxed{H_2} \quad + \quad \boxed{Cl_2} \quad \rightarrow \quad \boxed{HCl} \quad \boxed{HCl}$$

$$H_2 \quad + \quad Cl_2 \quad \rightarrow \quad 2HCl \qquad\qquad (16\text{-}10)$$

$$\boxed{H_2} \quad \boxed{H_2} \quad + \quad \boxed{O_2} \quad \rightarrow \quad \boxed{H_2O} \quad \boxed{H_2O}$$

$$\text{vapor}$$

$$2H_2 \qquad + \quad O_2 \quad \rightarrow \quad 2H_2O \qquad\qquad (16\text{-}11)$$

$$\boxed{N_2} \quad + \quad \boxed{H_2} \quad \boxed{H_2} \quad \boxed{H_2} \quad \rightarrow \quad \boxed{NH_3} \quad \boxed{NH_3}$$

$$N_2 \quad + \qquad 3H_2 \qquad\qquad \rightarrow \quad 2NH_3 \qquad\qquad (16\text{-}12)$$

Like all really great theories, Avogadro's not only explained the observations but also led to other predictions that were important and were found to be correct. If we weigh equal volumes of different gases (at the same temperature and pressure, of course) we can determine their relative weights. And since, according to Avogadro's law, these equal volumes contain equal numbers of molecules, the relative weights observed are actually in proportion to the *molecular* weights. If we know the molecular weight of one gas, we can determine the molecular weight of the other.

Example

If a flask that holds exactly 1 g of O_2 will hold 1.375 g of an unknown gas, a molecule of the unknown must weigh 1.375 times as much as an oxygen molecule. Since an oxygen molecule weighs 32 on our relative scale, the molecular weight of the unknown gas ($M.W._u$) is calculated as follows:

$$M.W._u = M.W._o \times \frac{W._u}{W._o} = 32.0 \text{ amu} \times \frac{1.375 \text{ g}}{1.000 \text{ g}} = 44.0 \text{ amu}$$

This value of 44.0 amu happens to be the molecular weight of CO_2. Such a determination does not, however, prove that the gas is CO_2; chemical tests are necessary for this.

16.9
THE KINETIC-
MOLECULAR
THEORY OF
GASES

So far in this chapter we have described the experimentally observed laws governing the relationships of pressure, temperature, and volume of a gas; the fact that gases diffuse; and the fact that volumes of reacting gases combine in small whole-number ratios. Two hundred years passed (1650–1850) while scientists discovered these facts and, as we have seen, even today many useful calculations can be made using these principles—all without any explanation for *why* gases behave as they do. At present, we do have a good understanding of the behavior of gases. Our modern theory makes it much easier to visualize the way gases act and to remember the laws mentioned in this chapter. The gas theories depend upon two simple ideas. First, *gases consist of individual particles, which are molecules*. (In the noble gases, the particles are atoms rather than molecules.) If we add the idea that *gas molecules are in constant motion*, as suggested by Graham's diffusion experiments, we have the basis for the modern kinetic-molecular theory of gases. (Kinetic refers to motion.) The essential features of the **kinetic-molecular theory of gases** are these:

1. Gases are composed of separate particles, called molecules.
2. Molecules are extremely small compared to the space between them, in the gas state. Except at very high pressures, the size of the molecules is insignificant compared with the volume occupied by the gas.
3. The molecules of a gas are in rapid motion, colliding with each other and with the walls of their container on the order of millions of times each second. The pressure produced *by* a gas is the result of these collisions.
4. Molecules are perfectly elastic; no energy is lost by their collisions. However, energy may be transferred from one molecule to another. Temperature is a measure of the *average* kinetic energy of molecules. (Kinetic energy is their energy of motion.) At a given temperature some molecules move faster and some slower, but the average velocity corresponds to that particular temperature.
5. The kinetic energy of all gases is the same, at the same temperature. According to Avogadro's law, equal volumes of different gases contain the same number of molecules. Since the kinetic energy of a particle (molecule) depends on both its mass and its velocity, this means that small molecules (hydrogen) must travel much faster, on the average, than large molecules (oxygen), if their kinetic energies are the same.
6. In the gas state, the attraction between molecules is so small that it may be considered negligible (except when we approach the conditions under which the gas is about to change to liquid). The examples discussed in the following paragraphs show how the kinetic-molecular theory explains the gas laws and illustrate how we can correctly predict the behavior of a gas in various situations.

Example 1   A tank contains one mole of oxygen. Suppose we add another mole, so that the tank contains two moles of oxygen (while maintaining the same temperature). How much will the pressure change? Since the pressure is due to the moving gas molecules bombarding the walls of the tank, if we have twice as many molecules in the same space, the pressure will be exactly twice as great.

Example 2   What is the explanation for Dalton's law of partial pressure? Suppose we have a liter flask containing enough oxygen molecules to exert a pressure of 400 torr, and in another liter flask we have nitrogen molecules, also exerting 400 torr pressure. Now suppose we force the nitrogen into the one-liter flask that holds the oxygen and keep the same temperature. The oxygen molecules will be hitting the walls of the flask as frequently as before and will still exert the same pressure, 400 torr, whether the nitrogen is present or not. However, in the same flask there are now the nitrogen molecules, which are colliding with the walls just as often and are also causing a pressure of 400 torr. So the total pressure in the flask is now 400 + 400 = 800 torr. At a certain temperature the pressure depends entirely on how many molecules are striking the surface. If there are twice as many molecules, there is twice as much pressure. It makes no difference whether the molecules are of two different kinds, as in this example, or are all the same, as in Example 1.

Example 3   Assume we have one mole of hydrogen in a strong metal tank, the volume of which does not change. If we raise the temperature from 200 K to 400 K, how much will the pressure change? The absolute temperature has been doubled, so the molecules are moving twice as fast, on the average. Even though the number of molecules is the same, if they are moving twice as fast they will be hitting the tank twice as often. Therefore they create twice the pressure.

Example 4   If we have a sample of gas which occupies a volume of 4 liters when the external pressure is 1 atm, why does the volume decrease to 2 liters when the external pressure is 2 atm? (see Fig. 16.1a, b). The pressure of the gas will exactly balance the external pressure. When external pressure is increased, the same number of gas molecules are forced to occupy a smaller and smaller space. Finally, when volume has been halved, the *concentration* of molecules will be twice as much as originally. As we have seen before, if there are twice as many molecules in a certain volume, the pressure they exert will be twice as much. So at this point the gas pressure is 2 atm; since this now balances the external pressure, the volume decreases no more. To say it in different terms, pressure is force per unit of area. If the same number of molecules occupy half the volume, there will be twice as many of them hitting the same area, thereby causing twice the force on that area.

Example 5   Imagine that we have one mole of hydrogen in a perfectly elastic balloon, surrounded by normal atmospheric pressure. What will be the change in volume if we raise the temperature from 200 K to 400 K? The volume must change so that the pressure caused by the gas molecules striking the inside of the balloon is still one atm, exactly balancing the surrounding pressure. When the absolute temperature changes from 200 K to 400 K, the molecules will be moving twice as fast. If each molecule is hitting the surface twice as frequently, and we have the same total number of molecules, the *concentration* of molecules will need to be half as great to maintain the same number of collisions on each unit area of the balloon. To make the concentration half as great, the volume must be doubled.

Some of the principles of the kinetic-molecular theory can be extended (with some modifications) to liquids and to solids. When we cool a gas and compress it we decrease the velocity of the molecules and bring them closer together. When the molecules are very close, the attractive forces between them do become significant, and at some stage become so great that the molecules condense to form a liquid. The molecules in a liquid still move, collide, and bounce apart enough that they maintain a random order. If the liquid is further cooled, the molecules move even more slowly, and the attractions between them become even greater. Then they may solidify into the orderly pattern of a crystal, in which the motion of the molecules is negligible.

16.10
STANDARD
CONDITIONS   Since the volume of a gas is affected so distinctly by both pressure and temperature, it means very little to report that we measured a gas volume and found it to be 1.5 liters. It is essential that the conditions of temperature and pressure during the measurement

also be reported. If we are to make meaningful comparisons of one gas volume with another, all observations should be based on the same conditions. Scientists have agreed upon a standard for this purpose. *Standard temperature and pressure (STP) are defined as a temperature of 273 K (or 0°C) and a pressure of 760 torr (or 1 atm)*. It is not necessary, of course, that we actually do an experiment under these conditions. If we know the volume of a gas at, say, 25°C and 745 torr, we can calculate the volume at STP. Or conversely, from someone else's report of a gas volume at STP, we could calculate what the volume would be if we decided to use a different temperature and pressure. The important point is that we must clearly state the temperature and pressure, and preferably use standard conditions.

**16.11
THE GENERAL
GAS LAW**

We have seen that the volume of a gas is inversely proportional to the pressure. From this we found that if the pressure applied to a gas is changed to a different pressure, the change in volume can be calculated from Eq. (16–2). Similarly, Eq. (16–4) allows us to predict the effect of temperature on volume, and Eq. (16–6) the relationship between temperature and pressure. All of these relationships can be combined into one equation, sometimes called the **general gas law**:

$$\frac{PV}{T} = \frac{P'V'}{T'} \qquad\qquad (16\text{–}13)$$

By means of this general equation we can calculate the effect on the volume of a gas if *both* pressure and temperature are changed; or, we can predict the change in pressure if both volume and temperature are changed, and so on. We may wish to predict the change in pressure of a gas when only one factor changes, for instance temperature, while the other factor, volume, remains constant. This means that $V = V'$, so that they cancel each other in Eq. (16–13) and the equation becomes identical to Eq. (16–6). In summary, the general gas law, Eq. (16–13), is the only equation we need for calculating any and all changes involving relationships of pressure, volume, and temperature of a gas. This is illustrated by the following examples.

Example 1

A chemical reference book reports that 1.00 g of CO_2 occupies 509 milliliters at STP (standard temperature and pressure; see Section 16.10). What volume would this CO_2 occupy in your laboratory, when the barometric pressure is 742 torr and the temperature is 24°C?

To solve this problem, let us first list all facts given. The "original" conditions, at STP, are these:

$V' = 509$ ml, $P' = 760$ torr, $T' = 273$ K

The "new" conditions are:

$V = ?$ $P - 742$ torr, $T - (24 + 273)$K $= 297$ K

Now all these values can be substituted into Eq. (16–13):

$$\frac{742 \text{ torr} \times V}{297 \text{ K}} = \frac{760 \text{ torr} \times 509 \text{ ml}}{273 \text{ K}}$$

Algebraic rearrangement gives:

$$V = \frac{760 \text{ torr} \times 297 \text{ K} \times 509 \text{ ml}}{742 \text{ torr} \times 273 \text{ K}} = 567 \text{ ml}$$

Example 2 You have a sealed, flexible plastic bag containing a loaf of bread. At home, when the temperature is 18°C (64°F) and the pressure is 758 torr, the air in the bag has a volume of 473 ml. You travel to a mountain top for a picnic and discover that afternoon that the plastic bag is considerably swollen. What is the volume of air in the bag if the pressure on the mountain is 608 torr and the temperature is 33°C (91°F)?

Solution It is best to carefully label the given facts:

$P' = 758$ torr, $V' = 473$ ml, $T' = (18 + 273)K = 291$ K
$P = 608$ torr, $V = ?$ $T = (33 + 273)K = 306$ K

Using Eq. (16–13) we have:

$$\frac{608 \text{ torr} \times V}{306 \text{ K}} = \frac{758 \text{ torr} \times 473 \text{ ml}}{291 \text{ K}}$$

$$V = \frac{758 \text{ torr} \times 306 \text{ K} \times 473 \text{ ml}}{608 \text{ torr} \times 291 \text{ K}} = 620 \text{ ml}$$

Example 3 At 25°C and 1.00 atm pressure, 127 ft³ of N_2 was compressed into a tank having a volume of 1.09 ft³. What will be the pressure in the tank at 15°C?

Solution Set up the calculation as in the previous examples.

$P' = 1.00$ atm, $V' = 127$ ft³, $T' = (25 + 273)K = 298$ K
$P = ?$ $V = 1.09$ ft³, $T = (15 + 273)K = 288$ K

$$\frac{P \times 1.09 \text{ ft}^3}{288 \text{ K}} = \frac{1.00 \text{ atm} \times 127 \text{ ft}^3}{298 \text{ K}}$$

$$P = \frac{1.00 \text{ atm} \times 127 \text{ ft}^3 \times 288 \text{ K}}{1.09 \text{ ft}^3 \times 298 \text{ K}} = 113 \text{ atm}$$

Example 4 At normal atmospheric pressure and 28°C we have 364 ml of hydrogen. What volume would this sample occupy at STP?

Solution We start with

$P' = 1.00$ atm, $V' = 364$ ml, $T' = (28 + 273)K = 301$ K

At STP we should have:

$P = 1.00$ atm, $V = ?$ $T = 273$ K

In this case we immediately notice that we already had standard pressure, that is, $P' = P$, so these factors drop out of Eq. (16–13) and we are left with:

$$\frac{V}{273 \text{ K}} = \frac{364 \text{ ml}}{301 \text{ K}}$$

$$V = \frac{273 \text{ K}}{301 \text{ K}} = 364 \text{ ml} = 330 \text{ ml}$$

16.12
THE MOLE
VOLUME OF
A GAS

Avogadro's law states that at the same temperature and pressure, equal volumes of gases always contain equal numbers of molecules (Section 16.8). From this principle we can also say that one mole (that is, 6.02×10^{23} molecules) of any gas will always occupy the same volume, at standard conditions. This is called the **mole volume**, V_m. One mole of hydrogen, or one mole of ammonia, or one mole of chlorine should each occupy exactly the same volume, V_m. If the numerical value of V_m is known, many useful calculations are possible. Therefore, the value of V_m has been carefully determined by experiments with a variety of gases. It is only necessary to measure the density of a gas.* From that information and the mole weight, V_m can be calculated, as shown by these experimental findings:

1. The observed density of oxygen, 1.429 0 g/liter, gives these results:

$$V_m = \frac{1.000\ 0\ \text{liter}}{1.429\ 0\ \text{g}\ \text{O}_2} \quad \frac{31.998\ 8\ \text{g}\ \text{O}_2}{\text{mol}\ \text{O}_2} = 22.392\ \text{liter/mol}$$

2. Using the observed density of hydrogen, 0.089 87 g/liter, gives:

$$V_m = \frac{1.000\ 0\ \text{liter}}{0.089\ 87\ \text{g}\ \text{H}_2} \times \frac{2.016\ \text{g}\ \text{H}_2}{\text{mol}\ \text{H}_2} = 22.43\ \text{liter/mol}$$

As expected, these two values of V_m are essentially the same, allowing for some experimental error. Considering many such experiments, the generally accepted value for V_m is:

mole volume of a gas $= V_m = 22.4$ liter/mol, at STP

The next three examples show some of the calculations made possible by the use of V_m.

Example 1   For a certain reaction we wish to use 0.200 mol of ammonia gas. What volume would this occupy at STP?

$$V = 0.200\ \text{mol} \times \frac{22.4\ \text{liter}}{\text{mol}} = 4.48\ \text{liter}$$

Example 2   If we have a two-liter flask filled with chlorine gas at STP, how much chlorine does it contain (a) in moles? (b) in grams?

$$\text{amount of Cl}_2 = 2.00\ \text{liter} \times \frac{1\ \text{mol}}{22.4\ \text{liter}} = 0.089\ 3\ \text{mol}$$

and

$$\text{amount of Cl}_2 = 0.089\ 3\ \text{mol} \times \frac{2(35.5)\text{g Cl}_2}{\text{mol}} = 6.34\ \text{g Cl}_2$$

Example 3   In the laboratory the mole weight of an unknown gas, U, was determined as follows. A small glass bulb weighed 27.495 5 g when empty. When filled with the unknown gas at 22°C and 756 torr, the bulb weighed 27.667 6 g. By previous measurement the volume of the bulb was found to be 94.3 ml. From these data calculate the mole weight of U.

---

* Since a gas is so diffuse at STP, its density is expressed as grams per *liter*, whereas the density of a solid or liquid is given in grams per *milliliter*.

Solution   As usual, we should begin by stating our known information and then see what other information or factors we need to use, to convert it to the desired answer. We know that 94.3 ml of U weighs a certain amount, which is:

wt. of bulb + U = 27.667 6 g
wt. of bulb   = 27.495 5 g
wt. of U   = 0.172 1 g

Thus we can write:

$$U = \frac{0.172\ 1\ g}{94.3\ ml}$$

Eventually we want the mole weight of U, or g/mol, not g/ml. We see that by using the mole volume V_m we could convert volume to moles. But when we look for the value of V_m we see it must be used at STP, so the original volume, 94.3 ml, measured in the lab must be converted to STP, using Eq. (16–13).

$$\frac{760\ torr \times V}{273\ K} = \frac{756\ torr \times 94.3\ ml}{295\ K}$$

$$V = \frac{756\ torr \times 273\ K \times 94.3\ ml}{760\ torr \times 295\ K} = 86.8\ ml$$

So at STP we have:

$$U = \frac{0.172\ 1\ g}{86.8\ ml}$$

and now applying V_m gives:

$$U = \frac{0.172\ 1\ g}{86.8\ ml} \times \frac{22.4\ liter}{mol} \times \frac{1\ 000\ ml}{liter} = 44.4\ g/mol$$

SUMMARY

In this chapter we have seen that laboratory experiments, beginning with those of Robert Boyle, culminated nearly 300 years later in our present kinetic-molecular theory of gases and the general gas law. These help us to explain, and thereby to understand, the behavior of molecules in the gaseous state, and to interpret chemical reactions in which products or reactants are gases. By using quantitative calculations, based on the gas law and the mole volume, it is relatively simple to find the pressure or volume of a gas under different conditions, or to find the molecular weight of a gas.

The kinetic-molecular theory of gases, which also helps us understand liquids and solids, illustrates many of the important aspects of science. When generally known observations become the object of careful experimentation, exact numerical relationships can often be seen, such as those discovered by Boyle, Charles, and Gay-Lussac. The mathematical relationships not only make it possible to predict other events, but indicate an order and regularity in nature. One is then challenged to learn and to understand more about that order. The discovery of facts about the universe is so intellectually exciting it provides one of the major incentives for study and research in the sciences.

IMPORTANT TERMS

absolute zero	diffusion	mole volume of gases
atmosphere	general gas law	standard conditions (gases)
Avogadro's law	kinetic-molecular theory	

WORK EXERCISES

1. The normal temperature of the human body is 98.6°F. What is this on the centigrade scale and on the Kelvin scale?
2. What is absolute zero on the Fahrenheit scale?
3. Convert $-40°C$ to Fahrenheit.
4. What is the meaning of the term "absolute zero"?
5. A sample of oxygen has a volume of 300 ml at a temperature of 30°C. What will be its volume at 0°C.
6. The pressure in an automobile tire reads 40 psi when the tire temperature is 100°C. What will be the gauge pressure in the tire when the tire cools to 25°C.
7. A steel cylinder with a capacity of 10 liters contains hydrogen gas at a pressure of 50 atm. What volume would the hydrogen occupy at 1 atm of pressure?
8. The heating of a sample of potassium chlorate liberated 250 ml of oxygen at a pressure of 730 torr. Assuming no change in temperature, find the volume of the sample at 760 torr.
9. The volume of a sample of nitrogen gas at 720 torr and 20°C is 160 ml. What will be its volume at STP?
10. A balloon holds 1 000 ml of oxygen at a temperature of 0°C. Assuming no change in pressure, find its volume when the temperature is 30°C.
11. A large balloon is filled with helium, which occupies 120 000 cu ft when the atmospheric pressure is 730 torr and the temperature 15°C. What will be the volume of the helium when the balloon climbs to an altitude such that the pressure is 640 torr and the temperature 8°C.
12. A liter of an unknown gas weighs 1.4 g/liter at STP. What is the molecular weight of the gas?
13. The molecular weight of butane is 58. What is the weight of 1 liter at STP?
14. At STP, 5 ml of a gas weigh 0.022 g. What is the molecular weight of the gas.
15. A liter of gas at a temperature of 25°C and a pressure of 730 torr weighs 1.52 g.
    a) What volume will the gas occupy at STP?
    b) What is the molecular weight of the gas?
16. Two balloons are filled to the same size, one with helium and the other with a gas having a molecular weight of 36. If the contents of both balloons are allowed to escape through the same size openings, which gas will escape more rapidly? What will be the comparative rates of escape?
17. In the equation $N_2 + 3H_2 \rightarrow 2NH_3$ how many liters of hydrogen are required to react with 3 liters of nitrogen?
18. What is the volume at STP of 4.2 g of oxygen?
19. Exactly $2.00(10^3)$ ml of nitrogen are collected over water at 25°C and at a pressure of 755 torr. What is the volume at STP?
20. We have $1.50(10^3)$ ml of a gas measured at 20°C and 730 torr; it weighs 5.05 g. What is the molecular weight of the gas?
21. An unknown gas was collected over water at 17°C and at a pressure of 752 torr. It was found that 209 ml of the gas weighed 0.736 g.
    a) What is the volume of the gas at STP?
    b) What is the molecular weight of the gas?
22. During a laboratory experiment 420 ml of hydrogen were collected over water at a temperature of 18°C and a pressure of 755 torr.
    a) What volume will it occupy at STP?
    b) What will be the volume of the hydrogen at 25°C and 730 torr?
23. A 684 ml sample of oxygen was collected over water at 15°C and a pressure of 742 torr. What volume will this amount of dry gas occupy at 1 000 torr and 60°C?
24. "Air pressure" keeps an automobile tire inflated. *How* does the air keep it from collapsing?
25. What volume of oxygen is required to completely burn:
    a) 10 liters of methane?
    b) 3 ft³ of hydrogen?
26. If a 20.0-liter flask contains 0.50 mol CO_2 at 0°C, what is the pressure in the flask?
27. If 0.10 mol O_2 was allowed to escape into a 10-liter container and then 0.20 mol N_2 was added to the same container,
    a) what percentage of the molecules in the container are oxygen?
    b) how many grams of O_2 are present?
    c) how many grams of N_2 are present?
    d) what is the pressure inside the container?

SUGGESTED READING

Davis, H. M., "Low-temperature physics," *Scientific American*, June 1949, pp. 30–39. The usual ceaseless motion of molecules would cease at absolute zero.

Hall, M. B., "Robert Boyle," *Scientific American*, Aug. 1967. Boyle was one of those most responsible for the establishment of the experimental method. This article describes his activities and reproduces illustrations of some of his apparatus.

Mason, E. A., and R. B. Evans III, "Gases in motion: Simple demonstrations of Graham's Law," *Journal of Chemical Education* **46** 358 (1969).

17

Radioactivity and Nuclear Chemistry

When we contemplate the history of chemistry, it is interesting to see how a few theories once widely held were completely overthrown only to emerge again, usually much altered, centuries later. The possibility of the transmutation of one element into another is an example of a theory with this history.

Alchemy, which developed as a fusion of Greek science with some aspects of religion, had many facets. Among them was the idea that baser metals such as lead and mercury could be transformed into gold. Both charlatans and honest, though naive, experimenters claimed some success in this endeavor. All hopes of changing one kind of element to a different one were dimmed when Dalton's atomic theory gained acceptance early in the nineteenth century. According to this theory, atoms of elements could enter into compounds, but the elements themselves were immutable. By the end of the nineteenth century, evidence was accumulating that atoms were not the simple, indivisible, unalterable particles Dalton had proposed, but that they were made of yet simpler units which came to be known as electrons, protons, and neutrons. Even more amazing was the observation that some atoms spontaneously gave off radiation, which was later identified as electrons or even larger particles. If an atom is made of a precise number of protons and electrons and owes its distinctive nature to the precise number of each it contains, what happens when it loses an alpha particle containing 2 protons and 2 neutrons? If mercury, which contains 80 protons, loses a proton, it must become element 79, gold. So the theory has come full cycle, and transmutation is again believable. The tools for this transformation are no longer the alembic and philosopher's stone, but cyclotrons and atomic piles. Modern alchemists, moreover, are interested in reactions more exciting and even more profitable than ones which change mercury to gold. This chapter tells the story of these latter-day alchemists—of their discoveries, their theories, their promise, and the threat to society in these discoveries.

**17.2
THE
DISCOVERY OF
RADIO-
ACTIVITY**

In 1896, Henry Becquerel, a French physicist, accidentally discovered that a uranium compound with which he was working darkened photographic films that were wrapped to protect them from light. It was evident to him that some type of radiation was coming from the uranium. On further investigation, it was found that all uranium compounds, regardless of source, produced similar radiations. Mme. Marie Curie named this phenomenon **radioactivity**.

Mme. Curie and her husband, Pierre, undertook a search for substances which showed radioactive properties, and to their surprise found that ore from which uranium was recovered was more radioactive than the pure uranium compound. As a result of their research, two new elements, polonium and radium, were found (see Section 13.10). With the production of relatively pure isotopes (nuclides) the details of radioactive decay could be studied in greater detail.

TABLE 17.1   Characteristic properties of alpha, beta, and gamma radiations

Ray	Character	Charge	Mass (amu)	Velocity	Penetration
α	He^{2+}	2^+	4.002 6	about 0.1 speed of light	stopped by paper or human skin
β	electron	$1-$	0.000 548	up to 0.9 speed of light	several mm in human tissue
γ	short x-rays	0	0	speed of light	human body or several feet of concrete

Naturally decaying nuclides emit three types of radiation: **alpha (α) particles** which are positively charged helium ions, **beta (β) particles**, which are electrons, and **gamma (γ) rays**, which are shortwave x-rays. Table 17.1 shows characteristic properties of alpha, beta, and gamma radiations. The presence of these radiations can be demonstrated by placing a small quantity of radium within a block of lead. A strong magnet is placed near the opening in the lead block so that escaping radiations must pass through the magnetic field. A photographic plate is placed a short distance away (Fig. 17.1). The positive alpha particles are deflected toward the negative field whereas the beta rays are attracted by the positive field. The gamma rays are unaffected. When developed, the photographic plate shows the areas of radiation impact.

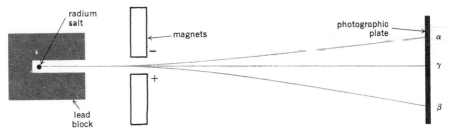

Fig. 17.1   The separation of alpha, beta, and gamma rays by a magnetic field.

It is now known that by losing alpha, beta, and gamma radiations, uranium disintegrates through an elaborate series of transmutations, terminating as the element lead. The decomposition of radium is only one step in this series.

17.3
DECAY BY
RADIOACTIVE
EMISSION

The first step in the disintegration of uranium-238 involves the emission of an alpha particle. This reaction can be represented as

$$^{238}_{92}\text{U} \rightarrow {}^{234}_{90}\text{Th} + {}^{4}_{2}\text{He} \tag{17-1}$$

uranium   thorium   alpha particle
(helium)

The alpha particle comes from the nucleus of the atom, and the reaction is called a nuclear reaction to distinguish it from a chemical reaction, which involves valence electrons. The thorium initially formed decomposes also, only in this instance, a beta particle is given off.

$$^{234}_{90}\text{Th} \rightarrow {}^{234}_{91}\text{Pa} + {}^{0}_{-1}\text{e} \tag{17-2}$$

thorium   protactinium   beta particle
(an electron)

We represent the formation of the electron that is emitted as the beta particle as follows.

$$^{1}_{0}\text{n} \rightarrow {}^{1}_{1}\text{p} + {}^{0}_{-1}\text{e} \tag{17-3}$$

neutron   proton   electron

This reaction indicates that a neutron (in the nucleus) decomposed into a proton, which stayed in the nucleus, and an electron (beta particle), which was emitted. The decay of uranium-238 proceeds through a series of reactions as shown in Table 17.2 and stops only when lead-206, a stable isotope, is formed. In this process some elements emit alpha particles and other beta particles. Emission of gamma radiation often accompanies emis-

TABLE 17.2   The steps in the disintegration of uranium-238

Element	Mass Number	Neutrons	Atomic Number	Particle Emitted	Half-Life*
uranium	238	146	92	α	4.498×10^9 yr
thorium	234	144	90	β	24.10 d
protactinium	234	143	91	β	69 s
uranium	234	142	92	α	2.48×10^5 yr
thorium	230	140	90	α	8.3×10^4 yr
radium	226	138	88	α	1 620 yr
radon	222	136	86	α	3.82 d
polonium	218	134	84	α	3.05 m
lead	214	132	82	β	26.8 m
bismuth	214	131	83	β	19.7 m
polonium	214	130	84	α	1.64×10^{-4} s
lead	210	128	82	β	22 yr
bismuth	210	127	83	β	4.85 d
polonium	210	126	84	α	138 d
lead	206	124	82	stable	stable

* Abbreviations used: yr = years, d = days, m = minutes, and s = seconds.

sion of other particles. Other types of radiation are also observed, particularly from nuclides that have been artificially produced (Section 17.5).

When writing nuclear equations, it is necessary to know the kind of radiation that is given off. Loss of an alpha particle produces a nuclide with an atomic number that is two less and an atomic weight that is four less than the original nuclide. Loss of a beta particle produces a nuclide with an atomic number that is one greater and an atomic weight that is the same as the original nuclide. Loss of gamma radiation does not change the atomic number or atomic weight but results in a nuclide with less energy. The mass units on both sides of a nuclear equation must be equal. This is not true of atomic numbers unless we assign the electron a number of -1, as was done in Eq. (17–3). Other types of nuclear equations will be presented and discussed in Section 17.5.

**17.4
HALF-LIFE OF
RADIOACTIVE
ELEMENTS**

The rates of nuclear reactions, unlike those of chemical reactions, are not altered by environmental changes. The disintegration of radium proceeds as rapidly at $-100°C$ as at high temperatures. Apparently the rates of decay are determined entirely by the stability of the nuclei.

The term used to indicate the rate of decay is the **half-life**, which is the interval of time required for one-half of a given quantity of a radioactive element to decompose. For example, if 1 kg of radium were observed for a period of 1 620 years, it would be found that one-half of it would remain as radium. The other portion would contain all the elements below radium in the uranium disintegration series (Table 17.2). At the end of the next 1 620 years, $\frac{1}{2} \times \frac{1}{2}$—or $\frac{1}{4}$ kg—of radium would remain. It can be seen that disintegration would eventually become minute. The situation is comparable to that in the old conundrum—if a frog is placed at one end of a yard stick and starts jumping toward the other end, each time jumping exactly one-half the distance to the end, how many jumps will he take to reach the end? Although the amount of a radioactive element may become very small, technically it is never all gone. Practically, we say that, after 20 half-lives have elapsed, the remaining radioactivity is negligible.

Table 17.2 reveals that half-life periods vary from millions of years to less than a millisecond. We cannot predict when one individual nucleus will decay. However, in an ordinary sample, containing approximately Avogadro's number of atoms, we can predict statistically that a definite number of the nuclei present will decay within a certain time.

Isotopes with short half-lives give off a large amount of radiation per unit of time and those with long half-lives may seem to hardly be radioactive. For instance, compare a mole of phosphorus-32, whose half-life is 14.3 days, with a mole of uranium-238, whose half-life is 4.5×10^9 years. In both cases, when one half-life has elapsed about 3×10^{23} radiations will have been given off, but these are emitted in 14.3 days by phosphorus-32, whereas it takes 4.5×10^9 years for uranium-238. Radiations from long-lived isotopes may be harder to detect since the isotopes decay so slowly, but they work well for long-term radioactive dating (Section 17.13). Short-lived isotopes are useful for destroying tumors and in other medical applications requiring short doses of radiation. They can be used to trace or destroy without concern that the isotope may stay around and continue to damage normal tissue for years (Section 17.12).

**17.5
TRANSMUTA-
TIONS**

The discovery of radioactivity offered new thresholds of research for chemists and physicists alike, and many of the world's most able scientists were attracted to this field. From a humanitarian standpoint, radium was of tremendous help in the treatment of cancer. From a strictly scientific standpoint, it proved to be an exciting source of minia-ture high-speed projectiles with which to bombard all sorts of matter.

The invention of the **cloud chamber** by Wilson, in 1912, made it possible to actually observe the effect of radiation with the unaided eye. In the cloud chamber a piece of radioactive material is surrounded by air which is supersaturated with moisture; each radiation particle leaves a minute vapor trail as it passes through the air. These vapor trails, or **fog-tracks** as they are commonly called, are produced when moisture condenses around the many ions caused when radiating particles collided with air molecules. These trails are similar to the condensation trails of high-flying jet aircraft. The fog-tracks can be photographed, and a trained scientist can identify the tracks of the different radiating particles (Fig. 17.2).

In 1919, the English physicist Ernest Rutherford tried bombarding various gases with alpha particles emitted by a radioactive substance. To his surprise, after nitrogen was

Fig. 17.2 A cloud chamber photograph showing tracks produced by subnuclear particles. [Courtesy of the Brookhaven National Laboratory.]

bombarded, he discovered that very small quantities of hydrogen were present. The other product, oxygen, was not identified until later. The equation for this reaction is

$$\tfrac{4}{2}\text{He} + \tfrac{14}{7}\text{N} \rightarrow \tfrac{17}{8}\text{O} + \tfrac{1}{1}\text{H} \tag{17-4}$$

Others had merely observed the formation of a new element as the result of natural radioactive decay; Rutherford had created a new element. He had accomplished what the alchemists had dreamed of—a transmutation of elements.

    This discovery so intensified research in radioactivity that within less than fifty years hundreds of transmutations have been carried out. The detection of many subatomic particles, including the neutron, resulted from such studies. A few such reactions are as follows:

$$\tfrac{4}{2}\text{He} + \tfrac{9}{4}\text{Be} \rightarrow \tfrac{12}{6}\text{C} + \tfrac{1}{0}\text{n} \tag{17-5}$$

$$\tfrac{4}{2}\text{He} + \tfrac{27}{13}\text{Al} \rightarrow \tfrac{30}{15}\text{P} + \tfrac{1}{0}\text{n} \tag{17-6}$$

$$\tfrac{4}{2}\text{He} + \tfrac{24}{12}\text{Mg} \rightarrow \tfrac{27}{14}\text{Si} + \tfrac{1}{0}\text{n} \tag{17-7}$$

$$\tfrac{12}{6}\text{C} + \tfrac{1}{1}\text{H} \rightarrow \tfrac{13}{7}\text{N} \tag{17-8}$$

$$\tfrac{40}{20}\text{Ca} + \tfrac{1}{0}\text{n} \rightarrow \tfrac{40}{19}\text{K} + \tfrac{1}{1}\text{H} \tag{17-9}$$

An alternative method of writing the nuclear reactions given above is to use the form: original nuclide (bombarding particles, emitted particle) resulting nuclide. In this system Eqs. (17–5) and (17–9) become

$$\tfrac{9}{4}\text{Be}(\alpha, \text{n})\tfrac{12}{6}\text{C} \tag{17-5a}$$

$$\tfrac{40}{20}\text{Ca}(\text{n}, \text{p})\tfrac{40}{19}\text{K} \tag{17-9a}$$

While this form of equation is simpler, the other form is easier to balance.

17.6 CHARGED-PARTICLE ACCELERATORS

Until 1930, all bombarding of atomic nuclei was brought about by high-speed alpha particles from naturally occurring radioactive isotopes, such as those used by Rutherford. In 1930, E. O. Lawrence, of the University of California, devised a method that involved the artificial acceleration of charged particles, such as protons and deuterium ions. These positively charged particles are repelled by the positively charged atomic nucleus. When they are accelerated they acquire energies much higher than that found in natural radio-active decay and the probability of a nuclear reaction resulting from bombardment is greatly increased.

    Two types of devices are used for bombardment, the cyclotron (which is circular in shape—see Fig. 17.3), and the linear accelerator (Fig. 17.4). The earlier devices were relatively small, being measured in inches, but the ones built more recently are of enormous proportions.

The function of a **cyclotron** may be described as follows. The two half-sections, B and C, are attached to a high-frequency alternating current which alternates the electric charge on each half of the device many times each second. A positively charged particle released at A moves outward and toward B, which at the moment is negative. It accelerates rapidly and, on reaching D, the fields are reversed so that the particle is literally pushed and pulled into

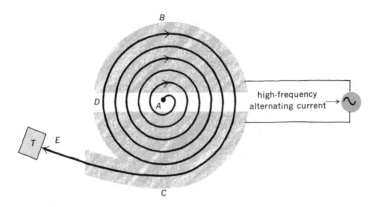

Fig. 17.3 The path of charged particles in a cyclotron.

the other section. This procedure happens millions of times per second until the particle is finally shot through E to a target, T. Because of the effect these particles produce, a device such as this is called an **atom smasher.**

The **linear accelerator** projects a charged particle in a straight line from its source to the target. In the diagram of a linear accelerator shown in Fig. 17.4, the odd-numbered sections will always have charges opposite to those of the even-numbered sections. Charges on sections will be alternating at a high frequency. Suppose a positively charged particle is released at S, when section 1 is negative. It will be attracted into the cylinder but, as it passes through, the electrical fields are changed so that section 1 is positive and section 2 is negative. The speed of the particle will be greatly increased since it is being pulled by section 2 and repelled or pushed by section 1. This process continues, with the particles increasing in speed with each step, until they reach the target, T. The phase of the oscillating field has to be adjusted over the length of the accelerator, of course, so that the field will keep in step with the particles.

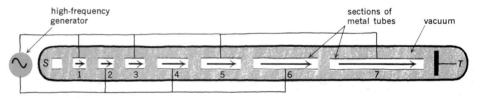

Fig. 17.4 Sketch of a linear accelerator. See also Fig. 17.5.

17.7
MASS–ENERGY
RELATIONSHIP

In the early years of this century, Einstein formulated the famous equation $E = mc^2$, in which E represents energy, m equals mass, and c is the speed of light. Since the speed of light is almost inconceivable (3×10^{10} cm/sec), E must have a still more fantastic value. The equation implies that there is a relationship between matter and energy, and that each may be converted into the other.

At first, only a few people were able to grasp the significance of the equation. It appeared, however, to answer heretofore unanswered questions regarding the source of

Fig. 17.5   An aerial view of the linear accelerator at Stanford University. This accelerator is two miles in length, the largest and longest one in the world. [Courtesy of the Stanford Linear Accelerator Center.]

the sun's energy. It was known that the sun was largely composed of hydrogen, but also contained some helium. On the basis of the Einstein equation, it was reasoned that four hydrogen atoms might in some manner combine to form one helium atom, with a small amount of the matter being converted into energy. In fact, if we compare the masses of four hydrogen atoms, accurate to three or four decimal places, with the accurate mass of a helium atom, we discover a discrepancy in the masses, which could be the source of energy. Equation (17–10) shows the reaction and the masses of the particles (the masses of the electrons have a very small effect and are not being considered). In this reaction, four protons can combine to form a helium nucleus if two of the protons lose their charge to become neutrons; thus, we are left with 2 protons and 2 neutrons, which makes a helium nucleus.

$$4\,{}^{1}_{1}\text{H} \quad \rightarrow \quad {}^{4}_{2}\text{He} + \text{energy} \tag{17–10}$$

$$\begin{array}{cc} 4.032 & 4.003 \\ (4 \times 1.008) & \end{array}$$

During reaction, Eq. (17–10), the mass changed from 4.032 to 4.003, a loss of 0.029 mass units. This decrease in weight has been called the **mass defect** of the helium atom. However, in a reaction of nuclear particles this lost mass is not expelled as particles or small fragments of matter, but instead is transformed to energy. Let us now see how much energy this would be, according to Einstein's equation

In order to have a reasonable quantity of material in our calculation we will, as usual, consider moles. Thus, 4.032 g of hydrogen would yield 4.003 g of helium, with 0.029 g transformed to energy. Using this value in Einstein's equation gives:

$$
\begin{aligned}
E &= mc^2 \\
&= 0.029 \text{ g } (3 \times 10^{10} \text{ cm/sec})^2 \\
&= 2.6\,(10^{19}) \text{ g cm}^2/\text{sec}^2 \\
&= 6.2\,(10^{8}) \text{ kcal} \\
&= 620 \text{ million kcal.}
\end{aligned}
$$

Although 0.029 g is a very small amount of material, it can be transformed to a tremendous amount of energy. The 620 million kilocalories released is enough to heat an average home for about twenty years. If one whole gram of mass, still a small amount, were all transformed to energy, it would heat an average home for hundreds of years! To say it yet another way, one mole of material participating in a nuclear reaction releases approximately one to ten million times as much energy as it does in an ordinary chemical reaction in which valence bonds are formed.

It is now known that energy loss of this type occurs during the formation of nuclei of all elements. Whenever protons and neutrons are packed together to form a nucleus, energy is released. The more energy that is lost, per nuclear particle, the more stable is the resulting nucleus. Therefore, this is called the **binding energy**. The most stable nuclei, that is, those that have lost the most energy during binding, are those of medium size (mass about 40 to 100). Large nuclei are somewhat less stable, and we see them breaking down during radioactive decay. Apparently if too many protons and neutrons are packed together in a nucleus, it cannot be as stable as a medium-sized nucleus. Binding energy is definitely released even during the formation of a large nucleus, but not so much per nuclear particle as is released in the building of nuclei of medium size.

<table>
<tr><td>

**17.8
NUCLEAR
FUSION AND
FISSION**

</td><td>

Since nuclei of medium size are the most stable, there are two ways that nuclei can release energy in the process of being converted to different nuclei. Small nuclei can engage in a **fusion reaction**, combining to form a medium-sized nucleus. Although considerable binding energy was lost when the small nuclei were formed from protons and neutrons, still more energy can be lost when the small nuclei unite to produce a more stable medium-sized nucleus. In the other process, a larger nucleus can release additional binding energy (beyond that lost during the original formation of the nucleus) if it breaks down into two or more medium-sized nuclei. This is called a **fission reaction**.

Both of these nuclear reactions have been studied experimentally and theoretically, especially since World War II. Experimentally, the fission reaction was the one first examined here on earth.

</td></tr>
</table>

Possibly the energy changes associated with nuclear fission and fusion will be understood better in terms of more familiar objects. Suppose it was discovered that the most desirable fruits were those of medium size, say the size of a grapefruit. As a result, a regulation was passed that, whenever possible, people should convert their fruit to the best size.

If we had a watermelon, we could cut it into smaller fragments of the optimum size. Of course, in the process a little juice would be lost. The important difference in nuclear conversions is that the loss could not come out as matter, but would have to emerge as energy.

Or, if we had four oranges, they could be pressed together to make one larger mass of fruit, again with some loss. As before, if this were a nuclear combination, the loss would be emitted as energy.

<table>
<tr><td>

**17.9
THE NUCLEAR
FISSION BOMB**

</td><td>

By 1935, attempts were being undertaken to bombard uranium atoms with neutrons in the hope that the neutrons would be captured in the nucleus, so that new elements, heavier than uranium, might be synthesized. The Italian physicist Enrico Fermi, one of the leaders in this research, produced isotopes which were difficult to identify. In Germany, Meitner and Hahn were obtaining comparable results from similar experiments. Their detection of barium in these radiation products indicated that instead of making larger, heavier atoms, uranium atoms were being broken down into smaller atoms by impact of a neutron.

</td></tr>
</table>

The scientific world was informed of this discovery in 1939, but its true significance was recognized at the time by only a few of the research workers who were studying nuclear structure.

The possibility of obtaining energy by splitting atoms was a new concept. When uranium atoms were bombarded by neutrons and split into two or more smaller atoms, not only was energy produced in great amounts, but there was also a production of excess neutrons. Since a neutron can cause an atom to fission or split, which in turn liberates two or more neutrons, it seemed logical that these freed neutrons could cause other atoms to split. If this continued, a so-called chain reaction would take place in which tremendous numbers of atoms would be split and tremendous amounts of energy released. Such energy, if controlled, would be a formidable weapon in the form of a bomb.

At the time of the publication of the findings of Meitner and Hahn in 1939, the United States was on the verge of entering World War II. Scientists in several countries soon realized the military possibilities of the energy from nuclear fission. There were, however, a great many difficulties in developing and handling such energy.

The American government established a highly secret project, staffed by top-rate scientists, to develop an atomic bomb. It was found that uranium-235 was the fissionable isotope of uranium. Since normally available uranium contained 140 uranium-238 atoms for every 1 of the uranium-235, a difficult task was involved in separating the two kinds of atoms that were identical chemically and very similar physically. Thousands of scientists collaborated to overcome this and many other technical problems involved in the production of the atomic bomb.

The mechanics of a fission bomb depend on the fact that for each fissioning atom of uranium-235, more than one neutron is released, and that these in turn trigger more fissions. If we have a small amount of uranium-235, the size of a ping pong ball, many of the neutrons will escape without striking a nucleus. As the size of the mass of uranium increases, however, the chance that neutrons will escape becomes less. Finally, at some point, known as the **critical mass**, the number of effective (i.e., captured) neutrons per fission is more than one and a chain reaction takes place which liberates literally billions of neutrons almost instantly. The size of a critical mass of uranium-235 is estimated to be approximately that of a baseball. The mechanism in the bomb provides that two or more pieces of uranium-235 be arranged in such a manner that they can be suddenly forced together. Any time the mass of the uranium exceeds the critical mass an explosion occurs. The energy released from each reacting atom in such a device is nearly ten million times the energy released from each atom in chemical explosions. There are sufficient stray neutrons in the atmosphere to provide the first neutron to initiate a chain reaction any time a sample of uranium-235 exceeds the critical mass.

Following the end of World War II, the United States began an extensive program of testing atomic bombs. Many of these tests occurred on or near remote Pacific islands, and later on the Nevada desert north of Las Vegas. In the Pacific, bombs were exploded underwater in the midst of a naval fleet of obsolete warships, to determine their effects on ships and the extent of radioactivity produced. In Nevada, numerous atomic devices were detonated underground, at ground level, and in the air to determine the resultant radiation and concussion intensities at different distances from the detonation point. The Soviet Union carried out similar experiments. Throughout the world, monitoring stations measured considerable increase in atmospheric radioactivity after these explosions. It was feared by some that continued release of such materials would increase the concentration of radioactive species enough to endanger life on earth. Consequently, the explosions of nuclear devices have been greatly curtailed by the United States and the Soviet Union. At the present time, the majority of such explosions are detonated underground where the radioactive debris does not reach the atmosphere.

The diplomatic situation has been complicated recently by the nuclear achievements of China and France. These nations have shown little or no inclination to agree to controlling radioactive pollution of the atmosphere.

17.10
THE NUCLEAR
FUSION BOMB

In Section 17.7, we noted that for some years prior to 1939 it was believed that the energy of the sun resulted from the fusion of hydrogen atoms into helium atoms. It was also believed that such reactions could take place only at the extremely high temperatures existing on the sun. Since no known method of producing such temperatures on earth existed, it was felt that the control of such forces was not available to humans. The development of the atomic bomb, however, created the means of providing temperatures comparable to those of the sun, and for the first time it seemed possible that fusion reactions could be produced on earth.

By including a quantity of deuterium or tritium,* or a mixture of the two, with the

---

* Deuterium occurs in ordinary water, and can be readily isolated (Section 14.4). Tritium is made synthetically by bombarding lithium-6 with a neutron: $^6_3Li + ^1_0n \rightarrow ^3_1H + ^4_2He$.

fissionable material in an atomic bomb, the first hydrogen or **thermonuclear bomb** was produced and exploded in 1952. The advantage of the thermonuclear bomb lies in the fact that the power of the bomb can be increased many times without an increase in the radioactive fallout. The greater power of the fusion bomb results from the higher percentage of material that is converted from matter to energy in the fusion reaction.

17.11
NUCLEAR
POWER FOR
PEACEFUL
PURPOSES

Fissionable materials are increasingly being used in the production of electrical power. Uranium (or some other fissionable material) is utilized in reactors (Fig. 17.6). The fission takes place in uranium metal (or other fissionable substance) which is often present in the shape of rods. The neutrons from fissions that occur in one rod trigger other disintegrations both in that rod and in other rods of the reactor. Low-speed neutrons have proved to be more effective than high-speed neutrons in causing fission. Graphite is one of the substances that tends to decrease the speed of neutrons flowing through it. Consequently, graphite, or a similar neutron-slowing substance, is incorporated in nuclear reactors.

Several other devices are used to control the rate of nuclear fission in a reactor. One of the dangers of a reactor is that it may become too efficient, or that an uncontrolled chain reaction may be initiated which might generate enough heat to melt the components of the reactor. To guard against this, cadmium metal, or some similar material that is a good absorber of neutrons, is also incorporated into a nuclear reactor. The cadmium is present in the form of rods that can be inserted into or withdrawn from the reactor. When inserted, these rods absorb neutrons and thus decrease the rate of nuclear reactions, and withdrawn they allow the reaction to proceed. The cadmium rods act as a control mechanism. Emergency cooling systems are also used to prevent accumulation of heat and a "melt-down."

The energy from the nuclear reactor is converted into heat and absorbed by a fluid of some sort which flows through a heat exchanger and converts water into steam. The steam is then used to generate electricity; it could be used for other types of power. Since the amount of heat generated by a reactor is an indication of the rate of fission in the reactor, automatic control devices have been developed to insert the cadmium rods farther into the reactor when the temperature of the reactor rises and withdraw them at lower temperatures. Devices such as this and various kinds of shielding make nuclear reactors relatively safe sources of power.

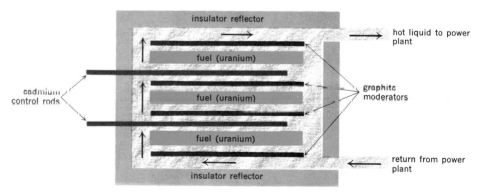

Fig. 17.6 Schematic diagram of the cross section of a nuclear reactor.

The United States, having good supplies of inexpensive fossil fuels (coal and petroleum), has been slow in commercializing large-scale nuclear reactors. However, there has been much research, and many types of nuclear reactors have been developed in the United States and in many other countries. The countries, such as Great Britain, which do not have low-priced fossil fuels, have moved faster than in the United States. Although numerous large nuclear plants are presently under construction in the United States, or are being planned by large power companies, the construction and operation of fission power plants has been postponed in many cases due to problems associated with processing and disposal of nuclear wastes. The safety of such plants has also been questioned.

Nuclear fission results in the formation of many different nuclides. Fuel rods eventually accumulate sufficient amounts of both radioactive and nonradioactive isotopes that their effectiveness is decreased and the reactor ceases to operate efficiently. Used fuel rods as well as other parts of the reactor that have become radioactive must be processed to separate wastes from usable nuclides. The waste nuclides must also be disposed of in some manner that will prevent radioactivity from being released into the environment. As this book is being written there are no commercial plants in the United States that are processing nuclear wastes, and the problem of permanent disposal of wastes is still unsolved. Some states have passed legislation limiting the construction of new nuclear facilities until problems of processing and disposal have been solved.

The **breeder nuclear reactor**, in which fissionable isotopes such as plutonium-239 are generated from nonfissionable isotopes (see Eq. (17–14)), shows some promise, but the problems of separating the fissionable isotopes and disposal of the long-lived isotopes (Pu-239 has a half-life of nearly 25 000 years) have kept such plants from being operated in the United States. European countries have some breeder reactors in operation even though they have not solved the processing or disposal problems at the present time.

The production of power by controlled nuclear fusion is the subject of a great deal of research. The basic fuels for nuclear fusion are the isotopes of hydrogen, which are much more abundant than fissionable isotopes such as uranium-235. The amount of radioactive waste products of fusion should be much, much less than with fission. The major problems that will have to be solved in order to make fusion reactors practical are that of achieving the very high temperatures (millions of degrees) that are needed to start the fusion reaction and that of condensing and confining the hydrogen isotopes so that particles will be close enough to react. The potentialities of nuclear fusion are very great and the finest minds and facilities are being used to solve the problems associated with the process.

While it seems likely that nuclear power will be used primarily for large installations rather than for automobiles and airplanes, miniature nuclear devices are currently being used in navigational satellites and weather stations. Enrico Fermi, who contributed so much to the development of nuclear physics, said, "I believe that the conquest of atomic energy may be readily used to produce not destruction, but an age of plenty for the human race."

**17.12
USES OF
RADIOACTIVE
ISOTOPES**

The presence of radioactive material can easily be detected by a Geiger counter, a device which may, according to the degree of sophistication, crackle, flash a light, or actually record the impulse of escaping rays. During the "big hunt" for uranium at the close of World War II, thousands of Geiger counters were bought by people to use on mountain vacations, as well as by professional prospectors hoping to strike it rich. Radiation detecting devices have also become necessary in industrial and medical research laboratories. With the hundreds of radioactive isotopes and much more sophisticated radiation detection devices now available, technicians and doctors have many ways of using the new techniques in the diagnosis and treatment of disease. Just as a car with its horn honking can readily be located and followed as it winds its way through city traffic, a radioactive isotope, which is continuously broadcasting its disintegration signals, can readily be located or followed in the body with the aid of a radiation detector.

Sodium-24 is a radioactive isotope with a half-life of 14.8 h. If radioactive ions are included in an isotonic salt solution and injected into the bloodstream, they can be used to detect faulty circulation. For example, if a radiation detector gives a higher count for one foot than from the other, we know there is poor circulation in the one with the low count.

Industrially, radioactive isotopes are used widely. Petroleum pipeline companies use plugs containing radioactive material to separate different grades of crude oil in their lines. As the plug moves through the line, Geiger counters monitor its progress. In the steel industry, the thickness of steel plates, which are being rolled, is controlled by a counter recording the rate at which the radiations pass through the plate; thin plates permit more radiation to pass than do thicker plates.

Carbon-14 atoms can be incorporated in organic compounds and then given to experimental animals. The "metabolic fate" of organic compounds such as the common sugar glucose is now well known. It has been found that after being converted to a number of different compounds, carbon-14 ends up as CO_2 which is exhaled from the lungs. (See Work Exercises, Chapter 35 for specific examples of this use of carbon-14.)

Iodine-131, an isotope which has a half-life of eight days, is used in checking the efficiency of thyroid glands. When the gland is functioning properly, it quickly removes iodine from the bloodstream. When a malfunctioning thyroid is suspected, the patient is given a drink of water containing a known amount of iodine-131. Since the normal rate at which the thyroid absorbs iodine is known, a radiation detector placed on the thyroid will indicate whether the thyroid is functioning properly. In cases of cancer of the thyroid glands, iodine-131 can be administered so that the patient will receive continuous radiation treatment. Pure iodine-131 goes almost exclusively to the thyroid gland and destroys the cancerous (along with some normal) tissue. If a thyroid cancer has spread to another part of the body, it will still take up iodine and thus I-131 controls even these scattered growths, called *metastases*. The short half-life of the isotope makes it possible to control radiation dosages. Within a month, less than 10% of the original activity remains.

The isotope technetium-99m (^{99m}Tc) is now widely used for diagnosis of tumors and diseased tissues. This isotope is readily produced from molybdenum-99. (The "m" in the designation indicates that ^{99m}Tc is a metastable isotope which decays to give another form of technetium-99 with a much longer half-life). ^{99m}Tc has a short half-life (about 6 hours) and produces a high-energy gamma radiation. This radiation is easily detected by devices known as scintillation cameras, which can be placed over various parts of the body to detect places where technetium-99m has accumulated. If the isotope is injected in the form of the pertechnate ion (TcO_4^-) it is absorbed in brain tumors and abscesses but not appreciably by normal brain tissue. Other compounds of technetium are taken up by the liver, spleen, or bone marrow. Thus a body scan using a scintillation camera may indicate a high uptake of the isotope and indicate cancerous growth in the brain, liver, spleen, or bone marrow. In other instances, the lack of radioactivity in the kidneys following injection of ^{99m}Tc indicates dead or abnormal tissue. The effectiveness of ^{99m}Tc is primarily due to its intense output of a penetrating radiation that is easily detected and its different absorption depending on the form in which it is administered. The fact that the isotope has a short half-life means that high doses can be given but that their radioactivity will disappear rapidly. Another advantage is that the isotope decays with the release of γ-radiation and that no α or β radiations (which are more likely to damage nearby tissue) are emitted.

The isotope phosphorus-32 is used to trace the movement of phosphate molecules in both plants and animals. Exposing parts of plants or animal bodies to photographic film following administration of ^{32}P gives an indication of the location of radioactive phosphorus, which is the same as the location of ordinary phosphorus. Higher levels of ^{32}P are used to treat chronic leukemia and polycythemia. The radioactive isotope is handled by the body much as is normal phosphorus and ^{32}P is concentrated in the bone. Since bone marrow is the place where blood cells are formed, ^{32}P can destroy the white blood cells which are produced in great quantities in leukemia or the red blood cells which are produced in great quantities in polycythemia. Phosphorus-32 emits a β radiation of moderate energy and has a half-life of 14.3 days. Thus it decays less rapidly than ^{99m}Tc, but more rapidly than many other isotopes. The advantage of using ^{32}P is that it has a relatively short half-life, emits a relatively intense radiation, and is handled the same as the normal nonradioactive isotope ^{31}P by plants and animals.

One of the first uses of radium was in the destruction of cancerous cells. It has now largely been replaced by cobalt-60, which is less expensive and emits radiation that is more intense and penetrating than that from radium. Both radium and cobalt-60 can be administered in the form of metal capsules or needles that are inserted into the body near the tumor that is to be destroyed. With ^{60}Co, the radioisotope is generally encased in a lead container (which absorbs radiation), with a small opening to direct the radiation at the tumor. Following treatment the needles or capsules are removed from the body.

Cobalt-60 is also used in instruments somewhat like x-ray machines which can direct radiation to certain parts of the body. Cobalt-60 has a half-life of 5.3 years, which means that it breaks down relatively rapidly, providing an intense source of radiation. However, this half-life is so short that the instruments that use this isotope have to be recharged periodically with newly formed cobalt-60. Fortunately, cobalt-60 is relatively easy to make by bombarding normal cobalt-59 with neutrons.

Isotopes with relatively short half-lives are used for tracer experiments and medical purposes. As discussed in Section 17.4, a given amount of these isotopes emits its radiation in a short time, making the isotope easy to detect and effective in destroying unwanted tissues. Once these isotopes have decayed, the remaining material is relatively harmless. Longer-lived isotopes are more useful for other purposes, particularly in radioactive dating.

17.13
RADIOCARBON
DATING

There are three isotopes of carbon: C-12, C-13, and C-14. Carbon-14 is a radioactive isotope produced in the higher atmosphere when neutrons from cosmic rays strike nitrogen-14:

$$^{14}_{7}N + ^{1}_{0}n \rightarrow ^{14}_{6}C + ^{1}_{1}H \tag{17-11}$$

It has a half-life of approximately 5 600 years, decaying to form nitrogen-14 and an electron:

$$^{14}_{6}C \rightarrow ^{14}_{7}N + ^{0}_{-1}e \tag{17-12}$$

Carbon dioxide molecules in the atmosphere exist in three isotopic forms in the ratio:

carbon-12: 99%,   carbon-13: 1%,   carbon-14: trace.

Carbon compounds in living tissues retain the same ratio of the three types of carbon atoms.

Dr. Willard Libby, formerly with the Atomic Energy Commission and now with the University of California, conceived the idea that the ages of carbon-containing substances, such as wood, could be determined by their degree of radioactivity which was due to their content of carbon-14. Carbon in recently grown plants undergoes approximately 16 disintegrations per gram per minute. One gram of carbon from wood grown 5 600 years ago (one half-life of C-14) would be expected to show only half as many disintegrations. Thus a determination of the radioactivity of the carbon in any organic substance can help in establishing the date the substance was synthesized. Figure 17.7 shows the correlation between ages indicated by carbon dating and ages established by other methods. Redwood trees serve well as a check for our theory. Even though the tree is alive, there is no exchange of carbon from the atmosphere into the wood within the tree, so we may count the annual rings and check the result with the age established by the carbon-dating method. The older the specimen, the less radioactivity is present, and dating becomes less accurate. Fairly accurate datings going back more than 10 000 years have been made on wood and charcoal.

Recently, discrepancies have been noted in the carbon-14 dating method. These discrepancies are minor on materials less than 3 000 years of age. Older specimens are now believed to be older than they were originally thought to be. The evidence was obtained by comparing the results found by carbon dating and those obtained from annual ring counts of old redwood trees and bristlecone pines. A difference of about 700 years was found in samples over 5 000 years old. We now believe that the amount of carbon-14 in the atmosphere has fluctuated from time to time. This may have been due to fluctuations in sunspots. Since atmospheric testing of nuclear devices began, the level of carbon-14 in the atmosphere has increased drastically.

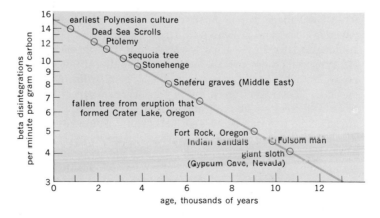

Fig. 17.7 Radiocarbon dating. The beta disintegration rate, noted at the left, fixes the approximate age of the carbon-containing specimens examined. [Adapted from Bruce H. Mahan, *College Chemistry*, Addison-Wesley, 1969.]

Longer-lived isotopes such as uranium-238 are used for dating rocks and other objects older than those that can be dated by carbon-14. The principles are the same as with carbon dating but the useful time span is much longer.

**17.14
SYNTHETIC
ELEMENTS**

By the beginning of 1937, all but four of the first 92 elements had been discovered. On the periodic tables of that time, numbers 43, 61, 85, and 87 were blank spaces. Now all these appear in the table, plus thirteen transuranium elements that have been prepared by synthesis in cyclotrons or linear accelerators. Some typical reactions for these preparations are:

$$\ce{^{98}_{42}Mo} \quad + \quad \ce{^{2}_{1}H} \quad \rightarrow \quad \ce{^{99}_{43}Tc} \quad + \quad \ce{^{1}_{0}n} \tag{17-13}$$

molybdenum + deuterium ion → technetium + a neutron

The name *technetium* is derived from the Greek word meaning artificial; it was the first artificially produced element. In 1940, neptunium and plutonium were prepared by neutron bombardment of the stable uranium-238 isotope, producing the unstable uranium-239 with a half-life of 23.5 minutes:

$$\ce{^{238}_{92}U} + \ce{^{1}_{0}n} \rightarrow \ce{^{239}_{92}U} + \gamma \tag{17-14}$$

Uranium-239 forms neptunium-239, which decomposes to form plutonium-239:

$$\ce{^{239}_{92}U} \xrightarrow[\substack{\text{half-life} \\ \text{23.5 m}}]{\beta} \ce{^{239}_{93}Np} \xrightarrow[\substack{\text{half-life} \\ \text{2.35 d}}]{\beta} \ce{^{239}_{94}Pu} \quad \substack{\text{half-life} \\ \text{24,920 yr}}$$

uranium   neptunium   plutonium

It was soon determined that plutonium-239 has fissionable characteristics similar to those of uranium-235. Plutonium-239 was produced in such quantities that it was used in the second atomic bomb exploded over Japan in 1945.

The elements having atomic numbers from 93 to 103 have been synthesized using various types of bombardment. Elements with numbers 104 and 105 have been reported but conflicting claims are made about their properties (see Section 4.10).

As with the elements discussed in Section 13.10, the discovery of most of the elements 93–103 was made by a comparatively small group of people, including Glenn Seaborg, Albert Ghiorso, and Edwin McMillan, working at the University of California at Berkeley. Although relatively new devices were used to confirm and detect the presence of the new elements, which are unstable and decay rapidly, the major new development was the machines that formed the new elements by bombardment of other elements with various particles. The names given to the new elements reflect both famous scientists and the place of discovery. Thus curium, mendelevium, fermium, einsteinium, and lawrencium are named in honor of Marie and Pierre Curie, Dmitri Mendeleev, Enrico Fermi, Albert Einstein, and Ernest Lawrence. Americium, berkelium, and californium reflect the place of their discovery. While these names might seem unusual, other elements such as germanium, francium, polonium, europium, erbium, terbium, ytterbium, and yttrium are all named for countries, cities, and other units. The last four elements were named for the small Swedish village of Ytterby, which had a quarry from which samples containing ores of these elements were obtained.

It is predicted that some of the even heavier elements may be relatively stable, and attempts are being made to extend the Periodic Table of the Elements to include more and more heavy elements.

17.15
THE DANGERS
OF RADIATION

Much has been said and written about the dangers of using radioactive isotopes and of fallout from nuclear testing. The problem is complex, and there are no simple answers to questions about these dangers. Unfortunately, much of what has been said has been influenced by emotion and by the attitude that either scare tactics or, on the other hand, false complacency is justified because the ends sought are good. The following discussion attempts to present the significant facts that must be considered in making an intelligent evaluation.

We are exposed to many kinds of radiation during our entire lifetime. The radiations of heat and visible light, for example, are usually not especially damaging. Ultraviolet radiation, however, can cause not only sunburn, but in some cases, skin cancer. When a person speaks of radiation damage he generally means damage from high-energy radiation. Alpha, beta, and gamma rays, along with x-rays, may have high energy and great penetrating power which would make them dangerous. X-rays and gamma radiation are particularly potent and likely to produce damage. Ordinarily, alpha and beta radiations have relatively low energies and are absorbed by the skin or clothing; however, the radiation may be dangerous if materials emitting such rays are eaten or inhaled, and thereby incorporated into the human body.

Any high-energy radiation can provide sufficient energy to break a chemical bond. Most of these breakages are harmful. While it may be possible to repair a watch by hitting it with a hammer, it is much more likely that its value as a watch will be permanently destroyed. If the chemical bond is broken in a vital molecule in a vital cell, the consequences may be serious. Cancer can be produced by high-energy irradiation, although the usual effects of exposure to x-rays or gamma radiation is destruction of parts of the tissues, followed by healing and regeneration of tissue.

We cannot avoid exposure to radiation. It reaches us continually from radioactive isotopes in the air we breathe (C-14), from the rocks of the earth's crust (K-40), or from the cosmic rays (ions such as H^+ and He^{2+} traveling at high speeds) of outer space which penetrate the earth's atmosphere. We are also occasionally exposed to x-rays during medical examinations for tuberculosis, broken bones, or dental decay; in these cases, we decide that the benefits outweigh the probable disadvantages.

Although radioactive fallout to date has increased the exposure of the general population by less than 1%, there are instances of concentration of fallout where exposures have been much greater, and the consequences serious. The fact that one of the isotopes produced by nuclear fallout, strontium-90, is found in milk and becomes concentrated in the bones of growing children leads to concern.

The effects of high-energy radiation are random. We cannot predict what any given ray will do. It can produce changes that lead to cancer or abnormal sperm or egg cells, or it may pass completely through the body with no effect. The question is one of probability. Each time we drive a car we face a certain probability of being involved in a fatal accident. Most of us believe the odds are such that we will take a chance. It is probable that the normal background radiation plus that from occasional x-rays and from the fallout up to the present time is not likely to produce serious damage. More research and study is needed, however, to establish the various probabilities involved in exposure to radiation.

On the other hand, there is no reason for complacency regarding the dangers of radioisotopes and fallout. Any exposure involves a potential hazard. Research experiments by high school students involving radioisotopes are probably not worth the potential dangers. Similar experiments by a trained researcher using proper shielding and other safeguards may be worth the risk. The advantages in treating tumors of the thyroid gland with radioactive iodine or in x-ray therapy for malignancies must be weighed against the possible dangers. The values of atomic testing and dangers from fallout are much harder to assess. We cannot predict what the long-range genetic effects of radiation will be.

The dangers of release of radioactivity from nuclear reactors is a subject of dispute. Reputable scientists have differing opinions about the possibility of release of radio-activity from reactors either in normal operation or in an accidental release. There is also disagreement about the effects of low levels of radiation on the general public. Due to the random nature of any damage produced by radiation, predictions are made on the basis of statistical data, and cause-and-effect relationships are hard to establish. As with other choices that a society or an individual makes, positive and negative factors must be carefully considered. There is one point, however, on which most scientists agree—a full-scale nuclear war would be disastrous, and would lead to the end of our civilization and perhaps of all humans.

IMPORTANT TERMS

alpha particle (α)   fog-tracks   nuclear fusion
beta particle (β)   gamma radiation (γ)   nuclear reactor
binding energy   half-life   radioactive isotopes
breeder reactor   linear accelerator   radioactivity
chain reaction   mass defect   radiocarbon dating
cloud chamber   mass-energy relationship   thermonuclear bomb
critical mass   nuclear fission

WORK EXERCISES

1. What discovery first caused doubt to be cast on Dalton's atomic theory?
2. Name the events which led to Madame Curie's discovery of radium.
3. Describe the rays emitted by radium.
4. How does radioactive disintegration differ from ordinary chemical reactions?
5. What is meant by the term *half-life*?
6. Would the whole life of a radioactive element be twice the half-life? Why?
7. Distinguish between the terms *fission* and *fusion*.
8. What does one actually see when fog-tracks are formed in a cloud chamber?
9. Who was the first person to produce the long-dreamed-of transmutation of an element?
10. What is the chief difference between a cyclotron and a linear accelerator?
11. In your own words, what is *mass defect*?
12. What physical condition prevented until 1945 the possibility of a nuclear fusion on the earth?
13. How is the rate of nuclear reactions controlled in a nuclear reactor used for the production of electrical power?

14. Define the term *critical mass* as used in nuclear reactions
15. What is the natural source of carbon-14, the isotope which makes carbon-dating possible?
16. If a sample of carbon from an ancient specimen produces 4 disintegrations/min/g, how old is the specimen?
17. Why are x-rays and gamma radiations more damaging than alpha and beta radiations?
18. Why is it impossible to find an environment free from radiation?
19. Why is strontium-90 a particularly dangerous isotope?
20. What is a major difficulty in harnessing a nuclear fission reaction?
21. What characteristics are desirable in a radioisotope
    a) that is to be used to destroy tumors?
    b) that is to be used as a tracer in the human body?
22. Where would you expect to find ^{90}Sr in the human body? Why?
23. Give the most important uses or applications of the following isotopes.
    a) ^{14}C b) ^{99m}Tc

c) ^{32}P   d) ^{131}I

e) ^{235}U   f) ^{3}H

24. Complete the following by supplying the particle or nuclide in the blank space

a) $^{40}_{20}$Ca (n, p) _____

b) $^9_4$Be (_____, n) $^{12}_6$C

c) $^6_3$Li (n, _____) $^3_1$H

d) _____ (n, γ) ^{60}Co

e) $^{10}_5$B (α, n) _____

f) _____ (n, p) $^{14}_6$C

g) $^{239}_{94}$Pu (α, n) _____

h) $^{238}_{92}$U (_____, n) $^{241}_{94}$Pu

25. Write the equations for three possible fission reactions of uranium-235.

26. Complete the following.

a) ^{238}U $\rightarrow \alpha +$ _____

b) ^{32}P $\rightarrow \beta +$ _____

c) $^{32}_{16}$S + _____ $\rightarrow ^{32}_{15}$P + ^{1}H

d) $^{234}_{90}$Th $\rightarrow \beta +$ _____

e) $^4_2$He + $^{24}_{12}$Mg $\rightarrow ^1_0$n + _____

f) ^{14}C $\rightarrow \beta +$ _____

g) $^{239}_{93}$Np $\rightarrow ^{239}_{94}$Pu + _____

h) $^{214}_{84}$Po $\rightarrow$ _____ + _____

i) $^{98}_{42}$Mo + $^2_1$H _____ + $^1_0$n

27. Give the application or significance of the following reactions that were given in Exercises 24 and 26.

a) 24 (c)   b) 24 (d)

c) 24 (f)   d) 24 (g)

e) 26 (a)   f) 26 (b)

g) 26 (c)   h) 26 (g)

i) 26 (h)   j) 26 (i)

SUGGESTED READING

Bebbington, W. P., "The reprocessing of nuclear fuels," *Scientific American*, December 1976, p. 30. A discussion of a process that is necessary if nuclear fission is to be practicable—the United States has no commercial facilities for this processing. Some of the reasons for this lack are discussed.

Bethe, H. A., "The necessity of fission power," *Scientific American*, January 1976, p. 21 (Offprint #348). The famous scientist says we must use nuclear fission to provide the energy we need.

Cohen, B. L., *Nuclear Science and Society* (pb), Anchor/Doubleday, 1974. This pocketbook provides a good introduction to nuclear reactions and their applications. His chapter on the nuclear controversy is moderate and less pessimistic than some views.

Cohen, B. L., "The disposal of radioactive wastes from fission reactors," *Scientific American*, June 1977, p. 21 (Offprint #364). Dr. Cohen believes there is no real problem involved in disposing of nuclear wastes in deep geological formations. This is a good article although somewhat more optimistic than others on the subject.

Gofman, J., and A. Tamplin, *Poisoned Power* (pb), Signet/New American Library, 1974. Drs. Gofman and Tamplin present a very negative criticism of the use of nuclear energy. Their views are not to be dismissed but differ from those of many scientists.

Kaplan, F. M., "Enhanced-radiation weapons," *Scientific American*, May 1978, p. 44 (Offprint #3007). This article discusses the controversial "neutron bomb." Both the scientific and political aspects of this weapon are discussed.

Valk, P., J. McRae, A. J. Bearden, and P. Hambright, "Technetium radiopharmaceuticals as diagnostic organ-imaging agents," *Journal of Chemical Education* **53**, 542, September 1976. An excellent short article on this important radioisotope.

PART TWO

ORGANIC CHEMISTRY

18

The Nature of Organic Compounds

Human existence on earth is, of course, utterly dependent on the air we breathe, the water in our lakes, rivers, and oceans, and the quality of our soil. But the chemical substances most intimately involved in our lives are organic compounds. The proteins and fats of our tissues, the hormones that regulate our bodily activities, and the carbohydrates we use for energy are all organic compounds. And so it is for all living organisms, be they plants or animals. The vegetables, meats, grains, and fish we eat consist predominantly of organic compounds, with only small amounts of inorganic substances (and, of course, a large quantity of water). The clothes we wear, the paper we write on, and nearly all the fuels we use (except those in nuclear reactors) are organic compounds.

Organic chemistry seems, indeed, to be a fascinating topic. But just what makes a substance an "organic" compound? Let us see, historically, how this question was first answered. Many of our ancestors were equally fascinated by the subject!

By 1800, chemistry had become firmly established as a science, so that during the next half-century there was keen interest in studying the composition of substances and the manner in which one might be changed to another. Because good methods of analysis had been developed, it was possible to determine the composition of a substance with considerable accuracy. The successful analyses stimulated two important results. First they provided the facts which led to the concepts of atomic weights and of atoms as discrete units which combined with one another to form larger particles called molecules. Second, chemists were encouraged to investigate the composition of almost any sort of material they discovered in nature.

As a result of these investigations, chemists began to distinguish two groups of substances. Those derived from plant or animal sources became known as **organic** compounds. All the others, which were obtained in one way or another from the mineral constituents of the earth, were called **inorganic** substances.

The concepts of atoms and molecules, mentioned above, were developed in terms of the inorganic compounds, which had relatively simple formulas. Formulas such as NaOH or $BaCl_2$, once assigned to particular compounds, seemed to represent their constitution very well. First of all, each inorganic compound had a unique formula; only one substance, for example, had formula $BaCl_2$. Second, the ratio of 1 barium to 2 chlorines in the compound was explained by the recently proposed concept of a valence or combining power for each atom. Finally, the Swedish chemist J. J. Berzelius proposed a logical explanation of why barium combined with a substance such as chlorine, rather than with sodium or magnesium. During this period of time, there was great interest in the newly discovered electric currents and in their effect on compounds in water solutions. Chemists had discovered that metals collected at one electrode, and nonmetals at the other electrode. Therefore, Berzelius theorized, a barium atom, because it was positive, was attracted to the negative chlorine atoms to make up the substance $BaCl_2$.

The chemists of the early 1800's were acquainted with a very large number of organic compounds. Some had been known for centuries, some had only recently been isolated, but all were fascinating. Among them were dyes, soap, vinegar, sugar, perfumes, gums, and rubber, to mention but a few. For a variety of reasons chemists were led to believe that these were a distinctly different type of compound than the inorganic substances. Not only were they produced by plants or animals, but more impressive was the knowledge that there was a tremendous number of different organic compounds and that they were made up from very few elements. The elements found were carbon, hydrogen, oxygen, often nitrogen, and occasionally phosphorus or sulfur. How could one imagine that hundreds or perhaps thousands of different organic compounds could be assembled from only three of four kinds of elements? Among the inorganic compounds it was possible to find FeO and Fe_2O_3, Cu_2O, and CuO. However, different combinations between oxygen and one particular metal were limited, and were explained by assuming that iron or copper had two different valences (or, in modern terms, *oxidation states*).

If one studied the formulas of even a small group of organic compounds, it seemed impossible to decide what the combining power of the atoms could be. For example, among compounds containing only carbon and hydrogen, the formulas C_2H_6, C_2H_4, C_3H_6, C_4H_{10}, and C_7H_8 were all discovered, and there were many more. Did this mean that the oxidation number of carbon in the first is 3, in the next two is 2, and in the last two, $2\frac{1}{2}$ or $1\frac{1}{8}$?

18.3 VITAL FORCE AND SYNTHETIC COMPOUNDS

Because of the profusion and complexity of organic compounds, many chemists despaired of ever understanding their nature. Furthermore, no organic compound had ever been made in a laboratory from simpler compounds or from the elements, although a number of inorganic compounds had been made synthetically. Consequently, many people believed that organic compounds were formed only under the influence of a "vital force" orginating in living plants or animals.

In 1828, Friedrich Wöhler made a remarkable discovery at the University of Göttingen in Germany. He attempted to prepare ammonium cyanate by means of a double decom-

(18–1)

position reaction in a solution of ammonium chloride and silver cyanate, both of which were regarded as inorganic substances. Instead of ammonium cyanate, Wöhler obtained crystals of urea! Previously, urea had been obtained only from the urine of animals. Now Wöhler had synthesized it without the presence of the "vital force." The synthesis was confirmed four years later when Wöhler and Justus Liebig succeeded in preparing ammonium cyanate by a different method. It proved to have properties entirely different from those of urea. However, when a solution of ammonium cyanate was boiled down to induce crystallization, the product isolated was always urea!

Within a few years, when acetic acid ($C_2H_4O_2$) and several other organic compounds were prepared by synthesis from inorganic materials, further doubt was cast on the need for a "vital force." As time passed, chemists also synthesized a number of *new* compounds; these were not merely copies of those produced by living plants or animals, but were compounds never before observed on earth. Because of their properties, however, the new compounds seemed clearly to belong in the same class as the organic compounds from natural sources. If it is not necessary for all organic compounds to be associated with living organisms or a "vital force," how then do we define "organic" compounds?

In the mid-1800's, it became apparent that the one factor common to all organic compounds was the element carbon. Therefore, we now say simply that **organic compounds** *are the compounds of carbon*. At the present time, more than three million organic compounds are known; that is, they have been isolated or synthesized, and their names, properties, and formulas have been described in some chemical journal.

18.4 **ISOMERS**	The formulas for urea and ammonium cyanate, shown in Eq. (18–1), reveal that both consist of the same amounts of identical elements, CH_4N_2O. Yet the properties of the two substances are different. So Wöhler's experiment had dispelled the need to believe in a "vital force," but had added a new complexity. Within a decade or two, many other groups of compounds related in this unusual way were discovered. It became clear that *compounds having different properties can have the same composition*. Such related compounds are called **isomers** (Gr. *iso*, equal, plus *meros*, part). This possibility contributes to the great number of different organic compounds. To consider just a few examples, there are three different compounds having formula C_3H_8O, and three compounds having formula C_5H_{12}. Corresponding to the formula $C_4H_{10}O$ there are seven compounds, each differing distinctly from the others in many of its properties.

18.5 **IMPORTANCE** **OF STRUCTURE**	We shall not be able to describe the many efforts made between 1840 and 1860 to solve the puzzle concerning the constitution of organic compounds. Suffice it to say, however, that there was much experimentation by the most skilled chemists of the era, and that many theories were proposed which later had to be discarded when it was found that they were not in harmony with all the observed facts. It turned out that the key to the whole puzzle was the idea of *structure*. We shall describe the concept as we now understand it.

If *covalent bonds* exist between the atoms of a molecule, the atoms are held in some particular arrangement; that is, *the molecule has a structure*. In a water molecule, for example, covalent bonds hold two hydrogen atoms in a specific way to an oxygen atom. Each hydrogen nucleus is 1.0 Ångstrom unit (0.1 nm) from the oxygen nucleus, and the angle between the two bonds is 105°, as shown in Fig. 18.1. The water molecule has a definite structure.

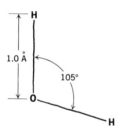

Fig. 18.1 Structure of the water molecule.

In contrast, an ionic compound such as barium chloride exists as ions in either a solution or a crystal. In a solution, two particular chloride ions are not held at definite distances and angles to one certain barium ion. All we know is that there are twice as many Cl^- ions as there are Ba^{2+} ions in the solution. It is true that a *crystal* of barium chloride has a definite structure because the ions are stacked together in a certain pattern, at regular distances from each other. But there is no *molecular* structure; we cannot find distinct groups of atoms we could designate as $BaCl_2$ molecules.

A carbon compound (an organic compound) *has a structure because carbon atoms can form stable* **covalent bonds** *to other carbon atoms and to a number of different atoms.* For instance, a molecule having composition C_3H_8O can have the structure shown in Fig. 18.2(a). In that structure, the covalence of each carbon is 4, each oxygen is 2, and each hydrogen is 1. A formula such as C_3H_8O is a **molecular formula**: it states the number and kinds of atoms in the molecule. A formula such as that in Fig. 18.2(a) is a **structural formula**: it shows which atom is bonded to which.

The fact that a carbon atom can form *covalent* bonds with other atoms leads to an interesting conclusion. It should be possible for three carbons, eight hydrogens, and one oxygen to be arranged into structures different from the one represented by (a) in Fig. 18.2. The other two possibilities are (b) and (c). These three different structures, defined on paper, explain the existence of three separate, real compounds having different properties. They are isomers, because all three have the same molecular formula, C_3H_8O.

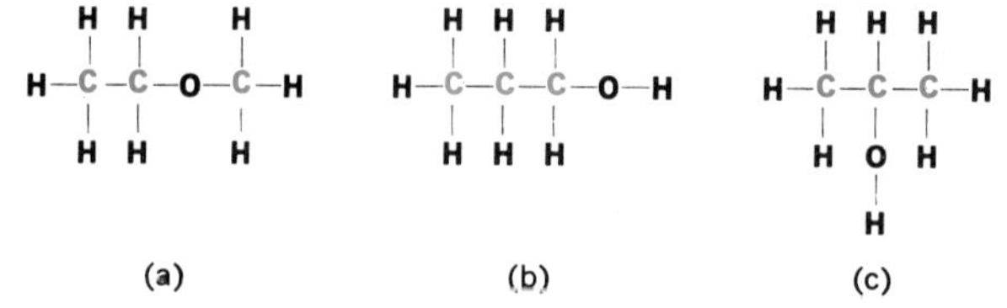

$$
\begin{array}{ccc}
\underset{\underset{H}{|}}{\overset{\overset{H}{|}}{H-C}}-\underset{\underset{H}{|}}{\overset{\overset{H}{|}}{C}}-O-\underset{\underset{H}{|}}{\overset{\overset{H}{|}}{C}}-H
&
H-C-C-C-O-H
&
H-C-C-C-H
\\
(a) & (b) & (c)
\end{array}
$$

Fig. 18.2 Structures of the C_3H_8O isomers.

The concept of structures based on covalent bonds also explains why carbon can combine with other elements in so many different ratios (Section 18.1). In the molecular formulas C_2H_6 and C_4H_{10}, the ratios of the hydrogens to the carbons are different. However, the possible structural formulas for these two substances show that all carbon atoms have a covalence of 4 in both compounds.

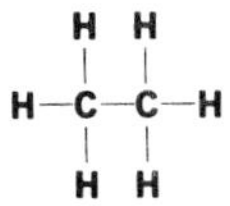

$$
\begin{array}{cc}
\underset{\displaystyle C_2H_6}{H\!-\!\overset{\displaystyle H}{\underset{\displaystyle H}{C}}\!-\!\overset{\displaystyle H}{\underset{\displaystyle H}{C}}\!-\!H}
&
\underset{\displaystyle C_4H_{10}}{H\!-\!\overset{\displaystyle H}{\underset{\displaystyle H}{C}}\!-\!\overset{\displaystyle H}{\underset{\displaystyle H}{C}}\!-\!\overset{\displaystyle H}{\underset{\displaystyle H}{C}}\!-\!\overset{\displaystyle H}{\underset{\displaystyle H}{C}}\!-\!H}
\end{array}
$$

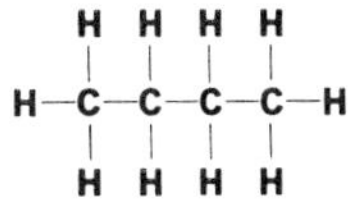

Quite a few elements can form covalent bonds. However, *carbon is unique in its ability to form long chains built up of carbon atoms bonded to one another.* An example is shown in Fig. 18.3(a). The chains may be of varying lengths or they may be branched, as in Fig. 18.3(b). Furthermore, a chain of carbon atoms can twist in such a way that carbons at the ends of the chain come close to each other and can link to form a ring. Fig. 18.3(c) is an example of a ring structure. For simplicity, the atoms occupying the other valences of the carbons in Fig. 18.3 are not shown. The simplified diagrams therefore represent the skeletons of the molecules.

Every organic molecule consists of some kind of carbon skeleton, to which are attached other atoms. Nearly always a large portion of the skeleton is filled out with hydrogen atoms. *The elements most often found in organic compounds, in addition to carbon and hydrogen, are nitrogen, oxygen, and occasionally sulfur.* Many synthetic compounds contain *halogens*, but natural compounds very rarely do. *Phosphorus* is found in certain natural compounds of great importance biologically.

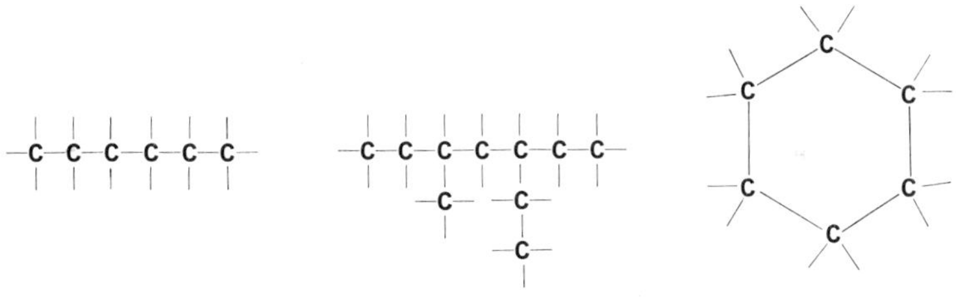

(a) (b) (c)

Fig. 18.3 Typical carbon skeletons in organic molecules: (a) a chain of carbon atoms; (b) a branched chain; (c) a ring.

**18.6
ASSIGNMENT
OF STRUCTURE**

In the previous section, it was shown that structural formulas can correctly account for the number of different compounds that actually exist. But how do we decide which structure, of several possible, represents a real substance that we have in a bottle? We do it by observing enough different properties of the substance so that we can logically say that the facts fit one, and only one, structure. In this way we can relate, or *assign* a structure to a real substance.

For illustration, let us use a simple case in which there are only two possible compounds having the same molecular formula. The first one comes from fermented fruit juice. It is a clear, colorless liquid called alcohol. Experimentation reveals that alcohol vaporizes easily, has a distinctive odor, has a stinging taste, affects the nervous system in a peculiar way, is very soluble in water, and burns with a hot, almost invisible flame. More careful investigation shows that the products of combustion are carbon dioxide and water. If a *weighed* amount of alcohol is burned, the CO_2 and H_2O produced can be captured (separately) and weighed. From these weights can be calculated the amounts of carbon and hydrogen originally present in the alcohol. Its molecular formula is found to be C_2H_6O. However, none of these facts tells us what the structure of alcohol is.

The second compound is obtained by heating wood alcohol (a different sort of alcohol) with concentrated sulfuric acid. The reaction produces a colorless gas, methyl ether. Besides being a gas rather than a liquid, this compound is decidedly different from alcohol in other ways. It is only slightly soluble in water, and its taste and odor are quite unlike those of alcohol. However, methyl ether, like alcohol, burns readily. A combustion analysis reveals that the molecular formula of methyl ether is C_2H_6O, the same as that of alcohol. In other words, the two compounds are isomers.

It is possible to write two, and only two, different structures for formula C_2H_6O; these are shown in Fig. 18.4. The theoretical possibility of two different structures agrees with the fact that two real compounds exist, but does alcohol have structure (a) or (b)? Clearly, more information is required.

Sodium metal provides some interesting facts. Methyl ether does not react with it. But a piece of sodium dropped into alcohol causes vigorous evolution of a colorless gas. The gas is highly flammable and thorough investigation proves that it is hydrogen. If all the alcohol left in the reaction mixture evaporates away, there remains a white, strongly basic solid. It has molecular formula C_2H_5ONa. Furthermore, even if an excess of sodium is added, the only products are hydrogen and the C_2H_5ONa compound. This must mean that one of the six hydrogens of alcohol is bonded differently than the other five. This suggests that the correct structure for alcohol must be (b) rather than (a) in Fig. 18.4.

In fact, this conclusion is especially appealing in view of the whole behavior of alcohol with sodium. It is surprisingly similar to the reaction of water with sodium. Both compounds evolve hydrogen gas and are converted to a basic product. In a water molecule, a hydrogen atom covalently bonded to the oxygen atom can be displaced by sodium, as shown in Eq. (18.2).

$$2\text{H}OH + 2Na \rightarrow 2NaOH + H_2 \tag{18-2}$$
$$\text{(a base)}$$

The fact that structure (b), but not (a), also has one hydrogen covalently bonded to an oxygen apparently explains why alcohol can react with sodium in an analogous way:

$$2\text{H}OC_2H_5 \quad + 2Na \rightarrow 2NaOC_2H_5 + H_2 \tag{18-3}$$
alcohol   (a base)
structure (b)

Several other experimental observations also provide a convincing case for assigning structure (b) to alcohol and (a) to methyl ether.

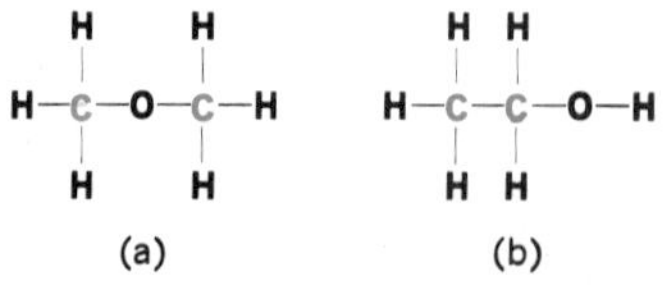

Fig. 18.4 Structures possible for C_2H_6O.

In another case involving a more complex molecule, a greater number of experiments would have to be observed in order to accumulate sufficient facts to assign a structure. Nevertheless, the process of logically comparing the properties of the real compound with a likely structure would be the same as that described above for alcohol and methyl ether. In one way or another, every structural formula you see written in this book, or elsewhere, has been deduced from experimental investigation.

**18.7
CARBON
COVALENT
BONDS**

So far in this chapter a covalent bond between two atoms has been represented by a line. Actually, each covalent bond consists of a pair of electrons shared between the two atoms, as shown for methane in Fig. 18.5. To better understand the properties of organic compounds, it is important to see in more detail how carbon atoms (or other atoms) form covalent bonds. Because covalent bonds arise from a sharing of electrons, it will be necessary to examine more closely the orbitals which the electrons occupy.

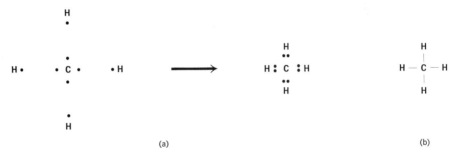

(a)   (b)

Fig. 18.5   Covalent bonding in methane, CH_4. (a) Electron-dot structures showing how the sharing of electrons creates four covalent bonds. (b) Valence-bond structure. Each line represents a pair of shared electrons.

**A. Shapes of
Atomic Orbitals**

Let us briefly review the concept of orbitals: (1) Each electron in an atom is found in a certain space called an orbital. (2) No more than two electrons can occupy one orbital. (3) There are different types of orbitals (regions of space), designated as **s**, **p**, **d**, and **f**. In nearly all organic compounds, only the **s** and **p** orbitals are of concern. Now it is useful to add to these ideas a picture of the *shape* of each kind of orbital space.

By means of some complex mathematical equations it is possible to calculate the limits within which an electron is confined. The calculations reveal the size and shape of the space (the orbital) which each electron inhabits. It is found that **s** electrons occupy a space that is a sphere; see Fig. 18.6(a). The space occupied by **p** electrons is shaped like an hourglass, Fig. 18.6(b). The three **p** orbitals which can exist within each main shell arrange themselves at 90° to each other, as shown in (c) of Fig. 18.6.

Although the calculations describe the space within which a certain electron can be found, they cannot tell *exactly* where the electron is at any particular moment because the electron moves at a fantastically high speed (approximately the speed of light). Therefore an orbital is often described as an "electron cloud." The one or two electrons in each orbital can be thought of as being blurred or dispersed throughout that entire space. This idea is illustrated by the single electron of a hydrogen atom shown in Fig. 18.7, which represents a cross section of the spherical **s** orbital. Note that the electron is not evenly dispersed throughout the space, but is concentrated in a certain region and fades

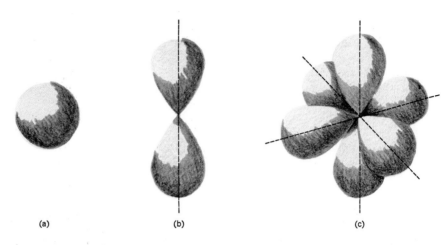

(a)   (b)   (c)

Fig. 18.6   The shape of orbitals (electron clouds). (a) An **s** orbital. (b) One **p** orbital. (c) Three **p** orbitals.

out toward the edge. Electrons in **p** orbitals are similarly dispersed, but form clouds of a different shape.

B. How Covalent Bonds Form

The theoretical calculations also suggest that covalent bonding results when an orbital from one atom overlaps, or fuses, with an orbital from another atom. The fusion creates a new orbital containing the two shared electrons. An example is the union of two hydrogen atoms to form an H_2 molecule, shown in Fig. 18.8. The resultant electron cloud, which surrounds both nuclei and binds them together, is a **molecular** orbital. A cross section of this cloud, perpendicular to the axis between the nuclei, is a circle. Any molecular orbital having a circular cross section is called a **sigma** (σ) molecular orbital.

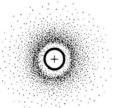

Fig. 18.7   Cross section of the spherical **s** orbital of a hydrogen atom. The electron is pictured as a "cloud" of negative charge dispersed around the nucleus.

(a)   H •   • H $\longrightarrow$ H **:** H

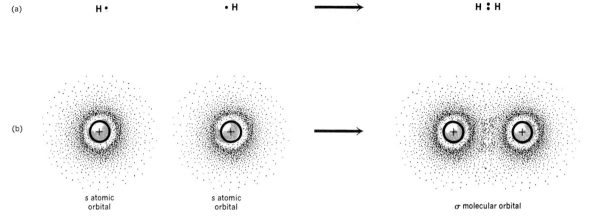

(b)

s atomic
orbital

s atomic
orbital

σ molecular orbital

Fig. 18.8 Formation of the covalent bond in an H_2 molecule. The electron sharing is represented by (a) electron-dot structures and (b) overlap of **s** atomic orbitals from each atom, producing a sigma molecular orbital.

C. Hybrid Orbitals for Bonding Now let us apply these principles of bond formation to a carbon compound, CH_4. Carbon is in the second period of the periodic table. Its first shell is filled with two electrons and it has four valence electrons in the second shell. Electrons in the second shell can occupy either the **s** orbital, which is spherical or one of the three **p** orbitals, which have two lobes (hourglass shape). However, *at the moment of bond formation*, when carbon is going to share its four valence electrons with other atoms, the **s** orbital and the three **p** orbitals

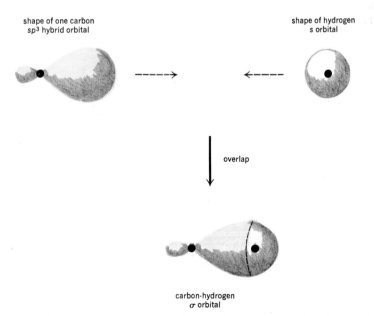

shape of one carbon
sp^3 hybrid orbital

shape of hydrogen
s orbital

overlap

carbon-hydrogen
σ orbital

Fig. 18.9 Formation of one carbon–hydrogen covalent bond. Overlap of atomic orbitals creates a sigma (σ) molecular orbital. (Nuclei are shown as small dots.)

are converted to four *equivalent* hybrid orbitals, called **sp**³ orbitals. Each of these **sp**³ hybrid orbitals contains one electron. Each **sp**³ orbital has two lobes, but one is much larger than the other (Fig. 18.9). For bonding, the larger lobe provides better overlap (and hence a stronger bond) with an orbital of some other atom, for example with the **s** orbital of hydrogen. Figure 18.9 also shows the hypothetical shape of *one* C—H sigma orbital, which represents one of the C—H covalent bonds of methane. Altogether, a carbon atom has four equivalent **sp**³ orbitals, which generate four sigma orbitals in a molecule of methane, shown in Fig. 18.10(a). This figure shows only the larger lobe of each **sp**³ orbital involved in the sigma molecular orbital; the small lobes have been omitted; compare Fig. 18.9.

D. The Tetra-hedral Arrange-ment of Carbon's Molecular Orbitals

One more idea must be included in this description of covalent bonding at a carbon atom. The electrons in one orbital strongly repel electrons in another orbital; therefore, the larger lobes of the four sigma orbitals of CH_4 spread out in space as far as possible from each other. This repulsion between the electron clouds determines the shape of the molecule. If we consider the axis of each sigma bond (that is, the line between the C nucleus and the H nucleus), the angle between any two bonds is 109° (more precisely, 109° 28′). This is shown on the ball-and-stick model in Fig. 18.10(b). All the bond angles and bond lengths of CH_4 are equivalent. If imaginary lines are drawn between the hydrogens, they outline an object that is **tetrahedral** (four-sided). For this reason, any carbon atom having four sigma bonds to other atoms is called a *tetrahedral carbon*; this term means that the four atoms bonded to carbon are spread out in space, at angles of approximately 109° from each other. Although electron-dot and valence-bond structures are often written for carbon compounds, as in Fig. 18.5, the true bond angles are *not* 90°, as those structural formulas might suggest. We must keep in mind that the real molecule has a three-dimensional, tetrahedral shape, as shown by Fig. 18.10, with bond angles of 109°. This idea is essential to understanding the nature of carbon compounds, as will become apparent in the next example.

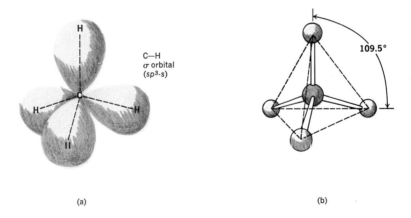

(a)   (b)

Fig. 18.10   Shape of a methane molecule, CH_4. (a) The four sigma orbitals spaced equally far apart. (b) Ball-and-stick model. Imaginary lines connecting ends of bonds outline a regular tetrahedron.

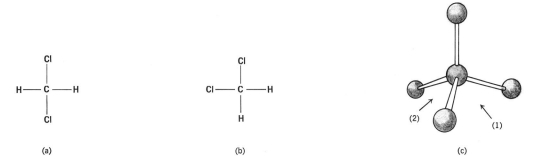

Fig. 18.11   (a, b) Two possible structural formulas for CH_2Cl_2. (c) Three-dimensional model. A view from position (1) is equivalent to (a); a view from position (2) is equivalent to (b).

E. Structural Formulas and Real Molecules

Dichloromethane has molecular formula CH_2Cl_2 and it might seem possible to write for it the two structures shown in Fig. 18.11; in (a) the two chlorines appear to be across from each other, while in (b) they seem to be next to each other and closer together. However, these two-dimensional formulas do not reveal the true arrangement in space of the atoms. Because of the actual tetrahedral arrangement of carbon's bonds, the angle between the two bonds holding the chlorines should always be the same, about 109°. The proper three-dimensional structure of CH_2Cl_2 is shown in Fig. 18.11(c). If we viewed this from position (1) and imagined the structure projected onto a flat surface, we would see formula (a). A view from position (2) would appear as formula (b). Therefore (a) and (b) are *not* two different structural formulas, because they both represent the same three-dimensional molecule. It is just as proper to use either (a) or (b) as a structural formula for dichloromethane. However, when using either one it is important to keep in mind a model of the three-dimensional shape, (c) of Fig. 18.11. There is only one substance having formula CH_2Cl_2 and all of the real molecules are identical.

18.8 ROTATION ON CARBON BONDS

Already we have seen several examples of compounds in which a carbon atom was covalently bonded to another carbon atom. Such a bond would be formed when an **sp**^{3} hybrid orbital from one carbon overlaps with an **sp**^{3} hybrid orbital of the other carbon, giving a molecular orbital often designated as σ (**sp**^{3}–**sp**^{3}). This C—C bond formation is illustrated in Fig. 18.12(a). Like other sigma molecular orbitals, the one between the two carbons has a circular cross section. As a result, in a molecule such as C_2H_6, Fig. 18.12(b), one carbon and its three hydrogens can rotate on the bond axis relative to the other carbon. This does not significantly change the strength of the C—C bond. The overlap of the two **sp**^{3} orbitals, which produced the sigma orbital, is the same in any rotational position because of the circular cross section of the sigma orbital. The possibility of rotation on a carbon–carbon sigma bond is an extremely important characteristic of all carbon compounds.

Some of the consequences of bond rotation are illustrated by the compound called 1,2-dichloroethane.* Three possible models of this compound are shown in (a), (b), and

---

* Dichloro in the name means there are two chlorines in the molecule, and 1,2- indicates that one chlorine is bonded to the first carbon and the other to the second carbon.

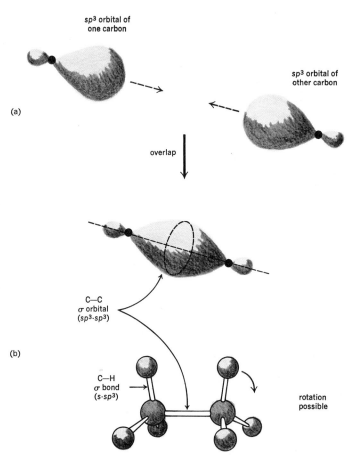

Fig. 18.12 Bond formation in ethane, C_2H_6. Rotation is possible on the C—C sigma bond because of its circular cross section. (a) Overlap of an **sp³** orbital from each carbon forms a sigma molecular orbital. (Each carbon also has three other **sp³** orbitals, not shown, which overlap with **s** orbitals of hydrogen atoms.) (b) Ball-and-stick model.

(c) of Fig. 18.13. Above the ball-and-stick models are space-filling models. The latter show, to scale, the space occupied by each atom; thus, a hydrogen atom is quite small and a chlorine atom is much larger. A space-filling model provides a more accurate view of the size and shape of a molecule, but the bond angles are more clearly seen on a ball-and-stick model. Now the important thing to be observed with these models is this: if rotation can occur on the C—C bond axis, a molecule in form (a) could change to form (b) or (c), without breaking any bonds. In fact, rotation is so easy that every time one molecule collides with another, enough force exists to make it rotate. This means that we cannot isolate, and keep in separate bottles, three different kinds of molecules, (a), (b), and (c), and call them three different compounds. At any given moment a bottle of 1,2-dichloroethane contains all these rotational forms (plus many other intermediate states of rotation), and one form is easily and frequently changed to another.

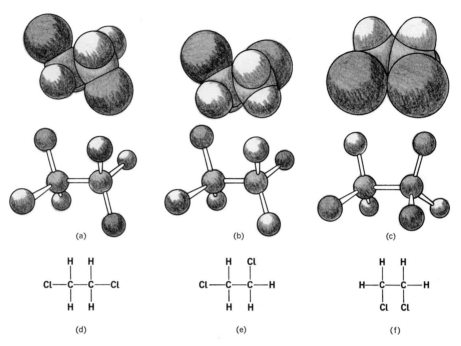

Fig. 18.13 Rotational forms of 1,2-dichloroethane. Three stages of rotation are shown in (a), (b), and (c), each represented by two types of models. Projection of these three forms yields the structural formulas (d), (e), and (f).

It is interesting to note that although various rotational forms are possible, they are not equally favorable. Form (c), Fig. 18.13, is least likely to exist because the bulky chlorine atoms are crowded when they are so close to one another. As a result, if the molecule gets flipped into form (c), it does not remain that way. It will immediately rotate into a less crowded form. Correspondingly, form (a) is the most favorable, and (b) is intermediate.

It is also important to understand how the real molecules and these three-dimensional models relate to structural formulas. On paper we might write formula (d), (e), or (f) shown in Fig. 18.13. At first glance it appears that these are three different structures, each having the chlorines held in different locations. However, a closer look reveals that the structural formulas (d), (e), and (f) actually represent the models (a), (b), and (c). That is, if we view model (a) from above and imagine it projected downward into a flat surface (the paper on which we write), it would be formula (d). So structural formulas (d), (e), and (f) do *not* represent three different compounds; instead, they merely correspond to different rotational forms of one kind of molecule, 1,2-dichloroethane. The significant condition here is that formulas (d), (e), and (f) all represent the same *structure,* which is described by the name 1,2-dichloroethane. In each one, each carbon holds two hydrogens and one chlorine, by the same types of sigma bonds. In contrast, the name 1,1-dichloroethane specifies a different compound having a different structure, Fig. 18.14.

When working with formulas of organic compounds it is essential to recognize which formulas represent true differences in structure and which are merely different rotational forms of the same structure. For example, (d) and (e) of Fig. 18.13 represent the same

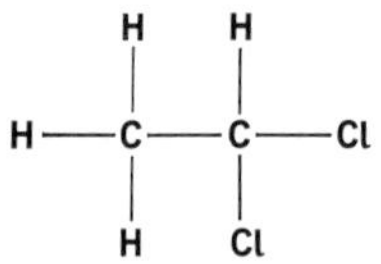

Fig. 18.14 Structural formula of 1,1-dichloroethane, an isomer of 1,2-dichloroethane.

compound, 1,2-dichloroethane. Both (d) and (e) have the same structure. One can be converted to the other by rotation, *without* breaking any bonds or making any new bonds. However, 1,1-dichloroethane (Fig. 18.14) has both chlorines bonded to the same carbon and therefore has a different structure than 1,2-dichloroethane. Because both compounds have molecular formula $C_2H_4Cl_2$ they are isomers. Rotation on the C—C bond does *not* change 1,1-dichloroethane to 1,2-dichloroethane. Instead, 1,1-dichloroethane could be changed to 1,2-dichloroethane only if bonds were broken and new ones formed, so that one of the chlorine atoms could be relocated on the other carbon. This is a reliable test for deciding whether two formulas really represent different structures (different real compounds): *If bonds must be broken to change one formula to the other they are different structures.* In the next section we shall see the important ideas of covalent bonds and bond rotations further applied to writing and understanding structural formulas of organic compounds.

18.9 **WRITING** **STRUCTURAL** **FORMULAS**	A structural formula written in the manner used in the previous sections is also called a **valence bond structure**, because all the bonds in the molecule are shown. Once we have had a little experience with the valence bond structure and understand its relation to a model or to the real molecule, we will often find it convenient to write a more abbreviated structural formula. Most of the bonds on the carbon skeleton of an organic molecule are occupied by hydrogen atoms. Since the hydrogens are so common and are also quite unreactive (as we will see in Section 19.4), a structural formula is often written in the style of Fig. 18.15(b). This is a **condensed structural formula**. In this form, all

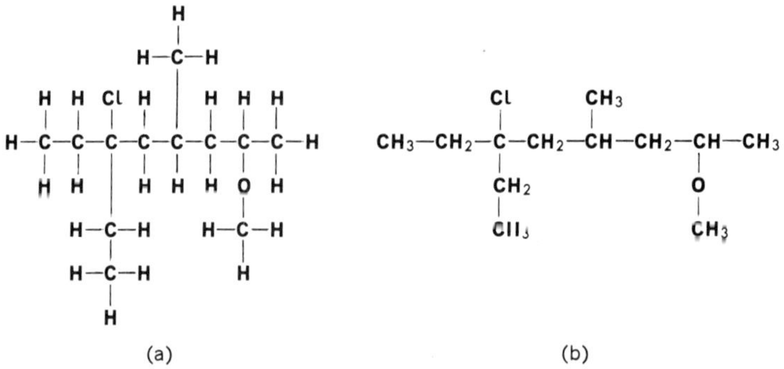

(a)   (b)

Fig. 18.15 Two types of structural formulas: (a) valence-bond structure; (b) condensed structural formula.

hydrogens bonded to a particular carbon are written beside the carbon, but their bonds are not shown; bonds to other atoms are shown however.

The condensed structural formula can be written more rapidly and in less space than the valence bond structure. Also, the form of the carbon skeleton and the location of important groups (those other than hydrogens) can be more easily visualized from the condensed structural formula. This is readily apparent if we compare the valence bond structure of the organic compound (a) with its equivalent condensed formula (b) in Fig. 18.15.

Now let us consider some compounds having the molecular formula C_5H_{12}. It turns out that more than one isomer of this composition is possible. That is, if we had five carbon atoms and twelve hydrogen atoms, each with its normal valence, we could put them together into more than one structure. Although we cannot handle individual atoms in this fashion, we can find the possible structures by using models or by drawing the possibilities on paper, representing each hydrogen with one covalent bond and each carbon with four.

Figure 18.16 shows a valence bond structure and a condensed structural formula for one isomer of C_5H_{12}, together with the corresponding models of the same molecule. Another structural arrangement for five carbons and twelve hydrogens is represented by Fig. 18.17; this isomer is said to have a **branched** chain carbon skeleton. In contrast, the compound in Fig. 18.16 has an **unbranched** structure, often called a **straight** chain. If we inspect a model of the latter compound, Fig. 18.16(b), we see that the chain is not really straight in the geometric sense, but is a zigzag. Therefore the term *straight* is used in a special way for describing organic compounds; it really means *not branched*.

One arrangement for C_5H_{12} that we might think of is diagrammed in Fig. 18.15. At first glance this seems to be different from the structures in either Fig. 18.16 or Fig. 18.17. It is indeed a different structure from that of Fig. 18.17. However, the molecule in Fig. 18.18 has the same structure as the one in Fig. 18.16. Every carbon atom in Fig. 18.18 is bonded to the same other atom as it is in Fig. 18.16. The two molecules are just different

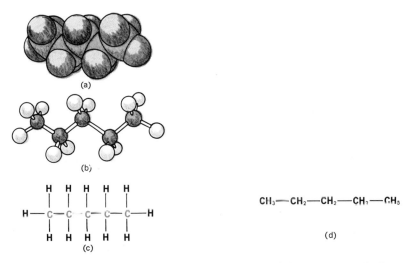

Fig. 18.16 An isomer of C_5H_{12} having a straight chain (an unbranched carbon skeleton): (a) space-filling model; (b) ball-and-stick model; (c) projection, a valence-bond structure; (d) condensed structural formula.

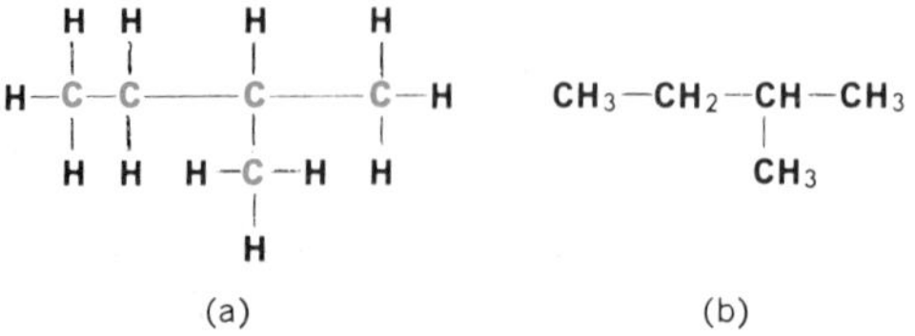

Fig. 18.17 An isomer of C_5H_{12} having a branched chain: (a) valence-bond structure; (b) condensed structural formula.

rotational forms of the same fundamental structure. (The situation is entirely analogous to the various rotational forms of 1,2-dichloroethane, described in Section 18.8.) When the C_5H_{12} molecule of Fig. 18.16 collides with another object, rotation may occur at one of the bonds. The second carbon of the chain may rotate relative to the third carbon, carrying the —CH$_3$ group into the downward location. The molecule would then have the arrangement of Fig. 18.18. To convince yourself that this change is possible merely by rotation on one of the carbon–carbon bonds, test the theory with a model and think about the relation between models and structural formulas.

The relationships between some possible isomers of C_5H_{12}, just described, can be summarized as follows: Molecules having the same molecular formula are isomers (different compounds) if their atoms are bonded differently (Figs. 18.16 and 18.17). On the other hand, if two molecules have only temporary differences in shape, due to rotation of groups on bonds, they are the same compounds (Figs. 18.16 and 18.18).

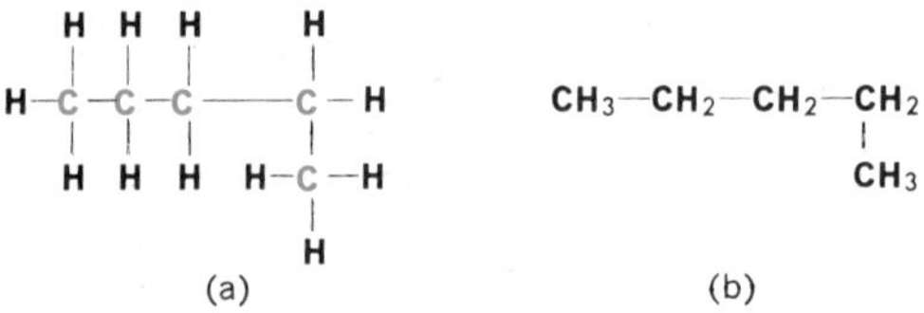

Fig. 18.18 A C_5H_{12} molecule having the same structure as that of Fig. 18.16, but in a different rotational form: (a) valence-bond structure; (b) condensed structural formula.

The existence of isomers creates a problem in naming the different compounds. Clearly, we need a distinctive name for each of the isomers of C_5H_{12}. This problem of nomenclature is discussed in Sections 19.2 and 19.3.

**18.10
RING
COMPOUNDS:
STRUCTURES
AND FORMULAS**

We mentioned in Section 18.4 that two carbon atoms within the same chain could bond together, thus closing a ring. Such ring structures are of frequent occurrence in both natural and synthetic compounds. Therefore, we should see how it is possible for ring compounds to exist if, as we have assumed (Section 18.6), the bonds around each carbon atom have a tetrahedral arrangement.

We will first examine the compound consisting of five carbon atoms in a ring, with ten hydrogen atoms to occupy the other carbon valences. A ball-and-stick model of the compound is shown in Fig. 18.19(a). In this arrangement all five carbons can be in one plane. If we look at the model from above, we see that the carbon skeleton forms a regular

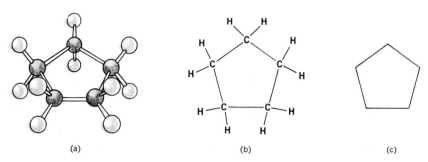

(a)   (b)   (c)

Fig. 18.19   The ring compound C_5H_{10}: (a) ball-and-stick model; (b) valence-bond structure; (c) abbreviated structural formula, showing outline of carbon skeleton.

pentagon, represented in Fig. 18.19(c). Mathematically, the angles within a regular pentagon are each 108°. This is very close to the value of 109° for a tetrahedral bond angle. Therefore the carbon atoms can adapt themselves quite readily to a five-membered ring. Furthermore, if we look again at the ball-and-stick model we see that the H—C—H bond angles around the edge of the ring can each be 109°. These possibilities apparently explain the fact that five-membered rings are quite stable and are easily and frequently formed.

Since it is not always convenient to make an accurate model or drawing of the compound C_5H_{10}, we can use the valence bond formula (Fig. 18.19(b)). As usual, the written valence bond formula does not adequately portray the true three-dimensional nature of the real molecule. In this case, it does not show that the hydrogen atoms are either above or below the plane formed by the carbon atoms. However, it is easy to draw and is satisfactory once we clearly understand what it represents.

Sometimes it is convenient to use the pentagon of Fig. 18.19(c) as an abbreviated formula to represent the compound. When we do, we must imagine that there is one carbon at each corner of the pentagon and that each carbon is holding two hydrogens. In other words, (c) is an abbreviated version of (b).

Figures 18.20 and 18.21 imply that the bond angles of three-and four-membered rings must be distinctly different from those of a five-membered ring. Since a four-membered ring produces a square, the carbon–carbon bond angles must be distorted away from their ideal value of 109°, down to 90°. In the case of the regular triangle formed by a three-membered ring, the carbon–carbon bond angles must be distorted way down to

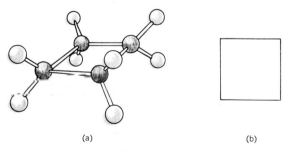

(a)   (b)

Fig. 18.20   The ring compound C_4H_8: (a) ball-and-stick model; (b) abbreviated structural formula.

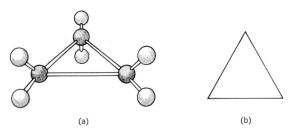

Fig. 18.21   The ring compound C_3H_6: (a) ball-and stick model; (b) abbreviated structural formula.

60°! Experimentally, we find that although such structures can exist, the bond distortion causes strains in the molecules, making them less stable.   A three-membered ring is sufficiently unstable to cause its chemical reactions to be modified.   The strain can cause the ring to break open and react in a way that is not possible in the case of a five-membered ring.

For the ring compound C_6H_{12} we can write the formulas shown in Fig. 18.22. Although we will frequently use these formulas, we find once again that they do not reveal the correct shape of the molecule very well.   If it is really a regular hexagon with all the carbon atoms lying in one plane, the carbon–carbon bond angles within the ring would have to be 120°.   This is far enough from 109° so that we might expect this compound to be strained also.

Actually, the six-membered ring does not have to be planar (that is, with all the carbon atoms in the same plane); it can instead be puckered, as shown by the ball-and-stick model in Fig. 18.23(a).   The rotations permitted by the carbon–carbon bonds allow a carbon skeleton to assume a zigzag shape (compare with Fig. 18.16(b)).   Consequently all the carbon bond angles can remain at their optimum value of 109° when the six-membered ring is formed.   The ring is quite stable, even a little more stable than a five-membered ring.

Figure 18.23(a) is drawn showing each carbon atom with two valence bonds unoccupied.   In the compound C_6H_{12} each of these bonds would be holding hydrogen atoms.   In other organic compounds some of the bonds may be occupied by different atoms or groups, such as —Cl, —OH, or —NH$_2$.   One of these could occupy either an

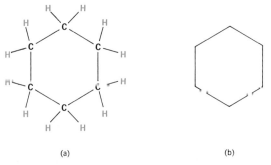

Fig. 18.22   The ring compound C_6H_{12}: (a) valence-bond structure; (b) abbreviated structural formula.

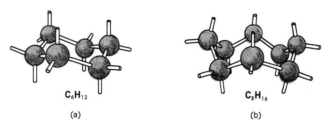

C₆H₁₂

(a)

C₈H₁₆

(b)

Fig. 18.23 Ball-and-stick models of puckered rings (hydrogen atoms not shown): (a) a six-membered ring; (b) an eight-membered ring.

axial or an equatorial position on the puckered carbon skeleton. These subtle differences in position can sometimes have an important effect on the chemical reactivity of six-membered ring compounds, which are common in nature and can be synthesized in the laboratory. Since a chemist or biochemist must sometimes be concerned with such effects, a skeletal formula of the type shown in Fig. 18.24 is often used. It shows the correct bond angles just as well as the model in Fig. 18.23(a).

Even larger rings can form because of the possibility of puckering. There can be some rotation at each carbon–carbon bond, permitting the skeleton to pucker into a shape that will finally allow ring closure without straining any of the bond angles. Figure 18.23(b) shows one possible shape for an eight-membered ring. Rings containing 15, 20, or even 30 atoms are known. There is no theoretical reason why rings of almost any size cannot form.

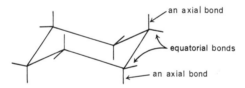

an axial bond

equatorial bonds

an axial bond

Fig. 18.24 Skeletal formula for a six-carbon ring compound. Only carbon valence bonds are shown. The unoccupied bonds could hold a variety of other atoms.

SUMMARY

It is amazing that well over three million organic compounds have been identified, even though the great majority of them are built up from only three or four different elements; carbon, hydrogen, oxygen, and nitrogen. This vast number is possible because all organic compounds contain carbon atoms, which are able to form strong covalent bonds to each other. The manner in which carbon atoms form covalent bonds has been described, as well as several important consequences of such bond formation: (a) three-dimensional structures with 109° bond angles, (b) the possibility of bond rotations producing many different shapes for one molecule, (c) the possibility of ring compounds, and (d) the possibility of isomers.

The most important general principle that has been discussed in this chapter is the idea that organic compounds have definite structures. In this connection it has been necessary to explore quite carefully what a structural formula really represents and how it relates to the actual molecules of a compound.

The discussions in the next chapter, and in all succeeding chapters, will be focused on the structural formula of each type of organic compound, because its structure dictates what all of its properties will be. For this reason, the study of organic chemistry must begin with structural formulas, how to write them, what they mean, and how to think about them. When each new class of compound is discussed we shall also tell how to name the structures.

IMPORTANT TERMS

branched chain	hybrid orbital	organic substances
bond rotation	inorganic substances	structural formula
carbon chain	isomers	sigma orbital
carbon ring	molecular formula	tetrahedral
condensed structural formula	molecular orbital	
covalent bond	orbital (atomic)	

WORK EXERCISES

1. Two isomers having composition C_5H_{12} are shown in Section 18.9. How many other isomers, if any, are possible for C_5H_{12}? Draw their structural formulas (either valence bond or condensed formulas). If you are not certain about the total number of isomers from writing structural formulas, try ball-and-stick models. Check your answer in Section 19.2.

2. From all the formulas shown below, pick out the pairs or groups of formulas which actually represent the same compound.

a) $CH_3CH_2CH_2CH_2CH_3$

b) 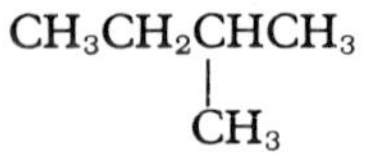 $CH_3CH_2CHCH_3$
   $\quad\quad\quad\;\; |$
   $\quad\quad\quad\;\; CH_3$

c) 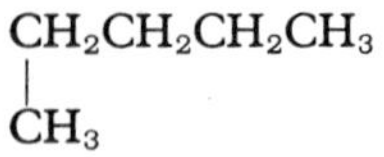 $CH_2CH_2CH_2CH_3$
   $\;\; |$
   $\;\; CH_3$

d)  $CH_3CHCH_2CH_3$
   $\quad\; |$
   $\quad\; CH_3$

e) 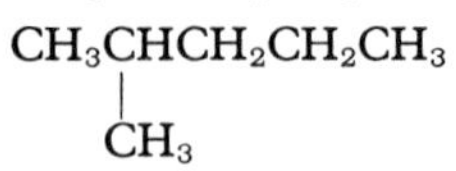 $CH_3CHCH_2CH_2CH_3$
   $\quad\; |$
   $\quad\; CH_3$

f) 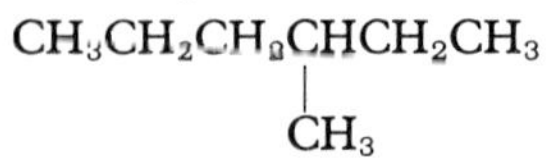 $CH_3CH_2CH_2CHCH_2CH_3$
   $\quad\quad\quad\quad\;\; |$
   $\quad\quad\quad\quad\;\; CH_3$

g) 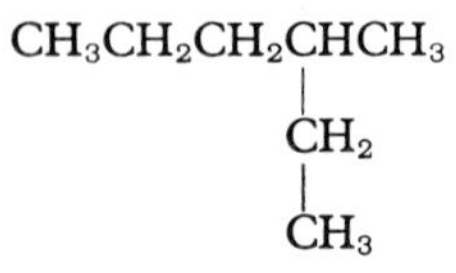 $CH_3CH_2CH_2CHCH_3$
   $\quad\quad\quad\;\; |$
   $\quad\quad\quad\;\; CH_2$
   $\quad\quad\quad\;\; |$
   $\quad\quad\quad\;\; CH_3$

h) 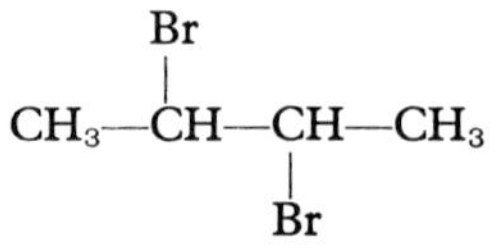 $CH_3\text{—}\overset{\displaystyle Br}{\overset{|}{CH}}\text{—}\underset{\displaystyle Br}{\underset{|}{CH}}\text{—}CH_3$

i) 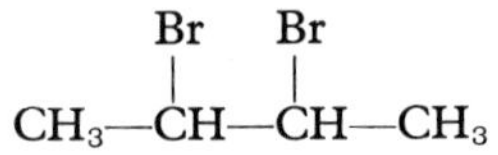 $CH_3\text{—}\overset{\displaystyle Br}{\overset{|}{CH}}\text{—}\overset{\displaystyle Br}{\overset{|}{CH}}\text{—}CH_3$

j) 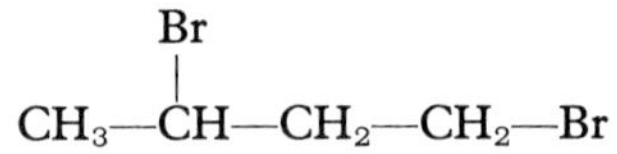 $CH_3\text{—}\overset{\displaystyle Br}{\overset{|}{CH}}\text{—}CH_2\text{—}CH_2\text{—}Br$

k) 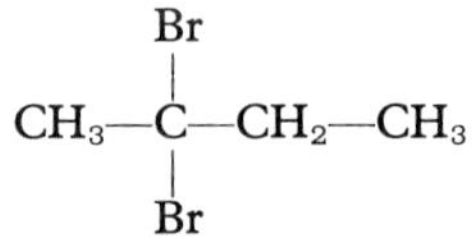 $CH_3\text{—}\overset{\displaystyle Br}{\underset{\displaystyle Br}{\overset{|}{\underset{|}{C}}}}\text{—}CH_2\text{—}CH_3$

l) 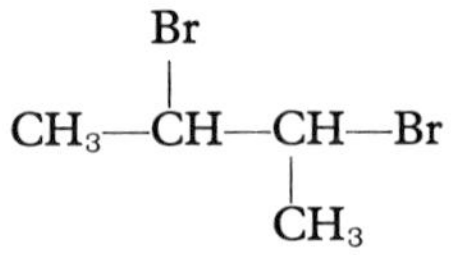 $CH_3\text{—}\overset{\displaystyle Br}{\overset{|}{CH}}\text{—}\underset{\displaystyle CH_3}{\underset{|}{CH}}\text{—}Br$

3. Write structural formulas to show how many different structures are possible for the molecular formulas:

a) C_4H_9Br b) $C_3H_6Cl_2$

c) $C_4H_{10}O$

SUGGESTED READING

Asimov, I., *The New Intelligent Man's Guide to Science*, Basic Books, New York, 1965, pp. 433–438. An excellent brief summary of early organic chemistry, from vitalism until the development of structural formulas.

Kurzer, F., and P. M. Sanderson, "Urea in the history of organic chemistry," *J. Chem. Educ.*, **33**, 452 (1956). The authors state "The early history of urea provides a vivid impression of the slow and difficult beginnings of chemical science . . ."

Lambert, J. B., "The shapes of organic molecules," *Scientific American*, Jan. 1970.

Lipman, T. O. "Wöhler's preparation of urea and the fate of vitalism," *J. Chem. Educ.* **41**, 452 (1964).

19

Saturated Hydrocarbons: The Alkanes

19.1
PETROLEUM,
A SOURCE OF
SATURATED
HYDRO-
CARBONS

Ancient civilizations in many areas of the world made use of the oily materials which were discovered oozing from cracks in the earth. In some localities the crude oil was quite fluid. In others, the seepage was chiefly asphalt (also called pitch or tar). In still other places natural gas escaped and frequently became ignited. These "eternal flames" gave rise to numerous myths and religious cults. Whether the materials were gas, oil, or pitch, they were all related and all arose from deposits of **petroleum** (L. *petra*, rock; *oleum*, oil). It is now generally believed that petroleum was produced millions of years ago by decay of marine organisms. Petroleum is nearly always found in sedimentary rock layers of marine origin.

By 5000 B.C., the Egyptians were already using asphalt as one ingredient for embalming. Asphalt was also well known in the region of the Euphrates river before 4000 B.C., in cities such as Babylon, Ur, and Nineveh. Archeologists have discovered that asphalt was used there as a paint, as mortar to hold bricks, and as waterproofing for jars, baths, and boats. The petroleum fields at Baku, on the Caspian Sea, have apparently been known since before recorded history and are still productive. Prior to 1000 B.C. the burning gas vents there attracted fire-worshipping cults.

The Chinese knew of petroleum more than 2000 years ago. They were perhaps the first to obtain it from drilled wells, although they may originally have been seeking water or salt. Both gas and oil were used for heat and light by the Chinese.

When Europeans began exploring the New World, they discovered that many Indian tribes were using petroleum products. Asphalt was an item of commerce among the Incas. Petroleum was known to Indians in widely scattered areas of North America. Oil was used for war paint, for softening leather and, especially in the northeast, for medication. The Seneca Indians had such a well-established trade with other tribes that their oil used for the treatment of coughs, sores, and bruises was called Seneca oil.

In modern times the first oil wells were drilled during the period 1858–1860 in the United States and Canada. Many others followed rapidly. From that time until 1900, the chief product isolated was kerosene, sold for use in lamps and cookstoves. Smaller amounts of lubricating oils were used. About the time that electric lamps might have put the oil companies out of business, automobiles appeared, causing a new spurt of growth in the oil industry (Fig. 19.1). Refinery practice had to be altered for the production of gasoline.

Fig. 19.1 A modern oil pump and crude-oil storage tanks. [Courtesy of the Shell Oil Company.]

During the period 1860–1900 chemists discovered that nearly all compounds in petroleum deposits were composed of nothing more than carbon and hydrogen. Whether from petroleum or elsewhere, compounds containing only carbon and hydrogen are called **hydrocarbons**. A compound is termed a **saturated hydrocarbon** if all its carbon atoms are saturated with hydrogen atoms; that is, hydrogens occupy all the carbon valence bonds, other than the minimum number of bonds required to link the carbons to one another in the chain. Because one chain can be composed of more carbons than are found in another chain, a whole series of saturated hydrocarbons is possible. Table 19.1 lists the first few members of the series. A complete list would include molecules that contain 10, 20, 50, or more carbon atoms linked together. Such a family of compounds can be found in deposits of petroleum. Saturated hydrocarbons usually constitute well over 90% of petroleum. The remainder often includes hydrogen sulfide and organic sulfur compounds, which have a strong odor. Small amounts of organic nitrogen or oxygen compounds may also be present. Petroleum from certain localities will contain several percent of aromatic compounds (Chapter 21), but many deposits have none. Some further remarks concerning petroleum products are found in Section 19–4.

A family of organic compounds, all of which have the same type of structure, is called a **homologous series** (Gr. *homos*, the same). In this book we will rarely have need to use the term again, but the *idea* of such a series of related compounds is important. Note (Table 19–1) that the molecular formula of any member differs from the one before it or after it by

CH_2. Inspection of the structural formulas reveals the reason for this. To convert the formula of ethane to propane we need to lengthen the chain by one carbon, and in the process need to add only two new hydrogens.

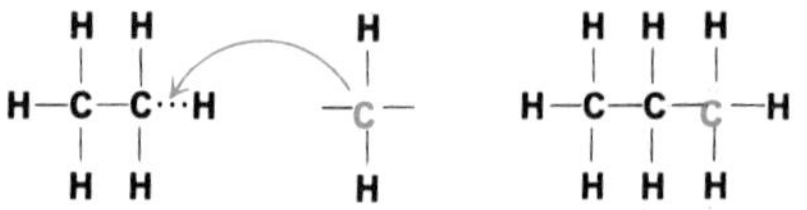

TABLE 19.1   Saturated hydrocarbons—the first four members in the series

Name	Molecular Formula	Structural Formula
methane	CH_4	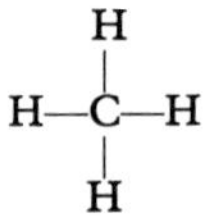
ethane	C_2H_6	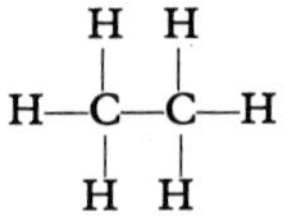
propane	C_3H_8	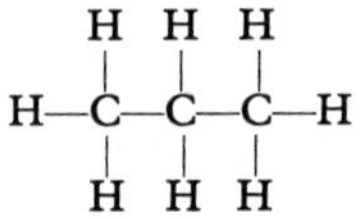
butane (*n*-butane)	C_4H_{10}	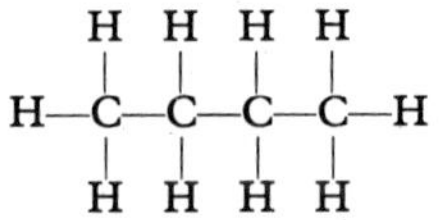

19.2 HISTORICAL DEVELOPMENT OF ORGANIC NAMES

Even as small a hydrocarbon as butane (Table 19.1) causes a problem of naming. A substance of molecular formula C_4H_{10} could also have this structure:

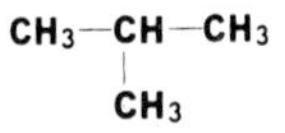

$$CH_3-CH-CH_3$$
$$|$$
$$CH_3$$

Since the latter is an isomer of butane, chemists years ago named it isobutane. To avoid any uncertainty, the unbranched butane (shown in Table 19.1) was designated *normal* butane, or *n*-butane.

The next member in the series is pentane. *Three* different structures, each having formula C_5H_{12}, are possible. All three are real compounds; they have been isolated (from petroleum or other sources) and found to differ in their melting points, boiling points, and so forth. Therefore different names are needed for the isomeric pentanes. The unbranched one was named *n*-pentane, and the one with the simpler branch was called isopentane.

Since in this case there was yet another isomer, a "new" one, it was called neopentane (Gr. *neos*, new).

$$CH_3CH_2CH_2CH_2CH_3 \qquad CH_3CHCH_2CH_3 \qquad CH_3-\overset{\displaystyle CH_3}{\underset{\displaystyle CH_3}{C}}-CH_3$$
$$\overset{|}{CH_3}$$

$$\text{\textit{n}-pentane} \qquad\qquad \text{isopentane} \qquad\qquad \text{neopentane}$$

The situation becomes rapidly more complicated. There are five different structures having formula C_6H_{14}, nine having formula C_7H_{16}, 75 having formula $C_{10}H_{22}$, and 4347 having formula $C_{15}H_{32}$! It would become increasingly difficult to think up appropriate little syllables to designate the various isomers, and even more difficult to remember which one should represent which structure. Clearly, what is needed is a relatively simple, easily remembered, logical system for naming organic compounds. It should be general and inclusive enough that it could apply to any possible organic compound, including those not yet discovered. Such a method for assigning names to organic compounds has been developed through the cooperation of chemists from many different countries. Hereafter we will call such names the **systematic names**.

By the 1880's this problem of nomenclature (naming) for organic compounds had become critical. Chemists had determined the structure of many natural products, including a few complex ones. Also, enough was known about organic reactions to permit the synthesis of completely new compounds, as well as the duplication of some of those discovered in nature. Consequently, the number of chemical structures known had surpassed any convenient method for naming them, and thereby readily communicating knowledge about them to others.

Work to fulfill the need for a systematic method of nomenclature was begun in 1889 by the International Congress of Chemists then meeting in Paris. A few members of the Congress were appointed to a study commission, to develop a set of rules for naming organic compounds. When the Congress next met in 1892 at Geneva, the recommendations of the commission were adopted by the approximately forty chemists present, and became known as the "Geneva rules." For the first time an adequate, internationally accepted system of nomenclature was available.

Thereafter, modifications had to be agreed upon occasionally, and a few new terms added to provide names for certain types of structure which had not been known previously. Some changes were accomplished at a meeting of the International Union of Chemistry (IUC) in 1930. The organization met again in 1949 under the name International Union of Pure and Applied Chemistry (IUPAC). At that time, and at subsequent meetings, the rules have been expanded. For a while the updated rules were called the IUC system; they are now known as the IUPAC system.

**19.3
SYSTEMATIC
NOMENCLA-
TURE OF
ALKANES**

In the systematic method of nomenclature, the saturated hydrocarbons, as a class, are called **alkanes**. Each individual member of the series has been assigned a name which represents the number of carbons in its chain and ends with -*ane*. Thus, pentane is an alk*ane* (saturated hydrocarbon) having five carbons. This name, used alone, designates only the *straight* chain (unbranched) isomer having formula C_5H_{12}. The systematic names for several alkanes are listed in Table 19.2. Except for the first four, nearly all the names were derived from the Greek names for numbers; *nonane* is of Latin origin.

TABLE 19.2   Names, boiling points (at 1 atm pressure), and melting points (if solid at 25°C) of several alkanes

Name	Number of Carbon Atoms	Formula	Boiling Point °C	Melting Point °C
methane	1	CH_4	−161	
ethane	2	CH_3CH_3	−89	
propane	3	$CH_3CH_2CH_3$	−42	
butane	4	$CH_3(CH_2)_2CH_3$	0	
pentane	5	$CH_3(CH_2)_3CH_3$	36	
hexane	6	$CH_3(CH_2)_4CH_3$	69	
heptane	7	$CH_3(CH_2)_5CH_3$	98	
octane	8	$CH_3(CH_2)_6CH_3$	126	
nonane	9	$CH_3(CH_2)_7CH_3$	151	
decane	10	$CH_3(CH_2)_8CH_3$	174	
dodecane	12	$CH_3(CH_2)_{10}CH_3$	216	
tetradecane	14	$CH_3(CH_2)_{12}CH_3$	254	
hexadecane	16	$CH_3(CH_2)_{14}CH_3$	287	
octadecane	18	$CH_3(CH_2)_{16}CH_3$	306	28
eicosane	20	$CH_3(CH_2)_{18}CH_3$	343	37
pentacosane	25	$CH_3(CH_2)_{23}CH_3$		53
triacontane	30	$CH_3(CH_2)_{28}CH_3$		66

For a saturated hydrocarbon *ring* compound the term **cyclo** (Gr. *kyklos*, circle) is included in the name. The general name is *cycloalkane*. Individual compounds have the usual name designating the number of carbons. For example:

is named cyclopentane

One more bit of terminology is necessary in order to name *branched* skeleton hydrocarbons by the systematic method. An **alkyl group** is a unit of structure derived by removing (imaginarily) one hydrogen from an alkane, so that the group has a bond available for attachment to a different atom. Illustrations are given in Table 19.3. Any

TABLE 19.3   Derivation of alkyl group names

particular group is named by using the ending -*yl* in place of the -*ane* ending in the name of the parent alkane. When we wish to generalize, the symbol R is used to designate any alkyl group.

The rules for writing systematic names of alkanes may be summarized as follows:

1. Name the longest carbon chain which can be found in the structure.

2. Preceding the longest-chain name, write down, in alphabetical order, the names of each of the kinds of alkyl groups which are attached to the main chain.

3. If there are several groups of the same kind attached to the main chain, list that group name only once, using the appropriate prefix such as *di, tri, tetra-*, etc., to indicate two, three, or four groups, respectively, of that kind.

4. Assign a number, as a prefix, to *each* of the alkyl groups in the name, to indicate the position of the group on the chain. For this purpose, start numbering the carbon atoms consecutively from either end of the chain, so that the attached groups will have the lowest numbers possible.

The rules are best illustrated by applying them, step by step, to some examples.

Example 1

1. 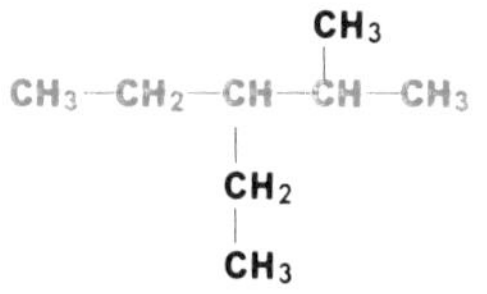
$$CH_3-CH_2-CH-\underset{\underset{CH_2}{\overset{\overset{CH_3}{|}}{|}}}{CH}-CH_3 \qquad \text{pentane}$$
$$\underset{CH_3}{}$$

2.
$$CH_3-CH_2-CH-\underset{\underset{CH_3}{\overset{\overset{CH_3}{|}}{|}}}{CH}-CH_3 \qquad \text{ethyl methylpentane}$$

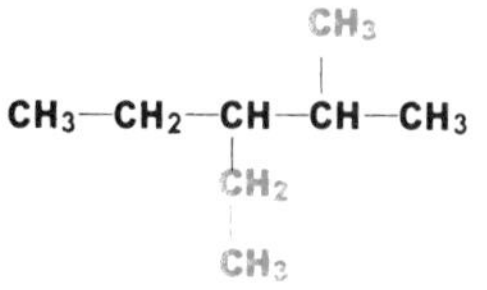

[Rule (3) is not required for this example.]

4. 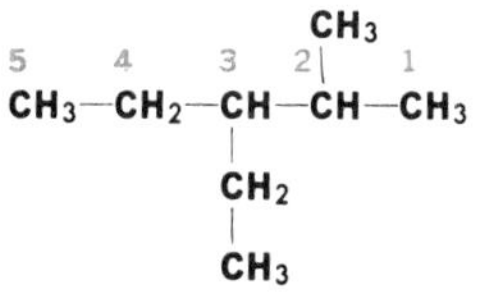
$$\overset{5}{C}H_3-\overset{4}{C}H_2-\overset{3}{C}H-\underset{\underset{CH_3}{\overset{\overset{CH_3}{|}}{|}}}{\overset{2}{C}H}-\overset{1}{C}H_3 \qquad \text{3-ethyl-2-methylpentane}$$

(*Not* 3-ethyl-4-methylpentane, which would result if we counted from the other end of the chain.)

Example 2

1. 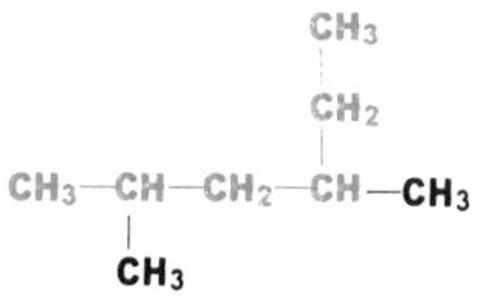
$$CH_3-CH-CH_2-\underset{\underset{CH_3}{}}{CH}-CH_3 \qquad \text{hexane}$$
with $\overset{CH_3}{\underset{|}{CH_2}}$ attached above the fourth carbon

2.
$$
\begin{array}{c}
\quad\quad\quad\quad CH_3 \\
\quad\quad\quad\quad | \\
\quad\quad\quad\quad CH_2 \\
\quad\quad\quad\quad | \\
CH_3-CH-CH_2-CH-CH_3 \\
\quad\quad | \\
\quad\quad CH_3
\end{array}
$$ methylhexane

3.   See preceding structure.   dimethylhexane

4.
$$
\begin{array}{c}
\quad\quad\quad\quad\quad\overset{6}{CH_3} \\
\quad\quad\quad\quad\quad{}^{5}| \\
\quad\quad\quad\quad\quad CH_2 \\
\overset{1}{\ }\quad\overset{2}{\ }\quad\overset{3}{\ }\quad{}^{4}| \\
CH_3-CH-CH_2-CH-CH_3 \\
\quad\quad | \\
\quad\quad CH_3
\end{array}
$$ 2,4-dimethylhexane

Example 3

1.
$$
\begin{array}{c}
\quad\quad\quad CH_3 \quad\quad\quad\quad CH_3 \\
\quad\quad\quad | \quad\quad\quad\quad\quad | \\
CH_3-CH_2-CH-CH-CH_2-C-CH_3 \\
\quad\quad\quad\quad\quad | \quad\quad\quad\quad | \\
\quad\quad\quad\quad\quad CH_2 \quad\quad\quad CH_3 \\
\quad\quad\quad\quad\quad | \\
\quad\quad\quad\quad\quad CH_2 \\
\quad\quad\quad\quad\quad | \\
\quad\quad\quad\quad\quad CH_3
\end{array}
$$ heptane

2.
$$
\begin{array}{c}
\quad\quad\quad CH_3 \quad\quad\quad\quad CH_3 \\
\quad\quad\quad | \quad\quad\quad\quad\quad | \\
CH_3-CH_2-CH-CH-CH_2-C-CH_3 \\
\quad\quad\quad\quad\quad | \quad\quad\quad\quad | \\
\quad\quad\quad\quad\quad CH_2 \quad\quad\quad CH_3 \\
\quad\quad\quad\quad\quad | \\
\quad\quad\quad\quad\quad CH_2 \\
\quad\quad\quad\quad\quad | \\
\quad\quad\quad\quad\quad CH_3
\end{array}
$$ methyl propylheptane

3.   See preceding structure.   trimethyl propylheptane

4.
$$
\begin{array}{c}
\quad\quad\quad CH_3 \quad\quad\quad\quad CH_3 \\
{}^{7}\quad{}^{6}\quad{}^{5}| \quad{}^{4}\quad{}^{3}\quad{}^{2}| \quad{}^{1} \\
CH_3-CH_2-CH-CH-CH_2-C-CH_3 \\
\quad\quad\quad\quad\quad | \quad\quad\quad\quad | \\
\quad\quad\quad\quad\quad CH_2 \quad\quad\quad CH_3 \\
\quad\quad\quad\quad\quad | \\
\quad\quad\quad\quad\quad CH_2 \\
\quad\quad\quad\quad\quad | \\
\quad\quad\quad\quad\quad CH_3
\end{array}
$$ 2,2,5-trimethyl-4-propylheptane

The purpose of the name is to describe accurately to another person the structure we have in mind. If the naming rules are satisfactory (and are properly used) it should be impossible to interpret them in such a way that the person would write a *different* structure than the one we had named.

Let us now apply the rules in the opposite manner, attempting to write a structure from a name.

Example 4

Given: the name 4,4-diethyl-2-methyloctane.

First, we see at the end of the name the word *octane*. Therefore we write out a chain of eight carbons:

C—C—C—C—C—C—C—C

Next, we start reading at the beginning of the name, and as we come to each group name we attach carbons to represent it onto the main chain, meanwhile counting down the chain to be sure we put each group in the correct position.

The *4,4-diethyl* means two ethyl groups, both on the fourth carbon:

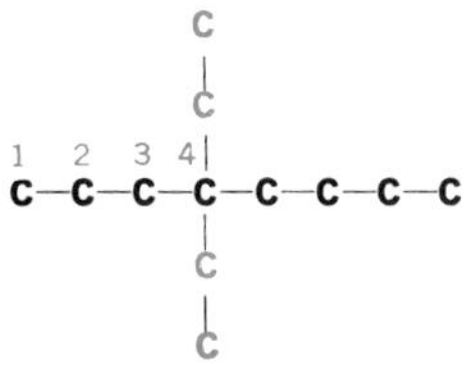

```
                 C
                 |
                 C
    1   2   3   4|
    C—C—C—C—C—C—C—C
                 |
                 C
                 |
                 C
```

Now, *2-methyl* requires a methyl group on the second carbon; we must be sure to count in the same direction in which we counted the first time:

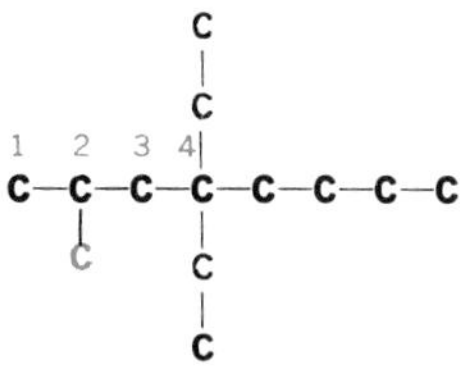

```
               C
               |
               C
   1   2   3  4|
   C—C—C—C—C—C—C—C
       |      |
       C      C
              |
              C
```

Finally, since we usually want to write a proper condensed structural formula, we can go back and put the required number of hydrogens at each carbon, keeping in mind that the covalence of carbon is 4:

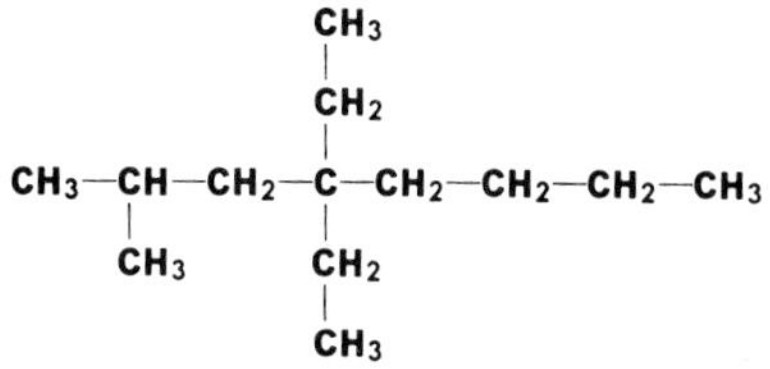

```
                 CH3
                 |
                 CH2
                 |
CH3—CH—CH2—C—CH2—CH2—CH2—CH3
      |        |
      CH3      CH2
               |
               CH3
```

Note in Example 4 that the group name *ethyl* was mentioned only once; the prefix *di-* told us that there were two such groups. However, even though both are at the same position on the chain, the number must be given for *both* of them, thus: 4,4-. Had it been written only as *4-diethyl*, we would have been uncertain when writing the structure. Did the author really mean both are at 4? Or is this a typographical omission; should it have been 3,4-diethyl?

If we count all the atoms in the structure shown in Example 1, we find the molecular formula is C_8H_{18}; the compound is one of the isomeric octanes. Although we first wrote the word *pentane* as the basis of a name for this structure, it does not mean the compound is one of the pentane isomers. If we examine the *complete* name we see that it does indeed say "eight carbons." Ethyl means two, methyl is one more, and pentane means five, for a total of eight. It is wise to check for errors by counting the total in both the name and the structure, to see whether they agree.

It is also important to realize that an alkyl group, such as any of those shown in Table 19.3, is a strictly imaginary device, convenient for naming purposes. It is not necessarily true that an alkyl group is a real compound that we might find in nature or that we might make one by some chemical reaction and attach it to some other group.

All organic compounds have a carbon skeleton of some sort. They may also have certain groups attached to this skeleton: —OH, —NH$_2$, —Cl, and so forth. In later chapters, the names used for these groups (and others) will be given. The group names together with the few simple rules for naming carbon skeletons, just described, provide the basis for naming organic compounds by the international system of nomenclature.

19.4 **GENERAL** **PHYSICAL** **AND CHEMICAL** **PROPERTIES**	Inorganic salts, being ionic compounds, are solids having quite high melting points and, of course, even higher boiling points. In sharp contrast, the covalent alkanes melt at very low temperatures, so that most are liquids at room temperature. A few are gases.

The boiling temperatures of several alkanes are shown in Table 19.2. These data indicate that there is a simple relationship between the size of alkane molecules and their boiling points. The smaller compounds are more volatile, which means they have lower boiling points. A larger molecule requires more energy to escape from the attraction of its neighbors in the liquid state and become a separate molecule in the vapor state. The temperature at which it boils is higher, corresponding to a greater energy requirement. Pentane, having a boiling point of 36°C, is the smallest alkane in the series that is liquid at room temperature and one atmosphere pressure. Those smaller than pentane are gases.

Since the many different alkanes present in petroleum have different boiling points, crude oil can be separated into various fractions by distillation. Several familiar petroleum products thus obtained are listed in Table 19.4, along with the size of alkane molecules normally found in each of the products. Although propane and butane are gases ordinarily, they can easily be liquefied if compressed and kept in a tank under pressure; they are often handled in this state and called liquefied petroleum gas (LPG). Either propane or butane is the type of "gas" carried in small tanks on recreational vehicles, to operate the lights, heaters, and cookstove. In very cold weather butane does not function well because it does not vaporize easily; note its boiling point of 0°C, shown in Table 19.2. Liquefied petroleum gas of some type is also kept in large tanks at rural homesites not having natural gas piped in.

Nearly all petroleum products are mixtures containing several different sizes of alkanes, and isomers of each. But if you compare the boiling ranges of some common materials listed in Table 19.4 (such as gasoline, diesel fuel, and mineral oil) with the sizes and boiling points of the alkanes in Table 19.2, it is clear that a higher boiling point corresponds to a larger size of molecule. Note also that alkane molecules having more than about 20 carbon atoms are solid at room temperature, as exemplified by petroleum jelly and paraffin wax.

Compounds consisting of larger molecules not only have higher boiling points but also are more viscous, that is, less fluid. Compare gasoline, kerosene, lubricating oil, and petroleum jelly.

At some time you have probably observed that oil floats on water; so do kerosene

TABLE 19.4   Some common petroleum products

Product	Main Alkanes Present	Boiling Range, °C
natural gas	C_1	below 0°
liquefied petroleum gas (LPG); propane, butane	C_3—C_4	below 0°
petroleum ether; light naphtha (solvents and cleaning fluids)	C_5—C_7	30–100°
gasoline	C_5—C_{10}	35–175°
kerosene; jet fuel	C_{10}—C_{18}	175–275°
diesel fuel	C_{12}—C_{20}	190–330°
fuel oil	C_{14}—C_{22}	230–360°
lubricating oil	C_{20}—C_{30}	above 350°
mineral oil (refined; decolorized)	C_{20}—C_{30}	above 350°
petroleum jelly (Vaseline)	C_{22}—C_{40}	(m.p. 40–60°)
paraffin wax	C_{25}—C_{50}	(m.p. 50–65°)

and gasoline. This simple observation reveals two other physical properties of the alkanes; they are insoluble in water and are less dense than water.

"Like dissolves like" is a general rule concerning solubilities. Water is a highly polar substance which generally dissolves other polar or ionic compounds. This suggests that *an alkane*, being insoluble in water, *is nonpolar*. Nonpolarity is an important characteristic of hydrocarbon skeletons, one which influences the behavior of all organic compounds. The reason an alkane is nonpolar can be understood if we examine a portion of its structure.

```
     H  H  H  H
     |  |  |  |
···— C— C— C— C —···
     |  |  |  |
     H  H  H  H
```

nonpolar bonds

We would expect the covalent bond between two carbon atoms to be nonpolar; since the atoms are identical, the electrons should be shared between them equally. It also happens that a hydrogen atom has just about the same tendency to attract electrons that a carbon atom does; that is, it has a similar electronegativity. Consequently, electrons shared between a carbon atom and a hydrogen atom also form a nonpolar bond.

The nonpolar nature of alkanes makes them good *solvents* for other nonpolar or only slightly polar substances. However, because of their flammability, extreme caution must be observed if volatile hydrocarbons are used as solvents. Year after year, hundreds of people are seriously burned and homes are destroyed because people were unwisely using gasoline to clean grease off a floor, an auto part, or a piece of clothing.

Alkanes are colorless. Recall that paraffin wax, kerosene, and the "white" gasoline used in camp stoves all lack color. The gasoline used in automobiles is usually orange because a little dye has been added. Common lubricating oils have a blue-green color because they still contain small amounts of colored substances originally present in the crude petroleum. The colored compounds need not be removed from lubricants. The mineral oil used medically is essentially the same mixture of hydrocarbons as lubricating oil, but is decolorized and refined.

Chemically, the distinctive feature of *an alkane* is that it *is generally inert.* There are only a very few chemical changes it will undergo. *Most* chemically reactive *compounds* (especially those studied in this book) are ionic or polar substances. Consequently, they do not react with the nonpolar alkanes. A few examples are listed below, for it is as important to know what will *not* happen as it is to know what will.

$$CH_3CH_2CH_2CH_3 + NaOH$$
$$CH_3CH_2CH_2CH_3 + HCl$$
$$CH_3CH_2CH_2CH_3 + H_2SO_4$$
$$CH_3CH_2CH_2CH_3 + Na \qquad \Big\} \rightarrow \text{no reaction}$$
$$CH_3CH_2CH_2CH_3 + KMnO_4$$
$$CH_3CH_2CH_2CH_3 + NaHCO_3$$
$$CH_3CH_2CH_2CH_3 + AgNO_3$$

One important reaction of an alkane is its conversion, at high temperature, to other hydrocarbons; this will be discussed in Sections 20.6 and 21.9.

Alkanes *do* react with oxygen (combustion) and with chlorine or bromine. These reactions will be discussed in the next two sections. It is interesting to note that even these reagents do not attack an alkane under ordinary conditions at room temperature. For instance, kerosene or gasoline remain unchanged when exposed to air for a very long time.

19.5 COMBUSTION

The complete combustion of methane (natural gas) is represented by Eq. (19–1). The burning, or combustion, occurs only if a flame or spark provides the initial burst of high temperature to start the reaction with oxygen. The reaction, once started, is exothermic, so the high temperature is maintained and promotes the reaction of other molecules.

$$CH_4 + 2O_2 \rightarrow CO_2 + 2H_2O + \text{heat} \tag{19-1}$$

Other alkanes burn in a similar manner. If an excess of oxygen is present, the final products will be carbon dioxide and water. For example, when heptane burns, a balanced equation can be obtained by first writing down $7CO_2$ to account for all the carbon atoms and then $8H_2O$ to account for the 16 hydrogen atoms from heptane. Finally, enough O_2 molecules must be provided to accomplish these changes:

$$C_7H_{16} + 11O_2 \rightarrow 7CO_2 + 8H_2O \tag{19-2}$$

If the amount of oxygen available to a burning hydrocarbon is not sufficient, carbon monoxide may be produced:

$$C_4H_8 + 4O_2 \rightarrow 4CO + 4H_2O \tag{19-3}$$

Or, if some carbon atoms are not oxidized at all, they form the fine black particles called soot.

Not all the details of exactly *how* a combustion reaction occurs are known, because it is not easy to study experimentally. However, it is certain that the requirement of a high temperature to start the reaction is due to the necessity of providing a high energy to break covalent bonds. Before the carbon atoms can be combined with oxygen there must be, at some stage in the process, a rupturing of the C—C and the C—H covalent bonds of the original alkane.

The largest consumption of alkanes is to make use of the energy released by combustion. Various sorts of engines have been designed to convert the heat energy to mechanical energy for doing work. The heat may be used directly for heating homes and public buildings or for cooking. Some furnaces and cooking ranges operate on kerosene, others on butane, and so on. One heating fuel widely used is natural gas. It is chiefly methane, with small amounts of other light hydrocarbons such as propane. Natural gas occurs in petroleum deposits, along with the mixture of liquid alkanes constituting the crude oil. The gas is carried by huge pipelines directly from the wells to major cities.

Since carbon chains form the skeletons of organic compounds, nearly all organic compounds will burn. Even solid organic substances are combustible; some familiar examples are wood, fats, cotton, and paraffin wax. How *easily* the compounds burn depends largely on their volatility. Gases and easily vaporized liquids, such as methane, butane, or octane, are readily ignited and burn vigorously.

19.6
SUBSTITUTION
BY CHLORINE

Under ordinary circumstances an alkane does not react with chlorine. A reaction will start, however, if a mixture of alkane and chlorine is exposed to a high temperature (400°–500°C) or to ultraviolet light from a lamp or from direct sunlight. Equation (19-4) shows the results with methane.

(19-4)

methane   chloromethane
                                     (methyl chloride)

This is a **substitution reaction**. One chlorine atom has been substituted in place of one of the hydrogens originally bonded to the carbon. The displaced hydrogen atom has paired off with another chlorine atom to form hydrogen chloride.

The need for bond-breaking once again explains the purpose of the energy required to initiate the reaction. Ultraviolet light, which provides more energetic rays than does visible light, can rupture the covalent bond of a Cl_2 molecule, producing Cl atoms. When a chlorine atom and a methane molecule collide (with enough energy), one of the C—H bonds of methane can be broken when the chlorine atom removes a hydrogen atom to form HCl. The remaining fragment of the methane structure (CH_3) can itself become a stable molecule by acquiring a chlorine atom to become $ClCH_3$.

The reaction represented by Eq. (19.4) could be carried out by mixing methane gas and chlorine gas in a tank having a window through which to shine ultraviolet light. As soon as a few molecules had reacted, some chloromethane (CH_3Cl) would be present in the tank. It could react with the remaining chlorine in much the same way that the methane did; the result, Eq. (19–5), would be the substitution of a second chlorine atom onto the same carbon.

$$
\underset{\substack{\\ H}}{\overset{\substack{H \\ |}}{H-\overset{|}{\underset{|}{C}}-Cl}} + Cl-Cl \xrightarrow{\text{UV light}} \underset{\substack{\\ Cl}}{\overset{\substack{H \\ |}}{H-\overset{|}{\underset{|}{C}}-Cl}} + H-Cl \tag{19-5}
$$

dichloromethane

Substitution of a third chlorine atom and a fourth is also possible.

$$CH_2Cl_2 + Cl_2 \xrightarrow{\text{UV light}} CHCl_3 + HCl \tag{19-6}$$

trichloromethane
(chloroform)

$$CHCl_3 + Cl_2 \xrightarrow{\text{UV light}} CCl_4 + HCl \tag{19-7}$$

tetrachloromethane
(carbon tetrachloride)

These reactions occur step by step. Each one is a repetition of the same fundamental substitution reaction, and at each step one molecule of hydrogen chloride is produced.

Consequently, when methane reacts with chlorine there results, in addition to the hydrogen chloride, a mixture of four different organic compounds. Fortunately the mixture can be separated by distillation, because the four compounds have distinctly different boiling points.

Alkanes other than methane can undergo substitution by chlorine. The chief difference is that a larger alkane yields a much larger number of different chlorinated products. For example, when butane reacts with chlorine, the first stage of substitution can yield *two* isomeric chlorobutanes.

$$
CH_3CH_2CH_2CH_3 + Cl_2 \xrightarrow{\text{UV light}}
\begin{array}{c}
CH_3-CH_2-CH_2-\underset{\substack{| \\ Cl}}{CH_2} + HCl \\[2mm]
\text{and} \\[2mm]
CH_3-CH_2-\underset{\substack{| \\ Cl}}{CH}-CH_3 + HCl
\end{array}
\tag{19-8}
$$

The chlorobutanes produced in Eq. (19–8) can react further with the chlorine in the reaction mixture to form *six* isomeric dichlorobutanes. Repeated substitutions can produce trichlorobutanes, tetrachlorobutanes, and so on up to decachlorobutane, C_4Cl_{10}.

Bromine reacts with alkanes in the same way that chlorine does, for example:

$$CH_3CH_3 + Br_2 \xrightarrow{\text{UV light}} CH_3CH_2Br + HBr \tag{19-9}$$

(and dibromoethanes, tribromoethanes, etc.)

<table>
<tr><td>

19.7
NAMES OF
ALKYL HALIDES

</td><td>

A compound having an alkane skeleton and a halogen in place of a hydrogen is called an **alkyl halide**; it has the general formula RX. An alkyl chloride is RCl; an alkyl bromide, RBr. With the equations in Section 19.6 are given the systematic names for the four chlorine derivatives of methane; the additional names shown there in parentheses are common names.

</td></tr>
</table>

To name an alkyl halide systematically we need only one new rule beyond those listed in Section 19.3: *an -o ending is used for a halogen group* (chloro, etc.), *and the group is given a smaller position number than is an alkyl group.*

$$CH_3\text{—}CH_2\text{—}CH_2\text{—}I$$

1-iodopropane

$$\overset{\displaystyle CH_3}{\underset{\displaystyle Cl}{CH_3\text{—}CH\text{—}CH\text{—}CH_3}}$$

2-chloro-3-methylbutane

$$\overset{\displaystyle C_2H_5}{CH_3\text{—}CH_2\text{—}CH\underset{\displaystyle Br}{\text{—}}CH\text{—}CH_2\underset{\displaystyle Br}{\text{—}}CH_2}$$

1,4-dibromo-3-ethylhexane

<table>
<tr><td>

19.8
RELATION OF
ALKANES TO
OTHER
ORGANIC
COMPOUNDS

</td><td>

There are three reasons why it is convenient and logical to begin our study of organic chemistry with the alkanes. First, an alkane is chemically quite inert. This fact simplifies the study of the reactions of other organic compounds. For example, an *alkene* molecule, which we will discuss in the next chapter, may be regarded as a double-bond unit of structure plus an alkane skeleton.

</td></tr>
</table>

an alkene $CH_3CH_2CH_2CH_2CH\!=\!CH_2$

alkane   functional
skeleton   group
(mainly inert) (reactive)

An alkene compound is very reactive due to the presence of a double bond between two of the carbons. *The double bond does something, or functions, and therefore that part of the structure is called a* **functional group**. While reactions are occurring at a functional group, the rest of the molecule, *the skeleton, survives unchanged*. This is a very important concept, and it simplifies the study of organic chemistry. For example, any compound having a C=C bond will have certain properties due to the reactivity of that bond. By learning what those properties are, we can predict the behavior of any of the thousands of different compounds in the series of alkenes. Similarly, the alcohols form another large series of compounds, each of which has an —OH group attached to an alkane skeleton. We can become acquainted with all alcohols by discovering what reactions to expect of an —OH group, and by keeping in mind that the alkane skeleton rarely reacts.

This approach to the study of organic chemistry is further illustrated with the structure of citronellol, Fig. 19.2. Each of the types of structure present (alkene, alkane,

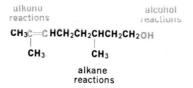

Fig. 19.2 Citronellol; 3,7-dimethyl-6-octene-1-ol.

alcohol) will behave in its own distinctive manner, and the chemical properties of citronellol will be the sum of all of these. This compound is found in the oil of roses and geraniums, as well as the citronella grasses of Asia.

Second, the rules for naming alkanes (Section 19.3) provide the basis for naming carbon skeletons in other compounds having functional groups.

Finally, the alkanes available from petroleum are important starting materials from which can be manufactured other, more valuable organic compounds. The carbon skeletons of alkanes can be converted to fibers for clothing, to medicines, to certain vitamins, to detergents, to dyes, to a great variety of plastics, and so on. The role of alkanes as raw materials is discussed further in Section 22.9.

19.9
CLASSES OF
ORGANIC
COMPOUNDS

It is very useful to classify organic compounds according to the type of structure, particularly the type of functional group, that they possess. Compounds having one particular kind of structural unit are grouped into one class or family. At the same time, it is important to recognize that one compound may have more than one functional group and will therefore react in several different ways, as illustrated by citronellol (Fig. 19.2).

In the next several chapters we will discuss, one by one, the various classes of organic compounds. This will include the distinctive properties, the uses, the names, and the occurrence in nature of the compounds in each class. Some idea of the types of structure that will be encountered is provided by Table 19.5, which lists the various classes of common organic compounds, with examples of each.

Most of the classes listed in Table 19.5 are distinguished by their functional groups.

TABLE 19.5 Classes of common organic compounds

Class of Compound	Functional Group	Representative Compounds and Names	References (Section)
alkane	—	RH $CH_3CH_2CH_2CH_3$ butane cyclohexane (C_6H_{12})	19.1, 19.3
alkene	$\diagdown C=C \diagup$	$CH_3CH=CH_2$ propene	20.1, 20.2
alkyne	$-C\equiv C-$	$CH_3C\equiv CCH_3$ 2-butyne	20.7
aromatic hydrocarbon	—	ArH benzene (C_6H_6) naphthalene ($C_{10}H_{14}$)	21.1, 21.2
alcohol	—OH	ROH CH_3CH_2OH ethanol (ethyl alcohol)	22.1, 22.2
ether	—O—	ROR' $CH_3CH_2OCH_3$ ethyl methyl ether ROAr $CH_3OC_6H_5$ methyl phenyl ether	22.1, 22.8
halide	—X	RX $CH_3CH_2CH_2Br$ 1-bromopropane ArX C_6H_5Cl chlorobenzene	19.7 21.2, 21.4

TABLE 19.5   (continued)

Class of Compound	Functional Group	Representative Compounds and Names	References (Section)
carboxylic acid		RCOOH CH$_3$CH$_2$CH$_2$CH$_2$—C (pentanoic acid) ArCOOH (benzoic acid)	23.1
ester		CH$_3$CH$_2$CH$_2$—C methyl butanoate ethyl benzoate	23.6
amide		CH$_3$—C acetamide benzamide	25.6
acid chloride		CH$_3$—C acetyl chloride benzoyl chloride	23.13
aldehyde	—C—H ‖ O	CH$_3$CH$_2$—C—H propanal benzaldehyde	26.1, 26.3
ketone	—C—R ‖ O	CH$_3$CH$_2$—C—CH$_2$CH$_3$ 3-pentanone 1-phenyl-2-propanone	26.1, 26.3

amine	—NH$_2$
	—NHR
	—NR$_2$
		CH$_3$CH$_2$—N—CH$_3$
		    |
		    H
		ethylmethylamine aniline
			25.1, 25.2

heterocycle	—	pyridine (C$_5$H$_5$N) thiazole (C$_3$H$_3$NS)	21.8, 25.7B

However, three of the classes—alkane, aromatic hydrocarbon, and heterocycle—represent the three fundamental types of carbon skeleton found in organic compounds. Some compounds consist of nothing more than one of these skeletons. In the great majority of compounds, however, the skeleton carries one or several of the functional groups.

As our study proceeds, you will find it helpful to refer frequently to Table 19.5 to check on names or symbols, or to find a reference for more details.

SUMMARY

Because alkanes are quite inert they undergo only a few kinds of chemical reactions. However, they are useful as fuels and lubricants, and, most important of all, as raw materials for manufacturing other organic compounds. Although there are still abundant supplies of alkanes on earth, in the form of petroleum, humans should develop other energy sources as quickly as possible, in order to save petroleum for raw material use.

The chains and rings of alkane molecules form the carbon skeletons of many other organic compounds. For this reason the systematic method for naming alkanes provides a basis for naming the other organic compounds as well.

Nearly every other organic compound has, in addition to its carbon skeleton, a unit of structure called a functional group. This is the portion of the molecule which is chemically reactive. Since each different functional group has a particular set of chemical and physical properties, organic compounds can be classified and studied in terms of the functional group present in each.

IMPORTANT TERMS

alkane	hydrocarbon
alkyl halide	saturated hydrocarbon
alkyl group	substitution reaction
cycloalkane	systematic name
functional group	unsaturated hydrocarbon

WORK EXERCISES

1. Show the condensed structural formula of the following.
   a) 3-ethylpentane
   b) 2,2-dimethylbutane
   c) 2-methyl-3-propylhexane
   d) 2,2,4-trimethylpentane
   e) ethylcyclopentane
   f) 1,3-dibromocyclohexane
   g) 4-ethyl-2,5-dimethyloctane
   h) 3,5-dimethylheptane
   i) 2,3-dichloro-4-ethylhexane
   j) 4,6-diethyl-3,4-dimethyl-6-propylnonane

2. Write condensed structural formulas for all isomers having the following molecular formulas.
   a) C_6H_{14} b) C_4H_9Cl
   c) $C_4H_8Cl_2$ d) $C_5H_{11}Br$

3. Write the systematic name for each structure in Exercise 2. Are any of the names the same? If

they are, look again to see whether the structures are really different. Is the naming system adequate—that is, does each structure have a unique name which applies to no other structure?

4. Name each structure.

   a) $CH_3CHCH_2CH_2CH_3$
           |
         CH_3

   b) $CH_3CH_2CH_2CHCHCH_3$
                       | |
                     CH_3 (with CH_3 above)

   c) $CH_3CH_2CHCH_2CHCH_3$
              | |
            CH_2 Cl
              |
            CH_3

d) $CH_3-CH_2-\underset{\underset{\displaystyle Br}{|}}{\overset{\overset{\displaystyle CH_3}{|}}{C}}-\underset{\underset{\displaystyle CH_3}{|}}{CH}-CH_3$

e)

f) —Cl

g) —CH$_3$

h) $CH_3CH_2-\underset{\underset{\displaystyle CH_3}{|}}{\overset{\overset{\displaystyle CH_3}{|}}{C}}-CH_2\underset{\underset{\displaystyle CH_2}{|}}{CH}CH_2CH_3$
$\underset{\underset{\displaystyle CH_3}{|}}{}$

i) $CH_3\underset{\underset{\displaystyle CH_3}{|}}{\overset{\overset{\displaystyle CH_3}{|}}{CH}}CHCH_2\underset{\underset{\displaystyle C_2H_5}{|}}{\overset{\overset{\displaystyle CH_3}{|}}{CH}}CHCH_2CH_3$

j) $CH_3\underset{\underset{\displaystyle CH_2}{|}}{\overset{\overset{\displaystyle Br}{|}}{CH}}CH\underset{\underset{\displaystyle CH_2}{|}}{\overset{\overset{\displaystyle CH_3}{|}}{CH}}CHCH_2CH_3$
$\underset{\underset{\displaystyle CH_3}{|}}{CH_2}\underset{\underset{\displaystyle CH_3}{|}}{CH_2}$

k) $CH_3CH_2\underset{\underset{\displaystyle CH_2}{|}}{\overset{\overset{\displaystyle CH_2CH_3}{|}}{CH}}CHCH_2\underset{\underset{\displaystyle CH_2}{|}}{\overset{\overset{\displaystyle CH_3}{|}}{CH}}CHCH_3$
$\underset{\underset{\displaystyle CH_3}{|}}{CH_3}\underset{\underset{\displaystyle CH_3}{|}}{CH_2}$

5. Give both a common name and a systematic name for the following.
a) $CHCl_3$
b) CCl_4
c) $CH_3-\underset{\underset{\displaystyle CH_3}{|}}{CH}-CH_3$

d) $CH_3-CH_2-\underset{\underset{\displaystyle CH_3}{|}}{CH}-CH_3$

e) CH_3Br
f) CH_3CH_2Cl
g) $CH_3\underset{\underset{\displaystyle CH_3}{|}}{CH}CH_2CH_2CH_3$

6. Write complete balanced equations for the reaction occurring in each mixture below, using structural formulas. If necessary, state "no reaction." If more than one reaction is possible in a given mixture, write equations for all of them.
a) propane burning in a camp stove
b) ethane and hydrogen chloride at 100°C
c) methane and chlorine at high temperature
d) isobutane and oxygen, heated
e) cyclopentane and chlorine in ultraviolet light
f) cyclohexane and air, heated
g) pentane and concentrated sodium hydroxide
h) methane and bromine in ultraviolet light
i) dichloromethane and bromine in ultraviolet light

7. Name the compounds shown in Figs. 18.20 and 18.21.

8. Write a name and an abbreviated structural formula for the compounds shown in Fig. 18.23.

9. List all the types of structural units (both carbon skeletons and functional groups) that you can find in the compounds below. Consult Table 19.5.
a) aspirin (Section 23.15E)
b) citral (Table 26.2)
c) coniine, the poison in hemlock which was used to kill Socrates (Section 25.9B)
d) the compound below, produced by the plant *coreopsis* (Compositae):

$-C\equiv C-C\equiv C-CH=CH-CH_2-OH$

e) thymol (Section 22.1)
f) vanillin (Section 22.1)
g) epinephrine (adrenalin):

$CH_3-\underset{\underset{\displaystyle H}{|}}{N}-CH_2-\underset{\underset{\displaystyle OH}{|}}{CH}-$ $-OH$

SUGGESTED READING

Drake, E., and R. C. Reid, "The importation of liquefied natural, gas" *Scientific American*, April, 1977. Is the transportation of LNG by ship safe ?

Eglinton, G., J. R. Maxwell, and C. T. Pillinger, "The carbon chemistry of the moon," *Scientific American*, Oct. 1972. The simple organic compounds found on the moon were not produced by living organisms, but they do provide additional clues to the origin of life.

Lawless, J. G., C. E. Folsome, and K. A. Kvenvolden, "Organic matter in meteorites," *Scientific American*, June 1972.

Nelson, T. W., "The origin of petroleum," *J. Chem. Educ.* **31**, 399 (1954). A thorough, documented review.

20

Unsaturated Hydrocarbons:
Alkenes and Alkynes

**20.1
OCCURRENCE
OF ALKENES**

In an **unsaturated hydrocarbon** the carbon skeleton holds *less* than the maximum number of hydrogens it might hold. An **alkene** is an unsaturated hydrocarbon having a carbon–carbon double bond; a good example is propene, Fig. 20.1.

The carbon–carbon double bond is very common in organic compounds. It is found in both chain and ring structures, in simple molecules and highly complex ones, in those produced within living cells, and in synthetic substances. Myrcene and limonene, Fig. 20.2, are examples of alkenes produced by plants. Myrcene is found in bay leaves. Limonene, along with certain other natural oils, occurs in the skin of oranges and limes. Both myrcene and limonene have ten-carbon skeletons with similar branching. Each is made up of isoprene carbon skeletons. Isoprene, shown in Table 20.1, has a five-carbon skeleton. Natural products built up from isoprene skeletons are called *terpenoids*. The structure of vitamin A, having four isoprene units, is shown in Section 32.7.

**20.2
ALKENE
NOMENCLATURE**

Table 20.1 shows the names of some representative alkenes. *Systematic names* have the ending *-ene* to designate the double bond. The position of the double bond in the carbon chain is indicated by a number placed just before the name for the chain; although the

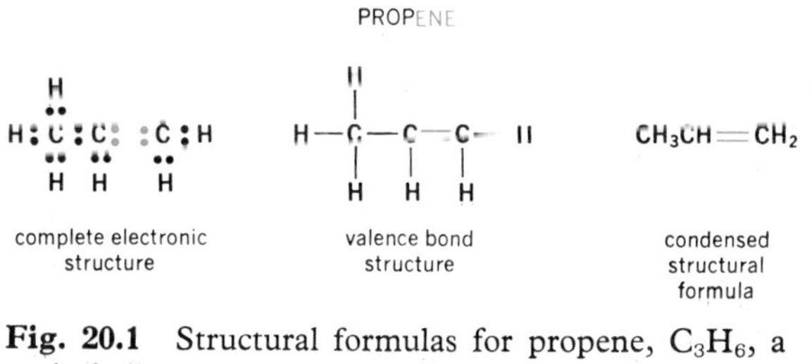

PROPENE

| complete electronic structure | valence bond structure | condensed structural formula |

Fig. 20.1 Structural formulas for propene, C_3H_6, a typical alkene.

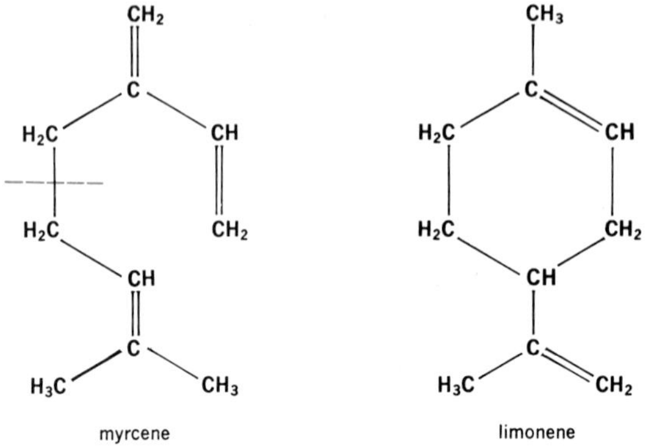

myrcene   limonene

Fig. 20.2   The structure of two naturally occurring alkenes built up from isoprene carbon skeletons.   The dotted line on the myrcene structure shows where its two isoprene units are linked together.

double bond is between two carbons, it is necessary to state only the number of the first of those carbons.

$CH_3CH=CHCH_2CH_3$ 2-pentene

If there are halogen atoms or alkyl groups attached to the chain, counting should be done from whichever end of the chain will allow *the lowest number for the double bond.* A

TABLE 20.1   Names of some alkenes

Structure	Systematic Name	Common Name
$CH_2=CH_2$	ethene	ethylene
$CH_3CH=CH_2$	propene	propylene
$CH_3CH_2CH=CH_2$	1-butene	α-butylene
$CH_3CH=CHCH_3$	2-butene	β-butylene
$CH_3-C=CH_2$ $\quad\ \ \mid$ $\quad\ \ CH_3$	methylpropene	isobutylene
$CH_3-C=CH-C-CH_3$ $\quad\ \ \mid\qquad\quad \mid$ $\quad\ \ CH_3\quad\ CH_3$	2,4,4-trimethyl-2-pentene	
$CH_2=C-CH=CH_2$ $\quad\ \ \mid$ $\quad\ \ CH_3$	2-methyl-1,3-butadiene	isoprene
 (cyclopentene ring structure)	cyclopentene	
$R_2C=CR_2$ (general)	alkene	olefin

systematic name for the structure below can be developed by applying the rules step by step.

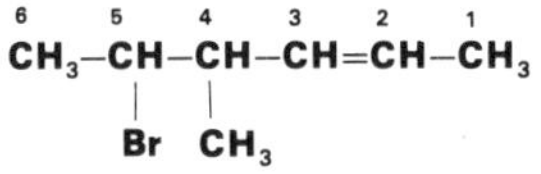

$$\overset{6}{CH_3}-\overset{5}{CH}-\overset{4}{CH}-\overset{3}{CH}=\overset{2}{CH}-\overset{1}{CH_3}$$
$$\underset{Br}{|} \quad \underset{CH_3}{|}$$

First, write the name of the longest carbon chain, including the *ene* ending to signify that a double bond is present:

hexene

Next, assign numbers to each carbon in the chain, giving the lowest possible number to the double bond:

2-hexene

Finally, write the names for the substituents and include their assigned position numbers.

5-bromo-4-methyl-2-hexene

Common names of several alkenes are shown in Table 20.1. These originated before systematic names were developed. They are still with us because it is difficult to persuade people to stop using common names for frequently encountered compounds. It is rather like trying to prevent a child's friends from calling him by a nickname. Unfortunately, some of the alkene common names are only slightly different from systematic names, which can lead to confusion and misspelling. The common names are given in the table only so that you will be aware of the problem and can look up a name here if you see it used elsewhere. As much as possible we will use proper systematic names in this book. In later chapters it will be necessary to learn a few common names because of their wide usage.

In common nomenclature, a double-bonded compound is called an **olefin**, as a general term.

That is, olefin is the common name equivalent to alkene. When "ethylene" gas was discovered in 1794, it was found to react with chlorine to form a water-insoluble oil. Hence it was called an olefiant gas (L. *oleum*, oil plus *-fiant*, making). Over the years the term evolved into the shorter olefin.

**20.3
STRUCTURE
OF THE C=C
BOND**

In order to understand the shape and the chemical properties of the C=C bond in an alkene, it is necessary to see a more detailed picture of its structure. We shall use ethene, the simplest alkene, to examine the bond formation. Figure 20.3 is a valence-bond structure of ethene. The sigma bonds, which form the skeleton of the molecule, are represented by the usual bond lines. The "extra" bond between the two carbons is called a **pi bond** (π). As we shall soon see, the pi bond is distinctly different from a sigma bond in both structure and reactivity; to emphasize this difference the pi bond in Fig. 20.3 is represented by a pair of electron dots.

It is apparent from Fig. 20.3 that each carbon participates in just three sigma bonds. For this purpose it hybridizes only three orbitals of its valence electrons, the s orbital and two **p** orbitals. This creates three equivalent **sp**^{2} hybrid orbitals. One of these overlaps with a similar **sp**^{2} hybrid orbital from the other carbon to form a C—C sigma bond. Each of the other two **sp**^{2} orbitals overlaps with an s orbital of a hydrogen to form a

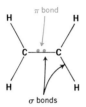

Fig. 20.3 Valence-bond formula of ethene, C_2H_4.

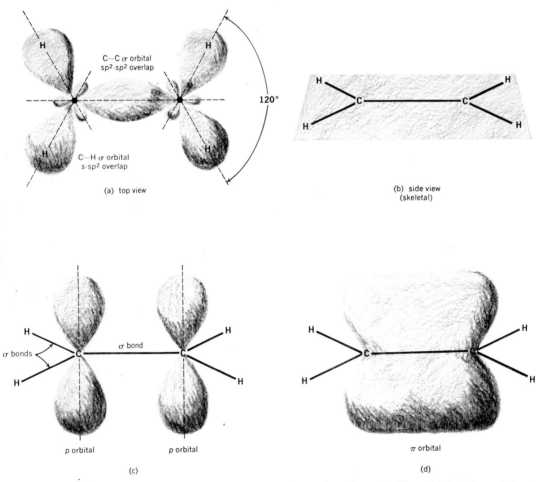

Fig. 20.4 Structure of a carbon–carbon double bond, shown for ethene, C_2H_4. (a) Top view of the sigma-bond framework. Each carbon forms three **sp²** orbitals for bonding. (Carbon nuclei shown as dots.) (b) Skeletal formula shows that all sigma bonds are in one plane. (c) Each carbon has one **p** orbital perpendicular to the plane of the sigma bonds. (d) Sideways overlap of **p** orbitals forms a pi orbital.

C—H sigma bond. As usual, the electron pair in one orbital repels those of another orbital, so the three sigma orbitals at each carbon spread out to give 120° bond angles, as shown in Fig. 20.4(a). The sigma bonds all lie in one plane, shown by the side view of the sigma framework in (b) of Fig. 20.4.

So far we have described only the sigma skeleton of the ethene molecule. But as seen in Fig. 20.4(c), each carbon still has one valence electron residing in a **p** orbital, which is not yet bonded (paired off) with another electron. The **p** orbital orients itself perpendicular to the plane of the sigma orbitals, as far away from them as it can be. Now the only way that the **p** orbitals on the two adjacent carbons can form a bond (a molecular orbital) is to overlap *sideways* with each other, as shown in (d) of Fig. 20.4. The resultant molecular orbital, occupied by the two bonded electrons, is called a *pi orbital*. The pi orbital has two lobes, like the **p** orbitals from which it originates. Because the adjacent **p** orbitals must overlap sideways to form a π bond, the depth of overlap is not great and the π bond is a weaker bond than the σ bond.

Thus it is evident that the two bonds of a carbon–carbon double bond are not really the same. Although C=C will often be used for simplicity in writing formulas, it is important to remember that one is a σ bond and the other is a π bond. We shall see in the next section that the shape of a C=C bond makes possible a particular type of isomer and in Section 20.5 that the π bond of an alkene reacts chemically quite unlike a σ bond of an alkane.

20.4
CIS–TRANS
ISOMERS

When two carbon atoms are joined by a double bond, one carbon cannot rotate in relation to the other, under ordinary circumstances. An interesting consequence of this property is shown in Fig. 20.5; there are *two* different compounds corresponding to the name 2-butene! If the molecule has the methyl groups held on the same side, the compound is named *cis*-2-butene; if the two methyl groups are across from each other, the compound is *trans*-2-butene. These are two distinctly different substances, each of which has its own physical properties. The double bond acts as a rigid barrier to rotation, thus restricting a methyl group in *cis*-2-butene from flipping over to produce a molecule of *trans*-2-butene.

A glance back at (c) and (d) of Fig. 20.4 reveals that the possibility of rotation at a C=C bond is restricted by the structure of the *pi* orbital. Rotation *is* possible on the C—C *sigma* bond without weakening it. However, if this occurred, the two **p** orbitals would no longer be directly lined up and would not be able to overlap to form the pi orbital. Or, to say it another way, to force one carbon to rotate on the sigma-bond axis, relative to the other carbon, would require breaking the pi bond. At a high temperature there would be enough energy to break a pi bond; even sigma bonds can be broken at very high temperatures. But at reasonable temperatures pi bonds are sufficiently stable

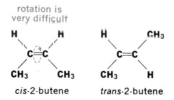

cis-2-butene   *trans*-2-butene

Fig. 20.5   The two isomers named 2-butene.

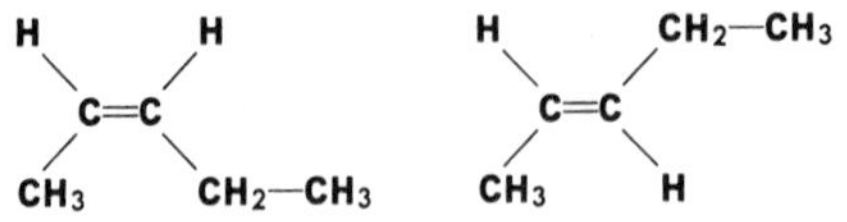

Fig. **20.6** Isomers of 2-pentene.

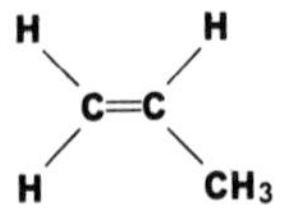

Fig. 20.7 Propene, no *cis-trans* isomers.

to exist and are very common. *Cis*-2-butene and *trans*-2-butene do exist as two different compounds.

Cis–trans isomerism is possible at a C=C bond as long as there are not two identical groups attached to one carbon. For example, there are *cis* and *trans* isomers of 1,2-dichloroethene or of 2-pentene (Fig. 20.6). However, *cis–trans* isomers of propene are not possible because one of the carbons at the double bond holds two hydrogens (Fig. 20.7). Consequently the methyl group, whichever side it is on, will always be across from a hydrogen and beside a hydrogen.

20.5
ALKENE
REACTIONS
A. Reactivity of
the Pi Bond

The electrons in the pi bond of an alkene make it very reactive for two reasons. First of all, the π bond is an "extra" bond; it is not part of the σ-bond skeleton required to hold the molecule together. The π electrons are available to make new covalent bonds to additional atoms, without disturbing the framework of σ bonds in the original molecule.

Not only are the π electrons available for bonding, but secondly they are *readily* available for reaction. The π bond is a weaker bond than a σ bond. As can be seen in Fig. 20.4, the *sideways* overlap of **p** orbitals to form a π bond is not nearly as effective as the deep overlap if an **sp**^{2} orbital *in the direction* of its axis, to form a σ bond. The σ electrons are tightly held, concentrated in a small space. In contrast, the π electrons occupy a large volume because they are loosely held. These π electrons are much more easily shared with some reacting substance that is seeking electrons, than are electrons in σ bonds.

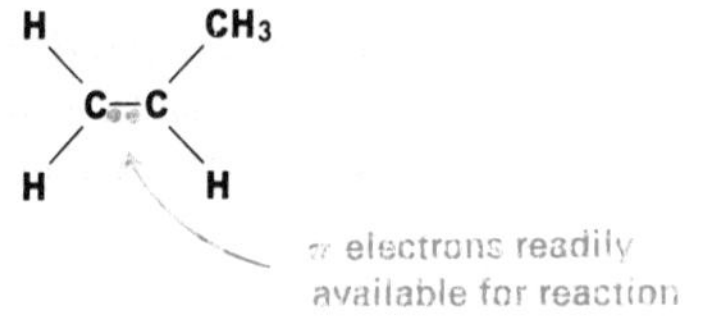

π electrons readily
available for reaction

Most reactions of alkenes are **addition reactions**. Reactants such as H_2 or Cl_2 simply *add* to the alkene to give one new molecule. Since the π electrons of the double bond can be used to hold other atoms, it is not necessary to displace atoms already present in order to make room for the new atoms. (Recall that, aside from combustion, an alkane reacts by *substitution* because its valence bonds are already saturated with hydrogens.)

Alkenes can also be attacked by common ionic oxidizing agents ($KMnO_4$, and so forth). This reaction, too, depends on the readily available electrons. Oxidation amounts to loss of electrons, or at least displacement of electrons toward a new atom.

In this chapter the properties of a carbon–carbon double bond will be illustrated with simple alkenes. But keep in mind that even when a large molecule contains other functional groups, a double bond reacts in its own characteristic way (Section 19.8).

B. Addition of Hydrogen (Hydrogenation)

When an alkene is treated with hydrogen gas, in the presence of certain metal catalysts, it is converted to an alkane; the process is called **hydrogenation**.

$$
\underset{\text{2-pentene}}{
\begin{array}{c}
\text{H} \;\; \text{H} \;\; \text{H} \;\; \text{H} \;\; \text{H} \\
| \quad | \quad | \quad | \quad | \\
\text{H}-\text{C}-\text{C}-\text{C}{=}\text{C}-\text{C}-\text{H} \\
| \quad | \quad\quad\quad | \\
\text{H} \;\; \text{H} \quad\quad \text{H}
\end{array}}
+ \underset{\text{hydrogen}}{\text{H}-\text{H}}
\xrightarrow{\text{Ni}}
\underset{\text{pentane}}{
\begin{array}{c}
\text{H} \;\; \text{H} \;\; \text{H} \;\; \text{H} \;\; \text{H} \\
| \quad | \quad | \quad | \quad | \\
\text{H}-\text{C}-\text{C}-\text{C}-\text{C}-\text{C}-\text{H} \\
| \quad | \quad | \quad | \quad | \\
\text{H} \;\; \text{H} \;\; \text{H} \;\; \text{H} \;\; \text{H}
\end{array}}
\tag{20-1}
$$

The metals most successful as catalysts for hydrogenation are nickel, palladium, and platinum. Reaction occurs when both the alkene and hydrogen are absorbed onto the surface of the metal. Since the amount of surface is important, the metal is used in a finely divided form; in this condition it is a black powder resembling charcoal.

Such a reaction would not be the most practical way to get a simple alkane such as pentane, which is easily obtained from petroleum. However, hydrogenation might be very useful to convert a segment of alkene structure in a complex molecule to a saturated segment. For instance, cottonseed oil can be converted to a fat (Section 24.4), and citronellol (see Section 19.8) to a saturated alcohol:

$$
\underset{\text{citronellol}}{
\begin{array}{c}
\text{CH}_3\text{C}{=}\text{CHCH}_2\text{CH}_2\text{CHCH}_2\text{CH}_2\text{OH} \\
| \quad\quad\quad\quad\quad\quad\quad\quad | \\
\text{CH}_3 \quad\quad\quad\quad \text{CH}_3
\end{array}}
+ \text{H}_2
\xrightarrow{\text{Pt}}
\underset{\text{3,7-dimethyl-1-octanol}}{
\begin{array}{c}
\text{CH}_3\text{CHCH}_2\text{CH}_2\text{CH}_2\text{CHCH}_2\text{CH}_2\text{OH} \\
| \quad\quad\quad\quad\quad\quad\quad | \\
\text{CH}_3 \quad\quad\quad\quad \text{CH}_3
\end{array}}
\tag{20-2}
$$

C. Addition of Chlorine or Bromine

Chlorine or bromine adds readily to a carbon–carbon double bond.

$$
\underset{\text{1-hexene}}{\text{CH}_3-\text{CH}_2-\text{CH}_2-\text{CH}_2-\text{CH}{=}\text{CH}_2} + \text{Cl}_2
\rightarrow
\underset{\text{1,2-dichlorohexane}}{
\begin{array}{c}
\text{CH}_3-\text{CH}_2-\text{CH}_2-\text{CH}_2-\text{CH}-\text{CH}_2 \\
| \quad\;\; | \\
\text{Cl} \quad \text{Cl}
\end{array}}
\tag{20-3}
$$

Reaction (20–3) can be carried out by simply bubbling chlorine gas into the liquid 1-hexene; no catalyst or heat is required.

Since chlorine is one of the few substances that can react with an *alkane*, it is especially important to note the differences in behavior of an alkene and an alkane with chlorine. *The alkene undergoes an addition reaction* (Eq. (20–3)) to give only one product. Ordinary conditions are sufficient. In contrast, *an alkane reacts by substitution*; besides the organic products, HCl is always formed (Eq. 19–4). Drastic conditions, either ultraviolet light or a high temperature, are required for the alkane reaction.

A reaction with bromine can be used as a simple laboratory test to detect the presence of unsaturation in a hydrocarbon.

$$
\underset{\substack{\text{an alkene}\\ \text{(colorless)}}}{\text{R}-\text{CH}{=}\text{CH}-\text{R}} + \underset{\substack{\text{bromine}\\ \text{(red-brown)}}}{\text{Br}_2}
\rightarrow
\underset{\substack{\text{a dibromoalkane}\\ \text{(colorless)}}}{
\begin{array}{c}
\text{R}-\text{CH}-\text{CH}-\text{R} \\
| \quad\;\; | \\
\text{Br} \quad \text{Br}
\end{array}}
\tag{20-4}
$$

Bromine itself has a dark red-brown color. If it is used up by addition to an alkene, the color disappears. However, if even one or two drops of bromine are mixed with an alkane, the solution takes on a red-brown color because the bromine remains unchanged. Consequently, an alkene can be distinguished from an alkane by testing with bromine. For a test, a very dilute solution of Br_2 in CCl_4 is most satisfactory and least dangerous. (Bromine is extremely irritating and poisonous.) Iodine has been used in a modification of this test, to detect double bonds in foods. The test results indicate the extent of unsaturation in an edible fat or oil.

Although chlorine, too, reacts easily with an unsaturated compound, for practical reasons it is not used in a test. Chlorine is such a pale yellow-green color that in dilute solutions it is impossible to detect a color change. Also, chlorine is troublesome to handle because it is a gas.

D. Addition of Hydrogen Chloride

A number of acidic compounds will add to an alkene. We will mention the most common examples, one of which is hydrogen chloride.

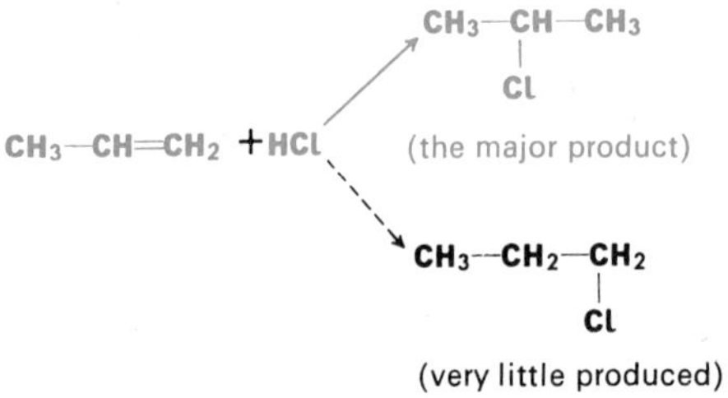

$$CH_3—CH=CH_2 \; +HCl$$

$$CH_3—CH—CH_3$$
$$|$$
$$Cl$$
(the major product)

$$CH_3—CH_2—CH_2$$
$$|$$
$$Cl$$
(very little produced)

(20-5)

It is apparent from Eq. (20-5) that hydrogen chloride could add to propene in two different ways. Curiously, one product greatly predominates. This fact was first reported in 1869 by the Russian chemist Markovnikov. Somewhat paraphrased, **Markovnikov's rule** states that *when an acid adds to an alkene, the hydrogen becomes attached to the carbon already bearing the most hydrogens.*

E. How an Acid Adds to an Alkene

Markovnikov originally formulated his rule after observing the behavior of several alkenes. He had no idea *why* the results were thus. In 1932, Professor Frank Whitmore, of Pennsylvania State University, proposed a general theory which explains how a reaction such as (20-5) occurs, and hence why Markovnikov's rule is correct.

Briefly, the explanation is as follows: A Cl^- ion would be repelled by the extra electrons in the alkene double bond. Instead, the alkene is first attacked by an H^+ ion which is seeking electrons. An approaching positive ion can distort the electron pair in the π bond of the alkene, pulling it into position to form a new covalent bond, a σ bond.

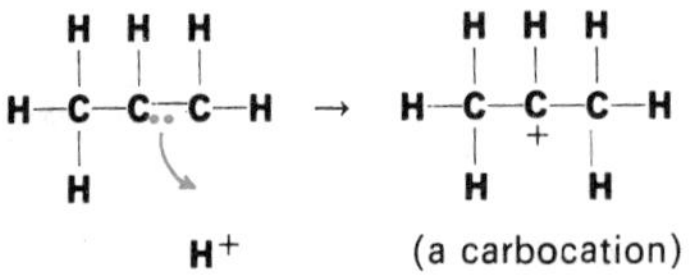

$$\begin{array}{c} H\ H\ H \\ |\ \ |\ \ | \\ H—C—C—C—H \\ |\ \ \ \ \ \ | \\ H \\ \\ H^+ \end{array} \rightarrow \begin{array}{c} H\ H\ H \\ |\ \ |\ \ | \\ H—C—C—C—H \\ |\ \ +\ \ | \\ H\ \ \ \ \ H \end{array}$$
(a carbocation)

(20-6)

The intermediate structure produced in Eq. (20-6) is a carbocation. The possible existence of such ions was first suggested in 1922 by the German chemist Hans Meerwein. A carbocation is a highly reactive ion which promptly combines with a chloride ion, as shown here:

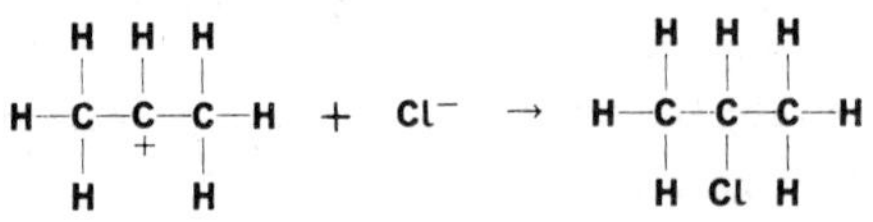

$$\begin{array}{c} H\ H\ H \\ |\ \ |\ \ | \\ H—C—C—C—H \\ |\ \ +\ \ | \\ H\ \ \ \ \ H \end{array} + Cl^- \rightarrow \begin{array}{c} H\ H\ H \\ |\ \ |\ \ | \\ H—C—C—C—H \\ |\ \ |\ \ | \\ H\ Cl\ H \end{array}$$

(20-7)

In the first step, the H^+ *might* have combined with propene as shown in Eq. (20–8), producing a different carbocation.

$$
\begin{array}{ccc}
\text{H H H} & & \text{H H H} \\
\text{H—C—C—C—H} & \rightarrow & \text{H—C—C—C—H} \\
\text{H} & & \text{H H} \\
\text{H}^+ & & \text{(a carbocation)}
\end{array}
\tag{20–8}
$$

The nature of carbocations is such that the type produced in Eq. (20–6) is more stable and more easily formed than that shown in Eq. (20–8). In the competition, reaction (20–6) wins and then gives, in Eq. (20–7), the results predicted by Markovnikov's rule.

F. Addition of Hydrogen Sulfate

Hydrogen sulfate (sulfuric acid) adds to an alkene, following Markovnikov's rule. In order to understand the structures resulting from this addition reaction, we should first examine the structure of sulfuric acid. The formula H_2SO_4 lists the atoms but tells nothing of the manner in which they are bonded. Actually, the four oxygens are clustered around the sulfur, and the hydrogens are bonded to oxygens.

SULFURIC ACID or HYDROGEN SULFATE

$$
\begin{array}{ccc}
\ddot{\text{O}}\ : & \text{O} & \\
\text{H}\!:\!\ddot{\text{O}}\!:\!\text{S}\!:\!\ddot{\text{O}}\!:\!\text{H} & \text{H—O—S—O—H} & \text{HOSO}_2\text{OH} \\
\ddot{\text{O}}\ : & \text{O} & \\
\text{electronic} & \text{valence bond} & \text{condensed} \\
\text{structure} & \text{structure} & \text{structural} \\
 & & \text{formula}
\end{array}
$$

Hereafter, in writing organic reactions, we will often use the condensed structural formula of sulfuric acid.

The addition of sulfuric acid to a double bond is illustrated by this reaction:

$$
\begin{array}{ccc}
\text{CH}_3\text{—CH}\!=\!\text{CH}_2 \ + \ \text{HOSO}_2\text{OH} & \rightarrow & \text{CH}_3\text{—CH—CH}_3 \\
 & & \text{OSO}_2\text{OH}
\end{array}
\tag{20–9}
$$

propene   isopropyl
                                            hydrogen sulfate

G. Addition of Water

In the presence of sulfuric or phosphoric acid, water will add to an alkene. In some cases heat is required, depending on the reactivity of the particular alkene.

$$
\text{CH}_3\text{—CH}\!=\!\text{CH}_2 \ + \ \text{HOH} \ \xrightarrow{\text{H}^+} \ \text{CH}_3\text{—CH—CH}_3 \atop \text{OH}
\tag{20–10}
$$

2-propanol
(isopropyl alcohol)

Note that this addition also obeys Markovnikov's rule. The isopropyl alcohol shown in this example is more commonly known as "rubbing alcohol."

If a different alkene is treated with water and acid, a different alcohol will be obtained. This reaction provides a good method for synthesizing several different kinds of alcohols on a factory scale.

The need for an acid catalyst in this reaction is understandable. In Section E we saw that this type of addition reaction begins when an H^+ attracts the electrons out of the π bond of the alkene. But water itself is so poorly ionized that it supplies almost no H^+ ions to a reaction mixture. We must add a strong acid, such as sulfuric, to provide H^+ ions and get the reaction started. Of course, in the presence of water the H^+ ions would actually be in the form of H_3O^+ ions. Step by step, this particular reaction would take place as follows:

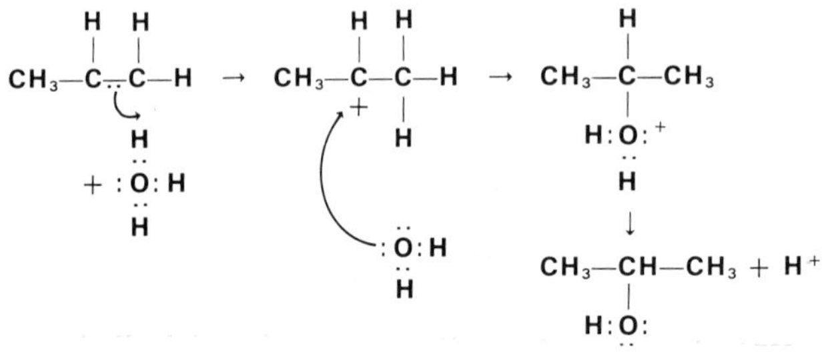

H. Oxidation of Alkenes

An alkene is easily oxidized under mild conditions. For example, hydrogen peroxide (usually in a solution of formic acid or acetic acid) oxidizes an alkene to a diol, a molecule having two alcohol groups:

$$CH_3-CH{=}CH_2 + H_2O_2 \rightarrow CH_3-\underset{\underset{OH}{|}}{CH}-\underset{\underset{OH}{|}}{CH} \qquad (20\text{--}11)$$

propene   1,2-propanediol
                     (propylene glycol)

A dilute solution of potassium permanganate at room temperature can also oxidize an alkene to a diol (a glycol):

$$3R{-}CH{=}CH{-}R + 2KMnO_4 + 4H_2O \rightarrow 3R{-}\underset{\underset{OH}{|}}{CH}{-}\underset{\underset{OH}{|}}{CH}{-}R + 2MnO_2\downarrow + 2KOH \qquad (20\text{--}12)$$

(purple)   (brown)

Reaction (20–12) is used chiefly as a test, which is possible because of the easily detected color changes. The permanganate is a *purple solution*, looking very much like grape juice. When it reacts, its color disappears, and we see in its place a *dark brown precipitate* of manganese dioxide. A positive reaction *suggests* the presence of unsaturation in the organic compound; as a check we should also try the bromine test (Section 20.5C). Certain organic compounds other than unsaturated hydrocarbons can react with potassium permanganate.

I. Combustion

If a mixture of an alkene and oxygen gas is ignited, combustion occurs.

$$CH_3CH_2CH{=}CH_2 + 6O_2 \rightarrow 4CO_2 + 4H_2O$$

This behavior is not unique to alkenes. In fact, alkanes and practically all other organic compounds will also burn (Section 19.5). Therefore, in the remaining chapters, when we discuss each new series of compounds, we will not bother to mention that each one is combustible.

20.6 **SOURCES AND** **USES OF** **ALKENES**	The most economical way to obtain simple alkenes in large quantity is to "crack" alkanes from petroleum. For instance, if butane is heated to about 500°C (with oxygen excluded, to prevent combustion) the covalent bonds begin to break. Since any of the bonds is susceptible to fracture at this temperature, all the imaginable fragments are formed.

$$
\begin{array}{c}
\text{H H H H}\\
| \ | \ | \ |\\
\text{H—C—C—C—C—H} \rightarrow\\
| \ | \ | \ |\\
\text{H H H H}\\
\text{butane}
\end{array}
\left\{
\begin{array}{l}
CH_3CH{=}CH_2 + CH_4\\
CH_2{=}CH_2 + CH_3CH_3\\
CH_3CH_2CH{=}CH_2 + H_2\\
CH_3CH{=}CHCH_3 + H_2
\end{array}
\right.
\qquad (20\text{-}13)
$$

Note in Eq. (20–13) that each time a segment is broken from the carbon chain there are insufficient hydrogens for all the carbon valences; consequently each break produces an alkene plus a saturated compound. The resultant mixture can be separated into its various components, each of which has a use. The cracking process is the industrial source of simple alkenes, because only small amounts exist originally in petroleum.

Alkenes which can be obtained inexpensively in this way are important intermediates for the synthesis of other organic compounds. For example, polyethylene, polypropylene, and butyl rubber are some of the materials obtained directly from small alkenes by a process called polymerization, which is discussed in Chapter 27. Also, alkenes can be converted to alcohols, and these in turn to a wide variety of other organic materials.

Another important use of the cracking reaction is in the manufacture of gasoline. The hydrocarbons suitable for gasoline are chiefly octanes and other alkanes slightly smaller or larger (see Table 19.4). The amount of these found naturally in petroleum is seldom great enough to satisfy the demand for gasoline. Therefore an important part of any petroleum refinery is a cracking unit in which larger alkanes found in kerosene or fuel oil are broken into smaller molecules. The fraction of alkanes and alkenes in the C_5 to C_{10} size range constitutes "cracked gas." It has a superior octane rating and is blended with the natural gasoline.

20.7 **ALKYNES**	Unsaturated hydrocarbons having a *triple* bond between two carbons are called **alkynes**. The ending *-yne* is used in a systematic name; the rest of the naming rules are the same as those for alkenes.

$$
\begin{array}{cc}
CH_3C{\equiv}CCH_3 &
\begin{array}{c}
\text{Br}\\
|\\
HC{\equiv}C—CH—CH—CH_3\\
|\\
CH_3
\end{array}\\
\text{2-butyne} & \text{4-bromo-3-methyl-1-pentyne}
\end{array}
$$

Only a very few of the carbon skeletons produced by living organisms have triple bonds. By comparison, alkene double bonds are very common in nature.

The structure of an alkyne triple bond is generally considered to involve the same principles of bonding envisioned for a double bond. As shown in Fig. 20.8, a carbon atom at a triple bond has only two sigma bonds. To provide for these sigma bonds, the carbon forms two **sp** hybrid orbitals and is left with *two* electrons in **p** orbitals. Each **p** orbital overlaps sideways with a similar **p** orbital from the adjacent carbon, which produces two pi orbitals.

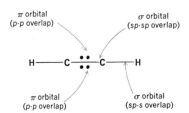

Fig. 20.8   Bond formation in acetylene, HC≡CH, showing the types of molecular orbitals.

The sigma bonds of an alkyne (like those of an alkane or an alkene) are oriented as far from each other as possible. This results in bond angles of 180° at the alkyne triple bond:

$$\overset{180°}{CH_3\overset{\frown}{-}C\equiv C-CH_3}$$

Chemical reactions of alkynes are similar to those of alkenes, the chief difference being that the alkyne has *two* pi bonds that can react.

$$CH_3-CH_2-C\equiv CH \;+\; 2H_2 \;\xrightarrow{Ni}\; CH_3-CH_2-CH_2-CH_3$$

$$CH_3-CH_2-C\equiv CH \;+\; HCl \;\longrightarrow\; CH_3-CH_2-\underset{\underset{Cl}{|}}{C}=CH_2$$

Markovnikov rule applies

(1 mole)   (1 mole)

$$CH_3-CH_2-C\equiv CH \;+\; 2HCl \;\to\; CH_3-CH_2-\overset{\overset{Cl}{|}}{\underset{\underset{Cl}{|}}{C}}-CH_3$$

(1 mole)   (excess)

Alkynes also give positive results with the bromine test (Section 20.5C) and the permanganate test (Section 20.5H). The tests, therefore, cannot be used to distinguish an alkene from an alkyne. Instead, they are general for any unsaturated compound.

20.8
ACETYLENE
(Ethyne)

Ethyne, HC≡CH, requires some extra attention because of its tremendous practical importance. It is used widely and therefore is nearly always called by its common name, acetylene. The production of acetylene in the United States is now well over one billion pounds a year. The oldest method of synthesis involves heating coke and lime in an electric furnace which converts them to calcium carbide. Treatment of the calcium carbide with water yields acetylene.

Slightly less than half of the acetylene manufactured in the United States now comes from a different raw material, natural gas (methane). Exposure to a very high temperature for less than one-tenth of a second converts the methane to acetylene. (Of course, oxygen must be excluded, to prevent combustion.)

$$2CH_4 \;\xrightarrow{1600°}\; HC\equiv CH + 3H_2 \qquad\qquad (20\text{-}14)$$

Combustion of acetylene, especially in the presence of oxygen rather than air, produces a very hot flame (near 2800°C). About 15% of the domestic production of acetylene is used in oxyacetylene torches for cutting and welding steel.

The rest of the acetylene goes into the synthesis of other organic compounds. The equations below show some important synthetic uses of acetylene. Note that each one starts with an addition to the unsaturated bond of acetylene.

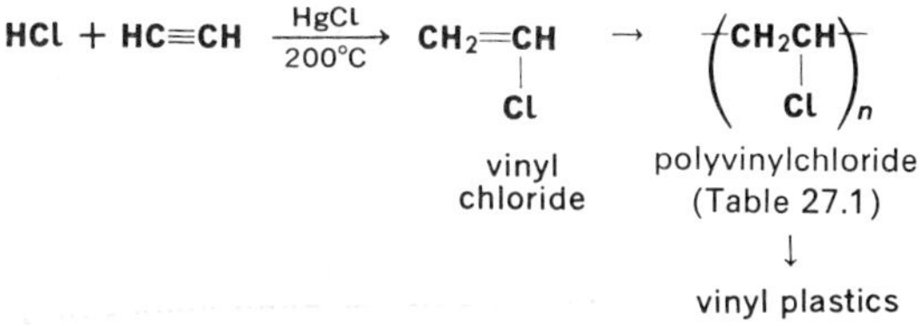

$$HCl + HC\equiv CH \xrightarrow[200°C]{HgCl} CH_2{=}CH \atop | \atop Cl \longrightarrow \left(CH_2CH \atop | \atop Cl \right)_n \tag{20-15}$$

vinyl
chloride

polyvinylchloride
(Table 27.1)
↓
vinyl plastics

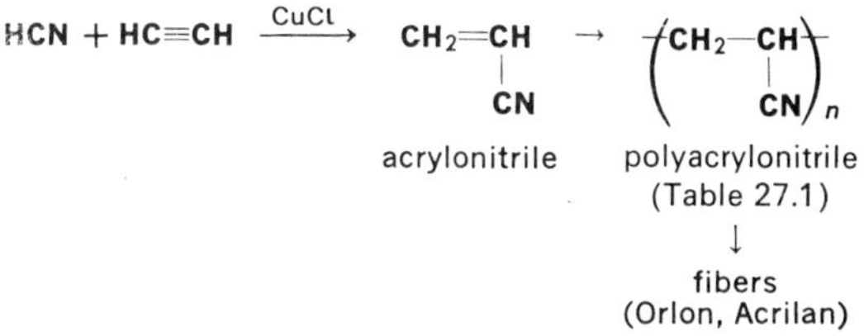

$$HCN + HC\equiv CH \xrightarrow{CuCl} CH_2{=}CH \atop | \atop CN \longrightarrow \left(CH_2{-}CH \atop | \atop CN \right)_n \tag{20-16}$$

acrylonitrile

polyacrylonitrile
(Table 27.1)
↓
fibers
(Orlon, Acrilan)

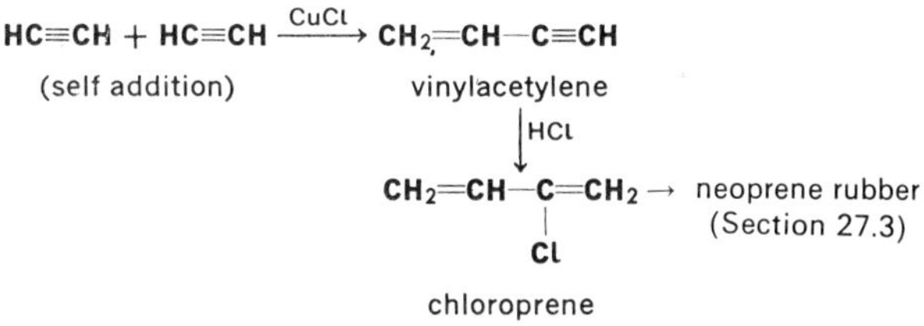

$$HC\equiv CH + HC\equiv CH \xrightarrow{CuCl} CH_2{=}CH{-}C\equiv CH \tag{20-17}$$

(self addition)   vinylacetylene
│HCl
$$CH_2{=}CH{-}C{=}CH_2 \atop | \atop Cl \longrightarrow neoprene\ rubber$$
(Section 27.3)

chloroprene

**20.9
PHYSICAL
PROPERTIES OF
UNSATURATED
HYDROCARBONS**

Because they have pi bonds, alkenes and alkynes are chemically much more reactive than alkanes. However, the pi bond causes only slight changes in physical properties. The boiling point of 1-hexene (63°C) or of 2-hexene (68°C) is about the same as that of hexane (69°C). Alkenes and alkynes are also like alkanes in solubility. All are insoluble in water but very soluble in nonpolar compounds.

SUMMARY

An alkene is a hydrocarbon having a carbon–carbon double bond. One of the bonds in the double bond is a skeletal bond, a *sigma* bond. The "extra" bond is a *pi* bond, which is reactive and constitutes the functional group of this class of compounds.

Many of the chemical reactions of the pi bond are *addition* reactions. The most common reagents which add to a pi bond are hydrogen, halogens (Cl_2, Br_2), acids (HCl, H_2SO_4), and water. Beside undergoing addition reactions, the alkene pi bond is easily oxidized. One useful oxidation product is a diol.

The structure of a pi bond also provides an interesting geometry to an alkene molecule: The bond angles at the C=C bond are 120° and *cis-trans* isomers are possible.

An alkyne has a carbon–carbon triple bond, two of which are pi bonds. Therefore the chemical properties of an alkyne are, for the most part, very similar to those of an alkene: additions and oxidation.

IMPORTANT TERMS

addition reaction	*cis-trans* isomers	olefin
alkene	cracking reaction	oxidation
alkyne	hydrogenation	pi bond
catalyst	Markovnikov rule	unsaturated hydrocarbon

WORK EXERCISES

1. Write the condensed structural formula and molecular formula for the following.
   a) hexane, 1-hexene, and cyclohexane
   b) heptane, 2-heptene, and cycloheptane
   c) 2-methylpentane;   2,3-dimethyl-2-butene; methylcyclopentane

2. Which compounds in Exercise 1 are isomers? What generalization can you make about the relation between alkenes and cycloalkanes?

3. Write the structural formulas of all possible open-chain compounds (i.e., those whose carbon skeletons are not rings) that have the following molecular formulas.
   a) C_4H_8 b) C_5H_{10}

4. Write the structural formulas of all possible ring compounds having these molecular formulas:
   a) C_4H_8 b) C_5H_{10}

5. In view of your answers to Exercises 3 and 4, give the total number of isomers of the following.
   a) C_4H_8 b) C_5H_{10}

6. Write a systematic name for each compound in Exercise 3.

7. Write the condensed structural formula for each of the following.
   a) 3-chloro-1-butene   b) *trans*-2 butene
   c) cyclohexene   d) 3-hexyne
   e) *cis*-4-methyl-2-pentene
   f) 1-bromo-2-butyne

8. Briefly describe the principal method for obtaining alkenes in large quantities. What is the raw material? What processes and reactions are used to obtain an alkene from it?

9. Why are alkenes more reactive than alkanes?

10. Write a complete, balanced equation for the reaction occurring in each mixture below, using structural formulas. If necessary, state "no reaction."
    a) 1-butene, hydrogen, platinum
    b) propene, bromine, 25°C
    c) 2-pentene burning in air
    d) 1-butene, hydrogen chloride
    e) cyclopentene, chlorine, 25°C
    f) cyclohexene, hydrogen, nickel
    g) 2-butene, concentrated sulfuric acid
    h) 1 mole of ethyne, 1 mole of hydrogen chloride
    i) propene, hydrogen chloride
    j) propene, chlorine
    k) propyne, excess bromine
    l) 1 mole propyne, 2 moles hydrogen chloride
    m) propyne burning in air
    n) 1-butene, hydrogen peroxide
    o) 2-pentene, dilute sodium hydroxide
    p) limonene (Fig. 20.2), excess chlorine
    q) myrcene (Fig. 20.2), excess hydrogen, nickel
    r) cyclohexene, hydrogen peroxide

11. Show the main organic product for each reaction mixture. (A complete, balanced equation is not necessary.)
    a) cyclohexene, dil. $KMnO_4$, 25°
    b) isoprene (Table 20.1), excess HCl
    c) 1-pentene, HI
    d) cyclopentene, conc. H_2SO_4
    e) 1 mol 2-butyne, 1 mol Br_2
    f) 1 mol propyne, 1 mol HCl
    g) 2-methyl-1-butene, HCl
    h) 2-butene, hydrogen peroxide
    i) 2-methyl-1-butene, conc. H_2SO_4
    j) 2,4-dimethyl-2-pentene, dil. $KMnO_4$
    k) 2,4-pentadiene, excess H_2, Pt

12. Which of the following would react with dilute permanganate solution at 25°C, causing its purple color to be replaced by a brown precipitate of MnO_2?
    a) propene   b) butane
    c) cyclopentane   d) butyne
    e) 2-butene   f) cyclohexene
    g) 3-methyl-1-pentene   h) 2-chlorobutane

13. Show structural formulas of at least six compounds, not counting *cis-trans* isomers, which would result from the cracking of pentane at 500°C.

14. Consider the synthesis of chloroethane from ethane, on an industrial scale.

a) What would be the source of the ethane?

b) Write an equation to show how ethane could be converted to chloroethane in one step; i.e., by one reaction.

c) Write equations to show how ethane could be converted to chloroethane by two steps.

d) Suggest a reason why method (c) might be more satisfactory for producing pure chloroethane than method (b), despite the fact that (c) requires two steps.

15. Write an equation for the addition of water to:

a) 2-butene

b) 1-pentene

c) methylpropene

d) cyclopentene

e) 3-methyl-1-butene

f) 2-methyl-2-butene

SUGGESTED READING

Jones, J., "The Markovnikov Rule," *J. Chem. Educ.* **38**, 297 (1961). A fascinating account of the development of a scientific "law." Includes many details and special examples not described in this book.

Noller, C. R., *Textbook of Organic Chemistry*, 3rd ed., Chapter 6, W. B. Saunders, Philadelphia, 1966.

21

Aromatic Hydrocarbons:
Benzene and Related Compounds

Chemists long ago became intrigued by the fragrant oils produced by certain plants, such as clove oil, wintergreen oil, and almond oil. The compounds seemed to share so many similar properties that chemists thought of them as a particular class of substances. They became known as *aromatic* compounds because of their odors.

One simple aromatic compound is **toluene**, C_7H_8, obtained by heating the balsam formed in the bark of the South American *tolu* tree. Another is produced when the fragrant *benzoin gum* is decomposed to yield a volatile liquid, of formula C_6H_6, which was named **benzene**. Gradually chemists recognized that the compounds in their "aromatic" class all had a low ratio of hydrogens to carbons, had at least six carbons, and were apparently related to benzene. They also soon learned that "aromatic" was not a very accurate description. Many of the compounds which obviously belong in this class because of their structures and properties do *not* have odors; conversely, many substances having entirely different types of structure *are* fragrant. Although the old name has persisted, we now define an **aromatic compound** as *one having a characteristic structure like that of benzene.*

Clearly, it is important to learn *what* about the structure of benzene is so distinctive. For several decades chemists were baffled about how to write the structure of this compound. We shall not have time to review all the experimental observations and the numerous attempts to write structures consistent with the facts. Instead, we shall present the formulas which now seem to be the best representation of the structure of benzene.

Benzene, C_6H_6, has its six carbons joined in a ring, with one hydrogen on each carbon. This would require only three of the four valence electrons on each carbon, leaving one electron on each carbon unshared, as shown in formula (a). It might seem that an unshared electron could pair off with a similar one from an adjacent carbon, forming a structure having three double bonds, such as (b) or (c).

[structural formula (a)]   [structural formula (b)]   [structural formula (c)]

    (a)   (b)   (c)

Formulas (b) and (c) raise some annoying questions, the most puzzling of which is the fact that *benzene does not have the chemical properties of an alkene*. We said that when a compound has a carbon–carbon double bond, and when we write this in its formula, we expect it to behave like a typical alkene. Since benzene does not behave in this way, neither formula (b) nor (c) seems to represent its true structure.

The bonding in benzene is best understood in terms of the orbitals, diagrammed in Fig. 21.1. It is necessary for each carbon to form only three σ bonds to build the skeleton of the benzene molecule. For this purpose the carbon uses **sp**^{2} hybrid orbitals. All the bond angles are 120° and all the carbon and hydrogen atoms lie in one plane. Perpendicular to the plane of the σ bonds, each carbon has one more electron in a **p** orbital. Figure 21.1 is a top view of these. From the side, each **p** orbital would have a two-lobed shape, like the **p** orbitals shown in Fig. 20.4, which were used to form an alkene π bond. In the case of benzene, each **p** orbital overlaps with the adjacent **p** orbitals on *both* sides of it. Since all the **p** orbitals do this, the result is one large, continuous ring containing six electrons, as shown in Fig. 21.2. This is often described as being a dispersed "cloud" of electrons. These are called π electrons because they are not part of the σ skeleton of the molecule. However, these benzene π electrons behave quite differently from alkene π electrons.

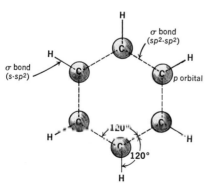

Fig. 21.1 Top view of the sigma skeleton of benzene. Each carbon still has one **p** orbital which is not yet bonded.

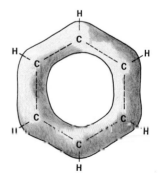

Fig. 21.2 Side-to-side overlap of all the **p** orbitals (from Fig. 21.1) produces one continuous "electron cloud" of six pi electrons.

The six electrons in the π cloud of benzene are said to be *delocalized*. This property makes the molecule especially stable. It is stable because the electron charges do not have to be concentrated in particular locations.

Since the benzene π electrons are not confined in *pairs* to a definite location, they are not accessible for the type of addition reaction that is so easy for an alkene. Also, benzene π electrons are not commonly attacked by an oxidizing agent. Benzene has its own unique chemical properties, which are quite unlike those of either an alkene or an alkane.

Although Fig. 21.2 gives a satisfactory picture of the distribution of the electrons in benzene, we are still left with the problem of how to write a simple formula using our standard symbols. It is inconvenient to draw a detailed picture such as Fig. 21.2 each time we want to write a chemical equation for benzene. Organic chemists now often use formula (d) or (e), in which the circle represents the continuous cloud of six π electrons.

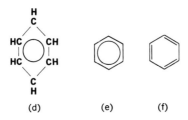

(d)   (e)   (f)

Formula (e) is the abbreviated style, in which we assume there is one carbon at each corner and each is holding one hydrogen. (Compare this formula with that for cyclohexane, Fig. 18.19.) In this book we shall use formula (e) for benzene. Another style, still found in many books, is to use a formula such as (c) or its abbreviated form (f). The disadvantage of formula (c) or (f) is that you must remember that benzene does not *really* have double bonds. The three double bonds shown in (f) must be thought of as representing a total of six π electrons which are delocalized and have the typical benzene properties.

21.2
NAMES

A compound having an alkyl group attached to a benzene ring is named by prefixing the alkyl group name to the word benzene. Methylbenzene, however, is ordinarily called toluene, even in systematic names.

benzene	CH_3	CH_2CH_3	$CH_2CH_2CH_2CH_3$
benzene	toluene (methylbenzene)	ethylbenzene	*n*-butylbenzene

The hydrocarbon names are used as the basis for the systematic names of other compounds. If two or more groups are attached to a ring, their relative positions are indicated by numbers. Since there is no "end" to the ring, we start counting at one of the groups.

bromobenzene   1,3-dichlorobenzene   2-bromo-5-chlorotoluene

In the last example, when the name toluene is used, counting must begin at the methyl group.

If *only* two groups are present, their relative positions can be designated by the terms *ortho*, *meta*, or *para*, abbreviated *o-*, *m-*, or *p-*. (These are acceptable systematic names, but originated from common names.)

o-bromotoluene   *m*-bromotoluene   *p*-bromotoluene

It is necessary to realize that, for benzene structures, the distance between groups is the important factor in assigning names. For example, all the formulas below represent the same compound, even though the molecules are shown in different positions.

1,3-dibromobenzene or *m*-dibromobenzene

It is possible for a compound to be built from two or more fused benzene rings.

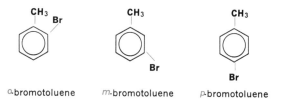

naphthalene, $C_{10}H_8$

The numbering system for naphthalene derivatives is shown below. In case only one group is present, the positions are sometimes designated by the Greek letters α and β, in common names.

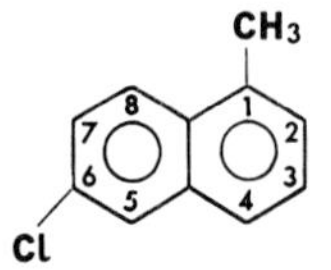

6-chloro-1-methylnaphthalene

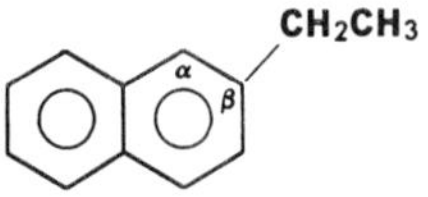

2-ethylnaphthalene
(β-ethylnaphthalene)

The skeletons of a few of the other possible fused-ring aromatic hydrocarbons are shown below.

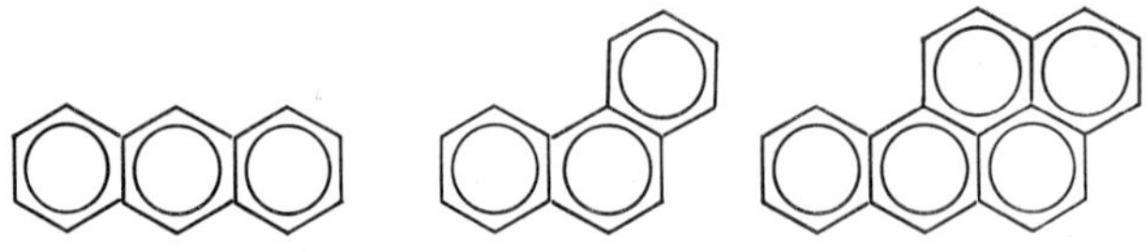

Some compounds composed of several rings induce the growth of cancer. One example is the last compound shown just above. Such compounds occur in coal tar and in cigarette tar.

In the examples of aromatic compounds shown so far, names have been written by naming the groups attached to the aromatic ring. Sometimes it is not convenient to do this, and it is easier to think of the ring as a group attached to some other carbon skeleton. In this situation, **phenyl** is the name used for the *group* obtained by removing one hydrogen from a benzene ring.

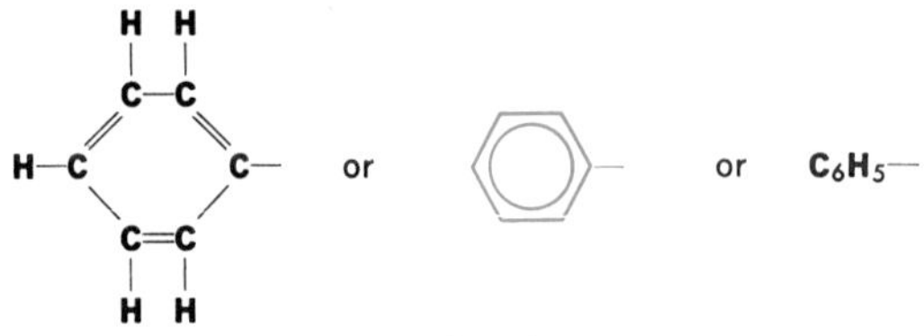

or   or C_6H_5-

phenyl group

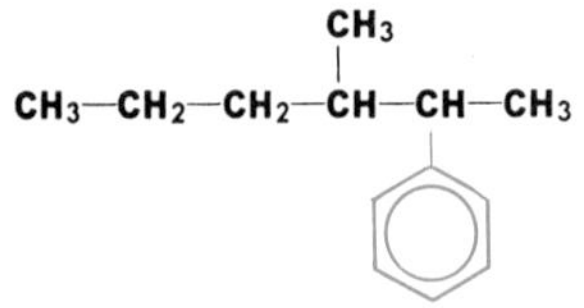

3-methyl-2-phenylhexane 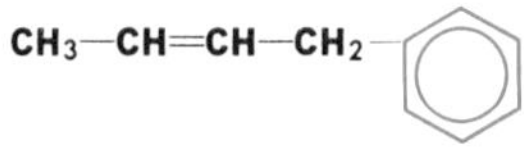   1-phenyl-2-butene

Here is a case of an illogical mixture of names surviving from the past. During the 1840's the compound now known as benzene was called *phene*. Hence it was reasonable to call the related group *phenyl*. The name benzene for the parent compound gradually became accepted, so phene died out, but unfortunately phenyl did not!

In general, any group of this type is called an **aryl** group; the following structures are illustrations.

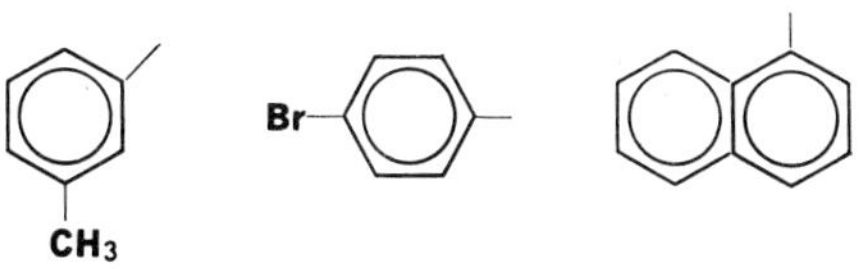

The symbol Ar- is used for an aryl group. Thus ArBr is an aryl bromide and ArH is an aromatic hydrocarbon.

**21.3
GENERAL
CHEMICAL
PROPERTIES**

We discovered in Section 21.1 that because benzene does not exhibit any of the customary alkene reactions it is necessary to develop a model, or picture, of benzene which truthfully represents these facts. Therefore we adopted the symbol (e) to portray benzene, with the understanding that the circle indicates the six π electrons which are held in a unique way. The concept to keep in mind is that the six electrons are very stable in the delocalized cloud around benzene. The cloud can rarely be broken up in such a way as to make the electrons available for addition reactions.

The following list shows several of the reactions in which benzene does *not* participate.

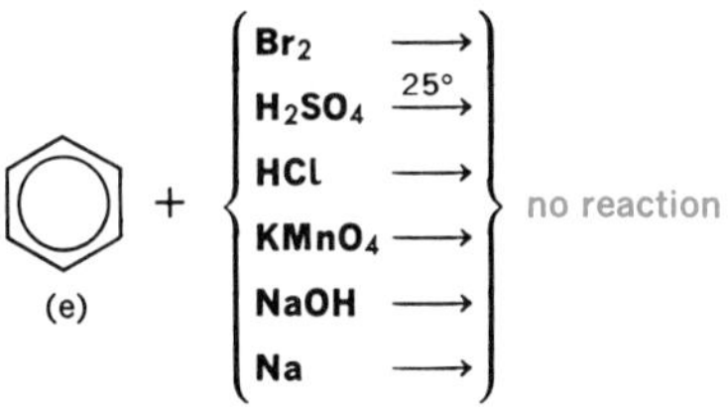

There are, however, a number of substances which react quite easily with benzene. Since addition reactions involving the electrons are rarely possible, the only other parts of the molecule available for reaction are the hydrogen atoms. *The reactions typical of a benzene ring are substitutions*, in which a new group replaces a hydrogen. These substitutions are distinctly different from alkane substitutions. Neither a high temperature nor unltraviolet light is necessary for benzene substitutions. If the proper catalysts are provided, the reactions occur easily at moderate temperatures. Three important examples are described in the following sections.

**21.4
REACTION
WITH
CHLORINE OR
BROMINE
(HALOGENA-
TION)**

If some bromine is added to benzene, the bromine dissolves in the liquid. However, no chemical change takes place; this was represented by the first equation of Section 21.3. However, if a small amount of iron (a tack will do, or a chip of iron from a machine shop) is added to the solution, some striking changes occur promptly. The solution becomes

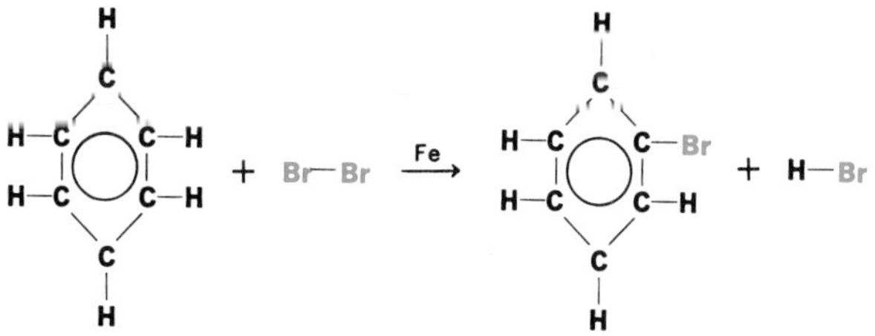

$$\text{benzene} + \text{Br-Br} \xrightarrow{\text{Fe}} \text{bromobenzene} + \text{H-Br} \tag{21-1}$$

warm. Gas bubbles out and turns to white fumes when it reaches the moist air outside the flask. These fumes have a sharp, acidic odor, and they turn litmus red; more thorough analysis proves that they consist of hydrogen bromide. If the liquid remaining in the flask is analyzed, one finds bromobenzene, C_6H_5Br. Clearly, a chemical reaction has occurred which was apparently catalyzed by the iron. The products suggest that a bromine atom was *substituted* in place of a hydrogen on the benzene ring.

Chlorine can react in the same way.

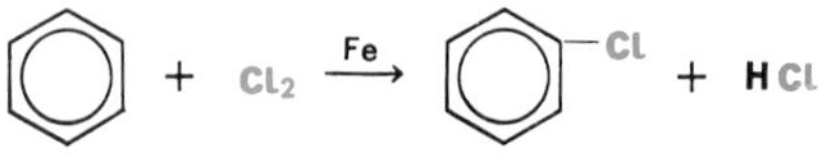

$$\text{(21-2)}$$

Equation (21–2) is actually balanced; remember that a hydrogen is present at each corner of the benzene ring even though it is not shown in the abbreviated formula.

Reactions of this type, called **halogenation**, provide a simple means to obtain aryl chlorides or bromides.

The unique behavior of benzene is well illustrated by its reaction with chlorine or bromine. The reactions occur by substitution, not addition. Furthermore, the substitution does not require ultraviolet light or high temperature (which is necessary for halogen substitution onto an alkane); instead, the reaction is catalyzed by another chemical substance, in this case, iron.

21.5
REACTION WITH
SULFURIC
ACID

Concentrated sulfuric acid does not react with benzene at room temperature; in fact, the two liquids are immiscible. However, if the mixture is heated to a moderate temperature (150°C) a substitution reaction, called **sulfonation**, occurs.

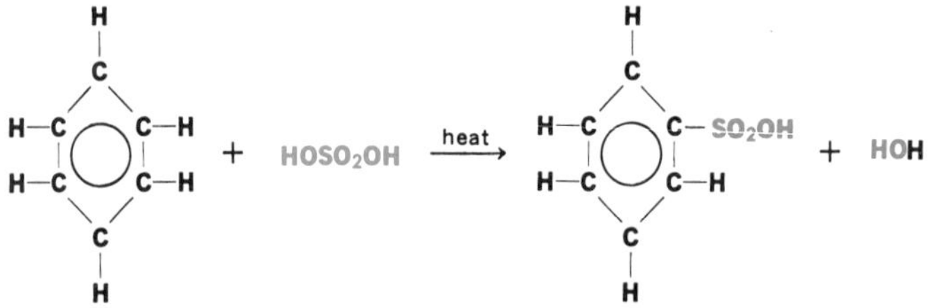

$$\text{(21-3)}$$

benzenesulfonic acid

Sulfonation, like the other benzene substitutions, requires a catalyst, despite the fact that in Eq. (21–3) none is apparent. Chemists have discovered that the reaction is self-catalyzed; one molecule of sulfuric acid promotes the reaction of another molecule.

Observe carefully the structure of the **benzenesulfonic acid**. There is a bond from a carbon to the sulfur (*not* to an oxygen). The three remaining oxygen atoms are all clustered around the sulfur. Review the structure of sulfuric acid discussed in Section 20.5F.

The structure of the sulfonic acid group is the same as that of sulfuric acid, except for the organic portion. Therefore, many of the properties of benzenesulfonic acid are similar to those of sulfuric acid. Benzenesulfonic acid is very soluble in water. It is a highly ionized, strong acid [Eq. (21–4)] and neutralizes bases [Eq. (21–5)].

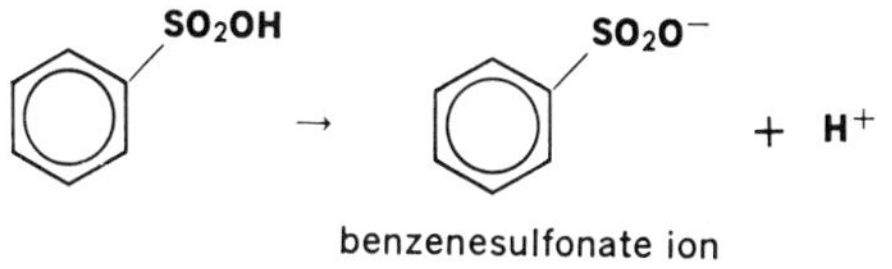

$$\text{(SO}_2\text{OH)} \rightarrow \text{(SO}_2\text{O}^-) + H^+ \qquad (21\text{-}4)$$

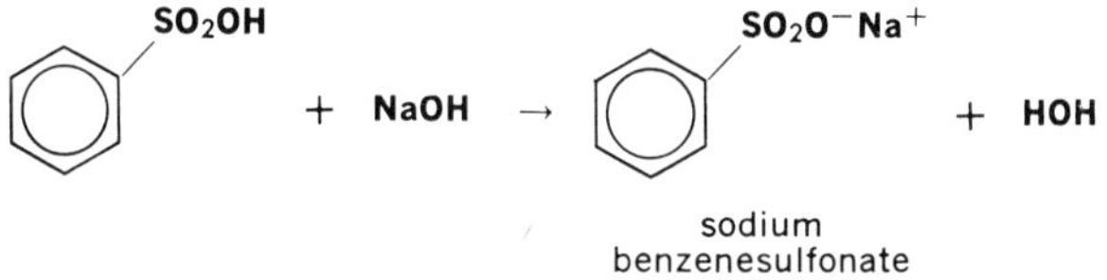

benzenesulfonate ion

$$\text{(SO}_2\text{OH)} + NaOH \rightarrow \text{(SO}_2\text{O}^-Na^+) + HOH \qquad (21\text{-}5)$$

sodium
benzenesulfonate

21.6
REACTION WITH
NITRIC ACID

If benzene is treated with a solution containing both concentrated nitric acid and concentrated sulfuric acid, substitution begins even at room temperature. A **nitro group** from nitric acid is substituted onto the ring; the process is termed **nitration**.

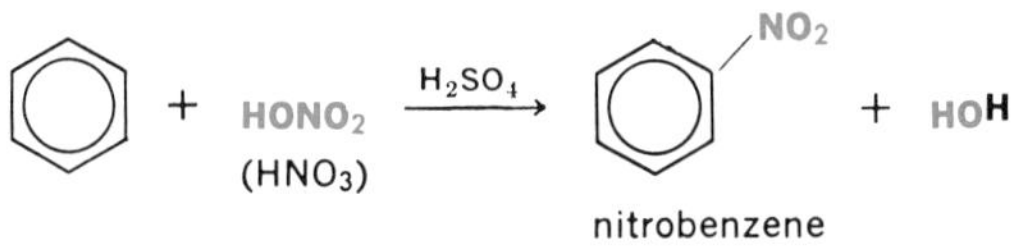

$$\text{(benzene)} + \underset{\text{(HNO}_3)}{HONO_2} \xrightarrow{H_2SO_4} \text{(NO}_2) + HOH \qquad (21\text{-}6)$$

nitrobenzene

The sulfuric acid is written over the arrow in Eq. (21–6) because it is the catalyst. Although sulfuric acid is important to promote the reaction, substitution by nitro groups is much easier than by sulfonic acid groups; the nitrobenzene will *not* be contaminated with any benzenesulfonic acid. In this connection, note that sulfonation [Eq. (21–3)] requires a higher temperature than does nitration [Eq. (21–6)].

The nitration reaction is used to prepare nitro compounds and is also one of the most frequently used methods for introducing other groups, indirectly, onto an aromatic ring. It will be shown in Section 25.5C that an aryl nitro compound $(ArNO_2)$ can easily be converted to an aryl amine $(ArNH_2)$. From the amine a great variety of other compounds can be made.

Compounds containing several nitro groups have explosive properties. Equation (21–7) shows how TNT can be made. Persons inexperienced in chemical procedures should not attempt this reaction because of the great danger involved.

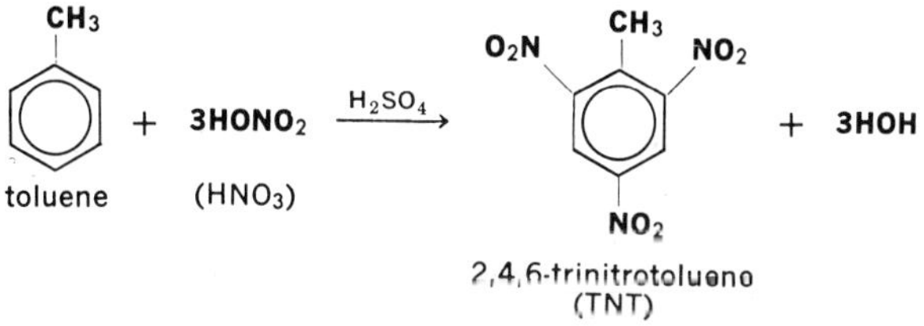

$$\underset{\text{toluene}}{\text{(CH}_3)} + \underset{\text{(HNO}_3)}{3HONO_2} \xrightarrow{H_2SO_4} \text{(O}_2N\text{, CH}_3\text{, NO}_2\text{, NO}_2) + 3HOH \qquad (21\text{-}7)$$

2,4,6-trinitrotoluene
(TNT)

21.7
OXIDATION OF
SIDE CHAINS

An alkyl group attached to an aromatic ring is often called a **side chain**. Although the saturated hydrocarbon structure found in an alkyl group normally resists oxidation, an aromatic ring affects a side chain in such a way that it *can* be oxidized.

It is often convenient to write an organic oxidation reaction as in Eq. (21–8). The

$$\text{C}_6\text{H}_5\text{—CH}_3 + 3[\text{O}] \xrightarrow{\text{KMnO}_4} \text{C}_6\text{H}_5\text{—C} \overset{\text{O}}{\underset{\text{OH}}{\diagdown}} + \text{H}_2\text{O} \qquad (21\text{–}8)$$

benzoic acid
(an organic acid)

oxygen in brackets indicates oxidation, but does not mean that O_2 is the oxidizing agent. In effect, the oxygen is supplied by the inorganic oxidizing agent, which is written above the arrow. We do not attempt to show the change in the oxidizing agent nor include it in the balanced part of the equation; instead, we focus our attention on the changes in the organic substances. The equation is balanced as far as the oxygen and organic compounds are concerned. Actually, the $KMnO_4$ taking part in reaction (21–8) would be reduced to MnO_2, since the toluene is being oxidized.

If there is more than one side chain attached to a benzene ring, all of them can be oxidized. Also, other oxidizing agents can be used, as shown in Eq. (21–9).

$$\text{(}o\text{-xylene)} + 6[\text{O}] \xrightarrow[\text{or KMnO}_4]{\text{Na}_2\text{Cr}_2\text{O}_7} \text{(phthalic acid)} + 2\text{H}_2\text{O} \qquad (21\text{–}9)$$

phthalic acid

Longer side chains can likewise be oxidized (Eq. 21–10). In each case the oxidation starts at the carbon attached to the ring, emphasizing the fact that it is the ring which makes the alkyl side chain sensitive to attack. Once the oxidation begins, the rest of the carbons in the side chain are, as a rule, completely oxidized to carbon dioxide.

$$\text{C}_6\text{H}_5\text{—CH}_2\text{CH}_2\text{CH}_3 + 9[\text{O}] \xrightarrow{\text{KMnO}_4} \text{C}_6\text{H}_5\text{—C} \overset{\text{O}}{\underset{\text{OH}}{\diagdown}} + 2\text{CO}_2 + 3\text{H}_2\text{O} \qquad (21\text{–}10)$$

Side-chain oxidation is useful for the preparation of an aromatic acid and for the identification of the original hydrocarbon.

**21.8
HETEROCYCLIC
COMPOUNDS**

Many important compounds, both natural and synthetic, have ring structures in which there are some atoms other than carbons; hence they are called **heterocyclic** compounds (Gr. *hetero-*, other, different). The *hetero* atoms commonly found are nitrogen, sulfur, or oxygen. The ring may be saturated, unsaturated, or aromatic. Page 409 gives examples. Heterocyclic compounds will be encountered occasionally later in the book because they occur in some very important compounds such as vitamins, enzymes, and nucleic acids. It will not be necessary to discuss their chemistry in detail, but it is important to be aware that structures of this kind exist.

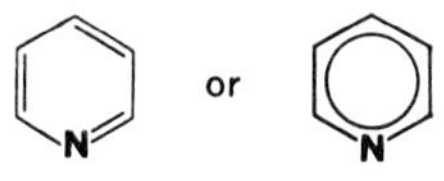

pyridine

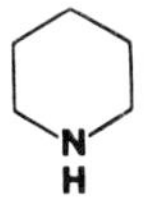

piperidine

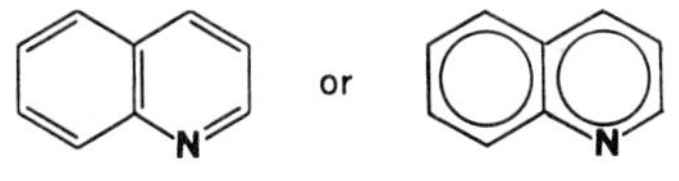

quinoline

furan

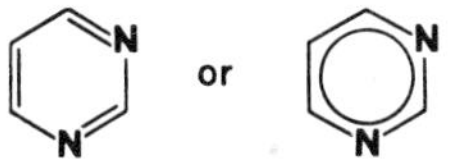

pyrimidine

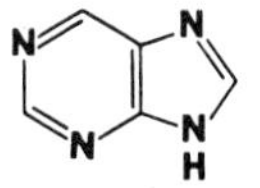

purine

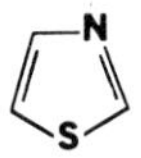

thiazole

**21.9
SOURCES OF
AROMATIC
COMPOUNDS**

If coal is heated in the absence of air (so that it is not just burned up), a portion of the material vaporizes. The residue, called **coke**, is carbon plus small amounts of mineral impurities, mainly clay.

$$\text{coal} \xrightarrow{\text{heat}} \underset{\text{(residue)}}{\text{coke}} + \underset{\text{(volatile)}}{\text{coal tar}}$$

The hot vapors that escape can be led to a cool chamber where they condense to become a thick liquid called **coal tar**. Coal tar is a valuable mixture composed of many different aromatic compounds, a few of which are listed in Table 21.1 to provide some idea of the variety present. Although living plants and animals produce aromatic substances, they rarely yield large quantities of them. Consequently coal tar is the richest natural source of aromatic compounds.* Of course, the coal tar mixture must be separated into various fractions before individual compounds can be obtained. Both chemical and physical methods of separation are used.

Conversion of 1 000 pounds of coal to coke yields only about 50 pounds of coal tar as a by-product. Since coke is used mainly for steel-making, the requirements of that industry determine the amount of coal tar available from this source. For many years the conversion of coal to coke has not supplied enough benzene and other simple aromatics to meet the great demand for their use in the manufacture of dyes, drugs, plastics, and fabrics. It has been necessary to seek additional sources of aromatic hydrocarbons.

Petroleum is a natural resource from which simple aromatic hydrocarbons can be *manufactured*. Most petroleum deposits contain only very small quantities of aromatic compounds. However, certain of the petroleum alkanes can be chemically transformed to aromatic hydrocarbons by the **catalytic reforming** process. Alkane vapors are passed over a platinum catalyst at about 500°C. For example, an alkane fraction containing primarily C_6 compounds is converted to benzene. In an analogous manner, toluene can be formed from a C_7 petroleum fraction. This reaction occurs because a six-membered ring closes more easily than does a seven-membered ring, and once it does, the very stable benzene ring structure can be formed.

---

* Recall that petroleum has been mentioned as the ultimate source of *alkanes* (Section 19.1).

TABLE 21.1   Types of compounds in coal tar

Hydrocarbons

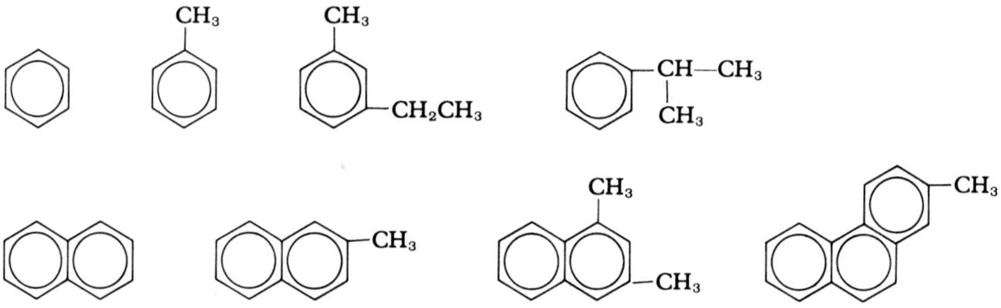

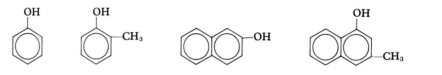

Phenols

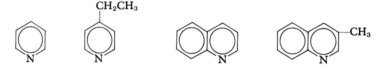

Nitrogen Heterocycles

21.10 ORGANIC HALOGEN COMPOUNDS

21.10
ORGANIC
HALOGEN
COMPOUNDS

In Section 19.6 we saw examples of alkyl halides and in Section 21.4 examples of aryl halides. A tremendous number of different organic halides have been synthesized in laboratories or factories. They have interesting and, in many cases, very useful properties. But because they are not natural, most are not readily biodegradable and often cause extensive damage to living organisms.

In nature, organic halogen compounds are exceedingly rare. Curiously, one of these rarities is absolutely essential to human life. The compound thyroxine is a part of the thyroid hormone.

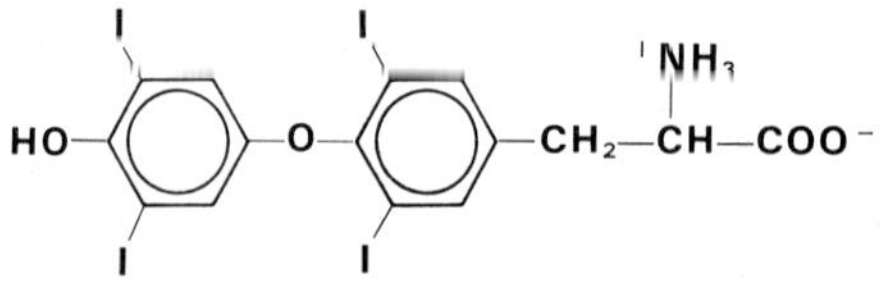

thyroxine

Thyroxine is in turn built up from the amino acid tyrosine (Section 31.2). Because of the iodine atoms present in the structure of thyroxine, humans must have a steady supply of iodine in their food. People who regularly eat seafood nearly always obtain sufficient iodine. For others, the easiest method is to use iodized salt. Two other natural organic halides are the antibiotics: Chloromycetin (chloramphenicol) and Aureomycin (chlorotetracycline).

Several synthetic organic halogen compounds are listed in Table 21.2; for each one the proper systematic name is shown first, followed by a common name or trademark names.

The highly halogenated compounds (such as CCl_4, $CHCl_3$, or CF_2Cl_2) are exceptions to the general rule that organic compounds are combustible; perhaps they are the only exceptions, aside from CO_2 and carbonates. In fact, carbon tetrachloride was formerly used in certain small fire extinguishers. The practice is extremely dangerous, however, because the vapors of carbon tetrachloride and its decomposition products can be lethal.

Nearly all of the organic halogen compounds are similar to each other in two characteristics. First, the liquid organic halides are good solvents for many *nonpolar* or moderately polar organic compounds, such as alkanes, fats, and waxes. Some organic halides will dissolve a greater variety of substances than the alkanes will. In water, the organic halides are insoluble. Second, their biological effects are pronounced. Depending on the particular halogen compound and its concentration, it may change the activity of individual cells, especially nerve cells. It may even kill the cells or kill the whole organism. The volatile halogen compounds, which can be inhaled, affect the central nervous system leading to unconsciousness if used in sufficient amounts. Large doses may cause permanent damage to certain tissues or may cause death due to suspension of vital functions.

The compound **d**ichloro**d**iphenyl**t**richloroethane, commonly called DDT, was first reported in the doctoral dissertation of a German chemist in 1874. In 1939, the Swiss chemist Paul Mueller discovered that it was an effective insecticide. DDT is a striking example of the dilemma so often created by organic halides. They can be both very beneficial and very dangerous.

The widespread use of DDT was stimulated by World War II. It was DDT that controlled an epidemic of typhus in Italy in 1943–44 by killing the lice that carried the disease. In fact, World War II was the first war in which disease did not kill more people than were killed in battle. This was due in part to DDT, but penicillin, the sulfa drugs, and better knowledge and medical care were also responsible. In 1953, it was estimated that on a worldwide basis DDT had saved five million lives and prevented 100 million illnesses, largely by control of malaria-carrying mosquitoes.

A number of chlorinated hydrocarbons such as DDT and hexachlorocyclohexane proved to be very effective insecticides initially. Unfortunately, many insect populations gradually developed tolerance to these pesticides at the dosage levels at which they were being applied. As a result it became necessary to use greater and greater concentrations to achieve the same degree of insect control. This in turn greatly increased the killing or harming of other organisms.

The threat to the environment is especially great because the halogenated compounds are persistent. They are destroyed only very slowly by the chemical, biochemical, and physical processes occurring in nature. It has been estimated that in the 25 years between 1945 and 1970, about 1.5 million tons of DDT were used and about 1 million tons were still present in the environment at the end of that period. Added to this is the fact that DDT is fat-soluble and tends to be stored in the bodies of animals that eat it. The amount of DDT increases in each member of a food chain. Water with as little as 0.000 001 parts per million of DDT may contain plants with 0.01 ppm. Fish that eat these plants contain 2 ppm, and birds of prey that eat fish contain as much as 10 ppm. Although there is still some controversy as to the exact relationship of DDT to their deaths, several birds such

TABLE 21.2 Some synthetic organic halides

Formula	Name	Applications
CCl_4	**tetrachloromethane**; carbon tetrachloride; Carbona contains CCl_4 and benzene	dry-cleaning fluid; solvent for oils, fats, waxes; some insecticide use; in veterinary medicine, used against worms and flukes
$CHCl_3$	**trichloromethane**; chloroform	solvent; general inhalation anesthetic (safety margin between anesthetic dose and lethal dose very small; can damage liver; may cause cancer)
CHI_3	**triiodomethane**; iodoform	antiseptic and local anesthetic applied to small wounds, especially in veterinary practice
CF_2Cl_2	**dichlorodifluoromethane**; Freon-12	fluid used in refrigerators; propellant in aerosols for cosmetics, paints, etc., but not for foods
CH_3CH_2Cl	**chloroethane**; ethyl chloride	local anesthetic for minor surgery (freeze technique)
Cl\ /Cl C=C Cl/ \Cl	**tetrachloroethene**	dry-cleaning fluid, especially in coin-operated machines; metal degreasing; medically, in both humans and animals, against worms and flukes
Cl\ /H C=C Cl/ \Cl	**trichloroethene**; Trilene	dry-cleaning fluid; metal degreasing; solvent; inhalation anesthetic for obstetrics and short operations
Br F | | H—C—C—F | | Cl F	**2-bromo-2-chloro-1,1,1-trifluoroethane**; halothane; Fluothane	inhalation anesthetic
Cl—⟨○⟩—Cl	**para-dichlorobenzene**; Dichloricide; Paramoth	kills moths in woolens (colorless crystals in flakes or lumps)

Cl   Cl
  | |
CH—CH
/   \\
Cl—CH   CH—Cl
\\   /
CH—CH
  | |
Cl   Cl
	hexachlorocyclohexane
	($C_6H_6Cl_6$); Lindane; 666;
	Gammexane; BHC, benzene-
	hexachloride (erroneous)
		effective insecticide; has had wide agricultural use; damages the environment

Cl—⬡—C(H)(—C(Cl)—Cl with Cl below)—⬡—Cl	**1,1,1-trichloro-2,2-di (p-chlorophenyl)ethane** dichlorodiphenyltrichloroethane; DDT	heavily used insecticide; creates many environmental problems
Cl, Cl—⬡(Cl)—O—CH$_2$—COOH	**2,4,5-trichlorophenoxyacetic acid**; 2,4,5-T	herbicide; enormous use as defoliant during Vietnam war

as the peregrine falcon, the brown pelican, and the bald eagle are rapidly decreasing in number. These birds and their eggs contain relatively large amounts of DDT. In contrast, a controlled experiment in which pheasants, quail, and chickens were fed DDT indicated that they were resistant to its effects.

Concern about the use of DDT and the availability of other insecticides led the Environmental Protection Agency of the United States to prohibit most uses of DDT after January 1973. But since, in general, other insecticides are both more toxic and more expensive than DDT, poorer countries and those less technically developed continue to use large amounts of DDT.

Other insecticides are continually being developed. Our experiences with DDT have shown us that we must be much more aware of the consequences of introducing any new substance into our environment, particularly if the new substance is introduced in large quantities. Our battle with the insects is not over, and methods other than the use of insecticides are being used. So far, biological control methods have been effective in a few cases. Crop rotation and control of winter breeding grounds of insects by selective burning are also effective.

SUMMARY

All aromatic hydrocarbons have distinctive properties which are like those of benzene, the most simple and most typical member of this class. Although benzene has six pi electrons, it does not undergo addition reactions and is highly resistant to oxidation. Instead, it participates in substitution reactions, most notably with halogens, with sulfuric acid, and with nitric acid.

Organic halogen compounds may be of either the alkyl or the aryl type. With rare exceptions, organic halides are synthetic rather than natural, and have characteristic effects on living organisms. Some are quite harmful if not handled with due precautions.

Heterocyclic compounds have rings containing at least one atom other than carbon. However, the heterocyclic rings may be saturated, unsaturated, or aromatic. Many compounds found in either plants or animals have heterocyclic rings.

All the types of carbon skeletons found in organic compounds have now been described. In the remaining chapters concerning organic chemistry we shall discuss functional groups containing oxygen or nitrogen atoms, and occasionally sulfur atoms.

IMPORTANT TERMS

aromatic compound	heterocyclic compound	*ortho*
aryl group	*meta*	*para*
benzenesulfonic acid	nitration	phenyl group
halogenation	nitro group	sulfonation

WORK EXERCISES

1. Write a structural formula for
   a) toluene
   b) ethylbenzene
   c) *para*-dichlorobenzene
   d) 1,2-dinitrobenzene
   e) 3-bromonitrobenzene
   f) pyridine
   g) sodium benzenesulfonate
   h) *o*-nitrotoluene
   i) *p*-bromobenzenesulfonic acid
   j) 2-bromo-4-nitropropylbenzene
   k) 1-chloro-2-phenylpropane
   l) *m*-diethylbenzene
   m) *trans*-1,2-diphenylethene
   n) 2-nitronaphthalene

2. Give the structure and name of the products formed by the reaction of benzene with the reagents shown. If necessary, state "no reaction."
   a) concentrated sulfuric acid at 150°C
   b) hot concentrated hydrochloric acid
   c) a mixture of concentrated nitric and sulfuric acids
   d) chlorine and iron
   e) bromine at 25°C

3. Write the structure of the organic acid produced by each reaction mixture.
   a) toluene with sodium dichromate
   b) ethylbenzene with potassium permanganate
   c) 1,2,4-trimethylbenzene with potassium permanganate
   d) *m*-bromotoluene with sodium dichromate

4. Describe briefly the following, and give an example of each.
   a) aromatic compound
   b) heterocyclic compound

5. Briefly discuss the two chief sources of aromatic hydrocarbons.

6. Write the structural formulas to show how many trichlorobenzene ($C_6H_3Cl_3$) isomers are possible.

7. What are the three principal types of compounds found in coal tar?

8. Name these compounds.

a)

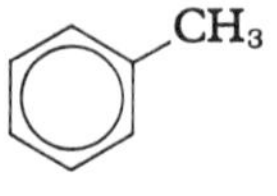

b)

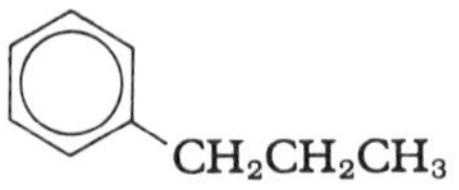

c)

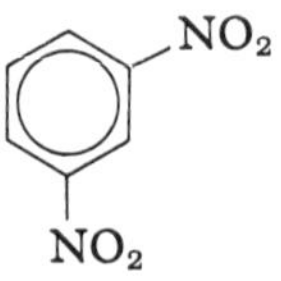

d)

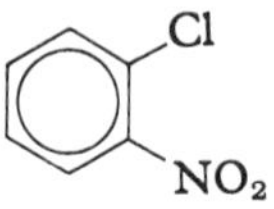

e)

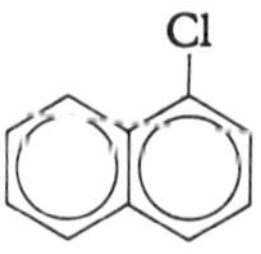

f)

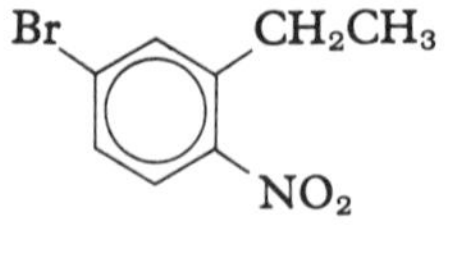

g)

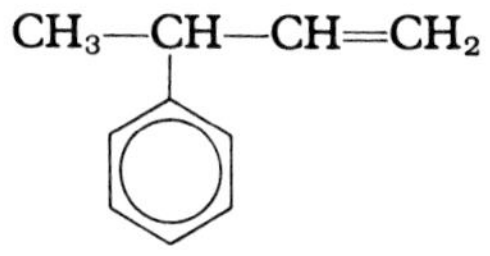

$$CH_3-CH-CH=CH_2$$

h)

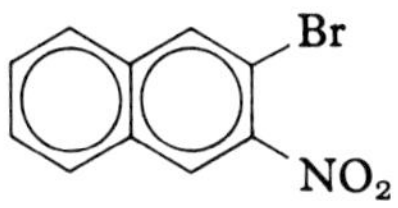

Br

NO_2

i)

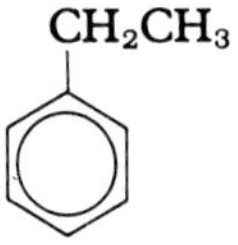

CH_2CH_3

j)

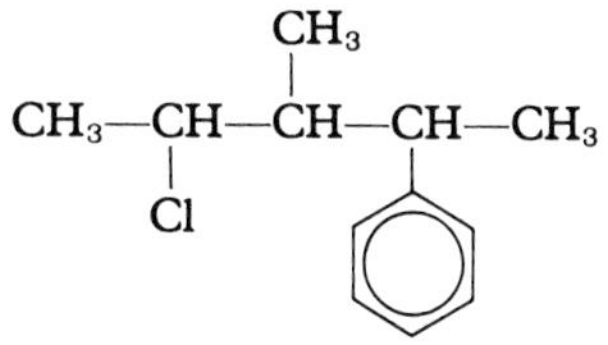

$$CH_3 \atop CH_3-CH-CH-CH-CH_3 \atop Cl$$

9. Write an equation for each reaction.
a) phenylethene, H_2, Pt
b) 3-phenylpropene, dil. $KMnO_4$, 25°
c) butylbenzene, Br_2, 25°
d) phenylethene, HCl
e) Trilene (Table 21.2), H_2, Ni
f) 2-phenylpropene, H_2O_2

SUGGESTED READING

Blumer, M., "Polycyclic aromatic compounds in nature," *Scientific American*, March 1976. These multiple-ring hydrocarbons have been found in soils and sediments around the world.

Breslow, R., "The nature of aromatic molecules," *Scientific American*, August 1972.

DiNardi, S. R., and A. M. Desmarais, "Polychlorinated biphenyls in the environment," *Chemistry* **49**, no. 4, p. 14, May 1976. An excellent discussion of the pollution produced by these persistent, poisonous substances.

Edwards, C. A., "Soil Pollutants and Soil Animals," *Scientific American*, April 1969 (Offprint #1138). The effects of DDT and other pesticides on various soil organisms are discussed and a warning given about bad effects noted.

Moore, J. W., "The Vinyl Chloride Story," *Chemistry*, **48**, no. 6, p. 12, June, 1975. Describes the discovery that this compound causes a rare form of cancer.

Peakall, D. B., "Pesticides and the Reproduction of Birds," *Scientific American*, April 1970 (Offprint #1174). The author tells about the effects of DDT and other pesticides on hawks and pelicans.

22

Alcohols, Phenols, and Ethers

22.1
INTRODUCTION

Nearly every primitive tribe has stumbled onto the fact that overripe fruits may undergo a pleasant type of spoilage (fermentation) which produces an intoxicating liquid. The more advanced societies discovered that by distillation they could concentrate the intoxicating substance from the original liquid. During the Middle Ages this essence or spirit became known as alcohol in some European languages. (The word was derived from the Arabic *al kuh'l*, but with considerable alteration in meaning.) Scientists eventually learned that this organic compound contained not only carbon and hydrogen, but also oxygen. Furthermore, many other oxygen-containing substances, having different amounts of carbon, were found to be structurally related to alcohol. Consequently the term alcohol was used to designate all compounds having this type of structure, and also to name the one particular compound. Two other classes of compounds, the ethers and phenols, proved to be structurally related, each in a different way, to the alcohols.

In the water molecule (H—O—H) the two covalent bonds of oxygen hold hydrogen atoms. Either one or both of these bonds may instead be linked to organic groups. In an **alcohol** (R—O—H), the oxygen holds one alkyl group. In a **phenol** (Ar—O—H), the oxygen holds one aryl group. In an **ether**, the oxygen holds two alkyl or aryl groups which may be the same or different (R—O—R, R—O—R′, R—O—Ar).

These oxygen-containing structures may be found in such diverse materials as sugars, pain-relievers, disinfectants, plastics, and solvents. Familiar examples of alcohols and ethers are these two compounds:

$$CH_3—CH_2—O—H \qquad\qquad CH_3—CH_2—O—CH_2—CH_3$$

ethanol
("alcohol," ethyl alcohol)

ethyl ether
(anesthetic "ether")

Vanillin, whose structure is shown on the next page, is the pleasant flavoring agent extracted from vanilla beans. Note that the compound has both an ether group and a phenol structure. Thymol, produced by the herb thyme, is used medically as a fungicide and also as a preservative for anatomical specimens.

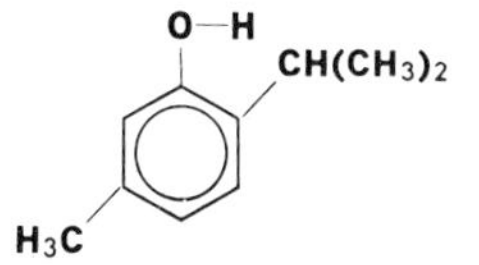

2-isopropyl-5-methylphenol
(thymol)

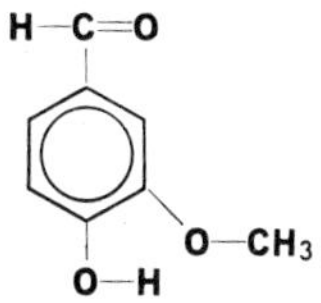

4-hydroxy-3-methoxybenzaldehyde
(vanillin)

**22.2
ALCOHOL
NOMENCLA-
TURE**

In *systematic nomenclature* an alcohol is named by using the ending *-ol* in place of the final *-e* in the hydrocarbon name. Thus, the two-carbon alcohol is called ethanol. In most cases it is necessary to include a number to indicate the location of the hydroxyl group on the carbon chain. The hydroxyl functional group is assigned the lowest number possible for it, in preference to any halogen or alkyl groups which may be present. These rules are illustrated in the examples below.

$CH_3CH_2CH_2OH$

1-propanol
(*n*-propyl alcohol)

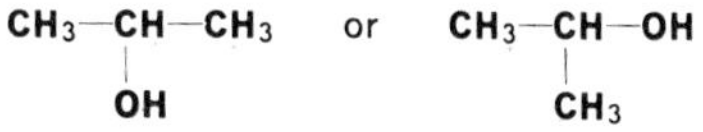

$CH_3-CH-CH_3$ or $CH_3-CH-OH$
  | |
  OH CH_3

2-propanol
(isopropyl alcohol)

$CH_3CH_2CH_2CH_2OH$

1-butanol
(*n*-butyl alcohol)

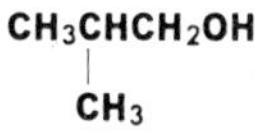

CH_3CHCH_2OH
  |
  CH_3

2-methyl-1-propanol
(isobutyl alcohol)

$CH_3CH_2CH_2CH_2CH_2CH_2OH$

1-hexanol
(*n*-hexyl alcohol)

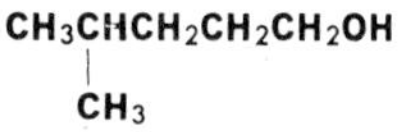

$CH_3CHCH_2CH_2CH_2OH$
  |
  CH_3

4-methyl-1-pentanol
(isohexyl alcohol)

Common names are also still used for many of the familiar alcohols. For these, a group name for the carbon skeleton is used, together with the word alcohol. In the preceding examples, the common name is shown in parentheses. Note that the term *iso* is used to indicate a methyl branch in the chain at the end farthest from the functional group; it must not be used for groups branched elsewhere.

If the reactions of the different butanol isomers are observed, one finds that their rates of reaction and kinds of products depend upon the position of the hydroxyl group in the molecule. Similar behavior of other isomeric alcohols leads to the generalization that their reactions depend not only on the presence of the hydroxyl group but also on the amount of branching in the carbon skeleton adjacent to the hydroxyl group. Consequently, an alcohol is classified as *primary*, *secondary*, or *tertiary* if there are, respectively, one, two, or three carbons directly bonded to the carbon holding the OH group.

$CH_3-CH_2-CH_2-OH$ $CH_3-CH-CH_2-CH_2-OH$
                                              |
                                            CH_3

a primary alcohol   a primary alcohol

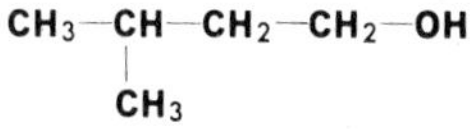

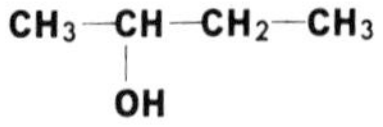

$$CH_3-CH-CH_2-CH_3$$
$$| $$
$$OH$$

a secondary alcohol

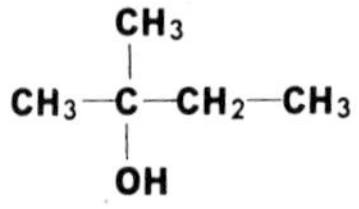

$$CH_3$$
$$|$$
$$CH_3-C-CH_2-CH_3$$
$$|$$
$$OH$$

a tertiary alcohol

Note that the total number of carbons in the molecule, or branching elsewhere in the carbon skeleton, is not considered when an alcohol is classified as primary, secondary, or tertiary.

Some common names for alkyl groups include the terms secondary (abbreviated *sec-*), or tertiary (abbreviated *tert-* or *t-*).

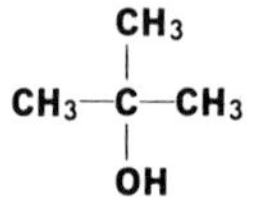

$$CH_3$$
$$|$$
$$CH_3-C-CH_3$$
$$|$$
$$OH$$

(tert-butyl alcohol)
2-methyl-2-propanol

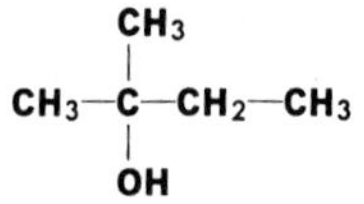

$$CH_3$$
$$|$$
$$CH_3-C-CH_2-CH_3$$
$$|$$
$$OH$$

(t-pentyl alcohol)
2-methyl-2-butanol

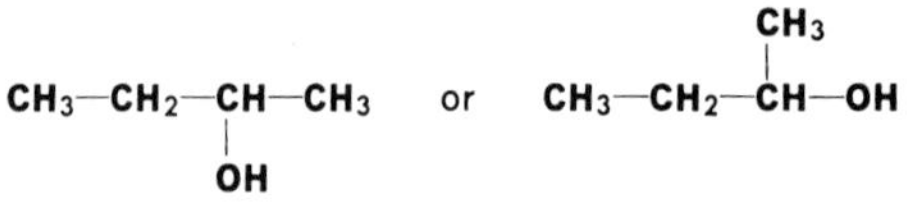

$$CH_3-CH_2-CH-CH_3$$
$$|$$
$$OH$$

or

$$CH_3$$
$$|$$
$$CH_3-CH_2-CH-OH$$

(sec-butyl alcohol)
2-butanol

Note carefully the difference between the sec-butyl group (shown just above) and the isobutyl group (shown on the preceding page). Also note that a common group name, such as isohexyl, designates in this one name the total number of carbons and that there are no additional names for alkyl substituents.

Some molecules may have more than one hydroxyl group. The most familiar of these are shown below. Their systematic names are included for illustration, but their common names are nearly always used. (The term *glycol* is a general name for those alcohols having two OH groups, usually on adjacent carbons. The name ethylene glycol is derived from the fact that this compound can be made from ethylene.)

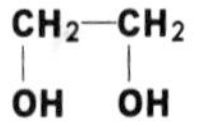

$$CH_2-CH_2$$
$$|\quad\ \ |$$
$$OH\ \ OH$$

(ethylene glycol)
1,2-ethanediol

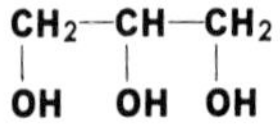

$$CH_2-CH-CH_2$$
$$|\quad\ \ |\quad\ \ |$$
$$OH\ \ OH\ \ OH$$

(glycerol)
1,2,3-propanetriol

22.3
SOURCES AND
USES OF
COMMON
ALCOHOLS

Methanol, also called *methyl alcohol,* or sometimes *wood alcohol,* was formerly obtained by destructive distillation* of wood. The product now manufactured synthetically is less expensive and so much purer that its industrial use has greatly increased. Each year tons of methanol are dehydrogenated (Section 22.5) to formaldehyde, which in turn is converted to phenol-formaldehyde, urea-formaldehyde, and melamine-formaldehyde polymers (Bakelite, Melmac, etc., Table 27.3) and to numerous other compounds.

---

* In destructive distillation a material is heated (in the absence of air) to a very high temperature, so that it decomposes and releases volatile substances.

Methanol is also used as an antifreeze, as a solvent, and as a starting material to build the structure of many dyes, drugs, and perfumes.

Methanol is a severe poison which can cause blindness and death. Even prolonged breathing of its vapor may be damaging; factory workers must be protected from the vapor by adequate ventilation.

Ethanol is the systematic chemical name for the substance popularly known as alcohol. Some other common names for it are spirits of wine, spirits, grain alcohol, and ethyl alcohol.

Physiologically, alcohol induces relaxation, a feeling of well-being, poor coordination, a dulled sense of judgment, and if consumed in sufficient quantities, causes drowsiness and unconsciousness. A large quantity consumed within a short time can even cause death. However, because of the initial effects of alcohol, a fatal concentration in the body is rarely established.

Grapes, honey, milk, cactus sap, berries, or almost any other imaginable source of sugar can be fermented to alcohol by yeasts. Furthermore, the starch in rice, corn, oats, potatoes, rye, etc., can be biochemically broken down to sugar and then fermented to alcohol. Fermentation rarely produces an alcohol content in excess of 13%. Natural table wines are 11–13% alcohol. The higher percentages in strong liquors can be obtained only by distilling to concentrate the alcohol. *Proof* is a term used in the United States to designate the strength of alcoholic beverages. Numerically, proof is twice the percentage; for example, a 90-proof gin contains 45% alcohol; the remainder is water and trace amounts of flavoring substances.

In almost every nation the production of alcohol is closely regulated and is taxed, often heavily. Enough beverage alcohol is consumed to make this tax a very important source of revenue. Unfortunately, alcohol also frequently creates medical and sociological problems.

Great quantities of alcohol are used industrially, in the form of **denatured alcohol**. This is alcohol to which a small amount of some other substance has been added to make it unfit for drinking. The additive has a disagreeable odor and taste and is often toxic. Some of the common denaturing agents are methanol, benzene, gasoline, pyridine, and isopropyl alcohol. Denatured alcohol is used to avoid the heavy beverage tax.

In both laboratory and industry, 95% alcohol is used extensively. This is the highest concentration which can be obtained by ordinary distillation; it is pure enough for many purposes. If, by special means, the last 5% of water is removed, the pure, water-free alcohol is called **absolute alcohol**.

Part of the industrial alcohol is obtained by fermentation of molasses, or sometimes of potatoes. Most is synthesized from ethene, obtained as usual from petroleum sources. The conversion from ethene to ethanol (Section 20.5G) is accomplished by passing steam and ethene through a hot tube containing phosphoric acid adsorbed on an inert material.

Ethanol is used for the synthesis of other organic compounds, as a component of lotions, perfumes, and cosmetics, and as a solvent.

Isopropyl alcohol is most familiar as the compound in "rubbing alcohol." Other applications are in hand lotions, after-shave lotions, cosmetics, quick-drying inks, and paints. The largest quantities of isopropyl alcohol are converted to acetone by the dehydrogenation reaction (Section 22.5). The isopropyl alcohol required for all these needs is synthesized from propene by the hydration process, utilizing sulfuric acid (Section 20.5G).

Ethylene glycol is an excellent automotive antifreeze because it is nonvolatile. It is also used in hydraulic brake fluids, printer's inks, ball-point pens, and as a solvent for certain paints, plastics, and other materials. Large quantities are converted to the polymers constituting Dacron fibers, Mylar film, and alkyd paints.

Glycerol, also called glycerin, is a constituent of all fatty foods and is readily used by the human body. This is interesting in view of the fact that other small alcohols, including ethylene glycol, are toxic.

Many countries obtain glycerol entirely as a by-product of soap-making. In the highly developed nations additional quantities are made synthetically because of the great demands for it. (One authority has reported more than 1 500 uses for glycerol!) The synthesis, developed by the Shell companies, uses propene from petroleum as the starting material.

Glycerol is a good humectant (moisture-retaining agent). This property makes it useful in the manufacture of tobacco, candy, cosmetics, skin lotions, inks, dentifrices, and pharmaceuticals. Eye corneas, blood cells, and other live tissues are often treated with glycerol for frozen storage. A few of the materials synthesized from glycerol are alkyd paints and other polymers, the explosive nitroglycerin, and monoglycerides and diglycerides used as emulsifiers and softening agents.

**22.4
PHYSICAL
PROPERTIES
OF ALCOHOLS,
PHENOLS, AND
ETHERS**

The volatility of an organic compound is directly related to its size. We stated previously that small alkanes are gases at room temperature, larger ones are liquids, and the largest are solids (Section 19.4). Similarly, within the group of alcohols a larger molecule has a higher boiling point than does a smaller one. It might seem, then, that methanol and ethanol would be gases, and would have very low boiling points. Instead, both are liquids. Evidently some factor other than size also affects the volatility of alcohols. It has been found that most physical properties of alcohols are influenced by hydrogen bonds.

Hydrogen bonds may be formed from one alcohol molecule to another, because of the highly polar covalent bond between oxygen and hydrogen (Fig. 22.1). These strong intermolecular attractions have a profound effect on the physical properties of **alcohols**, just as they do in the case of water (Section 9.7).

First, an alcohol has a *higher boiling point* than does an alkane hydrocarbon of about the same weight and size. This is illustrated by the data for 1-butanol and *n*-pentane, given in Table 22.1. Because of the hydrogen bond attractions between alcohol molecules, it is difficult for a molecule to break away from its neighbors in the liquid and exist as an individual molecule in the gas state. Therefore the alcohol requires more energy (higher temperature) for vaporization than does the hydrocarbon.

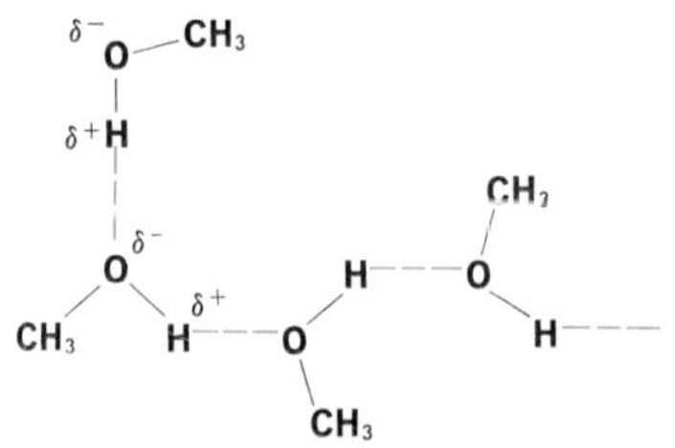

Fig. 22.1 Hydrogen bonds (color) create strong attractions between alcohol molecules.

TABLE 22.1   **Relation of boiling point to structure**

Compound	Molecular Weight	Structure	Boiling Point, °C
n-pentane	72	$CH_3CH_2CH_2CH_2CH_3$	36
1-butanol	74	$CH_3CH_2CH_2CH_2OH$	118
ethyl ether	74	$CH_3CH_2OCH_2CH_3$	35

Second, the simple alcohols are much *more soluble in water* than are the hydrocarbons of similar size. When an alcohol is mixed with water it can form hydrogen bonds to water molecules, and hence be pulled into the water, as indicated in Fig. 22.2.

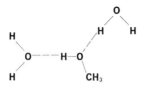

Fig. 22.2 Methanol dissolves in water because water molecules are attracted to it by hydrogen bonds (color).

However, a long-chain alcohol such as 1-octanol,

$$CH_3CH_2CH_2CH_2CH_2CH_2CH_2CH_2OH$$

is practically insoluble in water. The attraction of the polar OH group toward water molecules is not sufficient to pull such a large hydrocarbon chain into solution; the 1-octanol behaves nearly the same as does nonpolar octane. Alcohols of medium size, of course, have intermediate solubilities. The data in Table 22.2 show that the longer the carbon chain, the less soluble the alcohol is in water.

Although long-chain alcohols are insoluble in water, they do dissolve in octane or other hydrocarbons. The bulk of the long-chain alcohol molecule is the nonpolar hydrocarbon group, which readily associates with solvents similar to it in structure. Alcohol solubility is nicely summarized by the statement "like dissolves like."

TABLE 22.2   **Solubility of alcohols in water**

Alcohol	Solubility at 20°C (g ROH/100 g H_2O)
methanol	completely miscible
ethanol	completely miscible
1-propanol	completely miscible
1-butanol	8
1-pentanol	2.7
1-hexanol	0.6

In hot, dry locations tremendous amounts of water are lost from reservoirs through evaporation. Because of its solubility properties, octadecanol ($C_{18}H_{37}OH$) has been successfully to reduce this loss. When octadecanol is thrown on the water, the OH group is attracted to the water but is unable to pull such a long hydrocarbon group into solution. Consequently, the octadecanol molecules line up side by side on the surface, with the OH groups in the water and the long hydrocarbon "tails" sticking out of the water vertically. A continuous film (just one molecule thick!) forms on the surface, greatly hindering the evaporation of water molecules.

Alcohols can, in turn, act as solvents for *other* compounds; their ability to do this depends on the principles just discussed. Ethanol is a good solvent for alkanes and aromatic hydrocarbons as well as for polar compounds such as benzoic acid, acetone, and other alcohols. Methanol is a poorer solvent for alkanes, but dissolves the polar compounds and even some ionic salts.

Phenols follow the same solubility principles as do alcohols. Phenols are more water-soluble and have higher boiling points than do aromatic hydrocarbons of similar size. However, since even the smallest phenol has a skeleton of six carbons, phenols are only moderately soluble in water. Those having large carbon-skeletons are more soluble in hydrocarbons such as benzene. Most phenols are quite soluble in ethanol.

We note in Table 22.1 that the boiling point of **ethyl ether** is almost identical to that of *n*-pentane. This similarity is due to the fact that hydrogen bonding cannot exist between two ether molecules. In an ether, all the hydrogens are held on carbon atoms (Fig. 22.3). Since the carbon–hydrogen covalent bonds are *nonpolar*, a hydrogen atom of one molecule does not attract an oxygen atom in another molecule. Hydrogen bonding is important only when a hydrogen atom is caught between two very negative atoms, such as two oxygens (Fig. 22.1). Since hydrogen bonds do not exist between ethyl ether molecules, their volatility is very similar to that of *n*-pentane.

Ethyl ether, the most common ether, forms a separate layer when mixed with water. In a practical sense it is, therefore, often regarded as insoluble. Actually, some of the ether dissolves (about 7 g/100 g of water), but the amount is not easily observed by the eye. Ether is completely miscible with alkanes.

Ether can dissolve a wide variety of organic structures because of its own intermediate character; the alkyl portion of the molecule is typically nonpolar, but the oxygen bonds are moderately polar (Fig. 22.3). Consequently, ether associates with and dissolves

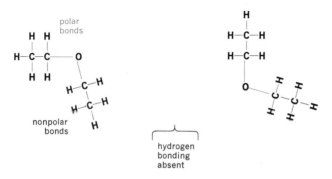

Fig. 22.3 There are no hydrogen-bond attractions between ether molecules because their hydrogen atoms are held by nonpolar bonds.

nonpolar compounds such as alkanes, benzene, and fats, as well as polar compounds such as phenol or acetic acid. The compounds which fail to dissolve in ether are chiefly those having several very polar groups, such as sugars, glycerol, or ethylene glycol. The chemical inertness of ether (Section 22.8) is another factor making it a desirable solvent for organic compounds. On the other hand, the great flammability of ether makes it dangerous, unless it is handled very carefully.

22.5 CHEMICAL PROPERTIES OF ALCOHOLS

A. Covalent Functional Group

First of all, it should be kept in mind that in an alcohol molecule such as CH_3—O—H, all the bonds, including those from oxygen to carbon and from oxygen to hydrogen, are *covalent*. Although the oxygen bonds are quite *polar* covalent bonds, they are not ionic. Therefore, the alcohol is *not* a strong base like NaOH which releases separate hydroxide ions, HO$^-$, into solution. The functional group of an alcohol may be called a hydrox*yl group*, but not a hydrox*ide ion*. Although the alcohol functional group does not ionize, in many other respects it is quite reactive, and hence an alcohol may be converted to a variety of other substances. Several important chemical properties of alcohols are discussed in the following paragraphs.

B. Dehydrogenation; Oxidation

When vapors of an appropriate alcohol are passed over a hot copper catalyst [Eqs. (22–1) and (22–2)], hydrogen is lost, and an aldehyde or ketone is produced.

$$CH_3-CH_2-\underset{\underset{H}{|}}{\overset{\overset{H}{|}}{C}}-O(H \xrightarrow[250°C]{Cu} CH_3-CH_2-\underset{\underset{O}{\|}}{C}{\overset{H}{}} + H_2 \tag{22-1}$$

a primary alcohol   an aldehyde

$$CH_3-CH_2-\underset{\underset{H}{|}}{\overset{\overset{CH_3}{|}}{C}}-O(H \xrightarrow[250°C]{Cu} CH_3-CH_2-\underset{\underset{O}{\|}}{C}{\overset{CH_3}{}} + H_2 \tag{22-2}$$

a secondary alcohol   a ketone

$$CH_3-\underset{\underset{CH_3}{|}}{\overset{\overset{CH_3}{|}}{C}}-O-H \xrightarrow[250°C]{Cu} \text{no reaction} \tag{22-3}$$

a tertiary alcohol

The reaction is termed **dehydrogenation** because it involves *removal of hydrogens* from the molecule; in this case, one from the oxygen and one from the carbon adjacent. Since the adjacent carbon in a tertiary alcohol does not hold a hydrogen, the reaction is not possible, Eq. (22–3).

The dehydrogenation reaction is important for at least three reasons:

1. It demonstrates the structural relations among the organic compounds involved. Note that a primary alcohol is converted to an aldehyde, a secondary alcohol becomes a ketone, and a tertiary alcohol is unchanged.

2. Dehydrogenation also occurs in vital biochemical reactions. In such cases, the reaction must take place at body temperature, and the catalyst is an enzyme. For

example, an enzyme called alcohol dehydrogenase can catalyze the conversion of ethanol to acetaldehyde.

3. This reaction provides a valuable synthetic method for making aldehydes or ketones from the readily available alcohols. It is adaptable to either laboratory or factory use. Aldehydes and ketones are discussed further in Chapter 26.

Dehydrogenation is in a sense equivalent to *oxidation*: the aldehyde or ketone product is in a state of higher oxidation than was the alcohol. This fact can be demonstrated experimentally by showing that the results are similar when alcohols are treated with the ionic oxidizing agents commonly used in the laboratory. In Eq. (22–4) below, the oxygen in brackets, [O], represents oxygen supplied by an ionic oxidizing agent such as $KMnO_4$ or $Na_2Cr_2O_7$. Note that this oxygen combines with the two hydrogens removed from the alcohol. (The actual mechanism for donation of the oxygen by the oxidizing agent is quite complex.)

$$CH_3-CH_2-\overset{\overset{\displaystyle CH_3}{|}}{\underset{\underset{\displaystyle H}{|}}{C}}-O-H + [O] \rightarrow CH_3-CH_2-C\overset{\displaystyle CH_3}{\underset{\displaystyle O}{\lessgtr}} + H_2O \tag{22-4}$$

Such a reagent is rarely useful for preparing an *aldehyde* from a primary alcohol, because the aldehyde is easily oxidized further to an organic acid:

$$CH_3-\overset{\overset{\displaystyle H}{|}}{\underset{\underset{\displaystyle H}{|}}{C}}-O-H \xrightarrow[[O]]{} CH_3-C\overset{\displaystyle H}{\underset{\displaystyle O}{\lessgtr}} + H_2O \xrightarrow[[O]]{more} CH_3-C\overset{\displaystyle O-H}{\underset{\displaystyle O}{\lessgtr}} \tag{22-5}$$

(difficult   organic acid
to isolate)

In contrast, the dehydrogenation reaction is generally an excellent method for synthesizing aldehydes as well as ketones.

Combustion occurs with alcohols, as it does with most other organic compounds, when they are exposed to oxygen at a sufficiently high temperature. The resultant complete oxidation yields the usual products, carbon dioxide and water:

$$CH_3CH_2OH + 3O_2 \xrightarrow{heat} 2CO_2 + 3H_2O$$

C. Reactions with Acids

1. Dehydration Alcohols are dehydrated when they are treated with strong acids such as sulfuric or phosphoric. The water lost may be split out from one alcohol molecule, in which case an alkene results (Eq. 22–6); or it may be split out between two molecules, so that an ether results (Eq. 22–7).

$$CH_3-\overset{\overset{\displaystyle H}{|}}{\underset{\underset{\displaystyle H}{|}}{C}}-CH_2 \xrightarrow[or\ H_2SO_4]{H_3PO_4} CH_3-CH-CH_2 + H_2O \tag{22-6}$$

1-propanol   propene

$$CH_3CH_2CH_2-O-H\ \ H-O-CH_2CH_2CH_3 \xrightarrow[or\ H_2SO_4]{H_3PO_4} CH_3CH_2CH_2-O-CH_2CH_2CH_3 + H_2O \tag{22-7}$$

1-propanol   propyl ether

Which product results depends on the reaction conditions and the particular alcohol used. A number of both alkenes and ethers have been synthesized in this fashion. Ethyl ether is manufactured by this reaction or a modification of it.

2. Halide formation   Concentrated halogen acids convert alcohols to organic halides.

$$ROH + HBr \rightarrow RBr + H_2O \qquad (22\text{-}8)$$

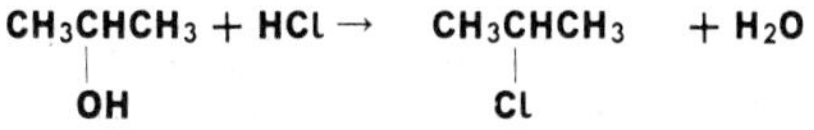

$$CH_3CHCH_3 + HCl \rightarrow \quad CH_3CHCH_3 \quad + H_2O$$
$$\quad\quad |\quad\quad\quad\quad\quad\quad\quad\quad\quad |$$
$$\quad\quad OH\quad\quad\quad\quad\quad\quad\quad\quad Cl$$

2-propanol   2-chloropropane

3. Ester formation   Alcohols react with organic acids to form compounds called esters.

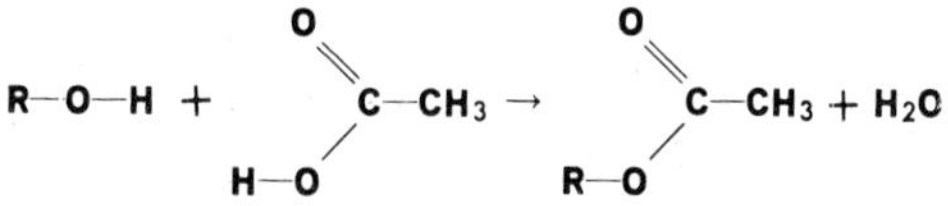

$$
R{-}O{-}H + \overset{O}{\underset{H-O}{\overset{\|}{C}}}{-}CH_3 \rightarrow \overset{O}{\underset{R-O}{\overset{\|}{C}}}{-}CH_3 + H_2O \qquad (22\text{-}9)
$$

an ester

The esters, of interest both chemically and biologically, will be discussed in Chapters 23 and 24.

22.6
THIOLS

Since oxygen and sulfur are both in the sixth group of the periodic table, both have a covalence of two. (Sulfur can also have other oxidation states.) In some organic compounds sulfur can form the same type of structure that oxygen does. Thus a **thiol** is like an alcohol but has sulfur in place of oxygen:

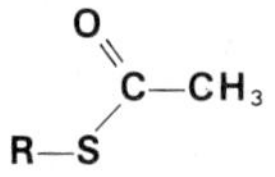

$$R{-}S{-}H \qquad\qquad \overset{O}{\underset{R-S}{\overset{\|}{C}}}{-}CH_3$$

a thiol   a thiol ester
(a mercaptan)

Several properties of a thiol are similar to those of an alcohol, although they may differ somewhat in degree of reactivity. For example, a thiol can be converted to a thiol ester. The latter structure is a key functional group in Coenzyme A and in certain other compounds of biochemical importance.

The most promptly noticed difference between a thiol and an alcohol is the difference in their odors. Like many divalent sulfur compounds, thiols have strong, offensive smells. They cause the distinctive scent of garlic and of skunk, for example.

Chemically, thiols and alcohols are remarkably different in the way they are dehydrogenated (oxidized). In Section 22.5 we stated that a primary alcohol loses two hydrogen atoms from *one* molecule and becomes an aldehyde. But a thiol loses hydrogen atoms from *two* molecules to form a **disulfide** structure:

$$RCH_2{-}S{-}H + H{-}S{-}CH_2R \underset{\text{reduction}}{\overset{\text{oxidation}}{\rightleftarrows}} RCH_2{-}S{-}S{-}CH_2R + 2[H] \qquad (22\text{-}10)$$

a disulfide

The loss of hydrogens from the thiol is, of course, equivalent to oxidation and the two hydrogens would be transferred to the oxidizing agent.

The arrow pointing to the left in Eq. (22–10) indicates that a disulfide is also easily converted to two thiol molecules by some compound that can donate two hydrogens (that is, some reducing agent). Several proteins and other biologically important compounds have thiol structures, and the conversion from thiol to disulfide structure, or the reverse, is essential to many life processes.

Either of these two reactions can also be conducted in the laboratory. For instance, a thiol can be oxidized with hydrogen peroxide, oxygen, iodine, or potassium permanganate. In reverse, a disulfide can be reduced to a thiol by a mixture of either zinc or tin and dilute acid.

Shown below are a few examples of names for individual thiols. Systematically, they are named the same as alcohols, except that the ending -*thiol* is used. In common nomenclature they are called mercaptans. If it is necessary to name the —SH structure as a group attached to a carbon skeleton, the prefix *mercapto* is used.

$HSCH_2CH_2CH_2CH_3$

1-butanethiol
(*n*-butyl mercaptan)

$$CH_3-\underset{\underset{SH}{|}}{CH}-CH_3$$

2-propanethiol
(isopropyl mercaptan)

$$CH_3\underset{\underset{CH_3}{|}}{CH}CH_2\overset{\overset{SH}{|}}{CH}CH_3$$

4-methyl-2-pentanethiol

$$\underset{\underset{SH}{|}}{CH_2}-\underset{\underset{SH}{|}}{CH}-\underset{\underset{OH}{|}}{CH_2}$$

2,3-dimercapto-1-propanol

22.7 PHENOLS

The term **phenol** is used to name a whole class of compounds, as well as the simplest example in the group. (The term phen*ol* implies a hydroxyl derivative of phene, the old name for benzene.)

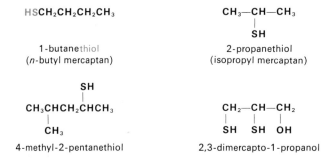

ArOH

general formula for a phenol

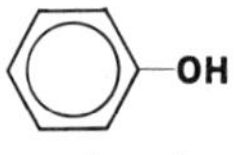

phenol

All phenols are weakly acidic and able to react with sodium hydroxide, Eq. (22–11). Most, however, are unable to react with sodium hydrogen carbonate. (In contrast, alcohols do not react even with sodium hydroxide.)

$$NaOH + H-O-\bigcirc \rightleftharpoons \overset{+}{Na}\overset{-}{O}-\bigcirc + H_2O \qquad (22-11)$$

phenol   sodium phenoxide

This acidic property is reflected in some of the names. An aqueous solution of phenol is still occasionally called "carbolic acid." Picric acid is an especially acidic phenol.

The structures and names of several phenols are shown on page 427, with common names in parentheses.

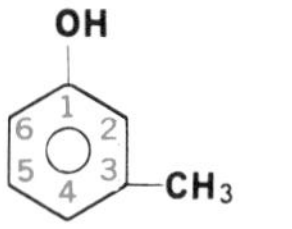

3-methylphenol
(m-cresol)

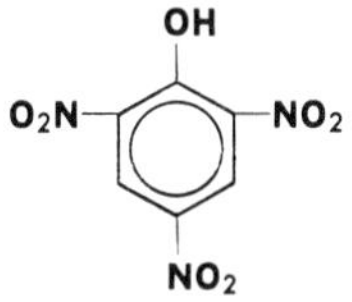

2,4,6-trinitrophenol
(picric acid)

2-naphthol
(β-naphthol)

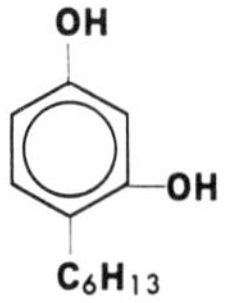

4-hexyl-1,3-dihydroxybenzene
(hexylresorcinol)

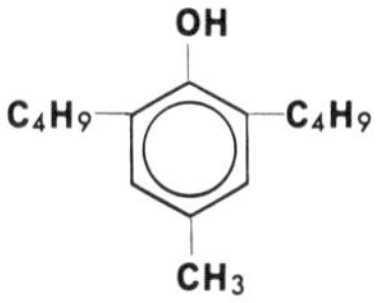

2,6-di-t-butyl-4-methylphenol
BHT

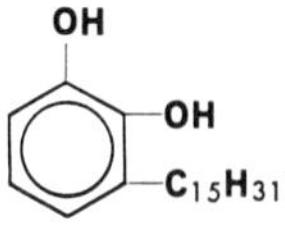

(urushiol)
the irritant in poison ivy
and poison oak*

Phenols have interesting physiological properties. Phenol itself is used to disinfect floors, walls, and apparatus. If strong solutions are spilled on the skin they will kill (burn) some of the tissue. Similarly, the germicidal properties of creosote (a crude mixture, obtained from coal tar, of aromatic hydrocarbons, cresols, and other phenols) make it a good wood preservative. Hexylresorcinol is used in humans and animals to combat intestinal worms and urinary infections. Butylated hydroxytoluene, BHT, is now used extensively as an antioxidant in foods, rubber, plastics, petroleum products, and soaps. It prevents deterioration caused by exposure to air.

Phenol is manufactured into important polymers, chiefly the phenol-formaldehyde and epoxy types. Thousands of pounds of dyes are made annually from naphthol. Alizarin, a phenolic compound found in madder root, has been used as a dye since ancient times. It is now made synthetically.

The phenols used in quantity are often obtained from coal tar (Section 21.9); for some, the supply is augmented by synthesis from petroleum materials.

22.8
ETHERS
A. Occurrence
and Chemical
Properties

An ether group occurs as a unit of structure in a variety of compounds produced by living organisms. However, the role, if any, that the ether unit plays in crucial biochemical reactions is seldom apparent. Possibly its chief contribution is to provide a molecule with the proper physical characteristics such as size or solubility. This situation may be related to the fact that ethers are chemically very inert, in much the same way that alkanes are. Hydrogen, strong bases, reactive metals, strong oxidizing agents, and strong

---

* Urushiol is actually a mixture of closely related compounds in which the C_{15} side-chains have varying numbers of double bonds.

reducing agents have no chemical effect on an ether. Some strong acids may attack an ether at high temperatures. Of course, combustion occurs with oxygen, and chlorine will substitute onto the alkyl portion of an ether in the presence of ultraviolet light.

Therefore, except when studying the structure of natural products, we are usually concerned with only a very few simple ethers. By far the most common of these is ethyl ether, sometimes called diethyl ether, but most often simply ether. It is used as a solvent (Section 22.3) and as an anesthetic.

B. Nomenclature   For simple ethers, common names are often used; the two groups attached to the oxygen are named, followed by the word ether. In a systematic name, the oxygen plus the smaller carbon skeleton is named as a group substituted onto the larger carbon skeleton; such a group has the name-ending *oxy*. In the following illustrations the common names are in parentheses.

$$CH_3—O—CH_2CH_2CH_3$$

1-methoxypropane
(methyl *n*-propyl ether)

$$CH_3CH_2—O—CH_2CH_3$$

ethoxyethane
(diethyl ether)

$$CH_3CH_2CHCH_3$$
$$|$$
$$O—CH_2CH_3$$

2-ethoxybutane
(ethyl sec-butyl ether)

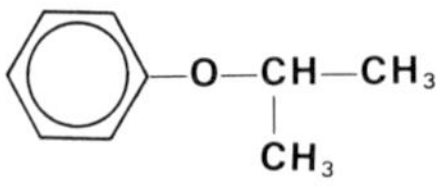

$$—O—CH—CH_3$$
$$|$$
$$CH_3$$

isopropoxybenzene
(isopropyl phenyl ether)

22.9
ANESTHETICS

In 1846, William Morton, a dentist, tested ether as an anesthetic while extracting a tooth. About two weeks later he was the first to make public its effectiveness, by administering it to a patient during surgery. The use of ether had been suggested to Morton by a chemist, Charles Jackson, who had experimented with inhalation of ether. Before this time surgery was an agonizing procedure. The patient was strapped to a table and (if he was lucky) soon lost consciousness due to pain. Ethanol had some value as a relaxant, but only by the use of ether and other anesthetics could surgery develop to its present state, involving kidney and heart transplants.

Ether is a very useful anesthetic. A safe level of unconsciousness can be achieved without depressing respiration or circulation. The amount of ether required is not toxic. Some disadvantages of using ether are that most patients experience prolonged and unpleasant recovery, and that ether is highly flammable. A number of other agents, or combinations of agents, have now replaced ether in many cases.

Ethene, cyclopropane, and vinyl ether ($CII_2—CII—O—CII—CII_2$) are also successful anesthetics, but share with ether a high flammability. Trichloromethane, commonly called chloroform (Table 21.5) is effective and nonflammable. It must be administered with great care to avoid overdosage or liver damage, but is used in many parts of the world. Another halogen compound, halothane (Table 21.5), was developed in 1956 specifically for anesthesia, and by 1966 had become widely accepted. The inorganic compound nitrous oxide (N_2O) is a weak anesthetic often used as a supplement to others. It is also effective in conjunction with intravenous administration of thiopental sodium, a heterocyclic compound (Section 21.8).

**22.10
CONSERVATION
OF ORGANIC
RAW
MATERIALS**

Synthesis is an important aspect of organic chemistry. By a chemical change some carbon compound which is quite abundant can be converted to a different, more valuable organic compound. We have seen reactions whereby an alkane can be converted to an alkene, the alkene to an alcohol, and the alcohol to an ester. There are also reactions for transforming an alkene to benzene, the benzene to nitrobenzene, the nitrobenzene to aniline, and so on. This means that nearly all the valuable synthetic organic compounds—polymers, medicines, dyes, etc.—are derived from two natural resources: coal and petroleum. At present the greatest quantities of both these resources are simply burned up to provide heat. Coal and petroleum are becoming so important as sources of material that we cannot afford much longer to waste them in combustion. Within a few generations humans must develop other sources of energy, before these critical sources of carbon compounds are completely exhausted. Coal and especially petroleum should be regarded as nonrenewable raw materials, like the ores of iron, tin, or chromium.

To reduce the consumption of carbon compounds, some plastics and other organic compounds can be recycled. Others cannot, and this fact poses a problem of disposal. The nonrecycled materials could probably be burned to provide some of the heat energy required by a complex civilization; this solution would be better than burning virgin petroleum. The compounds would have had at least one other use, perhaps many others, before being burned. Of course, any furnace for burning the disposed materials would have to be designed to provide complete combustion and the trapping of any noxious by-products, to avoid an unacceptable amount of air pollution.

SUMMARY

The covalently bonded OH group of either an alcohol or a phenol is very common among both biological and synthetic compounds. The OH functional group participates in a large variety of chemical reactions and consequently can be converted to numerous other organic compounds. For this reason alcohols are valuable starting materials for the manufacture of organic compounds, and both alcohols and phenols are biochemically very important. Thiols are sulfur compounds structurally related to alcohols; some of them are essential in certain biological molecules. Although ether structures are moderately common, they are not very reactive chemically and have few, if any, important functions in biochemical reactions.

The systematic names have the ending -ol for an alcohol and *phenol* for a phenol. If the OH structure must be named as a group, rather than in the name-ending, the term *hydroxy* is used. The common names of some of the most familiar alcohols should also be learned.

The possibility of hydrogen-bond attractions has a distinct effect upon the physical properties (solubility, boiling point) of alcohols, phenols, and ethers.

The behavior of an alcohol in the presence of an oxidizing agent relates the alcohol to other important organic compounds. A tertiary alcohol generally is not oxidized, a secondary alcohol is converted to a ketone, and a primary alcohol becomes first an aldehyde and then a carboxylic acid. In the presence of an acid, an alcohol can react to become an alkene, an alkyl halide, or an ester, depending on the particular acid present. However, a phenol does not undergo any of these reactions and also differs from an alcohol in being more acidic; the phenol can react with sodium hydroxide.

IMPORTANT TERMS

alcohol	ether	primary alcohol
alkoxide	hydrogen bond	secondary alcohol
dehydration	hydroxyl group	tertiary alcohol
dehydrogenation	oxidation	thiol
disulfide	phenol	

WORK EXERCISES

1. Show the structure of all possible alcohols having molecular formula $C_4H_{10}O$.
2. Write a systematic name for each isomer shown in Exercise 1.
3. Label each isomer in Exercise 1 as primary, secondary, or tertiary.
4. Show the product that each alcohol in Exercise 1 would give with copper at 250°C.
5. Show the structure of the following.
   a) 2-propanol
   b) phenol
   c) isobutyl alcohol
   d) 4-methyl-2-pentanol
   e) potassium phenoxide
   f) sec-butyl alcohol
   g) *m*-nitrophenol
   h) *n*-propyl mercaptan
   i) 5-bromo-2-methyl-3 hexanol
   j) two different secondary alcohols having formula $C_5H_{12}O$
   k) 3-ethyl-2-phenyl-1-pentanol
   l) 3-chloro-4-isopropyl-2-heptanol
   m) cyclohexanol
   n) 3-pentanethiol
6. Methyl ethyl ether, $CH_3OCH_2CH_3$, has mol. wt. 60 and is a gas (b.p. 10°C), whereas ethylene glycol, $HOCH_2CH_2OH$, has about the same mol. wt. (62), but has b.p. 197°C.
   a) Explain why ethylene glycol has such a high boiling point.
   b) Why is the high boiling point an advantage when ethylene glycol is used as antifreeze in an automobile engine?
7. Write complete equations for these reactions. If necessary, state "no reaction."
   a) 2-propanol, hydrogen bromide, heat
   b) dipropyl ether, sodium hydroxide
   c) 3-pentanol and copper at 250°C
   d) ethanol and copper at 250°C
   e) 2-butanol and warm potassium permanganate solution
   f) phenol and sodium hydroxide
8. Show how 2-propanol (rubbing alcohol) could be

manufactured, starting with petroleum as the raw material.

9. List all the types of structural units (both carbon skeletons and functional groups) which you can find in each of the mood-altering drugs shown below. (Consider the groups discussed in this chapter and consult Table 19.5.)
   a) mescaline

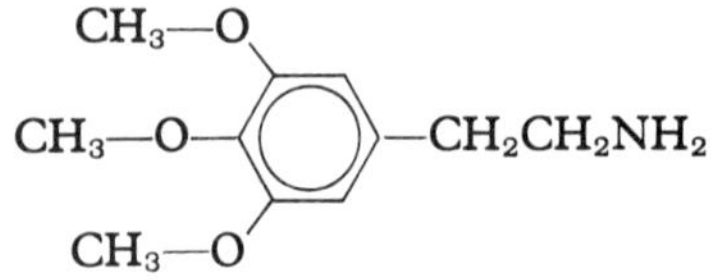

   b) tetrahydrocannabinol

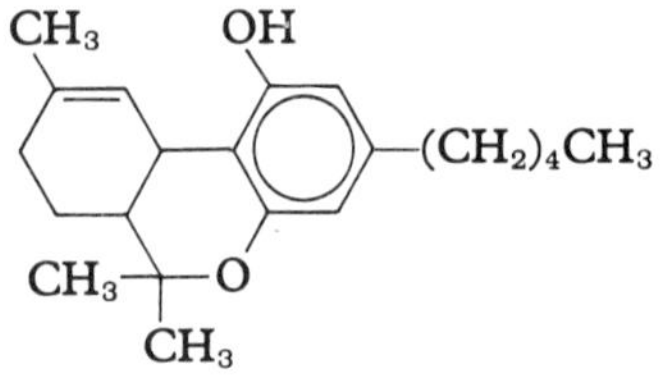

10. Write a structural formula for each of the following.
    a) 1-propanol
    b) 3-pentanol
    c) 2-methyl-2-butanol
    d) 3-methyl-2-butanol
11. Which alcohol in Exercise 10 is not dehydrogenated by hot copper?
12. Write an equation for the dehydrogenation of 1-butanol. In this reaction, is the alcohol reduced or oxidized? Discuss briefly.
13. Write an equation for one example of a biochemical dehydrogenation.
14. What type of compound is formed by the dehydrogenation of each of the following?
    a) a primary alcohol
    b) a secondary alcohol
    c) a tertiary alcohol
15. Write a systematic name for each of the following.
    a) $CH_3CH_2CH_2CH_2$
       |
       OH

b) HOCH$_2$CHCH$_3$
           |
           CH$_3$

c) CH$_3$CH$_2$CHOH
              |
              CH$_3$

         SH
         |
d) CH$_3$—C—CH$_3$
         |
         CH$_3$

e) CH$_3$CHCH$_2$CH$_3$
       |
       SH

f) CH$_3$CHCH$_2$CH$_2$OH
       |
       CH$_3$

g)

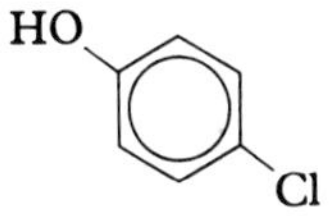

h) 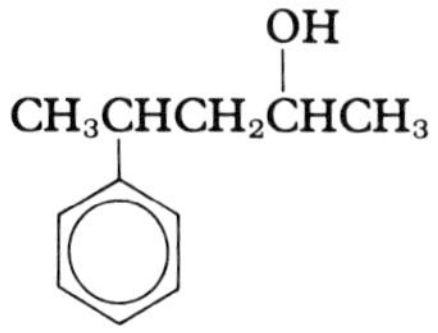 CH$_3$CHCH$_2$CHCH$_3$ (with OH substituents and phenyl ring)

i)

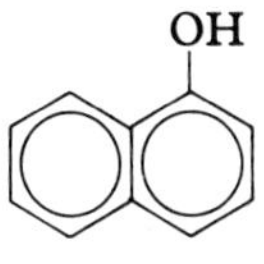

j)

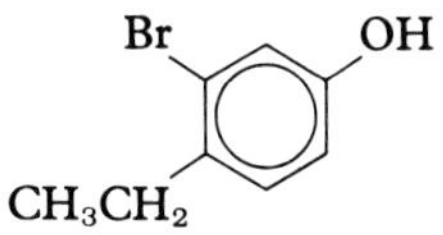

k) CH$_3$CHCH$_3$
      |
      OCH$_2$CH$_3$

l) 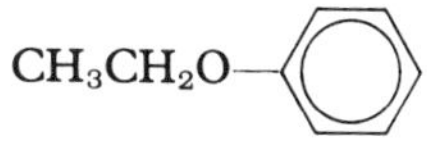 CH$_3$CH$_2$O—⬡

16. Which compound in each pair below is more soluble in water? Why?
   a) CH$_3$CH$_2$CH$_2$CH$_3$ or CH$_3$CH$_2$CH$_2$OH
   b) CH$_3$CH$_2$CHCH$_3$ or CH$_3$CH$_2$CH$_2$CHCH$_3$
               | |
               OH   OH
   c) CH$_3$CH$_2$CH$_2$CH$_2$ or CH$_2$CH$_2$CH$_2$CH$_2$
                   | | |
                   OH OH   OH
   d) ⬡—OH or ⬡—CH$_3$

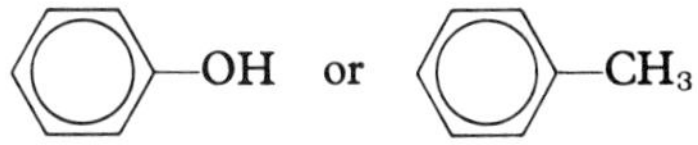

17. Which compound from each pair in Exercise 16 has a higher boiling point?

18. Which compound from each pair in Exercise 16 is more soluble in hexane?

19. Write an equation for each reaction or, if necessary, state "no reaction."
   a) cyclopentanol, conc. HBr, heat
   b) 2,4-dimethylphenol, NaHCO$_3$ solution
   c) ethanethiol, H$_2$O$_2$
   d) t-butyl alcohol, conc. H$_2$SO$_4$, heat → C$_4$H$_8$
   e) ethyl phenyl ether, dilute KMnO$_4$
   f) 2-naphthol, KOH solution
   g) isobutyl mercaptan, oxygen
   h) t-pentyl alcohol, conc. HCl
   i) cyclohexanol, conc. H$_3$PO$_4$, heat → C$_6$H$_{10}$
   j) cyclopentanol, warm KMnO$_4$ solution

20. Write an equation to show how each of the following alcohols could be prepared from an alkene. (If necessary, review Section 20.5G.)
   a) ethanol   b) isopropyl alcohol
   c) sec-butyl alcohol   d) tert-butyl alcohol
   e) 2-methyl-2-butanol   f) 2-pentanol

23

Organic Acids and their Derivatives

Some of the most abundant and well-known organic compounds are acids. Also familiar are the fats, which are made up of organic acids chemically combined with the alcohol glycerol. For centuries humans used tart fruits, sour milk, and vinegar, even though they were not acquainted with the individual compounds responsible for the sour tastes. When people did finally begin to isolate and identify separate acids (starting mostly in the eighteenth century), they usually named them after some familiar source, as shown in Table 23.1. Formic acid is the irritant in the sting of red ants, bees, and nettle plants. Butyric acid occurs in rancid butter, aged cheese, and human perspiration; it is the chief cause of their strong offensive odor.

As a class, the acids shown in Table 23.1 are most properly called **carboxylic acids**. The functional group characteristic of the class is a **carboxyl group**.

In most cases the structure should be carefully written out, as shown above. For the sake of brevity, the group is sometimes written —COOH (especially in the printing of books); when it is represented thus, be sure to remember the actual pattern of bonding. In abbreviated form, the acids themselves may be represented by RCOOH or ArCOOH.

The whole carboxyl group should be regarded as *one* functional group with its own distinctive properties. Most reactions of the carboxyl group are considerably different from those of a hydroxyl group, OH (Chapter 22), or a carbonyl group, C=O (Chapter 26).

The systematic names for carboxylic acids follow the usual pattern: To the name for a carbon skeleton add an ending for the carboxyl group by replacing the final *-e* with *-oic acid*. Thus, the five-carbon acid is called pentanoic acid. In the systematic nomenclature *benzoic acid* is used as a parent name for aromatic carboxylic acids (Table 23.1). The carbon in the carboxyl group is assigned the number 1 for counting down the chain

TABLE 23.1   Some common organic acids (carboxylic acids)

Common Name	Origin of Name	Structure	Systematic Name
formic acid	L. *formica*, ant	HCOOH	methanoic acid
acetic acid	L. *acetum*, vinegar	CH_3COOH	ethanoic acid
propionic acid	Gr. *pro(tos)*, first + *pion*, fat	CH_3CH_2COOH	propanoic acid
butyric acid	L. *butyrum*, butter	$CH_3CH_2CH_2COOH$	butanoic acid
caproic acid	L. *caper*, goat	$CH_3CH_2CH_2CH_2CH_2COOH$	hexanoic acid

lactic acid (α-hydroxy-propionic acid)	L. *lactis*, milk	CH_3—CH—COOH
		        |
		        OH
			2-hydroxy-propanoic acid

benzoic acid	obtained from benzoin, a plant gum	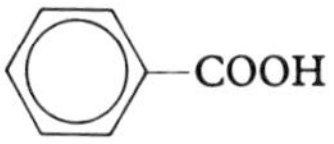 ⬡—COOH	benzoic acid
salicylic acid	obtained from the willow tree L. *salix*; Fr. *salicine*	⬡—COOH | OH	2-hydroxy-benzoic acid

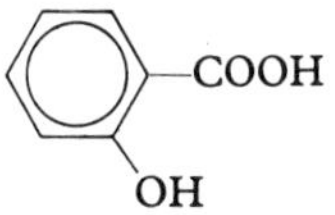

to locate other groups. In *common* names, positions are designated by the Greek letters α, β, γ, δ, starting with the carbon *next* to the carboxyl group:

$$CH_3CH_2CH_2CH_2CH_2COOH$$
$$\;\;\delta\;\;\;\gamma\;\;\;\beta\;\;\;\alpha$$

23.2 ACIDIC PROPERTIES

Acetic acid is a good example to use in beginning our discussion of chemical properties. By the mid-1700s chemists could isolate acetic acid in fairly pure condition from the distillation of vinegar. By the mid-1800s many of the properties of acetic acid had been carefully observed. The customary analysis indicated that the molecular formula was $C_2H_4O_2$. The substance had the sour taste thought to be typical of acids, and it neutralized caustic soda (sodium hydroxide). The resultant salt had the formula $C_2H_3O_2Na$. Even if an excess of caustic soda was used, only one of the four hydrogens of acetic acid could be removed by the hydroxide ions. Various other observations were made, including some which suggested that the two oxygens were bound to one carbon. To be consistent with all the experimental observations, and with the ideas of valence becoming accepted at that time (with C = 4, O = 2, H = 1), acetic acid was written this way:

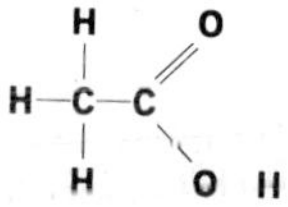

Over the years this has proven to be a satisfactory structural formula for acetic acid; that is, none of the more recent experimental facts have indicated that the atoms are bonded in some different pattern.

All the bonds of acetic acid are covalent. However, when acetic acid is dissolved in water an occasional acid molecule is able to ionize; the hydrogen escapes from the oxygen, to become an H^+ ion in solution. At any one time in a $0.1M$ solution, only about one acetic acid molecule out of every thousand exists in the ionized condition. The arrows in Eq. (23–1) are meant to show that most acetic acid molecules are in the nonionized form.

$$CH_3-C\underset{O-H}{\overset{O}{\Big\langle}} \quad \rightleftarrows \quad CH_3-C\underset{O^-}{\overset{O}{\Big\langle}} \quad + \quad H^+ \tag{23-1}$$

acetic acid   acetate ion

Recall that when HCl dissolves in water, *all* the molecules become ionized, yielding separate H^+ and Cl^- ions. By comparison, *acetic acid is termed a weak acid* because its water solution contains only a few hydrogen ions at any one time. However, the presence of even such a small number of hydrogen ions is enough to give the solution *typical acid properties*: it changes the color of litmus paper, etc.

The amount of ionization of acetic acid, indicated by Eq. (23–1), is expressed in more exact numerical terms by the ionization constant:

$$K_a = \frac{[CH_3COO^-][H^+]}{[CH_3COOH]}; \quad K_a = 1.8 \times 10^{-5}$$

The acidity of nearly all carboxylic acids is about the same, despite rather wide structural variations elsewhere in the molecule; for example, benzoic acid has almost the same acid strength as does acetic acid. For the great majority of carboxylic acids K_a is very close to 10^{-5}.

If a solution of sodium hydroxide is added to acetic acid, the hydroxide ions remove one hydrogen ion from *every* acetic acid molecule (rather than from just one out of a thousand, as water does). As usual, this acid-base reaction Eq. (23–2) produces a salt.

$$CH_3-C\underset{O-H}{\overset{O}{\Big\langle}} \quad + \quad Na^+OH^- \quad \rightarrow \quad CH_3-C\underset{O^-Na^+}{\overset{O}{\Big\langle}} \quad + \quad HOH \tag{23-2}$$

sodium acetate

The reaction occurs very rapidly, actually as fast as one can stir the solutions together. A second, very important feature of the reaction is that one gram molecular weight of acetic acid can convert one gram molecular weight of sodium hydroxide to a salt. In other words, molecule by molecule, a weak acid (e.g., acetic) has as much *capacity* to neutralize a base as does a strong acid (e.g., hydrochloric).

However, acetic acid, when alone in a dilute water solution, releases only a few hydrogen ions at any moment. Since the typical acidic properties are due to the hydrogen ions, the solution of acetic acid in water is not strongly acidic. This property of organic acids is important in biological systems. The tissues may need to have available a large supply of some acid, but at any given time a high concentration of hydrogen ions cannot be tolerated. The organic acids, being only slightly ionized, satisfy these conditions very nicely.

For the same reason, if we wish to halt the caustic effects of a strong base which has been spilled, vinegar or some other weak acid is a desirable neutralizing agent. If we used a strong acid, an excess might cause as much damage as the base we were trying to neutralize.

This whole situation can be dramatically summarized by considering another example. A $1M$ solution of hydrochloric acid is much more strongly acid that is a $10M$ solution of acetic acid. (In fact, the hydrochloric acid solution is about 100 times as acidic!) Yet, a given volume of the $10M$ acetic acid solution has the *capacity* to neutralize ten times as much sodium hydroxide as does an equal volume of the $1M$ hydrochloric acid solution.

To summarize, carboxylic acids have acidic properties for the same reason that inorganic acids do: they release hydrogen ions. The difference is one of degree; organic acids are weak acids because they are only slightly ionized. Like the inorganic acids, they form salts by reaction with metallic hydroxides, oxides, or carbonates.

23.3
REACTIVITY OF
THE CARBOXYL
GROUP

Why does a carboxyl group have acidic properties at all, compared to H—O—H or H—O—CH$_3$, which also have a hydrogen covalently bonded to oxygen? Clearly, the only thing different about the carboxylic acid is the C=O portion of structure; this is what affects the acidic property.

The C=O double bond is in many ways like a C=C double bond. One bond is a sigma bond and the other is a pi bond. The bond angles are 120° and the attached atoms lie in one plane. The one striking difference, however, is that the C=O bond is very polar, as indicated in Fig. 23.1. Because of the greater electronegativity of oxygen, electrons tend to be polarized toward it. But this polarization is even more intense than when oxygen is held by sigma bonds alone. The pi bond is a large, loosely held cloud of electrons, easily distorted toward oxygen. The total effect is that the carbon in a C=O group is very positive; it has much less than its share of electrons. This in turn means that any electrons in its vicinity will be strongly attracted toward the carbon. The C=O group is called an *electron-attracting* group.

The effect of the C=O group on acidity is illustrated by Fig. 23.1. If the carbon attracts electrons away from the OH group, the electrons will be less available there for the hydrogen, which therefore can escape more easily as an H$^+$ ion. Also, the RCOO$^-$ ion will be better able to exist as an independent ion because its minus charge will be partially absorbed by the electron-attracting C=O group.

We shall soon see that the electron-attracting property of the C=O group also affects other chemical reactions of a carboxylic acid or of its derivatives. Reactions of aldehydes and ketones, to be discussed in Chapter 26, are also determined by the polar C=O group.

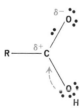

Fig. 23.1 Structure of the carboxyl group, showing that the positive carbon attracts electrons. (In the C=O portion of the structure above, the pi bond is represented by a pair of electron dots.)

23.4
SALT
FORMATION

The reaction of carboxylic acids with bases is quite general. Bases such as NaOH, KOH, $Mg(OH)_2$, or $Al(OH)_3$ could be involved.

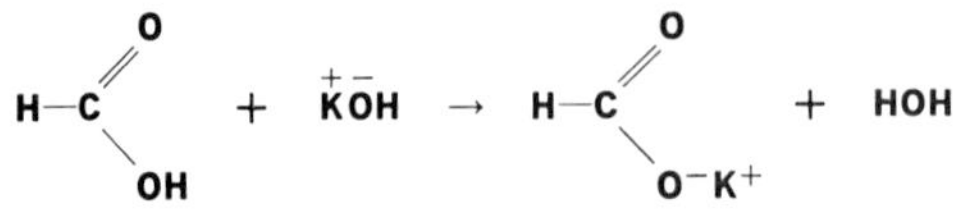

$$H-C\overset{O}{\underset{OH}{<}} \quad + \quad \overset{+}{K}\overset{-}{OH} \quad \rightarrow \quad H-C\overset{O}{\underset{O^-K^+}{<}} \quad + \quad HOH \tag{23-3}$$

form ic acid   potassium
                                      form ate

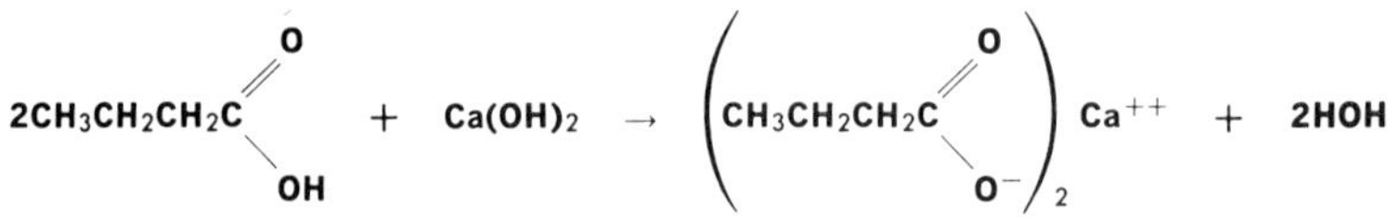

$$2CH_3CH_2CH_2C\overset{O}{\underset{OH}{<}} \quad + \quad Ca(OH)_2 \quad \rightarrow \quad \left(CH_3CH_2CH_2C\overset{O}{\underset{O^-}{<}}\right)_2 Ca^{++} \quad + \quad 2HOH \tag{23-4}$$

butano ic acid   calcium butano ate
(butyric acid)   (calcium butyrate)

As shown above, the salts of carboxylic acids are named by changing the *-ic* ending of the acid name to *-ate* for the salt. This applies to either common names or systematic names.

Still another type of base is ammonia, NH_3. It reacts with HCl to form the salt $NH_4^+Cl^-$. In a similar manner, ammonia reacts readily with a carboxylic acid to form a salt.

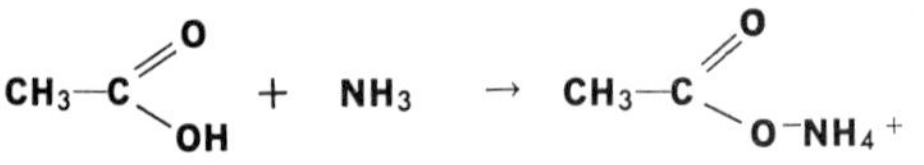

$$CH_3-C\overset{O}{\underset{OH}{<}} \quad + \quad NH_3 \quad \rightarrow \quad CH_3-C\overset{O}{\underset{O^-NH_4^+}{<}} \tag{23-5}$$

acetic acid   ammonia   ammonium acetate

In a carboxylate salt, such as $RCOO^-Na^+$, the functional group has an ionic bond. Therefore the compound has properties similar in many ways to those of an inorganic salt. The carboxylate salts are often quite soluble in water and in solution are excellent conductors of an electric current. They form relatively hard crystals and have high melting points.

When a piece of limestone ($CaCO_3$) or a little baking soda ($NaHCO_3$) is dropped into vinegar, a distinct fizzing occurs. Water solutions of other carboxylic acids give the same results. Proper testing reveals that the gas being evolved is carbon dioxide. Evidently a carboxylic acid is capable of donating a hydrogen ion to a carbonate or hydrogen carbonate ion, Eqs. (23–6) and (23–7). (Incidentally, this means that the carboxylic acid is a considerably stronger acid than is a phenol—see Section 22.7.)

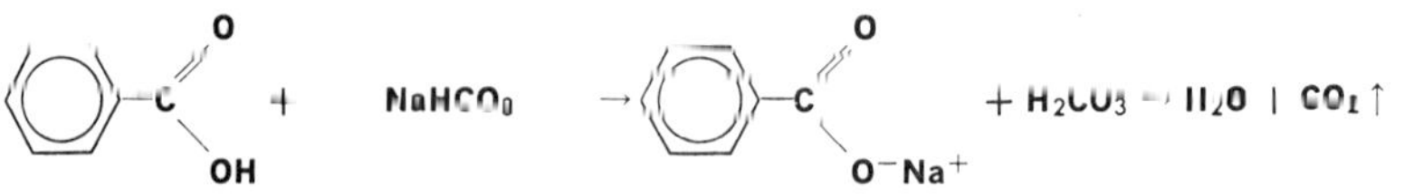

$$\underset{\text{benzoic acid}}{\text{C}_6\text{H}_5-C\overset{O}{\underset{OH}{<}}} \quad + \quad \underset{\substack{\text{sodium hydrogen} \\ \text{carbonate}}}{NaHCO_3} \quad \rightarrow \quad \underset{\text{sodium benzoate}}{\text{C}_6\text{H}_5-C\overset{O}{\underset{O^-Na^+}{<}}} \quad + \quad H_2CO_3 \rightarrow H_2O + CO_2\uparrow \tag{23-6}$$

$$2CH_3CH_2COOH + K_2CO_3 \rightarrow 2CH_3CH_2COO^-K^+ + H_2CO_3 \rightarrow H_2O + CO_2\uparrow \tag{23-7}$$

propanoic   potassium   potassium
   acid   carbonate   propanoate

Thus, the reaction of carboxylic acids with carbonates or hydrogen carbonates is another means by which salts are created.

If we add hydrochloric acid to an aqueous solution of sodium benzoate, a white precipitate of benzoic acid appears. (This acid happens to be a solid, and insoluble in water.) Or, if we add sulfuric acid to a solution of sodium butyrate, we promptly note the strong odor of butyric acid.

$$\text{C}_6\text{H}_5\text{—COO}^-\text{Na}^+ \; + \; \text{H}^+\text{Cl}^- \; \rightarrow \; \text{C}_6\text{H}_5\text{—COOH}\downarrow \; + \; \text{Na}^+\text{Cl}^- \tag{23-8}$$

sodium benzoate   benzoic acid

$$2\text{CH}_3\text{CH}_2\text{CH}_2\text{COO}^-\text{Na}^+ + (\text{H}^+)_2\text{SO}_4{}^{2-} \rightarrow 2\text{CH}_3\text{CH}_2\text{CH}_2\text{COOH} + (\text{Na}^+)_2\text{SO}_4{}^{2-} \tag{23-9}$$

sodium butyrate   butyric acid

In general, we find that mineral acids will convert carboxylate salts to the corresponding carboxylic acids. The explanation is simply that the carboxylic acids are weak acids, so that the strong mineral acids are capable of forcing hydrogen ions onto carboxylate ions, Eqs. (23–8) and (23–9).

**23.5
WATER
SOLUBILITY OF
CARBOXYLIC
ACIDS AND
SALTS**

In the preceding section we mentioned sodium benzoate and benzoic acid, Eq. (23–8). Sodium benzoate is quite soluble in water (61 g will dissolve in 100 g of water at 25°C). On the other hand, benzoic acid is only slightly soluble (about 0.25 g of benzoic acid will dissolve in 100 g of water at 25°C), so most of it is precipitated from the solution. In general, this relationship exists for all carboxylic acids and their alkali metal salts; although the salt may be either moderately soluble or very soluble in water, the acid is less soluble, often much less so. The alkali salts are more water-soluble because they are ionic; the fully developed charges on ions strongly attract polar water molecules, causing the salt to dissolve. The carboxylic acids, however, are chiefly nonionized. In order to dissolve, they must depend on hydrogen-bond attractions between their carboxyl groups, which are polar, and the water molecules. These attractions due to *partially* developed charges in polar molecules are less strong than the attractions due to ions (Fig. 23.2).

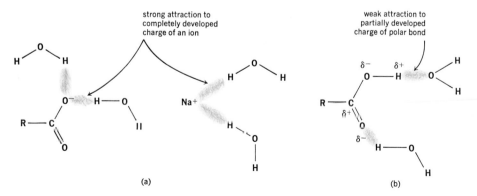

Fig. 23.2 Comparison of attraction of water molecules to (a) an ionic compound and (b) a polar compound. The greater the attraction, the greater the solubility in water.

The result is that the solubility of carboxylic acids in water is about like that of alcohols; the smallest acids are very soluble, those of five and six carbons are slightly soluble, and the larger molecules are "insoluble." In contrast, most sodium or potassium carboxylate salts are quite soluble in water.

These concepts of solubility have important applications to biology and medicine. Water is the principal liquid present in cells and body fluids, and the pH is usually near 7 (neutrality). Therefore the organic acids found in most tissues have been neutralized and exist in the form of their salts ($RCOO^-Na^+$ or $RCOO^-K^+$). This means most of them are quite soluble in the aqueous cell fluid. If a medicine happens to be an organic acid, it often is converted to a sodium salt. In that form it is more water-soluble; also, since it is neutral it is more easily tolerated by the body.

**23.6
REACTION
WITH
ALCOHOLS:
ESTERIFICA-
TION**

Before 1800, a few chemists had discovered that if they heated an organic acid with alcohol in the presence of sulfuric acid, two products resulted: water and an oily, water-insoluble liquid usually having a pleasant fruitlike odor. During the same era it was also observed that by proper treatment the fragrant oil could be converted back to the original acid and alcohol. Clearly, this new kind of compound seemed to be some combination of alcohol with an organic acid, achieved by splitting out water. During the 1800s, this reaction was found to be very generally applicable to a great variety of acids in combination with many different alcohols. Eventually the products were called **esters** (apparently a made-up name), and this method of preparing them was termed **esterification** (literally, ester-formation).

Equation (23–10) is a general example of esterification. In this kind of substitution reaction *the bond to the C=O group is broken*:

$$\text{(23-10)}$$

alcohol   carboxylic   ester   water
           acid

When —OH departs *with the electrons*, it picks up an H from ROH, forming a molecule of water. The RO— group from the alcohol bonds to the C=O group in place of —OH. Note carefully that this reaction involves the making and breaking of *covalent* bonds. It is *not* an acid–base reaction of H^+ *ions* with OH^- *ions*, and Eq. (23–10) should *not* be interpreted to mean than ROH is a stronger acid than R'COOH. The mechanism of this esterification reaction will be discussed in somewhat more detail in Section 23.7. Some specific examples of esterification reactions are shown in Eqs. (23–11) and (2312).

$$\text{(23-11)}$$

ethyl alcohol   butanoic acid   ethyl butanoate
                 (butyric acid)   (ethyl butyrate)

Note that the structure of an ester is described in its name. The group name for the alcohol portion is stated first, followed by the name of the acid component, with its *-ic* ending changed to *-ate*. This method of assigning ester names applies to both systematic and common names.

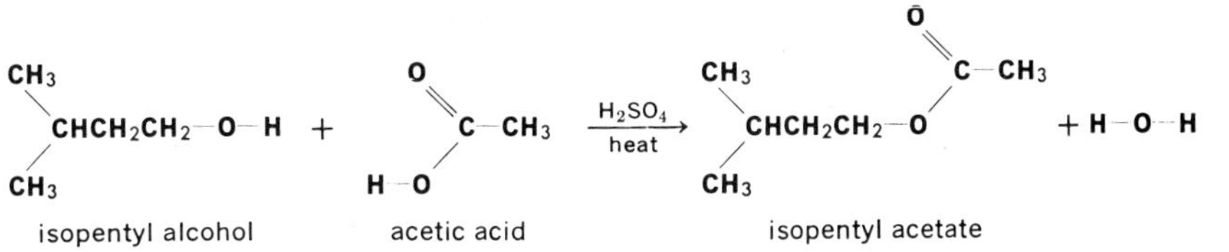

$$\underset{\text{isopentyl alcohol}}{\overset{\displaystyle CH_3}{\underset{\displaystyle CH_3}{\diagdown}}\text{CHCH}_2\text{CH}_2\!-\!\text{O}\!-\!\text{H}} \;+\; \underset{\text{acetic acid}}{\overset{\displaystyle O}{\overset{\displaystyle \|}{\underset{\displaystyle H\!-\!O}{\diagdown}}}\text{C}\!-\!\text{CH}_3} \;\xrightarrow[\text{heat}]{H_2SO_4}\; \underset{\text{isopentyl acetate}}{\overset{\displaystyle CH_3}{\underset{\displaystyle CH_3}{\diagdown}}\text{CHCH}_2\text{CH}_2\!-\!\text{O}\overset{\displaystyle \overset{O}{\|}}{\diagup}\text{C}\!-\!\text{CH}_3} \;+\; H\!-\!O\!-\!H \qquad (23\text{-}12)$$

Aromatic carboxylic acids undergo the esterification reaction as readily as do the alkanoic acids, for example,

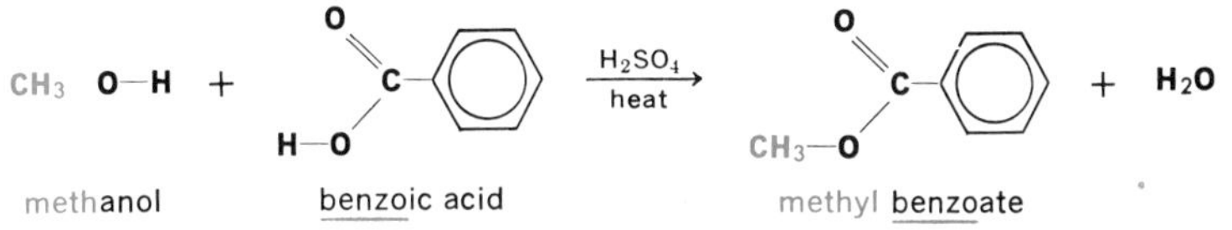

$$\underset{\text{methanol}}{CH_3\!-\!O\!-\!H} \;+\; \underset{\text{benzoic acid}}{\overset{O}{\overset{\|}{\underset{H-O}{C}}}\!\!-\!\!\bigcirc} \;\xrightarrow[\text{heat}]{H_2SO_4}\; \underset{\text{methyl benzoate}}{\overset{O}{\overset{\|}{\underset{CH_3-O}{C}}}\!\!-\!\!\bigcirc} \;+\; H_2O \qquad (23\text{-}13)$$

An ester is called a **derivative** of a carboxylic acid because the ester was derived from the acid. Amides and acid chlorides, which will be mentioned later, are also derivatives of carboxylic acids.

23.7 HOW ESTERIFICATION OCCURS

By following the pattern shown in Eq. (23–10) one can write a correct equation for an esterification reaction. But it is helpful to have a better understanding of *how* the reaction takes place. This is summarized in Fig. 23.3. The nonbonded electrons on the oxygen of $HOCH_3$ are attracted to the very positive carbon in the carboxyl group. When the two atoms come close together the electron pair from $HOCH_3$ forms a covalent bond to the carboxyl carbon. This then permits the OH portion of the carboxyl group to break its bond to C=O and depart with the electrons from that bond. Also, an H^+ drops off the $HOCH_3$ and combines with OH, producing an HOH molecule.

The mechanism of the esterification reaction, just described, has been thoroughly studied by chemists. They have found that the events summarized in Fig. 23.3 actually take place step by step. It is not necessary that we go into all the details; a few have been omitted. The important points to remember are these:

1. Once again we see that the polar C=O group attracts electrons; in this case it is the electrons on the oxygen of $HOCH_3$ or some other alcohol.
2. The bond is broken at the C=O group and the OH departs with the electrons.
3. An H is lost from $HOCH_3$ (or HOR), to be combined with the departing OH group.

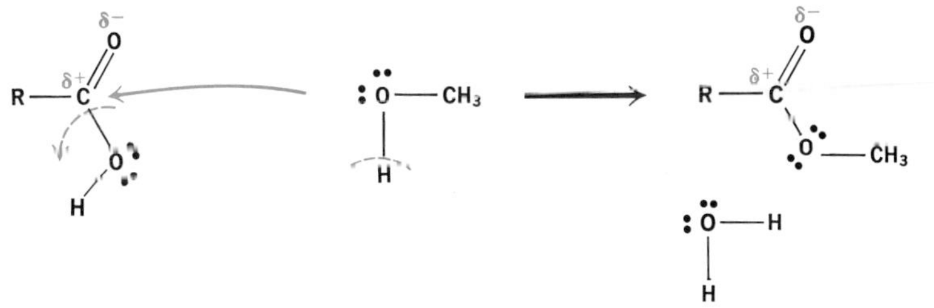

Fig. 23.3 Mechanism of the esterification reaction, summarizing the key events of bond-making and bond-breaking.

We shall soon see that these same events occur in other substitution reactions at a carboxyl group or a modified carboxyl group (a derivative). The substances which attack are things like ROH, HOH, and RNH_2, which have electron pairs available on oxygen or nitrogen. These are attracted to the positive carbon in the carboxyl group, and the bond holding the original group (—OH, —Cl, etc.) to the C=O is broken.

**23.8
LACTONES**

Many organic molecules have two (or more) functional groups. If the two are a carboxyl group and a hydroxyl group, an ester can form *within* one molecule:

$$CH_3{-}CH_2{-}CH{-}CH_2{-}CH_2{-}C \overset{O}{\underset{OH}{\diagup}} \quad \rightarrow \quad CH_3{-}CH_2{-}CH \underset{O}{\overset{CH_2{-}CH_2}{\diagup \diagdown}} C{=}O \quad + \quad HOH$$

a lactone

The product, called a **lactone**, is an intramolecular ester. The reaction occurs because during the continual twisting of the carbon skeleton (due to rotation on all the sigma bonds), sooner or later the OH group strikes the COOH group. Of course, this collision will occur most frequently if it leads to a ring of favorable size. In the example just shown, a five-atom ring resulted. If there were one more carbon between the OH and COOH groups, a six-membered ring, also favorable, would form:

$$R{-}CH \underset{CH_2{-}O}{\overset{CH_2{-}CH_2}{\diagup \diagdown}} C{=}O \qquad \text{a lactone}$$

Quite a few organic compounds, both natural and synthetic, have lactone structures. One interesting example is vitamin C:

$$\begin{array}{c} H_2C{-}OH \\ | \\ H{-}C{-}OH \\ \end{array}$$

vitamin C
(ascorbic acid)

**23.9
COMMON
ESTERS,
NATURAL AND
SYNTHETIC**

The odors of flowers and fruits are due to mixtures of several organic compounds. It was not until chemists developed skillful techniques that they were able to separate such mixtures into individual compounds and determine their structures. (Some of the more complicated mixtures are still being investigated.) Eventually, chemists found that some of the compounds in the plant materials were esters and that they were identical with those synthesized in the laboratory by the esterification reaction, Eq. (23–10). **Isopentyl acetate,** Eq. (23–12), was detected in bananas, apples, and some other fruits. It is often called *banana oil* (or *amyl* acetate, an old common name). Since it is a good solvent for certain lacquers, cements, and plastics, large quantities are made industrially by the esterification reaction. **Isopentyl butanoate** occurs in cocoa oil, along with several other compounds; by itself it has a distinctly sweet, fruity odor resembling pears.

Ethyl acetate is found in wine, pineapples, and other fruits. It is a valuable solvent used in nail polish removers and in the manufacture of perfumes, plastics, paints and photographic films.

Methyl benzoate, Eq. (23–13), occurs in a few plant oils; its odor is somewhat medicinal rather than sweet. Another example of an ester derived from an aromatic acid is methyl salicylate (Section 23.15D).

Several other simple esters, not known to occur in fruits or flowers, have nevertheless been prepared and used for decades by manufacturers of artificial flavors and scents. **Methyl butanoate** resembles the scent of apples. **Butyl butanoate** has a fine odor useful in pineapple essence. **Ethyl butanoate**, Eq. (23–11), is used in artificial peach flavor, as well as in pineapple and apricot. **Pentyl propanoate** resembles apricot essence.

The effects of vapors on the human nose are intriguing, but not well understood. For instance, the sweet, fruity odors of the simple esters present a marked contrast to the odors of the acids from which they are derived. Methanoic (formic) acid and ethanoic (acetic) acid have pungent, stinging odors. The larger acids, butanoic, hexanoic, and octanoic, have goatlike or rancid odors. Propanoic acid, of intermediate size, has some of both the pungent and the rancid qualities. Acids larger than about decanoic are so slightly volatile that we detect very little odor in their presence.

The esters discussed so far have been derived from alcohols. Esters of phenols are also of frequent occurrence in nature and in the laboratory. One example of a phenolic ester is **phenyl acetate**; note very carefully the structural difference between this ester and methyl benzoate, Eq. (23–13).

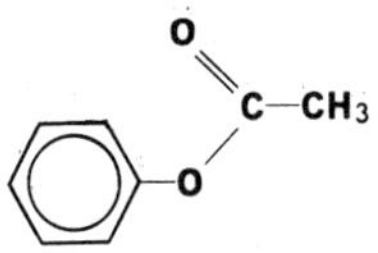

phenyl acetate

Many phenols do not react directly with a carboxylic acid to form an ester (another case in which phenols differ from alcohols). However, esters of phenols are easily synthesized by an indirect method discussed in Section 23–13.

All told, the ester group is very common in nature. It occurs in fats, to be discussed in Chapter 24, and in many simple compounds, as we have seen in this chapter. The ester group may also appear in a complex molecule, along with several other functional groups, as shown by the structure below.

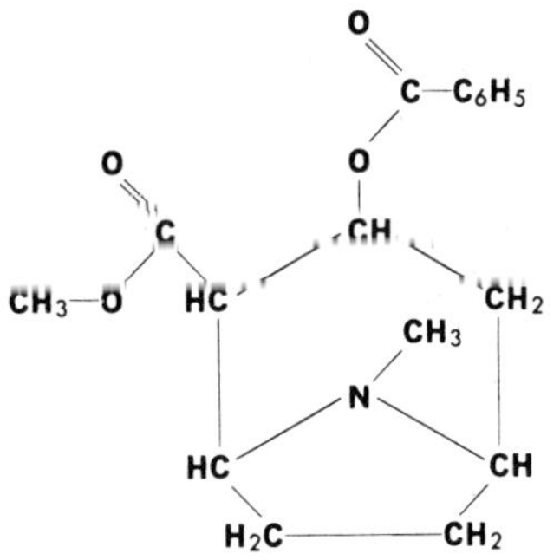

cocaine

Cocaine, produced by the coca bush of Bolivia and Peru, is used medically as a surface anesthetic. Note that cocaine has two different ester groups and a basic amino group (Section 25.2). Cocaine is also taken internally by many people for its stimulating effect. Because "street drugs" are always diluted and often mixed with other drugs, it is difficult to obtain reliable evidence as to what harmful effects, if any, cocaine may have. It does seem fairly certain, however, that it is not an addictive narcotic like heroin.

23.10 INORGANIC ESTERS

Esters of inorganic oxyacids are also known. Some examples are listed in Table 23.2, along with the acids from which they are derived. An organic ester is included in the list so that the structural comparison can be seen.

Glyceraldehyde-3-phosphate is one of the intermediates in the process of glycolysis (metabolism of carbohydrates), discussed in Section 35.6. Other important natural phosphate esters are adenosine monophosphate (AMP), mentioned in Section 33.2, and adenosine triphosphate (ATP), shown in Section 33.7.

Both isopentyl nitrite and glyceryl trinitrate (also called "nitroglycerin") are used to alleviate the painful effects of coronary attacks. Glyceryl trinitrate is also an explosive used for both military and peaceful purposes.

Another explosive, cellulose nitrate (nitrocellulose) is obtained when the cellulose of cotton or wood pulp is treated with nitric acid. The alcoholic OH groups in cellulose (Section 30.5) react with nitric acid. Working with either cellulose nitrate or glyceryl trinitrate is extremely

TABLE 23.2 Some esters of inorganic acids (compared to an organic ester)

Acid	Ester	Specific Example
H—O—C(=O)—R′ carboxylic acid	R—O—C(=O)—R′ carboxylate ester	CH_3CH_2—O—C(=O)—$CH_2CH_2CH_3$ ethyl butanoate
H—O—N(=O)→O nitric acid	R—O—N(=O)→O nitrate ester	CH_2—O—NO_2 | CH—O—NO_2 | CH_2—O—NO_2 glyceryl trinitrate ("nitroglycerin")
H—O—N=O nitrous acid	R—O—N=O nitrite ester	CH_3—CH(CH_3)CH_2CH_2—O—N=O isopentyl nitrite (isoamyl nitrite)
H—O—P(=O)(OH)—O—H phosphoric acid	R—O—P(=O)(O—H)—O—H phosphate ester	H—C=O | H—C—OH | CH_2—O—P(=O)(OH)—OH glyceraldehyde-3-phosphate

hazardous; early attempts to use them ended in disaster. Finally, in 1867, the Swedish chemist and industrialist Alfred Nobel patented dynamite, a mixture of glyceryl trinitrate with diatomaceous earth. He had found that glyceryl trinitrate could be safely handled in this condition. A decade later Nobel discovered that a mixture of glyceryl trinitrate and cellulose nitrate formed a colloidal mass which, curiously, stabilized both of them. This gelatin could be used for blasting purposes or, in a modified form, in guns.

Either cellulose nitrate alone or a mixture of glyceryl and cellulose nitrates is called *smokeless powder*. (Ancient gunpowder was a mixture of potassium nitrate, charcoal, and sulfur.) Nitrate materials of this kind are widely used in guns, rockets, and missiles. The explosives are tremendously useful in mining, road building, etc.; they are cruelly destructive in military use. Advances in technology very often present human beings with dilemmas.

Alfred Nobel acquired a large fortune from oil fields and from manufacturing explosives. He willed funds to the establishment of the **Nobel Prizes** for peace, chemistry, physics, literature, and physiology or medicine. The first prizes were awarded in 1901, five years after his death.

23.11 HYDROLYSIS OF ESTERS

In Section 23.6, we mentioned that after chemists discovered esters they also discovered how to split them into the component alcohol and acid. From the beginning, this has been a useful, frequently applied manipulation. An ester can be cleaved by heating it in water containing a little strong acid.

$$
\text{CH}_3\text{CH}_2\text{-O} \underset{}{\overset{O}{\diagdown}} \text{C-CH}_3 + \text{H-O-H} \xrightarrow[\text{heat}]{\text{H}^+} \text{CH}_3\text{CH}_2\text{-O-H} + \underset{\text{H-O}}{\overset{O}{\diagdown}} \text{C-CH}_3 \qquad (23\text{-}14)
$$

an ester   water   an alcohol   a carboxylic acid

A chemical decomposition, such as this one, due to reaction with water is termed **hydrolysis** (Gr. *hydro*, water plus *lysis*, loosen, break away). In this particular case, the strong acid serves as catalyst. Any of the common mineral acids are effective, for example, hydrochloric, sulfuric, of phosphoric acid.

The ester-hydrolysis reaction just written, Eq. (23–14), looks suspiciously like the equation of esterification written backward, Eq. (23–10). Furthermore, we find experimentally that *complete* hydrolysis is not achieved in the reaction mixture; some of the alcohol and the carboxylic acid molecules thus produced combine with each other to recreate some ester molecules and water. The relation between these two reversible reactions can be shown in one equation:

$$
\text{CH}_3\text{CH}_2\text{-O} \underset{}{\overset{O}{\diagdown}} \text{C-CH}_3 + \text{H-O-H} \underset{B}{\overset{A}{\underset{\longleftarrow}{\xrightarrow{\text{H}^+ \text{ heat}}}}} \text{CH}_3\text{CH}_2\text{-O-H} + \underset{\text{H O}}{\overset{O}{\diagdown}} \text{C-CH}_3 \qquad (23\text{-}15)
$$

As written here, the forward reaction, A, would be hydrolysis, the reverse reaction, B, would be esterification. The heat and catalyst have the same influence on both reactions, merely speeding up the action. This observation, and many others, causes us to believe that exactly the same events of bond-breaking and bond-making occur as the reactions proceed in either direction. Therefore the mechanism of the hydrolysis reaction would be like Fig. 23.3, except that in hydrolysis HOH would attack and HOCH$_3$ would be the group which leaves.

Fats are among the many kinds of esters produced by living organisms. During digestion, fats are hydrolyzed, giving acids and glycerol (the alcohol portion). The products pass into the body and, if not used for energy, are converted back to fats by an esterification reaction. In the body enzymes, rather than H^+ ions, are the catalysts for either hydrolysis or formation of fats.

In the laboratory either a hydrolysis or an esterification can be carried out, depending on which materials are put into the flask. However, because the products can react with each other to give back the starting materials, complete conversion never results. That is, if one attempts to hydrolyze an ester to an alcohol plus a carboxylic acid, *chiefly* those products will be obtained; but one must expect that there still will be *some* esters in the reaction mixture. Similarly, if one attempts an esterification reaction, Eqs. (23–15B) or (23–10), the conversion to ester will not be a complete one.

23.12 SAPONIFICATION OF ESTERS

A second method for cleaving an ester is to treat it with an aqueous solution of a metallic hydroxide, most often sodium hydroxide. The reaction can be conducted at room temperature within a reasonable time, but as usual can be hastened by heat.

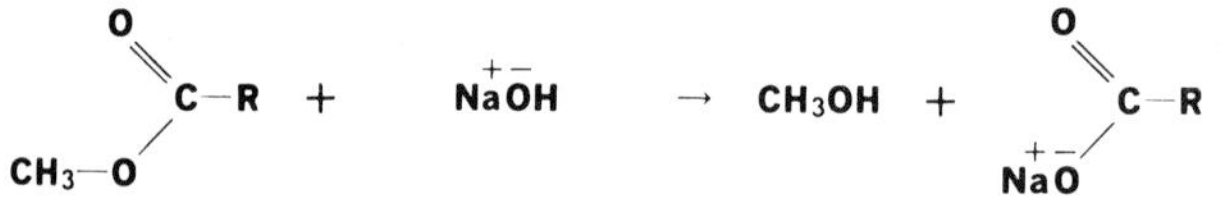

$$\text{(23-16)}$$

an ester	sodium hydroxide (in water)	an alcohol	sodium salt of an acid

Because of the base used, one of the products of this reaction is a salt of an acid rather than the acid itself; the other cleavage product is an alcohol, as in the hydrolysis reaction, Eq. (23–14).

If the R-group in this example happens to have a long straight chain (12 to 18 carbons), the salt produced is a soap. Indeed, we will see in Section 24.5 that treatment with sodium hydroxide solution is used to produce soap from fats (a special class of esters). The term **saponification** (literally, soap making) is now applied quite broadly to the reactions of hydroxides with any kind of ester (Eq. 23–16).

The reaction with base can be pictured as follows:

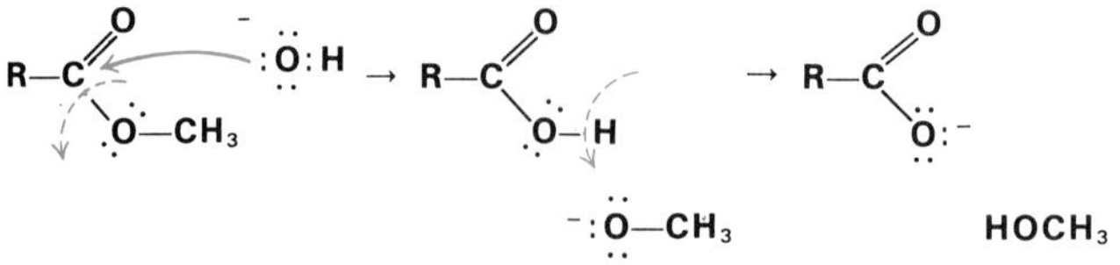

In stage 2, the CH_3O^- ion which was displaced is a strong base (about like HO^-) and immediately takes with it the H^+ from RCOOH, converting it to $RCOO^-$. (The Na^+ was omitted from the equations. It would be present in solution, balancing whichever negative ions were present at the moment.)

Although the *salt* of an acid is produced by saponification, mere acidification of the reaction mixture (for example, with hydrochloric acid) immediately converts the salt to the corresponding acid.

$$R-C\overset{\displaystyle O}{\underset{\displaystyle O^- Na^+}{\big\langle}} \;+\; H^+Cl^- \;\rightarrow\; R-C\overset{\displaystyle O}{\underset{\displaystyle O-H}{\big\langle}} \;+\; Na^+Cl^- \tag{23-17}$$

With this simple after-treatment, saponification can give the same net result as hydrolysis; an ester can be split into its component alcohol and carboxylic acid. Indeed, saponification may often be the preferred method because the cleavage is complete; the salt of the carboxylic acid cannot recombine with the alcohol to give back some of the ester.

**23.13
FORMATION
AND
REACTIONS OF
ACID
CHLORIDES**

A carboxylic acid can be converted to an acid chloride by treatment with phosphorus trichloride, as in this example:

$$3CH_3-C\overset{\displaystyle O}{\underset{\displaystyle OH}{\big\langle}} \;+\; PCl_3 \;\rightarrow\; 3CH_3-C\overset{\displaystyle O}{\underset{\displaystyle Cl}{\big\langle}} \;+\; H_3PO_3 \tag{23-18}$$

acetic acid   acetyl chloride   phosphorous
        (an acid chloride)   acid

The acid chloride is a highly reactive substance which can be converted to still other acid derivatives. For example, it reacts with an alcohol to produce an ester:

$$CH_3-C\overset{\displaystyle O}{\underset{\displaystyle Cl}{\big\langle}} \;+\; H-O-CH_2CH_3 \;\rightarrow\; CH_3-C\overset{\displaystyle O}{\underset{\displaystyle O-CH_2CH_3}{\big\langle}} \;+\; HCl \tag{23-19}$$

acetyl chloride   ethanol   ethyl acetate

An advantage of this indirect, two-step pathway from the acid to the ester, Eqs. (23–18) and (23–19), is that both steps give complete conversion to the desired products. This may be important if either the alcohol or the acid is expensive. Of course, the ester could also have been produced in one step by direct reaction of the alcohol with acetic acid (esterification—see Eq. 23–11). However, the conversion of ethyl acetate would not be very high because esterification is a reversible reaction (Section 23.11).

The reactive acid chloride also provides a method for synthesizing an ester of a phenol.

$$\bigcirc\!\!-O-H \;+\; \overset{\displaystyle O}{\underset{\displaystyle Cl}{C}}-CH_3 \;\rightarrow\; \bigcirc\!\!-O-\overset{\displaystyle O}{C}-CH_3 \;+\; HCl \tag{23-20}$$

phenol   acetyl chloride   phenyl acetate

Direct esterification of a phenol with a carboxylic acid is usually not possible.

An acid chloride reacts with ammonia in the same way it does with an alcohol. The reaction gives an amide, discussed in Section 25.6.

**23.14
SOURCES OF
ACIDS**

Carboxylic acids may be synthesized by a great variety of methods. A few of the common procedures will be discussed in this section. Other methods are of interest primarily to professional chemists.

A. By Oxidation   Often the acids may be obtained by oxidation of other organic compounds, on either a laboratory or an industrial scale. Primary alcohols, as well as aldehydes, may be oxidized to acids (Eq. (22–5)). In the laboratory these oxidations are usually accomplished by reagents such as sodium dichromate or potassium permanganate. For commercial production it is desirable to use an inexpensive oxidizing agent such as oxygen from air with a catalyst.

We stated in Section 21.7 that an alkylbenzene can be oxidized to benzoic acid. For industrial production, a vanadium pentoxide or manganese acetate catalyst makes possible the use of air for the oxidation. Nearly all benzoic acid is obtained in this way.

B. From Organic Cyanides   Hydrolysis of an organic cyanide provides an important laboratory method for the preparation of a carboxylic acid.

$$R\text{—}C\equiv N \ + \ 2H_2O \ + \ HCl \ \xrightarrow{\Delta} \ R\text{—}C\underset{O}{\overset{OH}{\diagup}} \ + \ NH_4^+Cl^- \tag{23–21}$$

an organic   a carboxylic
cyanide   acid
(a nitrile)

This reaction of an organic cyanide with water requires a catalyst. Either acid or base is an effective catalyst. However, if acid is used, as in Eq. (23–21), the carboxylic acid is obtained directly, without the necessity of acidifying a basic mixture after the hydrolysis reaction is complete.

The organic cyanide is also sometimes called a *nitrile*. It is important to note that the cyanide group is *covalently* bonded to the rest of the carbon skeleton in an organic compound, whereas in inorganic salts a cyanide group commonly exists as an ion.

The organic cyanide required for reaction (23–21) can be obtained in a variety of ways. One common starting material is an alkyl halide, which can undergo displacement by a cyanide ion:

$$\overset{+}{Na}\overset{-}{CN} \ + \ CH_3CH_2CH_2Br \ \rightarrow \ CH_3CH_2CH_2CN \ + \ \overset{+}{Na}\overset{-}{Br} \tag{23–22}$$

C. From Fats   By hydrolyzing a natural fat, one may obtain a mixture of octadecanoic acid (stearic acid) and certain other long-chain acids (Sections 24.1 and 24.2).

23.15 OTHER ACIDS, ESTERS, AND SALTS OF INTEREST

A. Acetic Acid   Dilute solutions of acetic acid (vinegar) have been used from antiquity and are still being used to preserve meat, fish, pickles, and other foods. Acetic acid arrests the growth of various microorganisms, preventing spoilage of the food to which it is added.

Acetic acid which is nearly pure, and undiluted by water, is called *glacial* acetic acid, so named because it freezes at about 16°C. Industrially, it is as important among organic compounds as sulfuric acid is among the inorganics. A weak acid, it is a useful acidulant in the dyeing and processing of textiles and in the coagulation of rubber latex. Acetic acid is one of the intermediates essential for the manufacture of cellulose acetate film, acetate textile fibers, and innumerable pharmaceuticals.

B. Propanoic Acid   This acid, commonly called propionic acid, occurs in small amounts in dairy products, probably some of it in esterified form. To a certain extent it acts as a natural preservative.

Propionic acid is also produced by some bacteria. Propionic acid and its salts are used medically to treat fungal infections such as "athletes' foot." Sodium or calcium propionate is added to some foods, especially to cheeses and breads, to inhibit the growth of microbes.

C. Benzoic Acid

Benzoic acid is an intermediate for the synthesis of many drugs and dyes. Increasing amounts are being used to improve the quality of alkyd enamels. Sodium benzoate, C_6H_5COONa, has long been used as a food preservative. It is most satisfactory in somewhat acid foods (pH 4 or lower), particularly in fruit juices, catsup, pickles, pie fillings, jams, and margarine. Sodium benzoate is an effective bactericide at the customary concentration of 0.1%, and is tasteless and nontoxic. It is also a preservative for drugs, cosmetics, toothpastes, gum, and starch.

D. Dicarboxylic Acids

In a number of organic compounds we may find two or more carboxyl groups in one molecule. The simplest example is **oxalic acid**, HOOC—COOH, which is just two carboxyl groups bonded together. Its name arises from the fact that it is the sour material in wood sorrel, botanically called oxalis, (Gr. *oxys*, acid). Since both the hydrogens in oxalic acid can be neutralized by a base, it can form either normal salts such as NaOOC—COONa, sodium oxalate, or acidic salts such as HOOC—COOK, potassium hydrogen oxalate. It is the latter form which occurs in sorrel plants, rhubarb, and spinach. The acid and its salt are toxic but are destroyed by cooking. Oxalic acid can be used in some circumstances to remove ink stains and rust.

Although many dicarboxylic acids are well known, we will mention just one more for now. **Succinic acid**, HOOC—CH$_2$—CH$_2$—COOH, apparently occurs in almost all plants and animals. It is one of the compounds participating in the citric acid cycle. By this process, discussed in Section 35.8, organisms obtain energy from their food. Succinic acid was first mentioned in 1550 by Agricola, who isolated some from amber (L. *succinum*). The acid is an important intermediate for the manufacture of numerous pharmaceuticals and polymers. The salt sodium succinate is an antidote for poisoning by heavy metals or by barbiturate drugs.

E. Hydroxy Acids

Another structural possibility is that a molecule may have both hydroxyl and carboxyl functional groups. We have already mentioned (Table 23.1) that sour milk contains **lactic acid** (2-hydroxypropanoic acid). This acid may also occur in fatigued muscle tissue.* Lactic acid is a common acidulant for food products because its mild acid taste does not overpower other flavors; a few examples of its use are in soups, olives, beer, soft drinks, cheese, and sherbets. The lactic acid for these purposes is commercially obtained by the action of *lactobacillus* organisms on molasses or starch.

Tartaric acid, a by-product of wine-making, is structurally dihydroxysuccinic acid:

$$\text{HOOC—CH—CH—COOH}$$
$$\qquad\;\; |\quad\; |$$
$$\qquad\;\; \text{OH}\;\; \text{OH}$$

During the aging of wine, a substance, originally present in the grape juice, crystallizes on the inside of the barrel. The medieval Greeks called this hard crust tartaron; in modern English it is tartar. The purified material, being white, is called *cream of tartar*. We now

---

* The breakdown of glucose, through a complex series of reactions, releases the energy needed for movement and body heat. Under certain circumstances, lactic acid may be one of the end products of glucose metabolism.

know that this is the potassium hydrogen salt of tartaric acid. Cream of tartar is widely used in baking powders. It is only slightly soluble at room temperature. However, at baking temperatures it dissolves and reacts with the sodium hydrogen carbonate also present in the baking powder, releasing bubbles of carbon dioxide. Tartaric acid itself may be obtained when desired by treating crude tartar with sulfuric acid. Tartaric acid is used in some foods, soft drinks, and metal polishes.

Citric acid (L. *citrus*) is a more complex substance having three carboxyl groups and one alcohol group.

$$\text{CH}_2\text{—COOH}$$
$$\text{HO—C—COOH} \qquad \text{citric acid}$$
$$\text{CH}_2\text{—COOH}$$

This acid occurs in several berries and other fruits, but especially in lemons, oranges, etc. In fact, it is present in very small amounts in all living cells which derive energy from the oxidation of carbon compounds. The oxidation process is called the citric acid cycle because of the key role of citric acid.

Although citric acid was formerly isolated from citrus wastes, since about 1925 it has been produced more economically by growing a fungus, *Aspergillus niger*, in a glucose solution. Citric acid is a metabolic product elaborated by the fungus as it consumes the glucose. The mat of fungus is then filtered off, and the citric acid is crystallized from the solution.

In food processing, citric acid is used more than any other solid organic acid because it is nontoxic, very soluble in water, and has a pleasant, mildly sour taste. A few of its applications are in fruit and vegetable juices, candies, desserts, jellies, frozen fruits, soft drinks, and effervescent tablets; others are in cosmetics, hair rinses, rust and scale removers, and bottle-washing mixtures. It is usually the most satisfactory acidulant for drug preparations.

The salt sodium citrate, in conjunction with citric acid, is valuable for its buffering ability in jellies, ice cream, candy, gelatin desserts, and whipping cream; the setting of these foods depends on the proper pH. The sodium citrate-citric acid mixture is likewise the most desirable buffer for medicines. In samples of human blood collected for transfusion, the mixture acts as a buffer and anticoagulant.

Salicylic acid and its derivatives are an interesting family. They constitute about half of all the "coal tar" drugs (i.e., benzenoid and related compounds) manufactured in the United States.

salicylic acid   methyl salicylate   acetyl salicylic acid
                        "oil of wintergreen"   "aspirin"

Salicylic acid, a better disinfectant than phenol, is a common ingredient of tropical ointments for skin diseases. It causes the outer layer of skin to flake off, but does not kill the underlying tissue.

The ester methyl salicylate is a fragrant oil occurring in numerous plants, especially wintergreen. It is a flavoring agent for candy, gum, foods, and dentifrices, and a constituent of some antiseptics and perfumes. Methyl salicylate is used in liniments as a counterirritant. The mild surface inflammation it creates affects the circulation in a manner which relieves sore muscles. Commercial quantities of methyl salicylate are produced by esterification of salicylic acid with methanol.

Aspirin is the most widely used synthetic drug because it is cheap, relatively safe, easily available, and quite effective in reducing fever and in relieving headaches or similar discomforts. Recently, enough aspirin has been produced in the United States to provide each person in the country with nearly 200 of the standard 5-grain tablets each year, this amounts to about 83 000 lb per day! Note in the structure of aspirin that the phenolic group has been converted to an acetate ester.

F. Waxes   Waxes are produced by a number of plants and animals. The wax in our ears and beeswax are typical of those produced by animals. Carnuba wax, from the leaves of certain Brazilian palm trees, makes an excellent polish for floors and automobiles. These waxes, although often mixtures of compounds, are chiefly esters in which both the acid portion and the alcohol portion have very long chains. A common example is $C_{27}H_{55}COOC_{30}H_{61}$. The simple term *wax* is usually reserved for these ester compounds. Modified names are used for other substances having waxy properties. For instance, recall that paraffin wax is a mixture of long-chain hydrocarbons.

SUMMARY

The organic acids (carboxylic acids) and their derivatives are another group of compounds found widely distributed in plants and animals. As with all classes of organic compounds, it is most important to learn the systematic names of carboxylic acids, esters, and salts. However, several of the most familiar acids, which have been known for centuries, have common names; these names are still so widely used that they should be learned. In common names α, β, γ, etc. are used to designate positions, rather than numbers.

A carboxylic acid is acidic enough to react with either hydroxide ion or hydrogen carbonate ion, forming a salt. A sodium or potassium carboxylate salt, being ionic, is always very soluble in water, whereas the carboxylic acid, if its skeleton has more than about six carbons, is not water-soluble. A knowledge of these solubility properties is valuable for understanding both the chemical and the biochemical behavior of carboxylic acids.

Carboxylic acids can react with alcohols to form esters, a reaction of considerable importance in living systems as well as in the laboratory. A lactone is fundamentally an ester which forms intramolecularly and therefore has a ring structure. Another variation in ester structure occurs when an inorganic acid reacts with an alcohol to form an ester. An ester can be broken down into its component alcohol and acid if it is heated with water in the presence of either a strong, inorganic acid (HCl, H_2SO_4) or a strong base (NaOH, KOH). Such a reacton is called hydrolysis.

Acid chlorides are carboxylic acid derivatives which provide another useful method for preparing esters, including esters from phenols. An acid chloride can also be converted to an amide, yet another type of acid derivative, which will be discussed in Section 25.6.

The C=O portion of structure in a carboxyl group strongly attracts electrons away from other atoms nearby in the molecule. This property is what makes a carboxylic acid much more acidic than water or an alcohol. The electron-attracting C=O structure is also responsible for the type of substitution reaction which transforms one kind of acid derivative to another, for instance, an ester to an acid, or an acid to an amide.

IMPORTANT TERMS

acid chloride

carboxylate salt

carboxylic acid

electron-attracting group

ester

esterification

hydrolysis

inorganic ester

lactone

saponification

WORK EXERCISES

1. Give the structural formula of the following.
   a) propanoic acid
   b) hexanoic acid
   c) benzoic acid
   d) sodium acetate
   e) potassium benzoate
   f) ethyl benzoate
   g) isopropyl acetate
   h) 3-chlorobenzoic acid
   i) a wax
   j) *n*-butyl nitrate
   k) isopentyl nitrite
   l) *t*-butyl propionate
   m) isopropyl α-ethyl-butyrate

2. Write complete equations for the following reactions and name the organic products.
   a) formic acid, sodium hydroxide
   b) acetic acid, sodium hydrogen carbonate
   c) benzoic acid, ethanol, H_2SO_4, heat
   d) sodium benzoate, HBr
   e) ethyl butanoate, H_2O, H_2SO_4, heat
   f) butanoic acid, isopentyl alcohol, H_2SO_4, heat
   g) methyl benzoate, dilute NaOH, heat
   h) CH_3CH_2CN, H_2O, heat, HCl

3. Show how each of the following could be obtained.
   a) hexanoic acid from 1-hexanol
   b) benzoic acid from toluene
   c) butanoic acid from 1-bromopropane

4. Write the equation for the reaction of methanol with *p*-bromobenzoic acid to produce an ester.
   a) Which would you estimate would be less expensive, the methanol or the *p*-bromobenzoic acid?
   b) If you were to prepare this ester in the laboratory, what conditions, concentrations, etc, would you use to make maximum use of the more expensive component?

5. Write a systemative name for each compound.

   a) 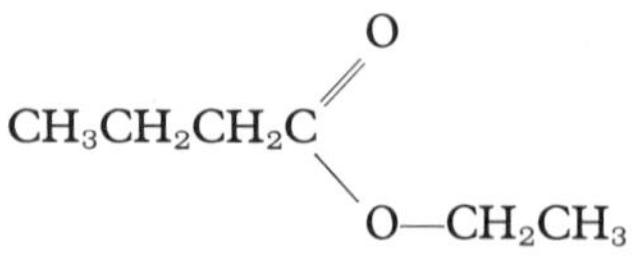 $CH_3CH_2CH_2C \overset{\displaystyle O}{\underset{\displaystyle O-CH_2CH_3}{\big<}}$

b) 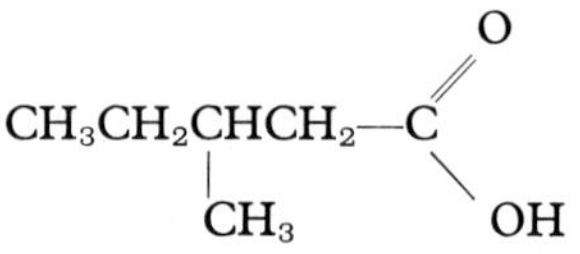 $CH_3CH_2CHCH_2-C \overset{\displaystyle O}{\underset{\displaystyle OH}{\big<}}$ with CH_3 branch

c) $CH_3CH_2CH-C \overset{\displaystyle O}{\underset{\displaystyle OH}{\big<}}$ with OH branch

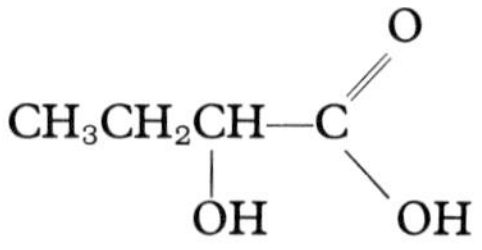

d) 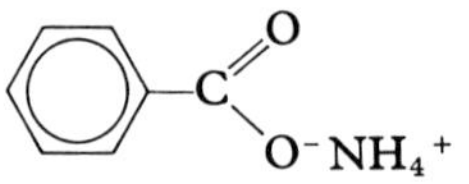 $C_6H_5-C \overset{\displaystyle O}{\underset{\displaystyle O^- NH_4^+}{\big<}}$

e) 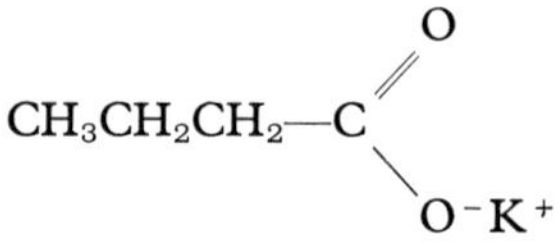 $CH_3CH_2CH_2-C \overset{\displaystyle O}{\underset{\displaystyle O^-K^+}{\big<}}$

f) 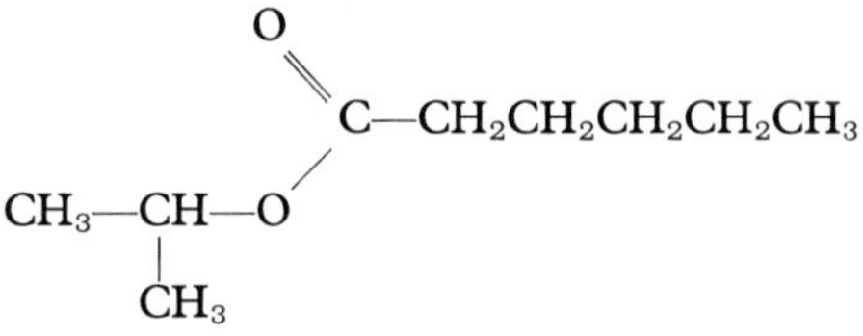 $CH_3-CH-O \overset{\displaystyle O}{\underset{}{\big<}} C-CH_2CH_2CH_2CH_2CH_3$ with CH_3 branch

6. Large amounts of phthalic acid are required for the manufacture of glyptal enamels (Section 28.8) and a variety of other substances. Show an industrial method for economically converting *o*-xylene (1,2-dimethylbenzene) to phthalic acid (benzene-1,2-dicarboxylic acid).

7. Show reactions that could be used to convert 1-butanol to the following.
   a) butanoic acid
   b) pentanoic acid. [Hint: This will require more than one step and is a more difficult problem than part (a). If you get stuck, consult Eq. (22–8) for a reaction that might give a useful intermediate.]

8. List all the types of structural units (both carbon

skeletons and functional groups) which you can
find in the compounds below.
a) oil of wintergreen (Section 23.15E)
b) ferulic acid, an intermediate formed during the
biosynthetic processes of many plants:

$$HO-C_6H_4-CH{=}CH-COOH$$
$$CH_3-O$$

SUGGESTED READING

Collier, H. O. J., "Aspirin," *Scientific American*, November 1963 (Offprint #169). There is now a better understanding of the reasons for the dramatic effectiveness of the most widely used drug.

Jacobson, M., and M. Beroza, "Insect Attractants," *Scientific American*, August 1964 (Offprint #189). The sex attractants excreted by insects are being isolated so that the chemical structure can be determined. If an attractant can be synthesized, it can be used to lure insects to traps. The structures discovered so far include alcohols, esters, and other types.

Noller, C. R., *Textbook of Organic Chemistry*, 3rd ed., Chapters 11 and 27, W. B. Saunders, Philadelphia, 1966.

24

Fats, Oils, Soaps, and Detergents

24.1
STRUCTURE
AND
HYDROLYSIS
OF FATS
AND OILS

Typical fats and oils, such as bacon fat or olive oil, can be hydrolyzed to glycerol and a mixture of fatty acids. ("Fatty acids" are defined in the next section.) From detailed studies of the hydrolysis fragments and of their other properties, it is certain that fats *are esters of glycerol with fatty acids.*

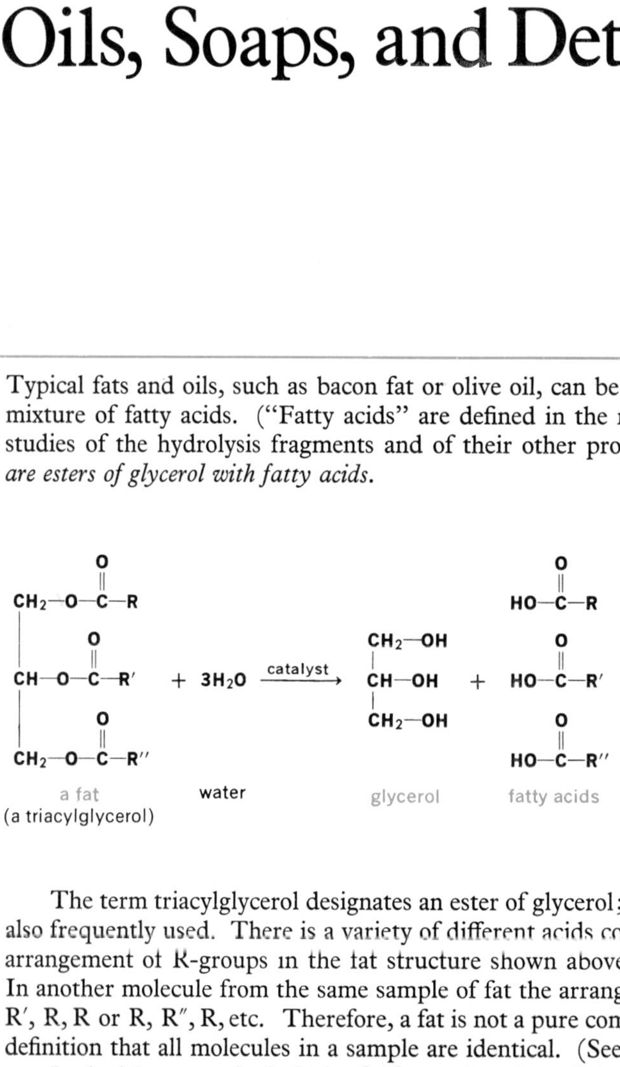

$$
\begin{array}{l}
\text{CH}_2\text{—O—}\overset{\overset{\text{O}}{\|}}{\text{C}}\text{—R} \\[4pt]
\text{CH—O—}\overset{\overset{\text{O}}{\|}}{\text{C}}\text{—R}' \\[4pt]
\text{CH}_2\text{—O—}\overset{\overset{\text{O}}{\|}}{\text{C}}\text{—R}''
\end{array}
\;+\; 3\text{H}_2\text{O} \;\xrightarrow{\text{catalyst}}\;
\begin{array}{l}
\text{CH}_2\text{—OH} \\[4pt]
\text{CH—OH} \\[4pt]
\text{CH}_2\text{—OH}
\end{array}
\;+\;
\begin{array}{l}
\text{HO—}\overset{\overset{\text{O}}{\|}}{\text{C}}\text{—R} \\[4pt]
\text{HO—}\overset{\overset{\text{O}}{\|}}{\text{C}}\text{—R}' \\[4pt]
\text{HO—}\overset{\overset{\text{O}}{\|}}{\text{C}}\text{—R}''
\end{array}
\qquad (24\text{–}1)
$$

a fat   water   glycerol   fatty acids
(a triacylglycerol)

The term triacylglycerol designates an ester of glycerol; the older term triglyceride is also frequently used. There is a variety of different acids combined in a natural fat. The arrangement of R-groups in the fat structure shown above is just one typical example. In another molecule from the same sample of fat the arrangement might be R′, R, R″ or R′, R, R or R, R″, R, etc. Therefore, a fat is not a pure compound by the usual chemical definition that all molecules in a sample are identical. (See Table 24.2.)

In the laboratory, hydrolysis of a fat can be accomplished with a mineral acid, such as hydrochloric or phosphoric, acting as catalyst. The chemical reaction taking place during the digestion of fatty food in the intestines of humans or animals is another example of the hydrolysis reaction, Eq. (24–1). The catalyst for the reaction is an enzyme.

In certain tissues of plants and animals, *biosynthesis* of fats occurs, catalyzed by a different enzyme:

$$\text{fatty acids} + \text{glycerol} \xrightarrow{\text{enzyme}} \text{fat} + \text{water} \qquad (24\text{–}2)$$

Plants first synthesize fatty acids and glycerol from very simple molecules and then combine them, Eq. (24–2), to produce a fat. Animals can do the same, but they also obtain large amounts of fatty acids and glycerol from digested food.

24.2
FATTY ACID
CONSTITUENTS

Since the glyceryl ester structure is common to all fats, the variations in properties from one type of fat to another must be due to differences in their fatty-acid components. It is therefore convenient to describe fats in terms of the fatty acids present.

In general, *a natural fatty acid has a skeleton composed of an even number of carbon atoms in a long, unbranched chain.* A chain length of sixteen or eighteen carbons is predominant. Carbon–carbon double bonds are common. The most typical acids found in fats are listed in Table 24.1. Common names, rather than systematic names, are often used for these acids. Acids containing fewer than twelve carbons, or more than twenty, occur in some fats but are less common.

The composition of a particular fat can be determined by hydrolyzing it, Eq. (24–1), and then analyzing the resultant mixture of fatty acids. The percent, by weight, of each acid present in various fats is listed in Table 24.2.

TABLE 24.1 Common fatty acids

Structural Formula and Common Name	Abbreviated Formula
$CH_3(CH_2)_{10}COOH$ lauric acid	$C_{11}H_{23}COOH$
$CH_3(CH_2)_{12}COOH$ myristic acid	$C_{13}H_{27}COOH$
$CH_3(CH_2)_{14}COOH$ palmitic acid	$C_{15}H_{31}COOH$
$CH_3(CH_2)_5CH{=}CH(CH_2)_7COOH$ palmitoleic acid	$C_{15}H_{29}COOH$
$CH_3(CH_2)_{16}COOH$ stearic acid	$C_{17}H_{35}COOH$
$CH_3(CH_2)_7CH{=}CH(CH_2)_7COOH$ oleic acid	$C_{17}H_{33}COOH$
$CH_3(CH_2)_4CH{=}CHCH_2CH{=}CH(CH_2)_7COOH$ linoleic acid	$C_{17}H_{31}COOH$
$CH_3CH_2CH{=}CHCH_2CH{=}CHCH_2CH{=}CH(CH_2)_7COOH$ linolenic acid	$C_{17}H_{29}COOH$

TABLE 24.2   Composition of some fats and oils

Fat or Oil	lauric C_{12} sat'd	myristic C_{14} sat'd	palmitic C_{16} sat'd	stearic C_{18} sat'd	oleic C_{18} one C=C	linoleic C_{18} two C=C	linolenic C_{18} three C=C
				Fatty Acid Present, % by Weight[a]			
lard		1–2	25–30	12–16	40–50	5–10	1
beef tallow		3–5	25–30	20–30	40–50	1–5	
mutton tallow		1–5	20–25	25–30	35–45	3–6	
butterfat (cow)[b]	2–5	8–14	25–30	9–12	25–35	2–5	
coconut fat[c]	45–48	16–18	8–10	2–4	5–8	1–2	
palm kernel fat[d]	43–47	15–20	8–9	2–5	10–18	1–3	
palm oil or fat		1–3	35–45	4–6	40–50	8–11	
peanut oil			8–10	3–5	55–60	25–30	
sardine oil		5–6	12–16	2–3	(75–82)[e]		
olive oil			8–16	2–3	70–85	5–15	
cottonseed oil		1	20–25	1–2	20–30	45–50	
soybean oil			10	3	25–30	50–55	4–8
safflower oil			6	3	13–15	75–78	
linseed oil					20–35	15–25	40–60
tung oil			4–6		5–10	9–12	(76–82)[f]

[a] The approximate range of typical values for each fat is indicated by these figures which were compiled from various sources.
[b] Also 3–4% butyric acid and 1–3% each of C_6, C_8, C_{10} acids.
[c] Also 5–9% each of C_8 and C_{10} acids.
[d] Also 4% C_8 and 4–8% C_{10} acids.
[e] Total unsaturated acids, 75–82%, including 10–15% palmitoleic; some as large as C_{24} with up to six C=C.
[f] Tung oil contains eleostearic acid rather than linolenic; the three C=C are in different locations.

24.3
FATS AND OILS

It has become customary to call a triacylglycerol a fat if it is solid, or an oil if it is liquid, at ordinary temperatures. This differentiation of course is arbitrary and depends on the climate. The differences in melting points of the triacylglycerols are due chiefly to the varying numbers of double bonds present; the more of these there are, the lower the melting point. For example, olive oil and cottonseed oil have considerably higher percentages of unsaturated acids than do lard and tallow (Table 24.2).

Palm oil or fat represents an interesting borderline case. It is usually called an oil because it is a liquid as it originates in the tropics. However, in the temperate regions it may be a semisolid. In palm oil the total percentage of unsaturates (oleic plus linoleic) is only slightly higher than that of lard or tallow. The data in Table 24.2 suggest that triacylglycerols from animals are usually fats, while those from plants are usually oils.

When the term *oil* is used in connection with plant products, it refers to triacylglycerols which are structurally esters. Petroleum oil, however, consists of hydrocarbons (Section 19.4).

24.4
HYDROGENA-
TION OF OILS

For some purposes, solid fats rather than liquid oils are preferred. Oils, which are more unsaturated, can be "hardened" to higher-melting solids by the addition of hydrogen to their double bonds. The hydrogenation reaction, first described for alkenes (Section 20.5B), can be applied to oils:

An oil
(lower melting;
liquid at room temperature)

A fat
(higher melting;
solid at room temperature)

$$
\begin{array}{l}
\text{CH}_2\text{—O—C—(CH}_2)_7\text{CH}=\text{CHCH}_2\text{CH}=\text{CH(CH}_2)_4\text{CH}_3 \\[4pt]
\text{CH—O—C—(CH}_2)_7\text{CH}=\text{CH(CH}_2)_7\text{CH}_3 \quad + 3\text{H}_2 \xrightarrow[\text{heat}]{\text{Ni}} \\[4pt]
\text{CH}_2\text{—O—C—(CH}_2)_{16}\text{CH}_3
\end{array}
\qquad
\begin{array}{l}
\text{CH}_2\text{—O—C—(CH}_2)_{16}\text{CH}_3 \\[4pt]
\text{CH—O—C—(CH}_2)_{16}\text{CH}_3 \\[4pt]
\text{CH}_2\text{—O—C—(CH}_2)_{16}\text{CH}_3
\end{array}
\qquad (24\text{-}3)
$$

glyceryl tristearate

Note that in this example of an oil molecule there are two unsaturated side chains, one oleic and one linoleic. In the resultant fat, all three groups are stearic, so the product is called glyceryl tristearate.

In this manner cheap, abundant vegetable oils, such as cottonseed or soybean oils, can be converted to oleomargarine, cooking greases (Crisco, Spry, etc.), or stocks to be further processed into soap (Section 24.5). The hydrogenation can be controlled to provide the degree of firmness most suitable for the product desired. "Soft" margarine is obtained by partial hydrogenation, leaving considerable unsaturation in it. A certain amount of milk, vitamins A and D, emulsifying agents, flavors, and yellow food colors are added to margarine. It should be noted that the same coloring agents are frequently added to butter to improve its appearance. When cows are not on fresh pasture the color of the butter is pale and the vitamin A content decreases by about half.

**24.5
SAPONIFICA-
TION**

It was shown in Chapter 23 that a typical ester can be cleaved either by water or by a sodium hydroxide solution. Since fats are esters, they display the same behavior. Cleavage of a fat by water was described in Section 24.1; cleavage by hydroxide, also called saponification, is illustrated in Eq. (24-4).

$$
\begin{array}{l}
\text{CH}_2\text{—O—C—C}_{17}\text{H}_{35} \\[4pt]
\text{CH—O—C—C}_{17}\text{H}_{35} \quad + \ 3\text{NaOH} \ \rightarrow \\[4pt]
\text{CH}_2\text{—O—C—C}_{17}\text{H}_{35}
\end{array}
\quad
\begin{array}{l}
\text{CH}_2\text{—OH} \\[4pt]
\text{CH—OH} \quad + \ 3 \\[4pt]
\text{CH}_2\text{—OH}
\end{array}
\quad
\begin{array}{l}
\text{O} \\
\text{C—C}_{17}\text{H}_{35} \\
\text{NaO}
\end{array}
\qquad (24\text{-}4)
$$

a fat   glycerol   (sodium stearate)
                                                    a soap

An alkali metal carboxylate salt having a chain of 12 to 18 carbons has cleansing properties and is called a **soap**; the most common example is sodium stearate.

**24.6
THE SOLU-
BILITY AND
CLEANSING
ACTION OF
SOAPS**

A sodium carboxylate salt (R—COO^- Na^+) is highly ionic and usually quite water-soluble because of the strong attraction of water molecules to the charges on the ions (Section 23–5). Let us now consider how this behavior might be modified in the case of a soap, in which the R-group is a *long* hydrocarbon chain. The structure of sodium stearate, a typical soap, is written out in Fig. 24.1. Beneath it is a diagram symbolizing the two important features of the structure. The circle represents the ionic carboxylate end of

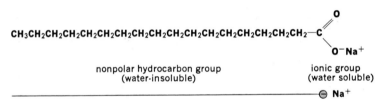

CH$_3$CH$_2$CH$_2$CH$_2$CH$_2$CH$_2$CH$_2$CH$_2$CH$_2$CH$_2$CH$_2$CH$_2$CH$_2$CH$_2$CH$_2$CH$_2$CH$_2$—C

O

O$^-$Na$^+$

nonpolar hydrocarbon group   ionic group
(water-insoluble)   (water soluble)

⊖ Na$^+$

Fig. 24.1 Structural features of a soap molecule.

the molecule and the long line represents the nonpolar hydrocarbon chain. The nonpolar hydrocarbon group is not soluble in water, but water would be attracted to the ionic groups and hence *tend* to dissolve the molecules.

The outcome of these conflicting tendencies is illustrated in Fig. 24.2. When soap molecules are placed in water, the hydrocarbon portions will not permit themselves to be exposed to water. Instead, they are attracted to each other, forming a cluster in which they are literally dissolved in each other. This grouping allows the ionic groups at the ends of the soap molecules to be attracted to the surrounding water molecules. The result is that the soap molecules are, in a sense, able to "dissolve" in the water. However, this is a colloidal solution, not a true solution.

The droplet of soap molecules surrounded by water, illustrated in Fig. 24.2, is called a *soap micelle*. Experimental measurements indicate that there are about 100 soap molecules in each cluster. The alignment of the hydrocarbon groups is random and frequently changing, as is typical in a liquid. In particular, the hydrocarbon chains must be bent, because of the crowding that would result at the center of the cluster if all were straight. The liquid soap micelles dissolved in liquid water constitute one example of an emulsion, one of the types of colloidal solutions.

Figure 24.3 shows how soap can pull into solution (emulsify) oil and dirt particles which alone would not be soluble in water. The interior of the soap micelle is essentially

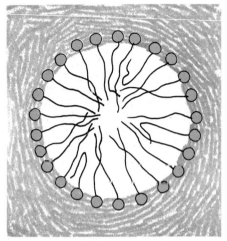

Fig. 24.2 A soap micelle—about 100 soap molecules clustered together and surrounded by water.

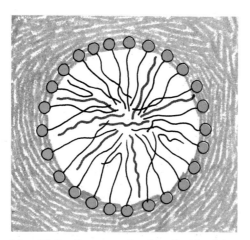

Fig. 24.3 Grease molecules dissolved inside a soap micelle.

the same as a liquid alkane mixture, such as kerosene. It is therefore a good solvent for materials similar to it in structure, namely nonpolar or slightly polar substances. Particles of dirt often cling to an object because of an oily film. Once the oil is emulsified and removed by the soap, the dirt particles are also loosened.

**24.7
SYNTHETIC
DETERGENTS**

It was shown in the previous section that a soap has cleansing properties because it can emulsify an oil in water. This action depends on the structural features shown in Fig. 24.1. In more general terms, any molecule will show cleansing and emulsifying behavior if it possesses the necessary combination of two different structural units: a large nonpolar, hydrocarbon group and a water-soluble end group. In addition to the soaps, there are numerous other compounds, both natural and synthetic, which fulfill these structural requirements.

The term **detergent** (L. *detergere*) means simply cleansing agent. Soap is just one of many compounds in this group. However, in everyday language when a person speaks of a detergent he is usually thinking of a modern synthetic detergent. The old familiar soap is still often regarded as being in a separate class.

The sulfonate salt shown below is a synthetic detergent which has excellent cleansing ability and is therefore widely distributed for both household and commercial use. Some common brand names are Tide, Fab, Cheer, and Wisk. (In general, synthetic detergents of this type are called alkylbenzenesulfonates.)

$$CH_3CH_2CH_2CH_2CH_2CH_2CH_2CH_2CH_2CH_2CH_2CH_2—\langle\bigcirc\rangle—SO_3^-Na^+ \qquad \text{a common synthetic detergent}$$

sodium *n*-dodecylbenzenesulfonate

This sulfonate salt is synthesized commercially by the route outlined in Fig. 24.4, starting from petroleum raw materials.

Another useful detergent is the sodium alkyl sulfate shown below. Note that it has the two features of structure previously described as being essential for cleansing action:

$$CH_3(CH_2)_{10}CH_2OSO_2O^-Na^+$$

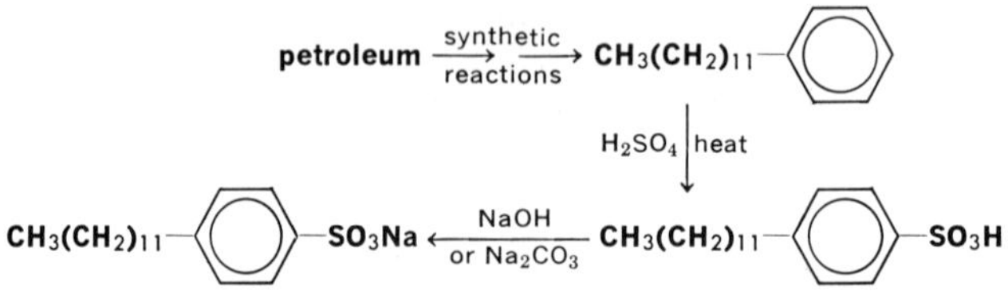

$$\text{petroleum} \xrightarrow[\text{reactions}]{\text{synthetic}} CH_3(CH_2)_{11}-\bigcirc$$

$$\downarrow H_2SO_4 \mid \text{heat}$$

$$CH_3(CH_2)_{11}-\bigcirc-SO_3Na \xleftarrow[\text{or } Na_2CO_3]{NaOH} CH_3(CH_2)_{11}-\bigcirc-SO_3H$$

Fig. 24.4   Synthetic route to a typical sulfonate detergent.

It is also possible for a detergent molecule to have a water-soluble end with a *positive* charge (rather than a negative charge as in the previous examples). Such a compound is therefore sometimes called an *invert soap*. Here is one example:

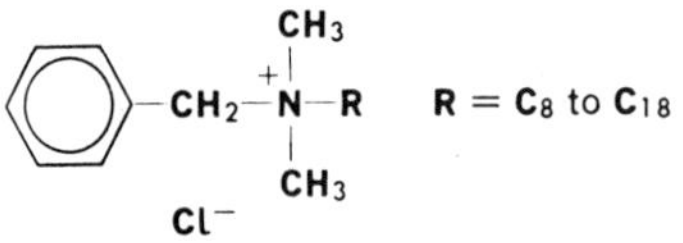

$$\bigcirc-CH_2\overset{+}{\underset{\underset{Cl^-}{CH_3}}{\overset{CH_3}{\underset{|}{N}}}}-R \qquad R = C_8 \text{ to } C_{18}$$

This type of compound is known as a quaternary ammonium salt (Section 25.5A). Since it is not only a good washing agent but also has germicidal properties, it is especially suitable for use in hospitals, etc.

It is not even necessary that the end group of a detergent be ionic, so long as it is water-soluble. Polar covalent structures, if there are enough of them, can provide the required solubility. This is illustrated in the examples below.

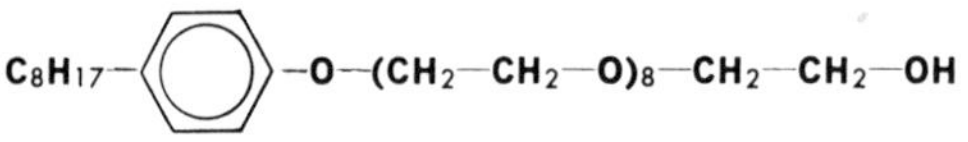

$$C_8H_{17}-\bigcirc-O-(CH_2-CH_2-O)_8-CH_2-CH_2-OH$$

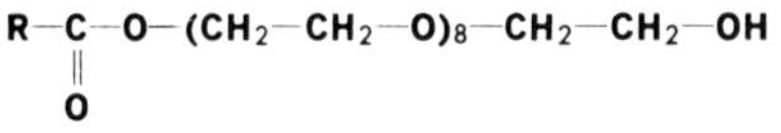

$$R-\overset{\overset{O}{\|}}{C}-O-(CH_2-CH_2-O)_8-CH_2-CH_2-OH \qquad \text{the R group is from a typical fatty acid}$$

Compounds of this type, known as *nonionic detergents*, are used extensively in dishwashing liquids, in low-sudsing products for clothes washers, and in special applications in which the absence of inorganic ions is desirable. One brand of nonionic detergent is named All.

SUMMARY

Fats and oils are natural products which are nutritionally important. The main distinctions between the two are that oils are chiefly plant products and are more unsaturated; that is, oils have more alkene pi bonds. Fats and oils are alike in that each is a triacylglycerol, that is, an ester of fatty acids with glycerol. The "fatty acids" found in natural fats and oils are distinguished by having unbranched, even-numbered, relatively long carbon chains (mostly C_{16} and C_{18}).

The most general and important reaction of a triacylglycerol is hydrolysis, which converts it to the component fatty acids and glycerol. Biochemically this hydrolysis is catalyzed by an enzyme and occurs during digestion, followed by assimilation and metabolism of the fatty acids and glycerol.

Chemically, the hydrolysis can be promoted by either acid or base catalyst. (The latter causes "saponification.") The sodium or potassium salts of the resultant fatty acids are soaps, which act as cleansing agents because they have both a large carbon skeleton and a water-soluble end group (the carboxylate group). This structural principle has been applied to synthetic detergents, which also have large carbon skeletons but different water-soluble end groups.

IMPORTANT TERMS

detergent	hydrolysis	triacylglycerol
fats	micelle	triglyceride
fatty acid	oils	unsaturated chains
glycerol	saponification	
hydrogenation	soap	

WORK EXERCISES

1. Write the structure of the following:
   a) an ester containing oleic, palmitic, and stearic acids combined with glycerol. With these components is more than one structure possible?
   b) a fat
   c) an oil
   d) a soap
   e) a synthetic detergent
   f) a drying oil
2. Write a complete equation, using structural formulas, for each reaction and tell the name used for that type of reaction.
   a) glyceryl tripalmitate, sodium hydroxide solution, heat
   b) glyceryl trioleate, hydrogen, nickel, heat
   c) glyceryl tristearate, water, hydrochloric acid, heat
3. Which of these compounds would you expect to be solid, and which liquid, at room temperature? Check your answers in a chemical handbook.
   a) glyceryl trioleate   b) glyceryl tristearate
   c) glyceryl tributyrate
4. What type of structure must an organic compound have to be a good cleansing agent?
5. Discuss briefly the process for manufacturing oleomargarine. In general, how do you think its nutritional value compares with that of butter?

SUGGESTED READING

Levey, M., "The early history of detergent substances," *J. Chem. Educ.* **31**, 521 (1954).

Meloan, C. E., "Detergents—soaps and syndets," *Chemistry*, **49**, no. 7, p. 6, Sept., 1976. The chemistry and physics of detergents are discussed, as well as additives such as builders, bleaches, and softeners.

Noller, C. R., *Textbook of Organic Chemistry*, 3rd ed., Chapter 12, W. B. Saunders, Philadelphia, 1966.

Snell, F. D., and C. T. Snell, "Syndets and surfactants," *J. Chem. Educ.* **35**, 271 (1958). A good survey of the types, properties, and manufacturing of synthetic detergents.

25

Amines, Amides, and Amino Acids

The amines are structurally related to ammonia. Whereas in ammonia the nitrogen holds three hydrogens, in an amine it is bonded to at least one organic group.

H—N—H   H—N—CH$_2$CH$_3$ CH$_3$—N—CH—CH$_3$ C$_2$H$_5$—N—C$_2$H$_5$
| | | CH$_3$ CH$_3$
H   H

ammonia   ethylamine   methylisopropylamine   diethylmethylamine

Amines and their derivatives are found in many different compounds of biological origin, some of the most abundant being amino acids and proteins. In many other compounds the amine structure is part of a heterocyclic ring (Section 21.8); these nitrogen heterocycles occur in nucleic acids, in several enzymes, and so on.

As might be expected, several of the properties of amines are similar to those of ammonia. The smaller amines, such as methylamine, dimethylamine, or ethylamine, are gases which are quite soluble in water and have a pungent odor very much like the odor of ammonia. Medium-sized amines (butyl, pentyl, etc.) have fishy odors. Two of the decomposition products found in decaying flesh are amines having foul, putrid odors; they have been given the descriptive names putrescine and cadaverine.

H$_2$NCH$_2$CH$_2$CH$_2$CH$_2$NH$_2$ H$_2$NCH$_2$CH$_2$CH$_2$CH$_2$CH$_2$NH$_2$

putrescine   cadaverine

Fairly small amounts of these compounds, formed by bacterial action, can cause "ptomaine" poisoning if one eats spoiled meat, fish, or eggs.

The easiest method of naming simple amines, shown in the preceding section, is to name the organic groups bonded to the nitrogen atom. These are common names. This method may not be possible if the organic group is complex. In such cases, illustrated below, an —NH$_2$ can be regarded as a *group* attached to a carbon skeleton. The name **amino** is used for an —NH$_2$ group, *methylamino* for —NHCH$_3$, and so on.

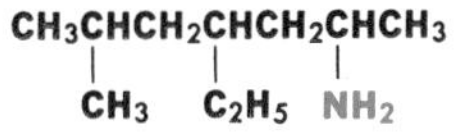

$$CH_3CHCH_2CHCH_2CHCH_3$$
$$\quad\quad |\quad\quad\quad |\quad\quad |$$
$$\quad\quad CH_3\quad C_2H_5\quad NH_2$$

2-amino-4-ethyl-6-methylheptane

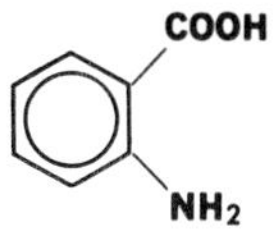

o-aminobenzoic acid

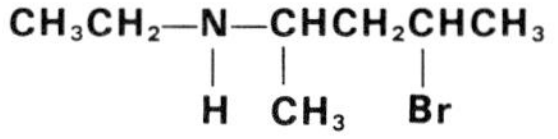

$$CH_3CH_2{-}N{-}CHCH_2CHCH_3$$
$$\quad\quad\quad\quad |\quad\quad |\quad\quad |$$
$$\quad\quad\quad\quad H\quad CH_3\quad Br$$

4-bromo-2-(ethylamino)pentane

The aryl amines are named as derivatives of aniline, the simplest amine in the aromatic class.

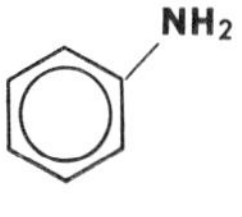

aniline 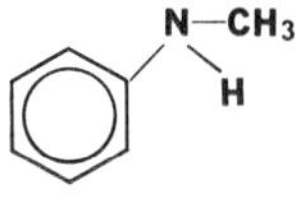   methylaniline* 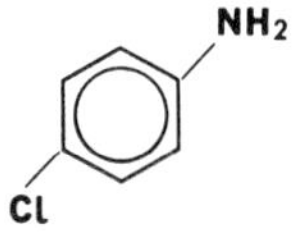   p-chloroaniline

25.3
BASIC
PROPERTIES

An ammonia molecule behaves as a base because it has an electron pair which it readily shares with a hydrogen ion. Either an organic or an inorganic acid may provide the hydrogen ion. These acid–base reactions occur rapidly at room temperature.

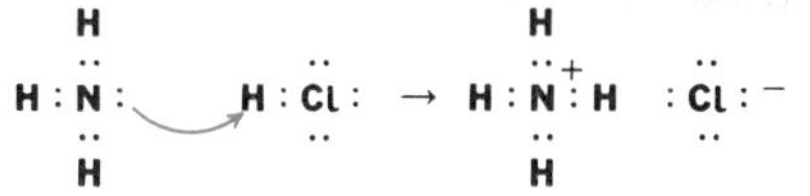

$$\quad\quad\quad\quad\quad\quad\quad\quad\quad\quad\quad\quad\quad\quad (25\text{-}1)$$

ammonia   hydrogen   ammonium chloride
                    chloride

$$\quad\quad\quad\quad\quad\quad\quad\quad\quad\quad\quad\quad\quad\quad (25\text{-}2)$$

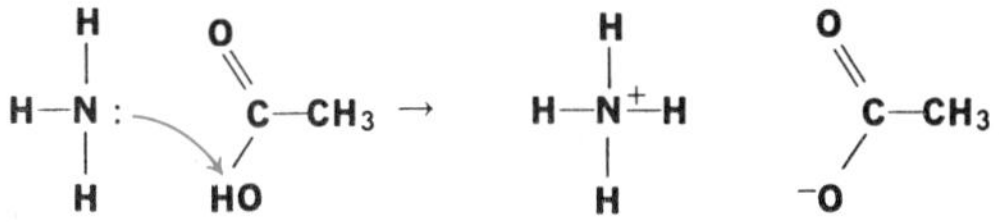

ammonia   acetic acid   ammonium acetate

Any amine also is a base; it, too, has an electron pair available which can attract a hydrogen ion:

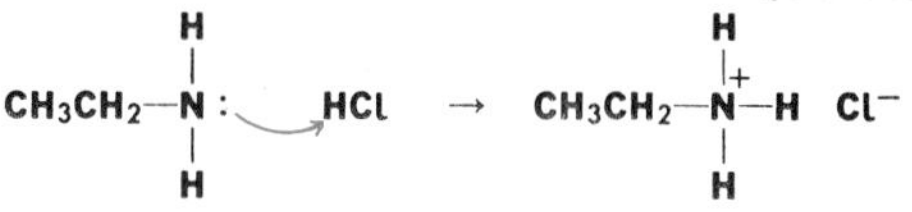

$$CH_3CH_2{-}N:\quad HCl \rightarrow CH_3CH_2{-}N{-}H\quad Cl^- \quad\quad (25\text{-}3)$$

ethylamine   hydrogen   ethylammonium chloride
                    chloride

$$\quad\quad\quad\quad\quad\quad\quad\quad\quad\quad\quad\quad\quad\quad (25\text{-}4)$$

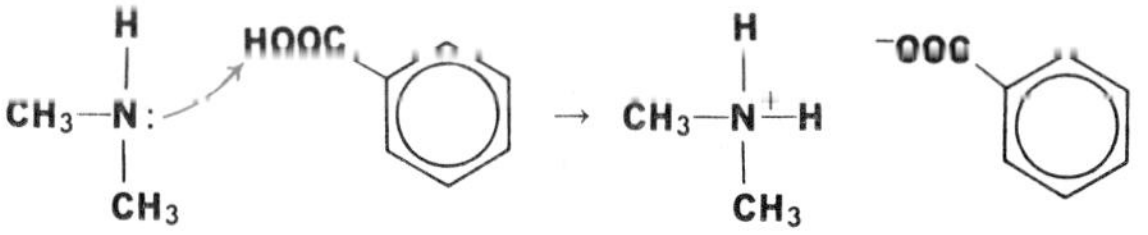

dimethylamine   benzoic acid   dimethylammonium benzoate

---

* Another, more precise, name for this compound is N-methylaniline. The capital N indicates that the methyl group is definitely on the nitrogen atom, not on the ring.

The salts which result from these acid-base reactions are named as **substituted ammonium salts,** as shown on p. 461. Since the basis property of an amine is due to the pair of available electrons, an amine of the type R_3N: reacts as a base, too.

$$\begin{array}{cc}
\overset{\displaystyle CH_3}{\underset{\displaystyle CH_3}{CH_3-\overset{|}{\underset{|}{N}}:}}\ \ \ \ \ HBr & \rightarrow\ \ CH_3-\overset{\displaystyle CH_3}{\underset{\displaystyle CH_3}{\overset{|}{\underset{|}{N}}}}\!\!\overset{+}{-}H\ \ Br^-
\end{array} \tag{25-5}$$

trimethylamine   hydrogen   trimethylammonium
                   bromide   bromide

For the same reason, the heterocyclic nitrogen compounds, Section 21.8, are likewise basic, despite the fact that the nitrogen is held in the ring.

$$+\ HCl \rightarrow \qquad\qquad Cl^- \tag{25-6}$$

pyridine   pyridinium chloride

The basic strength of an alkylamine is similar to that of ammonia. The ionization constant (K_b) for many amines is about 10^{-5} or 10^{-4}. This of course means that an amine is a much weaker base than sodium hydroxide, which is completely ionized. An arylamine, such as aniline, is even weaker, having a K_b of about 10^{-10}. Even so, the amines definitely form salts with strong acids.

**25.4
SOLUBILITY
OF AMINES**

The N—H bonds and C—N bonds of an amine are polar. (But not as strongly polar as the bonds at oxygen in an alcohol. Why?) Because amines are polar, they are attracted to water. Their solubility is similar to that of many other organic compounds. Small amines are very soluble in water, those having five or six carbons are only slightly soluble, the larger amines are insoluble in water. Most amines are quite soluble in the common organic liquids.

However, the solubility of an amine is changed if it is converted to a salt, which is ionic. An amine salt is very soluble in water, even if the carbon skeleton is rather large. But because of its ionic nature, the salt is not soluble in organic compounds. This behavior is illustrated by the next example:

$$-\overset{\displaystyle H}{\underset{\displaystyle\,}{\overset{|}{N}}}-CH_2CH_3\ +\ HCl\ \rightarrow\quad -\overset{\displaystyle H}{\underset{\displaystyle H}{\overset{|}{\underset{|}{N}}}}\!\!\overset{+}{-}CH_2CH_3\ \ \ Cl^- \tag{25-7}$$

N-ethylaniline   N-ethylanilinium chloride
insoluble in water,   soluble in water,
soluble in organics   insoluble in organics

The salt readily gives up the hydrogen ion in the presence of a stronger base and reverts to a water-insoluble amine:

$$-\overset{\displaystyle H}{\underset{\displaystyle H}{\overset{|}{\underset{|}{N}}}}\!\!\overset{+}{-}CH_2CH_3\ \ \ Cl^-\ +\ Na^+OH^-\ \rightarrow\quad -\overset{\displaystyle H}{\underset{\displaystyle\,}{\overset{|}{N}}}-CH_2CH_3\ +\ Na^+Cl^-\ +\ HOH \tag{25-8}$$

insoluble in water

In summary, if an amine is changed to a salt, the solubility changes also. This has an important bearing on whether an amine will be more soluble in one of the organic membranes of a cell or in the aqueous fluid bathing it. Similar solubility behavior for the carboxylic acids and their salts was mentioned in Section 23.5.

The solubility properties of amines must be considered when medications are administered. Many drugs, both natural and synthetic, are nitrogen compounds which are insoluble in water but will become water-soluble if converted to their salts. Lidocaine, a local anesthetic used in dentistry, is injected as an aqueous solution. For this purpose the manufacturer treats lidocaine with HCl to change it to a chloride salt.

Many of the natural alkaloids used in medicine are similarly handled in salt form; examples are morphine sulfate and codeine phosphate. Even though the pharmaceutical company has converted a drug to its salt, to make it water-soluble, the nurse or doctor who handles the drug must be aware of its chemical properties. For example, if morphine sulfate is administered along with a drug which is a salt of a weak acid (such as a barbiturate), both drugs may precipitate.

25.5
PREPARATION
OF AMINES
A. Alkyl Amines,
From Halides

A common method for the synthesis of an *alkyl* amine involves the reaction of an alkyl halide with ammonia, often in the presence of another base such as sodium hydroxide.

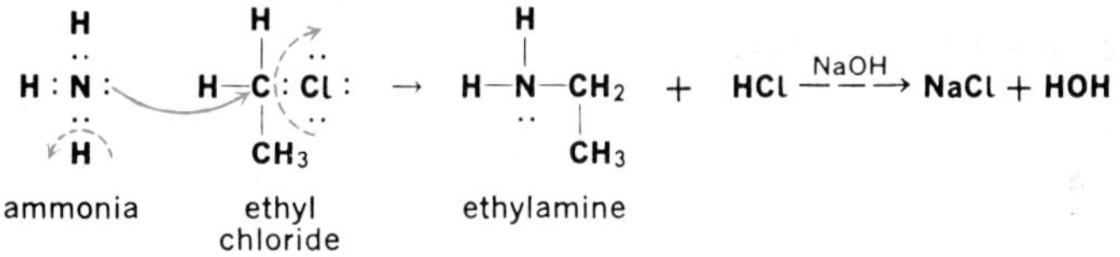

$$\text{ammonia} \qquad \text{ethyl chloride} \qquad \text{ethylamine} \qquad\qquad (25\text{-}9)$$

This is another reaction which depends on the electron pair of nitrogen. When the pair of electrons attacks the alkyl group, a halide ion (for example, Cl^-) is displaced off the other side of the carbon being attacked [Eq. (25–9)]. Following the substitution of the alkyl group onto nitrogen, a hydrogen ion can be lost from the nitrogen. Indeed, the reaction is promoted if a base is present to help the hydrogen ion leave the nitrogen. From this method of formation we see that an amine can truly be regarded as a substitution product of ammonia.

Further substitution occurs if an amine is allowed to react with more alkyl halide [Eqs. (25–10) and (25–11)]. The nitrogen of an amine still has a pair of electrons available, so it can react in the same way that ammonia does.

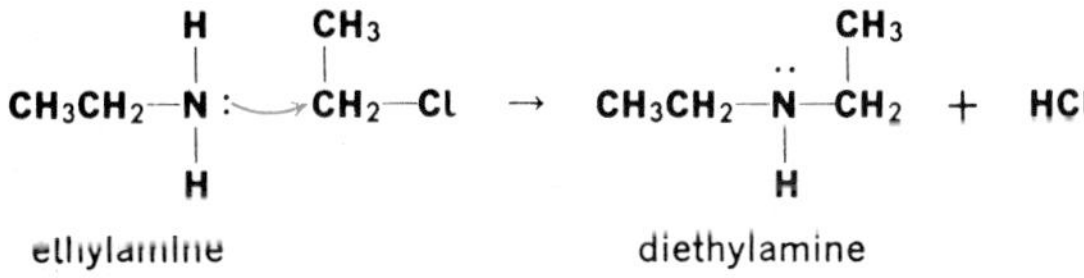

$$\text{ethylamine} \qquad\qquad \text{diethylamine} \qquad\qquad (25\text{-}10)$$

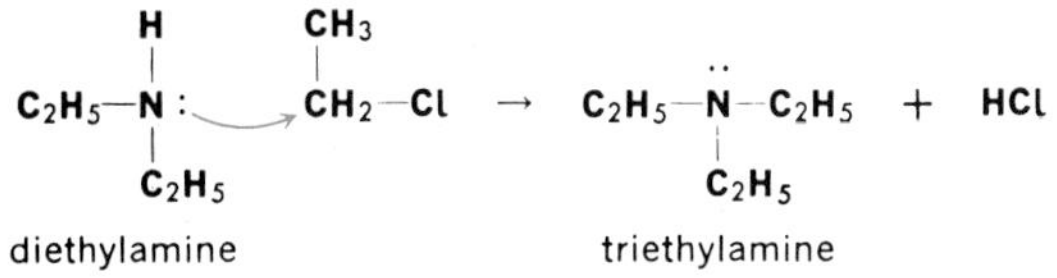

$$\text{diethylamine} \qquad\qquad \text{triethylamine} \qquad\qquad (25\text{-}11)$$

In fact, even a trialkylamine can react, because it too has an electron pair available [Eq. (25–12)]. However, in this case, there is no hydrogen to be lost from the nitrogen so the product is an ionic salt:

$$
\begin{array}{ll}
\underset{\text{triethylamine}}{\overset{\displaystyle \overset{\textstyle C_2H_5}{\underset{\textstyle C_2H_5}{|}}}{C_2H_5\!-\!N:}}\quad \overset{CH_3}{\underset{}{CH_2\!-\!\ddot{C}l:}} \;\longrightarrow\; \underset{\text{tetraethylammonium chloride}}{\overset{\displaystyle \overset{\textstyle C_2H_5}{\underset{\textstyle C_2H_5}{|}}}{C_2H_5\!-\!\overset{+}{N}\!-\!C_2H_5}}\; :\ddot{\underset{..}{C}l:^{-} & (25\text{–}12)
\end{array}
$$

Since this final product $(NR_4^+Cl^-)$ is a completely substituted ammonium salt (compare $NH_4^+Cl^-$), it is named accordingly. A compound of the type $NR_4^+Cl^-$ is also called, in general, a **quaternary ammonium salt** because of the four organic groups present.

Another variation of this synthetic method is to treat some amine with an alkyl halide having a different group.

$$
\underset{\text{aniline}}{\overset{\displaystyle \overset{N-H}{\underset{H}{|}}}{\bigcirc}} + 2CH_3I \xrightarrow{\text{NaOH}} \underset{\text{dimethylaniline}}{\overset{\displaystyle \overset{N-CH_3}{\underset{CH_3}{|}}}{\bigcirc}} + 2HI \qquad (25\text{–}13)
$$

B. Alkyl Amines, From Carbonyl Compounds

Another quite general method for preparing an alkyl amine consists of treating some carbonyl compound (either aldehyde or ketone) with ammonia and hydrogen in the presence of a nickel catalyst. The process is called **reductive amination.**

$$
NH_3 + \underset{R}{\overset{O}{\overset{\|}{C}}}\!\!-\!R' + H_2 \xrightarrow{\text{Ni}} H_2N\!-\!\underset{R}{\overset{H}{\overset{|}{C}}}\!\!-\!R' + H_2O \qquad (25\text{–}14)
$$

$$
NH_3 + \underset{H}{\overset{O}{\overset{\|}{C}}}\!\!-\!R' + H_2 \xrightarrow{\text{Ni}} H_2N\!-\!\underset{H}{\overset{H}{\overset{|}{C}}}\!\!-\!R' + H_2O \qquad (25\text{–}15)
$$

Once an amine has been synthesized, for example by Eq. (25–15), it can be made to react with another carbonyl compound (either the same or different from the first) in the same manner that ammonia did, thus producing a *di*alkyl amine.

$$
R'CH_2NH_2 + \underset{R}{\overset{O}{\overset{\|}{C}}}\!\!-\!R + H_2 \xrightarrow{\text{Ni}} R'CH_2\!-\!NH\!-\!\underset{R}{\overset{H}{\overset{|}{C}}}\!\!-\!R + H_2O \qquad (25\text{–}16)
$$

Further details of these reductive amination reactions are discussed in Section 26.9, including some related biological reactions.

C. Aryl Amines, from Nitro Compounds

A different synthetic method is usually employed to prepare *aryl* amines. A nitro group can be reduced to an amino group by hydrogens produced from the action of an acid on a metal. Iron or tin are commonly used.

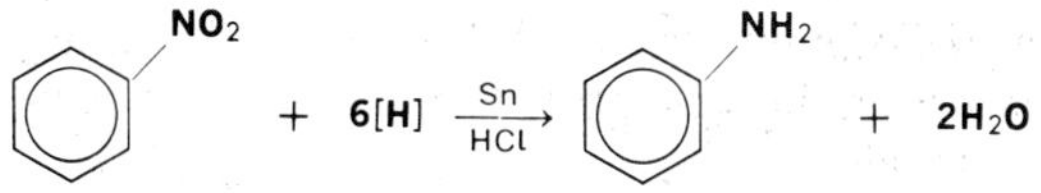

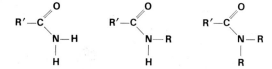

nitrobenzene   aniline

$$+ \quad 6[H] \quad \xrightarrow[\text{HCl}]{\text{Sn}} \quad \qquad + \quad 2H_2O \tag{25-17}$$

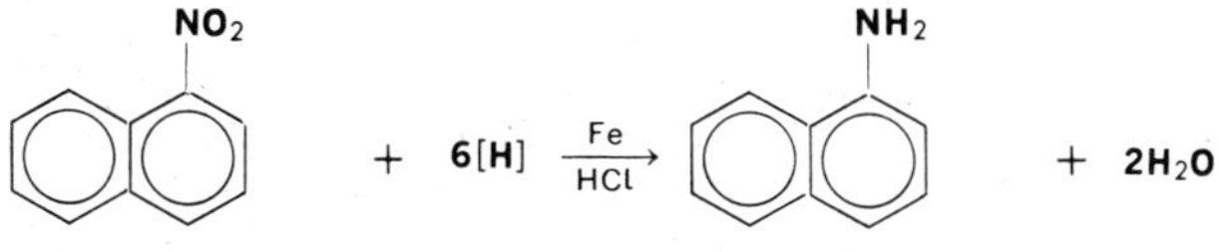

1-nitronaphthalene   1-aminonaphthalene

$$+ \quad 6[H] \quad \xrightarrow[\text{HCl}]{\text{Fe}} \quad \qquad + \quad 2H_2O \tag{25-18}$$

The reactions shown in Eqs. (25–17) and (25–18) are particularly useful because the nitro group is so easily substituted onto an aromatic ring (Section 21.6). This in turn means that a great variety of aryl amines can be synthesized. For example, consider the following synthetic route:

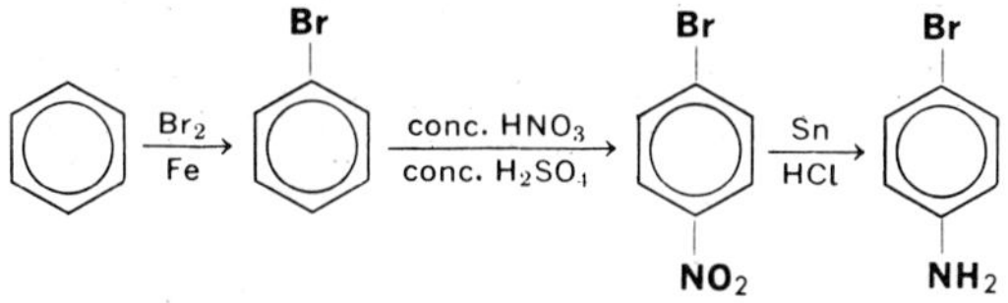

$$\xrightarrow[\text{Fe}]{\text{Br}_2} \qquad \xrightarrow[\text{conc. H}_2\text{SO}_4]{\text{conc. HNO}_3} \qquad \xrightarrow[\text{HCl}]{\text{Sn}}$$

25.6 AMIDES

Amides are derivatives of carboxylic acids and have structures such as these:

The latter two compounds can be thought of also as derivatives of amines. That is, amides are derivatives of amines with carboxylic acids, comparable to esters being derived from alcohols with acids.

A. Formation from an Acid Chloride

An acid chloride reacts rapidly with ammonia to form an amide. A particular amide is named according to the acid group from which it is derived.

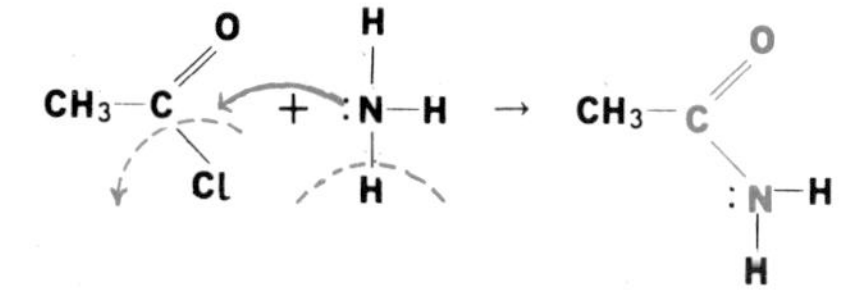

$$CH_3-C \quad + \quad :N-H \quad \rightarrow \quad CH_3-C \quad + \quad HCl \tag{25-19}$$

acetyl chloride   acet amide

$$+ \quad H-N-H \quad \rightarrow \quad + \quad HCl \tag{25-20}$$

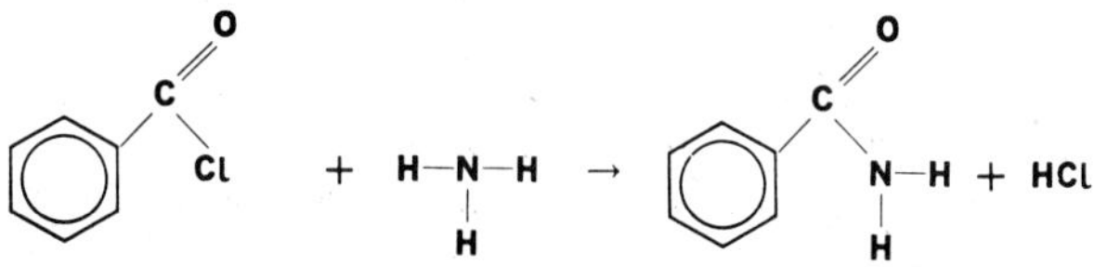

benzoyl chloride   benzamide

Acid chlorides react also with amines, provided at least one hydrogen is available for displacement; the trialkylamines therefore do not react. When amines react with acid chlorides, the products are N-substituted amides.

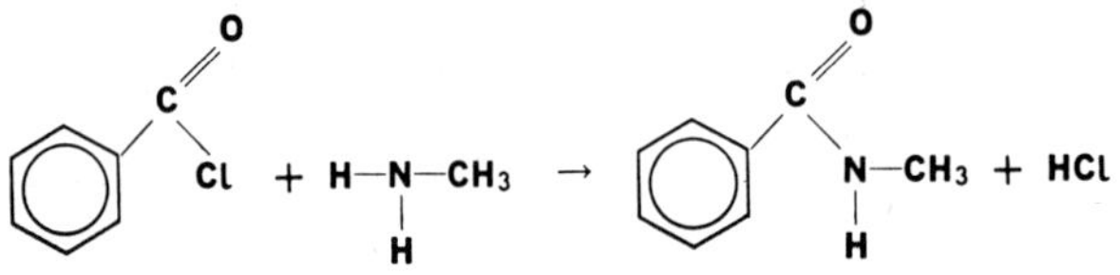

$$\text{(25–21)}$$

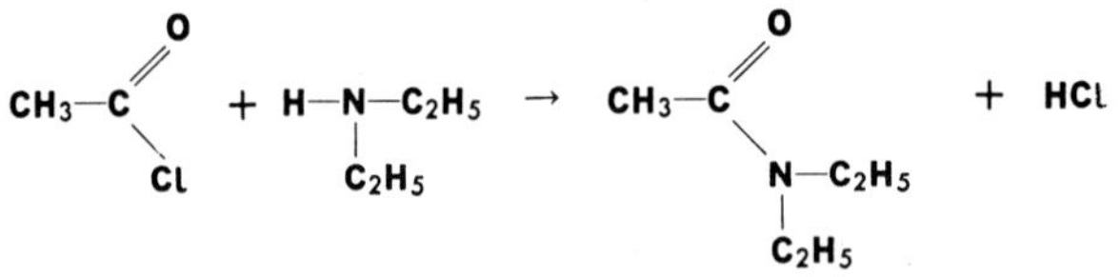

N-methylbenzamide

$$\text{(25–22)}$$

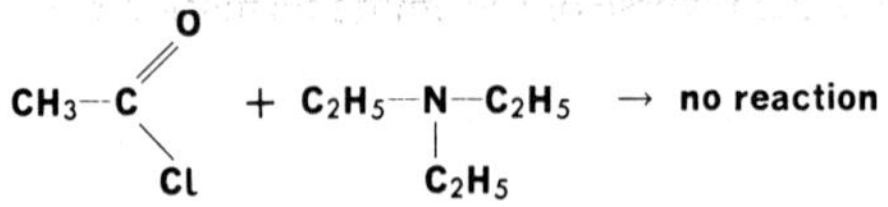

N,N-diethylacetamide

$$CH_3-C \underset{CL}{\overset{O}{\diagdown}} + C_2H_5-\underset{C_2H_5}{\overset{}{N}}-C_2H_5 \rightarrow \text{ no reaction} \qquad \text{(25–23)}$$

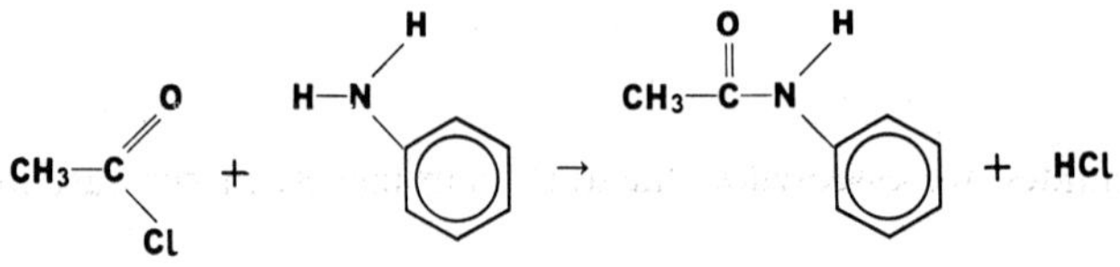

$$\text{(25–24)}$$

acetyl   aniline   acetanilide
chloride

These reactions of amines with acid chlorides are analogous to the reactions of alcohols with an acid chloride, Eq. (23–19).

B. Formation from an Acid

Amides can also be formed from the carboxylic acids themselves (Eq. 25–25). However, heat is required because a carboxylic acid is less reactive than an acid chloride.

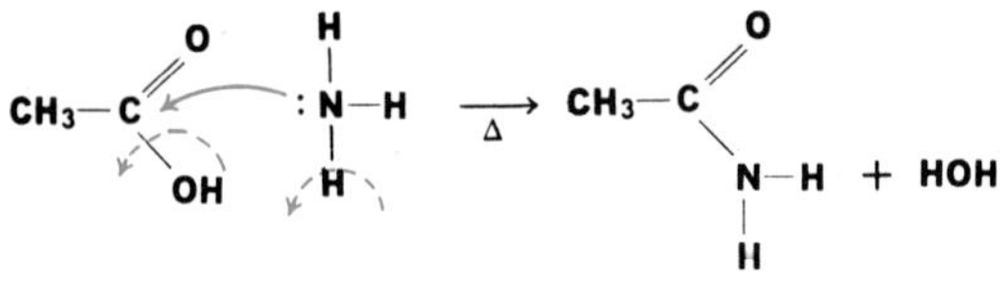

$$\text{(25–25)}$$

acetic acid   ammonia   acetamide

$$R-C \underset{OH}{\overset{O}{\diagdown}} + H-\underset{H}{\overset{}{N}}-R' \xrightarrow{\Delta} R-C \underset{N-R'}{\overset{O}{\diagdown}} + HOH \qquad \text{(25–26)}$$

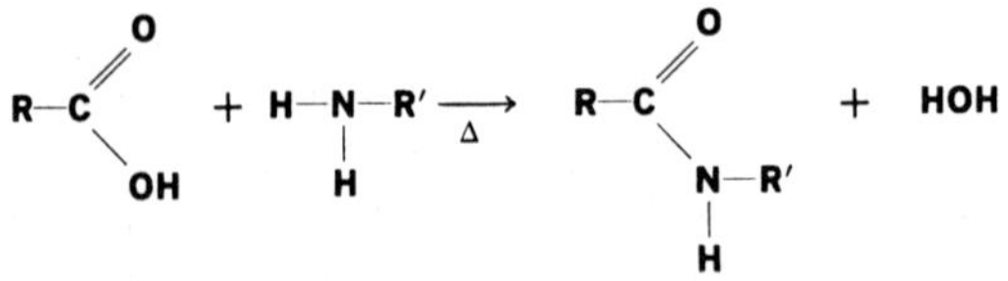

carboxylic   amine   N-substituted
acid   amide

Reaction 25–26 is comparable to the esterification reaction of an acid with an alcohol (Section 23.6) and, like it, is very slow unless heat is used.

The alert reader may have noticed that in equations such as (25–2) and (25–4) we showed ammonia, or an amine, reacting with a carboxylic acid to produce an ammonium salt, whereas in Eq. (25–25) we showed the same reactants producing an amide. The product obtained depends on the temperature. Reaction (25–2), which leads to ammonium salt, occurs rapidly at room temperature or below. In contrast, the amide-formation reaction (Eq. 25–25) is slow and difficult. It takes place only if the reactants are heated.

C. Hydrolysis of an Amide   We have seen, in Section 23.11, that an ester can be hydrolyzed to its component acid and alcohol. In similar fashion an amide can be hydrolyzed; in this case, too, an acid catalyst greatly hastens the reaction.

$$\underset{\underset{\underset{H}{|}}{\underset{N-C_2H_5}{|}}{\overset{\overset{O}{\|}}{CH_3-C}} + HOH \xrightarrow[\Delta]{H^+} \underset{OH}{\overset{\overset{O}{\|}}{CH_3-C}} + \underset{H}{\overset{}{H-N-C_2H_5}} \qquad (25\text{–}27)$$

This hydrolysis reaction is the reverse of the amide formation reaction seen in Eq. (25–26).

D. Amides in Nature   The formation and hydrolysis of amide bonds are not only frequent laboratory procedures but are also essential to the chemistry of proteins and numerous other natural compounds. But in living organisms there is one very important difference: enzymes are present to catalyze the reaction so that they occur easily at body temperature (provided energy sources are available).

E. Neutrality of Amides   The amides are essentially "neutral" substances, in contrast to the basic amines. A typical amide, such as acetanilide, does not form a salt when placed in dilute hydrochloric acid solution. Although the nitrogen in an amide still has a pair of nonbonded electrons, they are displaced toward the electron-attracting C=O group. Therefore the electrons are not sufficiently "available" to attract an H^+ ion from an acid.

$$\underset{\underset{H}{|}}{\underset{\overset{\delta+}{N}}{\underset{|}{\overset{\delta+}{R-C}}}}\overset{\overset{\delta-}{O}}{\diagup}\diagdown H$$

Electrons attracted toward C=O group are not available to hold an H^+.

25.7 THE STRUCTURE OF PROTEINS AND AMINO ACIDS   The tissues of animal bodies are composed chiefly of protein material, aside from the water present within the tissues or the bones supporting them. Muscle, skin, hair, cartilage, nails, and hooves all consist of proteins. In addition, a number of specialized compounds such as enzymes and certain hormones are composed mainly of protein. Experiments show that proteins are very large molecules. Many have molecular weights between about fifty thousand and several hundred thousand. Some have molecular weights as high as a few million. A clue to the nature of these gigantic molecules is the observation that they can be broken up into much smaller organic molecules. One way to do this is to boil a sample of protein in water containing some hydrochloric acid. After several hours, one finds that this dilute acid solution converts the protein to a mixture of amino acids.

protein + H$_2$O $\xrightarrow[\Delta]{\text{HCl}}$ mixture of amino acids   (25–28)

This result suggests that the amino acid molecules are building-units which can somehow be linked together to form a very large protein molecule.

If one separates the mixture from reaction (25–28) into the different amino acids, each type present can be identified. (Prior to about 1965, this was a tedious and difficult job! Since then, the availability of "amino acid analyzer" instruments has greatly reduced the time and labor required for such a study.) It turns out that there are as many as 20 to 25 different kinds of amino acids in the mixture and that because of the size of the original protein, many molecules of each of the different kinds of amino acids are present. Despite the variety of amino acids in the mixture, one finds that all are α-amino acids, having the structure shown in Fig. 25.1(a). They differ from each other only in the nature of their R groups. Just two of the common amino acids are shown in Fig. 25.1; others are listed in Section 31.2. The symbol α in the name α-amino acid specifies that the amino group is bonded to the carbon adjacent to the carboxyl group. It is possible, of course, for the amino group to be at some other position in the molecule. However, since all natural amino acids derived from proteins are α-amino acids, the remainder of our discussion will be confined to this type.

Greek letters are sometimes used to designate locations on a carbon skeleton, relative to a certain functional group (see Section 23.1). The letter α is assigned to the carbon nearest the functional group, β to the next carbon, and so on. Thus in leucine (Fig. 25.1b) the amino group is at the α position and the methyl group is at the γ position relative to the carboxyl group. The name "leucine" is a common name. To write a correct systematic name for this compound one should, as usual, use numbers to designate positions in the carbon skeleton, rather than $\alpha, \beta, \gamma, \delta$, and so on. Thus, the systematic name for leucine would be 2-amino-4-methylpentanoic acid.

Reaction (25–28) indicated that a protein is composed of amino acid units. Other experiments show further that the protein is built up when the amino acids link to each other by means of amide bonds. In Eq. (25–26) we saw that an amine can react with a carboxylic acid by splitting out water to form an amide. Now, an amino acid has a very interesting type of structure which we have not previously encountered; it has, within one molecule, both an amino group and a carboxylic acid group. Consequently, if the carboxyl group of one amino acid molecule encounters the amino group of some other amino acid molecule, water could be split out and an amide bond could be formed, thus linking together the two molecules. Furthermore, since one of the units would still have a carboxyl group available, it could form another amide bond with the amino group of yet a third molecule of amino acid. Since the newly attached amino acid would also have a carboxyl group to react, it could bond to a fourth unit, and so on until many amino acids

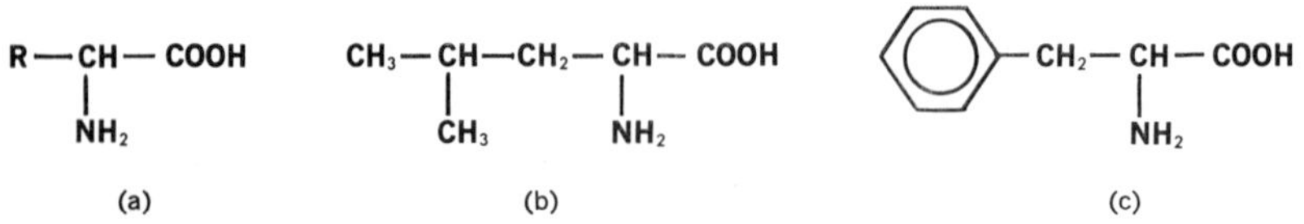

R—CH—COOH   CH$_3$—CH—CH$_2$—CH—COOH   ⬡—CH$_2$—CH—COOH
| | | |
NH$_2$ CH$_3$ NH$_2$ NH$_2$

(a)   (b)   (c)

Fig. 25.1 The structure of natural amino acids. (a) General formula for any α-amino acid. (b) Leucine. (c) Phenylalanine.

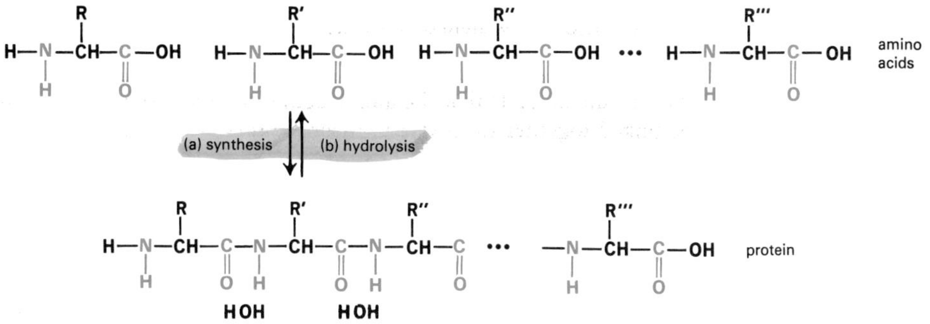

Fig. 25.2 Formation of a protein from amino acids, and the reverse. Reaction (a) represents the build-up of protein from amino acids. This complex biosynthesis occurs in living organisms. In the reverse reaction (b), water can break up the protein, provided a catalyst is present. In living organisms, enzymes are the catalysts; in the laboratory, strong acid can catalyze the hydrolysis.

were linked to each other. Each time an amide linkage is formed, one molecule of water is split out.

This process of combining many amino acid units, via amide bonds, to form a protein is represented in Fig. 25.2. Each kind of protein is unique, depending on the particular amino acids it contains and the sequence in which they are arranged. Therefore, the build-up or synthesis of a protein is exceedingly complex. Biologically it is accomplished by enzymes acting as catalysts, and it is controlled by ribonucleic acids, so that only the exact kind of protein required for a particular function is assembled. The protein of a hormone will be quite different than that found in a muscle fiber or in the membrane of a nerve cell. Also, the protein of beef muscle differs from that of human muscle. To duplicate such a synthesis in the laboratory is extremely difficult and has been accomplished only for relatively small proteins containing about one hundred amino acids or less. Some of the details concerning the biosynthesis of proteins are discussed in Chapter 36.

Hydrolysis of a protein is represented by reaction (b) in Fig. 25.2. Cleavage of each amide linkage requires one molecule of water and the aid of a catalyst. In the laboratory, hydrochloric acid is a satisfactory catalyst. Since hydrolysis involves breakdown of the protein molecule, it is much easier to carry out in the laboratory than is the synthesis of the protein. Even large, complex proteins can be hydrolyzed if heated for a few hours in the presence of hydrochloric acid. Figure 25.2(b) shows in greater detail the protein hydrolysis reaction that was first represented by Eq. (25–28).

Biologically, the hydrolysis reaction of Fig. 25.2(b) can occur in a variety of different tissues which have appropriate enzymes to catalyze the reaction. For example, the chemical reaction taking place during digestion is simply hydrolysis. In the digestive tract a piece of beefsteak reacts with water, under the influence of enzymes, and is broken up into its constituent amino acids. The resultant amino acid molecules are relatively small and water-soluble and, therefore, can be absorbed through the intestinal wall, dissolved in the blood, and carried to various tissues of the body. There, other enzymes can promote a synthesis reaction (Fig. 25.2a), reassembling the amino acids into new protein molecules. Because of the very specific nature of the enzymes and ribonucleic acids controlling the biosynthesis, the collection of amino acids which originated from the beefsteak will be arranged differently in the protein of human muscle or cartilage.

We have seen that it is the formation of amide groups which holds amino acids together in a protein and that hydrolysis of the amide bonds causes the protein to revert to amino acids. *The amide bonds in a protein are called* **peptide bonds**. This is merely a special term used to designate the particular kind of amide group that results from α-amino acids linking to each other to form a protein. However, as we have seen, the chemical properties of a peptide linkage are the same as those of any other amide group. Each peptide group in the protein structure of Fig. 25.2 is shown in color.

**25.8
CHEMICAL
REACTIONS OF
AMINO ACIDS
A. Amide
Formation**

Both the amino group and the carboxyl group of an amino acid can be involved in the formation of amide structures. In the preceding section we observed that when the amide structure links α-amino acids together to yield a protein, the amide group is called a peptide group. The formation of peptide bonds is undoubtedly the most common type of amide bond produced by a natural α-amino acid. However, the amino and carboxyl groups will always display their characteristic properties, and so there is always the possibility that an amide bond could be formed with some other molecule which was not an amino acid. One example of amide formation between an amino acid and some different type of compound will be discussed.

In the laboratory, hippuric acid can be made by treating glycine with benzoyl chloride:

benzoyl chloride   glycine   hippuric acid   (25–29)

Benzoic acid (or its salt) is sometimes ingested as a constituent of our meals. It occurs naturally in certain foods. Other foods may contain larger molecules which are broken down to benzoic acid during metabolism. Rather large amounts of such compounds are found in prunes and cranberries. Some foods contain benzoic acid or sodium benzoate which have been added as preservatives. Since benzoic acid from any of these sources is not further broken down, one way the body eliminates it is by combining it with glycine, which is the simplest amino acid and one usually present in the body in good supply. The benzoic acid and the glycine become linked together by an amide bond.

benzoic   glycine   hippuric acid   (25–30)
acid

The hippuric acid thus formed is subsequently excreted via the urine. Certain related compounds can be eliminated by the same route. If benzaldehyde is ingested, for instance in almond flavoring, the body can oxidize it to benzoic acid, which is then converted to hippuric acid.

**B. Salt
Formation**

Thus far most of our discussion of amino acids has been, quite properly, devoted to their ability to form peptide bonds and thus build proteins. However, it is important to consider also the properties of an individual amino acid as it might exist in a solution in the laboratory, in intestinal fluid, or in the bloodstream. We should expect that the amino group and the carboxyl group will have the typical properties we have come to expect of such groups. The only question will be to see how these groups might affect each other when both are in the same molecule.

So far we have written the structure of an amino acid as shown in Fig. 25.3(a) because this portrays the amino and carboxyl groups in the same way that they exist in many other compounds. Let us now consider acid and base properties. An amine is a fairly basic compound (Section 25.3), and it readily combines with a proton released from either a strong inorganic acid or from an organic carboxylic acid; the result is the formation of a substituted ammonium salt. An example would be the transfer of a proton that occurs when methylamine and propanoic acid are mixed:

$$CH_3\text{---}NH_2 \; + \; CH_3CH_2COOH \; \rightarrow \; CH_3\text{---}\overset{+}{N}H_3 \quad CH_3CH_2COO^- \tag{25-31}$$

methylamine   propanoic   methylammonium
              acid   propanoate

An amino acid will behave in the same way; a proton will leave the carboxyl group and reside instead on the amino group. The amino acid will actually exist as shown in Fig. 25.3(b) rather than as the simplified structure in Fig. 25.3(a).

$$\begin{array}{cc} R\text{---}CH\text{---}COOH & R\text{---}CH\text{---}COO^- \\ \mid & \mid \\ NH_2 & ^+NH_3 \\ \text{(a)} & \text{(b)} \end{array}$$

Fig. 25.3 The structure of an amino acid. (a) Simplified structure. (b) Dipolar ion form, the more accurate representation.

In Fig. 25.3(b) the organic skeleton is held together by covalent bonds as usual but in addition, part of the structure of (b) is ionic; it is sometimes described as being an **inner salt**. This ionic structure is most often called a **dipolar ion**, since there are two poles, positive and negative, within the same molecule.

When an amino acid is dissolved in water, nearly all the molecules will exist in the dipolar-ion form (b). In fact, even without being in water solution, an amino acid by itself will exist in the dipolar-ion form. Crystals of a pure amino acid actually contain the dipolar ions. When the solid dissolves in water these dipolar ions, already formed, are merely released into the solution.

We have just described the condition of an amino acid in a water solution—that is, near neutrality. However, if a moderately strong acid or base is present, the amino acid can react with either, which means that an amino acid is *amphoteric*. An amino acid would react with a strong acid as follows:

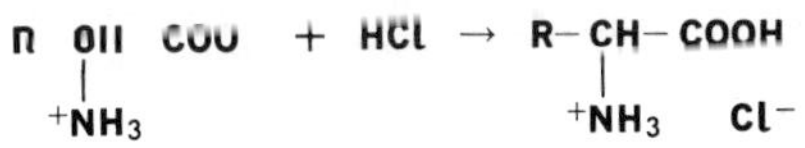

$$\begin{array}{ccccc} R\text{---}CH\text{---}COO^- & + & HCl & \rightarrow & R\text{---}CH\text{---}COOH \\ \mid & & & & \mid \\ ^+NH_3 & & & & ^+NH_3 \quad Cl^- \end{array} \tag{25-32}$$

In this reaction the hydrochloric acid is able to donate a proton to the negative carboxylate group. Since the solution is acidic, the amino group remains as a positive ion; the only difference is that the negative ion associated with it is now a chloride ion rather than the original carboxylate ion.

In a basic solution the following reaction of a dipolar ion would occur:

$$R-CH-COO^- + NaOH \rightarrow R-CH-COO^-Na^+ + HOH \qquad (25-33)$$
$$\overset{|}{\underset{^+NH_3}{}} \qquad\qquad \overset{|}{\underset{NH_2}{}}$$

Here the strongly basic hydroxide ion removes the proton from the amino group, a weaker base, and forms a molecule of water. The carboxylate group remains a negative ion, unchanged. However, its charge is now balanced by the sodium ion rather than by the ammonium ion group originally present.

25.9
MORE
NITROGEN
COMPOUNDS
OF INTEREST
A. Vitamins

Several (but not all) of the vitamins have nitrogen atoms in either amide groups or amine groups. Quite often the basic amino nitrogen is part of a heterocyclic ring. **Niacinamide** is a very important vitamin but is the simplest in structure. It has one basic nitrogen in a pyridine ring, plus a neutral nitrogen in an amide group. Most vitamins have a more complex structure, such as that of **thiamine** (vitamin B_1).

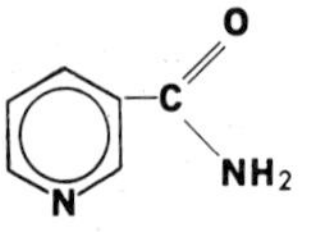

niacinamide
(vitamin B_3)

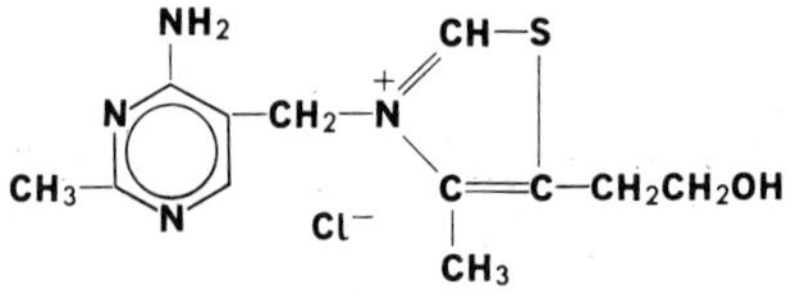

thiamine (vitamin B_1)

Note in thiamine that the five-membered heterocyclic ring contains both a nitrogen and a sulfur, with the nitrogen in the form of a quaternary ammonium salt (Section 25.5A).

B. Alkaloids A number of plants produce nitrogen-containing substances which are moderately basic and are therefore called alkaloids (alkali-like). Most of these compounds have striking effects on the nervous system of animals; some may even be fatal. Structurally, the nitrogens present in alkaloids are most often in heterocyclic rings. The simplest alkaloid is **coniine**, which occurs in the poison hemlock plant (see Fig. 25.4). **Cocaine** is also an alkaloid. Two other familiar examples, caffeine and nicotine, are shown in Fig. 25.4. **Caffeine** is present in tea, coffee, cola nuts, and a number of other plants. The compound is a stimulant. Apparently large amounts are necessary to cause death; few fatalities from caffeine have been reported.

More than ten different alkaloids have been found in tobacco leaves, but **nicotine** constitutes three-fourths of the weight of alkaloids present. Nicotine is highly toxic to

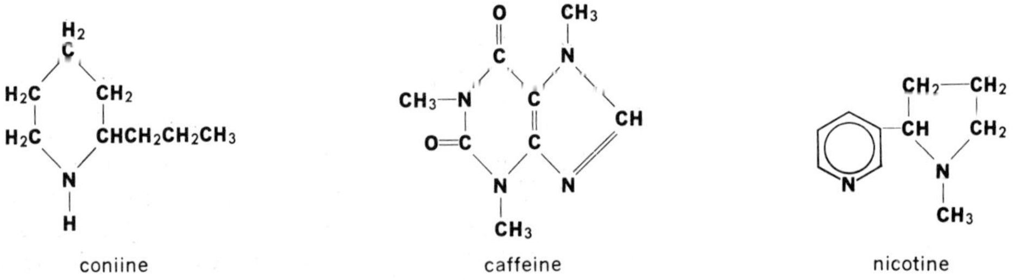

coniine

caffeine

nicotine

Fig. 25.4 Some common alkaloids.

animals and has considerable use as an insecticide. In very small smounts, as might be inhaled from smoking tobacco, it causes brief stimulation.

Ergot, a fungus disease of rye, produces several different amides of **lysergic acid.**

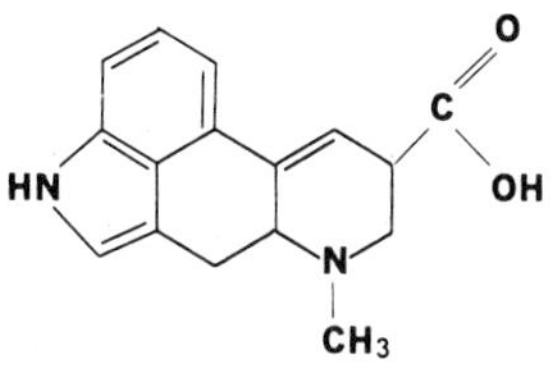

lysergic acid

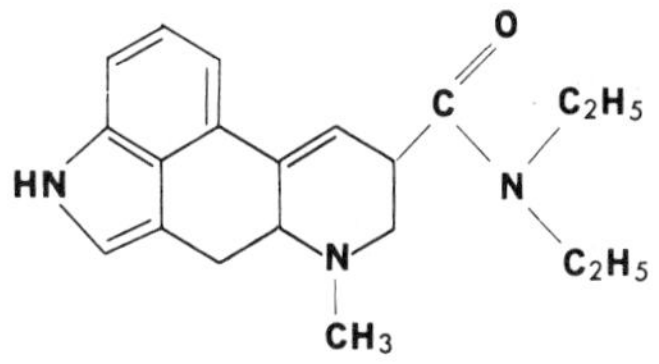

LSD

A synthetic modification of this acid is a hallucinogenic drug called LSD, which represents *lysergic acid diethylamide*. Appropriate treatment of the natural materials can yield lysergic acid. Its carboxyl group, by standard procedures, can react with diethylamine to become a diethylamide structure.

Many alkaloids have even more complicated structures than those shown above. One example is **quinine**, which is moderately effective in suppressing malaria infections. **Strychnine** is a poison causing convulsions and death. A certain species of poppy plant produces the drug **morphine**. **Codeine** is a slight modification of morphine. Both are habit-forming, but when properly administered in controlled amounts are medically useful as pain relievers and sedatives. Another modification of morphine is **heroin**, which is so addicting that its use even for medical purposes is illegal in the United States.

C. Urea Economically and biologically, urea is important. It is a colorless, odorless, crystalline, water-soluble compound playing a significant part in the balance of life between plants and animals. Chemically, it is the amide of carbonic acid; the structural relations are shown in the scheme below.

$$H_2O + CO_2$$
$$\updownarrow$$
$$HO-\underset{\underset{O}{\|}}{C}-OH + 2NH_3 \rightleftarrows H_2N-\underset{\underset{O}{\|}}{C}-NH_2 + 2H_2O \qquad (25\text{-}34)$$

carbonic acid   ammonia   carbamide
urea

Like other amides, urea can be *hydrolyzed* to yield ammonia plus the parent acid; the hydrolysis reaction is represented by Eq. (25-34) if it is read from right to left. Since carbonic acid is unstable, it decomposes to carbon dioxide and water. In the opposite direction, urea can be *formed* from carbon dioxide and ammonia, as represented by Eq. (25-34) when read from left to right.

Metabolism of proteins by an animal could potentially release ammonia into his body. The ammonia would be harmful if it accumulated. To avoid this, many animals dispose of the amino groups from proteins or amino acids by combining them with carbon dioxide to form urea. [They carry out Eq. (25-34), in effect, in several steps promoted by enzymes. See Section 36.5 for details of the conversion.] The urea is then excreted. In the soil the enzymes of certain bacteria hydrolyze the urea, thus reversing Eq. (25-34) and releasing the ammonia, which is then available for plants to rebuild into new proteins.

Urea can also be obtained by conducting in a factory the reaction shown in Eq. (25-34). Ammonia, carbon dioxide, and a little water are mixed under pressure at about 180°C. The resultant urea can be isolated as a granular solid by evaporating the water present.

Recent production of urea in the United States has exceeded one million tons per year. About 80% of this is used for fertilizers; the rest goes into urea-formaldehyde plastics and adhesives, a few pharmaceuticals, and miscellaneous other products.

D. Barbiturate Drugs

Urea is one of the essential ingredients for the synthesis of the **barbiturate** drugs. The other component is an ethyl malonate ester bearing two substituents. (Ethyl malonate itself is $C_2H_5OOC-CH_2-COOC_2H_5$, the ester of a diacid, malonic acid.) During the reaction [Eq. (25–35)], two new amide bonds are formed, closing a ring.

$$\text{(25–35)}$$

substituted ethyl malonate   urea   a barbiturate

In general, the barbiturates are effective sedatives and soporifics. They can be used to induce sleep, calm epileptic patients, provide sufficient unconsciousness for minor surgery, etc. Many variations of the drugs are possible because different substituents (R, R′) can be built into the compounds. These variations provide differences in depth of relaxation, effective time, and so forth. Of the dozens which have been synthesized and tested, a few of the most common are listed in Table 25.1.

TABLE 25.1 Some common barbiturates

Substituents R	R′	Generic Name	Some Brand Names
CH_3CH_2-	CH_3CH_2-	barbital	Veronal, Barbitone, Dormonal
(phenyl ring)	CH_3CH_2-	phenobarbital	Luminal, Gardenal, Barbenyl
$CH_3CHCH_2CH_2-$ | CH_3	CH_3CH_2-	amobarbital	Amytal, Somnal, Isomytal

E. Aniline and Related Compounds

Thousands of tons of aniline and aminonaphthalene are manufactured annually by the type of reduction shown in Eqs. (25–17) and (25–18). For commercial purposes, scrap cast iron is the least expensive metal to use as the source of hydrogen. Aniline and its derivatives are used largely for the manufacture of dyes and compounds for rubber processing. Smaller quantities are converted to photographic chemicals and important pharmaceuticals.

SUMMARY

The amines, which occur in many living organisms and in laboratory products as well, are organic derivatives of ammonia. Like ammonia, the amines are weak bases which can react with acids to form substituted ammonium salts. The ionic salts are usually very soluble in water, whereas an amine having a large carbon skeleton is not soluble in water. This change in solubility upon salt formation is the same type of behavior observed for the carboxylic acids and their salts. For both the amines and the acids, solubilities have a significant effect on their behavior in living cells.

One of the most important chemical properties of an amine is reaction with a carboxylic acid to split out water and form an *amide*. The amide-formation reaction occurs with both simple and complex amines and with amino acids. A protein consists of a great many amino acid molecules linked to each other by amide groups, which in this case have the special name, *peptide* group. If an amide is heated in water containing acid or base, it is hydrolyzed to yield its constituent amine and carboxylic acid. This hydrolysis reaction is comparable to that of another acid derivative, an ester.

An amino acid, having both a carboxylic acid group and a basic amino group, is *amphoteric* and can react with either an acid or a base to form a salt. An amino acid itself exists largely in *dipolar ion form*.

Although an amine is quite basic, an amide does not show any basic properties in water solution, because of the electron-attracting effect of the C=O portion of the amide structure.

IMPORTANT TERMS

amide	peptide
amine	protein
α-amino acid	quaternary ammonium salt
amino group	reductive amination
dipolar ion	substituted ammonium salt

WORK EXERCISES

1. Name these compounds.
   a) $CH_3CH_2NHCH_2CH_3$

   b) 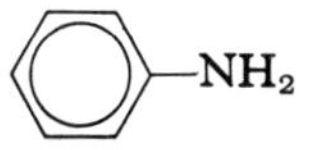 ⬡—NH_2

   c) $CH_3CHCH_2NH_2$
      |
      CH_3

   d) $CH_3NH_3{}^+Br^-$
   e) CH_3CH_2—C—NH_2
              ‖
              O

   f) 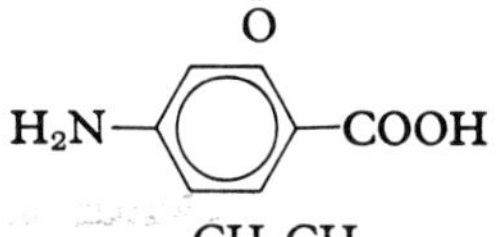 H_2N—⬡—$COOH$

   g) $CH_3CH_2CHCHCH_2CH_3$
           |
        CH_2CH_3 |
            $NHCH_3$

   h) 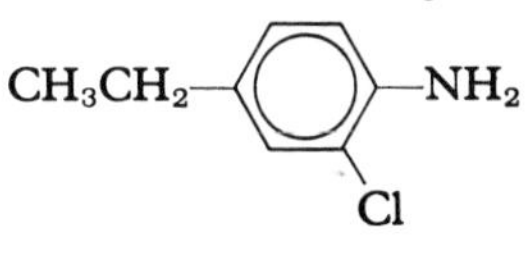 CH_3CH_2—⬡—NH_2
              |
              Cl

2. Give the structural formula of the following.
   a) trimethylamine
   b) ammonia
   c) ethyl-*n*-propylamine
   d) *o*-nitroaniline
   e) 3-amino-2,4-dimethylpentane
   f) *p*-aminophenol
   g) triethylammonium chloride
   h) benzamide
   i) N-ethyl-N-methylbutyramide
   j) methylisopropylammonium chloride
   k) N,N-dimethylanilinium bromide

3. Using structural formulas, show clearly how methylamine can behave as a base in reacting with hydrogen bromide.

4. Write complete equations for the following reaction mixtures. If necessary, state "no reaction."
   a) one mole of methyl bromide, one mole of ammonia
   b) trimethylamine, hydrogen chloride
   c) propanoic acid, ammonia, heat
   d) methyl chloride, trimethylamine
   e) *p*-nitrotoluene, tin, hydrochloric acid
   f) acetyl chloride, trimethylamine

g) benzoic acid, ethylamine, heat

h) methylamine, acetyl chloride

i) butyramide, water, enzyme catalyst

j) N-ethylbenzamide, water, acid catalyst, heat

5. Using RCOOH to represent lysergic acid, show the structure of the diethylamide derived from it.

6. Write a systematic name for each of the following amino acids. (See Section 25.2 and Table 23.1, and for structures, see Fig. 25.1.)

a) leucine

b) phenylalanine

7. Show the structure of the amino acid phenylalanine as it would exist:

a) in water

b) in stomach fluid, which is quite acidic (pH 2).

c) in intestinal fluid, which is quite basic (pH 9).

8. List all the types of structural units (both carbon skeletons and functional groups) which you can find in the compounds below. (Consider the groups discussed in this chapter, as well as your previous study. Check Table 19.5).)

a) procaine (Novocaine, Neocaine, etc.), a local and spinal anesthetic:

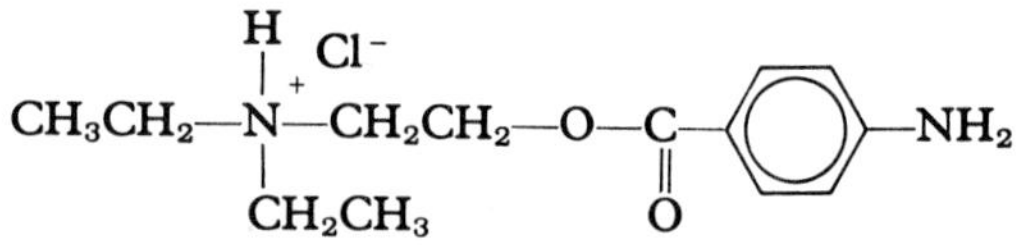

$$CH_3CH_2-\overset{\overset{\displaystyle H}{|}}{\underset{\underset{\displaystyle CH_2CH_3}{|}}{N^+}}-CH_2CH_2-O-\overset{O}{\underset{\displaystyle ||}{C}}-\bigcirc-NH_2 \quad Cl^-$$

b) The insect repellent called Delphene or Off:

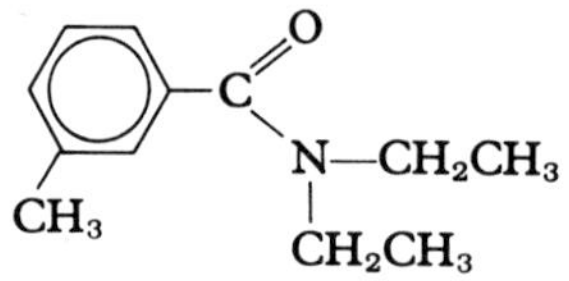

$$\overset{}{\underset{CH_3}{\bigcirc}}-\overset{O}{\underset{}{C}}-\overset{}{\underset{CH_2CH_3}{N-CH_2CH_3}}$$

SUGGESTED READING

Asimov, I., *The New Intelligent Man's Guide to Science*, Basic Books, New York, 1965, pp. 447–462. A fascinating account of the contributions from organic synthesis.

Asimov, I., *A Short History of Chemistry*, Anchor Books, Doubleday, New York, 1965, pp. 168–182. Description of the rise of synthetic and structural organic chemistry, including dyes, drugs, explosives, and proteins.

Dickerson, R. E., "The structure and history of an ancient protein," *Scientific American*, April 1972.

Gates, M., "Analgesic drugs," *Scientific American*, Nov. 1966 (Offprint #304). A discussion of morphine and the search for related drugs which would provide the same benefits without the undesirable effects.

Guillemin, R., and R. Burgess, "The hormones of the hypothalamus," *Scientific American*, Nov. 1972. Recently two of these hormones have been isolated and characterized. Both are peptides, one being composed of three amino acid units and the other of ten amino acid units.

Robinson, T., "Alkaloids," *Scientific American*, July 1959. Most alkaloids have a profound effect on animals and humans, but their function in plants is obscure.

26

Carbonyl Compounds

26.1
NAMES AND STRUCTURES

A carbonyl group has a carbon with a double bond to oxygen.

$$-\underset{\underset{O}{\|}}{C}-$$ carbonyl group

If both of the other bonds from the carbon hold organic groups, the compound is a **ketone**; if one of the bonds holds a hydrogen, the compound is an **aldehyde**.

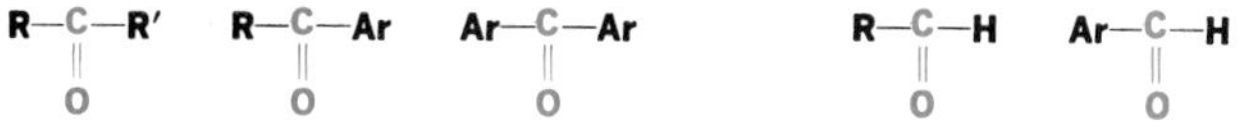

$$\underset{\underset{O}{\|}}{R-C-R'} \quad \underset{\underset{O}{\|}}{R-C-Ar} \quad \underset{\underset{O}{\|}}{Ar-C-Ar} \qquad \underset{\underset{O}{\|}}{R-C-H} \quad \underset{\underset{O}{\|}}{Ar-C-H}$$

ketones   aldehydes

Names for typical carbonyl compounds are shown in Table 26.1. In a systematic name, the ending *-one* is used for a ketone. As usual, a number is assigned to indicate where in the carbon skeleton this functional group is located. (Note that numbers are not really necessary for propanone or butanone.) The systematic ending for aldehyde is *-al*; it is unnecessary to assign the number *1* to the aldehyde group because if the carbonyl group were at a different position in the chain (say, 2 or 3), the compound would be a ketone, not an aldehyde.

26.2
SYNTHESIS FROM ALCOHOLS

Chemists have devised a great number of methods for building aldehyde or ketone groups into a variety of molecules. We will consider the one most general method of synthesis. Primary or secondary alcohols can be dehydrogenated to aldehydes and ketones, respectively. This reaction is summarized in Eq. (26–1) and was discussed in detail in Section 22–5B.

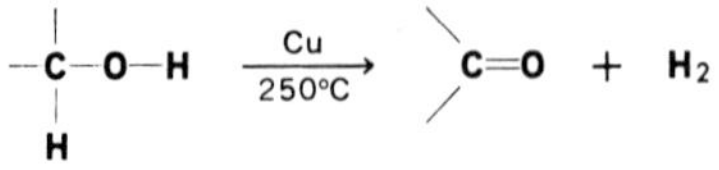

$$\underset{\underset{H}{|}}{\overset{|}{-}}C-O-H \xrightarrow[250°C]{Cu} C=O + H_2 \tag{26-1}$$

A specific example of the dehydrogenation reaction is the following:

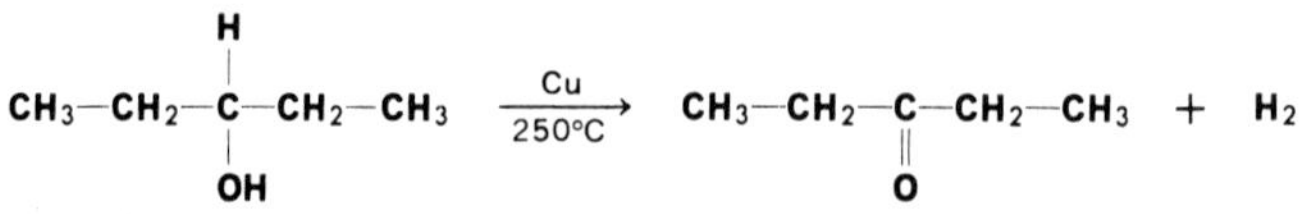

$$CH_3-CH_2-\underset{\underset{OH}{|}}{\overset{\overset{H}{|}}{C}}-CH_2-CH_3 \xrightarrow[250°C]{Cu} CH_3-CH_2-\underset{\underset{O}{\|}}{C}-CH_2-CH_3 + H_2 \tag{26-2}$$

3-pentanol   3-pentanone
                                            (diethyl ketone)

Many of the aldehydes and ketones which are manufactured in large amounts are made by this procedure, or a slight variation of it (e.g., silver catalyst is sometimes used in place of copper). Examples of compounds prepared in this fashion are acetone, methyl ethyl ketone, formaldehyde, and acetaldehyde.

TABLE 26.1   Nomenclature of aldehydes and ketones

Structure	Common Name	Systematic Name
$CH_3-\underset{\underset{O}{\|}}{C}-CH_3$	acetone	propanone
$CH_3-\underset{\underset{O}{\|}}{C}-CH_2-CH_3$	methyl ethyl ketone	butanone
$CH_3-CH_2-\underset{\underset{O}{\|}}{C}-\underset{\underset{CH_3}{\|}}{C}H-CH_2-CH_3$	ethyl *sec*-butyl ketone	4-methyl-3-hexanone
$H-\underset{\underset{O}{\|}}{C}-H$	formaldehyde	methanal
$CH_3-\underset{\underset{O}{\|}}{C}-H$	acetaldehyde	ethanal
$CH_3-CH_2-\underset{\underset{O}{\|}}{C}-H$	propionaldehyde	propanal
$CH_3-\underset{\overset{\|}{CH_3}}{\underset{}{C}}H-\underset{\underset{O}{\|}}{C}-H$	isobutyraldehyde	2-methylpropanal

26.3
NATURAL AND
SYNTHETIC
CARBONYL
COMPOUNDS

Many compounds produced by plants or animals have aldehyde or ketone functional groups. The examples listed in Table 26–2 give some idea of the wide variety possible and the complexity of some of them.

Many aldehydes and ketones can be manufactured inexpensively by the method shown in the previous section. They are, in turn, frequently used to synthesize still other organic materials. Thus, **acetaldehyde** is converted to acetic acid, polymers, chemicals

TABLE 26.2   Some carbonyl compounds found in nature

Structure	Name	Comments
	cinnamaldehyde	in cinnamon bark; causes the odor and flavor
	vanillin	in vanilla beans
	citral	in the rinds of citrus fruits and several other plants; characteristic lemon odor
	glucose	a sugar found in nearly all living organisms
	testosterone	a male sex hormone
	estrone	a female sex hormone

for the processing of rubber, and baits to kill snails and slugs. Approximately a million tons of **formaldehyde** are manufactured in the United States annually; nearly all is made into polymers for adhesives or plastic molding materials (Bakelite, Melmac). A solution of formaldehyde in water, called *formalin*, is used to preserve biological specimens. It reacts with the protein in such a way that decay is inhibited, and the tissue becomes firmer. Formaldehyde is also used in embalming. **Acetone** is the starting material for a variety of organic intermediates and solvents, including those manufactured into epoxy polymers and polymethacrylates (Lucite, Plexiglass). Another ketone, **cyclohexanone**, is an intermediate in the synthesis of nylon.

In addition, numerous ketones are effective solvents for materials such as lacquers, fingernail polish, waxes, adhesives, and plastics. **Acetone** (propanone) and **methyl ethyl ketone** (*MEK*; butanone) are the two used most extensively as solvents.

Acetone is also biologically important. Abnormal metabolism in individuals having diabetes causes the production of acetone; it is then excreted in the urine, or in severe cases even exhaled in the breath.

26.4 GENERAL CHEMICAL PROPERTIES

Aldehydes and ketones are reactive compounds which can be transformed into a wide variety of other substances having either practical or theoretical importance. Many of the possible carbonyl reactions are of interest primarily to the professional chemist. In this chapter we consider three general types of reactions which are of the greatest importance in both chemistry and biology: hydrogenation, oxidation, and addition of polar molecules. Following that, we briefly examine the carbohydrates, which constitute a unique group of carbonyl compounds.

26.5 HYDROGENA- TION

Hydrogen will add to a carbonyl group in much the same fashion that it adds to an alkene double bond (Section 20.5B). Catalysts such as nickel, palladium, or platinum are necessary to promote the addition reaction, another example of hydrogenation. If we consider specific examples, we note once again the structural relation of aldehydes to primary alcohols and of ketones to secondary alcohols:

$$\begin{matrix} \diagdown \\ \diagup \end{matrix}C=O \;+\; H_2 \;\xrightarrow{\text{Ni}}\; -\overset{\displaystyle |}{\underset{\displaystyle H}{C}}-O-H \tag{26-3}$$

$$CH_3CH_2\overset{\displaystyle }{\underset{\displaystyle \underset{O}{\|}}{C}}H \;+\; H_2 \;\xrightarrow{\text{Ni}}\; CH_3CH_2\overset{\displaystyle }{\underset{\displaystyle \underset{OH}{|}}{C}}H_2 \tag{26-4}$$

$$CH_3\overset{\displaystyle }{\underset{\displaystyle \underset{O}{\|}}{C}}CH_3 \;+\; H_2 \;\xrightarrow{\text{Ni}}\; CH_3\overset{\displaystyle }{\underset{\displaystyle \underset{OH}{|}}{C}}HCH_3 \tag{26-5}$$

Further, we see that the hydrogenation reaction (Eq. 26–3) is just the reverse of dehydrogenation (Eq. 26–1). In Section 22.5B we saw that dehydrogenation is equivalent to oxidation. Analogously, a hydrogenation reaction, such as that shown in Eq. (26–3), is equivalent to *reduction*. Indeed, we sometimes refer to the addition of two hydrogen atoms to a carbonyl group or to some similar structure as a reduction reaction. The reducing agent may not always be hydrogen gas. Certain other compounds can supply the two hydrogen atoms and thus act as reducing agents. This is always the case in living organisms.

For example, the coenzyme nicotinamide adenine dinucleotide (NAD) can be the source of hydrogen to convert ethanal (acetaldehyde) to ethanol:

$$
\underset{\underset{O}{\|}}{CH_3-C-H} + 2[H] \rightarrow \underset{\underset{O-H}{|}}{CH_3-\overset{\overset{H}{|}}{C}-H} \tag{26-6}
$$

(from NAD)

In the laboratory, hydrogenation of carbonyl compounds (Eq. 26–3) is seldom used to synthesize common alcohols. However, the hydrogenation reaction may often be important in preparing a more unusual alcohol or in establishing structural relations. An example of the latter is the hydrogenation of menthone to menthol (Eq. 26–7). The fragrant, flavorful oil produced by mint leaves contains both menthol and menthone.

$$
\text{menthone} + H_2 \xrightarrow{Ni} \text{menthol} \tag{26-7}
$$

menthone   menthol

Note that menthone, menthol, and citral (Table 26.2) are all *terpenoids* (Section 20.1), built up from isoprene units.

26.6
OXIDATION

Under ordinary conditions ketones resist attack by the common ionic oxidizing agents.

$$
\underset{\underset{O}{\|}}{RCH_2-C-CH_2R} + [O] \xrightarrow{KMnO_4} \text{no reaction}
$$

Aldehydes, on the other hand, are very easily oxidized to carboxylic acids.

$$
\underset{\underset{O}{\|}}{RCH_2-C-H} + [O] \xrightarrow{KMnO_4} \underset{\underset{O}{\|}}{RCH_2-C-OH}
$$

(aldehyde)   (carboxylic acid)

The oxidation involves attack at the *hydrogen* bonded to the carbonyl group. However, *carbons* bonded to the carbonyl group are resistant to attack, which explains why the ketone is unchanged.

This difference in behavior has been used as the basis of qualitative tests to distinguish an aldehyde from a ketone. One such test is provided by **Tollens' reagent,** a solution of silver nitrate in aqueous ammonia:

$$
\underset{\underset{O}{\|}}{R-C-H} + 2Ag^+ + 3NH_3 + H_2O \rightarrow \underset{\underset{O}{\|}}{R-C-O^-} + 2Ag\downarrow + 3NH_4^+ \tag{26-8}
$$

While the aldehyde group is being oxidized to a carboxyl group, the silver ion is reduced to silver metal, thus providing a change which can be seen. The metallic silver precipitates in very fine black particles or coats the surface of the glass vessel with a bright silver mirror. A carbonyl compound which behaves in this way is presumed to be an aldehyde; if it does not cause the formation of metallic silver, a ketone is suspected.

Two similar reagents, originally developed to test for sugars, are **Fehling's solution** and **Benedict's solution**. Both contain copper(II) ions (bright blue solution) which are reduced to copper(I) oxide (brick red precipitate).

$$R-\underset{\underset{O}{\|}}{C}-H \ + \ 2Cu^{2+} \ + \ 5OH^- \ \rightarrow \ R-\underset{\underset{O}{\|}}{C}-O^- \ + \ Cu_2O \ + \ 3H_2O \qquad\qquad (26\text{-}9)$$

The reader may wonder why OH^- ions in the same solution with Cu^{2+} ions do not cause precipitation of insoluble $Cu(OH)_2$. This reaction is prevented by including citrate ions or tartrate ions (not shown in Eq. 26–9) in the solution. These ions form a "complex" with Cu^{2+} which protects it from combination with OH^-.

**26.7
ADDITION
OF HYDROGEN
CYANIDE
A. Cyanohydrin
Formation**

In Section 26.5 the hydrogenation of carbonyl compounds was discussed; hydrogen, the substance being added, is nonpolar. There is an even greater variety of *polar* compounds which will add to a carbonyl group. Reaction with these polar molecules occurs quite readily, because the carbonyl group itself is polar. This polarity largely controls the reactions of carbonyl compounds, as it did for carboxylic acids (Section 23.3). However, since carbonyl compounds have no groups which can be displaced (such as Cl or OH), their reactions are *additions*. A good example of addition to a carbonyl group by a polar molecule, HCN, will be shown in this section. Other important examples will appear in the succeeding sections.

Hydrogen cyanide will add to ethanal as shown in Eq. (26–10). The reaction goes readily to completion and the product, a cyanohydrin, is stable and easily isolated.

$$CH_3-\underset{\underset{O}{\underset{\delta-}{\|}}}{\overset{\overset{\delta+}{H}}{C}} \quad HCN \ \rightarrow \ CH_3-\underset{\underset{OH}{|}}{\overset{\overset{H}{|}}{C}}-CN \qquad\qquad (26\text{-}10)$$

a cyanohydrin

This manner of addition is simple and logical. The positive fragment, the H^+, is attracted to the oxygen, the negative end of the carbonyl group. The other portion adding, the CN^-, bonds to the carbon, which is the positive end of the carbonyl group. All other reactions involving addition of polar molecules to a carbonyl compound follow this same pattern. In each case, the hydrogen from the adding molecule becomes bonded to the oxygen, and the other fragment adding, the negative portion, becomes bonded to the carbon.

In Eq. (26–10) hydrogen cyanide was shown adding to an aldehyde. However, the reaction is general for all carbonyl compounds; ketones react in a similar fashion. The resultant cyanohydrin has two functional groups, a hydroxyl group (an alcohol) and an organic cyanide. Each will display the reactions typical for such a group.

**B. Conversion of
Cyanohydrin to
Hydroxy Acid**

The cyanide functional group in a cyanohydrin can be hydrolyzed to yield a carboxylic acid group, just as can a simple organic cyanide (Eq. 23–21). Applying this hydrolysis reaction to the cyanohydrin obtained from reaction (26–10) results in the formation of lactic acid:

$$CH_3-CH-C\equiv N + 2H_2O + HCl \xrightarrow{\Delta} CH_3-CH-COOH + NH_4Cl$$
$$\quad\quad\quad | \quad\quad\quad\quad\quad\quad\quad\quad\quad\quad\quad\quad\quad | $$
$$\quad\quad OH \quad\quad\quad\quad\quad\quad\quad\quad\quad\quad OH$$
$$\quad\quad\quad\quad\quad\quad\quad\quad\quad\quad\quad\quad\quad\quad \text{lactic acid}$$

$$(26\text{-}11)$$

This series of reactions is of considerable interest as a synthetic route, because quite a few α-hydroxy acids occur in nature. Lactic acid, discussed in Section 23–15E, is just one example.

**26.8
ADDITION OF
AN ALCOHOL**   In the presence of an acid catalyst, an alcohol will add to a carbonyl group, with the oxygen from the alcohol bonding to the carbon of the carbonyl group and the carbonyl oxygen acquiring a hydrogen:

                                        a hemiacetal

$$(26\text{-}12)$$

This mode of addition by the alcohol is clearly similar to the pattern followed during the addition of hydrogen cyanide, Eq. (26–10). However, reaction (26–12) is readily reversible. The equilibrium does not especially favor the product, a hemiacetal, and the hemiacetal is not very stable. In fact, with most of the ordinary carbonyl compounds it is not possible to get the hemiacetal out of the reaction mixture. In Eq. (26–12), brackets are placed around the formula of the hemiacetal to indicate that it is unstable.

Although the hemiacetal is usually too unstable to be isolated, it can react further in this same solution to form an acetal, a product which *is* stable:

     a hemiacetal   an **acetal**

$$(26\text{-}13)$$

This second stage of reaction, Eq. (26–13), will be favored if we make it a point to include an excess of the alcohol in the reaction mixture. This stage is not an addition reaction; rather, it is a condensation, with water being split out. This reaction is also reversible.

Because of the reversibility of both Eq. (26–13) and Eq. (26–12), it is possible to *hydrolyze* an acetal. By treating an acetal with a large excess of water and an acid catalyst, we can convert it back to an aldehyde plus two molecules of alcohol. Of course, the conversion takes place by passing through the intermediate hemiacetal stage. The hydrolysis of an acetal functional group facilitates the study of those natural compounds which happen to contain such a structure; quite a few important compounds do. Hydrolysis will reveal which aldehyde and alcohol components were originally combined in the compound.

Since reactions (26–12) and (26–13) are both reversible, in order to get good conversion of the aldehyde to the acetal, we need to do whatever is possible to shift the equilibrium. We should use a large excess of alcohol. We notice that water is a *product* of the reactions. Therefore, we should be sure none is present to start with; we should use alcohol which is free of water, and for catalyst we should dissolve in the alcohol some dry HCl gas, *not* an aqueous solution of HCl. With these precautions, it is possible to carry the reaction through both stages, Eqs. (26–12) and (26–13), and obtain a good yield of the acetal.

By a similar line of reasoning, we can see how to convert the acetal back to the aldehyde by reversing the equilibrium in the two reactions. To promote reaction in the reverse direction, the acetal should be treated with a large excess of water. Acid catalyst should again be included, because any substance which catalyzes a forward reaction will similarly catalyze the reverse reaction. The reverse reaction, starting with attack by water, will first split one molecule of alcohol off the acetal, converting it to a hemiacetal. Subsequent loss of a second molecule of alcohol will finally produce the aldehyde. Altogether the reverse reaction would appear as follows:

$$\underset{\text{acetal}}{\overset{\text{H}}{\underset{\text{OCH}_3}{\text{R}-\text{C}-\text{OCH}_3}}} + \text{HOH} \underset{\text{HCl}}{\rightleftharpoons} \left[\underset{\text{hemiacetal}}{\overset{\text{H}}{\underset{\text{OH}}{\text{R}-\text{C}-\text{OCH}_3}}}\right] + \text{HOCH}_3 \underset{\text{HCl}}{\rightleftharpoons} \underset{\text{aldehyde}}{\overset{\text{H}}{\text{R}-\text{C}}\diagdown_{\text{O}}} + 2\text{HOCH}_3 \qquad (26\text{–}14)$$

Since this conversion of an acetal back to an aldehyde, Eq. (26–14), involves reaction with water, it would be called a hydrolysis reaction.

In summary, it should be emphasized that the reaction of a carbonyl compound with an alcohol begins, as shown in Eq. (26–12), with an *addition* reaction which is very similar to many other additions to carbonyls. Although an aldehyde was used for illustration in Eq. (26–12), ketones will react in a similar manner with alcohols, in which case the final product is called a **ketal**.

The formation and hydrolysis of hemiacetals and acetals are frequently observed in the laboratory and are important as well among natural products. In particular, a knowledge of these reactions is essential for an understanding of the properties of carbohydrates, which will be discussed in Chapter 30. In this connection, an interesting variation of hemiacetal formation is possible when the hydroxyl group and the carbonyl group are on the same carbon skeleton. An *intra*molecular reaction leads to a hemiacetal having a ring structure:

$$\underset{\text{OH}}{\text{R}-\text{CHCH}_2\text{CH}_2\text{CH}_2-\text{C}-\text{H}} \quad \rightarrow \quad \text{R}-\text{CH}\underset{\text{O}-\text{CH}}{\overset{\text{CH}_2-\text{CH}_2}{\diagup \diagdown}}\text{CH}_2 \qquad (26\text{–}15)$$

Carbohydrates have hemiacetal ring structures such as this.

The addition of ammonia to a carbonyl group takes place as follows:

$$\underset{}{\text{R}-\text{C}}\diagup^{\text{R}}_{\diagdown\text{O}} + \text{H}-\underset{\text{H}}{\text{N}}-\text{H} \rightleftharpoons \left[\underset{\underset{\text{AA}}{\text{OH}}}{\overset{\text{R}}{\text{R}-\text{C}-\underset{\text{H}}{\text{N}}}\diagdown\text{H}}\right] \rightleftharpoons \left[\underset{\underset{\text{an imine}}{\text{NH}}}{\overset{\text{R}}{\text{R}-\text{C}}}\right] + \text{HOH} \qquad (26\text{–}16)$$

In several ways, this reaction is similar to the addition of an alcohol to a carbonyl group.

Reaction (26–16) is reversible and the addition product, AA, which forms in the solution is unstable, just like a hemiacetal. Again we enclose in brackets the formulas of unstable compounds. As in the previous examples of carbonyl additions, when ammonia reacts the hydrogen will bond to the oxygen, and the electronegative atom, in this case nitrogen, will bond to the carbon of the carbonyl group. An aldehyde participates in a reaction such as (26–16) just as well as the ketone illustrated.

Although a substantial amount of the ammonia addition product, structure AA, is formed, it cannot be isolated. It easily breaks down by splitting out *ammonia* and reverting to the carbonyl compound. Alternatively, it can split out *water* to give an imine. This reaction is also reversible and the imine too is unstable in most cases.

Equation (26–16) is an important reaction despite the fact that neither structure AA nor the imine can be isolated. There is ample evidence that they *are* present in the reaction mixture and under just slightly different conditions stable compounds *can* result from such a reaction. This can happen if the compound adding is a derivative of ammonia, G—NH$_2$, bearing a group G in place of one of the hydrogens. There are several kinds of structure which G may be that will make the resultant imine, R$_2$C=N—G a stable, easily isolated substance. Products of this type will be discussed in Section 26.10.

B. Reductive Amination

Another way in which a significant product can be isolated from a reaction such as (26–16) is by reducing (hydrogenating) the imine existing in the mixture. This process converts it to an amine, a stable and useful compound:

$$\left[R-C \underset{NH}{\overset{R}{<}} \right] + H_2 \xrightarrow{Ni} R-\underset{H}{\overset{R}{\underset{|}{\overset{|}{C}}}}-NH_2 \qquad (26\text{–}17)$$

an imine   an amine

It is evident that this reaction is structurally similar to the addition of hydrogen to C=C or C=O groups. To transform the carbonyl compound all the way through to the amine simply requires placing in the same vessel all the necessary reactants: the carbonyl compound, ammonia, hydrogen, and nickel catalyst. The net reaction occurring in this mixture may be represented thus:

$$R-C \underset{O}{\overset{R}{<}} + NH_3 + H_2 \xrightarrow{Ni} R-\underset{H}{\overset{R}{\underset{|}{\overset{|}{C}}}}-NH_2 + H_2O \qquad (26\text{–}18)$$

What takes place, of course, is that reaction (26–16) forms the imine. As fast as any imine appears in the mixture, it is converted by hydrogen, as in reaction (26–17), to the amine. In fact, by adding Eqs. (26–16) and (26–17) algebraically, one arrives at the net equation (26–18). The process represented by Eq. (26–18) is called **reductive amination**. It was first shown in Section 25.5B as a general method for the synthesis of amines. In our present discussion, a ketone has been used in the examples, Eq. (26–16) and so on. However, these are very general reactions of carbonyl compounds; aldehydes react every bit as well.

C. Reductive Amination in Living Organisms

Reductive amination has just been discussed as a chemical process. It is also biologically important. In living systems, all the same steps occur, the chief differences being that the catalysts are enzymes and that the pair of hydrogen atoms required for the reduction step comparable to Eq. (26–17) come not from hydrogen gas but from some organic reducing agent in the cell.

One biochemical example of reductive amination is the production of the important amino acid, glutamic acid. A great variety of plants and animals are capable of carrying out this biosynthesis, represented by Eq. (26–19).

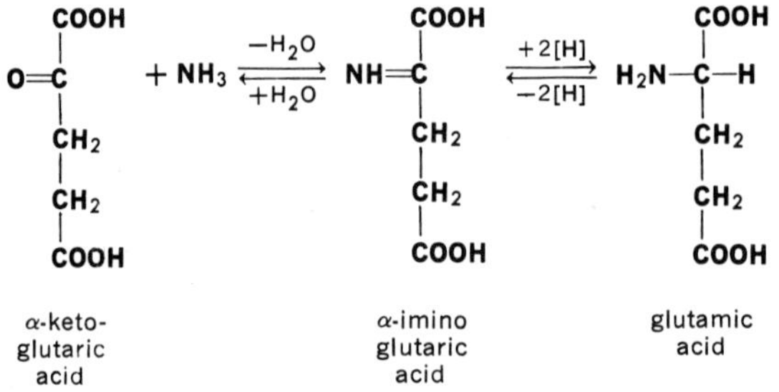

$$
\begin{array}{ccc}
\underset{\substack{\text{CH}_2 \\ | \\ \text{CH}_2 \\ | \\ \text{COOH}}}{\overset{\substack{\text{COOH} \\ | \\ \text{O}=\text{C} \\ |}}{}} + \text{NH}_3 \underset{+\text{H}_2\text{O}}{\overset{-\text{H}_2\text{O}}{\rightleftharpoons}}
\underset{\substack{\text{CH}_2 \\ | \\ \text{CH}_2 \\ | \\ \text{COOH}}}{\overset{\substack{\text{COOH} \\ | \\ \text{NH}=\text{C} \\ |}}{}} \underset{-2[\text{H}]}{\overset{+2[\text{H}]}{\rightleftharpoons}}
\underset{\substack{\text{CH}_2 \\ | \\ \text{CH}_2 \\ | \\ \text{COOH}}}{\overset{\substack{\text{COOH} \\ | \\ \text{H}_2\text{N}-\text{C}-\text{H} \\ |}}{}}
\end{array}
\qquad (26\text{--}19)
$$

α-ketoglutaric acid   α-iminoglutaric acid   glutamic acid

An enzyme called glutamate dehydrogenase is required as a catalyst for reaction (26–19). The necessary ammonia is obtained from cellular constituents. The two hydrogen atoms required for addition to the imine structure in the second stage are provided by a coenzyme, nicotinamide adenine dinucleotide. The α-ketoglutaric acid required for the carbon skeleton in reaction (26–19) can be derived from carbohydrates, through a complex series of metabolic reactions.

26.10 ADDITION OF AMMONIA DERIVATIVES

Certain ammonia derivatives will react with a carbonyl compound to give a *stable* product. For this reaction to be successful, the substituted group (which will be shown in a general way as G) must be of particular types. The reaction of these ammonia derivatives occurs as follows:

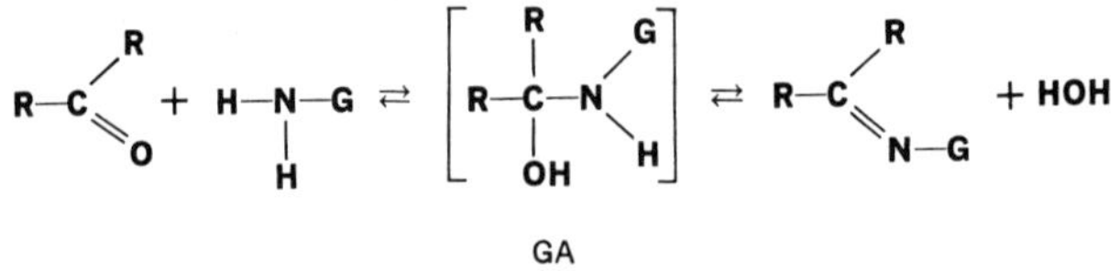

$$
\text{R}-\overset{\text{R}}{\underset{\text{O}}{\text{C}}} + \text{H}-\overset{\text{H}}{\underset{\text{H}}{\text{N}}}-\text{G} \rightleftharpoons
\left[\text{R}-\overset{\text{R}}{\underset{\text{OH}}{\text{C}}}-\overset{\text{G}}{\underset{\text{H}}{\text{N}}} \right] \rightleftharpoons
\text{R}-\overset{\text{R}}{\text{C}}\underset{\text{N}-\text{G}}{\diagdown} + \text{HOH}
\qquad (26\text{--}20)
$$

GA

Notice that the initial reaction involving *addition* to the carbonyl group once again gives an unstable product, just as it did with ammonia; compare structure AA in Eq. (26–16). The important difference in Eq. (26–20) is that when the intermediate structure, GA, splits out water in the second stage of the reaction, the final product is stable.

Shown below are the reactions of some of the important types of GNH_2 compounds. In each case only a net equation has been written, showing the final, stable product which one obtains. However, in each case the reaction will take place as shown in Eq. (26–20), going by way of the intermediate, GA.

$$R-C\overset{R}{\underset{O}{\diagup}} + H_2N-NH-\bigcirc \rightarrow R-C\overset{R}{\underset{N-NH-\bigcirc}{\diagup}} + H_2O \tag{26-21}$$

phenylhydrazine   a **phenylhydrazone**

$$R-C\overset{R}{\underset{O}{\diagup}} + H_2N-NH-\bigcirc-NO_2 \rightarrow R-C\overset{R}{\underset{N-NH-\bigcirc-NO_2}{\diagup}} + H_2O \tag{26-22}$$

NO$_2$ NO$_2$

2,4-dinitro-   a **2,4-dinitrophenyl-**
phenylhydrazine   **hydrazone**

$$R-C\overset{R}{\underset{O}{\diagup}} + H_2N-OH \rightarrow R-C\overset{R}{\underset{N-OH}{\diagup}} + H_2O \tag{26-23}$$

hydroxylamine   an **oxime**

Each of the reactions just shown is *very general for all kinds of aldehydes and ketones.*
The products, often called carbonyl *derivatives*, are nearly always solids which can be
filtered off and purified by crystallization.   Since each solid derivative has its own
characteristic melting point, it provides important information for identifying the original
carbonyl compound. Very often three or four such bits of information are sufficient for
positive identification of a carbonyl compound.   Since many carbonyl compounds are
liquids, the solid derivatives such as phenylhydrazones and oximes provide a great
advantage. Solids can be purified with more certainty and greater ease than can liquids.
And once the solid is obtained, its melting point is a much more reliable property for
identification than is the boiling point of a liquid.

SUMMARY

The carbonyl functional group is present in both aldehydes and ketones, which are very
common among either natural or synthetic organic compounds. Because the carbonyl group
has a pi bond, many of its reactions are additions. However, because it is a *polar* pi bond,
many of the reagents which add (HCN, ROH, GNH$_2$) are ones which do not add to an
alkene pi bond. Carbonyl compounds also add hydrogen, in the presence of a metal catalyst,
forming alcohols. All of these carbonyl addition reactions are common to both ketones and
aldehydes.

One reaction in which aldehydes and ketones differ is oxidation. The hydrogen attached
to the carbonyl group of an aldehyde is readily attacked and a carboxyl group results.
Ketones give no reaction with most typical laboratory oxidizing mixtures. This different
behavior with oxidizing agents is the basis for the Tollens' and the Benedict's tests to differentiate between aldehydes and ketones.

The systematic name-ending for a ketone is *-one*, and for an aldehyde is *-al*. Several of
the simple aldehydes have common names which originate from the common names of the
corresponding carboxylic acid; (for example, CH$_3$COOH is acetic acid; CH$_3$CHO is acetaldehyde).

IMPORTANT TERMS

acetal	cyanohydrin	ketone
aldehyde	dehydrogenation	oxidation
aldose	hemiacetal	reduction
carbonyl group	hydrogenation	reductive amination

WORK EXERCISES

1. Name each structure.
   a) CH_3CH_2CH
   $$\overset{\|}{O}$$

   b) $CH_3CH_2CCH_3$
   $$\overset{\|}{O}$$

   c) ⬡—C—CH$_2$CH$_3$
   $$\overset{\|}{O}$$

   d) $CH_3CHCH_2CHCH_2CH$
   with Cl, CH$_3$, O substituents

   e) cyclopentane ring =O

   f) $CH_3CH_2CCH_2CHCH_3$
   with O and CH$_3$

2. Write an equation to show how each of the following could be synthesized from the appropriate alcohol.
   a) 2-pentanone   b) butanal
   c) formaldehyde   d) propanal
   e) cyclohexanone   f) cinnamaldehyde

3. Show the structure that would result when each of these compounds is treated with hydrogen and nickel catalyst. [Consult Table 26.2 for structures.]
   a) 3-pentanone   b) vanillin
   c) citral   d) testosterone

4. Show how butanone (methyl ethyl ketone) could be synthesized from an alkene. [More than one step is required.]

5. Benzaldehyde, C_6H_5CHO, is a liquid. After it has been exposed to air for some time, a white solid appears at the mouth of the bottle. As time goes on, more and more of the solid settles to the bottom of the bottle. The solid is not soluble in water, but does dissolve readily in dilute sodium hydroxide solution. Suggest a probable structure for the white solid and a reason for its formation.

6. Write an equation for each reaction.
   a) the formation of a hemiacetal from an aldehyde
   b) the conversion of a hemiketal to a ketal
   c) the reductive amination of 3-hexanone
   d) the conversion of butanal to a cyanohydrin
   e) the hydrolysis of an acetal
   f) the reaction of acetaldehyde with Tollens' reagent
   g) the reduction of Benedict's solution by benzaldehyde
   h) the reaction of butanal with KMnO$_4$ solution

7. For what purpose might one wish to prepare the 2,4-dinitrophenylhydrazone of butanone? Write an equation for the reaction.

8. Lactic acid (Table 23.1) can be synthesized in three steps from ethanol
   a) First, the ethanol is converted to ethanal (acetaldehyde). Write an equation.
   b) Now, show how the acetaldehyde can be converted to lactic acid in two more steps.

9. Write an equation for a biological example of reductive amination.

10. List all the types of structural units (both carbon skeletons and functional groups) that you can find in each compound below (see Table 26.2).
    a) cinnamaldehyde
    b) vanillin
    c) testosterone
    d) estrone
    e) glucose

11. Tell which of the compounds below would reduce Tollens' reagent and what you would observe in case of a positive test.
    a) 2-butanone
    b) cyclopentanone
    c) benzaldehyde
    d) lactic acid, Eq. (26–11)
    e) ethyl phenyl ketone
    f) cinnamaldehyde (Table 26.2)

12. Tell which of the compounds below would reduce Benedict's solution and what you would observe in a positive test.
    a) ethanal

b) 2-pentanone
c) menthone (Eq. 26–7)
d) formaldehyde
e) glucose (Table 26.2)
f) cyclohexanone
g) citral (Table 26.2)

13. Show the structure of the organic product(s) from each reaction mixture.

a) *m*-chlorobenzaldehyde, hydroxylamine
b) 3-pentanone, H_2, Ni
c) 2-butanol, Cu, 250°

d) benzaldehyde, excess ethanol, dry HCl
e) butanone, NH_3, H_2, Ni
f) cyclohexanol, Cu, 250°
g) cyclopentanone, dil. $KMnO_4$, 25°
h) 2-methylbutanal, H_2, Ni
i) propanal, HCN
j) product from (i), H_2O, HCl, heat
k) vanillin, phenylhydrazine
l) pentanal, NH_3, H_2, Ni
m) estrone, hydroxylamine

SUGGESTED READING

Guild, W., Jr., "Theory of Sweet Taste," *J. Chem. Educ.* **49**, 171 (1972).

Johnson, R. D., "Organic Chemistry in Space," *Chemistry* **50**, no. 8, p. 16, October 1977. Discusses the organic molecules found in space, the instruments used to look for life on Mars, and related topics.

Roberts, J. D., "Organic Chemical Reactions," *Scientific American*, November 1957 (Offprint #85). This article is very good at this stage of your study. It will provide both some new insights and a review of some of the important ways in which organic molecules react.

27

Stereoisomers

27.1
INTRODUCTION This chapter deals with the fascinating isomers, called *stereoisomers*, which can exist because of particular spatial arrangements within the molecules. Since we have already encountered a few examples of one type, the *cis-trans* isomers, we will begin by investigating additional examples of *cis-trans* isomers. Then it will be possible to define some important terms and describe a very different type of isomer within the large category of stereoisomers.

27.2
CIS-TRANS
ISOMERS:
ADDITIONAL
EXAMPLES We discovered, in Section 20.4, that there could be both a *cis* and a *trans* isomer of an alkene such as 2-butene. The two isomers can exist because the rotation of groups about the double-bond is restricted. A similar situation arises in substituted cycloalkanes, whose carbon skeletons fall approximately in a plane, with attached groups held either above or below the plane of the ring. For example, *cis*-1,3-dichlorocyclopentane and *trans*-1,3-dichlorocyclopentane (Fig. 27.1) are two different compounds. The two chlorine atoms can both be below the plane of the ring, or one can be above the plane and one below. In this case, it is the ring itself which acts as a more or less rigid structure, preventing part

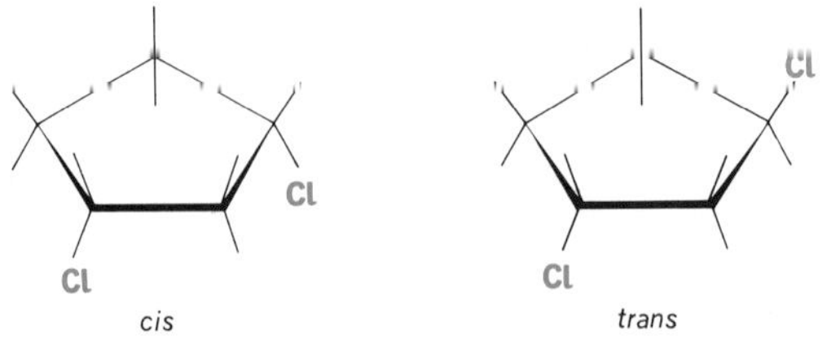

Fig. 27.1 Two different molecules of 1,3-dichlorocyclopentane.

of the skeleton in one isomer from rotating to produce the other isomer. Although each bond in the skeleton of the compound is a single bond, the carbon atoms are linked in a ring. Therefore, one carbon cannot rotate relative to the others without breaking a bond and tearing the ring open. So the situation is the same as it is with the *cis-trans* isomers of alkenes: a *cis* compound can be changed to its *trans* isomer only by way of a chemical reaction involving bond-breaking and bond-formation.

27.3 **STEREO-** **ISOMERS:** **DIFFERENCES** **IN CONFIGU-** **RATION**	For any given pair of *cis-trans* isomers (for example, Fig. 27.1), the *cis* compound has the same *structure* as the *trans* compound. *The* **structure** *of a molecule refers to which atom is bonded to which other atom, and to the type of bonds involved.* A *trans* isomer differs from a *cis* isomer only in the arrangement in space of its atoms. We say it is their con-figurations which differ. **Configuration** *refers to spatial arrangement.* Compounds differing in this fashion constitute a particular class of isomers: **stereoisomers** *have the same structure, but different configurations.* The *cis-trans* isomers represent just one type of stereoisomers, whose differences in configuration are due to some rigid structure in the molecule (a double bond or a ring). The other type of stereoisomers, to be described below, are called *optical isomers*; their configurations depend on a different factor.

27.4 **THE STRANGE** **ISOMERS OF** **LACTIC ACID**	Lactic acid occurs in a great variety of natural materials, including sour milk and sore muscles. It has a sour taste much like that of vinegar and other organic acids. Analysis of lactic acid reveals that its molecular formula is $C_3H_6O_3$; two of the oxygens are in a carboxyl group and the third is in a hydroxyl group. Furthermore, there is abundant evidence to prove that the hydroxyl group is on the carbon adjacent to the carboxyl group. Therefore, we are required to write the structural formula as in Fig. 27.2.

For many compounds this would appear to be a satisfactory formula. Not so in this case. Lactic acid seems to have two different isomers, both of which, however, must have the same *structure* (Fig. 27.2). Although there are many sources for the different kinds of lactic acid, one kind can be obtained conveniently from certain bacteria, another kind from muscle tissue. The properties of the two are listed in Table 27.1.

In *nearly* all respects the two kinds of lactic acid are identical. Are we justified in suggesting that they are different compounds? In one unusual property, their effect on the rotation of polarized light, the two kinds of lactic acid are curiously similar and yet different. Each affects the polarized light to the same extent, rotating it 3.8°. However, the results are exactly opposite. In one case it is a positive value (rotation to the right), while in the other case it is a negative value (rotation to the left). Indeed, it is remarkable that lactic acid has *any* effect on polarized light; many organic compounds do not. For this reason we have never before bothered to mention this particular physical property, as we have properties such as boiling point, density, or solubility. The nature of polarized light will be discussed in the next section. For now, we will say that because the distinctive

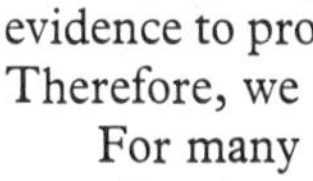

$$CH_3-CH-COOH$$
$$|$$
$$OH$$

Fig. 27.2 The structure of lac-tic acid.

TABLE 27.1   Properties of the two isomeric lactic acids

Property	Isomer A	Isomer B
source	muscle	bacteria
melting point	26°C	26°C
rotation of polarized light	+3.8° (to the right)	−3.8° (to the left)
other physical properties	A and B behave identically	
chemical properties	reactions correspond to the structural formula (Fig. 27.2); A and B behave identically	
assigned name	(+)-lactic acid	(−)-lactic acid

feature of the lactic acids involves light they were called **optical isomers**.* We shall soon see that numerous other organic compounds exhibit the same behavior.

In order to have names to designate each of the optical isomers of lactic acid, chemists have called them simply (+)-lactic acid and (−)-lactic, referring to their effect on light.

The idea that the two lactic acids really are different compounds is further supported by their behavior upon melting. Once again they are curiously similar yet different. Each type of lactic acid, when pure, melts (or freezes) at 26°C. However, if we mix a small amount of (+)-lactic acid with (−)-lactic acid, it melts at a temperature *lower* than 26°C. This suggests that the (+)-isomer is behaving as a foreign substance toward the (−)-isomer; that is, the molecules of the (+)-isomer must be different that those of the (−)-isomer. Recall that, whereas pure water freezes at 0°C, if we dissolve in it a different compound, such as salt or sugar, the water freezes at a lower temperature (Section 10.8).

Although the two varieties of lactic acid are exactly the same in most properties, there are these few differences which suggest that they are not the same compound. At least we will assume they are different compounds until we find out more about them. Even more important, we should try to discover what is causing this behavior. But before we do, we should learn a little more about polarized light.

27.5 POLARIZED LIGHT AND OPTICAL ACTIVITY

Many of the properties of light may be described by regarding it as wave motion (Fig. 27.3). That is, as a ray of light moves along in one direction, there are vibrations of an electromagnetic field in a direction perpendicular to the direction of travel. In the diagram only one wave is shown; an end view of its plane vibration would appear as a line. In ordinary light the vibrations may be in *any* direction perpendicular to the line of travel. Therefore, we must imagine that the planes of vibration of other waves could be tilted at various angles, so that in the end view there are many lines.

When ordinary light, vibrating in many different planes, passes through certain materials, the only light which gets through is confined to vibrating entirely in one plane. Light having only one plane of vibration is called plane-polarized light, or more briefly, **polarized light**. One material which has polarizing ability is calcite, a clear, crystalline form of calcium carbonate. The arrangement of atoms in the calcite crystal is such that light passing through it is forced to align its vibrations into one plane.

* Optics is a branch of physics concerned with light. The term was derived from the Greek word *optikos*, relating to the eye or vision.

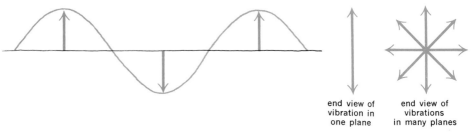

end view of   end view of
vibration in   vibrations
one plane   in many planes

Fig. 27.3   Ordinary light as wave motion.

A **polarimeter** is an instrument designed to measure the effect of substances on polarized light. It is a long cylinder holding various bits of apparatus; from the outside it usually looks like a telescope. Figure 27.4 shows the functioning parts of a polarimeter, in diagram form. The front end of the cylinder holds a calcite crystal, called a *polarizer*. This converts the ordinary light entering the instrument to polarized light. The light then passes through a tube containing the sample being studied. If the sample is a compound such as lactic acid, it can rotate the plane of vibration of the polarized light. The light will still be vibrating entirely in one plane, but the plane will be tilted so that it lies at a different angle than it did originally. At the back end of the instrument is another calcite crystal, the same as the first, but called an *analyzer*. Before the sample is placed in the instrument, the analyzer is carefully lined up with the polarizer, so that the polarized light vibrating in a vertical plane can pass on through the analyzer. However, if polarized light which passes through the sample is no longer aligned in a vertical direction, it is blocked by the analyzer. In order to allow the light to pass, we must rotate the analyzer until it is tilted at the same angle as the polarized light from the sample. The analyzer is attached to a round dial marked off in degrees, like a protractor. Therefore we can read on the dial the number of degrees through which we need to turn the analyzer to get it properly aligned with the polarized light.

The ability to rotate polarized light is a property of individual molecules. Therefore the amount of rotation observed in the polarimeter will be greater if the polarized light encounters more solute molecules by passing through a more concentrated solution or through a longer tube. Meaningful rotation values can be obtained only if we compare one with another under a standard set of conditions. The value used, called the **specific rotation** and symbolized by $[\alpha]$, is the *amount of rotation given by a pure liquid or a solution at a con-*

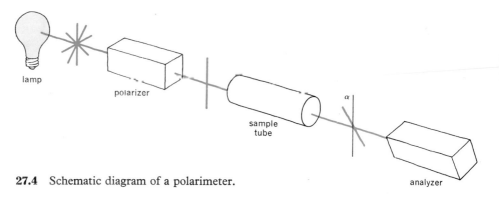

lamp

polarizer

sample
tube

α

27.4   Schematic diagram of a polarimeter.   analyzer *centration of one gram per milliliter, in a sample tube one decimeter long.* Very often the measurement will be observed using a tube of a different size or a solution of different concentration. In this case [α] can easily be calculated by the following equation:

$$[\alpha] = \frac{\alpha}{lc},$$

where [α] = the specific rotation,
      α = the observed rotation in degrees,
      l = the tube length in decimeters (dm).
      c = the concentration in g/ml.

Thus if a 2-dm tube is used for a certain sample, α (the degrees of rotation actually observed on the polarimeter dial) would be twice as great as it would have been for the same sample in a 1-dm tube. But when one substitutes 2 for l in the equation, α is divided and brought back to the standard conditions, giving [α].

We say that lactic acid is *optically active* because it affects the polarized light. **Optical activity** *is the ability to rotate the plane of vibration of polarized light.* Acetic acid, ethanol, acetone, and many other organic compounds do not rotate polarized light at all; these we say are optically inactive. If the rotation caused by an optically active compound is to the right (clockwise), the compound is said to be *dextrorotatory* (L. *dexter*, right). When writing the numerical value, we give it a positive sign, thus: +3.8°. Conversely, a *levorotatory* substance (Fr. *lévo*, L. *laevus*, left) rotates the light to the left (counterclockwise) and its numerical value is given a negative sign.

One characteristic of an optically active compound is that somewhere, somehow, there can always exist a companion which behaves exactly like it in all respects but one: the companion rotates polarized light the same number of degrees but in the *opposite*

TABLE 27.2 Comparison of optically active and inactive organic compounds

Optically Active	Optically Inactive
CH_3—CH—$COOH$ 　　　| 　　　OH 2-hydroxypropanoic acid (lactic acid)	CH_3—CH_2—$COOH$ propanoic acid
CH_3—CH_2—CH—CH_3 2-butanol　　OH	CH_3—CH_2—CH_2—CH_2 1-butanol　　　　　OH
CH_3—CH_2—CH_2—CH—CH_3 2-pentanol　　　　OH	CH_3—CH_2—CH—CH_2—CH_3 3-pentanol　　OH
CH_3—CH—$COOH$ 　　　| 　　　NH_2 2-aminopropanoic acid	CH_2—CH_2—$COOH$ 　| NH_2 3-aminopropanoic acid
CH_3—CH—CH_2—$COOH$ 　　　| 　　　Cl 3-chlorobutanoic acid	CH_3—CH—CH_2—$COOH$ 　　　| 　　　CH_3 3-methylbutanoic acid

direction. Thus we think of optical isomers as existing in pairs which somehow are opposites, like a right hand and a left hand. Even though we may first encounter only the right hand, sooner or later we will locate the left hand.

Quite a variety of organic compounds can display optical activity; a few examples are listed in Table 27.2 Surprisingly, many other organic compounds are optically inactive, despite the fact that structurally they are very similar to some of the active ones.

27.6
TETRAHEDRAL
CARBON TO
THE RESCUE

So far all we have done is to describe the observed properties of optical isomers. The big question is: *Why* are they that way? If they exist as pairs of different substances, yet have the same structure, how can we write two different formulas? Or, more properly, can the atoms somehow be assembled in two ways which will make most properties the same yet have opposite effects on light? Furthermore, why are other oganic compounds, apparently rather similar, *not* optically active?

Louis Pasteur was aware of this problem as long ago as 1860; he had been studying optically active salts of tartaric acid, obtained from the sediment in wine kegs. Despite his genius, there was not sufficient chemical knowledge at that time to make it possible for him to see an answer to the puzzle. He did surmise, however, that an optically active compound must have its atoms arranged in some kind of nonsymmetric pattern, and that there would be a companion pattern in which the nonsymmetric arrangement was laid out in the opposite order. Examples of such nonsymmetric objects are a right hand and a left hand or a clockwise and a counterclockwise helix (a coil). These pairs have the same structure; they differ only in the arrangement of their parts in space.

The answer to the problem was proposed by two young men independently and almost simultaneously in 1874. One was a Dutch chemist, Jacobus van't Hoff, age twenty-two; the other was a Frenchman, Joseph le Bel, age twenty-seven. They perceived that *if* a carbon atom holds four different groups and *if* its bonds have a tetrahedral structure (Section 18.7), then the arrangement of atoms around the carbon will automatically be nonsymmetric. And precisely because it is nonsymmetric, one other (and *only* one other) arrangement is possible, which is nonsymmetric in the opposite sense. The three-dimensional models pictured in Fig. 27.5 illustrate this principle.

The two possible spatial arrangements for lactic acid (Fig. 27.5) are related to each other in the way that an object is related to its mirror image. That is, if we hold one of the models up to a mirror, the image we see in the mirror looks like the *other* model; the image is the opposite or reflected arrangement. Equally important, these two arrangements (configurations) are *different*. Regardless of how one model is turned in space it will

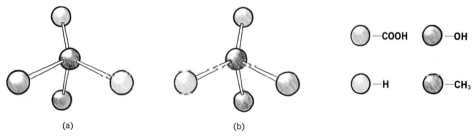

Fig. 27.5 Two nonsymmetric arrangements are possible when tetrahedral carbon holds four different groups. If lactic acid is an example, the groups would be: —COOH, —OH, —H, —CH$_3$.

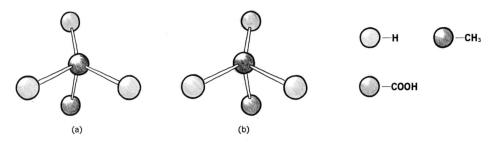

Fig. 27.6   A molecule of propanoic acid and its mirror image, which is identical.

never have the same configuration as its mirror-image companion. That is, one model cannot be *superposed* (placed in the identical position) on the other model. This is a good test to decide whether two objects are different. If one is superposable on the other, the two are identical.

A pair of hands nicely illustrates this concept of different (nonsuperposable) objects. A right hand has a nonsymmetric arrangement in space. A left hand has the same *structure*: it has fingers of the same length attached to the hand by the same kind of bones at the same angles. The left hand differs only in its spatial arrangement, which is an opposite nonsymmetric pattern. If we hold a right hand up to a mirror, the image we see there is a left hand! Furthermore, the hands are nonidentical. We cannot make a right hand become a left hand by merely turning it upside down or backward.

An optically active molecule has this "handed" property of being the mirror image of its partner. Therefore, an optically active molecule is called a chiral molecule, from the Greek word *cheir* meaning hand. A carbon atom which holds four different groups is called a **chiral carbon,*** since it is the source of the "handed" behavior.

A pair of isomers which differ only in the mirror image, or chiral, property are called enantiomers. The lactic acid isomers are an example. **Enantiomers** *are compounds which are nonidentical mirror images.*

The fact that four different groups are important to make enantiomers possible is further emphasized if we examine a molecule of propanoic acid, CH_3CH_2COOH. This compound is the same as lactic acid except that it lacks a hydroxyl and therefore has *two* hydrogens bonded to the central carbon. We can make a model of the propanoic acid molecule. Like any object, this model has a mirror image (Fig. 27.6). However, in this case the object and its mirror image are *identical*. That is, there can be only one kind of object having this structure. This uniqueness is due to the fact that both the object and its mirror image are symmetric; this symmetry, in turn, depends on the presence of two groups of the same kind (in this case, two hydrogens) attached to the central carbon atom.

Soon after van't Hoff grasped the idea that a tetrahedral carbon could possibly account for optical isomers, such as the lactic acids, he checked his hypothesis with many of the organic compounds that had been described in the chemical journals of his time. (For a few selected illustrations, see Table 27.2.) He found that compounds having one carbon holding four different groups were indeed optically active. However, a few of the compounds then listed as being optically active did *not* have four different groups at one

---

* Since the 1960s chemists have adopted the term chiral carbon. However, in older publications and in some current ones in other fields, you will see the names asymmetric carbon or dissymmetric carbon. All of these terms refer to a carbon holding four different groups.

carbon. Van't Hoff was confident enough to predict that these samples were contaminated with other compounds which were optically active. Because of the keen interest in the new concept, various chemists reexamined these "exceptions" and before long proved that van't Hoff's predictions were true. There is now no doubt that the explanation of optical isomerism offered by van't Hoff and le Bel is correct. It is supported by experimental evidence from thousands of compounds.

We say then that optical isomers have the same structure but different configurations. Their differences in configuration depend upon chiral patterns within the molecules. Since optical isomers differ only in configuration, they fall within the class of stereoisomers (Section 27.3).

27.7
FORMULAS
FOR OPTICAL
ISOMERS

Once again we are confronted by the problem of developing a written formula which will adequately represent a three-dimensional molecule. In fact, for optical isomers the problem is especially acute because their very nature depends entirely upon the subtle differences in spatial arrangement. Certain methods for writing formulas of optical isomers, shown below, have proved to be quite satisfactory and are widely used by chemists.

A perspective formula (Fig. 27.7) is easy to draw and effectively portrays the three-dimensional arrangement of groups in a molecule. It is evident that this formula is a close approximation to a picture of a three-dimensional model; compare Figs. 27.5 and 27.7. In a perspective formula the tapered bond lines imply that the two groups at the side are out in front of the central (chiral) carbon atom, closer to the viewer. Also, it is understood that the other two groups, attached by broken lines, are behind the central carbon.

It is even possible to write a satisfactory valence bond formula for an optical isomer (Fig. 27.8), *provided* we understand exactly what it represents and how to use it properly. By arbitrary convention chemists have agreed that in a valence bond formula the groups written at the sides of the chiral carbon project forward toward the viewer, whereas the other two groups extend backward behind the chiral carbon.* In other words, the lactic acid formula (a) in Fig. 27.8 represents the same configuration as those in Figs. 27.7(a) and 27.5(a).

A perspective formula is drawn in such a way that we can easily see the three-dimensional arrangement of groups. But when we look at a valence bond formula we must *imagine* the spatial arrangement; it is very important that we visualize the arrangement correctly, following the rule agreed upon. For example, one might be tempted to

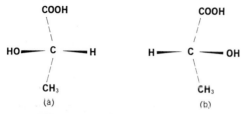

Fig. 27.7 Perspective formulas of the lactic acid enantiomers.

* The real molecules, or models of them, can of course be moved anywhere in space. However, when we wish to compare one with another and write their formulas, it is just a matter of convenience to view them always from the same direction.

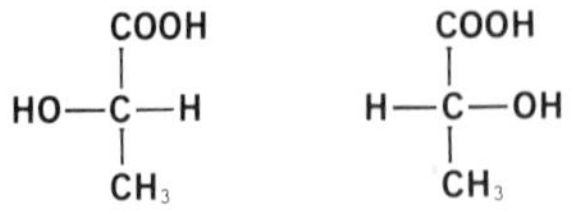

$$
\begin{array}{cc}
\text{COOH} & \text{COOH} \\
| & | \\
\text{HO} - \text{C} - \text{H} \qquad & \text{H} - \text{C} - \text{OH} \\
| & | \\
\text{CH}_3 & \text{CH}_3 \\
\\
\text{L-lactic} & \text{D-lactic} \\
\text{acid} & \text{acid} \\
\\
\text{(a)} & \text{(b)}
\end{array}
$$

Fig. 27.8 Valence-bond formulas of the lactic acid enantiomers, corresponding to the perspective formulas shown in Fig. 27.7.

argue that configuration (b) in Fig. 27.8 is really identical to that in (a); it just looks different because of the way it was written. After all, if (b) were just lifted off the paper and flipped over it would look the same as (a). The error in this argument results from ignoring the rule that one must assume the groups written at the sides are projecting *forward*. If (b) were flipped over, the groups at the side would then project to the *rear* and (b) would *not* coincide with (a). To say it another way, although a valence bond formula *appears* to be flat, it represents a molecule which actually has three dimensions. When we inspect the models (Fig. 27.5) or the perspective formulas (Fig. 27.7) of the two lactic acid configurations we realize that (b) is really different from (a).

**27.8
ABSOLUTE
CONFIGURA-
TION**

The formulas of the two enantiomers of lactic acid are often identified by the prefixes D and L, as shown in Fig. 27.8. The symbol D, derived from the Latin *dextro*, right, applies to the enantiomer having the hydroxyl group oriented to the right, in a written formula. The designation L, from the Latin *laevus*, left, designates the enantiomer having the hydroxyl group on the left.

Now if we stop to think about the *real* compound we are faced with a sudden surprise. One of the real samples of lactic acid that we might have in a bottle is labeled (+)-lactic acid because it rotates polarized light clockwise (it is dextrorotatory); this is a reliable physical property that can always be measured. But the question is this: is (+)-lactic acid the same as D-lactic acid or L-lactic acid? In other words, how do we know which *shape* of molecule, D or L (shown in Fig. 27.8), has the property of (+) rotation of polarized light?

This question can be answered only by experimental observations. The fundamental method is to study a pure sample of one of the enantiomers (say the (+) isomer) in an x-ray diffraction instrument. By this means, the location in space of the four different groups attached to the central carbon can be determined. It has been determined that (+) lactic has the L configuration; that is, it has the shape represented by formula (a) in either Fig. 27.7 or Fig. 27.8. This procedure of relating an actual sample of the compound (identified by (+) or (−)) to its configuration (identified by D or L) specifies the *absolute configuration* of the compound. Finally, all of this information can be given in the name; the compound just discussed has the complete name L-(+)-lactic acid. Its enantiomer is D-(−)-lactic acid.

The absolute configurations of quite a few optically active compounds have been determined by the x-ray method. However, many others have been determined because they could be related through chemical changes to one of the compounds of *known* configuration. For instance, (+)-glyceraldehyde has been proven to be D-glyceraldehyde. By mild oxidation D-glyceraldehyde can be converted to glyceric acid. This affects only the aldehyde group; since none of the other bonds is changed, we can be sure the glyceric acid will still have the same shape as the glyceraldehyde:

$$\text{D-(+)-glyceraldehyde} \qquad \qquad \text{D-glyceric acid} \tag{27-1}$$

After the oxidation has been carried out, the sample of D-glyceric acid obtained can be placed in a polarimeter. The optical rotation is found to be a *negative* value. This experiment has established that the (−) enantiomer of glyceric acid has the configuration we call D. Its complete name is D-(−)-glyceric acid. It is important to note that a (+) rotation value does not necessarily accompany a D configuration. Although both D-glyceraldehyde and D-glyceric acid have the same configuration (which is named D), since one has an aldehyde group and the other has a carboxylic acid group they affect the light differently, even to the extent of one being (+) and the other (−).

The study of the optically active carbohydrates first required (around 1900) that the symbols D and L be defined. A number of other related compounds, including the amino acids, were similarly specified with a D or L term. The D/L symbols are still almost universally used for these kinds of naturally occurring compounds, and we shall follow the custom in the remainder of this book. However, we should mention that during the 1960s another system, using the symbols R and S, was devised for use with *any* type of chiral compound. The R/S symbols are now widely used in conjunction with IUPAC names and are discussed in more comprehensive texts on organic chemistry.

**27.9
RACEMATES**

In Section 27.4 we found that lactic acid could be obtained from a variety of natural sources. Lactic acid can also be made synthetically [Eq. 27-2] by catalytic hydrogenation of pyruvic acid. The product thus obtained is called *racemic* lactic acid.

$$\text{pyruvic acid} \qquad \qquad \text{racemic lactic acid} \tag{27-2}$$

Is this yet a third type of lactic acid, differing from either the (+)-lactic acid or the (−)-lactic acid described in Table 27.1? Yes and no. The chemical properties of racemic lactic acid are identical to those of (+)- or (−)-lactic acid. However, it has a different solubility, its melting point is 18°C, and it is optically inactive ([α] = 0).

Chemists have discovered that *racemic* lactic acid is composed of exactly 50% (+)-lactic acid and 50% (−)-lactic acid. Such a mixture, called a racemate, can be formed by any pair of optical isomers. *A* **racemate** *is composed of exactly equal amounts of a pair of enantiomers.* A racemate is optically inactive because there are just as many molecules rotating the polarized light to the left as there are to the right; hence the net rotation observed in a polarimeter is zero.

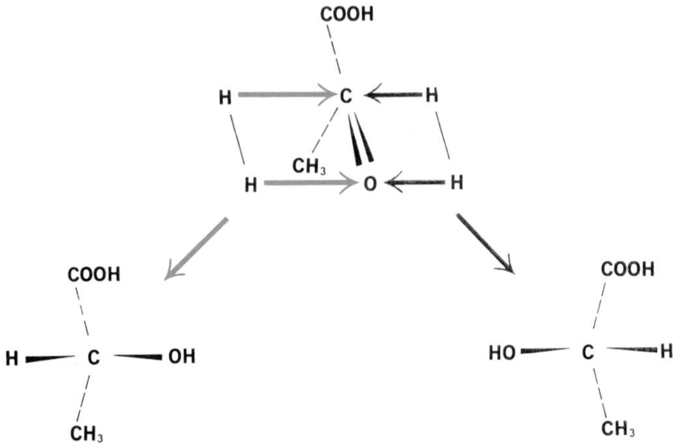

Fig. 27.9 A hydrogen molecule can add to either side of a pyruvic acid molecule, yielding either enantiomer of lactic acid with equal probability. The product thus obtained is a racemate.

When lactic acid is *synthesized* in the laboratory from pyruvic acid, *why* is a racemate produced? In contrast, some living organisms can build exclusively (+)-lactic acid and others, the (−) enantiomer. The laboratory reaction can be explained by considering the spatial arrangement of the molecules involved (Fig. 27.9). The starting material, pyruvic acid, is a symmetric molecule and hence is optically inactive. The molecule is flat; that is, the three groups attached to the central carbon all lie in one plane. In the diagram this plane is perpendicular to the page; the oxygen is nearest the viewer and the methyl and carboxyl groups are to the rear. Now when a hydrogen molecule adds to the double bond, it has an exactly equal chance to approach the flat pyruvic acid molecule from either side. Thus of the billions of molecules reacting, sheer chance will dictate that half of them will produce one enantiomer of lactic acid and the other half will produce the opposite enantiomer. Consequently, the product *as obtained in the reaction flask* is optically inactive; it is a racemate. This will always be the case, regardless of the type of reaction used. Whenever optically inactive materials are converted by chemical synthesis to new compounds, the resultant products are still optically inactive. Even though chiral carbons are created in the new molecules, the mixture as a whole will be a racemate.

If the two enantiomers constituting the racemate are separated from each other, then each will be optically active when alone. However, such a separation can *not* be accomplished by ordinary methods such as distillation or crystallization. Since the enantiomers have the same chemical and physical properties, both will show identical behavior during such treatment. A racemate can be separated into its component enantiomers only by the use of some other compound, which is already optically active, as a reagent or catalyst to influence the process.* In other words, we can generate new optically active material only if we use material which is already optically active! This is similar to the chicken-and-egg dilemma and the answer to it is similar: this is part of the mystery of life. The optically active material necessary to start the process must be obtained from a living organism.

_____
* The exact details of such a process are described in more comprehensive texts on organic chemistry.

To summarize, enantiomers behave differently from each other in the presence of another compound which is chiral (optically active). This is the second way in which enantiomers differ from each other. Their effect on polarized light is the other property in which they differ (Section 27.4).

We have just seen that an optically active reagent can be used to *separate* a racemate into two enantiomers, each of which will then be optically active. New optically active material can also be produced if an optically active catalyst or reagent is present at the time that new, chiral molecules are being formed. This optically active agent, being chiral itself, favors the formation of one enantiomer in preference to the other. A laboratory synthesis can be controlled by an optically active substance; the great majority of biochemical syntheses occur in this manner. We will discuss a biochemical example in the next section.

**27.10
OPTICAL
ISOMERS IN
LIVING
ORGANISMS**

Muscle tissue can convert pyruvic acid to lactic acid. This reaction, Eq. (27–3),

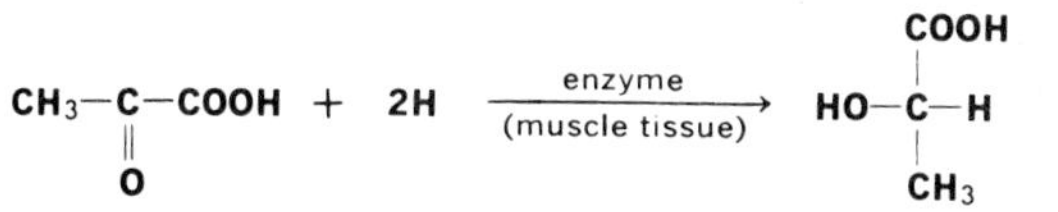

$$CH_3-\underset{\underset{O}{\|}}{C}-COOH \ + \ 2H \ \xrightarrow[\text{(muscle tissue)}]{\text{enzyme}} \ HO-\underset{\underset{CH_3}{|}}{\overset{\overset{COOH}{|}}{C}}-H \qquad\qquad (27-3)$$

pyruvic acid   L-(+)-lactic acid

is a biosynthesis. The crucial difference between it and the chemical synthesis [Eq. (27–2)] is that the biosynthesis produces only *one* enantiomer, L-(+)-lactic acid. In Eq. (27–3) the two hydrogen atoms added to the double bond come not from H_2 gas, but from an organic compound in the tissue. This difference, however, happens to be unimportant in determining which optical isomers will be produced. The important factor in Eq. (27–3) is the catalyst, an enzyme.

The enzyme catalyzes the reaction because it binds the pyruvic acid to itself and weakens the double bond, thereby facilitating the addition of hydrogen. The enzyme is a chiral molecule. (Incidentally, this of course makes it optically active.) Consequently, the enzyme acts as a pattern; a pyruvic acid molecule can fit onto its surface aligned in only one way (Fig. 27–10). Hydrogen atoms can then be added to the pyruvic acid from just *one* direction, thus yielding only one enantiomer of lactic acid.

As we have seen, certain other organisms may produce the opposite enantiomer of lactic acid. This just means that they have a different enzyme to promote the biosynthesis of lactic acid. Still different organisms may form both enantiomers, i.e., an optically inactive racemate. The significant point is that living organisms *can* produce single enantiomers (optically active) by means of their enzymes which are optically active molecules. In fact, in those biochemical reactions which are most crucial in sustaining life, organisms nearly always produce specifically one optical isomer. For example, we stated in Section 25.7 that amino acids are the essential units from which all proteins are built. Every amino acid molecule used for this purpose must have the L configuration shown below; the opposite enantiomer will not suffice.

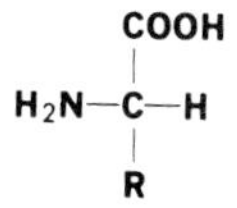

$$H_2N-\underset{\underset{R}{|}}{\overset{\overset{COOH}{|}}{C}}-H$$

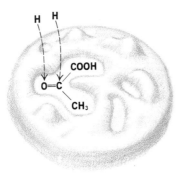

Fig. 27.10 An enzyme holding a pyruvic acid molecule, only one side of which is available for reactions. (The enzyme, represented schematically in color, is a complex organic molecule which would be much larger, in proportion, than suggested here.)

27.11
MORE COM-
PLEX OPTICAL
ISOMERS

One of the optically active compounds listed in Table 27.2 is 2-butanol; it has one chiral carbon, which is shown in color in Fig. 27.11. Because of its chiral carbon, 2-butanol can assume two different configurations. In a written valence bond formula, when the carbon skeleton is stretched out vertically, only two arrangements for the H and OH groups are possible: the OH can be either to the right or to the left.

Now let us imagine adding, to the 2-butanol molecules, one more chiral carbon, making 2,3-pentanediol (Fig. 27.12). Starting with one of the 2-butanol configurations, (a) from Fig. 27.11, one could add to it, at the upper end, another chiral carbon in either of two ways, with the OH to the right or to the left. This would give (a) and (b) of Fig. 27.12. Similarly, a new chiral carbon can be added in two ways to the other configuration of 2-butanol (Fig. 27.11(b)) giving Fig. 27.12(c, d). In this fashion, four different configurations of 2,3-pentanediol are possible.

In Fig. 27.12, compounds (a), (b), (c), and (d) are all optical isomers of each other. Furthermore, (a) and (d) are enantiomers; they are mirror images but are not identical. Compounds (b) and (c) make up another pair of enantiomers. A fifty-fifty mixture of (b) and (c) would constitute a racemate. A fifty-fifty mixture of (a) and (d) would yield another racemate having solubility, melting point, etc., different from the racemate of (b) and (c).

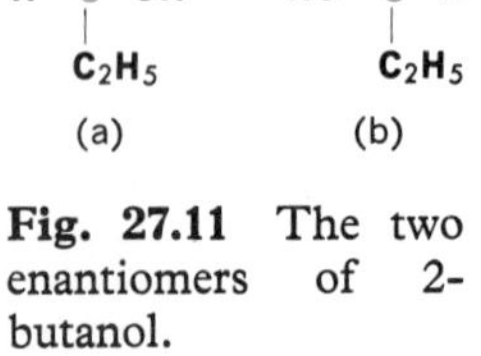

$$\overset{*}{C}H_3 \qquad\qquad OH_5$$
$$H-\overset{|}{\underset{|}{C}}-OH \qquad HO-\overset{|}{\underset{|}{C}}-H$$
$$C_2H_5 \qquad\qquad C_2H_5$$
$$\text{(a)} \qquad\qquad \text{(b)}$$

Fig. 27.11 The two enantiomers of 2-butanol.

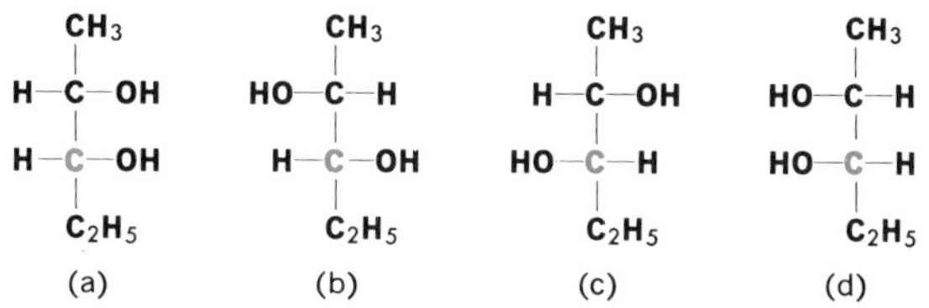

Fig. 27.12 The four configurations of 2,3-pentane-diol.

A further examination of the configurations in Fig. 27.12 reveals some additional facets of optical isomerism. Since (b) and (c) are enantiomers, they have identical melting points, boiling points, solubilities, etc. Even their rotation of polarized light is the same in amount, although of opposite sign. But what is the relation between (b) and (a)? They are **diastereomers**; *they are optical isomers but are not mirror images.* Compounds (b) and (a) have *different* melting points, boiling points, solubilities, etc. Each is optically active, but (b) rotates polarized light to a very different extent than does (a). Since (a) and (b) both have OH functional groups, they will have the same *type* of chemical properties. However, the *rate* at which they react may be different. To summarize, each configuration in Fig. 27.12 has one companion which is its enantiomer, its mirror image. But either of the other two configurations will be diastereomers of it. Thus (a) and (b) are diastereomers of each other; so also are (a) and (c), (b) and (d), or (c) and (d).

Many organic compounds, both natural and synthetic, have *several* chiral carbon atoms. How many optical isomers can we expect in such cases? We know that if there is 1 chiral carbon, 2 configurations are possible. Then if there are 2 chiral carbons, each of the first two possibilities is multiplied by 2 (Fig. 27.12). Hence, if there are three different chiral carbons, there would be $2 \times 2 \times 2$, or 2^3, or 8 possible configurations. In general, the number of different configurations will be 2^n, where n is the number of different chiral carbons. (This is often called *van't Hoff's rule*, for he was the first one to state the principle.) This simple mathematical relation leads to some astounding results. Careful inspection of the formula of glucose shown in Table 26.2, reveals that it has four chiral carbons. This means there are 2^4 or 16 compounds having the same structure but different configurations. One of these is glucose; another is its enantiomer. The other 14 compounds in this family of optical isomers are all diastereomers of glucose.

The important compound cholesterol has molecular formula $C_{27}H_{46}O$. Of these 27 carbon atoms, only 8 are chiral. However, $2^8 = 256$! The enzymes controlling the formation of cholesterol are so specific that only the one compound, out of 256 possible isomers, is formed by living organisms!

The number of different configurations predicted by the 2^n rule is the *maximum* number possible. When all the chiral carbons in a molecule are of different type, the maximum will be realized. Occasionally two of the chiral carbons will be similar, in the sense that the four groups bound to one carbon are the same as the four groups bound to another carbon. In this case not so many combinations will be possible, and the total number of configurations will be less than 2^n.

27.12
TYPES OF
ISOMERS

The various kinds of isomers have been introduced in different chapters. Since each of the fundamental types have now been mentioned, we will summarize them here:

Isomers are different compounds having the same molecular formula.

a) **Structural isomers** have different structures.

$$CH_3-CH_2-CH{=}CH_2 \text{ and } CH_3-CH{=}CH-CH_3$$

$$CH_3-O-CH_3 \quad \text{and} \quad CH_3-CH_2-OH$$

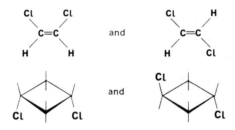

b) **Stereoisomers** have different configurations, but the same structure.
    1. *cis-trans* **isomers** have different configurations because of rigid structures in the molecules. (These are also called *geometric isomers*.)

2. **Optical isomers** have different configurations because of chiral patterns in the molecules.
    Enantiomers are nonidentical mirror images.

A **racemate** is composed of exactly equal amounts of a pair of enantiomers.
Diastereomers are optical isomers which are not mirror images of each other.

SUMMARY

Stereoisomers differ in their configurations—that is, in the arrangement in space of their groups. There are two types of stereoisomers, *cis-trans* isomers and optical isomers. In *cis-trans* isomers the different configurations result from a rigid structure in the molecule, either a C=C bond or a ring. In optical isomers the different configurations depend on chiral carbons. A chiral carbon has four *different* groups attached.

A chiral molecule is always different from its mirror image, like a right hand and a left hand. Such a pair of different mirror images are called enantiomers. Chiral molecules are always optically active. The two enantiomers rotate polarized light the same amount, but one enantiomer rotates the light in one direction, $(+)$, and the other enantiomer rotates it the opposite direction, $(-)$. These subtle differences in molecular shape have profound effect on the biochemical properties of a compound. In many cases, one enantiomer is absolutely required by a living organism, while the opposite enantiomer of the same compound is totally useless, or may even be toxic.

IMPORTANT TERMS

absolute configuration
chiral
configuration
D, L symbols
enantiomers

optical activity
polarimeter
polarized light
superposable
tetrahedral carbon atom

[See also the terms in Section 27.12.]

WORK EXERCISES

1. Which of these structures could exist in an optically active form?

a) 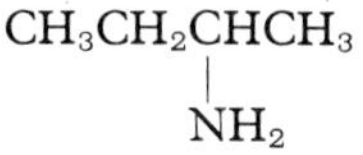 CH$_3$CH$_2$CHCH$_3$
$\quad\quad\quad$ |
$\quad\quad\quad$ NH$_2$

b) 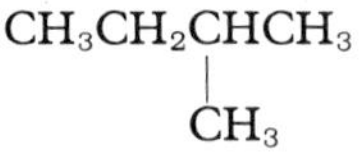 CH$_3$CH$_2$CHCH$_3$
$\quad\quad\quad$ |
$\quad\quad\quad$ CH$_3$

c) 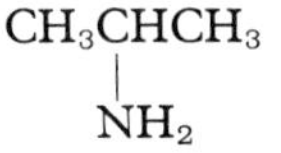 CH$_3$CHCH$_3$
$\quad\quad$ |
$\quad\quad$ NH$_2$

d) 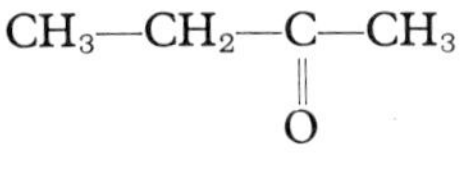 CH$_3$—CH$_2$—C—CH$_3$
$\quad\quad\quad\quad\quad$ ‖
$\quad\quad\quad\quad\quad$ O

e) 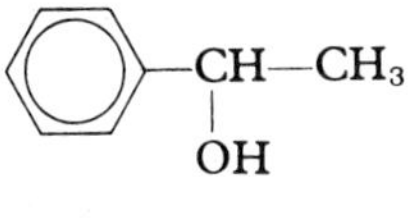 ⬡—CH—CH$_3$
$\quad\quad\quad$ |
$\quad\quad\quad$ OH

f) 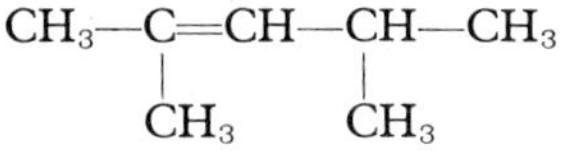 CH$_3$—C=CH—CH—CH$_3$
$\quad\quad$ |$\quad\quad\quad$ |
$\quad\quad$ CH$_3\quad\quad$ CH$_3$

2. The biologically important sugar ribose has the structure shown below. How many optical isomers (different configurations) are possible for this structure?

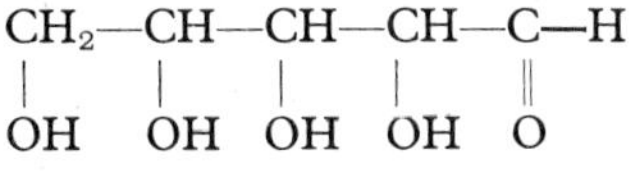

CH$_2$—CH—CH—CH—C—H
| $\quad$ | $\quad$ | $\quad$ | $\quad$ ‖
OH $\quad$ OH $\quad$ OH $\quad$ OH $\quad$ O

3. For each pair of molecules below, tell what type of isomerism is involved, *if any*.

a) 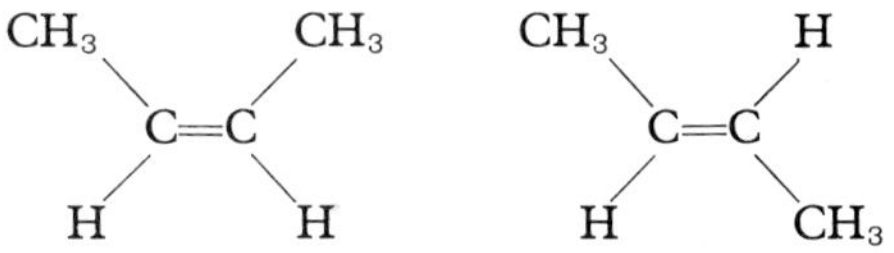 $\quad$ CH$_3\quad\quad$ CH$_3\quad\quad\quad$ CH$_3\quad\quad\quad$ H
$\quad\quad\quad$ \ $\quad\quad$ / $\quad\quad\quad\quad\quad$ \ $\quad\quad$ /
$\quad\quad\quad\quad$ C=C $\quad\quad\quad\quad\quad\quad$ C=C
$\quad\quad\quad$ / $\quad\quad$ \ $\quad\quad\quad\quad\quad$ / $\quad\quad$ \
$\quad\quad\quad$ H $\quad\quad\quad$ H $\quad\quad\quad\quad$ H $\quad\quad\quad$ CH$_3$

b) CH$_3$CH$_2$CH$_2$CH $\quad\quad\quad\quad$ CH$_3$CH$_2$CCH$_3$
$\quad\quad\quad\quad\quad\quad$ ‖ $\quad\quad\quad\quad\quad\quad\quad\quad\quad$ ‖
$\quad\quad\quad\quad\quad\quad$ O $\quad\quad\quad\quad\quad\quad\quad\quad\quad$ O

c) $\quad\quad\quad$ CH$_3\quad\quad\quad\quad\quad\quad\quad$ CH$_3$
$\quad\quad\quad\quad\quad$ | $\quad\quad\quad\quad\quad\quad\quad\quad$ |
$\quad\quad$ H—C—Br $\quad\quad\quad\quad\quad$ Br—C—H
$\quad\quad\quad\quad\quad$ | $\quad\quad\quad\quad\quad\quad\quad\quad$ |
$\quad\quad\quad\quad$ CH$_2\quad\quad\quad\quad\quad\quad\quad\quad$ CH$_2$
$\quad\quad\quad\quad$ | $\quad\quad\quad\quad\quad\quad\quad\quad\quad$ |
$\quad\quad\quad\quad$ CH$_2\quad\quad\quad\quad\quad\quad\quad\quad$ CH$_3$

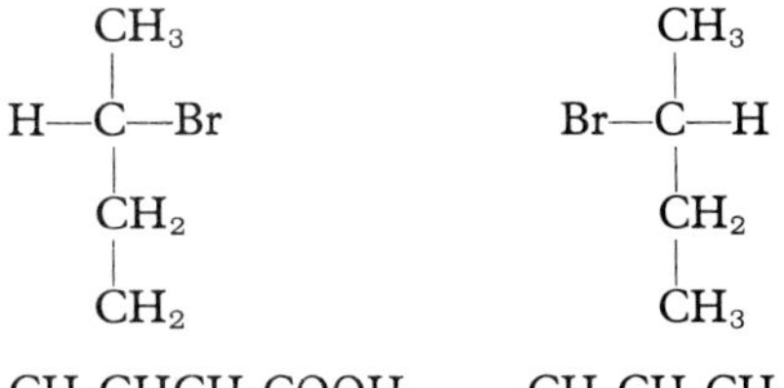

d)  CH$_3$CHCH$_2$COOH $\quad\quad\quad$ CH$_3$CH$_2$CH$_2$COOH
$\quad\quad$ |
$\quad\quad$ CH$_3$

e) $\quad\quad\quad$ CH$_3\quad\quad\quad\quad\quad\quad\quad$ CH$_3$
$\quad\quad\quad\quad\quad$ | $\quad\quad\quad\quad\quad\quad\quad\quad$ |
$\quad\quad$ H—C $\quad$ Cl $\quad\quad\quad\quad\quad$ Cl—C—H
$\quad\quad\quad\quad\quad$ | $\quad\quad\quad\quad\quad\quad\quad\quad$ |
$\quad\quad\quad\quad$ CH$_3\quad\quad\quad\quad\quad\quad\quad\quad$ CH$_3$

f) 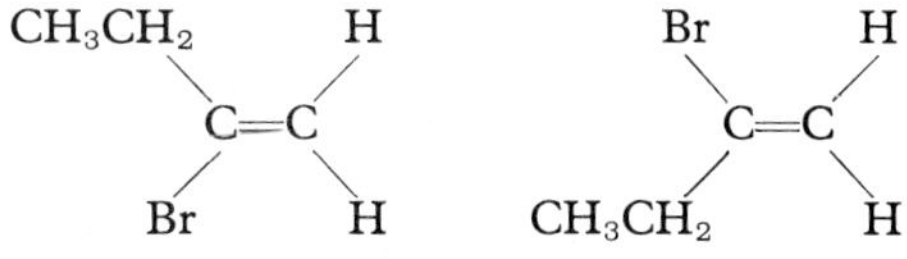 CH$_3$CH$_2\quad\quad\quad$ H $\quad\quad\quad\quad$ Br $\quad\quad\quad$ H
$\quad\quad\quad\quad$ \ $\quad\quad$ / $\quad\quad\quad\quad\quad$ \ $\quad\quad$ /
$\quad\quad\quad\quad\quad$ C=C $\quad\quad\quad\quad\quad\quad$ C=C
$\quad\quad\quad\quad$ / $\quad\quad$ \ $\quad\quad\quad\quad\quad$ / $\quad\quad$ \
$\quad\quad\quad$ Br $\quad\quad\quad$ H $\quad\quad$ CH$_3$CH$_2\quad\quad$ H

g) 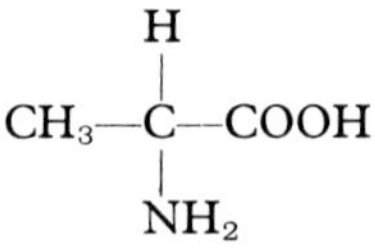 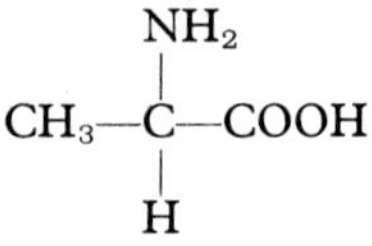 CH_3—$\overset{\displaystyle H}{\underset{\displaystyle NH_2}{C}}$—COOH

CH_3—$\overset{\displaystyle NH_2}{\underset{\displaystyle H}{C}}$—COOH

h)

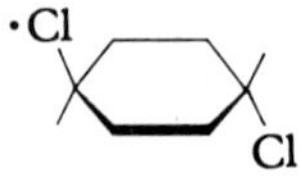

4. Review Problem 3, Chapter 18. In the light of your present knowledge of stereoisomers, what is the *total* number of isomers possible for each molecular formula?

SUGGESTED READING

Elias, W. E., "The natural origin of optically active compounds," *J. Chem. Educ.*, **49**, 448 (1972).

Kauffman, G. B., "Left-handed and right-handed molecules," *Chemistry*, **50**, no. 3, p. 14, April, 1977. An interesting introduction to chiral molecules and an account of Pasteur's separation of the enantiomers of tartaric acid.

Lawless, J. G., C. E. Folsome, and K. A. Kvenvolden, "Organic matter in meteorites," *Scientific American*, June 1972.

Perutz, M. F., "Hemoglobin structure and respiratory transport," *Scientific American*, Dec., 1978. Hemoglobin fulfills its role in transportation of oxygen and carbon dioxide by clicking back and forth between two alternative structures.

28

Polymers

**28.1
POLYMERS,
NATURAL
AND
SYNTHETIC**

As we have seen, the concept of structure in covalent compounds led to a vigorous development of organic chemistry from 1860 onward. Chemists became skilled at synthesizing compounds because, as a rule, they could reliably predict the outcome of reactions. They also became adept at proving the structures of the molecules, both natural and synthetic, with which they worked. Some of these molecules were highly complex and moderately large. However, it was not until the 1920s that chemists began to realize that the covalent structures so characteristic of organic substances could lead to the formation of *extremely* large molecules called polymers (meaning *many parts*). A single polymer molecule contains not thirty or one hundred or two hundred atoms, but *thousands* of atoms linked together covalently.

Most of the very stuff of life is polymeric material. Aside from water, a large amount of an animal body consists of proteins, which are polymers. The proteins make up not only the soft tissues but also such structures as hair, feathers, cartilage, horns, fingernails, and turtle shells. Starch, the chief food storage material of plants, is a polymer. Cellulose, the structural material of many plants, is a polymer. In wood the cellulose fibers are further strengthened by being cemented together by lignin, another polymer. Plant cells, like animal cells, contain protein. Nucleic acids, which play one of the most critical roles in living material, are also polymers. The nucleic acids carry the genetic information from one generation to the next and dictate the construction of the protein polymers which carry on many of the other functions of life.

In addition to consuming plants and animals for food, humans have for ages adapted many of the natural polymers for other useful purposes. Some notable examples are wood, cotton, linen, hemp, wool, and silk. Rubber has found widespread application only within the past hundred years. Not until 1930 were chemists able to make completely new *synthetic* polymers. The resultant plastics, elastomers (synthetic rubbers), adhesives, fibers, and coatings (paints and enamels) have profoundly affected our technology and standard of living.

In this chapter we will discuss a number of the common synthetic polymers. The biochemical functions of natural polymers will be considered in detail later in the book.

28.2
ADDITION
POLYMERS

Under the influence of the proper catalyst, one molecule of a substituted alkene, such as vinyl chloride, Eq. (28–1), can use its pi electrons to form a covalent bond to a second molecule of vinyl chloride, which can in turn link to a third molecule, and this to a fourth, and so on, to form an immensely long chain of such units. The gigantic molecule thus produced is a polymer. In Eq. (28–1) the n represents a very large number, from several hundred up to a few thousand.

$$n\ \text{CH}=\text{CH}_2 \xrightarrow{\text{peroxide}} \left(\text{CH}-\text{CH}_2\right)_n \qquad (28\text{–}1)$$
$$\underset{\text{vinyl chloride}}{\overset{|}{\text{Cl}}} \qquad \underset{\text{poly(vinyl chloride)}}{\overset{|}{\text{Cl}}}$$

vinyl chloride
a monomer

poly(vinyl chloride)
a polymer

The number n is called the degree of polymerization. Each of the small molecules from which the polymer is built is called a monomer (meaning *one part*). A structural formula for the polymer is written by drawing one portion of the structure in parentheses. The n beside the parentheses indicates that this portion of structure, called the repeating unit, is repeated n times. Note that there is a definite structural relation between the repeating unit and the monomer from which it is derived. For one particular polymer molecule the value of n may differ from that of another molecule in the same sample. It is not possible to determine n accurately, but experiments can tell us that for a certain polymer, n has a value of 1 000–1 400, for example.

Poly(vinyl chloride) is one example of an addition polymer, one of the two fundamental types of polymers. An **addition polymer** is so named because it is formed by an addition reaction. The structural characteristic of an addition polymer is that its backbone is a chain of carbon atoms built up during the polymerization reaction. One addition polymer differs from another in the nature of the groups—alkyl, aryl, or functional groups—which may be attached at alternate carbons along the chain.

It is possible to synthesize a great variety of polymers, each having different properties, by starting with different alkene monomers. For example, polystyrene is a stiff, relatively brittle plastic used to mold toys, model trains, knobs for radios, and inexpensive picnic spoons and cups. In contrast, polypropylene, although rigid, is extremely tough, impact-resistant, tear-resistant, and in thin sections is quite flexible. A contoured chair having the seat and back all in one piece can be fashioned from polypropylene. A complete box with hinge and lid can be molded in one operation from polypropylene. At the hinge the polypropylene is simply molded paper-thin so that it will be flexible. The hinge is so tough that it can be bent back and forth a great many times (probably more than a million) without breaking. Despite the flexibility of the hinge, the box and lid, being thicker, are rigid.

Several other common varieties of addition polymers are described in Table 28.1.

28.3
THE MECHAN-
ISM OF
POLYMERIZATION

The manner in which a polymerization reaction occurs is now reasonably well understood. The formation of poly(vinyl chloride) and many other polymers can be initiated by a peroxide catalyst. Figure 28.1 summarizes the stages in such a polymerization. First, the peroxide decomposes to form radicals. A radical, R·, is highly reactive; it can become

TABLE 28.1 Some common addition polymers

Structure	Chemical Name	Common Name or Brand Name	Uses
$\left(\begin{array}{c} \text{CH}-\text{CH}_2 \\ \mid \\ \text{Cl} \end{array}\right)_n$	poly(vinyl chloride)	PVC, "vinyl"	floor tile; pipes; wire covering; clear film for food packaging
$\left(\begin{array}{c} \text{CH}-\text{CH}_2 \\ \mid \\ \text{C}_6\text{H}_5 \end{array}\right)_n$	polystyrene	"styrene"	toys and models; household articles; instrument panels and knobs; styrofoam insulating material; hot drink cups, picnic jugs, etc.
$\left(\begin{array}{c} \text{CH}-\text{CH}_2 \\ \mid \\ \text{CH}_3 \end{array}\right)_n$	polypropylene		pipes; valves; packaging film; sterilizable bottles for babies and hospitals; utility boxes; fish nets; waterproof carpeting
$(\text{CH}_2-\text{CH}_2)_n$	polyethylene		squeeze bottles; flexible cups and ice cube trays; clear, soft film for food bags and clothes packages; weather balloons; moisture barriers in buildings
$(\text{CF}_2-\text{CF}_2)_n$	polytetrafluoroethylene	Teflon	linings for cooking pans; electrical insulation; gaskets, valves, and machine parts; heat- and chemical-resistant pipes and sheets
$\left(\begin{array}{c} \text{CH}-\text{CH}_2 \\ \mid \\ \text{CN} \end{array}\right)_n$	polyacrylonitrile	Acrilan, Orlon	fabrics and knitted goods
$\left(\begin{array}{c} \text{CH}_3 \\ \mid \\ \text{C}-\text{CH}_2 \\ \mid \\ \text{COOCH}_3 \end{array}\right)_n$	poly(methyl methacrylate)	"acrylic," Lucite, Plexiglass	airplane windows; models; light fixtures; costume jewelry; dental fillings and false teeth; tooth brush and hair brush handles; reflectors; automobile tail-lights

more stable if it finds one more electron to pair with its one odd electron. In the presence of an alkene, the radical can make use of one of the two electrons in the pi bond of the alkene. This, however, leaves one odd electron at the other end of the former alkene bond. Consequently, the newly formed particle is a radical also. As soon as it collides with another alkene molecule it binds one electron, once again leaving one unpaired. Thus, the stage is set for an indefinite number of repeated reactions, whereby the growing polymer chain adds to itself one monomer molecule after another.

It is apparent from Fig. 28.1 that at each stage of the polymerization a reactive site (in this example, a radical) is produced at the end of the growing chain. In fact, it might appear that the polymer chain would stop growing only when all the monomers had been used. If so, it would contain not a thousand or two but billions and billions of repeating units. The reason the polymer chain stops growing is that occasionally the odd electron

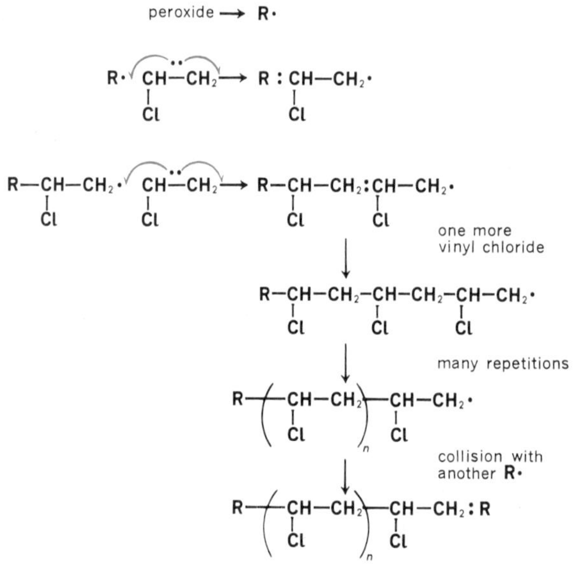

peroxide ⟶ R·

R·⌢CH—CH₂ ⟶ R : CH—CH₂·
      | |
      Cl   Cl

R—CH—CH₂·⌢CH—CH₂ ⟶ R—CH—CH₂:CH—CH₂·
   | | | |
  Cl   Cl   Cl   Cl

one more vinyl chloride

R—CH—CH₂–CH–CH₂–CH–CH₂·
   | | |
  Cl   Cl   Cl

many repetitions

R—(CH–CH₂)—CH–CH₂·
      | |
      Cl   ₙ Cl

collision with another R·

R—(CH–CH₂)—CH–CH₂:R
      | |
      Cl   ₙ Cl

Fig. 28.1 Mechanism of a typical addition polymerization catalyzed by peroxide; vinyl chloride converted to poly-(vinyl)chloride.

at the end of the growing polymer chain will encounter another radical, $R\cdot$, with which it can form a covalent bond *without* generating a new reactive site. Because only a trace of peroxide is added to the alkene monomer, only a few radicals are present at any given time. Therefore the growing polymer radical may typically collide with, and add, about a thousand alkene molecules before a collision with another radical causes termination. Still other radicals, released by the decomposing peroxide, simultaneously initiate the growth of numerous other polymer chains. In general, the greater the number of radicals supplied to the mixture, the faster will be the rate of polymerization, because more polymer chains will be growing at any given moment. The greater concentration of radicals will also cause the formation of shorter polymer chains (that is, n will be smaller); chain growth will terminate sooner because there will be a greater chance that the reactive site will encounter another radical.

Polymerization of an alkene can also be initiated by either a positive ion (Eq. 28–2) or a negative ion (Eq. 28–3), rather than by a radical.

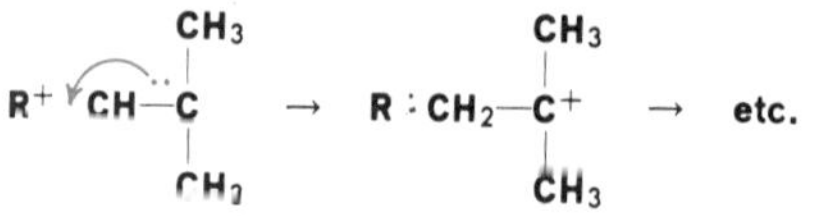

$$R^+ \,\widehat{\ }\, CH\!-\!\!\underset{\underset{CH_3}{|}}{\overset{\overset{CH_3}{|}}{C}} \; \rightarrow \; R:CH_2\!-\!\!\underset{\underset{CH_3}{|}}{\overset{\overset{CH_3}{|}}{C}}{}^+ \; \rightarrow \; etc. \qquad (28\text{–}2)$$

$$R:\!\underset{\underset{CH_3}{|}}{\overset{}{\widehat{\ }\,CH}}\!-\!CH_2 \; \rightarrow \; R:\underset{\underset{CH_3}{|}}{\overset{}{CH}}\!-\!CH_2:{}^- \; \rightarrow \; etc. \qquad (28\text{–}3)$$

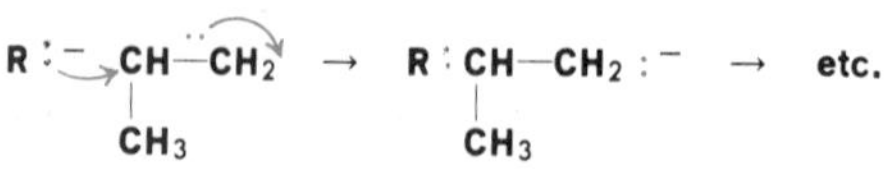

However, in every case the attack of the reactive particle on an alkene generates a particle having the same kind of reactive site. The new particle then tries to stabilize itself by attacking another alkene molecule, but in the process only generates a new reactive site, and so on.

**28.4
DIENE
POLYMERS**

A **diene** is an organic compound having *two* carbon–carbon double bonds somewhere in its structure. Of particular value in polymerization reactions are compounds such as 1,3-butadiene,

$$CH_2{=}CH{-}CH{=}CH_2$$

in which the two double bonds are separated from each other by one single bond. A polymer prepared from such a diene is just another example of an addition polymer.

Neoprene (Eq. (28–4)) was the first diene polymer successfully commercialized and the first synthetic rubber developed in the United States. Neoprene has been manufactured since the late 1930s for use in shoe soles, coverings for electric wires, and for rubber hoses, gaskets, and fittings which must withstand attack by oil or gasoline. Neoprene is superior to natural rubber in its toughness and its resistance to weathering and to oil. This polymer would make excellent tires, but is too expensive for this purpose.

$$nCH_2{=}\underset{\underset{Cl}{|}}{C}{-}CH{=}CH_2 \xrightarrow{\text{peroxide}} {-}\left(CH_2{-}\underset{\underset{Cl}{|}}{C}{=}CH{-}CH_2\right)_n{-} \tag{28-4}$$

chloroprene
2-chloro-1,3-butadiene

polychloroprene
neoprene

It is important to note that when chloroprene or a similar diene polymerizes, each monomer is linked into the polymer chain by the opposite ends of the original diene unit, and that a double bond remains in each repeating unit of the polymer.

The manner in which the diene becomes bonded at its opposite ends is shown in Eq. (28–5). There is a shift of electrons on all four carbons of the diene skeleton.

$$R{\cdot} \; CH_2{-}\underset{\underset{Cl}{|}}{C}{-}CH{-}CH_2 \;\rightarrow\; R{:}CH_2{-}\underset{\underset{Cl}{|}}{C}{-}CH{-}CH_2{\cdot} \;\rightarrow\; \text{etc.} \tag{28-5}$$

Consequently, the new reactive site (an odd electron) appears at the end of the chain. However, this process is just a variation of the fundamental addition reaction (compare with Fig. 28.1).

Natural rubber is a polymer of isoprene (see also Section 20.1):

$$nCH_2{=}\underset{\underset{CH_3}{|}}{C}{-}CH{=}CH_2 \xrightarrow{\text{catalyst}} {-}\left(CH_2{-}\underset{\underset{CH_3}{|}}{C}{=}CH{-}CH_2\right)_n{-} \tag{28-6}$$

isoprene
2-methyl-1,3-butadiene

polyisoprene
natural rubber

A number of plants, especially *Hevea* trees, can produce rubber, using enzymes to catalyze the synthesis. For years all attempts to form rubber in the laboratory, by polymerizing isoprene, failed. Then during the early 1950s, Karl Ziegler in Germany and Guilio Natta in Italy conducted extensive research on new types of catalysts which regulate reactions in a special way. Because of the fundamental principles they established during these and subsequent studies, Ziegler and Natta shared the Nobel Prize for chemistry in 1963. By applying their principles, chemists at the Firestone Rubber Company and at Goodrich-

Fig. 28.2 Clear, flexible film of polyethylene, an addition polymer, is widely used for food packaging. [Courtesy of Eastman Chemical Products, Inc.]

Fig. 28.3 A molded medicine bottle cap of polypropylene, an addition polymer. [Courtesy of Eastman Chemical Products, Inc.]

Fig. 28.4 Sturdy, attractive luggage can be fashioned from tough polypropylene. [Courtesy of Eastman Chemical Products, Inc.]

Gulf developed methods which would, in the laboratory or factory, polymerize isoprene into a material virtually identical to natural rubber.*

**28.5
COPOLYMERS**

When a mixture of *different* monomers is polymerized the product is a copolymer. By varying the kinds of monomers used and the percentage of each in the mixture, a chemist can often create a polymer having some property, such as toughness or heat stability, required for a certain application. Table 28.2 lists some of the copolymers presently being manufactured. These are further examples of addition polymers. Figure 28.5 illustrates how such materials can be utilized.

TABLE 28.2 Examples of (addition) copolymers

Monomer Components	Polymer Name	Uses
vinyl chloride, vinyl acetate	"vinyl," vinylite	phonograph records; shower curtains; rain wear
vinyl chloride, vinylidene chloride	Saran	"Saran wrap" for food; fibers for auto seat covers; pipes
vinyl chloride, acrylonitrile	Dynel, Vinyon	fibers for clothing
ethylene, propylene	ethylene-propylene rubber, EPR	tires
styrene, butadiene	styrene-butadiene rubber, SBR	tires
acrylonitrile, butadiene, styrene	ABS	crash helmets; women's spike heels; luggage; pipes; parts and cases for batteries, telephones, and other instruments; knobs and handles

**28.6
CONDENSATION
POLYMERS**

The second fundamental type of polymer is a condensation polymer, formed by a **condensation reaction**. A small molecule such as water or methanol is split out as each unit is added to the polymer chain. Structurally, condensation polymers are distinguished by the fact that there are almost always functional groups *within* the polymer backbone. (In contrast, the backbone of an addition polymer consists of carbon atoms—see Section 28.2).

Nylon is a good illustration of a condensation polymer. It is formed by the reaction of two monomers, a diamino compound and a dicarboxylic acid (Eq. (28–7)):

$$NH_2(CH_2)_6NH_2 \; + \; HOC(CH_2)_4COH \; \rightarrow \; \cdots NH(CH_2)_6NHC(CH_2)_4CNH(CH_2)_6NHC(CH_2)_4C \cdots$$

1,6-diaminohexane   1,6 hexanedioic acid (adipic acid)   nylon   (28–7)

---

* The synthetic polyisoprene described above is truly a synthetic *rubber*, in that it is a manufactured material which actually duplicates the substance produced by rubber trees. The other materials commonly called "synthetic rubber" are really different compounds which have elastic, rubbery properties. It is less ambiguous to call them *synthetic elastomers*.

Fig. 28.5 Since ABS plastic can be chrome-plated, it is used for decorative parts in cars and homes. Shown here is a lavatory faucet. [Courtesy of the Marbon Chemical Division, Borg-Warner Corporation.]

The amino groups react with the carboxylic acid groups in the usual way (compare Eq. (25–26)), forming amide groups. Note that the amide functional groups become part of the polymer backbone, and that for each amide link formed one molecule of water is split out.

It is apparent from Eq. (28–7) that for a condensation reaction of this type to lead to a polymer, each participating monomer must have *two* functional groups. Thus, after one end of the diamine has reacted with an acid molecule, the amino group at the opposite end is available for reaction with another acid molecule. Since the acid molecule also has another functional group at the opposite end, it can react with yet another diamine molecule. Consequently, innumerable repetitions of the reaction are possible. It would be good practice for the reader to rewrite Eq. (28–7) step by step, to see how the polymer chain grows.

In Eq. (28–7) a moderately long section of the polymer chain was written out in order to show the nature of the reaction. Figure 28.6 is a structural formula for the nylon polymer written in the usual more abbreviated form. Since two different monomers go into the structure of nylon, a repeating unit in the polymer formula must show each of

$$\left(\mathrm{NH(CH_2)_6NHC(CH_2)_4C}\atop\quad\underset{O}{\|}\qquad\underset{O}{\|}\right)_n$$

Fig. 28.6 The structural formula of nylon (a polyamide).

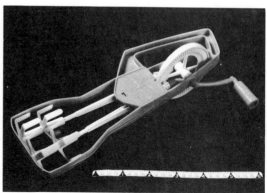

Fig. 28.7 Articles molded of nylon: left, machine gears; right, gears and blades of a beater. The frame and handle of the beater are of ABS. [Courtesy of du Pont de Nemours.]

them. In general, nylon is called a **polyamide** because amide functional groups are the links holding the polymer together.

Nylon makes strong, long-wearing fibers particularly suitable for hosiery, sweaters, and other articles of clothing. It was the first completely synthetic fiber marketed,* and to a very large extent has replaced silk, which it resembles in many of its properties. Nylon is superior to silk in most respects, and it is cheaper to produce. It is not entirely coincidental that silk and nylon have similar properties. Silk is a protein and all proteins are polyamides (Section 25.7). Nylon can be molded into solid articles (see Fig. 28.7). It is quite strong and has a slippery surface; gears made of nylon need no lubrication. It is also frequently used for glides on drawers and zippers on clothing, for combs and cups, and for knobs and handles of various sorts. In the United States the production of nylon now exceeds the astounding total of one billion pounds annually.

The most common type of nylon has the structure shown in Fig. 28.6. Other nylon polymers having somewhat different properties can be made by using a diamine or a dicarboxylic acid having a different carbon skeleton.

Another type of condensation polymer is a **polyester**; as the name implies, the functional groups repeated in this polymer chain are esters. The most familiar polyester is poly(ethylene terephthalate), fibers of which are known as Dacron or Terylene. The polymer is formed [Eq. (28-8)] by the reaction of ethylene glycol with the methyl ester of

$$n\text{HOCH}_2\text{CH}_2\text{OH} + n\text{CH}_3\text{OC} - \!\!\bigcirc\!\!- \text{COCH}_3 \rightarrow \left(-\text{OCH}_2\text{CH}_2\text{OC} - \!\!\bigcirc\!\!- \text{C} \right)_n + 2n\text{CH}_3\text{OH} \qquad (28\text{-}8)$$

ethylene glycol   methyl terephthalate   poly(ethylene terephthalate)
                                                (Dacron, Mylar)

---

* The manufacture of rayon began at an earlier date. Rayon is natural cellulose which has been *modified* by chemical reactions. The process depends on the fact that the polymer backbone of rayon was present originally in the cellulose. The structural changes are made only in the groups attached to the backbone.

terephthalic acid.   In the process, a molecule of methanol is split out for each new ester group formed; the net result is that terephthalic acid (a *di*carboxylic acid) is transformed from one type of ester to another.   A polymer results because ethylene glycol, the alcohol forming the new ester groups, is also a *di*functional molecule.

Polyester fibers, such as Dacron, are used extensively in fabrics for clothing and other goods.   The material can also be extruded into clear, tough films having excellent tear resistance and dimensional stability.   One application is in Cronar motion-picture film. Another is the Mylar film and sheets used for recording tapes and other purposes.

**28.7
THERMOPLAS-
TIC versus
THERMOSET
POLYMERS**

In the previous sections, polymers were classified on the basis of chemical structure as either addition or condensation types.   It is also convenient to classify polymers into two different groups on the basis of their behavior when exposed to heat, which profoundly affects their uses.   **Thermoplastic** polymers become plastic (that is, moldable) when heated.   They can be repeatedly heated until soft and remolded.   In contrast, a **thermoset** polymer hardens or sets when heated.   It can be molded only once; thereafter it is rigid and heat will not soften it.

Polyethylene is a good example of a thermoplastic polymer.   It can be heated and molded over and over again.   A Melmac dish is a thermoset polymer.   It remains rigid even when placed directly in a flame, but would soon char and decompose at such a high temperature.

All the polymers, both addition and condensation, described in the previous sections are *thermoplastic*.   However, it is also possible to make a *thermoset* polymer utilizing either the addition or the condensation type of linkage.   The essential difference is that a thermoset polymer is cross-linked, whereas a thermoplastic polymer is not.   The phenomenon of cross-linking is described in the next section.

**28.8
CROSS-LINKED
POLYMERS**

We discovered in Section 28.6 that a condensation reaction can lead to a polymer only if each monomer component has *two* functional groups.   Now let us see what can happen if at least one of the components has *three* functional groups on each molecule.   One common mixture of this type consists of phthalic acid and glycerol.   When the mixture is heated, the acid and the glycerol react with each other to form ester bonds and release water (Fig. 28.8).   Long polymer molecules result.   Although this product is similar in many ways to the polyester molecule shown in Eq. (28–8), it differs in one essential feature.   Each glycerol molecule has three hydroxyl groups, only two of which are needed to form the ester links holding the polymer backbone together.   The third hydroxyl group on each glycerol unit is available to react with phthalic acid, and frequently does so.

Figure 28.8 represents the initial stage of reaction between phthalic acid and glycerol, and shows two polymer chains.   This stage can be achieved by proper control of reaction conditions; for example, moderate heat can be maintained for a limited time.   We note however, that in each polymer chain carboxyl and hydroxyl groups are still available.   If the mixture is heated further, these remaining functional groups react to form additional ester bonds, linking one polymer chain to another.   The result, shown in Fig. 28.9, is a vast network in which each polymer chain is covalently bonded to many others.   Indeed, since each chain is bonded to others, and these in turn to others, the whole mass of material could be considered as one giant molecule.   After the final stage of reaction the polymer is said to be highly **cross-linked**.   Physically, it has become rigid.

The initial polymer prepared from glycerol and phthalic acid has very few cross links (Fig. 28.8).   It is a slightly viscous liquid which can be poured into a mold or treated in

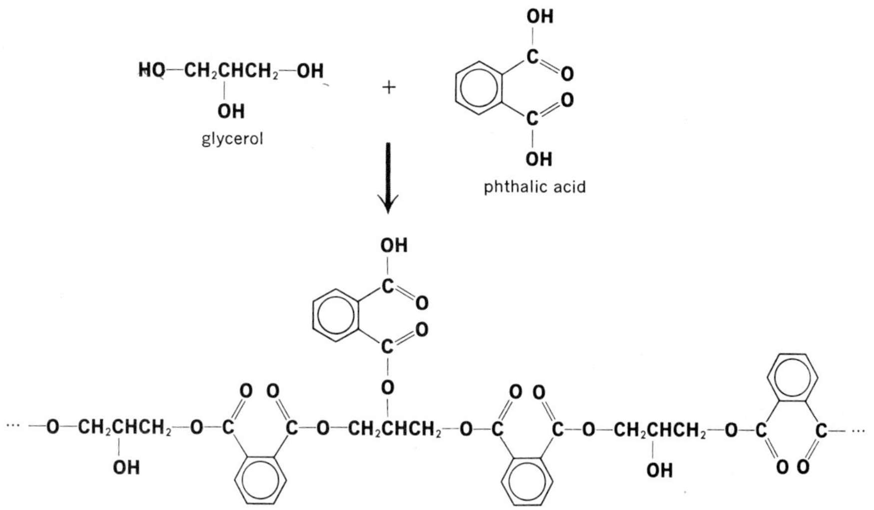

Fig. 28.8   The initial stage of reaction between glycerol and phthalic acid.

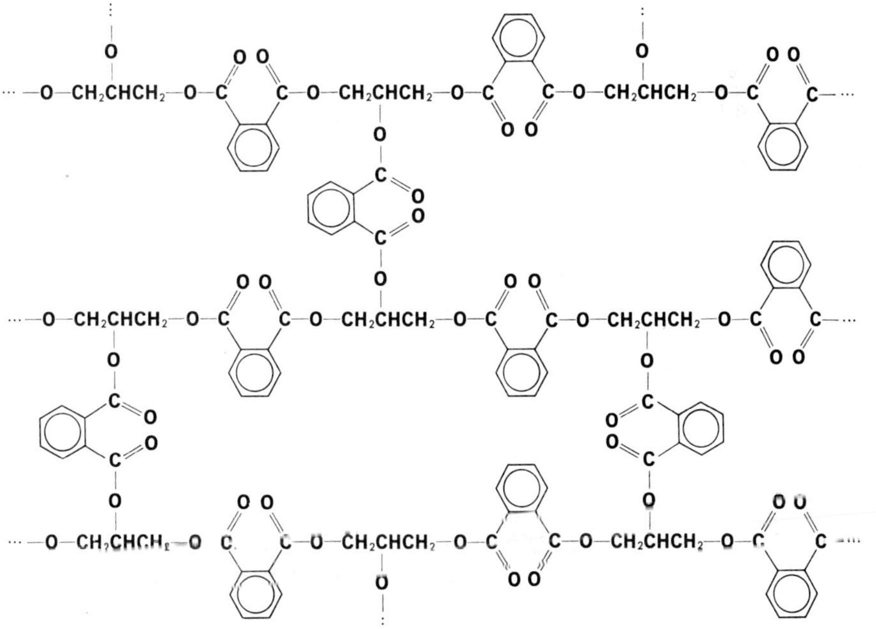

Fig. 28.9   The structure of glyptal, a thermoset polymer.   The final stage of reaction between glycerol and phthalic acid produces a highly cross-linked, rigid material.

TABLE 28.3   Some common thermoset (cross-linked) polymers

Monomer Components	Polymer Name	Uses
glycerol, phthalic acid	glyptal	baked enamels; molded articles
phenol, formaldehyde	phenol-formaldehyde resin, phenolic, Bakelite, Formica, Micarta	table tops; wall panels; resin bond for plywood; radio and appliance cabinets; washing machine agitators; handles; dials; drawer pulls
urea, formaldehyde	urea-formaldehyde resin	translucent light panels in homes and cars; decorative wall panels; bottle caps; adhesives; enamels; housings for radios and appliances
melamine, formaldehyde	melamine-formaldehyde resin, Melmac	dishes; panel boards; buttons; cases for hearing aids

some other way. After one thorough heating the material becomes a rigid, cross-linked polymer which can never again be melted or remolded. We see now that cross-linking creates a thermoset polymer. In contrast, a thermoplastic polymer is said to have a linear structure, because each molecule is just one long chain.

Glyptal is the common name for the glycerol-phthalic acid polyester (Fig. 28.9). Vast quantities of the material are used for baked enamels on automobiles and household appliances, and much smaller amounts for molded articles. For enameling, the first-stage glyptal polymer is mixed with pigments and sprayed onto a metal surface. Exposure to a bank of infrared heat lamps then causes the cross-linking leading to the thermoset polymer.

A variety of well-known thermoset polymers are listed in Table 28.3. All have one feature in common: they are cross-linked polymers.

28.9 THE SOCIAL IMPORTANCE OF SYNTHETIC POLYMERS

One of the distinctly human traits is the ability to construct useful and beautiful objects. Centuries ago people started using the materials they found in nature, such as wood, silver, gold, clay, leather, horn, and plant fibers. Gradually they developed more complex processes for obtaining useful materials. They discovered new metals that could be isolated by heating certain rocks (ores), and found that a mixture of sand and alkali subjected to very high temperatures yielded glass, an *inorganic* polymer. Only within the last few decades have humans learned to synthesize organic polymers, thereby providing a host of new structural materials.

With synthetic polymers, people have an immense new field for creative expression, both artistic and utilitarian. The synthetic polymers have properties different from those of any natural material previously employed; in many cases they are profoundly different. Moreover, the polymerization reaction is a very general process. By the proper choice of monomers one can create a new material having, within reason, almost any property desired.

This flood of new materials has had a tremendous effect upon our culture within a short period of time. As recently as 1940 the only synthetic polymer often encountered by the average citizen was Bakelite. Nylon hosiery, neoprene rubber soles, and Plexiglas, were relatively new. We now take for granted a wide variety of polymers used in fabrics, toys, gadgets, machinery, construction materials, and art objects—paintings, jewelry dishes, and sculptures.

SUMMARY

Polymers are gigantic molecules formed when a few hundred or a few thousand "ordinary" size molecules bond to each other. The bonding may result from (a) an addition reaction of some kind of alkene molecule, forming new C—C bonds, or (b) a condensation reaction of some other functional group, forming links such as ester or amide groups.

Some examples of natural polymers are rubber, starch, cellulose, proteins, and nucleic acids. Chemists have also learned how to promote the formation of polymers in a laboratory or factory, thus making available a large number of useful materials previously unknown. Some examples are fibers (Orlon, nylon, Dacron), plastics (vinyl, Lucite, polyethylene), and elastomers or synthetic rubbers (neoprene, butyl rubber).

The properties, and therefore the uses, of a *synthetic* polymer depend not only on the monomer units from which it was formed but also upon whether it is a thermoplastic or a thermoset (cross-linked) polymer.

You will find, when studying biochemistry that the properties of a *natural* polymer, such as a protein, also depend on its monomer units and the presence of cross-linking in the polymer.

CONCLUSION: THE IMPORTANCE OF STRUCTURE AND CONFIGURATION

From our first encounter with organic substances, in Chapter 18, to these final discussions of organic chemistry it has been apparent that covalent bonds give organic compounds specific structures and configurations, which in turn establish their chemical and physical properties. In fact, we can go one step further. Subtle differences in the shape and structure of organic molecules determine how they function in living cells. The reactions of organic molecules in living organisms fall within the field of *biochemistry*, our next broad topic for study.

IMPORTANT TERMS

addition polymer	**diene**	**polymer**
condensation polymer	**monomer**	**repeating unit**
copolymer	**natural rubber**	**synthetic rubber**
cross-linked	**polyamide**	**thermoplastic**
degree of polymerization	**polyester**	**thermoset**

WORK EXERCISES

1. Give the name and structure of the monomer from which each of the polymers in Table 28.1 can be prepared.

2. Using any one of the examples from Table 28.1, write an equation for the formation of the polymer showing the catalyst, the degree of polymerization, and the repeating unit.

3. All the double bonds in natural rubber (polyisoprene) have the *cis* configuration. Draw a portion of a rubber molecule in the correct *cis* arrangement. See Eq. (28–6).

4. Write a fairly long section of structure for styrene-butadiene rubber, keeping in mind that there will not necessarily be a regular alternation of monomer components as the polymer chain develops. (See Table 28.2.)

5. Write out, step by step, a fairly long segment of Dacron structure [Eq. (28–8)], so that you see how the polymer chain grows. What small molecule is eliminated as each new covalent bond forms? What type of polymer is Dacron?

SUGGESTED READING

Frazer, A. H., "High-temperature plastics," *Scientific American*, July 1969.

Mark, H. F., "The nature of polymeric materials," *Scientific American*, Sept. 1967. An excellent discussion of the properties of both natural and synthetic polymers and the possibilities of tailoring these molecules for new uses.

Meloan, C. E., "Fibers, natural and synthetic," *Chemistry*, **51,** no. 3, p. 8, April, 1978. The structure and properties of wool, silk, cotton, and synthetics such as polyesters are discussed.

"A close-up look at polymers," *Chemistry*, **51**, no. 5, June, 1978. This excellent issue of *Chemistry* is devoted entirely to polymers. Various articles describe both synthetic and biopolymers, their role in medicine and surgery, and future polymers.

PART THREE

BIOCHEMISTRY

29

Introduction to Biochemistry

29.1
DEVELOPMENT
OF BIO-
CHEMISTRY It is difficult to set a date for the beginning of the study of biochemistry. Discoveries of facts, proposals of classifications, and the development of the theories we now recognize as comprising the body of knowledge of modern biochemistry were made by many persons. Fundamental discoveries were made by persons we would now call chemists, physiologists, physicians, and many others who would be difficult to classify. For a long time all these people were held back by the doctrine that special vital forces were needed for biological reactions, and that these reactions could not be studied and understood by humans. However, some people, either from bravery, foolhardiness, or perhaps because they did not realize that what they were doing was supposed to be impossible, began to study biochemical questions. Even before the Renaissance many studies were made of the composition of various parts of plants and animals, of the chemistry of the soil, fertilizers, and the requirements for plant growth.

Theophrastus Bombastus von Hohenheim, better known as Paracelsus, who lived from 1493 to 1541, brought together much of the old alchemical mystique and added new ideas of his own. He believed, among other things, that food contained both poisonous and wholesome parts and that digestion separated the two parts. His theory was that when the digesting agent, the *Archaeus*, was not working properly, the poisonous elements were not separated and eliminated, so the person became ill. Medicines were given to restore the *Archaeus* to its normal state. Some of the medicines used were poisonous salts (mercury II chloride and lead acetate); others were extracts from plants. Among the latter were extracts made by distillation, sometimes a distillate from fermented material which contained ethanol. The euphoria produced by ethanol no doubt led to widespread acceptance of the remedies of Paracelsus. Historically more important was the fact that the ideas of Paracelsus and his followers were interesting and stimulated much experimentation. Probably many of their patients died not so much from their illnesses as from the "cures" prescribed. However, the ways of approaching and looking at the chemistry of living things proposed by Paracelsus led to better theories.

Alchemy and other ancient "sciences" emphasized the work of the past. Alchemists believed that people in previous ages had known the secrets of long life and of making gold. The best that living people could do was to rediscover things. When the emphasis shifted from the past to the present and the future, alchemy changed into science. This change in emphasis was very important for the growth of biochemistry as well as the other sciences.

One of the early scientists, we might even call him a biochemist, was the Frenchman Louis Pasteur. He was especially interested in the individual reactions taking place in living organisms. He helped to develop the idea that fermentation of grapes (we know now that it was actually the sugar in the grapes) to produce alcohol really involved a series of chemical reactions, and that these reactions occurred in living yeast cells. He further realized that the products formed in these reactions could be changed if different types of cells (either different yeasts or bacteria) were present. A further step was taken when the Buchner brothers showed in 1896 that fermentation did not require whole cells, as Pasteur had believed, but that cell-free yeast juices which contained a "ferment" could bring about certain reactions. It was later realized that the "ferment" was not a single substance but was really a mixture of many biological catalysts which came to be called enzymes. When these catalysts were extracted and purified, and as highly purified forms of the reactants became available to scientists, the road was opened for the investigation of the reactions that take place in living organisms.

Fig. 29.1   Methods of separation and analysis of mixtures by chromatography and electrophoresis.

a) Paper chromatography. In paper chromatography a mixture of substances is applied to a spot on a piece of filter paper. The paper is then placed in a container with a solvent mixture which either rises on the paper by capillary action or is allowed to flow down the paper from a solvent reservoir. If this procedure does not separate all components (as with C and D above) a second solvent mixture can be used in a direction at right angles to that of the first solvent mixture. This gives a two-dimensional chromatogram. Thin-layer chromatography is similar to one-dimensional paper chromatography except that a plate of glass coated with a thin layer of absorbent material is used.

(b) Column chromatography. In column chromatography a mixture is applied to the top of a glass column filled with a powdered adsorbent substance, and a solvent mixture is allowed to flow through the column. This produces a separation of substances in the mixture. If test tubes are moved under the column at given intervals by the use of a turntable, the different components can be separated.

(c) Gas (gas-liquid) chromatography. In gas chromatography a mixture of substances is injected into the instrument. A current of helium or other inert gas carries the mixture through a small tube packed with an adsorbent material. The tube is heated to help in separation of the components of the mixture. As the components leave the instrument they are detected by some type of device and a curve is plotted. From such curves both the number of components and the relative amount of each may be seen. Some gas chromatographs have arrangements for trapping the individual components as they leave the instrument. These trapped materials can then be analyzed by other methods.

(d) Paper electrophoresis. In paper electrophoresis the mixture is applied to a filter paper or other similar material which is then placed in an electrophoresis cell having an anode at one end and a cathode at the other. Electricity is allowed to flow, and the components of the mixture migrate in the electric field. In the illustration, substances A and B are anions (they migrate toward the anode) and D and E are cations. C has not migrated.

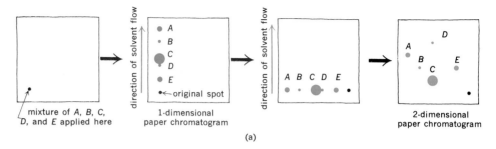

mixture of *A*, *B*, *C*, *D*, and *E* applied here

direction of solvent flow

1-dimensional paper chromatogram

direction of solvent flow

2-dimensional paper chromatogram

(a)

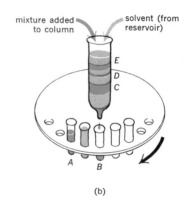

mixture added to column

solvent (from reservoir)

(b)

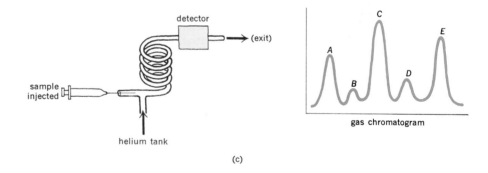

detector

(exit)

sample injected

helium tank

gas chromatogram

(c)

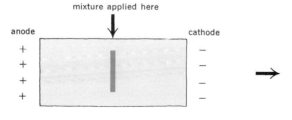

mixture applied here

anode

cathode

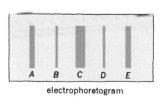

electrophoretogram

(d)

As investigation led to greater understanding, the early biochemists were nagged by a fear that the reactions observed when parts of plants or animals were placed in a glass container (*in vitro*) might not be the same as those that actually occurred in a living plant or animal (*in vivo*). Another difficulty some of the clearer thinkers recognized was that the substances isolated and purified from natural products might have been modified by the processes they used and thus might no longer be exactly the same as they were in the starting material. The possibility that they may be producing **"isolation artifacts"** and the *"in vitro–in vivo"* problem still plague modern biochemists.

The past few decades have seen an extremely rapid increase in biochemical knowledge, and predictions for the future make the older science fiction seem mild. Why is our knowledge of biochemistry growing so rapidly ? There are many answers. One is that the United States government has been, and is, supporting biochemical research. Another reason for the rapid growth is the discovery and development of new techniques, particularly those for separating and analyzing biochemical compounds. Researchers are purifying and separating substances of importance to biochemistry by using the techniques of **chromatography** and **electrophoresis**. Either solid material packed in columns, paper, or the process of gas-liquid chromatography (popularly called "gas" chromatography) has been especially useful. Chromatographic techniques have made it possible to separate individual amino acids from complex mixtures—a task that previously had been almost impossible. **Electrophoresis**—the movement of substances in an electrical field—is also an important technique. Some instruments use both chromatographic and electrophoretic processes for separating components of mixtures. Some of these analytical techniques are shown in Fig. 29.1.

Separation of components of cells by centrifugation is also an important technique (see Section 29.4). In this process, tubes containing complex mixtures are placed in a device known as a centrifuge. The tubes are spun rapidly and the most dense components of the mixture settle out. In some cases the biochemical substances to be separated are suspended in a salt solution to help in the separation. In salt solutions, as in ordinary liquid mixtures, some components may rise to the top of the centrifuge tube since they are less dense. Others will sediment and perhaps some will remain in suspension. If we use an ultracentrifuge—one that produces forces hundreds of thousands times gravity— and use centrifuge tubes containing solutions that are more dense at one part than another, it is possible to separate large molecules from one another. These separations have led to knowledge of the composition and structure of many of the compounds most important to life, including RNA and DNA.

Analytical methods which involve the absorption of infrared and ultraviolet radiation, as well as other techniques, have made it possible to analyze a sample consisting of only a few milligrams of material. The availability of radioactive isotopes, particularly the radioisotope of carbon, carbon-14 (C-14), has led to amazing progress in the study of biochemical reactions. Now specific molecules can be labeled with C-14 and their "fate" in an animal or plant shown by determining which excretion products are labeled. Thus, finding labeled carbon dioxide in the air expired by a rat given glucose which contained carbon-14 is evidence that glucose is converted, *in vivo*, to carbon dioxide by the rat (Fig. 29.2). By administering very small quantities of radioisotopes to humans, medical researchers can sometimes diagnose, and in other instances (usually with larger doses) treat, various diseases.

Many of the early and some of our modern advances in biochemistry have resulted from accidental discoveries in a research laboratory. Discovery of things not searched for, which is called serendipity, has been responsible for countless biochemical developments. For instance, the important discovery by the Buchners that cell extracts could carry out the

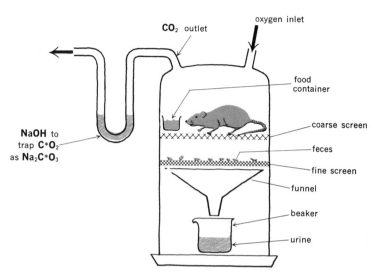

Fig. 29.2 Diagram of an apparatus used to measure metabolism of radioactive substances in an animal. When an animal that has been fed or injected with a radioactive substance is placed in the apparatus, the gases it exhales as well as its feces and urine can be collected and analyzed. (The asterisk is used to indicate radioactive atoms.)

reactions of fermentation was not made as the result of an experiment planned for this purpose, but came about as the result of their attempts to produce a "yeast juice" that they hoped would have medicinal value. When Eduard Buchner added a solution of sugar to the yeast juice with the hope that it would act as a preservative, he observed that fermentation was occurring, and concluded that units smaller than the whole cells (which Louis Pasteur insisted were necessary) could carry out fermentation. The realization that smaller units—the "ferments" that we now call enzymes—could serve as catalysts was a significant biochemical advance. Discoveries such as this require a trained, perceptive scientist, accurate observations of unexpected results, and the correct interpretation of these observations. It is interesting to speculate as to when—or whether—antibiotics would have been developed if Alexander Fleming had merely discarded his culture of staphylococcus that had been "contaminated" by a penicillin-producing mold.

Although such accidental discoveries are important to the progress of science, many discoveries and most of the development of new products have been made as the result of well-planned experiments. While researchers are always looking for the unexpected event in their routine work, this prospect does not provide the only motive for biochemical research. Biochemists are usually driven by at least two other motives in their research. For some, the desire to know how living organisms function—just for the pleasure of knowing—is the major motive. Others are more concerned with practical results—curing or preventing diseases. To most researchers, of course, both motives are important.

Although it is interesting to know how biochemistry developed and to realize some of the difficulties it faces, the major question, "What is biochemistry?" has not yet been discussed.

The simplest definition of biochemistry is that it is the chemistry of living plants and animals. Chemistry, as we have learned, is concerned with the composition, structure, and reactions of various types of matter. Thus biochemistry is the study of the composition and structure of those materials found in living organisms and the substances derived from them. Further, it is concerned with the *reactions* which occur in living organisms. The goal of biochemistry is to explain the processes of life in physical and chemical (biochemical) terms.

**29.2
BIOCHEMICAL
COMPOSITION**

Life exists in many varieties from the unicellular algae through the simpler animals and plants to the giant redwood trees and whales. In spite of their obvious differences, there is a remarkable similarity in the compounds which are found in these varied types of organisms. All living matter, as well as all inanimate matter, is composed of the approximately one hundred naturally occurring chemical elements. About half of these elements are found in measurable amounts in the human body and in most other organisms that have been studied. Of these fifty elements, less than half have some known function in the human body. Those elements for which no function is known may be present simply because they became incorporated in the organism passively—just because they happened to be in the environment. The other possibility is that further research will discover their function.

Only eleven chemical elements make up any appreciable percentage of the tissues of plants and animals. These elements are carbon, hydrogen, oxygen, nitrogen, sulfur, phosphorus, sodium, potassium, calcium, magnesium, and chlorine. The first three of these elements, carbon, hydrogen, and oxygen, are present in virtually all the compounds found in living plants or animals. Nitrogen is also a major component of biochemical compounds; sulfur and phosphorus are present in many instances. Sodium, potassium, calcium, and magnesium are often referred to as mineral elements, and are usually found either as ions or in relatively simple, "inorganic" compounds, although they may be loosely bound to a protein or other biochemical substance. Table 29.1 shows the percentages of the different chemical elements found in the human body.

TABLE 29.1   Chemical elements found in the human body

Element	Percent by Weight	Element	Percent by Weight
oxygen	65	sodium	0.15
carbon	18	chlorine	0.15
hydrogen	10	magnesium	0.05
nitrogen	3.0	iron	0.004
calcium	2.0	iodine	trace
phosphorus	1.0	fluorine	trace
potassium	0.35	silicon	trace
sulfur	0.25		

Although one must begin by discussing the elements found in biochemical substances, these elements are present both as ions and in compounds but primarily in the form of compounds. A real understanding of biochemistry, therefore, must involve a description of compounds. If one were to attempt to answer the question, "What compounds are found in living organisms?", the list of such compounds would be extremely long.

When scientists are confronted with such a long list, they usually look for similarities and differences and, on the basis of these, they develop a system of classification. While no simple system of classification can include all the compounds found in living matter, biochemists have developed a system which includes the large majority of these compounds, with the usual qualifications that some compounds may belong to two or more classifications, and that there be a classification of "miscellaneous."

The most abundant compound in all living organisms is water. The percentage of water varies from well over 90% in some of the simpler aquatic animals to about 50% in a severely dehydrated camel. Various parts of any plant or animal will contain different amounts of water. Muscle tissue is about 70% water, human blood is 80%, and even the proverbially dry bony tissue is about 40% water. The importance and functions of this simple but remarkable compound were discussed in Chapter 9.

If we remove the water from any biochemical substance, we find that the dehydrated material contains three major types of compounds. The relative amounts of these three types of compounds may vary widely, but all plants and animals contain at least small amounts of each, and together these three account for almost all the "dry weight" of a plant or animal.

One of the early attempts to describe these three types of compounds was made in 1827 when William Prout classified food substances as either saccharine, albuminous, or oily. With increasing knowledge came the realization that not only foods, but all animal and plant tissues contained these three types of compounds. We now use the terms **carbohydrates**, **proteins**, and **lipids** to describe these compounds which were originally differentiated so long ago. A fourth type of compound—the **nucleic acids**—is present in much lesser quantities, but is a vital component of living organisms.

Table 29.2 gives the approximate composition of several substances of interest to biochemists.

TABLE 29.2 Approximate composition of materials from plants and animals

	Water	Carbo-hydrate	Protein	Lipid	Nucleic* Acid	Ash†
mammalian muscle (lean)	72–80%	1%	17–21%	5%	0.3%	1%
brain	79	1	8–10	10	0.2	—
hen's egg	74	1	13	11	—	1
cow's milk	87	5	3	4	—	0.7
rat liver	69	4	16	5	0.3	1.4
fish, cod	83	0.4	16	0.4	—	1.2
bananas	75	23	1.2	0.2	—	0.8
rice	12	80	7.6	0.3	—	0.4
peanuts (roasted)	2.6	24	27	44	—	2.7
honey	20	80	0.3	0	—	0.2
potatoes, white	78	19	2	0.1	—	1.0
potatoes, sweet	68	28	2	0.7	—	1.1
spinach	93	3.2	2.3	0.3	—	1.5
beer (4% alcohol)	90	4.4	0.6	0	—	0.2
soybeans, dry	7.5	35	35	18	—	4.7

* Figures for nucleic acid content are, in general, not available for foods and other animal products. It is interesting to note that some viruses, the bacteriophages, may contain from 12 to 61% DNA, and that tobacco mosaic virus has from 5 to 6% RNA.

† The value given as ash indicates the amount of material remaining when all organic matter has been oxidized. It is an indication of the amount of mineral elements present.

Since the carbohydrates, proteins, lipids, and nucleic acids are complex compounds, although made of simple building blocks, it is important to know their structures. Most of the naturally occurring representatives of these compounds, with the exception of the lipids, may be considered to be polymers. The way the basic units of carbohydrates, lipids, proteins, and nucleic acids are linked together has been known for some time. The current emphasis is on the ways the carbon chains or polymer chains are arranged in space—the three-dimensional structure. In this text separate chapters are devoted to the description of the **composition** and **structure** of the **carbohydrates**, the **proteins**, the **lipids**, and the **nucleic acids**.

In addition to substances in these four groups, there are other compounds of major importance in biochemistry. Although these compounds make up less than 1% of most organisms, they may be extremely important for the continued life of the organism. Some of these groups of compounds are the vitamins, the porphyrin-containing compounds such as hemoglobin and chlorophyll, various plant pigments, and the medically important substances known as the alkaloids.

29.3 BIOCHEMICAL REACTIONS

When we understand the composition and structure of the basic biochemicals, we are ready to consider their reactions. The similarities of reactions in various types of organisms are much more striking than are their differences. The term used by biochemists to describe the reactions which take place in living organisms is **metabolism**. Metabolism is subdivided into a "building" aspect called **anabolism** (anabolic reactions), and a process of degradation or "tearing down" of compounds called **catabolism** (catabolic reactions).

Since the basic reactants in metabolic reactions are organic chemical compounds, biochemical reactions are much like organic reactions. There are, however, some important differences. If organic reactions are to proceed at an appreciable rate, it is often necessary that the reactants be heated or that acidic or basic catalysts be added to the reaction mixture. Since high temperatures and acidic or basic conditions almost always result in the death of any living thing, another milder method is utilized to bring about biochemical reactions.

Almost all biochemical reactions, both catabolic and anabolic, are catalyzed by enzymes. An enzyme is itself a protein, and may be defined as a catalyst of biological origin or alternately as a catalytically active protein. A further discussion of enzymes and of the energy relationships of biochemical reactions is given in Chapter 34, "Biochemical Reactions."

29.4 THE UNIT OF BIOCHEMICAL ORGANIZATION —THE CELL

All living things are made of cells. While there is also some extracellular material in most organisms, it is the organization of carbohydrates, proteins, lipids, nucleic acids, and other biochemical compounds in cells that makes life possible. All cells are surrounded by a cell membrane, which contains the cells and isolates them from other cells but still allows communication with the external environment of the cell. The mechanisms by which the cell membrane allows cells to admit certain substances and exclude others are fascinating and the subject of a great deal of research. Cell membranes are discussed in greater detail in Section 32.10.

We generally distinguish two types of cells, the **procaryotic** and the **eucaryotic** cells. Procaryotic cells are found in the simpler organisms, the bacteria and the blue-green algae. There is only a little organization inside the procaryotic cell. Organisms of this type, which are sometimes called **procaryotes**, have no nucleus. Although there is some organization of hereditary and synthetic materials into units, these units are not contained within individual membranes as in eucaryotes. The cell walls of procaryotic cells are much more complex than those of eucaryotic cells. We classify bacteria as being Gram positive $(G+)$ if they become stained when exposed to a mixture of dyes known as Gram's stain. This difference in staining correlates with differences in the biochemical composition of cell walls in $G+$ and $G-$ cell walls. One reason that penicillin and similar antibiotics are effective against certain bacteria is that these antibiotics interfere with the synthesis of bacterial cell walls. Since human cells are contained only in membranes with only minor amounts of cell wall materials, these antibiotics do not harm human cells. We also find that antibiotics that work against Gram positive bacteria are not effective against Gram negative bacteria, and attribute this to the difference in their cell walls.

Most living things are made of **eucaryotic cells**. These cells contain hereditary material within a nuclear membrane in a structure we call the nucleus. Other structural units that we call organelles are also contained within membranes in eucaryotic cells. The various organelles are the sites of different metabolic processes. For instance, the same cell may either synthesize a molecule of glucose into glycogen or catabolize it into carbon dioxide and water. Both reactions can occur in the same cell at the same time because the two processes occur in different parts of the cell separated by membranes. Only a well-organized unit having enzymes that catalyze one series of reactions separated by a semipermeable membrane from the enzymes that catalyze competing reactions is capable of the reactions necessary for life. Pictures taken with an electron microscope, which we call electron micrographs, such as that shown in Fig. 29.3 show the various organelles in a typical eucaryotic cell. Figure 29.4 shows idealized diagrams of our modern concepts of procaryotic and eucaryotic cells.

We can separate subcellular organelles of eucaryotic cells by centrifugation. If we remove the liver from an animal, cut it, and then grind it with sand or disrupt the tissue in a blender or homogenizer, we get a thick semisolid much like a milkshake in consistency. This mixture is called a *homogenate*, and when it is spun in a centrifuge, we can separate the different units or organelles which make up the cells. This separation is possible because the subcellular units are of different densities. Figure 29.5 shows the way such particles would be distributed after a short period of centrifugation. By centrifuging at definite speeds and pouring off the unsedimented fraction (the supernatant fraction), we can obtain relatively pure preparations of cell nuclei, the mitochondria and microsomes. The clear supernatant liquid from which all organelles have been removed is a rich source of enzymes—as are all the other fractions.

The **mitochondrion**, Fig. 29.6, even though extremely small, is a complex unit, the site of many of the metabolic reactions that release and store the energy from foods. It has been postulated that the enzymes which catalyze a series of reactions are arranged in sequence on the cristae (inner membranes) of mitochondria. Thus a molecule may be metabolized by passing from one enzyme to another on a mitochondrial assembly or, perhaps more correctly, a disassembly line.

The name **microsomes** was given to a preparation isolated by centrifuging a cell homogenate at a high speed after the nuclei and mitochondria had been removed. The microsome fraction contains the ribosomes and some of the threads of material that tie these small spherical bodies together into the endoplasmic reticulum. The *ribosomes*, so named because they are rich in ribonucleic acids (RNA), are important in the synthesis of proteins by the cell.

By other techniques we can isolate **lysosomes**, which contain enzymes that catalyze

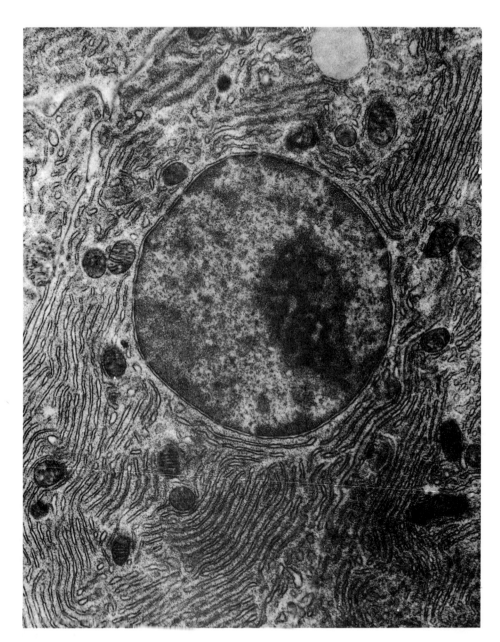

Fig. 29.3 The nucleus and part of the cytoplasm of a cell from the pancreas of a bat (magnified 18 000 times). The photograph, called an electron micrograph, was taken using an electron microscope. It shows clearly the double-layered nuclear membrane, ribosomes lining channels that probably serve for secretion, and many mitochondria. [Photograph courtesy of Dr. Don W. Fawcett.]

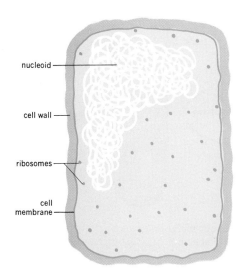

nucleoid

cell wall

ribosomes

cell
membrane

(a) cross-section of a typical procaryotic cell

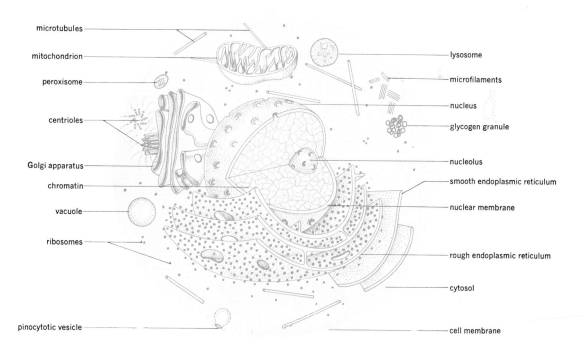

microtubules

mitochondrion

peroxisome

centrioles

Golgi apparatus

chromatin

vacuole

ribosomes

pinocytotic vesicle

lysosome

microfilaments

nucleus

glycogen granule

nucleolus

smooth endoplasmic reticulum

nuclear membrane

rough endoplasmic reticulum

cytosol

cell membrane

(b) three-dimensional diagram of a typical eucaryotic cell

Fig. 29.4 Drawings of typical cells. (a) Typical cell of a procaryote such as a bacterium. Note the absence of a nucleus and the thickened wall. (b) Idealized drawing of the cell of a eucaryote. Note the various organelles, not all of which are observed in every cell. See Fig. 32.3 for a detailed drawing of a cell membrane.

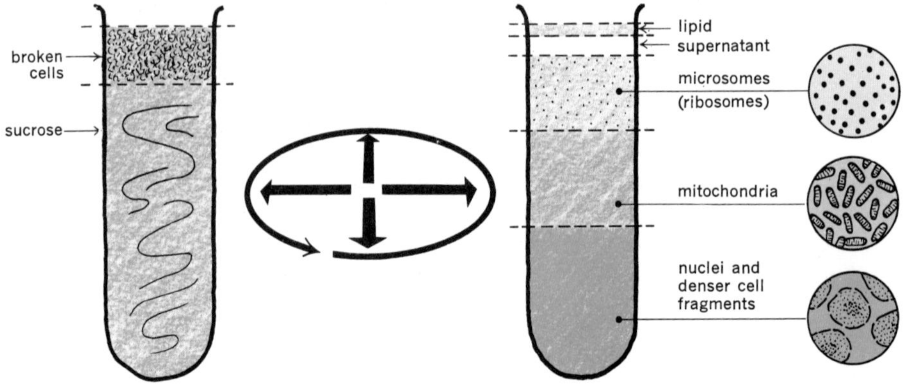

broken cells

sucrose

lipid
supernatant
microsomes
(ribosomes)

mitochondria

nuclei and
denser cell
fragments

Fig. 29.5 Diagram showing separation of subcellular fractions by centrifugation. [Adapted from an illustration of a centrifuge by the Fisher Scientific Company.]

the decomposition of many important compounds in the cell. It is believed that these "suicide bags" are ruptured after the death of a cell and are responsible for the process of self-destruction—*autolysis*—of dead cells.

The cells of green plants contain very complex organelles called chloroplasts. These organelles, which contain the green pigment chlorophyll, are the site of many of the reactions of photosynthesis. Although the cells of green plants are eucaryotic, many have heavy, complex cell walls like procaryotic cells.

Many of the differences in reactions carried on by different types of cells and particularly the reactions that distinguish whether an organism is a plant or an animal depend on the organelles present. The major differences in procaryotes and eucaryotes involves the presence or absence of organelles and the differences in cell walls. In the following chapters, both the structure of biochemical compounds and the reactions they undergo will be correlated with the type of cell and the organelles involved whenever possible.

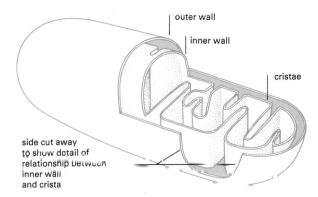

outer wall

inner wall

cristae

side cut away
to show detail of
relationship between
inner wall
and crista

Fig. 29.6 Cross section of a typical mitochondrion. This idealized view is based on electron micrographs similar to that shown in Fig. 29.3. There are many particles attached to the inner membrane (the cristae). These particles may play a role in formation of ATP.

All types of cells, though they differ in structure, carry on many of the same reactions. In many instances different eucaryotic cells work together to coordinate a series of reactions or a process. In more complex organisms such as the human, cells are combined with other similar cells having a similar function in an organization called a **tissue**. Tissues make up **organs** and a collection of organs is necessary for a complex **organism**. Just as a group of communicating individual persons make a society, cells having differing degrees of communication make up a living plant or animal. Even the unicellular bacteria communicate with other organisms through the release of substances into their environment, although the communications are often in the form of substances that destroy other organisms. Communication between eucaryotic cells by means of hormones and other substances helps to coordinate the actions of cells or tissues. This communication is easier because such cells lack cell walls. Failure of normal communications between eucaryotic cells may have drastic consequences. It might be mentioned that similar failures in communication between individual humans may also be disruptive. Some human communication seems to resemble that of procaryotes which release into the environment substances that harm or destroy other organisms.

SUMMARY

In this chapter we have traced briefly the beginnings of biochemistry and have given some answers to the question, "What is biochemistry?" We have said that biochemistry is the study of the composition, structure, and reactions of substances found in living systems. We have also said that the purpose of biochemistry is the explanation of living processes in physical and chemical terms. Another possible definition of biochemistry might be, "Biochemistry is the things a biochemist does." While this is less satisfactory as a formal definition of biochemistry, perhaps it is the best one. For biochemistry is still a growing, developing science. What it will be a few years from now depends on what the biochemists and their colleagues who call themselves biophysicists, molecular biologists, or cytochemists find interesting enough to experiment with and to speculate about. Like the living things with which it deals, biochemistry is changing. If we would really understand it, we must know not only what it seems to be today, but what it has been and the direction in which it seems to be going.

Biochemistry is a fascinating science because it is the chemistry of living things—plants and animals. Most of us are primarily concerned about one particular kind of animal—the human. There are, of course, many ways of answering the question, "What is a human?" The biochemical answer presents a picture in which some areas are well defined and often highly complex, while others are still hidden by the mists of ignorance. Nevertheless, the whole picture makes a beautiful kind of sense. In the next several chapters we will try to show you this biochemical way of looking at life. It is not a view that can be really understood after a brief, superficial glimpse. Some concentration on details is demanded if you would eventually see at least a bit of the grand design of life as seen by the biochemist.

IMPORTANT TERMS

anabolism	enzyme	microsome
biochemistry	in vitro	mitochondrion
catabolism	in vivo	organ
centrifugation	isolation artifacts	organism
chromatography	lysosome	ribosome
electrophoresis	metabolism	tissue

SUGGESTED READING

Biochemistry Texts

The reader may wish further details on many of the items discussed throughout the section on biochemistry. The following list represents a selection of basic biochemistry texts containing different degrees of detail.

Conn, E. E., and P. K. Stumpf, *Outlines of Biochemistry*, 4th ed., John Wiley, New York, 1976. An excellent textbook. The presentation is more comprehensive than this book, but less than that in the books by Lehninger or White *et al.*

Lehninger, *Biochemistry*, 2nd ed., Worth Publishers, New York, 1975. A major biochemistry text (1104 large size pages). Full of facts, well presented.

White, A., P. Handler, E. L. Smith, R. L. Hill, and I. R. Lehman, *Principles of Biochemistry*, 6th ed., McGraw-Hill, New York, 1978. A good, extensive biochemistry text written primarily for medical students.

Biology Texts

Two books that may be used for obtaining additional information for the study of biochemistry are given below. The book by Baker and Allen is highly oriented toward biochemistry. Kimball's book provides a somewhat more traditional approach to biology without neglecting biochemical aspects. Both are excellent, well-illustrated texts.

Baker, J. J. W., and G. E. Allen, *The Study of Biology*, 3rd ed., Addison-Wesley, Reading, Mass., 1977.

Kimball, J. W., *Biology*, 4th ed., Addison-Wesley, Reading, Mass., 1978.

Other Reading

The following references are both more general and, in some instances, more specific.

Brachet, J., "The living cell," *Scientific American*, September 1961, p. 50 (Offprint #90). The key article in an issue devoted to the topic of cells.

deDuve, G., "The lysosome," *Scientific American*, May 1963, p. 64 (Offprint #156). A discussion of a cell organelle which sometimes digests the cell itself.

Green, D., "The mitochondrion," *Scientific American*, January 1964, p. 63 (Offprint #175). A discussion of the molecular architecture and function of this important cellular component.

Kennedy, D., *The Living Cell*, W. H. Freeman, San Francisco, 1965. The book includes twenty-four articles reprinted from the *Scientific American*, nine of them from the Sept., 1961 issue devoted to the cell.

30

Chemistry of the Carbohydrates

When we heat a sugar in a test tube we observe several successive changes. The sugar melts and then begins to darken until finally, after prolonged heating, there is a black residue, which seems to be carbon, in the tube. Water condenses on the upper walls of the tube during the heating. From this experiment we might deduce that sugar can be decomposed into carbon and water, so that we could classify sugar as a hydrate of carbon or a carbohydrate. If we found a variety of compounds which behave similarly when heated, we could classify these compounds as carbohydrates. Chemical analysis of sugar and similar compounds would show that sugar and the other carbohydrates are composed only of carbon, hydrogen, and oxygen, and that their formulas can be expressed as $C_x \cdot (H_2O)_y$. In many cases x would be equal to y ($C_6H_{12}O_6$) while in others ($C_{12}H_{22}O_{11}$) accurate analysis would be required to show that x and y were not really equal.

Relatively early in the development of biochemistry, as the result of experiments like those listed above, certain compounds were classified as carbohydrates. Further observations have shown, however, that it is inaccurate to describe those compounds we know as carbohydrates as "hydrates of carbon." Not only do the hydrogen and oxygen atoms of carbohydrates fail to exhibit the properties of water of hydration, but the way these compounds react leads to the conclusion that there are alcohol and either aldehyde or ketone functional groups in the carbohydrate molecule. The modern definition for the classification known as **carbohydrates** is that the compounds are polyhydroxy aldehydes or ketones. This definition is usually expanded to include polymers and other compounds derived from, or closely related to, the polyhydroxy aldehydes and ketones.

A nutritionist often describes the carbohydrates as the sugars and starches. This description is incomplete, however, since it does not include the many polysaccharides other than starch. Sugars, which biochemists usually define as sweet-tasting carbohydrates, are the simplest carbohydrates. Their names usually contain the suffix -*ose*. The starches, which may be defined as the complex, granular or powdery carbohydrates found in seeds, bulbs, and tubers of plants, represent only one group of complex carbohydrates—the polysaccharides—which includes glycogen, cellulose, the pectins, and many other compounds. Carbohydrates are usually classified as follows: (1) monosaccharides, (2) derived or related compounds, (3) oligosaccharides, and (4) polysaccharides.

**30.2
MONO-
SACCHARIDES**

The **monosaccharides** may be described as those carbohydrates that cannot be further degraded by hydrolysis. Compounds containing three carbon atoms, the *trioses*, are the simplest monosaccharides.

**Names and
Structures**

The general names for the monosaccharides are formed from a prefix indicating the number of carbon atoms and the suffix *-ose* which we have already mentioned. Thus we have trioses, tetroses, pentoses, hexoses, and heptoses containing 3, 4, 5, 6, and 7 carbon atoms, respectively. Those monosaccharides which contain an aldehyde group are **aldoses**, those with a ketone group, **ketoses**. The indication that a sugar contains a ketone group may also be made by including the letters *-ul-* in the name of the compound. Thus we may speak of *aldopentoses*, *aldohexoses*, *ketoheptoses* and *heptuloses*, the latter two terms being different ways of naming a 7-carbon, keto sugar.

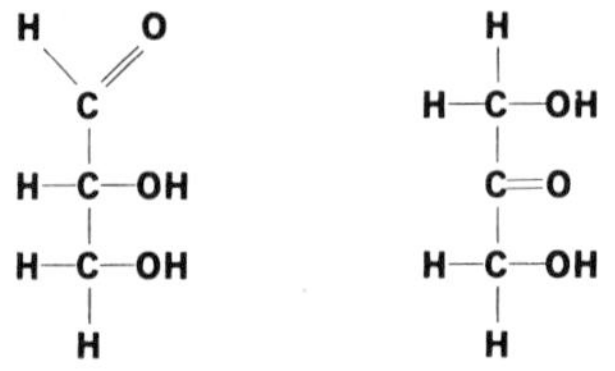

D-glyceraldehyde   dihydroxyacetone

The two trioses are glyceraldehyde, which may exist in either of two optically active forms, and dihydroxyacetone. Glyceraldehyde is especially important because most of the common sugars may be considered related to it by addition of carbon atoms to lengthen the three-carbon chain, although biosynthesis of sugars does not follow this course. Sugars are designated as D- or L-sugars, depending upon whether the chiral (asymmetric) carbon atoms farthest from the aldehyde or ketone group are similar to the D-form or the L-form of glyceraldehyde. In the structures shown below, chiral carbons are shown in color. (Optical activity and chiral carbon atoms were discussed in Sections 27.5 and 27.7.)

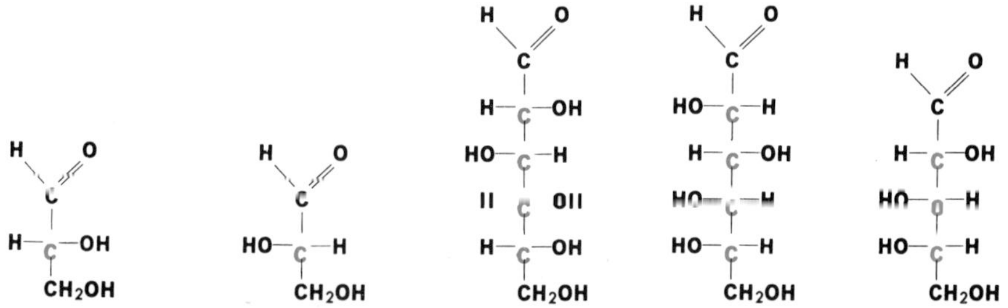

D-glyceraldehyde   L-glyceraldehyde   D-glucose   L-glucose   L-arabinose

Plane-polarized light is rotated to the right in solutions of D-glyceraldehyde and to the left in solutions of L-glyceraldehyde; however, compounds which are related to these parent compounds may have different rotations since they usually contain other chiral carbon atoms. For instance, *glucose* has four chiral carbon atoms and arabinose has three. Solutions of the ketohexose, *fructose*, which is related to D-glyceraldehyde and is thus D-fructose, rotate light to the left. To designate the actual rotation of polarized light we use a plus (+) sign to describe those solutions in which light is rotated to the right and a minus (−) sign for solutions in which the rotation is to the left. The terms dextrorotatory and levorotatory are also used, but + and − are preferable. Thus we describe the common isomer of fructose as D(−)fructose, the D indicating that its configuration is related to D-glyceraldehyde and the (−) indicating that plane-polarized light is rotated to the left in its solutions.

The monosaccharides that are related to D-glyceraldehyde are shown in Fig. 30.1. The most abundant monosaccharides are the hexoses and the pentoses. Glucose is the most widely found hexose, and there are smaller amounts of galactose and mannose present in natural products. The three pentoses xylose, arabinose, and ribose are also found, the first two principally in plant products. Ribose is found in both plants and animals. In most cases the monosaccharides are not found as such but as components of disaccharides, polysaccharides, or other compounds.

The ketoses which are related to dihydroxy acetone are not so widely found as the aldoses. The formulas for the most abundant ketoses are shown in Fig. 30.2.

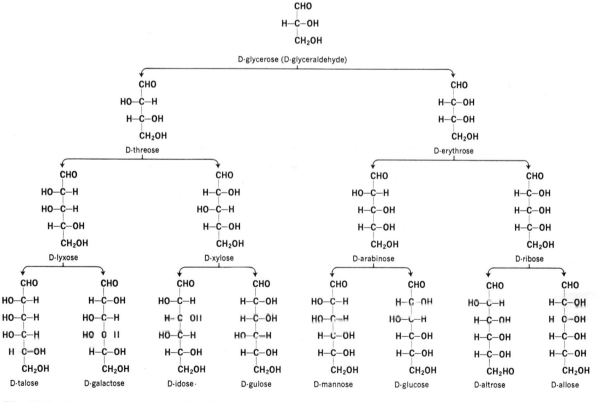

Fig. 30.1 Structures and names of the D-aldoses.

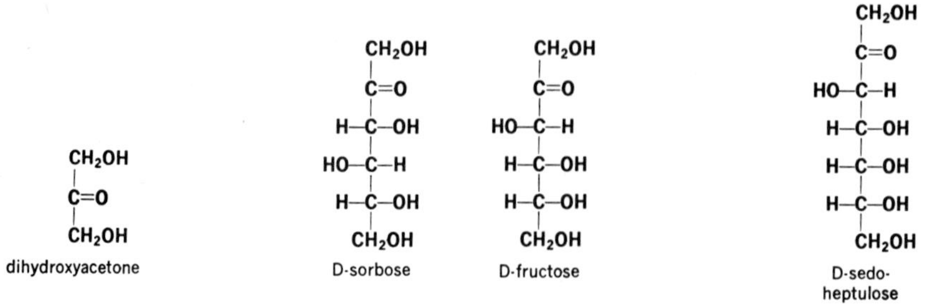

Structures in the figure:

dihydroxyacetone:
CH₂OH
C=O
CH₂OH

D-sorbose:
CH₂OH
C=O
HO—C—H
H—C—OH
H—C—OH
CH₂OH

D-fructose:
CH₂OH
C=O
HO—C—H
H—C—OH
H—C—OH
CH₂OH

D-sedoheptulose:
CH₂OH
C=O
HO—C—H
H—C—OH
H—C—OH
H—C—OH
CH₂OH

Fig. 30.2   Structures and names of some D-ketoses.

In discussing the structure and reactions of the monosaccharides we often need to designate a specific carbon in the chain. As with other organic compounds we give each carbon a number, with the carbonyl carbon being assigned the lowest possible number. Thus the carbonyl carbon is C-1 in glucose but C-2 in fructose. The chiral carbon that establishes whether the monosaccharide (saccharide) is D or L in both molecules is C-5, but it is C-4 in pentoses such as ribose.

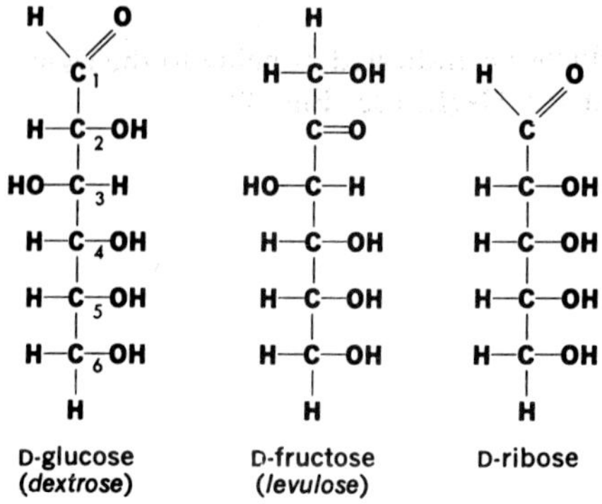

D-glucose
(*dextrose*)

D-fructose
(*levulose*)

D-ribose

The monosaccharides are given names which generally reflect the source or a property of the sugar.

The name *glucose*, which is derived from a Greek word for sweet wine, reflects the fact that this sugar was originally isolated from grapes. The alternative name, *dextrose*, resulted from the observation that solutions of this sugar rotate polarized light to the right. The name *fructose* is derived from the Latin word for fruit, *fructus*; and the other common name for this sugar, *levulose*, indicates its levorotatory property. *Galactose* (a component of the disaccharide lactose) is found in milk and owes its name to the Greek prefix *galact-*, meaning milk. *Lactose* is derived from the Latin prefix *lact-*, meaning milk. The name *mannose* reflects the fact that its related hexahydric alcohol, *mannitol*, was found in the dried juice of the manna tree. *Arabinose* is produced from gum arabic, an exudate from certain types of acacia trees. Apparently the name *ribose* was coined by selection and rearrangement of the letters in arabinose, which is appropriate, since the two are isomers. *Xylose* is found in woody parts of plants; its name is derived from the Greek word *xylon*, meaning wood. The name *sucrose* is derived from the French word for sugar, *sucre*, and *maltose* is named for its source, malt, which is formed in germinating cereals. The word *saccharide* is derived from words for sugar that are similar in Arabic, medieval Latin, and Greek.

The formulas given so far for the monosaccharides are called *open-chain formulas*. They are also known as Fischer formulas since they were proposed by the great German chemist Emil Fischer, who was responsible for establishing the configuration of many carbohydrates. Further studies of the monosaccharides, especially the hexoses and pentoses, have indicated that they do not have all of the properties of aldehydes. In particular, the change in optical rotation observed when we make solutions of glucose and similar sugars indicates that they are not just open-chain aldehydes. Freshly made solutions of glucose that have been crystallized from different solvents or solvent mixtures may have specific optical rotations ranging from $+18.7°$ to $+112.2°$. If we observe the optical rotation of these solutions we find that it changes and finally reaches an equilibrium value of $52.7°$. This change in rotation is called mutarotation and it is observed for most sugars. As a result of mutarotation and other properties, we believe that the carbonyl group of a carbohydrate reacts with one of the alcohol groups to give a hemiacetal (Section 26.8). In this reaction, the H from the alcohol goes to the carbonyl oxygen, converting it to an —OH group and forming an oxygen bond between the carbonyl carbon and the carbon that contained the alcohol group. Since the carbonyl group and the alcohol which react are on the same chain, we get cyclic compounds as shown in Fig. 30.3. In this reaction the carbonyl carbon becomes a chiral carbon, and two possible orientations of the newly formed —OH group are possible. Instead of giving these two forms totally new names, we call them the alpha and beta (α and β) forms of the sugar. By convention, if the —OH at C-1 of D-glucose is indicated as being to the right, we have α-D-glucose, and if it is to the left we have β-D-glucose (Fig. 30.3).

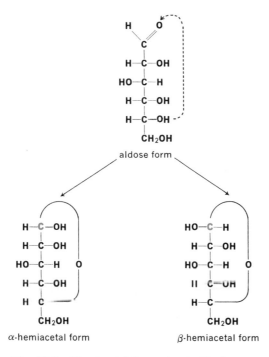

Fig. 30.3 Hemiacetal formation in D-glucose, resulting in ring closure.

It is possible to write hemiacetals in the Fischer representation, but the extremely distorted long bond, shown between carbons 1 and 5, makes us prefer writing these structures in Haworth representations, which are shown below.

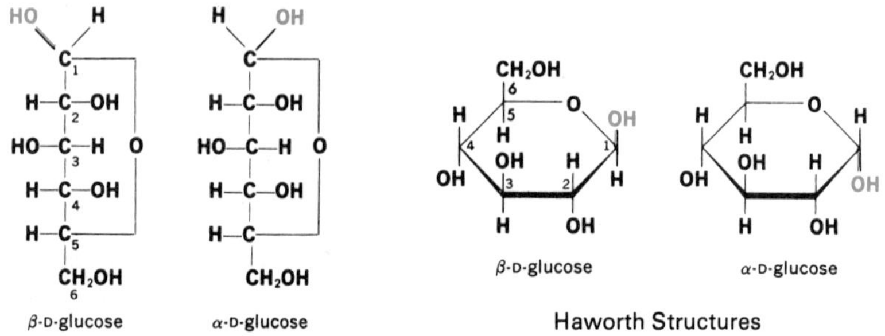

β-D-glucose   α-D-glucose   β-D-glucose   α-D-glucose

Haworth Structures

In the Haworth representations, the substituents written to the right in Fischer formulas are written as being below the ring, and those to the left are written as being above the ring. When the oxygen of the alcohol group becomes involved in ring formation, the ring is twisted so that the CH$_2$OH group is written above the ring for D-glucose and other D-isomers. (See Fig. 30.4.)

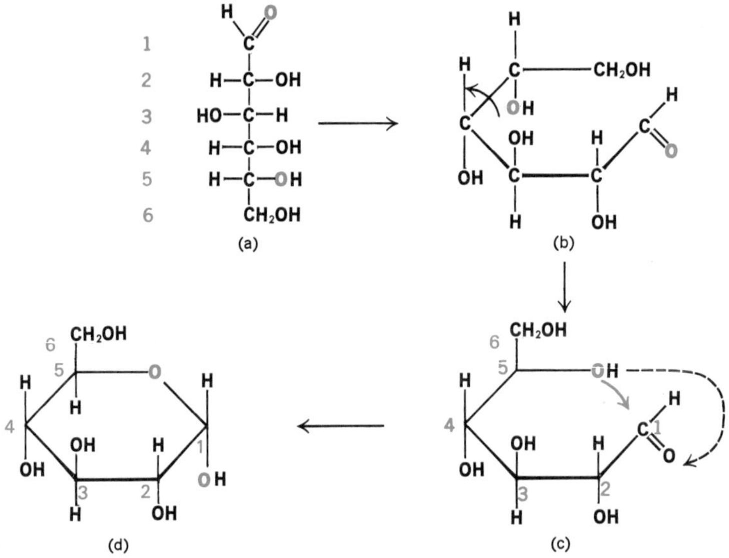

Fig. 30.4 Ring formation in D-glucose, shown in perspective for α-D-glucose.

The Haworth representations for the most common monosaccharides are given on page 543. Note that fructose and ribose are written as five-membered rings. They are found in these forms in natural products. Actually, glucose and other hexoses can also form five-membered rings, and fructose can form a six-membered ring if the alcohol attached to the C-6 atom is involved. The forms shown are those found in the greatest amount in natural products. In solutions of the respective monosaccharides, six-membered rings predominate even for fructose and ribose.

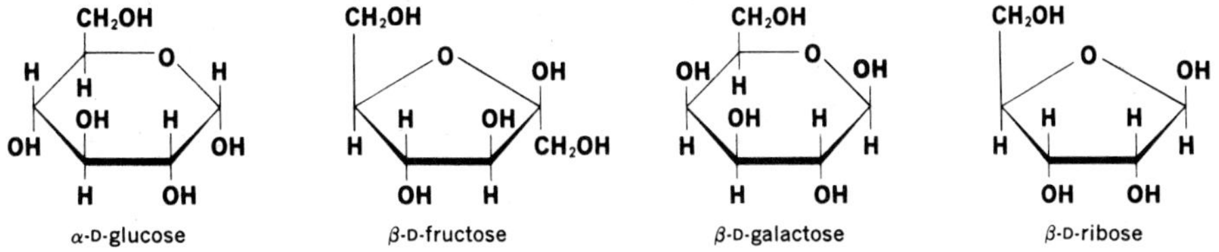

α-D-glucose   β-D-fructose   β-D-galactose   β-D-ribose

Hemiacetals of monosaccharides are more stable than the open-chain forms; however, in solutions they exist in equilibrium with small amounts of the open-chain aldehyde. Any solution of glucose is a mixture containing α and β forms as well as some of the open-chain aldehyde, as shown in Fig. 30.5. At equilibrium there is about 64% of β-D-glucose and 36% α-D-glucose, with less than 1% in the open-chain form. Crystals of glucose also have the hemiacetal form. In fact, it has been impossible to get crystals of glucose that contain the open-chain aldehyde. Solutions made from crystals of β-D-glucose have a specific rotation of +18.7°, and freshly made solutions of α-D-glucose crystals have a rotation of +112.2°. The equilibrium mixture has a rotation of +52.7°. It is the conversion of either the α- or β-form to the equilibrium mixture that is responsible for the phenomenon of mutarotation. (See Fig. 30.5.)

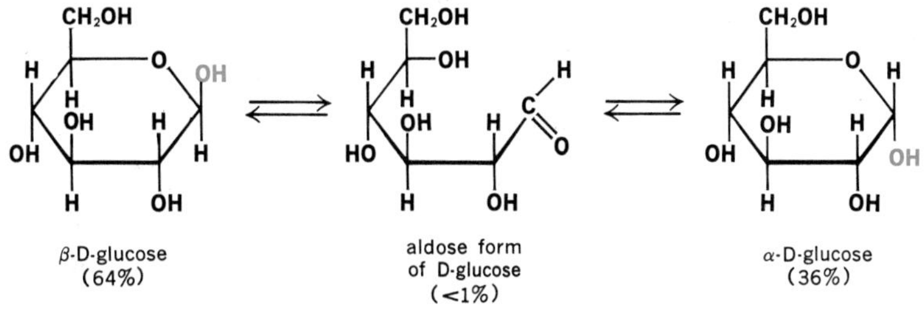

β-D-glucose   aldose form   α-D-glucose
(64%)   of D-glucose   (36%)
(<1%)

Fig. 30.5 Equilibrium of different forms of D-glucose.

We have chosen, for reasons of simplicity, to use Fischer and Haworth formulas for carbohydrates. We can also write such formulas in conformational structures. This method takes into account the fact that the structures of the carbohydrate ring are not flat. The formulas for some common sugars using these structures are given below.

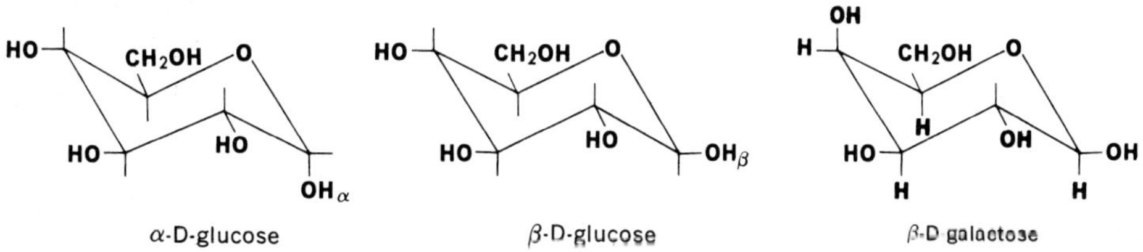

α-D-glucose   β-D-glucose   β-D galactose

Although conformational structures are more accurate representations than other forms, they are not commonly used in textbooks, because of their complexity.

These conformational structures show that most of the larger groups that are attached to the ring (—OH and —CH₂OH groups) are equatorial. This distribution means that there is more space for these large groups than if they were axial. The —OH at carbon-1 is equatorial in β-D-glucose and axial in the α form. Since molecules tend to assume the form in which large components have the maximum space surrounding them, the β form of D-glucose is the molecular form present in the greatest percentage at equilibrium (see Fig. 30.5).

It should be understood that all written formulas are representations of molecules, and the representation chosen depends on the situation. For the purposes of this book the Haworth formulas are the best representation, but occasionally we will use open-chain, or Fischer, formulas when the exact ring structure is not specified, or to explain chemical reactions. For instance, many of the reactions of monosaccharides are the reactions of free aldehyde groups and are best shown with Fischer formulas. Although the amount of the open-chain form is small in solutions of glucose, the law of mass action tells us that the removal of a molecule of any form from an equilibrium mixture will result in formation of more of that form. Continued reaction with the open-chain form, as in oxidation of an aldehyde group, can result in the oxidation of all of the glucose.

Occurrence and Preparation of Monosaccharides

The monosaccharides as such are not found in any appreciable quantity in nature. A monosaccharide is usually found combined with other monosaccharides in the form of polymers called *polysaccharides*, such as starch or cellulose. In some cases, monosaccharides are combined with compounds other than carbohydrates, as ribose is in ribonucleic acid (RNA).

Many fruits, berries, and plant juices do, however, contain some free monosaccharides. Glucose was first found in grapes, which may contain as much as 27% glucose. However, most fruits contain less glucose, and animals contain very little. The proportion of glucose in human blood, for example, is approximately 100 mg/100 ml. Even though the concentration is low, it must be maintained very near this level; persons with levels below normal go into *hypoglycemic* (low blood sugar) shock, and *hyperglycemia* (excess sugar in the blood) can result in coma.

Fructose, like glucose, is found in small quantities in many fruits. Honey contains relatively large amounts of both glucose and fructose. The other monosaccharides are generally found as components of polysaccharides.

Monosaccharides may be isolated from their natural sources by various physical methods. They are more commonly prepared by the hydrolysis of polysaccharides. Commercial preparation of glucose, for instance, is accomplished by the hydrolysis of starch. Although most of the monosaccharides have been synthesized in small amounts for theoretical purposes, such as the need to prove a certain structure or to establish the relationship of a given monosaccharide to other compounds, chemical synthesis is not commercially profitable.

Properties of Monosaccharides

Monosaccharides are colorless, crystalline solids. Crystallization is often difficult because of the presence of small amounts of impurities in the original solution. Since sugars are so similar to one another in chemical and physical properties, it is hard to separate them, and a small amount of fructose present as an impurity in a solution of glucose prevents formation of glucose crystals.

Both the carbonyl group and the alcohol groups on monosaccharides makes them soluble in water and insoluble in nonpolar solvents. Solutions of sugars are viscous and have a sweet taste. While most of the common monosaccharides belong to the D category, the actual (observed) rotation may be either to the right ($+$) or to the left ($-$). Measuring the degree of rotation of plane-polarized light by a solution is a technique that may be used to find the amount of a sugar in the solution, provided only one sugar (or only one optically active compound) is present.

The chemical properties of monosaccharides are those of the aldehydes and the alcohols. Heat, catalysts, and strong acids, bases, or other reagents are usually required if we wish to observe these chemical properties in the laboratory.

Oxidation of
Monosaccharides

Like simple aldehydes (Section 26.6), monosaccharides can be oxidized. In the laboratory hot alkaline solutions of metallic ions are often used as oxidizing agents, and in the process the metallic ions are reduced. Since the sugars are the reducing agents, they are often referred to as **reducing sugars.** Such reactions are used to identify monosaccharides. Fehling's and Benedict's tests, for example, use an alkaline solution of copper(II). In both cases the test is positive if it results in a yellow or red compound of copper(I). Tollens' test uses a solution containing silver ions (Ag^+). A test is considered positive if a silver mirror forms on the wall of the test tube. In fact, before aluminum was used as the reflective substance for mirrors, a reaction similar to Tollens' test was used in the manufacture of mirrors.

Although ordinary ketones are not oxidized by Tollens', Benedict's, and Fehling's reagents, ketones with an —OH on the carbon next to the ketone carbon are oxidized. Thus keto sugars such as fructose give a positive test with the reagents listed. The reactions of aldehydes with ions of copper and silver are shown in Eqs. (26–8) and (26–9). These reactions, particularly that with copper ions, are used to test for the presence of glucose in either the urine or blood.

Oxidations of monosaccharides in living systems are catalyzed by enzymes. Products in which the C-1, the C-6, and both the C-1 and C-6 are oxidized to an acid form are known. The oxidation products of glucose are shown below. Glucaric acid has the common name saccharic acid.

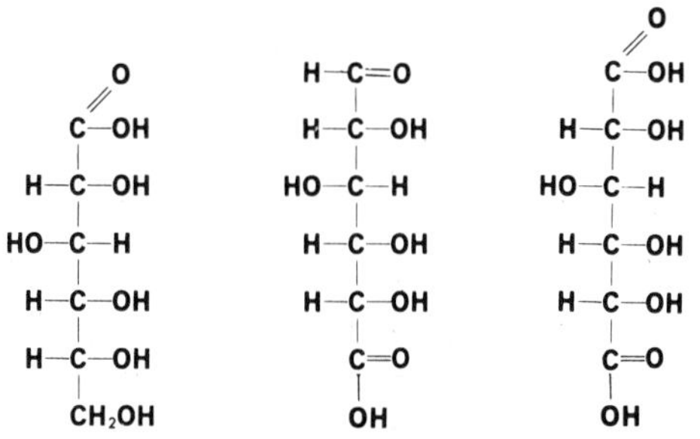

gluconic acid   glucuronic acid   glucaric acid

Glucuronic acid is found in a hemiacetal form. Gluconic and glucaric acids can be found in lactone forms which result from the reaction of the acid group with an alcohol group, forming an intramolecular ester group. As with hemiacetals, different ring structures may be formed; representative ones are shown below.

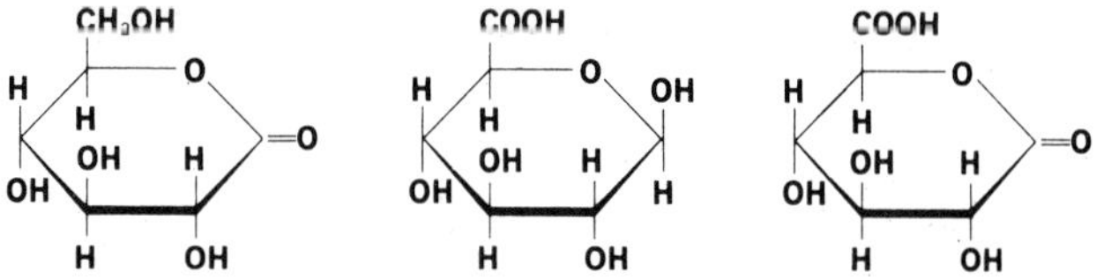

(lactone form of   glucuronic acid   glucaric
gluconic acid)   β-form   (saccharic) acid
                                                   (lactone form)

Similar compounds are formed by oxidation of other monosaccharides. Galactose gives galactonic, galacturonic, and galactaric (mucic) acid. As with glucose, the dicarboxy compound has a widely used common name—mucic acid. As with other organic compounds, complete oxidation of monosaccharides, whether by burning in the laboratory or in being metabolized by an animal, yields carbon dioxide and water.

Reduction of Monosaccharides The aldehyde or ketone group of monosaccharides can be reduced to give a hexahydric alcohol. Reduction of either glucose gives glucitol, which is primarily known as its common name, sorbitol. Reduction of the pentoses xylose or xylulose gives xylitol.

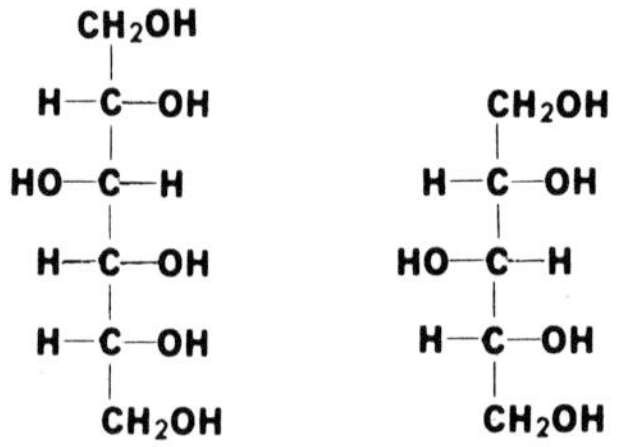

glucitol
(sorbitol)

xylitol

Glucitol and the other sugar alcohols such as mannitol (from mannose) and dulcitol (from galactose) are found in natural products and are also synthesized from sugars commercially. Sorbitol (glucitol) has many uses. Nearly 1 000 tons of ascorbic acid (vitamin C) are made each year via processes that start with sorbitol. It is also used as a humectant or moisturizing agent for leather, tobacco, and in printing processes. Sorbitol is about 60% as sweet as sugar and is used as a sweetening agent. Although it is metabolized by the body to yield energy, only a small part of sorbitol is actually converted to glucose in the process. This means that sorbitol can be used by diabetics as a sweetening agent.

Xylitol is also a sweetening agent and has recently been used in chewing gum, since mouth bacteria that cause tooth decay cannot use it as a source of energy. Preliminary reports that xylitol may cause cancer have, however, led to a decrease in its use as a sweetener.

Formation of Glycosides As we have seen, the reaction of a carbonyl group of a monosaccharide with one of the alcohol groups on the chain gives a hemiacetal. Reaction of the hemiacetal with a second alcohol group, either of a simple alcohol or one of another monosaccharide, gives an acetal and a molecule of water. These acetals, which are known as **glycosides**, are much more stable than the hemiacetals. Those glycosides that are made of two hexoses are important compounds—the disaccharides. The combination of two molecules of glucose forms the disaccharide maltose.

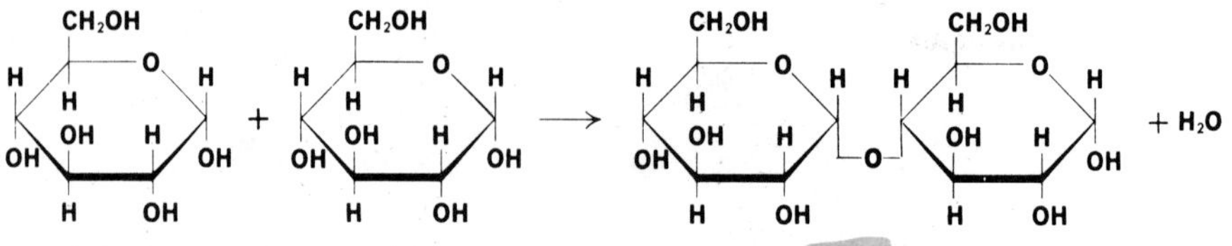

α-D-glucose   α-D-glucose   α-maltose

In this reaction, it is the C-1 of one molecule of glucose that has reacted with the —OH on the C-4 of the other. We call this a 1–4 linkage. We also find that the C-1 that has reacted was in either α or β form. Once the **glycosidic linkage** has been formed, it is fixed as either α or β, since the acetal linkage is stable. It can be hydrolyzed in water and acid, of course, but this linkage will not rearrange in water solution.

If the C-1 of the glucose molecule that formed the disaccharide linkage is in the β form, the disaccharide cellobiose is formed.

β-cellobiose

When speaking of the structure of disaccharides and other glycosides we need to specify the type of linkage and the carbon atoms that form the linkage. The linkage is alpha (α) in maltose and beta (β) in cellobiose. *In vitro* synthesis of glycosides gives a mixture of both α and β linkages. In living systems the formation of glycosides is catalyzed by enzymes and usually only one form (either α or β) is made. While the bond between monosaccharide units is fixed, the free hemiacetal OH group at position 1 of the disaccharide is not. Thus both maltose and cellobiose exist in both α and β hemiacetal forms.

The most common linkage between monosaccharide units is from the C-1 of one unit to the C-4 of the other. Linkages that are 1–6, 1–3, and 1–2 are also found. These linkages can be either α or β, depending on the configuration of the C-1 that formed the linkage.

We sometimes use the more specific terms *glucoside* or *galactoside* to indicate glycosides of these sugars. Maltose is a glucoside and lactose, which is a disaccharide containing a galactose joined β 1–4 to glucose, is a galactoside or, more specifically, a β-galactoside. Enzymes that catalyze the hydrolysis of lactose and similar compounds are called β-galactosidases.

Since the formation of glycosidic bonds involves the loss of a molecule of water and the combination of two large (monosaccharide) units, it is often called a condensation reaction. Continuation of the condensation process involving monosaccharides results in the formation of trisaccharides, tetrasaccharides, and eventually polysaccharides.

While the most abundant glycosides are produced by the reaction of monosaccharide units to give the disaccharides and polysaccharides, monosaccharides can also react with other alcohols. The reaction of glucose with methanol gives a methyl glucoside. If the reaction is performed *in vitro* both the α and β isomers are formed.

D-glucose   methyl α-D-glucoside

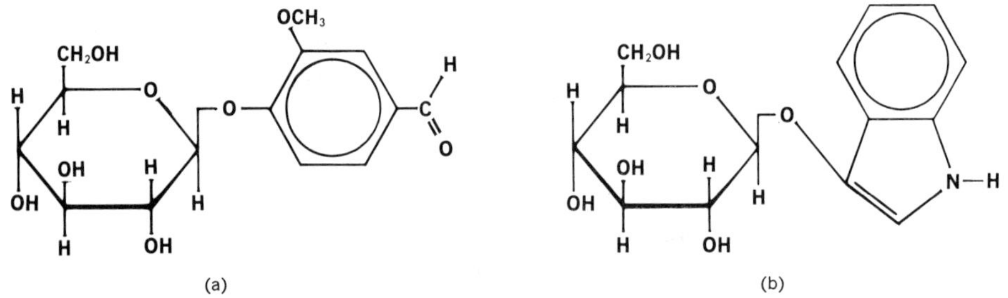

Fig. 30.6 Two naturally occurring glycosides. (a) The β-D-glucoside of vanillin. (b) Indican, the β-D-glucoside of 3-hydroxyindole.

There are many naturally occurring glycosides. Vanillin is an aldehyde containing a phenolic-OH group (Table 26.6) and in vanilla beans it is linked through its —OH group to glucose (Fig. 30.6). Preparation of vanilla extract involves the enzyme-catalyzed hydrolysis of the glucoside to yield free vanillin, which is then extracted with ethanol. The blue dye indigo is produced from a naturally occurring glucoside, indican (Fig. 30.6). The glucoside is hydrolyzed, and oxidation of the 3-hydroxyindole formed produces the bright blue dye, indigo.

Amygdalin, which can be obtained from almonds and the pits of peaches and apricots, is commonly called laetrile. It is a glycoside containing two glucose units linked β 1–6 which are then linked to a mandelonitrile group, as shown here.

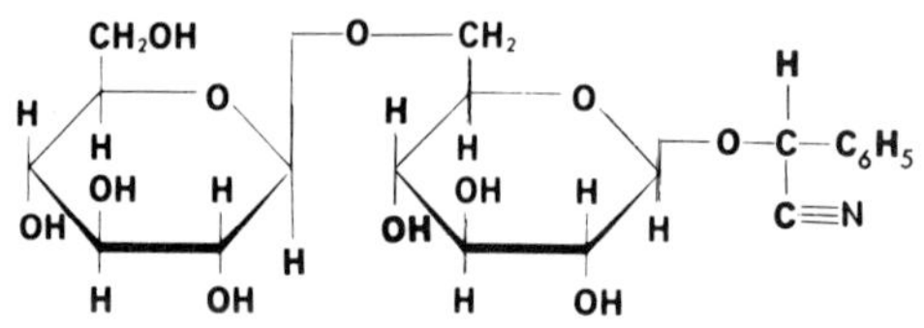

Some people believe that this compound is absorbed by cancerous tumors and that the —CN group which is then released by enzymes within the tumors will result in the death of the tumor. Most scientists question these beliefs, and as this book is being written, clinical studies are being initiated to test the effectiveness of this glycoside. It is known that people who consume too much amygdalin suffer from cyanide poisoning.

The drug digitalis, which is derived from the leaves of the purple foxglove plant, contains a complex glycoside made of five monosaccharide units containing 1–2, 1–3, and 1–4 linkages. The alcohol component is a sterol, digitogenin. Small amounts of digitalis are effective in treating heart conditions. Larger amounts, such as those consumed when a child eats the leaves of the foxglove plant, can cause death.

The human body forms compounds of *glucuronic acid,* known as **glucuronides**, with the —OH or —COOH groups of many compounds. Since many of the compounds that form glucuronides are foreign to the body and are toxic, we often called this process detoxication. In these reactions with glucuronic acid a phenolic compound gives a glycosidic or ether type of linkage, and benzoic acid or other similar acids give esters.

In addition to tying up a potentially toxic molecule, glucuronides are generally more soluble in body fluids and are easily excreted from the body. Since glucuronides are also formed with nontoxic compounds, some people object to the term detoxication and

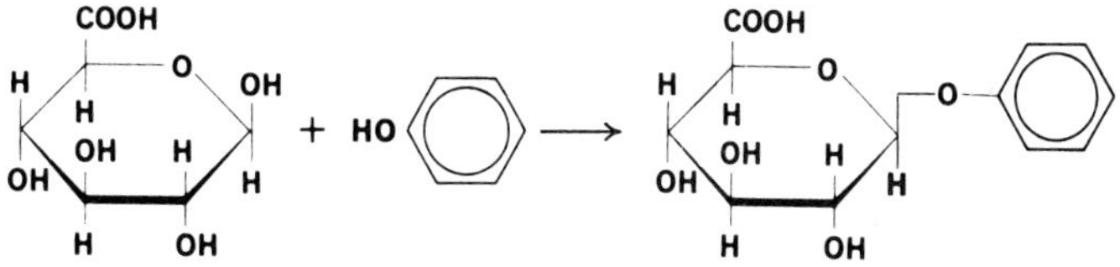

β-glucuronic acid   phenol

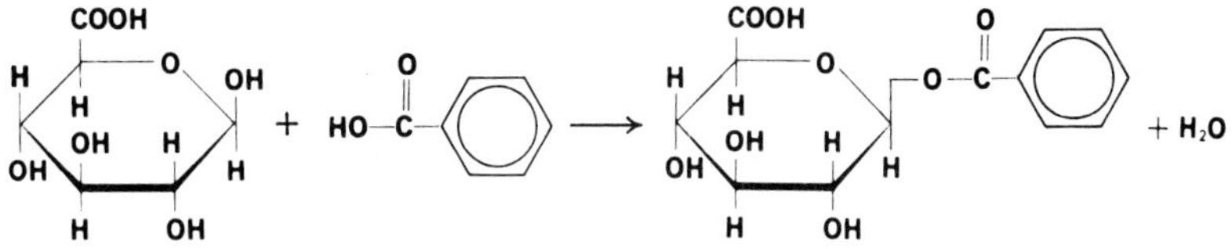

β-glucuronic acid   benzoic acid

merely say that this type of reaction that occurs in the body has definite advantages and that organisms that use this reaction may be better equipped to survive in a world of poisonous substances.

Ribose and deoxyribose are linked to nitrogen-containing compounds in the nucleic acids DNA and RNA and in nucleotides such as ATP. Here the linkage is from carbon to nitrogen rather than to oxygen.

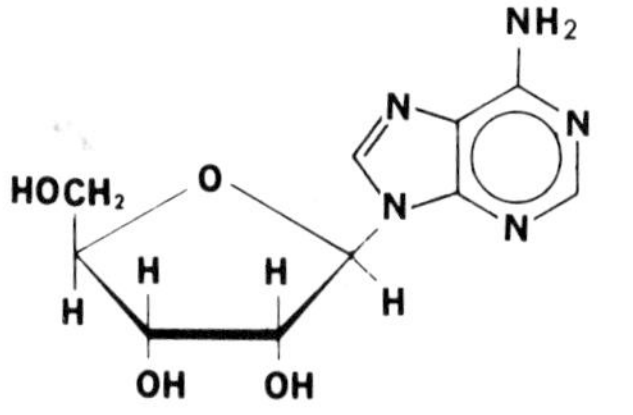

adenosine

Formation of Phosphate Esters   A structure of great biochemical importance is formed when either an alcohol or aldehyde group of a sugar is converted to a phosphate ester.

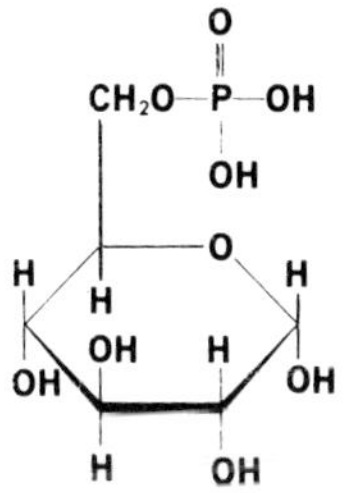

glucose-6-phosphate 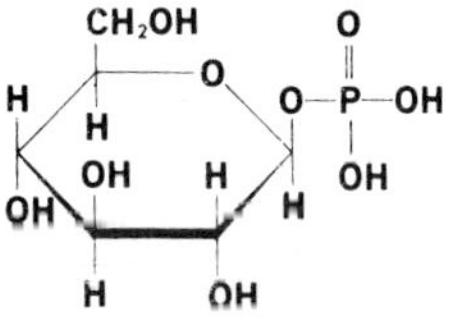   glucose-1-phosphate

Although the H's are written attached to the phosphate groups above, some of them dissociate in water solutions and we have phosphate ions. For simplicity, the phosphate group is often indicated by the symbol Ⓟ as shown on page 550.

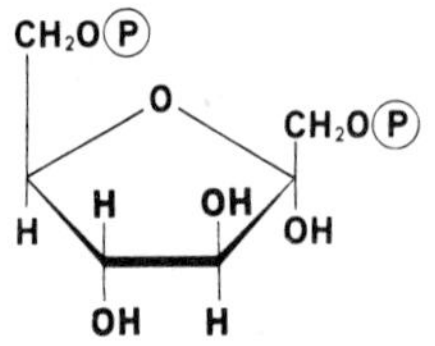

fructose-1,6-diphosphate

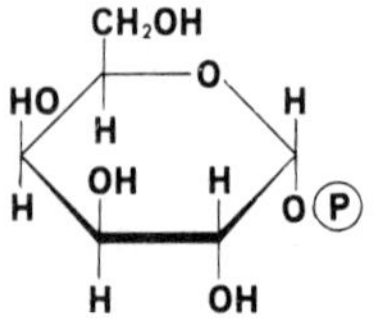

galactose-1-phosphate

In living organisms the source of the phosphate group in most such reactions is adenosine triphosphate (ATP) rather than phosphoric acid. In such reactions the ATP is converted to ADP (adenosine diphosphate).

glucose + ATP → glucose-6-phosphate + ADP

(See Fig. 33.12 for the formula for ATP.) The phosphorylation of glucose with ATP is probably the only important reaction of glucose in the human body. Other reactions proceed from phosphates of glucose.

**30.3
COMPOUNDS
RELATED TO
MONOSACCHA-
RIDES**

Some compounds do not fit the precise definition of carbohydrates since they lack certain groups (aldehydes) or contain others (amines). Since these compounds are similar in structure to the monosaccharides and may occur as units in polymeric carbohydrates, they are usually included in a discussion of carbohydrates.

The deoxymonosaccharides fit the definition of carbohydrates but differ from most others in that they have a CH$_3$ group in place of a CH$_2$OH group. The compound 2-deoxyribose is found as a constituent of the deoxyribonucleic acids (DNA), which are important in the transmission of inherited characteristics. L-rhamnose and L-fucose, which are deoxy forms of mannose and galactose, are found in plants and animals, particularly in the cell walls of microorganisms. They are usually present as units of a polysaccharide. These two compounds differ from most carbohydrates in being found naturally as L-isomers.

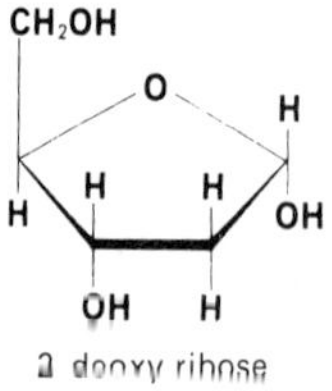

2-deoxy ribose

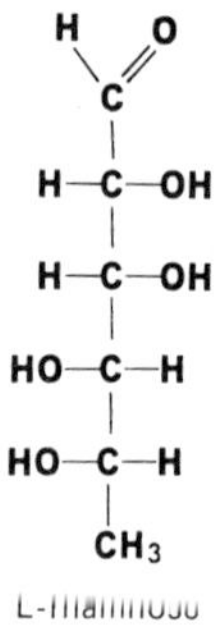

L-rhamnose
(6-deoxy-L-mannose)

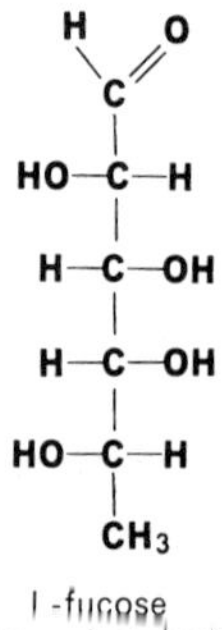

l-fucose
(6-deoxy-L-galactose)

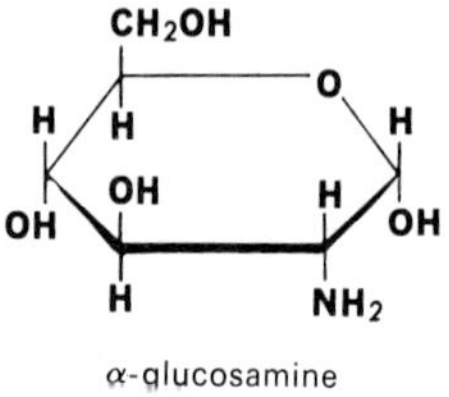

α-glucosamine

Glucosamine is a compound having an amino group in place of the alcohol group on carbon 2 of glucose. It is an important component of the polymer that makes up the tough outer covering (exoskeleton) of insects. Gluconic and galacturonic acids are components of polymers which will be discussed in greater detail in the section on polysaccharides.

30.4
OLIGO-
SACCHARIDES

Oligosaccharides are those carbohydrates which, when hydrolyzed, give two or more monosaccharide units. The prefix *oligo-* means few, and the upper limit of this classification is indefinite, but most authorities indicate that not more than ten or twelve monosaccharide units are found in oligosaccharides. While the number of possible oligosaccharides is great, only a few have been found in plants and animals. The most common oligosaccharides are the disaccharides shown below.

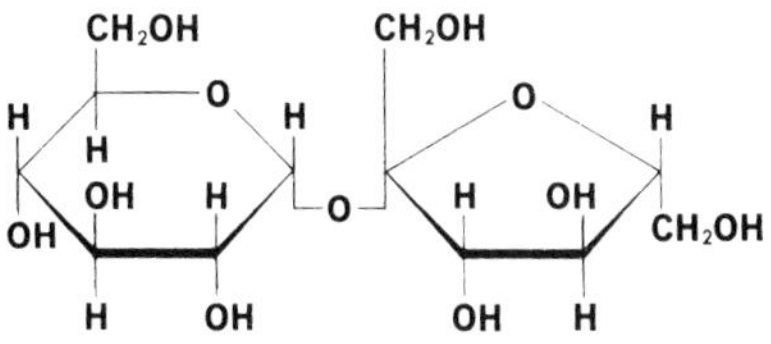

CH₂OH   CH₂OH   CH₂OH   CH₂OH

sucrose 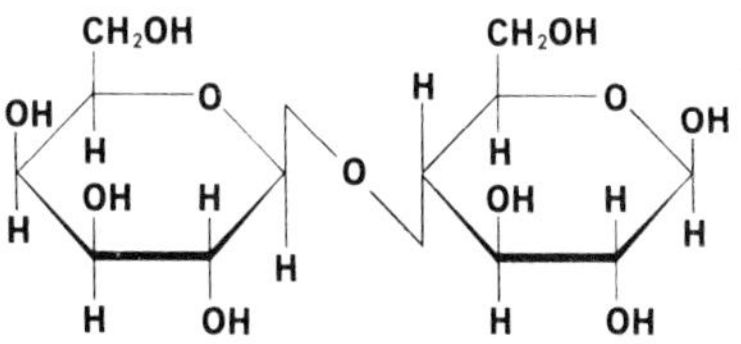   β-lactose

CH₂OH   CH₂OH

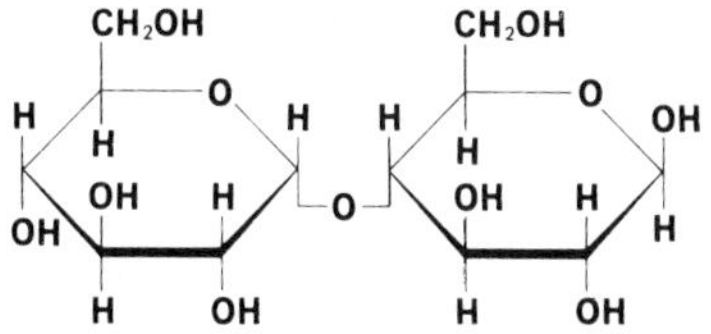

β-maltose

Sucrose contains one glucose unit and one fructose unit. The linkage involves the carbon-1 of glucose (α form) and the carbon-2 (β form) of fructose. In the formula above, the fructose has been flipped over with carbon-2 to the left. Since the hemiacetal OH of both monosaccharides has been used in the glycosidic linkage there are no α and β forms of sucrose and it does not show mutarotation.

Lactose contains one unit of galactose linked β 1–4 to a glucose, and maltose contains two units of glucose linked α 1–4. Both lactose and maltose exist in α and β forms and both show mutarotation.

Sucrose is especially abundant in sugar beets and in sugar and sorghum canes. It is also found in most common fruits and vegetables. Lactose is found primarily in milk. The milk from cows and goats is about 5% lactose; human milk contains from 6 to 7% of this sugar. Maltose is relatively rare. The other disaccharides are not found in appreciable amounts in natural products. While tri- and tetrasaccharides are apparently present in many plants, they have not been studied extensively, and the higher oligosaccharides are apparently quite rare.

The disaccharides are obtained commercially by isolation from natural products. Sucrose is commercially important and has the distinction of being one of the few chemically pure substances that can be obtained in a grocery store. Maltose is present, along with other substances, in corn syrup, which is made by the hydrolysis of corn starch. Maltose is also found in germinating seeds, where enzymes have produced it from the polysaccharide starch.

Biosynthesis of the disaccharides involves the synthesis of monosaccharides and finally the coupling of two monosaccharides (or their phosphate esters) to give the disaccharides. Only small amounts of disaccharides have been made this way in the laboratory; as with monosaccharides, the synthesis is not commercially feasible.

Both the chemical and physical properties of the disaccharides are similar to those of the monosaccharides. Table 30.1 shows the solubility and relative sweetness of common

TABLE 30.1   Solubility and relative sweetness of sugars

	Solubility (g/100 ml of H_2O)	Sweetness
sucrose	179	100
glucose	83	74
galactose	10	32
fructose	very soluble	173
xylose	117	40
lactose	17	16
maltose	108	33
saccharin*	—	55 000

* Saccharin, which is not a sugar, is included for comparison.

sugars. The measurement of sweetness involves the use of human tasters who determine the dilution that can be made before a solution no longer tastes sweet. Because they tend to form supersaturated solutions, the exact solubility of sugars is difficult to measure. It is interesting to note that there is a correlation between solubility and sweetness.

Disaccharides can be hydrolyzed to give monosaccharides. This process is easily catalyzed by either strong acid or enzymes. The hydrolysis of sucrose is especially interesting. Since the aldehyde group of glucose and the ketone group of fructose are combined to make the bond between the two molecules in sucrose, there is no reducing group present, and sucrose does not reduce Fehling's or Benedict's reagents. After hydrolysis, however, molecules of glucose and fructose are present and the hydrolyzed solution is now "reducing." Since a mixture of fructose and glucose is actually sweeter than one of sucrose alone, hydrolysis of a solution of sucrose increases its sweetness.

Sucrose has an optical rotation of $+66.5°$; glucose has a rotation of $+52.7°$; and fructose, $-93°$. Consequently, when one molecule of sucrose is hydrolyzed to yield one each of glucose and fructose, the net rotation of the mixture is to the left. This change in rotation has been called inversion, and hydrolysis of sucrose is sometimes called **inversion of sucrose**. An equimolar mixture of glucose and fructose is called *invert sugar*; the enzyme which catalyzes this hydrolysis is often referred to as *invertase* instead of by its systematic name, *sucrase*.

Food preparation or cooking that involves exposing sucrose solutions to high temperatures under acid conditions may produce this hydrolysis. While this may make the food concerned a bit sweeter, it is more likely that the real purpose of this cooking, as when we make jelly and some forms of candy, is to produce a mixture of sugars. Such a mixture is less likely to crystallize (as when a jelly "sugars") than a pure solution.

Sucrose is commercially the most important of the disaccharides. It is an important item of world trade because of its desirability as a sweetening agent. In spite of the use of synthetic noncaloric sweeteners, each person in the United States consumes an average of about 170 pounds of sucrose per year. Although other countries consume less sucrose, more than 35 million tons are produced in the world each year. Approximately one-third of this sugar comes from beets and two-thirds from cane.

Lactose is found in the milk of mammals. It can be purified and has some commercial uses but has never been used as widely as a sweetener because of its low solubility and because it is not nearly so sweet as sucrose.

The polysaccharides, which contain many monosaccharide units linked together, are the most abundant carbohydrates. It is difficult to determine experimentally exactly how many such units there are in a polysaccharide chain. Molecular weights ranging from several thousand to values in the millions have been reported for various polysaccharides. It is generally assumed that the higher molecular weights represent the true molecular weight of the polysaccharide as it occurs in the plant, and that the lower weights are caused by degradation of the polymers during purification. However, we cannot totally disregard the possibility that some polymerization occurs during the purification process. Probably the best conclusion is that the number of units may vary somewhat without affecting the overall properties of the polysaccharide. There is a possibility that the molecular weight of a polysaccharide is characteristic of the species of plant or animal in which it is found.

One difference in structure which is of great biological significance is the one between the alpha- and beta-linked polysaccharides. Those polysaccharides containing the alpha-linked monosaccharide units are much more digestible than β-linked polymers. Animals use α-linked starch as a major source of food, but only a few animals can digest the β-linked cellulose. Actually, the cellulose-utilizing animals, such as cattle and sheep and even the lowly termite, are able to use this β-linked polymer only because of micro-organisms in their digestive tract which actually perform the hydrolysis of cellulose to glucose for them. Compare the structures of starch and cellulose, shown below.

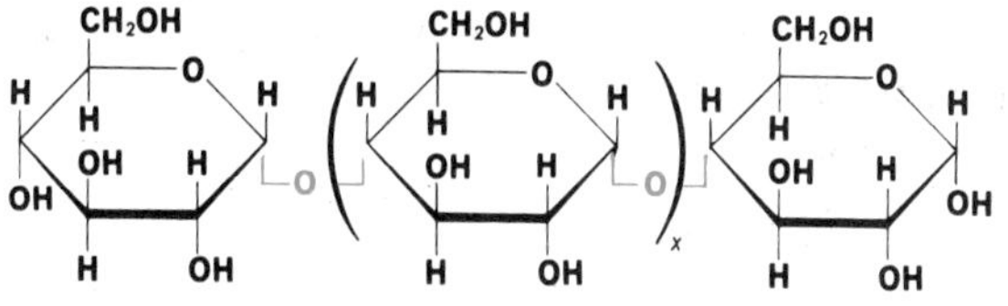

starch (α-linked)

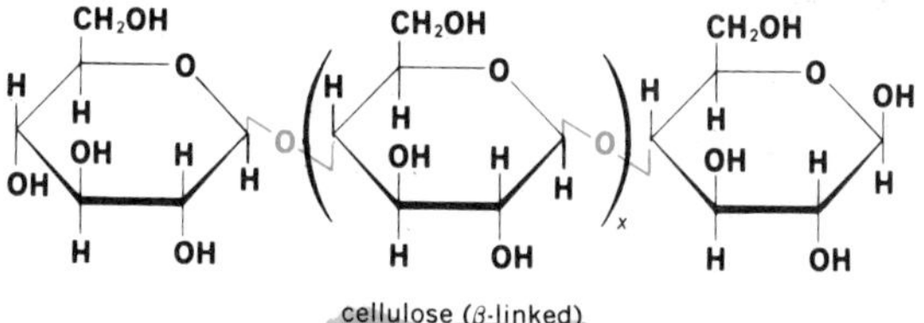

cellulose (β-linked)

The polysaccharides make up between 60% and 90% of the dry weight of plants, where they serve both as structural materials and as reserve food supplies. They are found in much smaller amounts, usually less than 1% of the total weight, in animals. Typical polysaccharides are starch, which may be obtained from the seeds and tubers of a variety of plants; cellulose, the structural material of plants; and glycogen, which is found in small amounts in animals.

The polysaccharides are white solids which are only slightly soluble in water. In contrast to the mono- and oligosaccharides, the polysaccharides are not sweet-tasting. Nor are the polysaccharides very reactive chemically, both because of their insolubility and because of the fact that their most active groups (the aldehyde and ketone groups) are tied up in the linkage between monosaccharide units. Strong acids or enzymes are required to catalyze the hydrolysis of polysaccharides to give monosaccharide units.

Polysaccharides that contain only one type of monosaccharide unit, such as cellulose, which contains only glucose, are called **homopolysaccharides**. Those polysaccharides that contain more than one type of monosaccharide unit, such as hyaluronic acid, are **heteropolysaccharides**. Polysaccharides composed of pentose units are called **pentosans**. They are found in wood, straw, and leaves of plants. The gums from some trees are the richest sources of relatively pure pentosans. Pentosans are not very useful commercially since most animals cannot digest them; however, chemical degradation can give furfural, which is used commercially for the manufacture of nylon and in the refining of petroleum.

The most important and abundant polysaccharides are polymers of the hexoses and are called **hexosans**. Glucose is the only monomeric unit of most of the abundant and important polysaccharides.

Starch   Various kinds of starch are found in the cereal grains and in many kinds of tubers. The starch from each type of plant has a characteristic form of granule. Experienced persons can tell the source of the starch by examining a sample microscopically (see Fig. 30.7). Chemically, all forms of starch are quite similar; all are composed totally of glucose. Most sources of starch yield two fractions, amylose and amylopectin (Fig. 30.8). These two forms are very similar and separation is difficult. While both kinds of starch contain alpha 1–4 links between monosaccharide units, the amylopectin molecules contain many more glucose units, are highly branched (1–6 linkages), and are somewhat more soluble in water.

The starches in the form of rice, potatoes, and wheat or cereal products supply about 70% of the world's food. In many cultures starch makes up much more than 70% of the diet, since most highly starchy foods are relatively inexpensive to buy or relatively easy to grow.

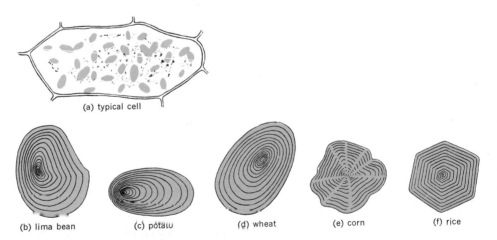

(a) typical cell

(b) lima bean   (c) potato   (d) wheat   (e) corn   (f) rice

Fig. 30.7   Characteristic starch grains. Starch grains are scattered throughout the cytoplasm of many plant cells. (a) In this sketch of a typical cell, the starch grains are indicated as colored bodies. Drawings of enlarged starch grains of typical plants are shown in (b) through (f). [Adapted from James Bonner and Arthur W. Galston, *Principles of Plant Physiology*, W. H. Freeman and Company. Copyright © 1952.]

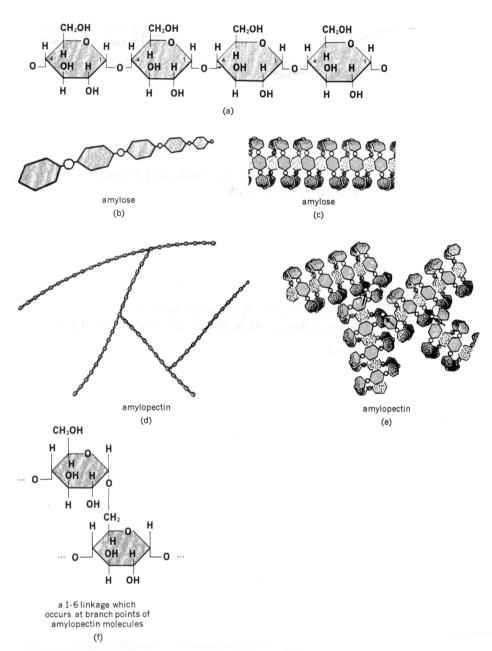

Fig. 30.8 The structure of the amylose and amylopectin components of starch. The structure of amylose can be represented as in (a) or simplified as in (b). The three-dimensional structure of amylose is given in (c). A simplified representation of a part of a molecule of amylopectin is shown in (d), and a three-dimensional diagram of amylopectin is represented in (e). The 1–6 linkage shown in (f) occurs at branch points of amylopectin molecules. [Adapted from James Bonner and Arthur W. Galston, *Principles of Plant Physiology*, W. H. Freeman and Company. Copyright © 1952.]

Glycogen Glycogen is found in the liver and muscles of animals, and it is often called "animal starch." Glycogen is similar to the amylopectin fraction of starch in that it is made totally of glucose and is highly branched. Since animals do not synthesize starch, glycogen represents the only form in which they can store excess glucose. The total amount of glycogen in an animal is small; about 1% of the muscle and between 1.5 and 4% of the liver is glycogen. However, this small amount is extremely important in maintaining the correct level of glucose in the blood and in supplying the muscles with a source of energy.

Cellulose Cellulose is the most abundant organic compound. It has been estimated that 50% of all carbon atoms in the vegetation on the earth is tied up in cellulose molecules. Like starch and glycogen, cellulose is made exclusively of glucose units, but unlike them, the glucose units are β-linked. This β-linked polymer of glucose is extremely tough and indigestible, and for this reason is an effective structural material both within plants and, in turn, for humans, who use many materials which contain cellulose. The woody part of plants contain large amounts of cellulose. Cotton is 98% cellulose. Cellulose-containing materials such as lumber, paper, cotton, and rayon (which is a chemically modified cellulose) are important commercially.

Chitin The tough outer covering of many insects and of crustaceans like crabs and lobsters contains appreciable amounts of another β-linked polymer. In this instance, however, the monomeric unit is not glucose but glucosamine, and the substance is called chitin. Complete hydrolysis of chitin yields both glucosamine and acetic acid, and apparently the monomeric structural unit is N-acetyl glucosamine. (See Fig. 30.9.)

30.6 HETERO-POLYSAC-CHARIDES Those polysaccharides discussed so far have been made of a single kind of monosaccharide. There are other polysaccharides which usually contain at least two different kinds of monomeric units. While we often call these substances **heteropolysaccharides,** they are also called **heteropolymers,** since some contain peptides or other nonsaccharide units. In many cases the linkage between monomers is from the number 1 carbon of one unit to the 2, 3, or 6 carbon of the other unit, as well as the more common 1–4 linkage found in starch, glycogen, and cellulose. Some of the more important heteropolysaccharides are discussed below.

Pectins Apples and milk are both about 87% water. The solid structure of apples and other fruits is due to the fact that they contain pectins. We use pectins in making semisolid jam and jelly from fruit juices and sucrose. Pectins contain pectic acid, which is a homopolymer containing units of the methyl ester of D-galacturonic acid (Fig. 30.9). Pectins also give arabinose and galactose when hydrolyzed; presumably they are present in polymeric forms also.

Mucopolysaccharides The mucopolysaccharides are a group of heteropolymers that have important functions in cells. They generally contain uronic acids and an amino sugar. There is often an acetyl group on the amino sugar, and many of the alcohol groups have been converted to sulfate esters.
  Hyaluronic acid is found in the vitreous humor of the eye, in the coat of human egg cells (ova), and as the ground substance between cells. Hyaluronic acid is a heteropolymer containing D-glucuronic acid linked β 1–3 to a unit of N-acetyl-D-glucosamine. The latter is linked β 1–4 to another unit of glucuronic acid, and the units then repeat them-

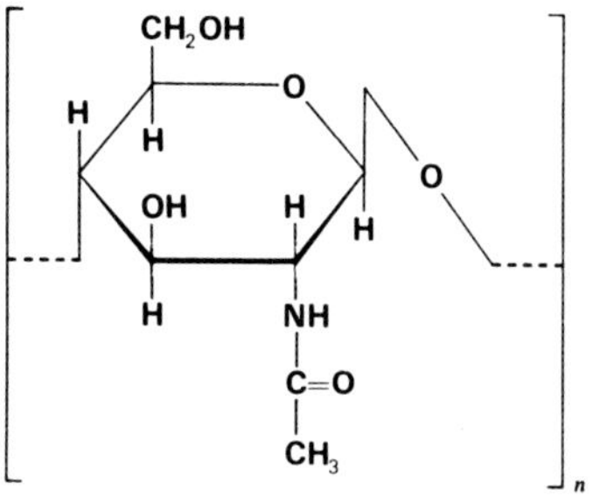

N-acetylglucosamine, the
monomer of chitin

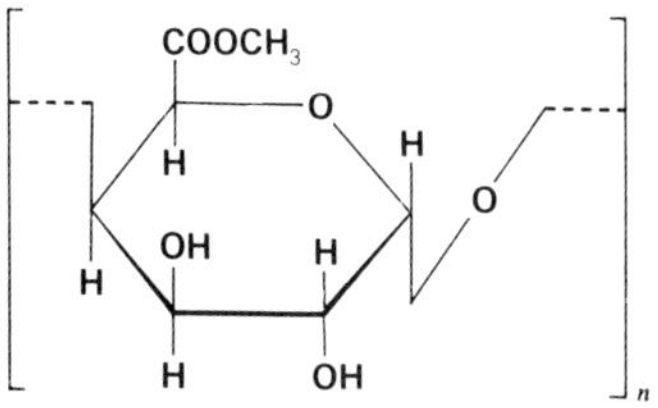

the methyl ester of
galacturonic acid, the
monomer of pectic acid

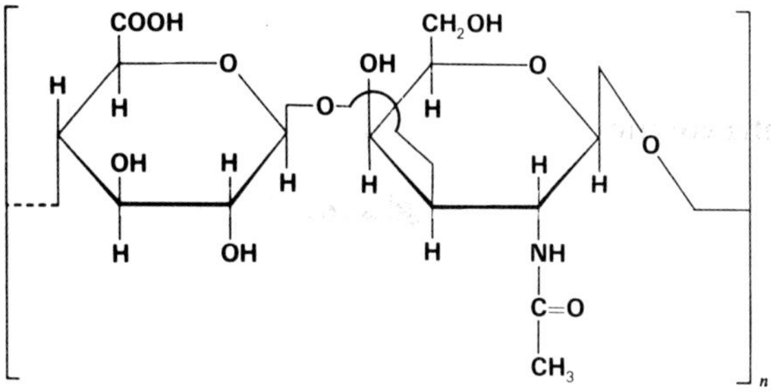

glucuronic acid and
N-acetylglucosamine,
which make up the
repeating unit of hyaluronic
acid

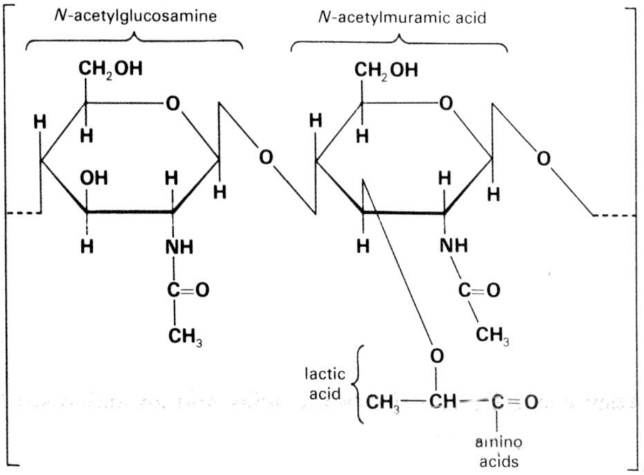

the repeating unit
of peptidoglycans

Fig. 30.9 The structural units of some important biological polymers.

selves (Fig. 30.9). The enzyme hyaluronidase catalyzes the hydrolysis of hyaluronic acid. Bacteria having hyaluronidase are more invasive than other bacteria, and the fact that human sperm cells contain the enzyme helps in penetration and fertilization of the human ovum.

Chondroitin is a heteropolymer like hyaluronic acid but contains N-acetylgalactosamine instead of N-acetylglucosamine. It is found at the surface of cells. Chondroitin sulfates containing a sulfate ester at the C-4 and C-6 positions of N-acetylgalactosamine are important structural components of cartilage, tendons, and bones.

Heparin, the naturally occurring anticoagulant found in the liver, lungs, and spleen, as well as in the blood, contains D-glucosamine and D-glucuronic acid monomers. It also contains sulfate groups esterified at C-6 of glucosamine and C-2 of glucuronic acid. Heparin is often given to people who have blood clots to prevent further clots from forming.

Peptidoglycans are found in the cell walls of Gram-positive bacteria. These heteropolymers have a backbone of N-acetylglucosamine and N-acetylmuramic acid. The linkages are β 1–4. N-acetylmuramic acid contains lactic acid joined by an ether link to C-3 of an acetylglucosamine unit. The carboxyl group of lactic acid is linked to a chain of amino acids which is known as a peptide (see Fig. 30.9). The enzyme lysozyme, which is found not only in egg white and in many bacteria but also in human saliva and tears, catalyzes the hydrolysis of the β 1–4 linkage of peptidoglycans. This hydrolysis weakens the cell walls of bacteria and makes them more likely to rupture, with the subsequent destruction of bacteria. The presence of lysozyme explains the antiseptic properties of tears and saliva. It is interesting to note that Alexander Fleming, who first discovered the effects of penicillin, which interferes with the synthesis of cell walls in growing bacteria, also spent a great deal of time in research on the enzyme lysozyme.

30.7 GLYCOPROTEINS AND GLYCOLIPIDS

The compounds discussed above are primarily carbohydrate with the obvious exception of the peptidoglycans, which contain peptides made of amino acids. Other biochemical substances contain appreciable quantities of proteins and lipids in addition to polysaccharides and, using the prefix glyco- to indicate the carbohydrate content, are called **glycoproteins** and **glycolipids**. Glycoproteins are important components of blood plasma and of the outer coats of human cells. It is these outer coats, sometimes called fuzzy coats because of their appearance in electron microscopic pictures, that influence the cell type, which is important in tissue transplants as well as in blood transfusions. Another glycoprotein found in the blood of Antarctic fish enables them to survive temperatures of $-1.9°C$ without their blood freezing.

The cell walls of Gram-negative bacteria contain appreciable amounts of a combination of lipid and polysaccharides known as **lipopolysaccharides**. The presence of certain types of lipopolysaccharides in the blood of animals, such as would result from the destruction of Gram-negative bacteria, can cause fever, shock, and tissue damage. Because of this, we refer to these lipopolysaccharides as endotoxins. The name lipopolysaccharides, which uses the combining form for lipids first, indicates that they contain a greater amount of lipids than polysaccharides.

Ongoing research concerning the structure and functions of glycolipids and glycoproteins will probably tell us much more about how the human body functions in health and disease. These mixed compounds are probably of greater significance to human health than are the simple homopolysaccharides such as starch and cellulose, which are found primarily in plants.

SUMMARY

The carbohydrates are found primarily in polymeric form in the structural parts of plants. The chemical properties of monosaccharides are due to their carbonyl (aldehyde and ketone) groups, to their alcohol groups, and especially to the interaction of their carbonyl and alcohol groups. When monosaccharides react to form polysaccharides, the reactive carbonyl groups are tied up in the glycosidic linkages. Thus the polysaccharides are not only insoluble in water, due to their high molecular weight, but also much less active chemically than are the monosaccharides and disaccharides. This insolubility and low reactivity make the polysaccharides suitable for structural components and food reserves.

The types of linkage (α or β) between monosaccharide units in polysaccharides are also very important in determining whether the polysaccharide serves a structural function (cellulose and chitin) or a storage function (starch and glycogen). Most organisms contain enzymes that can result in the rupture of α linkages, and these animals then use the products of hydrolysis for food. On the other hand, those organisms that contain enzymes capable of catalyzing the hydrolysis of β linkages have an added source of food. Sometimes the ability to hydrolyze β-linkages enables organisms to invade and destroy other organisms.

While the monosaccharides are not widely distributed in nature as are the polysaccharides, they are formed when polysaccharides are digested. The monosaccharides serve as the major source of energy for all animals. Without the monosaccharide glucose, life as we know it would not be possible.

IMPORTANT TERMS

aldose	homopolysaccharide
carbohydrate	heteropolysaccharide
cellulose	hyperglycemia
glucuronide	hypoglycemia
glucoside	ketose
glucosidase	lipopolysaccharide
glycogen	mutarotation
glycolipid	oligosaccharide
glycoprotein	pentose
glycosidic linkage	pentosan
hemiacetal	polysaccharide
hexose	reducing sugar
hexosan	starch

WORK EXERCISES

1. Define the following terms: ketotetrose, triose, aldopentose, octulose, deoxy sugar.
2. Draw open-chain structural formulas for the following. Indicate chiral carbons with asterisks.
   a) L-glucose   b) D-altrose
   c) a ketotetrose
3. Explain what is meant by each of the following.
   a) a D-compound
   b) an L-compound
   c) D(+)galactose
   d) L(−)glucose
   e) D(−)fructose
   f) a reducing sugar

g) a β linkage between monosaccharide units
h) a 1–4 linkage between monosaccharide units

4. Explain what happens when mutarotation occurs.
5. What are the advantages and disadvantages of using the following ways for writing formulas for carbohydrates?
   a) open-chain (Fischer) formulas
   b) Haworth representations
   c) conformational (chair and boat) representations
6. Using Haworth representations, draw the following.
   a) β-D-glucose
   b) α-D-Idose

c) L-glucose (either form)

d) the β form of a sugar containing an α 1–4 linkage between two glucose units

e) the α form of a disaccharide having an α 1–6 linkage between two glucose units

f) two possible β forms of a disaccharide containing one unit of glucose and one of galactose with a β 1–4 linkage between units

g) the α form of a disaccharide containing a β linkage between two ribose units

h) β-maltose

i) α-cellobiose

j) sucrose

k) a trisaccharide containing two glucose and one galactose units

7. How many possible isomers would there be for the trisaccharide in Exercise 6(k)? Give a general representation for each, without drawing each in detail.

8. Write both open-chain and Haworth structures for the following.
a) glucuronic acid
b) glucaric acid
c) galacturonic acid
d) galactaric (mucic) acid
e) mannuronic acid
f) the α form of glucose-1-phosphate
g) galactose-6-phosphate

9. Use Haworth formulas to write the following.
a) methyl α-D-glucoside
b) ethyl β-D-galactoside
c) the β form of N-acetyl glucosamine

d) β-galactosamine

10. Using structural formulas write equations for the following.
a) the hydrolysis of sucrose
b) the complete oxidation of glucose, of gluconic acid, of glucaric (saccharic) acid

11. How is a fruit jelly that has been boiled for ten minutes different from the uncooked mixture of ingredients?

12. What is the nutritional significance of the β linkage between the units of a polysaccharide?

13. What reason might be deduced for the fact that starch does not taste sweet?

14. Benedict's test is used to see whether glucose is present in the urine of persons with *diabetes mellitus*. During the period of time a mother produces milk after the birth of a child, she may excrete lactose or galactose in the urine. Would urine containing either lactose or galactose give a positive Benedict's test?

15. Give the names and importances of three common naturally occurring glucosides other than the disaccharides and polysaccharides.

16. List three important mucopolysaccharides and give the major function(s) of each.

17. What is the significance of the following enzymes?
a) amylase
b) β-galactosidase
c) hyaluronidase
d) lysozyme
e) an enzyme that catalyzes the hydrolysis of β linkages between glucose units

SUGGESTED READING

Sharon, N., "Glycoproteins," *Scientific American*, May 1974, p. 78 (Offprint #1295). The structures and some functions of those proteins that have carbohydrates bound to them are discussed in this well-illustrated article.

Strobel, G. A., "A mechanism of disease resistance in plants," *Scientific American*, January 1975, p. 80 (Offprint #1513). Disease resistance in sugar cane is discussed in terms of the toxin, a galactoside, and possible effects on cell membranes.

31

Chemistry of the Proteins

For humans as well as other animals the protein molecules are at the center of most important life processes. Our bodies are supported by bones which are made of a protein framework with minerals deposited in it. Muscles are primarily protein, and the ligaments and tendons that tie muscles and bone together are made of protein. The blood contains many proteins, one of which—hemoglobin—carries oxygen and carbon dioxide. Other proteins in the blood are important in blood clotting, in protecting against disease, as hormones, and as molecules that transport other molecules of food or wastes. The enzymes, which are necessary catalysts for almost every biochemical reaction, are primarily proteins. While carbohydrates, lipids, and nucleic acids play important roles, it is proteins that constitute a large percentage of the biochemicals in the body. In the form of enzymes, proteins control the body's biochemical reactions. About one-half of the dry weight of the human body is protein; much of the rest is depot fat which is not metabolically active but is used for energy as needed.

Although he could have had no idea of the real importance of proteins when he proposed their name in the 1840s, the Dutch chemist Gerardus Mulder derived the word protein from the Greek *proteios*, meaning of the first rank or order. As with many other biochemical substances, protein was originally thought to be one substance. When it was found that there were many different proteins, they were subdivided into various categories. The original classification was based on the one property that could be measured simply—solubility. We still classify proteins by solubility, but also classify them by structure, composition, and function.

About the time they were named, it was recognized that proteins contain nitrogen as well as carbon, hydrogen, and oxygen. While most proteins also contain sulfur and phosphorus, it is the presence of nitrogen that is the basis of the classification of a biochemical substance as a protein. In fact, since proteins are usually about 16% nitrogen, we often calculate the amount of protein in a biochemical specimen by determining the amount of nitrogen and multiplying by 6.25 (100%/16%).

Further studies of proteins led to the realization that it was not the nitrogen itself that was important in protein, but nitrogen in an amine group that was part of an amino acid. A **protein** is now defined as *a polymer which yields amino acids upon hydrolysis.* When hydrolyzed, proteins may give small amounts of other substances in addition to amino acids but the amino acids are always the major hydrolysis product of any protein.

Our discussion of the proteins will begin with a description of the amino acids. Only when we understand the properties of amino acids can we describe the properties and functions of protein. We will also discuss the peptides, those compounds intermediate in size between amino acids and proteins, before returning to proteins.

31.2
AMINO ACIDS

The **amino acids**, as their name implies, are organic compounds which contain both an acid and an amino group. The fundamental structure and reactions of the amino acids were discussed in Sections 25.5 and 25.6. Here we will extend that discussion to indicate more of the properties and reactions that are important to biochemistry. Those amino acids commonly found in proteins are given in Fig. 31.1 (pages 564–565). The amino acid group is shown in black and the R group—the part that makes each amino acid different— in color. Each amino acid is given a three-letter abbreviation which is very useful when writing structures of peptides and proteins. These abbreviations are also shown in Fig. 31.1 In addition to the commonly found amino acids, about 200 others have been detected in natural sources other than protein. While twenty amino acids might seem like a large number of compounds, on further reflection it is really amazing that with the millions of organic compounds, and thousands of possible amino acids only 20 of them are commonly found in proteins and only 200 more in other natural products.

Except for those amino acids containing an additional acid group, which are named as acids, the names of amino acids end in the suffix -ine.

Cystine was first found in bladder stones in 1810. Its name is derived from the Greek *kystis*, meaning a pouch or bladder. Cystine is one of the least soluble amino acids and is still occasionally found in the form of stones in the urinary bladder.

The simplest amino acid, glycine, was one of the first to be isolated by hydrolysis of proteins. Since it has a slightly sweet taste, it was named from the Greek word for sweet, *glykys*. When leucine was isolated as white, glistening plates, its name was chosen as a derivative of *leukos*, the Greek word for white. Isoleucine was named for its structural analogy to leucine.

The silk protein sericin (from the Latin word for silk) yields large amounts of serine. Cheese (*tyros* in Greek) was the first source of tyrosine. The amide asparagine is found in asparagus, and its hydrolysis product is aspartic acid. Arginine was named because if forms silver (L. *argentum*) salts. Proline derives its name from its chemical similarity to pyrrolidine; valine is named for its similarity to valeric (pentanoic) acid, and glutamic acid for its similarity to the dicarboxylic acid, glutaric acid.

The properties of amino acids are due to their amine and acid functional groups and to R groups. The meaning of an R group in protein chemistry is slightly different from that in organic chemistry. In organic chemistry an R group represents a hydrocarbon unit. The R groups of amino acids may be merely a CH_3 or similar hydrocarbon group, but in other instances the R group contains a functional group such as —OH or —SH. A listing of amino acids classified by their R group, which is also known as a side chain, is given in Table 31.1.

TABLE 31.1   Classification of amino acids according to side chains (R groups).

The properties of proteins depend on the state of charge and solubility of side chains of amino acids. Although glycine and alanine have nonpolar side chains, the amine and acid groups make them quite soluble in water.

---

1. Amino acids with nonpolar side chains
     alanine, cystine, glycine, isoleucine, leucine, methionine, phenylalanine, proline, tryptophan, valine
2. Amino acids with polar or potentially ionic side chains that are not charged at physiological pH values. Some of these form hydrogen bonds.
     asparagine, cysteine, glutamine, serine, threonine, tyrosine
3. Amino acids with an additional acid group on the side chain. These amino acids usually have a negative charge on the side chain at physiological pH values.
     aspartic acid, glutamic acid
4. Amino acids with an additional amine group on the side chain. These amino acids usually have a positive charge at physiological pH values.
     arginine, histidine, lysine

Although small quantities of amino acids are found in the blood after the digestion of a protein-containing meal, they are found primarily in the form of their polymers, the proteins. It is from the naturally occurring proteins that commercial quantities of amino acids are usually produced. Plants can synthesize amino acids from ammonia and carbon dioxide. Animals, however, have to use larger units for making amino acids, and usually take an amino group from one compound and transfer it to a keto-acid. Green plants can synthesize all the amino acids they require for the production of protein, but most animals can make only approximately half of the amino acids they require. Consequently, they must get the others from their diet to build the proteins of muscles, enzymes, and blood components.

An interesting, but commercially unfeasible, process for synthesizing amino acids was discovered in the mid 1950s by Miller and Urey, working at the University of Chicago. They were able to find very small quantities of amino acids after passing an electric spark through a mixture of methane, ammonia, water, and hydrogen. Other experimenters have proved that the gases used as starting materials can be varied somewhat and that very small amounts of amino acids are relatively easy to produce. These experiments are interesting because the gases used, often called the *primitive* gases, are the ones known to be present in the atmospheres of the other planets, and were the ones probably present on the primitive earth. These experiments indicate a way amino acids may have been synthesized on a lifeless earth. Even more important, these discoveries led to the realization that amino acids are stable compounds and that any reactions that supply large amounts of energy to hydrogen compounds of carbon, nitrogen, and oxygen are likely to synthesize amino acids.

By adding hydrocyanic acid (HCN) to mixtures of primitive gases, pyrimidine and purine bases like those of the nucleic acids can be produced. When phosphates are also present, ATP can be produced. Other researchers have produced protein-like substances by heating mixtures of amino acids with phosphate or other catalysts. Methane, carbon dioxide, hydrogen, and ammonia are prevalent in the atmospheres of the planets of our solar system and HCN has been detected in the tails of comets. Thus, we could predict that we should find most of the building blocks of life in materials from the surface of Mars and other planets. However, we have found only nanogram quantities of the amino acids in rocks from the moon. Perhaps their absence is due to the absence of water required for their synthesis. There are probably slightly higher amounts of amino acids on Mars.

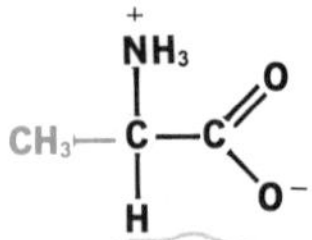

alanine (Ala)

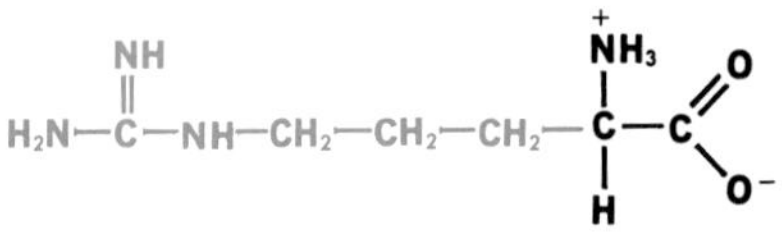

arginine (Arg)

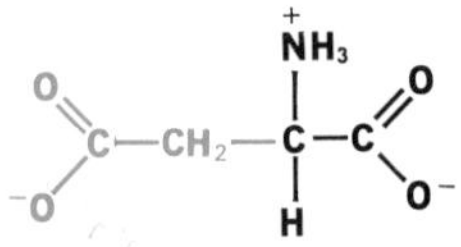

aspartic acid (Asp)

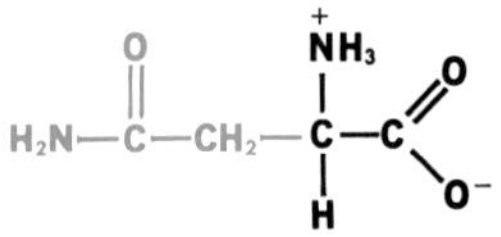

asparagine (Asn)

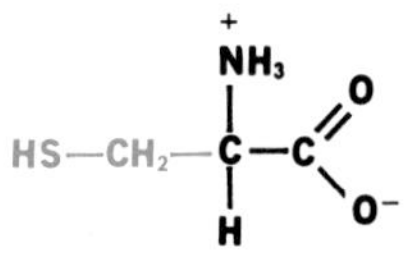

cysteine (CySH)

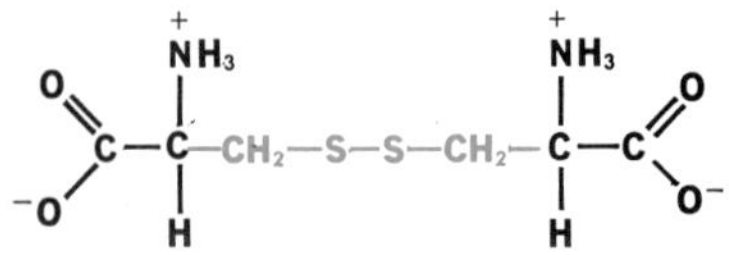

cystine (CyS·SCy)

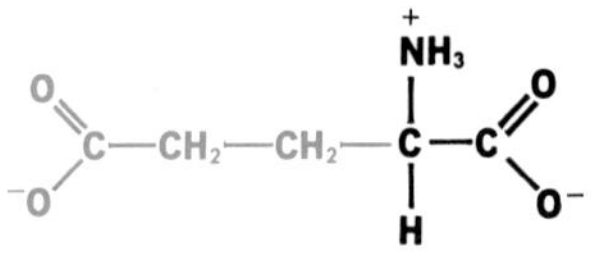

glutamic acid (Glu)

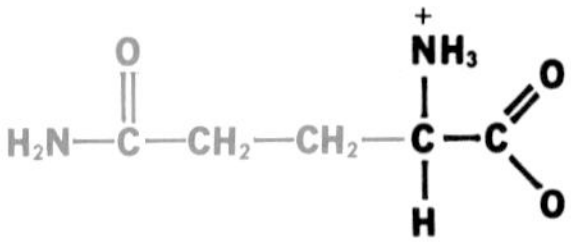

glutamine (Gln)

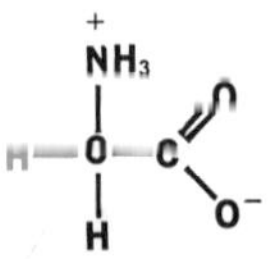

glycine (Gly)

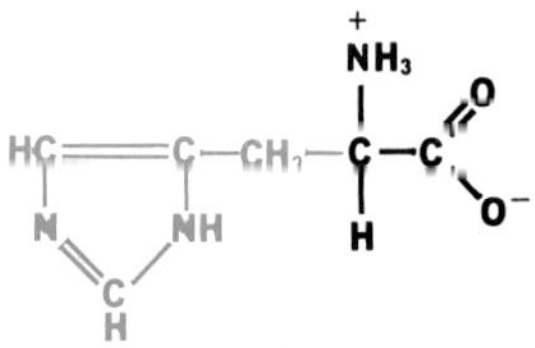

histidine (His)

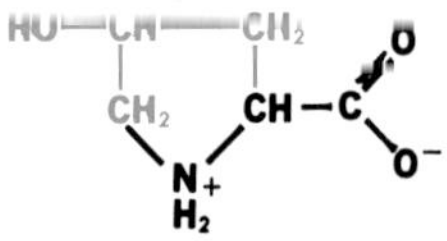

hydroxyproline (Hyp)

Fig. 31.1 The common amino acids. The formulas, names, and symbols used for each amino acid are given. The side chains (R groups) are shown in color.

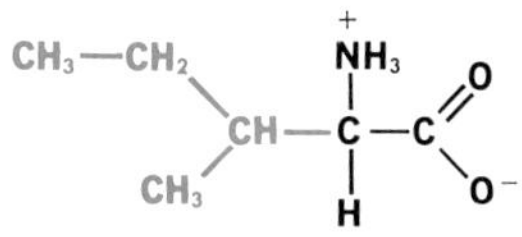

isoleucine (Ile)

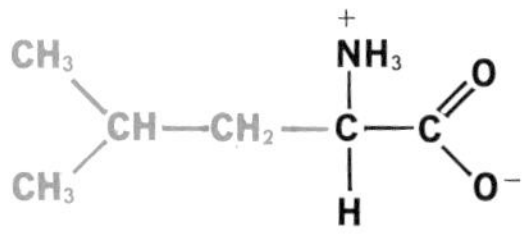

leucine (Leu)

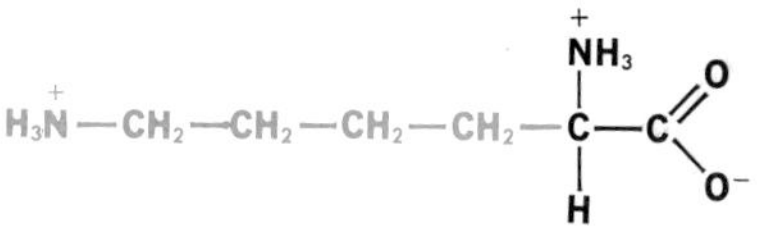

lysine (Lys)

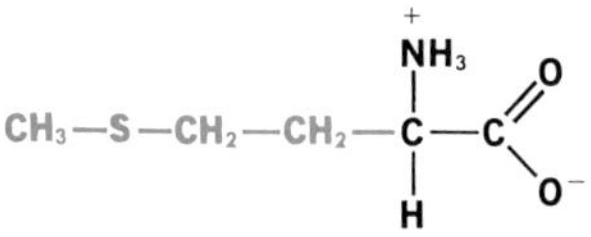

methionine (Met)

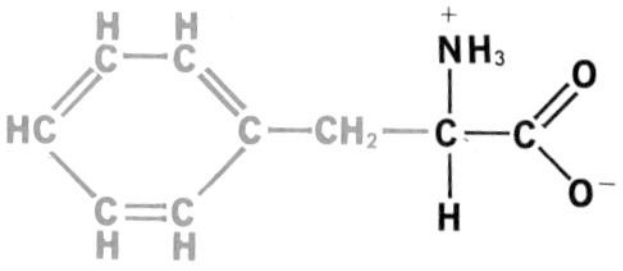

phenylalanine (Phe)

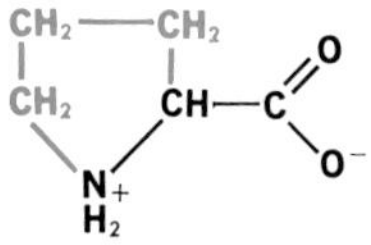

proline (Pro)

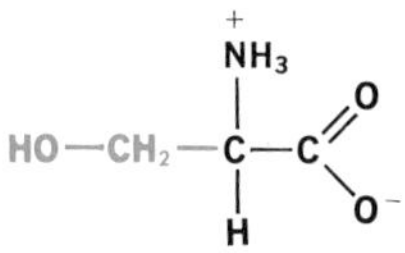

serine (Ser)

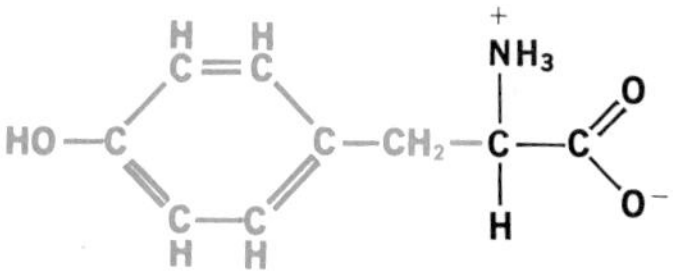

tyrosine (Tyr)

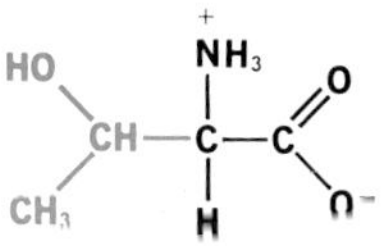

threonine (Thr)

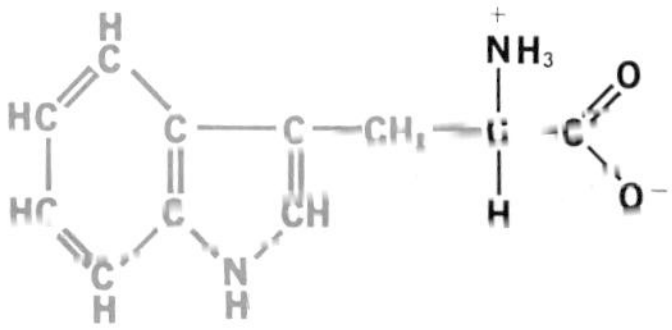

tryptophan (Try)

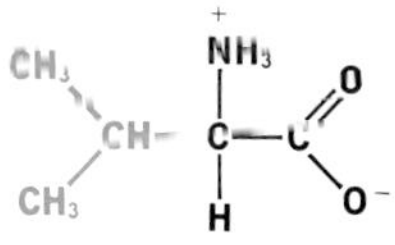

valine (Val)

Physical Properties of the Amino Acids

The amino acids are white, crystalline solids, and most are quite soluble in water. They have relatively high melting points for compounds of low molecular weight. These properties suggest that the amino acids, unlike simple acids and amines, resemble the inorganic salts and exist as ionic molecules. Measurement of the properties of amino acids in solutions confirms the fact that they are indeed charged molecules and also that their charge may change from negative to positive or vice versa with changes in the acidity of the solution. This change in charge is accompanied by a change in chemical and physical properties. At pH values near neutral, most amino acids have both a negative and a positive charge, and thus are dipolar ions. In acid solutions they tend to have a proton attached to their negative group and are cations, while in basic solutions they are usually anions. The charge on the side chain (R group) is also important in determining the charge of an amino acid. The pH at which a protein, amino acid, or other molecule will not migrate when placed in an electrical field is called the *isoelectric point*. (See Fig. 31.2). The isoelectric point is the pH at which the negative charges on the molecule exactly balance its positive charges. This point varies for the different amino acids and provides a method of separating one amino acid from all others, or at least all others having different isoelectric points.

ionic forms of the amino acid alanine

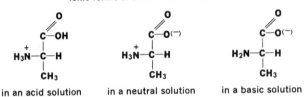

in an acid solution   in a neutral solution   in a basic solution

Since amino acids take up hydrogen ions (protons) in acid solutions and release them in basic solutions, they resist changes in the pH of a solution and are good buffers. The alpha carbon of amino acids usually has four different groups attached to it. We could predict, therefore, that the amino acids contain chiral carbon atoms and are optically active. This is indeed true with most of the naturally occurring amino acids having the L-configuration. We find D-amino acids in some antibiotics and in bacterial cell walls.

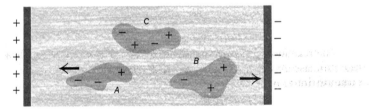

Fig. 31.2 Migration of charged particles in an electric field. This drawing shows how various amino acids or protein molecules can be separated by electrophoresis. Substance *A*, which has an excess of negative charges, migrates toward the anode; substance *B* has an excess of positive charges and thus is a cation. Substance *C*, although it has charged groups, has equal numbers of positive and negative charges and will not migrate when placed in an electrical field. Substance *C* is at its isoelectric point.

This unusual form prevents hydrolysis of cell walls by enzymes that act only on the more common L-isomers.

Chemical Properties of the Amino Acids   Several biochemical reactions are common to all amino acids. Among these are formation of peptide bonds, transamination, oxidative deamination, and decarboxylation.

When the amino group of one amino acid reacts with the acid group of another amino acid, a compound called a **peptide** is formed. The group $-\overset{\overset{\displaystyle O}{\|}}{C}-NH-$ is called the *peptide group*. It is a substituted amide group. The C—N bond is sometimes known as a peptide bond. The reaction of alanine and glycine to form a peptide is shown below.

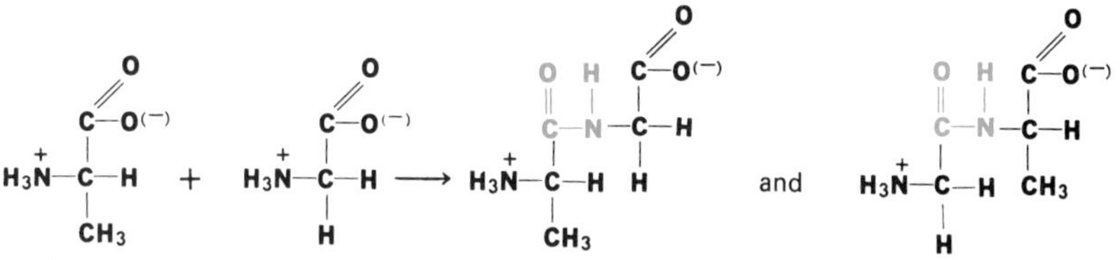

alanine   glycine   alanylglycine   and   glycylalanine

A compound containing two amino acid residues (but only one peptide bond) is known as a **dipeptide**. In addition to the two dipeptides given above, alanylalanine, glycylglycine, and tri, tetra, and polypeptides may be produced whenever these two amino acids react chemically to produce peptides. Therefore, the *chemical* synthesis of a specific peptide is difficult; it usually requires that the amino group of one amino acid and the carboxyl group of the other one be "blocked" so that only one reaction takes place. The *biological* synthesis of peptides is more specific and often gives only one product because of those remarkable catalysts, the enzymes. Three amino acids, when combined, give a **tripeptide**. The combination of many amino acids gives first polypeptides, and finally the extremely high-molecular-weight compounds, the proteins.

The structure of peptides may be indicated by names, as was done for alanylglycine. A shorter method is routinely used, however, whereby the three-letter symbols for each amino acid, Fig. 31.1, are strung together. It is understood that the free amino group is at the left end and the free carboxyl group at the right. However, the symbols (NH_2) and (COOH) can be used for unambiguous presentation. Thus alanylglycine is shown as ala·gly or (H_2N) ala·gly (COOH).

The reaction of an amino acid with a keto acid in the presence of an enzyme called a transaminase gives a different amino acid and a different keto acid. The process is called **transamination** since the amino group is transferred from one carbon chain to another one.

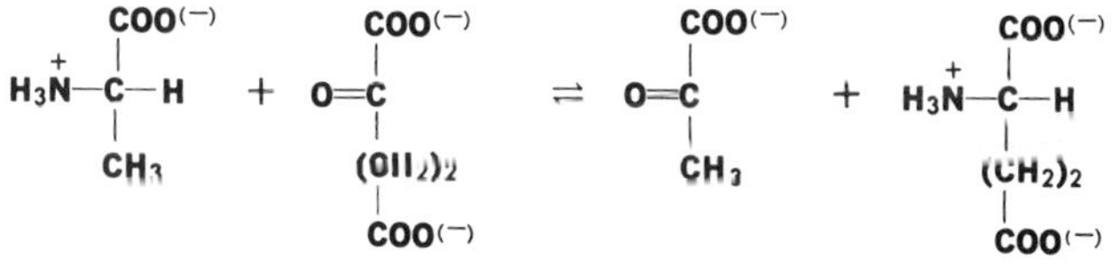

alanine   α-ketoglutaric acid   pyruvic acid   glutamic acid

The reaction is easily reversible, and at equilibrium there are appreciable quantities of both reactants and products. This reaction provides a method for synthesis of amino acids when the corresponding keto acid is available. Although it is generally said that certain amino acids are essential, i.e., they cannot be synthesized in a given animal, it is really the keto acid or the carbon skeleton which is essential. The reaction given above shows that either glutamic acid or alanine can be synthesized when α-ketoglutaric acid or pyruvic acid is present along with a source of amino groups. Since α-ketoglutaric acid, pyruvic acid, and sources of amino groups are common in the bodies of animals, glutamic acid and alanine are synthesized by most animals and, therefore, are not essential amino acids.

Amino acids may also be converted to keto acids by the process of **oxidative deamination**. This is a two-step process which involves first a loss of two hydrogen atoms, followed by the addition of a molecule of water. (For simplicity, the formulas for the amino acids are written in a nonionic form.)

$$\underset{\underset{R}{|}}{\overset{\overset{H_2N}{|}}{H-C-COOH}} \xrightarrow{2H} \underset{\underset{R}{|}}{H-N=C-COOH} \xrightarrow{HOH} \underset{\underset{R}{|}}{O=C-COOH} + NH_3$$

Unlike transamination, this series of reactions is not readily reversible. The overall reaction releases appreciable amounts of energy from the amino acid, and the reverse reaction does not serve as an important source of amino acids in animals. A process essentially the reverse of this, **reductive amination**, is used by plants for synthesis of amino acids

Amino acids may react to produce carbon dioxide and an amine by the process of **decarboxylation.**

$$\underset{\underset{R}{|}}{\overset{\overset{H_2N}{|}}{H-C-}}\overset{\overset{O}{\parallel}}{C-OH} \rightarrow CO_2 + \underset{\underset{R}{|}}{\overset{\overset{H_2N}{|}}{H-C-H}}$$

The extent to which such reactions occur in humans is probably slight, but histamine may be synthesized from histidine in this way. Histamine produces allergic reactions and promotes the motility of the stomach. The decarboxylation products of lysine and ornithine give the vile-smelling amines which have been given the expressive names of putrescine and cadaverine.

In addition to the general reactions which can occur with all amino acids, there are specific reactions. These are reactions of the side chain of the amino acid. Some illustrations are given below.

The hydrolysis of arginine to give urea and ornithine is a major source of urea in animals. Urea, which is found in the urine, is the major nitrogen-containing excretion product of the metabolism of amino acids. (See Section 36.5.)

$$\underset{\underset{\underset{\underset{NH}{\parallel}}{\underset{H-N-C-NH_2}{|}}}{\underset{(CH_2)_3}{|}}}{\overset{\overset{COO^{(-)}}{|}}{\overset{+}{H_3N}-C-H}} + HOH \rightarrow \underset{\underset{\underset{NH_2}{|}}{\underset{(CH_2)_3}{|}}}{\overset{\overset{COO^{(-)}}{|}}{\overset{+}{H_3N}-C-H}} + \underset{}{\overset{\overset{O}{\parallel}}{H_2N-C-NH_2}}$$

arginine   ornithine   urea

Two molecules of cysteine can be oxidized to form the disulfide cystine. This reaction is important in forming cross linkages between polypeptide strands in proteins. Mild reducing conditions can convert cystine to cysteine.

$$2 \ \underset{\substack{\text{cysteine}}}{\overset{\substack{COO^{(-)} \\ | \\ H_3\overset{+}{N}-C-H \\ | \\ CH_2 \\ | \\ SH}}{}} \ \rightleftharpoons \ \underset{\substack{\text{cystine}}}{\overset{\substack{COO^{(-)} \\ | \\ H_3\overset{+}{N}-C-H \\ | \\ CH_2 \\ | \\ S}}{} \ \ \ \overset{\substack{COO^{(-)} \\ | \\ H_3\overset{+}{N}-C-H \\ | \\ CH_2 \\ | \\ S}}{}} \ + \ 2H$$

A combination of general and specific reactions occurs in the conversion of the amino acid phenylalanine to the hormone epinephrine, which is also known as adrenalin. This conversion involves a series of reactions and illustrates how a normal dietary component is converted to a substance of vital importance to the continued life of the animal. A somewhat similar series of reactions which also includes the incorporation of iodine atoms leads to the synthesis of the hormone thyroxine.

Those amino acids (glutamic and aspartic acids) that contain an additional carboxyl group on the side chain can form amides with this second group. In some cases it is ammonia itself that reacts, in others the amine group comes from some other compound. The reaction with ammonia is an important way of removing this toxic compound from the bloodstream. A reverse of this action produces ammonia in the kidney cells, where it is rapidly converted to ammonium ions and excreted.

$$\underset{\substack{}}{\overset{\substack{O \\ \| \\ C-OH \\ | \\ (CH_2)_2 \\ | \\ H_3N^+-C-H \\ | \\ COO^-}}{}} \ + \ NH_3 \ \rightarrow \ \overset{\substack{O \\ \| \\ C-NH_2 \\ | \\ (CH_2)_2 \\ | \\ H_3N^+-C-H \\ | \\ COO^-}}{} \ + \ HOH$$

The amides formed from glutamic acid and aspartic acid are called glutamine and asparagine. They are often found in proteins instead of the amino acids with free carboxyl groups.

Although amino acids can be converted to many relatively simple compounds such as urea and epinephrine which are of vital importance, their major function is to serve as building blocks for peptides and proteins.

**31.3
PEPTIDES**

The amino acids and proteins have been the subject of study for many years. Except for the one tripeptide, glutathione, few peptides were known until comparatively recently. The structure of a representative pentapeptide is given in Fig. 31.3. Partly because of newer experimental techniques and partly due to other factors, research concerning peptides is now flourishing. A list of some of the peptides and their functions is given in Table 31.2. While it is difficult to study the functions of smaller peptides because they

TABLE 31.2   Some peptides found in the human body, their structure and functions

Name	Structure	Function
Glutathione*	-Glu-Cys-Gly	Widely distributed, probably functions as an oxidizing-reducing agent.
Oxytocin	Cys-Tyr-Ile | | S | | | S | | | Cys-Asn-Gln | Pro-Leu-Gly-NH$_2$	Contraction of smooth muscle. Used to induce labor in pregnant women.
Vasopressin	Cys-Tyr-Phe | | S | | | S | | | Cys-Asn-Gln | Pro-Arg-Gly-NH$_2$	Increases blood pressure. Promotes reabsorption of water by kidney cells.
Angiotensin I	(H$_2$N)Asp-Arg-Val-Tyr-Ile-His-Pro | (COOH)Leu-His-Phe	The angiotensins are produced in the kidneys and increase blood pressure throughout the body. They are the most potent vaso- constrictors known and are involved in some types of hypertension. Angiotensin II is more active than Angiotensin I, from which it is formed.
Angiotensin II	(H$_2$N)-Asp-Arg-Val-Tyr-Ile-His | (COOH)-Phe-Pro	

(continued)

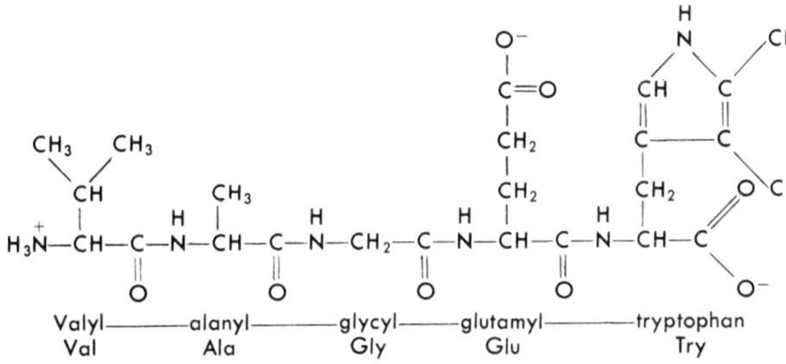

Fig. 31.3 A pentapeptide. The structure of this peptide is shown above, and the names and symbols of its components are shown below.

TABLE 31.2 (*continued*)

Name	Structure	Function
Thyrotropic Releasing Factor	Pyroglu†-His-Pro-NH$_2$	Promotes release of the pituitary hormone that stimulates the thyroid.
Follicle Stimulating and Luteinizing Hormone Releasing Factor	Pyroglu-His-Try-Ser | Arg-Leu-Gly-Tyr | Pro-Gly-NH$_2$	Promotes release of the hormones that stimulate ovary and testis.
β-endorphin	(H$_2$N)-Tyr-Gly-Gly-Phe-Met-Thr-Ser | (COOH)Thr-Val-Leu-Pro-Thr-Gln-Ser-Lys-Glu	β-endorphin and the enkephalins have a morphine-like activity in relief of pain.
Methionine 5-enkephalin	(H$_2$N)-Tyr-Gly-Gly-Phe-Met(COOH)	
Gramicidin‡	L-Leu L-Orn D-Phe | | L-Val L-Pro | | L-Pro L-Val | | D-Phe L-Orn L-Leu	Gramicidin is an antibiotic. It is not normally found in the human body, as are others in this list. It is unusual in that it contains D-amino acids.

The symbols (NH$_2$) and (COOH) serve to locate the amino and acid end of a peptide chain. The symbol —NH$_2$ indicates that the free COOH end of the chain has an amine group combined to give an amide structure.

* In glutathione the γ-carboxyl group of glutamic acid (the one on the side chain) is involved in forming the peptide bond. The α-carboxyl group remains free.

† Pyroglu is the symbol for pyroglutamic acid. In this form the amino group has reacted with the carboxy group on the side chain to give a cyclic structure

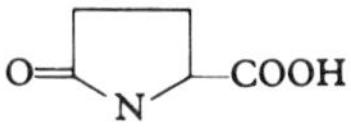

O=◯—COOH which leaves the α-carboxyl group

free to combine in peptide bonds.

‡ The symbol Orn stands for ornithine, an amino acid that is derived from arginine. Ornithine is not normally found in proteins.

are rapidly metabolized, it seems that processes such as the release of stored hormones, regulation of pain, and control of blood pressure may depend on small peptides. In fact, their rapid metabolism may be the property that makes them fit for producing a rapid but short-acting effect.

31.4 THE SYNTHESIS OF PROTEINS

Because of their great complexity and the difficulties involved in synthesizing specific optical isomers of individual dipeptides and tripeptides, only a few proteins have been chemically synthesized. Insulin, which contains 51 amino acids, has been synthesized, as has ACTH, which contains 39 amino acids, but these substances are near the polypeptide-protein border of classification. Relatively new techniques of synthesis are, however, beginning to make more complex syntheses possible. The enzyme ribonuclease, which is a small protein, has already been synthesized *in vitro*. Recent experiments involve altering bacteria so they will produce proteins such as insulin (Section 39.6). Techniques like these that are less time-consuming and less expensive will be required before we can expect commercially feasible *in vitro* synthesis of proteins in large amounts.

Proteins are, of course, synthesized by all living things. This process again is complex since it involves the linking of hundreds of specific units in a specific way. It can be calculated that there are approximately 1.35×10^{167} (135 followed by 165 zeros) possible arrangements of the amino acids found in the relatively small protein molecule of hemoglobin. This number is much greater than the number of atoms in the universe. The ability of a cell to synthesize just the right kind of proteins is truly remarkable.

The process of protein synthesis *in vivo* is under hereditary control. Although some of the details remain to be discovered, it now appears that the master plan for protein synthesis is found in the DNA (deoxyribonucleic acid) in the nucleus of the cell. This plan or code is transferred to the RNA (ribonucleic acid), which actually serves as the pattern for the extremely complex task of making just the right kind of protein. Protein synthesis is discussed in detail in Section 36.4. Several hereditary disorders are now known in which an individual is unable to synthesize one kind of protein. In some anemias there is a change in only one amino acid in the hemoglobin molecule. A similar "inborn error of metabolism" is responsible for the condition known as phenylketonuria which, if untreated, produces mental retardation in persons with the defect. The deficiency in this case is in a specific enzyme required for the metabolism of the amino acid phenylalanine. (See Section 39.4 for a further discussion of these hereditary defects.)

31.5 THE STRUCTURE OF PROTEINS

Most of the physical, chemical, and biological properties of proteins depend on their structure. We generally discuss protein structure in four categories: primary, secondary, tertiary, and quaternary structures. The **primary structure** of proteins involves the order of linkage of amino acid residues in the protein chain. Once formed, these peptide bonds between amino acids are not easily ruptured. The other aspects of protein structure (secondary, tertiary, and quaternary) are more subject to change. As an illustration of a biological property that is influenced by a difference in primary structure, we can consider normal and sickle-cell hemoglobin molecules. The difference of a single amino acid in the 150 amino acids in the structure of the β chain of the blood protein hemoglobin changes normal hemoglobin to sickle-cell hemoglobin.

Normal hemoglobin (H_2N)-Val-His-Leu-Thr-Pro-Glu-Glu-Lys- - -
Sickle cell hemoglobin (H_2N)-Val-His-Leu-Thr-Pro-Val-Glu-Lys- - -

The differences in properties of these two hemoglobins as well as the biochemical consequences are discussed in Section 39.4. For many people, having only sickle-cell hemoglobin means a life of pain and an early death.

Protein chains are not normally merely stretched out into individual straight chains, as might be expected from looking at the primary structures. There is almost always some **secondary** structure due to interaction between two chains or between the amino acids of a single chain. Many times both kinds of interaction occur. The most common bonds responsible for secondary structure are hydrogen bonds (Section 9.4). Hydrogen bonds form between the carbonyl portion of one peptide group and the amino portion of another. Two such bonds are shown below.

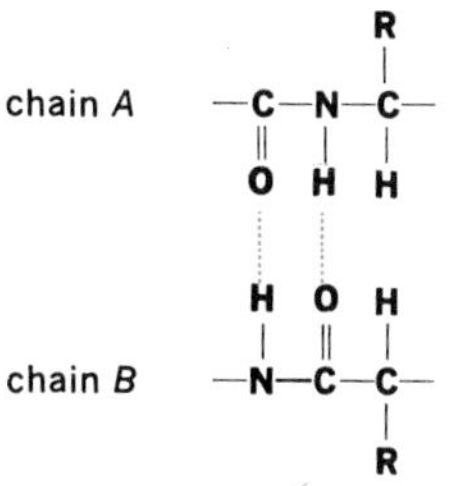

```
                         R
                         |
chain A       —C—N—C—
               ‖   |   |
               O   H   H


               H   O   H
               |   ‖   |
chain B       —N—C—C—
                         |
                         R
```

hydrogen bonds between amino acid chains

When hydrogen bonds occur between different chains, a pleated sheet structure results (Fig. 31.4). Bonding between units of a single polypeptide chain gives a spiral structure known as a helix (Fig. 31.4). Because hydrogen bonds are relatively weak and can be broken relatively easily, secondary structures are much easier to change than primary structures.

Tertiary structures produce the complex folding of protein chains. Bonds responsible for tertiary structures involve interactions of the side chains of amino acids. The amino acid side chains that react may be far from each other in a chain but are brought close together by folding of the chain. Tertiary bonding may be of several types. In some cases, covalent bonds between two —SH groups of two cysteine molecules may be found. These covalent disulfide bonds are much harder to break than hydrogen bonds.

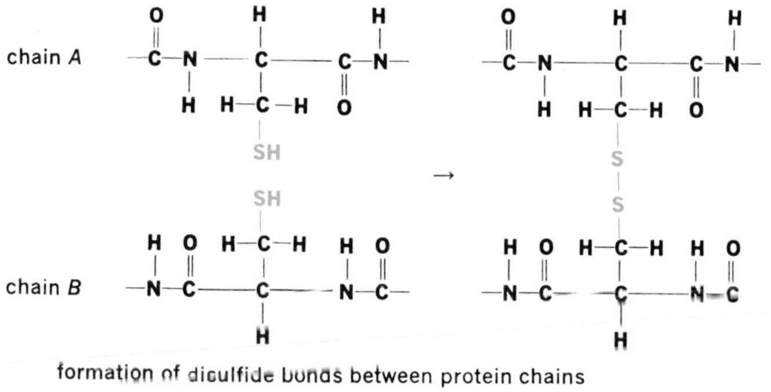

formation of disulfide bonds between protein chains

We break some disulfide bonds and establish new ones when we have a "permanent" wave put in our hair, or have it straightened.

Other tertiary bonds may involve hydrogen bonding between side chains, as between the —OH of a tyrosine or serine group and an acid from a glutamic or aspartic acid. In

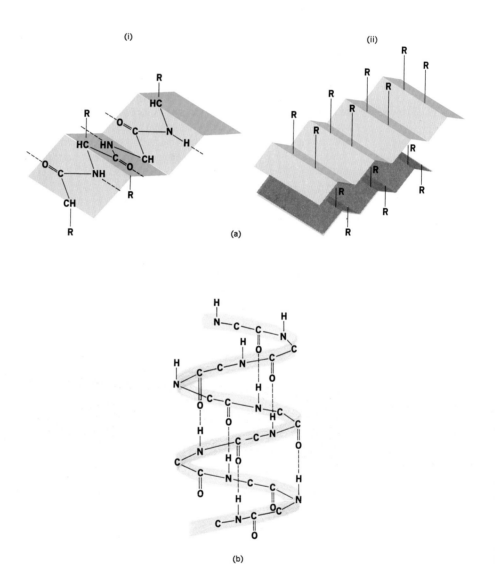

Fig. 31.4 Three-dimensional forms of polypeptides. Peptide chains in proteins are held together by secondary bonds, especially hydrogen bonds. Two forms of protein structures that are known to exist are the pleated sheet shown in (a) and the alpha-helix shown in (b). In (a), the pleated sheet form, the chains are linked together by the hydrogen bonds between the carbonyl groups of one chain and the amino groups of the other. Because of the natural bond angles, a structure such as this forms a pleated sheet. Several sheets may be held together as shown on the right. The alpha-helix shown in (b) represents another possible configuration of a polypeptide chain. The dotted lines between amino acid units represent hydrogen bonds. X-ray diffraction studies show that many proteins contain the alpha-helix in parts of their structure.

other instances we have an electrostatic attraction between a positively charged group ($-NH_3^+$) and a negative group ($-COO^-$) which are relatively far apart. If conditions are right, covalent bonds can be formed instead of the hydrogen bonds and electrostatic bonds described above. If such covalent bonds form, they are part of the tertiary bonding.

Another type of attraction that is important in establishing the tertiary structure of proteins is the attraction of nonpolar side chains for each other (see Table 31.1). It may be that this is really a mutual repulsion of these groups by water and polar parts of the molecule. It is true that in many proteins there is a sequence of amino acids with non-polar side chains, which come together and give a three-dimensional shape to the protein molecules. Such nonpolar areas are attracted to the similarly nonpolar lipids in combinations of lipids and proteins known as lipoproteins.

Thus there are many different kinds of bonds responsible for the tertiary structure of proteins. Along with the hydrogen bonds of the secondary structure, they give a protein a three-dimensional shape such as that seen for myoglobin (Fig. 31.5).

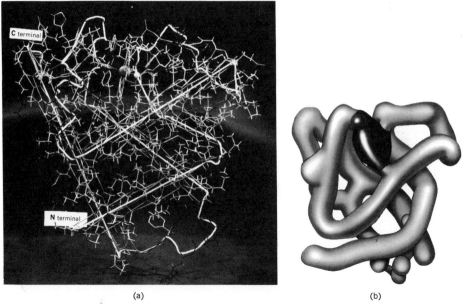

(a)   (b)

Fig. 31.5 The protein myoglobin: (a) photograph of a model of myoglobin from the sperm whale; (b) drawing of a molecule of myoglobin. In (b) the dark, disc-shaped portion is the heme group. Much of the structure of myoglobin is due to the attraction of non-polar side chains of amino acids within the molecule. [Photograph courtesy of Professor John C. Kendrew.]

Many proteins contain more than one subunit. For example, the protein hemo-globin contains four subunits (Fig. 31.6). Two of these are α chains and two are β chains. Only when we have all four chains (along with heme groups) do we have a hemoglobin molecule. This combination of subunits produces the fourth type or quaternary structure of proteins. There are many enzymes which require the combination of two or more subunits in order to be active. The units may be made of identical subunits or different subunits as in hemoglobin. The bonds formed to give quaternary structures are relatively easy to make or break.

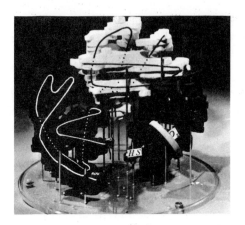

Fig. 31.6   The protein hemoglobin. Each of the four subunits of hemoglobin contains one heme group. The *a*-chains are represented in white; the *b*-chains in black. The model on the left shows two black *b*-chains and one white *a*-chain. The basic peptide backbone is shown by the white and black lines. The model on the right shows the complete hemoglobin molecule containing two *a*-chains and two *b*-chains. Two heme groups are shown; the other two are on the opposite side of the molecule. In this model oxygen is shown attached to the heme groups. [From *Nature* **185**, 416 (1960). Photographs courtesy of Dr. M. F. Perutz.]

In addition to amino acids, many proteins contain a nonprotein unit. In hemoglobin, this subunit is called heme and contains iron plus a complex ring structure (Fig. 40.2). Enzymes often contain some unit called a coenzyme attached to the protein. An example of such a unit is given in Fig. 34.9.

**31.6
CLASSIFICA-
TION
OF PROTEINS**

One of the first classifications of proteins was proposed in 1907 by the British Physiological Society. This classification was based on solubility. From it we retain the categories of **albumins** and **globulins**. The albumins are soluble in water and are precipitated only by relatively concentrated solutions of salts. The globulins generally require dilute salt concentrations for solubility but are precipitated out of solution by lower concentrations of salt than are albumins. If we have a solution containing albumins and globulins, we can separate them by making the solution 50% saturated with ammonium sulfate. At this point the globulins are precipitated while the albumins remain in solution. We can also separate the albumins and globulins by electrophoresis, since the albumins travel farther on an electrophoretogram than the globulins. Other groups from the 1907 classification, such as the prolamines, glutelins, and scleroproteins, are not commonly used now.

We often classify proteins as being **fibrous or globular**. The fibrous proteins are generally insoluble and the globular proteins (which include the albumins and globulins) are soluble in water solutions. The globular proteins are generally classified in terms of the roles they play in living tissues. Most enzymes, the blood proteins, and many proteins that help confer immunity to diseases are globular proteins. Fibrous proteins are the structural proteins, which may be further classified as follows.

Keratins are the tough proteins of fur, hair, hooves, claws, and feathers. Normally these proteins exist in a helical structure, often with disulfide linkages between coils of the helix. When stretched under heat and pressure, keratins may assume a pleated sheet form. Silk exists as a pleated sheet even without stretching. It is made of only three amino acids, glycine, serine, and alanine.

Collagens are fibrous proteins found in bone, teeth, cartilage, and skin. In bones and teeth, mineral elements are added to collagen. Normally collagen has a complex helical structure involving three helixes twined together. Collagen contains about 25% glycine, and another 25% is made of proline and hydroxyproline. Boiling collagen converts it to gelatin.

Closely related to collagens are the **elastins**, which are proteins that can be repeatedly stretched and will still regain their former shape. These are important in the lungs and blood vessels. Loss of elasticity by skin proteins is one of the aspects of getting old. The **myosins** of muscle, which can contract and relax under proper stimuli, and the **fibrin** that forms when the blood clots, are also fibrous proteins.

We also find that many proteins are closely associated with components other than amino acids. If lipids are closely bound to the protein molecule we have lipoproteins. Carbohydrates are a part of glycoproteins. Metalloproteins and phosphoproteins contain metals and phosphates. Nucleoproteins, which are rich in the basic amino acids lysine and arginine, are associated with nucleic acids in cells.

**31.7
PHYSICAL
PROPERTIES
OF PROTEINS**

From the large molecular weights of the proteins, which range from about twelve thousand to hundreds of thousands, we would expect these molecules to be insoluble in water. Many proteins are, in fact, insoluble in water, but many very important ones are quite water-soluble, probably because of the attraction of water molecules to charged groups in the protein molecule. Although most amino and acid groups are tied up in the peptide bond, other groups of amino acids (their side chains) are often in a charged state. The additional acid groups in glutamic acid and aspartic acid and the amino groups in lysine and arginine, as well as other side chains, can be charged. As with the primary amino and acid groups, the state of charge of the side chains depends upon the pH of the solution. When the net charges on a protein are at a minimum, i.e., when the negative charges are equal to the positive charges, the protein will not migrate when placed in an electrical field. As we have already seen, the process of placing a protein solution on paper or some other relatively inert substance and allowing the proteins present to migrate toward the anode or cathode is called electrophoresis (see Figs. 31.2 and 31.7). This technique is used to separate individual proteins from mixtures. It is now finding wide application for studying the various proteins in blood.

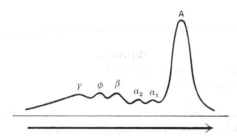

Fig. 31.7 Electrophoretic pattern of the proteins in human blood plasma. Albumin is identified by **A**, fibrinogen by ϕ, and the various globulins by the Greek letters α, β, and γ. See Fig. 40.3 for other electrophoretograms of human plasma. The exact shape of the curve obtained varies with the pH and salt concentration of the liquid used for electrophoresis. The arrow shows direction of migration.

Recall (Section 31.2) that the pH value at which a protein will not migrate when placed in an electrical field is called its **isoelectric point**. This is the point of minimal solubility for that protein. We can make use of such knowledge in separating it from other proteins, or in helping us decide which protein (or proteins) are present in a solution.

**31.8
CHEMICAL
PROPERTIES
OF PROTEINS**

The chemical reactions of the proteins are difficult to study. In most cases only a relatively small part of the huge protein molecule is actually changed or undergoes a reaction. For example, it is now believed that only a small part of the molecule of an enzyme, called its active site, is involved when the enzyme catalyzes a reaction. The combination of the iron-containing protein hemoglobin with oxygen is believed to involve the iron atom and one of the histidine molecules in the protein chain. Structural studies of the enzyme ribonuclease show an opening in the molecule just the right size for molecules of the substrate upon which the enzyme acts. Biochemists are only beginning to understand the true nature of the chemical reactions of proteins. More knowledge in this fascinating area will bring us much closer to understanding the chemical nature of living processes.

Other, *in vitro*, chemical reactions of proteins are used for their identification and characterization. In some cases a colored product is produced, as in the xanthoproteic reaction, in which proteins are treated with nitric acid, or in the biuret test, which uses a copper salt in conjunction with a basic solution of the protein. Other tests involve the precipitation of the protein by heat or salts.

Most of these reactions used to identify proteins result in the gross alteration of the protein. The term **denaturation** has been used to describe these changes. While there is some disagreement as to the use of the term denaturation, it is best defined as *any change in any property of a protein that does not involve rupture of the basic peptide bonds*. Some of these changes can be reversed, as when a globulin that has been "salted out" of solution is redissolved. Others, such as the heat treatment of egg albumin, cannot be reversed, i.e., you can't "unboil" an egg. (See Fig. 31.8.) Some biochemists restrict the definition of denaturation to those processes which are irreversible.

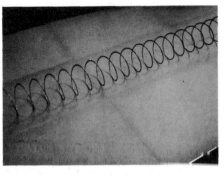
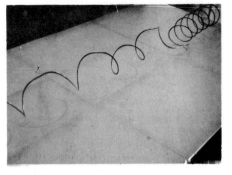

(a)   (b)

Fig. 31.8 Denaturation of proteins. (a) The coiled spring toy known as a "slinky" is used to represent the helical structure of a protein. When the elastic limit of the helix is exceeded, the shape is irreversibly altered as shown in (b). This is analogous to irreversible denaturation of a protein. [From J. J. W. Baker and Garland E. Allen, *Matter, Energy, and Life*, 3rd Ed., Addison-Wesley, 1977.]

Many denaturation reactions are used in biology and medicine. Ethanol and other organic solvents are used to destroy bacteria by denaturing their proteins. Similarly, heat and ultraviolet radiation can destroy bacteria by denaturation. Both heat and ultraviolet radiation can also denature some of the proteins of human skin.

Other substances such as tannic and picric acids denature proteins and in the process make them less soluble and tougher. Tannic acid (derived from oak bark) is used in making leather from the protein in hides of animals. Both tannic and picric acids are used to treat burns. In this case, the softer proteins exposed by a burn are toughened and prevent entrance of bacteria into the wound produced by the burn.

Heavy metal ions such as those of mercury, lead, and silver react with —SH groups in proteins to denature them. This explains why these metals are poisonous to humans. Solutions of these three metals have also been used to destroy bacteria, although such uses are now largely discontinued due to the possible toxicity of the metals to humans. Some hospitals still use a dilute solution of silver nitrate in the eyes of newborn babies. This is done to destroy bacteria (particularly the gonococci which can cause blindness) that the baby might have acquired from the birth canal. It is important that the silver nitrate solution be dilute so it does not significantly denature the proteins of the eye.

**31.9
PURIFICATION
AND
FRACTIONA-
TION OF
PROTEINS**

The purification of proteins has played an important role both in the study of these vital compounds and in the uses of the purified substances. Before determining the composition of a protein, the chemist must be sure that it is pure, i.e., that the sample consists of only one specific protein. The study of the exact nature of catalysis by enzymes also requires pure substances. Since any foreign protein that is injected into an organism usually produces antibodies to that protein, it is essential that any material that is to be injected into a human for medical purposes be highly purified. For these reasons it is necessary that insulin, which is used to treat diabetes, and the polypeptide ACTH, which is sometimes used to treat local inflammations, as well as many other proteins be as pure as possible.

Protein purification is based on a knowledge of the physical properties of the proteins. The proteins may be separated from lipids and carbohydrates since they have different solubilities and also because of their charged state. If relatively crude materials are used, the material is first fractionated by spinning it in a centrifuge. Proteins have different solubilities. This differential solubility of the various proteins in salt solutions or in solutions containing organic solvents such as ethanol or acetone forms the basis for many separations. Other purifications can be made by passing the material through a column of a chromatographic absorber, which utilizes the differences in affinity of the various proteins for the adsorbent. Electrophoresis uses the difference in migration rate of proteins as influenced by their charged state, which is, in turn, influenced by the pH of the solution. This method is useful in purifying relatively small amounts of proteins. (See Fig. 29.1.) All purification procedures, particularly those with organic solvents, are preferably done at low temperatures and with the use of other precautions to prevent irreversible denaturation of the protein. Since proteins are large molecules, they can be separated from salts and other small soluble molecules by dialysis. In this process a sac of semipermeable membrane retains the protein while the smaller particles escape into a surrounding liquid. Water may be removed from protein solutions under vacuum at low temperatures in a process called freeze drying. The dry powders so produced are much more stable in storage than are solutions of proteins. By using a variety of these techniques it is possible to separate the proteins of blood so that dried serum or gamma globulin becomes available for medical uses.

31.10
THE
IMPORTANCE
OF PROTEINS

Proteins play many vital roles in the living organism. Their importance in enzymatic catalysis, as oxygen carriers, and as structural materials has already been mentioned. While it is impossible in an elementary text to discuss all the important functions of proteins, a few more significant ones are given below.

In addition to hemoglobin, the blood contains many other proteins. There are albumins which serve to maintain sufficient osmotic pressure to help water and other nutrients pass freely through the walls of the capillaries. The particular mixture of proteins that has been called gamma globulin constitutes one of our major supplies of antibodies for resisting some diseases. The proteins fibrinogen and prothrombin are essential for the formation of the clump of red blood cells called a clot. Clotting is useful when it prevents excessive loss of blood, but clots can also form in the blood vessels of the heart or brain, causing a heart attack or a stroke.

Many of the hormones which serve as important regulators of the processes of the body are proteins. Most of the hormones of the master gland, the pituitary, are protein-containing. Two of them, vasopressin and oxytocin, are octapeptides; others are either glycoproteins (compounds containing both protein and carbohydrate) or pure proteins. The widely used hormone insulin contains 51 amino acids and may be classed as either a large polypeptide or a very small protein (see Fig. 31.9). The discovery and extraction

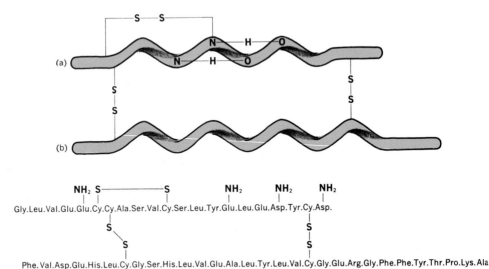

Fig. 31.9 Two representations of the structure of insulin. The lower diagram shows the sequence of amino acids in two protein chains called the a- and b-chains. These two chains are linked together with disulfide bonds. The upper representation is a drawing of the insulin molecule showing both the disulfide bridges and possible hydrogen bonds. The disulfide bridge in the a-chain links two parts of that chain and does not serve to bind two polypeptide chains together as do the other disulfide bonds. [Adapted from C. B. Anfinsen, *The Molecular Basis of Evolution*, 2nd. ed., New York, Wiley, 1963.]

of insulin and its use in the treatment of diabetes makes a dramatic chapter in the story of the human conquest of disease. Thousands of persons whose days would have been limited now live normal lives because large quantities of highly purified insulin are available. The use of the protein growth hormone has met with only limited success. Further study and purification of other hormones can be expected to lead to longer and better lives for some of those persons who today suffer from hormone insufficiency.

The importance of proteins in our lives begins when a sperm cell (made up of nucleic acids and proteins) unites with an ovum. In animals, the fertilized egg develops through stages of differentiation, which are largely influenced by a succession of enzymes, to the birth of the individual. Birds' eggs, which provide nutrition for the embryo birds, are especially rich in proteins. The gradual growth of a person is due to the increase in bone and muscle under the control of the proteinaceous hormones and with the help of the enzymes. The production of new sperm and ova under the influence of other protein-containing hormones leads to the perpetuation of the species. The continued life and health of the individual depends on sufficient nutrition by proteins (along with other foods) and on the ability to manufacture sufficient enzymes to catalyze the essential reactions and sufficient antibodies to resist infections. The failure of any of these protein-influenced systems or, in some cases, the overactivity of some of these processes leads to the death of the individual.

SUMMARY

The proteins are important compounds in both plant and animal life. The different properties of various proteins are due largely to the side chains of the amino acids of which they are made, since the amino and acid groups are tied up in the peptide bonds that hold proteins together. The order in which amino acids are linked in a protein molecule provides innumerable different proteins which can serve a wide variety of functions. This variety is further extended by the different shapes that may be assumed by a given sequence of amino acids due to secondary, tertiary, and quaternary bonds. These differences in structure and shape and the possibility of changes in configuration mean that the properties of proteins may change with external conditions. Such changes are needed for an organism to respond to a changing environment—both the external environment of wind and weather and the internal one of sugar, salt, and stress.

While some proteins are extremely stable, such as those found in animal hides, which can be converted to a tough durable leather, others are very delicate. Merely passing a current of air through a solution of some enzymes is sufficient to alter them so much that they can no longer function as catalysts.

The amino acids which humans derive from digestion of food play many important roles in the body. One of the major roles is to serve as building blocks for the proteins and peptides synthesized by humans. Another role of amino acids is the formation of relatively small molecules such as urea, epinephrine, and thyroxine, which are important to the normal functioning of the human body. Transamination of amino acids produces keto-acids which can be metabolized to give energy.

Peptides containing relatively few amino acids are being found to play important roles in the human and other animals.

The functions of proteins are many and varied, and while the importance of other classes of compounds cannot be forgotten, Mulder's statement that protein "is without doubt the most important of the known components of living matter, and it would appear that, without it, life would not be possible" is still true. Research studies of the nature and reactions of proteins are one of the most important ways we can achieve the fundamental objective of biochemistry—the explanation of life in chemical and physical terms.

IMPORTANT TERMS

albumin	electrophoresis	isoelectric point
amino acid	essential amino acid	peptide
deamination	fibrous protein	peptide linkage
decarboxylation	gamma globulin	polypeptide
denaturation	globular protein	protein
dipeptide	globulin	salting out
dipolar ion	hemoglobin	transamination

WORK EXERCISES

1. List several natural products that are primarily or largely composed of protein.

2. List the common dicarboxylic amino acids.

3. List the common amino acids containing more than one amino group.

4. What functional groups other than the carboxyl and amino groups are found in the side chains of proteins?

5. What element is always present in protein and not generally present in carbohydrates?

6. A sample of egg white is 1.70% nitrogen. What percentage of the egg white is protein?

7. Egg yolk is 16.3% protein. Assuming that all of the nitrogen in the yolk is due to protein, what is the percentage of nitrogen in egg yolk? Which contains the higher percentage of protein, the white or the yolk of egg?

8. Describe the following physical properties of amino acids: solubility; optical activity; and migration in an electrical field.

9. Write all possible charged states of cystine, of glutamic acid, of lysine.

10. List the chemical reactions that are common to all amino acids.

11. Write equations for each of the following reactions and name all the products of the reactions.
   a) the reaction of glycine and serine to form a dipeptide
   b) the transamination of alanine with alpha ketoglutaric acid
   c) the decarboxylation of histidine, of ornithine
   d) the complete hydrolysis of lysylphenylalanyl-asparagine
   e) the oxidative deamination of aspartic acid, of alanine

12. Write the structures for glycylcysteinylalanine and for threonylcysteinylproline, and show possible cross-linkages between these peptides.

13. Write the complete structure for the peptides indicated below.

   a) (H_2N)-Tyr-Ala-Ser(COOH)
   b) (H_2N)-Gly-Gly-Met-Phe(COOH)

14. List the different kinds of bonds that are involved in each of the basic structures of proteins (primary, secondary, and tertiary).

15. List some classifications of proteins that are based on solubility of the protein.

16. List some functions of peptides containing less than 40 amino acid residues.

17. You have a solution of egg white that contains both globulins and albumins. How would you separate these two types of proteins (two different methods)?

18. Give a major function of each of the following proteins or types of proteins.

   a) hemoglobin   b) keratins
   c) collagens   d) elastins
   e) myosins   f) fibrin

SUGGESTED READING

Fraser, R. D. B., "Keratins," *Scientific American*, August 1969, p. 86 (Offprint #1155). The type of protein known as keratin is found in skin, feathers, hair, and other external parts of many animals. This articles describes methods used to establish the three-dimensional structure of this type of protein.

Patton, S., "Milk," *Scientific American*, July 1969, p. 58 (Offprint #1147). The production and composition of a basic life-giving food are discussed at length in this well-written article.

32

Chemistry of the Lipids

If we separate the water, carbohydrates, and proteins from a cell or from a total organism, the major component left is **lipid**. In practice the procedure used is just the opposite of that given above—the lipids are separated from all other biochemical materials by extracting them in ethyl ether, ethanol, chloroform, or similar nonpolar solvents or mixtures of these solvents. This procedure is the basis for our definition of lipids as *biochemical substances that are soluble in nonpolar organic solvents*. The lipids are commonly known as fats; biochemists, however, reserve the name "fat" for one particular kind of lipid.

Since lipids are defined in terms of solubility, and since many types of compounds are soluble in the lipid solvents, we find the lipids include many heterogeneous compounds. They are not made of one particular kind of building block as are the proteins and carbohydrates, and almost no lipids are really polymers.

Since the lipids lack a really basic building block, the only way to study the lipids systematically is by a classification based on solubility and hydrolysis products. As you may expect, there are different systems of classifications. It is especially difficult to classify those lipids that yield two or more characteristic hydrolysis products. The following classification is widely accepted and is sufficient for a basic understanding of the lipids:

a) triacylglycerols (triglycerides)   b) phospholipids (phosphatides)
c) waxes   d) steroids
e) terpenoids   f) miscellaneous

The placement of a lipid in one of the categories above depends primarily on its hydrolysis products. Although there are many kinds of molecules in these hydrolysis products, as would be expected in a group of compounds defined only in terms of solubility, there is more similarity than one might expect. In fact, we find present in lipids relatively few of

TABLE 32.1 Common fatty acids

SATURATED ACIDS

Name	Formula	Common Source
acetic acid	CH_3COOH	vinegar
butyric acid	C_3H_7COOH	butter
caproic acid	$C_5H_{11}COOH$	butter
caprylic acid	$C_7H_{15}COOH$	butter
capric acid	$C_9H_{19}COOH$	butter, coconut oil
lauric acid	$C_{11}H_{23}COOH$	coconut oil
myristic acid	$C_{13}H_{27}COOH$	coconut oil
*palmitic acid	$C_{15}H_{31}COOH$	animal and vegetable fats
*stearic acid	$C_{17}H_{35}COOH$	animal and vegetable fats
arachidic acid	$C_{19}H_{39}COOH$	peanut oil
lignoceric acid	$C_{23}H_{47}COOH$	brain and nervous tissue
cerotic acid	$C_{25}H_{51}COOH$	beeswax, wool fat

UNSATURATED ACIDS

Name	Formula	Number of Double Bonds	Common Source
*palmitoleic acid	$C_{15}H_{29}COOH$	1	animal and vegetable fats
*oleic acid	$C_{17}H_{33}COOH$	1	animal and vegetable fats and oils
linoleic acid	$C_{17}H_{31}COOH$	2	linseed oil, vegetable oils
linolenic acid	$C_{17}H_{29}COOH$	3	linseed oil
arachidonic	$C_{19}H_{31}COOH$	4	brain and nervous tissue

* Acids indicated with an asterisk are among the most widely distributed.

the thousands of possible organic compounds. This natural economy comes about because of the great similarity in synthetic pathways in different organisms. We must not forget that the composition of biochemical substances is the result of the synthetic processes in plants and animals.

The most common hydrolysis products or building blocks of the lipids are the monocarboxylic organic acids, which are commonly called the fatty acids. Of all possible fatty acids, we find that only those having an even number of carbon atoms occur to any appreciable extent. Although there is some variation as to the number of double bonds (degree of unsaturation) in these acids, branched chains and ring compounds are rare in the fatty acids. The commonly found fatty acids are given in Table 32.1. Some fatty acids are found loosely bound to albumins in the blood. The level of these "free fatty acids" (FFA level) may be used in diagnosis of cardiovascular disease (see Section 40.7).

In addition to the fatty acids, certain other components are found as hydrolysis products of lipids. The more common are glycerol, choline, and aminoethanol.

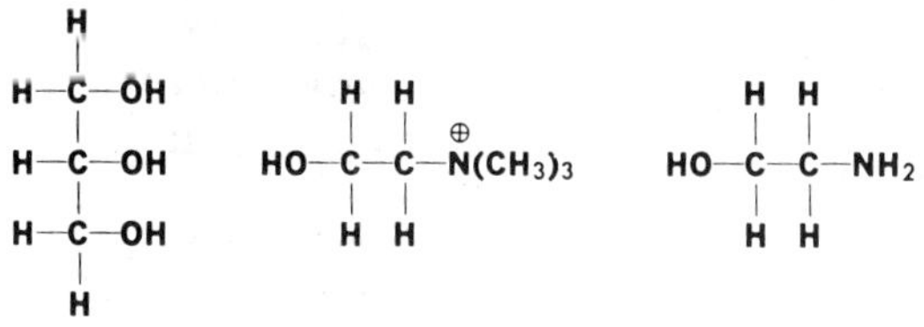

glycerol   choline   aminoethanol

Other lipids contain structural units which are related to isoprene.

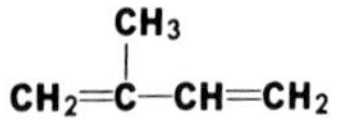

$$CH_2{=}\overset{\displaystyle CH_3}{\underset{\displaystyle |}{C}}{-}CH{=}CH_2 \quad \text{isoprene}$$

The isoprene units exist both in chain form and in ring structures, as in β-carotene.

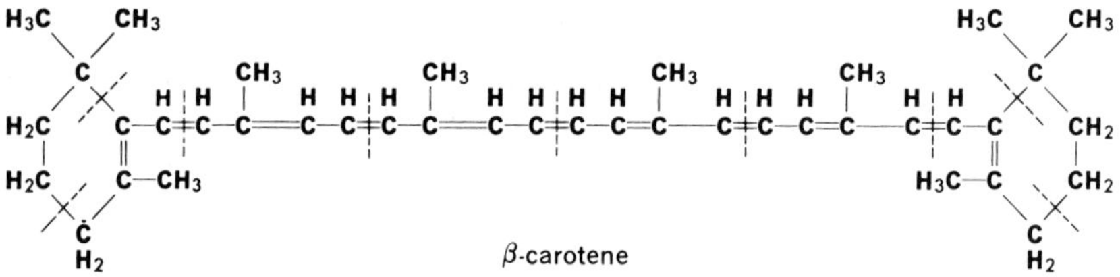

β-carotene

The dotted lines indicate isoprene units in this complex structure. β-carotene, which is a brilliant orange color, gives carrots their color. The carotenes are also present in grass and leaves of plants, but the color is masked by chlorophyll. Rubber is a polymer made of isoprene units [Eq. (27–6)]. Although it is more difficult to see, the complex structure known as the steroid nucleus contains several isoprene units. The biological synthesis of cholesterol involves the combination of several such units.

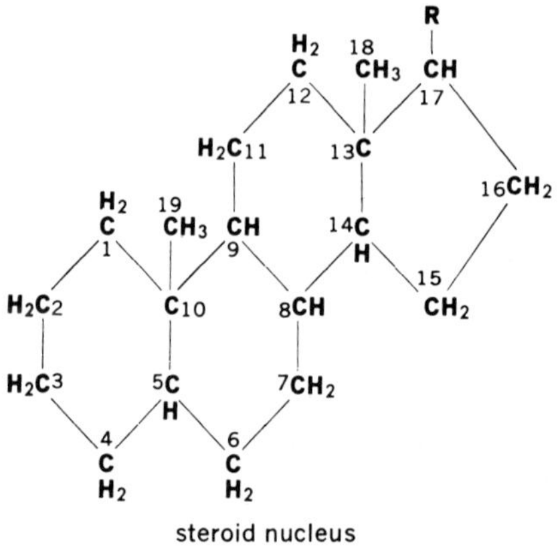

steroid nucleus 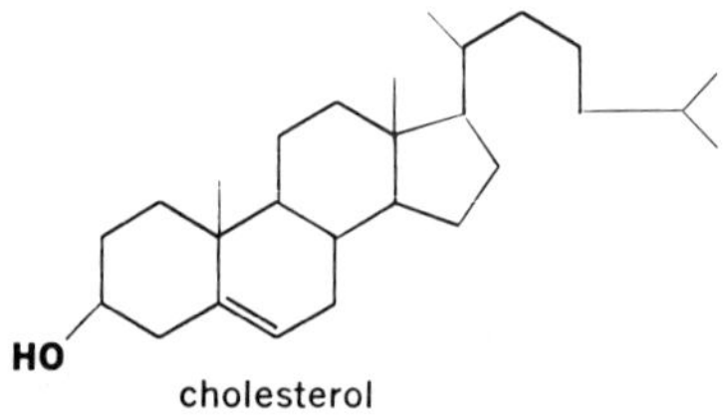   cholesterol

Although isoprene units are involved in the synthesis of the compounds given above, hydrolysis will not normally give these units.

Another class of compounds that are classified as lipids because of their solubility are the **prostaglandins**. They generally contain 20-carbon carboxylic acids with a five-membered ring, hydroxyl groups, and double bonds. The formulas for two prostaglandins are given in Fig. 32.1. These compounds are discussed further in Section 32.8.

**32.3
TRIACYLGLY-
CEROLS (TRI-
GLYCERIDES)**

The lipids which, upon hydrolysis, yield one molecule of glycerol and three of fatty acids are known as **triacylglycerols**; however, the common name, **triglycerides**, is also widely used in discussing these compounds. The basic chemistry of these compounds was discussed in Chapter 24. Here we will add details that are of special interest to biochemists.

The formula for a typical triacylglycerol is shown in Fig. 32.2, both in a structural

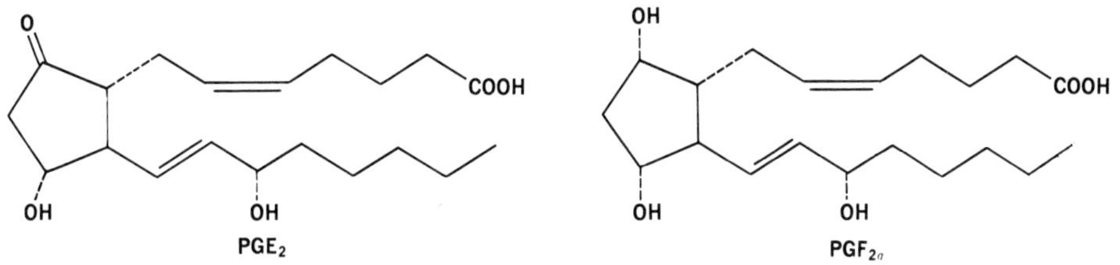

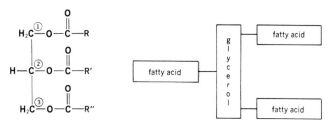

Fig. 32.1 Prostaglandins. The structure of Prostaglandin E_2 (PGE$_2$) and Prostaglandin F_{2a} (PGF$_{2a}$) are shown above. Bonds indicated with a dotted line (----) are considered to be below or in back of the five-membered ring. Solid bonds indicate groups above or in front of the ring.

formula and in a "block" formula that indicates the building blocks of these compounds. The fatty-acid units (sometimes called fatty-acid residues) at positions 1 and 3 are more likely to be saturated and those found at position 2 are usually unsaturated, often containing fatty-acid residues with more than one double bond. Naturally occurring triacylglycerols have a chiral carbon at position 2.

Fig. 32.2 The structure of a typical triacylglycerol (triglyceride). The R groups are those of the fatty acids. The block diagram at the right indicates units of the triacylglycerols and also indicates the products derived by hydrolysis.

We find triacylglycerols in both plants and animals. Although plants tend to store excess energy in the form of polysaccharides, they often contain appreciable amounts of triacylglycerols in their seeds. Animals accumulate triacylglycerols which we commonly call fat when their food intake exceeds energy output. Triacylglycerols from animal sources are generally more saturated than those from plant sources. As explained in Chapter 24, saturated triacylglycerols tend to be solids at room temperature and are commonly known as **fats**, while triacylglycerols containing a few double bonds per molecule are liquids and are called **oils**.

The separation of most fats and oils from their natural sources usually involves only a physical process. Olive and cottonseed oils are separated by pressing, and butter by churning cream which contains the butterfat. Lard is separated from various fatty tissues of the swine by heating to melt the fat, which is then filtered while hot to remove it from proteins and carbohydrates. In biochemical preparations, fats are usually extracted, along with the other lipids, by adding a mixture of organic solvents (usually ethanol and ethyl ether) to a natural product. Triacylglycerols can be separated from phospholipids by their differences in solubility. Triacylglycerols are separated from sterols and other lipids by saponification. This process converts triacylglycerols into water-soluble soaps and glycerols, while other lipids (the nonsaponifiables) remain soluble in nonpolar solvents.

Pure triacylglycerols are colorless, odorless, and tasteless. Characteristic tastes, odors, and colors of butter, olive oil, and cod liver oil are caused by compounds other than triacyl-glycerols. Many of the physical properties of the triacylglycerols depend on length or degree of unsaturation of the carbon chain of the esterified fatty acids; see Section 24.3.

The major biochemical reaction of triacylglycerols is hydrolysis. Enzymes which catalyze the hydrolysis are commonly called **lipases**. Other reactions may occur at double bonds in the fatty-acid chain. These reactions include oxidation, which occurs naturally, and addition of halogens, which does not normally occur in living systems. Halogenations are used primarily to identify and characterize lipids.

Butter and other triacylglycerols that have been left at warm temperatures for long periods of time tend to develop obnoxious odors and tastes. We say they have become **rancid.** The odors are due to free fatty acids (butyric acid in butter) and aldehydes. These free acids and aldehydes are formed as a result of two of the characteristic reactions of the triacylglycerols, *oxidation* of double bonds and *hydrolysis* of the ester bond. The *hydrolysis*, which releases odorless glycerol in addition to the vile-smelling acids, is often catalyzed by enzymes present in bacteria. *Oxidation* of fatty acids occurs principally at the double bonds. Exposure to the oxygen of air at warm temperatures induces oxidation, which results in the rupture of double bonds to form free aldehyde groups. These aldehydes have odors similar to those of the fatty acids and, in fact, are readily converted to fatty acids by further oxidation. Oxidative processes are the primary causes of rancidity in most foods.

Triacylglycerols are of major importance because they represent the primary energy storage compounds in animals. They provide insulation from both heat loss and mechanical shock. An insulating layer of fatty tissues just below the skin helps the human body maintain its required temperature of 98.6°F. Many vital internal organs, such as the kidneys, are encased in a thick layer of fat which helps to protect the animal against mechanical shock. Subcutaneous fat also performs this function. While fats serve many useful functions, persons with too much fat are less healthy and often die younger than average persons. High levels of triacylglycerols in blood are associated with heart attacks.

32.4 PHOSPHOLIPIDS

Many lipids contain the element phosphorus in addition to carbon, hydrogen, and oxygen. Those phospholipids are found in many parts of living organisms, but are especially prevalent in the brain and nervous tissue. Liver and egg yolks also contain appreciable amounts of a particular type of phospholipid—lecithin. The membranes of most cells contain phospholipids.

Many phospholipids are considered derivatives of phosphatidic acid. The R-groups are those of relatively long-chain fatty acids.

phosphatidic acid

Phosphatidic acid does not occur as such in living animals. In derivatives of phosphatidic acid the phosphate group is generally esterified with an OH group of a nitrogen-containing alcohol. Substances most commonly esterified here are choline, ethanolamine, or the amino acid serine.

a lecithin

Further classification of phospholipids that are derivatives of phosphatidic acid depends on the basic alcohol present. **Lecithins** contain choline; **cephalins** contain either ethanolamine or serine; and the **plasmalogens** contain an unsaturated long-chain ether in place of a fatty acid and either choline or ethanolamine. Phospholipids which are not derivatives of phosphatidic acid are also known.

The phospholipids, particularly the lecithins, are large molecules having both a non-polar component and a highly polar one. This property is utilized in the structure of membranes. Also, we would expect them to be good detergents (see Section 24.7). A major biochemical function they perform is in bringing together the water-insoluble lipids and the water-soluble components of any living organism. It has been proposed that phospholipids help in the transport of lipids in the bloodstream (which is basically water). As a matter of fact, tiny fat globules are present in the blood after digestion of a meal containing fat. It is interesting to speculate that perhaps the phospholipids are responsible for keeping these globules small enough so that they do not block the small blood vessels. Less vital uses of lecithins are as emulsifying agents to keep the fat, chocolate, and other components of many candies from separating. Egg yolks, which are rich in lecithins, are used to emulsify vinegar and salad oil into a mayonnaise or salad dressing.

The prevalence of phospholipids in brain and nervous tissue leads us to wonder about their significance in these important tissues. While it may be they are present only inadvertently, some biochemists are strongly influenced by the belief that "nature probably does nothing without a purpose," and they are trying to find the purpose or function of the phospholipids in these areas. At present, we can say only that their function is not yet fully understood.

**32.5
WAXES** The esterification of a fatty acid with a simple long-chain alcohol (i.e., one having only a single OH group rather than three as in glycerol) gives a wax. Waxes are present in the leaves and fruits of many plants. They provide protective coatings for these structures. Beeswax is secreted by bees and provides a structure for storing honey. Whale oil (which is not truly an oil, but a wax) contains cetyl palmitate, and beeswax contains myricyl palmitate.

$$CH_3(CH_2)_{14}-\overset{\overset{\displaystyle O}{\|}}{C}-O-(CH_2)_{15}-CH_3 \qquad CH_3(CH_2)_{14}-\overset{\overset{\displaystyle O}{\|}}{C}-O-(CH_2)_{29}-CH_3$$

cetyl palmitate   myricyl palmitate

A block diagram for waxes would be:

fatty acid	—	long-chain alcohol

**32.6
STEROIDS**   The term **sterol** means solid alcohol. Cholesterol was isolated from gallstones in 1775, and the prefix *chole* indicates its relation to the bile (*chole* is the Greek word for bile). We now know that cholesterol is present in many tissues of the body, especially in brain and nervous tissue. Cholesterol was only the first of a great variety of compounds having a steroid nucleus. Steroids having an —OH group are called *sterols*. (See Section 32.2 for the formula for the steroid nucleus and cholesterol.)

The steroid nucleus is essentially a flat structure. If we consider the nucleus to be in the plane of the page of this book, substituent groups will be either above or below the ring. When we draw a steroid structure, we use a solid line to indicate that the substituent is above or in front of the ring, and a dotted line to show it is below or to the back.

Since sterols are primarily hydrocarbons with a high molecular weight, they are soluble in the nonpolar organic solvents, and are classified as lipids. Modern biochemistry has found other relationships between the fatty acids and cholesterol that indicate the propriety of their being classified together. Acetate, in its active form (known as acetyl Coenzyme A), is the starting material not only for the synthesis of fatty acids and cholesterol, but for many other lipids. The physiological function of cholesterol is not known, but there seems to be some correlation between the level of cholesterol in the blood and a tendency toward certain types of heart attacks. The presence of cholesterol in gallstones represents an instance of an unnatural concentration of a normal body constituent.

There are many steroids that have well-known specific functions. Testosterone, the male hormone, is reponsible for secondary sexual development in the male, and estradiol and other related compounds perform a similar (although different) function in the female.

testosterone   estradiol

The hormones of the adrenal gland, such as cortisol and aldosterone, as well as the hormone progesterone, are steroids. Progesterone helps to maintain pregnancy in the female, and is known to inhibit the release of ova. Cholesterol is the precursor of many of these sterols.

cortisol   aldosterone   progesterone

A group of steroids which has received a great deal of attention comprises the oral contraceptives. Several compounds are used for this purpose, but they fall into two classes—the analogs of progesterone and those of the natural estrogens such as estradiol. Norethindrone and mestranol are widely used as oral contraceptives. Comparison of their formulas to those of progesterone and estradiol shows the obvious similarities.

norethindrone   mestranol

When these two steroids are given either in a single pill or in separate pills at different times during the menstrual cycle they result in essentially normal menstruation with the exception that no ova (eggs) are released from the ovary. Similar results are produced by administering progesterone and estradiol; however, these hormones are relatively ineffective when given orally and must be injected for maximal effects. Thus we can see that a slight difference in chemical structure allows a steroid taken orally to function in a manner similar to one normally produced by the body. While estrogens are produced regularly following puberty, progesterone is normally produced only during pregnancy. The combination of the two steroids inhibits ovulation and prevents conception in a manner similar to that occurring during normal pregnancy. While the oral contraceptives are widely used, harmful side-effects have been reported. These drugs are currently available only upon prescription and should be taken under the supervision of trained medical personnel.

Vitamin D is not truly a steroid, since one of the rings of this compound has been ruptured, but it is closely related to the steroids. It is produced by the action of ultraviolet radiation on 7-dehydrocholesterol.

7-dehydrocholesterol   vitamin D

(If R is C_9H_{17}, the compound is vitamin D_2. If R is C_8H_{17}, the compound is vitamin D_3.)

The bile acids, which are important for the digestion of lipids, also contain a steroid nucleus. These acids are usually combined with an amino acid derivative when functioning biochemically.

cholic acid   glycocholic acid (glycine derivative of cholic acid)

<table>
<tr><td>

**32.7
TERPENOIDS**

</td><td>

The substance we call turpentine can be obtained from the resin or gum of pine and fir trees. A great number of other compounds are related to turpentine in terms of both odor and chemical structure. As a group, these compounds are called terpenoids, a name derived from an older spelling for turpentine.

</td></tr>
</table>

The terpenoids have a distinctive fragrance and can be isolated from natural sources by steam distillation. They are responsible for the odors of pine trees, citrus fruits, geraniums, and many other plants. Chemically, they can be considered as being derived from iso-prene units (which can, in turn, be derived from acetate); terpenes contain 10 carbon atoms. Other similar compounds containing 15, 20, and 30 carbon atoms are called the sesquiterpenes, diterpenes, and triterpenes. Typical examples of terpenoids are shown in the following structures:

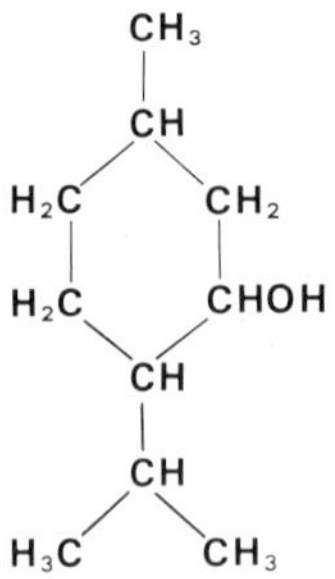

menthol 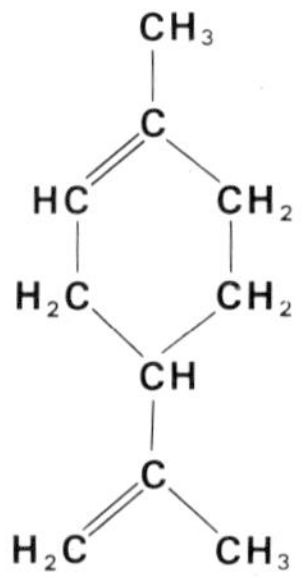   limonene   α-pinene 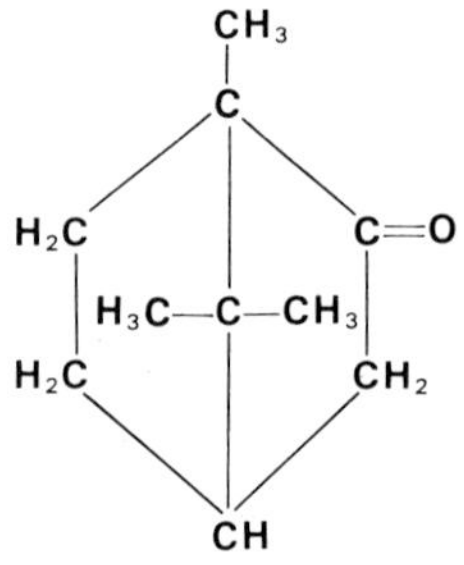   camphor
               (from citrus oils)

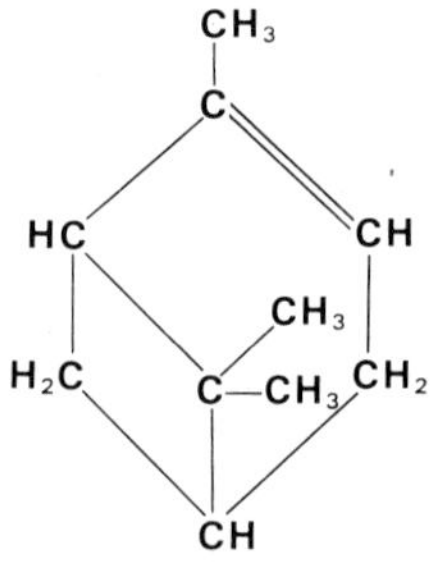

The carotenes, the orange pigment found in many green plants as well as in carrots, are tetraterpenes. Vitamin A is a diterpene; its molecular weight is sufficiently high that it does not have an odor as do the terpenes. (Only substances of relatively low molecular weight are sufficiently volatile at normal temperatures to produce odors.)

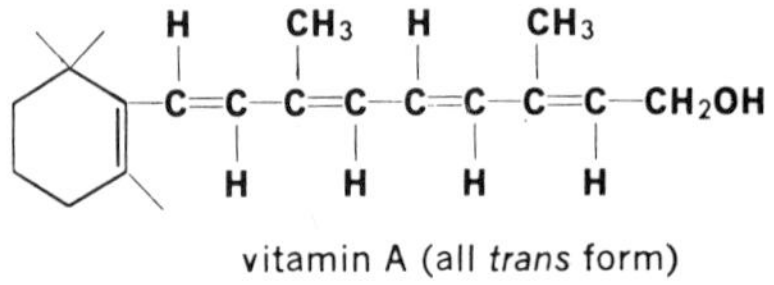

vitamin A (all *trans* form)

The odorous terpenes apparently attract insects and thus aid the fertilization of plants. Some insects use compounds similar to terpenes as pheromones. Pheromones are used to communicate alarm and also function as sexual attractants. In photosynthesis the carotenes probably play some role in the absorption of light energy. Vitamin A has several functions in the human. It is present, for example, in rhodopsin (visual purple) and plays an important role in vision (see Chapter 38). Terpenoids are used commercially in flavorings and in odorants (perfumes).

<table>
<tr><td>

**32.8
SPHINGOLIPIDS**

</td><td>

The sphingolipids contain the 4-sphingenine (formerly called sphingosine) group:

</td></tr>
</table>

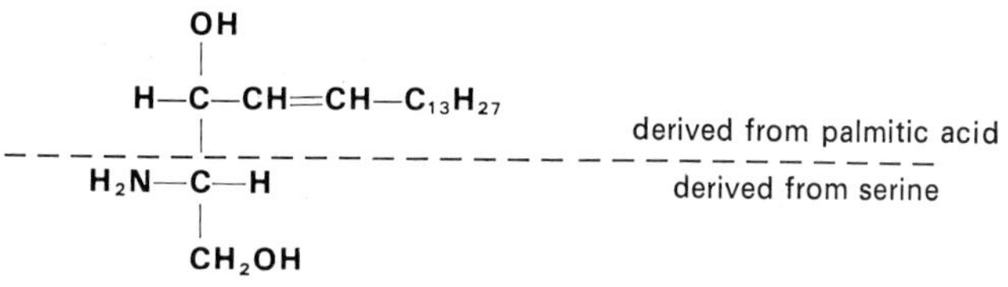

OH
|
H—C—CH=CH—C$_{13}$H$_{27}$
| ------------------------------- derived from palmitic acid
H$_2$N—C—H   derived from serine
|
CH$_2$OH

4-sphingenine

The upper part of the molecule is derived from palmitic acid and the lower from the amino acid serine. We find a variety of substances attached to the sphingenine unit. Some of the sphingolipids are also phospholipids. The sphingolipids are found in brain tissue and are associated with membranes. Some representative sphingolipids are shown below.

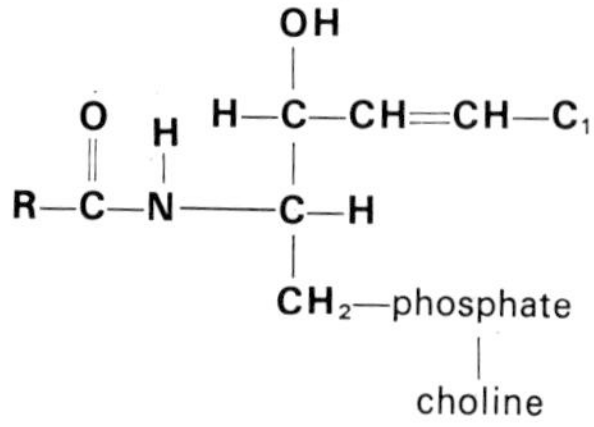

OH
|
O H H—C—CH=CH—C$_{13}$H$_{27}$
‖ | |
R—C—N——C—H
|
CH$_2$—phosphate
|
choline

a sphingomyelin

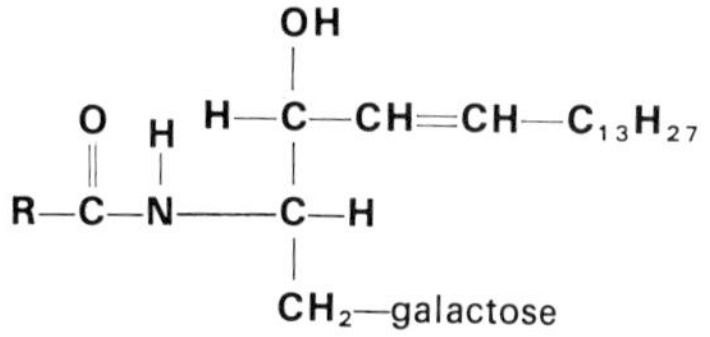

OH
|
O H H—C—CH=CH—C$_{13}$H$_{27}$
‖ | |
R—C—N——C—H
|
CH$_2$—galactose

a cerebroside

OH
|
O H H—C—CH=CH—C$_{13}$H$_{27}$
‖ | |
R—C—N——C—H
|
CH$_2$—O—

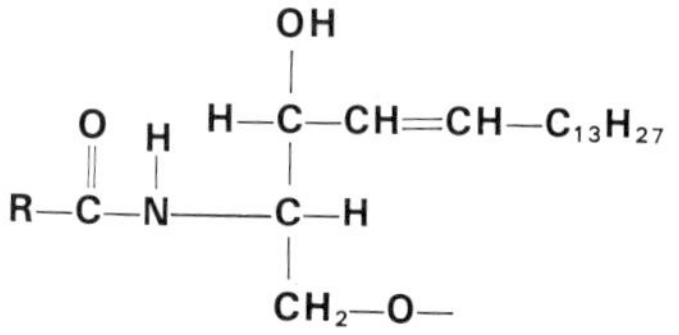

a ceramide unit

In the above, the R groups are those of the fatty acids and may vary. Cerebrosides contain galactose as the sugar unit. Gangliosides contain a ceramide unit and one or more carbohydrate residues. Glucose, galactose, the aminosugars, and N-acetylmuramic acid groups are all found in these compounds. Sphingolipids and phospholipids are lost from the nerve tissue in multiple sclerosis. There are several hereditary diseases that involve the metabolism of ceramides, particularly the gangliosides (Section 37.8).

**32.9
PROSTA-
GLANDINS**

The prostaglandins are a group of compounds derived from 20-carbon unsaturated fatty acids (Fig. 32.1). They were originally isolated from semen, and their name reflects the fact that they were believed to come from the prostate gland. We now know that these substances—about 20 different ones have been isolated—are present in almost all parts of the human body and in many other animals, also. The prostaglandins are synthesized from unsaturated fatty acids by a number of different tissues in the body. Many effects of the prostaglandins have been shown, but we do not know in any detail their function in a normal human body. They are effective in nanogram quantities and are involved in control of blood pressure and in the production of inflammation and pain. Derivatives of the prostaglandins, the thromboxanes, influence clotting of blood. Aspirin and other medications inhibit both the synthesis and metabolism of prostaglandins. In many cases the prostaglandins influence, both positively and negatively, the action of hormones. Preparations of prostaglandins have been used to induce labor in pregnant women and can also produce abortions. The functions of the prostaglandins are the subject of a great deal of current research that should help us understand these compounds, which seem to be involved in so many vital body processes.

32.10
LIPIDS AND
MEMBRANE
STRUCTURE

Cells as well as organelles within cells, such as mitochondria, are enclosed in membranes. Although membranes contain proteins, phospholipids, and glycolipids, their basic structure is a double layer of phospholipid molecules with the other components inserted in the double layer.

In the phospholipid molecule, the two fatty-acid chains are nonpolar and the phosphate and the groups attached to it are polar. We can say that the fatty acid chains are hydrophobic (water "fearing") and the phosphate and attached groups are hydrophilic (water "loving"). We can draw a phospholipid molecule in much simplified forms such as those shown below.

These drawings show how the two hydrophobic chains tend to associate and to be as far as possible from the hydrophilic part of the molecule. Similar structures for soaps were shown in Section 24.6. Whenever we place phospholipids or similar molecules in water they tend to form bilayers with the hydrophilic parts on the outside and a hydrophobic area in the middle.

Similar structures can be formed with glycolipids, which have a water-soluble part (the carbohydrate) and a hydrophobic (lipid) part. Similarly, proteins often have a section of their amino acid structure that contains amino acids with nonpolar side chains. We find these hydrophobic side chains associating with each other to form the tertiary structure of protein molecules.

If we put all the units together in a matrix of phospholipids, we have a double-layered membrane containing proteins and glycolipids. The glycolipids give the outer part of the membrane a surface that can bind substances such as antibodies and is probably responsible for the different types of blood cells as well as other cells. Some proteins, the integral proteins, seem to penetrate the lipid layers, and others, the peripheral proteins, are primarily on the outer surface. A model of this protein-lipid structure is shown in Fig. 32.3. This model of a membrane helps explain many aspects of cell structure and function, including the binding of hormones, antibodies, and enzymes to membranes.

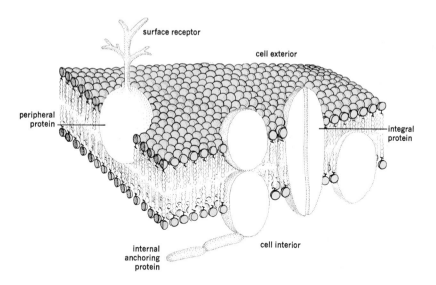

Fig. 32.3 A model of a cell membrane, showing the phospholipid double layer with proteins embedded in it. The peripheral proteins are primarily on the outer surface, and integral proteins penetrate the phospholipid bilayer. Some peripheral proteins contain surface receptors, and internal proteins serve to anchor the membrane to intracellular components.

It also explains the transport of some substances across membranes and the exclusion of others. Such a structure would tend to flow back together if pricked by a very small pin or otherwise disrupted slightly.

SUMMARY

From the foregoing we can see that the classification lipid includes many different types of compounds. The triacylglycerols are probably of greatest importance, from a quantitative aspect, to most animals. These compounds provide energy when eaten, and serve as compounds in which energy can be stored for future use. The phospholipids are also widely distributed in the animal body. Although we know many of their functions, the exact role they play is not yet fully understood. Cholesterol is found in many places in animal tissues and is especially prevalent in the brain. Although found in only relatively small quantities, the steroid-derived vitamin D and the steroid hormones such as testosterone, estradiol, and cortisol are vital. When there is a deficiency or excessive amounts of them, the consequences are severe and may even be fatal.

Other lipids—particularly the terpenoids—are of significance to plants, but only a few of them, such as vitamin A, are of biochemical importance to humans. However, the social and economic importance of these compounds especially of rubber, is great.

The phospholipids seem to form the matrix of membranes which incorporate proteins and glycolipids. The special properties of membranes are important in many vital processes. It is to be expected that further research on membrane structure and function will help explain how organisms function in health and how abnormal function is related to disease.

IMPORTANT TERMS

fat	prostaglandin	steroid
lecithin	phospholipid	terpenoid
lipase	rancid	triacylglycerol
lipid	saponification	unsaturated lipid
oil	sphingolipid	wax

WORK EXERCISES

Before you do the following exercises, you may wish to review Chapter 24.

1. List the hydrolysis products of the following.
   a) triacylglycerols   b) phospholipids
   c) waxes

2. Draw the structure of the following.
   a) a triacylglycerol containing stearic, palmitic, and lauric acids (Is there more than one possible answer to this question?)
   b) a highly unsaturated triacylglycerol
   c) fat
   d) a lecithin containing lignoceric and arachadonic acids
   e) cetyl stearate

3. Write equations for the following.
   a) hydrolysis of myristyloleyllaurin
   b) addition of bromine to palmitoleic acid
   c) saponification of a triacylglycerol

4. Indicate whether you think each of the following triacylglycerols would be a fat or an oil.
   a) tripalmatin   b) triolein
   c) tributyrin
   (You may wish to check your predictions by consulting a handbook of chemistry.)

5. Would you expect fats or oils to have the greater tendency to become rancid?

6. On the basis of the two causes of rancidity, how might you protect lipids from developing this undesirable characteristic?

7. Cow's milk produced when cattle are grazing on green pastures is usually appreciably yellower than that produced when they are fed dried alfalfa or other dry food. Can you give a reason for this?

8. What information can you find on labels of various food products that are high in lipids, such as cooking oils, oleomargarines, and butter, about the source and content of lipids in these foods?

9. Give the names of natural products that can be considered to be made of isoprene units.

10. List the commercial (including medical) uses of the following.
    a) lecithins   b) waxes   c) sterols

11. List the important biological functions of the following.
    a) triacylglycerols   b) terpenoids
    c) prostaglandins   d) steroids

12. What would be some of the effects on a human if there were no cholesterol in the diet and all body synthesis of cholesterol was prevented?

13. Write a short description of the structure of cell membranes.

14. Consult a general biochemistry text to answer the following.
    a) Find the formulas and significance of steroids other than those discussed in this chapter.
    b) List the formulas and explain the significance of terpenoids.

SUGGESTED READING

Baumann, G., "The cell membrane," a series of three articles in *Chemistry* **51**, No. 9, November 1978; **51**, No. 10, December 1978; and **52**, No. 1, January–February 1979. These well-illustrated articles discuss the formation and structure of cell membranes as well as their function in nerve conduction.

Pike, J. E., "Prostaglandins," *Scientific American*, November 1971, p. 84 (Offprint #1235). A good discussion of these hormone-like lipids.

33

Chemistry of the Nucleic Acids

<table>
<tr><td>

**33.1
INTRODUCTION**

</td><td>

The carbohydrates, proteins, and lipids, along with the mineral elements and water, constitute about 99% of most living organisms. The remaining 1%, however, includes some of the most important biochemical compounds. Without these compounds, cells would be unable to exist for any appreciable length of time and could never reproduce.

</td></tr>
</table>

Probably the most important of these minor constituents are the nucleic acids. These compounds were so named because they were originally found in the nuclei of cells. We now know that they are also found outside the nucleus and that there is some difference both in location and function of the two basic kinds of nucleic acids. In eucaryotic cells such as those of humans, deoxyribonucleic acid (DNA) is found primarily in the nucleus and ribonucleic acid (RNA) is found both in the nucleus and in the cytoplasm. In procaryotic cells, which have no nucleus, much of the DNA is associated with the inner membrane of the cell wall. Bacteria also contain plasmids, which are small rings of DNA. In the nucleus of eucaryotic cells, DNA is almost always associated with basic proteins known as histones, while in procaryotic cells there is no protein associated with DNA. From the names it is obvious that the presence or absence of an oxygen in a ribose component of the nucleic acid molecule is a chemical difference between DNA and RNA. There are other differences, as will be pointed out later. While the functions of RNA and, to some extent, of DNA still are not totally known, it seems that DNA plays a primary role in transmission of hereditary information from one cell to its daughter cells, and that the functions of RNA are secondary. The principal known function of RNA is its involvement in protein synthesis.

**33.2
COMPOSITION
AND STRUCTURE
OF THE
NUCLEIC ACIDS**

The hydrolysis of a pure sample of a nucleic acid yields phosphoric acid, a pentose (either ribose or 2-deoxyribose), and a mixture of organic compounds that have the properties of bases. (See Figs. 33.1 and 33.2.) The five most common bases are adenine, guanine, thymine, cytosine, and uracil. Adenine and guanine are called **purine bases** because of their relation to purine, and similarly thymine, cytosine, and uracil are **pyrimidine bases**.

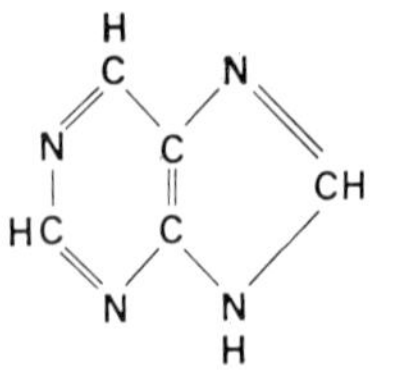

purine 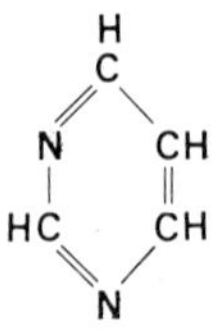   pyrimidine

Other bases such as 5-methylcytosine and 5-hydroxymethylcytosine are found in small amounts in some nucleic acid hydrolysates. Transfer RNAs, especially, have many unusual bases in their structures. Although the amounts of the different components vary some-what, there is always one phosphate group and one pentose for each base. In most kinds of DNA, the amount of purines is equal to the amount of pyrimidines, the amount of adenine equal to that of thymine, and the amount of guanine to that of cytosine. This ratio of bases is not usually found in RNA. **Nucleic acids** are defined as those complex biochemical substances which yield, on hydrolysis, purine and pyrimidine bases, ribose or deoxyribose, and phosphate.

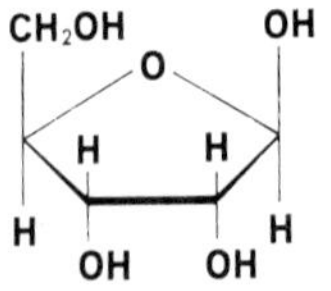

β-D-ribose β-2-deoxy-D-ribose

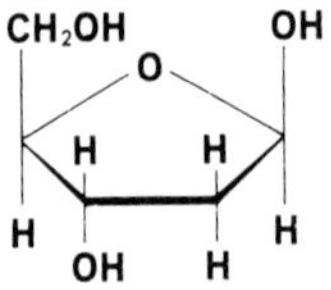

Fig. 33.1 Structures of the pentoses found in nucleic acids.

Although some false assumptions were originally made concerning the structure of DNA, when it was determined that the molecular weight of DNA was very high, it became obvious that DNA was a polymer. The monomeric unit was found to contain a base, deoxyribose, and a phosphoric acid group. This monomeric unit is a **nucleotide.** The compounds containing only the base and the sugar are called **nucleosides.** The structure of a nucleotide and a nucleoside are shown below. See also Table 33.1

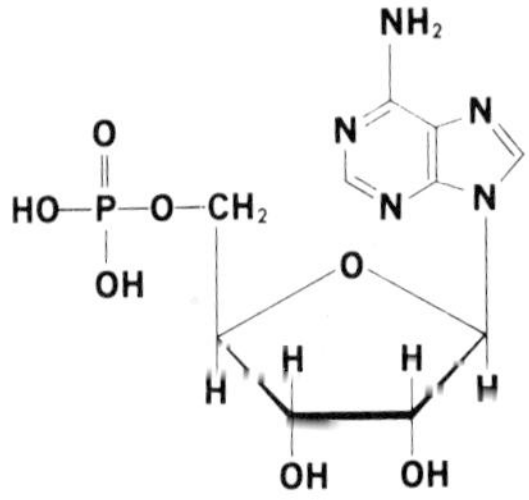

adenosine 5'-monophosphate
or adenylic acid
(a nucleotide)

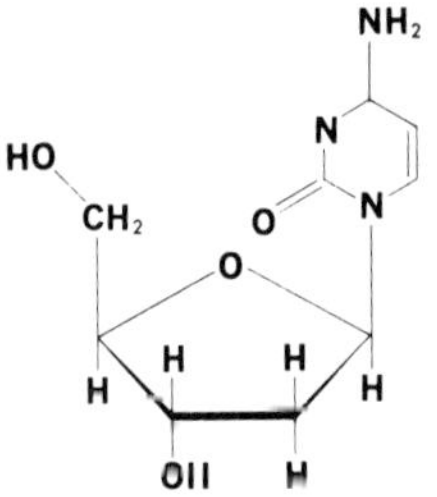

deoxycytidine
(a nucleoside)

In general, DNA exists as a double-stranded polymer with hydrogen bonds holding adenine of one strand to a thymine in another strand, and with cytosine and guanine

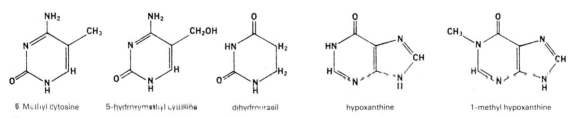

Fig. 33.2 Structures of purine, pyrimidine, and the bases related to them that are found in nucleic acids. The purines adenine and guanine are found in all forms of DNA and RNA, while the other purines listed are found less frequently, primarily in transfer RNA. Similarly, cytosine, uracil, and thymine are widely found and the other pyrimidines less frequently. Those bases that contain oxygen can exist in keto and enol forms as shown above. The lactam form predominates at physiological pH values.

TABLE 33.1   Names of nucleotides and nucleosides

Base	Nucleoside	Nucleotide
cytosine	cytidine or deoxycytidine	cytidylic or deoxycytidylic acid
uracil	uridine* (deoxyuridine)	uridylic* (deoxyuridylic acid)
thymine	thymidine†	thymidylic acid†
adenine	adenosine or deoxyadenosine	adenylic or deoxyadenylic acid (adenosine phosphate)
guanine	guanosine or deoxyguanosine	guanylic or deoxyguanylic acid

* In nature uracil is found combined with ribose, not deoxyribose. The deoxy compounds can be made synthetically.
† In nature thymine is always combined with deoxyribose, not ribose. Thymidine and thymidylic acid contain deoxyribose, not ribose.

similarly bonded. (See Fig. 33.3.) Single-stranded DNA is found in some viruses; however, such single-stranded forms now seem to be a minor exception to the rule that DNA is double-stranded. Three hydrogen bonds are possible between cytosine and guanine, while only two are present in the adenine-thymine pair. This finding correlates well with the fact that more energy is required to separate the two strands of DNA (a process called *melting*) when the guanine-cytosine content of the DNA is high. See Fig. 33.4. When DNA solutions are heated, the hydrogen bonds between strands are disrupted and the strands separate. Slow recooling permits essentially normal double-stranded DNA. Rapid chilling produces some intrastrand bonds as well as some interstrand bonds.

The three-dimensional structure of the nucleic acids was established on the basis of X-ray diffraction studies. We know now that DNA is a double-stranded, spiral (helical) molecule. For convenience, adenine, thymine, guanine, and cytosine are designated by the first letters of their names in the representation of the structure of DNA shown in Fig. 33.5. Other visualizations of the DNA structure are shown in Fig. 33.6. The two strands of DNA run in opposite directions and they are said to be antiparallel chains. This means that the upper end of the left half of the molecule contains a free OH group attached to the carbon 5 of ribose while the upper end of the right chain is an OH at the carbon 3 of ribose.

In addition to the obvious difference in having ribose instead of deoxyribose, RNA contains uracil and does not contain thymine. Since its molecular weight is usually lower than that of DNA, and since there seems to be no uniform ratio of bases, we could postulate that RNA is only single-stranded. On the basis of research, it seems that most forms of RNA are single-stranded and have a double-stranded helical structure in only part of their molecules. There seem to be at least three types of RNA that can be separated by essentially physical means: messenger RNA, ribosomal RNA, and soluble or transfer RNA. The names given these forms of RNA reflect the function that are believed to play in protein synthesis. The role of nucleic acids in protein synthesis is presented in Chapter 36.

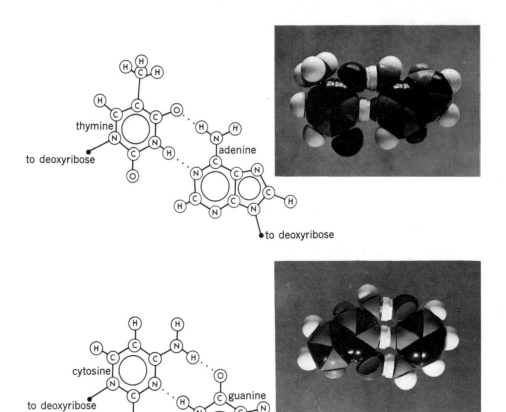

Fig. 33.3 Hydrogen bonds between the adenine:thymine and cytosine:guanine pairs of bases. Note that there are two bonds between adenine and thymine and three between guanine and cytosine. (Photographs from Jane M. Cram and Donald J. Cram, *The Essence of Organic Chemistry*, Addison-Wesley, 1978.)

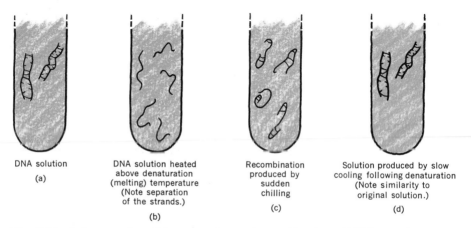

DNA solution	DNA solution heated above denaturation (melting) temperature (Note separation of the strands.)	Recombination produced by sudden chilling	Solution produced by slow cooling following denaturation (Note similarity to original solution.)
(a)	(b)	(c)	(d)

Fig. 33.4 Representation of the separation and recombination of DNA strands.

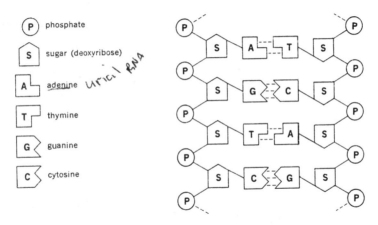

Legend:

- (P) phosphate
- (S) sugar (deoxyribose)
- (A) adenine *uracil RNA*
- (T) thymine
- (G) guanine
- (C) cytosine

Fig. 33.5 Representation of the structure of a segment of DNA. The sugar molecules in the right-hand chain are reversed to indicate that it is antiparallel.

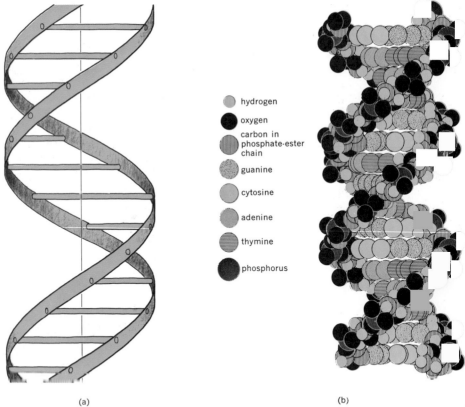

Legend:

- hydrogen
- oxygen
- carbon in phosphate-ester chain
- guanine
- cytosine
- adenine
- thymine
- phosphorus

(a)   (b)

Fig. 33.6 Schematic representations of the molecule of DNA: (a) a simplified model showing the two strands linked by ladderlike steps formed by the base pairs; (b) a more complex representation of the same structure. [(b) is reprinted by permission from *CA, A Cancer Journal for Clinicians* and from Dr. L. D. Hamilton; *CA, A Bulletin of Cancer Progress* **5**, 163 (1955) from material provided by Dr. W. E. Seeds and Dr. H. R. Wilson, Wheatstone Physics Laboratory, King's College, London.]

**33.3
OCCURRENCE
OF NUCLEIC
ACIDS**

Nucleic acids are found in all living cells, where they are usually present at a level of less than 1% of the total net weight of the cell. Yeast cells and some bacterial cells may contain up to 5% nucleic acid. The bacteriophages may contain as much as 60% DNA. It has been estimated that if all the DNA in an average adult human were uncoiled and laid out in a straight line it would reach five billion miles. This long string would, however, be only 0.1 to 0.2 nanometers wide.

In almost all instances the nucleic acids of eucaryotic cells are found in combination with proteins in the form of nucleoproteins. Many of the functions of nucleic acids are performed only poorly, if at all, when the specific protein they ordinarily are associated with is missing. Sperm cells are composed of a DNA nucleoprotein. (See Fig. 33.7.) Viruses contain either DNA or RNA surrounded by a protein coating. (See Figs. 33.8 and 33.9.) In some cases it has been possible to separate the protein outer coat from the inner thread of the virus and then to recombine the two fractions into a virus which is able to infect living cells. The proteins associated with nucleic acids contain high percentages of the basic amino acids such as lysine and arginine.

**33.4
SYNTHESIS OF
NUCLEIC ACIDS**

The nucleic acids can be synthesized from simple materials by most organisms. Some bacteria require adenine or other bases in order to synthesize their nucleic acids, but most organisms can synthesize the pentoses and bases, and combine them with phosphate from their diet to assemble the nucleic acids required. While each kind of organism makes its own specific kind of DNA, the RNA from different species seems to be quite similar. Some polymers that can function in place of natural RNA have been synthesized *in vitro*. The synthesis, however, requires an enzyme as a catalyst. Other enzymes can catalyze the synthesis of DNA when provided with the proper starting materials and a small amount of already-formed DNA as a **primer**. The synthesized DNA is similar in physical properties to the DNA of the primer, but until recently such synthetic DNA's had not been shown to

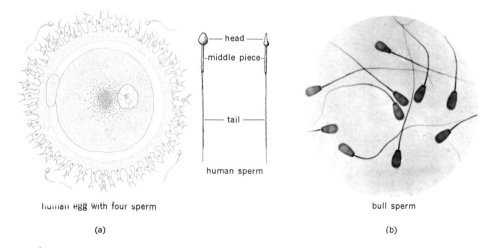

human egg with four sperm

(a)

human sperm

bull sperm

(b)

Fig. 33.7 Sperm cells. (a) shows a drawing of a human egg and sperm cells and (b) is reproduced from a microscopic slide of the sperm of a bull. [Adapted from J. J. W. Baker and Garland E. Allen, *The Study of Biology*, 3rd ed., Addison-Wesley, 1977.]

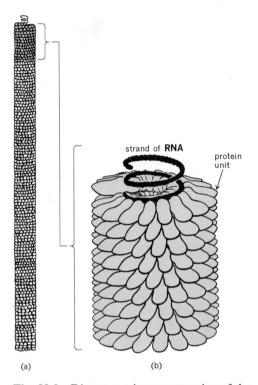

(a)   (b)

Fig. 33.8 Diagrammatic representation of the structure of the tobacco mosaic virus (TMV). (a) Drawing of a complete virus particle (note similarity to photomicrograph of TMV in Fig. 33.9). (b) Enlarged diagram of part of the TMV particle with a single coiled strand of RNA emerging from the stack of spirally arranged protein units that surround it. Each protein unit is identical to all others and contains 158 amino acids in a sequence that is known. The complete TMV particle has 2 200 of these protein units. [Adapted from a model in W. M. Stanley and E. G. Valens, *Viruses and the Nature of Life*, Dutton, New York.

have biological activity. Late in 1967, Kornberg and his co-workers reported the synthesis of a viral DNA that could infect microorganisms. The *in vivo* synthesis, or replication, of DNA is illustrated in Fig 33.10 (page 600).

It has recently become possible to synthesize relatively long chains of nucleic acids *in vitro*. Synthesis that formerly took months can now be done in days. While it is probably not correct to describe these and other experiments as the creation of life in a test tube, synthetic nucleic acids hold great promise for understanding life processes. In other instances (Section 39.6) practical developments are being derived from *in vitro* syntheses of DNA.

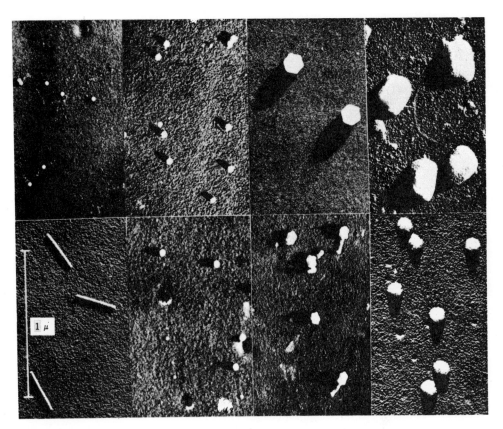

Fig. 33.9 Electron micrographs of eight typical viruses at the same magnification. Top (left to right): poliomyelitis, small bacteriophage, *Tipula irridescens* (an insect virus showing individual particles, not crystals), vaccinia (the virus that causes smallpox). Bottom: tobacco mosaic virus, rabbit papilloma, large bacteriophage, influenza virus. [Courtesy of the Virus Laboratory, University of California, Berkeley.]

33.5
PHYSICAL
PROPERTIES
OF THE
NUCLEIC ACIDS

Samples of DNA have been found to have an extremely high molecular weight. Values in the millions have been found, and it is accurate to say that the nucleic acids are probably the largest molecules known. Although some lower values for molecular weights have also been reported, it is likely that the samples were torn apart in the process of separating them from other cell components. The alternate possibility, that the high-molecular-weight components were aggregations of smaller molecules present in the intact cell, seems less likely, but cannot be ignored. It is probable that all the DNA in some bacterial and viral cells is represented by one single molecule. There are many forms of RNA, of which some have molecular weights as low as 75 000 while others have weights probably extending into the million range.

Because of their many polar groups and in spite of their high molecular weight, nucleic acids are water-soluble. The bases of the nucleic acids absorb ultraviolet radiation in the 250- to 290-millimicron range. This property is commonly used to identify and measure the amount of nucleic acids in a sample.

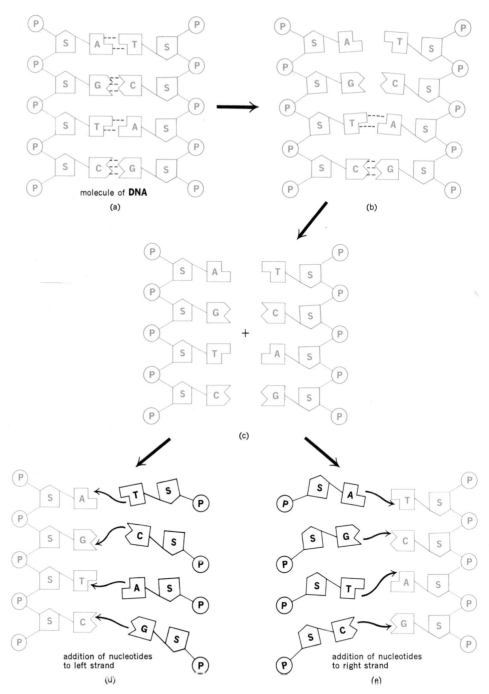

Fig. 33.10 The process of duplication (replication) of DNA. (a) shows part of the molecule of DNA (see Fig. 33.5 for identification of components). In (b) the two strands have begun to separate and in (c) they are completely apart. (d) shows how individual nucleotides join with the left strand of the DNA to form another double-stranded molecule, and (e) shows a similar process for the right strand of the original DNA. Since each half of the original DNA molecule has now produced a molecule identical to the original DNA molecule, duplication has occurred. This process occurs in the chromosomes of the cell nucleus prior to cell division.

<table>
<tr><td>

33.6
CHEMICAL
PROPERTIES
OF THE
NUCLEIC ACIDS

</td><td>

The hydrolysis of nucleic acids is catalyzed by acids, by bases, or by enzymes. Enzymes which catalyze the hydrolysis of only RNA (RNA-ases) or DNA (DNA-ases) are widely used in studies of nucleic acids. The first products of hydrolysis of the nucleic acids are nucleotides, but these can be further hydrolyzed to give nucleosides, and finally to give the bases, pentose, and phosphate. Each step in the hydrolysis requires a specific type of enzyme.

</td></tr>
</table>

In recent years it has become possible to find the exact sequence of nucleotides and to establish the structure of nucleic acids. The techniques involve using enzymes to break chains next to specific bases, followed by separation of fragments by electrophoresis and chromatography. The properties of nucleic acids, in particular their functions in living cells, are the direct consequence of the sequence of nucleotides (bases) in the polymeric chain. A change of just one base (cytosine instead of thymine) in a polymer containing thousands of bases may have drastic effects.

An important chemical property of nucleic acids is that the amino-containing bases of the nucleic acids can react with nitrous acid with the removal of the amino group. It is interesting to correlate this with the finding that nitrous acid can produce mutations when it reacts with DNA.

33.7
THE
IMPORTANCE
OF
NUCLEIC ACIDS

The role nucleic acids play in heredity and in protein synthesis is discussed in Chapters 36 and 39. A third important role played by nucleic acids is in the action of viruses. We tend to think of viruses as causing diseases, which they do, but they are of great theoretical interest also. A virus represents a kind of bridge between living and nonliving materials. Many viruses have been isolated as crystallized nucleoproteins. In this form they can be stored in bottles on shelves for long periods of time. Yet, when added to living organisms or to cultures containing living cells, the virus can take over the function of the living cell so that it produces new virus particles instead of new cells. So far scientists have not been able to get viruses to reproduce, even in a medium containing all known nutritional requirements of cells, unless living cells are also present. They seem to function only when they assume control of the reactions of a cell.

One type of virus, the bacteriophages, which are viruses that attack bacteria, has been studied extensively. These viruses consist of an inner core of nucleic acid surrounded by an outer coat of protein. They attach themselves to the wall of a bacterium and inject their nucleic acid into the bacterium. (See Fig. 33.11.) The injected nucleic acid then takes over the metabolism of the cell and so directs it that thousands of new virus particles, each capable of infecting other bacteria, are produced. In other instances the virus is apparently present in the bacterial cell in an inactive or latent form and the virus is, in fact, reproduced along with other cell components for several generations. When this infected cell is exposed to some kind of shock, such as ultraviolet irradiation or certain chemical agents, the virus becomes active and directs the cell toward its own destruction by production of virus particles.

Some forms of cancer are caused by viruses. In fact, the Nobel prize-winning scientist Dr. Wendell Stanley stated many years ago that he believed all cancer was caused by viruses. Other scientists have speculated that many humans carry inactive viruses which can cause cancer when irradiated or otherwise stimulated. While recent cancer research emphasizes the importance of chemical carcinogens in the environment, it does not exclude the involvement of viruses. The development of new research techniques coupled with a

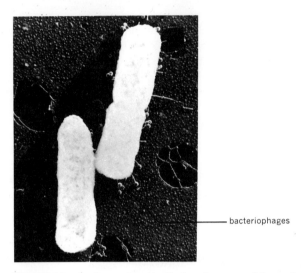

bacteriophages

Fig. 33.11 A photomicrograph showing several bacteriophages attached to one of the two cells of the bacterium *E. coli*. The bacteriophages are the pin-shaped objects sticking to the bacterium on the right. [Courtesy of Thomas F. Anderson, François Jacob, and Elie Wollman.]

better knowledge of the biochemistry of nucleic acids may lead to the understanding of cancer and to cures or controls for this disease.

Nucleotides have been discussed as the monomeric units from which the polymeric nucleic acids were made. Nucleotides which are present as such, not as polymers, also play an important role in biochemical reactions. Adenosine triphosphate (ATP), a nucleotide with two additional phosphate groups, serves as the energy source for hundreds of biochemical reactions (Fig. 33.12). When either of the two end phosphate groups is removed by hydrolysis, a large amount of energy is made available for driving energy-requiring reactions. Such diverse reactions as synthesis of proteins, muscular contraction,

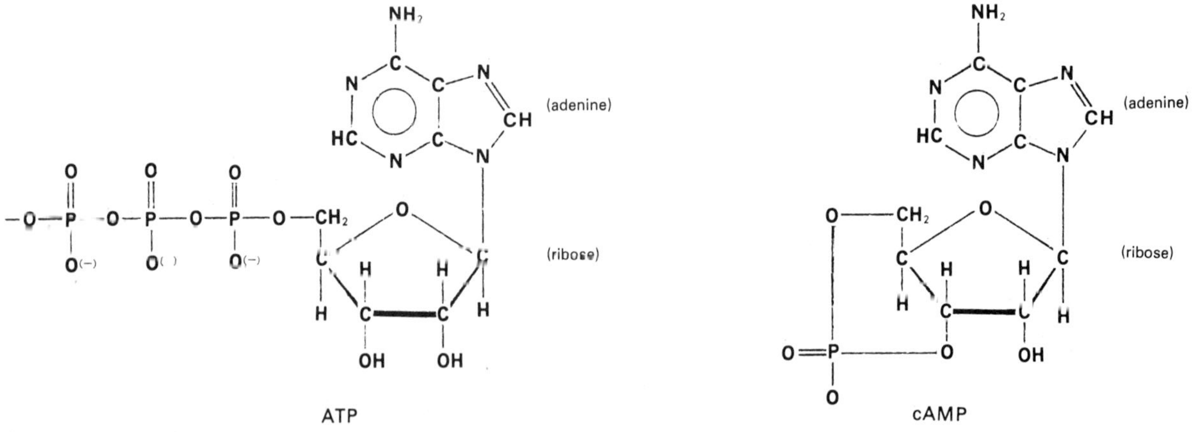

ATP

cAMP

Fig. 33.12 The structure of ATP (adenosine triphosphate) and cAMP (cyclic adenosine monophosphate).

and the production of light by fireflies all require ATP as an energy source. Triphosphates of guanine and other bases serve similar roles.

The form of adenosine monophosphate in which the phosphate is bound to both the 3 and 5 carbons of the ribose is called cyclic AMP. (See Fig. 33.12.) This substance is an important regulator of metabolic reactions. It is involved in the action of many hormones.

Block diagrams can be used to illustrate the structure of these important nucleotides. (HO— (P) is used to indicate a phosphate group.)

Adenine—Ribose—O—(P)—O—(P)—O—(P)—OH   Adenine—Ribose—O—(P)—O—(P)—OH
        ATP (adenosine triphosphate)   ADP (adenosine diphosphate)

Adenine—Ribose—O—(P)—OH   Adenine—Ribose—O—(P)
  AMP (adenosine monophosphate)   cAMP (cyclic AMP)

The active form of many of the vitamins is a nucleotide. In some cases the vitamin itself is the base, combined with a ribose and phosphate unit; in others, adenine or some other base is required. (See Figs. 33.13 and 33.14.)

The reason why a nucleotide "handle" is so widely found in the structure of extremely important biochemical compounds is not known. While it may be only accidental or coincidental that the base-pentose-phosphate combination is so widespread, this apparent coincidence is the kind of observation that makes a biochemist want to track down the significance of this chemical structure.

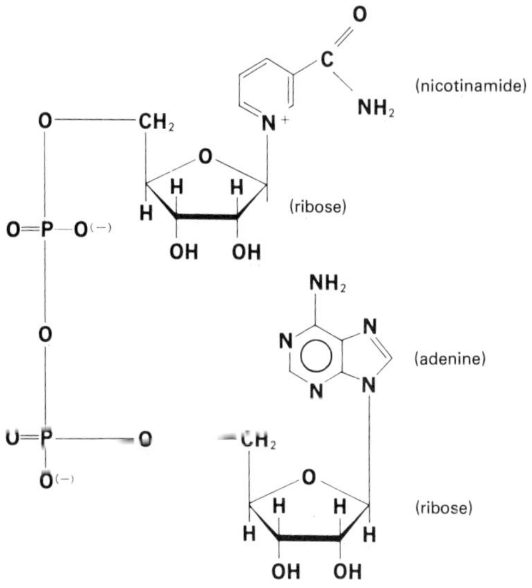

Fig. 33.13 The structure of NAD (nicotinamide adenine dinucleotide).

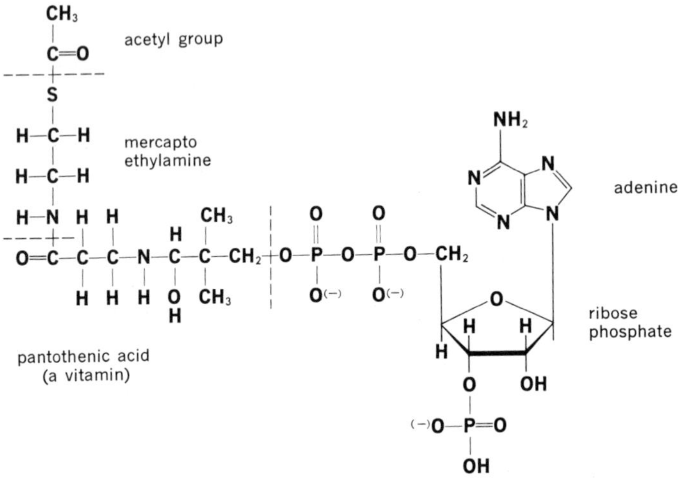

Fig. 33.14 The structure of Coenzyme A (CoA). An acetyl group is shown attached to the CoA to give acetyl CoA.

SUMMARY

The nucleic acids are high-molecular-weight polymers containing purine and pyrimidine bases, ribose or dixoyribose, and phosphate. The monomeric units of these polymers, known as nucleotides, are important both as building blocks of nucleic acids and in their own right. The nucleotides ATP, cAMP, and the coenzyme forms of vitamins such as NAD and Coenzyme A play important roles in metabolism.

Study of the properties and functions of nucleic acids brings us closer to the basic object of biochemistry, to explain life in physical and chemical terms. The double helix of DNA is held together by hydrogen bonds which break more easily than the covalent bonds between the bases, ribose, and phosphates. We know that the sequence of bases in both kinds of nucleic acids is important in determining not only their physical and chemical properties but also their function in a living system. In fact, we can say that their function in a living system is due directly to chemical and physical properties.

Studies not only showing the mechanism but also isolating the enzymes responsible for replication of a chain of DNA explain the basic process of cell division to give two essentially identical cells. The ways nucleic acids function in protein synthesis are discussed in Chapter 36 and their involvement in the chemistry of heredity and in hereditary diseases is described in Chapter 39. The involvement of nucleic acids in the basic life processes of a cell helps explain how viruses can "take over" the functions of a cell and why they need a living cell in order to reproduce.

IMPORTANT TERMS

bacteriophage
deoxyribonuclease (DNA-ase)
dexoyribonucleic acid (DNA)
nucleic acid
nucleoprotein

nucleoside
nucleotide
primer DNA
purine base

pyrimidine base
ribonuclease (RNA-ase)
ribonucleic acid (RNA)
virus

WORK EXERCISES

1. Draw the formulas for the following.
   a) deoxycytidine   b) uridylic acid
   c) thymidylic acid   d) deoxyguanosine
   e) adenosine diphosphate (ADP)
2. List the major hydrolysis products of each of the following.
   a) DNA   b) RNA
   c) deoxyuridine   d) adenylic acid
   e) guanosine
3. Draw the structure for the trinucleotides specified by the symbols CGA and its complementary nucleotide. Locate possible hydrogen bonds between these nucleotides. What is the maximum number of hydrogen bonds possible between these two trinucleotides?
4. What would be the expected products if

   a) DNA-ase were added to a sample of nucleic acids purified from a specimen of liver?
   b) RNA-ase were added to a purified sample of DNA?
5. What are the major differences between DNA and RNA?
6. How do the histones and other proteins associated with nucleic acids differ from other proteins?

Consult a standard biochemistry text for answers to the following.

7. What bases other than guanine, adenine, cytosine, thymine, and uracil are found in the nucleic acids? What percentage of total bases is represented by these "other" bases?
8. What techniques are used to separate and analyze the components of the nucleic acids?

SUGGESTED READING

Crick, F. H. C., "Nucleic acids," *Scientific American*, September 1957, p. 188 (Offprint #54). A basic article by one of the pioneers in establishing the structure of nucleic acids.

Fiddes, J. C., "The nucleotide sequence of a viral DNA," *Scientific American*, December 1977, p. 54 (Offprint #1374). The techniques to unravel the sequence of nucleotides are discussed and illustrated in this article.

Sayre, A., *Rosalind Franklin and DNA*, W. W. Norton and Co., New York, 1975. Although she did much of the work, Ms. Franklin has received almost no credit for the discovery of the structure of DNA. This book presents a different perspective from that in Watson's book, listed below.

Watson, J. D., *The Double Helix, A Personal Account of the Discovery of the Structure of DNA*, The New American Library, Signet Books, New York, 1968. Dr. Watson gives his version of the discovery. It's witty and well written. For a slightly different aspect, see the book by Sayre listed above.

34

Biochemical Reactions

With a few exceptions, all energy used on earth comes from the sun. The energy released from petroleum and coal was trapped long ago; that from wood and from food we eat left the sun only a relatively short time ago. Devices for making direct use of the sun's energy through solar cells are now being developed.

For millions of years plants have used chlorophyll and other pigments to trap radiation from the sun and convert this energy into the potential energy of chemical compounds. This ancient process of energy conversion has only recently been scientifically investigated and there are still many unanswered questions concerning the nature of photosynthesis. As the name implies, **photosynthesis** involves synthesis of all the different compounds found in plants utilizing the energy of sunlight. The synthesis of glucose is often summarized as

$$6CO_2 + 6H_2O \rightarrow C_6H_{12}O_6 + 6O_2$$

In this chapter ways are discussed in which animals use the energy stored in food for the many processes that are required for life.

Although animals absorb some radiation from the sun, energy for the real work of the body is provided by the food the animal eats. The green plant is the original source of this useful energy; for flesh-eating organisms, the energy of the plant has been converted to energy stored in meat of the animal which is eaten. (See Fig. 34.1.) The processes by which an animal utilizes food have been categorized as **ingestion, digestion, assimilation**, followed by the **elimination** of unused material and waste products. Biochemists are concerned with digestion and especially with assimilation of digested materials. In some cases, they are interested in the excretions, particularly the urine, for evidence of the abnormal functioning of an animal.

Following ingestion of food there is often some physical separation of the food particles. Whereas mammals and insects can chew their food, other animals must search for small morsels of food, since they have poorly developed mouth parts. In most animals that are able to chew, chewing not only separates the food into small particles but also moistens

plants make food   cow eats plants,   man drinks milk
                          produces milk   or eats meat

Fig. 34.1 A biological food chain. Energy from the sun is trapped by a
plant and converted into the potential energy of its leaves, seeds, and
fruit. An animal, such as a cow, eats the plants and converts the energy
stored in plants to that in its body or in the milk which it produces. A
man may derive energy either by drinking the milk produced by the cow
or by eating the flesh of the animal as steak or hamburger. At each step
some of the energy is converted to heat or is lost to the surroundings by
other means.

and lubricates the food with saliva. In the human, as well as in many other animals, the
saliva contains enzymes which catalyze the first steps in the process of digestion.

The **digestive system**, or **gastrointestinal tract**, is essentially a partially filled
tube extending between the mouth and the anus. (See Fig. 34.2.) Food in the tract,
which is called the **alimentary canal**, is not really "inside" the body. It enters and
becomes part of the body only when it is absorbed through the intestinal wall. We can
describe digestion and assimilation briefly by saying that as the food passes through this
hollow tube, various secretions are added, some material enters into the body via either
the blood or lymph, some waste products are added to the indigestible food residue,
bacteria act upon the contents of the intestine, and finally the remaining material, the
feces, is excreted.

Chemically, **digestion** is the hydrolysis of ingested food into smaller molecules.
Polysaccharides are converted to monosaccharides, proteins to amino acids, and the
triacylglycerols are broken down into fatty acids and glycerol. The function of digestion
seems to be the production of molecules small enough to be absorbed through the
intestinal wall.

The hydrolyses of digestion are catalyzed by enzymes which are secreted by a variety
of organs and are added to the food at strategic points in its passage through the digestive
tract. The common names for enzymes are formed by adding the suffix -ase to the name
of the substance whose transformation is catalyzed by the enzyme. Thus a peptidase acts
on a peptide, an amylase acts on starch (both amylose and amylopectin parts), and sucrase
acts on sucrose.

It is true, in one respect, that "we are what we eat"—that is, we use the amino acids
and monosaccharides and other components of our food to build our own bodies. However, human fat and human proteins are different from those of the milk, butter, and
steak we eat. Since only simple units of the proteins, carbohydrates, and lipids are
absorbed through the intestinal wall and since we are able to assemble the units only in
our own way, a good analogy to digestion is that of a complex brick building being
demolished—the mortar being dissolved, and the simplest units, the bricks, small pieces of
wood, or steel being reused for construction of another building.

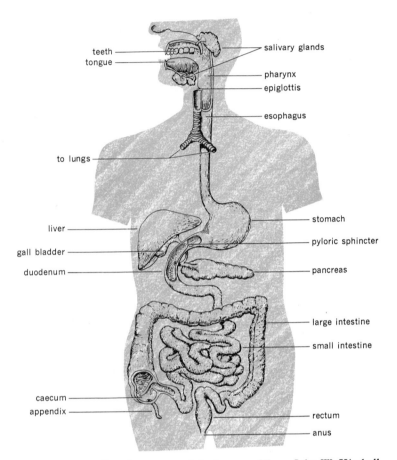

teeth
tongue

salivary glands

pharynx
epiglottis

esophagus

to lungs

liver

gall bladder

duodenum

stomach

pyloric sphincter

pancreas

large intestine

small intestine

caecum
appendix

rectum

anus

Fig. 34.2   The digestive system of the human. [From John W. Kimball, *Biology*, 4th ed., Addison-Wesley, 1978.]

The digestive process in the human is complex, but it can be simply described to provide a background for the biochemistry of this important series of events. Detailed discussions of the digestion of specific biochemical compounds are given in Chapters 35, 36, and 37. As we have already noted, when food is chewed it is mixed with saliva, which both lubricates the food and contains the enzyme (salivary amylase) which initiates digestion of starch. The food then enters the esophagus, which serves mainly as a connecting tube between the mouth and the stomach. It produces no digestive enzymes. When food reaches the stomach, action of salivary amylase is stopped by acidity which comes from the hydrochloric acid secreted by the stomach.

Food normally stays in the stomach for a few hours. Here it is exposed to a pH between 1.0 and 2.5. At these pH values there is some hydrolysis of carbohydrates and the enzymes pepsin and gastricsin, which require an acid pH for their action, begin the digestion of proteins. Pepsin, which is found in birds, fish, and reptiles as well as in mammals, catalyzes the hydrolysis of proteins to yield a mixture of smaller polypeptides. Gastricsin acts in a similar manner. Other digestive enzymes have been found in the stomach, but they are not effective at the pH of the stomach, and it has been speculated that perhaps these enzymes are holdovers from infancy when their secretion into a less acid stomach (characteristic of infants) may have been effective. Alternatively, they may become effective in the lower tract, where the pH is neutral or slightly alkaline.

As the partially digested food enters the small intestine, it is met with a large number of catalysts for the digestive process. From the gall bladder come the nonenzymatic **bile** acids, which aid in emulsification of lipids; from the pancreas comes a mixture of **amylases**, the peptidases **trypsin** and **chymotrypsin**, at least one **lipase, cholesterol esterase**, which catalyzes the hydrolysis of esters of cholesterol, and perhaps other, yet to be discovered, enzymes. Finally, the intestine itself secretes several enzymes. Among these are various **disaccharidases, peptidases, nucleosidases**, and **nucleotidases**. In addition to the enzymes and bile acids, the various secretions contain sufficient alkaline materials to neutralize the acid of the stomach. It is believed that the secretions added to the food in the intestine are sufficient to do a complete job of digestion and in fact many persons digest food reasonably well following removal of large portions of their stomachs. However, they must eat more frequently and follow some dietary restrictions for adequate digestion. Digestive enzymes are listed in Table 34.1.

Digested foods pass through the intestinal walls into the blood or the **lymph**. Lipids, in particular, are absorbed by the lymph. Undigested foods, along with bacteria which find the intestine a good home and breeding place, are passed into the large intestine or colon, where water and other substances are absorbed. The contents of the large intestine are finally discharged as the feces.

The other major way the body rids itself of waste products is through the urine, but

TABLE 34.1 Major digestive enzymes

Site of Secretion	Enzyme	Site of Action	Substrate Acted Upon	Products of Action
mouth	salivary amylase	mouth	starches	maltose, dextrins
stomach	pepsin	stomach	proteins	polypeptides (peptone)
stomach	gastricsin	stomach	proteins	polypeptides
pancreas	pancreatic amylases	small intestine	starches	maltose
pancreas	trypsin	small intestine	proteins, polypeptides	small peptides
pancreas	chymotrypsin	small intestine	proteins, polypeptides	small peptides
pancreas	carboxypeptidase	small intestine	polypeptides	peptides, amino acids
pancreas	pancreatic lipase	small intestine	triacylglycerols	fatty acids, glycerol
small intestine	disaccharidases	small intestine	disaccharides	monosaccharides
small intestine	nucleotidases	small intestine	nucleotides	nucleosides, phosphate
small intestine	aminopeptidase	small intestine	polypeptides	peptides, amino acids
small intestine	dipeptidases	small intestine	dipeptides	amino acids

this represents a different type of excretion. Urine is produced by the kidneys, which may be thought of as a filter for the blood. In addition to water, urine contains waste products produced by the further reactions of materials absorbed through the intestinal wall. The urine is a much better indicator of the kinds of reactions going on in the body than are feces, since it contains only the end products of the vital reactions of the body, and does not contain undigestible substances which might be present merely by chance. Consequently urinalysis is much more important than analysis of feces in diagnosis of the state of health of a person.

**34.3
FURTHER
METABOLISM**

The word **metabolism** is used to describe the sum total of all reactions that occur in a living organism. The study of metabolism emphasizes those reactions which follow the absorption of digested food; strictly speaking, however, digestion is also a part of metabolism. Those molecules which are absorbed into the body may be either built into more complex molecules in reactions described as **anabolic,** or degraded into smaller molecules in **catabolic** reactions. Anabolic reactions require energy and are characterized as **endergonic,** and catabolic reactions liberate energy and are **exergonic** reactions. Any chemical substance that is involved in metabolic reaction is a **metabolite.** Detailed discussions of the metabolism of carbohydrates, proteins, and lipids are the subject matter of Chapters 35, 36, and 37.

**34.4
ENERGETICS**

There are many ways of measuring the amount of energy released in or required for a chemical reaction. The most meaningful expression of biochemical energy is **free energy.** Free energy, also known as the Gibbs energy, is given the symbol G. The term is used to describe the energy of reactions that is available for doing useful work under the conditions of most living systems, i.e., under constant temperature and pressure. When energy is released in a reaction the change in free energy, ΔG, is negative. Since such energy-releasing reactions can occur without an external source of energy, they are called spontaneous reactions. Conversely, when the ΔG is positive, energy must be supplied for a reaction to proceed.

Although we cannot measure the free energy "contained" in substances, but can measure only the free-energy change that takes place in reactions, it is convenient to think of some compounds as containing more energy than others. Such compounds are often referred to as high-energy compounds, but what is really high is the negative ΔG observed when such compounds react. Since high-energy compounds are more likely to react at normal temperatures, they are relatively unstable, while those compounds formed in a reaction in which a great deal of energy was released are more stable. A spontaneous chemical reaction may be defined as a process in which matter proceeds to more stable states.

Since chemical reactions tend to proceed spontaneously when energy is released, the obvious question arises, "Why doesn't a potentially combustible substance, such as wood, start to burn spontaneously in air?" In this case, we know that the wood must first reach its kindling temperature. Speaking in more general terms, an **energy of activation** must be supplied before even a potentially spontaneous reaction will proceed. For some reactions, such as the burning of phosphorus in air, normal room temperature is sufficient to initiate the reaction.

A useful analogy to the energy of activation is provided by the water in a lake high on a mountain. Normally the water would run downhill until it reached sea level, but some barrier, perhaps a beaver dam or a landslide, prevents the water from running down-

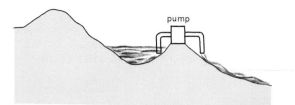

Fig. 34.3 Release of stored energy. The water in a mountain lake is analogous to chemical compounds which contain stored energy. Some of the energy can be released when the water runs downhill; however, the dam, which represents an energy barrier similar to the energy of activation in a chemical reaction, may prevent the water from flowing toward sea level. If a pump is installed to first lift the water over the dam, however, then release of energy can then take place spontaneously.

hill and releasing the energy—energy which might be used to drive turbines and give rise to electrical energy. If we supply energy by using a pump, we can raise the water over the barrier and let it flow downhill. The pump is analogous to a device for providing the energy of activation. Diagrams showing ways energy can be stored, converted and released are given in Figs. 34.3, 34.4, and 34.5.

Although glucose does not burn at body temperature even when placed in pure

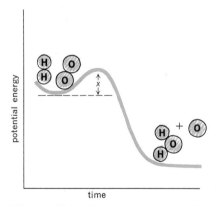

Fig. 34.4 Energy relationships between hydrogen and oxygen gases and water. The end products have less energy than the reactants; thus the reaction releases energy after the initial energy of activation (represented by x) is supplied. The energy barrier is relatively small. [Adapted from J. J. W. Baker and Garland E. Allen, *Matter, Energy, and Life*, 3rd ed., Addison-Wesley, 1977.]

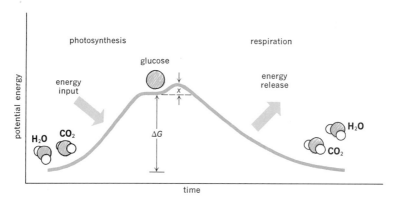

Fig. 34.5   An energy diagram showing the conversion of water and carbon dioxide to glucose, followed by the oxidation of glucose. Energy from the sun is "stored" in the "uphill" process of photosynthesis. This energy is released in respiration. The difference in free energy between glucose and the two substances, water and carbon dioxide, is represented by ΔG. The value x represents the energy of activation necessary for the oxidation of glucose. This energy is supplied by ATP. [Adapted from J. J. W. Baker and Garland E. Allen, *Matter, Energy, and Life*, 3rd ed., Addison-Wesley, 1977.]

oxygen, animals have systems for converting glucose, as well as other foods, into carbon dioxide and water. The oxidation of glucose, which initially requires an energy of activation, takes place in many steps. At some of these steps, small amounts of energy are released. Trying to use the energy of a molecule of glucose in a single reaction would be similar to the effect of placing a simple waterwheel at the base of Niagara Falls. Living organisms cannot utilize such large amounts of energy. Using the waterfall analogy, we can say that organisms divert the flow of water around a precipitous waterfall, and by using several small waterwheels at the bottom of small waterfalls, they are capable of trapping the energy in small, useful packets. (See Fig. 34.6.)

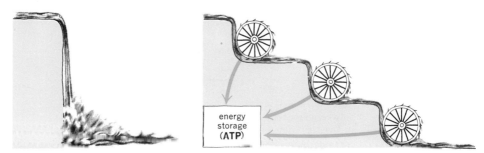

Fig. 34.6   Waterfall analogy to biochemical energetics. When water falls over a high waterfall it generates tremendous amounts of energy. A waterwheel at the bottom of such a fall would be smashed. If, instead of one precipitous waterfall, the stream is diverted into a series of smaller falls, each one having a water wheel, the energy the water loses can be converted into useful energy. When biochemical reactions that yield large amounts of energy proceed in a series of reactions, each of which can synthesize ATP, the same principle is shown.

Some energy so released is used to drive the anabolic reactions of the organisms. Other energy is used to keep the heart beating, move food through the digestive tract, and move the animal from one place to another. The problem that living organisms have solved, that of release of large amounts of energy in small units, is similar to the problem still faced by nuclear scientists who are trying to make the tremendous energy from the fusion of hydrogen useful to drive generators or automobiles instead of functional only as a bomb blast.

Plants and animals are able to store and use the energy of metabolic reactions largely because of the compound **adenosine triphosphate (ATP)**. This compound, which was discussed in Chapter 33, represents the principal "currency" of most living organisms.

ATP + HOH → ADP + phosphate + 7 to 12 kcal

When ATP is converted to ADP (adenosine diphosphate) and phosphate, a relatively large amount of energy is released. Conversely, when ADP combines with a phosphate group to give ATP, energy is required. Most living things use the synthesis of ATP to store energy from the oxidation of food; later, this ATP is hydrolyzed to provide energy for reactions such as the synthesis of proteins or for muscle contraction. (See Fig. 34.7.) Even the light given off by a firefly and the electricity produced by an electric eel use ATP as their energy source.

The exact amount of energy given off when a mole of ATP is hydrolyzed to ADP and phosphate in living systems is not known. The standard change in the free energy,

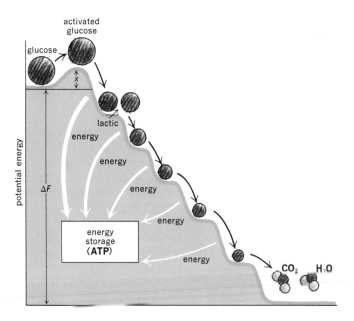

Fig. 34.7 The energetics of the oxidation of glucose. The diagram shows how glucose must first be activated (by ATP) before its energy is released. The release of this energy is shown as occurring in a series of steps, several of which generate ATP from ADP and phosphate.

ΔG, for this reaction is $-7\,300$ calories. The negative sign indicates that energy is released. The conditions under which ΔG is determined are a temperature of 25°C, a pH of 7, and a concentration of 1 mole/liter of ATP. Reactions in the human body take place at about 37°C, the normal body pH is 7.4, and concentrations of ATP are usually in the millimolar range—never so high as 1 molar. While it is difficult to determine the amount of energy released when ATP is hydrolyzed in living systems, various estimates place the value at about 10 000 to 12 000 calories (10–12 kilocalories). For general purposes it is sufficient to know that a relatively large amount of energy is given off when ATP is hydrolyzed or a relatively large amount can be stored by the synthesis of ATP. Purine and pyrimidine bases other than adenine may also form "high-energy" compounds, and we find that guanosine triphosphate (GTP) and uridine and cytidine triphosphates (UTP and CTP) are also used to store or provide energy.

Some compounds can release even more energy per mole when hydrolyzed or otherwise involved in a reaction than ATP does. Phosphoenolpyruvate (PEP), which will be discussed in future chapters, yields about 15 kcal/mole when it is converted to pyruvic acid under standard conditions. PEP is used as an energy source in many reactions. Creatine phosphate is used in muscle tissue to store and release energy. Thus ATP, PEP, and creatine phosphate are all known as high-energy compounds, although, as we have seen, it is more accurate to say that they are compounds that can release a relatively large amount of energy in specific reactions, rather than just containing the energy. Many other compounds "contain" a large amount of energy; the trick is to have a reaction that releases or stores large amounts of energy. Such "tricks" are very important to life.

34.5
ELECTRON
TRANSPORT

The energy contained in foods is released primarily in the process of oxidation of the hydrogen atoms contained in the food. Hydrogen atoms or, speaking more precisely, the electrons of these atoms are passed from one compound to another in the series of reactions known as electron transport. The enzymes and coenzymes necessary for electron transport are localized in the mitochondria of cells.

Electrons may take several different routes, and certain intermediates are not really known, but the scheme involving **nicotinamide adenine dinucleotide (NAD), flavine adenine dinucleotide (FAD),** and the **cytochromes** is probably the most important such system. The formulas for these compounds are given in Fig. 34.8. The substrate nicotinamide adenine dinucleotide phosphate (NADP), which functions much as NAD does, contains an additional phosphate on the ribose unit attached to nicotinamide. There are several cytochromes all having structures similar to cytochrome c (see Fig. 34.9).

In the system shown in Fig. 34.10, two hydrogen atoms (two protons and two electrons) are removed from an energy-rich molecule. Two electrons along with one proton are passed to NAD, which becomes NADH. The other proton (a hydrogen ion) goes into the surrounding medium. NADH then passes the electrons to FAD, which oxidizes the NADH, reduces the FAD, and releases energy. A hydrogen proton (hydrogen ion) from the solution combines with the FAD and we get $FADH_2$. $FADH_2$ then passes the electrons to the cytochromes. The cytochromes are complex molecules which contain iron in the $3+$ state which is reversibly reduced to the $2+$ state by the electrons it receives from $FADH_2$. One of the cytochromes eventually passes electrons to an oxygen atom which also acquires two protons, and the end product of the series of reactions is water. In this series of reactions the combination of hydrogen with oxygen to produce water, a process that releases large amounts of energy, has been broken into several steps, and the potentially

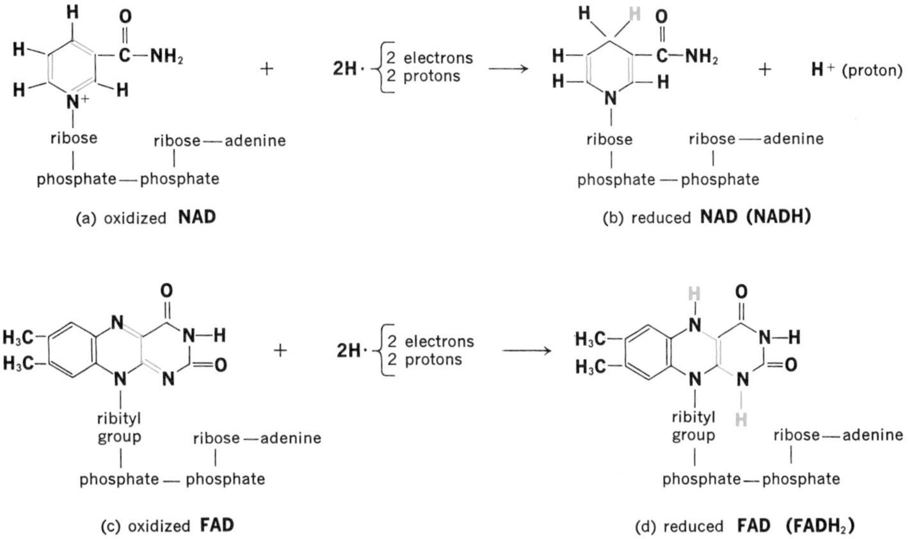

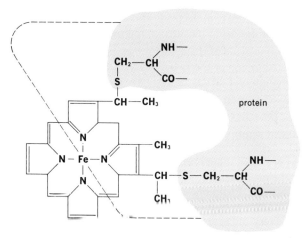

Fig. 34.8 Formulas of electron acceptors in the electron transport system: (a) shows oxidized NAD (see Fig. 33.13 for complete structure), (b) reduced NAD (NADH), (c) oxidized FAD, and (d) reduced FAD ($FADH_2$).

destructive or unusable amounts of energy are made available in useful-size packets. There are more intermediate compounds than those given above and in Figs. 34.8 and 34.10. However, the exact role played by each is not so firmly established as is that of

Fig. 34.9 The structure of cytochrome c. The heme group is bound to the protein molecule by sulfide linkages to the amino acid cysteine, which is part of the protein.

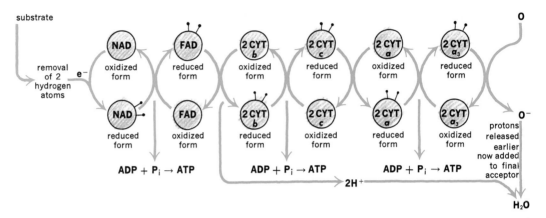

Fig. 34.10 A summary of electron transport. The diagram shows the removal of the two hydrogen atoms from an oxidizable substrate, the transport of the electrons, and the final acceptance of the electrons and two protons by an atom of oxygen to give water. The energy released is trapped in the synthesis of ATP from ADP and inorganic phosphate (P_i).

NAD, FAD, and the cytochromes. For this reason and for simplification, these other compounds have not been included in this presentation.

We can summarize the process of electron transport as follows:

1) The substrate loses two electrons and one proton to NAD, converting it to NADH and releasing a hydrogen ion (proton) to the surroundings.

2) The NADH gives two electrons and one proton to FAD. In addition, the FAD acquires a proton from its surroundings and becomes $FADH_2$.

3) Since each molecule of the cytochromes can accept only one electron, two molecules are required to accept the two electrons from $FADH_2$. The two protons lost in the transformation of $FADH_2$ to FAD are released into the surroundings.

4) An atom of oxygen acquires two electrons from two molecules of a cytochrome, oxidizing the cytochrome and reducing the oxygen.

5) The oxygen acquires two protons to form water.

This process is shown diagrammatically in Fig. 34.10.

During the process of hydrogen transport, in a way still not fully understood, ATP is synthesized from ADP and phosphate. If NAD is the original electron acceptor there are usually three molecules of ATP synthesized for every two electron pairs (or two hydrogen atoms) or for every molecule of water formed. If FAD is the original electron acceptor, only two molecules of ATP are synthesized in the formation of water. Thus, by the process of electron transport, any chemical substance that can yield electrons (which are almost always released in combination with protons as hydrogen atoms) can result in the synthesis of ATP and the storage of energy. The synthesis of ATP, as well as electron transport, occurs primarily in the mitochondria of cells.

To illustrate, ethanol is oxidized to give acetic acid (probably in the form of acetyl Coenzyme A) by two reactions, both of which involve NAD.

ethanol + NAD → NADH + acetaldehyde

acetaldehyde + H₂O + NAD → NADH + acetic acid

An alternative pathway involves the oxidation of acetaldehyde, with FAD as the hydrogen acceptor. Thus ethanol can yield a molecule of acetic acid with the resultant synthesis of either six molecules of ATP (if both reactions use NAD) or five molecules of ATP (if one reaction uses NAD and the other FAD).

The complete oxidation of acetic acid involves a more complex process and can yield a total of twelve molecules of ATP.

CH₃COOH + 2O₂ → 2CO₂ + 2H₂O

The energy *stored* as the result of the oxidation of a compound in a living system can be compared with the amount of energy *released* in the process. If one mole of ethanol is oxidized to give CO_2 and H_2O in a bomb calorimeter, the energy released is 312 kcal/mole (the ΔG is −312 kcal/mole). This is the same as the energy released when ethanol is oxidized to CO_2 and H_2O in a living system.

The oxidation of ethanol *in vivo* proceeds as shown in the equations given above. The first two reactions involve the conversion of NAD to NADH, and the oxidation of each of the two NADH molecules can result in the synthesis of three molecules of ATP from ADP and phosphate. In another series of reactions the complete oxidation of the acetic acid, which is formed from ethanol, results in the synthesis of 12 molecules of ATP per molecule of acetate. Thus the total oxidation of ethanol to CO_2 and H_2O gives 3 + 3 + 12 or 18 moles of ATP synthesized per mole of ethanol oxidized. The energy stored in this process is found by multiplying by 18 the amount of energy stored when one mole of ADP and one of phosphate are converted to ATP. If we use the standard free energy change of 7.3 kcal for this reaction, the result is 131.4 kcal.

Thus, of the possible energy (312 kcal/mole of ethanol completely oxidized), 131.4 kcal has been stored by the synthesis of ATP. This represents an efficiency of 131.4/312 or 42%. If we use a value of 10 kcal for the energy stored when a mole of ATP is synthesized from ADP and phosphate, the efficiency of storage becomes 18 × 10 kcal divided by 312 kcal, or 58%.

**34.6
ENZYMES**

In many respects biochemical reactions are similar to the chemical reactions occurring *in vitro*. They follow the laws of conservation of mass and energy and most require an energy of activation. The products formed are similar to those formed in similar organic reactions. Like many nonbiochemical reactions, biochemical reactions require catalysts in order to proceed at any appreciable rate. Although a few reactions that occur in living organisms proceed without catalysts, the greatest number require the biochemical catalysts, enzymes. An **enzyme** is usually defined in just these terms, as a catalyst of biological origin. All known enzymes contain protein. Some are composed only of protein; others require some cofactor other than protein in order to function. FAD and NAD are cofactors, usually called **coenzymes** and function primarily in conjunction with particular enzymes. (See Fig. 34.11.)

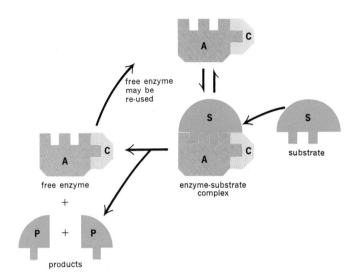

Fig. 34.11 A diagram showing the involvement of a coenzyme in an enzymatic reaction. The protein part of an enzyme (apoenzyme) is combined with its coenzyme, as shown upper center. The substrate molecule then attaches to the completed enzyme and is converted, first into an unstable enzyme-substrate complex, and then into the products of the reaction with the regeneration of the free enzyme. Not all enzymes require coenzymes.

Enzymes are remarkable catalysts in that they can accelerate the rate of reactions without either an increase in temperature or a major change in pH. They differ from most other catalysts since usually only one reaction is promoted and by-products are not formed. The exact way enzymes function is still the subject of active research. It is generally believed that the **substrate**, the substance reacting, is bound in some way to the surface of an enzyme. The binding site is apparently quite specific and can bind only one or only one kind of substrate molecule. This specificity has been compared to a lock and key. The enzyme (the key) functions only when a certain specific lock (substrate) or perhaps a certain type of lock is available. (See Fig. 34.12.)

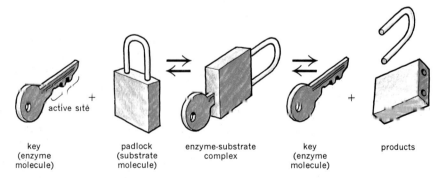

Fig. 34.12 The lock-and-key model for picturing the interaction of substrate and enzyme.

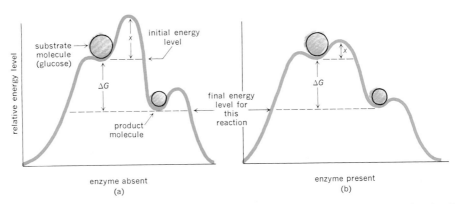

Fig. 34.13 The effect of an enzyme on the activation energy. The amount of activation energy is represented by *x* and the change in free energy by ΔG. The uncatalyzed reaction (on the left) requires a greater energy activation. In reactions that proceed in a series of stepwise reactions, each reaction has its energy of activation. [Adapted from J. J. W. Baker and Garland E. Allen, *Matter, Energy, and Life*, 3rd ed., Addison-Wesley, 1977.]

One way of answering the question, "What is it that enzymes really do?" is to say that they influence the course of a reaction so that less energy of activation is required. They *do not*, however, provide this energy. Enzymes cannot make *impossible* reactions proceed; the energy driving the reaction is still the difference in energy content between reactants and products. Enzymes, in common with other catalysts, only hasten the rate of attainment of equilibrium in a chemical reaction. (See Fig. 34.13.)

Another way of answering the question of what enzymes do is to say that they hold reactants at sites on the enzyme surface so that the probability of the two substances reacting is greatly increased. It has also been postulated that when enzymes bind substrates, they actually stretch the molecule, somewhat as the rack stretched the victims on this instrument of torture. This tension makes it easier for specific bonds to be broken. Perhaps this stretching exposes a specific bond (e.g., an ester bond) so that it is more easily attacked by another molecule (e.g., a water molecule) to bring about a reaction e.g., the hydrolysis of an ester). (See 34.14.)

To discuss enzymes we must have methods for naming them. Originally enzymes were given trivial names, and we still use the names pepsin, catalase, and steapsin for enzymes that catalyze the hydrolysis of proteins, the decomposition of H_2O_2, and a mixture of lipid-hydrolyzing enzymes. A better nomenclature consists of adding the suffix *-ase* to the name of the substrate upon which an enzyme acts, and we still use the names sucrase, urease, and arginase for enzymes that catalyze reactions involving sucrose, urea, and arginine. Since one substrate may undergo any one of several possible reactions, however, we now find it better to include not only the name of the substrate but also the type of reaction catalyzed when naming the enzyme.

The most recent system for naming enzymes is in fact, based on the type of reaction they catalyze. The categories used are given in Table 34.2. The complete name of the enzyme in this system includes the name of the substrate (or the general type of compound to which it belongs), as well as the type of reaction catalyzed. Under this system urease belongs to class 3, the hydrolases, and is called *urea amidohydrolase*, which indicates that the substrate urea is hydrolyzed and that the reaction is the hydrolysis of an amide.

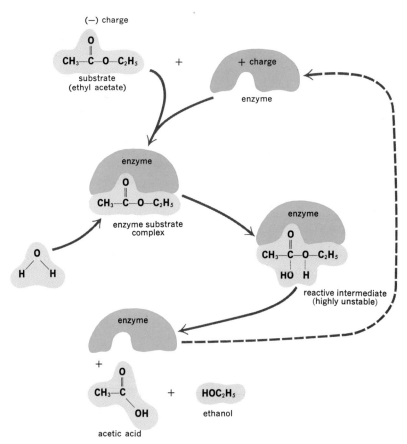

Fig. 34.14 A diagrammatic representation of an enzymatic reaction in which the enzyme makes the substrate more vulnerable to attack by the reacting molecule. In this reaction, the substrate (ethyl acetate) is bound to the enzyme both by complementary shapes and by opposite charges. This binding facilitates the hydrolysis of the ester bond by a molecule of water.

Sucrase is named as a β-D-*fructofuranoside fructohydrolase*, that is, an enzyme that hydrolyzes a fructose group from a compound which contains D-fructose in a beta linkage and in the furanose form. The basic classifications are broken down into subgroups and

TABLE 34.2   Types of enzymes

Enzymes	Reaction Catalyzed
1. oxidoreductases	oxidation and reduction
2. transferases	transfer of a chemical grouping from one compound to another
3. hydrolases	hydrolytic reactions
4. lyases	nonhydrolytic addition or removal of groups
5. isomerases	conversion of a substance into one of its isomers
6. ligases	syntheses or, more specifically, the formation of various types of bonds

we can refer to an enzyme by a series of numbers. Urease, for instance, is 3.5.1.5. For the purpose of this book, many aspects of the formal nomenclature are too specific, so we will use trivial names as a general rule, sometimes with explanations as to the exact reaction catalyzed.

34.7
CONTROL OF
ENZYMATIC
REACTIONS

Most cells in the human body are capable of carrying out hundreds of enzyme-catalyzed reactions. We can illustrate some types of reactions in the diagram below.

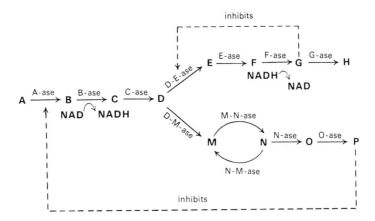

In this series of reactions, substance A is converted to B under the influence of an enzyme we will call A-ase. B is then converted to C and eventually either to H or P if all reactions proceed. A may be a specific type of food, perhaps the amino acid methionine, and H and P represent important metabolites such as ATP, a hormone such as insulin, or some other vital compound. The enzymes that influence these reactions are catalysts, and they only hasten the rate at which the normal equilibrium is reached. They do not change the concentrations of substances at equilibrium. In spite of this, there are many ways that enzymes can be controlled so that the processes in the body may be controlled. In our example we will show how these processes can be controlled to give the best amounts of H and P.

Law of Mass
Action

Since enzymes catalyze the attainment of equilibrium, the law of mass action (Section 12.2) indicates that the rate of conversion of A to B will be decreased as the amount of B (the product of the reaction) increases. However, if B is continually being removed by being further metabolized, the reaction of A → B will proceed as long as the enzyme A-ase is functioning and some of A is present.

Inhibitors

The rate of many enzyme-catalyzed reactions is decreased or may be completely stopped by inhibitors. Cyanide ions, dimethyl mercury, and the fluorophosphate nerve gases (which are also used in insecticides) irreversibly inhibit enzymatic reactions. If a sufficient number of vital enzyme molecules are removed from functioning by these irreversible inhibitors, death can occur. Other inhibitors are classified as reversible inhibitors. This type of inhibitor decreases the rate of a reaction but, since it is not bound permanently to the enzyme, when it is removed (by dissociation or other process) the enzyme is again effective. Atropine, which is used to dilate the pupils of the eyes in optical examinations, is a reversible inhibitor. It reversibly inhibits the same enzyme (cholinesterase) that is permanently inhibited by the fluorophosphate nerve gases. Many drugs and medications work by inhibiting enzymes and thus decreasing the rate of some reactions.

Sometimes the product of a reaction inhibits another reaction. For instance, in our diagram we have shown that P is an inhibitor of A-ase. Thus when P is produced, the whole series of reactions shown will be slowed or maybe completely stopped. This is an excellent means of control of reactions, since when sufficient P has been produced it would only be wasteful to keep synthesizing it. Inhibition can also be used to control the relative amounts of H and P by acting on one of the enzymes at the branching point of a series of reactions. In the diagram we have indicated that G inhibits D-E-ase. Thus as G is formed, we find that the synthesis of E, F, G, and H is decreased while that of M, N, O, and P continues.

Coenzymes   In the diagram, the coenzyme NAD is indicated as being required for the conversion of B to C and F to G. If there is a deficiency of NAD (as would result from a deficiency of the vitamin nicotinamide) these reactions would not proceed at an optimum rate. Another way that coenzymes control reactions is that they must not only be present, but must be present in a specific form. NAD is reduced in the reaction $B \rightarrow C$. If all the NAD present is in the reduced form, NADH, the conversion of B to C stops. We need a reaction such as $F \rightarrow G$, which requires NADH and regenerates NAD, for B-ase to continue to function. Bacteria which produce lactic acid apparently do so because they must use NADH produced in an earlier part of the metabolism of glucose so the metabolism will proceed.

$$CH_3{-}\overset{\overset{\displaystyle O}{\|}}{C}{-}COOH + NADH + H^+ \rightarrow CH_3{-}\overset{\overset{\displaystyle OH}{|}}{\underset{\underset{\displaystyle H}{|}}{C}}{-}COOH + NAD^+$$

pyruvic acid

Other organisms do not produce lactic acid from glucose since they oxidize NADH via the scheme shown in Fig. 34.10.

Allosteric Enzymes   So far we have talked primarily about enzymes whose actions are dependent on one site where a substrate is bound. Many enzymes that catalyze simple hydrolyses are of this type. Many more enzymes have their action modified by some process taking place at a site other than the active site. These are called **allosteric enzymes** (from *allo*, meaning other, and *steric*, meaning space). Most allosteric enzymes are made of several subunits and are sometimes known as oligomers (*oligo*, few; *mer*, body). The aggregation of several units of enzymes of this sort changes the activity of the individual units.

Allosteric enzymes are often found at the beginning of a sequence (such as A-ase in the diagram) or at a branching point (D) in a metabolic chain. They can be either activated or inhibited by a variety of factors, but these factors are never associated directly with the active site, the place where the substrate binds. All factors influencing allosteric enzymes are not fully understood, but there is general agreement that they represent one of the most important aspects of normal metabolic control.

Protein Synthesis   Since enzymes are primarily protein (although many require cofactors also), the rate of an enzyme-catalyzed reaction may be controlled by the amount of enzyme present. While many enzymes—those we call **constitutive enzymes**—seem to be synthesized at a relatively constant rate, others are synthesized only or primarily under special conditions. Since we can induce the formation of these enzymes, we call them **inducible enzymes**. For instance, many bacteria can use either glucose or lactose as the major source of energy. When supplied with glucose, the bacteria do not produce the enzyme β-galactosidase which is necessary for the hydrolysis of the disaccharide lactose. However, when lactose is present in the media in which the bacteria are being grown, β-galactosidase is produced. We now know the mechanism by which an inducer such as lactose or a substance similar to it chemically "turns on" the hereditary apparatus that lets the DNA pattern for the synthesis of the protein β-galactosidase be expressed. The enzyme alcohol dehydrogenase is present in greater quantities in the liver and kidneys of people who consume relatively large amounts of ethanol.

Some people lack the genetic information (DNA molecules) for the synthesis of a given enzyme. We can illustrate this in our metabolic diagram by saying that there is no C-ase in the bodies of such people. This leads to the accumulation of A, B, and C and to a deficiency of D and all metabolites that are normally derived from it. In some cases, as in phenylketonuria (Section 39.4), we find that the accumulation of a normal metabolite (phenylalanine in phenylketonuria; A, B, or C in our diagram) causes bad effects. These effects may be due to the fact that the metabolite which accumulates in large amounts, or a product (or products) derived from the normal metabolites, inhibits some reaction in another metabolic sequence. This is apparently the cause of the mental retardation of children with phenylketonuria. Another result of the inability to synthesize an enzyme is that one or more products of the metabolic series of reactions is not present. Albinos lack the pigment melanin in their hair and skin because they lack an enzyme necessary for the synthesis of the pigment. Hereditary defects of this sort are discussed in Sections 35.11 and 39.4.

Location of Enzymes   In the metabolic scheme we have indicated that there are two enzymes involved in the interrelationship of substances M and N.

Ordinarily we would expect that M → N if the equilibrium favors the production of N. However, if N-M-ase is also present, we have what some call futile reactions, or we can say there is a "vicious" circle. Just as soon as some N is formed, it is immediately changed back into M. Such reactions potentially exist with glucose and glucose-6-phosphate.

glucose + ATP → glucose-6-phosphate + ADP

glucose-6-phosphate → glucose + phosphate

These reactions are somewhat different since ATP is involved in one and only an ordinary phosphate is produced in the other; however, the net effect of the two reactions is a futile process (vicious circle). Now, the conversion of glucose to glucose-6-phosphate is the beginning of one of the most important sequences of reactions in all living things. If the glucose-6-phosphate were immediately converted back to glucose, there would be no energy derived from glucose, or starches which give glucose. Fortunately, this disastrous consequence is avoided because glucose-6-phosphate is found in the endoplasmic reticulum of the cell and hexokinase in the soluble parts of the cytoplasm. In other cases, enzymes that would produce futile reactions are located in different compartments of the cell, such as the mitochondria and lysosomes, or in the cell membranes. Such separation of enzymes is vitally important if a cell is to both catabolize compounds for energy and synthesize (anabolize) important polymers such as proteins, polysaccharides, nucleic acids, and the complex lipids. Some control of these "vicious cycle" reactions is also provided by allosteric enzymes, which can make a reaction proceed primarily in one direction, even when competing enzymes are present.

Another case in which the location of enzymes is important involves the manufacture of enzymes such as those that catalyze the hydrolysis (digestion) of proteins. Since all cells, including those that manufacture the protein-splitting enzymes (proteases), are composed largely of proteins, there is a problem. This problem is solved partly by having the enzymes removed from the cell by secretion through a duct. In fact, the final stages of synthesis may take place in part of the cell directly connected to the duct which is less subject to attack by the enzyme. The pancreatic enzyme trypsin is secreted from the pancreas through ducts into the small intestine, where it catalyzes the digestion of proteins.

Another important factor in the synthesis of proteases is that they are synthesized and secreted in an inactive form. These inactive forms, which are known as **proenzymes** or **zymogens** (*zyme*, enzyme; *-gen*, generator of), have a peptide unit covering the active site of the enzyme. Only when the proenzymes are safely out of the cell that made them and in a place such as the stomach or intestine, where the walls are less subject to being digested, is the protective group removed. Examples of conversions of many different proteases are given in Chapter 36.

Procaryotes, since they have no intracellular organelles, have much greater difficulty in separating enzymes from each other. It is their internal organization, which includes the separation of enzymes, that makes eucaryotic cells capable of more complex processes.

The methods of control of enzymatic reactions discussed in the previous paragraphs are some of the major controls both in normal metabolism and those brought about by drugs or poisons. There are others that are known and probably others still to be discovered. Some people believe that all disease can be explained by abnormalities of enzymatic reactions.

**34.8
HORMONES**

While the enzymes are very important in metabolic control as discussed above, a further role is played by hormones. A **hormone** *may be defined as a product of an endocrine gland.* There are several such glands, such as the pituitary, the thyroid, parts of the pancreas, the adrenals, and parts of the ovaries or testes. (See Fig. 41.4.) The products of these glands are secreted directly into the bloodstream and distributed to all parts of the body of an animal. Some of the glands listed above are also exocrine glands—they produce some secretions that flow through ducts to a definite site of action.

Hormones of higher animals are usually found to be peptides, proteins, glycoproteins, or steroids.   In many cases nonprotein matter is associated with the proteinaceous hormones.   Insulin, the hormone of the pancreas, is a polypeptide which normally contains small amounts of the metal zinc.   The hormones of the pituitary, such as the growth hormone, the thryroid-stimulating hormone, the hormones which influence the ovary and testes, as well as others from this source, are primarily proteins or polypeptides. The sex hormones produced by the ovary or testes as well as most of the hormones of the adrenal gland are steroid hormones. (See Section 32.6.)   The hormones of the thyroid gland and of the medulla of the adrenal gland are derivatives of amino acids.

Although the physiological effects of the hormones, such as the increase of the basal metabolic rate by thyroxine or the effect of testosterone on the secondary sexual characteristics in the male, are well known, the exact manner in which hormones exert their control on body processes is not known.   Recent research indicates that hormones function by increasing the rate of synthesis of certain proteins, perhaps by first increasing the rate of synthesis of ribonucleic acid (RNA), which in turn influences the rate of protein synthesis.   If the protein synthesized is an enzyme, the effects on metabolism are obvious. Other hormones, such as insulin, affect the permeability of cells.

Another theory has been developed which correlates many experimental observations concerning hormone action.   The hormone, which does not penetrate the cell, binds to the outer membrane of the cell.   In some instances this binding stimulates the release of a prostaglandin which activates the enzyme adenyl cyclase.   This enzyme catalyzes the synthesis of cyclic AMP from ATP within the cell.   The cyclic AMP then activates one or more enzymes within the cell.   For instance, adrenaline, which is released from the adrenal glands under stressful situations, results in the stimulation of an enzyme that catalyzes the breakdown of glycogen to glucose.   Other enzymes catalyzing the release of fatty acids from adipose tissue cells are also stimulated by adrenaline and the action of cyclic AMP.   Each molecule of adrenaline has its effect amplified at each step of the reactions, and one molecule of adrenaline can result in the release of 30 000 molecules of glucose.   This glucose (as well as the readily metabolizable fatty acids) can then be used by the body in the "fight or flight" reactions that are generally listed as one of the major effects of adrenaline. See Fig. 34.15 for a diagram of this process.

A great deal of interrelation exists between the various hormones.   One of the simplest examples is that between the thyroid-stimulating hormone of the pituitary and the thyroid hormone thyroxine and/or triidothyronine.   The thyroid-stimulating hormone induces the thyroid gland to produce its hormone thyroxine.   Thyroxine, in turn, influences the pituitary gland and inhibits the production of the stimulating hormone.   This kind of process has been called "feedback." (See Fig. 15.3 for an illustration of feedback.) If the thyroid gland cannot produce sufficient thyroxine because of a lack of iodine in the diet, excessive amounts of thyroid-stimulating hormones are produced, causing an enlargement of the thyroid gland, which is called a goiter.

There are many other feedback relationships between various endocrine glands, some of them extremely complex.   For example, the sexual cycles in females, the production of inflammation and wound healing, and the control of the level of glucose in the blood are all influenced by feedback relationships of endocrine glands and their hormones. Other hormones, particularly those of the pituitary gland, depend on releasing factors from the brain.   Only when releasing factors, which are peptides, are secreted is the hormone released into the blood system.

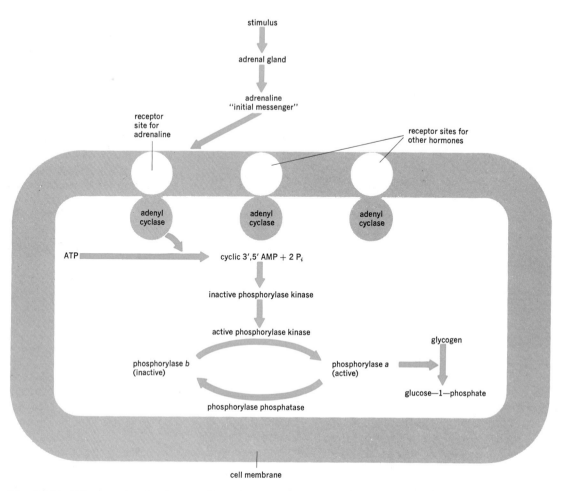

Fig. 34.15 Diagram showing how cyclic AMP functions in the action of adrenaline. The adrenaline is released from the adrenal gland and is carried by the blood to the cell where it is bound. This stimulates the activity of the membrane-bound enzyme adenyl cyclase, which catalyzes the conversion of ATP to cyclic AMP. Cyclic AMP in turn activates enzymes that convert glycogen to glucose. Other hormones also involve cAMP in their actions. [Diagram from J. J. W. Baker and Garland E. Allen, *The Study of Biology*, 3rd ed., Addison-Wesley, 1977.]

SUMMARY

Metabolic reactions are remarkably similar in all plants and animals so far studied. Not only are the same types of molecules involved as reactants, products, and catalysts, but in all cases the reactions occur in a number of small, sequential steps. The presence of a number of steps in an overall reaction not only aids in trapping energy from foods, but also transforms anabolic reactions requiring large amounts of energy to a series of small reactions, each of which can be driven by available energy. The introduction of a number of small steps also makes possible more sensitive control of reactions.

Each metabolic reaction is catalyzed by an enzyme. Failure of an enzyme to function normally usually causes a deficiency of some substance or substances and an accumulation of other substances. These deficiencies or excesses may have serious or fatal consequences for living organisms.

Anabolic reactions require energy which is usually provided by catabolic reactions, particularly oxidations. The ATP helps transfer energy to energy-requiring (endergonic) reactions from energy-releasing (exergonic) reactions. Any substance that can be oxidized by NAD or FAD usually brings about the synthesis of ATP from ADP and phosphate in living systems. A molecule of reduced NAD normally yields three molecules of ATP, while one of FAD yields only two molecules of ATP when complete oxidation has taken place.

Both anabolic and catabolic reactions are going on all the time in all living organisms. Although an animal may maintain a given weight for a long time, the molecules of which it is composed are constantly being changed. When animals are given isotopically labeled foods, we soon find these isotopes incorporated in their body tissues. They then begin to be eliminated slowly. We characterize this condition as a dynamic equilibrium. It is not like a chemical equilibrium (where the same atoms merely move from one compound to another), because new molecules are constantly being added and eliminated. The situation is illustrated (Fig. 34.16) where a vessel of water maintains a constant level even though water is constantly entering and leaving. We often speak of a **metabolic half-life**. This is the time it takes to replace half of a given body component. Half of the protein in the liver of a rat is replaced in 7 days, so we say that rat-liver protein has a metabolic half-life of 7 days.

Fig. 34.16 An illustration of dynamic equilibrium. Water flows from a faucet into a container that has an overflow outlet. Although the amount of water in the container remains constant, individual molecules are constantly being added from the faucet and removed from the overflow. If we added a dye to the water, we would see that it would gradually be washed away. Essentially, this is what we see when we give radioactive isotopes to a plant or animal and note their disappearance even though the plant or animal is neither growing nor shrinking in size.

The next several chapters are concerned with the metabolism of various types of compounds. We will discuss specific instances that led to the generalizations we have stated in this chapter. We will examine normal catabolic and anabolic reactions and trace the effects of blockages of such pathways or abnormal reactions, and the relation of these abnormalities to disease.

IMPORTANT TERMS

ATP	coenzyme	free energy	NAD
alimentary canal	constitutive enzyme	futile reactions	nucleosidase
allosteric enzyme	cytochromes	gastricsin	nucleotidase
amylase	digestion	gastrointestinal tract	pepsin
anabolic	disaccharidase	inducible enzyme	peptidase
assimilation	endergonic	lipase	photosynthesis
bile	energy of activation	lymph	proenzyme
catabolic	enzyme	metabolic half-life	saliva
chlorophyll	exergonic	metabolite	substrate
chymotrypsin	FAD	metabolism	trypsin

WORK EXERCISES

1. Write equations for the digestion of each substance.
a) sucrose   b) a triacylglycerol
c) ethyl acetate

2. List the enzymes necessary for the complete digestion of the following.
a) protein   b) a triacylglycerol
c) a polysaccharide

3. What type of organic compound is being hydrolyzed in the digestion of each of the following?
a) a protein   b) a triacylglycerol
c) a polysaccharide

4. Substance A gives 200 kcal/mole when completely oxidized *in vitro*. Its metabolism to the same end products results in the synthesis of 15 moles of ATP from ADP and phosphate. What is the efficiency of conversion of the energy of compound A to the energy stored in ATP? (Assume that ATP $\rightarrow$ ADP $+ \textcircled{P}$ gives 10 kcal/mole.)

5. How do enzymes differ from other catalysts?

6. Name the enzyme catalyzing each of these reactions.
a) triacylglycerol + water $\rightarrow$ fatty acids + glycerol
b) maltose + water $\rightarrow$ glucose
c) RNA + water $\rightarrow$ organic bases + ribose + phosphate
d) adenylic acid + water $\rightarrow$ adenosine + phosphate
e) adenosine + water $\rightarrow$ adenine + ribose

7. Some persons secrete little or no hydrochloric acid in their stomachs. How would their digestion differ from that of a normal person? What treatment might be used for these persons?

8. Why must the growth hormone be administered by injection rather than by mouth in order to be effective?

9. How does an enzyme affect the equilibrium of a reaction?

10. Using electron dot formulas, show the actual number of electrons in the oxidized and reduced part of the riboflavin molecule that changes when FAD $\rightarrow$ FADH.

11. Why are ATP, PEP (phosphoenolpyruvate) and creatine phosphate important?

12. List six means by which the rate of enzyme catalyzed reactions may be controlled. Give a specific example of each method of control.

Consult a basic biochemistry text to find the answers to the following.

13. Coenzyme Q, which is known as ubiquinone, is involved with NAD and FAD in electron transport. Find its function and the place it acts in the scheme.

14. What are the digestive hormones, and how do they function?

SUGGESTED READING

Changeux, J. P., "The control of biochemical reactions," *Scientific American*, April 1965, p. 36 (Offprint #1008). The author discusses the control of enzymes which in turn control other syntheses. A complex theory is involved, but is well explained in this article.

Hinkle, P. C., and R. E. McCarty, "How cells make ATP," *Scientific American*, March 1978, p. 104 (Offprint #1383). The latest details of how ATP is synthesized in mitochondria and in the chloroplasts of green plants is discussed and well illustrated. The role of membranes in this process is well presented.

Koshland, D. E., Jr., "Protein shape and biological control," *Scientific American*, October 1973, p. 52 (Offprint #1280). A discussion of the shape and changes in shape of proteins as they play important biological roles.

Pastan, I., "Cyclic AMP," *Scientific American*, August 1972, p. 97 (Offprint #1356). Cyclic AMP appears to control the rates of many metabolic processes. This control is complex but is well described in this article. The discovery of cyclic AMP is discussed, as well as theories of its action.

35

Metabolism of Carbohydrates

35.1
INTRODUCTION

The carbohydrates are the most important energy sources in the diets of most animals—including humans. Carbohydrates are generally ingested in the form of the high-molecular-weight polymers starch and cellulose. The disaccharides sucrose and lactose and the monosaccharides glucose and fructose are also present in many diets. The metabolism of carbohydrates occurs in several stages. First, they must be degraded by digestion into molecules small enough (monosaccharides) to be absorbed through the intestinal wall. Then, either the absorbed carbohydrates are converted to a polysaccharide (glycogen) and stored or they are further broken down and oxidized to yield energy. If food intake exceeds the energy expended, partially degraded carbohydrates can be converted to fat and stored. Complete breakdown of carbohydrates with maximum energy release results in the formation of carbon dioxide and water. Certain microorganisms, such as bacteria and yeasts, produce some carbon dioxide and water, but their catabolism of carbohydrates ends with lactic acid or ethanol.

While all animals ingest the polysaccharide cellulose, only a few are able, through the presence of microorganisms in their digestive tracts, to utilize the cellulose as a source of energy. For most animals, cellulose is not digested and is excreted essentially unchanged. It contributes to the bulk of intestinal contents and feces.

Plants metabolize carbohydrates quite differently. Instead of ingesting polymeric carbohydrates, green plants make their own carbohydrates through the process of photosynthesis. It is true that some of this carbohydrate may be oxidized to yield energy for energy-requiring reactions in the plant; however, the remainder of the saccharide units are polymerized to yield starches and cellulose.

35.2
DIGESTION OF CARBO-HYDRATES

The digestion of starch begins in the mouth, where salivary amylase is added to food as it is being chewed. **Amylases** catalyze the hydrolysis of starch to remove disaccharide units (maltose), leaving smaller starchlike molecules called **dextrins**. (See Fig. 35.1 for a representation of the action of an amylase.) The action of salivary amylase continues as

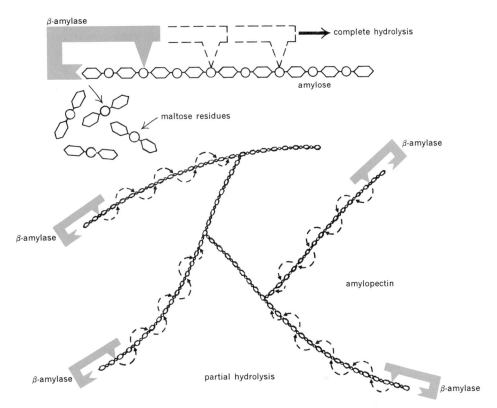

Fig. 35.1　The action of an amylase on starch.　The enzyme β-amylase aids in the digestion of starch by catalyzing the hydrolysis of every second α 1–4 linkage in a starch molecule. This reaction yields maltose which must be further digested under the influence of maltase to yield glucose.　Since β-amylase catalyzes the hydrolysis of only α 1–4 linkages, its action stops when a 1–6 branch is reached.　A starch molecule in which degradation has stopped at the 1–6 branches is called a dextrin.　[Adapted from James Bonner and Arthur W. Galston, *Principles of Plant Physiology*, W. H. Freeman and Company.　Copyright © 1952. Reprinted from an original article by W. Z. Hassid and R. M. McCready, *J. Am. Chem. Soc.* **65**, 1943, p. 1159.]

food passes through the esophagus but stops when the food is exposed to the acidity of the stomach.　Although the low pH of the stomach stops the action of salivary amylase, the hydrogen ions themselves may serve as catalysts for hydrolysis of bonds between saccharide units.　These acid-catalyzed hydrolyses tend to act at random sites in the starch molecules and produce a mixture of mono-, di-, tri-, and higher saccharides.　Following most normal meals, only part of the starches are converted to the monosaccharide level by the time the food leaves the stomach and passes into the small intestine.

　　In the small intestine, pancreatic amylase attacks the remaining starch and dextrin molecules, and the **disaccharidase** maltase completes the digestion by converting maltose to glucose.　Disaccharides such as sucrose and lactose are not digested by enzymes until they reach the small intestine, where the disaccharidases convert them to monosaccharides.　Monosaccharides present in food are not digested but are absorbed in the same form in which they were eaten.

As indicated in Section 35.1, the carbohydrate cellulose is not digested by higher animals. In most animals, cellulose is excreted as such; it makes up part of the so-called roughage of the diet and adds to the bulk of the feces. Cattle, sheep, and similar animals, however, are able to use some of the cellulose of their diet. Those animals have several food pouches or stomachs; the first of them contains various microorganisms, some of which can convert cellulose to glucose. Since this conversion occurs long before the food enters the intestine, the glucose produced by the hydrolysis of cellulose can be absorbed and utilized by ruminants.

The name *ruminant* is derived from the name given to the first food pouch—the rumen. In addition to sheep and cattle, goats, deer, antelope, and other grazing animals, giraffes, and camels have such a food pouch and are known as ruminants. These animals swallow their food without really chewing it well and later, while resting, regurgitate the food and chew it in a process described as chewing their cud. The process promotes mixing of saliva and microorganisms from the rumen with the food prior to its being carried to another food pouch, the abomasum, where it is acted upon by the acid gastric juice.

Another type of animal—the insect called the termite—can also make use of cellulose. This is, of course, why termites are so destructive. They eat their way through wooden structures using the cellulose of the wood for energy. Again it is not the termite itself, but a microorganism, a protozoan, living in the gut of the termite that contains the enzymes necessary to degrade cellulose into a useful form—glucose.

**35.3
ABSORPTION
AND FURTHER
METABOLISM
OF CARBO-
HYDRATES**

After complete digestion has converted carbohydrates into monosaccharides, they are absorbed through the wall of the intestine into the bloodstream. The portal vein carries the absorbed monosaccharides to the liver, where they may be synthesized into glycogen, oxidized to CO_2 and H_2O, or perhaps released as monosaccharides and allowed to circulate to other parts of the body. Although other organs are capable of metabolizing glucose, the liver is responsible for much of the carbohydrate metabolism. Under the influence of the pancreatic hormone insulin and other factors, the liver regulates the amount of glucose in the circulating blood. When the level of glucose in the blood is high, as it is after digestion of a meal rich in carbohydrates, the liver synthesizes large amounts of glycogen. When the amount of blood glucose is low, as it would be following intensive exercise, glycogen is depolymerized to yield glucose. Thus the liver, under the influence of insulin and other hormones, keeps the level of glucose in the blood relatively constant. The digestion and further metabolism of carbohydrates can be summarized as follows:

$$\text{polysaccharides} \xrightarrow{\text{digestion}} \text{monosaccharides} \xrightarrow[\substack{\text{intestinal} \\ \text{wall}}]{\substack{\text{absorption} \\ \text{through}}} \text{monosaccharides} \longrightarrow CO_2 + H_2O$$
$$\downarrow$$
$$\textbf{glycogen}$$

The glucose released by the liver travels throughout the body, where cells absorb it and use it for energy. In addition to catabolizing glucose, muscle cells can also (like liver cells) convert glucose to glycogen. While the liver may contain up to 8% glycogen, muscle cells normally contain only 0.5 to 1.0% of this polymer.

The concentration of glucose in the blood is extremely important for proper functioning of the human body. The normal fasting level of glucose is between 70 and 90 mg/100 ml. (Some methods of analysis which measure other compounds in addition to glucose give slightly higher values.) A concentration of glucose below the normal range is called

hypoglycemia and one higher than the normal level is called **hyperglycemia.** (See Fig. 35.2.)

Extreme hypoglycemia produces a series of reactions known as shock. The symptoms may include the trembling of muscle, a feeling of weakness, and a whitening of the skin. Serious hypoglycemia can cause unconsciousness and lowered blood pressure, and may result in death. Loss of consciousness is probably due to the lack of glucose in the brain, which depends on this sugar for most of its energy. Extreme hypoglycemia is usually due to the presence of excessive amounts of insulin. Some individuals who suffer from hyper-insulinism (overproduction of insulin) are subject to hypoglycemia after they eat a meal rich in carbohydrates. The most common cause of severe hypoglycemia is probably the injection of excessive amounts of insulin by diabetics; most diabetics carry candy or other glucose-rich substances with them to eat if they feel the onset of hypoglycemia. It is interesting to note that insulin shock has been produced by injection of insulin in treatment of mental diseases.

When the level of glucose in the blood rises, as it does following the digestion and absorption of a carbohydrate-rich meal, more insulin is secreted and this catalyzes the conversion of glucose to glycogen. At higher levels of blood glucose (Fig. 35.2) the synthesis of fatty acids and cholesterol increases. This synthesis is also influenced by insulin. At some level, between 140 and 170 mg/100 ml for most people, glucose is excreted in the urine. The concentration of glucose necessary for the appearance of **glucosuria** (glucose in the urine) is called the **renal threshold.** Actually, glucose is always filtered into the urine in the kidneys but is reabsorbed before the urine leaves the kidney. Thus glucosuria represents not so much an abnormal filtering out of glucose but rather an inability of the kidney tubule cells to reabsorb excessive amounts of the glucose. Extremely high levels of blood glucose are observed in cases of diabetic coma; factors other than hyperglycemia are probably responsible for the coma, however.

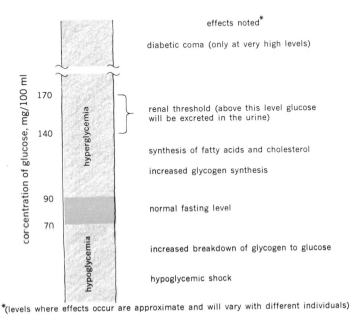

*(levels where effects occur are approximate and will vary with different individuals)

Fig. 35.2 Effects noted at various levels of blood glucose.

A variety of hormones controls the level of blood glucose. Insulin (from the pancreas) lowers blood glucose and enhances the conversion of glucose to glycogen. Adrenaline (epinephrine), one or more pituitary hormones, and a second pancreatic hormone—glucagon—all tend to raise the concentration of glucose in the blood. All these factors, acting together, keep the blood glucose level within certain limits. If these limits are exceeded, animals do not function normally. While moderate deviations in the level of blood glucose are not serious, extreme hypoglycemia may be fatal.

35.4 **GLYCOGEN** **SYNTHESIS** **AND** **BREAKDOWN**	The conversion of glucose to glycogen (**glycogenesis**) involves two intermediates, glucose 6-phosphate and glucose 1-phosphate. The same two intermediates are involved in the reverse process, the conversion of glycogen to glucose, or **glycogenolysis**. Although the intermediates in both processes are the same, the enzymes involved in the synthesis and gradation of glycogen are different. We might compare the processes to a round trip from city A to D, passing through cities B and C, but taking different roads for at least part of the two trips. (See Fig. 35.3.)

When glucose is converted to glucose 6-phosphate, the phosphate as well as the energy required to drive the reaction is provided by ATP. The reaction can be catalyzed by either glucokinase or a less specific enzyme, hexokinase. Another enzyme then catalyzes the conversion of glucose 6-phosphate to glucose 1-phosphate. This compound reacts with uridine triphosphate (UTP) to give the compound UDP-glucose. UDP-glucose then reacts with glycogen, forming an α 1–4 linkage between glucose units to lengthen the glycogen chain, and the UDP is set free. The synthesis of glycogen also requires another enzyme or set of enzymes to make the 1–6 branches which are part of the glycogen structure.

When glycogen is converted to glucose 1-phosphate in the process of glycogenolysis, the enzyme catalyzing this reaction is different from the glycogen synthetase which catalyzed the formation of starch from glucose 1-phosphate. The enzyme catalyzing the degradation is a **phosphorylase**, i.e., an enzyme that uses phosphoric acid to cleave a larger molecule. Although this enzyme can catalyze the synthesis of glycogen, equilibrium lies in the direction of the formation of glucose 1-phosphate. This glucose phosphate is then isomerized to glucose 6-phosphate, which can then be converted to glucose or be degraded as discussed in Section 35.5.

The transformation of glucose 6-phosphate to glucose involves an enzyme called a phosphatase. The reaction gives glucose and phosphoric acid. It does not result in the synthesis

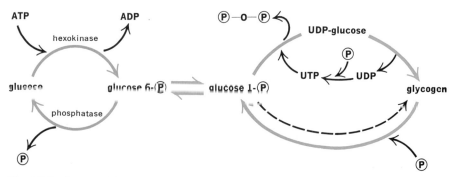

Fig. 35.3 Intermediate compounds and enzymes involved in the reversible conversion of glucose to glycogen.

of ATP from ADP and phosphate, so the reaction is not the reverse of that catalyzed by hexokinase.

The effects of deficiencies of the enzymes that catalyze these reactions are discussed in Section 35.12.

The irreversibility of parts of the scheme for the reversible conversion of glucose to glycogen is an illustration of the generalization that synthetic and degradative pathways may use the same general intermediates but do not follow the same paths. A further illustration and reference to this is made in Section 37.5.

35.5 DEGRADATION OF GLUCOSE

The conversion of glucose to carbon dioxide and water, with the release of much of the energy contained in the glucose molecule, involves first the splitting of glucose into three-carbon units and then a series of oxidations. The amount of free energy released in the complete oxidation of one mole of glucose is about 690 kcal.

glucose + oxygen → carbon dioxide + water + 690 kcal

Since this amount of energy is much greater than the amount that can be stored effectively in a molecule of ATP, we would expect that the degradation and oxidation of glucose proceeds in a series of steps, some of which are capable of yielding energy that can be stored in ATP. This is true, and we know a great number of intermediates involved in the process. For convenience, we generally think of one series of reactions as leading from glucose to either pyruvic* or lactic acid. We call this series of reactions **glycolysis**. The second series of reactions involves the conversion of pyruvic acid or lactic acid to acetic acid, and finally to carbon dioxide and water. This series is referred to as **aerobic metabolism** and involves the **citric acid cycle** or the *Krebs cycle*. (See Fig. 35.4.)

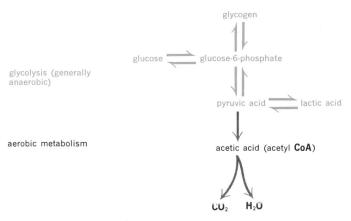

Fig. 35.4 A summary of the metabolism of glucose.

---

* Lactic, pyruvic, and acetic acids are weak acids and ionize only slightly; however, at the pH of the body, these substances are present as lactate, pyruvate, and acetate ions. Acetate is present in most biological systems as acetyl coenzyme A. Throughout the biochemistry section of this book, the term acid should be taken to mean an equilibrium mixture of an acid and its anion. Often the anion is present to a greater extent than the undissociated acid.

The term *aerobic* means in the presence of air. Actually, it is the oxygen of the air that is required for aerobic reactions. Anaerobic reactions do not require oxygen. In some cases, oxygen inhibits anaerobic processes, while in others the presence of oxygen has no effect on the reactions.

35.6
GLYCOLYSIS

Glycolysis is the term used to describe the breakdown of glycogen, glucose, or other common monosaccharides to give pyruvate. If the series of reactions proceeds in the absence of oxygen, the process is called **anaerobic glycolysis** and the end product is lactate or lactic acid.

The reactions of glycolysis are summarized in Fig. 35.5 and are discussed in detail in the following paragraphs. This summary is given in small type, indicating that it can be omitted if only a general knowledge of glycolysis is required.

We may consider glycolysis as starting with either glucose or glycogen. The first step in either case involves addition of a phosphate group. Ordinary phosphate is sufficient for phosphorylation of glycogen, but ATP is required to phosphorylate glucose. Glucose 6-phosphate is formed in one step from glucose, and after two steps when glycogen is the starting material.

By an isomerization, glucose 6-phosphate is converted to fructose 6-phosphate. This reaction proceeds via the aldehyde form and is best shown using straight-chain or Fischer formulas.

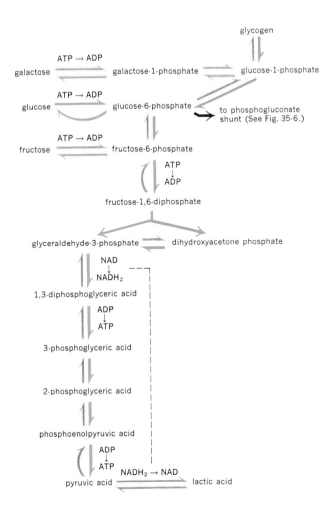

Fig. 35.5 Glycolysis. This series of reactions shows how glycogen or the monosaccharides glucose, galactose, and fructose can be converted to pyruvic or lactic acid. The series of reactions is reversible in the liver and kidney, where enzymes are able to bypass those steps that are essentially irreversible. The bypass reactions that are involved in reversal of the process are shown as curved arrows.

At this point another molecule of ATP is required to transform fructose 6-phosphate to fructose 1,6-diphosphate. This diphosphofructose is then split to give two three-carbon compounds, glyceraldehyde 3-phosphate and dihydroxyacetone phosphate. See diagram at the top of page 644.

Dihydroxyacetone phosphate represents a dead end in this metabolic pathway; only the fact that the enzyme triosephosphate isomerase can catalyze the conversion of dihydroxyacetone phosphate to glyceraldehyde 3-phosphate prevents the loss of half the energy of the glucose molecule. We can say that the net effect of the splitting of one molecule of fructose diphosphate is the formation of two molecules of glycerladehyde 3-phosphate. So

CH$_2$—O (P)
|
C=O
|
HO—C—H
|
- - -
|
H—C—OH
|
H—C—OH
|
CH$_2$O (P)

CH$_2$O (P)
|
C=O
|
HO—C—H
|
H

dihydroxyacetone phosphate

H—C=O
|
H—C—OH
|
CH$_2$O (P)

fructose 1,6-diphosphate   glyceraldehyde 3-phosphate

far the glycolytic scheme has required an energy input in the form of two molecules of ATP, but has not yet released any of the energy of glucose.

The next step in glycolysis, the conversion of glyceraldehyde 3-phosphate to 1,3-diphosphoglyceric acid, is a complex reaction. It involves the addition of a phosphate and the oxidation of the molecule from an aldehyde to an acid. This oxidation results in the formation of reduced NAD (NADH).

H—C=O
|
H—C—OH   + (P)OH $\xrightarrow{\text{NAD}\rightarrow\text{NADH}}$
|
H$_2$C—O(P)

O
‖
C—O (P)
|
H—C—OH
|
H$_2$C—O (P)

glyceraldehyde   phosphoric   1,3-diphospho-
3-phosphate   acid   glyceric acid

We will discuss the fate of this reduced NAD later, since we wish to follow the general course of glycolysis, which involves the further reaction of 1,3-diphosphoglyceric acid.

One bond between the phosphoric acid and glyceric acid in 1,3-diphosphoglyceric acid is an anhydride linkage, and as such its rupture provides sufficient energy to synthesize a molecule of ATP from ADP and phosphate. The other product of the hydrolysis of 1,3-diphosphoglyceric acid is 3-phosphoglyceric acid.

O
‖
C—O (P)
|
H—C—OH   + ADP $\longrightarrow$
|
H$_2$C—O (P)

O
‖
C—OH
|
H—C—OH   +   ATP
|
H$_2$C—O (P)

1,3-diphospho-   3-phospho-
glyceric acid   glyceric acid

The compound 3-phosphoglyceric acid is converted to its isomer, 2-phosphoglyceric acid. Removal of a molecule of water from the latter compound gives the enol form of phosphopyruvic acid.

O
‖
C—OH
|
H—C—O (P) $\xrightarrow{\text{H}_2\text{O}}$
|
H$_2$C—OH

O
‖
C—OH
|
C—O (P)
‖
H$_2$C

2-phospho-   phosphoenol-
glyceric acid   pyruvic acid

Hydrolysis of phosphoenolpyruvic acid again gives both the phosphate and sufficient energy to result in the synthesis of ATP from ADP.

With the removal of the phosphate we are left with the nonphosphorylated compound pyruvic acid. It is of interest to note that phosphorylated carbohydrates are found almost always totally within cells, and the nonphosphorylated compounds, glucose, pyruvic acid, and lactic acid, are found both in cells and in the bloodstream.

The difference in end products of aerobic and anaerobic glycolysis (pyruvate and lactate) involves the reduced form of NAD which we represent as NADH. It is formed in one of the earlier reactions in the scheme (the conversion of glyceraldehyde 3-phosphate to 1,3-diphosphoglyceric acid). The oxidized form (NAD) must be present if glycolysis is to proceed and not stop with the formation of 1,3-diphosphoglyceric acid, so NADH must be oxidized to NAD. When oxygen and the necessary enzymes are present, this oxidation occurs by the electron transport system as shown in Fig. 34.10. Most anaerobic organisms lack the enzymes for this process, and they oxidize NADH by reducing pyruvic acid to give lactic acid.

Although human muscles normally function aerobically, they may revert to an anaerobic mode when they work excessively. Overworked muscles accumulate lactic acid, which may cause soreness and cramps. If the lactic acid increases to about 10 milli-moles/liter, the muscles can no longer function. When a supply of oxygen is restored to the muscle, it can convert the lactic acid back to pyruvic acid and oxidize the NADH with oxygen. We say that a muscle that has formed lactic acid has an "oxygen debt." We keep breathing rapidly after exercise in order to repay this debt. Some forms of cancer cells seem to lack the enzymes for aerobic glycolysis and use the anaerobic process.

Monosaccharides other than glucose are also metabolized in glycolysis. Fructose makes up half of the sucrose molecule, and is a product of digestion of any sucrose-containing foods. Fructose is phosphorylated to give fructose 6-phosphate, which is a normal glycolytic intermediate. Lactose, commonly known as milk sugar, produces one molecule of glucose and one of galactose when hydrolyzed in the digestive process. Full utilization of the energy of lactose requires that galactose be metabolized.

The route by which galactose enters the glycolytic pathway involves phosphorylation to give galactose 1-phosphate. This compound is then converted to its isomer glucose 1-phosphate by a change in configuration of the alcoholic OH group on carbon number 4.

galactose 1-phosphate   glucose 1-phosphate

The glucose 1-phosphate so formed cannot be distinguished from that formed from glycogen or glucose, and it is metabolized via the glycolytic reactions.

The formation of phosphate compounds in glycolysis not only helps in the release of energy and the synthesis of ATP, but also "traps" those phosphorylated molecules within a cell. While glucose, pyruvate, and lactate pass through cell membranes, phosphorylated compounds do not, and thus they are kept near the site of their continuing metabolism. The ability of glucose, pyruvate, and lactate to travel has many advantages. We have already discussed the distribution of glucose from the digestion of food to the different cells of the body. As we shall see in Section 35.7, lactic acid may be removed from fatigued muscles, travel through the bloodstream, and be resynthesized to glucose in the liver.

Another function of glycolysis is that it provides compounds that can be used in other metabolic sequences. Pyruvic acid can be transaminated to give the amino acid alanine, and phosphoglyceric acid is readily converted to the amino acid serine, which can then be further metabolized to give cysteine or glycine. Dihydroxyacetone phosphate can be converted to glycerol 3-phosphate, which is used in the synthesis of lipids. UDP-glucose and UDP-galactose are used in the synthesis of cell wall components of bacteria.

The glycolysis of one molecule of glucose requires the investment of two molecules of ATP, and results in the synthesis of four molecules of ATP. (There are two reactions which yield ATP, but one molecule of glucose gives two molecules of each of the reactants, and hence two molecules of ATP are formed at each of two steps.) **Thus the net energy gain in anaerobic glycolysis is the formation of two molecules of ATP per molecule of glucose.** The NADH which is generated is oxidized to NAD by the conversion of pyruvate to lactate.

Aerobic glycolysis gives more ATP synthesis since the oxidation of each NADH, with the formation of H_2O, gives an average of three ATP's (Section 34.5); and since there are two reduced NAD molecules per molecule of glucose, we get six ATP's from this source. When we add these to the two that are obtained whether or not oxygen is present, **the net energy trapping in aerobic glycolysis is eight molecules of ATP per molecule of glucose.**

We can understand more of the energetics of glycolysis if we look at the energy in glucose, pyruvic acid, and lactic acid. If these compounds are completely oxidized, the following amounts of energy can be released:

glucose   690 kcal/mole
pyruvic acid   275 kcal/mole
lactic acid   315 kcal/mole

As we might expect, lactic acid, which is less oxidized than pyruvic acid, has a greater energy content. We can find the amount of the energy stored in glucose that has been released by first multiplying the energy content of pyruvic and lactic acids by 2, since one molecule of glucose gives two of pyruvic acid or lactic acid. We then subtract the values obtained from the energy in glucose.

1 mole of glucose	= 690 kcal		1 mole of glucose = 690 kcal	
2 moles of pyruvic acid (2 × 275)	= 550 kcal		2 moles of lactic acid (2 × 315)	= 630 kcal
energy released	140 kcal		energy released	60 kcal

From this calculation we see that 140/690 or about 20% of the energy of glucose is released when glycolysis is aerobic and the end products are pyruvic acid and water. Only 60/690 or about 9% of the energy is released when glycolysis is anaerobic and lactic acid is formed as the only end product.

We can calculate the efficiency with which the released energy is trapped in ATP as follows. The formation of pyruvic acid and water results in the conversion of 8 molecules of ADP to ATP. If we multiply 8 by 12 kcal, which is the energy stored when a mole of ADP is converted to one of ATP, we get 96 kcal of energy stored. Since 140 kcal of energy is released and 96 are trapped, the efficiency is about 96/140 or 70% for aerobic glycolysis. Anaerobic glycolysis results in the synthesis of only 2 ATP's per molecule of glucose, so we have a storage of only 24 kcal of energy. The amount of the energy of glucose released is 60 kcal, so the efficiency of trapping of the released energy in anaerobic glycolysis is only 24/60 or 40%. Thus we can see that not only does anaerobic glycolysis release less of the energy of glucose than does anaerobic glycolysis, but it also traps the released energy much less efficiently.

Glycolysis results in the conversion of some of the energy of glucose to ATP. Only a small part of the energy of glucose is released; most of it is still in the lactic acid or pyruvic acid which is left at the end of the process. However, this amount of energy is apparently sufficient to support life and growth of anaerobic bacteria which produce lactic acid. Some of these bacteria are characterized as lactobacilli because of the lactic acid they produce. Human muscles, particularly those working with inadequate amounts of oxygen, can use anaerobic glycolysis as a source of energy and they, too, produce lactic acid. A more beneficial scheme would result in the release of a greater percentage of the energy of the glucose and would end with products, such as carbon dioxide and water, that were nontoxic and easy to eliminate. Such a system exists and will be discussed, after a few digressions, in the section on aerobic metabolism.

**35.7
GLUCONEO-
GENESIS**

The reversal of the glycolytic scheme of reactions to give glycogen or glucose is called **gluconeogenesis**. This process, which requires "bypasses" for those reactions of glycolysis that are virtually irreversible,

glucose + ATP → glucose 6-phosphate + ADP

fructose 6-phosphate + ATP → fructose 1,6-diphosphate ADP

phosphoenolpyruvate + ADP → pyruvate + ATP

takes place primarily in the liver. (See Fig. 35.5.)

The kidney also contains enzymes necessary for gluconeogenesis. While the conversion of glucose to two molecules of lactate yields only two ATP, six energy-rich molecules, two of them guanosine triphosphate (GTP), are required for gluconeogenesis.

2 lactate + 4 ATP + 2 GTP + 6H$_2$O → glucose + 4 ADP + 2 GDT + 6 phosphates

Since they lack the enzymes for gluconeogenesis, muscle cells cannot convert lactic acid back to glucose, and lactic acid which is removed from fatigued muscles travels to the liver. Here, 15–20% of the lactic acid is oxidized in the tricarboxylic acid cycle to provide energy for the conversion of the rest into glucose. While it does require energy, in this process lactic acid, which may be regarded as a waste product of overworked muscle, is converted to glucose, which not only serves as the source of stored energy (glycogen) in muscles, but is the major energy source of the brain. Gluconeogenesis can serve as a source of glucose when none is available from the diet. Compounds such as the amino acid alanine, which can be converted to pyruvate, and glycerol from lipid metabolism can also enter into the gluconeogenesis scheme.

35.8
THE PHOSPHO-GLUCONIC ACID SHUNT

Not all glucose 6-phosphate is metabolized by being converted to fructose 6-phosphate. An alternative pathway involves the oxidation of glucose 6-phosphate to 6-phosphogluconic acid. The series of reactions resulting from this conversion is called the *phosphogluconate shunt*. The term shunt indicates a bypass, and in some ways this pathway is a bypass to glycolysis. The phosphogluconate pathway is shown in Fig. 35.6.

The pathway provides oxidation in the first two reactions and thus provides energy. The energy here is transferred to nicotinamide adenine dinucleotide phosphate (NADP), a compound much like NAD but containing an additional phosphate group. Reduced NADP is an important energy source for many biosynthetic reactions, particularly for lipid synthesis. Thus it is not surprising that enzymes for the phosphogluconate pathway are found in tissues that synthesize lipids. In addition to producing NADPH, this pathway results in the synthesis of pentoses, and is probably the major source of the ribose and deoxyribose needed for nucleic acid synthesis.

Many tissues, particularly the liver, have enzymes for both glycolysis and the phosphogluconate series of reactions. These tissues can metabolize glucose by either pathway; the pathway chosen depends on allosteric enzymes and conditions such as the availability of oxygen and the level of NADPH or of pentoses in the cells or in the circulating blood that

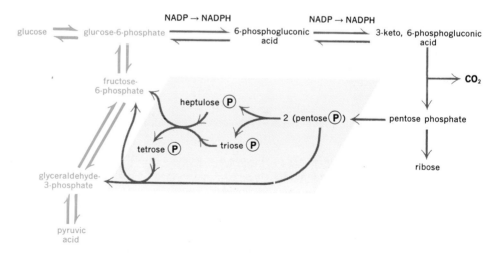

Fig. 35.6 The phosphogluconate shunt. This series of reactions is an alternative to those of glycolysis. The shaded area represents various possible condensations that may be used to convert pentose phosphate molecules to glyceraldehyde phosphate or fructose 6-phosphate. Glycolytic reactions are also shown in color for reference.

moves through the tissues. Muscle and brain tissue apparently use the glycolytic pathway exclusively for metabolism of carbohydrates, since they do not contain the enzymes necessary for the phosphogluconate pathway. Some bacteria use reactions similar to the phosphogluconate shunt. It is interesting to note that the photosynthesis of carbohydrates involves many of the intermediate compounds found in the phosphogluconate reaction sequence. It also involves NADP.

<table>
<tr><td>35.9
METABOLISM
OF
PYRUVIC ACID</td><td>Pyruvic acid may undergo any of several alternate reactions, and thus represents a major branching point in metabolism. Some of these reactions lead to other series of reactions, and some represent metabolic dead ends. Figure 35.7 summarizes these reactions.</td></tr>
</table>

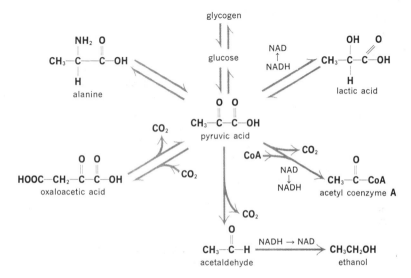

Fig. 35.7 Metabolism of pyruvic acid. Pyruvic acid represents a major branching point in metabolism. Some of the important products derived directly from pyruvic acid are shown.

Addition of an amino group transforms pyruvic acid to the amino acid alanine. The reverse reaction also occurs and represents one method for conversion of amino acids into carbohydrates. Addition of carbon dioxide, perhaps not to pyruvic acid itself but to phosphoenolpyruvic acid, gives the dicarboxylic acid oxaloacetic acid, whose significance will be shown later. The reversible conversion of pyruvic acid to lactic acid has been discussed previously. This conversion represents a metabolic dead end, and lactic acid, if it is to be further metabolized, must first be changed back to pyruvic acid. In normal muscle tissue when there is an adequate supply of oxygen, it is doubtful that any lactic acid is formed. Pyruvic acid may also be converted to glucose or glycogen in the liver or kidney by gluconeogenesis.

Pyruvic acid may lose a molecule of carbon dioxide in either of two ways. These two different methods lead to quite different pathways and represent a major difference in metabolism in different kinds of organisms. In the simpler process, carbon dioxide is lost and acetaldehyde is formed:

$$CH_3-\overset{\overset{\displaystyle O}{\|}}{C}-COOH \rightarrow CH_3-\overset{\overset{\displaystyle O}{\|}}{C}-H + CO_2$$

pyruvic acid   acetaldehyde

The acetaldehyde may then be reduced to yield ethanol. This pathway is found primarily in yeasts, and is the basis of alcoholic fermentation. In this process yeasts metabolize glucose through the anaerobic glycolytic pathway with the formation of pyruvic acid. They do not, however, oxidize the reduced NAD by transforming pyruvic acid to lactic acid. Instead, they use the reduced NAD to convert acetaldehyde to ethanol:

$$CH_3-\overset{\displaystyle O}{C}-H + NADH + H^+ \rightarrow CH_3-\underset{\underset{\displaystyle H}{|}}{\overset{\displaystyle OH}{C}}-H + NAD^+$$

acetaldehyde   ethanol

Thus the alcoholic fermentation of carbohydrates yields as end products carbon dioxide and ethanol. The series of reactions may be summarized as

$$C_6H_{12}O_6 \rightarrow 2C_2H_5OH + 2CO_2$$

The process does not involve oxygen, and the amount of energy trapped by the yeasts is small (two ATP molecules per molecule of glucose), but is sufficient to support their growth and reproduction.

The accumulation of ethanol in yeasts as well as the accumulation of lactic acid by bacteria eventually slows and stops their growth. Both organisms are poisoned by the products of their metabolism. Water and carbon dioxide, which are the products of aerobic glycolysis, are much less toxic.

Alcoholic fermentations have been known and widely used since prehistoric times. The widespread availability of both carbohydrates and yeasts has led to a variety of alcoholic beverages. Starches of wheat and barley yield beer, sugar of grapes gives wine, fermented honey gives mead—the list of conversions is almost endless.

Although fermentation stops when the alcohol content is about 15%, we can concentrate the alcohol by distillation to produce beverages with higher alcohol content. Some of these distilled products are vodka, saki, brandy, whiskey, and similar beverages. While all these beverages contain high percentages of alcohol, the differences between them are due to compounds other than alcohol which either distill over with the alcohol or are added later.

The process of fermentation has been explained in many ways. The Greeks described it as the god Dionysus acting upon grape juice; Pasteur realized that the fermenting agents were yeasts; we now describe fermentation as being the enzymatically catalyzed conversion of carbohydrates to ethanol and carbon dioxide. The process of alcohol production has always fascinated people. However, they have undoubtedly been more concerned with the effects achieved by consumption of the products than with understanding the process of the formation of alcohol.

An interesting abnormality, reported in the press, involved a Japanese man who was repeatedly accused of drunkenness even though he maintained that he had not consumed any alcoholic beverages. He was found to have, as the result of a previous injury, an extra pouch in his stomach. This pouch, which was protected from the effects of the acid gastric juice, was harboring a colony of yeasts which were efficiently producing ethanol. He then, quite unconsciously, absorbed the alcohol with the usual results. It is not likely, however, that similar pouches are present in the stomachs of others who protest in the face of overwhelming evidence to the contrary that they have not consumed alcohol.

The other way pyruvate may lose carbon dioxide is by a complex reaction. In this reaction or series of reactions, carbon dioxide is formed. The resulting acetaldehyde-like intermediate is indirectly oxidized by an enzyme using NAD as a hydrogen acceptor, and acetic acid in the form of acetyl coenzyme A (acetyl CoA) is formed. This series of reactions is virtually irreversible: in other words, acetyl CoA is not carboxylated to yield pyruvic acid. There are several possible fates for acetyl CoA; the most common one is the series of reactions known as the citric acid cycle.

**35.10
AEROBIC
METABOLISM—
THE CITRIC
ACID CYCLE**

The series of aerobic reactions in animals is variously called the *Krebs cycle*, the *tricarboxylic acid cycle*, and the *citric acid cycle*. We will use the last name since it seems to be the most descriptive. This series of reactions is not restricted to the metabolism of carbohydrates; it is also important in the metabolism of both lipids and proteins. Any substance that can be converted to acetyl CoA may enter the cycle. In this cyclic series of reactions a two-carbon unit, the acetyl group, is completely converted to carbon dioxide and water. The cycle requires oxygen to oxidize (through the electron transport-system) the NADH, NADPH, and $FADH_2$ which accept hydrogen atoms (electrons) from the intermediates in the cycle. Although there are minor variations, the cycle is used by almost all plants and animals that have been studied. Anaerobic organisms, of course, do not use it.

The citric acid cycle is summarized in Fig. 35.8, and discussed in detail in the following paragraphs.

In the first reaction of the cycle an acetyl CoA unit is condensed with an oxaloacetate molecule to give citric acid.

$$
\begin{array}{c}
\text{O}=\text{C}-\text{COOH} \\
|\\
\text{H}-\text{C}-\text{COOH} \\
|\\
\text{H}
\end{array}
+
\begin{array}{c}
\text{O} \\
\|\\
\text{CH}_3-\text{C}-\text{CoA}
\end{array}
\rightarrow
\begin{array}{c}
\text{H} \\
|\\
\text{H}-\text{C}-\text{COOH} \\
|\\
\text{HO}-\text{C}-\text{COOH} \\
|\\
\text{H}-\text{C}-\text{COOH} \\
|\\
\text{H}
\end{array}
+ \text{CoA}
$$

oxaloacetic   acetyl CoA   citric acid
acid

Actually, the citric acid formed is in equilibrium with *cis*-aconitic acid and isocitric acid, but the equilibrium favors the formation of citric acid.

$$
\begin{array}{c}
\text{H} \\
|\\
\text{H}-\text{C}-\text{COOH} \\
|\\
\text{HO}-\text{C}-\text{COOH} \\
|\\
\text{H}-\text{C}-\text{COOH} \\
|\\
\text{H}
\end{array}
\underset{\text{H}_2\text{O}}{\overset{\text{H}_2\text{O}}{\rightleftharpoons}}
\begin{array}{c}
\text{H} \\
|\\
\text{H}-\text{C}-\text{COOH} \\
|\\
\text{C}-\text{COOH} \\
\|\\
\text{H}-\text{C}-\text{COOH}
\end{array}
\underset{\text{H}_2\text{O}}{\overset{\text{H}_2\text{O}}{\rightleftharpoons}}
\begin{array}{c}
\text{H} \\
|\\
\text{H}-\text{C}-\text{COOH} \\
|\\
\text{H}-\text{C}-\text{COOH} \\
|\\
\text{HO}-\text{C}-\text{COOH} \\
|\\
\text{H}
\end{array}
$$

citric   *cis*-aconitic   isocitric
acid   acid   acid

Because of this reversibility citric acid would seem to be a poor substance for the initiation of a series of reactions. However, the reaction by which it is formed from oxaloacetic acid

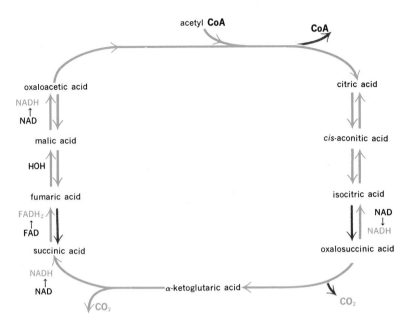

Fig. 35.8   The citric acid cycle.   This series of reactions, which is also known as the Krebs cycle, converts an acetate group to carbon dioxide and water.   The water is formed indirectly by the oxidation of NADH and $FADH_2$ through the electron transport system.   The isomers of citric acid and oxalosuccinic acid contain 6 carbon atoms each, α-ketoglutaric has 5, and succinic, fumaric, malic, and oxaloacetic contain 4 carbon atoms each.

and acetyl CoA is virtually irreversible, and there are no other significant biological reactions of citric acid.   The dehydrogenation of isocitric acid to give oxalosuccinic acid causes an imbalance in the equilibrium.   As would be expected from the law of mass action, the net effect is the conversion of citric acid to oxalosuccinic acid.

In the formation of oxalosuccinic acid, two hydrogen atoms are removed from isocitric acid to give reduced NAD (or NADP in some organisms) and oxalosuccinic acid:

In the next reaction of the citric acid cycle, <u>oxalosuccinic</u> acid loses carbon dioxide to give α-ketoglutaric acid:

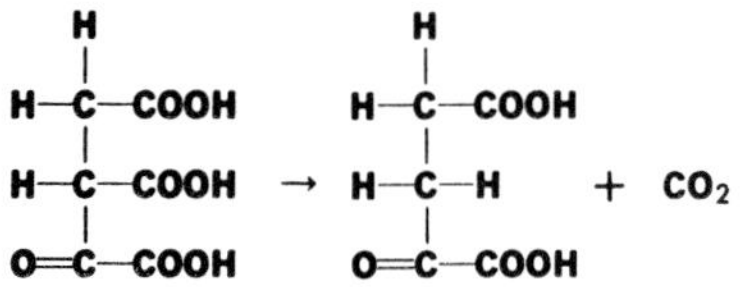

$$
\begin{array}{l}
\overset{\displaystyle H}{\underset{\displaystyle |}{H-C-COOH}} \\
\overset{\displaystyle |}{H-C-COOH} \quad \rightarrow \\
\overset{\displaystyle |}{O=C-COOH}
\end{array}
\qquad
\begin{array}{l}
\overset{\displaystyle H}{\underset{\displaystyle |}{H-C-COOH}} \\
\overset{\displaystyle |}{H-C-H} \quad + \quad CO_2 \\
\overset{\displaystyle |}{O=C-COOH}
\end{array}
$$

oxalosuccinic   α-ketoglutaric
acid   acid

The following reaction, the conversion of α-ketoglutaric acid to succinic acid, is complex and includes both an oxidation and a decarboxylation. It is similar to the conversion of pyruvic acid to acetyl CoA, in that it requires both NAD and coenzyme A. The succinic acid is actually succinyl CoA, and in its conversion to succinic acid, GTP (not ATP) is generated. This reaction, or perhaps to be completely accurate we should call it a series of reactions, is virtually irreversible.

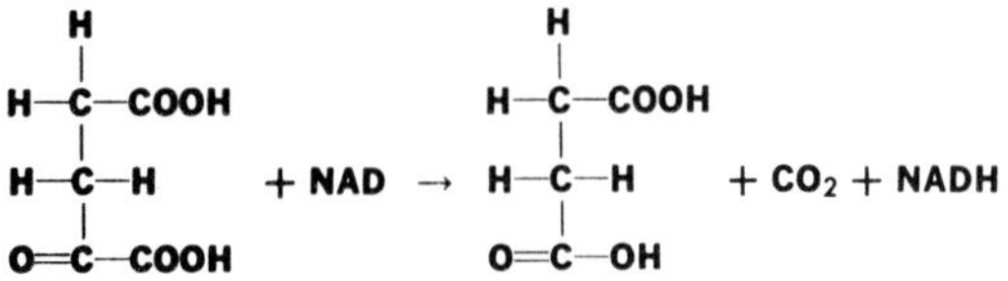

$$
\begin{array}{l}
\overset{\displaystyle H}{\underset{\displaystyle |}{H-C-COOH}} \\
\overset{\displaystyle |}{H-C-H} \quad + NAD \rightarrow \\
\overset{\displaystyle |}{O=C-COOH}
\end{array}
\qquad
\begin{array}{l}
\overset{\displaystyle H}{\underset{\displaystyle |}{H-C-COOH}} \\
\overset{\displaystyle |}{H-C-H} \quad + CO_2 + NADH \\
\overset{\displaystyle |}{O=C-OH}
\end{array}
$$

α-ketoglutaric   succinic
acid   acid

Succinic acid is dehydrogenated under the influence of a FAD-containing enzyme to give fumaric acid. Fumaric acid is then hydrated to give malic acid, which is dehydrogenated, in a reaction which requires NAD, to yield oxaloacetic acid.

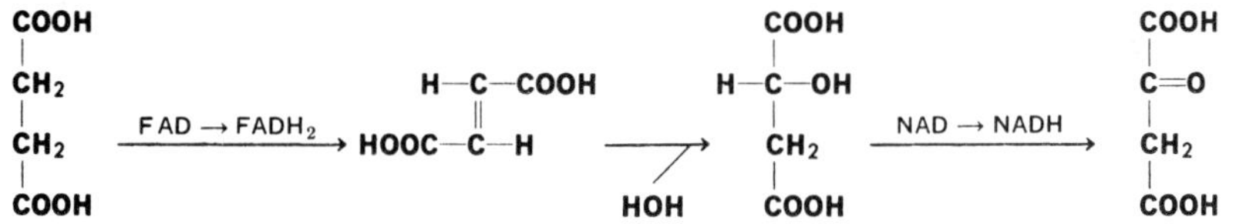

$$
\begin{array}{l}
COOH \\
| \\
CH_2 \\
| \\
CH_2 \\
| \\
COOH
\end{array}
\xrightarrow{FAD \rightarrow FADH_2}
\begin{array}{l}
H-C-COOH \\
\quad\quad \| \\
HOOC-C-H
\end{array}
\xrightarrow[HOH]{}
\begin{array}{l}
H-C-OH \\
| \\
CH_2 \\
| \\
COOH
\end{array}
\xrightarrow{NAD \rightarrow NADH}
\begin{array}{l}
COOH \\
| \\
C=O \\
| \\
CH_2 \\
| \\
COOH
\end{array}
$$

succinic   fumaric   malic   oxaloacetic
acid   acid   acid   acid

The oxaloacetic acid unites with an acetate unit to give citric acid, and the cycle begins again. Of course, with a cyclic series of reactions there is really no beginning and no end, but the general availability of acetyl CoA and the significance of having sufficient oxaloacetate to react with it makes this an especially important reaction in the cycle, and a good point to start and stop any discussion of the cycle.

In the citric acid cycle a molecule of acetyl CoA, or more precisely of a two-carbon unit, is converted to two molecules of CO_2 and four of H_2O. The energy contained in the acetyl CoA is converted to ATP via the intermediates NAD (or NADP), FAD, and the cytochrome system.

One turn of the citric acid cycle results in the formation of three molecules of NADH and one of $FADH_2$. This results in the synthesis of eleven molecules of ATP from ADP and phosphate (three from each NADH and two from the $FADH_2$). In addition, the energy released in the conversion of α-ketoglutaric acid to succinic acid through the involvement of coenzyme A is sufficient to synthesize a molecule of GTP, which is a high-energy compound like ATP. The number of high-energy molecules synthesized per acetate unit is twelve.

We can summarize the amount of ATP (and GTP) synthesized in the complete oxidation of glucose to carbon dioxide and water as follows:

aerobic glycolysis	8 ATP
conversion of 2 moles of pyruvic acid to 2 moles of acetyl CoA	6 ATP
oxidation of 2 moles of acetyl CoA	22 ATP
	2 GTP
total	38 ATP equivalents

Using 12 kcal as the energy stored in an ATP molecule, we have 12×38 or 456 kcal of energy stored in ATP. Since the total free energy change in the conversion of glucose to carbon dioxide and water is 690 kcal (Section 35.5), the series of reactions results in trapping 456/690 or 66% of the energy of glucose.

These calculations are based on an energy storage of 12 kcal per mole of ATP. If we use the standard free energy change of 7.3 kcal/mole, the efficiency of storage is only 40% rather than 66%.

35.11 SUMMARY OF CARBOHYDRATE METABOLISM

Carbohydrates are usually ingested in the form of starch and cellulose. Small amounts of sucrose, glucose, and fructose are also present in many foods. Lactose (milk sugar) is an important component in the diet of babies, children, and many adults. The process of digestion degrades all digestible carbohydrates to monosaccharides, and the indigestible carbohydrates such as cellulose are excreted unchanged. The products of digestion, the monosaccharides, are absorbed into the bloodstream and carried to various organs where they are utilized. Utilization may involve anabolic reactions leading to glycogen in animals and starch or cellulose in plants. For the purpose of simplification, catabolism of carbohydrates is usually divided into several series of reactions: glycolysis, the phosphogluconate pathway, and the citric acid cycle. The latter serves for the metabolism of proteins and lipids as well as for carbohydrates.

Other series of reactions are known to involve metabolites of the carbohydrates, but the three metabolic schemes listed seem to be the most important ones. It is possible to list more intermediate compounds in each of the series of reactions given in the glycolytic, phosphogluconate, and citric acid pathways. The degree of detail given in Figs. 35.5, 35.6, and 35.8, however, is sufficient and perhaps extravagant for a basic understanding of the process.

Although remarkable similarity is observed in the reactions in organisms ranging from the yeasts and bacteria to humans, there are some small differences. Many microorganisms use anaerobic glycolysis and release only small amounts of the potential energy of the carbohydrates they ingest. Since they are essentially nonmotile, perhaps they need less energy. It is possible that formation of excess ATP might even be deleterious for them. The growth of these microorganisms is, in fact, inhibited by the products of their own metabolism—ethanol and lactic acid. Other microorganisms, the aerobes, not only get more energy from each molecule of glucose, but also produce carbon dioxide and water, which are nontoxic and easy to eliminate as waste products.

Multicellular organisms, particularly those which move from place to place, often at rapid rates, use aerobic pathways. Certain of their tissues, such as the muscle tissues which in order to insure survival must often work under conditions of low oxygen concentration, retain the glycolytic system. This system allows oxidation reactions to proceed even in the absence of oxygen. When this happens, pyruvic acid is converted to lactic acid, and we say that the animal has incurred an "oxygen debt." The "debt" is repaid by rapid breathing, even after violent exercise has stopped. The rapid breathing results in (1) increased oxygenation of the blood flowing to the tissues, (2) the conversion of lactic acid to pyruvic acid and NAD to NADH, and (3) the oxidation of NADH through the electron transport system.

35.12 ABNOR- MALITIES OF CARBO- HYDRATE METABOLISM	Several diseases are caused by abnormalities of the carbohydrate metabolism. We will discuss briefly those that are best understood—*diabetes mellitus,* galactosemia, glycogen-storage disease, and glucose 6-phosphate dehydrogenase abnormalities. Another but less serious abnormality is due to the fact that some adults have almost none of the enzyme lactase, which is required to digest lactose. Apparently they lose this enzyme with maturity, because almost all children have it. For further discussion of this condition see the article by Kretchmer listed in the Suggested Reading.

Diabetes Mellitus

The word *diabetes* means excretion of large amounts of urine; *mellitus* means sweet. Therefore, the term *diabetes mellitus* signifies that a sweet urine—actually a urine containing glucose—is excreted in large amounts. Other forms of diabetes, such as *diabetes insipidus,* are known, and it is not totally wrong to describe the increased excretion of urine following ingestion of large amounts of beer as a temporary diabetes.

The condition we know as *diabetes mellitus* was probably known to the ancient Egyptians. The Greek scientist Celsus, who lived from 30 B.C. to 50 A.D., described a disease producing "polyuria without pain but with emaciation and danger," and a Chinese physician, about 200 A.D., described the condition as a "disease of thirst." About 1500, Paracelsus evaporated the urine of a diabetic and recovered "salt," which must have been a mixture of urea, salts, and sugar. Even in primitive societies it was known that ants were attracted to the urine of diabetics because of its sweetness.

From the dawn of civilization until the 1920's *diabetes mellitus* was fatal. No effective treatment was known. In severe cases, the patient could expect to live less than a year after discovery of the condition. The outlook then for diabetics is analogous to the prognosis today for people who develop some forms of cancer.

Diabetes mellitus is characterized by the presence of glucose and three other abnormal substances—acetone, acetoacetic acid, and β-hydroxybutyric acid in the urine. The glucose found in the urine is a reflection of the high level of glucose in the bloodstream. The high level of blood glucose (hyperglucosemia) is due to the fact that the normal metabolism is severely impaired in diabetics. They apparently cannot convert glucose either to glycogen or to carbon dioxide and water at anything like the normal rate. Since glucose cannot be utilized, it accumulates following digestion of a carbohydrate-containing meal, and is only slowly excreted. The variation in normal metabolism of glucose can be shown by a glucose tolerance test. The person undergoing the test fasts for at least 12 hours, and then drinks a solution containing a large amount of glucose. Blood samples are withdrawn after specific time intervals and analyzed for glucose. While both normal and diabetic persons will show a rise in blood glucose under the conditions of the test, the concentration of glucose in the blood will be higher and the rate of decrease will be slower in the diabetic than in the normal person. Typical glucose tolerance curves are shown in Fig. 35.9.

The other unusual urinary components of diabetics, acetoacetic acid, β-hydroxybutyric acid, and acetone, are believed to accumulate because the diabetic is unable to oxidize acetyl CoA. The acetyl CoA is produced primarily by the metabolism of lipids.

$$CH_3\overset{O}{\overset{\|}{C}}-CoA \rightarrow CH_3-\overset{O}{\overset{\|}{C}}-CH_2-COOH \quad CH_3-\overset{OH}{\underset{H}{\overset{|}{C}}}-CH_2COOH$$

acetyl CoA   acetoacetic acid   β-hydroxybutyric acid

$$CH_3-\overset{O}{\overset{\|}{C}}-CH_3 \quad CO_2$$

acetone

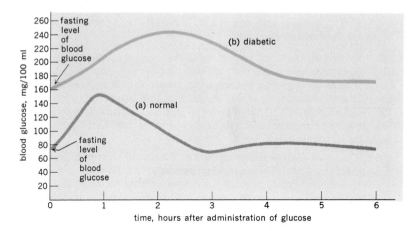

Fig. 35.9 Glucose tolerance curves. A glucose tolerance curve shows the amount of blood glucose at various intervals after an individual has consumed a large amount of glucose. Curve (a) is a normal curve. It shows a rise followed by a decrease and sometimes reaches a level slightly below the normal fasting level before returning to normal. Curve (b) represents results obtained when the test is made on a diabetic person. His fasting level is higher than normal and administration of glucose increases the blood glucose level to values that usually exceed the renal threshold. The typical "diabetic" curve falls off only slowly, and abnormally high values are found several hours after administration of the glucose.

Thus we say that, in the absence of glucose metabolism, acetyl CoA tends to accumulate. It has been postulated that the real deficiency here is in oxaloacetic acid, which normally condenses with acetyl CoA to give citric acid. Another view is that this deficiency is caused by the lack of pyruvic acid which can lead to the formation of oxaloacetic acid by a carboxylation reaction. Perhaps it is best merely to say that three abnormal metabolites, all probably derived from acetyl CoA, are present in the urine of diabetics. Acetone is also excreted with the air expired from the lungs of persons with severe, untreated diabetes.

In 1921, the Canadian scientists Frederick Banting and Charles Best (who was only a medical student at the time) proved that a substance that alleviated the symptoms of *diabetes mellitus* could be extracted from the pancreas. This substance, now known to be a hormone, is called *insulin*. Normally it is produced by special cells in the pancreas and released into the bloodstream and from the bloodstream absorbed by many tissues. Injections of insulin lead to the disappearance of all symptoms of *diabetes mellitus*.

How does insulin work? We really don't know. It has the net effect of promoting the metabolism of glucose. The best theory is that insulin promotes the absorption of glucose into cells. Until glucose is inside a cell, it cannot be metabolized. Insulin also promotes the synthesis of glycogen.

Insulin is now extracted from pancreases of beef, swine, and sheep, and made available in a highly purified form. Recent experiments have shown that the bacterium *E. Coli* can be induced to make insulin (see Section 39.6). Insulin must be injected into the diabetic, since ingestion as a pill would result in the digestion of insulin and reduce it to its component amino acids. Insulin is a treatment, not a cure, for diabetes. A cure would have to increase the synthesis or prevent the inactivation of insulin.

Mild cases of diabetes, those that develop in later life, can be treated with various chemical substances which can be administered orally. These drugs do not contain insulin, and are effective only in mild cases—those in which the patient is able to produce some but an insufficient amount of insulin. These drugs, whose trade names are Orinase® and Diabinese® apparently function by stimulating the secretion of the insulin that the diabetic person can still make or decreasing its destruction. However, the fact that these drugs can produce bad side effects has led many physicians to stop using them for the treatment of diabetes.

$$CH_3 - \bigcirc - SO_2 - \overset{H}{\underset{|}{N}} - \overset{O}{\underset{\|}{C}} - NH - (CH_2)_3 - CH_3$$

tolbutamide (Orinase)®

$$Cl - \bigcirc - SO_2 - \overset{H}{\underset{|}{N}} - \overset{O}{\underset{\|}{C}} - \overset{H}{\underset{|}{N}} - (CH_2)_2 - CH_3$$

chlorpropamide (Diabinese)®

Diabetes mellitus can be characterized as a disorder of carbohydrate metabolism. The exact nature of the defect is not known. The treatment of diabetes did not come from an understanding of the chemistry of the disease, but as the result of careful observation and a hunch on the part of Dr. Banting. In other disorders, treatment has been suggested only after the biochemical nature of the disease was understood.

Galactosemia   Galactosemia is a disease of infants. It is characterized by a high level of galactose in the blood. Galactose is also found in the urine. This condition becomes evident soon after the birth of affected children. They vomit when milk is fed to them, they fail to gain weight, their liver enlarges, and in severe cases they may die. Untreated survivors of this condition tend to be dwarfed, mentally retarded, and may have cataracts in their eyes. Fortunately, this disease is relatively rare and can be treated effectively.

When it was realized that the disease was caused by the inability to metabolize the galactose which was derived from lactose present in milk, the treatment was obvious— restriction of the intake of milk and milk products which contain lactose. If this treatment is begun during the first few weeks of life, all symptoms disappear and there are no long-term effects.

It is now known that galactosemia is caused by a hereditary deficiency of the enzyme responsible for the conversion of galactose 1-phosphate to glucose 1-phosphate. As the children grow older they usually develop an alternate pathway for metabolizing galactose, and thus the need to restrict milk is not permanent.

With the understanding of the cause and treatment of this condition, the problem becomes one of detection. Unfortunately, the early symptoms of galactosemia are similar to those of many other diseases of infants. Careful observation and examination of the urine and blood for the presence of galactose can lead to prompt diagnosis and treatment. Ignorance is the major reason for the prevalence of this disorder.

Glycogen Storage   A much more serious situation is involved in glycogen storage disease. Actually, several
Disease   forms of this disorder are known, each probably involving the failure of a specific reaction in the reversible conversion of glucose to glycogen. The disease is characterized by an

accumulation of glycogen in the liver, heart, or skeletal muscle. These conditions are hereditary and rare, but are usually fatal, with death often the result of decreased resistance to infections.

One form of glycogen storage abnormality, known as von Gierke's disease, involves the absence of the enzyme necessary to convert glucose 6-phosphate to glucose. No treatment has been found to be really effective. Other forms of these diseases involve the enzyme responsible for forming branches or for debranching the glycogen molecule—that is, a lack of the enzyme involved in the formation or disruption of 1–6 bonds between glucose units. A summary of the blocks involved in various types of glycogen storage disease is given in Fig. 35.10.

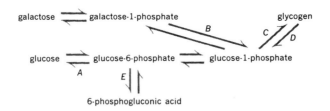

Fig. 35.10 Site of enzymatic deficiencies in disease of carbohydrate metabolism: *A*, site of defect in von Gierke's disease; *B*, site of defect in galactosemia; *C*, site of defect in diffuse glycogenolysis (Type III); *D*, site of defect in glycogen storage disease of liver and muscle (Type IV); *E*, site of defect in drug-induced hemolytic anemia and favism.

Disorders Involving Glucose 6-phosphate Dehydrogenase

The administration of certain drugs, namely phenacetin and some of the sulfonamides, causes the rupture of blood cells (hemolysis) and a subsequent anemia in some persons. In severe cases, half of the hemoglobin may be destroyed in this way. Although hemoglobin is normally synthesized by these persons, and blood tranfusions may be given to persons with the defect, this disorder is still a serious one. A related condition is caused when sensitive persons eat fava beans. The defect in these disorders involves the enzyme glucose 6-phosphate dehydrogenase, the enzyme involved in the entrance of glucose 6-phosphate into the phosphogluconate shunt. It is quite likely that persons having these disorders are somewhat more resistant to malaria than are normal persons, so the condition is not without its advantages. Since this trait is widely found in Africans, it has apparently helped them to survive in a malarial area and has consequently been transmitted to their offspring. We say such traits have "positive survival value."

The hereditary diseases discussed above seem to be rare. There is some tendency for *diabetes mellitus* to be inherited, although the mode of inheritance is not completely understood. It is highly likely that other hereditary diseases involving irregularities in carbohydrate metabolism will be found. In some cases, such as in galactosemia, understanding the disease has led to an obvious and effective treatment. In others, as in the glycogen storage diseases, no effective treatment has yet been found. The study of defects in metabolism, or "metabolic" diseases, has only begun, and represents a real challenge for the medical biochemist. This type of disease is discussed further in Chapter 39, "The Chemistry of Heredity."

IMPORTANT TERMS

adrenaline
aerobic
amylase
anaerobic
citric acid cycle (Krebs cycle)
diabetes mellitus
disaccharidase
galactosemia
glucagon
gluconeogenesis
glucosuria

glycogenesis
glycogen storage diseases
glycolysis
hyperglycemia
hypoglycemia
insulin
oxygen debt
phosphogluconate shunt
phosphorylase
phosphorylation
renal threshold

WORK EXERCISES

Part A

1. How does the metabolic oxidation of glucose differ from the reaction when a small amount of sugar is burned in a flame? In what ways are the reactions similar?

2. Which glycolytic intermediates have no phosphorus in their structure? What is the significance of the phosphorus-containing intermediates?

3. Describe how the level of glucose in the blood is kept within a narrow range.

4. What is the source of blood glucose in an animal that has eaten no carbohydrates for a long period of time?

5. Summarize the digestion of a molecule of a highly branched polysaccharide.

6. Does glucosuria always mean that a person has *diabetes mellitus*?

7. Compare and contrast the three major metabolic pathways by which a molecule of glucose can be metabolized—glycolysis, phosphogluconate shunt, and citric acid cycle.

In solving Exercises 8, 9, and 10, assume that the free energy "stored" when a mole of ADP is converted to one of ATP is 12 kcal.

8. The complete oxidation of acetic acid to carbon dioxide and water using a bomb calorimeter yields approximately 200 kcal of free energy per mole of acetic acid. Compare this energy with the amount of energy stored as ATP when acetic acid is oxidized via the citric acid cycle.

9. Fluoroacetic acid occurs in the leaves of a poisonous plant (*Dichapetalum cymosum*). The toxicity of this substance is due to the fact that it combines with oxaloacetic acid to form fluorocitric acid, which acts as an inhibitor of the dehydration of citric acid in the citric acid cycle. Assuming that the further reactions of citric acid are completely

blocked, but that no other reactions are blocked, how many ATP molecules would you estimate could be produced per molecule of glucose fed to an animal poisoned with fluoroacetic acid? How does this amount of ATP compare to that normally obtained in an unpoisoned animal?

10. Malonic acid is poisonous. Apparently the toxicity is due to the fact that it inhibits the dehydrogenation of succinic acid in the citric acid cycle. Assuming (for ease of calculation) that the free energy of the conversion of glucose to CO_2 and H_2O is 700 kcal/mole, estimate what percentage of the free energy of the glucose would normally be trapped as ATP in an animal poisoned with malonate. How does this figure compare with the amount of energy produced by glucose in an unpoisoned animal?

The Use of Isotopes in Biochemistry

The reactions by which the carbohydrates are metabolized have been studied by using isotopes. In these studies, a labeled compound (one containing the isotope) is administered to an animal either by injection or by feeding, and later some part of the animal or its waste products is analyzed for the isotope. Individual enzymatic reactions may be studied by adding isotopic substrates to purified enzyme systems.

The radioactive isotope carbon-14 (^{14}C) is probably the most widely used, but the non-radioactive isotopes deuterium (^{2}H), nitrogen-15, and oxygen-18 are often used instead. The radioactive isotopes tritium (^{3}H), phosphorus-32, and iodine-131 can also be used.

Some isotopic compounds are uniformly labeled (for example, all the carbon atoms may contain ^{14}C) while in other compounds only one carbon atom is isotopic. Using an asterisk to indicate labeled carbon atoms (which should not be confused with the use of an asterisk for indicating a chiral carbon in optically active compounds), we can differentiate between uniformly and specifically labeled pyruvic acid as follows:

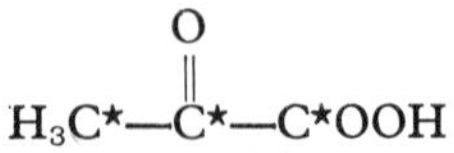

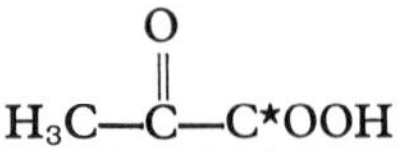

$$
\begin{array}{cc}
\overset{\text{O}}{\overset{\|}{H_3C^*-C^*-C^*OOH}} & \overset{\text{O}}{\overset{\|}{H_3C-C-C^*OOH}} \\
\text{uniformly labeled} & \text{carboxyl-labeled} \\
\text{pyruvic acid} & \text{pyruvic acid}
\end{array}
$$

In performing and interpreting experiments involving isotopes, we make the following assumptions:

1) The labeled compound is metabolized in the same manner as a nonlabeled compound.
2) The label stays with the atom which was originally labeled, and this atom maintains its position in a complex molecule. Care must be taken when labeling hydrogen atoms since the fact that the hydrogen of acids dissociates from its original molecule makes interpretation of results difficult.
3) When symmetrical compounds are formed, the results are the same as if the isotopic label were distributed between similar atoms. Thus if succinic acid is labeled in only one of the non-carboxylic carbons

$$HOOC-CH_2-C^*H_2-COOH$$

it behaves as if the label were distributed over the two similar carbons

$$HOOC-C^*H_2-C^*H_2-COOH$$

Metabolism of this succinic acid to give the non-symmetrical malic acid (in the citric acid cycle) would produce both

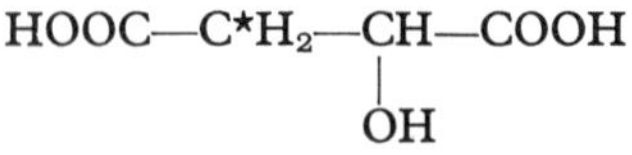

$$HOOC-C^*H_2-\underset{\underset{OH}{|}}{CH}-COOH$$

and

$$HOOC-CH_2-\underset{\underset{OH}{|}}{C^*H}-COOH$$

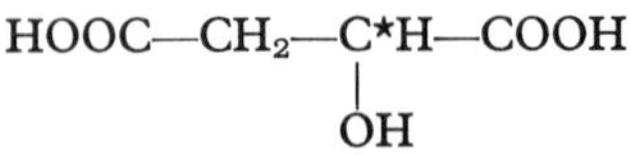

and analysis of such a mixture would give results indicating that the labeling was spread over the noncarboxyl carbons:

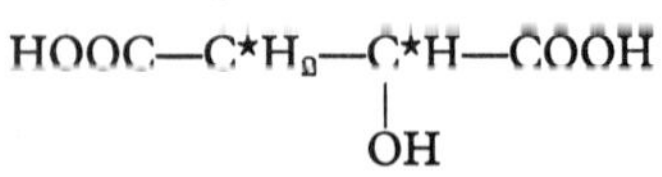

$$HOOC-C^*H_2-\underset{\underset{OH}{|}}{C^*H}-COOH$$

The carboxyl carbons, however, would not be labeled. There is one important exception to this rule about labeling as it applies to symmetrical compounds. Although citric acid is symmetrical, research indicates that the enzyme that catalyzes its metabolism in the citric acid cycle treats the citric acids as if it were nonsymmetrical, and

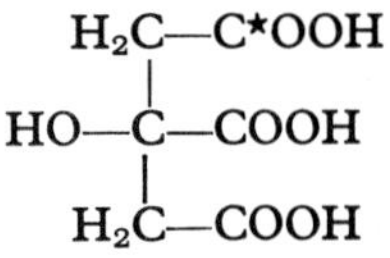

$$
\begin{array}{l}
H_2C-C^*OOH \\
\quad | \\
HO-C-COOH \\
\quad | \\
H_2C-COOH
\end{array}
$$

gives α-keto glutaric acid labeled only in one carboxyl group,

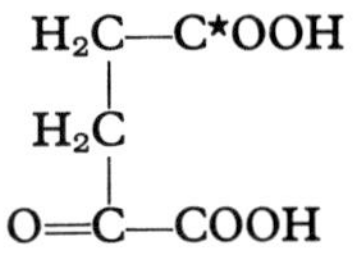

$$
\begin{array}{l}
H_2C-C^*OOH \\
\quad | \\
H_2C \\
\quad | \\
O=C-COOH
\end{array}
$$

Part B.

On the basis of the foregoing discussion, do the following work exercises.

11. List five radioactive compounds that might be isolated from a rat or its excretion products following the injection of pyruvic acid with the carboxyl carbon labeled with carbon-14. Locate the position of the radioactive carbon in each.

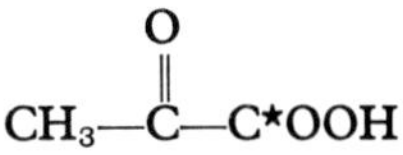

$$\overset{\text{O}}{\overset{\|}{CH_3-C-C^*OOH}}$$

12. What compounds which are normal metabolites of pyruvic acid would you expect to have no radioactivity when carboxyl-labeled pyruvic acid was administered to rats?

13. If carboxyl-labeled pyruvic acid had been added to a culture of yeast, which products of yeast fermentation would be radioactive and which products would not contain carbon-14?

14. One of the major reactions of pyruvate is its conversion to acetyl CoA. This reaction could be studied by the following experiments:

The enzymes and cofactors known to be necessary for the conversion of pyruvic acid to acetyl coenzyme A are used in both experiments. In Experiment 1, the addition of pyruvate labeled in the methyl or carbonyl carbon gave acetyl CoA containing radioactivity. In Experiment 2 the addition of acetate labeled in both carbon atoms did not result in the formation of labeled pyruvic acid.

What can you conclude about the reversibility of this reaction?

15. Glucose was synthesized to contain carbon-14 in the fourth carbon and later given to a white rat. The following compounds were later isolated from the animal: glucose, fructose-1,6-diphosphate, pyruvic acid, acetyl CoA, α-ketoglutaric acid. Indicate the position of the carbon-14 (if present) in each of these compounds.

SUGGESTED READING

Kretchmer, N., "Lactose and lactase," *Scientific American*, October 1972, p. 70 (Offprint #1259). Whereas children have the enzyme lactase, which is necessary to digest milk sugar, most adults don't. The significance of this finding and its racial and genetic aspects are discussed in this interesting article.

Levine, R., and M. S. Goldstein, "The action of insulin," *Scientific American*, May 1958, p. 99. This article discusses glucose metabolism and the function of insulin.

Wolf, G., *Isotopes in Biology*, Academic Press, New York, 1964. This paperback book gives a wealth of information on the use of isotopes for the study of biochemical processes.

36

Metabolism of Proteins

36.1
INTRODUCTION The digestion of proteins is necessary in order to convert them into amino acids which can be absorbed across the intestinal wall. Undigested proteins are potent antigens; that is, they bring about the producton of antibodies (that are themselves proteins) which produce allergic reactions. These body defenses against foreign proteins mean that absorption of proteins without digestion would produce severe allergic shock reactions following each meal. However, it is possible that some proteins are absorbed through the intestine without such effects during the first few days of a baby's life. In this way the mother's milk probably provides the child with preformed antibodies which will serve until he can manufacture his own. The permeability of the intestine is rapidly lost within a few days after birth.

Following digestion and absorption, the amino acid of proteins may either be catabolized to yield energy or be built into one of the thousands of proteins necessary for the life and well-being of the organism. The process of protein synthesis is extremely complex. Each protein contains at least a hundred amino acids in a highly specific sequence. The regular misplacement of just one amino acid can lead to serious consequences. At the same time that the exact structure is being formed, energy must be provided for this anabolic process.

This chapter discusses first the digestion of proteins and then the reactions of the amino acids that result either in their degradation to yield energy or in their synthesis into larger molecules, especially the proteins which play a vital role in the life of any organism.

**36.2
PROTEIN-
DIGESTING
ENZYMES** The digestion of proteins to yield amino acids is catalyzed by a variety of enzymes, all of which may be called **peptidases,** since the bond hydrolyzed is the peptide linkage between the amino acids. The word **proteolytic** is also used to describe these enzymes. The reaction may be written as follows:

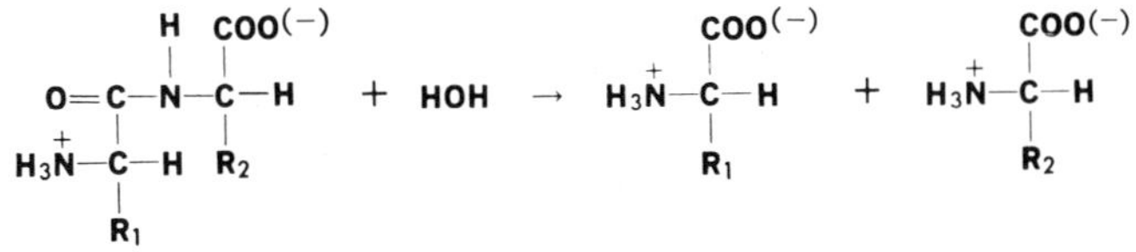

$$
\underset{\substack{\text{a dipeptide}}}{\overset{\substack{H \quad COO^{(-)} \\ | \qquad | \\ O{=}C{-}N{-}C{-}H \\ \overset{+}{\underset{|}{}} \qquad | \\ H_3\overset{+}{N}{-}C{-}H \quad R_2 \\ | \\ R_1}}{}}
\quad + \quad \underset{\substack{\text{water}}}{HOH}
\quad \rightarrow \quad
\underset{\substack{\text{amino acid}}}{\overset{\substack{COO^{(-)} \\ | \\ H_3\overset{+}{N}{-}C{-}H \\ | \\ R_1}}{}}
\quad + \quad
\underset{\substack{\text{amino acid}}}{\overset{\substack{COO^{(-)} \\ | \\ H_3\overset{+}{N}{-}C{-}H \\ | \\ R_2}}{}}
$$

The usual substrates for proteolytic enzymes are proteins, not dipeptides, but the basic reaction is the hydrolysis of a peptide linkage as shown above.

There are many varieties of peptidases. **Exopeptidases** act on peptide bonds nearest to the end of a long protein chain. They may be either **aminopeptidases** or **carboxypeptidases**, depending on whether or not they exert their action nearest the end of the protein (or polypeptide) chain having the free amino or carboxyl group. **Endopeptidases** catalyze the hydrolysis of peptide bonds that are not near the end of a chain. **Dipeptidases** act only on dipeptides. Many of the peptidases are specific. For example, trypsin catalyzes the hydrolysis of peptide bonds in which the carbonyl group was from lysine or arginine, and other peptidases have similar specificity.

All the proteolytic enzymes that have been carefully studied to date are secreted in inactive forms. Pepsin, for instance, is produced as an inactive protein **pre-pepsin** (also called pepsinogen★). Its conversion to pepsin involves the hydrolysis of some of its own peptide bonds. This hydrolysis is catalyzed by hydrogen ions or by pepsin itself. Two questions are immediately suggested: "How does hydrolysis of peptide bonds convert an inactive protein into an active enzyme?" and "Why are these enzymes secreted in an inactive form?" "How" questions are much easier to answer than "Why" questions, but answers have been proposed for both. Conversion of inactive pre-pepsin to pepsin probably involves the removal of an inhibitor and perhaps a "masking" group. The "masking" group is a part of the pre-pepsin molecule that covers or masks the active site of the enzyme much as a sheath masks a knife—although in this case the masking group is much smaller than the active agent. Pre-pepsin has a molecular weight of 42 600 and pepsin of 34 500, indicating that the inhibitor and other peptides removed have a weight of approximately 8 000.

The answer to the question concerning the reason for production of inactive precursors (sometimes called **zymogens**, or enzyme generators) is suggested by the realization that the cells that manufacture these enzymes are themselves proteins. It would be inefficient or even disastrous if these proteolytic enzymes destroyed the very tissues that made them. The basic question "Why doesn't the stomach or intestine digest itself?" has plagued scientists for some time. Apparently the answer is that most of the stomach and intestinal wall is protected by a lining of polysaccharide-protein complex (a mucopolysaccharide) called mucus which is relatively indigestible but which does, however, slough off continuously—much as do the outer layers of our skin. The glands that manufacture pepsin lack this lining but are protected because they make an inactive form of the enzyme. The same explanation can be used to describe the manufacture of pre-trypsin and pre-chymotrypsin by the pancreas.

36.3
PROTEIN
DIGESTION

Protein digestion does not start until the food reaches the stomach. Here the combined action of HCl and the endopeptidases, **pepsin** and **gastricsin** initiate protein digestion.

★ The name pepsinogen is most commonly used. The International Union of Biochemistry recommends pre-pepsin. They make similar recommendations for the precursors of trypsin and chymotrypsin.

Complete action by pepsin results in the hydrolysis of about 10% of the bonds of typical proteins and leaves particles having molecular weights of between 600 and 3 000. Such a mixture is also produced *in vitro* (using pepsin, of course) and sold as *peptone* for use in bacterial growth media. Gastricsin, a comparatively recently isolated enzyme, is similar to pepsin but is sufficiently different that it can be separated from pepsin and studied.

In the small intestine, the partially digested proteins are exposed to the action of the endopeptidases **trypsin** and **chymotrypsin**, and carboxypeptidases from the pancreas. Aminopeptidases and dipeptidases are probably secreted from the intestinal wall. Trypsin and chymotrypsin are secreted as pre-trypsin and pre-chymotrypsin, and it is probable that the exopeptidases are also secreted as inactive zymogens. As the result of the action of the various peptidases, proteins are converted into amino acids and these are absorbed through the intestinal wall into the bloodstream.

Some proteolytic enzymes are extracted from their natural sources and used for various purposes. Papain, which comes from the papaya fruit, is used in meat tenderizers, where its action partially digests the meat. Other proteolytic enzymes have been used to aid in freeing the lens of the eye prior to its removal in surgery for cataract and to facilitate regrowth of nerve cells that have been cut.

36.4 METABOLISM OF AMINO ACIDS

The digestion and the eventual fate of amino acids absorbed into the bloodstream may be summarized as follows:

proteins → amino acids
1. synthesis into protein,
2. removal of amino group to give a keto acid, which may:
   a) be converted to glucose or glycogen,
   b) be converted to CO_2 and water,
3. participation in a pathway peculiar to a given amino acid

Protein Synthesis

The synthesis of protein is an extremely rapid process when one considers the complexity of the protein being synthesized. Only minutes after the injection of radioactive amino acids into an animal, radioactive protein can be found in that animal, indicating that the synthesis occurs rapidly. We also know from other studies with isotopes that there is a continual degradation and resynthesis of all the proteins in any organism. The white rat replaces half of its proteins in 17 days. The human takes 80 days to do the same. These figures are somewhat misleading, however, since not all proteins are degraded and re-synthesized (the term protein turnover is used to describe the process) at the same rate. Half of the proteins in the blood serum are turned over in 10 days, liver proteins require only 20 to 25 days, but the turnover in the protein of bone (collagen) is slow.

The problem of exactly how an organism synthesizes protein has intrigued scientists for many years. How does a cell put together just the right amino acid, in the right sequence to give a protein that may serve as an enzyme, an antibody against bacteria, or perhaps a hormone? Since the number of possible arrangements of the 20 amino acids in even a small protein is greater than the number of atoms in the known universe, protein synthesis cannot be just a random synthesis, but must be under strict control from some kind of "director."

The realization that the kinds of proteins an organism synthesizes depends on its heredity led to speculation that DNA was probably involved in some way and might serve as the "director" or source of information. This assumption we now know to be true. In fact, both DNA and RNA are involved in the synthesis of proteins. Three kinds of RNA, each having different physical characteristics, are involved in protein synthesis. These three forms are called (1) **soluble (transfer) RNA**, (2) **ribosomal RNA**, and (3) **messenger (template) RNA.** Each plays a specific role in protein synthesis.

The general scheme for protein synthesis has been worked out only during the past twenty years. There are still unanswered questions about this highly detailed process, and research is continuing at a voluminous rate. In the discussion that follows, we have attempted to give a broad description of the process. It is beyond the scope of this book to discuss all the details of, and exceptions to, the general process.

We can say that the information (message) for making protein is contained in DNA. By processes shown in Fig. 33.10 the DNA duplicates itself (replicates). The DNA also serves as the pattern for synthesizing RNA in a process that is called **transcription.** The various kinds of RNA then help "translate" the message into protein structure.

$$\text{DNA} \xrightarrow{\text{transcription}} \text{RNA} \xrightarrow{\text{translation}} \text{Protein}$$

The first step in the synthesis of proteins involves a reaction of a given amino acid with ATP. This reaction produces a compound having a structure that can be summarized as adenine-ribose-phosphate-amino acid. The amino acid is said to be "activated" in this process. (See Fig. 36.1). This activated amino acid is then transferred to a specific kind of soluble RNA. This kind of RNA is known as transfer RNA. It is now certain that there are several kinds of transfer RNA for each amino acid but they all have certain characteristics in common. See Fig. 36.2 for the structure of a molecule of transfer RNA that binds to the amino acid alanine.

The transfer-RNA-amino-acid complex then travels to the ribosome (a small organelle within the cell) where the amino acid is added to other amino acids to form polypeptide chains and eventually proteins. The transfer RNA then leaves the ribosome and picks up another molecule of its particular kind of amino acid. The order in which the amino acids,

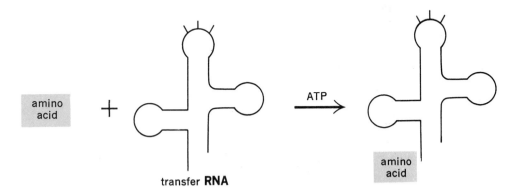

Fig. 36.1 The activation and binding of an amino acid to transfer RNA. In the reaction shown above, a single enzyme is responsible for both activating an amino acid and binding it to the specific transfer RNA that will carry it to the ribosome. The structures given here are only rough diagrammatic approximations of the molecules they represent.

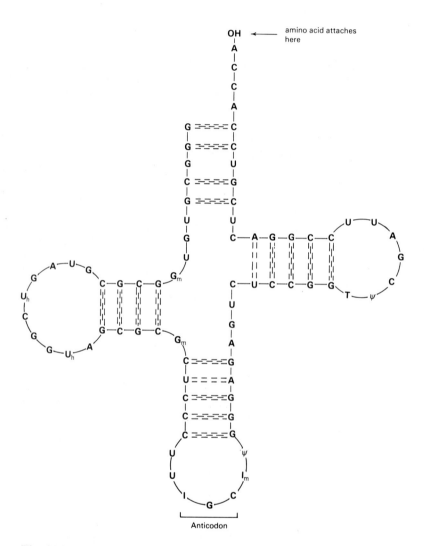

Fig. 36.2 Drawing of the structure of the soluble RNA that transfers the amino acid alanine. The amino acid attaches to the upper right portion of this molecule. The nucleotide triplet IGC is the anticodon that matches with the codon for alanine in messenger RNA. The symbols Ψ, I_m, I, G_m, and U_h indicate unusual bases found in this and many transfer RNAs. Transfer RNAs for other amino acids are similar to the one shown above, although some have slightly different forms as well as having different anticodons.

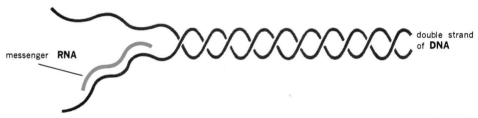

Fig. 36.3 Synthesis of messenger RNA. It is believed that DNA strands in the nucleus of the cell unwind sufficiently for a given strand to serve as a pattern for the synthesis of messenger RNA. The messenger RNA then carries the instructions contained in the DNA to the ribosomes.

which are brought to the ribosome by transfer RNA, are linked together in a peptide chain is specified by messenger RNA.

Messenger RNA, as well as other known kinds of RNA, is synthesized in the nucleus of the cell under the influence of DNA. The DNA not only duplicates itself (as described in Chapter 33) but also serves as a pattern for the synthesis of a specific kind of messenger RNA. In DNA duplication a guanine in one strand of DNA serves as a guide for adding a cytosine to a forming chain of DNA, and adenine is the guide for thymine. Since RNA contains uracil instead of thymine, an adenine in a DNA molecule directs the placement of a uracil (actually a uridylic acid molecule) in a chain of messenger RNA. (See Fig. 36.3.) Thus, we can summarize the synthesis transcription of RNA from DNA as follows:

DNA chain: —G—C—A—G—C—T—A—·····

RNA chain: —C—G—U—C—C—G—A—U—·····

Messenger RNA then travels from the nucleus to the ribosome, carrying in its structure the message describing, in a type of code, a particular protein molecule. Messenger RNA and the transfer-RNA-amino-acid combination meet at the ribosome (which itself is made of RNA and protein). Messenger RNA specifies the kind of amino acid to be placed at each part of a forming protein chain by matching some specific part of the transfer RNA. In the illustration shown below only a small area of the soluble RNA exactly matches the messenger RNA.

ribosome

messenger RNA: —C—G—U—(C—C—G)—A—U—·····

transfer RNA: —A—A—U—(G—G—C)—A—G—·····

From the experimental evidence, we are quite certain that the code requires three nucleotides (bases) to specify an amino acid. We call each of these three base sequences in messenger RNA a codon. In the previous illustration, the codon —C—C—G— is being matched with the complementary anticodon —G—G—C— located on the transfer RNA which is carrying the amino acid. Since we believe that the codon C—C—G is specific for the amino acid proline (see Table 36.1), a molecule of this amino acid is added to the growing protein chain whenever its codon is specified in messenger RNA. Figure 36.4 shows this process diagrammatically.

The establishment of the list of codons that specify each amino acid is one of the triumphs of modern biochemistry. It came about when experiments were carried out in

TABLE 36.1 The genetic code

This table gives the trinucleotides (codons) that are believed to contain the information for incorporating a given amino acid into a polypeptide chain. Although the code is degenerate in that more than one trinucleotide codes for a given amino acid, there is a general similarity in two of the bases in the codons for an amino acid. The codons UAA, UAG, and UGA serve the purpose of stopping peptide synthesis, and thus are called *terminator* codons.

Amino Acid	Codons	Amino Acid	Codons
alanine	GCA, GCC, GCG, GCU	leucine	CUA, CUC, CUG, CUU,
arginine	CGA, CGC, CGG, CGU,		UUA, UUG
	AGA, AGG	lysine	AAA, AAG
asparagine	AAC, AAU	methionine	AUG
aspartic acid	GAC, GAU	phenylalanine	UUC, UUU
cysteine	UGC, UGU	proline	CCA, CCC, CCG, CCU
glutamic acid	GAA, GAG	serine	UCA, UCC, UCG, UCU,
glutamine	CAA, CAG		AGC, AGU
glycine	GGA, GGC, GGG, GGU	threonine	ACA, ACC, ACG, ACU
histidine	CAC, CAU	tryptophan	UGG
isoleucine	AUC, AUU, AUA	tyrosine	UAC, UAU
		valine	GUA, GUC, GUG, GUU

which synthetic RNA molecules were added to a mixture of ribosomes, transfer RNA, amino acids, and other factors. It was found, for instance, that when polyuridylic acid [represented as —U—(U—U)$_x$—U—] was added to the mixture, only polyphenylalanine was formed. It was surprising to find that there is also another codon for phenylalanine, U—U—C, but this repetition proved to be the general rule. That is, there is more than one codon for most of the amino acids, but each known codon is specific; it governs the addition of only one kind of amino acid molecule to the growing peptide chain. Other codons serve to terminate the growing protein molecule.

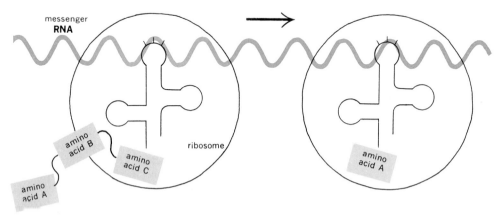

Fig. 36.4 Synthesis of protein at the ribosome. The ribosome apparently has two major active areas. In one, messenger RNA is held so that transfer RNA having the correct sequence of nucleotide bases can match it. When the matching takes place, the activated amino acid is positioned over the area where peptide bonds are generated between amino acids to give the protein chain.

Evidence indicates that messenger RNA from one kind of organism can be used with ribosomes and transfer RNA amino acids from another, and that the kind of protein synthesized is characteristic of the messenger RNA. That is, the process of protein synthesis is universal. This may indicate the process by which nucleic acids from viruses can "take over" the synthetic systems of the cell. It would also explain why viruses require whole cells—not just an assortment of nutrients—if they are to grow and reproduce.

Messenger RNA molecules are depolymerized after being used only once or, at most, a few times. The messenger RNA of higher plants and animals seems to be more stable than that of bacteria. While this destruction of messenger RNA seems wasteful, it provides a sensitive means of controlling the kind of protein synthesized. Bacteria usually need to respond to more environmental changes than do the cells of higher plants or animals. Bacteria that stop making a protein they no longer need are more efficient and tend to be successful in their never-ending struggle to survive and reproduce.

The synthesis of some peptides, such as glutathione and some antibiotic peptides, does not involve nucleic acids. These seem to be minor exceptions to the rule that peptide bond synthesis is directed by DNA and RNA. However, protein synthesis has been studied primarily in bacteria and this leads some scientists to question whether the process in humans and other animals follows exactly the same processes.

Amino Acid Degradation
The basic biochemical reactions of amino acids were described in Chapter 31 as peptide synthesis, transamination, deamination, and decarboxylation. Since these reactions provide different ways of degrading amino acids, we need to perform experiments with animals to decide which of the possible reactions are most important.

To determine the extent to which transamination actually occurs in an animal, we can feed it any one amino acid that has been labeled with the nonradioactive isotope nitrogen-15 in the amino group. Later when we examine the various tissues of the animal we find that nitrogen-15 is present in the amino acids, both free and in proteins, with the exception of lysine. If we make our examination only a short time after feeding the animal the isotopic amino acid, we find that the greatest amount of ^{15}N is still in the amino acid administered, with lesser amounts in glutamic acid and aspartic acids, and still lesser amounts in the other amino acids, except lysine, of course, which contains no ^{15}N. These findings lead to the conclusion that the keto acids corresponding to the amino acids glutamic acid and aspartic acid are the most important "amino acceptors" in transamination reactions. It is not surprising to find that these keto acids, α-ketoglutaric and oxaloacetic acids, are found as reactants in the citric acid cycle.

Another approach to finding the extent of transamination is made by feeding an animal the keto analogs of amino acids. We know that a supply of certain amino acids is essential for a given animal, a white rat, for example, because the animal cannot synthesize these amino acids at a sufficient rate to allow it to grow normally. When we feed the keto acids corresponding to the **essential amino acids** to this animal, we find that all the keto analogs can be substituted for the essential amino acids with the exception of those of lysine (which the previously described experiments had shown not to be transaminated) and threonine. Since our other experiments indicated that threonine could be transaminated, we must conclude that the failure of the keto analog to substitute for dietary threonine may be due to the fact that it is metabolized so quickly that it cannot be transaminated. A typical transamination reaction is shown in Fig. 36.5.

These experiments show one of the important uses which an animal makes of transamination reactions. If the basic carbon skeleton is present in the body from some other source, the amino acid can be synthesized. Thus a requirement of an animal for an essential

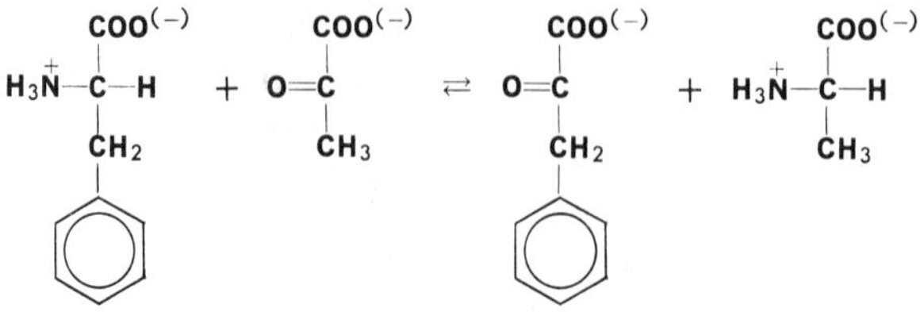

Fig. 36.5 A transamination reaction in which the amino group of the amino acid phenylalanine is transferred to pyruvic acid to yield phenylpyruvic acid and alanine. The reverse of this reaction would result in the synthesis of phenylalanine if sufficient phenylpyruvic acid were present from another metabolic pathway.

amino acid is really only a requirement for a certain type of carbon chain with a keto acid on it. A list of essential amino acids is given in Chapter 38, and it is interesting to see that they represent complex or unusual compounds whose carbon chains or benzene rings are not present in metabolites of the carbohydrates or lipids.

The conversion of an amino acid to a keto acid may also occur by a process of **deamination**. Many organisms contain the keto analogs of the amino acids; however, there is little evidence that oxidative deamination occurs to any extent in the bodies of humans and other higher animals.

With certain important exceptions, the amines that would be produced by the **decarboxylation** of amino acids are not found in the human; consequently, we conclude that decarboxylation of amino acids is not an important process in humans. Bacteria in the intestine are responsible for some decarboxylations, and some of the odor of normal feces is due to these vile-smelling amines.

From the above, we can conclude that those amino acids that are not used in protein synthesis but are degraded for energy are first transaminated. This transamination produces keto acids that can enter either the glycolytic or the citric acid pathway. The transamination of alanine, glutamic acid, and aspartic acid yields pyruvic acid, α-ketoglutaric acid, and oxaloacetic acid, respectively. Reactions in addition to transamination are required to convert other amino acids into compounds of one of the major metabolic schemes. (See Fig. 37.4.) Thus we can say that the carbon chains of all amino acids enter the glycolytic scheme or the tricarboxylic acid cycle. Via the reactions of these cycles, they either can follow the gluconeogenic route to give glucose and glycogen or can be converted to CO_2 and H_2O. The former sequence requires energy and the latter yields large amounts of energy.

36.5
EXCRETION OF
NITROGENOUS
WASTE
PRODUCTS

The conversion of amino acids to keto acids leaves an amino group. Although the amino group may be passed from compound to compound, the net utilization of amino acids for energy or glucose and glycogen synthesis means that the amino group must eventually be eliminated from the body. The metabolism of this group is extremely important since very low levels of ammonia in the blood are extremely toxic to all animals. A chronic alcoholic often dies from ammonia toxicity; the liver, which is the source of the enzymes that catalyze conversion of the amino group to the less toxic urea, has been so damaged by cirrhosis that the person is unable to get rid of the poisonous ammonia. Other animals, such as fish, can apparently tolerate slightly higher levels of ammonia in their blood, but they too are poisoned by relatively low levels of this substance. One of the specific biochemical differences in different species involves the manner in which they excrete the nitrogen from proteins.

There are two major sources of nitrogenous waste products—amino acid metabolism and the metabolism of the purine and pyrimidine bases which are either consumed in the diet as nucleic acids or synthesized in the body. In the human, approximately 95% of the nitrogenous excretion products come from amino acid metabolism. Nitrogenous waste products are excreted primarily via the kidneys in humans and other mammals. (See Section 40.9.) Not all animals do so; for instance, fish use their gills for excretion of ammonia. Most animals excrete more than one nitrogen-containing product, but there is often a predominance of one substance.

Amino Acid Excretion   Some simple animals excrete amino acids. This loss is essentially wasteful, since there is still a great deal of energy in amino acids. Humans excrete minor amounts of some amino acids in the urine, and some persons having hereditary abnormalities excrete relatively large amounts of specific amino acids. **Cystinuria** is the condition in which cystine is excreted in relatively large amounts in the urine. Since cystine is quite insoluble, it may form "stones" in the bladder or kidneys of persons who are cystinurics. In fact, the first chemical characterization of cystine was made from a bladder stone of a person who probably was a cystinuric individual.

Ammonia Excretion   Levels of only 5 mg of ammonia per 100 ml of blood are toxic to humans, and the normal concentration of this metabolite is only 1 to 3 micrograms per 100 ml of human blood. Animals that live in water, from simple one-celled organisms to the fish, excrete free ammonia. Even fish, however, do not have concentrations of ammonia greater than 0.1 mg (100 micrograms) per 100 ml of blood. How do these animals excrete ammonia when there is so little of it in the blood? Apparently the ammonia is usually combined with glutamic acid to give glutamine. Glutamine is then carried to a membrane next to the surrounding water (the gills for fish and certain other animals), where the ammonia is released.

Humans excrete ammonium ions (NH_4^+) in the urine, and presumably the ammonia is formed from glutamine in the kidneys. The amount of ammonium ions excreted is not very large, and there is no detectable excretion of ammonia by normal healthy humans. The reaction can be described as follows:

$$\underset{\text{glutamine}}{\begin{array}{c} COO^{(-)} \\ | \\ H_3\overset{+}{N}-C-H \\ | \\ (CH_2)_2 \\ | \\ C-NH_2 \\ \nparallel \\ O \end{array}} \;+\; \underset{\text{water}}{HOH} \;\rightleftharpoons\; \underset{\substack{\text{glutamic} \\ \text{acid}}}{\begin{array}{c} COO^{(-)} \\ | \\ H_3\overset{+}{N}-C-H \\ | \\ (CH_2)_2 \\ | \\ O=C \\ \searrow \\ OH \end{array}} \;+\; \underset{\text{ammonia}}{NH_3}$$

Urea Excretion   Many animals, including humans, excrete urea as the major nitrogen-containing metabolite derived from amino acid metabolism. Urea is much less toxic than ammonia. Levels between 18 and 38 mg/100 ml of blood are considered normal in a human. Sharks have

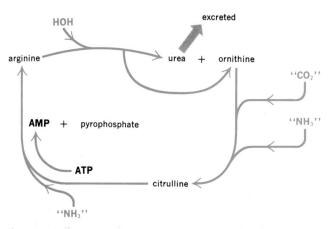

Fig. 36.6 The urea cycle. By the series of reactions shown above, urea is produced by the hydrolysis of the guanido group of arginine. Arginine is then regenerated in a series of reactions so that the cycle can continue.

as much as 2% (2 g/100 ml of blood) of urea in their blood. In fact, shark hearts will not beat if urea is not present. The fact that freshwater sharks have much less urea in their blood (0.6% instead of 2%) leads us to the conclusion that one function of the urea in sea sharks is that of balancing the osmotic pressure of the salty sea water.

Energy is required to synthesize urea. Apparently the production of the relatively non-toxic urea is sufficiently advantageous to compensate for the extra energy required for the synthesis. The series of reactions by which urea is synthesized in mammals is called the urea cycle. This series of reactions is summarized in Fig. 36-6 and discussed in detail in the following section.

In this series of reactions, arginine is hydrolyzed to give urea and a molecule of the amino acid ornithine.

$$\begin{matrix} \text{COO}^{(-)} \\ | \\ \overset{+}{\text{H}_3\text{N}}-\text{C}-\text{H} \\ | \\ (\text{CH}_2)_3 \\ | \\ \text{N}-\text{H} \\ | \\ \text{C}=\text{NH} \\ | \\ \text{NH}_2 \end{matrix} + \text{HOH} \rightarrow \underset{\text{urea}}{O=\overset{\overset{\displaystyle NH_2}{|}}{C}-NH_2} + \begin{matrix} \text{COO}^{(-)} \\ | \\ \overset{+}{\text{H}_3\text{N}}-\text{C}-\text{H} \\ | \\ (\text{CH}_2)_3 \\ | \\ \text{NH}_2 \end{matrix}$$

arginine   + water   urea + ornithine

Ornithine then combines with one unit of CO_2 and one of NH_3. Actually, neither CO_2 nor ammonia is present as such, but they are added in the form of carbamyl phosphate. Carbamyl phosphate is synthesized in the liver of mammals by the following reaction:

$$CO_2 + NH_3 + 2ATP \rightarrow H_2N-\overset{\overset{\displaystyle O}{\|}}{C}-O\,\text{(P)} + 2ADP + HO\,\text{(P)}$$

The addition of carbamyl phosphate to ornithine gives citrulline.

ornithine   +   carbamyl phosphate   →   citrulline   +   phosphoric acid

Citrulline acquires an amino group from aspartic acid. The final product of the reaction is arginine and fumaric acid.

citrulline   +   aspartic acid   arginine   fumaric acid

The arginine is then hydrolyzed to yield another molecule of urea and one of ornithine, and the cycle repeats itself. The fumaric acid can be oxidized via the citric acid cycle.

In this series of reactions we can see that ATP is used to provide the energy for the synthesis of urea. In mammals, this series of reactions takes place only in the liver. Extensive liver damage leads to the accumulation of ammonia, and can lead to death.

Uric Acid Excretion   In animals that excrete urea, the urea is dissolved in the urine. It can be said that the excretion of urea actually requires water. Animals which must conserve water (at least during parts of their life cycle), such as the reptiles, birds, and many insects, excrete uric acid. This substance is excreted in a solid form mixed with, but not dissolved in, small amounts of water. The synthesis of uric acid is a complex, energy-requiring process and we must assume, as we did in the case of urea synthesis, that the advantages of conservation of water are sufficient to justify the excretion of a relatively energy-rich compound which the animal must synthesize.

uric acid

It is difficult to determine what is cause and what is effect, but it is true that many reptiles are found in deserts, where conservation of water is of primary importance to

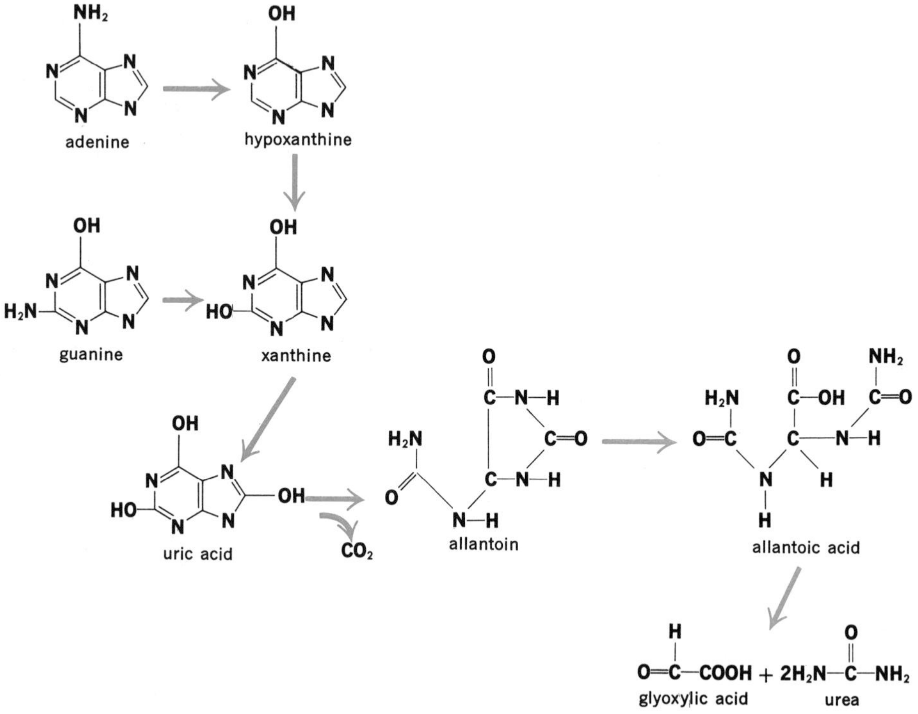

Fig. 36.7 Metabolism of purine bases. In this series of reactions adenine and guanine are oxidized to uric acid. Man and other primates excrete uric acid, since they lack the enzyme necessary for the formation of allantoin. Excretion products of other animals indicate that they carry the series of reactions further.

survival. The condition which is shared by birds, reptiles, and some insects is that the embryo develops within an egg which is enclosed in a shell that is impervious to water. Thus all the water the developing embryo will be able to use is within the egg at the time it is deposited by the female. The sequence of events observed in the development of the chicken embryo is interesting. During the first four days of incubation of the egg, ammonia is produced. From the fifth to the fourteenth day, urea is produced instead of ammonia, and indeed some of the previously produced ammonia is converted to urea. From the fourteenth day until hatching and throughout the rest of its life, uric acid is synthesized by the chicken. Animals that develop from eggs laid in water generally excrete ammonia— as do the adults of these species. However, when amphibians such as frogs begin to grow legs, their metabolism switches, and they begin to synthesize urea—apparently in anticipation of their coming conversion to land-dwelling animals. Philosophically, it is probably preferable to interpret the change in excretion products as being the factor that makes life on dry land possible, rather than to make the teleological interpretation that a change in excretion anticipates a change in environment.

Humans also excrete uric acid; however, it is not synthesized as a method of excreting ammonia but represents the end product of the metabolism of the purine bases adenine and guanine. With minor exceptions, only humans and other primates (apes) excrete uric acid as the normal end product of purine metabolism. Most mammals excrete allantoin, and simpler animals excrete allantoic acid or urea as the end products of purine metabolism (see Fig. 36.7 for this series of reactions). In this series of reactions, animals generally regarded as simpler than humans are actually more complex in that they have more enzymes and can tap more of the energy of the purines. The "loss" of these enzymes, especially of uricase, which catalyzes the conversion of uric acid to allantoin, can be blamed for some of the ills that humans suffer. Some types of painful arthritis and gout result from deposits of uric acid in the joints.

The "loss" of enzymes by higher species leads us to wonder about the advantages of losing the enzyme uricase and the attendant accumulation of uric acid in the blood and other parts of the body. Some scientists have speculated that perhaps the superior intelligence of humans is due to the high level of uric acid in their blood. While this is a tempting hypothesis, the fact that men normally have a slightly higher level of uric acid in their blood than women (4.5 mg/100 ml in men as against 3.5 mg/100 ml in women) has led to the rejection of this hypothesis by approximately half of the population. The fact that some fowls (chickens and turkeys) have an even slightly higher level of uric acid in their blood (5 mg/100 ml), and that insects have as much as 20 mg/100 ml again argues against the hypothesis!

Miscellaneous Nitrogenous Excretion Products

Ammonia, urea, and uric acid are major nitrogen-containing excretion products of amino acid metabolism. Allantoin and allantoic acid are also excreted by many animals, but these come primarily from purine metabolism. Trimethylamine oxide is excreted by some marine animals, but the source of this compound has not been definitely established. Swine and spiders both excrete guanine, the swine because they have no guanase. Presumably the same is true of spiders, whose metabolism has not yet been extensively investigated. In spite of the fact that it has the enzyme uricase, the Dalmatian dog excretes some uric acid. Apparently this results from a slightly different kidney function and not from an intent to imitate human metabolism!

36.6 SPECIFIC METABOLIC PATHWAYS FOR AMINO ACIDS

In addition to the general reactions of amino acids which have been previously discussed, each amino acid has a series of specific reactions in which it participates. In some cases, similar amino acids share at least parts of a specific metabolic pathway. As a result of these special pathways, either the amino acid is converted into a compound that can enter the glycolytic pathway or citric acid cycle, or it is converted into a special metabolite. There are many of these special metabolites. Some of the more important ones are summarized in Table 36.2. The specific pathways of phenylalanine and tyrosine are described in Chapter 39 in connection with a discussion of abnormalities in these pathways. A complete discussion of these special metabolic pathways is beyond the scope of this book, and a general biochemistry textbook should be consulted by those wishing further details.

36.7 DISORDERS OF PROTEIN METABOLISM

Several instances are known in which a lack of the correct enzyme has led to an abnormal accumulation of metabolites. From our knowledge of protein synthesis, we believe that failures in synthesis of proteins, especially where the condition is hereditary, are due to defects in nucleic acids. It is probably incorrect to describe these enzyme deficiencies as disorders of protein metabolism, even though they represent abnormalities in synthetic pathways. Numerous other conditions are known in which lack of enzymes leads to abnormal metabolism of amino acids. The most striking example involves the metabolism of phenylalanine and tyrosine. These defects are discussed in Section 39.4.

TABLE 36.2 Important products of the metabolism of specific amino acids

Amino Acid	Product	Importance of Product
alanine	pyruvate	glycolytic intermediate
aspartic acid	oxaloacetic acid	important intermediate of citric acid cycle
arginine	ornithine, citrulline	intermediates in urea synthesis
arginine	urea	important nitrogen-containing excretion product
glutamic acid	α-ketoglutaric acid	important intermediate in citric acid cycle
methionine	S-adenosyl methionine	important source of methyl group in biosynthetic reactions
cysteine	taurine	combines with cholic acid to give bile salts
histidine	histamine	important in allergic reactions
tryptophane	nicotinic acid	vitamin
phenylalanine, tyrosine	thyroxine, adrenaline	important hormones
tyrosine	melanin	pigment of skin and hair

SUMMARY

Proteins are digested under the influence of a variety of enzymes to yield amino acids. These amino acids can be used to synthesize new protein or other simpler molecules such as the hormones thyroxine and adrenalin. The process of protein synthesis is not only complex but amazing, because it causes complex molecules to have a precisely specific sequence of amino acids. Directions for the synthesis are found in DNA, but various types of RNA are required to carry out the process. Synthetic reactions are remarkably similar whether the organism is as simple as the bacterium *E. Coli* or as complex as a human. The energy of these anabolic reactions is provided by ATP. Amino acids not used for synthetic purposes can be degraded to yield energy. When this happens, the carbon, hydrogen, and oxygen are converted to carbon dioxide and water by reactions of the citric acid cycle and other common metabolic sequences. The nitrogen may be eliminated in a variety of compounds; different species use different reaction sequences to convert the amino acid nitrogen into a waste product, depending on the habitat and state of evolution of the animal. Thus throughout the animal kingdom, the metabolism of amino acids follows some pathways that are similar and others that are distinctive.

IMPORTANT TERMS

aminopeptidase
anticodon
carboxypeptidase
chymotrypsin
codon
cystinuria
dipeptidase
endopeptidase

essential amino acid
exopeptidase
gastricsin
messenger (template) RNA
pepsin
peptidase
pre-pepsin (pepsinogen)
proteolytic

replication
ribosomal RNA
soluble (transfer) RNA
transaminase
transcription
translation
trypsin
zymogen

WORK EXERCISES

1. Cite the experimental evidence that transamination occurs in an animal's body.
2. Write the reaction by which each of the following could be formed from an amino acid.
   a) glyoxylic acid (O=CH—COOH)
   b) alpha ketoglutarate
   c) pyruvic acid
   d) ethanolamine
   e) mercaptoethanolamine (H_2N—CH_2—CH_2SH)
   f) histamine
   g) oxaloacetic acid
3. What is the major difference in the sources of the carbon skeletons of essential and nonessential amino acids?
4. List the kinds of RNA involved in protein synthesis, and indicate the role of each in this process.
5. Tell what is meant by and the significance of each of the following in protein synthesis.
   a) replication   b) transcription
   c) translation
6. The peptide alanylserylmethionylglutamyltyrosine was incubated with several different enzymes. The products of each specific reaction are given below. List the type or classification of the enzyme that catalyzed each reaction.

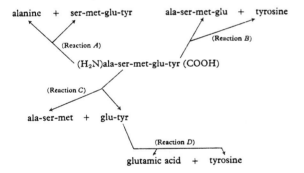

The following exercises involve radioactive tracers. Refer to the Work Exercises in Chapter 35 if necessary for a discussion of this type of exercise.

7. The metabolism of the amino acid histidine can be summarized as follows:

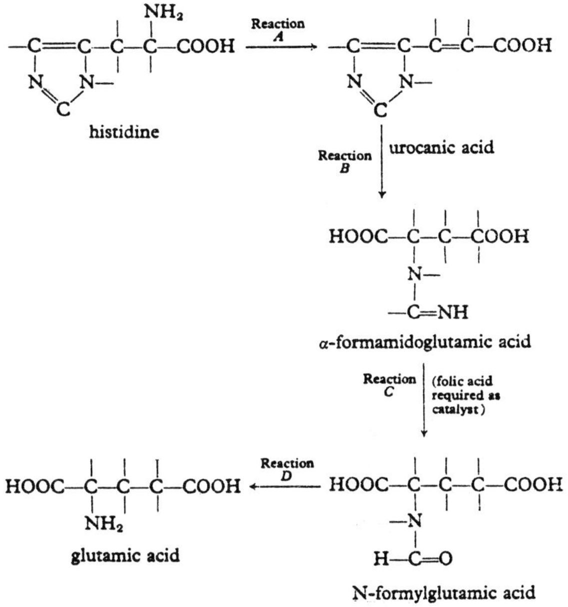

Reaction C has been shown to require the vitamin folic acid as a coenzyme. Answer the following questions with respect to the metabolism of histidine.
   a) If histidine containing ^{14}C in the carboxyl carbon is given to an animal, what compound in the citric acid cycle would you expect to be labeled first? (Give the structural formula and show the position of the label.)
   b) What clinical test might be developed to indicate whether an individual had a deficiency of folic acid?
8. Citrulline containing ^{15}N in the position indicated below was given to a dog. What two compounds (other than citrulline) in the animal should contain the greatest amount of ^{15}N in a few minutes after administration of the labeled compound?

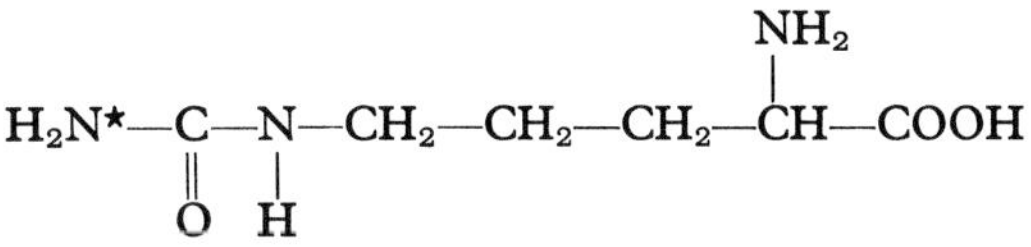

$$H_2N^\star\text{—}\underset{\underset{O}{\|}}{C}\text{—}\underset{\underset{H}{|}}{N}\text{—}CH_2\text{—}CH_2\text{—}CH_2\text{—}\underset{\overset{|}{NH_2}}{CH}\text{—}COOH$$

SUGGESTED READING

Asimov, I., *The Genetic Code*, a Signet Science Library Book (P2250), New American Library, New York, 1963.

The following articles from *Scientific American* are excellent references on the subject of protein synthesis. The illustrations make a complex process quite understandable.

Clark, B. F. C., and K. A. Marcker, "How proteins start," *Scientific American*, January 1968, p. 36 (Offprint #1092).

Crick, F. H. C., "The Genetic Code," *Scientific American*, October 1962, p. 66 (Offprint #123).

Crick, F. H. C., "The Genetic Code III," *Scientific American*, October 1966, p. 55 (Offprint #1052).

Nirenberg, M. "The Genetic Code II," *Scientific American*, March 1963, p. 80 (Offprint #153).

Nomura, M., "Ribosomes," *Scientific American*, October 1969, p. 28 (Offprint #1157). A well-illustrated discussion of these organelles and their role in protein synthesis.

Rich, A., and S. H. Kim, "The three-dimensional structure of Transfer RNA," *Scientific American*, January 1978, p. 52 (Offprint #1377). An excellent, well-illustrated article on this important RNA. If you're interested in the mechanism of protein synthesis this article is highly recommended.

37

Metabolism of Lipids

<table>
<tr>
<td>37.1
INTRODUCTION</td>
<td>Although lipids are found in both plants and animals, there are significant differences in the ways these two types of organisms use lipids. Animals contain greater percentages of lipids than plants do. Animals use the lipids in their diets as a source of energy and as a means of storing energy. When an animal undergoes even brief periods of fasting or starvation, the relatively meager supplies of stored carbohydrate (glycogen) are soon exhausted and fat stores are called on. These stores, which we sometimes call adipose tissue, consist of cells that, while having a nucleus containing all the organelles, contain up to 99% triacylglycerols (triglycerides). Removal of triacylglycerols from storage depots and the degradation and oxidation of these lipids (which are made primarily of carbon and hydrocarbon atoms) releases a great deal of energy.</td>
</tr>
</table>

Plants use simple compounds such as carbon dioxide, water, nitrates, and other mineral compounds as "food" building blocks. These compounds are relatively low in energy, and plants use sunlight rather than lipids (or carbohydrates) as a source of energy. Lipid stores of plants are used not by the plant, but by its seeds, as a source of energy. Animals, of course, eat plants and their seeds and use the lipids as well as other components of plants as a source of energy.

The reactions involved in the digestion, degradation, and oxidation of lipids, as well as their synthesis, will be discussed in this chapter.

<table>
<tr>
<td>37.2
DIGESTION OF LIPIDS</td>
<td>The enzymes required for the digestion of lipids are given the general name of lipases. There are recent reports of a lipase secreted in the pharynx, and another known as gastric lipase is found in the stomach. The latter is active only at pH values near 7, and thus it is ineffective in the stomach, where the pH is normally between 1.5 and 2.5. The principal digestion of lipids occurs in the small intestine, where the combination of bile and the pancreatic and intestinal lipases catalyzes the hydrolysis of the ester bonds of the</td>
</tr>
</table>

triacylglycerols. Complete digestion of triacylglycerols yields a mixture of free fatty acids and glycerol.

$$
\begin{array}{l}
\text{H}_2\text{C}-\text{O}-\overset{\text{O}}{\overset{\|}{\text{C}}}-\text{C}_{15}\text{H}_{31} \\
\overset{\text{O}}{\overset{\|}{\text{C}_{17}\text{H}_{33}-\text{C}}}-\text{O}-\text{C}-\text{H} \\
\text{H}_2\text{C}-\text{O}-\overset{\text{O}}{\overset{\|}{\text{C}}}-\text{C}_{17}\text{H}_{35}
\end{array}
\;+\; \text{HOH} \;\longleftrightarrow\;
\begin{array}{l}
\text{C}_{15}\text{H}_{31}\text{COOH} \quad \text{palmitic acid} \\
\text{C}_{17}\text{H}_{33}\text{COOH} \quad \text{oleic acid} \\
\text{C}_{17}\text{H}_{35}\text{COOH} \quad \text{stearic acid}
\end{array}
\;+\; \text{C}_3\text{H}_5(\text{OH})_3
$$

1-palmityl-2-oleyl-
3-stearyl-Sn-glycerol
(a triacylglycerol)   water   stearic acid   glycerol

   A major problem faced in the digestion of triacylglycerols is that of attachment of the water-soluble lipases to water-insoluble lipids. The process is more efficient when the lipids are broken into small globules. Since this breakdown increases their surface area, and consequently the area available for interaction with lipases, lipids in an emulsified form (broken into small globules) are much more readily digested. The **bile** does not contain enzymes, but the bile salts are effective emulsifying agents. Thus, while bile is not absolutely necessary for digestion of lipids, it helps greatly in the process.
   Although it seems that hydrolysis of two fatty-acid residues of triacylglycerols occurs readily, the last fatty acid (on the middle carbon) is often not removed in digestion, and monoacylglycerols (monoglycerides) are absorbed through the intestinal wall. The digested acylglycerols are resynthesized almost immediately (probably in the intestinal wall) to triacylglycerols. Thus only low levels of the free fatty acids are ever found in the blood. Lipids that are liquid tend to be more easily digested and absorbed than are solid lipids.
   Lecithinases (enzymes that catalyze the hydrolysis of the phospholipid lecithins) are present in the small intestine. Cholesterol esterases are also present. These latter enzymes catalyze the hydrolysis of cholesterol from the fatty acids with which it is often combined in natural products.

**37.3
ABSORPTION
OF LIPIDS**

Glycerol, fatty acids, possibly the mono- or diesters of glycerol and fatty acids, and cholesterol are absorbed across the intestinal wall into the blood and lymph. The **lymph** is a clear fluid much like blood, but without red corpuscles. It has a circulatory system similar to but separate from that of the blood (see Fig. 37.1). Following the digestion of a fatty meal, the lymph becomes cloudy from the presence of tiny globules of lipids which are called **chylomicrons**. They contain primarily triacylglycerols. Since the lymph flows into the blood, principally at the thoracic duct, the blood also becomes somewhat cloudy following the digestion of a fatty meal (this change can be observed only with serum or plasma, since the red blood cells tend to obscure the effect in whole blood). Thus the net effect of absorption of lipids into the lymph is about the same as if the lipids were absorbed into the bloodstream. Cholesterol is the only common sterol that is readily absorbed. Small amounts of the sterol-like compound vitamin D are also absorbed. Many of the substances classified as lipids are neither digested nor absorbed. Since these make up only a small part of the diet of animals, the indigestibility poses no real problems.

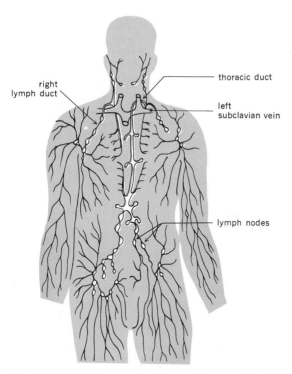

Fig. 37.1 The lymphatic system. The vessels which help in the circulation of the lymph and its connection with the blood system are shown here. In addition to its function in lipid absorption, the lymph returns valuable proteins to the circulatory system and is important in the response of the body to infections.

**37.4
FURTHER
METABOLISM
OF LIPIDS**

The major products of lipid digestion are the fatty acids and glycerol. As we indicated in Section 37.2, these recombine immediately after absorption to form triacylglycerols. They then circulate either as the small globules known as chylomicrons or as components of lipoproteins, until they reach an organ which can rehydrolyze them. There, either the fatty acids and glycerol are converted to a triacylglycerol characteristic of the species and of the specific organ doing the synthesis, or else the glycerol and fatty acids are oxidized to carbon dioxide and water with the attendant release of energy.

We can summarize the metabolism of triacylglycerols as follows:

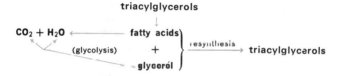

Glycerol is metabolized by addition of phosphate from ATP and oxidation to yield glyceraldehyde-3-phosphate, which is an intermediate in the glycolytic series of reactions (see Fig. 35.4).

Oxidation of Fatty Acids

The problem faced in the degradation of fatty acids is essentially that of breaking a carbon-to-carbon bond. In this case it is accomplished by dehydrogenation followed by a hydration, another dehydrogenation, and finally the rupture of the bond adjacent to the keto group formed in the process described above. In the major scheme of degradation, alternate carbon bonds are broken, giving two-carbon units. Coenzyme A (Fig. 33.14) is essential to the process. A schematic summary of the process is given in Fig. 37.2 and discussed in the following paragraphs. Degradation and oxidation of fatty acids takes place primarily in the mitochondria, which are also the site of the further metabolism of acetyl CoA.

In the series of reactions shown in Fig. 37.2, we can see that the product of the oxidation of fatty acids is acetyl CoA. This product is then further metabolized to carbon dioxide and water via the citric acid cycle. The removal of an acetyl CoA unit leaves a fatty acid which is shorter by two carbon atoms. This shortened fatty acid goes through the cycle again and again until it is all converted to acetyl CoA. Since the starting material differs each time, the series of reactions is sometimes called a spiral—the **fatty acid spiral**. Each time a two-carbon unit of the fatty acid is converted to a unit of acetyl CoA, 1 FAD is reduced to $FADH_2$ and 1 NAD is reduced to NADH. Thus a total of 5 ATP molecules can be synthesized each time a two-carbon unit is released. The oxidation of one molecule of palmitic acid ($C_{15}H_{31}COOH$) requires that it go through the spiral 7 times to yield 8 molecules of acetyl CoA. Each turn of the spiral yields 5 ATP, so we get 5×7 or 35 ATP's. Remembering that oxidation of acetyl CoA via the citric acid cycle gives 12 ATP molecules per acetate unit, we get 12×8 or 96 ATP's. Of course, to be precise, we should subtract 1 ATP since it is required for the attachment of the original fatty acid to coenzyme A.

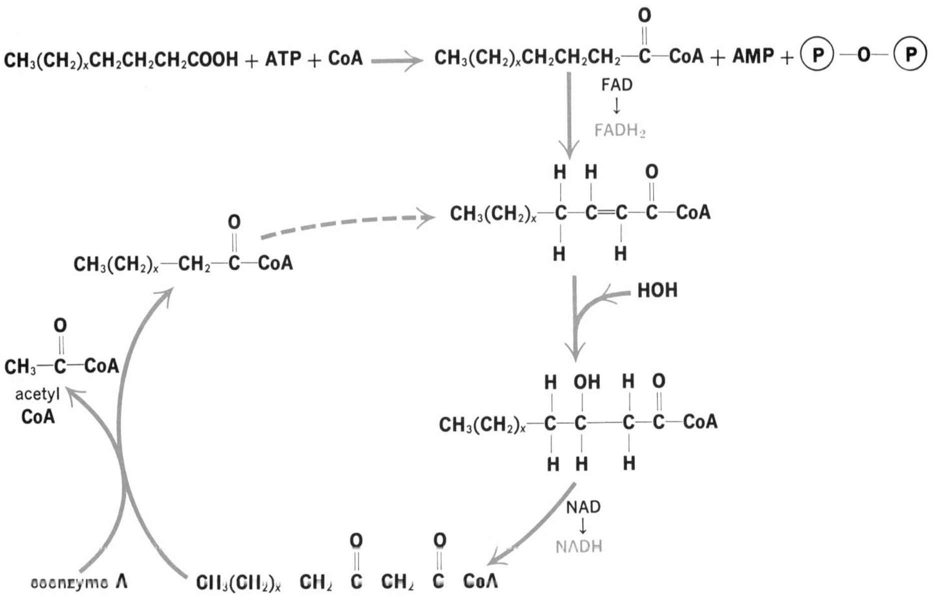

Fig. 37.2 The oxidation of fatty acids. The series of reactions given above shows how a two-carbon unit may be removed from a long-chain fatty acid. It should be noted that this is not truly a cyclic series of reactions since each reactant contains two fewer carbon atoms than it did in its previous trip through the series of reactions.

(The attachment of the CoA to the acetate unit that is released at the end of the spiral does not require the energy of ATP.) Thus we get:

$$5 \times 7 = 35$$
$$12 \times 8 = 96$$
$$\overline{131}$$
$$\underline{-1}$$
$$130 \text{ ATP's/mole of palmitic acid}$$

The process described in Fig. 37.2, in which a two-carbon unit is repeatedly removed from the carboxyl end of a fatty acid, is sometimes known as beta (β) oxidation. Other schemes of degradation also exist. In one, known as omega (ω) oxidation, the carbon farthest from the carboxyl carbon is oxidized before the chain is degraded by β-oxidation. Similar enzymes are found in some microorganisms that are capable of oxidizing hydrocarbons to acids and then metabolizing the acid formed. These microorganisms are probably important in oxidation of petroleum resulting from oil spills from tankers or off-shore oil wells.

Other metabolic schemes for the metabolism of fatty acids involve degradation into one-carbon units instead of two-carbon units and the addition of a carbon dioxide unit to acids with an uneven number of carbon atoms to give a dicarboxylic acid. This process converts propionic acid to succinic acid, which can then be metabolized in the tricarboxylic acid cycle.

**37.5
SYNTHESIS OF
FATTY ACIDS**

Most plants and animals can synthesize fatty acids. In some systems, synthesis seems to be primarily a reversal of the catabolic process; however, it now seems that this process is used primarily to add short units to fatty acid chains that are already formed. The major process for synthesizing fatty acids is shown in Fig. 37.3. The hydrogen atoms (electrons) and the energy for fatty acid synthesis is provided by reduced NADP, in contrast to the degradation, where FAD and NAD serve as hydrogen acceptors. It is interesting that most human tissues that synthesize fatty acids contain the enzymes for the phosphogluconate shunt (see Fig. 35.6), which produces reduced NADP.

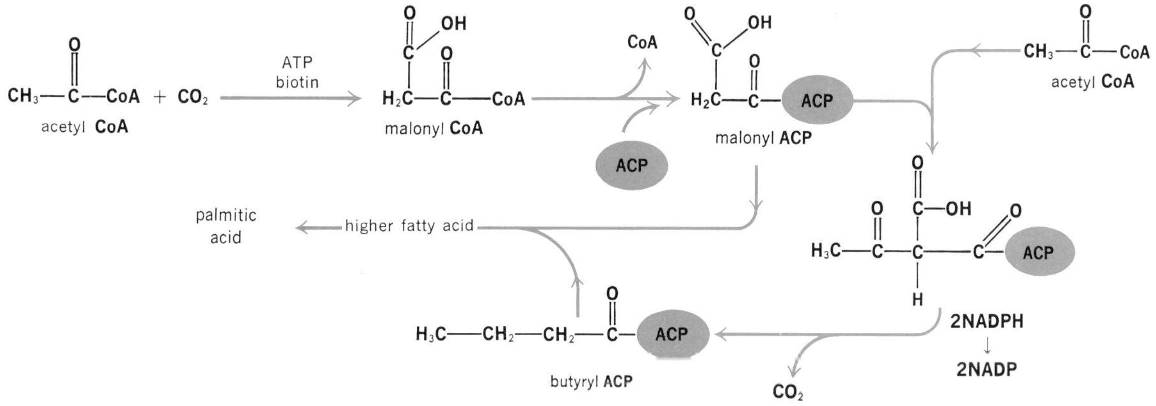

Fig. 37.3 The synthesis of fatty acids. The series of reactions represented above shows how a fatty acid is synthesized from acetyl CoA units. In the process CO_2 is first added and then eliminated. Also, during most of the synthesis the growing fatty acid chain is attached to an acyl carrier protein (ACP) which is in turn attached to a protein complex.

Although acetyl CoA is used as the starting material, the process of synthesis is not just the addition of these two-carbon units to one another. A carbon dioxide molecule is added to acetyl CoA to form malonyl CoA, but the CO_2 unit is subsequently removed, and all the carbon atoms of the synthetic fatty acid are derived from acetate, none from carbon dioxide. During the synthetic process, the growing fatty acid chain is bound to a complex of proteins through an acyl carrier protein (ACP). The end product of fatty acid synthesis by this process is usually palmitic acid. A different process is used for addition of a two-carbon unit to give 18 carbon fatty acids.

The process for synthesis of fatty acid varies in different kind of organisms and is significantly different in procaryotes and eucaryotes. In the human, enzymes for fatty acid synthesis are found primarily in the cytoplasm of the cell rather than being associated with organelles.

The minor differences between the processes of synthesis and degradation are apparently significant. In connection with these processes, Dr. David Green has stated, "Evidence is multiplying that life processes almost never follow the same pathway in building up and degrading a complex molecule. Readily reversible processes are rarely suitable for synthesis. The trick is to pick reactions that go almost exclusively in one direction." * It is important to note that synthetic and degradative reactions occur in different parts of the eucaryotic cell. Also, the involvement of ACP in synthesis but not degradation of fatty acids is significant.

Some organisms, especially plants, synthesize fatty acids having double bonds. Humans are able to saturate or unsaturate one carbon-to-carbon bond in the C_{18} fatty acid molecule. Thus they can convert oleic acid to stearic acid or vice versa. Since humans require fatty acids containing more than one double bond, a source of these fatty acids must be consumed in the diet, and we characterize them as essential fatty acids.

An animal or plant tends to synthesize triacylglycerols whose melting point is near the skin temperature of the animal or environmental temperature of the plant. Flax plants will grow at various temperatures, but the linseed oil from plants grown in colder temperatures is more highly unsaturated; that is, it will have a lower melting (freezing) point. Since liquid lipids are more easily metabolized by animals, there is a need for fat deposits that are solid enough to be structurally sound but sufficiently soft to be readily metabolized. It is illogical to believe that a seal living in Arctic waters would deposit a triacylglycerol like that found in beef, because the latter would be quite solid at the temperatures in which the seal lives.

We often divide the lipid stores of an animal into two categories, working or **tissue lipids** and storage or **depot lipids**. In any animal, tissue lipids are always present in about the same amount and are believed to be essential components of cells. Many of these tissue lipids are phospholipids, which are known to form an important part of membranes. Since, in humans, much of the lipid of the brain and nervous system is phospholipid, presumably the brain contains working and not just storage lipid. The amount of depot or storage lipid, which consists of triacylglycerols, is variable and depends on the nutritional state of the animal. The amount of depot lipid is high during periods of adequate or excessive food intake, and low following starvation.

There is a constant turnover of both kinds of lipids in the body, i.e., they are constantly being degraded and resynthesized. The metabolic half-life of the fatty acids in rat liver (mostly tissue lipids) is about 2 days; that of those in the brain and of the depot fats is

---

* D. E. Green, *Scientific American*, February 1960, p. 51.

about 10 days. Thus the depot lipids, while less active metabolically, are still being constantly degraded and resynthesized. In an experiment in which mice were maintained on a restricted diet so that they lost weight, they were given small amounts of isotopically labeled linseed oil. Their fats were soon found to contain the isotope. In fact, in 4 days, these animals made approximately 120 mg of depot fat and 130 mg of tissue lipid. Since they were losing weight, they were apparently catabolizing fats even faster than they were synthesizing them, but this did not mean that all synthesis stopped. This is only one example of the facts that lead us to believe that all components of the bodies of all organisms are in a state that has been described as **dynamic equilibrium**. Only when isotopes became available could scientists show that, even when a person's weight remains the same, the molecules in the body are constantly changing.

As we have said previously, an animal tends to accumulate lipids, particularly depot lipids that have a melting point near that of its skin temperature. Exceptions to this generalization are known. In an experiment designed to study the effect of diet on depot fat, dogs whose depot fat normally melted at 20°C were fed either mutton tallow (highly saturated, high melting point) or linseed oil (highly unsaturated, low melting point). The melting point of the body fat of the dogs fed mutton tallow was found to increase to 40°C and that of those fed linseed oil was reduced to 0°C. Where large amounts of a particular type of lipid are ingested and when almost none of other types is available, an animal is not always able to convert fatty acids just as it would on a varied diet. Since plants synthesize their own fatty acids (starting originally with CO_2 and H_2O), they tend to produce more consistent types of fats or oils, with the exception noted previously that the degree of unsaturation may depend on the climate in which the plant is grown.

Farmers who fatten hogs sometimes receive a lower price when they sell the animals if the fat is too soft—a condition called "soft pork." The condition is remedied if the diet is changed to include less unsaturated fat and more corn. Although the corn oil is relatively unsaturated, the starch in the corn is converted to fatty acids that are saturated, and consequently "solid" pork is produced and a higher market price can be obtained.

Fatty acids are joined with a glycerol unit derived from carbohydrate metabolism in the synthesis of triacylglycerols. This process is summarized below

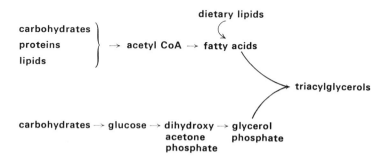

Cholesterol from foods is poorly absorbed from the small intestine; however, sufficient quantities can be absorbed to be of concern in the diets of some people. Cholesterol is also synthesized by almost all animals. The synthesis is a complex process which starts with acetyl CoA. Although we often think of cholesterol only in terms of its relationship to cardiovascular diseases (see Section 37.8), we should realize that it is the precursor of the steroid hormones (Section 32.6), which are important in many body processes.

The steroid nucleus is not degraded in the human body but is metabolized to a variety of compounds which are excreted both in the urine and in the feces. Urine levels of 17-keto steroids are significant in diagnosing several hormonal abnormalities. Much of the body's cholesterol is eliminated in the bile, which flows into the small intestine. Bile and compounds derived from it are responsible for some of the color of the feces, and lack of color in stool specimens indicates impaired bile secretion.

37.7
LIPIDS IN THE
BLOOD

The various types of lipids found in the blood are the subject of a great deal of study at the present time. The **chylomicrons** found immediately after digestion of a fatty meal are primarily triacylglycerols. Other lipids found in the blood are incorporated in **lipoproteins.** We can indicate the nature of these particles in the diagram below. It should be understood that this diagram represents only composition without implying structural relationships.

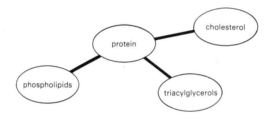

If we place blood serum in a centrifuge we can separate various lipid fractions differing in density. Some of the fractions and the relative percentages of the various components they contain are given in Table 37.1. Note that the triacylglycerols, which have a low density, make up a greater percentage of the low density lipoproteins (LDL). Protein, being denser, is found in greater amounts in the high density components (HDL). Analyses of the amounts of each of these fractions are now being made in many laboratories. Attempts to correlate concentrations of blood lipids with cardiovascular disease are the

TABLE 37.1   Characteristics and composition of some lipids of human blood serum

Fraction	Composition (%)				Molecular Weight (daltons)
	Triacyl-glycerols	Cholesterol	Phospholipids	Protein	
chylomicrons	80	10	8	2	10^9–10^{10}
very low density lipoproteins (VLDL)	55	18	19	8	5–10×10^6
low density lipoproteins (LDL)	10	27	28	21	2×10^6
high density lipoproteins (HDL)	6	12	25	58	1–4×10^5

object of a great deal of research. See Section 40.7 for a further discussion of blood lipids and lipoproteins.

**37.8
DISORDERS OF
LIPID
METABOLISM
Cardiovascular
Disease**

Cardiovascular disease is the leading cause of death in the United States. Actually, several conditions are included under this broad term, which really means diseases of the heart and blood vessels. The conditions we commonly call heart attacks, strokes, heart failure, and hypertension (high blood pressure) are all forms of cardiovascular disease. The underlying cause of cases of cardiovascular disease is **atherosclerosis**—a condition in which fatty deposits are formed in the inner lining of blood vessels. These deposits interfere with normal blood flow and may result in depriving some areas of an adequate blood supply, in the formation of a blood clot, or in the rupture of blood vessels.

The deposits formed in the blood vessels are composed largely of lipids in which cholesterol is found in the greatest quantity. There is no generally accepted theory as to the reason for formation of these fatty deposits in the blood vessels; however, it has been established that persons who have high levels of cholesterol in their blood are more likely to have heart attacks. This does not necessarily mean that the high level of cholesterol causes the attacks. Since the body receives some of its cholesterol from dietary sources and makes the rest, the extent to which cholesterol in the diet should be restricted is also a subject on which opinion is divided. It appears that cholesterol synthesis is decreased when diets are rich in cholesterol and that diets low in cholesterol stimulate cholesterol synthesis in the body. (But fasting and starvation *decrease* the rate of cholesterol synthesis.) On the other hand, diets high in saturated fats do tend to increase levels of blood cholesterol, and diets high in unsaturated fats tend to decrease blood cholesterol. It is also known that persons who are overweight are more likely to suffer from cardiovascular diseases. The explanation here probably depends on at least two factors—excess weight puts an extra strain on the heart, and overweight persons probably have high blood lipid levels, particularly high cholesterol levels.

The problems involved in understanding the cause and prevention of cardiovascular disease are much too complex to be adequately discussed in this text. They are problems about which the experts still disagree. However, we might summarize what we do know about lipid metabolism by saying that anything that can be metabolized to yield acetyl CoA can be synthesized to either cholesterol or fatty acids. Therefore, excessive intake of either carbohydrates or lipids is not beneficial to health. While it has not definitely been established that unsaturated lipids, such as those found in various vegetable oils, are valuable in preventing cardiovascular disease, some of the normal dietary lipids should probably be in this form. There is insufficient evidence at this time to justify excluding from one's diet the animal fats, which include butter and eggs as well as the fat found mixed in all meats. Indeed, it is difficult to design an adequate diet without the inclusion of milk and eggs. Probably the best policy a person can follow is to restrict the total number of calories eaten and to decrease the percentage of calories that are in the form of lipids, particularly the saturated lipids derived from animal fats. Americans derive from 40 to 45% of their calories from fats, whereas the Japanese, for example, derive only 10% of their calories from fats (and 78% from starches). When groups of men between the ages of 45 and 64 were compared, research showed that the mortality rate from cardiovascular disease for Japanese men is only 25% of that for American men. When Japanese move to the United States, however, and change their dietary habits to those of the Americans, their rate for heart attacks is similar to that for native Americans. We cannot, therefore, disregard the importance of the amount and type of lipid in the diet even though such factors as personality, sex, stress, and heredity are involved in cardiovascular disease. Much more research is necessary before we can give definite answers to problems associated with cardiovascular disease, and in particular, determine the part played by lipids.

Diabetes Mellitus   Although *diabetes mellitus* results from a defect in the metabolism of carbohydrates that is due to an insufficiency of insulin, it also involves changes in lipid metabolism. The decreased metabolism of carbohydrates means that more lipids are catabolized to provide energy for body processes. This leads to an increased level of fatty acids in the blood (hyperlipidemia), which in turn inhibits many important metabolic reactions.

In the absence of carbohydrate metabolism, fatty acids are not fully oxidized and are excreted as acetoacetic acid, β-hydroxy butyric acid, and acetone, which are sometimes called the ketone bodies. The ketone bodies produce the acidosis which is associated with *diabetes mellitus*. Diabetics therefore test their urine not only for glucose but also for the ketone bodies that are derived from lipid metabolism.

It has been noted that many people who develop diabetes late in life are obese, and that many diabetics who are being treated still have excessive amounts of fats in their liver. There has therefore been some speculation about the influence of obesity on the development of *diabetes mellitus*.

Hereditary Diseases of Lipid Metabolism   At least ten hereditary diseases have been classified as lipid metabolism diseases. (See the article by Brady in Suggested Reading for further information.) The best known— Gaucher's disease, Tay-Sachs disease, and Niemann-Pick disease—are named for people who originally described the condition. These diseases are usually fatal. In general, the diseases involve the accumulation of a specific type of lipid, due to a lack of the enzyme required for degradation of the lipid but not of the enzyme responsible for its synthesis. Symptoms of the diseases vary, but many involve enlargement of the liver or spleen and mental retardation. Tay-Sachs disease, in which the lipid accumulates in the brain cells, is particularly tragic. A normal-appearing baby who has the disease soon starts developing muscular weakness and mental retardation. This is followed by seizures and blindness. Death usually occurs at about three years.

While the specific type of lipid that accumulates varies with the disease, most are glycolipids containing one or more monosaccharide units attached to a sphingol residue and a fatty acid. In general, the factor that is deficient is the enzyme necessary for catalyzing the hydrolysis of the bond linking a monosaccharide to other components.

The hereditary enzyme deficiency diseases are further discussed in Chapter 39. As with most of these diseases, the diseases of lipid metabolisms may be detected in a fetus before birth. There have also been extensive programs designed to detect adult carriers of Tay-Sachs disease. Unfortunately, attempts to cure this and other lipid-storage diseases by supplying a missing enzyme have not been successful.

37.9 SPECIAL FUNCTIONS OF LIPIDS   In spite of the evidence pointing to lipids as a possible cause of cardiovascular disease, lipids play many vital roles in the normal functioning of the body. One gram of lipid gives an average of 9 kcal of energy while 1 g of carbohydrate or protein provides only 4 kcal/gram. Thus persons who work hard—who have a large demand for energy—must eat reasonable amounts of lipids. Lipids are stored in a water-free state, so a maximum amount of energy is stored with a minimum increase in body weight. The advantage of economic storage of energy was probably of great importance to our primitive ancestors, whose survival depended on both having energy stores and being mobile, especially if they had to run from sabre-toothed tigers. It is interesting to note that the plants, which are essentially not mobile, accumulate carbohydrates in their tubers (potatoes) or seeds (wheat) instead of accumulating lipids. The plant lipids, which are mostly oils, however, are located in the most mobile part of the plant—the seeds.

Fats are stored by animals both subcutaneously and around delicate organs. Subcutaneous fat helps insulate the body from extremes of temperature and cushions it from the blows of external objects. Fatty deposits around the kidneys and other vital organs probably help protect them from physical shock.

In the instances given above, it is tempting to assume that the human body "knows" where to store fat and that it is better to store fat than protein or carbohydrate. It is certainly much more scientific to say that those animals who stored fats had a better chance for survival than did similar animals who accumulated carbohydrates or proteins—or maybe even some other kind of compound. We could conclude that since those who survived were, of course, the ones who reproduced, the most advantageous traits have persisted. Whether the prehistoric requirements for survival are still most important for our way of life may be open to question.

There is, however, a lot of nonsense in the currently preferred image of the American, especially the American female. The standard of leanness and absence of depot fat that is set by many fashion models probably represents a condition only slightly less sound from a physiological point of view than the picture of the overweight housewife. The mental and physical harm done both by the dread of being fat and by the attempts to diet is tragic. Reasonable amounts of depot fat are not so bad as we might be led to believe. A sensible approach to the problem of body weight requires a knowledge of nutrition—a topic which is discussed in the following chapter.

**37.10
INTERRELA-
TIONSHIP OF
THE
METABOLISM
OF
CARBO-
HYDRATES,
PROTEINS, AND
LIPIDS**

Although proteins, carbohydrates, and lipids have distinctive metabolic pathways, there are several compounds that are common to two or more pathways. These represent important interconversion points. The citric acid cycle is used for the final oxidation of all three major types of compound, and both carbohydrates and lipids enter the cycle via acetyl CoA. A summary of intermediary metabolism is given in Fig. 37.4. Although the figure presents an abbreviated summary, it represents the major known pathways and shows the interrelationship of the metabolism of carbohydrates, proteins, and lipids.

From Fig. 37.4 we can see that carbohydrates can be converted to fatty acids and the lipids via acetyl CoA. The reverse is not true because the reaction of pyruvic acid to acetyl CoA is not reversible. The conversion of some amino acids to intermediates in the citric acid cycle and to pyruvic acid is also shown. Other amino acids can be converted to acetyl CoA or other intermediates by a relatively complex series of reactions. Many substances have been classified as being ketogenic or glucogenic. The ketogenic compounds will produce ketosis (excretion of acetoacetate, β-hydroxybutyrate, and acetone) when given to a diabetic animal. This reaction indicates that ketogenic substances are metabolized to yield acetate or acetoacetate. Glucogenic (antiketogenic) substances are those that are metabolized to yield a citric acid cycle intermediate, pyruvic acid, or some other glycolytic intermediate. A list of some common ketogenic and glucogenic substances is given in Table 37.2.

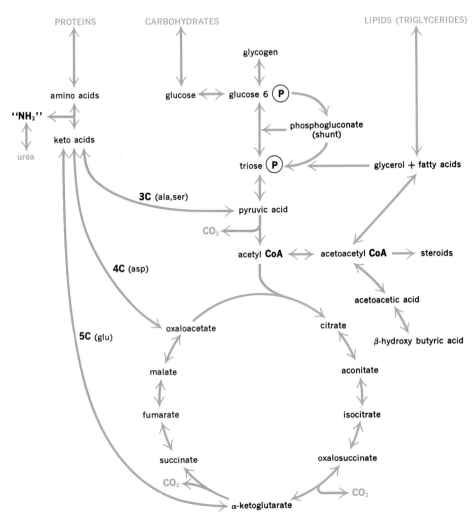

Fig. 37.4 Summary of intermediary metabolism. The reactions shown above represent the major metabolism of the basic biochemical compounds, the proteins, the carbohydrates, and the triacylglycerols.

TABLE 37.2 Ketogenic and glucogenic food substances

The lists given below contain the major nutrients that are classified as ketogenic or glucogenic food substances.

Ketogenic Substances	Glucogenic Substances
the amino acids leucine, isoleucine, phenylalanine, and tyrosine	the amino acids glycine, alanine, serine, cystine, methionine, aspartic acid, glutamic acid, proline, arginine, and histidine
all fatty acids	all carbohydrates, glycerol

SUMMARY

While animals consume a variety of lipids, it is the triacylglycerols and their digestion products, the fatty acids, that play a major role in human metabolism. The fatty acids can be degraded to acetyl CoA. Both the degradation and the oxidation of acetyl CoA via the tricarboxylic acid cycle yield energy. In anabolic reactions, acetyl CoA (which can be derived from metabolism of carbohydrates and proteins as well as lipids) can be synthesized by the body to give fatty acids and other products such as cholesterol. Synthetic fatty acids can be combined with a glycerol unit from carbohydrate metabolism to give triacylglycerols.

The human (and many other animals) requires some fatty acids which contain more than one double bond, but animals generally cannot make such acids. They need to eat lipids from plant sources. Plants can make a variety of highly unsaturated fatty acids and products derived from them.

IMPORTANT TERMS

adipose tissue	depot lipids	lipase
atherosclerosis	dynamic equilibrium	lipoproteins
bile	fatty acid spiral	lymph
cardiovascular disease	lecithinase	tissue lipids
chylomicron		

WORK EXERCISES

1. Summarize the digestion of a molecule of a triacylglycerol; of a lecithin.

2. List the series of reactions that is used to break a carbon–carbon bond in the metabolism of fatty acids

3. What is the metabolic fate of the glycerol from the digestion of a fat?

4. Why is the system for metabolism of fatty acids called a spiral rather than a cycle?

5. A table of food composition gives the following caloric values for 100 g of the edible portion of some sea foods: clams, 81; crabs, 104; oysters, 84; shrimp, 127; scallops, 78; salmon, 223; herring, 191. (All values are for the raw food; there is no significant difference in water content of these foods.) On the basis of the material in this chapter, can you give a logical explanation and interpretation of the difference in caloric value of these foods?

6. Calculate the number of ATP's formed from ADP and phosphate in the complete oxidation of one mole of capric acid ($C_9H_{19}COOH$). If the theoretical free energy change in the complete oxidation of this compound *in vitro* is 1,500 kcal/

mole, and the change in free energy for the synthesis of ATP from ADP and phosphate is 12 kcal/mole, what is the percentage efficiency of this biological oxidation?

7. Alanine containing ^{14}C was given to a diabetic dog. Later β-hydroxybutyric acid containing ^{14}C was found in the urine of this animal. List the probable intermediates in the conversion of alanine to β-hydroxybutyric acid.

8. On the basis of the metabolic scheme for the amino acid histidine given in Exercise 7 at the end of Chapter 36, explain why histidine is glucogenic.

9. It is known that a diet virtually free of lipids but high in carbohydrates can result in a person's accumulating large deposits of depot fat. Trace (in a general way) the pathway by which this conversion of carbohydrates to lipids can occur.

10. What major metabolic reaction of carbohydrates is irreversible? What is the significance of this fact?

11. Consult a general biochemistry text for the reactions involved in the synthesis of cholesterol by animals.

SUGGESTED READING

Brady, R. O., "Hereditary fat-metabolism diseases," *Scientific American*, August 1973, p. 88. Tests are described for adult carriers and for developing fetuses to detect hereditary defects in fat metabolism such as Tay-Sachs disease. Details of ten such diseases are given.

Dawkins, M. J. R., and D. Hull, "The production of heat by fat," *Scientific American*, August 1965, p. 62 (Offprint #1018). This article discusses the significance of "brown" fat deposits in heat production.

Green, D. E., "The metabolism of fats," *Scientific American*, January 1954, p. 32 (Offprint #16).

Green D. E., "The synthesis of fats," *Scientific American*, February 1960, p. 46 (Offprint #67). This article and the one above, covering the degradation and synthesis of fatty acids, are excellent.

Spain, D. M., "Atherosclerosis," *Scientific American*, August 1966, p. 48. A discussion of this leading cause of death and the relationship of diet to the disease.

38

Nutrition

<table>
<tr>
<td>38.1
INTRODUCTION</td>
<td>There are two reasons why a plant or an animal must receive nourishment. The first is that each organism must have a supply of energy for the many processes that must occur if life is to continue. In green plants most of the energy requirements are provided by the plant's ability to capture and use the radiation from the sun. Other plants, the fungi, not only cannot use the sun's radiation, but are killed or have their growth slowed by sunlight. This is especially true of bacteria, which are similar to fungi. Although humans like to acquire a suntan, they, along with other animals, do not utilize the sun's radiation as a source of energy; in fact, the tanning reaction probably serves to protect humans from the harmful effects of overexposure to the sun. The energy an animal uses to maintain basic life processes and to perform work comes from the many different kinds of food it eats.</td>
</tr>
</table>

The second reason a living organism must receive nutrition is to obtain necessary chemical substances which the organism itself cannot synthesize. For most plants, the nutritional requirements are merely a source of carbon dioxide, some nitrogen other than the elemental nitrogen found in the air, and a variety of mineral elements—the most important of which are phosphorus, potassium, and sulfur. Animals cannot use CO_2 as a source of carbon but use the carbohydrates, lipids, and proteins, which contain hydrogen as well as carbon. Usually nitrogen must be provided in the form of amino acids or proteins. Mineral elements are necessary for animals just as they are for plants. Animals also have a special requirement for small amounts of chemical substances called vitamins. Green plants do not require preformed vitamins because they can synthesize these substances; many bacteria, however, have definite requirements for vitamins. This chapter will discuss the nutrition of animals, principally of humans, in terms of both energy requirements and essential nutrients. Although we may tend to forget it, neither plants nor animals can exist for any length of time without water, so it also is a nutrient.

The discussions in this chapter are intended only to give a general introduction to nutrition. In a chapter of this sort we have had to make statements which are generally true, but there are many exceptions. Drastic diets or diets for people with conditions such as *diabetes mellitus* require professional advice from a dietician or physician. With these reservations, however, this chapter should help you understand a great deal about nutrition.

The energy requirements of an animal as well as the energy content of foods are commonly expressed in terms of the kilocalorie—the amount of energy required to raise the temperature of one kilogram of water one degree Celsius. As we mentioned in Section 2.9, in nutrition the term Calorie, abbreviated Cal, is usually used instead of the term kilocalorie. Both units are, of course, equal to 1 000 calories. The SI recommends the use of the unit known as the joule instead of the calorie, and some conversion to joules and kilojoules (kJ) is being made. However, in this text we have retained use of the calorie. To convert from Calories (kilocalories) to kilojoules we multiply by 4.2; conversely, to convert from kilojoules to kilocalories, we divide by 4.2.

The total number of calories people require depends a great deal on their actvity. People doing heavy labor expend more energy and consequently require more energy from their food than do desk workers. Even a person who spends the day in a hospital bed is doing some work—his heart is beating, there are peristaltic movements of the digestive tract, and nerves are sending messages. Since the amount of work depends on the amount of muscles, the size of the heart, and other factors, caloric requirements also depend on body weight. One reasonable estimate of the energy expended just to keep a person alive is that it requires 1 Cal/kg of body weight per hour. Thus a person weighing 50 kg (about 110 lb) requires 50×24 or 1 200 Cal/day as a minimum. For this person doing average work the energy requirements are probably from 2 000 to 2 400 Cal/day. A large person doing heavy work may require up to 5 000 or 6 000 Cal/day. These figures are only a general indication of caloric requirements. The efficiency of digestion and the basal metabolic rate of a person have a great deal to do with the number of calories required.

Carbohydrates and proteins provide 4 Cal/g and lipids (the triacylglycerols) give 9 Cal/g, although for specific foods the number of calories differs slightly from the averages given. These figures are for the digestible members of the groups listed. For example, although cellulose contains energy, cellulose is not digested by humans and therefore provides no calories for them. The opposite is true for ruminants, who can utilize cellulose (see Section 35.2). Actually, proteins and amino acids give slightly more energy than 4 Cal/g when they are oxidized *in vitro*. However, in humans their energy is not completely used, since urea is excreted rather than oxides of nitrogen. As a result, the average energy actually available from the metabolism of proteins is reduced to 4 Cal/g.

Most persons are aware that they feel somewhat warmer after they have eaten a meal. Those who are aware of different kinds of foods may have noticed that the increase in heat production is greater following a meal rich in protein-containing foods. This increased heat production which follows eating is called the **specific dynamic action** of foods. It represents the conversion of some of the caloric content of food into heat. The amount of this conversion is about 13% for lipids, 5% for carbohydrates, and an amazing 30% for proteins.

The conversion of the stored energy of foods into heat over and above that required for the maintenance of body temperature represents a "waste" of energy. This loss may be due to the failure of the body to trap the food energy which is released in the digestion and further metabolism of the food. The untrapped energy is then dissipated as excess heat. In practical terms, then, 30% of the caloric value of protein foods may be dissipated as heat. People who calculate caloric content of diets very accurately usually add about 10% of calories to compensate for the specific dynamic action.

38.3 THE ESSENTIAL NUTRIENTS

An essential food is defined as a *chemical substance which cannot be synthesized by a given animal at a rate equal to its need*. Strictly speaking, carbon dioxide is an essential nutrient for a plant. When we use the term **essential nutrient**, especially with respect to humans, we mean that the omission of this one nutrient would cause deficiency symptoms even though the individual had a varied diet containing sufficient energy sources. The deficiency symptoms might be indefinite, perhaps shown only in a failure to grow normally, or they might be specific as they are in the diseases known as rickets and pellagra.

The recommended daily dietary allowances of the essential nutrients are given in Table 38.1. It should be realized that these are *recommended allowances and not the minimum daily requirements*. It is also true that these allowances are for healthy people and that diseases and other conditions will alter a person's requirements. It should also be noted that the requirements of women who are pregnant or nursing a baby (lactating) are higher.

The recommended allowances for carbohydrates, proteins, and lipids (sometimes known as the macronutrients) are discussed in the material which follows and are summarized in Section 38.11.

38.4 CARBO-HYDRATES IN NUTRITION

None of the carbohydrates are essential, in the terminology of the nutritionist. However, for the majority of the people on earth more than half of the diet consists of carbohydrates. Rice, wheat and wheat flour, bread, potatoes, macaroni, and similar products are over 75% carbohydrate. In the more prosperous countries, proteins and lipids usually account for a large portion of the total food intake, but in poorer countries only small quantities of protein are available to most people. In these countries the land is used to produce cereal grains. Because of the expense involved and the continuous demand for grain, it is not fed to cattle or swine, which would convert the calories contained in the carbohydrate grains to protein. Animals are poor calorie converters—the food they supply as meat or milk represents only 10 to 25% of the calories they consume.

Recommendations about the amount and types of carbohydrates that should be consumed are given in Section 38.11. Other studies indicate that it is desirable to include appreciable amounts of foods that contain cellulose and similar substances that are indigestible but that contribute bulk ("fiber") to the diet. There is a correlation between high-bulk diets and a low incidence of cancer of the colon. And, although experimental confirmation is not yet available, it seems logical that having reasonably large amounts of materials going through the colon would prevent the accumulation of possibly carcino-genic substances in pockets (diverticula) in the large intestine. The indigestible materials that contribute to bulk are primarily carbohydrates, although other substances may be included also. This dietary component is listed as "crude fiber" in tables giving nutritional data.

38.5 PROTEINS IN NUTRITION

The proteins are significant in nutrition because of the amino acids they contain. Ten of the common amino acids are not synthesized by children at a rate sufficient to maintain normal growth. Adults probably require only eight amino acids. These essential amino

Table 38.1 Food and Nutrition Board, National Academy of Sciences–National Research Council—Recommended daily dietary allowances,[a] revised 1980

Designed for the maintenance of good nutrition of practically all healthy people in the U.S.A.

Age (years)	Weight (kg)	Weight (lb)	Height (cm)	Height (in)	Protein (g)	Fat-Soluble Vitamins Vitamin A (μg RE)[b]	Vitamin D (μg)[c]	Vitamin E (mg α-TE)[d]	Water-Soluble Vitamins Vitamin C (mg)	Thiamin (mg)	Riboflavin (mg)	Niacin (mg NE)[e]	Vitamin B-6 (mg)	Folacin (μg)[f]	Vitamin B-12 (μg)	Minerals Calcium (mg)	Phosphorus (mg)	Magnesium (mg)	Iron (mg)	Zinc (mg)	Iodine (μg)
Infants																					
0.0–0.5	6	13	60	24	kg × 2.2	420	10	3	35	0.3	0.4	6	0.3	30	0.5[g]	360	240	50	10	3	40
0.5–1.0	9	20	71	28	kg × 2.0	400	10	4	35	0.5	0.6	8	0.6	45	1.5	540	360	70	15	5	50
Children																					
1–3	13	29	90	35	23	400	10	5	45	0.7	0.8	9	0.9	100	2.0	800	800	150	15	10	70
4–6	20	44	112	44	30	500	10	6	45	0.9	1.0	11	1.3	200	2.5	800	800	200	10	10	90
7–10	28	62	132	52	34	700	10	7	45	1.2	1.4	16	1.6	300	3.0	800	800	250	10	10	120
Males																					
11–14	45	99	157	62	45	1000	10	8	50	1.4	1.6	18	1.8	400	3.0	1200	1200	350	18	15	150
15–18	66	145	176	69	56	1000	10	10	60	1.4	1.7	18	2.0	400	3.0	1200	1200	400	18	15	150
19–22	70	154	177	70	56	1000	7.5	10	60	1.5	1.7	19	2.2	400	3.0	800	800	350	10	15	150
23–50	70	154	178	70	56	1000	5	10	60	1.4	1.6	18	2.2	400	3.0	800	800	350	10	15	150
51+	70	154	178	70	56	1000	5	10	60	1.2	1.4	16	2.2	400	3.0	800	800	350	10	15	150
Females																					
11–14	46	101	157	62	46	800	10	8	50	1.1	1.3	15	1.8	400	3.0	1200	1200	300	18	15	150
15–18	55	120	163	64	46	800	10	8	60	1.1	1.3	14	2.0	400	3.0	1200	1200	300	18	15	150
19–22	55	120	163	64	44	800	7.5	8	60	1.1	1.3	14	2.0	400	3.0	800	800	300	18	15	150
23–50	55	120	163	64	44	800	5	8	60	1.0	1.2	13	2.0	400	3.0	800	800	300	18	15	150
51+	55	120	163	64	44	800	5	8	60	1.0	1.2	13	2.0	400	3.0	800	800	300	10	15	150
Pregnant					+30	+200	+5	+2	+20	+0.4	+0.3	+2	+0.6	+400	+1.0	+400	+400	+150	h	+5	+25
Lactating					+20	+400	+5	+3	+40	+0.5	+0.5	+5	+0.5	+100	+1.0	+400	+400	+150	h	+10	+50

[a] The allowances are intended to provide for individual variations among most normal persons as they live in the United States under usual environmental stresses. Diets should be based on a variety of common foods in order to provide other nutrients for which human requirements have been less well defined.

[b] Retinol equivalents. 1 retinol equivalent = 1 μg retinol or 6 μg β carotene.

[c] As cholecalciferol. 10 μg cholecalciferol = 400 IU of vitamin D.

[d] α-tocopherol equivalents. 1 mg d-α tocopherol = 1 α-TE.

[e] 1 NE (niacin equivalent) is equal to 1 mg of niacin or 60 mg of dietary tryptophan.

[f] The folacin allowances refer to dietary sources as determined by Lactobacillus casei assay after treatment with enzymes (conjugases) to make polyglutamyl forms of the vitamin available to the test organism.

[g] The recommended dietary allowance for vitamin B-12 in infants is based on average concentration of the vitamin in human milk. The allowances after weaning are based on energy intake (as recommended by the American Academy of Pediatrics) and consideration of other factors, such as intestinal absorption.

[h] The increased requirement during pregnancy cannot be met by the iron content of habitual American diets nor by the existing iron stores of many women; therefore the use of 30–60 mg of supplemental iron is recommended. Iron needs during lactation are not substantially different from those of nonpregnant women, but continued supplementation of the mother for 2–3 months after parturition is advisable in order to replenish stores depleted by pregnancy.

TABLE 38.2 Essential amino acids (for the human and the white rat)

arginine	leucine	phenylalanine
histidine	lysine	threonine
isoleucine	methionine	tryptophan
		valine

acids are given in Table 38.2 It will be seen that these amino acids are the ones having unusual carbon chains.

Most proteins derived from animals—such as beef and pork products, milk, and eggs —are good sources of the essential amino acids. Although some plants contain appreciable amounts of protein, many plant proteins lack one or more of the essential amino acids. Soybean protein contains less than desirable quantities of methionine, and corn protein is deficient in both lysine and tryptophan. Proteins having low amounts of one or more essential amino acids are considered poor quality proteins. For those whose diet consists primarily of plant products, and this means for much of the world's population, a variety of different plant proteins is necessary if they are to receive a balanced diet. In many instances one plant protein can supply amino acids that are lacking in another. These are known as complementary proteins. See *Diet For A Small Planet*, listed in the Suggested Reading for more information on complementary proteins. Adequate nutrition for vegetarians is easier to achieve if eggs or milk products are eaten. Fowl and eggs are good protein sources, and in countries located near the sea, fish, which is a good source of protein, is available.

A good general rule is that a person should eat about one gram of good quality protein each day for each kilogram of body weight. Since most protein foods are about 70% water, this means that a 70-kg (154-lb) person should eat a minimum of approximately 230 g (about half a pound) of high-protein foods each day. This rule of thumb assumes that the food is 30% protein, which is true only of high-protein foods like meat, eggs, and cheese. If the protein is provided by a plant that contains only 5% protein, the intake must be increased, of course. In spite of their low content of methionine, soybeans and products made from them are a good source of protein.

38.6
LIPIDS IN
NUTRITION

Lipids are good sources of food energy. In most instances an animal can make all the lipids it requires, provided it is given a diet containing a sufficient amount of carbohydrates.

It has been shown experimentally that white rats require small amounts of the highly unsaturated fatty acids. The formulas for these fatty acids are given in Table 38.3. Linoleic acid can be converted to the other two acids (linolenic and arachidonic acids) and therefore, strictly speaking, it is the truly essential fatty acid.

It is likely that humans also require highly unsaturated fatty acids in their diets, although definite symptoms of deficiency due to a lack of these acids are rarely seen in adults. Infants require small amounts of these fatty acids and develop deficiency symptoms if they are not supplied. We know that the highly unsaturated (polyunsaturated) fatty acids are converted to prostaglandins (Section 32.8) by the human. Since prostaglandins are important compounds and humans cannot synthesize polyunsaturated fatty acids, they should be included in human diets. Recent recommendations regarding the total amount of lipids as well as the type of lipids are given in Section 38.11 and Figure 38.2.

TABLE 38.3   The essential fatty acids

Name	Formula	Location of Double Bonds
Linoleic acid (octadecadienoic acid)	$C_{17}H_{31}COOH$	9–10, 12–13
Linolenic acid (octadecatrienoic acid)	$C_{17}H_{29}COOH$	9–10, 12–13, 15–16
Arachidonic acid (eicosatetraenoic acid)	$C_{19}H_{31}COOH$	5–6, 8–9, 11–12, 14–15

**38.7
MINERAL
ELEMENTS IN
NUTRITION**

There is no simple, unequivocal listing of the mineral elements required for adequate human nutrition. In the first place, there is some question as to just which elements are mineral elements. The generally accepted definition is that **mineral elements** *are those elements not normally found in carbohydrates, proteins or lipids.* This conception eliminates the elements carbon, hydrogen, oxygen, nitrogen, sulfur, and phosphorus, since these are commonly found in the three basic nutrients. However, adequate amounts of phosphorus for the human body's requirements cannot always be supplied from the three major sources of nutrients. Some inorganic phosphate must also be included in the diet. For this reason some authorities regard phosphorus as a mineral element. An adult requires about one gram of phosphorus per day either from inorganic phosphates or other compounds.

With the elimination of carbon, hydrogen, oxygen, nitrogen, and sulfur from the list of mineral elements, the major question remaining is, "Which of the other chemical elements are essential for normal health and growth?" The question is not easily answered. More than 50 elements are found in the human body. Some are present in large quantities, and a deficiency produces specific deficiency symptoms. A lack of iron, for example, produces anemia. But other elements, such as aluminum and silicon, are probably present primarily because they are so abundant in the soil of the earth that it would be extremely difficult not to eat them along with normal food. It is, in fact, virtually impossible to place a human on a completely aluminum- or silicon-free diet. Consequently, we have no experimental evidence as to whether these elements are essential. A summary of the mineral elements which are known to affect humans is given in Table 38.4.

We can list several general functions of the mineral elements. Many are found in the structure of parts of the body, as are calcium and phosphates in bone. Others are probably

TABLE 38.4   The mineral elements in nutrition

elements needed daily in large quantities	sodium, calcium, magnesium, potassium (phosphorus)
elements needed in smaller quantities	iron, copper, chlorine, iodine, cobalt, manganese, zinc, molybdenum
mineral elements whose need is questionable	bromine, fluorine, selenium, tin, silicon, arsenic, boron, vanadium, chromium
mineral elements which are toxic in relatively small quantities	tin, arsenic, barium, beryllium, bismuth, cadmium, lead, mercury, selenium, silver, tellurium, thorium

less obviously structural; iron and iodine are parts of the molecules of hemoglobin and thyroxine, respectively. The mineral elements are also important in maintaining the correct osmotic pressure of body fluids. The maintenance of a definite osmotic pressure is necessary to prevent dehydration or overhydration of body cells and to keep proteins in their active forms. Nerves and muscle cells also require certain of the mineral elements for normal function. In cases in which either of two or perhaps three elements has an effect, it is logical to assume that the effect is due to osmotic regulation. A requirement for a specific mineral element indicates a specific effect.

One of the most important functions of mineral elements is the part they play in enzymatic catalysis. When a mineral element (probably in an ionic form) is required for the maximum function of an enzyme, we call the mineral element an "activator." Almost all enzymatic reactions have been shown to require inorganic ions. In some cases a specific ion is required; in others similar ions may be effective in promoting the enzyme-catalyzed reaction.

38.8 IMPORTANCE OF SPECIFIC MINERAL ELEMENTS

The following paragraphs discuss briefly the specific functions of some of the more important mineral elements. For recommended daily allowances of mineral elements, see Table 38.1.

Calcium

Calcium is found in the form of a complex calcium phosphate in bones and teeth. It is also necessary for blood clotting. The removal of calcium ions by addition of citrate or other chelating agents keeps blood from clotting. Calcium activates some lipases and ATP-ase. Most persons require about 800 mg of calcium per day. Milk is one of the best sources of calcium. For pregnant women the calcium requirement is appreciably higher than normal because of the amount of calcium being deposited in the bones of the growing child before birth. Pregnant women who have insufficient calcium in their diets often deplete the calcium in their own bones, with resulting fragility of bones. Milk production requires large amounts of calcium in the diet. The old saying that a mother loses a tooth for every child she bears may have been true at one time, but there is no need for such damage with our modern knowledge of nutrition and the availability of milk and calcium supplements.

Studies of the diets of people in the United States indicate that many people, particularly women, normally consume amounts of calcium that are significantly below the recommended allowances. Increased consumption of milk and dairy products as well as green leafy vegetables can increase calcium intake. Extreme weight-loss diets, particularly the hydrolyzed protein diet, are very low in calcium. Deaths of people on these diets have been attributed to low blood calcium levels.

Magnesium

Magnesium is present in small quantities in bones, but is found primarily in body fluids. It activates many enzymes and is found as part of the chlorophyll molecule. Low levels of magnesium in the diet (a condition which is hard to produce experimentally, since magnesium is present in almost all natural water) can cause hyperirritability. Mice suffering from magnesium deficiency have running fits when given stimuli as simple as being exposed to a blast of air. High levels of magnesium in the blood produce anesthesia; however, many anesthetics are much more effective. Humans need about 250–500 mg of magnesium per day.

Sodium and Potassium Sodium ions account for 93% of the cations of the blood. Sodium ions are found in all body fluids, but most cells have little sodium inside the cell wall. Instead, there is a relatively high concentration of potassium ions within cells. While in a resting state, nerve cells contain potassium ions. When a nerve impulse passes, these potassium ions flow out into the surrounding medium and sodium ions flow into the nerve cell. In order for another impulse to be conducted by that nerve cell, the resting conditions must be reestablished, sodium must be pumped out, and potassium ions must enter. The whole process happens in milliseconds.

Potassium is present in relatively large quantities in plants, but sodium is not. Normally, humans consume from 2 to 4 g of potassium per day (in an ionic form, of course). Sodium ions are consumed primarily in the form of NaCl. While consumption varies widely, 5 g/day of sodium ions is a good average value. Herbivores (animals that eat only plants) often crave sodium chloride and will travel long distances to find a source of salt. Anthropologists have pointed to the fact that ancient civilizations developed in areas near the sea or near dried up seas and lakes, where good supplies of salts, particularly NaCl, were available. When the human body contains abnormally large amounts of sodium ions it tends to retain water, and we see the condition known as edema. Persons who develop edema—often as a result of certain heart conditions—are restricted from consuming NaCl, and are given diuretics to aid them in excreting sodium ions.

People who are taking diuretics may also excrete more potassium than usual. It is recommended that these people eat a banana or drink a glass of orange juice each day, since these foods are relatively high in potassium and low in sodium. Prolonged administration of intravenous solutions to hospital patients requires that potassium be added to prevent low potassium levels in the blood. The term for this condition, hypokalemia, reflects the Latin name for potassium, *kalium*. Symptoms of hypokalemia are irritability, muscular weakness, and tiredness, and may include paralysis and changes in heart action. Similar symptoms are often seen in people on some of the extreme weight-loss diets which are deficient in potassium.

Chlorine Chlorine is the principal anion of the body. No specific function is known for chlorine, and we assume that it is merely the most readily available and convenient anion to balance the cations Na^+ and K^+ in body fluids.

Iron Iron is part of the important body substances hemoglobin, myoglobin, and the cytochromes. Although the body contains from 3 to 5 g of iron, humans usually need only small amounts, perhaps 10 mg/day, in their diet. The fact that females lose some blood at each menstrual cycle means that they require a slightly higher intake of iron than do males. The body reuses iron, however, and although the iron-containing hemoglobin is turned over relatively rapidly in the body, iron is not excreted as are other parts of the hemoglobin molecule. It is retained and reused in the synthesis of a new molecule of hemoglobin. In spite of this, many women suffer from iron deficiency anemia. In fact some nutritionists say that anemia is the most common disease of malnutrition. Iron from red meats is well absorbed by the body. Supplements of iron-containing compounds are needed during pregnancy (Table 38.1). Since many forms of iron are poorly absorbed, care should be taken in choosing iron supplements. Iron from compounds where it is in the 2+ (ferrous) form is more readily absorbed than is 3+ (ferric) iron. The anion or negative component of the compound can also influence absorption.

Manganese Manganese is known to activate many enzymes, among them arginase and many phosphatases. A deficiency has been shown to cause sterility in male rats.

Copper   Copper serves as an activator for many enzymes, particularly oxidative enzymes. Copper is part of the pigment of the respiratory system of crabs, shrimp, and some other animals and serves much the same function that iron serves in the hemoglobin of humans. This pigment is blue and has the name *hemocyanin*.

Cobalt   Cobalt is part of the structure of vitamin B_{12}, and the name of the vitamin, cobalamin, reflects this fact. Deficiencies of cobalt in the food of ruminants leads to deficiencies of cobalamin and to anemia. Humans cannot synthesize cobalamin but must ingest the vitamin, which already contains the cobalt.

Fluorine   It is known that large amounts of fluoride ions are poisonous. It is also true that small amounts may be incorporated into teeth, and that these small amounts make teeth less subject to decay. The addition of fluoride ions to water supplies has led to a great deal of controversy.

Iodine   Iodine is part of the thyroid hormones thyroxine and triiodothyronine. Lack of iodine in the diet is one of the causes of an enlargement of the thyroid gland which is called goiter. The use of iodized salt, which contains some sodium iodide in addition to sodium chloride, eliminates goiters caused by iodine deficiency.

Other Mineral Elements   Molybdenum and zinc serve as enzyme activators. In addition, zinc is found as part of the structure of the hormone insulin. It also helps in the process of wound healing. Bromide ions have a sedative effect when present in relatively high concentrations in the blood. An appreciable amount of aluminum in the diet may lead to poor bone structure because aluminum prevents the absorption of the phosphates normally present in the diet and necessary for bone formation. This condition apparently comes about because aluminum phosphate is quite insoluble. Studies on animals maintained in germ-free conditions indicate that they require fluorine, silicon, tin, and vanadium for growth. Boron and zinc are required for normal plant growth although high levels, especially of zinc, may be toxic to plants. Very low levels of chromium may be necessary for diabetics.

38.9 VITAMINS   At the beginning of the twentieth century is was generally believed that an animal given a diet containing sufficient calories, adequate amounts of carbohydrates, proteins, lipids, and mineral elements would be well nourished. Additional research, however, with dietary components that were purified further indicated that something more than these nutrients was needed. Animals given highly purified diets soon got sick and, if not treated with less highly purified foods, they died. Gradually nutritionists realized that the animals on these purified but inadequate diets had symptoms similar to diseases seen in humans. The addition of certain substances to the purified diets cured the "deficiency" diseases in experimental animals, and it was not long until similar additions were made to diets of humans.

Placing experimental animals on given diets whose composition was well established has developed into an important experimental procedure for establishing exact nutritional requirements. The first new food substance to be identified was an amine—an important or vital amine. From these two words the name vitamine, which we now spell *vitamin*, was coined.

We now know that these vital food substances are not all amines. The current definition of a **vitamin** is that it is *an organic compound occurring in natural foods, either as* *such or as a utilizable precursor, which is required for normal growth, maintenance, or reproduction.* The lack of some, but not all, vitamins leads to a **deficiency disease**. Vitamins are catalysts and are not sources of body energy.

The vitamins were originally named for letters of the alphabet. When a vitamin which was thought to be only one substance was found to be two or more different substances, both letters and numbers were used. Now that chemical structues are known for the vitamins, it is better to use descriptive names, although the old letters and numbers are still widely used. Another classification of vitamins is based on their solubility. Vitamins A, D, E, and K are insoluble in water and are often called the fat-soluble vitamins. It is better to call them lipid-soluble vitamins. Substances classified as one of the vitamins B or vitamin C (ascorbic acid) are water soluble. Although these two classifications (lipid- and water-soluble) were developed merely for convenience in classifying vitamins, there is an underlying significance in the classification. The lipid-soluble vitamins are stored in the fats of the body; the water-soluble vitamins are not stored to any extent. Thus the water-soluble vitamins must be ingested daily, while stores of vitamins A, D, E, and K may be sufficient to keep deficiencies from developing even after long periods on vitamin-deficient diets. Now that vitamins are readily available in supplements as well as in foods, physicians are beginning to see cases of **hypervitaminoses**—an excessive amount of a vitamin. The only vitamins for which true hypervitaminoses are known are the lipid-soluble vitamins—the ones the body can accumulate, sometimes to its own detriment. The recommended daily allowances for vitamins are given in Table 38.1.

A vitamin must always be defined with respect to a given type of animal. Ascorbic acid (vitamin C) is a necessary substance for all animals. Most animals synthesize this compound. Humans, other primates, and the guinea pig, however, cannot synthesize ascorbic acid. For these animals ascorbic acid is a vitamin—for others it is not.

Any real understanding of the vitamins must involve some idea of their structure and their general functions.

Vitamin A   The formula for this vitamin is given below:

vitamin A (retinol)

It can be seen that vitamin A (chemically known as retinol) is a hydrocarbon with one —OH group; the compound is a lipid and is classified as lipid-soluble. Vitamin A is similar to β-carotene (see Chapter 32), and can be derived from β-carotene in the diet. The vitamin itself is yellow and its precursor, β-carotene, is orange, so that small amounts of it give a yellow color. Thus many foods that are good sources of the vitamin are yellow or orange (carrots, squash). Many green foods also contain vitamin A, but its color is masked by the more brilliant chlorophyll. The vitamin can be destroyed by heating and oxidation.

Vitamin A is necessary for growth, maintenance of epithelial tissues, and proper vision. Lack of the vitamin causes night blindness, poor dark adaptation, and a disease of the eyes called xerophthalmia. Although vitamin A undoubtedly plays many roles in the animal body, the one best understood is its function in vision. The visual cycle can be summarized as follows:

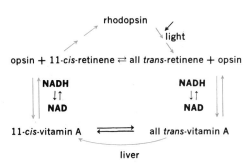

This cycle involves rhodopsin (visual purple), which is a combination of a protein called opsin and retinene (the aldehyde corresponding to vitamin A). Since there are five double bonds in the side chain of vitamin A, each one could exist in either the *cis* or *trans* configuration. Vitamin A is usually found in the all-*trans* form; however, in rhodopsin there is the *cis* configuration between carbon atoms 11 and 12. This 11-*cis* configuration is necessary for the binding of retinene to the protein opsin. When a quantum of light hits a molecule of rhodopsin it converts the 11-*cis* form of the vitamin to the all-*trans* form. This change of configuration apparently changes the shape of the retinene so much that it cannot stay bound to opsin. The other reactions of the series serve to regenerate the 11-*cis*-retinene so that it can recondense with opsin and again react with light.

Rhodopsin is not the only visual pigment. It is most important in vision in dim light; lack of rhodopsin due to a dietary deficiency of vitamin A (or carotene) causes a condition known as "night blindness." Eating liver, which is a good source of the vitamin, and carrots, which contain β-carotene, prevents or cures night blindness.

In the United States today there may be more excesses of vitamin A than deficiencies. The observation that Eskimos generally do not eat polar bear's liver correlates with the fact that it contains extremely high levels of vitamin A. When eaten, it produces symptoms of hypervitaminosis A, a condition characterized by skin irritations and a lack of appetite. In many areas of the world, however, vitamin A deficiencies create serious problems, particularly since they cause the disease xerophthalmia. It was recently estimated that, in India, at least one million persons are blind because of vitamin A deficiencies. Yet the materials to prevent these deficiencies would cost only a few cents a year per person.

The B Vitamins At one time it was believed that there was a substance called vitamin B. However, further study indicated that this substance was really a mixture of more than one chemical compound. Further fractionation proved that there were many compounds, all water-soluble, most containing at least one nitrogen in a ring structure, and having somewhat similar functions. These vitamins were given names such as B_1, B_2, and B_3. Some of those originally given a numerical classification proved to be either one of the other vitamins or else compounds that were not vitamins; only those called B_1, B_2, B_6, and B_{12} have retained the letter number designation. These designations are gradually being replaced with names which reflect the chemical structures. Other substances now classified as B vitamins are known primarily by their chemical names. For purposes of discussion we will group these various substances under the general heading of B vitamins.

All the B vitamins are necessary components of all living cells. They usually function as coenzymes, usually with a nucleotide, or at least a phosphate, attached. Since many organisms can synthesize these chemical compounds, they are not truly vitamins for these organisms.   In many animals, bacteria in the gastrointestinal tract synthesize appreciable amounts of the B vitamins, and the host organism may use the products of intestinal synthesis.   In fact, in some animals certain vitamin deficiencies cannot be produced without destroying their gastrointestinal bacteria.

All B vitamins are involved in promotion of the appetite, growth, and general well-being of the animal which requires them.   Foods that are good sources of one B vitamin as a rule are rich in the others also.   The classical deficiency diseases which have been attributed to a deficiency of a specific B vitamin were probably multiple deficiency diseases.   Actually, we know much more about the specific deficiency diseases in the white rat and other experimental animals than we do in humans, since our knowledge of human deficiencies has come primarily from reports of doctors who were much more interested in curing the disease than prolonging the condition in order to study it.   We may feel some displeasure about the lack of precise knowledge concerning human nutrition, but when we realize that definitive knowledge could come only from experimentation on humans, we are willing to accept the limitations.

Thiamine (Vitamin B₁) — The disease beriberi has been characterized as a thiamine-deficiency disease.   Beriberi was once prevalent in the Orient, where a large part of the diet is composed of polished rice.   The simple way to cure this disease is to restore to the diet the materials from rice hulls which are removed in polishing the rice.   While the classical human beriberi may have been a multiple deficiency disease, many of the symptoms are similar to those shown in rats on diets deficient only in vitamin B₁.

thiamine chloride

Thiamine deficiency causes abnormalities in the nervous system.   An abnormality in the nerves in muscles leads to muscular weakness.   Anxiety and mental confusion are also seen in humans.   In some cases the heart does not function normally.

Thiamine functions as a coenzyme in decarboxylation reactions.   A biochemical deficiency symptom is the finding of high levels of pyruvic and lactic acids in the blood, apparently due to the failure of a thiamine-deficient animal to decarboxylate pyruvic acid to acetyl CoA.   We do not yet know how the biochemical abnormality—lack of decarboxylation—leads to the physiological symptoms.

Thiamine was the "vital amine" which resulted in the name vitamin.   It is one of the most important vitamins; the complete absence of this substance from the diet will produce death in experimental animals in a week or two.   A human requires about 1 mg of thiamine per day.   Yeast, pork, fresh fruits, vegetables, and whole grains are good sources of thiamine.   Thiamine is now added to white bread and polished rice.   The vitamin is hydrolyzed to an inactive form when heated in neutral or alkaline solutions.   Consequently, prolonged cooking of foods containing thiamine, particularly if sodium bicarbonate has been added, as it often is to retain the fresh green color of vegetables, destroys the vitamin.

Some interesting experiments he did with white rats led Dr. R. J. Williams to reach some conclusions regarding alcoholism and thiamine. When white rats were given a choice of either water or a dilute solution of ethanol, each rat established an individual drinking pattern. Some preferred water, others drank both water and alcohol. When the diet of the rats was altered so that the thiamine was decreased, all the rats started drinking large amounts of alcohol. This kind of alcoholism could be cured by adding thiamine to the deficient diets. It is known that human alcoholics often suffer from thiamine deficiencies. Apparently the intake of alcohol diminishes the desire for food so that these people develop a slight thiamine deficiency, which leads to further alcoholism, more serious vitamin deficiency, and so on. The administration of thiamine is effective in treating some cases of alcoholism; however, on the basis of current evidence we cannot say that adequate nutrition is a cure for all cases of alcoholism, since many psychological factors are also involved. Relatively large doses of thiamine have been touted as a preventive or cure for a "hangover." This prescription is probably as effective as most cures, other than abstinence, that have been proposed. To our knowledge this area has not been subjected to rigorous scientific experimentation.

Riboflavin (Vitamin B$_2$) The deficiency of riboflavin can be said to produce generally poor health. White rats given diets deficient in vitamin B$_2$ fail to grow normally, have a poor coat of hair, a tendency toward skin irritations, and can be described as "looking sick." Human deficiencies of riboflavin that are not complicated by other deficiencies are not well characterized. The vitamin in the form of FAD (flavine adenine dinucleotide) and FMN (flavine mononucleotide) serves as an important coenzyme (see Fig. 34.8 for the formula of FAD).

riboflavin

The vitamin is found especially in yeast, liver, and wheat germ. Milk and eggs, although containing smaller amounts, probably represent the major sources of the vitamin in the average American diet. Green leafy vegetables are also good sources of riboflavin.

Nicotinamide This vitamin is related to nicotinic acid (which is only remotely related to nicotine, see Section 25.6A and B). Nicotinic acid can be converted to the vitamin by the human. The name niacin is also used for nicotinic acid, and niacinamide is nicotinamide.

nicotinamide

The classical deficiency disease of this vitamin is pellagra. Until 40 or 50 years ago many of the residents of the southeastern United States and of Spain suffered from this condition. It was characterized by a dermatitis, mental symptoms, and diarrhea. The addition of nicotinamide to the diets of persons suffering from pellagra cures the disease; so does the consumption of a diet rich in good quality protein. This originally led to confusion as to the real cause and cure of pellagra. When it was found that the amino acid tryptophan could be converted to nicotinamide by a series of reactions, the link between the two cures for pellagra was found. Those areas in which pellagra was widespread were areas in which corn was a major part of the diet. In fact, the diet of a typical pellagrin consisted principally of the fatty part of pork or "side meat," cornmeal mush, and molasses. Thus the caloric needs were satisfied, but no good source of either nicotinamide or tryptophan was ingested. The protein present in corn is notably lacking in the amino acid tryptophan. The story of the discovery of the cause and the treatment of pellagra by Dr. Goldberger of the United States Public Health Service is a striking example of nutritional detective work. One dramatic account of this story is in the book *Hunger Fighters* by Paul DeKruif. (See the Suggested Reading at the end of the chapter.)

Nicotinamide is found in the coenzymes NAD and NADP. (For formulas see Fig. 33.13.) These coenzymes function in a great number of oxidation reactions and provide an important link in the conversion of the energy of foods to the energy stored in ATP.

The requirement for this vitamin depends on the intake of tryptophan in the diet. The vitamin is found in a variety of plant and animal foods. Meat products, especially liver, are good sources of nicotinamide.

Pyridoxal (Vitamin B₆)

Pyridoxal is the functional form of the substance also known as vitamin B_6. The closely related compounds pyridoxine and pyridoxamine can also serve as sources of the vitamin.

pyridoxal   pyridoxine   pyridoxamine

No human diseases due to a deficiency of this vitamin are known, although convulsions have been reported in infants who had been fed a formula deficient in sources of pyridoxal. Rats on a diet deficient in vitamin B_6 develop a dermatitis and fail to grow normally. The fact that the amounts of urinary excretion products of pyridoxal are greater than the amount known to be ingested has led to the conclusion that pyridoxal is synthesized in the human body. Whether this synthesis is due to bacteria in the intestinal tract or whether some part of the body synthesizes the compound is not known. If the latter is found to be true, pyridoxal may no longer be considered a vitamin. Whole grains, fresh meats, vegetables, and milk are good sources of pyridoxal.

Pyridoxal phosphate is the coenzyme for many of the reactions of amino acids. Transaminations are among the most important reactions catalyzed by pyridoxal-containing enzymes. Decarboxylation and racemization (conversion from the D- or L-form to a DL mixture) of amino acids are also catalyzed by B_6-containing enzymes.

Pantothenic Acid

Pantothenic acid is part of the structure of coenzyme A. (See Fig. 33.14 for the formula.)

pantothenic acid

Human deficiency diseases are not known. In rats deficiencies produce a decreased rate of growth, impaired reproduction, and graying of hair. The latter is not observed, of course, unless the rat used is a dark-haired one instead of the usual albino white rat used in nutritional research. No relationship has been found between pantothenic acid and graying of human hair. As coenzyme A, the vitamin functions in a great number of reactions, especially those involving acetic or other fatty acids. Because of its widespread occurrence in foods, pantothenic acid deficiencies are rare. Liver and yeast are rich sources of the vitamin.

Biotin Biotin is widely distributed in foods. In most animals the excretion of metabolites of biotin exceeds the intake, and we assume that intestinal synthesis provides sufficient amounts of this compound.

Biotin serves as a coenzyme for reactions in which carbon dioxide is added to a substrate. The synthesis of fatty acids is one of the more important of such reactions. The conversion of pyruvate to oxaloacetate requires biotin, as does the incorporation of CO_2 into urea via the urea cycle.

Folic Acid Deficiencies of folic acid produce various kinds of anemias. Such deficiencies in the human are often due to a malfunction of the digestive tract rather than to an actual dietary insufficiency. The vitamin is widespread in nature, and most good diets contain appreciable quantities of it. It is also synthesized by bacteria in the intestine.

Folic acid functions in the metabolism of one-carbon compounds, particularly formyl and hydroxymethyl groups. It is certainly a necessary component of the body, although the daily requirements have not been established. Some clinical nutritionists believe folic acid deficiency is relatively common. This is due to poor diets and the lack of this vitamin in some vitamin pills, particularly those used for pregnant women.

Cobalamin (Vitamin B$_{12}$) This vitamin was one of the last ones to be studied and characterized. Deficiency of vitamin B_{12} causes the condition known as pernicious anemia. As with folic acid, most deficiencies are due to failure to absorb the vitamin which is present in food. The vitamin is widespread in animal products, but is not found in most green plants. Apparently the human body can store some amounts of this vitamin and since the daily requirement is small, perhaps only 1 μg (microgram)/day, B_{12} is not especially significant in normal nutrition. The formula for B_{12} is given in Fig. 38.1.

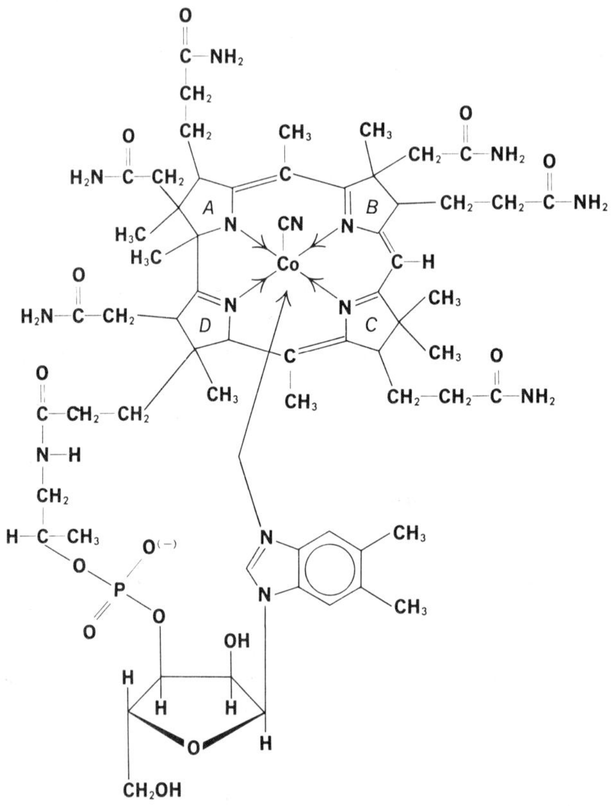

Fig. 38.1 Structure of vitamin B_{12} (cobalamin). Note the cobalt atom in the center of the pyrrolidine rings. These rings are in a plane with the cobalt, with cyanide group, and the dimethylbenzimidazole groups above and below the ring. Note also that rings *A* and *D* do not have a carbon bridge between them as was found in hemoglobin and chlorophyll.

Ascorbic Acid (Vitamin C)

Scurvy has been known for hundreds of years. It was especially seen among sailors. Scurvy was characterized by sore gums and loss of teeth, swollen and tender joints, small hemorrhages located just below the skin, and anemia. The British learned that lemons or limes which could be kept on long voyages would not only cure but prevent this condition. From this discovery came the practice of carrying limes and the name "Limey" to describe a British sailor. Later developments led to the identification of ascorbic acid as vitamin C. Two ways of writing the structure of ascorbic acid are shown on the following page. It belongs to the L-group of compounds and is related to simple carbohydrates.

Most animals can synthesize ascorbic acid from glucose, but humans, the apes, and the guinea pig cannot. The human requirement has not been definitely established, but is apparently in the range between 30–50 mg/day. The vitamin is present not only in citrus fruits, but in most fresh fruits and even in potatoes. Ascorbic acid is relatively easily oxidized, however, and consequently its effectiveness as a vitamin is lost when foods containing it are cooked.

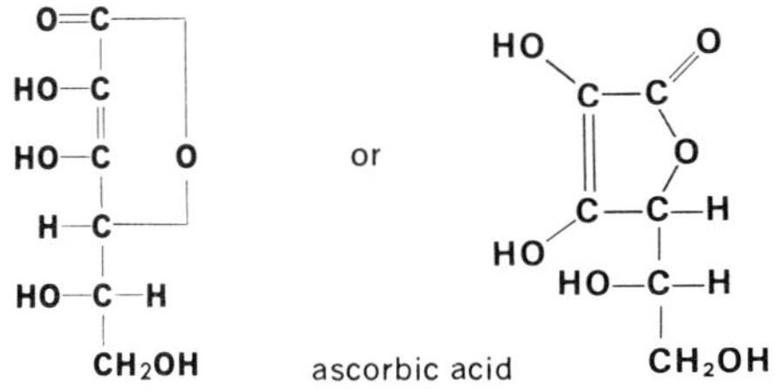

ascorbic acid

The exact function of ascorbic acid in human metabolism is not known. It presumably functions in oxidation-reduction reactions, and many such reactions have been shown to be aided by ascorbic acid. For most of these systems other easily oxidizable and reducible substances also function. No reaction is known for which ascorbic acid is a specific coenzyme. Despite much publicity, there is little scientifically valid evidence that vitamin C cures or prevents colds. Such evidence may be found, but large doses of the vitamin seem not to be warranted at present. On the other hand, vitamin C is water-soluble and is not stored, so overdoses are not so bad as those of lipid-soluble vitamins.

Vitamin D   Several substances have the function of Vitamin D. They are steroids differing only in the side chain (R group) at position 17. Children who have inadequate amounts of the vitamin in their diet develop rickets. The disease is now rapidly disappearing, but cases are prevalent in the poorer countries of the world. Rickets is characterized by the failure of the bones to harden normally. Children with rickets usually have bowed legs and are stunted in growth. The vitamin promotes the absorption of calcium and phosphorus through the intestinal wall and functions in depositing these two elements in the protein matrix of bone.

Vitamin D is sometimes called the sunshine vitamin because the ultraviolet radiation of the sun can convert an inactive precursor in the skin (7-dehydrocholesterol) to vitamin D. (See Section 32.6.)

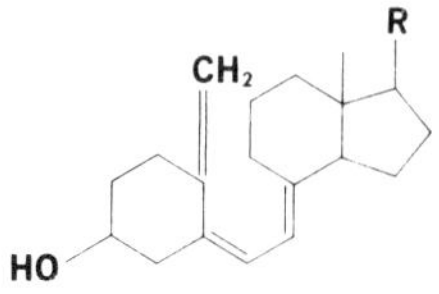

vitamin D

Children need vitamin D during the time they are growing and, to a lesser extent, mothers require the vitamin during pregnancy and lactation. The vitamin D requirement for adult males can probably be met by exposure to sunlight. Vitamin D is readily available, and since it is lipid-soluble and can be stored, cases of hypervitaminoses have been reported. In these cases bones are actually demineralized, and the calcium and phosphate deposited in many soft tissues. Deposits may also form in the kidney with serious consequences. Deaths have been reported due to kidney failure brought on by excessive intake of vitamin D in infants. Considerations such as these make one consider whether the widespread practice of adding vitamin D to milk, which has proved to be an excellent preventive for rickets, may not also have unfavorable consequences.

Vitamin E   The substance α-tocopherol and related compounds are necessary for normal reproduction in the rat.

α-tocopherol

Also, the absence of these compounds, which are collectively known as vitamin E, causes a serious muscular dystrophy in rabbits. Although there are other indications that the vitamin is necessary for several species of animals, there is no convincing evidence of deficiency symptoms in humans. It is ineffective in treating human muscular dystrophy and sterility. On the other hand, since other species do require the vitamin, it is possible that there is sufficient dietary intake and storage of vitamin E in humans so that symptoms of deficiency do not appear. The best natural sources of vitamin E are the oils derived from plants; wheat germ oil and cottonseed oil are good sources. The lipids of the leaves of green plants also contain vitamin E.

Excessive claims are often made for the effectiveness of vitamin E in helping certain heart conditions and in improving muscle action. Some athletic coaches advise the use of plant oils that are rich in vitamin E. To the best of our knowledge, at the present time, such uses are unwarranted. The ever-present danger of hypervitaminoses of the lipid-soluble vitamins should keep one from indiscriminate use of this substance. The 1974 revision of recommended daily allowances decreased its recommendation (by about 50%) for this vitamin. The 1980 Revision (Table 38.1) further decreased the recommended allowances.

Vitamin K   Vitamin K is necessary for proper clotting of blood; deficiencies are characterized by hemorrhages. Since the vitamin is widely found in foods and is probably synthesized by intestinal bacteria, deficiencies are usually due to poor absorption.

vitamin K

There are several forms of vitamin K, each differing in the length of the side chain. The widespread occurrence of the vitamin and its production by intestinal synthesis mean that normal persons need not be concerned with the amount of this nutrient in their diets.

In cases where clotting of the blood is damaging to an individual, extensive use is now made of substances which oppose the action of the vitamin, reducing the tendency of the person's blood to clot. These substances are structurally similar to vitamin K, but sufficiently different so that they interfere with the action of the vitamin. Dicoumarol was originally isolated from spoiled clover hay which caused the death of cattle due to internal hemorrhages. Now both dicoumarol and a synthetic analog, warfarin, are given to persons who have had heart attacks caused by blood clots. Warfarin is also used now as a rat poison. It causes death by producing internal hemorrhages. Continued consumption of the poison is required, however, before it is lethal.

dicoumarol   warfarin

Do Other Vitamins Exist?

The vitamins that have been discussed are the only ones accepted by most nutritionists, although they would not rule out the possibility that other vitamins might be discovered. In general, nutritionists recommend that a variety of natural foods be eaten so that any vitamins not yet discovered (and these certainly would be needed in extremely small amounts) would be present in the diet.

Many people claim that substances other than those we have discussed are vitamins. In these cases insufficient evidence is available to classify the substances as vitamins. Some are necessary substances but are normally synthesized by the body from the macronutrients. For others, the effects claimed have not been validated in controlled experiments.

For a substance to be accepted as an essential nutrient several tests must be met. It is generally required that experiments be performed with animals (other than humans) which show that there are deficiency symptoms when the nutrient is omitted from the diet and that these symptoms are decreased when the nutrient is administered. Even this does not necessarily mean that the substance is required by humans. A requirement for humans is accepted if a deficiency disease exists which can be cured by administration of the nutrient.

One method used to evaluate both nutrients and medications in humans is the double-blind experiment. In such experiments some of the people being studied are given the substance to be tested and others are given placebos. (Placebos are capsules or pills that look like the test substance but contain an inactive substance.) The term double-blind reflects the fact that neither the person being tested nor the doctor or scientist who observes the effects of the test substance knows who is getting the placebo and who the test substance. This procedure is used to rule out the desire of both the experimental subjects and the observer to have the experiment "succeed." Only when there is clear statistical evidence, based on a large number of cases, that the substance being tested is effective do scientists accept the new substance.

Many reports in the press of nutritional findings are based on a few cases in which both the patient and the observer knew when the new substance was being given. In other cases the reports concern persons with abnormal metabolic conditions. In general, scientists are reluctant to accept such claims until a double-blind experiment has shown effectiveness. In this text, we have taken a conservative position in cases where we believe there is insufficient evidence concerning reports on nutrition. Some of these reports will probably prove to be true, but a scientist cannot tell others that something is true until the evidence is very strong.

38.10 ON BEING WELL FED

An adequate diet should contain foods that provide sufficient energy. Either carbohydrates, proteins, or lipids can provide this energy. Evidence from studies of persons having cardiovascular disease indicates that in the normal American diet the percentage of calories derived from lipids is probably too high. Many nutritionists recommend that lipids should provide no more than 15% of the calories a person consumes. (See Section 38.11 for other recommendations.)

When the consumption of energy-producing foods exceeds the energy requirements of an individual, the body usually converts the energy of the food to energy stored in triacylglycerols, and the person gets fat. The simplest way to lose weight or to keep from gaining weight is to decrease the number of calories consumed.

A good diet, however, must also include all essential nutrients. It should contain the essential amino acids which are derived from proteins, the essential fatty acids, the mineral elements, and the vitamins. To meet these requirements, a varied diet should include high quality protein, vegetables, and fruits, some of which should be uncooked, and some plant oils. Vitamin pills are not necessary for normal persons who are able to have a varied diet as outlined above. Large amounts of the water-soluble vitamins are probably not harmful, but serious consideration should be given before taking pills containing large amounts of the lipid-soluble vitamins. While there is no doubt that these substances are required, there have been many reports of serious consequences of overdoses and subsequent hypervitaminoses of vitamins A and D in children. Other overdoses have been reported in adults who take excessive amounts of vitamin pills containing vitamins A and D. Legislation to make these vitamins (in large amounts) subject to prescription has been proposed but not accepted as of this writing.

Eggs, milk, and other dairy products are excellent foods. Both eggs and milk are "specifically designed" for the nutrition of a developing individual—the designer, of course, being the process of evolution, which assures the nonsurvival of the inadequately nourished. It is also true that egg yolks contain high levels of cholesterol (250 mg per egg yolk). For adult males, it is probably wise to limit the consumption of egg yolks.

Many Americans are overweight, and many others are obsessed with the idea that they must lose weight. They keep hoping that there is some easy way to eat all they want and still be thin. Well-meaning but ignorant persons and charlatans are only too ready to take advantage of this situation. Many good types of diet are known, but others, even some that are widely publicized, are so grossly unbalanced that continued adherence to them would have serious consequences.

People who want to lose weight should decrease the number of calories consumed while maintaining an adequate intake of protein. Decreasing the consumption of lipids and carbohydrates is desirable, although the lipid consumption should still include linoleic or other polyunsaturated fatty acids. It might be well to use vitamin and mineral supplements when cutting consumption of calories. For serious dieters as well as those interested in nutrition, charts such as those listed in the Suggested Reading are valuable.

To be well fed, a person should realize that nutritional requirements change with different body conditions. During pregnancy and lactation, requirements for a variety of nutrients are increased, and women who are either pregnant or nursing a child should receive good nutritional information. Babies and children have different nutritional requirements than adolescents and adults. Premature babies, in particular, have special nutritional requirements. Illnesses of both children and adults may produce changes in nutritional requirements.

Many medications are known to affect the requirements for nutrients. Prolonged use of antibiotics may produce deficiencies of folic acid and vitamin K, among others. These are probably due to the changes the antibiotic brings about in the microorganisms in the intestinal tract which normally synthesize these vitamins. There are reports that requirements for vitamin B_6 are increased when a woman is taking oral contraceptives. To be well nourished, any person taking medications for a prolonged period of time should obtain advice about possible effects of the medication on nutritional requirements.

Although they are not normally considered as nutrients, food additives are also consumed by most people. A discussion of the advantages and disadvantages of these compounds is given in the Suggested Reading at the end of this chapter.

The discussions in this chapter are intended to serve as a general introduction to the science of nutrition. There are exceptions to most of the generalizations presented. This chapter is intended to give an understanding of normal nutrition, not to provide sufficient information for those who are ill, taking medications, or otherwise different from the "average." Any prolonged dietary restrictions or drastic supplementation of the diet should be undertaken only upon the advice of a doctor or other competent nutritionist.

**38.11
DIETARY
GOALS FOR
THE
UNITED STATES**

In February, 1977, a Committee of the United States Senate issued a report establishing dietary goals for the United States. This report was the result of hearings and other information from many scientists. As a result of criticism, the goals were somewhat modified and issued again in December, 1977. The major recommendations of the committee as revised are given below:

1. To avoid overweight, consume only as much energy (calories) as is expended; if overweight, decrease energy intake and increase energy expenditure.

2. Increase the consumption of complex carbohydrates and "naturally occurring"* sugars from about 28% of energy intake to about 48% of energy intake.

3. Reduce the consumption of refined and processed sugars by about 45% to account for about 10% of total energy intake.

4. Reduce overall fat consumption from approximately 40% to about 30% of energy intake.

5. Reduce saturated fat consumption to account for about 10% of total energy intake; and balance that with polyunsaturated and monounsaturated fats, which should account for about 10% of energy intake each.

6. Reduce cholesterol consumption to about 300 mg a day.

7. Limit the intake of sodium by reducing the intake of salt to about 5 grams a day.

The effects of these goals on the diets as well as the best estimates of the current diets of Americans are given in Fig. 38.2.

The committee made further recommendations for changes in food selection and preparation:

1. Increase consumption of fruits and vegetables and whole grains.

2. Decrease consumption of refined and other processed sugars and foods high in such sugars.

3. Decrease consumption of foods high in total fat, and partially replace saturated fats, whether obtained from animal or vegetable sources, with polyunsaturated fats.

---

* "Naturally occurring" sugars are those found in a natural food as opposed to refined (cane and sugar beet) and processed (corn sugar, syrups, molasses, and honey) that are added to foods either in preparation or just before eating.

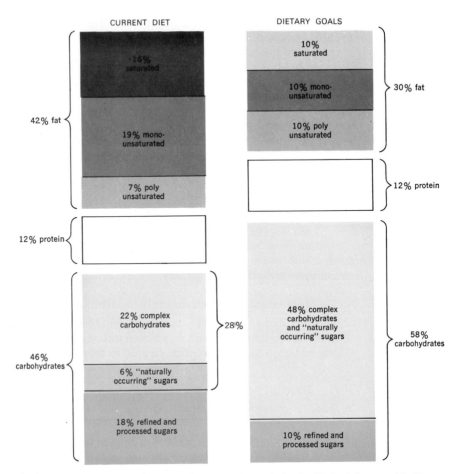

Fig. 38.2 Comparison of the current diet of people in the United States with dietary goals. [From "Dietary Goals for the United States" (second ed.), The Select Committee on Nutrition and Human Needs, United States Senate, December 1977.]

4. Decrease consumption of animal fat, and choose meats, poultry, and fish which will reduce saturated fat intake.

5. Except for young children, substitute low-fat and nonfat dairy products for high fat dairy products.

6. Decrease consumption of butterfat, eggs, and other high cholesterol sources. Some consideration should be given to easing the cholesterol goal for premenopausal women, young children, and the elderly in order to obtain the nutritional benefits of eggs in the diet.

7. Decrease consumption of salt and foods high in salt content.
   The committee also made the following statements:

Persons with physical and/or mental ailments who have reason to believe that they should not follow guidelines for the general population should consult with a health professional having expertise in nutrition, regarding their individual cases.

Although the Dietary Goals are stated in terms of specific levels, each specific level should be considered as the center of a range.

The report of the committee as well as other similar reports appear in some of the Suggested Readings at the end of the chapter. The committee recommended that its complete 83-page report of December, 1977, be read for further details. This is an excellent recommendation.

The report stresses that it is concerned only with the macronutrients and notes the recommendations of the National Research Council—National Academy of Sciences (given in Table 38.1) for micronutrients such as the vitamins and minerals.

In spite of questions that have been raised about the Dietary Goals, they seem to be excellent goals. It is difficult to see how following them would be harmful to the normal person. Many of these suggested modifications in the American diet make sense in terms of world food problems. If Americans ate more whole grains and less meat, as well as consuming fewer calories, it could alleviate some of the starvation and hunger in the world.

38.12 WORLDWIDE NUTRITIONAL PROBLEMS

It is often said that half the world's people go to bed hungry every night. Although the truth of such a claim is hard to establish, it is a fact that far too many people suffer from malnutrition, either from lack of calories or from lack of a specific essential nutrient. This problem exists because fewer people are dying from infectious diseases and more of them are living to help aggravate the rapid rate of population growth in many countries.

Lack of sufficient protein containing the essential amino acids is probably second only to general lack of food as a worldwide problem. Although most plant proteins are poor sources of several of the essential amino acids, many people cannot afford to feed plant materials to animals in order to secure the higher quality, hence more desirable, animal protein. Since animals convert only a relatively small percentage of the protein in their food into body protein, it is highly questionable whether we should continue the wasteful practice of feeding high-protein foods to cattle. The situation is different when cattle or sheep eat grass or other food unavailable for direct human use.

There are ways of making a diet of plant protein more nutritious. One is by carefully mixing plant proteins so that the deficiences of one food are met by the amino acids supplied by another. Instructions for doing this are given in the book *Diet For a Small Planet* by Lappé. Another procedure for increasing the quality of protein food is by breeding new kinds of plants that have either higher protein content or a higher content of an important amino acid. The article by Harpstead in the list of Suggested Reading tells of the development of a high lysine corn. This and similar breeding experiments combined with better methods of production are generally known as the "Green Revolution." It shows promise of meeting some of the world's food problems.

We will probably never solve our worldwide nutrition problems until we find ways of dealing with population growth and with the excessive consumption of food by the developed countries. It is tragic to contemplate that dog and cat food, so much seen in every supermarket, contains fish protein that children in many parts of the world badly need. Disputes over territorial fishing rights show how countries value the fish caught off their shores.

While science can help solve some of the worldwide food problems, social problems are involved also. Only when awareness and concern are coupled with scientific and social knowledge will we have a chance of solving the crucial problems of the world.

IMPORTANT TERMS

Calorie	hypervitaminosis	nutrition
deficiency disease	mineral element	specific dynamic action
essential nutrient	nutrient	vitamin

WORK EXERCISES

1. Why is tryptophan an essential amino acid?

2. Why must vegetarians eat a variety of types of plant food?

3. Calculate your basic daily requirements for calories and protein.

4. Keep a record of all foods eaten for a day. Using a dietary chart, compute your intake of carbohydrates, proteins, lipids, and total calories.

5. Why should you include nonhydrogenated vegetable oils in your diet?

6. What are the basic functions of mineral elements in the human body?

7. What are the biological functions of each of the following mineral elements?
a) iron   b) chlorine
c) calcium   d) cobalt
e) iodine

8. What vitamin deficiency is the primary cause of each of the following diseases?
a) beriberi   b) rickets
c) scurvy   d) pellagra

9. What vitamin is found in each of the following coenzymes?
a) NAD   b) FAD
c) coenzyme A

10. Examine a carton of milk to find the amount of vitamin D that has been added. Check in a grocery or dairy store to see whether adding vitamin D is widespread.

11. Examine the label on a loaf of bread to determine the amount of vitamins added to the bread. Most breakfast cereals also have a listing of their vitamin content. How do they compare with bread as a source of vitamins?

12. Which is the "sunshine" vitamin? Why is it so named?

13. How do the requirements of certain animals for vitamin C correlate with the generalization (Chapter 36) that animals higher in the evolutionary scale often lack enzymes that simpler animals have?

14. List the vitamins which contain the following elements: N, S, Co.

15. Describe the changes in nutrition that should accompany pregnancy.

SUGGESTED READING

Burton, B. T., *Human Nutrition—Formerly the Heinz Handbook of Nutrition,* 3rd Ed., McGraw-Hill (Blakiston), New York, 1976. This excellent book has basic discussions of food elements, utilization of food, nutrition in health, and nutrition in disease. It also contains brief tables of nutrients in various foods.

DeKruif, P., *Microbe Hunters,* Harcourt, Brace and World, New York, 1932. A classic popularization of the lives of many great scientific microbe hunters. Also available in paperback.

Deutsch, R. M., *The New Nuts Among the Berries,* Bull Publishing Co., Palo Alto, California, 1977. This book is a delightful presentation of the food faddists—past and present. The author is well informed and critical.

Fernstrom, J. D., and R. J. Wurtman, "Nutrition and the brain," *Scientific American,* February 1974, p. 84 (Offprint #1291). The changes in chemicals in the brain following eating are discussed.

Frieden, E., "The chemical elements of life," *Scientific American* July 1972, p. 52. An excellent discussion of the functions of mineral elements in life processes. Particularly interesting is its report on the elements fluorine, silicon, tin, and vanadium, which have recently been shown to be essential for normal growth of some animals.

Harpstead, D. D., "High-lysine corn," *Scientific American,* August 1971, p. 34 (Offprint #1229).

This article tells of the breeding of strains of corn that contain this essential amino acid, so important in world nutrition.

Hoffman, L., *The Great American Nutrition Hassle*, Mayfield Publishing Co., Palo Alto, California, 1978. This is a collection of recent articles which Ms. Hoffman has annotated and organized under headings such as: Dietary Trends, Prevention and Therapy Through Nutrition, The Health Food Controversy, The Politics and Economics of Food, and What Can Be Done? The articles come from a variety of sources and represent different viewpoints. Specific topics include "Professors on the Take," "Alcohol and Nutrition," "Diet Books that Poison Your Mind . . . and Harm Your Body," and articles on Megavitamin Therapy, Sugar and Hypoglycemia, and Cholesterol. This is a fascinating, readable book.

Jacobson, M. F., *Eater's Digest: The Consumer's Factbook of Food Additives* (pb), Anchor/Doubleday, Garden City, New York, 1972. This is a good middle-of-the-road presentation on a controversial subject. Dr. Jacobson has a Ph.D. in Microbiology from Massachusetts Institute of Technology. He lists and discusses many additives as well as writing some general chapters on testing of additives, nutritional labeling, and similar topics.

Lappé, F. M., *Diet for a Small Planet*, Revised (pb), Friends of the Earth/Ballantine, New York, 1975. This book discusses world protein problems and gives specific instructions and recipes for making plant protein diets more nourishing.

Majtenyi, J., "Food additives—Food for thought," *Chemistry* **47**, No. 5, p. 6, May 1974. This article is an excellent introduction to the subject of food additives. The reasons various substances are added as well as a safety rating of almost 50 substances are given in an excellent table which accompanies the article.

"Dietary Goals for the United States," 2nd edition. Prepared by the staff of the Select Committee on Nutrition and Human Needs—United States Senate, December 1977, Stock No. 052-070-04376-8. This 83-page report is recommended for those who are concerned with improving their own diets and those of others.

"Dietary Goals for the United States—Supplemental Views," Prepared by the staff of the Select Committee on Nutrition and Human Needs, United States Senate, November 1977, United States Government Printing Office, Washington, D.C. This 869-page book gives the arguments of those who objected to the Dietary Goals. There are reports by scientists, the National Dairy Council, Salt Institute, and others. A good reference for those interested in politics and science. It shows why the goals were modified by the Senate Committee.

Recommended Dietary Allowances, 9th edition. National Academy of Science, Washington, D.C., 1980.

For those interested in finding the nutritional content of foods, the following are excellent, inexpensive source books. The first is sufficient for most people; the second contains much more information and is recommended for those conducting dietary surveys and similar studies. It contains 291 large sized pages, each full of data.

Nutritive Value of Foods, United States Department of Agriculture, Home and Garden Bulletin, No. 72, April 1977. Stock No. 001-000-03667-0.

Nutritive Value of American Foods in Common Units, United States Department of Agriculture, Agriculture Handbook No. 456, November 1975. Stock No. 001-000-03184-8.

39

The Chemistry of Heredity

Since the times of the ancient Greeks, and probably long before, people have been aware of the fact that a young animal resembles its parents. Characteristics regarded as hereditary have included eye color, stature, and even a fiery temper. Occasionally heredity seemed to fail when, for example, normal parents produced an albino offspring. As with many biological theories, those concerning heredity have become more and more detailed in order to explain observations. Underlying the details, however, we can now see a few simple generalizations which help us to explain heredity and to predict what will happen in many cases.

When it was realized that all living things were produced by parents similar to them, and that lice, frogs, and mice were not generated spontaneously from decaying matter, mud, or old rags, theories of heredity became important. Part of the way in which characteristics are transmitted was deduced by the Austrian monk Gregor Mendel, whose studies of pea plants were the basis for adding to the theory of heredity the concepts of specific unit characteristics which do not blend. Figure 39.1 illustrates one of Mendel's experiments involving tall and dwarf peas.* The development of Mendel's theories led to the belief that there were at least a pair of factors called **genes** for each hereditary characteristic. Some characteristics, such as skin color, are now believed to be controlled by more than one pair of genes. For each hereditary characteristic or for each pair of genes, both dominant and recessive varieties are known.

With sweet peas, a plant produced by crossing one having red flowers with one having white flowers produces only red flowers, not pink. Since the new plant presumably has genes for both red flowers and white flowers, we conclude that the gene for red flowers is dominant. If we cross the red-flowering pea plants from mixed parents with

---

* Recently some scientists have noted that Mendel's results were too "good." While not questioning his conclusions, his critics do question whether some of his data were "manipulated" to make them look better.

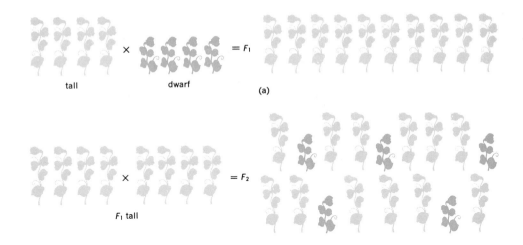

tall   dwarf   = F_1

(a)

F_1 tall   = F_2

Fig. 39.1 Mendel's experiments with tallness in pea plants. Diagram (a) represents Mendel's first cross of tall and dwarf peas. All the offspring (F_1 generation) were tall. Diagram (b) shows the results of matings between the members of the F_1 generation. When Mendel performed the experiment he found 787 tall plants and 277 dwarf plants. [Adapted from J. J. W. Baker and Garland E. Allen, *The Study of Biology*, 3rd ed., Addison-Wesley, 1977.]

each other, their offspring produces red or white flowers, but no pink ones. In the illustration of the tall and dwarf peas (Fig. 39.1) we see that plants were either tall or dwarf, and that tallness is dominant. An individual having two different genes for a single trait is called a *hybrid* or a **heterozygote**. A pure-bred individual, one having a pair of identical genes, is homozygous—or is a **homozygote**—for that trait.

Our present theory about inheritance can be summarized as follows. There is a pair of genes for each hereditary characteristic in most cells of each individual. These genes may be identical, in which case the individual is homozygous for that characteristic, or they may be different, in which case the individual is heterozygous for that characteristic. When an individual forms reproductive cells, whether in the parts of a flower or in the testes or ovary of a mammal, only one member of the pair of genes goes into an individual reproductive cell. (The general name given this reproductive cell is *gamete*.) A homozygote produces only one kind of gametes; a heterozygote, two. When the gametes from a male (sperm cells) meet with and fertilize the gametes (eggs or ova) from a female, various combinations are possible. If the number of offspring of the mating of a single male and a female is large, the proportions predicted by laws of probability are found. In cases where the number of offspring is small, the lack of an adequate sample may result in some deviation from predictions. The situation is analogous to tossing a coin only a few times and having the percentage of "heads" not equal to that of "tails."

An analysis of the hereditary factors involved in the tall and dwarf pea plants of Mendel is given in Fig. 39.2. (We use T to represent the dominant characteristic—tallness—and t to represent the dwarf characteristic.) The male ($\male$) produces gametes containing

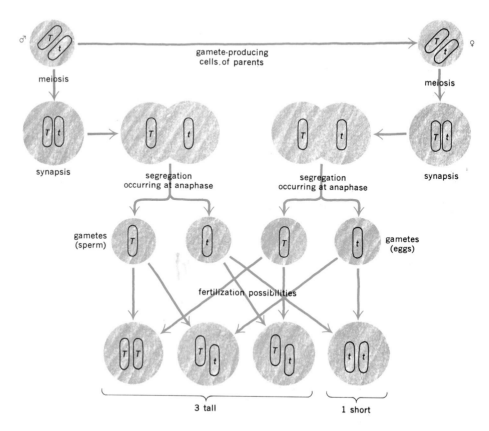

Fig. 39.2 Diagram showing the mechanism of inheritance of tallness in pea plants. *T* indicates the gene for tallness and *t* the gene for the dwarf condition. [Reprinted from J. J. W. Baker and Garland E. Allen, *The Study of Biology*, 3rd ed., Addison-Wesley, 1977.]

either *T* or *t*, and so does the female (♀). Chances are 1 in 4 that the offspring will receive *TT* or *tt*, and 2 in 4 that the new individual will be heterozygous (*Tt*). If *T* is completely dominant, 3 or 4 individuals (those having *TT* and *Tt*) will show the dominant characteristic, and only 1 in 4 will show the recessive character. Similarly if *B* represents brown hair color in rats, and *b* represents an albino condition, the mating of two heterozygotes, both of whom had brown hair, would be expected to produce 25% albino offspring and 75% normal-appearing brown-haired offspring. However, 50% of the total offspring (the heterozygotes) of such matings would be normal-appearing carriers of the albino trait.

The theory and examples given above are derived from observation and statistical interpretation. No special instruments or chemical techniques are required for these observations and their interpretation. Other observations about heredity involve the use of the microscope, biochemical information, and specialized techniques.

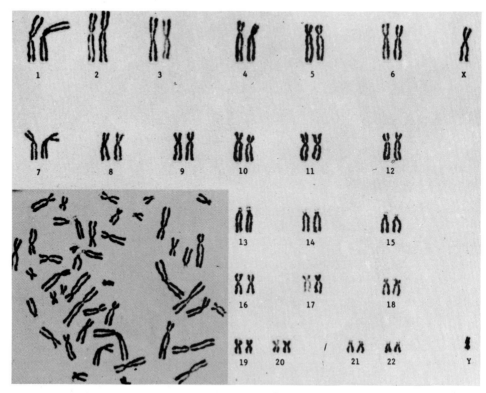

Fig. 39.3 Human chromosomes. When the contents of a human cell are treated in certain procedures and mounted on a microscope slide, we can take a picture such as that shown in the lower left segment shown above. If the individual chromosomes are cut out and arranged in the special form shown in the photograph, we have what is known as a karyotype. The karyotype given here is that of a normal human male. It contains one X and one Y chromosome. Females have two X chromosomes and no Y chromosome in each cell. The cells of some people contain more or less than 46 (23 pairs) of chromosomes. In most cases, these abnormalities in chromosome number have serious consequences. Persons with Down's syndrome, who are commonly called Mongoloids, have three chromosomes 21. [Photograph courtesy of James German, M.D.]

**39.2
GENETICS
FROM A
CELLULAR
LEVEL OF
OBSERVATION**

Information about the cellular mechanism of heredity is based on microscopic observation of the behavior of individual cells. During part of their life cycle, cells are seen to contain long threadlike bodies called chromosomes in their nuclei (see Fig. 39.3). These chromosomes are normally present in pairs. During normal cell division, each member of the pair of chromosomes splits to form two chromosomes, one of which goes to each daughter cell. In this form of cell division, which is called **mitosis** and represents an asexual form of reproduction of cells, the parent and offspring have exactly the same kind of chromosomes. Since the chromosomes contain the hereditary material—the units a geneticist would call genes—each offspring has exactly the same genes as its parents.

In the process of sexual reproduction, **gametes** (sperm and ova) are formed which contain half the number of chromosomes contained in the normal body cells of the parents. This is accomplished by means of **meiosis** (Fig. 39.4), a type of cell division in which

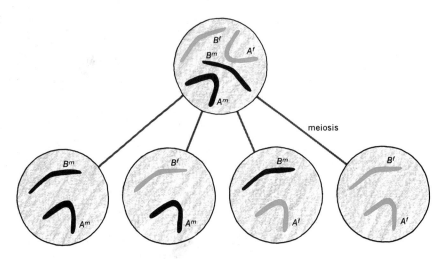

Fig. 39.4 Meiosis. The diagram above represents the random assortment of chromosomes from the mother (A^m and B^m) and the father (A^f and B^f) during meiosis. Note that each of the cells formed contains only half as many chromosomes as the parent cell.

only one member of the pair of chromosomes goes to each developing gamete. Thus, on a cellular level, the observations correlate nicely with the theory derived from observation of individual organisms. The genetic information is contained in the chromosomes. The production of a new individual involves first the separation and then the recombination of chromosomes. We postulate that each chromosome contains the units we call genes. While we cannot see the genes, we picture them as being specific areas of a chromosome.

**39.3
GENETICS
FROM A
BIOCHEMICAL
VIEWPOINT**

The **chromosomes** seen in the microscopic examination of cells are found, upon chemical analysis, to be composed of deoxyribonucleic acid (DNA) and protein. While either the protein or DNA could be the chemical carrier of hereditary information, most evidence supports the fact that DNA is the hereditary substance. Some of the evidence for this conclusion is given below.

1. ·While the amounts of all other constituents of a cell vary widely, the amount of DNA per cell is generally the same in all cells of all animals in any given species.

2. The amount of DNA per cell in different species is different.

3. When DNA from certain bacteria having one characteristic is added to similar organisms lacking the characteristic, the characteristic is acquired by the deficient organism. For example, resistance to certain antibiotics can be transferred from a culture of bacteria that *is* resistant to another culture of the same organism that is *not* resistant merely by adding DNA of the resistant culture to the growth medium of the nonresistant bacteria. This type of transfer of DNA from one organism to another appears to occur only in microorganisms.

4. Viruses, which can be said to alter the heredity of the cells they attack, are known to inject only DNA (in some cases RNA), not protein, into the cell they infect.

When the structure of DNA was shown to be a double-stranded helix which had the capability of replicating itself (see Chapter 33), a mechanism for the duplication of the chromosomes was suggested.

Thus DNA is the basic hereditary substance—the molecules that are transferred from parent to offspring. It is found primarily in the nucleus of the cell. However, the major synthetic work of the cell is done outside the nucleus, in the cytoplasm. This means that the transmission of the information contained in DNA requires a messenger. This function is fulfilled by the kind of RNA called messenger RNA see (Chapter 36). DNA controls the synthesis of messenger RNA to transfer the information to the cytoplasm of the cell concerning types of protein to be synthesized. At the ribosomes where protein synthesis occurs, the message of DNA is actually put into action. The theory assumes that all hereditary characteristics are somehow dependent on the synthesis of a specific kind of protein. Thus DNA has the specifications for synthesis of the proteins of muscle, hemoglobin, the protein hormones, and most significantly, the enzymes. Since biochemical reactons do not proceed at any appreciable rate without enzymatic catalysis, the kind of enzymes present in a cell probably controls the general course of metabolism within that cell. For example, hereditary differences between species with respect to the kind of waste products made from amino acids (which was described in Chapter 36) are dependent not on the presence of the amino acids but on the presence of specific enzymes.

Since the kind and amount of DNA in any cell of an organism is identical with that of every other cell in the organism, we are faced with a major problem in explaining differences in cell function. DNA that has the information for making the enzymes necessary for urea synthesis is in every cell of the human body, yet only liver cells actually carry out the synthesis of urea. If we compare the total information in each cell to a large book of instructions, the question becomes, "How does each cell know just which page of instructions to follow?" Answers to this question are given by the operon theory, which is presented in the article by Changeux in the list of references at the end of this chapter.

We can summarize the biochemical or molecular interpretation of heredity as follows: Males produce sperm cells, which consist largely of a head which contains DNA surrounded by protein and a long tail-like structure called a flagellum. Egg cells contain lipids and proteins, in addition to DNA (see Fig. 39.5). When a sperm cell fertilizes an egg cell, a chromosome from the sperm cell pairs up with a similar chromosome from the egg cell. In order for these chromosomes to form matched pairs, the male and female must have the same kind of chromosomes—that is, be of the same species.

At the moment of fertilization (conception), the zygote, or fertilized egg cell, has all the hereditary information it will ever receive. As the zygote divides to form 2, 4, 8, 16 cells, and finally the millions or trillions of cells found in the new individual, some cells begin to assume different functions in a process appropriately known as differentiation. Differentiation is believed to be the result of directions from repressors and de-repressors which control exactly which molecules will be synthesized of the many thousand possible kinds for which instructions are available in the DNA. As the organism grows and matures, it will produce sperm or egg cells containing only the kinds of DNA (genes) that it received from its parents. Occasionally some agent, perhaps some form of radiation, changes the DNA in a gamete, and this gamete produces a **mutant organism** having different hereditary characteristics. Such **mutations** are rare. The probability of mutations is expressed as mutations per gene per generation. In humans this probability is between 1 in 100 000 and 1 in 1 000 000, with different genes having different rates of mutation. Thus the long-range preservation of a species is assured by the transfer of potentially "immortal" DNA from parent to offspring. Sufficient variety is introduced by

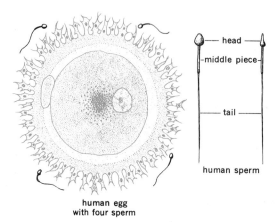

- head -
- middle piece -
- tail -

human sperm

human egg
with four sperm

Fig. 39.5 A human egg with four sperm cells. The drawing shows the relative size of human sperm and egg cells. In the process of fertilization one sperm cell penetrates the egg. [From J. J. W. Baker and Garland E. Allen, *The Study of Biology*, (Third Ed.), Addison-Wesley, 1977.]

formation of hybrid individuals and by mutations. Most mutants are probably less able to survive than are the normal parents, but on extremely rare occasions a mutant is able to do things its parents could not do, and in the struggle for survival the mutant wins and becomes the first of a new kind of individual. A change in environmental conditions may help a mutant which is in competition with unmutated individuals.

The development of our understanding of the process of heredity is a striking example of the way in which science functions. Earlier civilizations could talk about heredity only in terms of individuals, and could offer no real explanation of the processes involved. This general observation was followed by refinement to the cellular and finally the biochemical level of observation and explanation. Our present way of looking at and thinking about heredity has had two major consequences. First, we have, of course, a better understanding of just how the process works. While new knowledge is one of the motives for scientific research, the second major consequence of our progress is that we can predict, and in certain cases control, some of the effects of heredity. Control always brings problems, but the benefits are so great that we must find solutions to these problems. Some examples of biochemical understanding and control of hereditary abnormalities are given in the remainder of this chapter.

**39.4
INBORN
ERRORS OF
METABOLISM**

In 1906, the English physician Garrod wrote a book describing several abnormal patterns of excretion. He recognized that these conditions were inherited, and called them inborn errors of metabolism. We now know hundreds of such conditions. The following sections discuss abnormalities in protein structure and enzyme formation that are known to be hereditary.

**Sickle Cell
Anemia**

More biochemical information is available about sickle cell anemia than about any other inherited condition. The correlations of chemistry, physiology, and genetics are amazing. In fact, studies of sickle cell anemia were responsible for some of our current theories about the chemical nature of heredity.

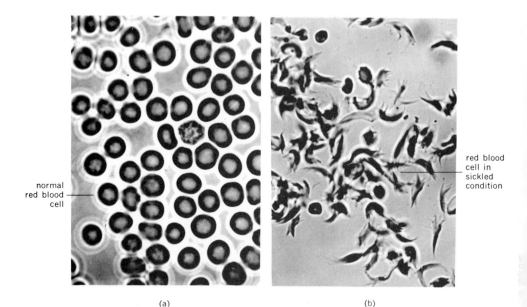

normal red blood cell

red blood cell in sickled condition

(a)   (b)

Fig. 39.6 Normal red blood cells and sickle cells: (a) normal human red blood cells; (b) a similar preparation that was made of the red blood cells of a patient with sickle cell anemia. The sickle-like form of the blood cells in (b) gives the disease its name. [Photograph courtesy of Dr. A. C. Allison.]

If we take blood samples from persons with this form of anemia, we find that the red blood cells, when placed under conditions where the oxygen concentration is low, form peculiar sickle shapes (see Fig. 39.6). We find that some of the red blood cells of certain other, apparently healthy persons will also sickle when similarly treated. If we study the hemoglobin from these persons we find that they have two different kinds. When the mixture is analyzed by electrophoresis at pH 6.9, normal hemoglobin migrates in one direction and the abnormal hemoglobin—called sickle cell hemoglobin—migrates in the other direction. Persons with sickle cell anemia, a serious condition, have only sickle cell hemoglobin (abbreviated Hgb S), and those who have two kinds of hemoglobin have both Hgb S and normal hemoblobin (abbreviated Hgb A). These latter persons are characterized as having the sickle cell trait.

The tendency to form sickle cell hemoglobin is hereditary. Those having sickle cell anemia are homozygous, and those with the sickle cell trait are heterozygous. Thus we have a hereditary condition that is directly related to the kind of protein that an individual makes.

Studies of Hgb A and Hgb S show that they have different solubilities in weak salt solutions (similar to the salt solution of the blood). When carrying oxygen, both forms of hemoglobin are about equally soluble. However, when the oxygen is removed (a process that occurs as blood cells pass from the lungs to other parts of the body) Hgb S is only 1/50 as soluble as Hgb A. Apparently this decreased solubility is the cause of the formation of the "sickles." Hgb S actually crystallizes out of solution, and the crystals then either stretch or break the membrane of the red blood cells. Such sickled cells cannot perform the function of normal red blood cells, and the person with sickle cell anemia must either manufacture red blood cells at a faster rate or suffer from the effects of a deficiency in red blood cells. The sickled cells also tend to block the small blood vessels, the capillaries, and this failure of blood circulation is the apparent cause of the pain these people suffer. Thus sickle cell anemia is a disease caused by a difference in solubility of an abnormal protein.

The next questions are: "Why is Hgb S less soluble than Hgb A, and why do the two forms migrate differently in an electrical field?" The answer to both questions is the same, and is found in the chemical structure of the protein hemoglobin. This protein contains two kinds of subunits, with two of each kind combined to give the hemoglobin molecule (Fig. 31.6). In one of the chains there is a glutamic acid residue in Hgb A and a valine in Hgb S. Glutamic acid has the side chain $-CH_2-CH_2-COO^-$ and valine has $-CH-(CH_3)_2$. This difference might seem to be minor, since both have three carbons in the side chain. The effect, however, is due to the charge on the carboxyl group of glutamic acid which makes that group much more soluble in water than the hydrocarbon-like valine. The fact that one has a negative charge and the other no charge also explains the difference in electrophoretic mobility. Thus the difference on only one amino acid in a chain of about 150 amino acids can result in a life of weakness, pain, intermittent illness, and the constant threat of death from those infections that a normal person handles easily.

Why do some people make Hgb S and others only Hgb A? Since this condition is a hereditary trait, we assume that one person produces DNA that has the pattern for making messenger RNA which makes normal hemoglobin (Hgb A), and that the other has DNA that leads to the production of Hgb S. How different are the "codes" for glutamic acid and valine? From Table 36.1 we find that the codes for glutamic acid are GAA and GAG, and two of the possible codes for valine are GUA and GUG. The difference between pairs is only in one nucleotide. It is logical to assume that a single mutation may be responsible for the change. Why has this mutation persisted? The answer is also fascinating.

Since persons with sickle cell anemia often do not live to an age where they can have children, why hasn't the condition died out? Why hasn't the evolutionary principle of survival of the fittest eliminated persons with the gene for sickle cell anemia? The answer is related to the disease malaria. While persons with sickle cell anemia (homozygotes) die young and are eliminated, those with the sickle cell trait (heterozygotes with both Hgb S and Hgb A) have an increased resistance to malaria. The parasite that causes malaria lives in red blood cells, and apparently it can't find as good a place to live and reproduce if its host's red blood cells contain half Hgb S. The gene for Hgb S is found almost totally among blacks. Approximately 8% of American blacks have some Hgb S. The occurrence of this abnormal hemoglobin is as high as 45% in some African tribes—which may lead us to ask whether we can really call this an "abnormal" hemoglobin.

Thus the story of sickle cell anemia and the sickle cell trait relates physics and chemistry (the solubility and electrophoretic behavior of Hgb S) to physiology (anemia and decreased supply of oxygen to the tissues) to genetics where it shows how recessive traits may be present but not obvious. This line of reasoning supports the theory that genes influence protein synthesis. The sickle cell anemia story further shows how a normally deleterious condition can persist in a given environment, and we have a picture of natural selection (evolution) in action. The difference in incidence of the sickle cell trait in Africa and America indicates that, in the absence of malaria, the trait is less common.

In no other instance are all aspects of a condition so well known and the interrelations so well established. Unfortunately, in the case of sickle cell anemia, understanding does not lead to curing the condition. Some medications seem to increase the survival rate of persons with sickle cell anemia, but in the long run, the only hope seems to lie in preventing the birth of children doomed to lives of suffering and to early death from sickle cell anemia. Control will require widespread examinations of the hemoglobin of prospective parents followed by genetic counselling of individuals who have this trait, so that they will seriously consider the consequences of bringing afflicted children into the world. One of the aims of science in general, and biochemistry in particular, is to reach a similar understanding of all diseases.

Hereditary Abnormalities in Phenylalanine Metabolism

The metabolism of the amino acid phenylalanine is summarized in Fig. 39.7. This summary indicates that phenylalanine from proteins in the diet may be metabolized in a variety of ways. It may be resynthesized into protein; it may be transformed into the thyroid hormone thyroxine, into the adrenal hormone epinephrine (adrenalin), or into the basic pigment of human hair and skin, melanin; it may also be oxidized completely into carbon dioxide and water with the release of an appreciable amount of energy. Tyrosine, which can be derived from phenylalanine, follows the same general pathways

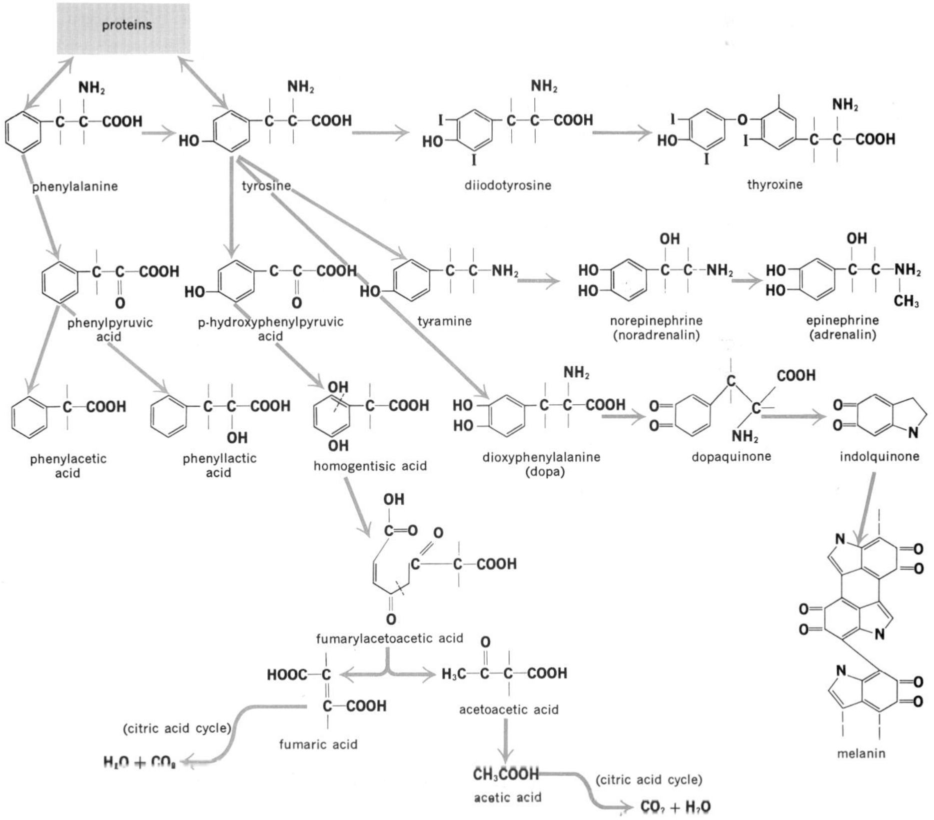

Fig. 39.7 The metabolism of phenylalanine. The series of reactions shown above represents some of the possible products and intermediates that can be formed from the amino acid phenylalanine. The consequences of failure to produce some of these reactions are discussed in the accompanying text.

except that it cannot be converted to phenylalanine. In terms of nutrition, phenylalanine is an essential amino acid; tyrosine is not.

The understanding of the metabolism of phenylalanine assumed a new importance when it was observed that a certain type of mental defectives—e.g., severely retarded children—excreted an unusual substance in their urine. This compound was identified as phenylpyruvic acid. It was found that about 1% of the retarded persons in institutions excreted this abnormal metabolite. The questions immediately suggested are: "Does this excretion have anything to do with the retardation?" and "If so, can an understanding of the condition lead to a cure?" Further observation and study indicated that the persons who excreted phenylpyruvic acid (they are called phenylketonurics) often came from marriages of cousins. This type of observation is a strong indication that the disease may be associated with the inheritance of recessive genes.

We now know that the condition is inherited and that the reason phenylketonurics excrete this abnormal compound is that they cannot convert phenylalanine to tyrosine. When this reaction is blocked, the phenylalanine which is not used for protein synthesis has only one pathway left—the conversion to phenylpyruvic, phenyllactic, and phenylacetic acids. Why is the conversion of phenylalanine to tyrosine blocked? It is blocked because the phenylketonuric does not make the necessary enzyme for conversion. In genetic terms, the DNA of the phenylketonuric lacks the information required for synthesis of phenylalanine oxidase. Whether it dictates the making of a similar protein that is just different enough to lack catalytic activity is not known, although our theory suggests that it does.

Since most parents of phenylketonurics have essentially normal intelligence and do not excrete phenylpyruvic acid, they must be heterozygotes having one gene (or type of DNA) that has the information for synthesis of phenylalanine oxidase and one that doesn't. The next question that presents itself is: "Can we find these hybrid carriers of the defect?" Previously, geneticists believed that the only way a recessive characteristic could be detected in a heterozygote was by letting the individual produce offspring and examining them for the trait. This is a tragic procedure when human lives are involved. Reasoning that perhaps the heterozygous carriers of the recessive gene for phenylketonuria (abbreviated as PKU) might have less than normal amounts of phenylalanine oxidase, research scientists tested parents of phenylketonuric children. While the test has not been entirely successful, it does show that these heterozygous carriers excrete larger amounts of phenylpyruvic acid when given a large dose of phenylalanine than do normal persons. Thus the understanding of the metabolism of phenylalanine and the biochemistry of a hereditary defect has led to the development of a test for a hidden recessive characteristic. Results of such tests can be used for counselling individuals related to phenylketonurics as to the possibility of their having children with this disorder. If both a husband and wife are given such tests the probability of having children with this genetic defect can be predicted. Procedures of this sort offer great promise in improving the genetic qualities of humans. Control would also pose a great threat in the hands of unscientific or inhumane persons.

The prediction of the possibility of having phenylketonuric children may be an interesting development, but what about the baby born with the disease? Is this infant doomed to develop no further than the level of severe retardation? Fortunately, the understanding of the biochemistry of the defect has led to an effective treatment—not to prevent the defect, but to help children who have the disease.

One treatment that might be proposed would be to give the phenylketonuric child some of the enzyme that is lacking. Since enzymes are digested into their component amino acids in the gastrointestinal tract, administration would presumably have to be made by injection. This technique works for the administration of insulin to diabetics, but would pose two more problems in the case of phenylketonuria. The first would be that of getting the large enzyme molecule from the blood into the cells where it must work. The second problem is even more formidable—the human body generally responds to the injection of a foreign protein by forming antibodies to the protein. These antibodies neutralize any effect of the protein.

Fortunately, an alternative procedure can be used and, in fact, is used successfully. This treatment consists of restricting the amount of phenylalanine in the diet of the phenylketonuric. It has been found that dietary restriction of this sort allows the infant with PKU to develop normally and that after the age of five or six, perhaps even earlier, the diet may be discontinued without bad effects. Thus the condition, while it can never be cured, can be treated.

The problem of phenylketonuria now becomes one of discovering the babies who are potential phenylketonurics. Most states now require that the urine, or better yet, the blood, of all newborn babies be tested for the presence of phenylpyruvic acid or phenylalanine. The blood tests require only a small sample of blood and are probably best, since the concentration of phenylalanine in the blood increases some time before urinary excretion of phenylpyruvic acid is readily detectable. Thus treatment can be initiated at the earliest possible moment. There is now no reason for children with this defect to grow up facing life mentally retarded.

The question, "Why does an excess of phenylalanine or phenylpyruvic acid in the blood cause mental defects?" has not yet been definitely answered. It is probable that either phenylpyruvic acid or phenylalanine interferes with the synthesis of serotonin—a compound involved in brain function.

Phenylketonuria is not the only hereditary defect related to the metabolism of phenylalanine. Albino individuals lack the ability to synthesize melanin. Apparently their systems lack the information for making one of the enzymes for the conversion of tyrosine to melanin. Since this condition represents the lack of a metabolic product, melanin, and not an excess of a metabolite (as in PKU), there is no known treatment for albinism. If tests for detection of normal carriers of the recessive gene are developed, control of the incidence, if not treatment, might become possible. The consequences of albinism, while serious, are not nearly so tragic as those of PKU.

Lack of an enzyme responsible for oxidizing homogentisic acid leads to the excretion of this product in the urine. This condition is hereditary and presents another inborn error in the metabolism of phenylalanine and tyrosine. The condition is usually not serious, and no treatment is given. Another abnormality, the excretion of p-hydroxyphenylpyruvic acid, is traced to a blockage in the oxidation of tyrosine. The incidence of this condition apparently is rare, and it has few, if any, bad consequences.

The metabolism of phenylalanine was chosen to illustrate inborn errors of metabolism. Although errors are known in the metabolism of other substances, more have been found in the metabolism of phenylalanine than in that of other substances. This fact may mean that there really are more such errors in phenylalanine metabolism, but it probably means only that the abnormal metabolic products of phenylalanine tend to be obvious because of their distinctive odors or color.

Other Genetic Diseases Diseases associated with the metabolism of carbohydrates (Section 35.12) and lipids (Section 37.8) have already been discussed. Many other diseases, such as sickle cell anemia and PKU, which are due to defects in protein and amino acid metabolism are known. These diseases are easy to describe biochemically, since they are caused by failure of only one gene and since we have been able to identify the protein molecule (hemoglobin or enzyme) that is abnormal. Cystic fibrosis is also definitely a genetic disease, but the exact biochemical defect is not yet known. Some other diseases such as *diabetes mellitus* seem to have a hereditary component, but the exact pattern of heredity and development of the diseases are not understood. It seems that they may involve more than one gene and may also require certain environmental (physiological) conditions to be expressed.

**39.5
THE ONE-GENE–
ONE-ENZYME
THEORY**

The principles of hereditary control of enzyme formation which are so well illustrated by the metabolism of phenylalanine and its abnormalities were first formulated on the basis of work with the red bread mold, *Neurospora*. From their studies with this organism, Doctors Beadle and Tatum formulated the *one-gene–one-enzyme theory*, which should perhaps now be characterized as the *one-gene–one-protein theory*. A good summary of this theory, which has led to a much better understanding of heredity, was given by Dr. E. L. Tatum in his Nobel Prize Lecture of 1958. He stated that:

1. All biochemical processes in all organisms are under genic control.
2. These overall biochemical processes are resolvable into a series of individual, stepwise reactions.
3. Each single reaction is controlled in a primary fashion by a single gene, or, in other terms, in every case a 1:1 correspondence of gene and biochemical reaction exists, such that:
4. Mutation of a single gene results only in an alteration in the ability of the cell to carry out a single primary chemical reaction. . . . As has been repeatedly stated, the underlying hypothesis, which in a number of cases has been supported by direct experimental evidence, is that each gene controls the production, function, and specificity of a particular enzyme.*

**39.6
GENETIC
ENGINEERING**

The understanding of any chemical process allows us to control it. Since we now understand a great deal about the biochemistry of hereditary processes, we are now beginning to control these processes in various ways that are commonly described as **genetic engineering**. Many people are seriously concerned about the dangers and ethical problems associated with genetic control.

In some respects, those who grow plants and raise animals have long been genetic engineers, since they have selected the best seeds to be planted or the best animal to breed and reproduce. Today these processes have been extended so that, for example, we not only produce hydrid corn but have deliberately sought out varieties of corn that produce more lysine (an amino acid normally deficient in corn protein) than normal. We then cross these "high-lysine" plants with others that have other desirable characteristics. The "green revolution" which has contributed so much to improving the world's food supply was based on the development of new varieties of plants, particularly rice and wheat. In some cases we now try to alter the genes of plants by inducing mutations

Other experiments demand much more knowledge of chemistry in that they attempt to alter genes by altering the DNA of a plant or animal—or microorganism. The most

---

* E. L. Tatum, "A case history in biological research," *Science* **129**: 1711–1715 (26 June 1959), copyright by the American Association for the Advancement of Science.

exciting experiment in this area at the time of writing is one in which the bacterium *E. coli* has had its DNA altered so that it produces the human form of the protein insulin. The development was very complex and involved a great number of experiments but can be summarized relatively simply. The DNA code for insulin synthesis was determined and this DNA was made synthetically. It was then incorporated into the DNA in a plasmid from *E. coli*, along with other genetic factors that would stimulate production of a large amount of insulin. In order to keep the insulin that was produced from being destroyed by proteases in the bacterial cells (since it was a foreign protein to them) the code for insulin was patched onto that for the normally produced enzyme β-galactosidase.

Although we can still obtain insulin from the readily available pancreases of cattle, sheep, and swine, and this insulin is effective in the human, the process of inducing a bacterium to make a protein will no doubt find other applications. Human growth hormone is a protein that is highly specific—humans do not respond to the growth hormone of cattle or other species. It now seems that bacteria can also be induced to produce human growth hormone. If the hormone becomes available, it will represent a great benefit for those people who are dwarfs because of a lack of growth hormone. Other proteins or peptides that are difficult to synthesize *in vitro* may also be made by bacteria in the future.

It should be emphasized that the techniques used in inducing *E. coli* to produce insulin did not exist even a few years ago. The development of new methods of chromatography and electrophoresis, as well as new techniques of synthesis, coupled with the operon theory, suggested lines of investigation that might be pursued. Seeing the fruits of such developments is one of the joys of being a scientist.

While it is exciting to contemplate using fast-growing bacteria to synthesize an important biochemical for us, many scientists have been worried about another possibility. What if a harmless (nonpathogenic) bacterium that is used in genetic experiments acquires traits that make it capable of causing widespread disease and death? We know that many bacteria have acquired immunity to common antibiotics, and in fact, many of these resistant strains have been used in genetic research. At one point, the possibility of producing a pathogenic bacterium or virus led scientists to impose a moratorium on genetic research until safeguards were developed. These include isolation of altered bacteria and their destruction by sterilization at the end of experiments. Also, the bacteria used are often "double mutants," that is, strains of bacteria that are unable to synthesize two different growth factors. While these can be grown in a laboratory where such growth factors are supplied, it would require a double mutation for them to be able to live outside the laboratory. Other safeguards are also used.

These are some of the concerns about genetic engineering that arise when we deal with bacteria. When we talk about genetic experiments conducted with humans, ethical concerns also arise. The knowledge of genetic processes has enabled physicians to fertilize a human ovum with sperm *in vitro* (commonly referred to as test-tube conception). The fertilized egg has then been implanted in the uterus and a normal baby produced. The possibility of producing abnormal children by this process, as well as the question whether we should "interfere" in a natural process such as conception, make this procedure controversial. Even more controversial are the consequences of amniocentesis.

Amniocentesis involves removing a small amount of the amniotic fluid which surrounds a developing child (fetus) in the mother's uterus. This fluid contains a few cells from the fetus which will grow and multiply in cell culture media until there are enough cells to give reliable tests. The first amniocentesis procedures routinely involved counting the number of chromosomes in each cell. By doing this, conditions such as Down's syndrome, which results from an extra chromosome number 21 (see Fig. 39.3), and other chromosomal abnormalities could be detected. We can now also test the cultured fetal cells for enzyme abnormalities. Thus it is possible to detect genetic disease before the birth of a defective child. The major problem involves the course to be followed after making a diagnosis of a genetic disease in the fetus. For many, an abortion is the natural consequence; others reject abortion for moral reasons. At the present time amniocentesis is not performed on women who disapprove of abortion, and the final choice concerning action to be taken following amniocentesis is always left to the woman carrying the defective child.

Less controversial genetic engineering involves finding ways to introduce a missing enzyme into a person with a genetic defect. The problem involves getting the enzyme—a protein molecule—into many cells so they may function normally. The most promising line of research involves incorporating the DNA having the instructions for the missing enzyme into a virus. This virus could then be used to "infect" cells with the missing DNA and let the cells make the enzyme from the DNA pattern. Experiments recently reported indicate that a gene has been successfully transferred from the cells of monkeys to rabbit cells using the virus SV-40.

Other kinds of genetic engineering (on a cellular or tissue level) involve transplanting cells having the ability to make a missing substance into a deficient animal. Such transplants involving tissues that make the thyroid hormone thyroxine and the hormone insulin have been attempted for many years. The use of drugs to control the rejection of foreign tissue has made such experiments more likely to succeed.

SUMMARY

The scientific explanation of hereditary processes has progressed from a whole organism level, through a cellular and subcellular (chromosome) level to a physical and chemical level. Most of the genetic processes can now be explained in terms of the chemicals DNA, RNA, and proteins. In some instances it has been possible to show that a genetic difference is due to a difference in only one amino acid in a protein chain. The one-gene–one-enzyme (protein) theory has proved to be very important. In some instances, however, the chemical or physical explanation for a genetic difference is not known.

The control of genetics by physical and chemical techniques, known as genetic engineering, is just beginning. Such experiments raise questions that are not only scientific but also moral, ethical, and social. The problems raised, such as "To what extent should we control genetics?" and "Who initiates and supervises controls?" are formidable and are beyond the scope of this book. They cannot and must not be discussed without a basic understanding of the biochemical, as well as the cellular and organismic, basis of heredity.

IMPORTANT TERMS

amniocentesis	genetic engineering	meiosis
chromosome	heterozygote	mitosis
gamete	homozygote	mutation
gene	inborn error of metabolism	ribosome
genetic code		

WORK EXERCISES

1. List the evidence which supports the conclusion that DNA is the hereditary material.
2. Explain what is meant by the one-gene–one-protein theory. What kind of evidence supports this theory?
3. List all examples of inborn errors of metabolism that have been discussed in this text. List the specific protein missing or altered in each case.
4. What might be the consequences of an inborn error of metabolism that prevented synthesis of thyroxine from tyrosine?
5. Differences in chemical structure in several kinds of hemoglobin are listed in Table 39.1. Check the genetic code as given in Table 36.1 to determine the differences in the code for the normal and abnormal amino acids in these structures.

TABLE 39.1 A summary of amino acid differences in some hemoglobin variants

Type of Hemoglobin	Position Number in B-Chain					
	6	7	26	63	67	121
normal	Glu	Glu	Glu	His	Val	Glu
Hgb S (sickle cell)	Val					
Hgb C	Lys					
Hgb G San José		Gly				
Hgb E			Lys			
Hgb M Saskatoon				Tyr		
Hgb M Milwaukee					Glu	
Hgb D Punjab						Glu·NH_2
Hgb Zurich				Arg		
Hgb O Arabia						Lys

SUGGESTED READING

Beadle, G. W., "The genes of men and molds," *Scientific American*, September 1948, p.30 (Offprint #1). Dr. Beadle discusses some of the experiments that led to the one-gene–one-enzyme theory and the Nobel Prize.

Changeux, J. P., "The control of biochemical reactions," *Scientific American*, April 1965, p. 36 (Offprint #1008). The control of enzyme activity and production by feedback is discussed.

Friedman, T., "Prenatal diagnosis of genetic disease," *Scientific American*, November 1971, p. 34 (Offprint #1234). A discussion of techniques used to determine, even before birth, whether a child will have a genetic defect. The social implications are discussed.

Grobstein, C., "The Recombinant DNA Debate," *Scientific American*, July 1977, p. 22 (Offprint #1362). While the article emphasizes the scientific and ethical problems of working with recombinant DNA, the illustrations show how it is done.

McKusick, V. A., "The royal hemophilia," *ScientificAmerican*, August 1965, p. 88. This article traces an inherited trait back to Queen Victoria of England.

Riesenberg, L. B., "Recombinant DNA—The containment debate," *Chemistry*, 50, No. 10, p. 13, December 1977. Although somewhat out of date in this rapidly developing area, this article gives an excellent introduction to the problems and methods of control involved in genetic engineering.

"Human insulin from bacteria," *Chemistry* 51, No. 10, p. 21, December 1978. This simply written and well illustrated excerpt from the magazine's Research Reporter Section describes this feat of genetic engineering. References to other articles are included also.

40

The Chemistry of Body Fluids

**40.1
INTRODUCTION**
To a simple organism such as an amoeba, many of the basic life processes are merely a matter of osmosis. Although finding something small enough to eat and engulfing it may provide problems, the ingesting of nutrients and eliminating of leftovers involves only the passage of molecules and ions from and into a surrounding watery medium, through a membrane and, perhaps, a cell wall. More complex animals, including humans, with their trillions of cells, still require a liquid medium to provide food and a means of elimination of wastes for each individual cell. Most multicellular plants and animals do not live in watery environments. They do, however, contain fluids which provide not only for nutrition and elimination but also for communication between cells. The internal watery medium has the further advantage of protecting the cells relatively well from environmental changes. But if the puddle that is home for an amoeba dries up or gets too hot, it means either death or, for a lucky individual, formation of a capsule to wall itself off from the hostile environment and survival in an inactive state until better conditions reappear.

Compared with an amoeba in a mud puddle, unicellular organisms that live in the sea have a much more constant environment, not only in terms of temperature and supply of water but also in terms of a ready supply of calcium, sodium, and other ions. Since this is a better environment it is not surprising that we find thousands of kinds of organisms living in the sea. The internal fluids of complex organisms, whether they live in the sea or on the land, provide a sealike environment in terms not only of water, but also of the inorganic ions we call electrolytes.

In order to provide each individual living cell with a fluid which contains the best kind and amount of electrolytes, a fluid from which nutrients may be obtained and into which wastes can be excreted, the organism must have a means of moving the fluid and of removing things from it. Without efficient movement of fluids, each cell would be forced to live in an environment from which other cells had exhausted the food and which they had contaminated with wastes. In most animals, pumps and filters keep the fluid that bathes the cells moving and purified. All complex animals have a major pump we generally

734

call a heart; they also use the action of other muscles to control the flow of blood and other fluids. While the kidneys of mammals are highly developed and selective filters, even the simplest animals have some sort of device for removing some things from the blood while leaving others. The substances removed are excreted as urine. The second major circulating fluid, the lymph, is filtered through lymph nodes, although they do not eliminate fluids from the body as the kidneys do.

This chapter will describe briefly the composition of major body fluids as well as some aspects of the pumping and filtering systems of the human. The implications of the amounts of various chemical substances in the body fluids will also be discussed. We should realize that all body components are being rapidly metabolized and that what we measure in blood and urine represents the difference between catabolic and anabolic processes as well as the ingestion, digestion, and excretion of nutrients. A low level of a substance in the blood usually means failure to ingest, digest, or synthesize the substance or else indicates increased catabolism and excretion of the substance. Low levels may also indicate that a substance is being stored abnormally or in an unusual body compartment.

Blood is like a highway on which the "traffic" changes from time to time throughout the day. We usually measure the amount of substances in the blood only after a person has been fasting for at least 12 hours, to avoid the "peak traffic" effects that occur just after that person has digested a large meal. This is necessary if we are to study basal or "normal" metabolism. However, few people actually fast for this long between their bedtime snack and breakfast toast and coffee, and what is really the "normal" state involves levels of compounds that are much above the fasting level. We study both fasting and "normal" aspects in tests such as the glucose tolerance test, in which we first measure the fasting level of blood glucose, then give a "meal" of glucose and measure the glucose concentrations in the blood at various times after ingestion of the glucose. Sometimes we test the function of the kidneys or liver by giving a test substance and seeing how rapidly it is metabolized. Other tests, such as those for the amount of an enzyme present in blood, are less dependent on previous food consumption.

In a text such as this we will not attempt to provide a complete reference for all chemical analyses done on blood and urine but will describe only the most common ones. We have chosen to concentrate on those analyses that involve chemical processes. There are many tests, such as counting the number and describing the types of white blood cells, which involve little chemistry, and we will not discuss those in detail. The tests we will describe, however, are ones that a person should know about and understand, whether that person is a patient or a person working in a health-related science. We will discuss how the tests reflect normal or abnormal metabolism and will indicate what interpretations are likely when abnormal levels of certain substances are detected.

| 40.2 |
| THE BLOOD |
| AND ITS |
| CIRCULATION |

Since the blood carries the products of digestion and metabolism of food, we can find out a great deal about metabolism in the body by studying the compounds found in the blood. Table 40.1 lists some of the substances for which blood is commonly analyzed, along with the normal concentration ranges of these substances. Since people vary in all characteristics we can measure, it is more useful to give the normal range than to list single values. Sometimes more than one method is used to analyze for a substance and it may be necessary to specify the method used when citing a value.

The general circulation of the blood and its relationship to other body fluids is shown in Fig. 40.1. Blood leaves the heart via an artery. Arteries branch into smaller vessels called arterioles and finally into very tiny, thin-walled vessels called capillaries. Here

TABLE 40.1   Normal constituents of blood (values obtained may vary with method of analysis)

I. Cells (formed elements)

Red blood cells

Males	$4.6–6.2 \times 10^6/mm^3$
Females	$4.2–5.4 \times 10^6/mm^3$
Children*	$4.5–5.1 \times 10^6/mm^3$
White blood cells	$5–10 \times 10^3/mm^3$
Thrombocytes (platelets)	$1.5–3.5 \times 10^5/mm^3$

Hematocrit

Males	40–54 ml/100 ml
Females	37–47 ml/100 ml
Newborn	49–54 ml/100 ml
Children*	35–49 ml/100 ml

II. Soluble substances in the serum or plasma

Determination	Normal Value	Substance Analyzed
Bilirubin (direct)	0.1–0.4 mg/100 ml	serum
Calcium	4.5–5.5 meq/liter	serum
Carbon dioxide	24–29 meq/liter	serum
Chloride	96–106 meq/liter	serum
Cholesterol, total	150–250 mg/100 ml	serum
Copper		
Males	70–140 μg/100 ml	serum
Females	85–155 μg/100 ml	serum
Fibrinogen	200–400 mg/100 ml	plasma
Glucose (fasting) (true value)	60–100 mg/100 ml	blood
Glucose (fasting) (Folin method)	80–120 mg/100 ml	blood
Glucose (fasting)	70–115 mg/100 ml	plasma
Hemoglobin		
Males	14–18 g/100 ml	blood
Females	12–16 g/100 ml	blood
Iodine, protein bound (PBI)	3.5–8.0 μg/100 ml	serum
Iron	75–175 μg/100 ml	serum
Magnesium	1.5–2.5 meq/liter	serum
Nitrogen, nonprotein (NPN)	15–35 mg/100 ml	serum
Oxygen		
Arterial	15–23 ml/100 ml	blood
Venous	10–16 ml/100 ml	blood
pH	7.35–7.45	blood (arterial)
Phosphates, inorganic	3.0–4.5 mg/100 ml	serum
Potassium	2.5–5.0 meq/liter	serum
Proteins, total†	6.0–8.0 g/100 ml	serum
Albumin	3.5–5.5 g/100 ml	serum
Globulin	2.5–3.5 g/100 ml	serum
Sodium	136–145 meq/liter	serum
Sulfate	0.5–1.5 meq/liter	serum
Triacylglycerols	< 165 mg/100 ml	serum
Urea	20–30 mg/100 ml	serum
Urea nitrogen (BUN)	8–20 mg/100 ml	blood
Uric acid		
Male	2.5–7.0 **mg**/100 **ml**	serum
Female	1.5–6.0 mg/100 ml	serum
Vitamin A	2.5–8.0 mg/100 ml	serum
Vitamin B$_{12}$	1.5–6.0 mg/100 ml	serum
Vitamin C (fasting)	0.1–1.5 mg/100 ml	plasma

* Varies with age

† Values are by salt precipitation. Electrophoresis gives different values.

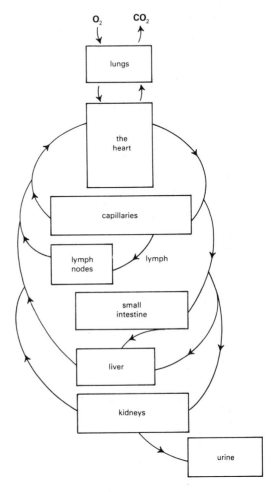

Fig. 40.1 The circulation of blood and other fluids. This diagram merely indicates relationships; it does not show anatomical detail. Vessels leaving the heart (shown on the right) are arteries and those returning to the heart (on the left) are veins. Several alternative pathways for blood circulation are shown. Although capillaries are indicated in only one instance, circulation of blood through all organs (lungs, small intestine, liver, kidneys, etc.) involves capillaries in those organs. The heart is the pump driving the blood; the capillaries of the lungs, small intestines, liver, and kidneys are filters, which add some substances to the blood as well as removing substances from it.

water and many other substances diffuse out of the capillaries into the space around the cells of the body. As the blood continues its flow, water and other substances are drawn back through capillary walls into the blood, which is collected into venules and veins and finally returned to the heart. Some fluids that left the capillaries are not returned to other capillaries but are collected in the lymphatic system (see Fig. 37.1) which eventually returns them to the blood. There are many alternative paths that blood may follow in its circulation. It may go to the lungs and pick up oxygen and lose carbon dioxide; it may go to the intestine and pick up digested food, then pass through the liver on its way back to the heart. Another pathway involves passing through the kidneys, where it is filtered and the urine is formed. Other pathways lead to the brain, the muscles of the arms or legs, even the big toe. In all cases the sequence of artery $\rightarrow$ capillary $\rightarrow$ vein is found, except that sometimes we have an extra vessel such as the vein going from the intestine to the liver before the vein from the liver goes back to the heart.

We use the suffix -*emia* to indicate "in the blood." Thus hypoglycemia means less (*hypo*) of a carbohydrate (*glyco*) in the blood. **Anemia** technically means without blood, but actually it indicates a decrease in the color (which is due to hemoglobin) in the blood. Sometimes we go back to Latin names for the elements and **hypokalemia** and **hypernatremia** indicate low potassium (kalium) and high sodium (natrium) levels in the blood, respectively.

As will be made clear in the following discussions, concentrations of substances in the blood often reflect the health or disease of a major metabolic organ such as the liver. Although many kidney defects are shown by analysis of the urine, others are detected by an increase in the concentration of a metabolite in the blood due to the failure of the kidney to remove that substance from the blood.

Males normally have about 85 ml of blood for each kilogram of body weight; thus an 80-kg (176-lb) man would have about 6 800 cm^3 or 6.8 liters of blood. Women normally have slightly less blood (about 70 ml/kg). The blood normally makes up about 7–8% of the body weight.

40.3
THE CELLS OF
THE BLOOD

If we allow blood to stand, it soon clots to leave a clear liquid known as **serum** and a semisolid clot which contains several kinds of cells trapped in a network of the protein fibrin. The serum may then be analyzed for the presence of various substances. For other analyses of blood, a substance (an anticoagulant) is added to prevent the blood from clotting. This sample is then placed in a special kind of test tube and spun in a centrifuge. After a period of time we find that the various blood cells (sometimes called the formed elements) have settled to the bottom of the tube. Normally the proportion of total blood volume that is accounted for by the blood cells (a value called the **hematocrit**) is 45% for males and 41% for females. The remaining 55–59% of the blood is called the **plasma**. The plasma contains many substances in solution. Its analysis is described in Section 40.4. Although we can analyze either serum or plasma, serum is often preferred since it has no anticoagulants in it.

The major substances that settle out of blood when it is centrifuged are of three types, the red blood cells (RBC's), which are also known as erythrocytes, the white blood cells (WBC's) or leucocytes, and the thrombocytes or platelets. Normal values for the number of cellular elements per volume of blood are given in Table 40.1.

Red Blood Cells
and Hemoglobulin

One of the major components of the red blood cells, or erythrocytes, is the protein hemoglobin. This protein, which is a deep red color, contains heme groups (Fig. 40.2)

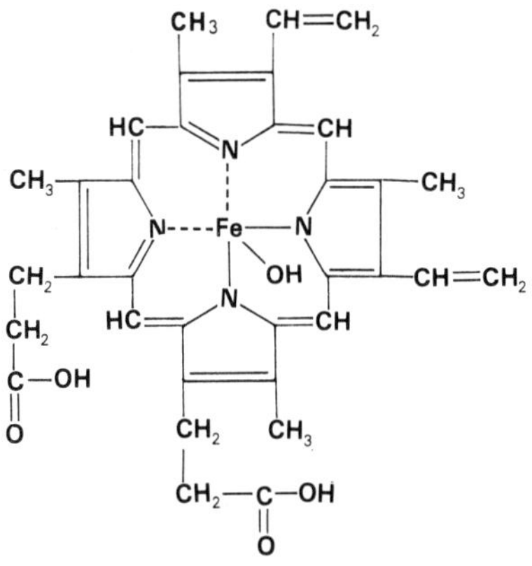

heme

Fig. 40.2 The structure of heme. This substance, containing an iron atom at the center of a complex cyclic structure, is attached to proteins in hemoglobin. The structure may be thought of as being composed of four five-membered, nitrogen-containing rings attached with carbon bridges. When heme is metabolized, one carbon bond is broken and the four rings are present in a chainlike structure.

as well as four protein chains (Fig. 31.6). Hemoglobin combines reversibly with oxygen, picking it up in the lungs and releasing it at other locations in the body where the oxygen concentration is low. We use the symbol Hb for hemoglobin and write equations for the union of oxygen and hemoglobin as follows:

$$Hb + O_2 \rightarrow HbO_2$$

The oxygen is loosely bound in this reaction, which is called an oxygenation. There is no oxidation involved; that is, the iron stays in the Fe^{2+} state, and no electrons pass from or to the oxygen. In other reactions hemoglobin is oxidized to form methemoglobin, with the iron in the Fe^{3+} state. This form of hemoglobin cannot carry oxygen.

Hemoglobin also carries carbon dioxide, picking it up in tissues where the concentration of CO_2 is high and releasing it in the lungs. When *carbon monoxide* binds to hemoglobin, it prevents the binding of oxygen, and when sufficient amounts of hemoglobin are converted to carbomonoxyhemoglobin, death occurs. (See Section 14.18 and Table 14.7.)

The normal amounts of oxygen and carbon dioxide in the blood are shown in Table 40.1. As we would expect, arterial blood (that drawn from arteries) contains more oxygen than venous blood. A low value of blood oxygen indicates failure of the blood to be oxygenated in the lungs, which is probably due to impaired function of the lungs (e.g., by pneumonia) or decreased circulation of the blood to the lungs (e.g., heart disease).

High levels of CO_2 in the blood indicate conditions similar to those of low O_2. Of course, the amount of oxygen carried by blood depends on the amount of hemoglobin per 100 ml of blood. Low levels of hemoglobin, which are found in the condition known as anemia, will give low blood oxygen. Many people, particularly women, have hemoglobin levels below normal. This is due to the slight loss of blood in menstruation, combined with failure to eat foods that provide iron and protein (amino acids) for the synthesis of hemoglobin.

We now know that there are many forms of hemoglobin which do not function in the same way as normal hemoglobin. Sickle cell hemoglobin and its effects were discussed in Section 39.4. At least 100 other kinds of abnormal hemoglobins are known, in which there is a substitution of one or another amino acid for that found in normal hemoglobin. Most of these produce less serious effects than sickle cell hemoglobin. Abnormal hemoglobins are detected by electrophoresis of the hemoglobin fractions from blood.

The red blood cells (which contain no nucleus) have a relatively short life, about 120 days. Normally humans destroy (and produce) about 90 mg of hemoglobin per kilogram of body weight per day. When a red blood cell dies, the iron from the hemoglobin is recycled within the body and the proteins are degraded and metabolized as amino acids. With the exception of the iron, the heme portion of the molecule (Fig. 40.2) is discarded. The liver metabolizes the heme (minus the iron) to a variety of colored compounds, with **bilirubin** probably being the most important. These colored compounds are secreted into the bile and urine, where they provide much of the yellow-orange color of both fluids. In cases where the turnover of RBC's is very rapid or when liver function is impaired (as in forms of hepatitis), bilirubin accumulates in the blood. In extreme cases, the body tissues acquire a yellow-orange color due to the bilirubin and similar compounds in the blood, and we say that the person has **jaundice**. The color may be seen especially in the white of the eye, where skin pigments do not obscure the color of the bilirubin. Feces also owe much of their color to the pigments in the bile, as well as to the hemoglobin in meats which were eaten. A fecal specimen (stool) that is whitish or clay-colored often indicates obstruction of the bile duct. Such conditions are usually accompanied by jaundice. When iron-containing mineral supplements are ingested, stool specimens may be very dark or almost black. A similar color, which is due to the metabolic products of hemoglobin, is seen when there is excessive bleeding in the upper gastrointestinal tract. The feces of vegetarians is usually lighter than that of people who eat meat.

White Blood Cells

There are several types of leucocytes or white blood cells (WBC's), and not only the number but the types of white blood cells are important in interpreting blood analysis. The major function of white blood cells is to act against bacteria and other foreign substances. Thus an increase in the WBC count usually indicates an infection such as appendicitis. Very high levels of WBC's are found in leukemia, while low levels are associated with inability to fight infections.

Thrombocytes and Blood Clotting

The **thrombocytes**, which are also known as blood platelets, are required for blood clotting, since their alteration on being exposed to the air or a rough surface starts the clotting process. The process is a complex one, involving at least 12 different factors. In the final phase of clot formation, long thin protein molecules of fibrinogen are converted into the netlike complex of the protein fibrin. Calcium ions also are required for blood clotting, and most substances used as anticoagulants when blood is drawn for analysis remove calcium ions from solution to prevent clotting. Hemophilia, the hereditary disease in which blood does not clot properly, is caused by a lack of one or more of the factors necessary for clotting. While there are different types of hemophilia with different factors missing, the cause of most forms of hemophilia seems to be a deficiency in DNA and a consequent deficiency in one or another protein for whose synthesis it codes. Patients with conditions caused by blood clots forming in the coronary artery of the heart or other vital places may be given medications such as warfarin or dicoumarol (Section 38.9) to act as antagonists to vitamin K, which is necessary for the synthesis of a factor required for blood clotting.

40.4	
BLOOD	
PLASMA	

40.4 BLOOD PLASMA

Many blood analyses are performed on blood plasma, the fluid obtained by centrifuging the blood cells out of a blood sample which has been prevented from clotting by addition of an anticoagulant. Plasma contains many substances either as solutes or in colloidal suspension. In some cases, before we can do the analysis we must prepare a protein-free filtrate (PFF) of the plasma. The purpose of this step is to precipitate the albumins, globulins, and other soluble proteins of plasma which could interfere with specific tests. The anticoagulants and the reagents used to precipitate proteins must be carefully selected so they themselves will not interfere with tests to be made on the blood. Obviously, a PFF is not prepared if the plasma is to be analyzed for a protein component. Since blood serum has not had anticoagulants added, it is often analyzed instead of plasma.

The only carbohydrate for which blood analyses are routinely made is glucose. We not only determine the fasting level of glucose but also may use a glucose tolerance test (Section 35.12) to study the metabolism of this important energy source. The fasting glucose level that is considered normal depends on the method used for analysis. Alternative values are given in Table 40.1.

Most lipids in the blood are present as lipoproteins, which are discussed in Section 40.7. Those relatively large globules of lipids, the chylomicrons, are usually present in blood only following digestion of a fatty meal. If they persist, it indicates an abnormality in fat metabolism.

The most abundant biochemical compounds found in blood (plasma) are the proteins, which are discussed in the following section.

40.5 THE PROTEINS OF BLOOD PLASMA

Human blood plasma contains from 6 to 8 grams of protein per milliliter. The normal ranges for the albumin, the globulin, and the fibrinogen content of plasma are given in Table 40.1. Although we formerly separated these three types of proteins by precipitation with salt solutions and occasionally do so today, it is now more common to use electrophoresis for such analyses. In this procedure a sample of blood plasma is placed on a strip of plastic and an electric current is applied (see Fig. 29.1). The different proteins, because they are of different sizes and have different charges, migrate at different rates. The proteins are then stained and the results are analyzed to give an electrophoretogram as shown in Fig. 40.3. On these electrophoretograms we find a single peak representing the protein that migrates the farthest—the albumin fraction. There are several peaks for the globulins, which represent different proteins, lipoproteins, and glycoproteins. The globulins are generally classified using the Greek letters α, β, and γ; the latter (γ-globulin) contains many of the factors used by the body to combat disease. Blood plasma also contains fibrinogen. This protein is absent in blood serum, since in the clotting process it has been converted to fibrin and is not present in the liquid serum. This is the only major difference between blood serum and blood plasma.

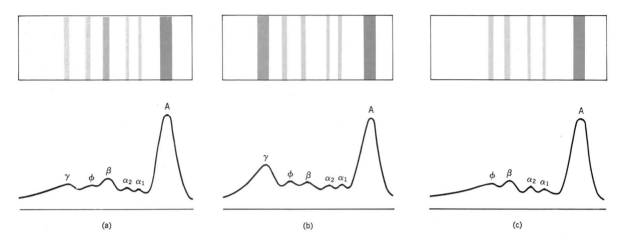

Fig. 40.3 Electrophoretograms of human plasma, with the corresponding stained electrophoresis strips shown above them. The curve of an electrophoretogram is plotted by running the strip through an instrument that is sensitive to the intensity of color developed by staining. The letter A indicates albumin, ϕ indicates fibrinogen, and the various globulins are indicated by the Greek letters $\alpha, \beta,$ and γ. (a) Pattern produced by normal blood plasma. (b) An abnormal pattern, showing increased gamma globulin such as that seen in infectious mononucleosis. Even greater increases in the gamma globulin fraction are seen in multiple myeloma. (c) Pattern seen in agammaglobulin-emia.

One of the major functions of plasma proteins, particularly the albumin, is the regulation of osmotic pressure of the blood. As blood circulates, water and other small molecules pass through the thin capillary walls into the tissues. They are forced out of the capillaries by the relatively high pressure given to the blood by the heartbeat. The difference in osmotic pressure due to the dissolved proteins then promotes the passage of substances, many of them waste products, back into the blood as it enters the capillaries near the veins. When the level of plasma proteins is lower than normal, some substances (primarily water) remain in the tissues, producing the condition known as **edema**. A person with edema has swollen tissues, and if they are poked with a finger, the depression formed will remain for a short time. Although edema also results from kidney abnormalities, it is lack of protein that accounts for the swollen, edematous condition we so often see in starving children.

The blood proteins are synthesized from amino acids of the diet, primarily in the liver. Some of the globulins are also synthesized elsewhere. Albumin in plasma is rapidly catabolized; therefore, if protein synthesis is less than normal, protein deficiency and edema soon appear. Inadequate albumin synthesis may be caused by lack of protein (containing the needed amino acids) in the diet or by a failure of liver tissue to synthesize the protein. Liver diseases such as cirrhosis—a condition due to hepatitis, alcoholism, or poisoning by nonpolar organic solvents—generally lead to decreased protein synthesis. As edema develops, people gain weight even though they are seriously ill—the weight gain is mostly water, of course.

Due to their size (about 610 amino acid residues per molecule and a molecular weight of 69 000 daltons), albumin molecules do not normally pass through capillary walls to any extent. Globulin molecules are even larger. However, surgical shock and some other traumas may suddenly increase capillary permeability, permitting large amounts of albumin to leave the bloodstream. This not only causes an edema but may seriously decrease the blood volume, since water is not drawn back into the blood due to its low osmotic pressure. Administration of either whole blood or plasma may be needed to combat the effects of the shock caused by a decrease in blood volume. Another type of malfunction occurs when capillaries are ruptured by a blow or by bacterial infection. If this happens to the capillaries in the kidney, we may find albumin and even red blood cells in the urine. If capillaries just beneath the skin are ruptured, the hemoglobin and its metabolic products produce a bruise or a "black eye."

Malnutrition and cirrhosis may not only produce edema but may also result in failure to synthesize normal amounts of fibrinogen, so that the blood cannot clot normally. A high level of albumins or fibrinogen in plasma is rare. It usually means that there has been a loss of liquid from the blood leaving a relatively greater concentration of the proteins.

Globulins The proteins in the globulin fraction are a much more complex group of substances than is the albumin fraction of plasma. Along with albumin, the globulins help transport steroid hormones, fatty acids, antibiotics, barbiturates, and various other substances in the blood. The lipoproteins, which migrate with the globulins, are described in Section 40.7.

Some people have a genetic disease in which **gamma globulin** (γ-globulin) is not synthesized. In this condition, known as agammaglobulinemia, victims are unable to synthesize certain types of protein that serve as antibodies to diseases, and they must be protected from infection. The condition can be detected in electrophoretograms of plasma proteins. Occasionally, slight increases in the globulins are seen. For example, as increase in γ-globulins is sometimes seen as a response to immunization or infections. In other conditions, particularly in multiple myeloma or other cancerous conditions, there are very high concentrations of a particular protein in the globulin fraction, which aids in diagnosis of the condition.

Enzymes in Blood Measuring the level of enzymes in blood is a well-established technique for detecting tissue damage. Given the normally low levels of many enzymes in blood, higher levels usually mean that the enzymes have been released from cells due to rupture or death of the cells. For example, high levels of the pancreatic enzyme amylase indicate an inflammation of the pancreas. Measurement of amylase levels in the blood is vital when determining whether the severe abdominal pains of a given patient are due to pancreatitis, in which case surgery is dangerous, or to appendicitis or gall bladder disease, in which case surgery is probably necessary. The transaminases SGOT (serum glutamic oxaloacetic transaminase) and SGPT (serum glutamic pyruvate transaminase) also called alanine transaminase, are released in certain conditions. Both transaminases are increased in acute liver disease (hepatitis). The SGOT in blood increases following certain types of heart attacks; therefore, determining the level of this enzyme in blood provides a sensitive indication not only of whether such an attack has occurred but also of the extent of the damage to heart tissue. Lactic dehydrogenase (LDH) is also released in heart attacks as well as in leukemia, general carcinomatosis, and sometimes in acute hepatitis.

Creatine phosphokinase measurements are used to study disorders of the muscles, including some kinds of muscular dystrophy. Analysis for phosphotases, both those that function best under alkaline and those that function under acid conditions, are also made. Alkaline phosphotase is elevated in many conditions and acid phosphotase is elevated in cancer of the prostate.

Cholinesterase levels in blood may be used to diagnose certain diseases as well as in determining the amount of exposure a person may have had to cholinesterase-inhibiting insecticides such as parathion. In this case, the enzyme levels are lower than normal, instead of being elevated as in most other abnormalities. With increased understanding of genetic metabolic defects (Section 39.4), more analyses for the missing or defective enzymes are being made. None of these is part of a routine blood analysis, but extensive screening for carriers of Tay-Sachs disease (Section 37.8) has been done utilizing a test for a hexosaminidase. Similarly, people suspected of carrying other genetic traits involving enzyme deficiencies can be given tests for those specific traits. Many states require tests for phenylketonuria (PKU) on all newborn babies.

40.6 NONPROTEIN NITROGEN-CONTAINING SUBSTANCES IN PLASMA

In addition to proteins, we find some substances related to the metabolism of the proteins and of the purines and pyrimidines of nucleic acids in blood plasma. Amino acids from digestion of foods are quickly removed from plasma, but substances such as urea, which represents the end product of protein metabolism, are found in blood. We express the measurement of urea as the blood urea nitrogen or BUN. Ammonia resulting from amino acid metabolism is extremely toxic and is converted to urea by the liver. Finding a low level of blood urea or a high level of ammonia, or both, indicates a failure of metabolism. It is often associated with cirrhosis of the liver.

Uric acid is the end product of purine metabolism in the human (Section 36.5). High levels of uric acid in the blood generally mean that some uric acid deposits will also be found in joints and other places, causing the painful diseases known as gout or gouty arthritis. Such high levels of uric acid (uricemia) are probably due either to increased purine synthesis or to decreased excretion by the kidney. The normal blood levels of uric acid are given in Table 40.1. Blood analyses of uric acid are valuable in diagnosing and in following the treatment of diseases involving high levels of uric acid in the blood.

Sometimes we just measure the total **nonprotein nitrogen** (NPN) rather than testing for urea and uric acid. High levels of NPN indicate failure of the kidneys to function normally in the excretion of urea and uric acid, which are the end products of protein and purine metabolism. A high NPN can also mean an excessive catabolism of either protein or purines.

40.7 LIPIDS IN THE BLOOD

Among the lipids found in blood are the triacylglycerols (triglycerides), phospholipids, cholesterol and cholesterol esters, and a small amount of free (not esterified) fatty acids. These lipids, especially the triacylglycerols, are present in the blood in large amounts following digestion of a fatty meal, and it is necessary for people to fast before blood is withdrawn for analysis of blood lipids. Even in fasting individuals, there are appreciable quantities of lipids being transported to and from fat deposits in the body.

Most lipids found in the blood are bound to each other and to proteins as lipoproteins. The lipoproteins, which migrate with the globulins in electrophoresis, have been classified into various categories. (See Table 37.1 and Section 37.7.) It now seems that persons with a high level of the very low-density lipoproteins (VLDL) are more likely to have cardiovascular disease. A greater proportional amount of the high density fraction (HDL) is correlated with a low incidence of cardiovascular disease.

Tables generally list a concentration of 500–600 mg of total lipids and 150–250 mg of cholesterol per 100 ml of serum as normal levels. Authorities disagree on the significance of the cholesterol level in the blood. It is probably more important to know the types of lipids and the way they are bound together than to know total levels in making diagnoses and predictions based on levels of blood lipids.

The analysis of plasma for mineral elements is very important. Those elements that are found primarily in ionic forms (sodium, potassium, calcium, and magnesium) are discussed in Section 40.12. Other mineral elements are discussed below.

While most of the iron of blood is found in hemoglobin, it is also found in serum. The normal values for it and for copper are given in Table 40.1. These metals are not found as ions but are attached to proteins. Sometimes we measure the amount of iron and the iron-binding capacity of a sample of serum or plasma. This analysis reveals how much iron is being mobilized and transported from iron storage in the liver, spleen, bone marrow, and other tissues. Hepatitis (inflammation of the liver) is characterized by high serum iron values, apparently due to tissue damage and a resultant release of iron stores. Copper, important in the synthesis of hemoglobin, is carried by a blood protein called ceruloplasm. Some diseases are associated with abnormal storage of copper. Plasma levels of copper are low in some anemias.

Although they are not done routinely, blood analyses for poisonous metals such as lead and mercury are performed in cases where poisoning due to industrial or other exposure is suspected. Analyses have shown that many children in large cities have relatively high levels of lead in their blood (see Section 13.13). Like iron and copper, these poisonous metals are found bound to proteins in plasma.

We often measure iodine in plasma in the form of protein-bound iodine (PBI). This component represents the thyroid hormones thyroxine and triiodothyronine that are being carried in the blood. Measurements of PBI are used to determine whether the thyroid gland is functioning normally. A high level of PBI indicates hyperthyroidism.

Analysis of plasma for vitamins gives a method for diagnosing vitamin deficiencies that is more sensitive than observing deficiency symptoms. It also allows the detection of low levels of some vitamins before a deficiency is so advanced that symptoms appear. Obviously, detection of a low level of a vitamin in plasma means that the vitamin should be ingested in greater amounts. Determinations of levels of vitamins A, B_{12}, and C (ascorbic acid) are the most commonly made; normal values for these substances are given in Table 40.1.

The human kidney is a remarkable organ. When functioning normally, it first *filters* the blood and then *reabsorbs* certain substances, such as glucose, that have been filtered out but are still useful to the body. Other substances, for example creatinine, are not only filtered out but are also *secreted* by kidney cells into the urine. Thus these substances are removed from circulation in larger amounts than if they were merely filtered out of the blood. In performing these functions, the kidney eliminates waste products and at the same time maintains the optimal concentration of dissolved substances in the blood. It is also the site of many metabolic reactions.

About one liter of blood flows through the kidneys each minute. Thus in a person with a blood volume of 5 liters an amount of blood equal to the total volume of blood in the body is filtered in about 5 minutes. About 25% of the output of the heart goes to the kidneys, and the pressure on the capillaries of the kidneys is higher than that on other capillaries. The anatomic filtering and reabsorbing unit is known as a **nephron**; there are about 10^6 such units in each kidney. A diagram of this basic unit is shown in Fig. 40.4. All together, the nephrons produce about 100 ml of filtrate (from the

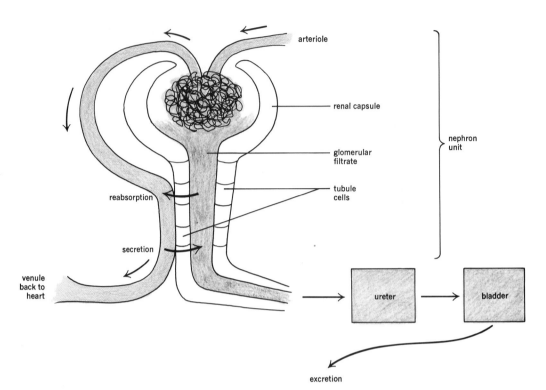

Fig. 40.4 Schematic diagram showing the filtration of the blood and formation of urine by the kidney. Blood enters from an arteriole (upper right) and is filtered in the renal capsule. The filtrate (glomerular filtrate or preurine) proceeds through various tubules where substances are absorbed from it or secreted into it before it leaves the kidney as urine. It then proceeds through the ureters to the bladder. Substances are reabsorbed through the walls of the tubules and returned to the capillaries that surround them. Some substances are not only filtered in the glomerulus but also secreted from the capillaries through the tubule cells into the forming urine. After having reabsorbed and secreted various substances, the capillaries collect the blood into veins and the blood returns to the heart.

one liter of blood filtered) per minute. This glomerular filtrate, sometimes called pre-urine, is reduced to about 1 ml of urine per minute. These values are very general ones and there are many normal variations.

As the blood is forced into the small knot of capillaries known as the glomerulus, small molecules, including water, are forced out of the blood. This fluid (the glomerular filtrate or preurine) contains all of the smaller molecules of the blood, including glucose, amino acids, and other nutrients being transported in the blood. It also contains many ions such as sodium, potassium, calcium, magnesium, chlorine, and the bicarbonate ion, as well as urea and uric acid, which can be regarded as waste products. Soon after the preurine has been forced into the collecting tubule, **reabsorption** of various substances begins. See Table 40.2 for a list of substances reabsorbed and for concentrations of glomerular filtrate and urine. The reabsorbed substances are transported across the walls of the tubule back into the capillary, and from there are returned to veins and to the heart.

The capacity of the tubule cells to reabsorb materials is limited. Normally about 300 mg of glucose can be reabsorbed per minute in the two kidneys of the human. If we assume that 100 ml of glomerular filtrate is formed per minute, it can be seen that when

TABLE 40.2 Comparison of components of blood plasma, glomerular filtrate (preurine), and urine in the human*

Component	Concentration (g/100 ml)		
	Plasma	Glomerular Filtrate	Urine
Urea	0.03	0.03	2.0
Uric acid	0.004	0.004	0.05
Glucose	0.10	0.10	trace
Amino acids	0.05	0.05	trace
Total inorganic ions	0.72	0.72	1.50
Proteins	8.00	trace	trace

Substances that are reabsorbed by the tubule cells of the kidney are water, glucose, amino acids, and inorganic ions. Substances that are secreted by the tubule cells are creatinine, potassium ions, uric acid, and foreign substances such as phenol red, diodrast, penicillin, and *p*-aminohippurate.

* Data adapted from a table by Charles K. Levy, *Biology*, 2nd edition, Addison-Wesley, Reading, Mass., 1978.

the blood contains more than 300 mg of glucose per ml, some of the glucose will not be reabsorbed and will appear in the urine. We say, then, that glucose has a **renal threshold** of 300 mg/100 ml, meaning that whenever the level of glucose in the blood is above that level some glucose will be found in the urine. Similar thresholds exist for other substances, but they are rarely exceeded. As in other characteristics, there is individual variation in renal thresholds—they are really a measure of the capacity for reabsorption and depend on rate of filtration as well as metabolic efficiency of the tubule cells.

If we consider the reabsorption of glucose in an area of the tubule where the concentration of glucose in the capillary is 100 mg/100 ml and the content of the glomerular filtrate has been reduced by reabsorption to 50 mg/100 ml, we can see that the reabsorption is taking a substance (glucose) from a less concentrated solution into one more concentrated. This requires the expenditure of energy, which is provided by ATP. The cells of the kidney tubules actively produce ATP by metabolism of carbohydrates and other substances. Interference with normal metabolism of the kidney cells has serious consequences.

The process of reabsorption is very important to us. If humans eliminated large amounts of glucose, amino acids, and inorganic ions in the urine, they would have to consume and metabolize large amounts of these substances. If no water were reabsorbed, the urine volume would be about 150 liters (37 gallons) per day instead of 1.5 liters. Not only would this increase problems of disposal of wastes, but we would have to consume enormous amounts of water. In fact, if we wanted to maintain the filtration rate in the human kidney without reabsorption we would probably have to live in water. People whose ability to reabsorb water is impaired have the condition known as *diabetes insipidus*. These people may excrete 10–15 liters of urine a day—and must consume this amount of liquid if they are to remain in water balance.

In a process opposite to reabsorption, some substances are not only filtered in the glomerulus, but are actively secreted from the capillaries through the cells of the tubules into the preurine. These represent waste products and include creatinine, uric acid, and hydrogen ions (although the process of elimination of hydrogen ions involves several reactions and is not merely the transfer of ions across the cells). Again, such processes work against a concentration gradient and therefore require ATP or some other energy source.

Many substances foreign to the body are secreted into the urine. The dye phenol red (phenolsulfonphthalein or PSP), diodrast, and penicillin are all secreted by the tubule cells. Phenol red is sometimes injected into the blood or muscle; the speed with which it is excreted into the urine then provides a measure of kidney function. Diodrast, a substance which shows up on x-rays, is sometimes injected so that x-ray pictures of the kidneys can be made. The secretion of penicillin makes it difficult to maintain a level of this antibiotic in the blood that is sufficient to be effective. In fact, when penicillin was first used experimentally, it was so rare and expensive that it was extracted from the urine of people receiving the antibiotic, purified, and reinjected.

40.10 **NORMAL** **CONSTITUENTS** **OF THE URINE**	By the processes of filtration, reabsorption, and secretion, the kidney forms the substance we know as urine. Urine contains more of the end products of metabolism and fewer of the still-usable metabolites than does the blood. Nevertheless, we can still tell a great deal about a person's metabolism and general health by analysis of the urine—urinalysis.

Although we can measure many substances accurately in a small sample of urine, for greater accuracy we often ask a person to collect for analysis all urine voided during a 24-hour period. Most of the values for normal contents of urine are specified in terms of a 24-hour specimen. Some of the values for normal constituents of urine are given in Table 40.3.

Even more than with blood, the compounds found in urine are influenced by diet. A single urinalysis should always be interpreted in terms of food and liquid consumption during the previous several hours. We use the suffix *-uria* to indicate presence in the urine. Thus glucosuria means that there is glucose in the urine. **Polyuria** (also called **diuresis**) means the excretion of a greater than normal volume of urine. Chronic polyuria is one of the symptoms of diabetes.

The volume of the urine varies with water intake and external temperature, as well

TABLE 40.3   Normal values for urine

specific gravity	1.003–1.030
pH	4.7–8.0 (6.0 average)
volume	600–2 500 ml
ammonia	0.3–1.0 g
calcium	0.1–0.2 g
chloride	5.5–10 g
creatinine	1.0–1.8 g
phosphorus	2.0–2.5 g
potassium	1.0–4.0 g
sodium	3.0–6.0 g
sulfur	0.3–1.4 g
urea	25–30 g
uric acid	0.5–0.8 g

Most of the values given above vary greatly with diet. Except for specific gravity and pH, all values are excretions for a 24-hour period.

as with physical and mental states. Aspects of water balance are discussed further in Section 40.11. The specific gravity of the urine depends on water intake. If the amount of solids (dissolved substances) in urine remains constant and water intake increases, a large volume of urine will be excreted and the urine will have a lower specific gravity. The specific gravity (S.G.) of urine increases when water intake decreases or when there is loss of body fluids by vomiting or diarrhea. See Table 40.3 for normal values of the specific gravity of the urine. The color of urine also varies with volume, since the rate of excretion of pigments (primarily urochrome, derived from hemoglobin metabolism) remains about the same. If the volume of urine produced in a given period is large, it may be a very faint yellow color, whereas if the volume is small, the urine is likely to be dark-colored. In some cases of abnormal metabolism of hemoglobin, we can observe green or red urine due to the unusual metabolites. Some medications also produce unusual colors in the urine. Normal urine has little odor, although one develops when microorganisms attack the components of urine. Some foods, such as asparagus and fish, may produce distinctive odors in the urine that are probably due to alkylamines or mercaptans (—SH-containing compounds). It was the odor of phenylpyruvic acid in urine that gave a clue to the cause of phenylketonuria.

Since most of the metabolic reactions of the body produce more acids than bases, most urine specimens are slightly acid. Here again, diet is important. For example, we find that many vegetarians excrete a neutral or alkaline urine. There are various conditions that alter the pH of the urine. High consumption of vitamin C (ascorbic acid) produces an acid urine. The fact that uric acid is more likely to precipitate (and form kidney or bladder "stones") in acid solutions has led many people to be concerned about the effect of high consumption of vitamin C if a person has a high level of uric acid in his or her blood—and consequently in the urine. On the other hand, alkaline urine tends to precipitate salts of calcium, and stones may form from this source.

Since carbohydrates and lipids are composed almost entirely of carbon, hydrogen, and oxygen, the end products of their metabolism are carbon dioxide and water. The carbon dioxide is excreted largely by the lungs, although some appears as bicarbonate ions in the urine. Proteins and nucleic acids, however, also contain nitrogen and some proteins contain sulfur. The end products of protein and of the purine and pyrimidine components of the nucleic acids are found in the urine. The major nitrogen-containing constituent in urine is urea, which normally makes up about half of the dissolved solids in a urine specimen. Urea is the end product of the metabolism of amino acids, pyrimidines, and various other compounds. Present in much smaller amounts, normally, is uric acid, the end product of metabolism of the purines.

High levels of urea generally indicate a high-protein diet, with low levels indicating the opposite. It is possible to find high levels of urea when the body is catabolizing its own proteins, and a low level may mean that some stage in the metabolism of proteins or amino acids is not proceeding normally, as when there is cirrhosis of the liver. Uric acid, which is filtered from the blood and probably also secreted by the kidney, may be high in some people. Examples of this were discussed in the section on NPN in the blood.

We find cations of many metals, such as sodium, potassium, and calcium, in the urine. Ammonium ions are also present. While the metal ions tend to reflect dietary consumption, they may also reflect diseased conditions. Diuretics are used to increase excretion of sodium ions. Ammonium ions are formed from ammonia (NH_3) produced from amides, and hydrogen ions from carbon dioxide (H_2CO_3). The excretion of NH_4^+ is a measure of the extent of production and elimination of hydrogen ions. High excretion indicates a compensation for acidosis.

The most abundant anions of the urine are chloride, phosphate, sulfate, and bicarbonate. Chloride and phosphate ions are largely derived from these substances in the diet, although phosphate may also be derived from the breakdown of some body tissues. Whether the phosphate is present as the monohydrogen phosphate (HPO_4^{2-}) or dihydrogen phosphate ($H_2PO_4^-$) depends on the amount of acid produced in the body. Variations in ratios of the two forms are a measure of acidosis or alkalosis. This is discussed further in Section 40.12, pH control. Sulfate ions are largely derived by metabolism of the sulfur-containing amino acids, cystine and methionine, and the subsequent oxidation of the sulfur to sulfate. The bicarbonate ion is an indicator of the acid-base balance of the body.

Creatinine is one of the most constant components of the urine. It is derived from creatine, which plays an important role in muscle action. In fact, if the amount of creatinine in a 24-hour urine specimen is low, clinical chemists always suspect that the provider of the specimen has not collected all urine voided during 24 hours.

Other substances are found in the urine in much smaller amounts. Normally we do not find appreciable amounts of amino acids, glucose, and protein in the urine, and these substances are considered to be abnormal constituents when present in more than trace quantities. However, very sensitive methods show that the kidney is not totally efficient and there are small amounts of even these substances in normal urine.

40.11 **ABNORMAL** **CONSTITUENTS** **OF THE URINE**	The abnormal component of urine that receives the most attention is glucose. Presence of glucose in urine is one symptom of *diabetes mellitus*. There are, however, other reasons why glucose or other sugars are found in the urine. Five-carbon sugars, the pentoses, may be found after eating large amounts of certain fruits. Galactose and lactose may be found in the urine of pregnant or lactating women. While small amounts of galactose may be found in the urine of normal infants, a high level is indicative of the hereditary disease galactosemia. The use of test strips containing glucose oxidase gives a urine test that is specific for glucose. Other tests give reactions with a variety of sugars. Any detection of glucose or another sugar in the urine requires further study before diagnosing the person as having diabetes or some other disease.

Since proteins do not penetrate capillary walls, the presence of more than traces of albumin in the urine indicates capillary damage, perhaps due to infection. The presence of red blood cells in urine indicates even more serious damage to the capillaries. The level of protein in urine may be slightly elevated following strenuous exercise or a very-high-protein meal. In many cases normal pregnancies are accompanied by some proteinuria. Since urine is stored in a urinary bladder before it is excreted, sometimes the reason for unusual substances being found in the urine is a bladder infection or the rupture of capillaries of the bladder or other parts of the urinary tract.

Bilirubin is present in the urine in jaundice, when it is filtered out of the blood in greater than normal amounts. We also find porphyrins (related to the heme part of hemoglobin) in the urine of some people; such a condition is called porphyuria. Some porphyurias are hereditary, and may be related to insanity or strange behavior. It is suspected that King George III of England, who was king during the American Revolution, may have suffered from porphyuria.

Some abnormal constituents of urine are due to hereditary deficiencies in the reabsorption processes. Sometimes a single metabolite is affected and in other cases, several. Other abnormal conditions—for example, cystinuria, in which not only the amino acid cystine, but also lysine, arginine, and ornithine are excreted in large amounts in the urine—are due to hereditary diseases that affect organs other than the kidney. These amino acids are found in the urine simply because they were present in the blood at extremely high levels. This is also true of the excretion of phenylalanine and phenyl-pyruvic acid in phenylketonuria. We can say that the renal threshold has been exceeded in these cases.

**40.12
WATER,
ELECTROLYTE,
AND pH
BALANCE**

The **electrolyte** (ionic) content and pH as well as the volume of water that bathes the cells of the body must remain relatively constant for life to continue. The human and many other animals have evolved a complex system involving the lungs, blood, and kidneys which controls pH and electrolyte composition as well as volume of body fluids. The major electrolytes of the body are Na^+, K^+, H^+, Ca^{2+}, Mg^{2+}, Cl^-, and HCO_3^-. These are involved in regulation of osmotic pressure and in irritability (ability to respond to a stimulus) of nerve and muscle, as well as in control of pH.

TABLE 40.4   Approximate water intake and excretion

Water Intake		Water Output	
Water, beverages	1200 ml	Urine	1500 ml
Water in food	1100	Lungs	700
Metabolic water	300	Skin (perspiration)	300
		Feces	100
Total	2600		2600

The above are representative intake and output values of water for a 24-hour period for a normal person. The intake of water and beverages and the output of urine are the most variable values.

Water Balance

While we are conscious of the water that we drink and that we eliminate as urine, we are generally much less aware of the water present in foods, the water produced in metabolic reactions, and that lost by perspiration or when breathing. A list of water input and output is given in Table 40.4. Body fluids which are produced and sometimes reabsorbed are listed in Table 40.5. Again, we are generally unconscious of the production of gastric

TABLE 40.5   Volumes and major electrolytes of body fluids

	Volume (24 hours)	Major Electrolytes	pH
Gastric juice	2500 ml	Na^+, K^+, H^+, Cl^-	1.0–1.5
Bile	700–1000	Na^+, K^+, Cl^-, HCO_3^-	slightly alkaline
Intestinal juice	3000	Na^+, K^+, Cl^-, HCO_3^-	slightly alkaline
Pancreatic juice	1000	Na^+, K^+, Cl^-, HCO_3^-	slightly alkaline
Perspiration	500–4000	Na^+, Cl^-	neutral

The volumes are average values for a 24-hour period. The electrolytes listed are those most important in terms of losses of these fluids from the body.

juice and of the digestive juices of the small intestine, but severe diarrhea or vomiting may result not only in the loss of the water but also in the loss of electrolytes and buffers of these fluids.

Although the amount of water ingested varies from day to day and at different times during a single day, the amount of water in the body is remarkably constant. Almost all water ingested is absorbed into the blood (some stays in the feces). The reason that the amount of water in the blood and other fluids, as well as intracellular water, remains so constant is that there are cells in the brain that are extremely sensitive to changes in osmotic pressure. Since the osmotic pressure of any liquid is due to the number of particles (ions or molecules) per volume of liquid and since the amount of substances dissolved in blood is relatively constant, ingestion and absorption of water lowers the osmotic pressure of the blood. Loss of water through the urine, perspiration, or a decrease of water consumption raises the osmotic pressure. When the osmotic pressure changes by as little as 2%, the osmoreceptor cells in the brain influence the pituitary gland (also known as the hypophysis) to change its rate of release of the octapeptide vasopressin. This peptide hormone, sometimes known as the antidiuretic hormone (ADH), is released from the pituitary and is carried by the blood to its site of action, the cells of the collecting tubule in the kidney. High levels of vasopressin (ADH) increase the reabsorption of water by the tubules. A decrease in ADH decreases the reabsorption. Thus high osmotic pressure of the blood (relatively low concentration of water) stimulates the release of ADH and more water is reabsorbed by the kidney. Conversely, low osmotic pressure of blood inhibits the release of ADH and excess water is excreted. High osmotic pressure also increases thirst and normally more water is consumed.

Some water is reabsorbed in the kidney tubules even without ADH, but this hormone is needed for proper kidney function. People whose pituitary function is not normal may fail to produce sufficient amounts of ADH. This results in *diabetes insipidus*. (Insipidus means "without taste"—thus distinguishing the disease from *diabetes mellitus*, in which the urine tastes sweet. Of course, we usually use methods other than taste to detect the latter form of the disease.) People with *diabetes insipidus* may excrete 10–15 liters of water a day. Of course, to compensate for this loss they must drink a similar amount of water or other beverages. The condition can be treated with synthetic vasopressin.

If there is an appreciable loss of water from the body by excessive perspiration, diarrhea, or in other ways and the water is not replaced, serious effects may result. In these cases the osmotic pressure of the blood and extracellular fluids is high, and there is a shift of water from cells into extracellular fluids. This causes severe thirst; it may also cause nausea and vomiting, and the volume of urine is drastically decreased. Without sufficient fluid the kidney does not excrete the products of metabolism, and therefore their level in the blood increases. This, in turn, further increases the osmotic pressure of the blood, and cells are further dehydrated. Unless the loss of water is stopped, more water (and probably electrolytes) are provided, and the flow of urine increases, people in this state lose coordination, and eventually death ensues. In heat prostration, perspiration and high temperatures lower the supply of water so much that there is not enough water for the formation of more perspiration and cooling of the body. In this case the body temperature may rise, and immediate measures to bring it back down are indicated. Severe diarrhea, with its loss of electrolytes as well as of water, produces dehydration and often causes death in infants and children.

The opposite condition, low osmotic pressure due to the increased consumption or decreased elimination of water, also has serious consequences. In this condition (sometimes called water intoxication), water passes into cells, the volume of extracellular fluids decreases, blood pressure falls, circulation is decreased, and the kidneys, whose function depends on adequate blood pressure and blood volume, do not work well. Since the kidneys do not respond by eliminating the excess water, the symptoms are similar to kidney failure due to high osmotic pressure of the blood. If we analyze the blood itself, we find an increase in plasma protein concentration along with lowered amounts of sodium and chloride ions.

Any time we excrete appreciable amounts of glucose, sodium ions, and some other substances, water is also excreted at higher rates. The diuresis (increase in urine volume) of *diabetes mellitus* is due to the fact that glucose is being excreted and taking water with it. The same is true of the excretion of sodium ions. Acidosis, with its increased secretion of some of the more acid ions, also generally causes a diuresis. The increased excretion happens because the greater concentration of solutes increases the osmotic pressure, so that water is drawn out of the kidney cells. Another factor is that many ions are "hydrated"; that is, they carry water of hydration with them.

Electrolytes and Water The effects of water imbalances are largely due to changes in osmotic pressure of the body fluids. Since the osmotic pressure is a function not only of the volume of water present but also of the amount of electrolytes present, changes in osmotic pressure can be caused by changes in electrolyte ingestion or excretion. Glucose, amino acids, urea, and other small molecules diffuse freely across most cell membranes and thus they have relatively little effect on osmotic pressure. The importance of the nondiffusible proteins in determining osmotic pressure has been discussed previously. The inorganic ions—sodium, potassium, calcium, magnesium, chloride, hydrogen phosphates, sulfates, and hydrogen-carbonate (bicarbonate)—can also play a major role in influencing osmotic pressure.

Although the relative amounts of these ions is less than that of the proteins, the osmotic pressure depends on the number and not the size of particles, and thus one mole of NaCl (58.5 g) produces 2 moles of ions (Na^+ and Cl^-). By contrast, it takes 69 000 g of serum albumin to give one mole of particles. We generally express the concentrations of the inorganic electrolytes in body fluids in terms not of moles but of milliequivalents per liter (see Section 11.10). The reason we use equivalents is to enable us to balance positive and negative charges. The weight of an equivalent is the weight that gives a charge of $+1$ or -1. We may convert moles to equivalents by multiplying by the charge on the particle. Since the values we obtain are so small, we multiply by 1 000 and express the values as milliequivalents. Millimoles and milliequivalents are equal for ions having a valence of 1.

We can illustrate the calculations involved in determining the concentration of calcium ions as follows. Suppose we have a solution containing 100 mg of calcium per liter. We know that the weight of a mole of calcium is 40 g and that each calcium ion has a charge of 2. Thus an equivalent of calcium is 40/2 or 20 g. If we divide by the equivalent weight, 20 g, we obtain 100 mg/20 g = 0.005 equivalents or 5 milliequivalents (meq) per liter. Alternatively, 100 mg divided by 40 g gives 0.002 5 moles or 2.5 millimoles of calcium. If we multiply this by the charge on the calcium ion, 2, we again get 5 meq/liter.

The concentrations of the principal electrolytes of extracellular fluid (plasma) and intracellular fluid are given in Table 40.6. From this table we can see that the ions inside the cell differ from those in the extracellular fluid. This is because the cell membranes are not as permeable to some of these small charged ions as they are to uncharged

TABLE 40.6   Average values of major ions of extracellular and intracellular fluids

Extracellular Fluid (Blood Plasma)				Intracellular Fluid			
Anions		Cations		Anions		Cations	
Na^+	142 meq	Cl^-	103 meq	Na^+	35 meq	$\left.\begin{array}{l}Cl^-\\ HPO_4{}^{2-}\end{array}\right\}$	75 meq
K^+	5	$HCO_3{}^-$	27	K^+	110		
Ca^{2+}	5	$HPO_4{}^{2-}$	2	Ca^{2+}	5	$HCO_3{}^-$	10
Mg^{2+}	3	$SO_4{}^{2-}$	6	Mg^{2+}	15	$SO_4{}^{2-}$	15
		Organic acids	6			Proteins	65
		Proteins	16				
Total	155 meq		155 meq		165 meq		165 meq

The electrolytes of extracellular fluid (blood plasma) are relatively constant and can be measured easily; those for intracellular fluids vary widely and are difficult to measure. Values for the intracellular fluid are approximations of average values. Concentrations are expressed in milliequivalents (meq).

glucose and other similar small molecules. Since water molecules pass both ways freely, it is the inorganic ions, especially the sodium and potassium ions, that influence osmotic pressure and control the passage of fluids into and out of cells. Sodium ions are the most abundant *cations* in extracellular fluids and potassium ions are the major *cations* inside the cell. It is these two electrolytes that are most often measured and administered intravenously to control electrolyte imbalance. *Anions* are always present to balance the *cations*, but they seem to pass through cell membranes more readily than sodium and potassium and have less influence on osmotic pressure. When giving electrolytes, chloride is usually used as the anion unless buffering is desired.

Imbalances of sodium and potassium ions are brought about by excessive ingestion, decreased elimination, or abnormal shifts from one place (intracellular or extracellular) to another within the body. About the only common increase in ingestion with serious consequences is found in those who are forced to drink sea water after a shipwreck or similar disaster. Sea water (3.5% NaCl) contains much more salt than human plasma (0.9%). The human kidney cannot excrete urine containing salt in excess of about 1.75% NaCl. Thus body fluids increase in osmotic pressure, cells become dehydrated, and death may follow. People who consume a great deal of salt in their diet generally compensate by consuming and excreting large amounts of water to "wash out" the salt. The compensation is more or less automatic, since salty foods usually make us thirsty. However, some people are concerned that we may overwork the kidneys or raise blood pressure by eating a high-salt diet and recommend (Section 38.11) a decrease in salt consumption.

People who lose a great deal of salt because of excessive perspiration when working hard under hot conditions need to add more NaCl to their diets. In most cases, it is the sodium ion that is the most important in salt consumption. The chloride ion is there to balance the charge on the sodium cation. People also lose electrolytes when they excrete large amounts of fluids. The kidneys have controls so that ions and water are eliminated by different mechanisms, and the electrolyte and water content of urine varies widely. On the other hand, the electrolyte concentrations of perspiration and the gastric and intestinal juices are fixed. We have already discussed the effects of excessive water loss by perspiration. Profuse or prolonged vomiting and diarrhea can produce serious shortages of water and some electrolytes, particularly sodium and potassium ions. The effects are particularly severe in infants and children, whose volume of body fluids is smaller and thus more subject to radical alteration than that of adults.

While details of electrolyte balance and the types and concentrations of electrolytes that should be administered are too complex to be presented in this text, the study of fluid and electrolyte balance is important in diagnosis and treatment of certain types of disease. A striking example is found in the disease cholera. This disease is caused by a microorganism which is transmitted through water, food, or other material contaminated by the feces of victims of the disease. The disease is characterized by severe diarrhea and vomiting. Often dehydration, acidosis, and circulatory collapse follow, and this sequence leads to death in 50% of untreated cases. If people with cholera are promptly given intravenous injections of solutions containing the proper electrolytes to correct the dehydration, acidosis, and hypokalemia, the fatality rate is decreased to less than 1%. However, since cholera epidemics now occur in areas where there are normally few doctors or others trained to administer the correct fluids, cholera still kills a great number of its victims.

In addition to their role in controlling water balance, the inorganic cations are important in controlling nerve and muscle action. High levels of calcium and magnesium ions in body fluids produce anesthesia by increasing the level of stimulus required for a nerve to fire or a muscle to react. On the other hand, low levels of these two ions result in convulsions because nerves fire and muscles contract with only slight stimuli. Although factors other than calcium and magnesium ions are also involved, it is low levels of these two ions (and particularly calcium ions) that are often the precipitating factors in hyperirritability of nerves and muscles.

pH Control   Most metabolic reactions are influenced by the pH. The normal pH of blood is 7.35–7.45, and values outside this range generally reflect serious conditions. If there is an excess of acid, the pH is lowered and a person is said to be in a state of **acidosis**. A high pH is called **alkalosis**. The pH of the body is maintained by the presence of **buffers**, which we define as those substances that resist changes in pH. Buffering substances, often a buffer pair of similar ions, have the ability to either combine with or release hydrogen ions. (See Section 12.12 for a further discussion of buffers.) The most important buffers of the human body are the proteins. It should be realized that buffers are only stopgap devices and that the real solution to problems of acidosis or alkalosis must involve elimination of the excess acid or base from the body. The elimination of the proteins (which require energy to synthesize and which have many functions other than buffering) would be wasteful. Protein excretion would also require different types of membranes than those found in the human.

The electrolytes and the gas CO_2 are small molecules and represent efficient methods of excreting excess acid or basic substances. The main routes of elimination are the air exhaled from the lungs and the urine produced by the kidney. The two major buffer systems involving electrolytes are the carbonic acid–bicarbonate system and the hydrogen phosphates. The components of these buffers are given below, with the more alkaline compound above the line and the more acid one below it.

$$\frac{HCO_3^-}{H_2CO_3(CO_2)} \qquad \frac{HPO_4^{2-}}{H_2PO_4^-}$$

The pH of solutions containing these ions depends on the ratio of the two components.

In each case the upper member of the buffer pair is converted to the lower member by addition of a hydrogen ion. The loss of a hydrogen ion results in the opposite conversion. When there is an excess of acid (acidosis) in the body, there is increased elimination of the more acid of the pair (the one on the bottom). This means that more CO_2 is exhaled by breathing faster (hyperventilation) and that the urine contains more $H_2PO_4^-$ ions when the body is acidotic. The respiratory control center in the brain is sensitive to slight changes in CO_2 and the resulting change in pH of the blood. When that center detects high carbon dioxide levels and acidosis, it stimulates the muscles used in breathing. We take advantage of this system when we administer a brief dose of carbon dioxide to start or stimulate the breathing of a person.

To compensate for alkalosis, a person may hypoventilate to conserve CO_2 and eliminate more HCO_3^- and HPO_4^{2-} in the urine. Another method of eliminating acids involves reactions in the kidney tubules, where amides—particularly of the dicarboxylic acids (such as glutamine)—release NH_3. At the same time the excess of CO_2 produces more H_2CO_3, which produces some H^+ ions, and the net product is NH_4^+ ions, which are then eliminated in the urine.

$$R-\overset{\overset{\displaystyle O}{\|}}{C}-NH_2 + HOH \rightarrow R-\overset{\overset{\displaystyle O}{\|}}{C}-OH + NH_3$$

$$H_2O + CO_2 \rightarrow HCO_3^- + H^+$$

$$NH_3 + H^+ \rightarrow NH_4^+$$

The HCO_3^- and the only slightly ionized organic acid are retained.

Because humans metabolize most foods to give CO_2 and water, the body has to deal with an excess of acids while still maintaining a slightly alkaline blood pH. This is done by excreting an acid urine and by exhaling CO_2. Additional acidosis can be produced by the loss of intestinal juices (which are slightly alkaline) in diarrhea, by *diabetes mellitus*, starvation, overdose of some medications such as aspirin (acetylsalicylic acid), and severe infections. In other instances, acidosis is produced when the normal rate of elimination of CO_2 by the lungs is decreased in emphysema or pneumonia, or by poor circulation of blood to the lungs, as when the heart is not functioning normally. Overdoses of drugs such as narcotics and barbiturates may also decrease the rate of breathing and promote acidosis.

Alkalosis can result from loss of the acid gastric juice by vomiting or from over-consumption of sodium bicarbonate in antacids. Excessive loss of CO_2 through the lungs due to hyperventilation associated with hysterics, crying, or improper use of a respirator can also lead to alkalosis. Mild alkalosis can result from increases in the rate of breathing in atmospheres deficient in oxygen.

Acidoses and alkaloses are classified as **respiratory** when the condition is brought about by abnormalities in elimination of gases by the lung. Other causes of these conditions are classified as **metabolic** acidoses or alkaloses. We diagnose the extent and type of alkalosis and acidosis by measuring not only the pH of the blood but also the CO_2 and HCO_3^- of blood.

Although the body compensates for acidosis and alkalosis, we can often aid the process. We can administer air containing a relatively high level of CO_2 to treat alkalosis. We can also treat this condition by administering solutions of lactic acid or NH_4Cl intravenously. Since ammonium chloride is the salt of a strong acid and a weak base, its hydrolysis yields a strong (highly ionized) acid and a weak (poorly ionized) base. Acidosis may be treated by administering solutions of HCO_3^- intravenously. These latter treatments are made with the assumption that the kidney is functioning and will eventually eliminate the substances administered. If the kidneys are not functioning, we may need to use hemodialysis with an artificial kidney. Any use of hemodialysis requires close attention to electrolyte concentrations as well as acid–base balance.

The only real treatment for acid–base imbalance is to remove the cause—for example, by injecting insulin for the acidosis of *diabetes mellitus,* by administering an antibiotic or other medication for bacterially caused diarrhea, or by calming a hysterical patient who is hyperventilating. Except for the last mentioned, these treatments should of course be instituted only by a professional. The discussion above is intended merely to help the reader understand conditions and treatments. It has not been extensive enough to give sufficient information for adequate diagnosis or prescription of injections for acidosis or alkalosis.

Many diets that promote rapid weight loss involve primarily the loss of water. Some of them accomplish this by producing an acidosis, with the attendant loss of some electrolyte as well as water. Others call for administration of a diuretic, which causes sodium ions to be eliminated. Diets that work this way can be dangerous if continued for any period of time.

Several hormones in the human body promote water retention or excretion—usually coupled with retention or excretion of sodium ions. It is the action of such hormones that causes variation in a woman's weight and water content during a normal menstrual cycle. Hormones that promote water retention are also involved in some kinds of hypertension (high blood pressure). In these cases, the blood volume and pressure are increased along with the retention of sodium ions. Treatment consists of reducing the amount of salt in the diet and administering diuretics, which work on the tubule cells of the kidneys to increase the excretion of sodium ions and water.

40.13
OTHER BODY
FLUIDS
Lymph

The fluid that is filtered out at the arterial end of the capillaries but not returned at the venous end is picked up by the lymphatic system (see Fig. 37.1). This system, which absorbs lipids following digestion, returns fluids to the blood at the thoracic duct. The lymph passes through lymph nodes, which remove some substances as well as adding antibodies to the lymph and, indirectly, to the blood. Because the lymph nodes filter bacteria, sore, swollen lymph nodes are indicative of infections.

Lymph is similar in composition to plasma except that it has a lower content of protein. The composition of lymph from different parts of the body varies; for example, that flowing away from the intestine obviously contains more chylomicrons of lipids and glucose than that from other parts of the body. Analysis of lymph is performed only in special circumstances.

The importance of the lymphatic system becomes clear when the lymph nodes are removed or are blocked by infection. The tropical disease *filariasis,* commonly known as elephantiasis, results in grossly swollen legs. Often the genital organs of the male also swell. The disease is caused by the blockage of lymph nodes by a parasite which is carried by mosquitoes. (See Fig. 40.5.) Similar swelling can follow surgery, particularly that for cancer, in which lymph nodes are removed or the lymphatic circulation is otherwise disturbed. Thus, as in many other situations, we become aware of an important body function only when it fails.

The Cerebrospinal
Fluid

The cerebrospinal fluid (CSF), samples of which may be obtained by doing a spinal tap, is derived from blood plasma by being filtered through various membranes in the brain.

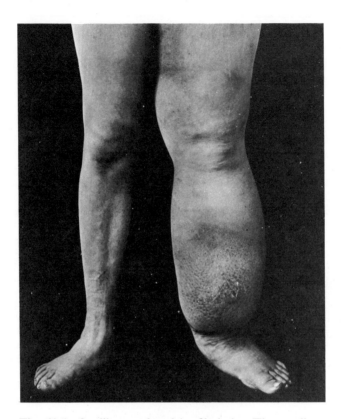

Fig. 40.5 Swelling produced by filariasis. The condition results from blockage of lymph nodes and is sometimes called elephantiasis because of the swelling and the similarity of the skin to that of an elephant.

It contains many of the substances present in the blood, but the lower concentrations of glucose (50–85 mg/100 ml) indicate that some glucose has been absorbed and metabolized in the brain. Also, since it is filtered at least twice (passes through at least two membranes), the CSF normally has a low protein content. The spinal tap is a delicate procedure; therefore, the CSF is not routinely analyzed. However, if brain diseases such as encephalitis or meningitis are suspected, an analysis is usually done. Bacteria or viruses are present in the CSF in meningitis. Other metabolites or electrolytes are altered in encephalitis, in syphilis of the brain, and in poliomyelitis.

Other Fluids That May Be Analyzed   Although they are not nearly so common as the analysis of blood and urine, a few other kinds of analysis are done often enough to deserve mention. The **gastric juice** can be removed from the stomach by passing a tube through the mouth or nose down into the stomach and withdrawing fluid. This technique is not used as frequently today as it was in the past since other procedures are available to measure stomach function. In special cases, semen from the male may be collected and examined. In other cases, fluids which accumulate in the abdominal cavity, around joints, and in other places in the body may be withdrawn and analyzed. And, although it is not a fluid, analysis of hair is also being used for various purposes.

SUMMARY

As animals evolved from creatures that needed a womb of water in a stream or sea, they carried the sea with them in their internal fluids. This made animals mobile and the use of homeostatic mechanisms gave them the ability to control their internal environment. With appropriate external aids, humans are now capable of living in all kinds of environments from the Arctic to the tropics and from deep beneath the sea to outer space. Devices such as the artificial kidney and heart-lung machines have been designed to help maintain homeostasis of body fluids when the normal pumps and filters fail or during surgery. More commonly, we help the body maintain a constant internal environment by administering fluids and electrolytes.

Study of the components of the body fluids, particularly the blood and urine, help us determine how well reactions are proceeding within the body. In many instances, we make these determinations only to confirm a diagnosis made on the basis of symptoms or to decide which of two or more possible diagnoses is the best. In other cases, routine examination of body fluids at a regularly scheduled check-up shows abnormalities before overt symptoms appear—a point at which treatment may be highly effective.

Many of the procedures involved in analysis of body fluids have now been automated; for example, machines can routinely analyze blood samples for as many as 10–12 components. All such analyses, however, must be interpreted on the basis of diet and external conditions as well as possible abnormalities of metabolism. The measurement of enzyme levels as part of blood examinations shows a great deal of promise for further understanding of how the body functions in health and disease. There have been speculations that technicians may soon perform a series of analyses routinely and then use a computer to integrate the results of such analyses. It is doubtful that we are yet ready to eliminate the physician who examines the patient and interprets computer output; however, the number of tests that can be run is becoming so large and the possible relations between different findings so complex that it may be necessary to provide the memory of a computer to assist in diagnosis.

With the availability of extensive data and the ability to interpret such data, it is becoming increasingly important that not only the health professionals but also that everyone become aware of how his or her body functions. The importance of water, electrolyte, and pH balance and the ways of preventing imbalance should be understood by all educated people. If we maintain the proper balance by sensible eating habits, we are likely to place less stress on the body's homeostatic mechanisms and thus prolong the period of healthy life we can expect.

IMPORTANT TERMS

acidosis
alkalosis
anemia
antidiuretic hormone (ADH)
bilirubin
buffer
cerebrospinal fluid (CSF)
diabetes
diabetes insipidus
diuresis

edema
electrolyte
erythrocyte (RBC)
γ-globulin
glomerular filtrate
hematocrit
hemoglobin
hypernatremia
hypokalemia
jaundice

leucocyte (WBC)
lymph
nephron
nonprotein nitrogen (NPN)
plasma
polyuria
renal threshold
serum
thrombocyte

WORK EXERCISES

1. List the major fluids of the human body, giving total volume or daily production and elimination wherever possible.
2. Why should people fast for at least 12 hours before blood is withdrawn for analysis?
3. List three major functions of hemoglobin.
4. List the most important proteins found in blood.
5. What is the difference between blood serum and blood plasma? Why is the former often preferred for blood analyses?
6. List the functions of globulins in the human bloodstream.
7. List and discuss the three major functions of the kidney.
8. Why is a 24-hour specimen of urine often used for analyses?
9. What makes an analyst suspect that a person returning a 24-hour urine specimen has probably not collected all of the urine voided during the past 24 hours?
10. What are some of the harmful effects that may result from urine that is more acid than normal? From urine that is more alkaline than normal?
11. List the important normal constituents of urine.
12. Explain how the blood volume is kept relatively constant.
13. Discuss the cause and treatment of *diabetes insipidus*.
14. Describe the causes and effects of excessive loss of water from the human body.
15. Describe the consequences of a human's drinking only sea water. Consult Chapter 9, Section 6, of this text and indicate how other animal species manage to drink sea water without harmful effects.
16. Why are vomiting and diarrhea so serious in children?
17. List some of the important causes of and ways the human body responds to acidosis; to alkalosis.
18. List the methods of treating acidosis; of treating alkalosis.
19. What is the relationship of body water content and some weight-loss diets?
20. What are the causes and effects of blockage of lymph nodes?

21. List the possible significance of the following results of blood analyses. The number in parentheses indicates the number of answers to be given.
    a) low level of oxygen in arterial blood (3)
    b) high level of CO_2 in arterial blood (3)
    c) high level of bilirubin in blood (2)
    d) high level of white blood cells (2)
    e) low level of white blood cells (1)
    f) low level of thrombocytes (1)
    g) high level of glucose (2)
    h) low level of serum albumin (2)
    i) high level of serum albumin (1)
    j) low level of fibrinogen (1)
    k) high level of globulins (2)
    l) high level of amylase (1)
    m) high level of SGOT (2)
    n) high level of lactic dehydrogenase (LDH) (3)
    o) high level of acid phosphotase (1)
    p) low level of cholinesterase (1)
    q) low level of urea (1)
    r) high level of uric acid (2)
    s) high level of nonprotein nitrogen (NPN) (2)
    t) high level of lead (1)
    u) high level of protein bound iodine (PBI) (1)
    v) high level of magnesium (1)
    w) low level of calcium (1)
22. List the possible significance of the following results of urine analysis. The number in parentheses indicates the number of answers to be given.
    a) a high volume of urine (2)
    b) a urine with a high specific gravity (1)
    c) a urine with a low specific gravity (1)
    d) phenylpyruvic acid in the urine (1)
    e) glucose in the urine (1)
    f) a urine with a pH of 8.0 (1)
    g) a high level of urea (1)
    h) a low level of urea (1)
    i) albumin in the urine (1)
23. What abnormal conditions are suggested by the following? The number in parentheses indicates the number of answers to be given.
    a) a whitish-colored fecal specimen (stool) (1)
    b) a very dark-colored fecal specimen (2)
    c) edema (1)
    d) extremely light-colored urine (1)

SUGGESTED READING

Asimov, I., *The Bloodstream*, Collier MacMillan Publishers, New York, 1961. Although it is not newly written, this book provides an interesting account of the blood and things in it.

Hirschorn, N., and W. B. Greenough III, "Cholera," *Scientific American*, August 1971, p. 15. Although cholera can be treated by replacing body fluids, it continues to kill, particularly in under-developed countries. Treatment and possible antidotes are discussed in this article.

Lott, J. A., "Hydrogen Ions in Blood," *Chemistry* **51**, No. 4, p. 6, May 1978. Many aspects of the subject are covered, including measurement of blood pH, the role of lungs and kidney in H^+ control, blood buffers, and abnormal values of blood pH. A complex subject is well presented in this article.

Maugh, T. H. II, "Hair: A Diagnostic Tool to Complement Blood Serum and Urine," *Science*, **202**, 1271, December 22, 1978. The analysis of hair is now being used to diagnose zinc and iron deficiencies; lead, mercury, and cadmium poisoning; and some diseases such as cystic fibrosis, celiac disease, and phenylketonuria. Even drugs such as barbiturates, morphine, and amphetamines have been found in hair, and reliable tests for these are being studied. Possible links of metal content of hair and learning disabilities are also being investigated.

Pinckney, C., and E. R. Pinckney, *The Encyclopedia of Medical Tests* (pb), Wallaby/Pocketbooks, New York, 1978. This 221-page paperback book is subtitled: "What you should know to prepare for and understand the medical tests you take. The comprehensive, authoritative, straight-forward guide that enables you to know—Why a test has been ordered, When it has been incorrectly administered or diagnosed, and What the results really mean." The book is written at a level understandable to readers of this text and represents a good "consumer's guide" to medical tests.

41

The Biochemistry of
Disease and Therapy

Attempting to discuss, in a single chapter, the biochemistry of disease and therapy (the treatment of disease) is a major task. Whole books are devoted to the subject, and in many cases they discuss only a single disease. Nevertheless, in this chapter we will present some generalizations about diseases and their relationship to biochemistry. We believe that despite the dangers of generalizations the information will give the reader a better understanding of some medical practices and will illustrate some practical applications of biochemical theories.

For the purposes of this discussion we will define **disease** as any departure from health or, in other words, any failure of the body to function normally. From the viewpoint of the biochemist the functioning of the body depends on chemical reactions. A breakdown in these reactions leads to abnormal functioning which we can call disease. To simplify our discussion, we have divided the diseases into several categories. These are arbitrary categories, of course, under which we attempt to describe processes which we do not completely understand. Some diseases may belong in two or more categories and others do not really belong in any of those listed.

The major division is between (1) infectious diseases (those caused by bacteria, viruses, and other organisms) and (2) noninfectious diseases. Noninfectious diseases are further classified as (a) nutritional diseases, (b) mental diseases, (c) diseases of the endocrine glands, (d) drug-induced diseases, and (e) the metabolic or degenerative diseases. Cancer and arthritis will be discussed under the last classification. Cardiovascular disease, which also belongs with the degenerative diseases, was discussed in Chapter 37. Those conditions or diseases known as inborn errors of metabolism were discussed in Chapter 39.

Forms of therapy have two major aims, to relieve the symptoms and to cure the disease. As will be evident from the following discussion, there is too little of the latter and far too much of the former. Our knowledge of the causes and cures of diseases is still so inadequate that often all we can do is to make the patient as comfortable as possible. In many cases this simple treatment is all that is needed because the natural defenses of the body can take care of the diseased condition. In others, treatment of symptoms only prolongs the time a patient must wait before dying from an incurable disease. Unfortunately, also, many drugs used to treat pain are addictive and may themselves induce diseases.

41.2
THE
INFECTIOUS
DISEASES

"At some early stage of evolution, possibly before plants and animals had become clearly differentiated from each other, one or more species must have made the discovery that a very convenient source of food consists of the tissues of other organisms. In such tissues may be found materials, already preformed, suitable as energy sources and for building one's own tissues . . . Among some of the species which had discovered the importance and desirability of eating other organisms as a source of food, the expedient of parasitism developed. After all, it is not economical to destroy the source of your food material completely, although some microorganisms, perversely enough, do this to their hosts. Instead, it is easier to live in or on the host, making use of the food supply which he provides directly or indirectly. This is parasitism of which infection is an example."*

As humans, we have a clear idea of what the parasites are: the bacteria, viruses, and simple animals such as lice. However, from the viewpoint of many plants, humans and indeed all animals are potential parasites. The animals that have survived have developed various forms of defense against their enemies. In addition to weapons and fortifications against large foes, humans have developed several defenses against the microscopic agents of infection.

41.3
DEFENSES
AGAINST
INFECTION

The human body has several natural defenses against infection. The tough, relatively impermeable skin is a major defense. For those invaders that do manage to get around or through this barrier, the **phagocytes**, those cells (such as some kinds of white blood cells) that engulf or "eat" invaders, are potent antagonists. Biochemists are most interested in two other natural defensive agents—the antibodies and some of the hormones. **Antibodies** are produced by the body in response to the invasion of various types of foreign molecules. The human can make antibodies against materials as varied as egg white and viruses. The process of antibody formation is not completely understood, but apparently we have a recognition system that can tell when a substance in our bodies is foreign. If the foreign particle is of sufficient size and proper character, the body manufactures antibodies that are apparently shaped in such a way as to combine with the antigen (invading foreign particle) and negate its effects. The antibodies are primarily protein, although many also contain carbohydrates in their structure. Some of the most important antibodies are present in the blood serum and can be separated in a group called the gamma globulins. Vaccination is effective because it stimulates the production of antibodies. It is assumed that the pooled gamma globulin from many individuals contains antibodies to many specific diseases, and thus it is often given to help a person fight a specific disease.

Some of the steroid hormones produced in the cortex of the adrenal gland also help the body fight infection. These substances enable the body to "wall off" the foreign invaders and thus keep them from spreading throughout the body and doing damage. A large percentage of the American population has had a mild infection of the tubercule

---

* B. S. Walker, W. C. Boyd, and I. Asimov, *Biochemistry and Human Metabolism*, 3rd ed., Williams and Wilkins, Baltimore, 1957.

bacillus, has developed antibodies to it (as shown by skin tests), and may also have "walled off" groups of these bacteria in small nodules in the lungs. Some adrenal cortical hormones help the body produce these inflammatory walls. The adrenal cortex also produces hormones which decrease the reaction of the body to foreign substances by suppressing antibody formation. These hormones, known as **anti-inflammatory corticosteroids,** are effective in diminishing inflammation in many parts of the body. They are widely used in the treatment of arthritis and other diseases. In the early days of corticosteroid therapy with cortisone, some persons who received this medication developed tuberculosis, apparently because the tuberculosis bacteria that had been "contained" were again set free. Physicians prescribing cortisone or other similar substances now take measures to prevent such consequences.

The transplanting of organs such as the heart and kidney from one human to another is made possible by the use of corticosteroids and similar substances that repress the responses of the body to foreign tissue. Ironically, after successfully undergoing a transplant operation many persons have died from pneumonia or other diseases. Their deaths were due, at least in part, to the fact that the drugs given to keep the body from rejecting the transplanted organ also keep the body from rejecting or fighting the bacteria that cause pneumonia and other diseases. In the healthy individual there is a delicate balance of the pro-inflammatory and anti-inflammatory substances.

| 41.4 CHEMO-THERAPEUTIC AGENTS | Against some diseases the natural body defenses can be aided by chemical substances either synthesized in the laboratory or extracted from other organisms. These substances are classified as synthetic **chemotherapeutic agents** and **antibiotics,** respectively. |

**41.4
CHEMO-
THERAPEUTIC
AGENTS**

Against some diseases the natural body defenses can be aided by chemical substances either synthesized in the laboratory or extracted from other organisms. These substances are classified as synthetic **chemotherapeutic agents** and **antibiotics,** respectively. Although antibiotics can now be synthesized in the laboratory, they were originally isolated from microorganisms, particularly the molds. Even in the laboratory, synthesis of antibiotics often relies on cultures of molds to perform certain steps in the synthesis. In most cases, laboratory synthesis is not economical; it is less expensive to grow the molds and let them do the entire job of producing the antibiotic.

A perfect chemotherapeutic agent would be totally effective against a kind of bacteria or other infectious agent and would not harm the host. The action has been characterized as being like a magic bullet that hits and affects only infectious agents. Since the biochemistry of all organisms from bacteria to humans is essentially similar, what is poisonous to one organism is usually poisonous to the other. However, sufficient exceptions to this general rule exist to make chemotherapeutic agents useful.

One of the first synthetic therapeutic agents was the compound called Salvarsan, or Dr. Erlich's compound 606, which was effective against syphilis. (The story of Dr. Erlich is dramatically told in the book *Microbe Hunters* by Paul de Kruif.) Probably the most important synthetic chemotherapeutic agents are sulfanilamide and its derivatives, known as the sulfonamides, or popularly as the sulfa drugs. The effectiveness of these drugs led to the proposal of a basic theory of chemotherapeutics. It was postulated that sulfanilamide was effective against bacteria because it was an antimetabolite to *para*-aminobenzoic acid (PABA). An **antimetabolite** is defined as a compound that is sufficiently similar in chemical structure to a normal metabolite so that it can replace it in an organism. However, the antimetabolite does not function as does the metabolite, a circumstance which often leads to failure in either growth or reproduction and thus to the eventual death of the organism. Bacteria use PABA in the synthesis of folic acid. It was argued that because of similarity in structure between PABA and sulfanilamide, organisms tend to incorporate sulfanilamide and compounds related to it into a molecule that is similar to folic acid but which cannot perform the functions of folic acid.

sulfanilamide   PABA   folic acid

This deficiency would lead to decreased ability to grow and reproduce. The test of a true antimetabolite is that its effects must be reversed by the addition of large amounts of the normal metabolite. Thus bacteria that are destroyed by sulfanilamide or another sulfonamide (actually perhaps just kept from reproducing, with the natural body defenses doing the destroying) should be able to live and reproduce in the presence of the sulfonamide when sufficient folic acid is supplied to them. While many sulfonamide-sensitive bacteria can meet this test, others cannot. It now seems most likely that the sulfa drugs are not true antimetabolites but that they do inhibit a specific step in folic acid synthesis. However, the antimetabolite theory has led to the design of many chemotherapeutic agents specifically designed to be antimetabolites for a chemical substance. The compound aminopterin, which is successful, although only for a limited time, in the treatment of leukemia, is another folic acid antimetabolite.

Why do sulfa drugs inhibit the growth of bacteria but have little effect on humans? The answer may be twofold. First, humans require folic acid as such, and do not synthesize it—it is a vitamin. Thus interruption of its synthesis or synthesis of a sulfonamide-containing folic acid is not a serious matter. Secondly, humans can survive even if deprived of all folic acid for a period of a day or two, but this period of time represents several lifetimes (times between divisions) to bacteria. Whenever the bacteria are prevented from reproducing at a normal rate, the natural defenses of the human body can usually effectively combat the invaders.

Sulfanilamide and some other sulfa drugs are relatively insoluble in acid solutions such as urine. Therefore, use of these chemotherapeutic agents can result in the formation of crystals in the kidneys—an undesirable action. In the past, the drugs were given along with sodium bicarbonate, and the patient was told to drink a great deal of water so that solutions in the urine would be dilute and less acid than normal. As a result of further work, the sulfa drugs that are now being synthesized are more soluble in acid solutions. There are many similar instances in which the structure of a drug has been modified to decrease or eliminate side effects; the problem that usually arises is that modifications in structure can also change the effectiveness of a drug.

A different form of chemotherapy is represented by certain drugs used in treatment of cancer. Here, rather than attacking an invading microorganism, the agent destroys the abnormal cells of the body itself. Some agents are not selective, but destroy both normal and cancerous cells; however, since the body can tolerate the loss of some normal cells, these agents are, on balance, useful weapons against cancer.

41.5
ANTIBIOTICS
The first antibiotic to be widely used was penicillin. Its discovery came about when there was an accidental contamination of a bacterial plate by a particular kind of blue-green mold. The experimenter, Dr. Alexander Fleming, observed that the colony of mold was surrounded by a zone in which no bacteria (a staphylococcus culture) were growing,

although the rest of the plate was covered with bacteria. He realized that the mold was secreting something which was inhibiting the growth of bacteria in the surrounding medium. To overcome the difficulties in extracting, testing, purifying, and producing in quantity the bacterial inhibiting substance required a great deal of research. It was many years before the substance which we call penicillin became available for general use.

Many other workers, notably Dr. Selman Waksman, have found hundreds of antibiotics. The general procedure is to make culture plates of the organism that we wish to control and expose them to extracts from other organisms—particularly the molds. An extensive search has been conducted for rare kinds of molds and yeasts that might produce antibiotics. Although hundreds of antibiotics are known, only a limited number are useful in treating diseases. Most of the antibiotics are toxic not only to the infectious agent but to the humans or other animals being treated. In the case of penicillin and similar antibiotics, their effectiveness is due to the fact that they inhibit the synthesis of materials in the cell walls of the procaryotic bacteria. Human cells are eucaryotic and do not have cell walls—only cell membranes.

Penicillin has the advantage of being both effective against certain infections and the least toxic of the common antibiotics; some individuals however, have acquired a sensitivity to penicillin so that administration causes serious consequences. Apparently what has happened is that these penicillin-sensitive persons have produced antibodies to penicillin itself, since it, too, is a foreign molecule. In cases of this kind the natural defenses of the body are acting in a way which is not good for its general health.

The formulas for some of the more important antibiotics are given in Fig. 41.1.

The widespread use of antibiotics has resulted in the prevalence of antibiotic-resistant microorganisms. Soon after penicillin came into common use, it became obvious that some types of bacteria that had been sensitive to penicillin could no longer be controlled by this antibiotic. Resistant bacteria were found to contain the enzyme penicilinase, which catalyzes the hydrolysis of penicillin and destroys its effectiveness. In fact, some bacteria could metabolize penicillin and use these reactions as a source of energy. Apparently, exposure to penicillin was inhibiting the growth of most bacteria of a certain type, but the few mutants which could grow in the presence of the antibiotic were then reproducing—rapidly. Among many kinds of bacteria, there now exist mutants that are resistant to almost every known antibiotic. In fact, in some cases it seems that bacteria can transfer resistance to an antibiotic to other bacteria by sexual processes. Many types of bacteria that were previously not thought to cause disease (they were nonpathogenic) are now both pathogenic and antibiotic-resistant.

The presence of antibiotic-resistant pathogens in hospitals is a major problem. In fact, people run some risk of acquiring a resistant pathogen just by being hospitalized for simple procedures. Obviously this danger is a great cause for concern among many people associated with the health professions. Although new antibiotics are being developed and it will probably require time for bacteria to become resistant to them, the standards of sanitation in hospitals must be raised, and increased emphasis placed on isolation of patients with resistant infections. Unnecessary prescription and use of antibiotics must be avoided.

Because of the wide use of vaccination and the development of effective chemotherapeutic agents, and perhaps also because of the relatively good state of nutrition of most persons in the United States, deaths caused by infectious diseases have greatly decreased in the past few decades. This is not true of all countries in the world, and infectious diseases still take a large toll, especially of children. Part of the solution to the problems of infectious diseases lies in education and part in the provision of vaccines, chemotherapeutic

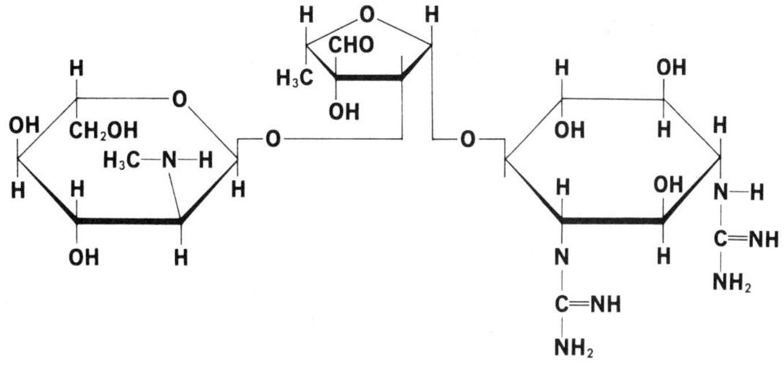

streptomycin

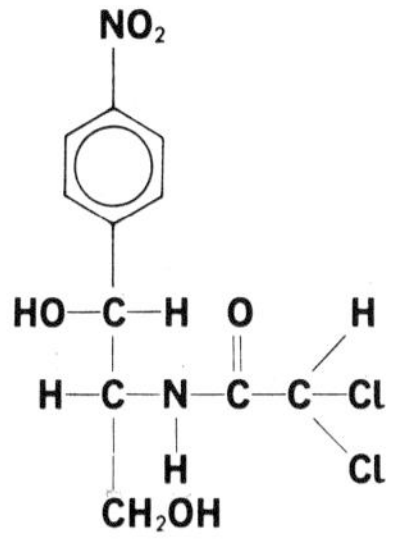

chloramphenicol 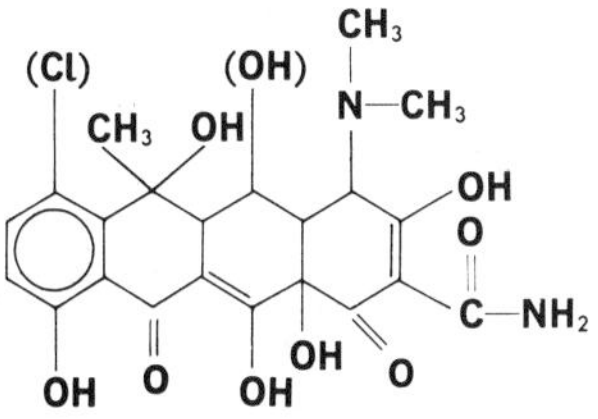   tetracycline 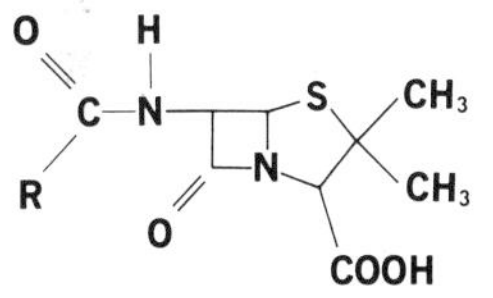   penicillin

Fig. 41.1 The structure of some common antibiotics. The form of penicillin known as penicillin G has a benzyl group at R. Other forms have the *p*-hydroxybenzyl, pentenyl, *n*-amyl, or *n*-heptyl R groups. The tetracycline molecule is known as Achromycin. Oxytetracycline which has an OH as indicated on ring number 3 is known as Terramycin, and chlorotetracycline, which has a chlorine atom on the first ring, is known as Aureomycin. The structure of the antibiotic gramacidin is given in Table 32.1.

agents, and nutritious foods. The role of adequate sanitation in the prevention and spread of infectious diseases is also important.

**41.6
ANALGESICS,
SEDATIVES,
HYPNOTICS,
AND
ANESTHETICS**

When a disease or injury produces pain or anxiety, chemical substances known as medications or drugs are usually given to relieve these symptoms. Unlike chemotherapeutic substances, such medications do not help cure the disease or injury but merely help make it more bearable. An **analgesic** brings about insensitivity to pain, a **sedative** tends to calm or tranquilize a person, a **hypnotic** produces sleep, and an **anesthetic** promotes the loss of sensation. Local anesthetics produce insensitivity in a localized part of the body, whereas general anesthetics produce sleep or general insensitivity to external stimuli. Many substances produce more than one effect, depending on the amount given (dose) and the route of administration (e.g., oral or by injection). In discussing these drugs it is important to understand drug tolerance and dependence.

When medications are given over a prolonged period of time, it is often necessary to increase the dosage to produce a given effect. We then say that the body has become **tolerant** to the action of the drug. In some instances tolerance develops because of the increased synthesis of an enzyme that metabolizes the drug, thus keeping the amount of the drug in the blood at a low level. In other instances the level of the drug in the blood increases but its effect on body cells does not increase proportionally. Doses that are lethal in the nontolerant individual produce only minor effects in tolerant individuals. It is thought that this happens because the drug inhibits an enzyme whose synthesis is controlled by the amount of product formed from the reaction (see Section 34.7). Inhibition of the enzyme results in decreased amounts of the product, which in turn results in the synthesis of more enzyme. Thus a vicious cycle is set in motion. Sometimes both forms of tolerance are found with a single drug.

Many drugs produce a **dependence**; that is, when use of the drug is discontinued, **withdrawal symptoms** are seen. In general these symptoms are the reverse of the effect of the drug. Drugs such as ethanol, barbiturates, and narcotic analgesics depress the central nervous system so that a normal amount of stimulation by pain or other stimulus no longer is effective in producing a response. Withdrawal of these drugs from a person dependent on them produces hyperexcitability, meaning that normal stimulation produces excessive responses. This is consistent with the theory that this kind of drug inhibits an enzyme that controls a reaction, with the result that more enzyme is made. Upon withdrawal of the drug, there is a lot of enzyme present with nothing to inhibit it, and the reaction it catalyzes proceeds at a very rapid rate, perhaps converting almost all of its substrate to product in a very short time. After a period of withdrawal most people return to normal, apparently because enzyme levels also return to normal.

Dependence is generally classified as being **psychological** (**habituation**) or **physical** (**addiction**). Withdrawal of a drug from an addicted person has serious consequences. If the addiction is severe, withdrawal can produce physiological reactions such as vomiting, severe diarrhea, muscle spasms, and increased heart rate and blood pressure, and it can result in death. Withdrawal of a drug from people who are only habituated to it has less violent physiological consequences, but depression or other psychological states can lead to suicide or to reckless actions.

Apirin   Acetylsalicylic acid, which is known as aspirin, is the most familiar of the analgesics. Aspirin not only relieves pain but also decreases fever (it is an antipyretic) and tends to decrease certain types of inflammation. Since aspirin and other salicylates can be obtained without a prescription, they are widely used. However, prolonged use may cause slight habituation and some degree of tolerance. Furthermore, although aspirin is one of the safest analgesics, it can produce stomach irritation and it decreases the clotting time for blood. It should be used with discretion. Indeed, some people classify aspirin as an abused drug. Methyl salicyclate (oil of wintergreen) is used in lotions and salves that decrease inflammation and pain. Aspirin and the other salicylates work by influencing the action of prostaglandins.

biturates   The barbiturates are general depressants derived from barbituric acid, which was synthesized in 1864 by von Baeyer. (The name of the compound derives from the name Barbara, plus urea, which was used in the synthesis of the compound. Whether the Barbara was a waitress in Munich or whether the compound was prepared on St. Barbara's day is a matter of conjecture.) Barbituric acid itself is relatively inactive medically. Its formula and those for some of the derivatives of barbituric acid which we call the barbiturates are given in Fig. 41.2. (See also Section 25.7 and Table 25.1.) While 2 500 barbiturates have been synthesized, only 50 have been used clinically and only 6–10 are commonly used.

The barbiturates depress a variety of functions in the cells of many vital organs. Certain barbiturates at low levels produce sedation. Higher levels produce sleep and still higher levels produce anesthesia. Excessive levels can depress the central nervous system

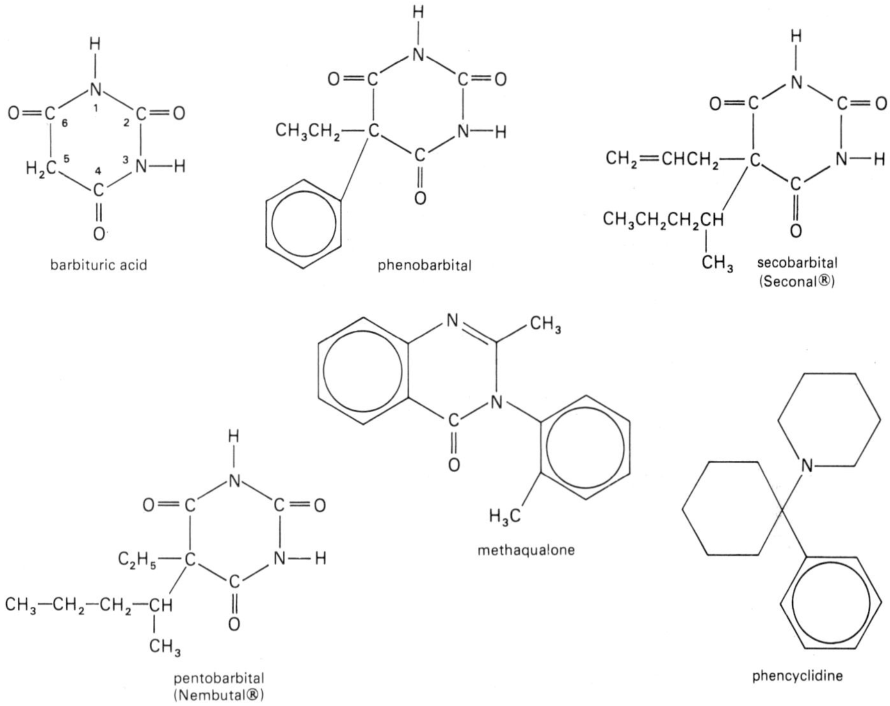

Fig. 41.2 The barbiturates and other sedative and anesthetic drugs. See Table 25.1 for other barbiturate formulas.

and cause coma and death. The barbiturates are obviously dangerous, because a person in a sedated state may forget how many pills have already been taken and continue taking the medication until death results. Barbiturates are often used by those who commit suicide.

The effect of barbiturates depends on the particular barbiturate used, the route of administration, the degree of excitability of the nervous system and the extent of tolerance that has been induced by previous use. The length of the hydrocarbon side chains at position 5 in the barbiturate molecule influences the solubility and perhaps the pK_a of these substances. These physical properties, in turn, influence the rate of absorption, the metabolism, and the length of time the barbiturate action lasts. Many barbiturates have been synthesized in an effort to alter biological properties by changing chemical structure.

People rapidly develop a tolerance to most barbiturates, and their regular use as a hypnotic (sleeping pill) results in dependence. In addition to sedation and sleep some barbiturates also produce a sense of well-being or euphoria, and since they have few unpleasant side effects, they are often abused to the point of addiction. Although barbiturates are legally obtained only with a prescription, many people obtain them illegally. Barbiturates are commonly encased in capsules of distinctive colors which lead to their common or "street" names. For example, doses of sodium amobarbital, which is dispensed in blue capsules, are known as "blues," "blue dolls," "blue devils," and by other similar names. Sodium pentobarbital, also known as Nembutal®, is dispensed in yellow capsules and secobarbital or Seconal® in red capsules. Thus the names "red dolls," "red devils," and "yellows," "nembies," and "yellow jackets" are commonly used to refer to these drugs. Phenobarbital is generally available as white tablets.

Tolerance of barbiturates develops, in part, from the synthesis of enzymes that metabolize the compounds in the microsomes of the liver. Apparently the enzymes promote the oxidation and removal of side chains at position 5. This leaves barbituric acid, which is not effective as a drug. Abrupt withdrawal from barbiturates without treatment results in death 5–7% of the time.

Morphine and Other Narcotic Analgesics Derived From Opium

Morphine, which is derived from the juice of poppy capsules, has been extracted and used since prehistoric times. A solution of opium in ethanol known as laudanum was used by Paracelsus. In 1803, the compound morphine was isolated from opium. Codeine, papaverine, and about 20 other alkaloids have also been extracted from crude opium. The formulas for some of these and related compounds are given in Fig. 41.3. Heroin, which is derived chemically from morphine by acetylating the —OH groups at positions 3 and 6 of morphine, was first synthesized in a search for a nonaddictive alkaloid. Preliminary tests indicated that it was not only more effective than morphine but less addictive. Unfortunately, it was later found to be highly addictive, and its use has been banned in the United States. Heroin's effectiveness is due to the fact that the acetylation results in more rapid uptake by the brain. Methadone (Fig. 41.3), which is a synthetic compound not derived from opium, has properties similar to the opiates. It is effective when taken by mouth and has a long duration of action. It can be used to treat people addicted to heroin; however, it is also addictive. The advantages of replacing a heroin addiction with one for methadone are that methadone is available legally (under controlled conditions) and that withdrawal from it has less drastic effects than heroin withdrawal.

Morphine and the other opium-derived narcotic analgesics depress the central nervous system and relieve intense pain. Perhaps their true mode of action is by altering perception of pain and/or the anxiety produced by pain. These drugs also constrict the pupils of the eye and inhibit the rhythmic action of the intestines. Although the major use of morphine is in treatment of intense pain, it and related compounds are sometimes given to treat diarrhea such as that brought on by food poisoning. Codeine is used in some cough syrups to suppress coughing and as an analgesic for moderate pain. It, like morphine, inhibits intestinal action; thus constipation is an undesired side effect of codeine use. Many of the effects of the opium alkaloids, such as cough suppressing (antitussive) and antidiarrheal actions, are produced by levels of the drugs much lower than those used to relieve pain.

As discussed in Section 25.9, alkaloids are weak bases. They form salts with acids, and these salts are relatively soluble in water. Alkaloids are often administered intravenously in water solutions. Addition of a base to a solution of an alkaloid can precipitate the alkaloid (as the free base) with a decrease in activity. This can happen if a basic drug such as sodium phenobarbital is administered along with an alkaloid.

All narcotic analgesics derived from opium are addictive, and all produce side effects that make them less than ideal drugs. Attempts to synthesize drugs with structures that will have the analgesic effect without the side effects have been unsuccessful. However, nonaddictive analgesics that can be used for severe pain are not available, and therefore morphine and other opium derivatives still have a place in medicine. Meanwhile, the search continues for nonaddictive analgesics that can relieve intense pain. As with other substances that relieve symptoms, it should be realized that treatment with an analgesic does not treat the disease but only the symptoms. Most physicians insist on diagnosing the cause of pain before administering morphine to relieve the pain.

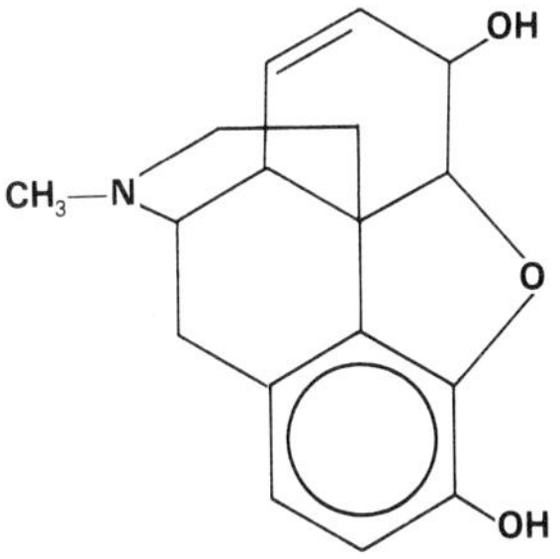

morphine

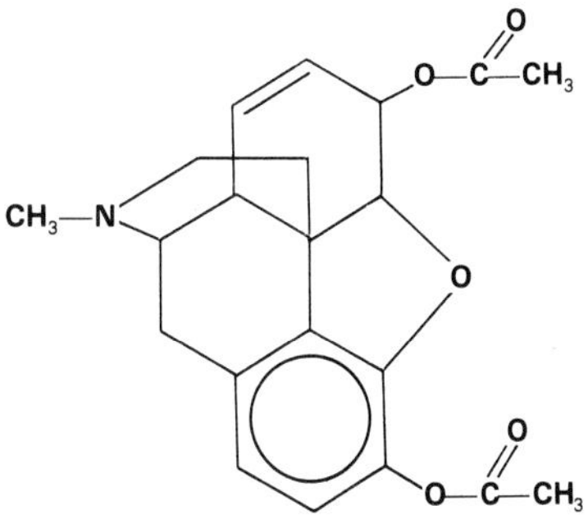

heroin

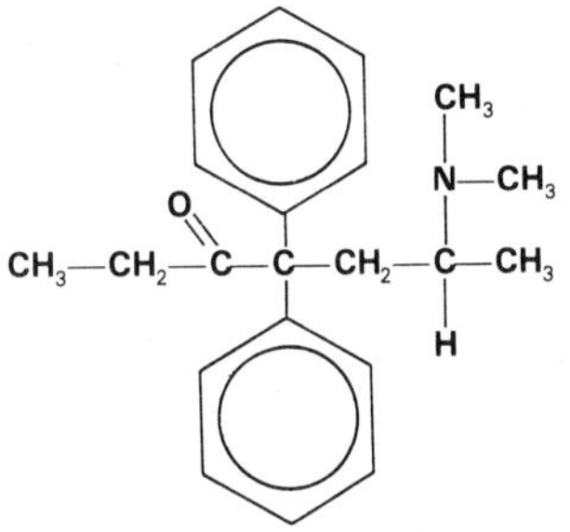

methadone

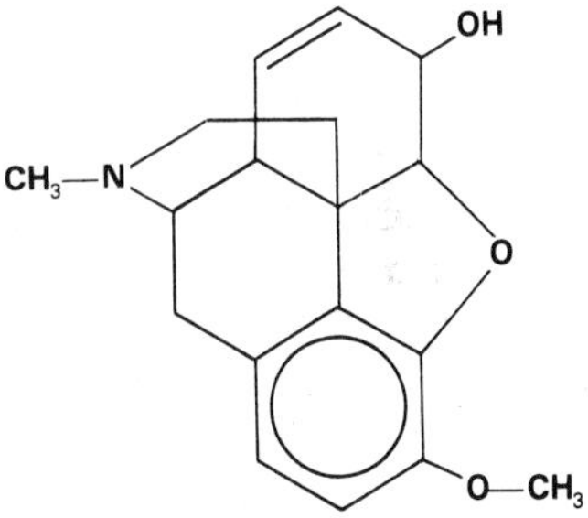

codeine

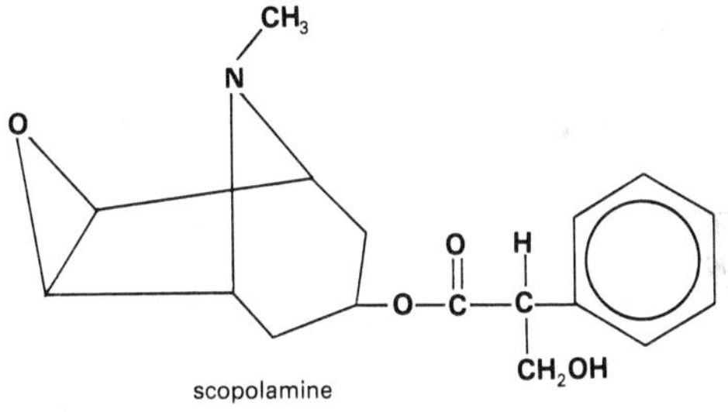

scopolamine

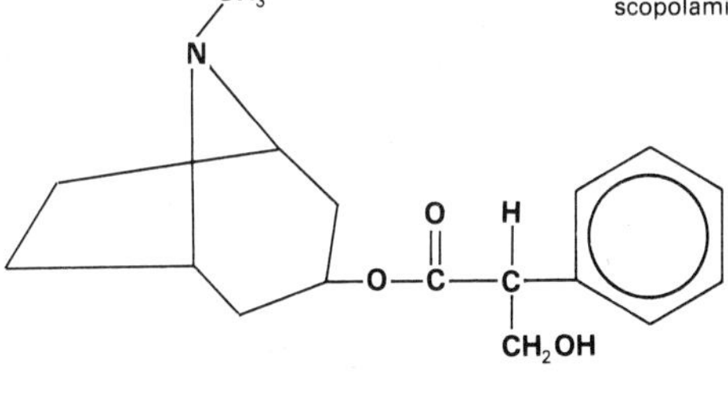

atropine

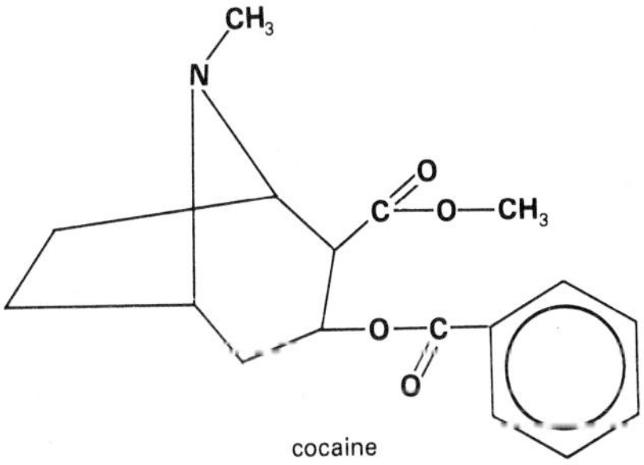

cocaine

Fig. 41.3 Alkaloidal drugs. Morphine, heroin, and codeine are derived from opium. Methadone, which is similar in action, is a synthetic substance. Atropine and scopolamine are derived from the deadly nightshade and cocaine is derived from the coca plant.

Cocaine   Cocaine, which is also an alkaloid, is derived not from the poppy but from the leaves of the coca shrub, which grows in the Andes mountains of South America. The people who live at those high elevations chew or suck the leaves of the coca plant, apparently for the euphoria and indifference to pain it produces. Sigmund Freud was one of the first to study the effects of cocaine. He used it to cure a morphine addiction in one of his colleagues. The colleague then became dependent on cocaine.

Cocaine is used medically as a local anesthetic, especially in eye examinations and procedures requiring local anesthesia of the upper respiratory tract. It was also used for dental procedures at one time, but has been replaced by other anesthetics. Cocaine, administered by injection, was the first agent used to produce local anesthesia by "blocking nerves." An early substitute for cocaine was the synthetic compound procaine, also known as Novocaine®. While procaine is much faster-acting and less toxic than cocaine, other substances such as lidocaine (Xylocaine®) and Carbocaine® are now more commonly used as local anesthetics.

Cocaine is widely used illegally at the present time as a drug to produce altered mental states. The dependence it produces appears to be primarily psychological, not physical. Long-term use can produce tolerance, and a toxic reaction that involves intense paranoia is often seen. Since users of cocaine are often stimulated rather than depressed, delusions of persecution sometimes occur, and the use of weapons on alleged persecutors has been reported. Overdoses of cocaine, particularly if administered by injection, can be fatal. Sniffing cocaine constricts the blood vessels of the nose and can destroy the tissues of the nose and produce perforation of the nasal septum.

Methaqualone   The search for better and/or nonaddictive drugs has produced many substances that origi-
and   nally seemed promising but were later found to have unexpected effects. One of the first
Phencyclidine   of these was heroin, which has been discussed previously. Two other drugs that have now assumed the status of "street drugs" are discussed in the material that follows. We will use the term "street drug" to designate a drug sold illegally and often, but not always, synthesized illegally. Some street drugs are also available by prescription. Drugs such as aspirin that are available without a prescription are commonly known as over-the-counter (OTC) drugs.

Methaqualone (Fig. 41.2) was originally developed to be used as a nonbarbiturate, nonaddictive sleeping pill (hypnotic). Further testing, however, showed that it was addictive and that it was capable of altering mental states. There is strong evidence that it can cause birth defects if taken by a pregnant woman. These drawbacks have not prevented its sale as a street drug under the name of qualudes, sopors, or wallbangers.

Phencyclidine (Fig. 41.2), which is 1-(1-phenylcyclohexyl)piperidine, and is called PCP, was developed and tested as an anesthetic for surgical procedures. It was patented in 1963 under the name Sernyl®. When trials in humans indicated agitation, disorientation, and delirium when those treated emerged from anesthesia, the company marketing the drug requested that human clinical investigation be stopped. It is still used legally as an anesthetic for animals.

Since 1967 the drug has been manufactured illegally and sold as a street drug. It is known as "angel dust" and has other similar names. The street products are usually grossly contaminated, and it is difficult to know which effects reported by those who have used PCP are due to the drug and which to contaminants. The effects include changes in body image, perceptual distortion, and feelings of apathy and estrangement. Reports of difficulty in thinking and concentrating, and preoccupation with death are also common.

Continued use of PCP produces symptoms of psychosis. Death due to overdoses or to actions taken while under the influence of the drug are common.

Atropine and Scopolamine

Atropine and scopolamine are known as the belladonna alkaloids. The name atropine was derived from that of the oldest of the Three Fates of Greek mythology, Atropos—the Fate that cut the threads of life. Belladonna is derived from Italian words for beautiful woman. These names reflect the fact that the belladonna alkaloids, which are derived from the plant known as the deadly nightshade, were used for centuries as poisons and to increase the size of the pupils of the eyes—making a woman more beautiful. Many other plants, such as the jimson weed, also contain these alkaloids. (See Fig. 41.3 for formulas.)

Atropine is still used, in its pure form, to dilate the pupils of eyes in an eye examination. While we classify these drugs in terms of their effects in blocking certain sections of the nervous system, rather than sedatives and tranquilizers, they are often used for the latter purposes. Scopolamine is used to prevent motion sickness and is present in some over-the-counter sleeping pills such as Sominex®. The belladonna alkaloids are also used as preanesthetic medications. In addition, atropine is used to treat poisoning by organophosphate insecticides. Atropine antagonizes the action of acetylcholine, which accumulates when the organophosphate insecticides inhibit cholinesterase.

All belladonna alkaloids are toxic and can cause death. They should be used with care. In some cases their mind-altering effects have led to their abuse. There have been efforts by regulatory agencies to have drug companies remove scopolamine from over-the-counter medications.

41.7 THE NON-INFECTIOUS DISEASES

Diseases in this classification, as far as we know, do not result from the presence of an infectious agent. It is possible, however, that some of them are caused by agents which have not yet been associated with the disease. This is especially true of certain kinds of cancer which may be caused by viruses. While many of the noninfectious diseases are being controlled as effectively as are the infectious diseases, others seem to be increasing. In some cases these diseases are more prevalent in the developed, industrialized countries and may be called diseases of industrialization. Some of the important noninfectious diseases are discussed in the sections that follow.

41.8 NUTRITIONAL DISEASES

Malnutrition is a minor problem in the United States today; however, among the poor it still prevails. But there are definite evidences that some diseases are aggravated, if not caused, by overeating or "excessive" nutrition, which some would classify as another form of malnutrition. It is estimated that 15% of the American population, some 30 000 000 persons, are overweight and that at least 5 000 000 of these are pathologically overweight. Quite the opposite is true of many other countries of the world. In Africa and in other underdeveloped countries throughout the world, kwashiorkor, a disease caused by inadequate protein nutrition, is prevalent. Pellagra, beriberi, and other deficiency diseases are still present in large areas of the world. (See Sections 38.11 and 38.12.)

Many nutritionists believe that the laziness, stupidity, and other characteristics which are often attributed to the natives of poor countries are really due to prolonged nutritional deficiencies. Controlled research on the effects of starvation on human volunteers supports this conclusion. Recent studies with experimental animals and of malnourished children indicate that malnutrition of infants produces mental defects that can never be overcome. Offspring of malnourished mothers show similar irreversible mental deficiencies.

Individuals vary in their utilization of food and in their needs for specific nutrients.

While not much is known about this subject, we do know that there is wide variability in all measurable human characteristics, including the requirements for vitamins. With respect to nutritional requirements, we are too often prone to remember the generalization and forget the individual. Screening of people on a massive scale and a great deal of research will be required to determine the individual variations. However, this lack of refinement should not keep us from acting upon the best available generalizations as to normal requirements. Malnutrition in the poorer countries of the world can be combated by education in nutrition and by increasing food supplies. The food problem is becoming more and more serious as a result of the rapid increase in population. This increase is due in a large part to the control of infectious disease and to better nutrition. Thus the solution to problems of disease and malnutrition may lead to even more complex problems. It may be said that in treating diseases humans have interfered with the balance of nature, and this is, of course, true. Few people, however, would approve the alternative of doing nothing to combat continued infant mortality, starvation, and disease. A second alternative is birth control. The problems raised by the population explosion are not insurmountable, as shown by some countries that have reversed their increase in population. Although many persons regard birth control as another form of interference with nature, oral contraceptives show great promise for those who can accept use of "the pill" (see Section 32.6).

**41.9
MENTAL
ILLNESS AND
ITS TREAT-
MENT BY
DRUGS**

Various forms of mental illness may affect as many as 10% of the people in the United States at some time during their lives. An appreciable percentage of all the patients in United States hospitals are victims of mental diseases. These figures lead many persons to believe that mental diseases are the primary health problem in the United States at the present time. It is true, of course, that the high percentage of mental patients in hospitals is due to the fact that the hospital stay is much longer for these patients than for patients in general hospitals. Other authorities consider the greatest health problem is in the control of cardiovascular disease or cancer, which rank highest as causes of death.

One difficulty in discussing mental diseases is the lack of a good definition of the diseases and objective reports of the effects of drugs used in their treatment. Mental illness is usually categorized as a **psychosis** or **neurosis**, and the victims of such conditions are known as psychotics or neurotics. Psychotics are generally thought of as persons who differ seriously from normal persons in their idea of reality and contact with their environment. They may be in a stupor, may be wildly irrational, may see and hear things no one else does, or may become uncontrollably violent. Neurotics are described as persons whose mental illness does not involve loss of contact with reality. They may be overwhelmed by persistent fears, haunted by anxiety, subject to unexplainable aches and pains, and perhaps unable to sleep. Definite borderlines between psychotics, neurotics, and normal persons are not easily drawn. It has been said that the mentally ill are just like normal persons except that some tendencies are exaggerated.

For many years biochemists have looked for an abnormal metabolite in the blood or urine of psychotics. This search has led to a number of compounds that can produce some of the symptoms of psychoses when given to experimental animals or humans. However, there are no clear indications that these substances are present in the blood or urine of psychotics in sufficient quantities to account for their illness. The search for abnormal metabolites is continuing and may yet be successful. There is evidence that some forms of schizophrenia are hereditary, and this lends support to the search for an inherited biochemical defect.

While the search for metabolic abnormalities in the mentally ill has been largely unsuccessful, therapy in treating mental illness has made major advances in the past thirty years. The use of tranquilizing drugs such as reserpine and chlorpromazine has transformed mental hospitals from madhouses or snake pits with straitjackets and padded cells to quieter, calmer places. These potent tranquilizers alleviate the symptoms, but do not cure mental diseases. They make it possible for psychiatrists to establish contact with a psychotic patient and thus help them treat the disease by psychotherapy. These drugs have also made possible the release of many patients from mental hospitals. Unfortunately, this release has often not led to a return of patients to supportive home care, but instead to their assignment to a "board and care" home where they are maintained on drugs with little therapy or chance for eventual recovery.

Reserpine can be isolated from an extract of the roots of the plant *Rauwolfia serpentina*. This plant has been used by medicine men of India for centuries for the treatment of a variety of conditions ranging from snakebite to madness. Purified extracts of *Rauwolfia* were first used in the United States in the treatment of high blood pressure. The tranquilizing effects were noted, and the most active of the many substances (more than 25 are known) in *Rauwolfia* extract was found to be reserpine (see Fig. 41.4). Chlorpromazine was synthesized to be used as an antihistamine, but its usefulness as a potent tranquilizer was soon observed. Since 1954, reserpine, chlorpromazine, and related compounds have made possible the closing of many mental hospitals and the treatment of mental patients in homes. However, the recently reported carcinogenicity of reserpine, which is also

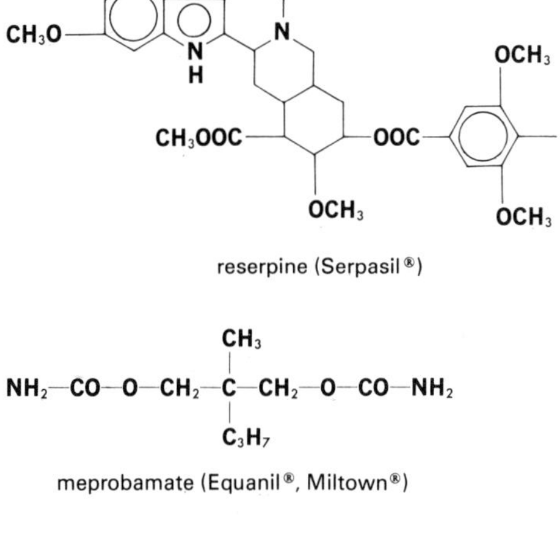

reserpine (Serpasil®)

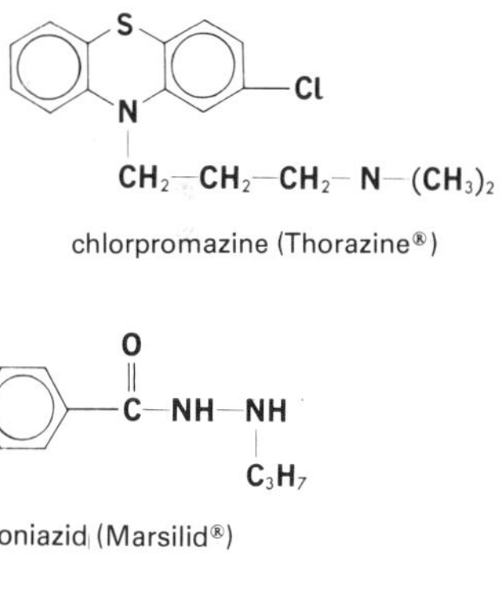

chlorpromazine (Thorazine®)

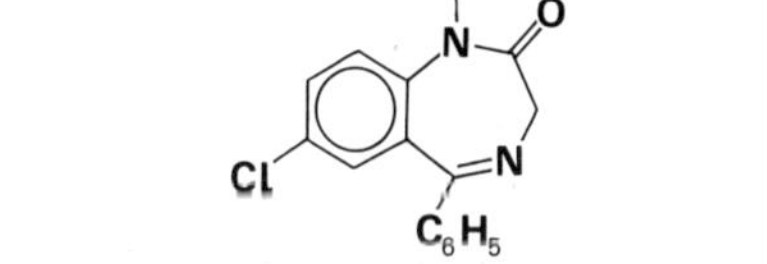

meprobamate (Equanil®, Miltown®)

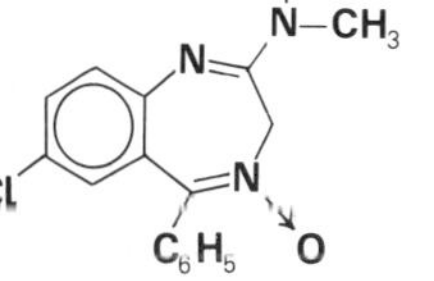

iproniazid (Marsilid®)

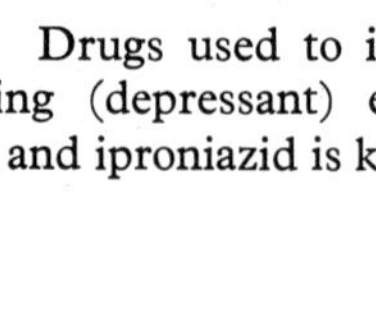

diazepam (Valium®)

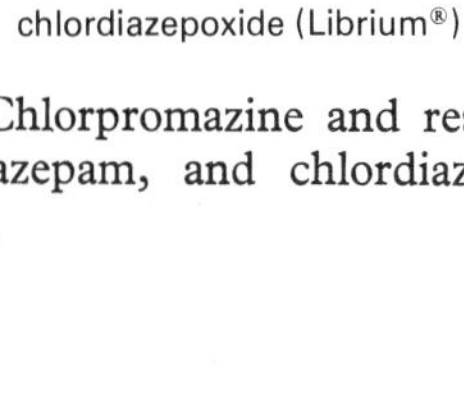

chlordiazepoxide (Librium®)

Fig. 41.4 Drugs used to influence mental states. Chlorpromazine and reserpine have tranquilizing (depressant) effects; meprobamate, diazepam, and chlordiazepoxide are sedatives; and iproniazid is known as an antidepressant.

widely prescribed to lower blood pressure, will no doubt lead to a decrease in the use of this drug.

Equanil® and Miltown® (meprobamate) have been replaced by Valium® (diazepam) and Librium® (chlordiazepoxide) as popular "tranquilizers." These drugs (Fig. 41.4) are primarily sedatives which produce euphoria at higher doses. Many health authorities believe that they are being grossly overused and that they pose a significant health problem. Habituation, addiction, and even death have been attributed to the use of Valium® and Librium®, sometimes in combination with each other and alcohol.

Other substances are also used to counteract depression—a sense of despair, fear, or guilt—without making a patient jittery or irritable. Iproniazid was originally used in the treatment of tuberculosis. When it was observed that patients taking the drug became elated and sometimes so happy they began dancing in the hospital rooms, the mood-elevating aspects were obvious. Unfortunately, many side effects of iproniazid became evident, and it was withdrawn from general use as an antidepressant (and more effective drugs were found for treating tuberculosis). Since iproniazid is a derivative of hydrazine (H_2NNH_2), chemical synthesis of similar compounds has produced a variety of effective and less toxic antidepressants. These drugs inhibit the enzyme monoamine oxidase (MAO) that is important in the oxidation of various amines (norepinephrine, for example, or serotonin) which are important in the functioning of the brain and the nervous system. Apparently the inhibition of MAO is responsible for the antidepressant properties of the hydrazine derivatives.

The amphetamines (Benzedrine and Dexedrine) which are similar in structure to the hormones epinephrine and norepinephrine, are used as stimulants. They have the general disadvantage of producing a nervous irritability and depressing the appetite. The latter effect, however, makes them useful for weight reduction. Since these are potent drugs, they are legally available only on a doctor's prescription. Although they have been used for many years by college students who want to stay up all night to finish a term paper or study for an examination, there is little evidence that such use is warranted. They can keep a student awake and may elevate spirits, but they probably do not elevate grades. Indeed, they may instead produce confusion or a false sense of security. In recent years the amphetamines, and especially methylamphetamine ("speed"), have been available as street drugs. See Fig. 41.5 for the formulas of these drugs.

Another class of mind-influencing substances which are variously described as hallucinogenic, psychotogenic, or psychedelic drugs are the object of much interest today. The first of these substances to be chemically characterized, lysergic acid diethylamide (LSD) (see Fig. 41.5) has been the object of much experimentation, both scientific and unscientific. The drug is relatively nontoxic in the dosages used, but it can have profound effects on mental states and on the personality. Psilocybin, mescaline, and tetrahydrocannabinol (from marijuana) are reported to have effects similar, but not identical, to those of LSD. The effects of these drugs are largely not describable in objective, scientific terms but are reported in subjective, personal language. As a result, scientific evaluation is difficult. It is to be hoped that further, well-controlled experimentation with these potent substances will lead to better evidence concerning both their usefulness and their potential danger.

An examination of the formulas of the compounds given in Fig. 41.5 shows that many of the substances affecting the mind are similar in structure to norepinephrine and serotonin. These latter substances, as we have already noted, are important in the functioning of the brain and nervous system. It is also known that reserpine and chlorpromazine influence the levels of serotonin in the brain. The best-established fact on the biochemical

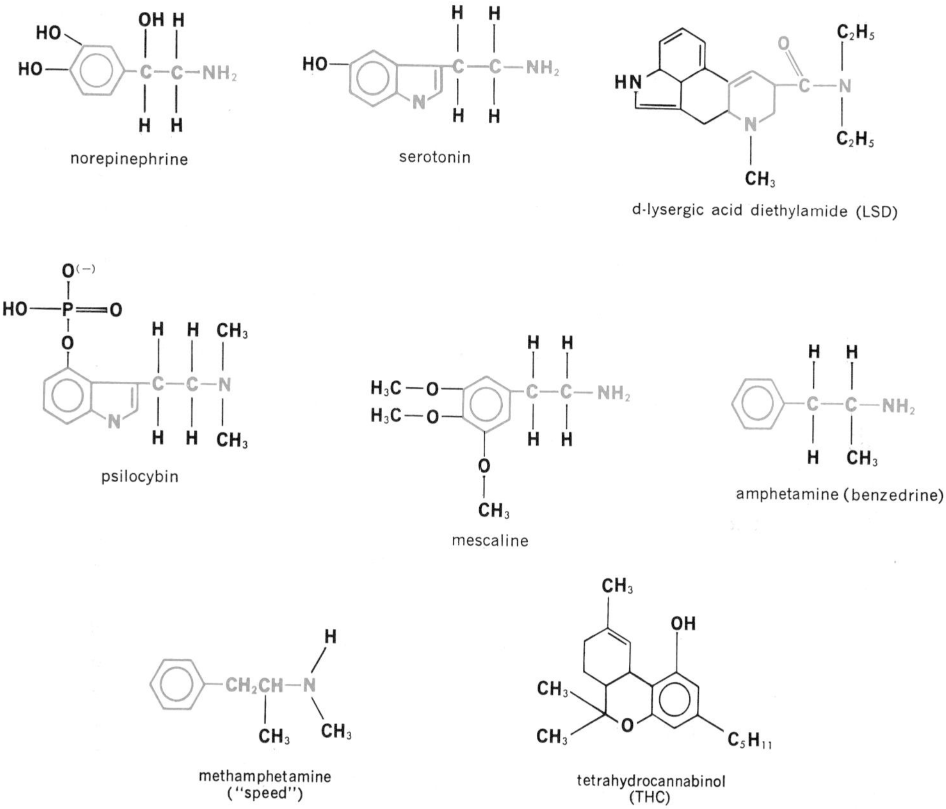

Fig. 41.5 Drugs that affect the central nervous system. Norepinephrine and serotonin are involved in transmission of nerve impulses. LSD, psilocybin, and mescaline produce altered mental states and may be called psychedelic or hallucinogenic substances. Tetrahydrocannabinol also alters mental as well as physical states. Amphetamine and methamphetamine are stimulants. The racemic (DL mixture) of amphetamine is known as Benzedrine and the D isomer or dextroamphetamine is known as Dexedrine. Structural elements that are similar are shown in color.

effect of any of these drugs is that the hydrazine derivatives inhibit the enzyme monoamine oxidase, which is important in the metabolism of norepinephrine and serotonin. Thus the evidence is mounting that these two naturally occurring amines are very important in influencing mental states. LSD is known to block the metabolism of serotonin; however, other similar substances that also block serotonin metabolism do not have the mind-altering effects of LSD. These observations are intriguing and suggest possible hypotheses about the action of mind-altering substances. However, no consistent theory has yet been able to stand the challenge of explaining all the experimental data.

Marijuana   One of the most widely sold illicit drugs in the United States today is tetrahydrocannabinol (THC) (see Fig. 41.5). It is usually obtained in an impure state as marijuana, hashish, or under a variety of names. Some street dealers claim to sell THC, but tests have shown many of such substances to be LSD or other drugs. At the time of this writing, the marijuana "industry" is said to be the third largest in the United States in terms of dollars, being surpassed only by Exxon and General Motors Corporation.

Although there has been a great deal of evidence both for and against the use of marijuana, some facts seem to be accepted by all. It generally produces relaxation, euphoria, increased heart rate, and some alteration of time perception, of judgment, and of coordination. Most people are not capable of intellectual activities when they consume appreciable amounts of THC. Other effects, such as the lowering of testosterone in males, have not been either established or disproved. Some investigators are studying the use of THC to treat glaucoma and to treat nausea accompanying radiation and some forms of chemotherapy for cancer.

There is insufficient evidence to give the effects of long-term use of drugs containing THC. References concerning marijuana and other psychoactive drugs are given in the Suggested Reading. It is beyond the scope of this text to discuss them in the detail that is necessary for understanding. It is obvious, however, that THC is a potent chemical, and its widespread use, especially by very young people, is cause for a great deal of concern.

41.10
DISEASES
CAUSED BY
MALFUNCTION
OF THE
ENDOCRINE
GLANDS

The endocrine glands secrete chemical substances called hormones directly into the bloodstream. The hormones travel throughout the circulatory system and eventually affect many parts of the body. The endocrine glands are the pituitary (or hypophysis), located just below the brain; the thyroid, located in the neck; the parathyroids, located in the thyroid gland; the pancreas, located in the abdominal region; the adrenals, located on top of the kidneys; and either the ovaries, located abdominally in the female, or the testes located in the scrotal sac in the male. (See Fig. 41.6.) Each endocrine gland produces at least one hormone, and many of them produce several chemically distinct hormones. The deficiency or oversupply of any of the hormones has serious consequences for the individual. We can give examples of only a few of these conditions for the purposes of illustration.

Diabetes mellitus, which was discussed in Chapter 35, is caused by a deficiency of the pancreatic hormone insulin. Treatment by insulin does not cure diabetes, but continued use of it allows the diabetic to lead an essentially normal life. Control of diabetes represents one of the most successful instances of correction of a hormone deficiency. Purified hormones from other animals or synthetic products have also been used successfully to treat endocrine insufficiencies. The administration of thryoxine for hypothyroidism (underactivity of the thyroid gland), the adrenocorticotropic hormone (ACTH) to compensate for underactivity of the adrenal cortex, and either male or female hormones to persons who fail to mature normally are other examples of treatment for hormone deficiencies.

Treatment is more difficult in cases of overproduction of a hormone or hyperactivity of an endocrine gland. In many cases, surgery is used to remove the overactive gland. This may be followed by replacement of the hormones—at a normal level—from a synthetic or natural source. In other instances, specific substances are known which can decrease the secretion of a gland. For instance, propylthiouracil decreases the output of thyroxine in a hyperthyroid person. In other cases, the action of one hormone is opposed to that of another, and administration of the antagonist may be helpful. An example of such antagonism exists between the male and female hormones and in the inflammation-promoting and inflammation-suppressing hormones of the adrenal cortex.

Hormones are sometimes used to treat diseases not directly attributable to the endocrine glands. Male hormones are effective in treating some forms of cancer of the breast in the female, and cortisone is widely used to treat various diseases characterized by inflammatory reactions.

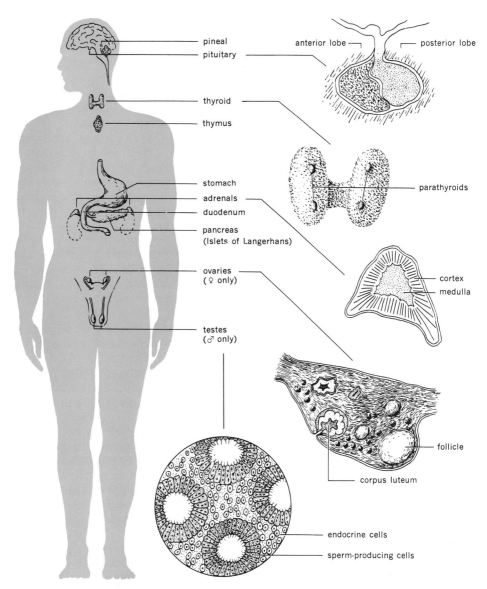

Fig. 41.6 The endocrine glands of the human. The location of the glands is shown at the left and drawings of the detailed structures at the right. [Reprinted from John W. Kimball, *Biology*, Addison-Wesley, 1965.]

There are no real cures for diseases of the endocrine other than surgical removal of overactive glands and possibly the surgical transplantation of glandular tissue to compensate for underactive glands. The purification and laboratory synthesis of hormones has, however, led to treatment of many endocrine disorders. Our knowledge of just how the hormones work is minimal. It is to be hoped that as we find out more about the reactions that hormones influence, other means of treatment and perhaps control of endocrine diseases will emerge. (See Section 34.8 for a discussion of hormones.)

**41.11
DRUG-INDUCED
DISEASES**

With the increased availability of drugs for the treatment of diseases, physicians are now observing many drug-induced diseases. These result from both drugs prescribed by a physician and drugs obtained illegally. We can classify these conditions as **iatrogenic** and **egogenic** diseases. The prefix *iatros* means physician and *ego* refers to the self, with *genic* referring to origin. Thus iatrogenic diseases are physician-caused and egogenic diseases are self-induced.

**Iatrogenic
Diseases**

The incidence of iatrogenic diseases began long ago. Many people died as a result of "bleeding" (deliberate letting of blood)—a practice that persisted for centuries. Moreover, early physicians such as Paracelsus (Section 29.1) were often ignorant and did not hesitate to administer substances that are now known to be poisonous. Although modern physicians are much more aware of which substances are poisons, the number of medications available (at least 10 000 are listed by the American Medical Association's Nomenclature) presents problems. Only 1 200 medications are widely prescribed, yet even this list presents a formidable challenge to the memory of the prescribing physician. When we add the number of possible interactions when two or more drugs are prescribed at the same time, the potential for **adverse drug reactions** (ADR's) is almost overwhelming. It is not surprising that ADR's are seen in about 10% of all patients receiving drug therapy (for further details, see the article by R. Mulroy in the Suggested Readings). The drugs for which adverse effects are most often seen are digoxin, the antibiotics, corticosteroids, anticoagulants, analgesics, tranquilizers, and antihypertensive drugs.

Digoxin is used to increase the force of heart contractions and to correct heart rhythm disorders. The effective and safe doses of this drug (and other digitalis preparations related to it) have a narrow range. Overdoses and underdoses can have serious consequences. The same is true for anticoagulants given to prevent further blood clotting after heart attacks and other incidences of abnormal blood clotting. People receiving these drugs must carefully observe directions for taking medications.

The antibiotics have been shown to produce allergic type reactions. Many people, for example, are sensitive to penicillin. In addition, antibiotics inhibit bacteria and alter the normal balance of microorganisms on the skin and mucous membranes, and thus many people acquire infections of a yeast type of microorganism as a result of taking antibiotics.

The corticosteroids—a group of drugs related to the steroids produced in the cortex of the adrenal gland—are given to prevent inflammation in many conditions. The two most commonly used corticosteroids are prednisone and prednisolone. They provide relief from inflammation and pain in rheumatoid arthritis and can be used for treatment of skin rashes such as those caused by poison oak or ivy. Most swelling and inflammation caused by injuries also responds to the corticosteroids. Unfortunately, these drugs can cause ulcers of the stomach, impaired resistance to bacterial and viral diseases such as those causing pneumonia, and retention of salt and water, as well as other adverse reactions.

Analgesics, tranquilizers, and some antihypertensive drugs (drugs given to lower blood pressure) may produce abnormal mental conditions including, but not limited to, depression. Some of these drugs also induce dependence. As we have seen, if a diuretic is administered as an antihypertensive drug it may promote excessive excretion of potassium and thus produce hypokalemia (see Section 38.8).

A variety of good references are now available to help a person understand the nature and possible adverse reactions (side effects) of the most common medications. Common sense dictates that people should become familiar with these publications. A listing of these books is given in the Suggested Readings. It is also highly desirable that people taking certain types of medication and people who are allergic to penicillin or other drugs should carry information, perhaps in the form of a warning bracelet, indicating such facts.

Other iatrogenic diseases may be due to improper procedures in inserting tubes, surgical operations, and other manipulations. While these are probably less common than the increased number of malpractice suits would indicate, they do exist. Many people in the health professions are seriously concerned about iatrogenic diseases, whether they are the fault of a physician or result from mistakes made by others in the health professions. The tragic death of babies who were given a diet in which salt (sodium chloride) had been substituted for sugar is an example of the latter sort of mistake.

Egogenic Diseases   Drug abuse is a problem of our civilization. It is practiced not only by the person who buys drugs on the street but also by the person who obtains prescriptions for a drug from several different physicians, each unaware of the other's prescriptions. Improper dosage (primarily overdosage) of prescribed medications is also a problem with many people. Librium® and Valium® are especially being overused at the present time.

The street drugs that are in use tend to change with the times and fashion. Most of them are illegally made and may be impure or may be adulterated with other substances. Some are obtained from reliable manufacturers through illegal processes. In almost all cases the drugs were first developed as analgesics, tranquilizers, or for other purposes but were withdrawn from the market because of undesirable effects. They then appeared as street drugs.

The uses and abuses of heroin, the barbiturates, methaqualone, phencyclidine, and other common street drugs have been discussed previously. All of these can cause disorders. Some of the disorders are best classified as mental diseases; others, especially when addiction has developed, are clearly physiological. In the next subsection we will discuss another chemical substance that is often abused—alcohol, or more specifically, ethanol.

Abuse of Ethanol   Although it may be classified as a nutrient, since we derive energy from its metabolism, ethanol must also be considered a drug, since it depresses the central nervous system. Consumption of alcohol promotes tolerance and addiction. Alcohol can be obtained legally and without prescription by people over a certain age, yet it is a drug that is abused and that is responsible for a number of drug-induced diseases.

Ethanol depresses the central nervous system. Some of the effects of this depression and their correlation with the blood levels of ethanol are given in Table 41.1. As indicated in the table many of the harmful effects of consumption of alcohol result from automobile accidents attributable to drivers who were intoxicated with ethanol. All too often, the harm done in these cases is inflicted on persons who have not themselves been drinking.

Even when it is not a killer, alcohol takes its toll. The unpleasant effects that are commonly known as a "hangover" may be due to alcohols other than ethanol (known as cogeners) that are present as impurities in the beverages consumed. Other effects, such as dehydration, stomach irritation, and acidosis, are probably due to the presence and metabolism of ethanol. The body normally oxidizes ethanol to acetic acid, which can produce acidosis.

TABLE 41.1   The relationship between the amount of alcohol consumed, blood alcohol level, and the effects noted

Amount Consumed (90 proof beverage*)	Blood Level (mg/100 ml)	Effect
3 oz	0.05	relaxation, euphoria, no significant impairment of driving skills
6 oz	0.10	slightly slurred speech, impaired judgment, and impaired driving skills (sixfold increase in driving fatalities and accidents)
9 oz	0.15	gross intoxication, impaired gait and judgment, irritability
18 oz	0.40	coma and possible death
30 oz		death

The relationship between amount consumed and blood level is approximate and depends on individual characteristics. There is a much greater correlation between blood level and effect. (Adapted from *Psychoactive Drugs*, by Vincent J. D'Andrea, Cummings Publishing Co., 1977).
* A 90 proof beverage is 45% ethanol. Most table wines contain about one-fourth as much ethanol. Consequently, about four times the volume of table wine would be required to produce the same effect as 90 proof beverages. Beer is commonly 3–6% ethanol.

Many of the effects of long-term overconsumption of alcohol (alcoholism) may be due to malnutrition. Since alcoholics derive energy from the metabolism of ethanol, they may eat little food. Conditions due to deficiencies of vitamin B_1 and nicotinamide are often seen in alcoholics. (See the discussion of thiamin in Section 38.9.) Cirrhosis of the liver, which often accompanies continued consumption of ethanol, may be due to a combination of malnutrition and the effects of ethanol. Similar defects in the liver are seen in cases of poisoning by carbon tetrachloride and related solvents. The replacement of normal cells of the liver by cirrhotic tissue means that many of the reactions of the body that are so essential to metabolism are depressed. In particular, ammonia is not converted into urea at a normal rate in the liver of people with severe cirrhosis.

Withdrawal from ethanol after prolonged use produces weakness, nausea, anxiety, and trembling. In some cases, the addicted person suffers hallucinations and other altered mental states (*delirium tremens*). Withdrawal is therefore a serious matter and should be carefully supervised. Although many people have been able to overcome an addiction to alcohol, psychologists believe that addiction involves psychological as well as physical factors. Thus, if the emotional environment is not changed following withdrawal, the alcoholic may soon become addicted again. This, of course, is true of many addictions.

The Uses of Drugs   Throughout history, nearly all societies have used some forms of drugs in their daily life and for religious festivals. Our society is different only in that it has available (and uses) a much greater number of alkaloids and other drugs. For some people, drug use includes only doses of the alkaloid caffeine in coffee or tea, the alkaloid nicotine in cigarettes, and perhaps some ethanol in wine or beer with a meal. Other people, however, depend on a continuous use of stimulants ("uppers") and sedatives or tranquilizers ("downers") to get them through the day. While a religious tone rarely prevails at a cocktail party, a beer bust, or a gathering where marijuana or other mind-altering drugs are used, these occasions may serve some of the same social purposes as primitive rituals. While some degree of drug use may be necessary for survival and happiness in our civilization, serious consideration should be given to the effects of drugs. The egogenic drug-induced diseases may be rapidly replacing other diseases as major sources of pain, anxiety, and death.

41.12
METABOLIC
OR
DEGENERATIVE
DISEASES

The diseases that we see more prominently in developed industrialized countries are sometimes called the metabolic or degenerative diseases. The terminology used to describe these diseases is not totally satisfactory. From one aspect, all diseases are metabolic since they influence reactions in the human body. The term degenerative diseases is perhaps better, yet that is not totally accurate, either.

Cancer

The term cancer is applied to a variety of conditions, all of which are characterized by a nonphysiological growth or multiplication of cells. Sometimes a growth may remain restricted in area and have little tendency to recur after removal. In a case like this we describe the growth, or neoplasm, as benign. In common usage, benign neoplasms are not cancerous. When a tumor grows unrestrictedly and spreads through the blood or lymph to other parts of the body to form new growths, it is described as malignant and called a cancer. Physicians use more precise terms and speak of *carcinoma*, a malignant tumor of epithelial cells; *sarcoma*, a malignant tumor of the muscle or connective tissue; or *leukemia*, which is characterized by excessive numbers of leucocytes or white blood cells.

Biochemists and other medical researchers have attempted for years to find some abnormal metabolite in cancerous cells, but so far the search has been unsuccessful. Cancerous cells seem to be similar to others except that they have lost the ability to stop reproducing. Malignant cells tend to lose their differentiation or specialization and revert to a more primitive type of cell, and in some cases this reversion involves the use of anaerobic glycolysis even in the presence of oxygen. These cells apparently do not use the reactions of the citric acid cycle.

More is known about the causes of cancer than about the changes it brings about in the metabolism at a cellular level. Ingestion of certain organic chemicals such as dimethylbenzanthracene and methylcholanthrene,

dimethylbenzanthracene   methylcholanthrene

can produce cancer in rats and other experimental animals. Closely related compounds found in cigarette smoke are also effective **carcinogens** (producers of cancer). Another cancer-causing agent is radiation. Ultraviolet irradiation of the skin can lead to skin cancer. X-rays and gamma radiation from radioisotopes penetrate deeper and produce cancer in other organs. The radioactive substances strontium-90 and radium, which the body tends to treat like their sister element, calcium, are found in bones and are likely to produce cancer of the bone or bone marrow.

Recently there has been a major emphasis on the role of chemical carcinogens. Many authorities say that 75–90% of cancer is caused by relatively simple substances in the

TABLE 41.2   Substances that produce cancer in humans

2-Acetylaminofluorene
Aflatoxins
Asbestos
Benzanthracenes
Benzidine (and its derivatives)
4-Biphenylamine
bis-Chloromethylether (and related compounds)
4-Dimethylaminoazobenzene
N-nitrosodimethylamine
Ethylenimine
Hydrazine (and derivatives)
1-Napthylamine (α-napthylamine)
2-Napthylamine (β-napthylamine)
Nickel carbonyl

The substances listed above are known to produce cancer in humans. Extensive lists of substances that are believed to cause cancer are given in *Chemical and Engineering News* **56**, No. 31, 20–22 (July 31, 1978) and **56**, No. 34, 23–24 (Aug 25, 1980). Radioactive isotopes, particularly radium-226 and strontium-90, have been shown to cause cancer.

---

environment. A short list of substances that are known to cause cancer in humans is given in Table 41.2. The Occupational Safety and Health Administration (OSHA) has compiled a list of over 200 substances that cause cancer in some species and that can, according to some evidence, cause cancer in humans. (See Table 41.2 for a reference to this list.) The recent banning of saccharin and the cyclamates because they caused cancer in experimental animals illustrates the current concern (which some people regard as excessive) with the problem of carcinogenesis.

Studies now indicate that the human body converts many substances that are not themselves carcinogens into cancer-causing substances. The reactions responsible for this transformation are those that the cell uses to "detoxify" foreign molecules. In this case, however, a substance that is more harmful than the original one is formed. It is also evident that some substances are **co-carcinogens**. Alone, they do not produce cancer, but when present with another substance they promote its carcinogenicity.

The problems involved in carcinogenesis cannot be treated fully in this text. The intelligent person, however, must be aware that many commonly used substances, including organic solvents, are carcinogens. Continued research and the dissemination of information to consumers is necessary to prevent increased incidence of cancer. The most reliably established carcinogens are those derived from smoking tobacco, especially in the form of cigarettes. All authorities agree that a decrease in smoking would drastically lower the incidence of cancer of the lung.

Evidence is accumulating that some forms of cancer are caused by viruses. In fact, some authorities believe that all cancer is caused by viruses. If this proves to be true, we will have to reclassify cancer as an infectious disease. However, cancer does not appear to be contagious in the sense that many other infectious diseases are.

One theory has been proposed that links all the bits of evidence concerning the causes of cancer. This theory must still be tested, but it presents an interesting way of regarding the cancerous conditions. According to this theory, viruses may infect cells and remain in the cells in an essentially inactive form. When the proper stimulus comes along, either radiation or a chemical carcinogen, the virus becomes active and produces cancer. A different theory has proposed that viruses are not necessary but that carcinogenenic chemicals or irradiation affect the normal DNA in a cell in such a way that it produces malignancy. (See Section 33.7.)

There are various treatments for cancer. Probably the best is surgical removal at an early stage of development of the neoplasm. In some cases irradiation with x-rays destroys the cancerous cells, along with some normal cells. Thus x-rays can both cause cancer and be used to treat it. The probability that they will cure a known neoplasm is greater than the probability of their producing a new one, so x-ray therapy is useful. Significant progress is being reported toward finding antibodies to cancerous cells. Presumably the antibodies have to come from someone other than the victim of the cancer, since the presence of cancer indicates that the person cannot make sufficient antibodies. The form of cancer therapy most closely related to biochemistry is the use of antimetabolites. Antimetabolites of many of the components of the nucleic acids have been effective in treating malignancies. Mercaptopurine and fluorouracil,

mercaptopurine   fluorouracil

have been used for this purpose. Treatment with antimetabolites often prolongs the life of the victim of cancer, but it does not offer a cure. Over a period of time cancerous cells can become resistant to most antimetabolites. The antimetabolites eventually begin to affect the normal cells of the body to such an extent that treatment must be stopped.

Some antibiotics which are too toxic for general use—for example, actinomycin D— are used in cancer therapy. Vincristine, an alkaloid from the periwinkle plant, is also used in chemotherapy of cancer. Both of these substances work by inhibiting protein synthesis. Actinomycin D combines with DNA and prevents formation of messenger RNA; vincristine prevents attachment of amino acids to transfer RNA.

It is to be hoped that further studies of the basic defect, causes, and treatment of cancer will lead to either a cure or an effective treatment of this dread disease.

Arthritis   Arthritis means, literally, inflammation of the joints. In any condition so generally described, there are probably many causes of the symptoms which characterize the condition. Two major types of arthritis have been distinguished, gouty arthritis and rheumatoid arthritis. In the former, deposits consisting largely of uric acid and urates form at various places in the body, most noticeably at the joints. Deposits at places other than joints are classified as gout. Gout and gouty arthritis represent an abnormal deposition of a normal metabolite, apparently due to the synthesis of excessive amounts of uric acid.

Rheumatoid arthritis is one of a group of diseases called the collagen diseases. In these there are inflammatory changes in connective tissue which can produce lesions in joints, skin, muscle, heart, and blood vessels. The cause of the inflammatory changes is not definitely known, but in some cases there are antibodies in the serum which react with other components of the host's own tissues. Therefore, one well-accepted theory states that rheumatoid arthritis is an auto-immune disease.

Both forms of arthritis represent abnormalities in essentially normal body constituents. Gout and gouty arthritis can be treated with colchicine and allopurinol, both of which apparently decrease the synthesis of uric acid. It is interesting that allopurinol, which is a purine antimetabolite, was first synthesized to be used as a chemotherapeutic agent against cancer. Both forms of arthritis, or perhaps it is best to say both diseases, can be treated with anti-inflammatory drugs such as cortisone.

Cardiovascular Disease

This disease, which was discussed in Chapter 37, is also included with the degenerative diseases. It is, in fact, probably the most important of these diseases, at least with respect to the number of persons affected.

41.13 AGING

It is wrong to classify the process of aging as a disease; however, the process has much in common with metabolic or degenerative diseases. We are all aware of the general symptoms of aging; we know that the skin wrinkles, the hair turns gray and may fall out, sight is impaired, the bones become brittle, and the body no longer can stand so much stress as previously.

The biochemist is interested in three basic aspects of aging: What is the biochemical nature of the changes accompanying aging? What is the cause of these changes? What treatments or procedures may offset or delay the onset of these changes? At the present time there are only fragmentary answers to these questions, but research is proceeding, and we will undoubtedly have better answers in the future.

The search for the nature of the biochemical changes in aging has centered on the connective tissue—the material between the cells that helps hold them together. The protein collagen is the most abundant component of connective tissue. As the body ages, the collagen becomes more rigid, less soluble in phosphate and citrate buffers, and assumes an almost crystalline form. This increased rigidity is probably due to increased cross linkage between collagen molecules. Other components of the connective tissue, elastin and various mucopolysaccharides (combinations of protein and polysaccharides), change in ways that may make nutrition and elimination of waste products of the cells, which are surrounded by the connective tissue, much more difficult. It is known that with aging many cells die and are replaced with connective tissue.

A brown lipoprotein pigment called lipofuscin accumulates in many cells with increasing age. Neither the source nor the significance of this pigment is known. However, in persons who live to be 100 years old, 6–10% of the volume of some heart cells is made up of this pigment.

While there is no lack of theories to explain why persons age, most of the theories have not been subject to any real verification. Since many of the cells of the body (nerve and muscle cells) never divide, they may just "wear out." We must ask, then, what wearing out means in biochemical terms. It has been suggested that the DNA is disrupted by chance irradiations so that it no longer has all the "information" a cell needs for survival. The amount of RNA found in the nucleus of cells decreases with age, as does its rate of turnover. Perhaps toxic products produced in the cell disrupt some part of the cellular machinery. Other theories of aging point to the fact that many older persons do not eat well-balanced, nutritious diets, and thus some of the effects of aging may be due to malnutrition.

Experiments have yielded the interesting fact that cells maintained in cell culture divide only a certain number of times and then die. If cells are derived from younger animals, they divide more times before they die. Thus it may be that aging and death happen because cells have only a definite life span. Only cancerous cells keep on dividing indefinitely in cell cultures.

Few valid treatments have been found to delay the onset of aging. In experiments with rats, a drastically reduced caloric intake, with maintenance of the levels of vitamins and proteins, increased the life span from a normal value of 1 000 days to 1 400 days. In severely starved animals the life span was doubled. In countries where starvation is common, the average life span is not increased, but these countries also have poorer nutrition and more infectious diseases, so perhaps the comparison is not valid. It is certainly true that obesity is not consistent with long life.

From time to time we hear reports that one or another ethnic group—often in some geographically isolated area—contains a high percentage of very old people. Close investigation usually reveals that birth records for these people are nonexistent or doubtful, and most scientists are therefore skeptical of claims concerning these groups. The groups do contain many healthy old people, however, and some of this is probably due to the fact that the old ones are very active and do not overeat.

Various people have advocated the consumption of yogurt or the implantation of monkey glands (especially the testes) as a means of preventing aging. Since heredity is important in longevity, others have flippantly suggested that the way to live to be old is to advertise for a pair of long-lived parents. And then, we should keep in mind the point attributed to George Bernard Shaw, to the effect that one comforting way to think about the prospect of growing old is to consider its alternative!

In a more serious vein—research into the cause of aging and possible ways of slowing down the process should prove fruitful in the next several years. Dr. Hans Selye has said: "Among all my autopsies (and I have performed quite a few), I have never seen a man who died of old age. In fact, I do not think anyone has ever died of old age yet . . . To die of old age would mean that all the organs of the body would be worn out proportionately, merely by having been used too long. This is never the case. We invariable die because one vital part has worn out too early in proportion to the rest of the body."[*]

Some may question the desirability of prolonging life for persons who feel alienated from society even while young. The prospect of a prolonged existence filled with mental anguish or boredom is not inviting. It would seem, then, that various areas must be explored for ways of making life more meaningful as well as longer.

SUMMARY

Increased chemical knowledge has led to the production of substances that can aid the human body in its fight against infectious diseases. Other chemical developments have led to new compounds for the relief of pain and anxiety caused by disease or injury. Our knowledge of nutrition and the functioning of the endocrine glands enables us to prevent or treat still other diseases that in the past had taken the lives of many people.

On the negative side, we are also using the results of chemical research and development to cause new diseases. The philosophy that "more is better" does not apply to the use of many chemical substances. The pollution of the environment and the widespread use of drugs cannot be listed as benefits of chemical development.

As with most problems, greater knowledge and awareness can lead to solutions. Understanding how our bodies function normally and in what ways we can help them regain

---

[*] H. Selye, *The Stress of Life*, New York, McGraw-Hill, 1956.

normal function when they are diseased is one of the benefits of the study of biochemistry. In an era of many medications available from a variety of sources, each individual must assume some responsibility for the things that go into his or her body. This requires, among other things, a willingness to go beyond the basics presented in one textbook and to read as much of the current reference material as possible. Such detailed understanding is necessary if one is to live a long and happy life in our present state of civilization.

IMPORTANT TERMS

addiction
adverse drug reaction (ADR)
analgesic
anesthetic
antibiotic
antibody
antigen
antimetabolite
barbiturate
carcinogen
chemotherapeutic agent
co-carcinogen
corticosteroid
dependence
disease

egogenic disease
endocrine gland
habituation
hypnotic
iatrogenic disease
infection
metabolic (degenerative) disease
morphine
neurosis
phagocyte
psychosis
sedative
therapy
tolerance (to drugs)
withdrawal

WORK EXERCISES

1. Distinguish between an antibiotic and an antibody. How are they different, and in what ways are they similar?
2. Distinguish between carcinoma and sarcoma, between a chemotherapeutic agent and an antibiotic, between a tranquilizer and an antidepressant, and between the effects of benzedrine and iproniazid.
3. What are the characteristics of a substance that is a good chemotherapeutic agent against bacteria? Against cancer?
4. The enzyme monoamine oxidase (MAO) probably catalyzes the metabolism of serotonin and norepinephrine. What would be the net chemical effects of inhibiting MAO?
5. What is the significance of the antibiotic-resistant bacteria?
6. What are the major differences between street drugs and over-the-counter (OTC) drugs?
7. Would you classify overeating and being overweight as an egogenic disease? Why?
8. Discuss the relationship of radiation to cancer.
9. List the chemical changes that accompany aging.
10. Discuss the types of diseases listed below in terms of cause, prevention, and treatment.
    a) infectious diseases
    b) malnutrition (undernutrition)
    c) malnutrition (overnutrition)
    d) mental diseases
    e) diseases of the endocrine glands
    f) *diabetes mellitus*
    g) drug-induced diseases
    h) cancer
    i) rheumatoid arthritis
11. List the major effects and uses of the following.
    a) aspirin
    b) morphine
    c) barbiturates
    d) methadone
    e) cocaine
    f) atropine
    g) digoxin
    h) corticosteroids (anti-inflammatory)
12. What are the dangers associated with each of the following?
    a) barbiturates
    b) morphine
    c) heroin
    d) methadone
    e) cocaine
    f) atropine
    g) anti-inflammatory corticosteroids
    h) ethanol
    i) marijuana or THC

SUGGESTED READING

Collier, H. O. J., "Aspirin," *Scientific American*, November 1963, p. 96 (Offprint #169). A discussion of ways in which this long-used "miracle-drug" works.

DeKruif, P., *Microbe Hunters*, Harcourt, Brace and World, New York, 1932. A classic popularization of the lives of many great scientific microbe hunters. Also available in paperback.

Harris, R. J. C., *Cancer*, 3rd Ed. (pb), Penguin Books, Baltimore Maryland, 1976. This book written by a medical researcher is an excellent explanation of sometimes complex material. Its 176 pages are packed with highly readable information.

Hughes, R., and R. Brewin, *The Tranquilizing of America* (pb), Warner Books, New York, 1979. This book has the subtitle "Pill Popping and the American Way of Life." It is dramatically written but based on good information about a serious problem.

Keller, M. (ed.), "Alcohol and Health—New Knowledge," United States Department of Health, Education, and Welfare, United States Government Printing Office, Stock No. 017-024-00399-9. This report to the United States Congress is authoritative and wide-ranging. Topics include heredity and congenital effects, alcohol and cancer, alcohol and highway safety, and trends in treatment of alcoholism.

Lehmann, P., "Food and drug interactions," *Chemistry* **51**, No. 7, 18, September 1978. The interactions between licorice and medicines to control high blood pressure, between the antibiotic tetracycline and milk, between cheese and antidepressants, as well as others, are discussed in this interesting article.

Majtenyi, J. Z., "Antibiotics—Drugs from the soil," Part I, *Chemistry* **48**, No. 1, 6, January 1975: Part II, *Chemistry* **48**, No. 3, 15, March 1975. These articles provide an excellent introduction to the subject of antibiotics. There is a brief history of the discovery of penicillin, a description of the methods used to find new antibiotics, and discussion of many of the newer antibiotics. The way penicillin damages cell walls is well illustrated.

Mellinkoff, S. M., "Chemical intervention," *Scientific American*, September 1973, p. 102. In an issue devoted to "Life and Death and Medicine," this article describes and discusses the vitamins, hormones, antimicrobial drugs, and other chemotherapeutic agents. An excellent summary that discusses adverse reactions as well as beneficial effects of medications.

Peterson, R. C. (ed.), "Marihuana and Health," Sixth Annual Report to the United States Congress from the Secretary of Health, Education, and Welfare, United States Government Printing Office, 1976. Stock No. 017-024-00570-3. This report discusses the chemistry and metabolism of cannabis, human effects, and the trends in marihuana use in the United States.

Peterson, R. C., (ed.), "Marihuana Research Findings: 1976," United States Department of Health, Education, and Welfare, United States Government Printing Office. Stock No. 017-024-00622-0. This book contains summaries of research on marihuana and a listing of the original reports. There is a lot of information in this 250 page book.

The following books discuss the many prescription drugs that are generally used. They differ in amount of detail, the first being the more complete. It lists a large number of drugs and talks about how the drug works, possible side effects, and many other subjects. The other book is less complete but still valuable.

Long, J. W., *The Essential Guide to Prescription Drugs*, (pb), Harper and Row, 1977. (752 pages).

Graedon, J., *The People's Pharmacy*–2 (pb), Avon Books, New York, 1980. (468 pages).

GLOSSARY

Absolute zero: The lowest attainable temperature; theoretically, the temperature at which molecular action would cease (approximately $-273°C$ or $-459°F$).

Acetal: A compound produced by reaction of two molecules of an alcohol with an aldehyde; it has the structure

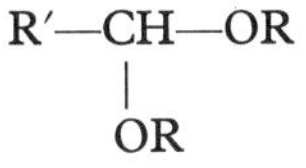

$$R'—CH—OR$$
$$|$$
$$OR$$

Acid: A proton donor. Common examples are HCl, H_2SO_4, and $HC_2H_3O_2$ (acetic acid).

Acid anhydride: A substance which on combining with water produces an acid. SO_3 and other soluble nonmetallic oxides are acidic anhydrides.

Acid chloride: An organic compound having the functional group:

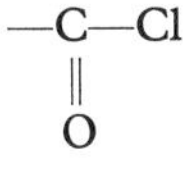

$$—C—Cl$$
$$\|$$
$$O$$

Acidosis: An excess of an acid or of the more acid component of a buffer pair in the body.

Addiction: Compulsive use of a substance beyond voluntary control. Failure to ingest the addicting substance causes physiological withdrawal symptoms.

Addition reaction: One in which new atoms bond with available electrons in a compound without displacing other atoms.

Adipose tissue: Parts of the body where fats are stored.

Adrenalin (epinephrine): A hormone of the adrenal gland that has a generally stimulating effect on the body.

Adsorption: The attraction toward or the adhesion of molecules or ions to solid surfaces.

Adverse drug reaction (ADR): An unexpected, negative response to administration of a medication; sometimes due to drug interactions.

Aerobic: Literally, in the presence of air. Aerobic reactions require oxygen in order to proceed.

Aerosol: Very small particles of a liquid or a solid suspended in a gas.

Albumin: A protein that is water soluble, coagulated by heat, and is not precipitated in a solution that is 50% saturated with ammonium sulfate.

Alcohol: An organic compound having an —OH functional group on an alkyl skeleton, ROH.

Aldehyde: A compound in which one hydrogen and an organic group are bonded to a carbonyl group:

$$R—C—H \quad \text{or} \quad Ar—C—H$$
$$\underset{O}{\overset{\|}{}} \qquad \qquad \underset{O}{\overset{\|}{}}$$

Compare with the definition of a *ketone*.

Aldose: A carbohydrate that contains an aldehyde group. Glucose, galactose, and ribose are simple aldoses.

Alimentary canal: The food tube that carries the food through the body and serves for both digestion and elimination.

Alkali metal: Any element in Group IA of the Periodic Table, e.g., sodium, potassium.

Alkaline earth: Any element in Group IIA of the Periodic Table, e.g., calcium, magnesium.

Alkalosis: An excess of a base (alkali) or of the more alkaline component of a buffer pair in the body.

Alkane: The systematic name for saturated hydrocarbons.

Alkene: The systematic name for a hydrocarbon having a double bond, such as butene, $CH_3CH_2CH=CH_2$.

Alkyl group: A unit of organic structure equivalent to an alkane minus one hydrogen atom (Section 19.3). For example, CH_3CH_2— is an ethyl group.

Alkyne: Systematic name for a hydrocarbon having a triple bond, such as propyne, $CH_3C\equiv CH$.

Allosteric enzyme: An enzyme whose activity is altered by action of substances that are bound to sites other than the active site.

Allotropic forms: Forms of a pure element having different physical and chemical properties. Diamond, graphite, and charcoal are allotropic forms of carbon.

Alloy: Mixture of metals or of metals with small amounts of nonmetals, usually combined to produce certain desirable properties not present in pure metals.

Alpha particle (α). A positively charged helium atom; a helium ion; the nucleus of a helium atom.

Amide: A carboxyl derivative of the type

$$R—C—NH_2 \quad \text{or} \quad R—C—NHR$$
$$\underset{O}{\overset{\|}{}} \qquad \qquad \underset{O}{\overset{\|}{}}$$

Amine: A derivative of ammonia in which one or more hydrogens have been replaced by organic groups; for example, $ArNH_2$ or R_2NH. Such compounds are weak bases.

Amino acid: An organic compound containing an amino (—NH$_2$) and an acid (—COOH) group. The majority of the naturally occurring amino acids have the carboxyl group and the amino group attached to the same carbon and thus are α-amino acids.

Aminopeptidase: A peptidase that catalyzes the hydrolysis of the peptide bond nearest the end of a peptide chain that contains the free amino group.

Amniocentesis: The withdrawal of fluid from the amniotic sac surrounding a fetus. The fluid is analyzed for abnormal cells or substances that indicate hereditary defects.

Amphoteric: Having the ability to react with either an acid or a base. Amino acids and the hydroxides of aluminum, tin, and lead, for example, are amphoteric.

Amylase: An enzyme that catalyzes the hydrolysis of starch.

Anabolic: A word used to describe reactions in living systems that combine simple molecules to make more complex molecules. Such reactions require energy in order to proceed.

Anabolism: Those reactions in a plant or animal that result in the synthesis of larger molecules from smaller ones. (See *Catabolism*.)

Anaerobic: A word used to describe reactions that occur in the absence of air. Actually, anaerobic reactions are those that do not require oxygen to proceed. Some anaerobic reactions are inhibited by the presence of oxygen.

Analgesic: A substance that produces insensitivity to pain without loss of consciousness. Aspirin is the most commonly used analgesic.

Anemia: Any condition in which the amount of hemoglobin in the blood is less than normal. There are many types of anemia.

Anesthetic: A substance that produces loss of sensation, either with or without loss of consciousness.

Anhydrous: Without water; e.g., hydrates from which the water of crystallization has been removed.

Anion: A negative ion.

Anode: The electrode at which electrons leave the solution; the electrode to which negative ions migrate.

Antibiotic: A substance produced by a microorganism that inhibits the growth of another organism.

Antibody: A macromolecule produced by an animal in response to an anitgen. The antibody reacts to overcome the effects of a specific antigen.

Anticodon: The section of a transfer RNA molecule that matches (by hydrogen bonding) a codon for protein synthesis in messenger RNA.

Antidiuretic hormone (ADH): A hormone from the pituitary gland that promotes reabsorption of water in cells of the kidney. Lack of ADH causes *diabetes insipidus*.

Antigen: A substance that initiates the production of antibodies.

Antimetabolite: A molecule that is similar to a normal metabolite but sufficiently different to interfere with reactions of the normal metabolite.

Aromatic compound: An organic substance whose carbon skeleton contains a benzene ring or closely related type of ring.

Aryl group: One derived from an aromatic hydrocarbon or derivative; for example, a phenyl group.

Assimilation: The process of distributing and using digested food.

Atherosclerosis: A condition in which fatty deposits are formed in the inner lining of blood vessels.

Atmosphere (the unit of pressure): The average atmospheric pressure at sea level, equaling 760 mm of mercury or 760 torr.

Atom: The smallest existing unit of an element.

Atomic mass unit (amu): A unit of mass equaling 1/12 the mass of a carbon-12 atom; approximately the mass of a proton or neutron.

Atomic number: The number of protons in the nucleus of an atom; therefore, it also equals the number of electrons in the shells of a complete atom.

Atomic weight: The average weights of the atoms of an element based on carbon-12 being equal to twelve.

ATP: Adenosine triphosphate. A compound whose hydrolysis provides energy to drive energy-requiring reactions. ATP represents an energy-storing compound. (See Fig. 33.12 for the formula for ATP.)

Avogadro's law: This law states that when equal volumes of different gases are under the same conditions of temperature and pressure they will contain the same numbers of molecules.

Bacteriophage: A type of virus which infects bacteria. Many bacteriophages that attack the bacteria *E. Coli* have been studied. They contain DNA and protein.

Barbiturate: Any derivative of barbituric acid. Most barbiturates are used as sedatives and hypnotics.

Barometer: Any of several types of mechanisms used to measure atmospheric pressure.

Base: A proton acceptor. Common examples are OH^- and NH_3.

Basic anhydride: A substance which on combining with water produces a base. Na_2O and other metallic oxides are basic anhydrides.

Beta rays (β): Streams of electrons traveling at speeds approaching the speed of light.

Bile: The fluid secreted from the liver which aids in digestion of lipids. It contains emulsifying agents but no enzymes.

Bilirubin: The organge-red pigment derived from breakdown of red blood cells. An excess of bilirubin in the body produces jaundice.

Binary acid: An acid composed of only two kinds of atoms; e.g., hydrochloric acid, HCl.

Binding energy: The energy holding protons together in an atomic nucleus. This energy is released in atomic fission.

Biochemical oxygen demand (BOD): The amount of oxygen required for the oxidation of a substance; normally used in referring to the oxidation of sewage or other waste products discharged into a body of water.

Biochemistry: The science of the chemical composition, structure, and reactions of plants and animals.

Biosphere: The area of the earth's atmosphere, hydrosphere, and crust in which life exists.

Boiling point: The temperature at which the vapor pressure of a liquid equals or just exceeds the pressure of the atmosphere above it.

Boyle's law: The volume of a gas is inversely proportional to the pressure exerted upon it if the temperature remains constant.

Brass: Any one of varying mixtures of copper with smaller amounts of zinc (Table 13.4).

Breeder reactor: A fission type of nuclear reactor that produces additional fissionable material as it operates.

Bronze: Mixtures of copper with varying amounts of zinc and tin. Bronzes are more resistant to corrosion and much stronger and harder than copper (Table 13.4).

Brownian movement: The continuous, erratic movements of colloidal particles which can be observed through a microscope.

Buffer: A solution whose pH changes only slightly with the addition of moderate quantities of either strong acids or strong bases. Stabilization of the pH results from the presence of a weak acid and the salt of the weak acid in the solution.

Calorie: The amount of heat required to raise the temperature of one gram of water by one degree centigrade. The "calorie" used to describe the energy content of food is actually a kilocalorie, which is 1 000 calories.

Calorimeter: A device in which the heat released in chemical reactions (particularly oxidations) can be measured.

Carbohydrate: Carbohydrates are polyhydroxy aldehydes or ketones or condensation products thereof (Sections 26.11 and 30.1).

Carbon black: A powdery form of carbon produced by incomplete burning of natural gas (primarily methane, CH_4), used as a pigment and in the manufacture of automobile tires.

Carbonyl group: the functional group

Carboxyhemoglobin: The compound formed when carbon monoxide reacts with hemoglobin. This compound does not combine with oxygen.

Carboxylate salt: The salt related to a carboxylic acid; for example, $RCOO^-Na^+$.

Carboxylic acid: An organic compound having a —COOH functional group. Such a compound is a weak acid compared to hydrochloric or sulfuric acid.

Carboxypeptidase: A peptidase that catalyzes the hydrolysis of the peptide bond nearest the end of a peptide chain that contains the free carboxyl group.

Carcinogen: A substance that causes cancer.

Cardiovascular disease: Any disease of the heart or blood vessels.

Carnivore: An animal whose major source of food is the flesh of other animals.

Catabolic: A word used to describe reactions in living systems that degrade or oxidize large molecules to smaller or more highly oxidized molecules. Such reactions result in the release of energy.

Catabolism: Those reactions in a plant or animal that result in the degradation or oxidation of molecules.

Catalyst: A substance which alters the rate of a chemical reaction without being consumed in the reaction; it does *not* change the position of equilibrium.

Cathode: The electrode at which electrons enter a solution; the electrode to which positive ions migrate.

Cation: A positive ion.

Cellulose: A polysaccharide found in the structural parts of plants. It is composed of glucose units linked β 1–4.

Celsius: A temperature scale on which 100° is the boiling point and 0° is the freezing point of water. Also known as the centigrade scale.

Centi-: A prefix denoting 1/100 part.

Centrifugation: A process by which particles can be sedimented and separated by rapid rotation of solutions in a device known as a centrifuge.

Cerebrospinal fluid: The fluid surrounding the brain and spinal cord.

Chain reaction: A reaction or series of reactions in which the products start further reactions. Nuclear fission as well as free radical reactions in polymer production are chain reactions.

Charles' law: The volume of a gas is directly proportional to the absolute temperature if the pressure remains constant.

Chemical change: Changes in a substance which involve composition and, often, the energy content.

Chemical formula: A group of symbols that indicates the kind and number of atoms present in a compound.

Chemical properties: Those properties which describe the composition and reactivity of substances, including energy changes and changes in composition.

Chemistry: The science concerned with the composition of the substances of which the universe is composed, the properties of these substances, and the changes they undergo. It is also concerned with the energy relationships involved in the changes.

Chemotherapeutic agent: A chemical compound that is used to treat disease.

Chiral: A chiral object is one that is different from its mirror image; for example, a pair of hands, or a tetrahedral carbon bonded to four different groups.

Chlorophyll: The green pigment found in plants. It is necessary for photosynthesis.

Chromatography: A process for separation of substances in a mixture by adsorption on paper or other materials. See Fig. 29.1 for illustrations of methods of chromatography.

Chromosome: Long, threadlike bodies that can be seen at certain stages in the life of a cell. The chromosomes are made of DNA and protein and are the site of hereditary information in the cell.

Chylomicron: Small fat particles (about 1 micrometer in diameter) found in lymph and blood.

Chymotrypsin: A proteolytic (protein-digesting) enzyme secreted by the pancreas.

Cirrhosis: The formation of abnormal amounts of connective tissue accompanied by the disappearance of the normal tissue of the liver or other organs.

***Cis*-trans** **isomers:** Those having different configurations because of a rigid structure, such as a $C=C$ group or a ring.

Citric acid cycle (Krebs cycle): A cyclic series of reactions by which acetic acid units are oxidized to carbon dioxide and water with the release of energy.

Cloud chamber: A device which makes visible the paths produced by high-speed particles passing through a supersaturated atmosphere. The particles collide with a very large number of gaseous molecules producing ions. Moisture collects on these ions forming minute droplets resulting in visible "vapor trails" which are called fog tracks.

Co-carcinogen: A substance that does not produce cancer by itself but increases the rate of cancer production by a carcinogen.

Codon: A three-nucleotide sequence of messenger RNA that carries the information for addition of an amino acid to a polypeptide chain.

Coenzyme: A nonprotein substance that is required for activity of an enzyme.

Colloids: Particles of such small size that they may stay suspended in gases or in liquids for an indefinite period of time.

Combustion: An exothermic reaction which proceeds so rapidly that flames or light are produced.

Compound: A substance containing two or more different elements combined chemically in a definite proportion by weight.

Concentrated solution: A solution containing a relatively high proportion of solute.

Condensation: The change of a gas into a liquid; also a type of reaction in which two or more molecules react to form a larger molecule with the elimination of water or another small molecule.

Configuration: The arrangement in space of atoms within a molecule having a given structure.

Constitutive enzyme: An enzyme that is present in normal quantities without the need for an inducer to stimulate its synthesis.

Corticosteroid: A steriod produced by the cortex of the adrenal gland. Most of the commonly used corticosteroids prevent or decrease inflammation.

Covalent bond: A pair of electrons shared between two atoms.

Critical mass: The least amount of fissionable material that can sustain a chain reaction.

Cyclo-, cyclic compound: A ring compound; for example, cyclopentane (Section 19.3).

Cyclotron: A device which accelerates electrically charged particles in a circular path to tremendous velocity and therefore to great energy. See also *Linear accelerator*, another device for a similar purpose.

Cystinuria: An abnormality in which large amounts of cystine are excreted in the urine. This condition may lead to formation of stones of cystine in the urinary bladder.

Cytochromes: A group of colored substances found in many cells. The cytochromes function in electron transport. See Fig. 34.9 for the structure of a cytochrome c.

Dark reaction: A series of the reactions of photosynthesis which can proceed without light.

Deamination: A process by which an amino group is removed from an organic compound. Deamination of an amino acid usually gives a keto acid.

Decant: To separate a liquid from a solid by allowing the solid to settle and pouring off the liquid.

Decarboxylation: The process of eliminating carbon dioxide from an organic molecule. Decarboxylation of an amino acid gives an amine.

Deci-: A prefix meaning one tenth; for example, a decimeter is 1/10 or 0.10 of a meter.

Deficiency disease: A disease whose symptoms can be cured by administration of a specific element or compound. The most well-known deficiency diseases result from vitamin deficiencies.

Dehydration: Reaction in which water is removed from a molecule.

Dehydrogenation: Reaction in which hydrogen atoms are removed from a molecule.

Deliquescence: The dissolving of solids in moisture absorbed from the air.

Delta H (ΔH): The change (Δ) in a quantity known as enthalpy (H). It is a measure of energy released or consumed in a reaction.

Denaturation: A change in the properties of a protein. Denaturation is believed to involve only the secondary linkages of proteins, not hydrolysis of peptide bonds. Denutaration may be reversible, in which case the starting properties are restored, or irreversible so that the protein in permanently altered.

Density: Mass per unit of volume; density may be expressed in a variety of units such as grams per milliliter or pounds per cubic foot.

Deoxyribonuclease (DNA-ase): An enzyme that catalyzes the hydrolysis of DNA.

Deoxyribonucleic acid (DNA): A nucleic acid that yields the pentose deoxyribose on hydrolysis. DNA is found primarily, but not exclusively, in the nuclei of cells and is a polymer of deoxynucleotides.

Dependence: A state in which a person needs or desires a particular substance. Two types are generally recognized: psychloogical dependence (habituation) and physiological dependence (addiction). See Section 41.6.

Depot lipids: The lipids that an animal accumulates or stores. The amount of these lipids may vary greatly as opposed to the tissue lipids which always make up a certain percentage of a cell or an animal.

Desalination: The removal of salt from a solution, especially used to describe removal of sodium chloride from sea water.

Desiccant: A substance that removes moisture (a liquid) from the air (a gas) by absorbing the liquid.

Detergent: An organic compound which has cleansing action in water. It has a large nonpolar hydrocarbon group and at least one ionic or highly polar group.

Deuterium: The naturally occurring isotope of hydrogen having mass number 2. It is found in water.

Diabetes: A condition in which the volume of urine is greater than normal. There are several types of diabetes.

Diabetes insipidus: Excretion of excessive volumes of urine due to lack of antidiuretic hormone. The urine does not contain glucose in this condition.

Diabetes mellitus: A disease characterized by a high level of glucose in the blood and excretion of glucose in the urine. It is caused by a lack of insulin.

Dialysis: The separation of dissolved crystalline material from colloidal materials by use of a semipermeable membrane.

Diamond: A clear crystalline form of the element carbon. Diamond is one of the hardest substances known.

Diastereomers: Stereoisomers that are not mirror images of each other.

Diene: An unsaturated organic compound having two double bonds somewhere in the molecule; for example, butadiene, $CH_2{=}CH{-}CH{=}CH_2$.

Diffusion: The migration of molecules among other molecules. Diffusion takes place rapidly within gases and slowly within liquids.

Digestion: The process of converting ingested food to simpler forms that can be absorbed.

Dilute solution: A solution containing a relatively small proportion of solute.

Dipeptidase: An enzyme that catalyzes the hydrolysis of dipeptides.

Dipeptide: A compound containing two amino acids joined by a single peptide linkage.

Dipolar ion: A molecule having both a positive charge and a negative charge; also sometimes called an "inner salt." The most common example is an amino acid:

$$R{-}CH{-}COO^-$$
$$|$$
$$^+NH_3$$

Disaccharidase: An enzyme that catalyzes the hydrolysis of a disaccharide to give two monosaccharide molecules. Important disaccharidases are sucrase, maltase, and lactase.

Disaccharide: A carbohydrate composed of two monosaccharides linked by a glycoside bond.

Disease: Any departure from health.

Distillation: A process in which a liquid is converted to a gas and then recondensed. Distillation is used to separate and purify liquids from mixtures.

Diuresis: The excretion of larger than normal volume of urine. Diuresis can be caused by ingestion of a diuretic which increases the excretion of ions such as sodium as well as increasing the excretion of water.

Dry ice: Solid carbon dioxide is called dry ice since it sublimes, passing directly from a solid to a gas. Therefore, no liquid is formed by the warming of the solid CO_2 at atmospheric pressure.

Ductile: Capable of being drawn into elongated form; e.g., ductile metals can be drawn into a wire, and semimolten glass into rods.

Duplication (replication) of DNA: The process in which a double strand of DNA separates and produces two identical copies of the original DNA.

Dynamic equilibrium: An equilibrium in which the total number of each of kind of molecules present remains constant although individual molecules are reacting or being changed in form.

Edema: A condition in which there is an excess of water in many tissues of the body.

Efflorescence: The loss of water from hydrates under natural circumstances; i.e., without placing the substances in artificially hot or dry places.

Egogenic disease: A disease that is self-induced.

Electrolysis: A chemical change brought about by the passing of an electric current through a solution of an electrolyte.

Electrolyte: A substance (acid, base, or salt) which conducts an electric current in a water solution.

Electron: A negatively charged subatomic particle with a mass 1/1837 that of a hydrogen atom.

Electron shell: A layer or group of electrons near the nucleus of an atom.

Electronegativity: The tendency of an atom to attract electrons to itself.

Electrophoresis: A process in which components of solutions may be separated by being allowed to migrate in an electrical field. The process is used in the purification of proteins and other charged molecules.

Electrovalent bond: See *Ionic bond*.

Element: A substance containing atoms of only one kind, i.e., having the same atomic number.

Empirical formula: The simplest formula representing the ratio of atoms in a molecule.

Emulsion: A finely divided suspension of two immiscible liquids.

Enantiomers: Compounds whose configurations are nonidentical mirror images. They are optically active.

Endergonic: A word used to describe an energy-requiring or energy-consuming reaction. Anabolic reactions are endergonic.

Endocrine gland: A gland which secretes directly into the blood stream. The secretions of the endocrine glands are called hormones.

Endopeptidase: A peptidase that catalyzes the hydrolysis of peptide bonds in the interior of a protein molecule.

Endothermic reaction: A chemical reaction in which heat is absorbed. (See also *Endergonic*.)

Energy: The ability or capacity to do work.

Energy of activation: The amount of energy required to start a reaction.

Enzyme: A protein that catalyzes a biochemical reaction. Some enzymes also contain factors in addition to the protein.

Epithelial: A word used to describe the cells and tissues on the surface and lining of the body. The skin and the lining of the digestive tract and of the lungs are made up of epithelial cells.

Equilibrium: A state of balance or equality between opposing forces; e.g., when opposing chemical reactions are proceeding at the same rate.

Equilibrium constant (K_{eq}): The ratio of the concentrations at equilibrium of the reaction products to the concentrations of the reactants. A measure of the tendency of a reaction to proceed. Those reactions with a large value of K_{eq} proceed to the right (as written).

Equivalent: The amount of material that contains one mole of reacting units, such as one mole of H^+ ions.

Equivalent weight: The number of grams per equivalent.

Erythrocyte (RBC): The red blood cell or corpuscle. RBC's have no nucleus but contain large amounts of hemolgobin, which serves to transport oxygen from the lungs to the rest of the body.

Essential amino acid: An amino acid that cannot be synthesized by an animal at a rate equal to its need for that amino acid.

Essential nutrient: A specific compound required by an animal which the animal cannot synthesize at a rate equal to its needs.

Ester: An organic compound having a functional group of the type

$$-\underset{\underset{O}{\|}}{C}-O-R \quad \text{or} \quad -\underset{\underset{O}{\|}}{C}-O-Ar$$

It is derived from a carboxylic group linked to an alcohol or phenol.

Esterification: The process of forming an ester group by direct reaction of a carboxylic acid with an alcohol.

Ether: An organic oxygen compound of the type ROR or ROAr.

Eucaryote (eukaryote): An organism containing cells each of which has a nucleus and various organelles such as mitochondria and ribosomes.

Eutrophication: Increase in growth of certain kinds of plants in response to increased amounts of nutrients. Generally used to describe increased growth of algae when amounts of nitrates, phosphates, and organic substances are increased due to pollution.

Evaporation (vaporization): The change of a substance from the liquid to the gaseous state.

Exergonic: A word used to describe a reaction that converts potential or stored energy to heat or other forms of energy. Catabolic reactions are exergonic.

Exopeptidase: A peptidase that catalyzes the hydrolysis of peptide bonds adjacent to the end of a peptide chain.

Exothermic reaction: A chemical reaction in which heat is released. (See also *Exergonic.*)

FAD: Flavin adenine dinucleotide. A substance that serves as an electron acceptor and donor in bioligical oxidations and electron transport. See Fig. 34.8 for the structure of this compound.

Fahrenheit: A temperature scale on which 212° is the boiling point and 32° the freezing point of water.

Fat: A triacylglycerol that is solid at room temperature.

Fatty acid spiral: The sequence of reactions by which a fatty acid is degraded into acetate units.

Feedback: A process in which a product of a reaction influences a step in that reaction. We speak of positive feedback when the product accelerates the reaction and negative feedback when the reaction is slowed by a product.

Fermentation: The anaerobic metabolism of carbohydrates. Fermentations are catalyzed by the enzymes of bacteria and yeasts, and may result in the formation of lactic acid or of carbon dioxide and ethanol.

Fibrous protein: An insoluble protein. Fibrous proteins have structures that are rigid and tough and often serve as structural elements. The proteins of hair, fingernails and claws, skin, and feathers are fibrous proteins.

Filtration: A process in which solids are separated from liquids or gases by passing the mixture through a filter such as filter paper.

Flocculation: A process in which small particles of a precipiate come together to form small clumps, which may settle out of a fluid and remove impurities.

Fluid: A word used to describe both liquids and gases—those states in which substances flow and assume the shape of the container.

Fixation: The chemical conversion of a gaseous substance to a solid or liquid. Nitrogen is fixed by being incorporated into nitrates, nitrites, or ammonia. Although ammonia is a gas, it is extremely soluble in water, and is generally considered to be a fixed form of nitrogen.

Fog tracks: The vapor trails produced by high-speed particles when they pass through a supersaturated atmosphere.

Formula: See *Chemical formula.*

Formula weight: The numerical equivalent of the sum of the masses of the atoms in the formula of a substance, but without units.

Fossil fuels: Substances that can be extracted from the earth and used for fuel. The most important fossil fuels are coal, petroleum, and natural gas.

Free energy: The energy available for doing useful work.

Freezing point: The temperature at which a liquid becomes a solid.

Freon: Any one of a number of compounds composed of carbon (one or two atoms), fluorine, and usually chlorine. They are used as refrigerants and as aerosol propellants.

Functional group: A group of atoms (a unit of structure) within a molecule which reacts or functions. For example, the —NH_2 group in ethylamine, $CH_3CH_2NH_2$.

Futile reactions: Two or more reactions in which the products(s) of one are converted to reactants(s) of the other; for example, the reactions $A \rightarrow B \rightarrow C$ $\searrow D \swarrow$

Galactosemia: The presence of galactose in appreciable amounts in the blood. A hereditary abnormality in which an individual cannot metabolize galactose.

Gamete: A cell capable of participating in fertilization and formation of a new individual. Ova and sperm are gametes.

Gamma globulin: A variety of globulin found in blood plasma. This fraction contains many antibodies against diseases.

Gamma rays (γ): Radiation possessing great penetrating ability, produced by radioactive material; similar to X-rays.

Gas: The physical state of matter in which a substance tends to fill completely any container in which it is located. Attractions between particles of a gas are minimal. Gases form when liquids evaporate.

Gastricsin: A protein-digesting enzyme found in the stomach.

Gastrointestinal (GI) tract: The stomach and intestines.

Gay-Lussac's law: In chemical reactions involving gases, the reacting volumes are always in the ratios of small whole numbers—one to one, one to two, etc.

Gel: A dispersion of a liquid within a solid.

Gene: The factor that influences the inheritance of a characteristic.

Genetic code: The sequence of organic bases in a nucleotide that specifies an amino acid sequence in a protein molecule.

Genetic engineering: Using a knowledge of biochemistry and genetics to alter hereditary characteristics.

Globular protein: A protein that is soluble or easily suspended in water solutions. Although they contain long polypeptide chains, globular proteins are folded to give a spherical form. Enzymes and many proteins of blood are globular proteins.

Globulin: A protein that can be coagulated by heat, is soluble in a dilute salt solution but precipitates in solutions that are 50% saturated with ammonium sulfate. There are several different globulins in the blood, one of which, gamma globulin, contains antibodies to diseases.

Glomerular filtrate: The fluid produced in the kidney by filtration of blood. This filtrate is modified by reabsorption of water and glucose and addition of other substances before it leaves the kidney as urine.

Glucagon: A hormone of the pancreas that increases the level of glucose in the blood. See also *Insulin.*

Gluconeogenesis: The formation of glucose from noncarbohydrate sources such as proteins and lipids.

Glucoside: A compound containing a glucose unit linked to another molecule. Many natural products other than carbohydrates are glucosides.

Glucosuria: The presence of glucose in the urine.

Glucuronide: A compound containing a glucuronic acid unit linked to another molecule. The excreted forms of many substances are glucuronides.

Glycogen: A polysaccharide found in the muscles and liver of animals. It is made of glucose units.

Glycogenesis: The formation of glycogen: The reverse of glycogenolysis or glycolysis.

Glycogen storage disease: A hereditary disease in which glycogen can be synthesized but not degraded.

Glycolipid: A complex substance containing both lipid and carbohydrate units.

Glycolysis: The process of degradation of glycogen and other carbohydrates. If the process is anaerobic the end product is lactic acid. In the presence of oxygen, glycolysis yields pyruvic acid.

Glycoprotein: A complex substance containing both carbohydrate and protein (or amino acid) units.

Glycoside: A compound formed by reaction of an alcoholic —OH group with the carbonyl group of a sugar. It is a particular example of an acetal.

Glycosidic linkage: The ether linkage between a monosaccharide and another unit. The second unit may be another monosaccharide unit or a different type of molecule.

Graham's law: The rates of diffusion of two gases are inversely proportional to the square roots of their densities.

Gram: A basic unit of mass or weight in the metric system. A gram is the weight of approximately one milliliter of water.

Gram-atomic weight: The atomic weight of an element expressed in grams; the weight in grams of Avogadro's number (6.022×10^{23}) of atoms of the element.

Graphite: A solid form of carbon composed of layers. Within each layer the carbon atoms are covalently bonded in hexagonal patterns. It is a soft, black, greasy-feeling substance which is often used as a lubricant.

Group (of the periodic table): The vertical columns of elements in the periodic table.

Haber process: The most commonly used synthetic method of nitrogen fixation.

Habituation: Dependence on a substance that is not addicting. Withdrawal from the substance produces psychological symptoms such as depression or anxiety rather than physiological symptoms.

Half-life: The time required for one-half of a given amount of radioactive material to disintegrate.

Halogen: Any one of the group of elements occurring in Group VIIA of the periodic table.

Hard water: Water containing minerals (usually calcium and magnesium ions) which interfere with the detergent action of soaps.

Heat of formation: The heat released or required for the formation of a compound from its elements.

Heat of fusion: The heat energy (calories) required to convert one gram of a solid to a liquid without a change in temperature of the substance being melted.

Heat of reaction: The heat released or required when a chemical reaction occurs.

Heat of vaporization: The heat energy (calories) required to convert one gram of a liquid to vapor without a change in temperature of the substance which is being vaporized.

Hematocrit (packed cell volume): The percentage of the volume of blood that is due to red blood cells. The normal value is 40–50%.

Hemiacetal: A compound resulting from addition of one molecule of an alcohol to an aldehyde, yielding the functional group

R′—CH—OR
  |
  OH

Hemodialysis: Dialysis of blood accomplished by passing the blood through an artificial kidney.

Hemoglobin: The red-colored protein which reversibly binds oxygen in the red blood cells.

Heterocyclic compound: A substance having a ring structure composed of carbon atoms and at least one different atom; for example, pyridine,

Heteropolysaccharide: A polysaccharide made of two or more different monosaccharide units.

Heterozygote: An individual having unlike genes for any given characteristic.

Hexosan: A polysaccharide made of hexose monomers.

Hexose: A monosaccharide containing six carbon atoms. Glucose, fructose, and galactose are important hexoses.

Homeostasis: Maintaining an equilibrium value of various components even though individual units of each of the components are being added and removed.

Homopolysaccharide: A polysaccharide made of only one specific monosaccharide.

Homozygote: An individual having identical genes for a given characteristic.

Hormone: The product of an endocrine or ductless gland. Hormones are carried throughout the body by the blood stream.

Humectant: A substance (generally a liquid) used to retain moisture in a mixture. Humectants such as glycerol are added to tobacco and cosmetics to keep them moist.

Hydrates: Compounds formed by the union of salts with water; for example, $CuSO_4 \cdot 5 H_2O$. Water of hydration may be driven from such compounds by heating.

Hydride: Any binary compound in which hydrogen is one of the elements. In one type, the metallic hydride, the valence of hydrogen is -1.

Hydrocarbon: A compound composed only of hydrogen and carbon.

Hydrogen bond: The special attraction that exists between molecules containing hydrogen bonded to oxygen, nitrogen, or fluorine. The attraction is due to the polar nature of such bonds. Water is an outstanding example of a compound containing hydrogen bonds.

Hydrogenation: Addition of hydrogen to an unsaturated molecule.

Hydrogen salt: A salt that has one or more acidic hydrogens.

Hydrolysis: The splitting (lysis) of a compound by a reaction with water. Examples are the reaction of salts with water to produce solutions which are not neutral, and the reaction of an ester with water.

Hydronium ion: A hydrated hydrogen ion, H_3O^+.

Hydrosphere: The general name given to the oceans, lakes, and other bodies of water on the earth's surface.

Hydroxyl group: The —OH functional group, covalently bonded, as in R—O—H. (*Not* the same as a hydro*xide* ion, OH⁻.)

Hydroxysalt: A salt that has one or more hydroxide groups.

Hygroscopic: Adjective describing a substance which absorbs moisture from the atmosphere.

Hyperglycemia: A concentration of glucose in the blood that is higher than normal.

Hypernatremia: A condition in which the concentration of sodium in the blood is higher than normal.

Hypertonic solution: A solution with a salt concentration greater than that of blood.

Hypervitaminosis: A disease due to excessive intake of a vitamin.

Hypnotic: A substance that tends to produce sleep.

Hypoglycemia: A concentration of glucose in the blood that is lower than normal.

Hypokalemia: A condition in which the concentration of potassium in the blood is lower than normal.

Hypotonic solution: A solution containing a percentage of salt less than that contained in human blood.

Iatrogenic disease: A disease caused by a doctor or other medical person.

Imine: A compound having the functional group

$$\text{>C=NH} \quad \text{or} \quad \text{>C=NR}$$

Immiscible: Two liquids which will not dissolve in each other are said to be immiscible; e.g., water and oil.

Inborn error of metabolism: An inherited abnormality which results in failure to produce a normal protein. Many inborn errors of metabolism involve failure of reactions due to lack of the necessary enzyme.

Inducible enzyme: An enzyme whose synthesis is stimulated by the presence of an inducer. Lactose promotes (induces) the formation of the enzyme β-galactosidase, which is necessary for the hydrolysis of lactose.

Infection: The invasion of the body of a plant or animal by a microorganism.

Ingestion: The act of taking in. In simple animals it involves surrounding a food particle. In more complex animals we use the term eating to describe ingestion.

Inhibitor: A substance that decreases the rate of an enzymatic reaction.

Inorganic substances: All the elements and their compounds, except the compounds of carbon. (See *Organic substances.*)

Insulin: A hormone from the pancreas that promotes the utilization of glucose. Deficiencies of insulin result in hyperglycemia and the disease *diabetes mellitus.*

In vitro Literally means "in glass." The term is used to describe reactions occurring in a reaction vessel as distinguished from those that occur *in vivo*, in the living animal.

In vivo Literally means "in the living." The term is used to describe reactions that occur in a living animal.

Ion exchange: A process in which a resin or similar substance containing one type of ions (a) is exposed to a second type (b) which it absorbs with the release of the first type of ion (a) into a solution.

Ionic bond. A bond resulting from the attraction of oppositely charged ions.

Ionization: The formation of ions in a solution.

Ionization constant (K_i): An equilibrium constant describing the ionization of a compound. The constant is calculated by dividing the concentration of ion(s) produced by the concentration of the nonionized compound.

Ions: Electrically charged atoms or groups of atoms. They may be positively or negatively charged, depending on whether the atoms from which they were formed lost or gained electrons.

Isoelectric (isoionic) point: The pH at which a protein or other charged molecule will not migrate when it is placed in an electric field (electrophoresis apparatus).

Isolation artifact: A chemical compound or structure that does not occur as such in a living organism but is produced during the process of extraction and purification.

Isomers: Different compounds having the same molecular formula. For example, methyl ether, a gas, and ethyl alcohol, a liquid, each have the formula C_2H_6O.

Isotonic solution: A solution with a salt concentration equal to that of blood.

Isotopes: Atoms of the same element containing different numbers of neutrons and therefore having different nuclear masses.

Jaundice: A yellowish color of the skin and some other tissue due to the presence of an excess of bilirubin.

K_a (acid dissociation constant): A value obtained by multiplying the ionization constant of a weak acid by the concentration of water. It is a measure of the strength of an acid and indicates the region in which a given acid and its salts will be a good buffer.

Kelvin: A temperature scale named after Lord Kelvin. It is also called the absolute temperature scale. A kelvin is one unit on the Kelvin scale.

Ketal: A compound produced by reaction of two molecules of an alcohol with a ketone; it has the structure

$$R_2'C{-}OR$$
$$|$$
$$OR$$

Ketone: A compound in which two organic groups are bonded to a carbonyl group:

$$R{-}\underset{\underset{O}{\|}}{C}{-}R', \qquad Ar{-}\underset{\underset{O}{\|}}{C}{-}R, \text{ etc.}$$

Compare with *Aldehyde.*

Ketose: A carbohydrate that contains a ketone group. The best known ketose is fructose.

Kilo-: A prefix meaning 1 000; for example, a kilogram is 1 000 grams.

Kilogram: The basic unit of mass in the SI. A kilogram (kg) is the mass of the standard kilogram maintained by the French Bureau of Standards.

Kindling temperature: The temperature at which combustible substances burst into flame as a result of combining with oxygen of the atmosphere.

Kinetic-molecular theory of gases: A series of statements describing the characteristic behavior of gases, based on the theory that molecules of gases are tiny, elastic bodies in constant motion.

Krebs cycle: See *Citric acid cycle.*

K_w: The dissociation or ionization constant for water. Its value is 10^{-14}, which is obtained by multiplying the K_i' for water by the concentration of undissociated water in a liter. (See Section 12.7.)

Lactone: An intramolecular ester; the ester group is part of a ring.

Lecithin: A phospholipid whose complete hydrolysis yields glycerol, two units of fatty acids, phosphate, and choline. Lecithins are dipolar molecules that are present in cell membranes.

Lecithinase: An enzyme that catalyzes the hydrolysis of a lecithin.

Leucocyte (white blood cell or WBC): A blood cell that does not contain hemoglobin. There are several types of leucocytes.

Light reaction: Those reactions of photosynthesis that stop when no light is present.

Linear accelerator: A device by which electrically charged particles are electrically accelerated in a straight line to tremendous velocities and therefore great energy. See also *Cyclotron*, another accelerating device.

Lipase: An enzyme that catalyzes the hydrolysis of a lipid. The most important lipases are those that catalyze the hydrolysis of triacylglycerols to give glycerol and fatty acids.

Lipid: A constituent of a plant or animal that is soluble in nonpolar solvents. A variety of solvents particularly diethyl ether, ethanol, and chloroform, are used to dissolve lipids.

Lipoprotein: A complex substance made of lipids and protein (peptide) units.

Liposaccharide: A complex substance made of lipid and carbohydrate units.

Liquid: The state of matter between a gas and a solid. Liquids assume the shape of the container in which they are present.

Liter: A unit of volume employed in the metric system. It is the volume of one kilogram of water at 4°C.

Lymph: A fluid similar to blood except that it contains no red blood cells. It is derived from the tissues of the body and conveyed to the bloodstream by the lymphatic vessels. It is important in absorption of lipids and in response of animals to infections.

Lysosome: A small body or organelle found in cells. Lysosomes contain hydrolytic enzymes.

Malleable: That which can be hammered into different shapes. Metals like gold, silver, copper, and iron are malleable.

Mass: A measure of the amount of matter in an object. Mass remains constant regardless of the location of the object or of the gravitational attraction.

Mass defect: The term used to describe a condition in which the mass of an atom is less than the sum of the masses of the individual particles of which it is made.

Mass-energy relationship: Theory conceived by Einstein that matter can be converted into energy and vice versa. (This necessitated a slight revision in older versions of the law which stated that neither matter nor energy could be created or destroyed.)

Mass number: The sum of the numbers of protons and neutrons in the nucleus of an atom.

Matter: That which has mass, inertia, and occupies space.

Mega-: A prefix meaning 10^6; for example, a Megagram is 1 000 000 grams.

Meiosis: The process of cell division in which daughter cells contain half as many chromosomes as the parent cell. Meiosis is important in the formation of gametes (ova and sperm).

Melting point: The temperature at which a solid becomes a liquid.

Messenger (template) RNA: A relatively high molecular weight RNA that carries information from DNA of the chromosomes in the nucleus of the cell to the ribosomes where protein is synthesized.

Metabolic (degenerative) disease: A noninfectious disease which results from the change in some aspect of metabolism. The causes and cures for most of these diseases are not well understood.

Metabolic half-life: The time required for half of any body component to be replaced by similar but different molecules. The half-life of many proteins is only a few days, while that of calcium in bones is much longer.

Metabolic water: Water that is produced as the result of reactions within an animal, specifically from the oxidation of hydrogen occurring in foods.

Metabolism: A term used to describe the total reactions taking place in any living organism.

Metabolite: Any chemical substance that participates either as a reactant or product in a reaction in a plant or an animal.

Metal: An element that loses electrons when it forms compounds. See Table 13.1.

Metallic bond: A type of bond in which electrons move freely between atoms in crystals of pure metals. This bonding explains the fact that metals are good conductors of electricity.

Metalloid: An element possessing some of the characteristic properties of both metals and nonmetals.

Meter: The basic unit of length in the SI, equivalent to approximately 39.37 in.

Micro-: A prefix meaning 10^{-6}; for example, a micrometer is $1/1\,000\,000$ or $0.000\,001$ meter.

Microsomes: Small bodies separated from broken cells by centrifugation. Microsomes are probably ribosomes plus some of the materials that tie ribosomes together.

Milli-: One-thousandth part, i.e., a millimeter is one-thousandth of a meter.

Milliequivalent: One-thousandth of an equivalent; the amount of acid or base in one milliliter of a 1 Normal solution of an acid or base.

Mineral element: In biochemistry, any chemical element other than C, H, O, N, and S. That is, the elements not commonly found in carbohydrates, proteins, and lipids are called the mineral elements.

Miscible: A term used to describe liquids that will dissolve partially or completely in each other; e.g., water and alcohol.

Mitochondrion: A subcellular particle or organelle. Mitochondria are the site of much of the energy production of cells. (See Section 29.4).

Mitosis: The process of cell division in which each daughter cell has the same number of chromosomes as the parent cell.

Mixture: A physical combination of two or more pure substances. Components of a mixture are not combined chemically and can generally be separated by physical means.

Molal solution: A solution in which concentration of solute is expressed in moles per kilogram of solvent.

Molar solution: A solution containing one mole of solute in one liter of solution.

Molarity: The strength of a solution expressed in moles per liter; usually abbreviated M.

Mole: The amount of substance that contains Avogadro's number (6.022×10^{23}) of particles (ions, atoms, or molecules).

Mole volume of gases: The volume (22.4 liters) of a molecular weight in grams (one mole) of any gas at standard conditions.

Mole weight: The weight in grams of one mole of substance.

Molecule: A particle containing two or more atoms.

Molecular formula: A formula which states the kind and number of all atoms in a molecule.

Molecular orbital: An orbital in a molecule, formed by overlap of atomic orbitals.

Monomer: A molecule that becomes the building unit of a polymer. (See Section 28.2 and *Polymer*.)

Monosaccharide: A polyhydroxy aldehyde or polyhydroxy ketone. The simplest type of carbohydrate.

Morphine: A narcotic analgesic derived from opium; it can produce addiction.

Mutarotation: The change in optical rotation observed after optically active compounds are dissolved.

Mutation: An abrupt change in an inheritable characteristic.

NAD: Nicotinamide adenine dinucleotide. A substance that serves as an electron acceptor and donor in biological oxidations. (See Fig. 33.13 for the formula of this compound.)

Nephron: The functional unit of the kidney in which the blood is filtered and urine formed.

Net ionic equation: An equation that shows only the particles participating in a reaction involving ions.

Neurosis: An emotional or psychological disorder in which thinking and judgment are impaired but there is no appreciable loss of contact with reality.

Neutralization: The reaction of an acid with a base to produce an approximately neutral salt.

Neutron: A neutral subatomic particle with a mass approximately that of a hydrogen atom. Neutrons account for one-half or more of the mass of all atoms except hydrogen.

Nitration: Substitution of a nitro group,—NO_2, onto a molecule (Section 21.6).

Nitrogen fixation: The combining of elementary nitrogen with other elements to form usable compounds.

Noble gases (also called rare gases): The six elements usually located at the extreme right in the periodic table. Since their valence shells are complete, they are relatively inert.

Nonelectrolyte: A substance that does not conduct an electric current in a water solution; a covalent compound.

Nonmetal: An element that gains electrons when it forms compounds. See Table 13.1.

Nonprotein nitrogen (NPN): A measure of the nitrogen-containing compounds other than proteins in blood. Urea is normally the largest component of NPN.

Normality: The number of equivalents per liter of solution.

Nuclear fission: The transmutation in which an atom is converted into smaller atoms. Such reactions are accompanied by liberation of energy.

Nuclear fusion: The transmutation occurring naturally in the sun, or synthetically in a hydrogen bomb, in which small atoms combine to form larger atoms. Such reactions liberate greater amounts of energy than do comparable fission reactions.

Nuclear reactor: The term commonly applied to the device which is used to control the rate of nuclear reactions and to produce useful nuclear power.

Nucleic acid: A high-molecular-weight substance from natural sources which, on hydrolysis, yields organic bases, pentoses, and phosphate. Although originally found in the nuclei of cells, it is present throughout the cell.

Nucleoprotein: A combination of a nucleic acid and a protein. The proteins in such combinations contain many basic amino acids. The bonds between the acidic nucleic acids and basic proteins are saltlike and relatively weak.

Nucleosidase: An enzyme that catalyzes the hydrolysis of nucleosides to yield an organic base and a pentose.

Nucleoside: A compound which, on hydrolysis, yields one molecule of an organic base and a pentose.

Nucleotidase: An enzyme that catalyzes the hydrolysis of nucleotides to give nucleosides and phosphate.

Nucleotide: A compound which, on hydrolysis, yields one molecule of an organic base, a pentose, and a phosphate group.

Nuclide: A general term used for any isotopic form of any element.

Nutrient: A chemical substance that provides either energy or a necessary component of an animal.

Nutrition: The science that is concerned with foods and their effect on organisms, particularly on humans.

Oil: In biochemistry, a triacylglycerol that is a liquid at room temperature.

Olefin: The old common name for an alkene.

Oligosaccharide: A molecule containing from two to ten monosaccharide units. The most important oligosaccharides are the disaccharides.

Optical activity: The ability to rotate the plane of vibration of polarized light.

Optical isomers: Those having different configurations because of chiral centers in the molecules.

Orbital: A region of space near an atomic nucleus which can be occupied by no more than two electrons.

Ore: A naturally occurring substance or mixture from which a metal can be extracted.

Organ: A group of tissues in a plant or animal that performs some specific function or functions.

Organic substances: The compounds of carbon. They may be produced synthetically or by living organisms.

Organism: Any living plant or animal.

Osmosis: The selective passage of liquids through a semipermeable membrane in a direction which tends to make concentrations of all substances on one side of the membrane equal to those on the other side.

Oxidation: The loss of electrons from an atom.

Oxidation number (also called *oxidation state*): Corresponds to the number of electrons gained or lost by an originally neutral atom. The oxidation number of sulfur in S^{2-} is $2-$ and in SO_2 is $4+$.

Oxyacid: An acid compound, such as HNO_3, which contains hydrogen, oxygen, and a non-metal; also called a *ternary acid*.

Oxygen debt: A condition in muscles that are not receiving sufficient oxygen. In this condition, excess hydrogen atoms (electrons) are used to convert pyruvic acid to lactic acid. Adequate supplies of oxygen reverse the process.

Ozone: An allotropic form of oxygen with a molecular formula of O_3.

Pascal: The unit of pressure in the SI. It is equal to 1 newton per square meter.

Pentosan: A polysaccharide composed of pentose monomers.

Pentose: A monosaccharide containing five carbon atoms. Ribose, arabinose, and xylose are pentoses.

Pepsin: A gastric enzyme that catalyzes the hydrolysis of peptide bonds of proteins. It is an endopeptidase.

Peptidase: Any enzyme that catalyzes the hydrolysis of the peptide linkage between amino acids.

Peptide: A compound formed by a combination of the amino group of one amino acid with the acid group of another. Peptide is also used to describe the linkage between the two amino acids.

Peptide linkage: A group containing the carbonyl group from one amino acid and the N—H from another, for example:

This linkage is found between the α-amino acid units in a protein. It is just a particular example of an amide linkage.

Periodic law: A law which states that when the elements are arranged in the order of their increasing atomic numbers, they exhibit a periodic recurrence of properties.

Periodic table: An arrangement of the chemical elements in table form to show the periodic recurrence of properties. See also *Periodic law*.

Periods: The horizontal rows of elements in the periodic table.

pH: A number denoting the hydrogen ion concentration in a solution.

Phagocyte: A cell of the body that can engulf foreign material.

Phenol: An organic compound having an —OH functional group on an aryl skeleton, ArOH. (The term is also used for the simplest compound of this class, C_6H_5OH.)

Phenyl group: The structural unit C_6H_5— or

Phosphogluconate shunt: A seies of reactions that provides for release of energy from glucose in fewer reactions than in glycolysis. This series of reactions may also result in the synthesis of ribose. (See Fig. 35.6).

Phospholipid: A lipid that contains a phosphate group. The most abundant phospholipids are derivatives of phosphatidic acid in which the phosphate is esterified with one of the —OH groups of glycerol.

Phosphorylase: An enzyme that catalyzes the addition of a phosphate group to a molecule. The phosphate group comes from ATP. It is better to call such enzymes phosphotransferases.

Phosphorylation: The addition of a phosphate group to a molecule. In many phosphorylation reactions phosphoric acid functions as water does in hydrolysis; such a reaction is called a phosphorolysis.

Photosynthesis: The process in which the energy of light is stored in chemical compounds of a plant.

Physical change: Changes in physical state or of shape with no change in composition. Change of ice to water and water to steam are physical changes.

Physical properties: Those properties of a substance (solubility, boiling point, freezing point, etc.) that can be measured without changing the composition of a substance.

Pig iron: The impure iron containing carbon, silicon, and sulfur which comes from the blast furnace. It is much too hard and brittle to be of much value until most of these impurities are removed.

pK_a: The negative of the logarithm of the acid ionization constant. This value indicates the effective buffering range of such acids and their salts.

Plasma (blood): The clear fluid of the blood. Plasma is prepared by addition of an anticoagulant and centrifugation to remove blood cells.

Polar covalent bond: A covalent bond is polar when it is formed between two atoms having a considerable difference in electronegativity. In such a bond, the pair of shared electrons is shifted toward the more negative atom.

Polarized light: Light that has only one plane of vibration. (See Section 27.5.)

Polymer: A large molecule formed by successive reactions that link together individual molecules (the monomers).

Polypeptide: A substance containing many amino acids linked by peptide bonds. Polypeptides contain fewer amino acid units than do the proteins.

Polyprotic acid: An acid, such as H_2SO_4, having more than one ionizable hydrogen.

Polysaccharide: A polymer of monosaccharide (i.e., glucose) units. Prominent members of this group are the starches, cellulose, and glycogen.

Polyuria: Excretion of a volume of urine greater than normal.

Potential energy: The energy a body possesses by virtue of its position. Also the stored energy present in fuels and foods.

Precipitate: A word used to describe a substance that settles out of a solution or suspension; also used to describe the process in which the substance settles out.

Precipitation: Either the process or the product of a process in which a substance settles out of a solution or suspension.

Pre-pepsin (pepsinogen): The inactive precursor of the enzyme pepsin.

Primate: A classification of animals that includes man, apes, and monkeys.

Primer DNA: A molecule of DNA that must be present in a system capable of synthesizing DNA before the synthesis can start. The kind of DNA synthesized resembles the primer DNA.

Procaryote (prokaryote): A simple type of organism characterized by lack of nuclei and other organelles.

Product: A substance formed by a chemical reaction.

Proenzyme: An inactive precursor of an enzyme.

Prostaglandin: A lipid with hormone-like properties which is derived from polyunsaturated fatty acids.

Protein: A high-molecular-weight substance which, on hydrolysis, yields amino acids. A polymer of amino acids.

Proteolytic: A word used to describe those enzymes that catalyze the breakdown of proteins.

Protium: The most common isotope of hydrogen. Its mass number is 1.

Proton: A hydrogen ion. The nucleus of a hydrogen atom. A positively charged subatomic particle with a mass approximating that of a neutron.

Protozoan: A member of a class of single-celled animals.

Psychosis: A severe mental disorder in which there is loss of contact with reality.

Purine bases: Important components of nucleic acids related to purine,

Pyrimidine bases: Important component of nucleic acids related to the organic base pyrimidine,

Quaternary ammonium salt: An ammonium salt having four organic groups (the same or different); for example, $R_4N^+Cl^-$.

Radioactive isotope: An isotope of an element so unstable that it decomposes to give α, β, γ or other radiations.

Radioactivity: The name suggested by Madame Curie for the disintegration of atomic nuclei in which α, β, or γ radiation occurs.

Radiocarbon dating: A process by which the age of ancient materials can be estimated by measuring the ratio of carbon-14 to carbon-12.

Rancid: A term used to describe foul-smelling lipids. The foul odors are due to aldehydes and acids formed by the hydrolysis or oxidation of triacylglycerols or unsaturated fatty acids.

Reactant: A substance that undergoes a chemical reaction.

Reactivity: A term used to describe the tendency of a chemical substance to react.

Reducing agent: A substance that removes oxygen or electrons from, or adds hydrogen or protons to, another substance.

Reducing sugar: A sugar that reduces metallic ions in an alkaline solution; e.g., glucose in Benedict's Reagent.

Reduction: The gain of electrons or hydrogen by an element or compound.

Reductive amination: The reaction of ammonia with a carbonyl group in the presence of a reducing agent (often hydrogen). The resultant product is an amine. Both chemical and biochemical examples are very important.

Renal threshold: The level of a substance in blood above which it will be excreted in urine.

Replication (duplication): The process in which a double strand of DNA separates and produces two identical copies of the original DNA segment.

Representative elements: Those in the eight main groups (A groups) of the periodic table.

Ribonuclease (RNA-ase): An enzyme that catalyzes the hydrolysis of RNA.

Ribonucleic acid (RNA): A polymer composed of nucleotides which contain ribose. Three types of RNA are important in protein synthesis: messenger RNA, ribosomal RNA, and soluble (transfer) RNA. Many viruses contain RNA and protein.

Ribosomal RNA: The RNA found in ribosomes.

Ribosome: A small spherical body or organelle found in cells. The ribosomes are an important site of protein synthesis in a cell.

Saliva: The fluid secreted in the mouth by the salivary glands. It contains complex polysaccharides and the enzyme amylase.

Salt: An ionic compound having positive ions other than hydrogen ions and negative ions other than hydroxide ions; for example, KBr.

Salting out: The precipitation of a protein by addition of a salt, usually as a salt solution.

Saponification: The reaction of an ester with a metallic hydroxide (or certain other bases) to produce a carboxylate salt and an alcohol. The name originates from the fact that when a fat undergoes such a reaction, the salt formed is soap. (See Sections 23.12 and 24.5.)

Saturated hydrocarbon: A carbon compound in which all carbon bonds, other than those required to hold the skeleton together, are occupied by hydrogen atoms.

Saturated solution: A solution in which the undissolved solute and the dissolved solute are in equilibrium. No greater amount of solute can be dissolved at the existing conditions.

Sedative: A substance that decreases nervous activity, tending to calm and tranquilize a person.

Serum: The clear fluid produced when blood cells are removed by the clotting of blood.

Shell: See *Electron shell*.

Smog: Originally used to describe a mixture of smoke and fog. Now generally used to describe visible air pollution.

Soap: An alkali metal carboxylate salt having a total of twelve to about eighteen carbon atoms; for example, $C_{17}H_{35}COONa$.

Solid: The state of matter in which particles are most closely packed. Solids are formed when liquids freeze.

Soluble (transfer) RNA: A relatively low-molecular-weight RNA that attaches to an amino acid during the process of protein synthesis. Transfer RNA is believed to match with a specific part of messenger RNA known as the code. Following this, the amino acid is added to the growing peptide chain of the protein being synthesized.

Solute: A dissolved substance. In solutions composed of a solid and a liquid, the solid is the solute. In solutions composed of two liquids, the one present in the lesser proportion is considered to be the solute.

Solution: A homogeneous mixture of molecules, atoms, or ions of two or more substances.

Solvent: The dissolving substance. In a solution composed of a solid and a liquid, the liquid is considered to be the solvent. In a solution composed of two liquids, the one present in the greater proportion is considered to be the solvent. Water is the most common solvent since it is so abundant and so many substances are soluble in it.

Specific dynamic action: The increased heat production which accompanies digestion. It is especially noticeable during digestion of proteins.

Specific gravity: The ratio of the density of a substance to the density of water.

Specific heat: The heat energy required to raise the temperature of one gram of a substance one degree centigrade.

Spectator ions: The ions that are left unchanged by a chemical reaction; they are partners of the reacting ions.

Sphingolipid: A lipid containing the amino alcohol *sphinginene*. Such lipids are found in the central nervous system.

Spontaneous combustion: An exothermic reaction that proceeds so rapidly that the kindling temperature is reached and the material is self-ignited.

Standard conditions (gases): Often abbreviated S.T.P. meaning Standard Temperature and Pressure. A temperature of zero degrees Celsius and a pressure of 760 torr.

Starch: The polysaccharide found in seeds, roots, and tubers of plants. It is composed of glucose units linked α 1–4 with some α 1–6 branches.

Steel: Iron, alloyed with varying quantities, usually smaller amounts, of other metals or nonmetals.

Stereoisomers: Isomers having the same structure but differing in configuration.

Steroid: A compound containing a complex hydrocarbon nucleus. See Section 32–2 for a diagram of this nucleus. Among the important compounds having a steroid skeleton are bile acids, sex hormones, and adrenal hormones.

Structural formula: Way of writing the structure of a molecule to show which atoms are bonded to which other atoms and the types of bonds used.

Sublimation (sublime): The transition from the solid to the gaseous state without passing through the liquid state. The reverse process is also sometiems referred to as sublimation.

Substance: A term used to describe both elements and compounds. A substance is considered to be pure, i.e., contains only one kind of atoms or one kind of molecules.

Substitution reaction: A reaction in which certain covalently bonded atoms of a compound are displaced by different atoms.

Substrate: The substance on which an enzyme acts.

Sugar: A sweet-tasting monosaccharide or disaccharide.

Sulfonation: Substitution of a sulfonic acid group, $—SO_2OH$, onto a molecule (Section 21–5).

Supersaturated solution: A solution that is holding more solute than it can normally dissolve at existing conditions.

Suspension: A mixture of one substance suspended in another. Many suspensions will settle out but colloidal suspensions will not unless conditions are changed.

Temperature inversion: Ordinarily the temperature of the air decreases with altitude. When there is a warm layer on top of a cooler layer, we call this a temperature inversion.

Template RNA: See *Messenger RNA*.

Ternary acid: See Oxyacid.

Terpenoid: A class of naturally occurring lipids derived from the 5-carbon isoprene unit. Terpenoids containing 10 carbon atoms are known as terpenes.

Tetrahedral: Four-sided. A term used to describe the regular arrangement of four covalent bonds at a carbon atom (Section 18.7 D).

Therapy: The treatment of disease.

Thermonuclear bomb: An explosive device that works because of nuclear fusion.

Thiol: A compound of the type RSH; a sulfur analog of an alcohol.

Thrombocyte: A blood cell involved in formation of blood clots; thrombocytes are also known as *platelets*.

Tissue: A group of similar cells along with the substances surrounding the cells.

Tissue lipids: Lipids that play an important role in the functioning of a cell and are found in all cells. The amount of tissue lipids in an animal does not vary appreciably as do the depot lipids.

Titration: The process of reacting a solution of unknown strength with one of known strength. This procedure is commonly used to determine concentrations of solutions of acids and bases.

Tolerance (to drugs): A situation or condition in which increasingly larger doses of drugs are needed to produce a given effect.

Ton: A large unit of weight. A metric ton is 1 000 kilograms (2 205 lb). A long ton is 2 240 lb and a short ton is 2 000 lb.

Torr: A unit of pressure equal to one millimeter of mercury. One torr is equal to about 133 pascals.

Transaminase: An enzyme that catalyzes the transfer of an amino group to a keto acid. Measurements of transaminases in blood are used in diagnosis of heart attacks and liver damage.

Transamination: A reaction in which an amino group is transferred from an amino acid to a keto acid. This results in the synthesis of a new amino acid and the conversion of the original amino acid to a keto acid.

Transcription: The process in which the information of DNA is incorporated into the structure of messenger RNA.

Transfer RNA: See *Soluble RNA*.

Transition elements: Elements in the B groups of the Periodic Table.

Translation: The process in which the information in messenger RNA is transferred to the amino sequence of a polypeptide.

Triacylglycerol: A lipid whose hydrolysis produces three moles of fatty acids for each mole of glycerol. Also known as a *triglyceride*.

Triglyceride: See *Triacylglycerol*.

Tritium: The heaviest isotope of hydrogen. It is produced in nuclear reactors; its mass number is 3.

Trypsin: A proteolytic enzyme secreted by the pancreas.

Unsaturated: A term used to describe an organic compound having double or triple bonds between some atoms. The most common examples are $C=C$, $C\equiv C$, $C=O$, $C=N$, and $C\equiv N$. These unsaturated units of structure can be saturated by addition of hydrogen.

Unsaturated hydrocarbon: A carbon compound having one or more double or triple bonds, so that not all bonds are occupied by hydrogens. (Compare with *Saturated hydrocarbon.*)

Unsaturated lipid: A lipid containing double bonds.

Valence: A term used to indicate the combining ability of an element. Valence may be measured in terms of the number of electrons lost or gained by an atom or group of atoms, or by the number of pairs of electrons shared.

Valence shell: The outermost shell of electrons around an atom.

Vapor: Another term for the gaseous state of matter.

Vaporization (evaporation): The change of state from a liquid to vapor (gas).

Virus: One type of agent that causes diseases in both plants and animals. Viruses contain large amounts of nucleic acids and protein. Viruses are generally smaller than bacteria and were originally distinguished from them because viruses could pass through filters.

Vitamin: An organic compound required in small amounts in the diet of an animal. A vitamin must be defined in terms of a given type of animal. Ascorbic acid is needed by all animals, but is synthesized by most animals and thus is not a vitamin for them. Ascorbic acid is a vitamin (vitamin C) for humans, other primates, and the guinea pig, since these animals do not synthesize it.

Volatile: A term used to describe a substance that is easily converted to a vapor (gas).

Wax: An ester of a fatty acid and a long-chain, monohydric alcohol.

Weight: Commonly used to indicate an amount or mass of material. Actually, weight is dependent upon the gravitational attraction between bodies, and therefore is not constant, as is mass.

Withdrawal (from drugs): A process characterized by symptoms such as vomiting and changes in heart rate, which is brought on by discontinuing the administration of a drug to which a person is addicted.

Zeolite: A silicate mineral used in water softeners.

Zymogen: The inactive precursor of an enzyme. Pre-pepsin is the zymogen of pepsin.

Answers to Selected Work Exercises

Chapter 1

none.

Chapter 2

1. a) 2.345×10^3 c) 1×10^0 e) 8.35×10^{-1} g) 1×10^6

2. a) 2 345   c) 689 000   e) 0.001   g) 70.0

3. a) 4   c) 1   e) 1   g) 3

4. a) 2 350   c) 6.9×10^5 e) 7 390 or 7.39×10^3 g) 30.00 or 3.000×10^1

5. a) 2 550 cm   c) 500 mm   e) 2.5 km   g) 75 cm   i) 30 cm   k) 90 mi

6. a) 0.5 kg   c) 6.893 mg   e) 1 400 g   g) 5.5 lb

7. a) 150 ml   c) 0.25 dm^3 e) 1.53 liter   g) 0.47 liter

8. a) 38°C   c) 14°F   e) -40°F   g) 310 K

10. 900 lb   12. 900 kg   14. 80 km/hr   16. 9 yd   18. 820 m

20. a) 4.0 km   c) 260 km/hr

22. 156 m^2 24. 320 cm^3 26. 9.0×10^3 liter, 2.4×10^3 gal   28. 1.1×10^3 kg, 2.4×10^3 lbs, 1.2 short tons

32. 3.5 oz butter, 7.0 oz sugar, 350°F

34. a) 6.8 fluid drams   c) 9.1 oz (avoir.)   e) 5 teaspoons

36. 59 ml   38. 6.6 × 10^{-6} lbs

Chapter 3

1. a) 480 cal   c) 37 kcal   e) 1.3 × 10^9 kcal

2. a) 370 cal   c) yes

3. 2.65 kcal   5. 13.5 g/ml   7. 19.3 g/ml   9. 0.71, 0.71 g/ml   11. 23 g (ice), 25 g (water)   13. 240 g

15. a) 61 kg   b) 134 lb

17. more dense, near the floor

19. a) cool and/or compress   c) cool

27. 16.7 g/cm^3 29. no; density should be 17.1 g/cm^3; it is 16 g/cm^3

Chapter 4

3. CaCl$_2$ 5. CO$_2$

15. a) 6.022 × 10^{23} atoms   c) 1.205 × 10^{24} atoms

16. a) 1.7 × 10^{-24} g   c) 5.3 × 10^{-23} g

17. a) 3 × 10^{23} atoms   c) 2.4 × 10^{22} atoms

18. a) $^{19}_{9}$F   c) $^{235}_{92}$U

19. a) 6   c) 8   e) 11   g) potassium-39   i) radium-226

20. 63.6 amu   22. 39.1 amu

Chapter 5

8. a) cesium   c) chlorine

9. a) Si $^{14p}_{14n}$ 2, 8, 4   c) K $^{19p}_{20n}$ 2, 8, 8, 1   g) Ne $^{10p}_{10n}$ 2, 8

11. a) O $^{8p}_{8n}$ 2, 6   c) Ca $^{20p}_{20n}$ 2, 8, 8, 2   e) F $^{9p}_{10n}$ 2, 7

12. a) lose one; metal   c) lose two; metal   e) gain two; nonmetal
    g) no loss or gain; a noble gas (nonmetal)

13. a) Rb$^+$ c) Ba^{2+} e) S^{2-}

17. a) $1s^2\, 2s^2 2p^3$ c) $1s^2\, 2s^2 2p^6\, 3s^2 3p^5$ e) $1s^2\, 2s^2 2p^6\, 3s^2 3p^6\, 4s^2 3d^4$

Chapter 6

1. a) ionic   c) ionic   e) covalent   g) ionic   i) covalent   k) covalent

2. e) nonpolar   i) polar   k) polar

4. a) 2−, 6+   c) 3−, 5+   e) 2−, 6+

6. a) NaBr   c) Ba$_2$O   e) CO$_2$ g) Al$_2$O$_3$

8. a) See Fig. 6.10   c) see Fig. 6.7   e) :Ï:Ï: g) see Fig. 6.8

9. a) H$_2$O   c) BF$_3$

10. a) 2+   c) 4+   e) 3−   g) 5+   i) 2+   k) 7+

11. a) strontium chloride   c) carbon dioxide   e) ammonia (hydrogen nitride)   g) sodium phosphate
    i) iron(II) sulfide (ferrous sulfide)

12. a) aluminum oxide   c) silver nitrate   e) sodium hydroxide   g) barium nitrite
    i) potassium cyanide   k) lead sulfide   m) calcium chloride   o) iron(III) carbonate (ferric carbonate)

13. a) KBr   c) Ag$_2$O   e) AlCl$_3$ g) ZnCO$_3$ i) PbSO$_4$ k) Ca(HCO$_3$)$_2$ m) KNO$_2$
    o) Na$_2$SO$_3$ q) NH$_4$I   s) Cu$_2$CO$_3$ u) Ba$_3$(PO$_4$)$_2$

15. a) PCl$_3$ c) SO$_2$ e) K

Chapter 7

1. a) 103   c) 97.4   e) 197   g) 160

2. a) 1.00   c) 0.724   e) 0.551   g) 0.841

3. a) 3.79 × 10^{24}

6. a) 0.450   b) 0.450   c) 0.90   d) 32.0   e) 2.71 × 10^{23} f) 36.0

8. a) 39.3% Na, 60.7% Cl   c) 88.9% O, 11.1% H   e) 62.2% Fe, 35.6% O, 2.22% H
   g) 82.8% C, 17.2% H

10. a) (1) carbon and sulfur are reactants; carbon disulfide is the product.
    (2) carbon and sulfur react to produce carbon disulfide.
    (3) one atom of carbon reacts with two atoms of sulfur to produce one molecule of carbon disulfide.

11. a) KCl + AgNO$_3$ → AgCl + KNO$_3$ c) 3NaOH + FeCl$_3$ → 3NaCl + Fe(OH)$_3$
    e) NaOH + HNO$_3$ → NaNO$_3$ + HOH   g) SO$_3$ + HOH → H$_2$SO$_4$

12. a) C + O$_2$ → CO$_2$ c) 2K + S → K$_2$S   e) CaO + CO$_2$ → CaCO$_3$ g) H$_2$O + SO$_2$ → H$_2$SO$_3$

13. a) 2HgO $\xrightarrow{\Delta}$ 2Hg + O$_2$↑   c) Ca(OH)$_2$ $\xrightarrow{\Delta}$ CaO + H$_2$O   e) Na$_2$CO$_3$ $\xrightarrow{\Delta}$ Na$_2$O + CO$_2$↑
    g) H$_2$SO$_3$ $\xrightarrow{\Delta}$ H$_2$O + SO$_2$↑

16. a) KCl + AgNO$_3$ → KNO$_3$ + AgCl↓   c) 3NaOH + FeCl$_3$ → Fe(OH)$_3$↓ + 3NaCl
    e) Mg + NiCl$_2$ → Ni + MgCl$_2$ g) ZnO + 2HCl → ZnCl$_2$ + H$_2$O
    i) K$_2$SO$_3$ + 2HBr → 2KBr + H$_2$SO$_3$

17. a) $Cl^- + Ag^+ \rightarrow AgCl\downarrow$ c) $3OH^- + Fe^{3+} \rightarrow Fe(OH)_3\downarrow$ e) $Mg + Ni^{2+} \rightarrow Ni + Mg^{2+}$
    g) $ZnO + 2H^+ \rightarrow Zn^{2+} + H_2O$ i) $SO_3^{2-} + 2H^+ \rightarrow H_2SO_3$

18. a) $Zn + S \xrightarrow{\Delta} ZnS$ c) $Zn + CuSO_4 \rightarrow Cu\downarrow + ZnSO_4$
    e) $3Ca(OH)_2 + 2H_3PO_4 \rightarrow Ca_3(PO_4)_2 + 6H_2O$ g) $2K + Cl_2 \rightarrow 2KCl$

19. a) 160   c) 197   e) 170   g) 84   i) 89

20. 81.3 g   22. 9.0 g

23. a) 22.0 g   c) 2.0 mol

25. a) 3, 7.5, 0.9   c) 1.96 g

Chapter 8

13. a) 1.5 mol   c) 3.9 g   e) 61 g

14. a) 88 g   c) 400 g

15. a) 350 lbs   c) 142 lbs

17. 59 kcal   19. 3.72 kcal; 670 kcal

Chapter 9

6. a) $Ca(HCO_3)_2 \xrightarrow{\Delta} CaCO_3 + H_2O + CO_2\uparrow$ c) $3MgSO_4 + 2Na_3PO_4 \rightarrow Mg_3(PO_4)_2 + 3Na_2SO_4$
    e) $2Na\cdot zeolite + Ca^{2+} \rightarrow Ca(zeolite)_2 + 2Na^+$

9. a) 1.40 tons   c) 47.2%

10. 56 g; 141 g

11. a) 20.9%   c) 51.2%   e) 14.0%   g) 26.3%   i) 43.9%

Chapter 10

14. a) 4.5 g of NaCl + water to make 500 ml of solution   c) 0.85 g of $CuSO_4 \cdot 5H_2O$ + 24.15 g of water
    e) 10 g of iodine + ethyl alcohol to make 100 ml of solution
    g) 10.0 g of NaOH + water to make 1.50 l of solution
    i) 1 gram of $AgNO_3$ + water to make 100 ml of solution

15. a) 160 g of NaOH   c) 290 g of NaCl   e) 1.8 g HCl

16. a) 1.50 M   c) 0.10 M

17. 57.6 g

19. a) 1.80 kg   b) 1.57 kg

20. a) 1.2 kg   b) 28%

22. a) 412 g   b) 389 g

24. a) 17.2 m   b) −32°C

26. a) 1500 ml   c) 12.5 ml

28. a) 3 mol   b) 3.65 M

30. a) 16.7 ml of 6.0 M H_2SO_4 + water to make 1.0 dm³ of solution
    c) 370 ml of 95% ethyl alcohol + water to make 500 ml of solution

31. a) Add 120 mg of thiamin and 150 mg of riboflavin to 100 g of sugar   c) 3.75 mg

Chapter 11

12. KI, $BaSO_4$, $CuBr_2$, MgS

15. in H_2SO_4: H^+, HSO_4^-, SO_4^{2-};   in HNO_3: H^+, NO_3^-;   in H_3PO_4: H^+, $H_2PO_4^-$, HPO_4^{2-}, PO_4^{3-}

21. a) $2NH_4^+ + 2Cl^- + Ca^{2+} + 2OH^- \rightarrow Ca^{2+} + 2Cl^- + 2NH_3\uparrow + 2HOH$

    $\qquad\qquad\qquad$ calcium $\qquad\qquad\qquad\qquad\qquad$ ammonia
    $\qquad\qquad\qquad$ chloride

    c) $2H^+ + SO_2^{4-} + 2K^+ + 2CN^- \rightarrow 2HCN\uparrow + 2K^+ + SO_4^{2-}$

    $\qquad\qquad$ hydrogen $\qquad\qquad\qquad$ potassium
    $\qquad\qquad$ cyanide $\qquad\qquad\qquad$ sulfate

    e) $Al^{3+} + 3Cl^- + 3Na^+ + 3OH^- \rightarrow Al(OH)_3\downarrow + 3Na^+ + 3Cl^-$

    $\qquad\qquad\qquad$ aluminum $\qquad\qquad\qquad$ sodium
    $\qquad\qquad\qquad$ hydroxide $\qquad\qquad\qquad$ chloride

    g) $3Ag^+ + 3NO_3^- + Al^{3+} + 3Cl^- \rightarrow 3AgCl\downarrow + Al^{3+} + 3NO_3^-$

    $\qquad\qquad\qquad$ silver $\qquad\qquad\qquad$ aluminum
    $\qquad\qquad\qquad$ chloride $\qquad\qquad\qquad$ nitrate

22. a) $NaOH + HCl \rightarrow NaCl + HOH$ $Na^+ + OH^- + H^+ + Cl^- \rightarrow Na^+ + Cl^- + HOH$

    $\qquad\qquad\qquad\qquad$ sodium
    $\qquad\qquad\qquad\qquad$ chloride

    c) $Ca(OH)_2 + 2HNO_3 \rightarrow Ca(NO_3)_2 + 2HOH$ $Ca^{2+} + 2OH^- + 2H^+ + 2NO_3^- \rightarrow Ca^2 + 2NO_3^-$

    $\qquad\qquad\qquad\qquad\qquad$ calcium $\qquad\qquad\qquad\qquad\qquad\qquad\qquad\qquad\qquad$ + 2HOH
    $\qquad\qquad\qquad\qquad\qquad$ nitrate

    e) $NH_3 + HNO_3 \rightarrow NH_4NO_3$ $NH_3 + H^+ + NO_3^- \rightarrow NH_4^+ + NO_3^-$

    $\qquad\qquad\qquad\qquad\qquad$ ammonium
    $\qquad\qquad\qquad\qquad\qquad$ nitrate

    g) $2Al(OH)_3 + 3H_3SO_4 \rightarrow Al_2(SO_4)_3 + 6HOH$

    $\qquad\qquad\qquad\qquad\qquad$ aluminum
    $\qquad\qquad\qquad\qquad\qquad$ sulfate

      $2Al^{3+} + 6OH^- + 6H^+ + 3SO_4^{2-} \rightarrow 2Al^{3+} + 3SO_4^{2-} + 6HOH$

23. a) $2HCl + Na_2CO_3 \rightarrow 2NaCl + H_2O + CO_2$ c) $H_2O + SO_3 \rightarrow H_2SO_4$

    $\qquad\qquad\qquad\qquad$ sodium $\qquad\qquad\qquad$ carbon $\qquad\qquad\qquad\qquad$ sulfuric
    $\qquad\qquad\qquad\qquad$ chloride $\qquad\qquad\qquad$ dioxide $\qquad\qquad\qquad\qquad$ acid

    e) $CaO + H_2O \rightarrow Ca(OH)_2$ g) $Al_2O_3 + 3H_2SO_4 \rightarrow Al_2(SO_4)_3 + 3HOH$ i) $H_2O + CO_2 \rightarrow H_2CO_3$

    $\qquad\qquad$ calcium $\qquad\qquad\qquad\qquad\qquad$ aluminum $\qquad\qquad\qquad\qquad\qquad\qquad$ carbonic
    $\qquad\qquad$ hydroxide $\qquad\qquad\qquad\qquad\qquad$ sulfate $\qquad\qquad\qquad\qquad\qquad\qquad$ acid

26. a) 40 g   c) 49 g   c) 14.8 g

27. a) $0.25N$ c) $1.N$ e) $0.5N$

28. $0.5N$

30. a) $0.4N$ c) $0.06N$

Chapter 12

11. acid with pK_a of 4.5; infinity

13. a) neutral   c) basic   e) neutral   g) basic

14. a) 0.25   c) 0.18

16. a) 3×10^{-4} b) 4.36×10^{-4}

17. a) 6.25×10^{-8}

18. a) 5   c) 9

19. a) 1×10^{-4} mol/liter   c) 4.47×10^{-8} mol/liter   e) 10^{-9} mol/liter   g) 1.0×10^{-1} mol/liter

20. a) 6   c) 4.25

21. a) 4.7   c) 3.7

23. 3×10^{-2} mol/liter

25. The pK_a is too far from body pH.

28. a) 2.000   c) 1.5866   e) -2.000
    g) -1.1267 i) 5.2695

29. a) 100   c) 2.06   e) 8.25   g) 526   i) 2.5×10^{-4}

Chapter 13

27. a) $2\,NaCl + \text{electricity} \rightarrow 2\,Na + Cl_2\uparrow$ c) no reaction   e) $Zn + S \rightarrow ZnS$
    g) $2ZnS + 3O_2 \rightarrow 2ZnO + 2SO_2\uparrow$ i) $CaCO_3 \xrightarrow{\Delta} CaO + CO_2\uparrow$ k) $2HgO \xrightarrow{\Delta} 2Hg + O_2\uparrow$
    m) $2Cu_2S + 3O_2 \rightarrow 2Cu_2O + SO_2\uparrow$ o) $2Au + 3S \xrightarrow{\Delta} Au_2S_3$ q) $Ag^+ + Cu^0 \rightarrow Cu^{2+} + 2Ag^0$

29. 840 lbs   31. 32 g   33. 50 g

34. a) $Zn + 2HCl \rightarrow ZnCl_2 + H_2\uparrow$ c) $Ca + 2H_2O \rightarrow Ca(OH)_2 + H_2\uparrow$ e) no reaction
    g) $Zn + Pb(NO_3)_2 \rightarrow Zn(NO_3)_2 + Pb$ i) $H_2O + C \xrightarrow{\Delta} CO\uparrow + H_2\uparrow$ k) no reaction
    m) $Mg + 2HCl \rightarrow MgCl_2 + H_2\uparrow$ o) no reaction   q) $Mg + H_2O(\text{steam}) \rightarrow MgO + H_2\uparrow$
    s) $CuO + H_2 \rightarrow Cu + H_2O$

36. 48.5 g

38. a) 11 g   c) 21 g

Chapter 14

10. a) $Ca(ClO)_2$ c) $HBrO_2$ e) $Mg(BrO_2)_2$ g) HFO_4

16. a) 4^+ c) 3^- e) 2^- g) 6^+ i) 7^+

17. a) $H_2 + Cl_2 \rightarrow 2HCl$ c) $2K^+ + 2Br^- + Cl_2 \rightarrow 2K^+ + 2Cl^- + Br_2$
    e) $Ca^{2+} + 2Cl^- + F_2 \rightarrow Ca^{2+} + 2F^- + Cl_2$ g) $Cu + Cl_2 \rightarrow Cu^{2+} + 2Cl^-$
    i) $2K^+ + 2I^- + Br_2 \rightarrow 2K^+ + 2Br^- + I_2$ k) $Ca + Cl_2 \rightarrow Ca^{2+} + 2Cl^-$
    m) $Cl_2 + Br^- \rightarrow 2Cl^- + Br_2$ o) $I_2 + Cl^- \rightarrow$ no reaction

18. a) $2Na^+ + 2Cl^- + 2H^+ + SO_4^{2-} \rightarrow 2HCl\uparrow + 2Na^+ + SO_4^{2-}$
    c) $2Na^+ + 2Cl^- + 2HOH \xrightarrow{\text{electrolysis}} 2Na^+ + 2OH^- + Cl_2\uparrow + H_2\uparrow$ e) $H_2 + X_2 \rightarrow 2HX$
    g) $KClO_3 \xrightarrow{\Delta} KCl + 1\frac{1}{2}O_2\uparrow$

19. a) 15 g   c) 9.0 g

20. a) $Na_2B_4O_7 + Mg^{2+} \rightarrow MgB_4O_7 + 2Na^+$ c) $C_3H_8 + 5O_2 \rightarrow 3CO_2 + 4H_2O$
    e) $P_2O_5 + 3H_2O \rightarrow 2H_3PO_4$ g) $FeS + H_2SO_4 \rightarrow FeSO_4 + H_2S\uparrow$
    i) $C_8H_{18} + 12\frac{1}{2}O_2 \rightarrow 8CO_2 + 9H_2O$ k) $H_2O + SO_3 \rightarrow H_2SO_4$

21. a) 21.5%   c) 32.7%   e) 36.8%

22. 3.1 tons   24. 32 lbs

Chapter 15

Chapter 16

1. $98.6°F = 37°C = 310°K$ 3. $-40°F$ 5. 270 ml   7. 500 liters   9. 141 ml

11. 133 500 cu ft   13. 2.59 g/liter

15. a) 880 ml   b) 38.1 = mol wt

17. 9 liters   19. $1.76(10^3)$ ml   21. a) 190 ml   23. 576 ml   26. 426 torr

Chapter 17

24. a) $^{40}_{19}K$ c) α particle   e) $^{13}_7N$ g) $^{242}_{96}Cm$

26. a) $^{234}_{90}Th$ c) n   e) $^{27}_{14}Si$ g) β particle   i) $^{99}_{43}Tc$

Chapter 18

3. a) four   c) seven

Chapter 19

1. a) 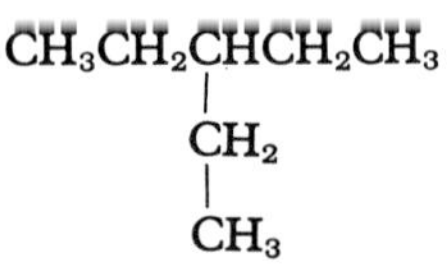 $CH_3CH_2CHCH_2CH_3$ c) 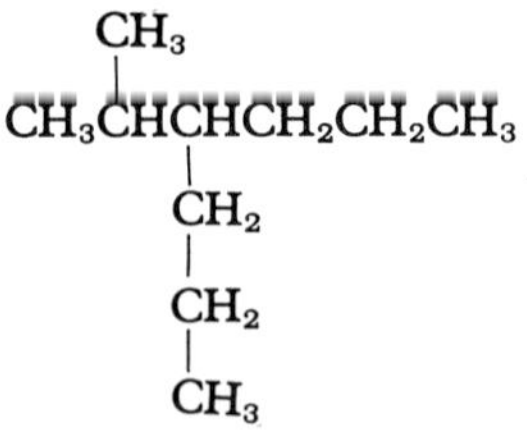 $CH_3\overset{\overset{\displaystyle CH_3}{|}}{C}HCHCH_2CH_2CH_3$ e) 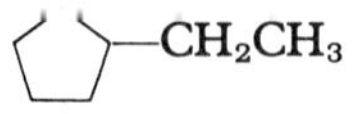 —CH_2CH_3

a) with branch
$$\begin{array}{c} | \\ CH_2 \\ | \\ CH_3 \end{array}$$

c) with branch
$$\begin{array}{c} | \\ CH_2 \\ | \\ CH_2 \\ | \\ CH_3 \end{array}$$

$$\text{g) } \underset{\underset{CH_3}{|} \quad \underset{C_2H_5}{|}}{CH_3CHCH_2CHCHCH_2CH_2CH_3}$$

$$\text{i) } \underset{\underset{C_2H_5}{|}}{CH_3CH-\overset{\overset{Cl \quad Cl}{|\quad|}}{C}CH_2CH_2CH_3}$$

2. and 3.

a) $CH_3CH_2CH_2CH_2CH_2CH_3$
hexane

$$\underset{\underset{CH_3}{|}}{CH_3CH_2CH_2CHCH_3}$$
2-methylpentane

$$\underset{\underset{CH_3}{|}}{CH_3CH_2CHCH_2CH_3}$$
3-methylpentane

$$\underset{\underset{CH_3}{|}}{CH_3CH_2-\overset{\overset{CH_3}{|}}{C}-CH_3}$$
2,2-dimethylbutane

$$\underset{\underset{CH_3 \ CH_3}{|\quad|}}{CH_3-CH-CH-CH_3}$$
2,3-dimethylbutane

c) $\underset{\underset{Cl}{|}}{CH_3CH_2CH_2\overset{\overset{Cl}{|}}{CH}}$
1,1-dichlorobutane

$$\underset{\underset{Cl \quad Cl}{|\quad|}}{CH_3CH_2CH-CH_2}$$
1,2-dichlorobutane

$$\underset{\underset{Cl \quad Cl}{|\quad|}}{CH_3CHCH_2CH_2}$$
1,3-dichlorobutane

$$\underset{\underset{Cl \qquad\quad Cl}{|\qquad\quad|}}{CH_2CH_2CH_2CH_2}$$
1,4-dichlorobutane

$$\underset{\underset{Cl}{|}}{CH_3CH_2-\overset{\overset{Cl}{|}}{C}-CH_3}$$
2,2-dichlorobutane

$\underset{\underset{Cl \ Cl}{|\ |}}{CH_3CH-CHCH_3}$
2.3-dichlorobutane

$$\underset{\underset{CH_3 \ Cl}{|\quad|}}{CH_3-CH-CH-Cl}$$
1,1-dichloro-2-methylpropane

$$\underset{\underset{CH_3}{|}}{CH_3-\overset{\overset{Cl}{|}}{C}-CH_2Cl}$$
1,2-dichloro-2-methylpropane

$$\underset{\underset{CH_3}{|}}{ClCH_2CHCH_2Cl}$$
1,3-dichloro-2-methylpropane

4. a) 2-methylpentane   c) 2-chloro-4-ethylhexane   e) cyclopentane   g) methylcylopentane
   i) 6-ethyl-2,3,5-trimethyloctane

5. a) chloroform: trichloromethane   c) isobutane; methylpropane   e) methyl bromide; bromomethane

6. a) $CH_3CH_2CH_3 + 5O_2 \rightarrow 3CO_2 + 4H_2O$ c) $CH_4 + Cl_2 \rightarrow CH_3Cl + HCl$
$CH_3Cl + Cl_2 \rightarrow CH_2Cl_2 + HCl$
$CH_2Cl_2 + Cl_2 \rightarrow CHCl_3 + HCl$
$CHCl_3 + Cl_2 \rightarrow CCl_4 + HCl$

e) ⬠ + $Cl_2 \rightarrow$ ⬠—Cl + HCl   ⬠—Cl + $Cl_2 \rightarrow$ ⬠(Cl)(Cl) + HCl   and so forth

g) no reaction   i) $CH_2Cl_2 + Br_2 \rightarrow CHCl_2Br + HBr$
$CHCl_2Br + Br_2 \rightarrow CCl_2Br_2 + HBr$

9. a) carboxylic acid, ester, alkane, aromatic   c) amine, alkane   e) alcohol, alkane, aromatic

Chapter 20

1. c) $\underset{\underset{CH_3}{|}}{CH_3CHCH_2CH_2CH_3}$ C_6H_{14}; $\underset{\underset{CH_3}{|}}{CH_3-\overset{\overset{CH_3}{|}}{C}=C-CH_3}$ C_6H_{12};   ⬠—CH_3 C_6H_{12}

3. a) $CH_3CH_2CH=CH_2$

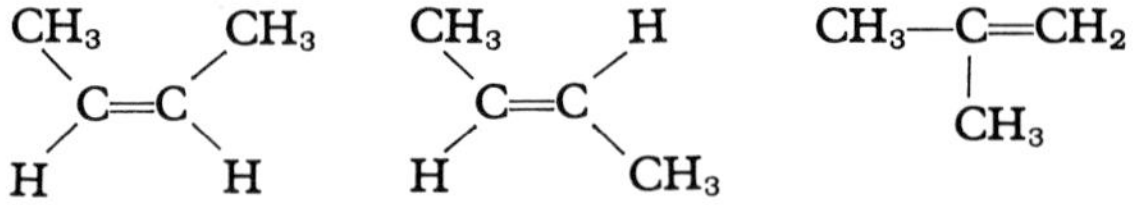

$CH_3-C=CH_2$
            |
           CH_3

4. a)

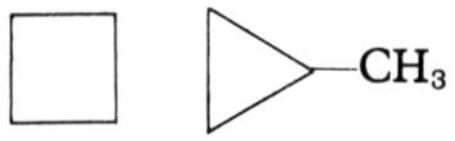

5. a) six

6. a) 1-butene; *cis*-2-butene; *trans*-2-butene; methylpropene

7. a) 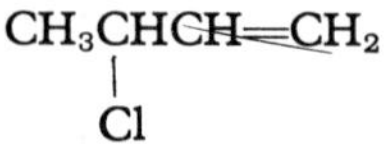 $CH_3CHCH=CH_2$ c)    e) 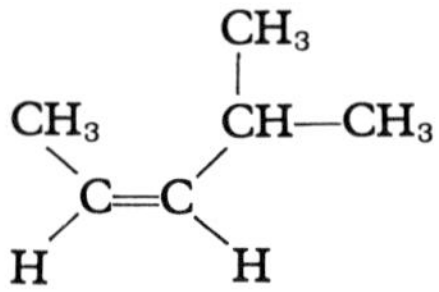
         |
         Cl

10 a) $CH_3CH_2CH=CH_2 + H_2 \xrightarrow{Pt} CH_3CH_2CH_2CH_3$

c) $2CH_3CH_2CH=CHCH_3 + 15O_2 \rightarrow 10CO_2 + 10H_2O$

e)

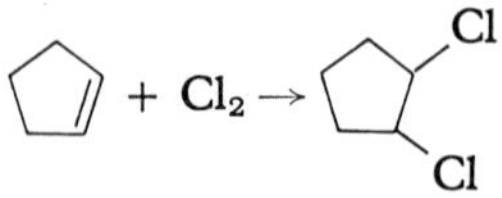

g) $CH_3CH=CHCH_3 + HOSO_2OH \rightarrow CH_3CH_2CHCH_3$
                                                    |
                                                 OSO_2OH

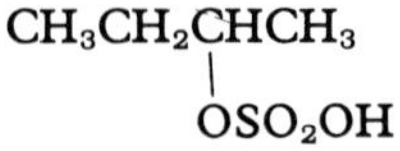

i) $CH_2=CHCH_3 + HCl \rightarrow CH_3CHCH_3$
                                         |
                                         Cl

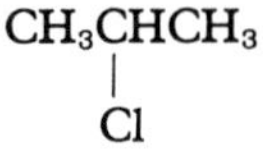

k) $CH_3C\equiv CH + 2Br_2 \rightarrow CH_3C-CH$ (with Br Br above and Br Br below)

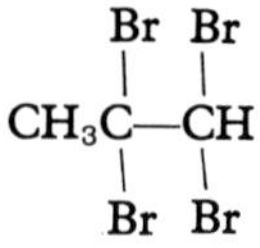

m) $CH_3C\equiv CH + 4O_2 \rightarrow 3CO_2 + 2H_2O$

12. a) yes   c) no   e) yes   g) yes

14. a) petroleum   b) $CH_3CH_3 + Cl_2 \xrightarrow{UV} CH_3CH_2Cl + HCl$ c) $CH_3CH_3 \xrightarrow{500°C} CH_2=CH_2 + H_2$

$CH_2=CH_2 + HCl \rightarrow CH_3CH_2Cl$

d) Method (c) gives only the desired product. Method (b) gives many other compounds, besides the desired product: dichloroethanes, trichloroethanes, etc. This result means that only a small amount of chloroethane is obtained, and it must be separated from all the other substances.

Chapter 21

1. a) 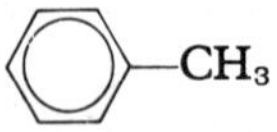   c) 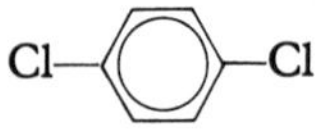   e) 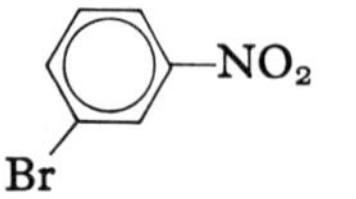   g)

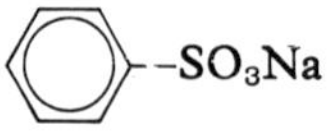

i) 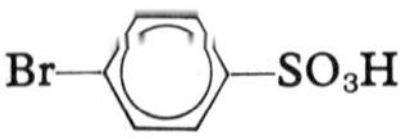   k)

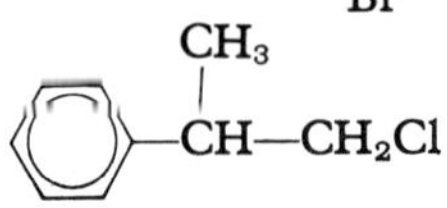

2. a)

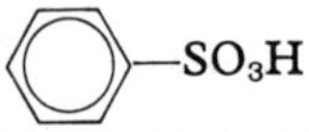

benzenesulfonic acid

c)

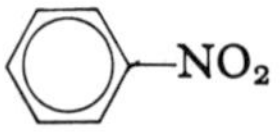

nitrobenzene

e) no reaction

3. a) [benzene ring]—COOH   c) [benzene ring with COOH at top, —COOH on right, COOH at bottom]

8. a) toluene   c) *m*-dinitrobenzene   e) 1-chloronaphthalene (or α)
   g) 3-phenyl-1-butene   i) ethylbenzene

Chapter 22

1. and 3.

$CH_3CH_2CH_2CH_2OH$ (primary) $CH_3CH_2CHCH_3$ | OH (secondary) CH_3CHCH_2OH | CH_3 (primary) $CH_3{-}\underset{\underset{OH}{|}}{\overset{\overset{CH_3}{|}}{C}}{-}CH_3$ (tertiary)

5. a) CH_3CHCH_3 | OH c) CH_3CHCH_2OH | CH_3 e) [benzene ring]OK   g) [benzene ring with OH and NO$_2$]

i) $CH_3CHCHCH_2CHCH_3$ with | OH and | Br and | CH$_3$ k) $HOCH_2CHCHCH_2CH_3$ with | C$_2$H$_5$ and phenyl substituent   m) [cyclohexane ring]OH

7. a) CH_3CHCH_3 | OH $+ HBr \xrightarrow{\Delta} CH_3CHCH_3$ | Br $+ HOH$

   c) $CH_3CH_2CHCH_2CH_3$ | OH $\xrightarrow[250°C]{Cu} CH_3CH_2CCH_2CH_3$ ‖ O $+ H_2$

   e) $CH_3CH_2CHCH_3$ | OH $+ [O] \rightarrow CH_3CH_2CCH_3$ ‖ O $+ H_2O$

9. a) amine, ether, alkane, aromatic

11. 2-methyl-2-butanol

14. a) aldehyde   b) ketone   c) no reaction

15. a) 1-butanol   c) 2-butanol   e) 2-butanethiol   g) *p*-chlorophenol   i) 1-naphthol
    k) 2-ethoxypropane

16. a) 1-propanol   c) 1,4-butanediol

17. a) 1-propanol   c) 1,4-butanediol

Chapter 23

1. a) CH_3CH_2COOH c) [benzene ring]COOH   e) [benzene ring]COOK

g) $CH_3CH-O-C-CH_3$ **i)** $C_{27}H_{55}-C-O-C_{30}H_{61}$
    $\underset{CH_3}{|}$ $\underset{O}{||}$ $\underset{O}{||}$

2. a) $H-C-O-H + Na^+OH^- \rightarrow H-C-O^-Na^+ + HOH$
    $\underset{O}{||}$ $\underset{O}{||}$

c) ⬡$-C-O-H$ + HOC_2H_5 $\xrightarrow[\text{heat}]{H_2SO_4}$ ⬡$-C-O-C_2H_5$ + HOH
    $\underset{O}{||}$ $\underset{O}{||}$

e) $CH_3CH_2CH_2C-O-C_2H_5 + HOH \xrightarrow[\text{heat}]{H_2SO_4} CH_3CH_2CH_2C-O-H + HOC_2H_5$
    $\underset{O}{||}$ $\underset{O}{||}$

g) ⬡$-C-O-CH_3$ + NaOH → ⬡$-C-O^-Na^+$ + $HOCH_3$
    $\underset{O}{||}$ $\underset{O}{||}$

3. a) $CH_3(CH_2)_4CH_2OH + 2[O] \rightarrow CH_3(CH_2)_4C-OH$
    $\underset{O}{||}$

5. a) ethyl butanoate   **c)** 2-hydroxybutanoic acid   **e)** potassium butanoate

7. a) oxidation ($KMnO_4$ or $Na_2Cr_2O_7$)

8. a) ester, phenol, alkane, aromatic

Chapter 24

1. a) More than one structure is possible.
   c) Two or more of the fatty acids chains would contain double bonds.   **e)** See Section 24.7.

2. a) saponification   **c)** hydrolysis

Chapter 25

1. a) diethylamine   **c)** isobutylamine; 1-amino-2-methylpropane   **e)** propionamide
   g) 4-ethyl-3-(methylamino)hexane

2. a) CH_3-N-CH_3 **c)** $CH_3CH_2NHCH_2CH_2CH_3$ **e)** $CH_3CHCHCHCH_3$ **g)** $(CH_3CH_2)_3\overset{+}{N}H$ Cl^-
    $\underset{CH_3}{|}$ $\overset{NH_2}{\underset{CH_3\ \ CH_3}{|}}$

   i) $CH_3CH_2CH_2C-N-C_2H_5$
    $\underset{O\ \ CH_3}{||\ |}$

4. a) $CH_3Br + NH_3 \rightarrow CH_3NH_2 + HBr$ **c)** $CH_3CH_2-C-OH + NH_3 \xrightarrow{\Delta} CH_3CH_2-C-NH_2 + HOH$
    $\underset{O}{||}$ $\underset{O}{||}$

e) CH_3-⬡$-NO_2 + 6[H] \xrightarrow[\text{HCl}]{Sn} CH_3-$⬡$-NH_2 + 2H_2O$

g) ⬡$-C-OH$ + $H_2NC_2H_5 \xrightarrow{\text{heat}}$ ⬡$-C-NHC_2H_5$ + HOH
    $\underset{O}{||}$ $\underset{O}{||}$

i) $CH_3CH_2CH_2-\underset{\underset{O}{\|}}{C}-NH_2 + HOH \rightarrow CH_3CH_2CH_2-\underset{\underset{O}{\|}}{C}-OH + NH_3$

5. $R-\underset{\underset{O}{\|}}{C}-\underset{\underset{CH_2CH_3}{|}}{N}-CH_2CH_3$

8. a) amine, ester, substituted ammonium salt, alkane, aromatic

Chapter 26

1. a) propanal   c) ethyl phenyl ketone

2. a) $CH_3CH_2CH_2CHCH_3 \xrightarrow[250°C]{Cu} CH_3CH_2CH_2\underset{\underset{O}{\|}}{C}CH_3 + H_2$ c) $CH_3OH \xrightarrow[250°C]{Cu} H-\underset{\underset{O}{\|}}{C}-H$
with OH under CHCH₃

3. a) $CH_3CH_2CHCH_2CH_3$ c)
with OH under

$$
\begin{array}{c}
CH_3 \\
| \\
CH \\
\diagup \quad \diagdown \\
H_2C \quad\quad CH_2 \\
| \quad\quad\quad | \\
H_2C \quad\quad CH_2OH \\
\diagdown \\
CH_2 \\
| \\
CH \\
\diagup \quad \diagdown \\
CH_3 \quad CH_3
\end{array}
$$

5. Benzoic acid is formed by exposure to air because the aldehyde is so easily oxidized.

7. For identification; see Eq. (26–23)

8. a) Cu, 250°   b) First, add HCN; second, heat in water with HCl.

11. a) aldehyde, alkene, aromatic   c) alcohol, ketone, alkene, alkane

12. a) no   c) yes, black precipitate

13. a) yes, red precipitate   c) no

14. a) ⟨benzene ring with Cl⟩—CH=NOH   c) $CH_3CH_2\underset{\underset{O}{\|}}{C}CH_3$ e) $CH_3CH_2CHCH_3$ with NH₂

g) no reaction   i) CH_3CH_2CHCN with OH

Chapter 27

1. a) active   c) inactive   e) active

3. a) *cis-trans* isomers   c) optical isomers   e) not isomers   g) optical isomers

Chapter 28

1. $\underset{Cl}{CH{=}CH_2}$ $\underset{C_6H_5}{CH{=}CH_2}$ $\underset{CH_3}{CH{=}CH_2}$ $\underset{\text{ethylene}}{CH_2{=}CH_2}$ $\underset{\text{tetrafluoroethylene}}{CF_2{=}CF_2}$

vinyl chloride   styrene   propylene

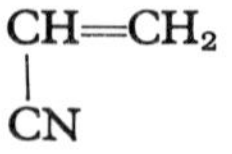

CH=CH₂
|
CN
acrylonitrile

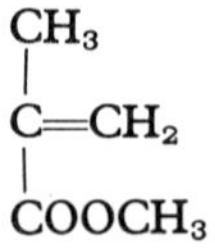

CH₃
|
C=CH₂
|
COOCH₃
methyl methacrylate

2. See Eq. 28–1.

3.

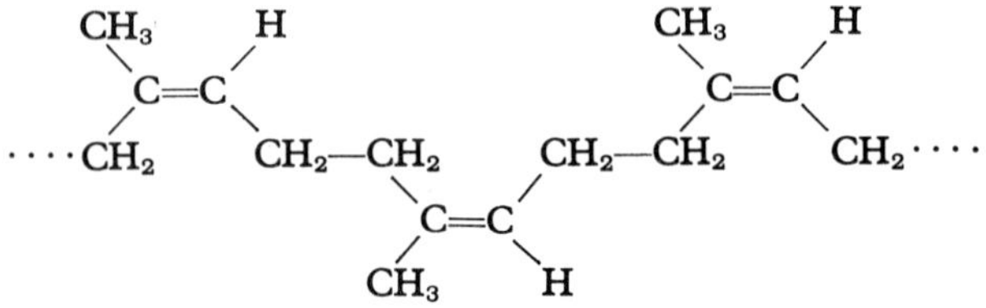

CH₃   H
  C=C
···CH₂   CH₂—CH₂   CH₂—CH₂   CH₂····
              C=C
            CH₃   H

CH₃   H
  C=C

Chapter 29

Chapter 30

2. a)

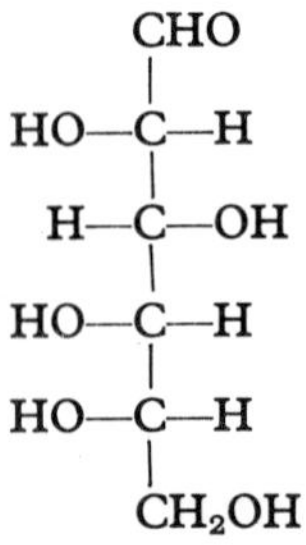

CHO
|
HO—C—H
|
H—C—OH
|
HO—C—H
|
HO—C—H
|
CH₂OH

c)

CHO
|
C=O
|
H—C—OH
|
H—C—OH
|
H

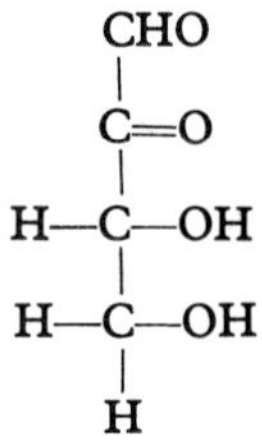

6. a)

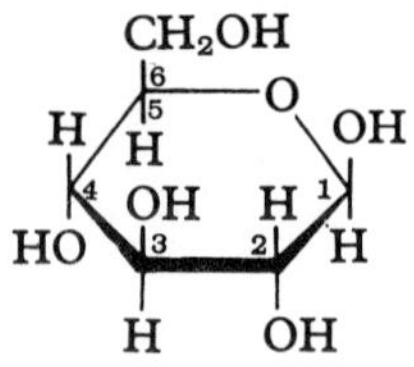

CH₂OH
       6
      5 O
H H   OH
  4 OH H 1
HO 3   2 H
   H   OH
β-D-glucose

c)

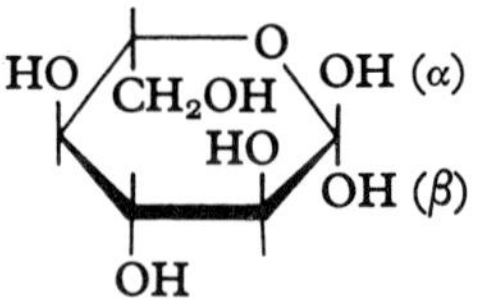

HO   O   OH (α)
   CH₂OH
      HO
            OH (β)
     OH

e)

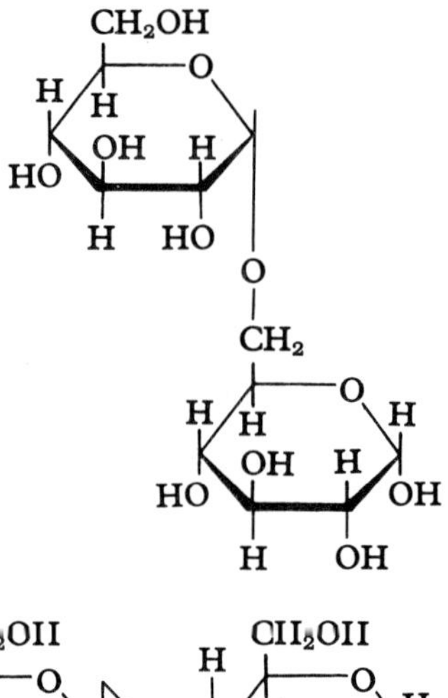

CH₂OH
       O
H H
   OH   H
HO
   H   HO
          O
          CH₂
        H H   H
          OH   H
        HO   OH
           H   OH

g) 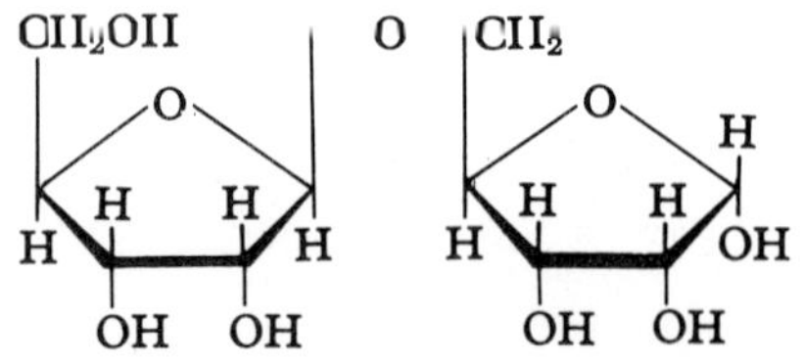 CH₂OH   O   CH₂
       O
   H   H   H   H
 H   H   H   OH
   OH OH   OH OH

i) 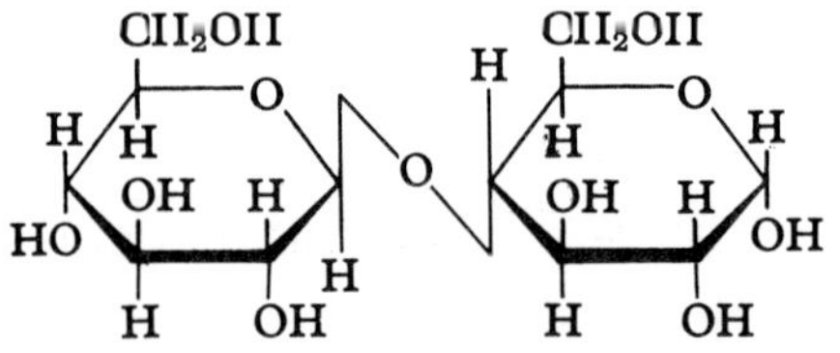 CH₂OH   CH₂OH
                 H
   H H   O   H   H   O
     OH H   OH H
 HO   O   HO   OH
     H   OH   H   H   OH

k)

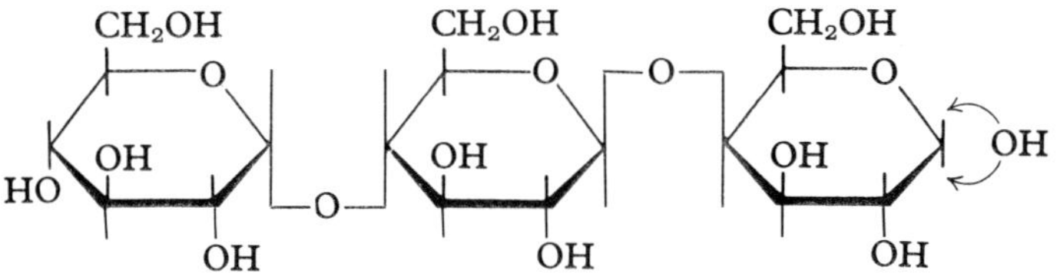

(One of many possible structures)

8. a)

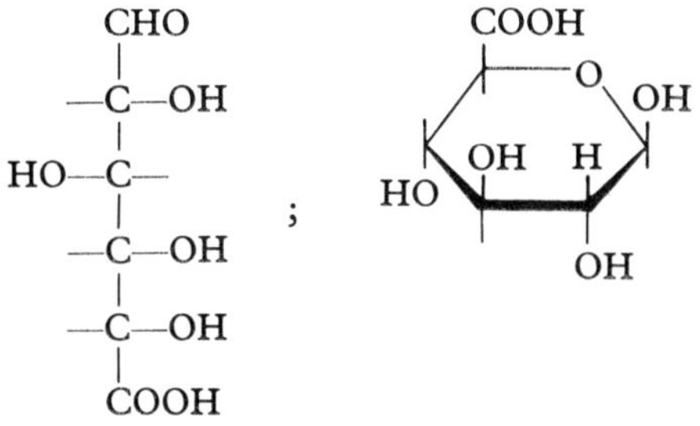

b)

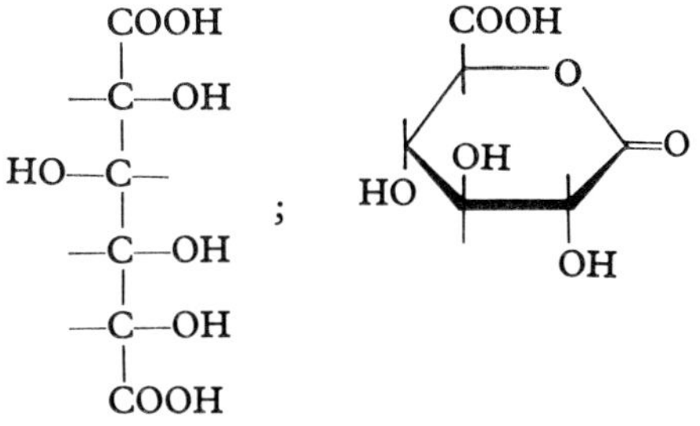

d)

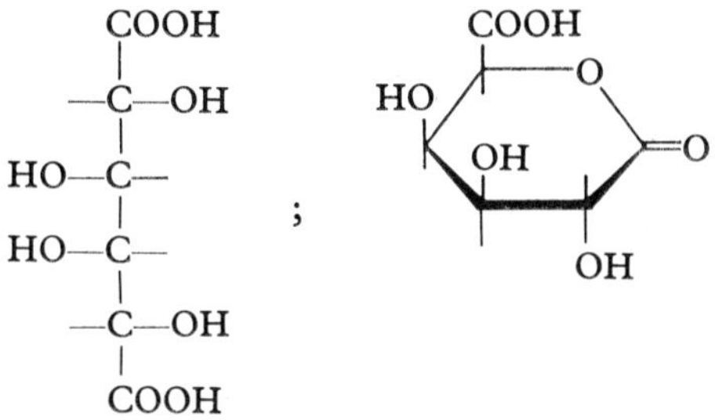

f)

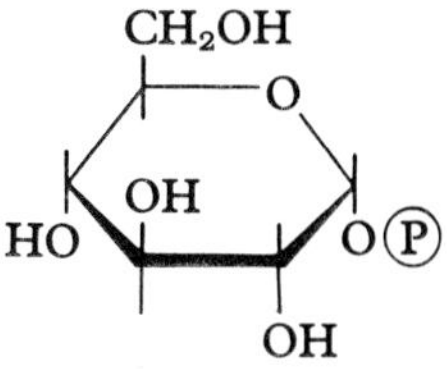

9. a)

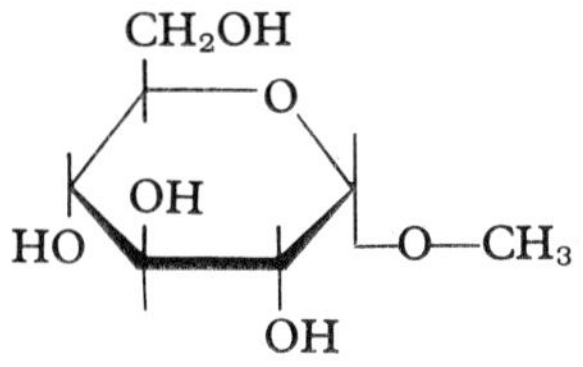

c)

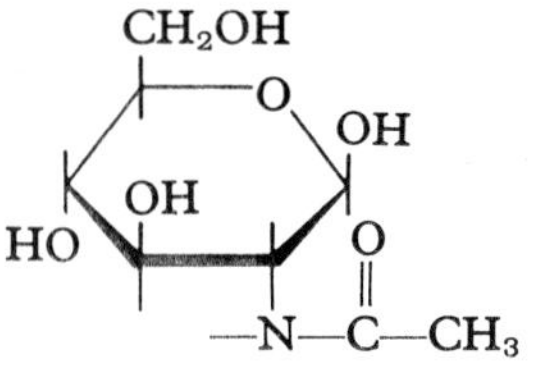

10. a)

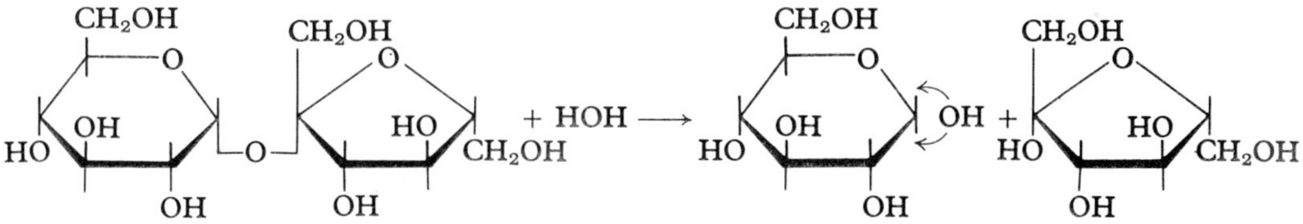

Chapter 31

6. 10.6%

9. cystine:

1) 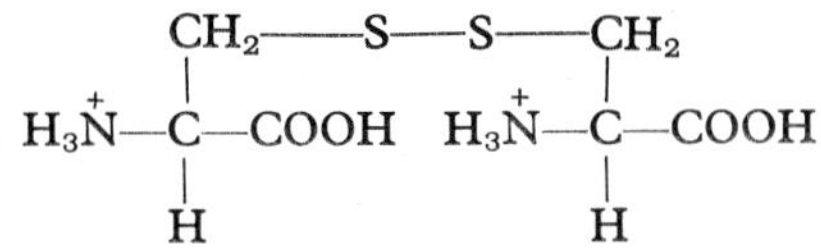 $H_3\overset{+}{N}$—C—COOH $H_3\overset{+}{N}$—C—COOH

2) 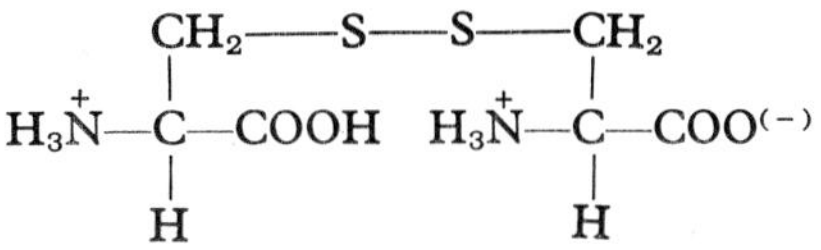 $H_3\overset{+}{N}$—C—COOH $H_3\overset{+}{N}$—C—COO$^{(-)}$

(Continued)

$$\text{3) } \underset{\underset{H}{|}}{\overset{\overset{\displaystyle CH_2—S—S—CH_2}{|}}{H_3\overset{+}{N}—C—COO^{(-)}}} \quad \underset{\underset{H}{|}}{H_3\overset{+}{N}—C—COO^{(-)}}$$

$$\text{4) } \underset{\underset{H}{|}}{\overset{\overset{\displaystyle CH_2—S—S—CH_2}{|}}{H_2N—C—COO^{(-)}}} \quad \underset{\underset{H}{|}}{H_3\overset{+}{N}—C—COO^{(-)}}$$

$$\text{5) } \underset{\underset{H}{|}}{\overset{\overset{\displaystyle CH_2—S—S—CH_2}{|}}{H_2N—C—COO^{(-)}}} \quad \underset{\underset{H}{|}}{H_2N—C—COO^{(-)}}$$

lysine:

$$\text{1) } \underset{\underset{H}{|}}{\overset{\overset{\displaystyle H_2C—\overset{+}{N}H_3}{\underset{(CH_2)_3}{|}}}{H_3\overset{+}{N}—C—COOH}}$$

$$\text{2) } \underset{\underset{H}{|}}{\overset{\overset{\displaystyle H_2C—\overset{+}{N}H_3}{\underset{(CH_2)_3}{|}}}{H_3\overset{+}{N}—C—COO^{(-)}}}$$

$$\text{3) } \underset{\underset{H}{|}}{\overset{\overset{\displaystyle H_2C—NH_2}{\underset{(CH_2)_3}{|}}}{H_3\overset{+}{N}—C—COO^{(-)}}}$$

$$\text{4) } \underset{\underset{H}{|}}{\overset{\overset{\displaystyle H_2C—\overset{+}{N}H_3}{\underset{(CH_2)_3}{|}}}{H_2N—C—COO^{(-)}}}$$

$$\text{5) } \underset{\underset{H}{|}}{\overset{\overset{\displaystyle H_2C—NH_2}{\underset{(CH_2)_3}{|}}}{H_2N—C—COO^{(-)}}}$$

It should be realized that all the formulas given above represent possible structures. In any given solution two or more of the forms will probably be in equilibrium with each other.

11. a) $\underset{\underset{H}{|}}{H_3\overset{+}{N}—\overset{\overset{\displaystyle H}{|}}{C}—\overset{\overset{\displaystyle O}{\|}}{C}—O^{(-)}} + \underset{\underset{H}{|}}{H_3\overset{+}{N}—\overset{\overset{\displaystyle CH_2OH}{|}}{C}———\overset{\overset{\displaystyle O}{\|}}{C}—O^{(-)} \rightarrow H_3\overset{+}{N}—\overset{\overset{\displaystyle H}{|}}{\underset{\underset{H}{|}}{C}}—\overset{\overset{\displaystyle O}{\|}}{C}—\overset{\overset{\displaystyle H}{|}}{\underset{\underset{H}{|}}{N}}—\overset{\overset{\displaystyle CH_2OH}{|}}{C}———\overset{\overset{\displaystyle O}{\|}}{C}—O^{(-)} + H_2O$

glycine   serine   glycylserine

Serylglycine is also formed.

b) $\underset{\underset{H}{|}}{H_3\overset{+}{N}—\overset{\overset{\displaystyle CH_3}{|}}{C}—\overset{\overset{\displaystyle O}{\|}}{C}—O^{(-)}} +$ $\underset{\underset{\displaystyle O=C—COO^{(-)}}{|}}{\overset{\overset{\displaystyle COO^{(-)}}{|}}{\underset{\underset{CH_2}{|}}{CH_2}}} \rightarrow O=\overset{\overset{\displaystyle CH_3}{|}}{C}—COO^{(-)} +$ $\underset{\underset{\displaystyle H}{|}}{\overset{\overset{\displaystyle COO^-}{|}}{\underset{\underset{H_3\overset{+}{N}—C—COO^{(-)}}{|}}{\overset{\overset{\displaystyle CH_2}{|}}{CH_2}}}}$

alanine   α-ketoglutaric acid   pyruvic acid   glutamic acid

c) $\underset{N\diagdown NH}{\boxed{}}—CH_2—\underset{\underset{\displaystyle \overset{+}{N}H_2}{|}}{CH}—COO^{(-)} \rightarrow CO_2 + \underset{N\diagdown NH}{\boxed{}}—CH_2—CH_2—NH_2$

histidine   histamine

$\underset{\underset{\displaystyle ornithine}{}}{\overset{\overset{\displaystyle CH_2—NH_2}{\underset{(CH_2)_2}{|}}}{H_3\overset{+}{N}—CH—COO^{(-)}}} \rightarrow CO_2 + H_2N—(CH_2)_4—NH_2$

1,4-diaminobutane
or putrescine

d) structure of lysylphenylalanylasparagine + 3H₂O →

$$\text{d) H}_3\overset{+}{\text{N}}-\text{C}-\text{C}-\text{N}-\text{C}-\text{C}-\text{N}-\text{C}-\text{C}-\text{O}^{(-)} + 3\text{H}_2\text{O} \longrightarrow$$

lysylphenylalanylasparagine

lysine + phenylalanine + aspartic acid + ammonium ion

$$\text{H}_3\overset{+}{\text{N}}-\text{C}-\text{COO}^{(-)} + \text{H}_3\overset{+}{\text{N}}-\text{C}-\text{COO}^{(-)} + \text{H}_3\overset{+}{\text{N}}-\text{C}-\text{COO}^{(-)} + \text{NH}_4^+$$

lysine · · · phenylalanine · · · aspartic acid · · · ammonium ion

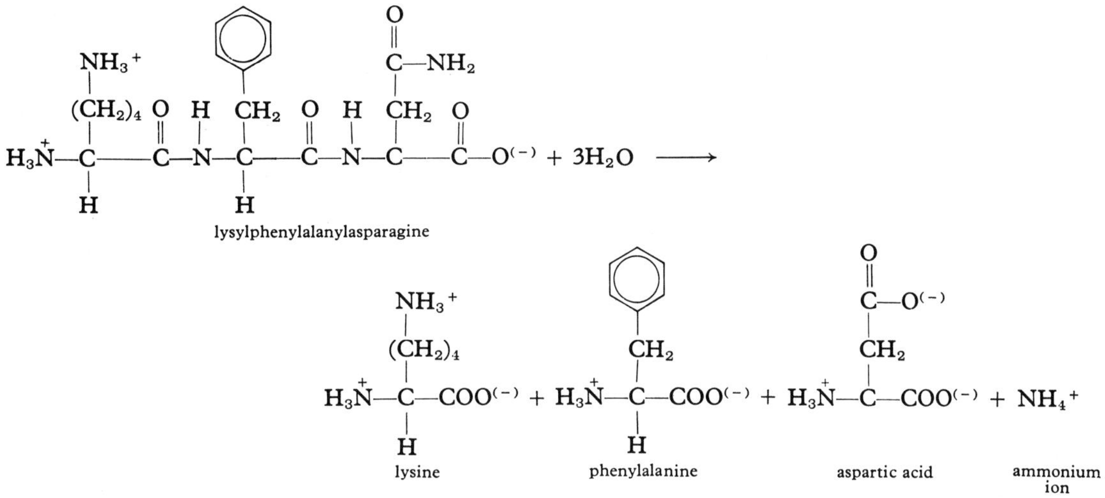

e) aspartic acid + [O] → NH₃ + oxaloacetic acid + H⁺

alanine + [O] → NH₃ + pyruvic acid + H⁺

$$\text{H}_3\overset{+}{\text{N}}-\text{C}-\text{COO}^{(-)} + [\text{O}] \rightarrow \text{NH}_3 + \text{O}{=}\text{C}-\text{COO}^{(-)} + \text{H}^+$$

alanine · · · pyruvic acid

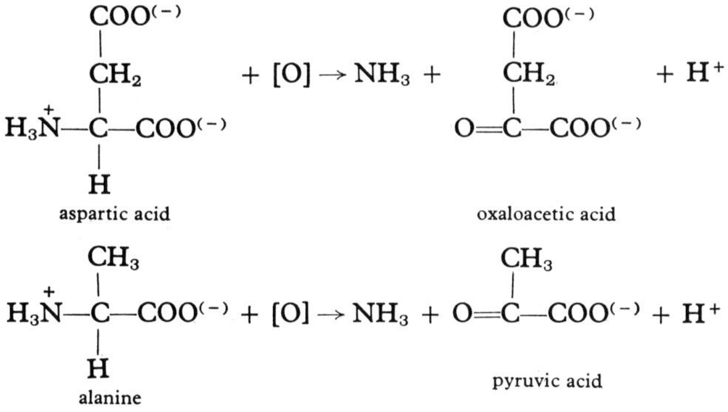

12.

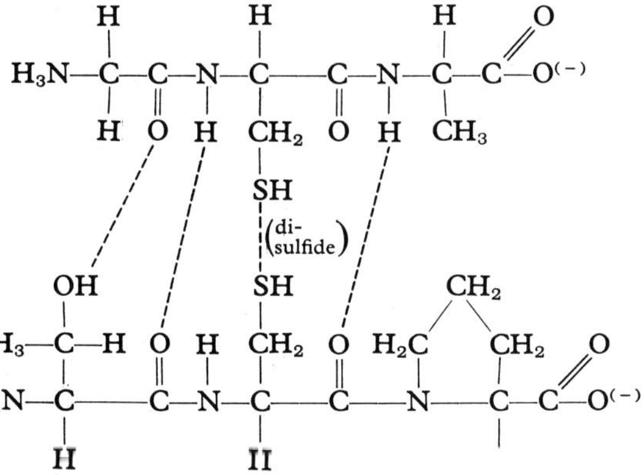

One disulfide and several hydrogen bonds are shown. Other hydrogen bonds are possible. Their formation would depend on the nearness of N—H and C—O groups to each other.

Chapter 32

1. a) glycerol, fatty acids

b) phosphate, glycerol, choline, ethanolamine, serine, fatty acids, possibly sphingenine

c) a fatty acid and a long chain alcohol

2. a)

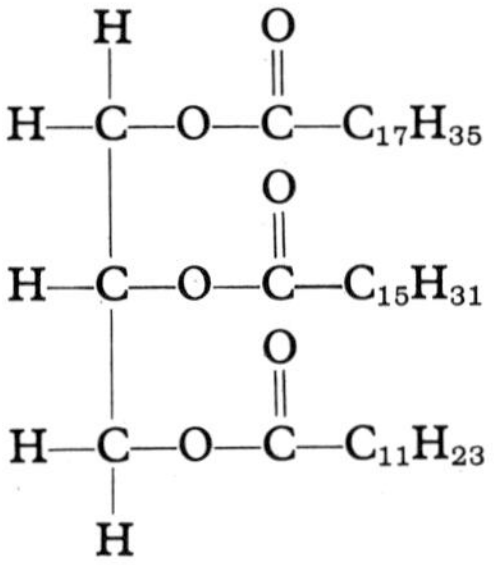

$$
\begin{array}{l}
\text{H}-\overset{\displaystyle \text{H}}{\underset{}{\text{C}}}-\text{O}-\overset{\displaystyle \text{O}}{\overset{\|}{\text{C}}}-\text{C}_{17}\text{H}_{35} \\[4pt]
\text{H}-\overset{}{\text{C}}-\text{O}-\overset{\displaystyle \text{O}}{\overset{\|}{\text{C}}}-\text{C}_{15}\text{H}_{31} \\[4pt]
\text{H}-\overset{}{\underset{\displaystyle \text{H}}{\text{C}}}-\text{O}-\overset{\displaystyle \text{O}}{\overset{\|}{\text{C}}}-\text{C}_{11}\text{H}_{23}
\end{array}
$$

(This is one of several possible structures)

d)

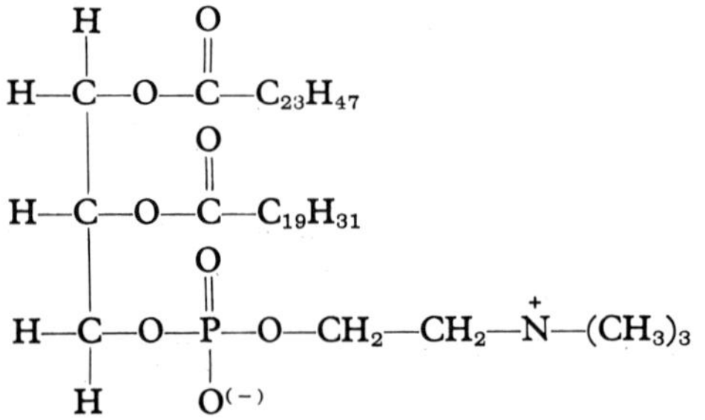

$$
\begin{array}{l}
\text{H}-\overset{\displaystyle \text{H}}{\text{C}}-\text{O}-\overset{\displaystyle \text{O}}{\overset{\|}{\text{C}}}-\text{C}_{23}\text{H}_{47} \\[4pt]
\text{H}-\text{C}-\text{O}-\overset{\displaystyle \text{O}}{\overset{\|}{\text{C}}}-\text{C}_{19}\text{H}_{31} \\[4pt]
\text{H}-\underset{\displaystyle \text{H}}{\text{C}}-\text{O}-\underset{\displaystyle \text{O}^{(-)}}{\overset{\displaystyle \text{O}}{\overset{\|}{\text{P}}}}-\text{O}-\text{CH}_2-\text{CH}_2-\overset{+}{\text{N}}-(\text{CH}_3)_3
\end{array}
$$

3. a)

$$
\begin{array}{l}
\text{H}-\overset{\displaystyle \text{H}}{\text{C}}-\text{O}-\overset{\displaystyle \text{P}}{\overset{\|}{\text{C}}}-\text{C}_{13}\text{H}_{27} \\[4pt]
\text{H}-\text{C}-\text{O}-\overset{\displaystyle \text{O}}{\overset{\|}{\text{C}}}-\text{C}_{17}\text{H}_{33}\;+\;3\text{HOH} \rightarrow \\[4pt]
\text{H}-\underset{\displaystyle \text{H}}{\text{C}}-\text{O}-\overset{\displaystyle \text{O}}{\overset{\|}{\text{C}}}-\text{C}_{11}\text{H}_{23}
\end{array}
$$

$$
\begin{array}{l}
\text{H}-\overset{\displaystyle \text{H}}{\text{C}}-\text{OH} + \text{C}_{13}\text{H}_{27}\text{COOH} \\[4pt]
\text{H}-\text{C}-\text{OH} + \text{C}_{17}\text{H}_{33}\text{COOH} \\[4pt]
\text{H}-\underset{\displaystyle \text{H}}{\text{C}}-\text{OH} + \text{C}_{11}\text{H}_{23}\text{COOH}
\end{array}
$$

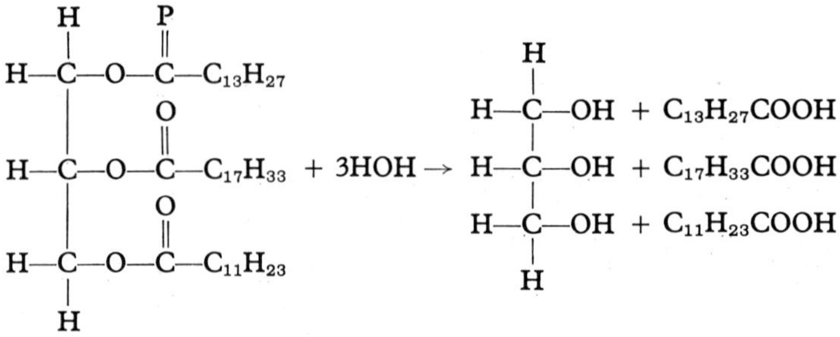

b) $\text{C}_{15}\text{H}_{29}\text{COOH} + \text{Br}_2 \rightarrow \text{C}_{15}\text{H}_{29}\text{Br}_2\text{COOH}$

Chapter 33

1. a)

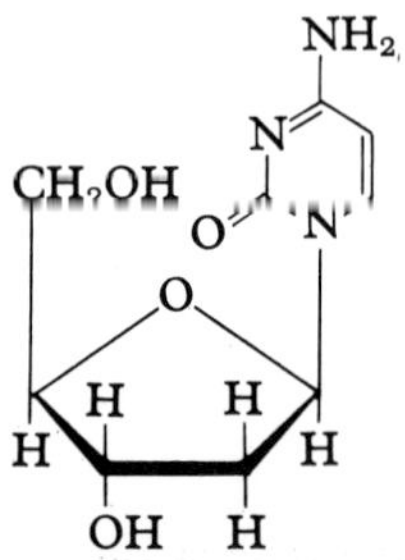

b)

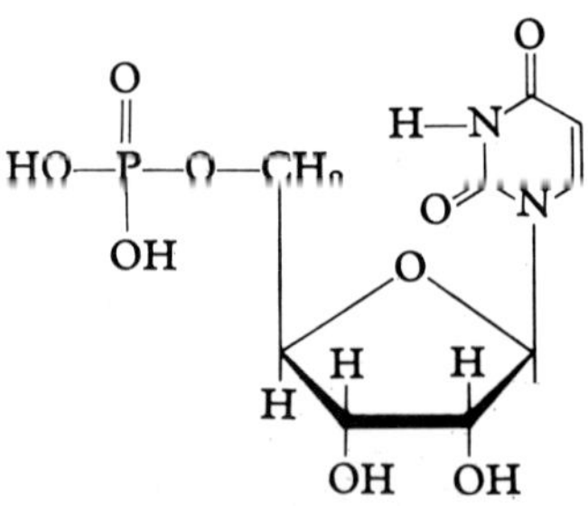

c)

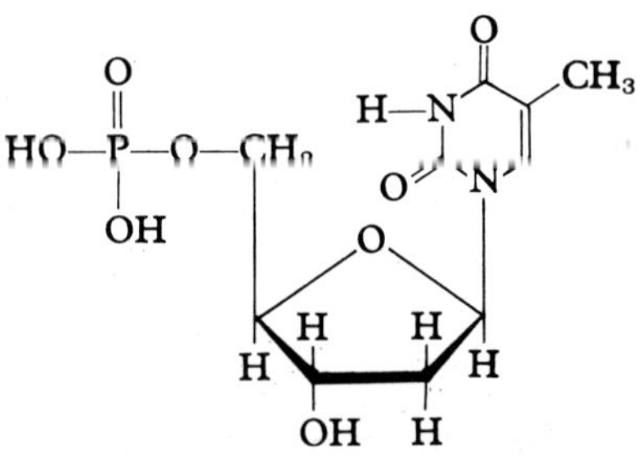

d)

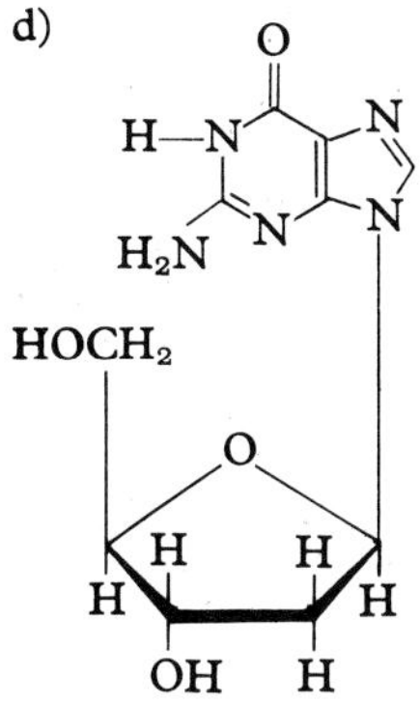

e)

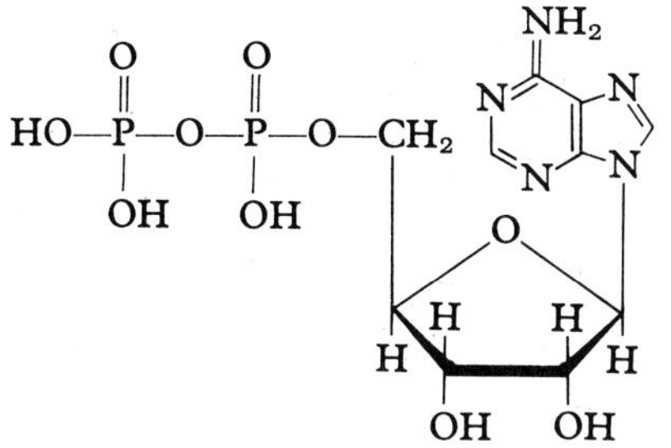

3.

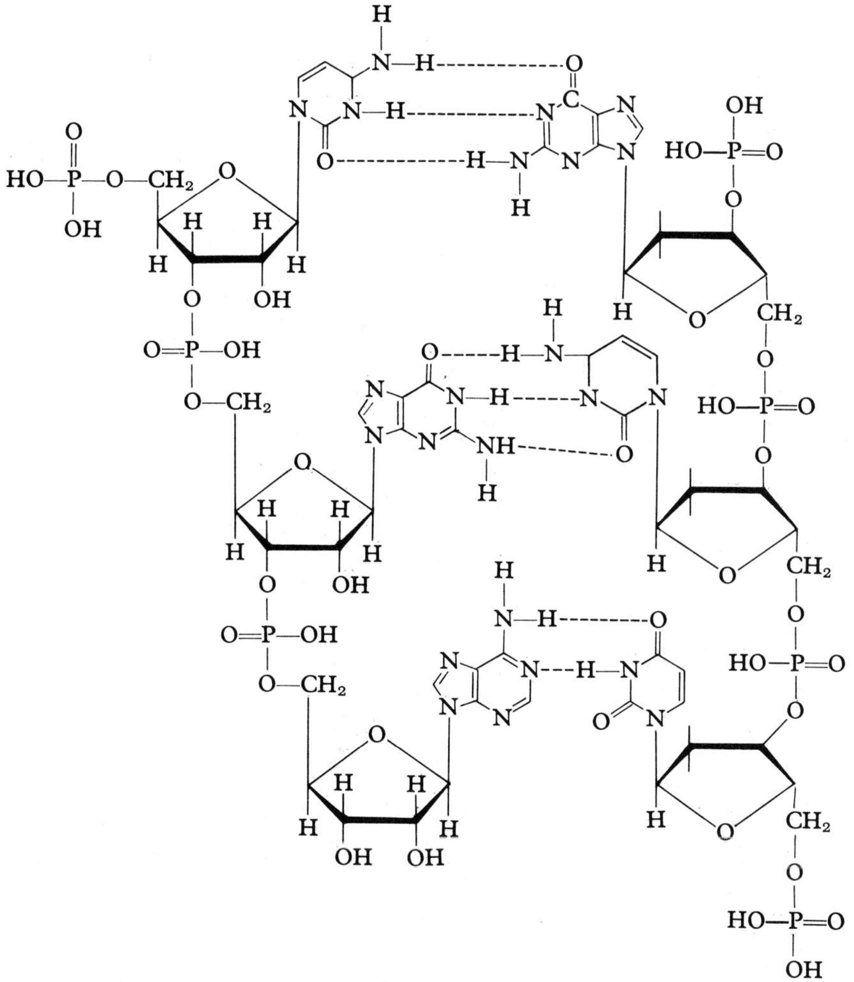

The nucleotide on the left contains ribose, the one on the right, deoxyribose. Other variations are possible.

Chapter 34

1.

b) triglyceride hydrolysis:

$$\text{HC—O—C—R'} + 3HOH \rightarrow C_3H_5(OH)_3 + RCOOH + R'COOH + R''COOH$$

a triglyceride   glycerol

c) $CH_3C\text{—O—}C_2H_5 + HOH \rightarrow CH_3COOH + C_2H_5OH$

ethyl acetate   acetic acid   ethanol

3. a) a substituted amide   b) an ester   c) a hemiacetal—a special kind of ether

Chapter 35

1. In metabolism the oxidation of glucose is catalyzed by enzymes and follows a definite pathway. Some of the energy of glucose is converted to the energy in ATP. When burned in a flame the oxidation of glucose is random. The energy of oxidation of glucose is released as heat. Both processes, however, give the same end products, CO_2 and H_2O, and both yield the same amount of energy.

3. Glucose can be metabolized by being converted to glycogen or by being oxidized. Insulin promotes the conversion of glucose to glycogen when the blood glucose is high. Other hormones, particularly glucagon and adrenalin, promote the formation of glucose when blood glucose levels are low.

4. Glycogen represents the most likely source. In severe carbohydrate deprivation, glucose can be formed from amino acids and the glycerol of fatty acids.

6. Other conditions can cause glucose in the urine, in addition to *diabetes mellitus*. A low kidney threshold for glucose and emotional stress may produce glucosuria.

9. Only 12 of the possible 38 molecules of ATP are produced in a poisoned animal.

11. CO_2, pyruvic acid, lactic acid, glucose, glycogen, or any intermediate in the glycolytic scheme. The pyruvate and lactate would be labeled in the carboxyl carbon, the glucose and glycogen in carbons 3 and 4.

Chapter 36

2. a) $H_3\overset{+}{N}\text{—}CH_2\text{—}COO^{(-)} \xrightarrow[\text{oxidative deamination}]{\text{transamination or}} O{=}CH\text{—}COOH$

glycine

c) 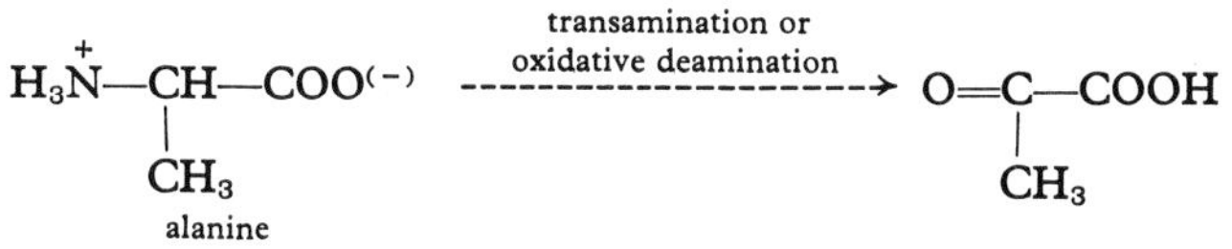 $\overset{+}{H_3N}-\underset{\underset{CH_3}{|}}{CH}-COO^{(-)}$ $\xrightarrow[\text{oxidative deamination}]{\text{transamination or}}$ $O=\underset{\underset{CH_3}{|}}{C}-COOH$

alanine

e) 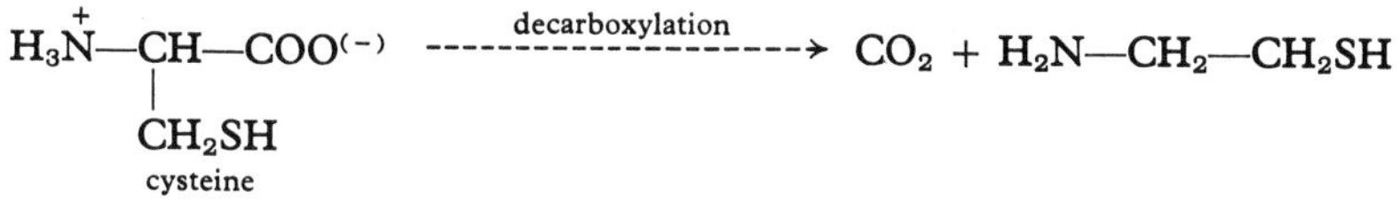 $\overset{+}{H_3N}-\underset{\underset{CH_2SH}{|}}{CH}-COO^{(-)}$ $\xrightarrow{\text{decarboxylation}}$ $CO_2 + H_2N-CH_2-CH_2SH$

cysteine

g) 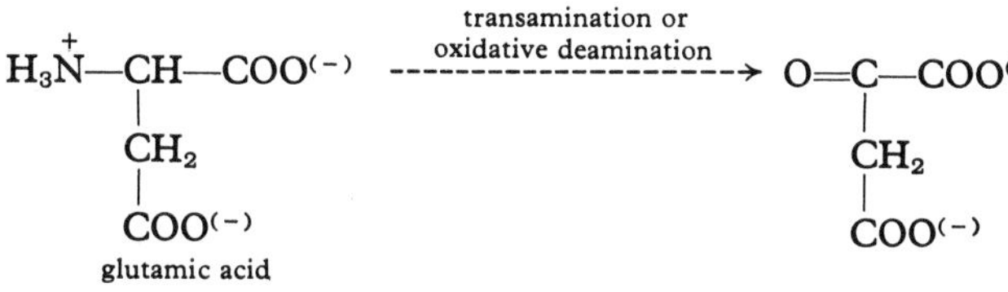 $\overset{+}{H_3N}-\underset{\underset{\underset{COO^{(-)}}{|}}{CH_2}}{CH}-COO^{(-)}$ $\xrightarrow[\text{oxidative deamination}]{\text{transamination or}}$ $O=\underset{\underset{\underset{COO^{(-)}}{|}}{CH_2}}{C}-COO^{(-)}$

glutamic acid

6. a) aminopeptidase   b) carboxypeptidase   c) an endopeptidase   d) a dipeptidase

8. arginine, urea

Chapter 37

2. dehydrogenation, hydration, dehydrogenation, reaction with coenzyme A.

3. It is phosphorylated with ATP to yield glucerol phosphate which can then be reduced to glyceraldehyde phosphate. The latter is an intermediate in glycolysis and can yield glucose, glycogen, or be oxidized to pyruvic acid and eventually to CO_2 and H_2O in the citric acid cycle.

4. Although the series of reactions is the same, degradation of a fatty acid results in a loss of two carbons each time the series of reactions is repeated. Thus each reactant is less by two carbons than the one preceding it, and the series is not truly a cycle.

5. Those animals that are the most mobile contain the highest amount of calories per gram. Mobile fish, for example, store fat, whereas clams and oysters store glycogen.

6. 79 moles of ATP/mole of capric acid. Efficiency is 948/1500 or 63%.

7. alanine $\rightarrow$ pyruvate $\rightarrow$ acetyl CoA $\rightarrow$ acetoacetate $\rightarrow$ β hydroxybutyric acid

9. glucose or starch $\rightarrow$ pyruvate $\rightarrow$ acetyl CoA $\rightarrow$ fatty acids $\rightarrow$ triglycerides

Chapter 38

2. Many plant proteins are deficient in some amino acids.

3. Minimal caloric requirements are found by multiplying weight in kilograms by number of hours in a day (24). A person should eat one gram of protein per kilogram of body weight per day. If foods are 70% water, the formula becomes: weight in kilograms $\times$ (100/30) = grams of protein food per day.

5. They supply energy and serve as a source of essential fatty acids. They may help prevent atherosclerosis.

9. a) nicotinamide   b) riboflavin   c) pantothenic acid

14. Nitrogen is found in thiamin, riboflavin nicotinamide, pyridoxal, pantothenic acid, biotin, folic acid, and vitamin B_{12}. Sulfur is found in thiamin and biotin. Cobalt is found in vitamin B_{12}.

Chapter 39

2. The theory states that a gene governs the synthesis of one protein and the effects of the gene are due to the action of the protein. The theory is based on evidence that hereditary defects are known to be caused by protein abnormalities.

4. The person would suffer from hypothyroidism. Such deficiencies could result in death of a fetus and a miscarriage.

5.

Hemoglobin type	Normal codes	Altered codes
Hgb S	GAA, GAG	GUU, GUC, GUA, GUG
Hgb C	GAA, GAG	AAA, AAG
Hgb G San José	GAA, GAG	GGU, GGC, GGA, GGG
Hgb E	GAA, GAG	AAA, AAG
Hgb M Saskatoon	CAU, CAC	UAU, UAC
Hgb M Milwaukee	GUU, GUC, GUA, GUG	GAA, GAG
Hgb D Punjab	GAA, GAG	CAA, CAG
Hgb Zurich	CAU, CAC	CGU, CGC, CGA, CGG, AGA, AGG
Hgb O Arabia	GAA, GAG	AAA, AAG

Chapter 40

21. a) Indicates decreased ability of the heart to pump blood to lungs; lung infection that prevents oxygenation of blood; anemia.
    c) Indicates impaired liver function; rapid turnover (increased destruction) of red blood cells.
    e) Indicates decreased ability to fight infections.
    g) May indicate *diabetes mellitus* or recent ingestion and assimilation of carbohydrates.
    i) Indicates decreased amount of water in the blood.
    k) Indicates an infection or multiple myeloma.
    m) Indicates acute liver disease or some types of heart attacks.
    o) Indicates possibility of cancer of the prostate.
    q) Indicates decreased urea synthesis which is generally due to impaired liver function, especially cirrhosis of the liver.
    s) Indicates impaired kidney function or excessive catabolism of protein or purines.
    u) Indicates hyperthyroidism.
    w) Hyperirritability of nerves and muscles.

22. a) Indicates increased liquid intake, may indicate *diabetes mellitus* or *diabetes insipidus*.
    c) High urine volume due to causes listed in (a).
    e) Indicates *diabetes mellitus* or low urinary threshold for glucose.
    g) Increased catabolism of protein.
    i) Indicates kidney infection.

23. a) Decreased excretion of bile, perhaps due to the presence of gallstones.
    c) Indicates failure of protein synthesis due to malnutrition or impaired liver function; water retention due to retention of sodium ions.

Chapter 41

1. Antibiotics and antibodies both serve to protect against infections.
   Antibiotics are produced by simple plants principally by microorganisms and are simpler in structure than antibodies. Apparently antibiotics are produced without the need for prior exposure to the infectious agent.

Antibodies are complex molecules (proteins with other additions possible) and are produced by animals. Antibodies are produced in response to an antigen.

3. Antibacterial chemotherapeutic agent: An antimetabolite or other substance that destroys or prevent reproduction of bacteria without affecting human cells. An anticancer agent should have the same effects on cancerous cells without affecting other human cells.

4. Increased levels of serotonin or epinephrine in tissues.

11. a) Used as an analgesic, antipyretic, antiinflammation agent; also decreases blood clotting.
    c) Are depressants, produce sedation and sleep; used legally primarily as a "sleeping pill".
    e) Is a local anesthetic. Used in anesthetizing eye and upper respiratory tract; also alters mental states and is used illegally for this purpose.
    g) Increases the force of heart contractions, corrects cardiac arrhythmias.

12. a) Addiction, accidental overdoses.   c) Addiction.   e) Produces habituation or dependence.
    g) Produces gastric ulcers, decreased ability to resist infections, retention of salt and water.
    i) Produces altered mental states, lack of ability to perform intellectual or precise work, and probably decreases motivation.

INDEX

period

○ Group IA   ○ IIA

Periodic Table of the Elements

Simplified atomic weights for practice in chemical calculations
(Not sufficiently accurate for quantitative work.)
Values in parentheses indicate mass numbers of the most stable isotope.

	Group IA	IIA	IIIB	IVB	VB	VIB	VIIB		VIII
1	1 1.0 **H** Hydrogen								
2	3 7.0 **Li** Lithium	4 9.0 **Be** Beryllium							
3	11 23.0 **Na** Sodium	12 24.3 **Mg** Magnesium							
4	19 39.0 **K** Potassium	20 40.0 **Ca** Calcium	21 45.0 **Sc** Scandium	22 48.0 **Ti** Titanium	23 51.0 **V** Vanadium	24 52.0 **Cr** Chromium	25 55.0 **Mn** Manganese	26 56.0 **Fe** Iron	27 **Co** Coba
5	37 85.5 **Rb** Rubidium	38 87.6 **Sr** Strontium	39 89.0 **Y** Yttrium	40 91.2 **Zr** Zirconium	41 93.0 **Nb** Niobium	42 96.0 **Mo** Molybdenum	43 (97) **Tc** Technetium	44 101 **Ru** Ruthenium	45 **Rh** Rhodi
6	55 133 **Cs** Cesium	56 137 **Ba** Barium	57-70* 71 175 **Lu** Lutetium	72 178 **Hf** Hafnium	73 181 **Ta** Tantalum	74 184 **W** Tungsten	75 186 **Re** Rhenium	76 190 **Os** Osmium	77 **Ir** Iridiu
7	87 (223) **Fr** Francium	88 226 **Ra** Radium	89-102† 103 (256) **Lr** Lawrencium	104 (257) ‡	105 (260) ‡	106 (?) ‡	107	108	109

□ Metals

□ Metalloids

□ Nonmetals

□ Noble gases

*LANTHANIDES

†ACTINIDES

57 139 **La** Lanthanum	58 140 **Ce** Cerium	59 141 **Pr** Praseodymium	60 144 **Nd** Neodymium	61 (147) **Pm** Promethium	62 **Sm** Samari
89 (227) **Ac** Actinium	90 232 **Th** Thorium	91 231 **Pa** Protactinium	92 238 **U** Uranium	93 237 **Np** Neptunium	94 **Pu** Pluton

‡ Names not yet official